Intermetallic Compounds

WILEY SERIES ON THE SCIENCE AND TECHNOLOGY OF MATERIALS

Advisory Editors: J. H. Hollomon, J. E. Burke, B. Chalmers, R. L. Sproull, A. T. Tobolsky

---

INTERMETALLIC COMPOUNDS
J. H. Westbrook, editor
ORGANIC SEMICONDUCTORS
Felix Gutmann and L. E. Lyons
THE PHYSICAL PRINCIPLES OF MAGNETISM
Allan H. Morrish
FRICTION AND WEAR OF MATERIALS
Ernest Rabinowicz
HANDBOOK OF ELECTRON BEAM WELDING
R. Bakish and S. S. White
PHYSICS OF MAGNETISM
Sōshin Chikazumi
PHYSICS OF III-V COMPOUNDS
Otfried Madelung (translation by D. Meyerhofer)
PRINCIPLES OF SOLIDIFICATION
Bruce Chalmers
APPLIED SUPERCONDUCTIVITY
Vernon L. Newhouse
THE MECHANICAL PROPERTIES OF MATTER
A. H. Cottrell
THE ART AND SCIENCE OF GROWING CRYSTALS
J. J. Gilman, editor
SELECTED VALUES OF THERMODYNAMIC PROPERTIES OF METALS AND ALLOYS
Ralph Hultgren, Raymond L. Orr, Philip D. Anderson and Kenneth K. Kelly
PROCESSES OF CREEP AND FATIGUE IN METALS
A. J. Kennedy
COLUMBIUM AND TANTALUM
Frank T. Sisco and Edward Epremian, editors
MECHANICAL PROPERTIES OF METALS
D. McLean
THE METALLURGY OF WELDING
D. Séférian (translation by E. Bishop)
THERMODYNAMICS OF SOLIDS
Richard A. Swalin
TRANSMISSION ELECTRON MICROSCOPY OF METALS
Gareth Thomas
PLASTICITY AND CREEP OF METALS
J. D. Lubahn and R. P. Felgar
INTRODUCTION TO CERAMICS
W. D. Kingery
PROPERTIES AND STRUCTURE OF POLYMERS
Arthur V. Tobolsky
PHYSICAL METALLURGY
Bruce Chalmers
FERRITES
J. Smit and H. P. J. Wijn
THE METALLURGY OF VANADIUM
William Rostoker

Intermetallic Compounds

Edited by J. H. Westbrook

Research and Development Center
General Electric Company
Schenectady, New York

John Wiley & Sons, Inc. New York · London · Sydney

Library of Congress Catalog Card Number: 66-21063
Printed in the United States of America

To

Cecil Henry Desch
Nikolai Semenovitch Kurnakov
Gustav Tamman

—— pioneers in the study of
intermetallic compounds

Contributors

AITKEN, EDWARD A.
Nuclear Materials and Propulsion Operation
General Electric Company
Cincinnati, Ohio

BAKISH, ROBERT
Bakish Materials Corporation
Englewood, New Jersey

BEATTIE, HARRY J., JR.
General Electric Company
Schenectady, New York

BEELER, JOE R., JR.
Nuclear Materials and Propulsion Operation
General Electric Company
Cincinnati, Ohio

BEVER, MICHAEL B.
Department of Metallurgy
Massachusetts Institute of Technology
Cambridge, Massachusetts

BROWN, ALLAN
University of Uppsala
Uppsala, Sweden

BROWN, NORMAN
School of Metallurgical Engineering and Laboratory for Research on the Structure of Matter
University of Pennsylvania
Philadelphia, Pennsylvania

CADOFF, IRVING
New York University
New York City, New York

CHRISTIAN, J. W.
Oxford University
Oxford, England

COLES, BRYAN R.
Department of Physics
Imperial College of Science and Technology
University of London
London, England

DUWEZ, POL
W. M. Keck Laboratory
of Engineering Materials
California Institute of Technology
Pasadena, California

DWIGHT, AUSTIN E.
Argonne National Laboratory
Argonne, Illinois

GUILLET, LÉON
École Centrale des Arts et Manufactures
Paris, France

HAGEL, WILLIAM C.
Engineering Recruiting Service
General Electric Company
Schenectady, New York

KORNILOV, I. I.
Baikov Institute of Metallurgy
Moscow, U.S.S.R.

KOUVEL, J. S.
General Electric
Research and Development Center
Schenectady, New York

LAVES, F.
Institut für Kristallographie und Petrographie
der Eidgenössischen Technischen Hochschule
Zurich, Switzerland

LAWLEY, ALAN
The Franklin Institute Research Laboratories
Philadelphia, Pennsylvania

Le ROUX, R.
Conservatoire National des Arts et Métiers
Paris, France

NEVITT, M. V.
Argonne National Laboratory
Argonne, Illinois

PARTHÉ, ERWIN
School of Metallurgical Engineering and
Laboratory for Research on the
Structure of Matter
University of Pennsylvania
Philadelphia, Pennsylvania

READ, THOMAS A.
University of Illinois
Urbana, Illinois

ROBERTS, B. W.
General Electric
Research and Development Center
Schenectady, New York

ROBERTS, C. SHELDON
Consultant
Los Altos, California

ROBINSON, PETER M.
Department of Metallurgy
Massachusetts Institute of Technology
Cambridge, Massachusetts

RUDMAN, PETER S.
Battelle Memorial Institute
Columbus, Ohio

RUNDLE, R. E.
Department of Chemistry and
Ames Laboratory of the Atomic Energy
Commission
Iowa State University
Ames, Iowa

SCHUBERT, K.
Max Planck Institut für Metallforschung
Stuttgart, Germany

SCHWAB, G.-M.
Institute of Physical Chemistry
University of Munich
Munich, Germany

WAYMAN, C. M.
University of Illinois
Urbana, Illinois

WERNICK, JACK HARRY
Bell Telephone Laboratories, Inc.
Murray Hill, New Jersey

WESTBROOK, J. H.
General Electric
Research and Development Center
Schenectady, New York

WOLFF, GUNTHER A.
Tyco Laboratories, Inc.
Waltham, Massachusetts

Preface

The first published paper dealing with intermetallic compounds appeared in 1839, and more than sixty years elapsed before the field was sufficiently developed to warrant the first review paper by Neville in 1900. However, new results were then appearing so rapidly that fifteen years later two books were printed, devoted exclusively to this subject, one by Desch in England and one by Giua and Giua in Italy. Curiously, however, although the literature of the field has since continued its exponential proliferation, no other book has been written over the past fifty years which treats intermetallic compounds comprehensively. The present book was initiated in an attempt to fill this evident void.

There are other compelling reasons, however, for attempting the production of a comprehensive work on intermetallics in addition to that of simply filling an obvious gap in the written record. Although the subject may at first seem to be a highly specialized bypath, well off the main highway of physical metallurgy, its centrality to the larger field is evidenced by the remarkable association of virtually all the big names of metallurgy with some phase of the development of knowledge of intermetallic compounds. Of these we may cite Matthiessen, Tamman, Kurnakov, and Desch of the early period and Bain, Hume-Rothery, and Westgren from more recent times. Intermetallic compounds are also of great technical importance. As minor constituents they play the critical role in obtaining the properties of many of our commercial alloys, for example, high-strength aluminum alloys ($CuAl_2$), nickel-base superalloys (Ni_3Al), alnico magnets (NiAl), and bearing alloys (SbSn). As the major phase, they perform in such diverse applications as band instrument keys (γ-brass), frogmen's demolition tools (NiTi), Invar low expansion parts (Ni_3Fe), crucibles (Mo_3Al), and tunnel diodes (GaAs). Intermetallics can also claim the material with the record superconducting transition temperature (Cb_3Sn), with the maximum magnetic coercive force (FePt), and with the unique capability for solid-state laser action (GaAs).

The book was planned neither as a student textbook nor as a highly erudite monograph for the narrow expert, but rather as a comprehensive treatise that would present the state-of-the-art in both theory and experiment for the materials scientist who has become concerned with intermetallics as a class of materials or the engineer who seeks to exploit their properties in new devices. To secure a high degree of authoritativeness, the book has been written by an international group of contributors, each an active worker in the subject of a particular chapter.

Just what is an intermetallic compound? Although the term is concise and gives an immediate connotation of the nature of the material, an exact definition is elusive and, as with most simple appellations, highly arbitrary. Some restrict the designation to a very narrow grouping of substances, and others choose not to distinguish intermetallics from a larger grouping of metallic compounds (for example, the Russian term *metallides*). Here we regard as intermetallic compounds all metal-metal compounds, both ordered and disordered, binary and multicomponent. Upon occasion, when it appears useful to do so, even the metal-metal aspect of the definition is relaxed somewhat in the consideration of some metal-metalloid compounds, such as the silicides or phosphides.

The book begins with a brief historical sketch of the development of the subject. The remainder of the first half of the book is concerned with the presentation of the bonding and crystal structure of intermetallic compounds. Each of these two parts is introduced by a survey chapter reviewing the main features of the topic. In the case of bonding, this survey is followed by a series of chapters presenting viewpoints of bonding as revealed by various experimental or theoretical techniques. The chapters on crystal structure are organized by structural families. The next three parts of the book consider, in turn, topics relating to microstructure, substructure, formation, stability, constitution, kinetics, and transformations. These chapters therefore have their greatest utility to the experimentalist synthesizing or studying intermetallic compounds, although much material of fundamental interest is also contained. The final part of the book consists of seven chapters treating specific classes of properties and the resultant applications. Throughout the book are included tabulations of structures, properties, materials constants, etc., which may be of aid either to the scientist designing a new experiment or testing a new theory or to the engineer selecting a potential material for a device.

Although every effort has been made to provide a comprehensive treatment of the subject, certain minor topics had to be eliminated for reasons of space or unavailability of qualified contributors. Among these topics are fracture, electron distribution, and liquid and vapor species.

Intermetallic studies began with the development and application of methods for the discovery and verification of the existence of intermetallic compounds. The subsequent period of amassing structural, constitutional, and property data together with the empirical systematization of such results is now coming to a close. In the future we can look to the application of increasingly sophisticated experimental and theoretical methods for incisive probing of fundamental questions of bonding, structural imperfections, and detailed examination of the influence of these factors and basic chemistry on the resulting properties. A further growth in technological uses of intermetallics is also expected, as well as the application of fundamental information to improvement of materials in which the intermetallic compound is only a minor constituent. It is hoped that this book will not only constitute a record of past progress and a useful reference in current research and application, but most importantly will serve as a stimulus to the further growth of the use and understanding of intermetallic compounds.

The plan of the book as well as the content of each of the chapters benefited greatly from the invaluable critical appraisal of colleagues and professional acquaintances too numerous to cite individually. The support and encouragement of the General Electric Research and Development Center is also greatly appreciated.

J. H. Westbrook

Schenectady,
New York
June 1966

Contents

part I

Introduction

chapter 1

Historical Sketch

J. H. WESTBROOK

General Electric Research and Development Center
Schenectady, New York

I may remind you, however, that much of what is both interesting and full of suggestion, even at the present day, is to be found buried in the treatises of the old writers whose work we inherit and continue.

Roberts-Austen

1. INTRODUCTION

The principal purpose of this sketch is to outline the earliest beginnings of some of the prominent concepts and experimental developments in the field of intermetallic compounds. No attempt has been made at completeness; the reader is directed to the subsequent chapters of the book for further details in each topical area. It is hoped, however, that these few pages will convey some of the human story of the first faltering efforts in the area and may inspire some readers to turn directly to the early literature for the historical drama which there unfolds and, indeed, as quoted above from Roberts-Austen,[1] for fruitful suggestions for today's research.

Near the beginning of the nineteenth century, when the first systematic studies of alloy systems were being made, a number of investigators noted behavior at certain alloy compositions strikingly like that of ordinary chemical compounds and began to speculate as to whether compounds might exist between metals. It appears that the first true observation of an intermetallic was that of the German chemist, Karl Karsten, who in 1839 noticed that a discontinuity occurred in the action of acids on alloys of copper and zinc at the equiatomic composition and suggested the formation of a compound.[2] In his words:

> From the behavior of all these alloys the conclusion can be drawn that they are true chemical compounds and not perchance mixtures of a certain alloy with an excess of one of the component metals.

The compound Karsten claimed does exist and is the now familiar beta brass, CuZn.

During the second half of the century

similar discontinuities in other properties—electrical, mechanical, and magnetic, as well as chemical—were detected in other alloy systems and led to the suggested formulation of still other intermetallic compounds. It is of interest to note that not all of the early studies of intermetallic compounds were motivated by sheer intellectual curiosity. For example, Calvert and Johnson[3] stated in 1855: "We . . . believe, that by producing alloys having a definite composition (i.e., intermetallics), we should point the way to much cheaper and better alloys than hitherto . . . ." Unfortunately, their hopes were not realized, and, except for some use as master alloys, application of intermetallic compounds as such has not been for reasons of convenience, economics, or incrementally improved properties but rather, for the most part, because of the *unique* properties they offer, frequently in the form of a dispersion in a metallic matrix or as single-crystal wafers in an electronic circuit.

2. CONSTITUTIONAL STUDIES

Acid treatment continued for some years to be a popular technique in the search for and identification of intermetallic compounds. Various alloys in a system would be subjected to attack by strong reagents in an attempt to isolate the compound from the matrix; subsequently, the residues were analyzed and, when possible, the results expressed in whole-number proportions of the components. Although many true compounds were found by this approach, it also led to many spurious results; for example, of the twenty-five compounds claimed in 1848 by Croockewit[4] only three were confirmed by subsequent investigations.[5] The method was fundamentally unsound in that it ignored the possibility of limited attack on solid solutions or mixtures and further presumed without basis that the compound would be more resistant to attack than its constituents.

Thermal analysis was also much used and often incorrectly interpreted. For example, Rudberg[6] in 1830, in studying the freezing points of certain low-melting alloys, observed a second thermal arrest on cooling, the temperature of which appeared to be independent of composition. He therefore concluded that a compound or "chemical alloy" was formed. This phenomenon is now, of course, recognized as a eutectic reaction in which two phases solidify simultaneously. However, Rudberg's view persisted for some time, and such "compounds" in aqueous systems even acquired a distinctive name, "cryohydrates." The true nature of such eutectic compositions was not generally recognized until many years later.

Many cases of intermetallic compound formation were straight-forward, as in systems

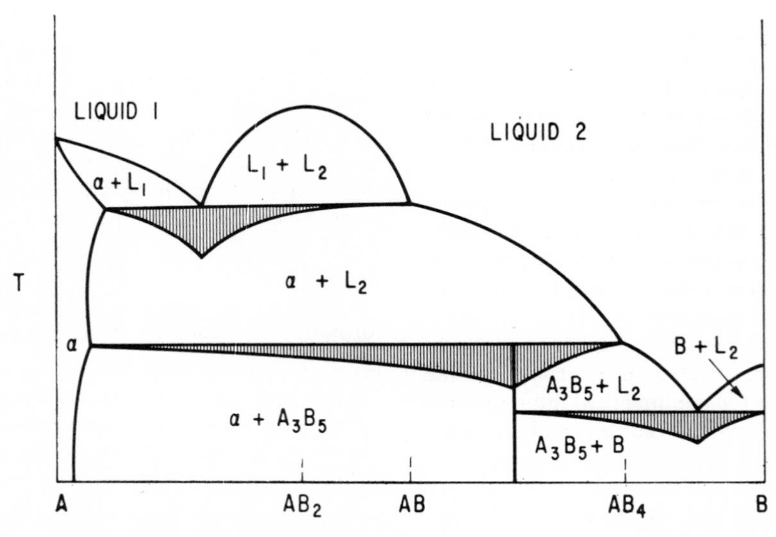

Fig. 1. Schematic phase diagram of a hypothetical system A–B with one intermetallic compound A_3B_5.

in which a single compound gave rise to a corresponding maximum in the melting curve at a composition corresponding to a simple atomic ratio of the component elements. For a considerable time only the freezing-point curves were determined, and formulas for supposed intermetallic compounds were derived for all maxima and kinks on the freezing-point—composition curves; minima were ascribed to eutectic formation. The imaginary system of Fig. 1 illustrates how wrong it was possible to be with this naïve approach. Working solely from the freezing-point data, the investigator might well have reported compounds corresponding to the maximum AB_2, and the two kinks, AB and AB_4, and missed the only real compound A_3B_5 completely.

The German metallurgist Tamman, whose portrait is shown in Fig. 2, made enormous contributions in many different fields of physical metallurgy over a period of almost fifty years following the turn of the century. Particularly prominent in this large volume of work were his many studies of the occurrence and properties of intermetallic compounds. He made an especially important contribution to the technique of thermal analysis by introducing the method of quantitative measurement of the times of thermal arrest for alloy samples of identical weight, cooled under identical conditions.[7,8] Since the time for a given arrest was a maximum at the composition corresponding to the pure constituent, the composition of any true compound could be determined quite unequivocally. These times were conventionally plotted by the Tamman school vertically downward from the invariant temperature in a phase diagram, as shown by the shaded sections in Fig. 1. Errors of the sort previously described, which might arise from cases like the hypothetical system of Fig. 1, were thus avoided. Tamman's constitutional work was widely criticized as being crude and fraught with errors, but his aim was a broad survey of a large number of systems rather than the careful study of a few, so that generalizations of binary constitution and the nature of intermetallic phases would be feasible. In this sense he was highly successful.

Fig. 2. Portrait of G. A. Tamman. (From *Z. Metallk.*)

Thermal analyses and analyses of residues after chemical attack as methods of compound identification gradually came to be supplemented by studies of the composition dependence of physical properties such as density, resistivity, thermal conductivity, hardness, and thermoelectric power. Although physical properties of intermetallics had been examined earlier by such men as Calvert and Johnson,[9] Matthiessen,[10,11] and LeChatelier,[12] the development of the methods and exploitation of this approach of "physicochemical analysis" must be credited to N. S. Kurnakov, the great Russian metallurgist whose portrait is shown in Fig. 3. Kurnakov, like his contemporary Tamman, was a man of great ability, productivity (more than 200 publications) and far-ranging interests—from geology to physical chemistry to metallurgy.[13] His contributions to our understanding of intermetallics began with phase-diagram determinations, extended to property-composition studies as just related, and also included theoretical discussions of the nature and stability of intermetallics. A typical early result which combined the methods of physicochemical analysis with conventional phase-diagram techniques is reproduced from the work of Smirnov and Kurnakov[14] on the Ag–Mg system in Fig. 4.

Departure by the experimentalists from dependence on a single technique as well as a growing application of the theoretical studies of phase equilibria by Gibbs and Roozeboom did much to place intermetallic compound identification on a firmer footing. Despite the aforementioned difficulties, by 1900 a published list[15] contained thirty-seven confirmed intermetallic compounds. Within two more decades the number was multiplied nearly tenfold, in large part as a result of the prodigious researches of Tamman and his students at Göttingen, in a long series of papers, the first of which was published in 1903.[16]

With the accumulation of data from the schools of Tamman and Kurnakov as well as others, problems began to arise which were to puzzle workers for some years to come. It appeared that some compounds could dissolve substantial amounts of either or both of their constituent elements and still retain their structure, homogeneity, and physical properties. These findings revived the classic controversy on the law of definite proportions between Berthollet on the one hand and Proust and Dalton on the other, and debate continued for some time. However, the existence of intermetallic compounds over ranges of composition was finally established beyond a doubt. One of the contributions made by Kurnakov[17] was to point out that singularities on composition-property curves may be used to identify compound formation:

> It is not the composition of the phase, which is generally variable, but the composition of the singular or Daltonian point which is characteristic on the diagrams showing the properties of a determinate compound . . . . A chemical individual represents a phase showing singular or Daltonian points on the curves of its properties. The composition which corresponds to these points is the same in all changes of the factors of equilibrium of the system.

It is now clear that the singularities to which Kurnakov called attention are a result of the strong effects of departures from stoichiometry on the properties of ordered compounds as in the case of AgMg in Fig. 4. Unfortunately, Kurnakov's definition has no general applicability, for singular points on property curves are not exhibited by many intermediate phases which are either not ordered, e.g., Ag_3Al, or for which the stoichiometric composition lies beyond the stability limits of the compound, e.g., βCuZn.

Another serious matter was the obvious failure of metals to obey the ordinary rules of chemical valence in their compound formation. Formulas, such as Cu_5Si, Ag_3Li_{10}, $FeZn_7$, seemed strange indeed to chemists accustomed to simple salts and other ionic or covalent compounds. Still more disturbing was the apparent ability of a given metal pair to exhibit variable valence as, for example, KHg, KHg_2, KHg_3, K_2Hg_9, and KHg_{10}. Yet some intermetallics did appear to exhibit normal valence, such as Mg_2Si or BaTe. Rationalization of the compounds of variable and abnormal valence remained a problem for some time. Still another perplexing observation was the lack of complete mutual solid solubility between Mg_2Sn and Mg_2Pb which were presumed to be isomorphous. On the other hand, the unlikely pair Cu_3Al and CuZn were found to form a continuous solid solution series.

Fig. 3. Portrait of N. S. Kurnakov. (From *Usp. Khim.*)

Most of the early work naturally concerned itself with binary systems, yet the first true ternary compounds CdHgNa and Hg_2KNa

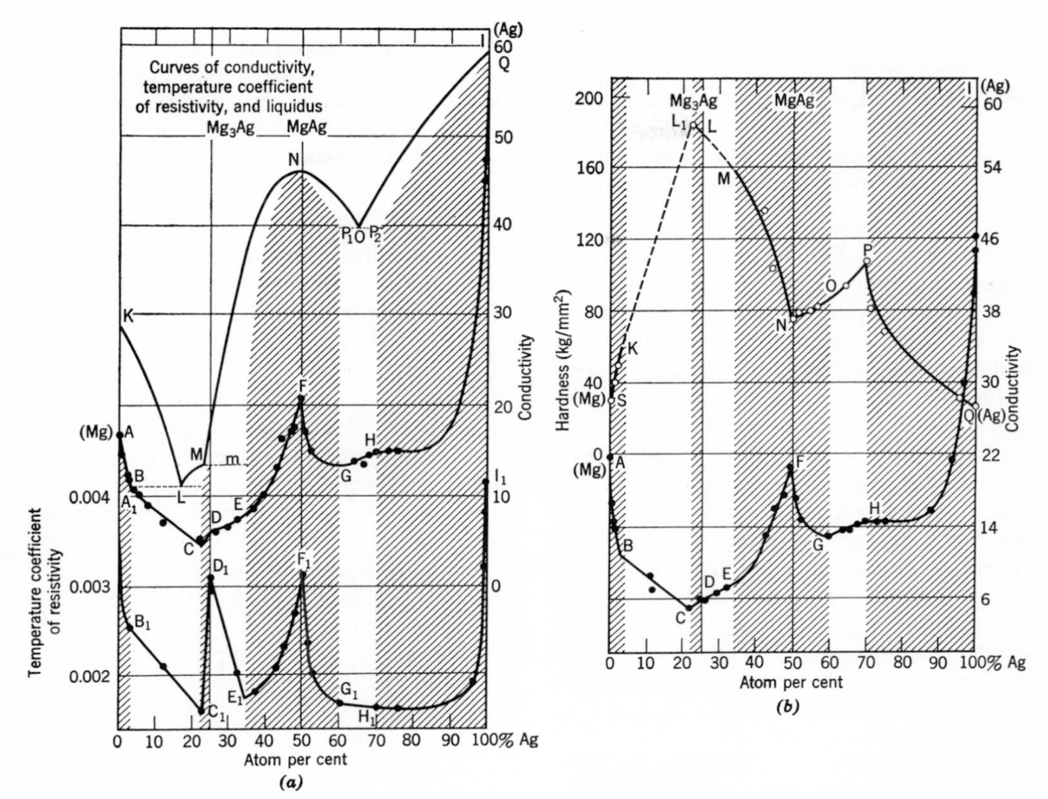

Fig. 4. Property-composition curves for the Mg–Ag system: (*a*) melting point, electrical conductivity, and temperature coefficient of resistivity; (*b*) hardness and electrical conductivity. (After Smirnov and Kurnakov.[14])

were reported by Jänecke[18] as early as 1906. Scores of ternary compounds, some quaternaries, and even a few quinary compounds have been identified in the subsequent sixty years.

3. CRYSTALLOGRAPHIC STUDIES

Crystallographic studies of intermetallics were impeded for many years by the difficulties attendant to isolating, from an alloy, crystals suitable for goniometric measurement.[19] Further, few intermetallic compounds appeared to occur naturally as minerals; these were limited principally to compounds of the metalloids, e.g., niccolite NiAs, dyscrasite Ag_3Sb, and tetradymite Bi_2Te_3. As a consequence, selected natural crystals were of little help in a general study of the crystallography of intermetallic compounds.

Following Laue's suggestion to Friedrich in 1912 that X-rays should be diffracted by crystals,[20] a powerful new tool was placed in the hands of the crystallographers. Although the elemental metals and natural crystals, such as the halides, were among the first materials studied, the introduction of powder methods by Debye and Scherrer[21] and by Hull[22] opened the way for much more widespread use of the X-ray diffraction technique. Apparently, credit for the first X-ray studies of intermetallic phases must go to Mary Andrews, an associate of Hull who, in 1921, published her studies of the Fe–Ni, Fe–Co, and Cu–Zn systems.[23] In the first two cases she found no new phases but only the expected change in crystal structure from that of one component to that of the other as the composition was varied across the system. In the third, however, the β-phase was found to be body-centered cubic and the ϵ-phase rhombohedral. She also examined a sample of Heusler's alloy (v.i.) which, she showed, was cubic. Continued application of these techniques in the hands

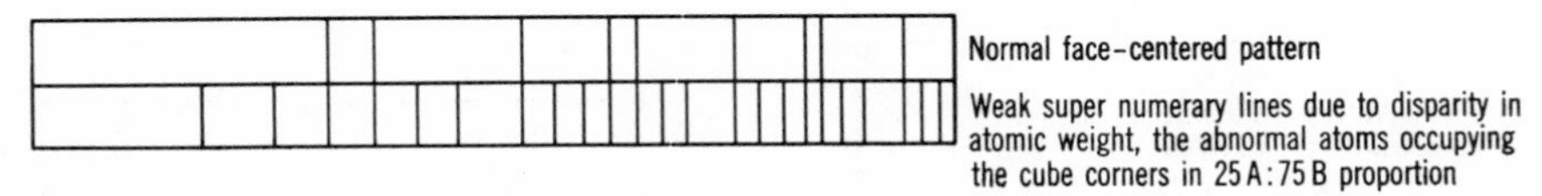

Fig. 5. Face-centered pattern with superposed lines from simple cube, resulting from the ordered arrangement of atoms in Cu_3Au. (After Bain.[26])

of such men as Goldschmidt in Norway,[24] Becker and Ebert in Germany,[25] Bain in the U.S.,[26] Westgren and Phragmén in Sweden,[27,28] and Bradley in England,[29,30] began rapidly to build a catalog of intermetallic structure types.

Tamman, in 1919, studying the effects of nitric acid on Cu-Au alloys, found that the equiatomic composition, when suitably heat-treated, was very resistant to attack.[31] However, if a slight excess of copper were present the alloy was readily attacked, the excess copper at the surface being dissolved leaving the gold behind. He concluded from these results that certain compositions at simple atomic proportions might consist of ordered arrangements of the two species. Although the resistance to acid attack was found to arise from another source, Tamman's surmise of ordered atomic arrangements in intermetallics was later shown by Bain[26] to be correct. This 1923 publication illustrated for the first time an X-ray diffraction pattern (the sketch reproduced in Fig. 5) for an ordered alloy containing, in Bain's words, "supernumerary" lines which were explained as a natural consequence of the ordered structure. Becker and Ebert,[25] in the same year, examined a number of other compounds and confirmed the existence of what are now called superlattice lines, but refuted Bain's suggestion that the structure of an intermediate phase might be related to those of the component elements

The X-ray structural studies proved so direct and informative that attempts were made to define intermetallics on a structural basis. Thus Westgren and Phragmén[32] suggested that "in an ideal intermetallic compound structurally equivalent atoms are chemically identical." However, many exceptions began to appear, such as Ag_3Al, which by its limited homogeneity range and simple formula seemed to belong to the intermetallic class, yet was demonstrated from its diffraction pattern to have a structure consisting of a random statistical distribution of silver and aluminum atoms in the β-Mn structure.[33]

The absence of superlattice reflections in the diffraction patterns of many other compounds, however, was explained by Bain[26] to be due to lack of sufficient difference in the scattering power of the component atomic species. With the advent of high-intensity neutron sources

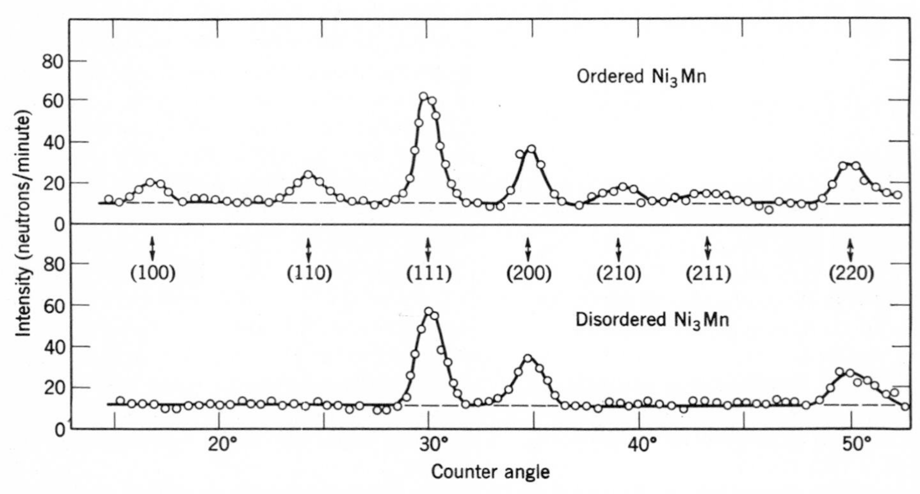

Fig. 6. Neutron diffraction patterns for ordered and disordered samples of Ni_3Mn. (After Shull and Siegel.[34])

from nuclear reactors, it was realized that a new tool was at hand for the determination of order in compounds by virtue of the very different scattering power of atoms for neutrons. Shull and Siegel[34] were the first to publish neutron diffraction patterns for intermetallic phases. Thus, as shown in Fig. 6, the expected extra reflections in the ordered state of the compound Ni_3Mn were readily found, although no suggestion of them can be obtained in the normal X-ray pattern due to the similar X-ray scattering power of Ni and Mn.

Although many intermetallic compounds were found by X-ray diffraction to possess simple structures, identical to those of the common metals, or simple ordered variants thereof, many others were fantastically complex—essentially indecipherable by powder pattern techniques and requiring single-crystal studies for solution. Rossi,[35] in 1932, attempted the first synthesis of single crystals of intermetallics. Although usually more difficult to grow than the elemental metals, sufficient success was frequently obtained to permit detailed structure studies, if not mechanical and physical property measurements. Crystallographic study of intermetallic phases continues to this day to be an important and active field that provides information useful for constitutional studies, interpretation of properties, and understanding of the basis for the occurrence of certain compounds and structures.

4. THEORY OF INTERMETALLIC PHASES

Desch,[36] in writing his monograph in 1914, enunciated a particularly prophetic statement:

> ... attempts to form a theory of the constitution of intermetallic compounds ... have been comparatively unsuccessful. The subject offers a promising field of research, and the results are likely to have an important bearing on the further development of atomic and molecular theory.

This has certainly proved to be the case, as the several books, scores of symposia, and thousands of research publications which have appeared since then testify. As Nevitt[37] has set forth, the goals of the effort have been threefold:

Fig. 7. Portrait of C. H. Desch. (From *J. Inst. Metals.*)

1. The description and classification of alloy phases in terms of composition and crystal structure.
2. The association of structures and structural elements with the character and distribution of bonds between the atoms.
3. An understanding of the factors which determine the relative stability of the phases.

Since, to the present day, not even the simplest binary system can be calculated, the effort has remained largely empirical and relies heavily on ingenious interpretations of large volumes of data.

For a long time the data at hand on the constitution and structure of intermetallics were so few and so fraught with errors that little progress was made in interpretation and generalization. Tamman[38,39] offered two general rules:

1. Neighboring compounds in the periodic system (excepting only the first two short periods) *do not* form compounds.

2. An element either forms compounds with all members of a natural group of the periodic system or with none.

Exceptions began to appear almost immediately, however, and Tamman's concepts died aborning. The earlier generalization of Abegg,[40] to the effect that the tendency to form compounds increases with the increasing heteropolarity of the two elements concerned, that is, with their separation in the periodic table (a sort of inverse corollary of Tamman's first rule), continues to this day as the *electronegativity principle*.

In 1926 Hume-Rothery, whose portrait is shown in Fig. 8, made an outstanding contribution to our understanding of a large group of intermetallics. He pointed out[41] that a great number of compounds, principally in the systems of Cu, Ag, and Au with other elements, could be rationalized by considering the ratio of valence electrons to atoms in each compound. For example, of the first intermediate phases or β-phases in copper-base systems, that in the Cu-Zn system is CuZn with three valence electrons and two atoms, or an electron-atom ratio of 3:2, whereas in the Cu–Sn system the first compound is Cu_5Sn with nine electrons and six atoms, again a ratio of 3:2. Succeeding compounds in these and other systems follow similar rules, the γ-phases with a ratio of 21:13 and the ϵ-phases with a ratio of 7:4.

Fig. 8. Portrait of W. Hume-Rothery. (From *Proc. ASTM.*)

It is surprising that what is now universally regarded as a classic paper was not received with unadulterated admiration at the time it was first read by Hume-Rothery to the British Institute of Metals. The discussions accompanying the published paper make interesting reading. One member described the paper as a "courageous, if not altogether successful attempt." Another expressed the hope that

> . . . the author would not write any more papers like the present until he had done a lot more experimental work, because there was so very much experimental work suggested by the theoretical part of such papers, and he (the discussor) thought that it would be a great mistake to let theories of that kind run too far ahead of practice. They would only get themselves into trouble by carrying such speculations too far when it was quite possible to verify them by experiment.

Evidently, Hume-Rothery was not very far ahead of practice, for in the same volume of the *Journal of the Institute of Metals* two other papers describe other phases with electron-atom ratios of 3:2. Even more striking confirmation was quickly afforded by the extensive X-ray diffraction studies by Westgren and co-workers[42–44] wherein it was shown that with few exceptions all the β-phases had one common crystal structure, the γ phases another, and the ϵ-phases a third.

Only a few years later Jones[45,46] was able to put forward a theory based on a knowledge of the crystal structure and the wave theory of electrons, which accounted for the empirical observation of critical electron-atom ratios. In essence, for structural stability the theory requires that all electrons be accommodated on the lowest portion of the energy-electron wavelength curve, which is to say that only certain electron-to-atom ratios are allowable.

One of the next attempts at systemization of intermetallic compound types to enjoy some success was that of Zintl[47] who proposed a rule for differentiating intermetallics whose bonding is essentially ionic or heteropolar and which possess ionic-type crystal structures from those whose bonding and structures are more typically metallic. He suggested that elements capable of entering into such heteropolar bonds are located in the periodic table no more than 1 to 4 places from the noble gases. Although this rule has had to be modified somewhat in later years, it served sufficiently well to characterize a large family of compounds such as Mg_2Sn, Mg_3Bi_2, and CaTe.

Another important principle of compound formation, that based on the relative sizes of the component atoms, was first set forth by Laves.[48,49] From Goldschmidt's work on ionic compounds, radius ratio was known to be an important structure determining parameter, but in this class of materials the proportions of the component elements are fixed or limited to a small number of possibilities. Laves's concept was that certain metallic structures might be understandable in terms of the creation of dense packings of atoms of different sizes, and this might then require both a limited range of radius ratio for the two species as well as a specific proportion of the components. Compounds of this sort with the type formula AB_2 have since been extensively studied and are known as Laves structures.

With the accumulation of a large volume of crystallochemical data on intermetallics, it was realized that only in limiting cases will the action of a single one of the three structure-determining factors—size ratio, electrochemical difference, or electron-atom ratio—be sufficient to determine structure and bonding type. In the usual case the observed structure is a result of the combined action of two or all three factors.

5. PROPERTIES

Space will permit the citation of only a few of the pioneer works in each of the most interesting areas of physical and chemical properties. For a more detailed historical review the reader is referred to the monographs by Desch[36] and by Giua and Giua–Lollini[50] as well as to the individual chapters of this book.

5.1 Mechanical Properties

The principal mechanical characteristics of intermetallics were recognized very early: their brittleness and their usual extreme hardness relative to the pure components. These characteristics were so outstanding that use was frequently made of them (e.g., by indentation hardness methods) in detecting the presence of compounds in unknown systems. One of the first systematic studies of the mechanical properties of intermetallics was that of Kurnakov and Zhemchuzhnii[51] in 1908. These authors compared the Mohs hardness of about twenty binary compounds with that of their components and found without exception that the hardness of the compound considerably exceeded that of the components. The first studies of the deformation process were by Tamman and Dahl[52] in 1928, who also were among the first to examine the temperature dependence of strength and ductility. They employed compression, ball indentation, and impact techniques and attempted a classification of the compounds studied on the basis of the temperature dependence of the ductile-brittle behavior.

Modern concepts of plastic behavior of crystalline solids are in terms of the structure and motion of dislocations. Koehler and Seitz[53] predicted that the ordered structure of most intermetallics would require a superdislocation, that is, a pair of partial dislocations separated by a strip of material with the ordering 180° out of phase with the normal structure. Such superdislocations were first revealed by Marcinkowski et al.[54] in electron transmission in 1960.

5.2 Magnetic Properties

The magnetic properties of para-, dia-, and ferromagnetic alloys known or suspected to contain intermetallic compounds attracted interest from the earliest times not only as a means of supporting the existence of unique compounds but also for the insights given to the fundamentals of magnetic behavior. The para- and diamagnetic behaviors were not so unusual, but particularly intriguing results were found for ferromagnets. Although intermetallic compounds, one component of which was a ferromagnet, were most frequently found to be nonmagnetic, the reverse case was found in 1903 by Heusler. He demonstrated that entirely nonmagnetic alloys of copper and manganese are rendered magnetic by alloying with a third element such as tin or aluminum.[55] Although it was quickly established that these properties were associated with the formation of intermetallic compounds, agreement on a specific model was not readily attained. Early hypotheses were based on a supposed enhancement of the magnetic properties of the compound MnAl or on raising the magnetic transformation temperature of elemental manganese. Only much later was it finally established that the unique properties are to be associated with the ternary compound Cu_2MnAl. Many other ferromagnetic compounds comprised of nonferromagnetic elements have since been discovered, both isomorphous with the Heusler alloy and with other structures. The ferromagnetism of MnBi was also established very early[56] and has continued to attract attention up to the present time because of its very high coercive force which can reach several thousand oersteds. Another important milestone to be marked is the suggestion of Néel[57] of the possibility of an antiferromagnetically ordered state. This and the related ferrimagnetic ordering have since been found to be especially prevalent in intermetallic compounds. The ability of neutrons to interact, because of their own magnetic moment, with the moments of individual atoms in crystals, hence, to reveal the magnetic structure, has opened up a whole new field of research.

5.3 Chemical Properties

Studies of the heats of formation of alloys were considered useful both for the determination of the existence of compounds and for the insight they gave into the strength of the interatomic bonding. In the first experiments in the late 1800's attempts were made to compare the heat evolved on dissolving a series of alloys in a suitable reagent. The results were poor because it was not fully recognized that the reactions in dissolving the alloy were not the same as those in dissolving the component metals. Apparently, the first successful measurement of the heat of formation of an intermetallic was that of Baker[58] on the CuZn system, using ferric ammonium chloride and cupric ammonium chloride as solvents.

Several important catalysts are prepared by first synthesizing an intermetallic compound and then subjecting it to attack by a strong reagent. Perhaps the most famous of these is the Raney nickel prepared from NiAl intermetallics by sodium hydroxide treatment and patented by Raney in 1927.[59] Although this is the most commonly used of nickel catalysts, and the most versatile of all catalysts, little is yet known of the true metallurgical state of the final product.

Although many intermetallic compounds are quite oxidation- and corrosion-resistant, it was early noted in attempts to isolate intermetallic compounds by reagent attack that many were *more* readily attacked than the constituent elements. Indeed, many magnesides, bismuthides, and aluminides were found to disintegrate spontaneously in air,[60,61] a phenomenon rediscovered by Fitzer[62] in $MoSi_2$ in 1955 and dubbed the "pest" effect.

5.4 Semiconducting Properties

The most common and probably the most important group of semiconducting intermetallic compounds are those found between elements of group III and V of the periodic table. Although the existence of some III-V's such as AlSb and AlP were known in the nineteenth century, it was not until the work of Goldschmidt[63] in 1929 that it was appreciated that a large group of isomorphous compounds existed in this class. Semiconducting properties were first reported for the compound InSb in 1950 by Blum et al.[64] and by Goryunova.[65] The field was really opened up, however, by Welker in a classic paper in 1952[66] in which he emphasized that the high carrier mobilities and wide range of band gaps available among the III-V compounds made them an important new family of semiconductors. Since then interest in this group of compounds has grown so much that well over one thousand papers on their preparation and properties have been published.

6. STATUS AND PROSPECT

It appears that a period of amassing structural, constitutional, and property data and the empirical systemization of the results is coming to a close. Experimental activity directed at understanding is, from this point on, likely to be increasingly devoted, on the one hand, to the application of sophisticated techniques, such as Mössbauer measurements or nuclear magnetic resonance which yield direct information on the nature of the bonding in compounds, and, on the other, to the development of refined metallurgical procedures to control structure and properties. Perhaps the day is not far distant when purely theoretical calculations of structure, bonding, and properties of intermetallics may be attempted. On the practical side we are already witnessing in compounds such as superconducting Cb_3Sn, ferromagnetic CoPt, and semiconducting GaAs the first large-scale commercial applications of this rich family of compounds. This trend is certain to continue and to accelerate.

REFERENCES

1. Roberts-Austen W. C., Fifth Report to the Alloys Research Committee; Steel (1899).
2. Karsten K., *Pogg. Ann.* **46** (1839), Series 2, 160.
3. Calvert F. C., and R. Johnson, *Phil Mag.* **10** (1855) [IV] 240.
4. Croockewit K., *Ann. Chem.* **68** (1848) 289–293.
5. Hansen M., and K. Anderko, *Constitution of Binary Alloys.* McGraw-Hill Book Co., New York, 1958.
6. Rudberg F., *Ann. Phys. Chem.* **18** (1830) 240. See also *ibid.* **21** (1831) 317.
7. Tamman G., *Z. Anorg. Chem.* **47** (1906) 296.
8. Tamman G., *Z. Anorg. Chem.* **45** (1905) 24.
9. Calvert F. C., and R., Johnson, *Phil. Trans.* **148** (1858) 349.
10. Matthiessen A., *Phil. Trans.* **148** (1858) 369.
11. Matthiessen A., *Phil. Trans.* **150** (1860) 177.
12. LeChatelier H., *Rev. Gen. Sciences* **6** (1895) 531.
13. Urasov G., and V. Nikolaev, *Usp. Khim.* **8** (1939) 812.
14. Smirnov V. I., and N. S. Kurnakov, *Z. Anorg. Allgem. Chem.* **72** (1911) 31, orig. in *Zh. Russ. Khim. O.* **43** (1911) 725.
15. Neville F. H., *British Assoc. for Advan. Sci. Reports* (1900) 131.
16. Tamman G., *Z. Anorg. Allgem. Chem.* **37** (1903) 303.
17. Kurnakov N. S., *Z. Anorg. Chem.* **88** (1914) 109.
18. Jänecke E., *Z. Physik. Chem.* **57** (1906) 507.
19. Barlow W., and W. J. Pope, *Trans. Chem. Soc.* **89** (1906) 1675.
20. Laue M., W. Friedrick, and P. Knipping, *Univ. Münich Sitzber. K. Akad. Wiss. München* (1912) 303.
21. Debye P., and P. Scherrer, *Physik Z.* **18** (1917) 291.
22. Hull A., *Phys. Rev.* **10** (1917) 661.
23. Andrews Mary R., *Phys. Rev.* **18** (1921) 245.
24. Goldschmidt V. M., *Z. Metallk.* **13** (1921) 449.
25. Becker K., and F. Ebert, *Z. Physik* **16** (1923) 168.
26. Bain E. C., *Chem. and Met. Eng.* **28** (1923) 21, 65.
27. Westgren A., and G. Phragmén, *Phil. Mag. J. Sci.* **50** (1925) 311.
28. Westgren A., and G. Phragmén, *Z. Metallk.* **18** (1926) 279.
29. Bradley A. J., and J. Thewlis, *Proc. Roy. Soc.* **A112** (1926) 678.
30. Bradley A. J., *Phil. Mag. J. Sci.* **6** (1928) 878.
31. Tamman G., *Z. Anorg. Chem.* **107** (1919) 1.
32. Westgren A., and G., Phragmén, *Metallwirtschaft* **7** (1928) 700.
33. Westgren A., and A. J. Bradley, *Phil. Mag. J. Sci.* **6** (1928) 280.
34. Shull C. G., and S. Siegel, *Phys. Rev.* **75** (1949) 1008.
35. Rossi C., *Z. Physik* **74** (1932) 707.
36. Desch C. H., *Intermetallic Compounds.* Longmans Green, London, 1914.
37. Nevitt M. V., "Alloy Chemistry of Transition Elements," in *Electronic and Alloy Chemistry of Transition Elements*, P. A. Beck, ed. Interscience, New York, 1963, p. 101.
38. Tamman G., *Z. Anorg. Chem.* **49** (1906) 113.
39. Tamman G., *Z. Anorg. Chem.* **55** (1907) 289.
40. Abegg R., *Z. Anorg. Chem.* **39** (1904) 330.
41. Hume-Rothery W., *J. Inst. Metals* **35** (1926) 307.
42. Westgren A., and G. Phragmén, *Trans. Faraday Soc.* **25** (1929) 379.
43. Westgren A., *Metallwirtschaft* **9** (1930) 1919.
44. Westgren A., *Z. Metallk.* **22** (1930) 368.
45. Jones H., *Proc. Roy. Soc.* **A144** (1934) 225.

46. Jones H., *Proc. Roy. Soc.* **A147** (1934) 396.
47. Zintl E., and H. Kaiser, *Z. Anorg. Allgem. Chem.* **221** (1933) 113.
48. Laves F., and H. Witte, *Metallwirtschaft* **14** (1935) 645.
49. Laves F., *Naturwiss.* **27** (1939) 65.
50. Giua M., and C. Giua-Lollini, *Chemical Combination among Metals* (English trans. from Italian). Blakiston's Son and Co., Philadelphia, 1918.
51. Kurnakov N. S., and S. F. Zhemchuzhnii, *Z. Anorg. Chem.* **60** (1908) 1.
52. Tamman G., and K. Dahl, *Z. Anorg. Allgem. Chem.* **126** (1923) 104.
53. Koehler J. S., and F. Seitz, *J. Appl. Mech.* **14** (1947) A217.
54. Marcinkowski M. J., R. M. Feiker, and N. Brown, *J. Appl. Phys.* **3** (1960) 1303.
55. Heusler F., *Verhandl. Deut. Phys. Ges.* **5** (1903) 219.
56. Heusler F., *Z. Angew. Chem.* **17** (1904) 260.
57. Néel L., *Ann. Phys.* **17** (1932) 5.
58. Baker T. J., *Phil. Trans.* **196A** (1901) 529.
59. Raney M., U.S. Patent 1,628,190 (1927).
60. Sperry E. S., *Trans. AIME* **29** (1899) 280. See also discussion by A. E. Hurt and by S. Peters, *ibid.* 1029.
61. Tamman G., and A. Ruhenbeck, *Z. Anorg. Chem.* **223** (1935) 288.
62. Fitzer E., "Molybdenum Disilicide as a High Temperature Material," in *Plansee Proceedings* (1955). F. Benesovsky, ed. Pergamon Press, London, 1956.
63. Goldschmidt V. M., *Trans. Faraday Soc.* **25** (1929) 253.
64. Blum A. N., N. P. Mokrovskii, and A. R. Regel, *7th All Union Conference on Properties of Semiconductors*, Kiev (1950).
65. Goryunova N. A., Ph.D. Thesis, Leningrad Polytechnic Institute (1951).
66. Welker H., *Z. Naturforsch.* **11** (1952) 744.

part II

Bonding and Related Properties

chapter 2

Theories of Bonding in Metals and Alloys

R. E. RUNDLE*

Department of Chemistry
and
Ames Laboratory of the Atomic Energy Commission
Iowa State University
Ames, Iowa

1. INTRODUCTION

There is as yet no real theory of metals and alloys. In this chapter we will explore, mostly qualitatively, the sense in which there is a theory, its limitations, and some approaches to other "theories" which may or may not be helpful in providing a theory of metals and alloys. This chapter contains some chemical ideas rather than a purely physical outlook on this subject. These ideas are still empirical and qualitative, but metals and alloys appear to have some surprising resemblances to other branches of chemistry, and perhaps a purely physical approach has caused generalizations to be made too early in the development of a theory of metals. In a few places some of these chemical resemblances are pointed out, and some suggestions are made as to where metal theory may need modification in directions similar to those necessary in chemical valence.

By far the simplest approach to metals has been free electron theory, and it has also been the most generally successful. Some conceptually simple modifications of this theory go far indeed toward providing a real theory of metals. This theory and some of its most promising ramifications are then a first and important part of this chapter.

Unfortunately, even modified free electron theories are almost structureless and have little to say about metal-metal distances, metal and alloy structure, compositions of alloy compounds, and other matters of considerable

* Deceased. Unfortunately Dr. Rundle died before he had the opportunity to revise his contribution. Editor.

interest to chemists and metallurgists. Hence it seems worthwhile to examine empirical theories and their foundations, since they have, without question, considerable power to give good values to measurable quantities of interest, though foundations for these empirical theories are weak.

2. FREE ELECTRON THEORY*

2.1 Simple Theory

The conductivity of metals suggests that some of the electrons of a metal are essentially free, moving under the slightest electrical potential, that is, without a potential barrier, even at the lowest temperatures. First, the results of assuming that the valence electrons of a metal are free but subject to the Pauli principle are reviewed briefly to show that the results are in surprisingly good agreement with many of the properties of metals. Then a section is devoted to explaining why and in what sense this is a reasonable approximation and how the worst features of the approximation can be eliminated to give a truer picture of the electronic state of a metal.

Completely free particles in the quantum theory[1] behave, in a sense, as waves, in which the wavelength and energies of the particles may take on any values as long as the particles are unconfined. If the electrons in a metal behaved in this manner, the wavelengths, momenta, and energies of the various electrons could differ infinitesimally from one another, and all the electrons could occupy very low energy states and still fulfill the Pauli principle, which requires that they occupy as many different states as there are electrons. If, however, the particles are confined, for example, to a rectangular box, the wave function ψ must go to zero at the edges of the box if the walls constitute an infinite barrier. (Otherwise, from the form of the Schroedinger equation it can be shown that ψ and the probability density $\psi\psi^*$ for a given particle will go to infinity outside the box, which is physically forbidden.) Out of the infinity of states allowed to the unconfined electron, confinement allows only a discrete number of states which fulfill this requirement.

For a single confined particle in a box with one corner at 000 and others at $a00$, $0b0$, $00c$, etc., the physical requirement that ψ be interpretable as a probability amplitude function demands that the wave function for this particle have the form

$$\psi = \sqrt{\frac{8}{V}} \sin \frac{n_x \pi x}{a} \sin \frac{n_y \pi y}{b} \sin \frac{n_z \pi z}{c} \tag{1}$$

where the coefficient $\sqrt{8/V}$ is a normalizing factor such that the total probability of finding the particle in the box is 1; n_x may have values of 1, 2, 3, etc. ψ, then, has the form of standing waves; ψ for one dimension is shown in Fig. 1*a*, and ψ^2, the probability density of the particle, is shown for one dimension in Fig. 1*b*. The energy of the particle is

$$E = \frac{h^2}{8m}\left(\frac{n_x^2}{a^2} + \frac{n_y^2}{b^2} + \frac{n_z^2}{c^2}\right) \tag{2}$$

E may be divided into components, $E_x = (h^2/8m)(n_x^2/a^2)$, etc.

The deBroglie wavelength of a particle is given by the familiar expression

$$\lambda_x = \frac{h}{m\dot{x}_x} \quad \text{or} \quad \lambda_x = \frac{h}{p_x} \tag{3}$$

where p is momentum. One may, then write classically

$$p_x = \sqrt{2mE_x} = \frac{hn_x}{2a} \quad \text{or} \quad \lambda_x = \frac{2a}{n_x} \tag{4}$$

This λ_x, then, is identical to the wavelength of the standing waves, as shown in Fig. 1*a*.

Other boundary conditions are permissible and more convenient for particles in three dimensions; they are the periodic boundary conditions with the periods a, b, and c, for the box and have the form

$$\psi = \left(\frac{1}{abc}\right)^{1/2} \exp\left[2\pi i\left(\frac{n_x x}{a} + \frac{n_y y}{b} + \frac{n_z z}{c}\right)\right] \tag{5}$$

as can be easily verified. It is convenient to deal with a cubic box, which does not change the character of the problem. Then

$$\psi_{(xyz)} = \left(\frac{1}{V}\right)^{1/2} \exp\left[\frac{2\pi i}{a}(n_x x + n_y y + n_z z)\right] \tag{6}$$

$$E = \frac{h^2}{2mV^{2/3}}(n_x^2 + n_y^2 + n_z^2) = \frac{h^2 n^2}{2mV^{2/3}} \tag{7}$$

* For more quantitative treatments see Seitz[1] or Kittel.[2]

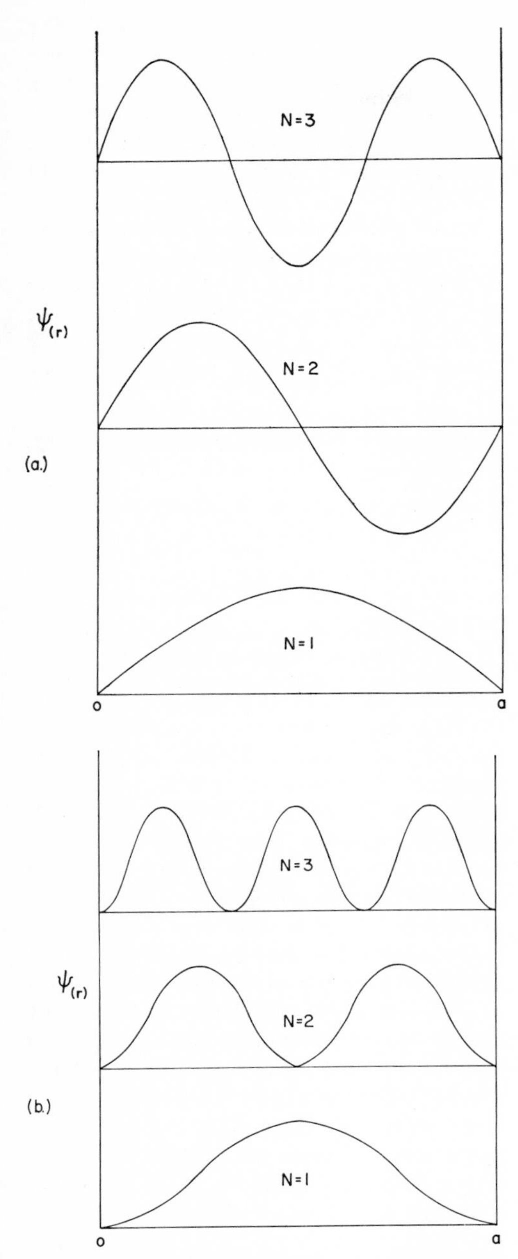

Fig. 1. Standing-wave solutions for the particle in a box problem: *(a)* ψ, the probability amplitude, *(b)* ψ^2, the probability density.

where $n^2 = n_x^2 + n_y^2 + n_z^2$ and $V = a^3$, the volume of the box. It is common and convenient to let $k_x = 2\pi n_x/a$, etc., so that (6) becomes

$$\psi_{(xyz)} = \left(\frac{1}{V}\right)^{1/2} e^{i(k_x x + k_y y + k_z z)} \tag{6a}$$

and (7) becomes

$$E = \frac{\hbar^2}{2m}(k_x^2 + k_y^2 + k_z^2) \tag{7a}$$

Since $p_x = \sqrt{2mE_x}$, etc., it is found that

$$k_x = p_x/\hbar \tag{8}$$

A state is now characterized by the three quantum numbers, n_x, n_y, n_z, plus a quantum number n_s, the spin quantum number, which may have values of $\pm\frac{1}{2}$. The Pauli principle requires a distinct set of quantum numbers for each electron, and, since the value E, the energy, rises with the values chosen for n_x, n_y, and n_z, it is clear that the kinetic energy of the free electrons now increases with the electron density.

In what follows enough quantitative details are given to show that for the "free" electron densities found in metals the Pauli principle requires that electron states be filled to surprisingly high kinetic energy values and that many of the physical properties of metals are determined by this fact alone. It is then shown that even in more complex theories, in which the potential energies of the "free" electrons with the positive cores and the "free" electrons with each other are taken into account, the kinetic energy of the electrons is still exactly or nearly exactly equal to the kinetic energy given by free electron theory and that, even though the electrons are not free, those properties dependent only or mostly on the kinetic energy terms are quantitatively or semi-quantitatively understandable by free electron theory. Of course, it must be expected that the cohesive energy and many other properties of metals depend also on potential energy terms, and for these properties free electron theory is no theory at all.

Thus k_x, k_y, and k_z are given in units of momenta/$\hbar$ and have the dimensions of cm^{-1}. For a cubic crystal the state of an electron can be given by a point in k or reciprocal space. These wave functions represent running waves, not standing waves. Because $\psi\psi^* = 1/V$, the electron density for such waves is constant throughout the box. In accordance with the Pauli principle, two electrons with $m_s = \pm\frac{1}{2}$ may occupy each separate level, as given by the three quantum numbers n_x, n_y, n_z or the values of k_x, k_y, and k_z. With $a = b = c$ there are many degeneracies, but for present purposes the problem is to find the top energy level necessary to hold all of the valence electrons of the metal.

The states for electrons may be considered as forming a cubic lattice (Fig. 2) in which the origin is at the state (000). Each point can be given a designation (n_x, n_y, n_z) or (k_x, k_y, k_z), and clearly the farther the point from the origin the higher the energy. Points at the same distance from the origin are degenerate, and degeneracies increase rapidly with energy. There is one point per unit volume in n-space, so that the number of points contained out to the value $n = n_F$ is

$$\int_0^{n_F} 4\pi n^2 \, dn = \frac{4\pi}{3} n_F{}^3 \tag{9}$$

Each point represents two electronic states, or there are $\frac{8}{3}\pi n_F{}^3$ states out to $n = n_F$.

Now for a metal of electron density ρ the number of electrons to be accommodated in the box is $a^3\rho$; hence we may write

$$\frac{8\pi n_F{}^3}{3} = \rho a^3 \tag{10}$$

The energy of the highest level which must be filled to accommodate all the electrons, after we eliminate n_F by the use of (10), is

$$E_F = \frac{h^2}{8m}\left(\frac{3\rho}{\pi}\right)^{2/3} \tag{11}$$

in which E_F is the Fermi level of the metal. At $T = 0°K$ all the electronic levels with $n \leqq n_F$ would be filled, but none with $n > n_F$. The Fermi energy for all metals is very high. Thus for potassium metal, in which there are two electrons/unit cell and the lattice constant is 5.33 Å, E_F is found to be about 2.4 eV by (9), corresponding to a temperature of about 28,000°K.

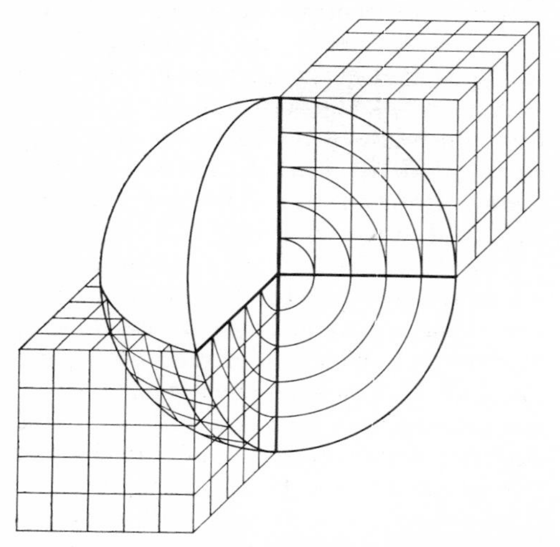

Fig. 2. States in n- or k-space. Spheres in this space are constant-energy surfaces. These states are for the periodic (plane-wave) solutions for the particle in a box.

It is of interest also to learn, the total energy of the electrons in the metal. This is determined by weighting each state with the energy of that state and summing over all states. As in the computation of the number of electrons, the summation is obtained by integration. Using (7)

$$E_T = 2\int_0^{n_F} \frac{h^2 n^2}{2ma^2} 4\pi n^2 \, dn = \frac{4\pi h^2 n_F{}^5}{5ma^2} \tag{12}$$

On substituting (9), n_F is eliminated and

$$E_T(0) = \frac{\pi h^2 a^3}{40m}\left(\frac{3\rho}{\pi}\right)^{5/3} \tag{13}$$

2.2 Electronic Contribution to the Specific Heat of Metals

Electrons in a metal are distributed over the available energy levels in accordance with Fermi-Dirac statistics. Because of the Pauli principle, the electrons at 0°K occupy all levels up to E_F and none above it. Above 0°K there is excitation into higher levels, but because the Fermi levels for metals are several electron volts, or equivalent to a temperature of almost 50,000°K for most metals, this excitation involves only the small fraction of electrons at the top of the Fermi level which may be excited only into levels approximately kT above the Fermi level; that is, the fraction of the electrons that can be excited is $\propto T/T_F$, and the excitation energy is also proportional to kT, so that the extra molal energy at some temperature T above the energy at 0°K is proportional to $(NkT^2)/T_F$. Thus the molar electronic heat capacity is expected to be

$$C_e \propto \frac{RT}{T_F} \tag{14}$$

Because T/T_F for most metals is of the order of 0.001 at room temperature, we expect the electronic heat capacity to be relatively unimportant at this temperature. However, since C_e decreases linearly with T, whereas at very low temperatures the lattice heat capacity decreases as T^3, C_e may become the predominant heat capacity at sufficiently low temperatures. A quantitative application of statistical mechanical methods to the molar electronic heat capacity for free electron theory has been made[1] and gives

$$C_e = \frac{\frac{1}{2}\pi^2 ZRT}{T_F} \tag{15}$$

in which Z is the valence of the metal.

2.3 Paramagnetism in Metals

For paramagnetic salts the Langevin equation holds quite well. In such cases each magnetic moment μ can be aligned parallel or antiparallel to an applied field H. The energy difference between the two states is $2\mu H$, and by the Boltzmann distribution law the fraction aligned is proportional to $\exp(-2\mu H/kT)$ or, if μH is small, to $2\mu H/kT$. The net magnetic moment is then proportional to $\mu^2 H/kT$.

For metals, as we have seen, all states of each spin are filled except for a region proportional to T/T_F at the top of the Fermi level, and only these electrons can orient in the direction of the field. Hence for a metal the net magnetic moment in a field H must be weighted by this factor or

$$M \propto \frac{T}{T_F}\frac{\mu^2 H}{kT} \propto \frac{\mu^2 H}{kT_F} \tag{16}$$

Hence a metal should have a slight, temperature-independent magnetism. The magnetic susceptibility $\chi = M/H \propto N\mu^2/kT_F$, as first noted by Pauli. Pauli's detailed calculation gives the proportionality constant[3]

$$\chi = \frac{3N\mu_B{}^2}{2kT_F} \tag{17}$$

for a free electron gas.

Many other physical properties of metals, including conductivity, may be treated rather well by free electron theory to get quantitative or good semiquantitative agreement with observations of the properties of metals,[1,2] so that there must be a close correspondence between simple free electron theory and a true theory of metals.

In another sense, simple free electron theory cannot be a true theory of metals, since all that has been treated is the kinetic energy of the electrons. To understand what holds a metal together we must recognize that the "free" electrons are held in a periodic potential field of the positive metal cores. A proper treatment of this problem is necessary to an understanding of the cohesive energy of metals.

3. CELLULAR THEORIES OF METALS

3.1 Wigner-Seitz Treatment

One of the most successful treatments of metals through a cellular model has been that of Wigner and Seitz.[1,2] The model is as follows: the valence electrons of the metal are treated as though they were confined to a cell or polyhedron about each metal core. This cell is formed by drawing lines from one atom to all its near-neighbors and constructing planes through the midpoints of these connecting lines and perpendicular to the line. These cells for body-centered and face-centered cubic structures are a cubically truncated octahedron and a dodecahedron with rhombic faces and O_h symmetry, respectively (Fig. 3).

Since these polyhedra approximate spheres, they are replaced by spheres of equal volume. (All treatments of metals by this model have used a spherical approximation.*) The best

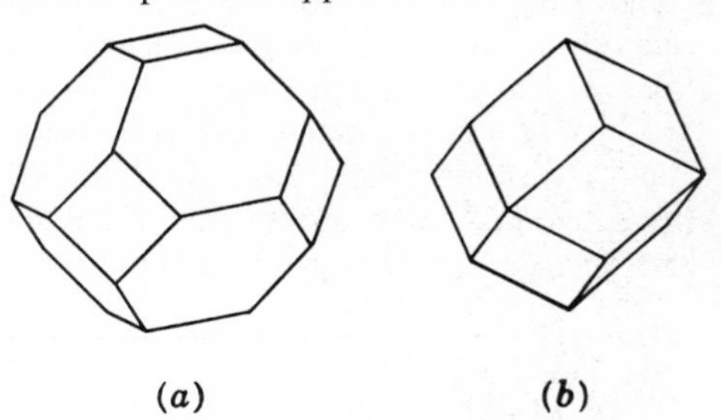

Fig. 3. Polyhedra enclosing metal atoms. Truncated octahedron encloses atom in body-centered cubic metal. Cubic dodecahedron encloses atom in cubic closest-packing.

* Some early works did attempt to use polyhedra, but they were generally too complex to prove useful. In the last ten or fifteen years many sophisticated refinements of the cellular model have been made and have been applied with some success to polyvalent metals and with lesser success to transition metals and copper. H. Brooks has presented two recent reviews of this subject.[4,5]

Before any of these methods become quantitative, rather formidable problems of electron interaction and correlation must be solved. The status of work in this field has been reviewed by Pines.[6]

It will be clear to anyone who examines these reviews that only the barest essentials of the cellular method stripped of its complexities are given here.

possible approximate potential function $V(r)$ is obtained from theoretical self-consistent field calculations for isolated atom cores. This $V(r)$ is the potential energy to which the valence electron is subject within the spherical cell as already defined. It is further assumed that $(\partial\psi/\partial r)_{r=r_0} = 0$ where ψ is the wave function for the valence electron and r_0 is the value of r at the boundary of the sphere. With these approximations the Schroedinger equation is solved for $\psi(r)$ for the electron, and r_0 is chosen to minimize the energy. The result of this treatment is that the wave function for the valence electron in the metal is nearly identical to the wave function for the valence electron in the free atom at values of r close to the nucleus, whereas for the metal electron ψ becomes very flat and nearly constant for larger values of r (Fig. 4). The minimization of the energy produces an r_0 much greater than the ionic radius of the core, so that throughout most of the volume of the cell $\psi(r)$ is a constant and therefore looks much like the electron distribution assumed in free electron theory. The reasons for this can be followed quite easily. Near the nucleus the valence electron distribution cannot differ greatly from the free atom because of the requirements of the Pauli principle and the strong potential function near the nucleus. That is, the wave function near the nucleus cannot be appreciably different from the atomic case and still be orthogonal to the core electrons and of low energy. Beyond this point,

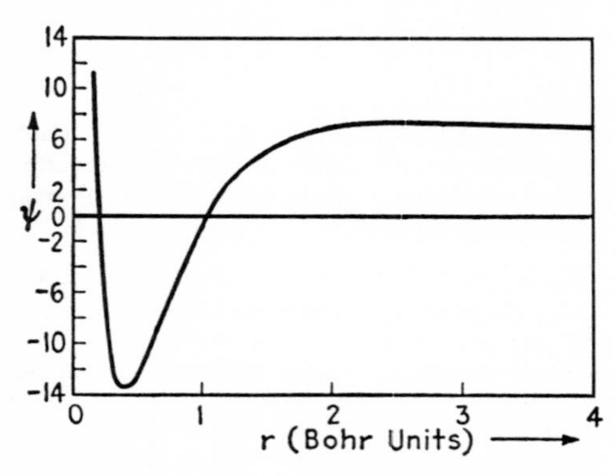

Fig. 4. Wave function, ψ, for valence electron in sodium metal, according to Wigner and Seitz. ψ is very similar to that for the free atom near the nucleus, but flattens out to become nearly constant throughout 90 per cent of the volume of the surrounding sphere. (From Seitz,[1] p. 351, by permission of author and publisher.)

where the core electron density is small and puts minor requirements on the valence electron, the kinetic energy is lowered by increasing r_0 and keeping $\psi(r)$ very flat, whereas the potential energy is increased the farther the electron is from the nucleus. The valence electron actually has a higher density near the nucleus in the metal than in the free atom to lower the potential energy, but the electron density does not fall off so rapidly at larger distances, for it is constrained to be finite at the cell boundary and leaves a very flat electron density which lowers the kinetic energy. In these respects the wave function for the valence electron behaves very much like that for a bonding electron in a covalent bond along the bond and for nearly identical reasons.[7] The value of r_0 arises from the competition between the potential energy, favoring low values of r_0, and the kinetic energy, favoring large values of r_0. The wave function thus found is joined continuously with those in other cells, hence it is correct only for one given electron with zero momentum for $k = 0$. The succeeding electrons placed in the metal must have greater momenta in just the way computed for free electron theory.

This treatment is quite successful for the alkali metals in predicting interatomic distances (and molar volumes), cohesive energies, and other physical properties. It is not so successful when applied to copper (assuming a valence of 1). The assumption of a spherical cell leaves the theory structureless, but there is good evidence that for many metals the energy of transformation from one form of close-packing to another (even bcc to ccp or hcp) leads to trivial energy and volume changes. So far the theory has not been extended to higher valences.

One of the real merits of the model is that the electron density derived from it justifies, to considerable extent, the type of valence electron distribution assumed in free electron theory and leaves the successful conclusions from free electron theory valid, since the behavior of the kinetic energy is just that obtained from free electron theory.

3.2 Simplified Cellular Model

It is quite difficult to apply the Wigner-Seitz model so that it will take care of more complex problems. For example, in free electron theory or the Wigner-Seitz treatment it has not been possible to treat metal valence. It is generally assumed that the number of valence electrons will correspond to the group number of the periodic table to which the metal belongs. This is unambiguous and almost certainly correct for the simpler metals, but it would be more satisfying if the valence of the metal arose naturally from the theory.

It is possible to remove most of the complex features of the Wigner-Seitz treatment without ruining its main successes, and the simplified model can be used to discuss certain more complex problems including valence.

This model, due to Dr. R. H. Good,[8] again takes a spherical cell for treatment. The result of the Wigner-Seitz treatment is used as justification for dividing the electrons into shell electrons, in which the electron distribution may be the same as that for the free ion core, and into valence electrons, which are distributed uniformly throughout the sphere. (It is not more troublesome to exclude the valence electrons from the volume occupied by the core electrons by taking the size of the positive core as equal to the size of the metal ion as found in crystals. This would improve the results to some extent, but for simplicity this refinement is not included here.) The potential energy of the valence electron is then calculated, taking into account the effect on one another of the coulombic field of the ionic core and the field of the valence electrons. The kinetic energy is computed as for the particle in a box satisfying a periodic potential, as described for free electrons. Both energies will depend on r_0 and Z, the number of valence electrons removed from the metal. The energy is minimized with respect to r_0 for any given Z, leaving r_0 a function of Z and the energy a function of Z alone. It is now possible, using known ionization energies of the metal, to calculate the cohesive energy for values of Z and to find the minimum energy for the condensed metal. It turns out naturally, now, that for simple metals Z equals the number of the group in the periodic table, for, as Z increases, the potential energy of the valence electrons decreases rapidly in the field of the higher-charged core. However, this energy must be paid for by ionization of the metal valence electrons. The ionization energies rise slowly enough with Z so that the extra lowering of the energy easily offsets increased ionization until the core is reduced to a rare gas core. At this point further ionization requires energy so great that it is energetically unfavorable to attain higher valences.

The model also shows how molar volume depends on valence and gives good compressibilities for metals. It is clear that since the treatment of kinetic energy is the same as that for free electron theory this simple cellular theory will reproduce the results obtained from free electron theory which depend primarily on kinetic energy.

Briefly, a straightforward classical treatment gives the potential energy for the coulombic interaction of the electron gas with the positive ion core and the electron gas on itself as

$$V = -\frac{9}{10}\frac{Z^2e^2}{r_0} \tag{18}$$

In computing the kinetic energy it is again assumed that the electrons occupy momentum states from 0 to p_F corresponding to E_F, where it is to be noted that E_F is the kinetic energy only. By altering (12) so that it is suitable for a spherical shell and letting $Z = 1$ and $a^3 = (4\pi/3)r_0^3$, we obtain

$$E_F = \frac{3}{10}\frac{\hbar^2}{m}\left(\frac{9\pi}{4}\right)^{2/3}\frac{1}{r_0^2}Z^{5/3} \tag{19}$$

$$E = E_F + V$$

$$= \frac{3}{10}\frac{\hbar^2}{m}\left(\frac{9\pi}{4}\right)^{2/3}\frac{1}{r_0^2}Z^{5/3} - \frac{9}{10}\frac{Z^2e^2}{r_0} \tag{20}$$

Minimization of E with respect to r_0 gives

$$r_0 = \frac{2}{3}\frac{\hbar^2}{me^2}\left(\frac{9\pi}{4}\right)^{2/3}Z^{-1/3} \tag{21}$$

whence

$$V\,(\text{molar}) = N\frac{4}{3}\pi r_0^3 = 3\pi/2Z \text{ cm}^3/\text{mole} \tag{22}$$

That the volume dependence on Z is approximately correct can be verified from the molal volumes

Element	Na	Mg	Al	
$\bar{V}m$	24	14	10	cm³/mole
$\bar{V}m$ calculated*	24	12	8	

* Taking $\bar{V}m$ for sodium as correct. Clearly, the constant term $3\pi/2Z$ is incorrect by a factor of about 5. This is improved considerably by excluding the valence electrons from the volume occupied by the Na^+, Mg^{2+}, and Al^{3+} cores, respectively.

Using the equilibrium value of r_0, we have

$$E = -\frac{27}{40}\left(\frac{4}{9\pi}\right)^{2/3}\frac{me^2}{\hbar^2}Z^{7/3} \cong -5Z^{7/3}\ \text{eV/mole} \tag{23}$$

This is the energy change when a mole of gaseous ions of charge $+Ze$ and Z moles of electrons condense to form a mole of metal. The cohesive energy of a metal/mole is the energy decrease when a mole of gaseous atoms condenses to form a metal. Hence

$$E_{coh} = -5Z^{7/3} + E_I \tag{24}$$

E_I (ionization energy) can be obtained very accurately from spectroscopic data.[9] In Table 1 the three terms in (24) are shown. It is clear that the cohesive energy would continue to rise with higher and higher ionizations if it

TABLE 1

Comparison of Ionization Energy and Observed Cohesive Energy With Cohesive Energy From a Simple Cellular Model (Energies in eV)

Element	Z	$-5Z^{7/3}$	E_I	E_{coh} (calculated)	E_{coh} (observed)
Na	**1**	**−5**	**5.1**	**+0.1**	**−1.12**
	2	−25	47.3	+22	
Mg	1	−5	7.6	+2.6	
	2	**−25**	**22.6**	**−2.4**	**−1.6**
	3	−65	102.7	+37.7	
Al	1	−5	6.0	+1.0	
	2	−25	24.8	−0.2	
	3	**−65**	**53.2**	**−11.8**	**−3.2**
	4	−127	173.2	+46.2	
Si	1	−5	8.15	+3.15	
	2	−25	24.5	−0.5	
	3	−65	58	−7.0	
	4	**−127**	**103**	**−24.0**	**−3.7**
	5	−214	270	+56.0	
P	1	−5	11	+6	
	2	−25	30.7	+5.7	
	3	−65	60.8	−4.2	
	4	−127	112.2	−15.8	
	5	**−214**	**177.2**	**−36.8**	**−3.3**
	6	−345	397.6	+52.6	
S	1	−5	10.4	+.54	
	2	−25	33.8	+5.8	
	3	−65	68.8	+3.8	
	4	−127	116.1	−10.9	
	5	−214	188.6	−25.4	
	6	**−345**	**276.6**	**−68.4**	**−2.3**
	7	−469	557.6	+88.6	
Cl	1	−5	13.0	+8.0	
	2	−25	36.8	+11.8	
	3	−65	76.7	+11.7	
	4	−127	130.2	+3.2	
	5	−214	198.0	−16.0	
	6	−345	294.7	−50.3	
	7	**−469**	**409.0**	**−60.0**	**−1.3**
	8	−640	757.3	+117.3	

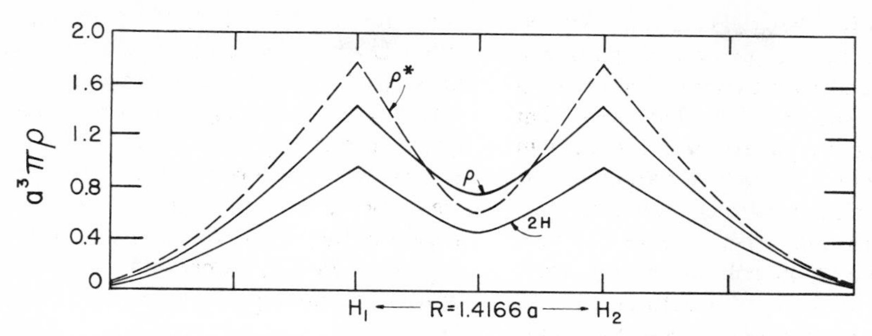

Fig. 5. Electron density in the H_2 bond, according to the Weinbaum function, ρ; for two hydrogen atoms with exponential coefficient as found for Weinbaum function, ρ^*; and for two normal hydrogen atoms, 2H. (Courtesy of K. Ruedenberg,[7] p. 370, and *Rev. Mod. Phys.*)

were not for E_I. Within one shell E_I rises more slowly than $5Z^{7/3}$, but when a rare gas core is reached further ionization greatly exceeds the increase in $5Z^{7/3}$ so that the stable valence for a metal arises quite naturally and corresponds to chemical valence for the same reason, namely the high energy of ionization of a rare gas core.

In Table 1 the valence predicted by this simple theory is in boldface. Clearly, the theory accounts satisfactorily for the valence of the elements in the first four groups and fails for those in the last four, although in the form given here the actual values for the cohesive energies are very poor.*

The reasons for this are not hard for a chemist to understand. In the cellular models which have been used so far it is assumed that the wave function of the electron in the cell has $\partial\psi/\partial r = 0$ at the boundary, giving a Wigner-Seitz type of wave function (Fig. 4). This is analogous to bonding molecular orbitals in chemistry. See, for example, the wave function for H_2 along the molecular axis in Fig. 5. In molecular problems in which it is simple to follow the requirements of the Pauli principle, orthogonality and symmetry, it is easily shown that only one such molecular orbital (MO) can be formed for each valence orbital of the atom considered if this AO overlaps AO's of neighboring atoms. This MO is filled by one electron from each atom of the bonded pair. For each bonding molecular orbital there is also an antibonding MO of higher energy which has a node between the two atoms bound, as in Fig. 5. Hence, by analogy with what is well established in molecular theory, it seems reasonable to conclude that wave functions with $(\partial\psi/\partial r)_{r=r_0} = 0$ at the cell boundary are limited to *half* the total number of states for the atom, whereas the remaining states have a node at the cell boundary, that is, $\psi(r)_{r=r_0} = 0$. These wave functions are now antibonding; that is, they will increase the energy of the system if they are occupied by electrons. Some other analogies between metal bonding and chemical bonding are given in Section 4 in an attempt to establish that these analogies deserve serious study.

In Table 1 Good's theory is tested for valence predictions only for those elements just above a rare gas in atomic number. If it is tested for the metals just beyond closed *d*-shells (Cu, Zn, Ga, etc., Ag, Cd, In, etc.), the results are more ambiguous, but generally the theory predicts valences higher than the group number, more nearly in line with Pauling's metal valences.[10] Thus this theory along with Pauling's empirical theory implies that the valence electrons are decreasing in the series Cu, Zn, Ga, for the volume (and bond distances) are increasing in this series. This simple theory also suggests very high valences for metals of the iron group, but until it can be made more fundamental in nature it is not to be trusted quantitatively in this complex area of metal theory. Qualitatively, (24) suggests that it is probably incorrect to speak of a very low valence for metals with low atomic volumes and short interatomic distances. This is again in agreement with chemical analogies used by Pauling.

* This agreement looks worse than it is for high valence states because it is the difference of two large numbers.

More recently the deHaas-van Alphen effect has been used to give the shape of the Fermi surface and to get conduction electron numbers in several metals.[11] This treatment leads to one conduction electron in copper in disagreement with the foregoing argument. It is not entirely clear that such numbers must be identical to the number of valence electrons used in determining the cohesive energy, since valence shells for certain structures may be filled by valence electrons which contribute to bonding without contributing to conduction (see §4.4). The extra electrons may form a conduction band that contributes little, if anything, to bonding and may indeed be antibonding. It is not inconceivable that this happens in elementary metals, especially in the transition-group elements. For the time being it seems best to consider metal valence for transition metals an open question and to restrict free electron and cellular models to the simple A-group metals of groups I, II, and III.

3.3 Brillouin Zones and Bands

In free electron theory the traveling waves of the electrons are expressed in terms of n_x or k_x [see (6) and (6a)]. For small values of n_x or k_x the wavelengths of the electrons are very large, comparable with the linear dimension of a macroscopic piece of metal, but they decrease with increase in n or k. In one dimension

$$\lambda = \frac{a}{n} = \frac{2\pi}{k} \tag{25}$$

where, as before, $k_x = (2\pi n_x)/(a)$.

From the previous section it can be seen that n_x is a large number, so that a great number of states are filled. For example, using (10), $8\pi/3 n_F^3 = \rho V$, whence, for a cubic crystal,

$$\left(\frac{n_F}{a}\right)^3 = \frac{3\rho}{8\pi} \tag{26}$$

For potassium metal $\rho = 1.3 \times 10^{22}$ electrons/cm^3. From this $a/n_F = \lambda_F = 4 \times 10^{-7}$ cm $=$ 40 Å. This is approaching the value of the lattice constant for potassium, which is 5.3 Å. (For a macroscopic piece of metal n_F is of the order of 10^7.) For metals of higher valences it is possible for λ_F to be less than the cell dimensions.

When wavelengths of the electrons are of the same order of magnitude as the cell constants of the metal, the traveling waves suffer Bragg reflection so that their k vectors are altered. In these circumstances the traveling waves are no longer eigenfunctions (stationary states) for the system. Instead, just at these points the stationary states must become standing waves with a period equal to some period of the lattice. It is easiest to follow this argument in one dimension. A particularly graphic account given by Kittel[2] is outlined here.

In one dimension the traveling waves travel along the one-dimensional axis. At certain wavelengths they will be Bragg-reflected in accordance with Bragg's equation

$$n\lambda = 2d \sin\theta \tag{27}$$

where n is the order of the diffraction maximum. For one dimension θ, half the scattering angle, must be $\pi/2$. Furthermore, $d = a_0$ the cell dimension. Hence Bragg's equation becomes, for one dimension, $n\lambda = 2a_0$. Substituting (25) for λ we find that (27) becomes

$$k = \frac{\pi}{a_0} n = \pm\frac{\pi}{a_0}, \pm\frac{2\pi}{a_0} \cdots \tag{28}$$

(This n should not be confused with n_x, the quantum number of the electron, and a_0 the lattice constant should not be confused with the macroscopic a, the length of the one-dimensional box. For example, for the first-order reflection $k = \pi/a_0$, and by substitution $\lambda = 2a_0$ and $n_x = ka/2\pi$ or $n_x = a/2a_0$. For a box of 1 cm and a lattice constant of 5 Å, then $n_x = 10^7$.)

Traveling waves satisfying the Bragg relation (28) would be $e^{\pm ikx}$, either $e^{in\pi x/a_0}$ or $e^{-in\pi x/a_0}$. Either of these waves would be rapidly diffracted by the lattice so as to become the other. Neither of them, then, is a stationary state, but their sums and differences

$$\psi_{1n} = \frac{e^{in\pi x/a_0} + e^{-in\pi x/a_0}}{2} \quad \text{and}$$

$$\psi_{2n} = \frac{e^{in\pi x/a_0} - e^{-in\pi x/a_0}}{2i}$$

are stationary states and, as is easily seen, are

$$\psi_{1n} = \cos\frac{n\pi x}{a} \qquad \psi_{2n} = \sin\frac{n\pi x}{a} \tag{29}$$

The pairs of equations ψ_{1n} and ψ_{2n}, though derived from the same k_x or n_x states, are no longer identical in energy for, as shown by Kittel and illustrated in Fig. 6, ψ_1 has its

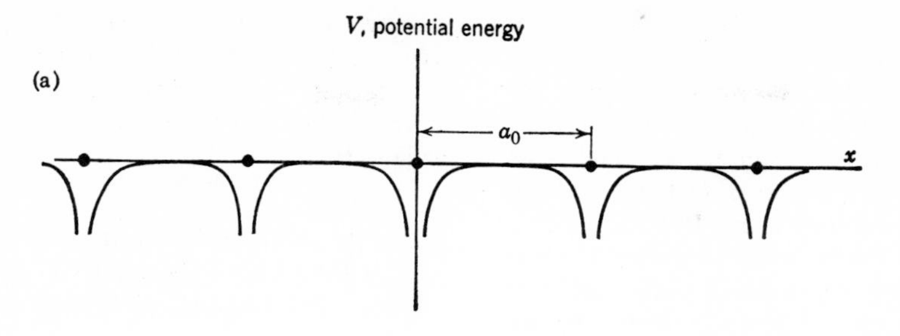

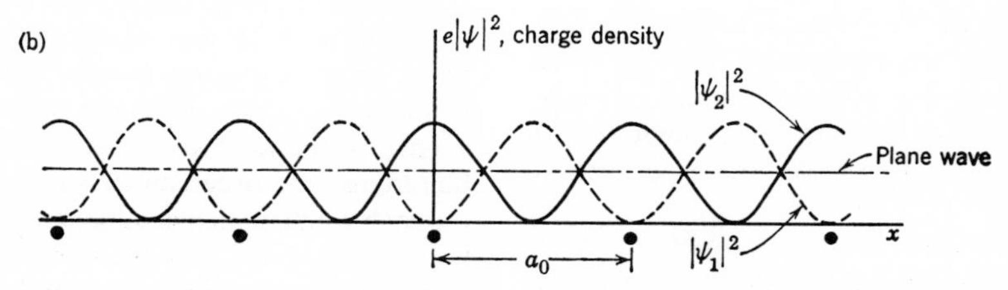

Fig. 6. Standing-wave functions for states cos kx and sin kx. It is to be noted that the latter has its maxima at the metal cores and will have a lower state, due to coulombic interactions, than the former which has nodes at the metal cores. (From Kittel,[2] p. 273, by permission of author and publisher.)

maxima between positive ion cores, its minima at these cores, whereas ψ_2 has its maxima at the cores, its minima between. Hence coulombically ψ_2 will have a lower *potential* energy than ψ_1.

It can be shown, then, that the change of energy with k (or n_x) is not so simple as the elementary picture given by free electron theory in which the energy increases quadratically with k (or n) as in (7) or (7a). Instead, as already shown, there are places in which the energy changes abruptly to give two energy levels with the same value of k. This leads to energy bands in which the energy may change with k nearly quadratically, separated by energy gaps in which the energy changes abruptly due to Bragg interaction of the electrons with the lattice. The zones leading to these energy bands have been called Brillouin zones and can be treated in three dimensions. Current treatments leave out details which must be part of a final theory. There is no doubt, however, that the basic ideas of the theory are correct.

A rather large body of work has gone into these zones and their implications for metals and alloys.[12] Direct experimental evidence for bands can be obtained. Insulators must have filled bands well separated from other bands energetically. Conductors must have only partly filled bands, and there are correlations of electron and hole conductors with the nature of the vacancies in such a band. Semiconductors, where impurity levels lie between bands, have been interpreted in terms of this concept. All of this is well known. There is no question that basically these deductions from band theory are valid and that the band concept has an enormous utility. No doubt it will form the basis for much that is contained in this book. But here we are concerned with Brillouin zones and the band theory of solids as a theory for metals, and here, unfortunately, very few things have been treated with sound theory to be certain of anything except rather general concepts. For example, it is commonly said that a band will contain two electron states, including spin, for each atomic orbital. As seen already in discussing cellular models, we suppose that good treatments will show that there is but one low state and one state in a higher band, which we shall call antibonding. Empirical evidence for metals shows that this view must be correct, and a detailed treatment for carbon shows that there are four, not eight, states in the lowest band for the diamond form of this element,[13] in agreement with experiment.

Computing at what point a discontinuity should occur in filling a Brillouin zone—whether at the inscribed sphere (energy increases rapidly with k beyond this point) or only when the zone is completely filled—has been, up to now, a matter of taste rather than a matter of science (see later). Recent work on the shapes of Fermi surfaces makes it clear that considerable theoretical work, much of it now in progress, will be necessary before there is enough detail in the theory to make it clear just what the true shapes of the zones are and just why the Fermi levels assume the odd shapes found experimentally. Much more detailed theoretical treatments must be made before theory can be useful as a guide for or even in the interpretation of experimental work.

The Hume-Rothery alloys[14] and their interpretation by Mott and Jones[15] once led to the hope that many intermetallic compounds owed their composition, structure, and stability to electron/atom ratios which could be simply interpreted in terms of filling Brillouin zones or filling these zones to the largest inscribed sphere. As time has gone on, it has become clear that composition boundaries are not so sharp as once believed, that valences of transition metals have been used as data fitting devices, that the theory is not well founded, especially for alloys, that Fermi surfaces are not simple spheres, and that, finally, the applications have degenerated into numerology.

This is not to say that there are no "electron compounds." There is increasing evidence for such phases beyond those discovered by Hume-Rothery. This is not to say that structure and composition are not to be interpreted by a theory arising from the concept of Brillouin zones or something close to it. It is just to say that there is still no really satisfactory theory. It appears that for a while theory in this area will wait on experiment.

Indeed, in the last several years a direct experimental approach to the shapes of Fermi surfaces has been made, and theorists are now concerned with understanding these experimentally determined and quite complex surfaces. Many of the transport and other physical properties can be obtained from the experimental Fermi shapes with far less demanding theories than are required for a complete understanding of metals. A particularly simple and readable account of recent experimental developments and their implications has just appeared.[16] For the much more sophisticated reader there are other reviews.[17]

4. ANALOGIES BETWEEN COVALENT BONDING AND METALLIC BONDING

4.1 Pauling's Theory of Metals

Pauling, in 1938,[18] noted that metallic radii of metals were similar to covalent radii, and more recently he has shown that a simple empirical equation for the dependence of bond length on the numbers of electron pairs in a bond could be found so that only one radius for each element was necessary to account for both covalent and metallic bond distances. In this view the electrons of a metal are too few to form electron pair bonds between the atoms of a metal and are delocalized (resonate) over the atoms pairs. For example, a metal of valence three and with a ccp or hcp structure may form only three electron pair bonds with its twelve closest neighbors, but these three electron pairs are delocalized over the twelve bonds, or resonate among them, so that each bond effectively contains $\frac{3}{12}$ of an electron pair, or the bond number is one fourth for these bonds.

Pauling's empirical relationship relating interatomic distance D to bond number n is

$$D(n) = D(1) - 0.6 \log n \qquad (30)$$

Use of this equation is straightforward. Bond distances for single bonds have been found to be additive, the sums of covalent radii of the atoms bonded[5] [i.e., $D_{AB}(1) = R_A(1) + R_B(1)$], so that if a metal M forms covalent single bonds with itself or some other element, X, whose covalent radius is known, the single-bond radius, $R_M(1)$, may be found for the metal and used in (30) to predict intermetallic distances in either metals or alloys. Consistent and accurate results are then obtained.

The empirical correlations of Pauling require that metallic bond distances decrease as metallic valences increase. As seen in the preceding section, this is in accord with experience. For the transition elements bond distances decrease and cohesive energies increase until chromium (group VI) is reached, but then remain constant. Pauling uses this observation and other data to argue that the metallic valence increases to six at chromium, parallel with the group number, and that it then remains constant for the next several elements. The increasing saturation moment observed from chromium through iron and the decreasing moment thereafter are interpreted as being due to essentially nonbonding magnetic levels which electrons occupy with spins unpaired until this band is half filled (at a point between iron and cobalt). Beyond that point electrons pair in this magnetic band until it is filled between nickel and copper. Throughout this interval the valence electrons remain constant at six. After filling the magnetic band, it is presumed that the added electrons pair with valence electrons on one atom to form core electrons no longer available for bonds.

More quantitatively, the way in which these bonding and magnetic orbitals are treated is as follows: for the transition elements of the first long period the five 3*d*-orbitals, the 4*s*-orbital, and the three 4*p*-orbitals are assumed to be states of low enough energy to be of consequence to the metallic states of these elements. Some six of these are, on empirical grounds, assumed to form bonding orbitals (hybrids of *s*, *p*, and *d*) which electrons occupy with paired spins because of the creation of electron pair bonds using these orbitals, until these bonding orbitals are filled. He assumes, further, that in order to have "unsynchronized resonance," which he believes is common to and stabilizes all metals, about 0.72 bonding orbitals/atom must be reserved and unoccupied. The remaining 2.28 orbitals form a nonbonding magnetic set of orbitals. Thus, up to group VI, all electrons outside the argon core are bonding electrons. Next the magnetic orbitals are half filled to get a maximum saturation moment of about 2.28 μ_B for, say, an Fe–Co alloy having 8.28 electrons/atom. Thereafter, 2.28 electrons, making a total of 10.56 electrons/atom, can be added, thus decreasing the saturation moment as the electrons pair in the magnetic orbitals but leaving the bonding orbitals unaffected. Ni–Cu alloy with 10.56 electrons/atom should have six bonding electrons and a completely filled magnetic core with no saturation moment. Further electrons must destroy bonding by pairing electrons on a single atom. Copper with eleven electrons outside the argon shell is reduced to a valence of 5.56 (6 − 0.44). The antibonding band is filled further for zinc and gallium, leading to metallic valences of 4.56 and 3.56. In this picture *d*-, *s*- and *p*-orbitals form a bonding band that persists far beyond nickel and indeed until sufficient antibonding electrons are added to cause instability of this bonding band.

Many features of his theory, and especially the 0.72 metallic orbitals/atom, seem entirely *ad hoc* and fanciful. It is improbable that these features are correct in detail or that they are necessary to the main argument. This interpretation is, nonetheless, in accord with the atomic distances and cohesive energies through the first long period. It also gives as good an interpretation of the magnetic behavior of the iron group and their alloys as any that has been suggested. The way in which it accounts for magnetic properties of transition elements and their alloys is shown in Fig. 7. But, as can be seen, this theory is wholly out of keeping with the usual physical interpretations of valence in this interval in which valences are usually given low values by physicists and in which physicists assume that the metallic valences of Cu, Sn, Ga, etc., follow the group number. A very different interpretation of Fig. 7, compatible with low valences, has been given by Slater[19] and others.[1,2] There is no fundamental basis for either valence assumption. The simple cellular model outlined in Section 3.2 seems to require, as does Pauling, that valence and cohesive energies go together and that short distances seem to require high valences. It has already been noted that the Fermi surface in copper seems to require one electron per atom, but, as shown in Section 4.4, this may not be inconsistent with a high valence for bonding electrons in copper in a theory having features similar to Pauling's theory.

Since, as noted earlier, there is no *metal theory* to present, the purposes of this chapter may be well served by trying to persuade physicists and physical metallurgists to look for empirical hints lying about in chemistry which may be useful as guides toward *a theory*. An attempt, however crude, will be made to translate Pauling's theory into a language more consistent with current physical thought in the hope that analogies will be clearer and more acceptable, so that some theoretical physicist may read this and re-examine the whole area of transition metal valence.

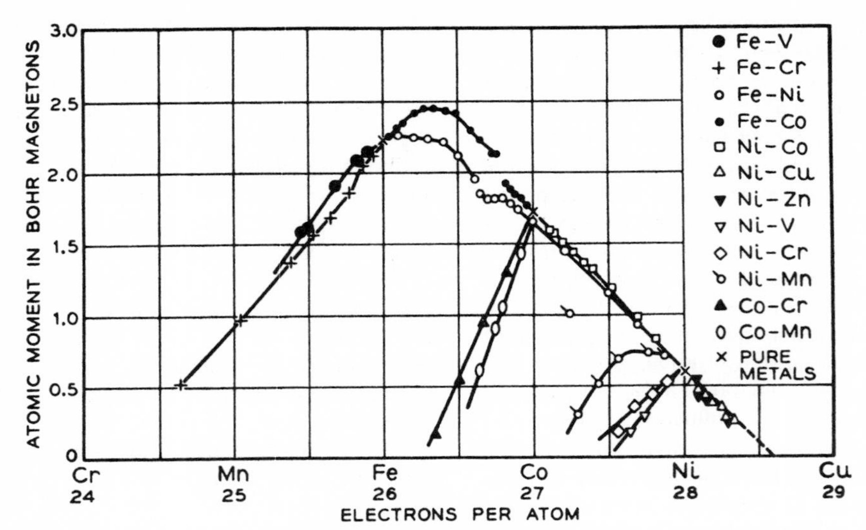

Fig. 7. Saturation moments of transition elements and their alloys. Pauling's theory accounts for the metals and alloys on the main curve. Several alloys fall off this curve, and Pauling has suggested that these may be antiferromagnetic. This point has not yet been well established. (After R. M. Bozorth, *Ferromagnetism*, D. Van Nostrand Co., New York, p. 441, by permission of author and publisher.)

The idea that *nd*-, $(n + 1)s$- and $(n + 1)p$-orbitals must all be considered in transition metals is supported in Section 4.2. The idea of bonding orbitals leading to electron pairing even if they are not localized pair bonds is essentially equivalent to saying that there is a bonding band where bonding is loose and that the kinetic energy causes pairing through the Pauli principle, as in the cellular theories of metals. Just as outlined in the examination of cellular theories, it seems likely that there will be a division into a bonding band, with $\partial\psi/\partial r = 0$ at some point between metallic neighbors and an antibonding band with nodes between neighbors. This would change Pauling's bonding orbitals to a bonding band, and the decrease in number of bonding orbitals or decrease in metallic valence between nickel and copper would then be due to filling an antibonding band. (This is analogous to the MO treatment of the number of bonds in, for example, the first row homonuclear molecules X_2).[20]

It remains to describe a nonbonding magnetic set of orbitals, for which no one has given a really satisfactory account. It seems clear, however, that in proceeding across the transition series, where the nuclear charge and the number of valence electrons are increasing, inner cores will shrink. It seems not unlikely that the coulombic attraction of the core may finally, at high valences, compete successfully with the tendency for electrons to go into a bonding band and that at some point a *d*-core of some number of orbitals may develop which does not involve the *d*-, *s*-, and *p*-orbitals most effective in forming the bonding band. Such a core would be rather tightly bound, and this band would not be spread widely by kinetic energy terms. It would then be a core in which electrons would tend to have parallel spins. Whether ferromagnetic interactions are then carried by the polarization of the electrons in the conduction or bonding band or by some more direct exchange mechanism is really an unsolved problem. But the proper ingredients for magnetic behavior in the transition metals have been set up. Indeed, it seems that the foregoing is a rather logical expectation (but no theory). It could happen only for the transition metals in which valence is high.

To a chemist this seems less fanciful than to have the valence electrons, which are the bonding electrons of the system, approach zero at just the point at which atomic volumes become least and at which cohesive energies are the highest for any metals in the periodic table. In view of the cellular treatment, it should be hard also for physicists to take such low valences seriously. This picture suggests that conduction electrons may well approach a very low value for the transition metals, for a bonding band has been filled and electrons are entering a tightly bound set of core states. Only some interactions between the two could cause holes to develop in the bonding band. At copper the bonding situation would be only slightly weakened by the addition of an electron or nearly an electron to the antibonding band. This addition of about one electron to an empty band capable of holding six electrons should dramatically influence conductivity, as it does. Similarly, the lower cohesive energy of zinc and gallium is accounted for in a natural way. This seems to be a model worthy of careful theoretical examination.

What will Pauling's empirical theory do? Unfortunately, an honest answer seems to be that the best it can do is correlate the interatomic distances in metals and alloys in known structures and relate them in a consistent way to bond distances of the same elements in nonmetallic compounds. The use of (30) leads, in nearly all cases, to valences that are consistent with Pauling's valences if covalent radii are taken from nonmetallic compounds or to single-bond radii that are consistent with other data if the valences are assumed. In some cases, after a structure is known, it is possible to look back and see features in the theory that favored some unusual structural aspect. So far it has not been useful in predicting compositions of alloys or structures of intermetallic compounds before they are known.

There is one other feature of Pauling's theory that should be mentioned, since it is a pertinent idea and is frequently misunderstood. This is the concept of electron transfer to increase metallic valence. It is based on a type of transfer well known in chemistry. For example, boron of group III has three valence electrons; nitrogen of group V has five valence electrons. Both elements have four valence orbitals. In forming electron pair bonds, boron is limited to three because of the number of valence electrons, and nitrogen is limited to three because one orbital must contain an unshared pair, leaving three orbitals with three electrons for forming bonds. Now, if nitrogen transfers an electron to boron, each element can form four electron bonds, for each of the four valence orbitals will now contain one electron. Compounds such as

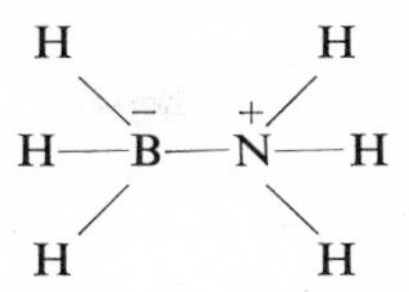

result from such electron transfer. (It is, of course, possible to describe this process in other terms, but the end result is the same.) This is not to say that the electron density will look like this. For the foregoing compound boron has a negative formal charge and nitrogen has a positive formal charge. All bonds to each will be changed in ionic character to minimize this unfavorable charge distribution, as Pauling has pointed out. But the existence of many such complexes shows that this type of charge transfer does occur. It is the key feature in such compounds as GaAs. In general, since it is metals which have fewer valence electrons than valence orbitals and nonmetals which have fewer valence orbitals than valence electrons, the transfer will usually be from the nonmetal toward the metal, in order to obtain more bonds. This is, of course, the wrong direction for electron transfer so far as electronegativities are concerned, and therefore electronegativity considerations will favor partly ionic bonds which operate in the opposite direction. Overall, the total charge transferred is likely to be very small.

How far this sort of charge transfer can go is illustrated by the fact that germanium with the diamond structure and GaSb, ZnSe, and CuBr are isostructural (the latter have the closely related sphalerite structure), have the same number of total valence electrons, and nearest bond distances vary only ± 0.1 Å throughout the series. There would be transfer of one, two, and three electrons to form four bonds in each of these compounds and of course the degree of ionic bonding would become greater through the series. Nonetheless, it is hard to escape the conclusion that covalent character is structure-determining throughout this series. In order to use (30) with success, it is necessary to take proper account of charge transfer.*

* For a detailed discussion of this topic see Parthé, Chapter 11.

4.2 Metal Carbonyls and Valence Orbitals

Evidence based on the composition of metal carbonyls supports Pauling's contention that 3*d*-, 4*s*-, and 4*p*-orbitals are valence orbitals for the transition elements of the first long period and that corresponding orbitals are valence orbitals of the second and third long periods. Table 2 lists the known carbonyls of the transition elements. Both compositions and structures are completely in accord with the view that in these carbonyls all the foregoing orbitals are filled by lone pairs of bonding electrons, so that the total number of electrons held or shared by all of these first-row transition elements is 36 (the number of electrons in the Kr atom) and is equal to the numbers 54 and 86 for the second and third transition series. However, remarkable derivatives of the

TABLE 2

Carbonyls of the Transition Metals and Their Structures[a]

Compound	Electrons/M	Configuration	References
$Ni(CO)_4$	36	Tetrahedral	(Brockway and Cross, 1935; Ladell *et al.*, 1952)
$Co_2(CO)_8$	36		(Cable *et al.*, 1954)
$Co_4(CO)_{12}$	36		(Corradini, 1957)
$Fe(CO)_5$	36		
$Ru(CO)_5$	54	Trigonal bipyramid	(Ewens and Lister, 1939)
$Os(CO)_5$	86		
$Fe_2(CO)_9$	36		(Powell and Evans, 1939)
$Ru_2(CO)_9$	54		
$Os_2(CO)_9$	86		
$Fe_3(CO)_{12}$	36		(Dahl and Rundle, 1957)
and			
$Ru_3(CO)_{12}$	54		(Dahl, in press)
$Mn_2(CO)_{10}$	36		(Dahl *et al.*, 1957)
$Tc_2(CO)_{10}$	54		(Trueblood and Wallach, 1961)
$Re_2(CO)_{10}$	72		(Dahl *et al.*, 1957)
$Cr(CO)_6$	36	Octahedral	(Brockway *et al.*, 1938)
$Mo(CO)_6$	54	Octahedral	(Brockway *et al.*, 1938)
$W(CO)_6$	86	Octahedral	(Brockway *et al.*, 1938)
$V(CO)_6$	35	Octahedral	(Natta *et al.*, 1959)

[a] Evidence that transition metals of first, second, and third long periods use 4*s*- and 4*p*-, 5*s*- and 5*p*-, and 6*p*-orbitals, respectively, in forming.

carbonyls are appearing so rapidly that it is impossible to review this literature in a small space. One generalization seems to be possible. Either the complex carbonyl also has an exactly filled Kr shell for the transition metal or the metal atom is so well shielded by stable ligands that no further ligands can approach it.[21] In the latter case it is possible to have uncompleted shells about the metal ion.

In conclusion, a considerable bulk of chemical data exists to suggest that not only are transition elements capable of using all valence orbitals out to the next rare gas but they insist on doing so unless sterically hindered from attaining a rare-gas configuration. In a good many cases attaining a rare-gas configuration involves forming metal-metal bonds. Hence the idea that they may form delocalized metal-metal bonds in metals using 3*d*-, 4*s*-, and 4*p*-orbitals appears to be chemically reasonable.

4.3 Delocalized Bonding in Electron-Deficient Compounds

Further evidence on this point comes from the compositions and structures of the boron hydrides and the organometallic compounds. It is common for metals to have more valence orbitals than valence electrons, counting all orbitals as valence orbitals out to the following rare-gas shell. Thus boron and aluminum of group III have three valence electrons but four valence orbitals, the *ns*- and three *np*-orbitals corresponding to their position in the periodic table. Similarly, beryllium and magnesium have two valence electrons but the same four valence orbitals. In these cases it is found that the number of valence electrons determines the gross composition but that all of the valence orbitals are used nonetheless.[21,22] The result generally is delocalized bonds in which some bonds have fractional bond numbers and longer distances, usually in good agreement with (30). Some examples of bond delocalization of this sort are shown in Fig. 8.

In boron hydrides[23] and in interstitial compounds[24] it is possible to have higher metal-to-nonmetal ratios than in the compounds shown in Fig. 8, in which case bonds are further delocalized and distances are lengthened. It seems quite clear that one can find any stage of bond delocalization from none in "normal" chemical compounds which generally contain many unshared electron pairs, on down through organometallic compounds, through the higher boron hydrides and interstitial compounds and simpler alloys, to metals, with the extent of bond delocalization progressing steadily as the ratio of valence orbitals to valence electrons increases. From this point of view metals are but the final extension of covalent bond theory to the most highly delocalized state.

Bond delocalization does not occur until there are atoms, always metals, in chemical combination with atoms or groups that have no unshared pairs. This and the increase of bond delocalization with excess orbitals is in keeping with the suggestion made in the first section that both in chemical compounds and in metals the number of bonding states, where $d\psi/dr = 0$ at some point between the bonded atoms, cannot exceed the number of valence orbitals of the metal. In consequence the valence of a metal is limited to the number of stable orbitals and is not double this number as is often assumed in band theories of metals.

4.4 Molecular Orbital Treatments

In several cases Longuet-Higgins and von Roberts[25] (hereafter LH-VR) have been able to treat alloys to show in detail how closed-shell, diamagnetic compounds may arise. They have treated certain forms of elementary boron, MB_6 compounds, B_4C, etc., with considerable success. This treatment is illustrated here by a treatment of MB_6 only.

The structure of MB_6 is shown in Fig. 9. Of first interest is the boron framework in which boron octahedra are bonded to six other octahedra along the cubic axes. LH-VR show that the bonds between octahedra are ordinary electron pair bonds and use group theory to find the proper linear combination of atomic orbitals for linking the boron octahedra to neighboring octahedra. These are a'_{1g}, e'_g, and t''_{1u} of Fig. 10. Since there are six such orbitals (*e*-orbitals are doubly degenerate, *t*-orbitals are triply degenerate), each octahedron of boron atoms will have to furnish six electrons to form bonds linking the octahedra. There remain six inward-directed *sp*-hybrids, some mixed with *p*-tangential orbitals, to give a total of nine orbitals belonging to the interior of the B_6 octahedron. Their proper linear combinations which form bases for irreducible representations of the group O_h are a_{1g}, e_g, t_{1u}, and t'_{1u} of Fig. 10. Of these a_{1g} is clearly

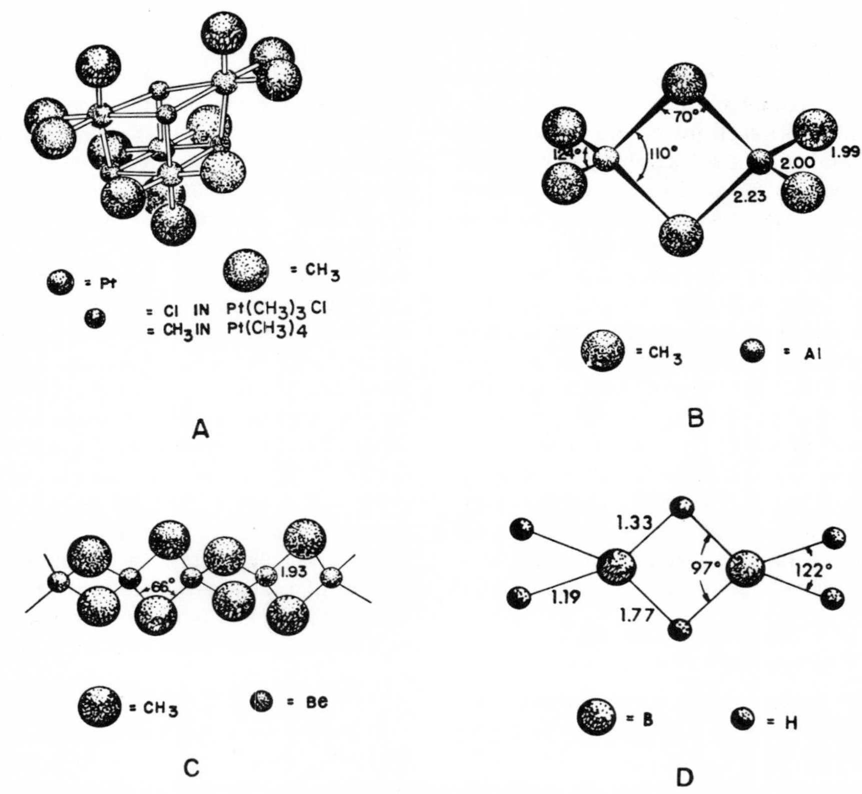

Fig. 8. Examples of structures of "electron-deficient compounds." In all of these compounds there are more bonds than electron pairs, so that some electron pairs are delocalized over more than one bond. All are diamagnetic. All valence orbitals of the metals are used in bond formation.

bonding, whereas e_g and t'_{1u} appear to be clearly antibonding and t_{1u} appears to be probably bonding. A further treatment is necessary to be sure about t_{1u}.

The remaining p-orbitals are tangential to the B_6 octahedron. Their linear combinations, which form basis sets for irreducible representations, are shown as t_{1g}, t_{2u}, and t_{2g} of Fig. 10. Of these t_{2g} is bonding, t_{2u} is antibonding, and t_{1g} needs closer attention. Making conventional assumptions about overlap and exchange integrals, LH-VR next computed an energy-level diagram for all states for B_6. T_{1u} is confirmed as a bonding state, so that there are seven inner-octahedron bonding states that are lower in energy than any of the other orbitals. These bonding states will be occupied by fourteen electrons. Six other electrons/B_6 are required for the bonds to other octahedra, or twenty electrons/B_6 in all. Six borons have only eighteen electrons, so that two are required from the metal ion to make a closed set. Furthermore, the overall bonding will now include the ionic bonding of M^{2+} and $B_6^{=}$.

As LH-VR note, CaB_6 is apparently a colorless insulator. But rare earths can be substituted for Ca, and as the substitution proceeds the compound becomes colored, gains a metallic luster, and becomes a conductor. Clearly an ion, RE^{3+}, is formed; B_6^{2-} picks up an electron in an antibonding state to become B_6^{3-}. This is energetically favored by the higher coulomb (Madelung) energy of the crystal, but in the antibonding orbitals of B_6 the electrons are held very loosely and at sufficiently high electron densities form a conduction band.

This is not the only such compound. The well-known sodium tungsten bronzes are similar. There is closely parallel behavior in CaC_2, a colorless salt with an acetylene ion,

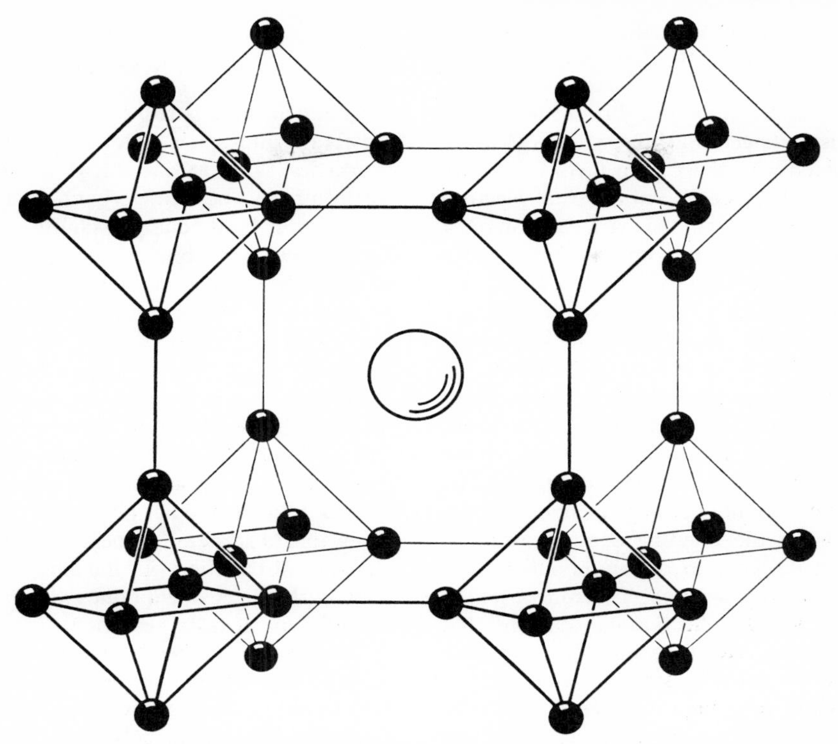

Fig. 9. CaB_6 structure. Trivalent metals may be substituted for calcium in this structure, leading to metallic conductivity and luster, although pure CaB_6 shows no such properties. (From Pauling,[10] p. 366, by permission.)

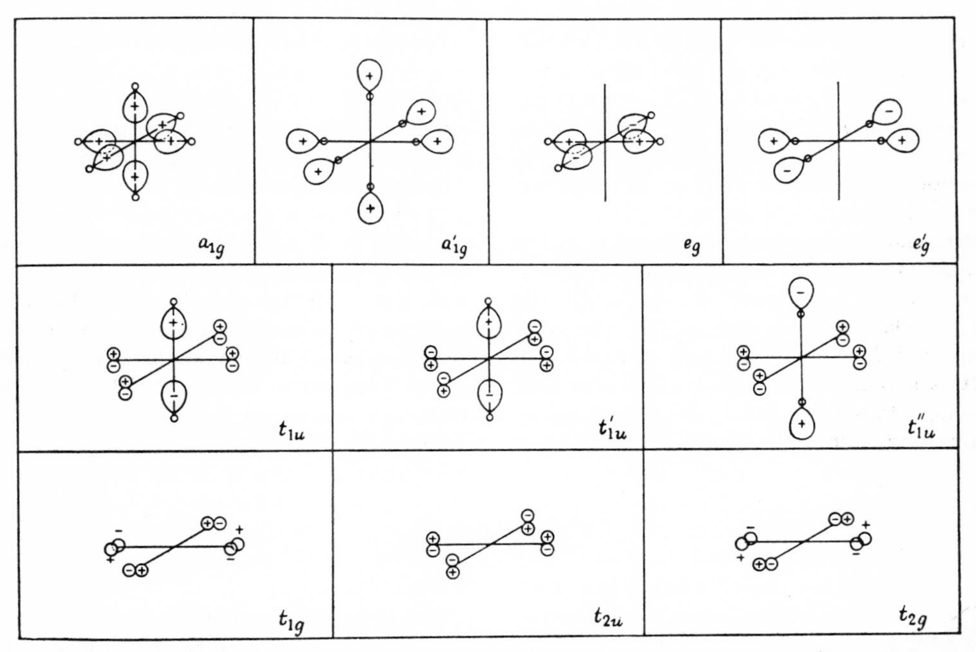

Fig. 10. Molecular orbitals derived from group theory for the B_6 framework in CaB_6. (From Longuet-Higgins and deV. Roberts,[25] p. 337, by permission.)

C_2^{2-}, which has the same C—C distance as in acetylene, 1.20 Å.[24] In this case, too, substitution by a rare earth leads to color, metallic luster, and metallic conductivity. Here it is quite certain that the extra electron is associated with the C_2^{2-} to form C_2^{3-}, since the C—C distance *increases* with electron concentration.[27] Again the extra electron must be loosely associated with the antibonding orbitals of the C_2^{2-} ion to cause this affect, and the crystal must be stabilized by a much larger Madelung energy. However, the loosely bound electrons again form a conduction band.

It is to be noted here that in both cases, and there are others, bonding bands are filled, contributing to the cohesive energy, but because a closed shell is reached the bonding electrons contribute nothing to conduction. It is only when the bonding band is exceeded and electrons go into an antibonding band that conductivity is found. This illustrates the point mentioned in Section 4.2, namely, that the conduction electrons and bonding electrons are the same in simple metals, but in metals with very high valences it may be possible to fill a bonding band and more, so that the conduction electrons are antibonding electrons. An attempt to show how this might be possible for copper has been given in Section 4.1.

Finally, the method of LH-VR deserves further study. No one has yet carried out a good similar MO treatment for a continuous structure, and the success with the alloy or metalloid structures so far handled by this method has depended on treating only finite parts of the problem. In recent years many intermetallic compounds have been found in which there are particular metal polyhedra, some of which are found in a whole class or several classes of compounds. Their stability may well depend on achieving a closed-shell MO configuration for this polyhedron, and attempts to use the method of LH-VR appear to be worth exploring.

5. SUMMARY

Although there is still no general theory of metals and alloys, it is clear that many physical properties of metals, such as conductivity, electronic-heat capacities, paramagnetic susceptibilities, and compressibilities, depend completely or mainly on the kinetic energy of the electrons, and the kinetic energy behaves very much as though the electrons were free electrons, subject to the Pauli principle.

Cellular models can account quantitatively for the cohesive energy of the simplest metals and furnish a reasonably adequate account of these metals. This model can and is being refined. Cellular models give some promise of being able to account for metallic valence. The electron distribution in metallic bonds resembles that in covalent bonds rather well and for the same physical reasons.

There is no adequate theory from which metallic valence and magnetic properties can be inferred in the transition metals. Chemical analogies suggest models that may prove profitable for theoretical study.

For alloys the promise held out by the Brillouin zone theory for dealing with "electron compounds" has been dimmed, and a thorough theoretical analysis is needed to see if a more satisfactory theory can be obtained. So far only empirical theories produce any useful correlations, and these do not offer substantial aid in predicting stable phases or physical properties.

A molecular orbital method has been used successfully to account for the stability of certain polyhedra sometimes found in alloys. This method is more rigorous than Brillouin zone theory, but it still seems more limited in scope. But since this method has received only minor attention it may be possible that more thorough study may lead to something of value.

REFERENCES

1. Seitz F., *Modern Theory of Solids*, McGraw-Hill, New York, 1940.
2. Kittel C., *Introduction to Solid State Physics*, 2nd edition, Wiley, New York, 1956.
3. Pauli W., *Z. Physik* **41** (1927) 81.
4. Brooks H., *Trans. AIME* **227** (1963) 546.
5. Brooks H., *Nuovo Cimento*, Series X, **7**, Supplement 2 (1958) 165.

6. Pines D., *ibid.*, 329.
7. Ruedenberg K., *Rev. Mod. Phys.* **34** (1962) 326.
8. Good R. H., unpublished.
9. Moore C. E., "Atomic Energy Levels," *N.B.S. Circular* **467,** U.S. Government Printing Office, Washington, D.C. 1949.
10. Pauling L., *Nature of the Chemical Bond*, 3rd edition, Cornell Univ. Press, Ithaca, N.Y., 1960.
11. Schoenberg D., *Nature* **183** (1959) 171.
12. Pippard A. B., "Experimental Analysis of the Electronic Structure of Metals," in *Reports on Progress in Physics*, The Physical Society London, 1960.
13. Kimball G. E., *J. Chem. Phys.* **3** (1935) 560.
14. Hume-Rothery W., and G. V. Raynor, *The Structures of Metals and Alloys*, 4th edition, The Institute of Metals, 1962. First account, Hume-Rothery, *J. Inst. Metals* **35** (1926) 295.
15. Mott N. F., and H. Jones, *The Theory of the Properties of Metals and Alloys*, The Clarendon Press, Oxford, 1936; Jones, *Proc. Roy. Soc.* (*L*) **A144** (1934) 225, **A137** (1934) 396.
16. Mackintosh A. R., *Sci. Am.* **209** (1963) 110.
17. Harrison W. A., and M. B. Webb, *The Fermi Surface*, Wiley, New York, 1960. See also Ref. 12.
18. Pauling L., *Phys. Rev.* **54** (1938) 899.
19. Slater J., *J. Appl. Phys.* **8** (1937) 385.
20. Eyring H., J. Walter, and G. Kimball, *Quantum Chemistry*, Wiley, New York, 1944, pp. 203–211.
21. Rundle R. E., *Record Chem. Prog.* **23** (1962) 195.
22. Rundle R. E., *J. Am. Chem. Soc.* **68** (1947) 115; *J. Chem. Phys.* **17** (1949) 67; *J. Phys. Chem.* **61** (1957) 45.
23. Lipscomb W., *J. Phys. Chem.* **61** (1957) 23.
24. Rundle R. E., *Acta Cryst.* **1** (1948) 180. For a translation into MO terms and additions, see H. Biltz, *Z. Physik* **153** (1958) 338.
25. Longuet-Higgins H. C., and M. deV. Roberts, *Proc. Roy. Soc.* (*L*) **A224** (1954) 336; **A230** (1955) 110.
26. Atoji M., and R. C. Medrud, *J. Chem. Phys.* **31** (1959) 332.
27. Atoji M., K. Gschneidner Jr., A. H. Daane, R. E. Rundle, and F. H. Spedding, *J. Am. Chem. Soc.* **80** (1958) 1804.

chapter 3

Thermodynamic Properties

PETER M. ROBINSON*

Department of Metallurgy
Massachusetts Institute of Technology
Cambridge, Massachusetts

MICHAEL B. BEVER

Department of Metallurgy
Massachusetts Institute of Technology
Cambridge, Massachusetts

1. INTRODUCTION

This chapter provides a summary and interpretation of the published thermodynamic data for intermetallic compounds. Most of the data presented are drawn from the collections of Kubaschewski and Evans[1] and Hultgren et al.[2] In addition, recent data have been taken from original papers.

Thermodynamic information on intermetallic compounds is still quite incomplete. For some classes of compounds, it is almost entirely lacking. For example, although more than 250 Laves phases have been identified, the heats of formation of only thirteen have been measured, and other thermodynamic information on Laves phases is even more limited. Thus some of the generalizations

* Presently associated with the Commonwealth Scientific and Industrial Research Organization, Division of Tribophysics, Melbourne, Australia.

concerning various types of compounds based on the information available at the present time may be influenced by historical accident; the compounds that have been investigated may not be representative of the group to which they belong.

Thermodynamic properties of intermetallic compounds can contribute to an understanding of their nature. The free energy of formation gives an indication of the stability, and the heat of formation is closely related to the type of bonding. The entropy of formation reflects changes in the vibrational behavior and configurational arrangement of the atoms on formation of the compounds. The heat capacity and the absolute entropy also reflect the vibrational behavior. The heat capacity of the solid just below the melting point sheds light on premelting phenomena. The heat of fusion, the entropy of fusion, and, to a lesser extent, the level of the melting point reveal some aspects of the nature of the transition from the crystalline to the liquid state. The activities of the components of intermetallic compounds give an indication of their bonding.

2. FREE ENERGIES OF FORMATION

2.1 General

A decrease in free energy accompanies the formation of an intermetallic compound from its component elements. A compound is stable with respect to competing neighboring phases in a multicomponent system if its free energy is lower than that of a mixture of these phases.

The free energy of formation of an intermetallic compound which exists over a homogeneity range changes with composition in such a way that the curve is concave toward the composition axis.[3] If the intermetallic compounds occurring in a binary system have narrow homogeneity ranges, their free energies of formation plotted as a function of composition lie at the corners of a polygon which is concave toward the composition axis. The straight lines of this polygon represent the two-phase regions between the compounds. Any reentrant angles indicate that the corresponding compound is metastable with respect to a mixture of the neighboring phases.

The minimum value of the free energy of an intermetallic compound existing over a homogeneity range has often been assumed to occur at the stoichiometric composition on which such a compound appears to be based.[4,5] Hume-Rothery, Christian, and Pearson[6] have pointed out that there is no fundamental reason for this assumption, except perhaps that whole-number ratios of atoms may be expected to pack together most conveniently and to have the lowest free energy. The limited data available for binary systems[2] show that the minimum free energy of formation of an intermetallic compound usually does not occur at the stoichiometric composition. As an example, the free energies of formation of the phases in the silver-zinc system are shown in Fig. 1. The full lines represent experimental

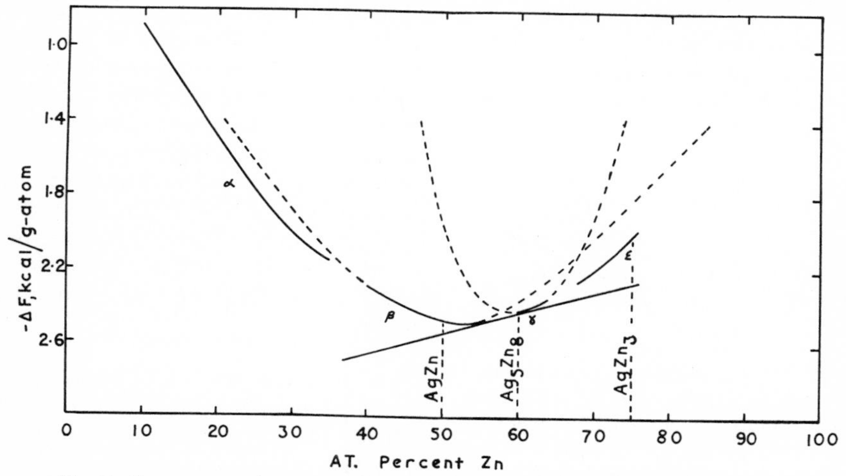

Fig. 1. Free energy changes on formation of the α-terminal solid solution and the β-, γ-, and ε-electron compounds in the Ag–Zn system.

values determined by vapor pressure and emf measurements[7,8] for the α terminal solid solution and the β-, γ-, and ϵ-electron compounds. The dotted lines represent extrapolations of the free energies of formation of the β and γ compounds. The minima for all three electron compounds occur at compositions other than the stoichiometric. It may be concluded that in compounds in which the bonding is mainly metallic, the stoichiometric composition is not necessarily the most stable. On the other hand, the compounds occurring in some systems have the lowest free energy at the stoichiometric composition. An example is the β-electron compound, AuCd, in which a small ionic addition to the metallic bonding makes the stoichiometric composition more stable than the neighboring compositions. (See §3.7.)

The stoichiometric composition on which an intermetallic compound appears to be based is not necessarily included in the range of composition over which the compound is stable. Rushbrooke[4] has demonstrated that the relative position of the free energy curves of the compound and the neighboring phases determines whether the stoichiometric composition of an intermetallic compound is included in the stable homogeneity range. In this demonstration it is assumed that the stoichiometric composition occurs at the minimum. The stoichiometric composition is included in the stable range if the free energy curve of the compound lies below that of both neighboring phases, but it is not included if the free energy curve occupies an intermediate position between the curves of the neighboring phases.[6] Figure 1 differs from this simple case inasmuch as the stoichiometric compositions do not occur at the minima of the free energies, although they lie within the homogeneity ranges. If the free energy of the β-phase in the silver-zinc system were lower, the common tangent of the β- and γ-phases would move in such a way as to make the stoichiometric composition of the γ-phase unstable.

The conditions for the stability of intermetallic compounds in ternary systems are analogous to those for binary systems. In the ternary free energy versus composition diagrams curved surfaces replace the simple curves of binary diagrams and ruled surfaces and tangent planes take the place of simple tangents. The resulting envelope of surfaces must be concave toward the composition plane.

Raynor has analyzed the following alternatives involving compounds in ternary systems.[5,9] If compounds exist in the three constituent binary systems, there are two possibilities: in the case illustrated by Fig. 2*a* the free energies of formation of the binary compounds are approximately of the same magnitude, whereas the phase relations in Fig. 2*b* result if one compound has an appreciably larger free energy of formation; this compound then coexists with the terminal solid solution based on the third component.

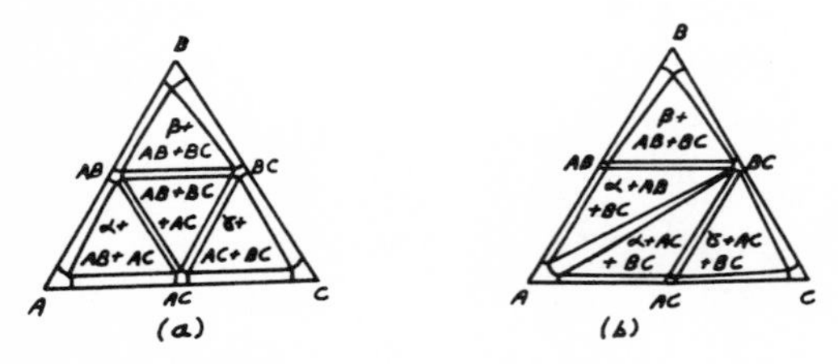

Fig. 2. Plan of the tangent planes at a given temperature for a ternary system, showing (*a*) binary compounds with approximately equal free energies of formation, and (*b*) a binary compound with an appreciably larger free energy of formation than that of any other compound in the system. (After Raynor.[5])

2.2 Determination of the Free Energies of Formation

The free energies of formation of intermetallic compounds can be obtained from equilibrium measurements. Since most equilibrium measurements involving intermetallic compounds are carried out at elevated temperatures, the experimental values of the free energies of formation obtained in this way are mainly for high temperatures, but some values are also available for room temperature.

Electromotive-force measurements provide equilibrium data from which free energies of formation can be obtained. If an intermetallic compound coexists with one of its constituent elements, the following relation holds for a suitable cell

$$\Delta F = -n\mathscr{F}\mathscr{E} \tag{1}$$

where n is the number of equivalents, $\mathscr{F}$, Faraday's constant, and, $\mathscr{E}$, the reversible cell voltage. The literature[1,2] should be consulted for details of the method and the experimental difficulties. This method can be used only with relatively few intermetallic compounds. If the compound does not coexist with one of the constituent elements, the activity coefficient of the components may be determined from emf values. The free energies of formation can then be calculated by the Gibbs-Duhem equation.

Equilibrium measurements involving the vapor pressure of one or both components of an intermetallic compound are possible in some cases. This method can be very useful, especially since its applicability has been extended to low pressures by the use of the mass spectrometer. In this manner Goldfinger[10] has measured the changes in thermodynamic properties associated with the formation of III–V compounds.

The free energy change on formation, ΔF, can also be evaluated from the equation

$$\Delta F = \Delta H - T\Delta S \tag{2}$$

where ΔH is the heat of formation, ΔS, the entropy change on formation, and T the absolute temperature. The heat of formation can be measured by calorimetry and the entropy of formation can be found from standard entropies of the compound and the elements. (See §§3 and 5.)

Wagner[11] proposed an analysis of phase relations which makes it possible to estimate the free energy of formation at the congruent melting point of a compound of single composition. The liquidus in the region of the compound and its heat of fusion and heat of formation must be known; if the elements are liquid at the melting point of the compound, their entropies of fusion must also be known. This method has been used to estimate the free energy of formation of III–V compounds.[12]

3. HEATS OF FORMATION

3.1 General

The heat of formation of an intermetallic compound is usually negative because the free energy change must be negative and the entropy-temperature product is small (2). In principle, the heat of formation of a stable phase may be positive if the free energy of formation is small and the entropy-temperature product is positive and sufficiently large.

If several intermetallic compounds with single compositions occur in a binary system, their heats of formation lie at the corners of a polygon similar to that formed by the free energies.[3] (See Fig. 3.) A reentrant angle in such a polygon means that the corresponding phase is either metastable or that the entropy-temperature product is such that it results in an appropriate value of the free energy.

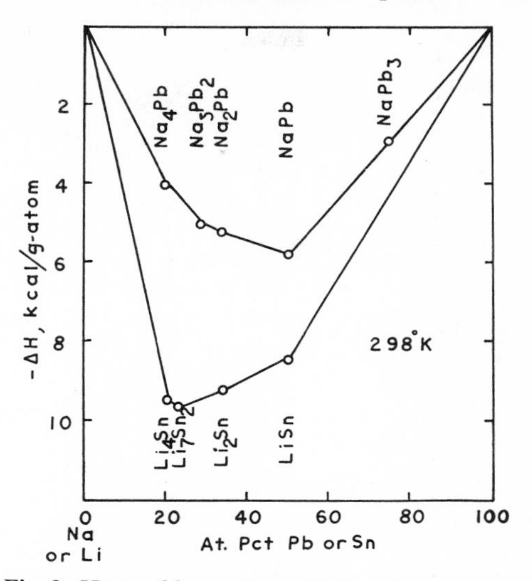

Fig. 3. Heats of formation at 298°K of compounds in the systems Na–Pb and Li–Sn.

The exothermic heat of formation of a phase which has a homogeneity range changes with composition in such a manner that the curve is concave toward the composition axis. This is illustrated in Fig. 4. The curve can be slightly convex, however, if the entropy-temperature product changes with composition in an appropriate manner. (See §4.4.) If the exothermic heat of formation of a phase with a homogeneity range goes through a maximum as a function of composition, this must be the largest heat of formation in the entire system and is likely to occur at or near the equiatomic composition.

3.2 Measurement of Heats of Formation

The heats of formation of intermetallic compounds are measured by calorimetry. The main methods are direct-reaction calorimetry, Oelsen's pouring technique, combustion calorimetry, and solution calorimetry.[1,13,14] Heats of formation can also be determined indirectly from equilibrium data for several temperatures. Such equilibrium data are usually obtained by emf measurements or in some cases by vapor pressure measurements.

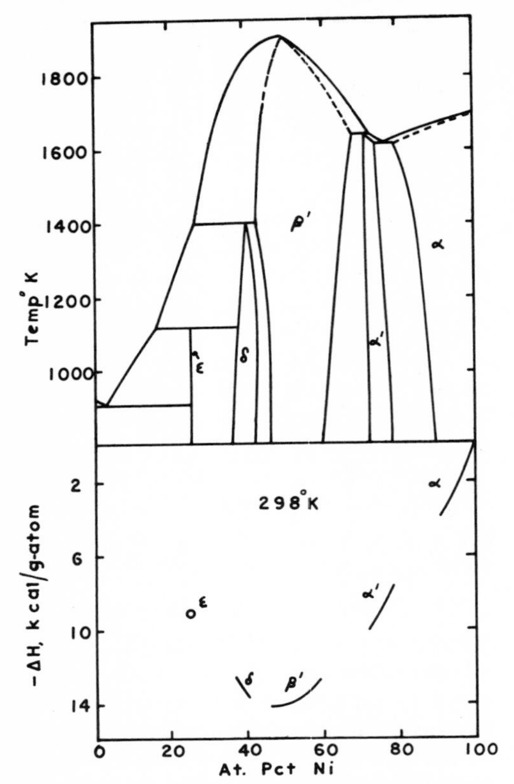

Fig. 4. Phase diagram of the system Al–Ni and the heats of formation at 298°K of phases in this system.

In the direct calorimetric methods of Kawakami,[15] Kubaschewski and Dench,[16] and Kleppa[17] the heat effect of the reaction of the solid components that form a compound is measured in the calorimeter vessel, which is kept at constant temperature. This temperature must be sufficiently high to ensure complete reaction within a reasonable time.

In Oelsen's pouring technique one component is poured over the second component at room temperature and the heat of formation is calculated from the heat effect during the cooling of the liquid alloy and the heat content of the molten component.[18] This method has been used extensively in the past, for example, to measure the heats of formation of silicides and aluminides of iron,[19] but experimental difficulties and the possibility of forming metastable compounds during cooling have limited its applicability.

Combustion calorimetry has been used for measuring the heats of formation of selenides and tellurides,[20] but it is not applicable to most intermetallic compounds because of the difficulty of attaining complete combustion. In solution calorimetry the heat of formation is determined from the difference in the heat effects produced by dissolving additions of the compound and a mechanical mixture of its components. In general, if an acid is used as a solvent, the large heat effects limit the accuracy, but the use of liquid metals such as tin, in which the samples have a low heat of solution, greatly improves the accuracy.[17,21,22]

The applicability of any calorimetric method depends on the nature of the system to be investigated. Application of a suitable method at the present time should result in an accuracy of ±5 per cent or better. The direct calorimetric method gives heats of formation for the temperature of reaction. Heats of formation obtained by solution calorimetry refer to the temperature of the addition, which may be the temperature of the bath or some other temperature, depending on the calorimeter and the method of operation.

The heats of formation can be determined from emf measurements by the equation

$$\Delta H = n\mathscr{F}\left(\frac{Td\mathscr{E}}{dT} - \mathscr{E}\right) \tag{3}$$

The use of this equation is subject to the same conditions as those discussed for (1). Heats of formation derived in this manner from emf data have been published for a number of systems.[1,2,23]

The heats of formation of intermetallic compounds containing a volatile element, such as zinc, tellurium, or arsenic, have been determined from vapor pressure measurements by an application of the van't Hoff equation.[2,7,10] The heats of formation derived from values of the emf or vapor pressure tend to be of low accuracy because small errors in the change of these values with temperature have a large adverse effect.[1,2] The heat of formation of a compound may also be found by (2) if the free energy and entropy of formation are known.

3.3 Heats of Formation and Bonding of Intermetallic Compounds

The heat of formation of a compound is the difference between the enthalpies of the compound and the component elements; the reference state of the latter is their stable state at the temperature and pressure of formation of the compound. The heat of formation is related to the bond energies in the compound, but individual bond energies cannot be calculated from the thermodynamic data unless the interactions of atoms other than single nearest-neighbor pairs are neglected. If only nearest-neighbor pairs are considered, a bond energy can be obtained by dividing the energy of the compound by the number of such pairs. This ignores the effect of other atoms on the bond energy of the nearest-neighbor pair and also ignores any bonding between non-nearest neighbors, which could be taken into account only if the dependence of the bond energy on distance and position in the crystal were known. Wagner[24] has pointed out that for fundamental reasons it may not be possible to calculate or measure individual bond energies in a crystal and that the concept of such energies may have to be discarded. Even without knowledge of individual bond energies, however, it is obvious that the heat of formation is closely related to the bonding of the compound.

Ionic, covalent, metallic, and van der Waals bonds and the transitions between these pure types must be considered in discussing the heats of formation of intermetallic compounds. Also, in a single compound, more than one type of bond may occur. For example, in compounds with layer structures the bonding within the layers differs from that between the layers.[25]

An intermetallic compound has a large exothermic heat of formation when the nature of its bonding is different from the bonding in the component elements, e.g. when metallic elements form a compound in which the bonding is ionic or covalent. The heats of formation of intermetallic compounds with ionic bonding range from approximately -10 to -40 kcal/g-atom; the numerical values of the negative heat of formation (i.e., the exothermic heat of formation) increase with the degree of ionicity of the bond. The exothermic heats of formation of covalent compounds are generally smaller than those of ionic compounds because the sharing of the electrons between the atoms in the covalent bond involves a smaller energy change than the electron transfer in the ionic bond. The heats of formation of most covalent intermetallic compounds range from approximately -3 to -8 kcal/g-atom. The heats of formation of compounds with metallic bonding are mainly due to the difference in coordination of the atoms in the component metals and the compounds. (See §3.8.) The heats of formation of most of these compounds are numerically smaller than -8 kcal/g-atom.

The crystal structure of intermetallic compounds is related to the nature of their bonding. The main factors governing both the structure and the bonding are the atomic or ionic radii of the components, the ratio of valence electrons to atoms in the compound, and the difference in the electrochemical characteristics of the components. One of these factors is usually dominant, although the others may also have a strong effect. In the following discussion groups of intermetallic compounds are considered in which the bonding is mainly ionic, covalent, or metallic or in which the bonding is intermediate between these types.

3.4 Compounds with Predominantly Ionic Bonding

Intermetallic compounds in which the bonding is predominantly ionic are formed in binary systems if the electrochemical characteristics of the components differ greatly. In general, the more purely ionic the bonding of a compound, the more restricted its homogeneity range with the limiting case of a "line compound." Predominantly ionic compounds of type AB usually have a sodium chloride (B1), zinc blende (B3), or wurtzite (B4) structure, and compounds of type A_2B_3 and AB_3 have the La_2O_3 (D5) and BiF_3 ($D0_3$) structures, respectively.

The heat of formation of an ionic compound from its component atoms at infinite distance can be derived from considerations based on the Haber-Born cycle,[26,27] provided the lattice energy of the compound is mainly electrostatic. The heat of formation of such a compound consists of the heat of sublimation and ionization energy of the electropositive component, the heat of sublimation and the electron affinity of the electronegative component, and the lattice energy of the compound. The lattice energy depends on the interatomic distance in the compound and the other factors are characteristic of the elements involved. If the electronegative component is a diatomic gaseous element, such as a halogen, its heat of dissociation replaces the heat of sublimation.

The heats of sublimation and ionization of

the electropositive component, together with an appropriate fraction of the lattice energy, give a characteristic value for the element, which is designated its electronegativity. The electronegativity of an electronegative component may be obtained by combining the heat of sublimation and electron affinity with a fraction of the lattice energy. Since the ionic diameter of an element is approximately the same in different compounds, the fraction of the lattice energy to be assigned to the electronegativity is assumed to be constant for a given element and independent of the compound.

Pauling[26] suggested the following equation for the heat of formation of an ionic compound:

$$\Delta H(\text{kcal/g-atom}) = -23.07Z(X_A - X_B)^2 \quad (4)$$

where Z is the number of valence links and X_A and X_B are the electronegativities of the component elements. Scales of electronegativity values have been proposed by Pauling,[26] Gordy and Thomas,[28] and Sanderson.[29] The values are generally determined from the heats of formation of the halides, since the availability of the energies of sublimation and ionization of the metals and the electron affinities and heats of dissociation of the halogens make it possible to estimate the lattice energy. The electronegativities determined from the heats of formation of halides, in which the metallic element is the cation, have been used with reasonable success to determine the heats of formation of compounds in which the metallic element is the anion.[30] Since electronegativities are estimated only to ± 0.1 eV, the accuracy of the heats of formation calculated from (4) cannot be better than ± 2.5 kcal/g-atom.

As several writers have pointed out,[30,31] heats of formation calculated by Pauling's equation agree reasonably well with experimental values for ionic and partly ionic intermetallic compounds. This is true, however, only if no single electronegativity scale is adhered to, as shown in Table 1. If a single

TABLE 1

Heats of Formation at 298°K of Typical Ionic and Partially Ionic Compounds—Experimental Values and Values Calculated from Different Electronegativity Scales

		Heat of Formation, $-\Delta H$, kcal/g-atom			
			Calculated from Electronegativity Scales		
Compound	Structure	Experimental[a]	Pauling	Gordy & Thomas	Sanderson
MgSe	NaCl(B1)	32.6 ± 2.0	33.2	33.2	55.4
MgTe	ZnS(B4)	25.0 ± 2.5	18.7	18.7	29.5
ZnSe	ZnS(B3)	17.0 ± 1.5	14.8	18.6	—
ZnTe	ZnS(B3)	14.4 ± 0.5	5.8	16.6	—
GaTe	—	14.3 ± 1.5	5.8	16.6	—
Li_3Bi	$BiF_3(D0_3)$	13.4 ± 0.08[b]	14.0	12.5	—
InTe	TlSe(B37)	11.5 ± 1.5	3.7	11.3	—
Li_3Sb	$BiF_3(D0_3)$	10.8 ± 0.8	14.0	12.5	—
PbSe	NaCl(B1)	9.0 ± 1.0	8.3	14.7	7.5
PbTe	NaCl(B1)	8.4 ± 0.2[b]	2.1	8.3	0.5
NaSb	$CuAu(L1_0)$	7.9 ± 0.5	13.5	9.3	—
NaBi	$CuAu(L1_0)$	7.8 ± 0.5	13.5	9.3	—
Mg_2Si	$CaF_2(C1)$	6.3 ± 0.3	11.1	11.1	8.6
Mg_2Sn	$CaF_2(C1)$	6.4 ± 0.05[c]	11.1	7.7	20.2

[a] All values in this column are taken from Kubaschewski and Evans,[1] except those indicated by (*b*) and (*c*).
[b] Hultgren et al.[2]
[c] Beardmore, Howlett, Lichter, and Bever.[34]

TABLE 2

Heats of Formation at 298°K, $-\Delta H$(kcal/g-atom), of Ionic and Partially Ionic Compounds as Functions of the Atomic Number of the Electropositive or Electronegative Component[a]

II–VI Compounds NaCl Structure (B1)		II–V Compounds La_2O_3 Structure ($D5_2$)		II–IV Compounds CaF_2 Structure (C1)	
MgS	41.5 ± 1.0	Mg_3P_2	(25.6)	Mg_2Si	6.3 ± 0.3
MgSe	32.6 ± 2.0	Mg_3As_2	—	Mg_2Ge	9.2 ± 0.02[d]
[b]MgTe	25.0 ± 2.5	Mg_3Sb_2	15.8 ± 0.8	Mg_2Sn	6.4 ± 0.05[d]
	—	Mg_3Bi_2	7.4 ± 0.2[c]	Mg_2Pb	4.0 ± 0.03[d]
CaS	55.0 ± 1.3	Ca_3P_2	24.0 ± 1.2	[e]Ca_2Si	16.7 ± 1.10
CaSe	37.4 ± 3.0	Ca_3As_2	—	[e]Ca_2Ge	—
CaTe	—	Ca_3Sb_2	34.8 ± 1.8	[e]Ca_2Sn	25.0 ± 1.5
	—	Ca_3Bi_2	21.4 ± 1.0[c]	[e]Ca_2Pb	16.9 ± 2.0[c]
BaS	53.0 ± 2.5	Ba_3P_2	23.6 ± 2.0	[e]Ba_2Si	—
BaSe	37.1 ± 2.5	Ba_3As_2	—	[e]Ba_2Ge	—
BaTe	31.8 ± 4.0	Ba_3Sb_2	35.0 ± 2.0	[e]Ba_2Sn	30.0 ± 2.7
		Ba_3Bi_2	29.0 ± 4.0[c]	[e]Ba_2Pb	23.3 ± 1.0

[a] All values of ΔH are taken from Kubaschewski and Evans,[1] except those marked (*c*) and (*d*). Values in parentheses are estimates.
[b] ZnS (B4) structure
[c] Hultgren et al.[2]
[d] Beardmore, Howlett, Lichter, and Bever.[34]
[e] The structure of these compounds has not been determined.

scale is used, the agreement between the calculated and experimental values for different compounds varies widely, and in most cases good agreement can be obtained only by choosing a scale by trial and error.

Regardless of the choice of an appropriate electronegativity scale, Pauling's equation becomes less applicable as ionic bonds are replaced by covalent or metallic bonds. Conversely, if this equation yields an approximately correct value of the heat of formation of an intermetallic compound, the bonding may be presumed to be at least partly ionic.

A general relation between the heats of formation of a homologous series of ionic compounds and the atomic number of one of the component elements can be deduced from the electronegativities. Within most groups of the periodic table the electronegativities decrease with increasing atomic number.[28] Since the electronegativities of the electronegative elements are larger than those of the electropositive elements,[26,28] the difference in the electronegativities of the components in a homologous series of compounds in which the electropositive element is kept constant decreases as the atomic number of the electronegative element increases within a given group. This is associated with a decrease in the exothermic heat of formation, as shown in Table 2. If the electronegative element is kept constant, however, the difference in electronegativity increases with increasing atomic number of the electropositive component. In this case the exothermic heat of formation increases with increasing atomic number of the electropositive element within the homologous series. (See Table 2.) These generalizations, however, hold only if the electronegativities of the elements which are being varied decrease with atomic number within the group. An example of an increase of the electronegativities on the Pauling scale is given by the elements of group III-B. Such behavior, however, is found only in exceptional cases throughout the periodic system. The occurrence of exceptions differs according to the different scales.

Although the magnitude of the heat of formation of ionic compounds depends to a large extent on the ionization energy of the component atoms, the packing density also affects the heat of formation. A normal valence compound in which the bonding is ionic and the stoichiometric proportion is 1:1 may

TABLE 3

Structures, Electronegativities and Heats of Formation of Group II-A and II-B Selenides as Functions of Ionic Radii

Compound	Difference in Goldschmidt Ionic Radii	Structure	Difference in Electronegativity, Pauling Scale	Heat of Formation,[a] $-\Delta H$, kcal/g-atom
CaSe	[b]1.91 − 1.06	NaCl	[b]2.4 − 1.0	37.4 ± 3.0
SrSe	1.91 − 1.25	NaCl	2.4 − 1.0	39.4 ± 2.0
BaSe	1.91 − 1.43	NaCl	2.4 − 0.9	37.1 ± 2.5
ZnSe	1.91 − 0.83	ZnS	2.4 − 1.6	17.0 ± 1.5
CdSe	1.91 − 1.03	ZnS	2.4 − 1.7	12.5 ± 2.0

[a] All values in this column are taken from Kubaschewski and Evans.[1]
[b] This column refers to selenium.

crystallize in the sodium chloride, zinc blende, or wurtzite structure. The choice between these structures appears to depend largely on the relative magnitude of the ionic radii. Small differences favor the cubic lattice of the sodium chloride type, as in the selenides of group II-A elements (Table 3). If the difference of the ionic radii is larger, there is a tendency to form the zinc blende structure in which the anions lie on a close-packed cubic sublattice while the smaller cations occupy one-half the tetrahedral interstices. Typical examples are the selenides of group II-B elements. The closer packing of the anions and cations in the zinc blende structure than in the sodium chloride structure leads to increased coordination and to a smaller exothermic heat of formation, as shown in Table 3.

The Haber-Born cycle takes the packing of the atoms in a compound into account in the lattice energy term. This energy is inversely proportional to the ionic radii of the components. For elements with approximately the same ionization energy and heat of sublimation, a decrease in ionic radii is associated with an increase in the lattice energy and in the electronegativity. If the elements are combined with the same electronegative component, such as selenium, the difference in the electronegativities of the components, and thus the magnitude of the heats of formation, are small when the electropositive component has a small ionic radius (Table 3).

The structures and heats of formation of other ionic and partially ionic compounds are listed in Table 4.

TABLE 4

Heats of Formation at 298°K of Ionic and Partially Ionic Compounds ($-\Delta H$, kcal/g-atom)[a]

NaCl Structure (B1)		ZnS Structure (B3, B4)		CaF_2 Structure (C1)	
SrTe	31.0 ± 4.0	Al_2Se_3	25.8 ± 0.7	Li_2Se	33.2 ± 2.3
SnSe	8.3 ± 1.0	Ga_2Se_3	21.0 ± 0.6	K_2Se	29.7 ± 2.0
SnTe	7.3 ± 0.2	Ga_2Te_3	13.0 ± 0.6	Na_2Se	27.3 ± 1.0
		CdTe	12.3 ± 0.3	Na_2Te	25.0 ± 3.0
		HgSe	2.6 ± 1.8	$CoSi_2$	8.2 ± 0.7
		Ag_2Se	1.7 ± 0.7	Cu_2Se	4.7 ± 0.1
		HgTe	1.5 ± 2.3		

[a] The heats of formation of additional compounds with these structures are to be found in Tables 1, 2, and 3. All values of ΔH are taken from Kubaschewski and Evans.[1]

3.5 Compounds with Transitional Ionic-Metallic Bonding

As a general rule, intermetallic compounds in which the bonding is intermediate between ionic and metallic possess a close-packed arrangement of the electronegative components.[32] Various partially ionic compounds crystallize in the antiisomorphous fluorite structure (C1), in which the atoms of the electronegative component are arranged on a cubic close-packed sublattice and the atoms of the electropositive component occupy the tetrahedral interstices. The structure of these compounds is the result of the difference between the electronegativities of the components, the difference between their valences, and the difference between their ionic radii.

The compounds of magnesium with group IV-B elements have the antiisomorphous fluorite structure. This series of compounds has been cited as an ideal example of the gradual replacement of ionic by metallic bonding as the differences in the electrochemical characteristics of the components decrease from Mg_2Si to Mg_2Pb.[33] A recent investigation[34] of the thermodynamic properties of these compounds shows that the heats of formation, and presumably the ionic contributions to the bonding, decrease from Mg_2Ge to Mg_2Pb, but an earlier published value[1] of the heat of formation of Mg_2Si is smaller than that of Mg_2Ge, as shown in Table 2. This discontinuity in the homologous series suggests that the bonding in Mg_2Si may be less ionic than in Mg_2Ge, although the difference in the electronegativities of the components of the two compounds predicts the opposite relation.

Raynor[32] has proposed that as the difference in electronegativity between the components decreases in a homologous series, the size and diffuseness of the ionic core increase and attractions of the van der Waals type, which are favored by close-packing, may develop between the electronegative components. According to Raynor, these attractive forces may change gradually to metallic bonds as the loosely bound electrons come under the control of several nuclei. According to another view,[35] covalent rather than metallic bonding may develop in place of ionic bonding. In particular, the predominance of covalent bonding in Mg(IV-B) compounds (with the exception of Mg_2Pb) and other semiconducting compounds has been emphasized.[36,37]

3.6 Compounds with Mixed Ionic, Covalent, and Metallic Bonding

In certain compounds the bonding is mixed ionic, covalent, and metallic. Many of these compounds have the nickel arsenide (B8) structure, in which the electronegative atoms form a close-packed hexagonal sublattice and the electropositive atoms are situated in the octahedral interstices. Compounds having this structure are based on the AB type where A is usually a transition metal and B belongs to group III-B, IV-B, V-B, or VI-B. This structure is adaptable to changes in axial ratio; attendant changes in composition extend the range over which the nickel arsenide structure can exist from A_2B to AB_2.[33] This results in mixed bonding and a range of heats of formation (Table 5).

Among the compounds having the nickel arsenide structure, the largest exothermic heats of formation and the largest axial ratios (see Fig. 5) are generally found in those in which the electropositive component is a transition element and has the smallest number of *d*-electrons, whereas the electron affinity (electronegativity) of the electronegative component is the largest. In such compounds there is a substantial ionic contribution to the bonding and their heats of formation range from approximately -8 to -11 kcal/g-atom. The ionic contribution to the bonding is also reflected in the restricted homogeneity range over which such compounds exist, although the nondirectional nature of ionic bonds allows deviations from the ideal composition and extensions of the homogeneity range sometimes occur.[32,38]

In a series of compounds having the nickel arsenide structure and a common electronegative component the ionic contribution to the bonding decreases as the number of electrons in the *d*-shell of the transition element increases. The difference between the electronegativities of the two components also decreases in such a series. As the ionic contribution to the bonding becomes smaller, the exothermic heat of formation changes from a level of approximately -11 kcal/g-atom for compounds in which the bonding is largely ionic to values that lie in a broad band ranging from -9 to -3 kcal/g-atom. A decrease to a

TABLE 5

Heats of Formation at 298°K of Compounds with the NiAs Structure (B8)[a]

Compound	$-\Delta H$, kcal/g-atom	Compound	$-\Delta H$, kcal/g-atom
FeS	11.4 ± 0.2	MnSb	6.0[d]
MnTe	(11.3)	CoSe	5.0 ± 1.3
NiS	11.1 ± 0.7	CoTe	4.5 ± 1.3
CoS	11.0 ± 0.6[b]	CoSb	5.0 ± 0.4
FeSe	9.0 ± 0.5	NiSe	5.0 ± 1.5
FeTe	(8.0)[b]	NiTe	4.5 ± 1.5
NiSb	7.9 ± 1.0[c]	CoSn	3.6 ± 0.3[c]
Ni_3Sn_2	7.5 ± 1.0[c]	AuSn	3.6 ± 0.05[e]
MnAs	6.9[d]	Co_3Sn_2	2.7 ± 0.2[b]
NiAs	(6.5)	MnBi	2.4[d]

[a] All values of ΔH are taken from Kubaschewski and Evans[1] except those marked (*b*), (*c*), (*d*), and (*e*). Values in parentheses are estimated values.
[b] Raynor.[32]
[c] Hultgren et al.[2]
[d] Shchukarev et al.[55]
[e] Misra, Howlett, and Bever.[53]

level of −9 to −6 kcal/g-atom may be assumed to be associated with increasingly covalent bonding. At lower levels of the exothermic heat of formation the bonding becomes predominantly metallic. (See Fig. 6.)

The transition from largely ionic to largely covalent bonding in compounds with the nickel arsenide structure is accompanied by a contraction of the homogeneity range over which they exist because of the requirements of valency and spatial orientation of the covalent bonds. On the other hand, the transition to largely metallic bonding is accompanied by a tendency toward more extensive homogeneity ranges.

Compounds with the nickel arsenide structure and a high degree of ionicity of the bond tend to have homogeneity ranges that extend in the direction of a deficit of the more metallic atoms by their omission from some of the octahedral sites until at the limiting composition AB_2, the cadmium iodide structure (C6) is formed.[33] The increase in the axial ratio due to such a change in composition is associated with an increase in the ionicity of the bond, which is reflected in an increase of the heat of formation. By contrast, if the bonding of a compound with the nickel arsenide structure is largely metallic, the interaction between the more metallic atoms A extends the homogeneity range in the direction of excess A atoms. These atoms occupy some of the trigonal interstices. The interaction of the metal atoms leads to decreases in the axial ratio and the exothermic heat of formation. Thus in a compound with a nickel arsenide structure the homogeneity range extends toward the AB_2 side of the AB composition when the bonding is largely ionic and to the A_2B side when the bonding is largely metallic. This is well illustrated by the series of

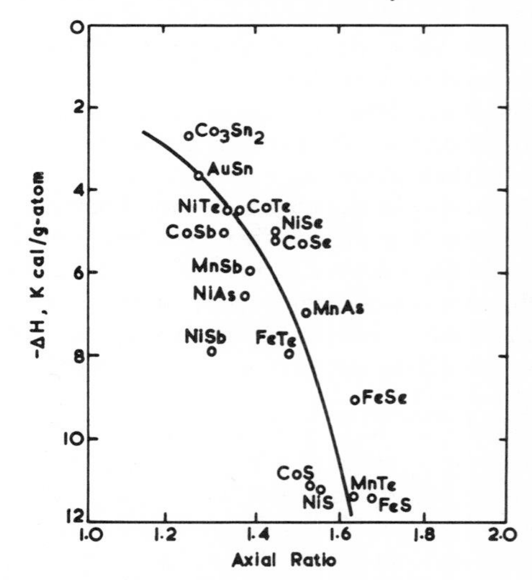

Fig. 5. Heats of formation at 298°K of intermetallic compounds with NiAs (B8) structure as a function of axial ratio.

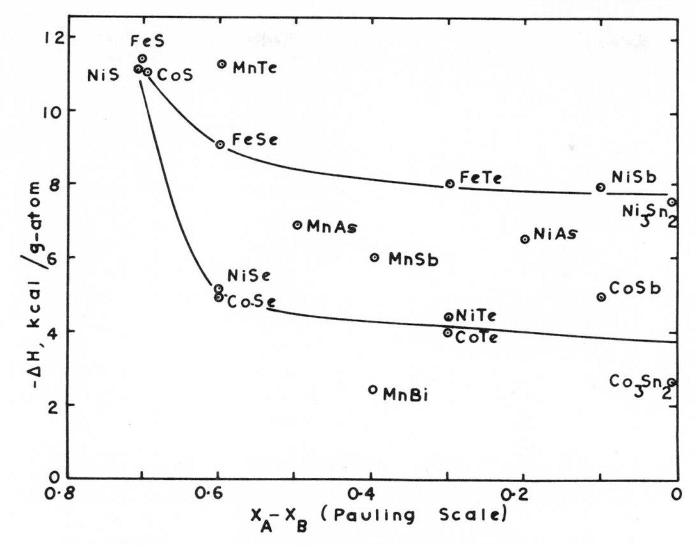

Fig. 6. Heats of formation at 298°K of intermetallic compounds with NiAs structure as a function of the difference in electronegativities between the components.

compounds NiSe, NiSb, NiBi, and NiSn, in which the center of the homogeneity range moves from 41.5 at.% Ni for NiSe to 62 at.% Ni for NiSn as the metallic contribution to the bonding increases.[39]

3.7 Compounds with Metallic Bonding—Electron Compounds

Compounds with metallic bonding are formed by metallic elements which have approximately the same atomic size and similar electronegativities. Such compounds are electron compounds if their components belong to different groups in the periodic table and their atomic sizes do not differ by more than approximately 15 per cent. These compounds are stable over ranges of composition. The main effect of substituting a solute for a solvent atom is a change of the electron-to-atom ratio. Electron compounds are based on compositions at which this ratio has the following values:

Electron-to-Atom Ratio	Crystal Structure
3/2	β-brass type
21/13	γ-brass type
7/4	ϵ-brass type

Typical examples of electron compounds are found in systems in which one component is copper, silver, or gold and the other zinc, aluminum, or tin. The heats of formation of most electron compounds are small in accord with the metallic nature of their bonding. (See Table 6.) If a system has several electron compounds, the largest exothermic heat of formation usually occurs in the γ-brass structure.

Minor ionic contributions to the bonding of electron compounds may arise as the difference between the electronegativities of the components increases. This is well illustrated by the β-phases in the systems Ag–Zn, Ag–Cd, and Ag–Mg; the electrochemical difference between the components increases in this order. The increased ionicity of the bonding is reflected in the increased heats of formation (Table 6) and an increased tendency to retain order; the compounds AgZn and AgCd have an ordered bcc structure at low temperatures, but a disordered bcc structure at high temperatures, whereas the compound AgMg retains its ordered bcc structure up to the melting point.[38]

The increasing ionic contribution to the bonding of the compounds AgZn, AgCd, and AgMg is also reflected by the solid-to-liquid transition (Fig. 7). The compound AgMg has a congruent melting point at the stoichiometric composition, whereas the compounds AgZn and AgCd decompose by a peritectic reaction. Other electron compounds with largely metallic character also undergo peritectic decomposition.[38]

TABLE 6

Heats of Formation of Electron Compounds
($-\Delta H$, kcal/g-atom)[a]

β-Brass Type (A2 and B2)			γ-Brass Type ($D8_2$)			ε-Brass Type (A3)		
Composition	T°K	$-\Delta H$	Composition	T°K	$-\Delta H$	Composition	T°K	$-\Delta H$
AgZn (0.50Zn)	873	0.75 ± 0.05	Ag_5Zn_8 (0.61Zn)	873	1.1 ± 0.05			
AgCd (0.49Cd)	723	1.6 ± 0.1	Ag_5Cd_8 (0.60Cd)	673	2.0 ± 0.1	$AgCd_3$ (0.75Cd)	673	1.4 ± 0.05
CuZn (0.48Zn)	773	2.3 ± 0.1	Cu_5Zn_8 (0.60Zn)	773	2.8 ± 0.1	$CuZn_3$ (0.78Zn)	773	1.6 ± 0.1
LaMg (0.50Mg)	292	2.9 ± 0.5	Cu_5Cd_8 (0.60Cd)	723	1.1 ± 0.07	Ag_3Sb (0.75Ag)	723	0.03 ± 0.04
AuCd (0.50Cd)	700	4.7 ± 0.1	Au_5Cd_8 (0.64Cd)	700	3.7 ± 0.1	$AuCd_3$ (0.75Cd)	700	3.7 ± 0.1
AgMg (0.50Mg)	273	4.4 ± 0.05[b]	Cu_9Al_4 (0.65Cu)	298	5.5 ± 0.8	Ag_3Sn (0.78Ag)	723	0.76 ± 0.2
MgTl (0.50Tl)	—	6.0 ± 1.0				Au_3Sn (0.84Au)	298	0.97 ± 0.05
LiTl (0.50Tl)	298	6.4 ± 0.5[c]				Cu_3Sn (0.75Cu)	723	1.9 ± 0.05
LiPb (0.50Pb)	298	7.3 ± 0.6[c]				Au_4In (0.80Au)	723	2.8 ± 0.1
HgLi (0.49Li)	292	10.0 ± 1.0						
CoAl (0.50Al)	298	13.2 ± 1.0[d]						
NiAl (0.50Al)	298	14.1 ± 1.5[d]						
CaTl (0.50Tl)	—	19.5 ± 1.5						

[a] All values of ΔH are taken from Hultgren et al.[2] except those marked (*b*), (*c*), and (*d*).
[b] Robinson and Bever.[54]
[c] Kubaschewski and Evans.[1]
[d] Kubaschewski and Heymer.[49]

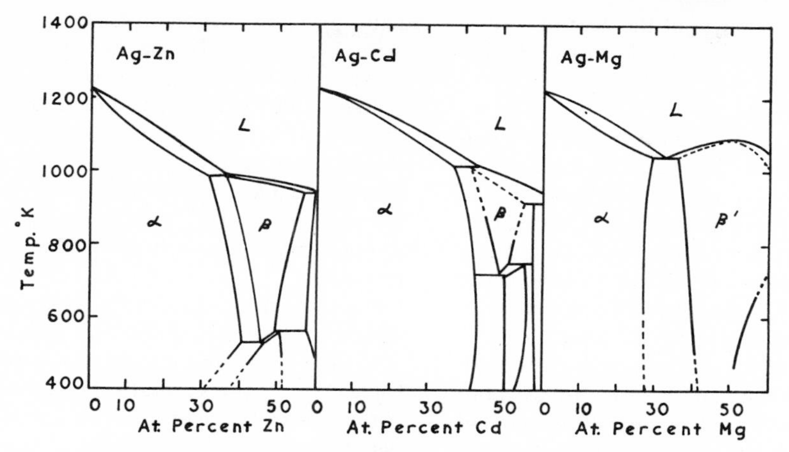

Fig. 7. Phase diagrams of Ag–Zn, Ag–Cd, and Ag–Mg systems, indicating the increasingly ionic nature of the bonding in the series of β-electron compounds.

Electron compounds exist over appreciable homogeneity ranges, but the heats of formation of most electron compounds are known only at a single composition. Some data are available for the heat of formation as a function of composition in the homogeneity range of the β-brass type of electron compounds.[2] In the system Cu–Zn, in which the conditions for the formation of electron compounds are ideal, the exothermic heat of formation increases with increasing zinc content across the homogeneity range.[2] (See Fig. 8*a*.) This increase with zinc concentration is due to an increase in the energy required by the electrons as they fill the Brillouin zone. At the higher electron concentrations, however, the density of states decreases rapidly as the Fermi surface touches the Brillouin zone boundary, and the additional energy required decreases as the phase boundary on the zinc-rich side is approached. The slope of the curve of the heat of formation as a function of composition thus decreases with increasing composition near this phase boundary.[40] This behavior is typical of electron compounds in which the bonding is metallic, as shown for the β-phase in the system Cu–Zn in Fig. 8*a*.

In the β-phases in the systems Ag–Mg and Au–Cd, and similar systems in which the electronegativities of the components differ

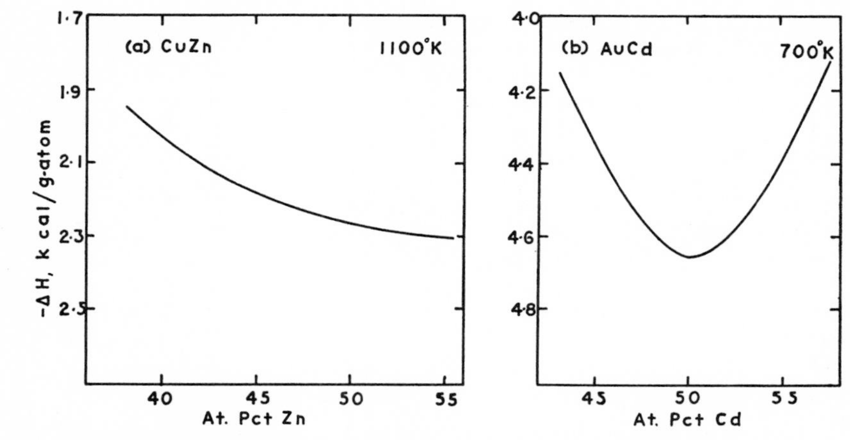

Fig. 8. Heats of formation of the β-electron compounds CuZn and AuCd as a function of composition, showing (*a*) effect of metallic bonding and (*b*) effect of partially ionic bonding.

appreciably, the change of the heat of formation with composition is modified by an ionic contribution to the bonding. Within the homogeneity range of the β-phases electron transfer is most probable at the stoichiometric composition. The partially ionic nature of the bonding causes an increase in the exothermic heat of formation. The heat of formation as a function of composition, therefore, has a maximum at the stoichiometric composition of those electron compounds in which the bonding is partly ionic. This is shown in Fig. 8*b* for the β-phase in the system Au–Cd. A maximum of the exothermic heat of formation is most likely in the central region of the phase diagram as mentioned in Section 3.1.

3.8 Compounds with Metallic Bonding—Laves Phases

Intermetallic compounds with structures characteristic of the Laves phases form if three conditions are fulfilled.[41] First, the difference between the atomic sizes of the components must be large enough to permit packing denser than that resulting from the close packing of spheres of equal size. Second, the electrochemical nature of the components must be similar. Third, the electron-to-atom ratio must be within certain limits. These conditions are sufficiently broad for Laves phases to form between elements in various parts of the periodic system. There are three main structural types of Laves phases: the cubic $MgCu_2$ (C15), the hexagonal $MgNi_2$ (C14), and the hexagonal $MgZn_2$ (C36).[33] Unlike most other types of metallic compounds, most Laves phases are line compounds, but a few, such as $MgCu_2$, are believed to have a relatively wide range of homogeneity.[38] The restriction to narrow homogeneity ranges is due to packing requirements.

The exothermic heats of formation of Laves phases tend to be numerically smaller than -5 kcal/g-atom, but a few are larger. (See Table 7.)[42–44] The generally small magnitude of the heats of formation is indicative of the metallic nature of the bonding. King and Kleppa[42] showed that there is no correlation between the heat of formation and the difference in electronegativity between the components of six compounds investigated by them, and the same may be assumed to hold for Laves phases generally. However, these investigators found that the exothermic heat of formation decreased with positive deviations from the ideal radius ratio of 1.225 for Laves phases. This decrease can be explained by the strain energy associated with nonideal ratios, which reduces the exothermic heat of formation.

Dehlinger and Schulze[45] suggested that the stability of Laves phases arises from a decrease in enthalpy due to the increase in the average

TABLE 7

Heats of Formation at 298°K of Laves Phases ($-\Delta H$, kcal/g-atom)

$MgCu_2$ Type (C15)		$MgZn_2$ Type (C14)		$MgNi_2$ Type (C36)	
$CaAl_2$	19.2 ± 0.1[a]	$ZrZn_2$	11.8 ± 0.2[f]	$MgNi_2$	4.4 ± 0.1[c]
$LaAl_2$	12.0 ± 0.5[b]	$CaCd_2$	10.0 ± 1.0[g]		
$MgCu_2$	2.7 ± 0.1[c]	$CaMg_2$	3.2 ± 0.1[c]		
UFe_2	2.6[d]	$MgZn_2$	2.6 ± 0.1[c]		
$PbAu_2$	0.1 ± 0.03[e]	$SrMg_2$	1.7 ± 0.1[c]		
		$BaMg_2$	0.5 ± 0.2[c]		
		$CdCu_2$	0.5 ± 0.1[h]		

[a] Kubaschewski and Heymer.[49]
[b] Hultgren et al.[2]
[c] King and Kleppa.[42]
[d] Akhachinski et al.[56]
[e] Kleppa.[50]
[f] Chiotti and Kilp.[51]
[g] Kubaschewski and Evans.[1]
[h] Kleppa.[52]

coordination number in these compounds compared with that in the component metals. Kubaschewski[46] has proposed a method for calculating the heats of formation of intermetallic compounds based on the heat of sublimation of the component metals and the change in coordination on formation and has calculated the heats of formation of various compounds, including the Laves phases. The agreement between the calculated and experimental values for Laves phases is satisfactory. This indicates that their heats of formation can be largely attributed to increases in coordination. The calculations involve the assumption that the energy of interaction between like atoms, calculated from their sublimation energies, is equally distributed over all like bonds and that the energy of interaction between two unlike atoms is the arithmetic mean of the A–A and B–B bond energies. The agreement obtained by Kubaschewski is of special interest in view of the doubts expressed concerning the "pair energy" approach to metallic systems.[24,42]

The stability of Laves phases containing a transition metal is increased by the transfer of an electron from the other metal to a partly filled inner shell of the transition metal.[42,47] This electron transfer introduces an ionic contribution to the bonding and is accompanied by an increase in the heat of formation.[42] An example of electron transfer occurs in the compound $MgNi_2$, which has a larger exothermic heat of formation than those of the compounds $MgCu_2$ and $MgZn_2$, which are otherwise similar to $MgNi_2$.[42] The energy effect of the electron transfer is opposed to the strain energy resulting from unfavorable atomic size ratios and thus helps to make the formation of Laves phases possible in systems with atomic size ratios that are not ideal. It is not surprising, therefore, that the largest deviations from the ideal ratio are found in Laves phases that contain a transition element.

3.9 Compounds with Covalent Bonding

Typical covalent bonding occurs in crystals in which an atom has four nearest neighbors, as in certain elements of group IV-B, in their solid solutions, and in compounds of elements of group III-B with elements of group V-B. The ratio of valence electrons to atoms in these crystals is 4:1, and each atom tends to form four covalent bonds. These phases have the zinc blende (B3) structure.

TABLE 8

Heats of Formation at 298°K of Covalent Compounds[a]

Compound	Structure	$-\Delta H$, kcal/g-atom
InSb	ZnS (B3)	3.5 ± 0.1
GaSb	ZnS (B3)	5.0 ± 0.2
InAs	ZnS (B3)	7.4 ± 0.6

[a] Values of ΔH are taken from Schottky and Bever.[12]

Intermetallic compounds in which the bonding is largely covalent form only if the difference in electrochemical characteristics of the components is small. As this difference increases, the electrons shared between the atoms in the covalent bond become increasingly polarized and this contributes an element of ionicity to the bond. As the ionic contribution to the bonding increases, the exothermic heat of formation also increases. The degree of polarization of the anion depends on the sizes and deformabilities of the anion and cation as well as on the difference between their electronegativities.

The heats of formation of the covalent III-V compounds range from -3 to -8 kcal/g-atom. Values are listed in Table 8.

Folberth[48] has deduced from some of their properties that III-V compounds have mixed covalent and ionic bonds and that the two types of bond are present in comparable amounts. This view, however, is not generally held. The heats of formation give little support to the assumption of an appreciable ionic contribution to the bonding of III-V compounds.

3.10 Comparison of the Heats of Formation of Various Types of Intermetallic Compounds

The preceding sections have dealt with the relation between the bonding and the heat of formation of certain classes of intermetallic compounds. Figure 9 summarizes how their occurrence depends on the difference in atomic size and the difference in electronegativities of the component elements. The contour lines for equal heats of formation are also plotted in this figure.

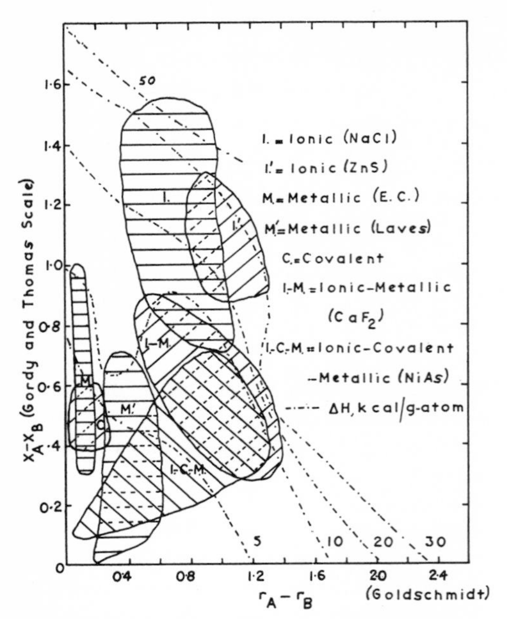

Fig. 9. The occurrence of intermetallic compounds in terms of the difference in electronegativities (X_A-X_B) and the difference in the atomic or ionic radii (r_A-r_B) of the components.

For metallic compounds the difference in atomic radii between the components is small and the difference in electronegativities is usually also small, but may reach surprisingly large values. As the ionic contribution to the bonding of compounds increases, both the size factor and the electronegativity difference increase; compounds with predominantly ionic bonding are characterized by large differences in electronegativity. In compounds with covalent bonding the differences in atomic size and electronegativities are small.

The published heats of formation of the structural types of intermetallic compounds discussed previously are summarized in Fig. 10. This illustration gives an indication of the magnitude and distribution of the heats of formation and permits a comparison between the heats of formation of compounds with different types of bonding. The distribution of the heats of formation indicated by Fig. 10 is likely to be affected by the possibly nonrepresentative distribution of compounds within any one class for which heats of formation have been measured. The heats of formation of metallic and covalent compounds tend to be relatively small, whereas the heats of formation of ionic compounds are larger and extend over a wide range (Fig. 10).

The classification of intermetallic compounds adopted in the preceding sections is based on the type of bonding as the primary criterion, but the structures of the compounds are also taken into account. Consequently, only compounds with certain structures have been considered. The heats of formation of compounds with complex structures have not yet been dealt with. Also, heats of formation have been published for some compounds, the structures of which are not known.

Figure 11 shows the heats of formation of compounds with complex or unknown structures, together with those of compounds considered in preceding sections, as a function of

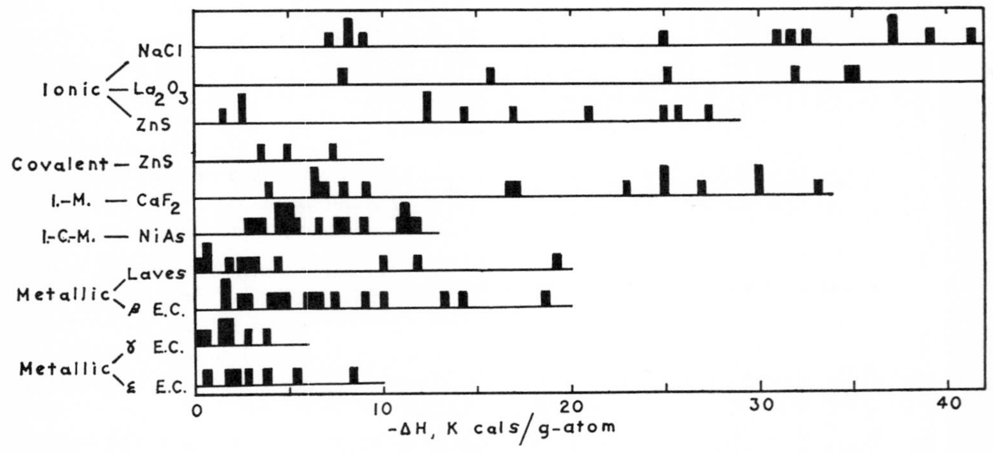

Fig. 10. Heats of formation at 298°K of intermetallic compounds with ionic, covalent, metallic, or transitional bonding.

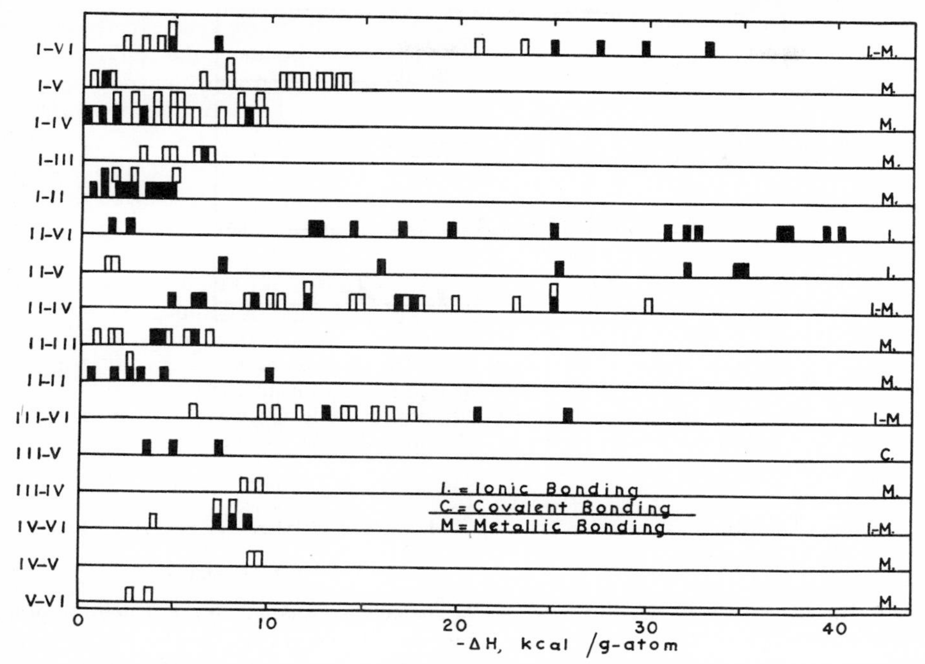

Fig. 11. Heats of formation at 298°K of intermetallic compounds as a function of the position of the elements in the periodic table. ▮ Compounds discussed in Sections 3.4 to 3.9. ▯ Compounds with complex or unknown structures not discussed in Sections 3.4 to 3.9.

the position of their component elements in the periodic table. This type of classification gives some indication of the heat of formation and type of bonding to be expected when elements from different parts of the periodic system combine to form a binary intermetallic compound.

The heats of formation of compounds of group I elements, with elements of groups II, III, IV, and V, are numerically smaller than −15 kcal/g-atom. The upper numerical limit of the heat of formation of each type of compound increases regularly in the sequence extending from I–II to I–V compounds. The compounds between these elements which were considered in Sections 3.7 and 3.8 have essentially metallic bonding and are electron compounds or Laves phases. It is reasonable to conclude that other compounds between elements in these groups with similar heats of formation are also metallic. The compounds between elements of group I and group VI classified by the magnitude of their heats of formation are concentrated in two ranges; one with small heats of formation, which suggest metallic bonding, and another with heats of formation in excess of −20 kcal/g-atom, which are ionic. The I–VI compounds with known structure have the antiisomorphous fluorite structure and transitional ionic-metallic bonding; this type of bonding may also be assumed for I–VI compounds of unknown structure, although some of them appear to be predominantly metallic.

The compounds of group II elements with group II or group III elements have small heats of formation suggesting phases with metallic bonding. The known compounds in this group are all Laves phases. The heats of formation of II–IV compounds with known structures range from −4 to −26 kcal/g-atom, and the heats of formation of other compounds between elements of these groups fall into the same range. They may be considered to be of intermediate bonding type—that is, transitional ionic-metallic bonding. The II–V and II–VI compounds of which the structures are known are ionic; their heats of formation have a wider range than the II–IV compounds.

The two measured heats of formation of III–IV compounds suggest that their bonding is metallic. The III–V compounds are known

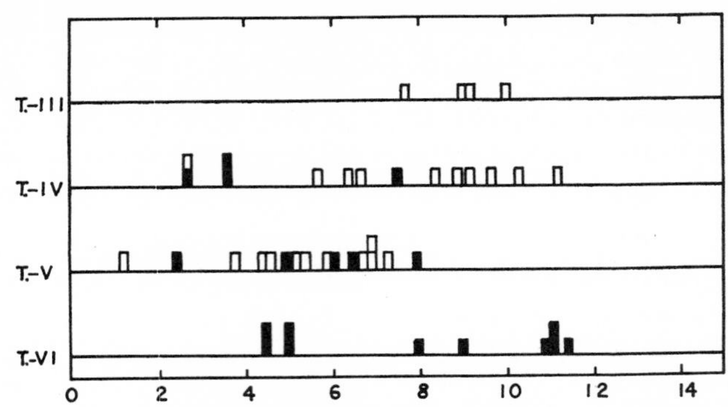

Fig. 12. Heats of formation at 298°K of intermetallic compounds containing a transition element as a function of position of the second element in the periodic table. ▮ Compounds with NiAs structure. ▯ Compounds with complex or unknown structures.

to have covalent bonding. The heats of formation of III–VI compounds range from −7 to −26 kcal/g-atom, and their bonding must be of transitional ionic-metallic type. Some III–VI compounds are known to have the zinc blende structure; their heats of formation range from −13 to −26 kcal/ g-atom and their bonding is predominantly ionic.

The compounds of group IV and group VI elements of which the structure is known have the sodium chloride structure and are ionic. The compounds of unknown structure between elements of these groups have heats of formation of the same magnitude. The few published heats of formation of IV–V and V–VI compounds are numerically smaller than −10 kcal/g-atom.

The heats of formation of compounds containing a transition metal are shown in Fig. 12. These compounds fall into three categories: those for which a nickel arsenide structure has been determined, those with complex structures, and those for which no structure determinations have been reported. The heats of formation of these compounds are numerically smaller than −12 kcal/g-atom. The properties of the compounds with a known nickel arsenide structure suggest that they have mixed ionic-covalent-metallic bonding, as discussed in Section 3.7. The heats of formation of the compounds containing a transition metal with other known or with unknown structures are similar to those of the compounds with nickel arsenide structures and do not give an indication of any difference in the bonding.

4. ENTROPIES OF FORMATION

4.1 General

The entropy change on the formation of an intermetallic compound can be treated as the sum of the changes in the terms which make up the total entropy

$$\Delta S = \Delta S_{\text{vib}} + \Delta S_{\text{conf}} + \Delta S_{\text{el}} + \Delta S_{\text{magn}} \quad (5)$$

The change in vibrational entropy ΔS_{vib} is due to the difference between the thermal vibrations of the atoms in the compound and in the component elements.[57] This difference is caused by changes in the bond strength and in the crystallographic arrangement of the atoms. A stronger bond in the compound than the average of the bonds in the elements reduces the vibrational entropy. Similarly, if the geometrical freedom of the atoms to vibrate is less in the compound than in the elements, the vibrational entropy decreases.

In the formation of a disordered intermetallic compound the uncertainty in the type of atom occupying a given lattice position leads to a positional or configurational entropy term. For the formation of a binary disordered compound this term is

$$\Delta S_{\text{conf}} = -R(N_{\text{A}} \ln N_{\text{A}} + N_{\text{B}} \ln N_{\text{B}}) \quad (6)$$

where N_{A} and N_{B} are the gram-atomic fractions of the components A and B, respectively. The configurational entropy term decreases with increasing degree of order and is zero for the formation of a perfectly ordered compound. Electronic and magnetic entropy effects may occur in the formation of a compound, but are usually not appreciable unless the compound contains a transition element.[42,57]

The entropy of formation can be obtained from equilibrium data in accordance with the equation

$$\Delta S = -\frac{d\Delta F}{dT} \tag{7}$$

The equilibrium data which have been used mainly for this purpose are values of the emf. Because most suitable galvanic cells operate at elevated temperatures, the resulting entropies of formation refer to high temperatures.[1,2] The entropy of formation can also be found as the difference between the standard entropies of the compound and the elements. Furthermore, the entropy of formation can be obtained by Eq. 2 if the heat and free energy of formation are known.

4.2 Ordered Intermetallic Compounds

The entropy change on formation of ordered intermetallic compounds, in the absence of appreciable electronic and magnetic entropy changes, is due to vibrational effects, since the configurational term is zero.[1,57] The entropy change therefore depends on the type of bonding of the compound and the change in crystallographic arrangement when it forms.

When a metallic compound is formed, there is little change in the type of bonding and any vibrational entropy change results from geometrical changes, in particular, the volume change. In the formation of electron compounds the volume change is generally small and, consequently, the entropy change is small. In such cases the entropy of the ordered compound may be estimated from the sum of the entropies of the elements.[1] In the formation of Laves phases, such as $MgCu_2$, $CaMg_2$, and $MgNi_2$, the difference in the atomic sizes of the components permits structures of high-packing density to form. The resulting decrease in volume is associated with a negative vibrational entropy change.[42] The entropy of formation of the compound $MgNi_2$ is significantly less negative than that of $MgCu_2$ and $CaMg_2$, although the volume contractions are approximately the same in the formation of these three compounds. This difference has been attributed to changes in the electronic and magnetic entropies on formation of $MgNi_2$, which oppose the entropy decrease resulting from the volume contraction.[42] In the formation of the compound AuSn, which has the nickel arsenide structure, there is a volume expansion and an increase in vibrational entropy.[53] (See Table 9.)

TABLE 9

The Entropy Change on Formation of Ordered Intermetallic Compounds

Compound	Structure	ΔS (cal/g-atom deg)	Compound	Structure	ΔS (cal/g-atom deg)
ZnS	ZnS (B3)	-1.9 ± 0.3[a]	$CaMg_2$	Laves (C14)	-2.0[d]
CdS	ZnS (B4)	-1.7 ± 0.3[a]	$MgCu_2$	Laves (C15)	-2.0[d]
ZnTe	ZnS (B3)	-1.4 ± 0.4[a]	$MgNi_2$	Laves (C36)	-0.2[d]
CdTe	ZnS (B3)	-1.0 ± 0.7[a]	KNa_2	Laves (C14)	-0.16[e]
PbS	NaCl (B1)	-0.7 ± 0.4[a]			
PbTe	NaCl (B1)	-0.5 ± 0.6[a]	AuSn	NiAs (B8)	$+0.55 \pm 0.3$[f]
			MnTe	NiAs (B8)	$+1.5 \pm 0.4$[a]
InAs	ZnS (B3)	-2.1[b]			
InSb	ZnS (B3)	-2.0 ± 0.3[c]	Ag_5Zn_8(0.61Zn)	γ-brass ($D8_2$)	$+1.53$[e]
GaSb	ZnS (B3)	-1.4 ± 0.3[c]	AgCd(0.50Cd)	CsCl (B2)	$+0.50 \pm 0.5$[a]
GaAs	ZnS (B3)	-1.4[b]	AgMg(0.50Mg)	CsCl (B2)	$+0.26$[e]
AlSb	ZnS (B3)	-1.2[b]	AuCd (0.50Cd)	CsCl (B2)	-0.07 ± 0.5[e]

[a] Calculated from standard entropies of compound and elements—Kubaschewski and Evans[1]: $\Delta S_{298°K}$.
[b] Piesbergen[58]: $\Delta S_{298°K}$.
[c] Schottky and Bever[12]: $\Delta S_{298°K}$.
[d] King and Kleppa[42]: $\Delta S_{298°K}$.
[e] Hultgren et al.[2]: Ag_5Zn_8, $\Delta S_{873°K}$; AgMg, $\Delta S_{773°K}$; AuCd, $\Delta S_{700°K}$; KNa_2, $\Delta S_{280°K}$.
[f] Misra, Howlett, and Bever[53]: $\Delta S_{298°K}$.

In the formation of ordered intermetallic compounds with largely ionic bonding a decrease in vibrational entropy tends to occur because of the pronounced increase in the strength of the bond. The magnitude of this decrease depends inversely on the strength of the bonding which is determined by the difference between the electronegativities of the elements. The exothermic heat of formation of ionic compounds increases with this difference and a correlation between the heat of formation and the entropy of formation may be expected, for there are no configurational effects. Such a correlation is shown in Fig. 13 for ionic compounds and ordered electron compounds.

The vibrational entropy term can be the determining factor in the formation of a stable intermetallic compound. The compound Au_2Bi has a positive heat of formation ($\Delta H_{623°K}$ = 0.328 kcal/g-atom[50]) and is stable only because the entropy of formation ($\Delta S_{646°K}$ = 0.86 cal/g-atom deg[50]) is large enough to cause a decrease in free energy, provided the temperature is sufficiently high. On this basis it has been calculated that the compound is stable over the temperature range 646 to 381°K, but at temperatures below 381°K the entropy-temperature product is too small to stabilize the compound.[2,50]

The entropies of formation of the covalent compounds GaAs, InAs, AlSb, GaSb, and InSb range from −1.2 to −2.1 cal/g-atom deg.[12,58] (See Table 9.) The values are of the same order of magnitude as the entropy change of −1.68 ± 0.17 cal/g-atom deg associated with the transformation of white tin, in which the bonding is metallic, to gray tin, in which the bonding is covalent. It is reasonable to assume that the entropy change on the formation of the covalent III–V compounds is due to a similar change from metallic to covalent bonding.[12]

4.3 Disordered Intermetallic Compounds

In a disordered intermetallic compound the component atoms do not occupy definite positions on the lattice. A random arrangement is facilitated by weak bonding. Most ionic compounds, therefore, are ordered and any lack of perfection in their structure is mainly due to point defects. Covalent compounds do not disorder (except to the minute extent necessitated by equilibrium requirements) because of their relatively strong bonding and the geometrical requirements of the covalent bond. Among compounds with metallic bonding only the Laves phases do not disorder appreciably because their stability depends on retaining the especially dense packing of their atoms. Electron compounds are able to disorder, for they have low bond

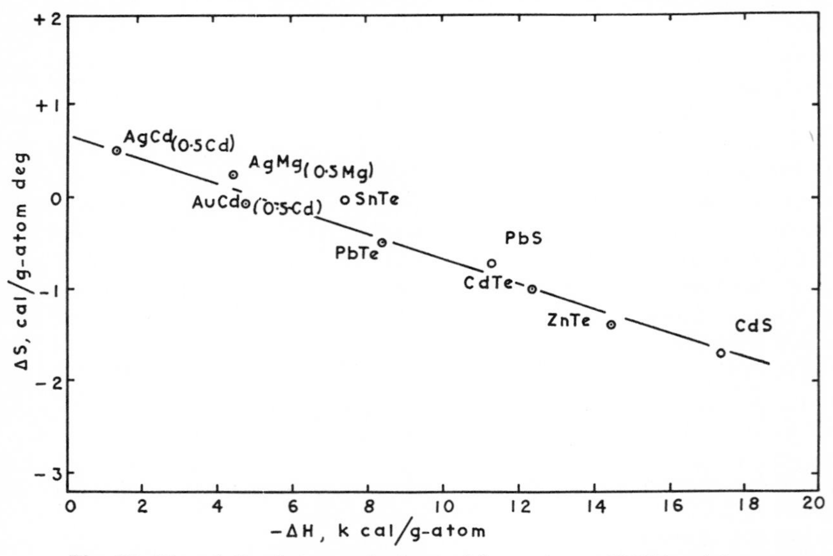

Fig. 13. The relation between the heat of formation at 298°K and the entropy of formation at 298°K, calculated from S°_{298} for compound and elements, for ionic compounds, and electron compounds with an ordered structure.

TABLE 10

The Entropy Change on Formation of Disordered Intermetallic Compounds[a]

Compound	Structure	T°K	ΔS (cal/g-atom deg)	S^E (cal/g-atom deg)
$AgZn_{(.50\,Zn)}$	bcc (A2)	873	$+1.97 \pm 0.6$	$+0.59$
$CuZn_{(.48\,Zn)}$	bcc (A2)	773	$+0.80 \pm 0.15$	-0.58
$AgCd_{3(.75\,Cd)}$	ε-brass (A3)	673	$+0.61 \pm 0.3$	-0.49
$AuCd_{3(.75\,Cd)}$	ε-brass (A3)	700	$+0.37 \pm 0.5$	-1.47
$CuZn_{3(.78\,Zn)}$	ε-brass (A3)	773	$+0.60$	-0.50
$Cu_5Zn_{8(.60\,Zn)}$	γ-brass ($D8_2$)	773	$+0.30$	-1.04

[a] All values of ΔS taken from Hultgren et al.[2]

strength and no rigorous geometrical requirements.

Equation 6 gives the configurational entropy change on formation of a completely disordered compound as a function of composition. The maximum configurational entropy of 1.38 cal/g-atom deg occurs at the equiatomic composition. The difference between the total entropy change and the configurational entropy change for complete disorder is the excess entropy S^E. It represents both the change in vibrational behavior and any deviation from complete disorder without separating their contributions.

The published entropies of formation of disordered intermetallic compounds are listed in Table 10. They are all positive. The excess entropies, with the exception of that of AgZn, are all negative.

In the formation of electron compounds the changes in vibrational entropy due to changes in bonding or in the crystallographic arrangement of the atoms are likely to be small. The total entropy change on formation of a fully disordered electron compound, therefore, consists primarily of the configurational entropy term and will have a positive value. The stability of disordered β-electron compounds is due to this relatively high entropy term rather than the small enthalpy change on formation.[57] The entropy-temperature product increases with temperature, and its stabilizing effect accounts for the widening of the homogeneity range in these compounds with increasing temperature.[33,59,60]

4.4 The Entropy of Formation as a Function of Composition

The entropies of formation as a function of composition are available for some ordered and some disordered electron compounds. The examples given in Fig. 14 refer to different temperatures, but the differences are not important in the present context. Electron compounds with an ordering tendency have the highest degree of order at the stoichiometric composition, on either side of which they disorder by the substitution of excess atoms on the lattice.

Figure 14 shows, as a function of composition, the entropy change on formation of the disordered β-electron compounds AgZn and CuZn (at temperatures above their critical temperatures for disordering) and of the ordered β-electron compounds AgMg and AuCd. The entropies of formation of the two disordered compounds increase slightly with zinc content across the homogeneity range. By contrast, the entropies of formation of AgMg and AuCd go through minima at the equiatomic composition.[2]

In disordered compounds the degree of disorder may be assumed to remain the same throughout the homogeneity range. If the compounds AgZn and CuZn are fully disordered, the configurational entropy changes with composition according to Eq. 6. It follows that the excess entropy increases slightly with increasing zinc content. This increase can be attributed to an increase in vibrational entropy. In the ordered compounds AuCd and AgMg there is no contribution to the entropy from configurational effects at the stoichiometric composition. The entropy change occurring on the formation of these stoichiometric compounds is therefore entirely due to vibrational effects. The marked increase in the entropy of formation with increasing deviations from stoichiometry may be attributed to the configurational effects

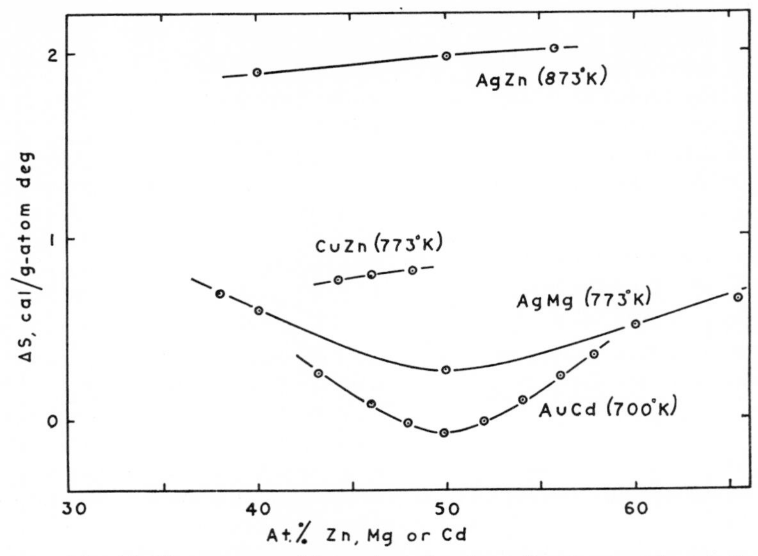

Fig. 14. The entropy of formation of ordered (AgMg, AuCd) and disordered (AgZn, CuZn) β-electron compounds as a function of composition.

arising from progressive disordering which occurs by the substitution of atoms of one component on the lattice.

The ζ-phase in the gold-tin system is a partly disordered electron compound which exists over a range of composition.[38] The exothermic heat of formation as a function of composition is convex toward the composition axis.[53] For the free energy of formation as a function of composition to be concave toward the composition axis the corresponding entropy function has to be positive and convex toward the composition axis with the degree of curvature depending on the temperature. If the configurational contribution to the entropy change were the larger, the curve would have the more usual concave form. The required convex shape, therefore, must be due to changes in the vibrational entropy term with composition. The lattice parameter of the phase increases slightly with composition across the homogeneity range until that composition is reached at which the interaction of the Fermi surface with the Brillouin zone causes a large increase in lattice parameter.[61] The resulting sharp change in volume expansion with composition will cause similar changes in the vibrational entropy and will lead to an entropy change-composition curve which is convex toward the composition axis.

5. HEAT CAPACITIES AND STANDARD ENTROPIES

5.1 General

The entropy of an ordered crystalline substance at any temperature has an absolute value, which is given by

$$S^\circ(T) = \int_0^T \left(\frac{c_p}{T}\right) dT \qquad (8)$$

The low-temperature heat capacities required for determining such standard entropies, for example, at 298°K, are measured by special calorimetric techniques.[62,63] Although the heat capacities of pure metals are vanishingly small at 0°K, those of intermetallic compounds may have a finite value because at low temperatures the compounds may fail to reach equilibrium with respect to configurational effects.[64] Experimental evidence on this point, however, appears to be still lacking.

According to the Debye relation, the heat capacity at low temperatures due to vibrational effects is proportional to the cube of the absolute temperature.[65] Deviations from this relation below approximately 10°K indicate that in this temperature range the electronic heat capacity is appreciable compared to the heat capacity associated with vibrational effects. At higher temperatures the electronic contribution is masked by other contributions, except in some compounds containing a transition metal. The contribution of vibrational effects to the entropy at any temperature can be determined if the Debye temperature of the compound is known or can be estimated.[57]

5.2 The Heat Capacities of Intermetallic Compounds

The heat capacity of an intermetallic compound, according to the Kopp-Neumann rule, is equal to the sum of the heat capacities of the component elements.[1] This rule can be expected to apply only to compounds in which the bonding and crystallographic arrangement are similar to those in the component elements.

Experimentally determined heat capacities at 298, 500, and 1000°K, together with values calculated by the Kopp-Neumann rule, are listed in Table 11. Typical deviations of the experimental from the calculated values are plotted for these temperatures in Fig. 15. It should be recognized that the accuracy of some of the experimental values may not be satisfactory. Also, the heat capacities of some

TABLE 11

Experimentally Determined and Calculated Heat Capacities of Intermetallic Compounds[a]

	Heat Capacity (cal/g-atom deg)					
	298°K		500°K		1000°K	
Compound	Expt.	Calc.	Expt.	Calc.	Expt.	Calc.
PbS	5.91	5.88	6.31	6.73	7.29	7.41
ZnS	5.49	5.75	6.12	6.61	6.63	7.30
CaS	5.67	5.87	6.05	6.82	7.00	7.75
Mg_3Sb_2	5.95	5.98	6.18	6.46	6.75	7.57
$MgZn_2$	5.90	6.03	6.38	6.54	7.58	7.76
$MgCu_2$	5.74	5.89	6.31	6.28	7.73	7.20
$MgNi_2$	5.86[b]	6.13	6.40[b]	7.09	—	—
Cu_2Sb	6.45	5.91	6.56	6.23	—	—
$AuPb_2$	6.70	6.23	7.90	6.62	—	—
AuZn	6.29	6.05	6.65	6.42	7.54	7.33
Co_2Sn	6.28	6.05	6.85	7.02	—	—
NiTe	6.28	6.15	6.61	7.40	—	—
$AuSb_2$	6.17	6.03	6.48	6.35	—	—
Ag_2Al	6.00	6.01	6.44	6.31	—	—
Ag_3Al	6.03	6.03	6.35	6.29	—	—
CuZn	5.89[c]	5.92	6.32[c]	6.32	—	—
Ag_3Sb	6.07	6.09	6.88	6.29	—	—
Cu_3Al	5.77	5.87	6.23	6.23	—	—
CuAl	5.75	5.84	6.18	6.29	—	—
AuSn	5.90	6.16	7.02	6.93	—	—
Cu_2Cd_3	5.58	6.06	6.81	6.50	—	—
CdSb	5.46	6.12	6.98	6.57	—	—

[a] The experimental values of the heat capacities tabulated above, except those marked (*b*) and (*c*), were obtained from equations given by Kelley.[86] The calculated values are the sum of the heat capacities of the elements obtained from equations given by Kelley.[86] For temperatures above the melting point of a component, an extrapolated value of the heat capacity of the solid component is used.

[b] Wollam and Wallace.[73]

[c] Hultgren et al.[2]

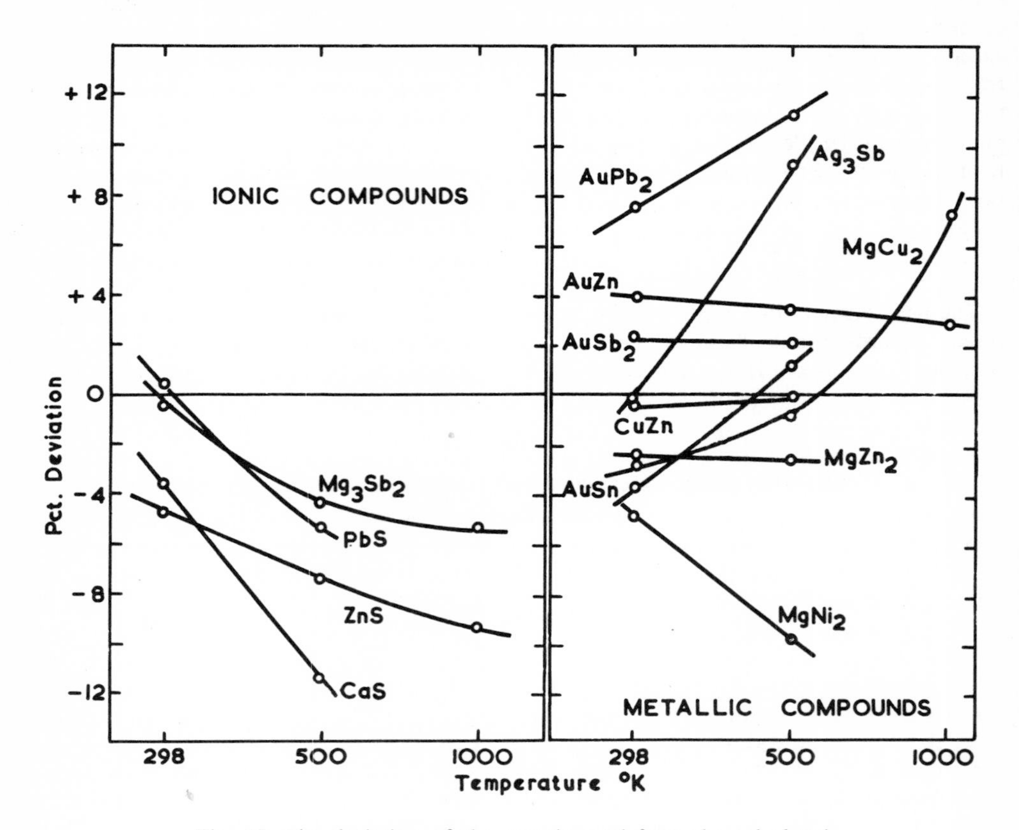

Fig. 15. The deviation of the experimental from the calculated heat capacities at 298, 500, and 1000°K for selected intermetallic compounds.

of the components used in the calculations may not be accurate.

The extent of agreement between the experimentally determined and the calculated heat capacities of various intermetallic compounds at 298°K varies widely. The heat capacities of the ionic compounds PbS and Mg_3Sb_2 and of the metallic compounds CuZn, Ag_3Sb, Ag_2Al, and Ag_3Al closely follow the Kopp-Neumann rule. The published heat capacities at 298°K of the ionic compounds ZnS and CaS, however, show a marked negative deviation from the Kopp-Neumann rule. This deviation may be due to two causes: the ionic bonding in these compounds may appreciably reduce vibrational effects below those in the component elements, and the electronic heat capacities in the compounds may be less than the sum of those of the component elements. The known heat capacities at 298°K of compounds with metallic bonding show both positive and negative deviations from the Kopp-Neumann rule (Table 11). The heat capacities of the Laves phases $MgCu_2$, $MgZn_2$, and $MgNi_2$ have negative deviations, which presumably are related to their high coordination number and dense packing.

The effect of temperature on the agreement between the experimental and calculated heat capacities can be seen in Table 11 and Fig. 15. The published high-temperature heat capacities of intermetallic compounds with ionic bonding, except PbS, show an increasing negative deviation from the Kopp-Neumann rule with increasing temperature above 500°K. This suggests that the vibrational effects and perhaps also the electronic heat capacities increase less rapidly with temperature in ionic compounds than in the component elements.

The heat capacities of many compounds with metallic bonding, on the other hand, show an increasing positive deviation (Table 11), which may be due to a weakening of the bonds with increasing temperature and to a configurational contribution to the heat capacity because of disordering. The heat capacities of the Laves phases $MgZn_2$ and $MgNi_2$ have an increasing negative deviation with temperature, although the deviation of the heat capacity of $MgCu_2$, which has a similar structure, becomes increasingly positive (Fig. 15).

Heat capacities at low temperatures are known for only a few intermetallic compounds. Piesbergen has measured the heat capacities of the compounds GaAs, InAs, AlSb, GaSb, and InSb between 10°K and room temperature.[58] Gul'tyaev and Petrov made similar measurements on the compounds AlSb, GaSb, and InSb in the temperature range 80 to 300°K.[66] The heat capacities of these III–V compounds show a negative deviation from the Kopp-Neumann rule, which results in a negative entropy of formation at 298°K (§ 4.2). The heat capacity of the compound Cu_2Sb is also smaller than the value given by the Kopp-Neumann rule in the temperature range 100 to 350°K.[67]

Deviations from the Kopp-Neumann rule of the heat capacities are expected when order-disorder or magnetic transitions occur in a compound. In such cases the heat capacity has a large value in the temperature interval of the transition. (See § 8.3.)

The heat capacity measurements of Tret'yakov and Khomyakov[68] on the compound AlCo indicate an anomaly near 1063°K which these authors interpreted as a disordering reaction. Kubaschewski[69] found a small negative deviation of the heat capacity of the β-electron compound AuCd from the Kopp-Neumann rule below 600°K. Near the melting point the deviation attained a large positive value which may be explained by a tendency toward disorder.[2] Similar pre-melting phenomena are discussed in Section 5.4.

Schröder and Cheng found anomalous behavior in the heat capacity of the compound FeTi between 1.5 and 4.0°K.[70] They interpreted their results, in conjunction with the magnetic measurements of Nevitt,[71] on the assumption that the compound is superparamagnetic because of ferromagnetic clustering at low temperatures; the magnetization vector of these clusters gives rise to an additional electronic contribution to the heat capacity.[70]

Westrum, Chou, and Grønvold measured the heat capacities of the β- and ϵ-phases in the iron-tellurium system at temperatures from 5 to 350°K.[72] They found a λ-type anomaly in the β-phase with a maximum value of the heat capacity at 63°K and ascribed this anomaly to magnetic or structural disordering of the iron atoms.

The heat-capacity measurements of Wollam and Wallace on the Laves phase $MgNi_2$ have a negative deviation from the Kopp-Neumann rule at temperatures above 50°K.[73] This behavior suggests that the vibrational effects decrease on formation of the compound because of the increased packing density and coordination. The heat capacity as a function of temperature has a small anomaly, similar to a λ-point, at 465°K; the associated heat effect is 1 cal/g-atom. Elliott has shown that with increasing temperature some Laves phases may transform from one structural type to another,[74] but X-ray analysis did not reveal structural changes in the compound $MgNi_2$.[73] Wollam and Wallace therefore attributed the anomalous heat capacity to a small magnetic effect due to an irregularity in the lattice.

The heat capacity of the δ-phase in the nickel-tellurium system has been measured for compositions from 52 to 67 per cent tellurium and temperatures from 5 to 350°K.[75] No order-disorder or magnetic transitions were observed. The heat capacities are of interest because they cover a range of compositions within the homogeneity field of the compound. The heat capacity changes gradually with composition and shows a positive deviation from the Kopp-Neumann rule at all compositions.

5.3 Standard Entropies

The standard entropies, $S°$, at 298°K are known for approximately 30 intermetallic compounds.[1,10] Their values range from 6 to 14 cal/g-atom deg. When they are plotted against the heat of formation at 298°K, they fall into two main groups (Fig. 16); one consists of the entropies of compounds with ionic bonding and the other of entropies of compounds in which the bonding is either transitional ionic-metallic or metallic. The

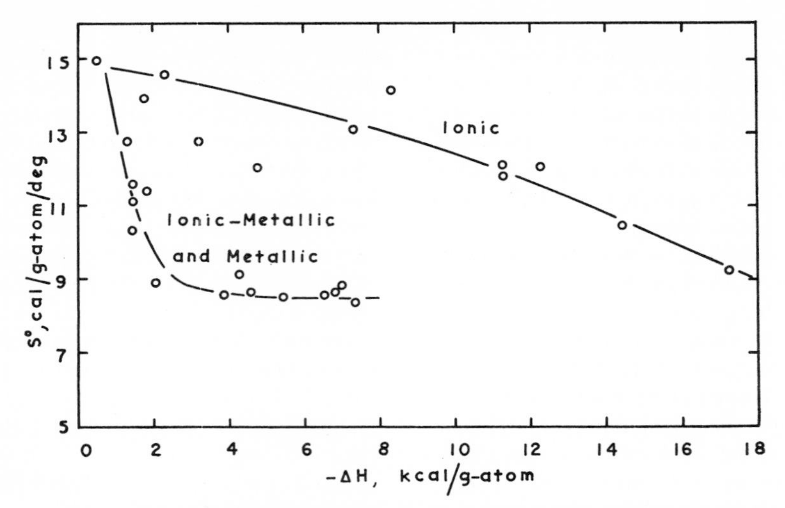

Fig. 16. Standard entropies at 298°K of intermetallic compounds as a function of their heats of formation. (Values of S° from Kubaschewski and Evans.[1])

standard entropies of ionic compounds tend to decrease and their heats of formation tend to increase with increasing bond strength. The second group of compounds shows a similar correlation, but the range of values of the heat of formation is smaller. For equal heats of formation, the standard entropies of ionic-metallic and metallic compounds are lower than those of ionic compounds. This suggests that the high coordination number and dense packing in the former reduce the vibrational freedom of the atoms more than does the strong bonding of the latter.

The standard entropies of some III–V compounds and the related group IV elements are listed in Table 12. The standard entropy of each III–V compound is approximately equal to that of the intermediate group IV element.[10] The bond strength and vibrational freedom of the atoms of the compound and the related element may be considered to be comparable.

5.4 High-Temperature Heat Capacities and Premelting Phenomena

A marked nonlinear increase in the heat capacity just below the melting point has been reported for a number of pure metals. Carpenter and Steward observed anomalous behavior of the heat capacity 50 to 100°K below the melting points of lithium, sodium, and potassium and attributed it to a temperature-dependent increase in the concentration of vacancies.[76] A similar increase in heat capacity just below the melting point has been observed for tin, cadmium, and bismuth.[77,78] Kubaschewski and Schrag explained the effect in bismuth by a loosening of the bonds in the lattice before melting.[78] Changes in other properties just below the melting point of pure

TABLE 12

Standard Entropies of III-V Compounds and Related Group IV Elements[a]

Compound	Temp. °K	Standard Entropy, S_T° (cal/g-atom deg) Compound	Element
GaAs	1080	17.35	S_T° (Ge) = 15.45
AlSb	298	8.25	S_T° (Ge) = 7.45
InAs	1000	15.75	$\frac{1}{2}S_T^\circ$ (Ge) + $\frac{1}{2}S_T^\circ$ (Sn) = 17.90
InSb	298	10.42	S_T° (Sn, grey) = 10.70

[a] Modified after Goldfinger.[10]

metals also indicate premelting phenomena. It is difficult to determine, however, to what extent these phenomena in pure metals are due to the presence of impurities or the method of measurement.[79] Premelting phenomena are more likely in intermetallic compounds, for changes in order and bond type may occur on melting.

A nonlinear increase in heat capacity with increasing temperature has been observed just below the melting points of the compounds Mg_2Sn, Mg_2Pb, AuSn, and various tellurides, including Bi_2Te_3 and Sb_2Te_3.[34,80,81] Because zone-refined samples were used in some of these investigations, contamination by impurities and segregation of the components were ruled out as explanations of the nonlinear increase in heat capacity. Measurements in the same calorimeter of the heat capacities of tin, lead, and bismuth near their melting points failed to show such a deviation.[82]

Schneider and co-workers have observed nonlinear behavior of several properties just below the melting points of NaTl and related compounds.[83,84] They discussed accelerated weakening of the bonds, accelerated generation of point defects, and accelerated disordering of an ordered compound as possible causes of such a premelting phenomenon. Drier, Craig and Wallace found a nonlinear increase in the heat capacity of the Laves phase KNa_2 just below the melting point of 280°K.[85] Because a disordering process is unlikely in a Laves phase, the rapid increase in heat capacity may be attributed to a weakening of the bonds as the melting point is approached. The layer structure of the compounds Bi_2Te_3 and Sb_2Te_3 makes accelerated weakening of the bonds near the melting point particularly probable because the van der Waals bonds between the layers of tellurium atoms are likely to weaken before the bonds between other layers do, thus contributing to premelting phenomena.[80] In general, the nonlinear increase in heat capacity below the melting points of the other compounds is probably due to one or a combination of the causes suggested by Schneider and Heymer.[84]

6. HEATS AND ENTROPIES OF FUSION

6.1 General

Fusion is a first-order transformation which takes place isothermally at the melting point with the absorption of heat. As discussed in Section 5.4, there is evidence that premelting phenomena occur in solid intermetallic compounds below the melting point. Also, some compounds may not dissociate completely on melting and clusters of atoms resembling the solid may persist over an appreciable range of temperature above the melting point.[87,88]

At the melting point T_f of a congruently melting compound, the free energies of the solid and liquid phases are equal and the entropy of fusion ΔS_f is related to the heat of fusion ΔH_f by

$$\Delta S_f = \frac{\Delta H_f}{T_f} \tag{9}$$

The heat of fusion of a metal is the energy required to change the periodic arrangement of atoms in the solid into the less rigid arrangement characteristic of the liquid state. In intermetallic compounds the heat of fusion includes the energy required for the destruction of any order present in the solid at the melting point and for changes in the type of bonding accompanying the fusion process, specifically a change from ionic or covalent bonding in the solid to metallic bonding in the liquid. The entropy of fusion is associated with the increase in vibrational freedom on melting and any disordering of an ordered compound.

Heats of fusion can be measured in various calorimeters, such as drop calorimeters and constant-temperature gradient calorimeters.[13,14] The entropies of fusion, hence the heats of fusion, of a disordered compound can be estimated to a first approximation by the addition of the entropies of fusion of the component metals.[1] If a solid compound is ordered at the melting point, a disordering term σ must be added to this value.[1] If the solid is completely ordered, the disordering term for binary compounds is given as a function of composition by the equation

$$\sigma = -R(N_A \ln N_A + N_B \ln N_B) \tag{10}$$

Complete disorder in the liquid is assumed in both cases. The values for compounds which are partly ordered in the solid state lie between the estimated entropies of fusion for completely ordered and completely disordered compounds.

The melting points of intermetallic compounds depend on the free energies of both the solid and liquid phases and therefore cannot generally be expected to give an unambiguous indication of their stability in the solid. For compounds that retain some association in the liquid, however, the level of the melting point is significant; for example, the melting point of certain intermetallic compounds with ionic bonding is a measure of their stability.

The shape of the solidus and liquidus curves in the region of the melting point of a compound gives some indication of the degree of order in the solid and liquid states.[89] The solidus curve of a compound existing over a range of composition approximates a hyperbola if the compound is partly disordered. The liquidus curve of such a compound has a similar shape if there is some association in the liquid, but it is parabolic if the liquid is completely dissociated. The solidus and liquidus curves meet at a point that may or may not be the stoichiometric composition of the compound depending on the energy of formation of the defects on either side of the stoichiometric composition.

6.2 Published Values of Heats and Entropies of Fusion

The heats and entropies of fusion have been determined for approximately twenty-five intermetallic compounds. The entropies of fusion range from 2 to 7.6 cal/g-atom deg.[1] The measured values for disordered and ordered intermetallic compounds are in reasonable agreement with the estimated values, as can be seen from the examples given in Table 13. Table 14 lists heats and entropies of fusion of compounds and the type of bonding in the solid, if known.

The largest heats and entropies of fusion tend to occur when complete disordering of an ordered compound on melting is accompanied by a change in the type of bonding. For example, the compounds GaSb and InSb, which have covalent bonding and are ordered in the solid, disorder completely on melting and have metallic bonding in the liquid.[12] Their entropies of fusion are 6.15 and 7.64 cal/g-atom deg, respectively.[12] In contrast, the entropy of fusion of the Laves phase $MgZn_2$, in which the bonding is metallic in both the solid and liquid state, is 3.24 cal/g-atom deg.

In a homologous series of compounds the entropy of fusion gives an indication of their relative stabilities. In such a series the entropies of fusion usually decrease with increasing atomic number of the component that is being varied. The heats of formation of compounds of magnesium with group IV–B elements indicate that in the solid state the ionic contribution to the bonding decreases and the bonding becomes more metallic in the order Mg_2Si, Mg_2Sn, and Mg_2Pb, as discussed in Section 3.5. It is probable that all three compounds are metallic conductors in the liquid state. It follows that the change in

TABLE 13

The Heats and Entropies of Fusion of Ordered and Disordered Intermetallic Compounds[a]

Compound		Heat of Fusion (kcal/g-atom)	Entropy of Fusion (cal/g-atom deg)	
Ordered	Disordered		Experimental	Calculated
CdSb		3.9 ± 0.2	5.3	5.3
AuSn		3.4[b] ± 0.1	4.9[b]	4.2
$AuPb_2$		1.9 ± 0.1	3.6	3.1
Na_5Pb_2		1.7 ± 0.1	2.5	2.9
NaPb		2.0 ± 0.2	3.1	3.1
Ni_3Sn_2		5.5 ± 0.2	3.6	3.5
	Bi_3Tl_2	1.7 ± 0.1	3.5	3.8
	Cu_2Cd_3	2.3 ± 0.1	2.8	2.4

[a] Experimental values are taken from Kubaschewski and Evans,[1] except those marked (*b*). Calculated values are determined from the entropies of fusion of the component metals (Lumsden[57]) and the disordering factor σ.
[b] Values from Misra, Howlett, and Bever.[53]

TABLE 14

The Heats and Entropies of Fusion of Intermetallic Compounds with Different Types of Bonding in the Solid

Compound	Heat of Fusion (kcal/g-atom)	Entropy of Fusion (cal/g-atom deg)	Type of Bonding in Solid
InSb	6.1 ± 0.37[a]	7.6	Covalent
GaSb	6.0 ± 0.37[a]	6.2	Covalent
Bi_2Te_3	5.7 ± 0.03[b]	6.6	—
Sb_2Te_3	4.7 ± 0.04[b]	5.3	—
Mg_3Sb_2	7.3 ± 0.8[c]	4.9	Ionic
Mg_3Bi_2	5.1 ± 0.5[c]	4.7	Ionic
Mg_2Si	6.8 ± 0.8[c]	5.0	Ionic-metallic
Mg_2Sn	4.5 ± 0.15[d]	4.3	Ionic-metallic
Mg_2Pb	3.2 ± 0.05[d]	3.9	Metallic
InTe	4.3 ± 0.09[e]	4.5	—
In_2Te_3	3.9 ± 0.02[e]	4.1	—
$MgZn_2$	3.4 ± 0.3[c]	3.9	Metallic
AuCd	2.2 ± 0.1[c]	2.4	Metallic
AuZn	3.0 ± 0.2[c]	2.9	Metallic
Ni_2Si	4.0 ± 0.2[c]	2.5	Metallic

[a] Schottky and Bever.[12]
[b] Howlett, Misra, and Bever.[80]
[c] Kubaschewski and Evans.[1]
[d] Beardmore, Howlett, Lichter, and Bever.[34]
[e] Robinson and Bever.[81]

bonding on fusion becomes less pronounced in the order Mg_2Si, Mg_2Sn, and Mg_2Pb and the entropies of fusion decrease in the same order. (Table 14).

The entropies of fusion of the compounds Sb_2Te_3 and Bi_2Te_3 are 5.31 and 6.58 cal/g-atom deg, respectively.[80] These compounds are semiconductors in the solid state and are probably metallic in the liquid state.

7. ACTIVITIES OF THE COMPONENTS OF INTERMETALLIC COMPOUNDS

7.1 General

The activity of a component of a solid solution or compound is defined as the ratio of its fugacity in the solid solution or compound to its fugacity in the standard state at the same temperature. In the case of a compound it is convenient to choose the pure component as the standard state. The activity of a component of a binary compound is equal to the gram-atomic fraction if its behavior across the homogeneity range is ideal. The activity coefficient f is the ratio of the activity to the gram-atomic fraction and is unity for ideal behavior.

Under suitable conditions the activity of a component of an intermetallic compound can be measured as a function of composition by vapor pressure or emf measurements. Various experimental techniques and results have been discussed by Wagner.[3] If a component of an intermetallic compound has an appreciable vapor pressure, values of the activity of the compound can be determined from the ratio of its vapor pressure over the compound to that over the pure component at the same temperature

$$a_{\mathrm{A}} = \frac{P_{\mathrm{A}}}{P_{\mathrm{A}}^{\circ}} \tag{11}$$

Mass spectrometer and radioactive tracer techniques are of special value in determining small vapor pressures.[10,90] The activity of a component can be obtained from emf measurements by the relation

$$\ln a_{\mathrm{A}} = \frac{n\mathscr{F}}{RT}\,\mathscr{E} \tag{12}$$

if the compound forms a reversible cell with a suitable electrolyte.

If the activity or activity coefficient of one component of a binary system is known as a function of composition, the activity or activity coefficient of the second component can be calculated by the Gibbs-Duhem equation. In general, the accuracy of this calculated value is not high, especially if the compound contains a large fraction of the other component. It would be desirable to measure the activities of both components, but this is not often possible.

The activity of a component is related to the partial molar free energy of formation by

$$\Delta\bar{F}_A = RT \ln a_A = RT \ln f_A N_A \quad (13)$$

When the activities of both components are known, the integral free energy of the compound can be calculated from this equation and the relation

$$\Delta F = N_A \Delta\bar{F}_A + N_B \Delta\bar{F}_B \quad (14)$$

If the activity of one component is known, the integral free energy of formation of the compound can be calculated by the Duhem-Margules equation.

7.2 Activity Coefficients of Components of Intermetallic Compounds

The components of a few intermetallic compounds exhibit nearly ideal behavior at certain compositions and temperatures. For example, the activity coefficient of cadmium in the compound $AgCd_3$ at a concentration of 79.0 at.% cadmium is 1.051 at 673°K, and the activity coefficient of thallium in the γ-phase at a concentration of 95.0 at.% thallium in the bismuth-thallium system is 1.044 at 423°K.[2] In general, however, the activity coefficients of components of intermetallic compounds are smaller than unity at all compositions and temperatures (Table 15).

The deviation of the activity coefficient from unity for a given type of compound increases with increased strength of the bonding as indicated by the free energy of formation (Table 15). The activity coefficient of silver in the β-electron compound AgZn at the stoichiometric composition is 0.57 at 873°K, whereas in the β-electron compound AgMg at the same composition the activity coefficient of silver

TABLE 15

The Free Energies of Formation of Intermetallic Compounds and the Activity Coefficients of Their Components[a]

Compound	T°K	f_A	f_B	$-\Delta F$ (kcal/g-atom)
AgZn (0.50Zn)	873	0.57	0.41	2.47
CuZn (0.48Zn)	773	0.40	0.21	3.00
AgMg (0.50Mg)	773	0.29	0.01	5.39
AuCd (0.50Cd)	700	0.11	0.05	4.61
NiAl (0.50Al)	1273	—	0.03[b]	—
Ag_5Zn_8 (0.60Zn)	873	0.39	0.56	2.41
Cu_5Zn_8 (0.61Zn)	773	0.33	0.24	3.04
Au_5Cd_8 (0.64Cd)	700	0.01	0.32	4.08
$AgZn_3$ (0.76Zn)	873	0.17	0.82	1.98
$CuZn_3$ (0.78Zn)	773	0.04	0.83	2.09
$AuCd_3$ (0.75Cd)	700	0.002	0.65	3.42
$AgCd_3$ (0.75Cd)	673	0.06	0.94	1.77
AgCd (0.50Cd)	673	0.31	0.38	2.37
Ag_5Cd_8 (0.60Cd)	673	0.20	0.52	2.30
Bi_2Tl (0.40Tl)	423	1.43	0.15	1.03
$BiTl_4$ (0.80Tl)	423	0.12	0.58	1.14
NiTe (0.60Te)	973	—	0.26	—
$FeAl_2$ (0.67Al)	1173	—	0.19[c]	6.20[b]
Fe_2Al_5 (0.70Al)	1173	—	0.18[c]	6.12[b]
$FeAl_3$ (0.75Al)	1173	—	0.52[c]	5.65[b]

[a] The values of f_A, f_B, and ΔF are taken from Hultgren et al.[2] except those marked (*b*) and (*c*).
[b] Steiner and Komarek.[91]
[c] Eldridge and Komarek.[92]

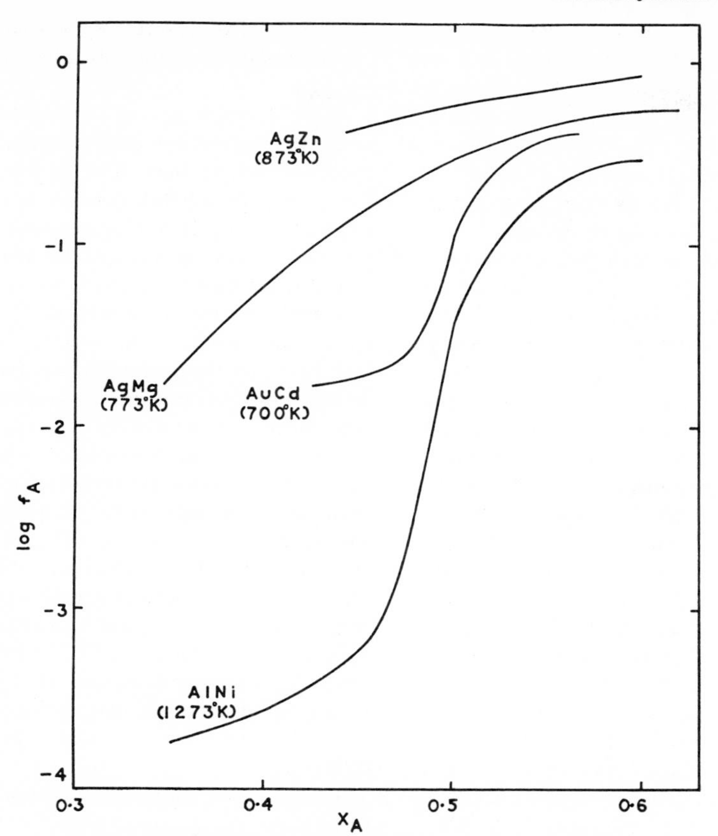

Fig. 17. The activity coefficients of components of β-electron compounds as a function of composition.

at 773°K is 0.290.[2] The bonding is stronger in the latter because of the partially ionic nature of the compound, as discussed in Section 3.7.

The activity coefficients of the components of most intermetallic compounds existing over a homogeneity range are less than unity over the entire range of composition. The activity and activity coefficient may change by a large amount across the homogeneity range. The activity coefficient of aluminum in the β-electron compound NiAl at 1273°K, for example, changes by more than three powers of ten.[91] Large changes of the activity are characteristic of stable intermetallic compounds. The concentration dependence of the activity coefficient can frequently be expressed by the formula

$$\ln f_A = \tfrac{1}{2}\alpha_2(1 - N_A)^2 + \tfrac{1}{3}\alpha_3(1 - N_A)^3 + \cdots \tag{15}$$

where α_2, $\alpha_3 \cdots$ are constants.[3] For certain compounds, however, such as the β-electron compounds NiAl and AuCd, the activity coefficients as a function of composition do not follow this general form, but have an inflection point at or near the stoichiometric composition, as shown in Fig. 17. This indicates that the stoichiometric composition is the most stable composition in the homogeneity range.[3]

8. TRANSFORMATIONS IN INTERMETALLIC COMPOUNDS

8.1 General

In this section changes in thermodynamic properties associated with polymorphic transformations and order-disorder transitions in intermetallic compounds are considered. Since solid solutions, in which order-disorder transitions occur, have some features in common with intermetallic compounds, they are also discussed.

Transformations can be classified according to the manner in which certain thermodynamic properties of the transforming phase change at the transformation temperature. In a first-order transformation the first derivatives of the free energy with respect to temperature and pressure, equal to the entropy and volume, respectively, are discontinuous at the transformation temperature. A latent heat, therefore, is associated with a first-order transformation. In a second-order transformation the first derivatives of the free energy are continuous and the entropy and volume do not change abruptly. However, the second derivatives of the free energy, equal to the specific heat, expansion coefficient and compressibility, are discontinuous at the transformation temperature.[93]

Polymorphic transformations in solids are first-order transformations. Some order-disorder transitions are second-order transformations, but others have characteristics of both first- and second-order transformations.

The heat effects associated with transformations are usually determined by heat-capacity measurements involving continuous heating of the specimen.[94,95] The heat effects of order-disorder transitions may be difficult to measure in this manner because there may not be sufficient time during continuous heating for the disordering transition to reach equilibrium. The heat effects of transformations can also be determined by measuring the difference in the enthalpy of the phases by difference methods such as solution calorimetry.[96,97]

8.2 Heat Effects of Polymorphic Transformations in Intermetallic Compounds

At the transformation temperature the heat of formation of the high-temperature phase is less exothermic than that of the low-temperature phase, but the larger entropy term associated with the former leads to a negative free energy change on transformation. Polymorphic transformations are known to occur in approximately thirty-three intermetallic compounds. The heats of formation of twenty-six of these compounds are less than 10 kcal/g-atom, and their bonding may be considered to be predominantly metallic. However, polymorphic transformations also occur in a few intermetallic compounds with predominantly ionic bonding, such as Mg_3Sb_2.

The heat effects associated with polymorphic transformations have been determined only for a few compounds. The published values, listed in Table 16, range from 0.26 to 0.59 kcal/g-atom. They are of the same order of magnitude as the heat effects of polymorphic transformations in elements.

Two transformations are possible in the disordered bcc β-electron compound AgZn, which is stable at temperatures above 258°C. On slow cooling below this temperature, the compound transforms to the disordered cph ϵ-phase with a heat effect of -0.585 ± 0.100 kcal/g-atom. If the disordered bcc β-compound is quenched, however, the polymorphic transformation is suppressed and the compound transforms to an ordered bcc

TABLE 16

The Heat Effects of Polymorphic Transformations in Intermetallic Compounds

Compound	T_t°C[a]	Low-Temperature Phase	High-Temperature Phase	ΔH_t, kcal/g-atom
AgCd	230	β', CsCl (B2)	ζ, cph (A3)	0.330 ± 0.150[b]
AgCd	440	ζ, cph (A3)	β, bcc (A2)	0.260 ± 0.150[c]
AgZn	258	ζ, disord. cph	β, bcc (A2)	0.585 ± 0.100[c]
Ag_2Se	128	α, orthorhombic	β, CaF_2 (Cl)	0.533 ± 0.133[d]
Cu_2Se	110	α, def. cubic	β, CaF_2 (Cl)	0.387 ± 0.33[d]

[a] Values of T_t°C are taken from Hansen.[38]
[b] Anderson.[98]
[c] Hultgren et al.[2]
[d] Kubaschewski and Evans.[1]

structure at a lower temperature. The heat effect of this ordering transition is -0.635 ± 0.100 kcal/g-atom.[2]

8.3 The Energies of Ordering in Intermetallic Compounds and Solid Solutions

Order-disorder transitions resemble polymorphic transformations in that at the transition temperature the high-temperature phase has a less exothermic heat of formation than the low-temperature phase. The disordered phase is stable at the higher temperatures because of the configurational entropy term.

The published values of the energies of ordering in intermetallic compounds and solid solutions range from -0.20 to -0.95 kcal/g-atom (Table 17). They are of the same order of magnitude as the energy effects associated with polymorphic transformations in intermetallic compounds and metallic elements.

Order-disorder transitions in some solid solutions have characteristics of both first- and second-order transformations. For example, in Cu_3Au, CuAu, and $CuAu_3$, a latent heat is associated with the order-disorder transitions but the long-range order parameter decreases gradually with increasing temperature before changing abruptly at the transition temperature. The heats of formation of ordered and disordered CuAu are shown in Fig. 18. The latent heat associated with the transition is -0.39 kcal/g-atom.[97] Approximately 0.54 kcal/g-atom is evolved over a temperature interval of 160°C below the transition temperature of 410°C. The total heat effect associated with the order-disorder transition in CuAu, therefore, is -0.93 kcal/g-atom.[97] The total heat effects associated with order-disorder transitions in other solid solutions are given in Table 17.

The order-disorder transitions in the solid solutions of composition CdMg, Cd_3Mg, $CdMg_3$, and CuPd and the intermetallic compounds CuZn, FeCo, and Fe_3Al are second-order transformations.[99] The heat capacity and heat of formation of the compound CuZn as a function of temperature in the region of the order-disorder transition are shown in Fig. 19 as an example of the change in these properties during a second-order transformation. The heat effect associated with the transition in CuZn is -0.55 kcal/g-atom[2] (Table 17).

TABLE 17

The Energies of Ordering in Intermetallic Compounds and Solid Solutions

Compound	Ordering Temperature, °C	Energy of Ordering kcal/g-atom
CuAu	410	-0.950[a]
		-0.930[b]
Cu_3Au	390	-0.530[c]
		-0.562[d]
		-0.650[e]
		-0.640[a]
$CuAu_3$	210	-0.200[a]
		-0.320[f]
		-0.270[g]
CdMg	252	-0.350[h]
Cd_3Mg	84	-0.315[h]
$CdMg_3$	151	-0.425[h]
CuPd	600	-0.660[h]
CuZn	468	-0.550[h]

[a] Orr.[97]
[b] Orr, Luciat-Labry, and Hultgren.[100] Combined measurements of Orr, Luciat-Labry, and Hultgren; Oriani and Murphy;[101] Hirabayashi;[102] Borelius, Larsson, and Selberg.[103]
[c] Sykes and Jones.[104]
[d] Rubin, Leach, and Bever.[96]
[e] Hirabayashi, Nagasaki, and Kono.[95]
[f] d'Heurle and Gordon.[105]
[g] Hirabayashi.[106]
[h] Hultgren et al.[2]

The experimental values of the energy of ordering are of interest because they provide a check for the various theories of ordering. (See Chapters 14 and 21.) For example, the measured energies of ordering in Cu_3Au range from -0.53 to -0.65 kcal/g-atom (Table 17). The Bragg and Williams theory of ordering predicts an energy change of -0.605 kcal/g-atom for the formation of a superlattice on the ordering of a completely random solid solution of Cu_3Au.[107] Peierls'[108] application of Bethe's nearest-neighbor theory to Cu_3Au gives a value of -0.56 kcal/g-atom for the formation of a superlattice from a matrix which initially contains short-range order. Cowley[109] extended the nearest-neighbor approach to include as many as five shells of neighbors; on this basis a change in energy of -0.50 kcal/g-atom is expected. Eguchi,[110]

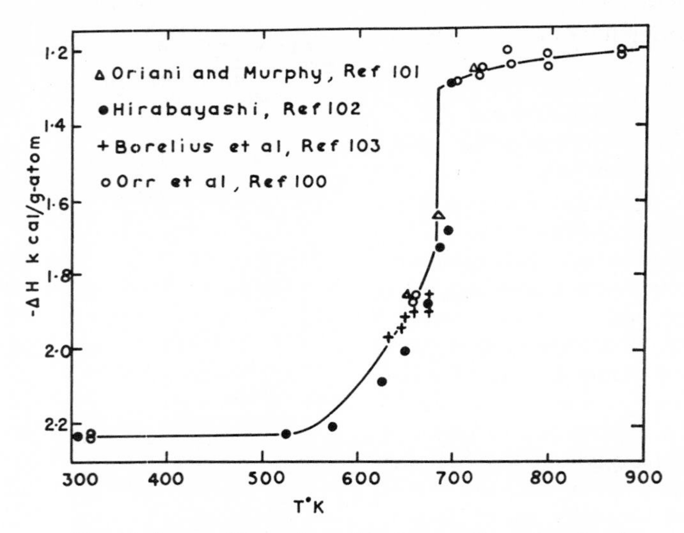

Fig. 18. Heats of formation of AuCu near the order-disorder transition temperature of 683°K. (After Orr et al.[100])

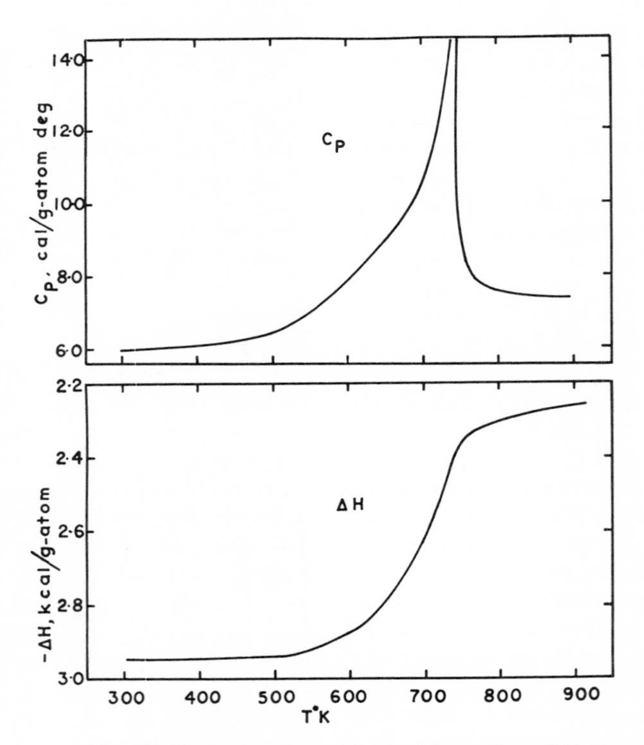

Fig. 19. Heat capacity and heat of formation of CuZn as a function of temperature near the order-disorder transition.

using a quantum-mechanical treatment, calculated a value of -2.26 kcal/g-atom for the difference in the energy of completely disordered and completely ordered Cu_3Au.

The results of the analyses of the second-order transformation in the compound CuZn by the Bragg and Williams and Bethe theories are in reasonable agreement with the heat capacity measurements of Moser[111] and Sykes and Wilkinson,[94] although the transition occurs in a narrower temperature range than is calculated.[112]

In the analysis of order-disorder transitions the experimentally determined activities of the components of compounds or solid solutions may be compared with those derived from the quasichemical theory.[99] The activity determinations of Oriani[113] showed that the transition of CuAu(I) to CuAu(II) is a first-order transformation. This was not shown by either the heat capacity measurements of Hirabayashi[106] or the enthalpy measurements of Borelius et al.[103] and Nystrom.[114] The failure of these investigations to find a first-order transformation may have been due to the difficulty in reaching equilibrium.[99] Oriani also used activity measurements to determine the degree of short-range order in the disordered CuAu.

Order-disorder phenomena in ternary systems will be discussed in Section 9.3.

9. TERNARY INTERMETALLIC COMPOUNDS AND SOLID SOLUTIONS OF BINARY COMPOUNDS

9.1 General

Structural investigations have established the existence of intermetallic compounds in various ternary systems.* Ternary intermetallic compounds are often isomorphous with those in binary systems. For example, Laves and Witte[115] found the existence of ternary Laves phases and Raynor et al.[116–118] and Gladyshevsky[119] carried out structural investigations of ternary electron compounds. Nowotny et al.[120] reviewed the conditions for the formation of ternary intermetallic compounds with partially ionic bonding such as LiMgSb, LiMgBi, CuMgSn, and CuMgAs.

---

* See the discussion by Kornilov in Chapter 19.

The thermodynamic requirements of the stability of an intermetallic compound in a ternary system are analogous to those of a compound in binary systems. They have been discussed in Section 2.1.

9.2 Heats of Formation of Ternary Intermetallic Compounds

The only investigation of the heats of formation of ternary intermetallic compounds was conducted by King and Kleppa[42] on the quasibinary system between the Laves phases $MgCu_2$ and $MgZn_2$. The phase diagram of this system is shown in Fig. 20*a*. The terminal solid solution with the $MgCu_2$ (C15) structure has a maximum homogeneity range of 0 to 69 mole % $MgZn_2$ at 725°C. The compound $MgCu_2$ has only limited solubility in $MgZn_2$, the maximum homogeneity range being 1 mole % $MgCu_2$ at 600°C. A ternary intermetallic compound with the $MgNi_2$ (C36) structure exists at intermediate compositions. At 600°C the homogeneity range of the compound extends from 73 to 89 mole % $MgZn_2$.

The heats of formation at 298°K in the quasibinary system $MgCu_2$–$MgZn_2$, based on the compounds, are shown in Fig. 20*b*. King and Kleppa explained these data by the new

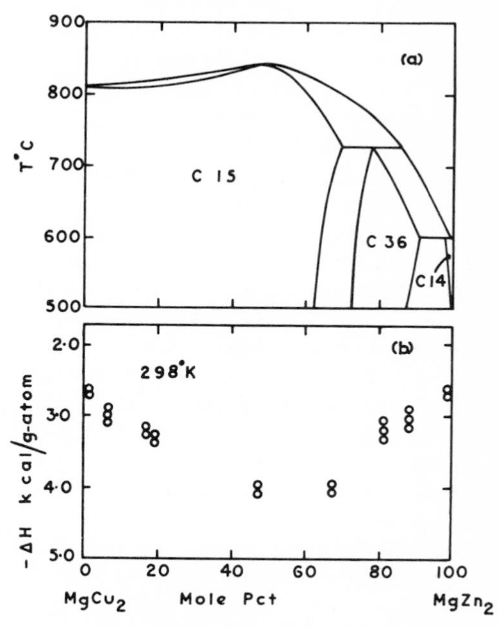

Fig. 20. (*a*) The phase diagram of the quasibinary system $MgCu_2$–$MgZn_2$; (*b*) the heats of formation at 298°K of alloys in the quasibinary system. (After King and Kleppa.[42])

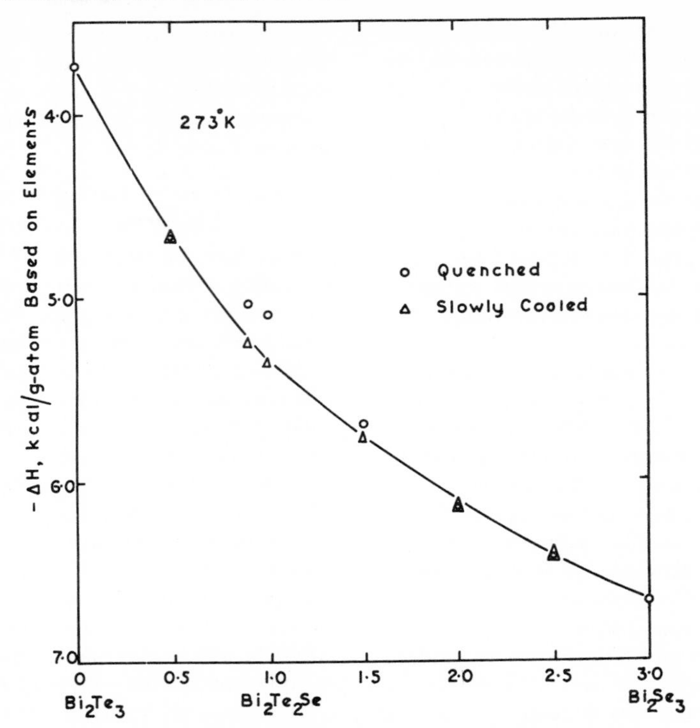

Fig. 21. Heats of formation at 273°K of $Bi_2Te_{3-x}Se_x$ alloys, based on the elements, as a function of the composition variable, x. (After Misra and Bever.[122])

type of nearest-neighbor bonds, namely Cu–Zn bonds, formed in the quasibinary system. The increase in the heat of formation up to approximately 60 mole % $MgZn_2$ is associated with the formation of Cu–Zn bonds which are not present in the binary compounds. At compositions in excess of 60 mole % $MgZn_2$, the heat of formation decreases as the number of the Cu–Zn bonds formed decreases. King and Kleppa stated that the heat of formation of the quasibinary alloys, based on the compounds, was approximately the same as that for the formation of $\frac{2}{3}$ of a gram-atom of the corresponding copper-zinc alloy.[42]

In some ternary systems a complete series of solid solutions is formed between two binary compounds. Such solid solutions require that the binary compounds be isomorphous, have one component in common, and have bonding of a similar type.[121] Various solid solutions between intermetallic compounds are described by Kornilov in Chapter 19.

The only investigation of the heats of formation of a complete series of solid solutions of compounds has been carried out by Misra and Bever[122] on the system Bi_2Te_3–Bi_2Se_3 (Fig. 21). The exothermic heats of formation of alloys in the vicinity of the composition Bi_2Te_2Se which had been slowly cooled from 20°C below the solidus temperature were larger than those of alloys which had been quenched from similar temperatures. This suggests that a transition, probably an order-disorder transition, occurs at this composition (see § 9.3).

9.3 Order-Disorder Transitions in Ternary Systems

Raynor and co-workers[116–118] have investigated the phase relations in the ternary silver-magnesium-tin and silver-magnesium-zinc systems. Henry and Raynor[123] found order-disorder reactions in the β-electron compounds AgMgSn and AgMgZn, which are based on the binary compound AgMg. The compound AgMg has an ordered CsCl (B2) structure; in dilute ternary solutions the tin or zinc atoms occupy the lattice sites randomly. When these solutions reach a critical concentration, however, a superlattice of the Fe_3Al type is formed.[123] Order-disorder transitions have also been observed in the copper-manganese-aluminum system.[124]

A number of attempts have been made to analyze the characteristics of order-disorder transitions in ternary systems. Hardy[125] has attempted to explain the order-disorder transition in FeNiAl; for this purpose he assumed the effects of second-nearest neighbor pairs of atoms. He showed that the temperature of ordering of a binary compound decreases linearly with the addition of a third component. Hasoya[126] used the Bragg and Williams theory[107] to predict the characteristics of an order-disorder transition in A_2BC ternary alloys with simple cubic or bcc structures. The calculations indicated that order-disorder transitions in such ternary alloys are second-order transformations. Matsuda[127] took into account the effect of the interaction between next-nearest as well as nearest-neighbor atoms on superlattice formation in ternary alloys with a bcc structure. By this analysis he predicted various types of superlattice formation in ternary alloys and compared his predictions with the experimental results for the copper-manganese-aluminum system.[124]

Wojceichowski[128,129] gave a statistical theory of order-disorder transitions in ternary alloys of various compositions and structures. He considered the long-range order on the basis of a generalization of the Bragg and Williams theory. Wojceichowski deduced a general equation for the configurational free energy and gave a method of calculating the order-disorder transition temperature for ternary systems. His results are in reasonable agreement with the experimental values of the transition temperature in the Heusler alloy Cu_2MnAl and in the β-electron compound AgMgSn.

Misra and Bever[122] investigated the ordering process in the alloy of composition Bi_2Te_2Se in the quasibinary system Bi_2Te_3–Bi_2Se_3. The continuous series of solid solutions in this system have the C33 crystal structure in which a unit of five basal layers is repeated. Atoms of the electropositive component (Bi) and of the electronegative components (Te, Se) occupy separate layers. If Y refers to layers of the electronegative atoms, the stacking order of two consecutive units may be represented as

—Y(2)—Bi—Y(1)—Bi—Y(2)—Y(2)—
Bi—Y(1)—Bi—Y(2)—

The layer of Y atoms in the center of a unit, designated Y(1), lies between two layers of bismuth atoms, whereas each layer at the outside of a unit, designated Y(2), has as neighbors a layer of bismuth atoms on one side and a layer of Y atoms on the other. In these solid solutions the selenium and tellurium atoms may occur randomly on the Y(1) and Y(2) layers, but an ordering reaction is possible by which selenium atoms preferentially occupy the Y(1) layers and tellurium atoms occupy the Y(2) layers. The energy of ordering in these solutions had a maximum value of 0.25 kcal/g-atom at the composition Bi_2Te_2Se.

ACKNOWLEDGMENTS

The authors wish to record their indebtedness to Professor Carl Wagner, Dr. Oswald Kubaschewski, and Dr. Werner F. Schottky for reading large portions of the manuscript and contributing helpful comments.

REFERENCES

1. Kubaschewski O., and E. Ll. Evans; *Metallurgical Thermochemistry*, 3rd edition. Pergamon Press, London, 1958.
2. Hultgren R., R. L. Orr, P. D. Anderson, and K. K. Kelley, *Selected Values of Thermodynamic Properties of Metals and Alloys*. Wiley, New York, 1963.
3. Wagner C., *Thermodynamics of Alloys*. Addison-Wesley, Reading, Mass., 1952.
4. Rushbrooke G. S., *Proc. Phys. Soc. (London)* **52** (1949) 701.
5. Raynor G. V., *Progress in Metal Physics*, Vol. 1. Butterworths, London, 1949.
6. Hume-Rothery W., J. W. Christian, and W. B. Pearson, *Metallurgical Equilibrium Diagrams*. Institute of Physics, London, 1952.
7. Scatchard G., and R. A. Westlund, *J. Am. Chem. Soc.* **75** (1953) 4189.
8. Trzbiatowski W., and J. Terpilowski, *Bull. Acad. Polon. Sci.* **3** (1955) 391.

9. Raynor G. V., *Theory of Alloy Phases*. ASM, Cleveland, 1956.
10. Goldfinger P., *Compound Semiconductors*, Vol. 1, *Preparation of III–V Compounds*, Chapter 52, R. K. Willardson and H. L. Goering, ed. Reinhold Publishing Corp., New York, 1962.
11. Wagner C., *Acta Met.* **6** (1958) 309.
12. Schottky W. F., and M. B. Bever, *Acta Met.* **6** (1958) 320.
13. Wittig F. E., *The Physical Chemistry of Metallic Solutions and Intermetallic Compounds*, N.P.L. Symposium No. 9, **1**, 1A. H.M.S.O., London, 1959.
14. Kubaschewski O., and R. Hultgren, *Experimental Thermochemistry*, H. A. Skinner, ed., Chapter 16. Interscience, New York, 1962.
15. Kawakami M., *Z. Anorg. Chem.* **167** (1927) 345.
16. Kubaschewski O., and W. A. Dench, *Acta Met.* **3** (1955) 339.
17. Kleppa O. J., *J. Phys. Chem.* **59** (1955) 175.
18. Körber F., W. Oelsen, W. Middel, and H. Lichterberg, *Stahl Eisen* **56** (1936) 1401.
19. Oelsen W., and W. Middel, *Mitt. K.W.I. Eisenforsch. Düsseldorf* **19** (1937) 1.
20. Hahn H., and F. Burow, *Angew. Chem.* **68** (1956) 382.
21. Orr R. L., A. Goldberg, and R. Hultgren, *Rev. Sci. Instr.* **28** (1957) 167.
22. Howlett B. W., J. S. Ll. Leach, L. B. Ticknor, and M. B. Bever, *Rev. Sci. Instr.* **33** (1962) 619.
23. McAteer J. H., and H. Seltz, *J. Am. Chem. Soc.* **58** (1936) 2081.
24. Wagner C., *The Physical Chemistry of Metallic Solutions and Intermetallic Compounds*, N.P.L. Symposium No. 9, **1**, 3K, 14. H.M.S.O., London, 1959.
25. Drabble J. R., and C. H. L. Goodman, *J. Phys. Chem. Solids* **5** (1958) 142.
26. Pauling L., *The Nature of the Chemical Bond*, 3rd edition. Cornell Univ. Press, Ithaca, New York, 1960.
27. Lewis G. N., M. Randall, K. S. Pitzer, and L. Brewer, *Thermodynamics*, 2nd edition. McGraw-Hill Book Co., New York, 1961.
28. Gordy W., and W. J. O. Thomas, *J. Chem. Phys.* **24** (1956) 439.
29. Sanderson R. T., *J. Chem. Phys.* **23** (1955) 2467.
30. Kubaschewski O., and H. A. Sloman, *The Physical Chemistry of Metallic Solutions and Intermetallic Compounds*, N.P.L. Symposium No. 9, **1**, 3B, 2. H.M.S.O., London, 1959.
31. Goodman C. H. L., *J. Electron.* **1**, (1955–56) 115.
32. Raynor G. V., *The Physical Chemistry of Metallic Solutions and Intermetallic Compounds*, N.P.L. Symposium No. 9, **1**, 3A, 2. H.M.S.O., London, 1959.
33. Hume-Rothery W., and G. V. Raynor, *The Structure of Metals and Alloys*, 4th edition. The Institute of Metals, London, 1962.
34. Beardmore P., B. W. Howlett, B. D. Lichter, and M. B. Bever, *Trans. Met. Soc. AIME* **236** (1966) 102.
35. Mooser E., and W. B. Pearson, *J. Electron.* **1** (1955–56) 629.
36. Welker H., *Ergeb. Exakt. Naturw.* **29** (1956) 231.
37. Krebs H., *Acta Cryst.* **9** (1956) 95.
38. Hansen M., and K. Anderko, *Constitution of Binary Alloys*, 2nd edition. McGraw-Hill Book Co., New York, 1958.
39. Collongues R., *The Physical Chemistry of Metallic Solutions and Intermetallic Compounds*, N.P.L. Symposium No. 9, **1**, 3K, 2. H.M.S.O., London, 1959.
40. Kleppa O. J., *J. Phys. Chem.* **60** (1956) 846.
41. Laves F., and H. Witte, *Metallwirtschaft* **14** (1935) 918.
42. King R. C., and O. J. Kleppa, *Acta Met.* **12** (1964) 87.
43. Smith J. F., and J. L. Christian, *Acta Met.* **8** (1960) 249.
44. Smith J. F., and R. L. Smythe, *Acta Met.* **7** (1959) 261.
45. Dehlinger U., and G. E. R. Schulze, *Z. Metallk.* **33** (1941) 157.
46. Kubaschewski O., *The Physical Chemistry of Metallic Solutions and Intermetallic Compounds*, N.P.L. Symposium No. 9, **1**, 3C, 2. H.M.S.O., London, 1959.
47. Wernick J. H., S. E. Haszko, and D. Dorsi, *J. Phys. Chem. Solids* **23** (1962) 567.
48. Folberth O. G., *Compound Semiconductors*, Vol. 1, *Preparation of III–V Compounds*, Chapter 2, R. K. Willardson and H. L. Goering, ed. Reinhold Publishing Corp., New York, 1962.
49. Kubaschewski O., and G. Heymer, *Trans. Farad. Soc.* **56** (1960) 473.
50. Kleppa O. J., *J. Phys. Chem.* **60** (1956) 446.
51. Chiotti P., and G. R. Kilp, *Trans. Met. Soc. AIME* **218** (1960) 41.
52. Kleppa O. J., *J. Phys. Chem.* **60** (1956) 852.
53. Misra S., B. W. Howlett, and M. B. Bever, *Trans. Met. Soc. AIME* **233** (1965) 749.
54. Robinson P. M., and M. B. Bever, *Trans. Met. Soc. AIME* **230** (1964) 1487.

55. Shchukarev, S. A., M. P. Morozova, and T. A. Stolyarova, *Zh. Obshch. Khim.* **31** (1961) 1773, quoted after *Chem. Abstracts* **55** (1961) 23026b.
56. Akhachinski V. V., L. M. Kopitin, M. I. Ivanov, and N. S. Podolskaya, *Thermodynamics of Nuclear Materials*, International Atomic Energy Agency, Vienna, 1962.
57. Lumsden J., *Thermodynamics of Alloys*, Institute of Metals, London, 1952.
58. Piesbergen U., *Z. Naturf.*, **18a** (1963) 141.
59. Zener C., *Z. Metallk.* **50** (1959) 3.
60. Dehlinger U., *Theoretische Metallkunde*, Springer-Verlag, Berlin, 1955.
61. Massalski T. B., and H. W. King, *Acta Met.* **8** (1960) 677.
62. Hill R. W., *Progress in Cryogenics*, Vol. 1 K. Mendelssohn, ed. Chapter 4. Academic Press, Inc., New York, 1959.
63. Keesom P. H., and N. Pearlman, *Handbuch der Physik*, S. Flügge, ed., Vol. 14/1, p. 282. Springer, Heidelberg, 1956.
64. Friedel J., *Perspectives in Materials Research*, Part 2, Appendix, L. Himmel, J. J. Harwood and W. J. Harris, eds. O.N.R., Washington, 1963.
65. Kittel C., *Introduction to Solid State Physics*, 2nd edition. Wiley, New York, 1956.
66. Gul'tyaev P. V., and A. V. Petrov, *Fiz. Tverd. Tela* **1** (1959) 368; *Soviet Phys.—Solid State* **1** (1959) 330.
67. Schimpff H., *Z. Phys. Chem.* **21** (1910) 257.
68. Tret'yakov Yu. D., and K. G. Khomyakov, *Zh. Neorgan. Khim.* **4** (1959) 13; *Russ. J. Inorg. Chem.* **4** (1959) 5.
69. Kubaschewski O., *Z. Phys. Chem.* **A192** (1943) 292.
70. Schröder K., and C. H. Cheng, *J. Appl. Phys.* **31** (1960) 2154.
71. Nevitt M. V., *J. Appl. Phys.* **31** (1960) 155.
72. Westrum E. F., C. Chou, and F. Grønvold, *J. Chem. Phys.* **30** (1959) 761.
73. Wollam J. S., and W. E. Wallace, *J. Phys. Chem. Solids.* **13** (1960) 212.
74. Elliott R. P., *ASTIA Document* No. AD-88033, 1956.
75. Westrum E. F., C. Chou, R. E. Machol, and F. Grønvold, *J. Chem. Phys.* **28** (1958) 497.
76. Carpenter L. G., and C. J. Steward, *Phil. Mag.* **27** (1939) 551.
77. Khomyakov K. G., V. A. Kholler, and S. A. Zhanko, *Vestn. Mosk. Univ.* **7,** No. 3; *Ser. Fiz. Mat.* **2** (1952) 41.
78. Kubaschewski O., and G. Schrag, *Z. Elektrochem.* **46** (1940) 675.
79. Schneider A., and G. Heymer, *The Physical Chemistry of Metallic Solutions and Intermetallic Compounds*, N.P.L. Symposium No. 9, vol. II, 4A, p. 2. H.M.S.O., London, 1959.
80. Howlett B. W., S. Misra, and M. B. Bever, *Trans. Met. Soc. AIME* **230** (1964) 1367.
81. Robinson P. M., and M. B. Bever, *Trans. Met. Soc. AIME* (in press).
82. Beardmore P., S. Misra, and M. B. Bever, unpublished work.
83. Schneider A., and O. Hilmer, *Z. Anorg. Allgem. Chem.* **286** (1956) 97.
84. Schneider A., and G. Heymer, *Z. Anorg. Allgem Chem.* **286** (1956) 118.
85. Drier C. A., R. S. Craig, and W. E. Wallace, *J. Phys. Chem.* **61** (1957) 522.
86. Kelley K. K., *Contributions to the Data on Theoretical Metallurgy, XIII. High-Temperature Heat Content, Heat Capacity and Entropy Data for the Elements and Inorganic Compounds*. Bureau of Mines, Bulletin 584, 1960.
87. Niwa K., and M. Shimoji, *The Physical Chemistry of Metallic Solutions and Intermetallic Compounds*, N.P.L. Symposium No. 9, vol. 1, 2B, p. 3. H.M.S.O., London, 1959.
88. Hendus H., *Z. Naturforsch.* **2A** (1947) 505.
89. Kröger F. A., *Chemistry of Imperfect Crystals*. North-Holland Publishing Company, Amsterdam, 1964.
90. Schadel H. M., and C. E. Birchenall, *Trans. AIME* **188** (1950) 1134.
91. Steiner A., and K. L. Komarek, *Trans. Met. Soc. AIME* **230** (1964) 786.
92. Eldridge J., and K. L. Komarek, *Trans. Met. Soc. AIME* **230** (1964) 226.
93. Pippard A. B., *The Elements of Classical Thermodynamics*. Cambridge University Press, 1957, p. 136.
94. Sykes C., and H. Wilkinson, *J. Inst. Metals* **61** (1937) 223.
95. Hirabayashi M., S. Nagasaki, and H. Kono, *J. Appl. Phys.* **28** (1957) 1070.
96. Rubin L. R., J. S. Ll. Leach, and M. B. Bever, *Trans. AIME* **203** (1955) 421.
97. Orr R. L., *Acta Met.* **8** (1960) 489.
98. Anderson P. D., *J. Am. Chem. Soc.* **80** (1958) 3171.
99. Guttman L., *Solid State Phys.* **3** (1956) 188.
100. Orr R. L., J. Luciat-Labry, and R. Hultgren, *Acta Met.* **8** (1960) 431.

101. Oriani R. A., and W. K. Murphy, *J. Phys. Chem. Solids* **6** (1958) 277.
102. Hirabayashi M., *J. Japan Inst. Met.* **15** (1951) 565.
103. Borelius G., L. E. Larsson, and H. Selberg, *Ark. Fys.* **2** (1950) 161.
104. Sykes C., and F. W. Jones, *Proc. Roy. Soc. (London)* **157** (1936) Ser. A, 213.
105. d'Heurle F., and P. Gordon, *Tech. Report No. 1*, Naval Res. Com., Nonr. 1958, 1406 (03) NR 031–50.
106. Hirabayashi M., *J. Phys. Soc. Japan* **6** (1951) 129.
107. Bragg W. L., and E. J. Williams, *Proc. Roy. Soc. (London)* **145A** (1934) 699.
108. Peierls R., *Proc. Roy. Soc. (London)* **150A** (1935) 552.
109. Cowley J. M., *Phys. Rev.* **77** (1950) 669.
110. Eguchi T., *Mem. Fac. Sci., Kyushu Univ.* **1B** (1951) 25; quoted after *Chem. Abstracts* **46** (1952) 3933*b*.
111. Moser H., *Physik. Z.* **37** (1936) 737.
112. Muto T., and Y. Takagi, *Solid State Phys.* **1** (1955) 194.
113. Oriani R. A., *Acta Met.* **2** (1954) 608.
114. Nystrom J., *Arkiv Fysik* **2** (1950) 151.
115. Laves F., and H. Witte, *Metallwirtschaft* **15** (1936) 840.
116. Frost B. R. T., and G. V. Raynor, *Proc. Roy. Soc.* 1950 (A) 203, (1072) 132.
117. Raynor G. V., and B. R. T. Frost, *J. Inst. Metals* **75** (1948–49) 777.
118. Hall W. H. *J. Inst. Metals* **75** (1948–49) 805.
119. Gladyshevsky E. I., *Dopovidi L'vivs'k. Derzh. Univ. im. I. Franko*, 1957, Vol. 8 (3) 190. Quoted after *Met. Abstr.* **26** (1958–59) 666.
120. Nowotny H., F. Holub, and A. Wittmann, *The Physical Chemistry of Metallic Solutions and Intermetallic Compounds*, N.P.L. Symposium No. 9, Vol. 1, 3E. H.M.S.O. London, 1959.
121. Kornilov I. I., *The Physical Chemistry of Metallic Solutions and Intermetallic Compounds*, N.P.L. Symposium No. 9, Vol. 1, 3J. H.M.S.O., London, 1959.
122. Misra S., and M. B. Bever, *J. Phys. Chem. Solids*, **25** (1964) 1233.
123. Henry W. G., and G. V. Raynor, *Can. J. Phys.* **30** (1952) 412.
124. Bradley A. J., and J. W. Rodgers, *Proc. Roy. Soc.* **114A** (1934) 340.
125. Hardy H. K., *Acta Met.* **1** (1953) 210.
126. Hosoya S., *J. Phys. Soc. Japan* **9** (1954) 489.
127. Matsuda S., *J. Phys. Soc. Japan* **8** (1953) 20.
128. Wojceichowski K. F., *Acta Met.* **6** (1958) 396.
129. Wojceichowski, K. F., *Acta Met.* **7** (1959) 376.

chapter 4

Bond Character as Revealed by Electronic Properties

Bryan R. Coles

BRYAN R. COLES

Department of Physics
Imperial College of Science and Technology
University of London
London, England

1. THE CONCEPT OF BOND TYPE: CHEMICAL AND PHYSICAL DESCRIPTIONS

In writing of "bond character" in solids containing one or more metals, an author's first duty is to declare his point of view or his prejudices. The subject cannot be raised in public discussion, it seems, without acrimonious arguments about valences, bond energies, etc. (and they arise from fundamentally different ways of looking at solids). I believe that these difficulties result mainly from the efforts of those with chemical backgrounds to apply the concept of a *bond*—something responsible for an attractive interaction between a *pair* of atoms—to discussions of solids.

On *physical* grounds I think one can give a description of the problem, but any question about a particular material is best answered by an experimental examination of the material. (But the answers, even so, can never be given in the form—"this material has x per cent ionic character, y per cent metallic character, etc.") The description of the general problem must be given in terms of the character of the wave functions of electrons involved in the bonding process, and it is worth emphasizing that as far as metals are concerned the cohesive energy (or bonding energy) is no guide at all to the number of electrons so involved. In titanium the number is 4, in nickel 10; the cohesive energies are closely similar.

The term "ionic bonding" is fairly straightforward and means that with two types of atom each has about it an integral number of electrons (different from its atomic number) which can accurately be described by wave functions like those of an isolated ion and which overlap very little with wave functions of the other type of atom.

The term "covalent bond" is best restricted to a description of two-center wave functions based on a pair of atoms and often capable of being written as a linear combination of atomic wave functions of the individual atoms. (Note that in these terms the presence of the expressions $\psi_a(1)\psi_a(2)$, and $\psi_b(1)\psi_b(2)$ in the molecular orbital wave function of the hydrogen molecule does not lessen its covalent nature, for they appear with equal weight.)

The remaining bond required (if we forget for the present van der Waals bonding) is normally called the metallic bond, but perhaps is better described as the extended wave-function bond. Bonding has this character when the electron wave function extends throughout the crystal and a polyhedron containing an individual atom is electrically neutral.

In these terms it is misleading to speak of germanium and lead as having different bond character simply because the former has only completely full and completely empty Brillouin zones. The fact that one can carry out a useful tight-binding calculation[1] for germanium, using as a starting point (sp^3) hybrid wave functions on the individual atoms, is irrelevant, for in the final crystal wave functions the individual character of particular bonds is completely lost. It is possible, however, to point to covalent bonds in solid hydrogen or solid methane. It should also be noted that arguments about the number of electrons in a pure metal which should be described by localized atomiclike wave functions are not arguments about the nature of the bonding; it is therefore not appropriate to describe models that use such wave functions for *d*-electrons as Heitler-London models.

In compounds it is at times useful to speak of mixed-bonding character, especially when marked differences in electronegativity (a more useful conceptual contribution from chemistry to solid state studies than "the bond") are present. There is a definite meaning to be attached to the statement that the HCl molecule or the Mg_2Sn crystal possesses ionic character, and in principle this could be expressed in terms of formal wave functions. It is therefore worth examining the extent to which information about bond character can be derived from the physical properties of a compound.

In quasichemical descriptions of solids special significance is attached to crystal structure and interatomic distances, and these are in fact used more often than physical properties as criteria of bond type. After what has already been said we need only remark that CsCl and CuZn have identical crystal structures and totally distinct bond characters (as do InSb and AgI).

It may be significant in a particular compound that an A atom has one or two especially close B neighbors, but general arguments from interatomic distances to bond type have little more significance than efforts to explain the interatomic distances of the alkali metals in terms of resonating covalent bonds.

Useful discussions of the factors affecting the crystal structures of intermetallic compounds (especially atomic size, electronegativity, and electron concentration) can be found in many places, (Chapter 8 of this book for example). Another review has been given by Raynor in the proceedings of a symposium[2] containing a number of papers relating to the properties and character of intermetallic compounds.

2. ELECTRICAL CONDUCTION IN INTERMETALLIC COMPOUNDS

2.1 Conductivity and its Temperature Coefficient

The electrical conductivity of solids varies more from material to material than any other easily measured property; experimental data, especially when combined with those for the temperature coefficient, provide very clear indications of the presence or absence of electrons of metallic character. In a simple formal treatment of the conductivity (σ) of electrons in solids we may write

$$\sigma = \frac{ne^2\tau}{m}$$

where n is the number of carriers, m, their mass, and τ, their time of relaxation. In a metal n is fairly insensitive to temperature or purity, and observed changes in σ can be discussed in terms of the sensitivity of τ to these variables. In a solid that is an insulator at low temperatures, however, large changes in n can be produced by variations of temperature and purity.

An important contribution of wave mechanics is the result that τ tends to infinity as the imperfections and lattice vibrations tend to zero, so that a very small number of metallic electrons per atom will confer a high conductivity on a well-ordered solid even when they give only a small contribution to the total bonding energy. This is very striking in high-frequency conductivity, as indicated by the optical properties; the metallic luster of selenium, many sulfides, and even solid iodine is well known. Conversely, large changes in conductivity (by a factor of 10^5) can be produced in a metal with negligible change in the number and nature of the conduction electrons.

For these reasons the stoichiometric perfection of an intermetallic compound can be determined directly by measurements of its electrical resistivity, especially at low temperatures, but whether we should look for a very high or a very low conductivity depends on the nature of the compound.

A clear example of the derivation of useful information from the conductivity of a compound is provided by the distinction between two magnesium alloys, each of which is of limited composition range in the phase diagram and possesses an atomic ratio that would seem to follow from "normal" chemical valences. $MgCu_2$ has a conductivity close to that of pure magnesium, whereas that of Mg_2Sn is a factor of at least 10^5 smaller. The semiconducting character of Mg_2Sn does not, however, of itself allow us to decide whether extended or ionic wave functions should be used to describe the bonding electrons.* It seems likely that a compound of elements differing so strongly in electronegativity would have more ionic character than, for example, InSb, another intermetallic semiconductor; but it is difficult to find any simple way of making a quantitative measure of this difference.

Ordinary electrical and magnetic measurements do not provide clear indications of polar character; and, although they can in principle be obtained from measurements of the infrared absorption, the optical effects associated with excitations of electrons often obscure behavior due to polar character in semiconductors. A careful study has recently been made[4] of the infrared reflection spectra of the III–V semiconducting compounds, and effective ionic charges have been derived; however, as pointed out by the authors and by Cochran[5], they do not necessarily indicate the presence of permanent charges on the two types of atom but perhaps only differences in their polarizabilities. Detailed analysis of the luminescent behavior under infrared stimulation of ZnS and ZnSe containing impurities[6] does, however, give strong indirect evidence of polar character.

Most simple interstitial metallic compounds have electrical properties dominated by the conduction electrons of the metallic matrix. Measurements of conductivity at low temperatures can show whether the interstices are occupied in an ordered or random fashion, but this is information about the geometry rather than the bonding character of the material. (It is interesting to note that it is possible in principle for small covalently bonded gas molecules to occupy interstices in some metals, but it should be remembered that, just as with localized atomic states, if the energies of the electrons in these "molecular" electronic states lay in the energy range of the conduction electrons they would cease to be real bound states.)

It is often stated that metallic and semiconducting materials can be clearly distinguished by the temperature coefficient of resistance in the absence of knowledge of the specific resistivity. This is by no means always true; Bi_2Te_3, which seems to have optical and other semiconducting properties, has a positive temperature coefficient of resistance between 4°K and room temperature, whereas certain palladium-silver alloys which are clearly metallic have negative temperature coefficients over some temperature ranges.

2.2 Hall Constant

Much of what has been said about the electrical resistivity applies also to the Hall effect, for the principal usefulness of Hall-effect data in the present problem is in distinguishing between metals and semiconductors. For semiconductors the Hall constant yields the number and sign of the current carriers with reasonable accuracy, provided

* A comparison of the properties of Mg_2Si, Mg_2Ge, Mg_2Sn, and Mg_2Pb is given by J. M. Whelan.[3]

that one type of carrier is predominant. Both parameters will normally depend strongly on the number and character of impurities present or on the departures from stoichiometry. By combining resistivity and Hall data the mobility of the carriers can be obtained and will vary with the character of the scattering processes. The behavior of the mobility in an ionic semiconductor has been calculated by Howarth and Sondheimer[7]; their results have been used by Ehrenreich[8] in a theoretical analysis of scattering mechanisms in InSb which seems to point to the presence of appreciable polar character. The presence of such character in SiC has also been deduced recently[9] from the temperature dependence of the mobility.

It is scarcely necessary now to issue a warning against attempts to "count" valence electrons in metallic alloys by measuring their Hall constants. The Hall effect in metals depends so strongly on anisotropies of the Fermi surface and relaxation time that such attempts have no meaning.

2.3 Superconductivity

In recent years superconductivity has been observed in a large number of intermetallic compounds (see Chapter 29), and suggestions have been made that certain structure types are particularly favorable for its appearance. It is therefore conceivable that detailed study of the variation of transition temperature within such structure types will yield some correlation with bonding character, although the most successful correlations up to the present have been with the total number of outer electrons.

The changes in magnitude and sign of the Hall constant on ordering in structures like Cu_3Au must, of course, be ascribed to changes in the Fermi surface produced by new Brillouin zone planes, and not to changes in bond character.

2.4 "Electrolytic" Transport in Alloys

Evidence has been available for many years that the application of electric fields to metallic alloys can transport atoms through the solid toward one electrode or the other. In earlier work relative transport numbers for the two components were derived; more recently application of marker techniques has given individual transport numbers. (Methods and data are reviewed by Heumann[10].) Rather surprisingly *both* constituent metals often move to the anode, although C in Fe and H in Pd move rather more rapidly to the cathode. It had been hoped to derive evidence for ionic character from such measurements,[11] but it is now clear that such hopes are illusory. The reason is clearly set out in a theoretical paper by Bosvieux and Friedel.[12] The presence of conduction electrons reduces the direct force to zero, and the observed transport is due only to momentum transfer from the current carriers during collisions; since these can be electrons or positive holes or both, no general rules can be laid down.

3. THE MAGNETIC PROPERTIES OF ELECTRONS IN SOLIDS

In all types of chemical bonding pairing of opposed electron spins is the general rule, and because the presence of unpaired spins is easily detected by magnetic measurements (modern resonance techniques are particularly sensitive) such experiments are often used to detect the presence and number of electrons *not* involved in bonding processes. The susceptibility contributions of unpaired spins (unless ferro- or antiferromagnetism sets in) show a strong decrease with increasing temperature, but even in the absence of such contributions the smaller, temperature-independent dia- or paramagnetic susceptibilities of solids are sometimes examined for clues to their electronic nature.

3.1 Magnetic Criterion for Bond Type

In studies of the coordination complexes formed by transition metals of the first long period Pauling[13] was able to show that the symmetry of the coordination bonds could be successfully ascribed to particular hybrid bonds (e.g., d^2sp^3 for octahedral complexes). If an atom forming such bonds possesses more electrons outside a closed (s^2p^6)-shell than are used in bonding, they will occupy the atomic *d*-levels not used in constructing the bonds and give, in accordance with Hund's rule, the maximum possible spin. Magnetic susceptibility measurements can give evidence of the total spin of the atom, for one may normally write

$$\chi = \frac{C}{T - \theta}$$

where

$$C = \frac{N\mu_{\text{eff}}^2}{3k}$$

and

$$\mu_{\text{eff}} = \sqrt{4S(S + 1)}\mu_B$$

if, as is often true for elements of the first transition group, any orbital contribution to the magnetic moment can be ignored; (χ is the susceptibility per gram, μ_B the Bohr magneton, and S and N are the spin and number per gram of the moment-bearing atoms.)

It is not possible, from the chemical formula, to decide the nature of the bonding to the transition metal atom of coordinated atoms or groups. Thus the $[Fe(CN)_6]^{3-}$ complex ion could be Fe^{3+}, with only ionic links with the $(CN)^-$ groups or with coordination links that use (d^2sp^3) orbitals to provide them. In the "ionic" situation the five nonbonding d-electrons could occupy all five of the atomic d-levels to give $S = \frac{5}{2}$, but if coordination (semicovalent) links are formed only three d-levels are available so that $S = \frac{1}{2}$. In a large number of complexes the indications provided by the magnetic moment have been in excellent agreement with those provided by the symmetry of the complex.

3.2 Localized Magnetic States in Metallic Crystals

Pauling's success with the coordination complexes of the transition elements led him to apply similar concepts to these elements in the metallic state; but for these the situation is more complicated, and his ideas have, in general, been rejected by *physicists* concerned with the transition metals. Pauling originally proposed[14] that, because the maximum spin moment found in ferromagnetic alloys of the first transition group metals is $2.44\mu_B$, this number represented the number of states that were atomic in character and could not be used in building up metallic bonds. It is now generally believed that (*a*) it is quite possible to have ferromagnetism with electrons described by extended (Bloch) wave functions, and (*b*) it is not possible to describe d-electrons whose energy levels lie in the range covered by the conduction band as strictly bound (atomic) levels.[15] In the rare-earth metals, on the other hand, the $4f$-electrons seem to be sufficiently strongly bound so that they may be given an integral number per atom and described by wave functions that fall to a very low amplitude between neighboring atoms.

In certain dilute alloys of transition metals with nontransition metals the susceptibility behavior can be interpreted in terms of magnetic moments similar to those the transition element manifests in inorganic compounds, and it is then probably fair to say that the electrons providing these moments make little contribution to the bonding. Indications of unpaired spins *spatially* localized on particular atoms are given by neutron diffraction studies of ferromagnetic and antiferromagnetic metals and alloys, but studies of some of them at high temperatures (above Curie or Néel temperatures) show that extended wave functions should be used to describe the magnetically active electrons and contributions to bonding from them cannot be ruled out.

In compounds of transition metals of the earlier groups (IV–VI) with semimetals or nonmetals—TiO, VSi_2, $MoSi_2$, for example—the absence of strong, temperature-dependent paramagnetism makes it certain that marked ionic character is absent, and, from a study of a number of disilicides, Robins[16] concludes that the general character of the energy bands of the pure metals persists in these compounds.

Manganese and chromium, in compounds with elements of Groups VB and VI, often show strongly magnetic behavior, but it does not seem possible to link this behavior in any direct way with ionic character in the bonding.

With nickel compounds, however, the well-known tendency of the $3d$-band of nickel to fill on forming metallic alloys with non-transition elements makes it reasonable to suggest that strong magnetic behavior (in NiO and NiS, for example) is associated with ionic character, a feature one would in any case have suspected in these particular materials.

Some interesting results have been obtained for the effects on saturation magnetization and Curie temperature of substitutions of nitrogen for carbon in Fe_3C, and they have been interpreted in terms of electron transfer effects.[17] It seems likely that changes of lattice spacing will also have appreciable effects, but more work of this type would be of great interest.

In some alloys and compounds of rare-earth metals (Y–Gd, GdB_6) the presence of Curie or Néel temperatures of more than a few degrees absolute can safely be ascribed to the role of conduction electrons in providing an interaction mechanism, but rather strong *direct* interactions seem to be present in the divalent europium compounds with S, Se, and Te, which must be predominantly ionic.

3.3 Temperature-Independent Dia- and Paramagnetism

In the absence of partly filled *d*- or *f*- shells the conduction electrons of metallic solids rarely give rise to striking effects in magnetic properties. They can make both para- and diamagnetic contributions to the susceptibility, and since ion-core diamagnetic contributions are always present neither the sign nor the magnitude of small temperature-dependent susceptibilities is significant. The rather large diamagnetic susceptibilities of antimony and bismuth have been ascribed[18] to a transitional stage between covalent and metallic bonding, but a correct theoretical treatment of this diamagnetism can be given[19] without recourse to such terminology, and a similar effect is present in the thoroughly metallic γ-phase of the Cu–Zn alloys.

The diamagnetic susceptibility of a simple closed-shell ion is a fairly well-defined quantity, and tabulated values can be found in many places. It is therefore possible in principle to detect any departures from pure ionic character in compounds, but this requires accurate absolute measurement of rather small susceptibilities. A more sensitive technique might be to look for shifts in the frequency of the nuclear magnetic resonance from its value (at the same field) for an isolated ion. Conduction electrons produce the well-known Knight shift, and "chemical" shifts have been observed in some nonmetallic solids.

There are, not surprisingly, difficulties in reaching firm conclusions about bond character from the resonance shifts or from the electric field gradients revealed by quadrupole effects.

4. CONCLUSION

In view of the rather cautious response given in this chapter to most simple suggestions for means of distinguishing different bonding mechanisms in the solid, it seems that the question—"What is the bonding in this material?"—is rarely sufficiently precise. However, electronic measurements can yield important information which can be correlated with other properties, and they are therefore often worthwhile for their own sake. It is only if one's aim is to "calculate" the energetics of intermetallic compounds from "bond energies" and "bond mechanisms" that one will suffer final disappointment.

REFERENCES

1. Leman G., and J. Friedel, *J. Appl. Phys.* **33** (1) (1962) 281S.
2. Physical Chemistry of Metallic Solutions and Intermetallic Compounds (*NPL Symposium No. 9*). H.M.S.O., London, 1959.
3. Whelan J. M., in *Semiconductors*, N. B. Hannay, ed. Reinhold Publ. Co., New York, 1960 p. 389.
4. Hass M., and B. W. Henvis, *J. Phys. Chem. Solids* **23** (1962) 1099.
5. Cochran W., *Nature* **191** (1961) 60.
6. Morehead F. F., *J. Phys. Chem. Solids* **24** (1963) 37.
7. Howarth D. J., and E. H. Sondheimer, *Proc. Roy. Soc.* **219** (1953A) 53.
8. Ehrenreich H., *J. Phys. Chem. Solids* **9** (1959) 1929.
9. Van Daal H. J. et al. *J. Phys. Chem. Solids* **24** (1963) 109.
10. Heumann T., Ref. 2, Paper 2C. (See also Verhoeven J., *Met. Rev.* **8** (1963) 311.
11. See, for example, Kubachewski O., and K. Reinartz, *Z. Elektrochem.* **52** (1948) 75.
12. Bosvieux C., and J. Friedel, *J. Phys. Chem. Solids* **23** (1952) 1923.
13. Pauling L., *The Nature of the Chemical Bond.* Cornell Univ. Press, Ithaca, N.Y., 1940.
14. Pauling L., *Phys. Rev.* **54** (1938) 899.
15. Friedel J., *Nuovo Cimento* (*Supp*) **7** (1958) 287.
16. Robins D. A., *Phil. Mag.* **3** (1958) 313.
17. Nicholson M. E., *Trans. AIME* **209** (1957) 1.
18. Klemm W., quoted by O. Kubachevski and H. A. Sloman, Ref. 2, Paper 3B.
19. See R. Peierls *Quantum Theory of Solids.* Oxford University Press, 1955.

chapter 5

Bonding as Revealed by Crystal Morphology

Gunther A. Wolff

GUNTHER A. WOLFF

Tyco Laboratories, Inc.
Waltham, Massachusetts

1. INTRODUCTION

The symmetrical appearance and polyhedral form of crystals has intrigued many people since ancient times. It was not until 1669 (N. Steno, law of constancy of interfacial angles), however, that a law was recognized concerning the formation of crystals. In 1784 R. J. Hauy[1] introduced the concept of rational indices of crystal planes and thus laid the foundation for modern crystallographic systematics. J. W. Gibbs[2] subsequently (1875–1878) derived the thermodynamic conditions governing the equilibrium morphology of single crystals in contact with their matrix. Other theoretical treatments[3–5] of the equilibrium form of crystals are based on the theorem of G. Wulff[3] which states that this form can be constructed geometrically in a simple manner: planes are made to intercept normally the (radial) plane vectors of the specific interface free energy values of all crystal planes; the resulting geometrical form is then congruent to the crystal equilibrium form (see Fig. 1).

For homopolar or covalent, metallic, and van der Waals crystals, I. N. Stranski[6] subsequently showed that the molecules or building units in the apex positions of Wulff's polyhedron are in equilibrium with the crystal matrix; they are bonded as strongly to the crystal and to as many neighbors as the molecules or building units in the regular growth or equilibrium positions of the crystal surface.

These units also make up the crystal edges of the equilibrium polyhedron where they are bonded together to form periodic bond chains (PBC vectors[7]); in these bond chains each unit is bonded to one more neighbor than are the units in the equilibrium position (Fig. 2). The bonding forces within the crystal surface thus directly determine the orientation of the crystal edges and, in turn, the overall crystal morphology. Inversely, from data on crystal morphology, we can obtain information on the type and relative strength of bonding in crystal surfaces. It is fortunate that this method can be extended to crystal-growth forms, for

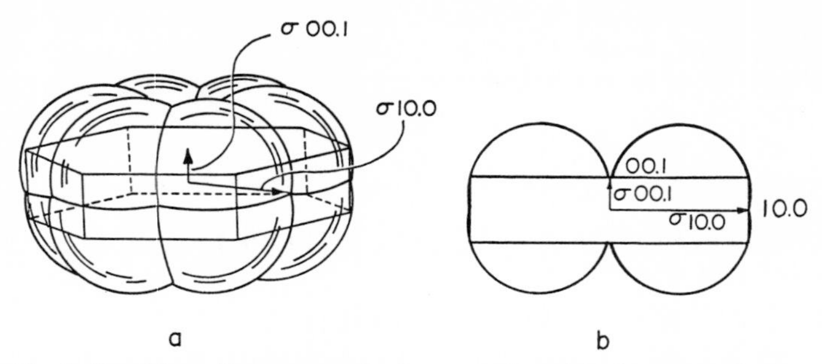

Fig. 1. Specific surface energy plot for normal graphite (structure shown in Fig. 9). The values were calculated as described in Section 4. (*a*) Plot in spherical coordinates (Wulff's plot or σ-plot). (*b*) Polar diagram representing cross section through spherical plot in (*a*).

growth planes are always equilibrium planes (although the reverse is not necessarily true). The term crystal morphology as used in this context is not applied to the shape of the crystal (as implied in expressions like whisker, platelet, needle-type, or dendritic habit); rather it refers specifically only to the idealized crystal habit or the "crystal form." The crystal morphology in this sense is then given by the crystal form composed of either observed or derived crystal planes; in this form the relative size of the crystal planes is comparatively unimportant, in contrast to crystal morphology in the regular sense in which this is not so.

In the following it is shown how a comparison of experimental and theoretical crystal forms will lead to information on

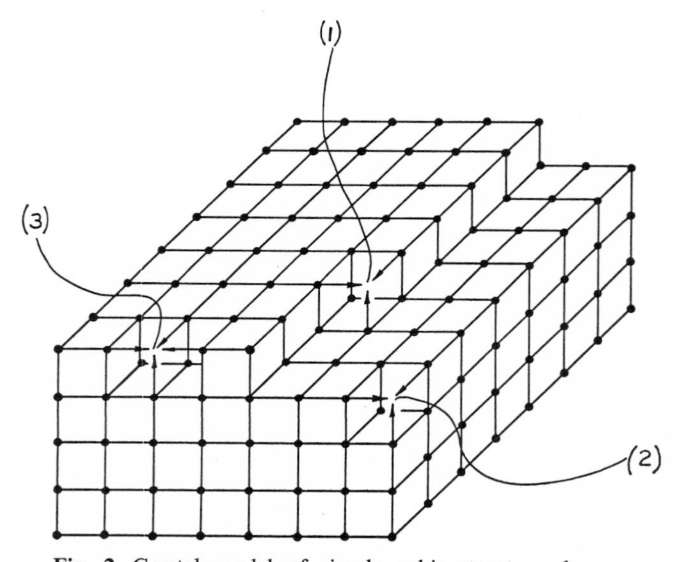

Fig. 2. Crystal model of simple cubic structure demonstrating bonding in the growth (equilibrium) position (1), apex position (2), and edge position (3). An atom in the equilibrium position (1) is held in place by three nearest-neighbor bonds. The same holds true for an atom in position (2). This atom is therefore likewise in an equilibrium position when only nearest neighbors interact. The atom in position (3) is part of the periodic bond chain (PBC) of the edge and is therefore bonded to one more neighbor, that is, to four neighbors in all.

bond-energy ratios, changes in the surface structure, mechanisms of reactions at the surface, and properties of covalent surface bonding.

2. CRYSTAL MORPHOLOGY

2.1 Growth, Equilibrium, Solution, and Cleavage Forms of Crystals

Crystals grow or dissolve when their surrounding matrix is sufficiently supersaturated or undersaturated. They are in equilibrium with their matrix when there is zero net transfer of molecules or crystal-building elements between matrix and crystals so that the crystals neither grow nor dissolve. In growth as well as in the equilibration process crystals will assume a polyhedral shape. However, when crystal growth takes place through a screw dislocation nucleation mechanism at low supersaturations rather than through two-dimensional nucleation at higher supersaturations, the growth of hillocks via vicinal planes may occur, thus resulting in growth forms that are not polyhedral. In general, it is found that the surfaces of these forms are more or less cylindrically curved and as such are parallel to the crystal edges of the polyhedral growth forms.

Crystal solution end forms, on the other hand, are bounded by curved surfaces which in turn are bounded by curved ridges. The ridges are terminated by protruding crystal corners which correspond to the planes most slowly attacked by the solvent in question. The axes of the curved ridges are identical in orientation with the edges of the polyhedral crystal growth forms.

The same principles which hold for "positive" crystals also hold for "negative" crystals. A negative crystal is defined as a void in the interior of a crystal. Just as growth and equilibrium forms of positive crystals are polyhedral (type a) and solution forms are of rounded shape (type b), so are the respective forms of negative crystals (Fig. 3). When crystal voids are in equilibrium with their growth matrix within the voids or are being increased in size, for example, by etching of the crystal surface constituting the walls of the cavity, the form of the resulting void will be polyhedral (type $\bar{a}$). On the other hand, when the void is being decreased in size, for example, by growing the crystal further into the void, the form of the resulting void (or negative crystal) will be of rounded shape of solution form type (type $\bar{b}$) just as the form of a positive crystal would be on etching.[8]

The matter becomes more complex, however, when "positive" crystals of convex surface curvature and "negative" crystals of concave curvature are physically associated, as in ordinary crystal surfaces on which concave etch pits with convexly curved rims and settings occur (Fig. 3). Here the underlying principles are still the same, although their interpretation by ordinary means may appear more difficult.[9,10]

The difficulty in the evaluation of such complex cases can easily be circumvented by the application of the light-figure method (Fig. 4). In the light figure, that is, in the projected figure obtained from the reflection of parallel light impinging on the crystal surface, crystallographic planes will appear as intense spot reflections or as the intersection of lines on the projection screen. The latter represent light reflections from at least two sets of vicinal planes tilted along particular zonal directions. When the surface is entirely composed of macroscopic flat or *F*-planes, spot reflections alone appear. The spot reflections then correspond to polyhedral *F*-planes, whereas the line reflections, if any, correspond to more or less rounded crystal edges (or stepped *S*-planes) joining the polyhedral *F*-planes. The line reflections would appear between the spot reflections or between the intersections of other line reflections. This follows from the fact that the line reflections intersect each other or meet at the points at which the spot reflections would normally appear. All other nonreflecting planes are kinked or *K*-planes. It is most fortunate that in the most common cases (as in preferential etching) the light figure is the same for the etched and grown concave and convex surfaces so that their interpretation, in general, poses little difficulty. This is not true for nonpreferential etching, however, in which the interpretation is different (see Fig. 1 of Reference 11).

In the past the study of growth and etch habit patterns has been most useful in the identification of crystal materials and their polarity. Unfortunately, however, the respective habit types observed frequently depend on the type of growth matrix or etchant and

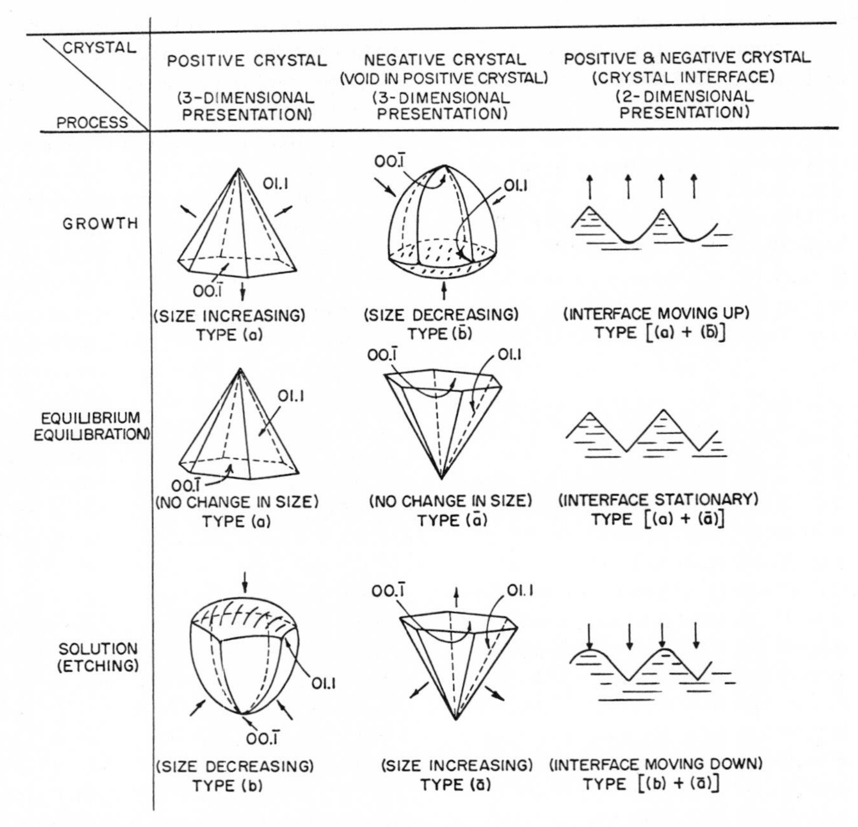

Fig. 3. Idealized presentation of growth, equilibrium, and solution forms of positive crystals, negative crystals, and combination of both when present in geometrically "flat" crystal surfaces. Idealized positive and negative crystals are of convex and concave curvature, respectively. The crystals selected are of 6 mm point group symmetry as in wurtzite. The type (a) and (ā) crystal forms are enclosed by *F*-planes (*S*- and *K*-planes can persist in some instances). The type (b) and (b̄) forms have PBC vectors parallel to the curved ridges. These forms are enclosed by vicinal *K*-planes.

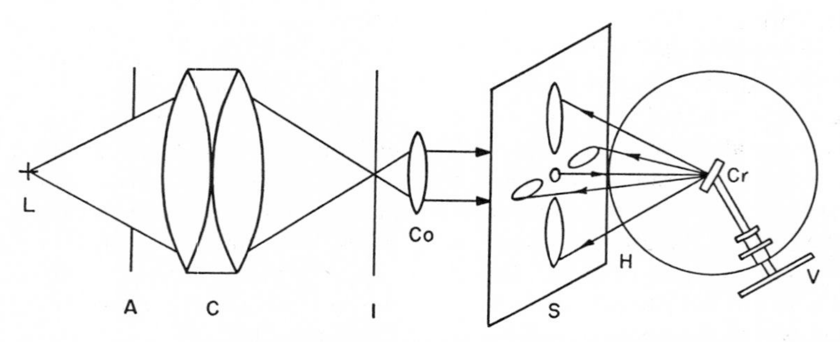

Fig. 4. Sideview schematic of light-figure apparatus. For better visualization the screen and an idealized light-figure reflection are shown in normal perspective. *L*, Light source; *A*, aperture; *C*, collimator; *Co*, condenser; *S*, screen; *H*, horizontal circle of two-circle goniometer; *V*, vertical circle; *Cr*, investigated crystal.

are often ambiguous. For a successful material identification, therefore, the various growth and etch habits must be known for each of the respective matrices and etchants. In crystal cleavage this is not necessary, for there is only one pattern for each material and therefore the aforementioned difficulty does not exist. It is unfortunate, therefore, that the application of cleavage and/or mechanical etching has been neglected. Cleavage and etching techniques are useful techniques to supplement X-ray methods in material identification.[12,13]

2.2 Derivations of the Crystal Equilibrium Form from the Crystal Structure, PBC Vector, and Bond Energy Ratio

The first to associate crystal habit and crystal structure was R. J. Hauy.[1] Then it was A. Bravais[14] who related the importance of crystal planes to their reticular density. Much later J. D. H. Donnay and D. Harker[15] showed the influence of screw axes and glide planes on crystal morphology. The significance of close-packed atomic or molecular arrays in determining crystal habit was extensively demonstrated by W. Kossel,[16] I. N. Stranski,[17] P. Hartman,[7] and others. The picture would certainly be incomplete without reiterating J. W. Gibbs's principle,[2] which requires that the total interface free energy of a crystal of constant volume in equilibrium with its matrix be at a minimum.

$$\sum_i \sigma_i A_i = \text{minimum} \tag{1}$$

where σ_i = specific interfacial free energy of the ith plane and A_i = area of the ith crystal plane. A useful interpretation of this principle appeared when G. Wulff[3] showed that

$$\frac{\sigma_i}{h_i} = \text{constant} \tag{2}$$

This leads to the geometrical construction of the crystal equilibrium form as mentioned in Section 1. Here h_i represents the central distance of the ith crystal plane.*

The derivation of the crystal equilibrium forms would be of minor importance if it were not for the fact that the crystal growth forms are also always comprised of the same crystal planes (though not necessarily all of them and not of the same dimensions) as are present in the equilibrium form. This statement is strictly correct only for crystals grown via the formation of two-dimensional nuclei. In a limited sense, however, this is also true if crystal growth takes place through the catalytic effect of screw dislocations; that is, if the concentration of the screw dislocations is low and if the formation of vicinal planes can be neglected.

A crystal equilibrium form may be derived in three ways as elaborated in the following paragraphs:

1. The first method consists of varying the central distances of all crystal planes until they are proportional to the specific surface free energy, the crystal volume remaining constant. This is Wulff's method which follows from Eq. 2.

In a first approximation the specific surface free energy may be replaced by the specific surface free energy at 0°K, that is, the specific surface energy. It is most important to emphasize that only the lowest value of the specific surface free energy σ of a crystal plane which any particular configuration of its orientation can assume is considered. In {011} planes of NaCl structure materials the configuration given in Fig. 5*a* is thus of a lower σ-value than that which is shown in Fig. 5*b*. The former configuration therefore, is preferred. Such an arrangement also could constitute a puckered configuration or one which is composed of a multitude of "sub-pyramids" of equilibrium shape and any suitable size (Stranski: "Vergröberung") and a more or less symmetrical orientation, with the pyramid base parallel to the plane in question.

2. The second method consists of removing from a simple crystal form (e.g. cube, octahedron, etc.), all crystal building units, that is, atoms, molecules, ionic groups, and the like, which are bonded less strongly to the crystal than the building units in the equilibrium position. The evaporation probability of such loosely bonded crystal building units in various corner positions will be higher than the evaporation probability of a building unit in the equilibrium position. The apex building units will thus evaporate and leave the corner positions empty more often than not. In this

* Wulff's point and the crystal center coincide only in crystals with a symmetry center.

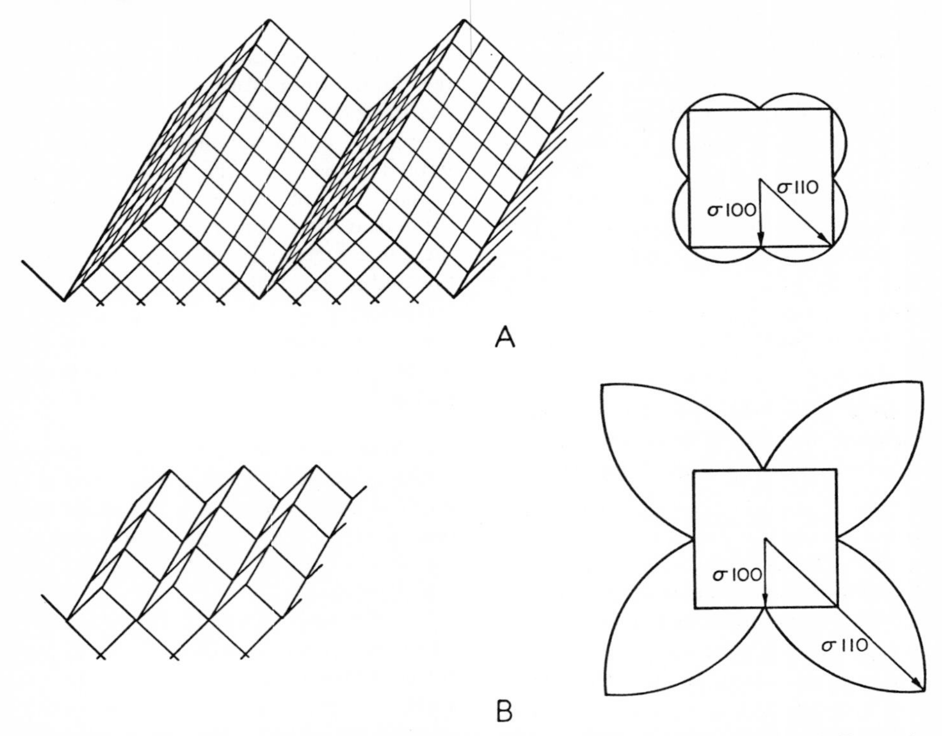

Fig. 5. (*a*) "Faceted" (or Vergröbert")[49] (110) plane of NaCl structure. Corresponding polar σ-plot for the [001] zone is also shown. The picture presented holds true also for (110) of simple cubic structure. (*b*) Atomically smooth (110) surface of NaCl of nonequilibrium configuration and related polar σ-plot of the atomically smooth configurations of the [001] zone. The smooth, unfaceted (110) surface shown is enlarged with respect to the faceted (110) surface in (*a*). In both figures the elementary cubic crystal building blocks are of the same size.

way, after the adjacent layers are removed, another crystal plane is then formed which belongs to the equilibrium form. This process is repeated until all crystal building units are removed from the crystal which are less strongly bonded to the crystal than the building units in the equilibrium positions. Once all crystal equilibrium planes have been derived in this way, the resulting crystal form is varied until the mean detachment energy of all building units for all crystal planes is equal (I. N. Stranski[6]).

3. Determine all "periodic bond chain vectors" (after P. Hartman[7]) of the equilibrium surface. In the periodic bond chain (PBC) in the crystal surface the building unit is bonded more strongly (by one or more neighbors) to the crystal than the unit in the equilibrium position. Add all PBC vectors (expressed in bond strength and geometrical length) which are not parallel to the plane in question. The end points of the resulting vectors then represent the apices of the corresponding equilibrium form.[18,19] The derivation of the equilibrium form in this way is quite simple when there are only straight atomic or molecular PBC vectors, as in face-centered cubic crystals. In the case in which there are PBC vectors of zigzag (geometrical or bonding) arrangement, as in the diamond structure, care must be taken that only those PBC vectors are considered that fulfill the previously stated requirements for the planes in question (i.e., the building units of the periodic bond chains must be bonded more strongly to the crystal than the building units in the equilibrium position). Zigzag arrays of building units are thus not PBC vectors for all planes to which they are parallel.

As under (2), the central distances of the derived crystal equilibrium planes are varied until the mean detachment energy of all units is identical for all crystal planes, or, as under

(1), until for all crystal equilibrium planes the central distances are proportional to their specific surface free energy.

When there are two types of chemical bond, which are of different strength, the expected equilibrium form is abruptly changed when certain critical values of their bond-strength ratio are reached and/or exceeded. Also the equilibrium form of crystals without a symmetry center changes in like manner with changing ratio of the surface energy contribution of unlike surface atoms. The difference in surface energy of unlike surface atoms can be caused by their differences in their ionic polarization or by the differences in their more or less completely filled electron orbital structure (as in surface anions and cations of wurtzite and sphalerite structure crystals).[11,18,19]

2.3 Differences Between Surface Structure and Bulk Structure

Materials of diamond, sphalerite, and wurtzite structure at high pressures generally transform to modifications of greater density and higher coordination. The corresponding types and critical pressures at which the transformations for the various materials take place are given in Table 1. In these transitions the fourfold or tetrahedral s^2p^6 coordination of these materials changes to the denser sixfold or octahedral p^6 packing of NaCl (B1) structure or to the closely related β-Sn type and cinnabar (B9) structure of (4 + 2) and (2 + 2 + 2) coordination, respectively. This type of change to a sixfold coordination has also been observed in Ge and InSb on melting.[31] It is therefore not surprising that the surfaces of these materials also experience a similar change in crystal structure. The structural properties of the crystal surfaces are expected to be intermediate between the respective structural properties of crystal bulk and crystal melt. Likewise, in materials of covalent bonding and low coordination in the bulk the atomic coordination in the crystal suface is expected to be higher than in the crystal bulk.

In particular, this effect is likely to be expected in materials of tetrahedral coordination and s^2p^6 bonding. At the surface of these materials the lack of nearest neighbors causes the dehybridization of the s^2p^6 states, thus favoring an octahedral coordination.

Table 1

Selected Pressure Transitions of Group IV-B Elements, III-V, II-VI, and I-VII Compounds

Material	Structure of High-Pressure Modification	Critical Pressure in Kilobars
Ge[20]	β-Sn	120
Si[20]	β-Sn	160–200
InSb[21–24]	β-Sn-type	23
GaSb[23]	β-Sn-type	90
AlSb[23]	β-Sn-type	125
InAs[23]	β-Sn-type	100
GaAs[25]	(not known)	245
InP[23]	NaCl (B1)	133
HgTe[26]	Cinnabar (B9)	20
HgSe[26]	Cinnabar (B9)	15
CdTe[26,27]	NaCl (B1); β-Sn-type	36; 90
CdSe[26]	NaCl (B1)	32
CdS[26,28]	NaCl (B1)	33
ZnTe[25]	(not known)	140
ZnS[25]	(not known)	245
ZnO[29]	NaCl (B1)	100
AgI[30]	NaCl (B1)	3.5

Anomalies in the cleavage habit of HgSe, HgTe, CdTe, Si, Ge, and of α-Sn have been interpreted in this way.[12] In these cases new PBC vectors are being formed at the surface which had not been present in the ideal surface.[18,19] These deviations from the ideal surface configuration have also been found in Si, Ge,[32,33] and CdS[34] by low-energy electron diffraction and atom ejection experiments.[35] In addition, differences between the surface structure of diamond, on the one hand, and of silicon and germanium, on the other, have been shown by habit studies[12] and slow-electron diffraction studies.[36] These results have been further substantiated by the dual behavior of the crystal surface of silicon carbide and its polymorphic modifications.[10]

Morphological studies of vapor-phase growth and solution forms of group IV elements have revealed that {111} and {001} are the stable planes in diamond, whereas {111}, {001}, and {113} appear on Si, Ge, and α-Sn (Fig. 6). Theoretical equilibrium form calculations taking into account first and second nearest-neighbor interactions account for the appearance of {111} and {001}, respectively, in the diamond structure; but the appearance of {113} could be accounted for

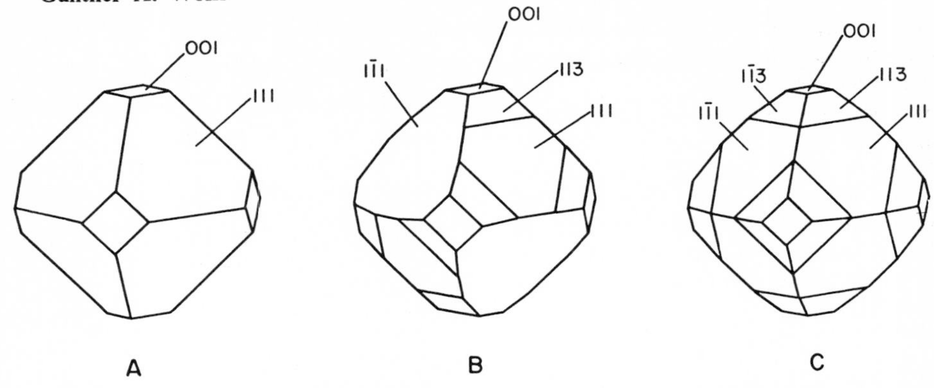

Fig. 6. Common habits of (*A*) diamond {111}, {001}; (*B*) β-silicon carbide {111}, {1$\bar{1}$1}, {001}, {113}; (*C*) silicon, germanium, gray tin {111}, {001}, {113}. Note that in β-silicon carbide {1$\bar{1}$3}, i.e., the faces of the negative trigonal tristetrahedron, are not observed.

only after a surface deformation was assumed in which surface atoms moved from tetrahedral into energetically more favorable positions, presumably of the octahedral-like type. In like manner the morphology of β-SiC of cubic sphalerite structure reported by N. W. Thibault[37] reveals the presence of {113} but not of {$\bar{1}\bar{1}\bar{3}$} (Fig. 6). A study of this structure reveals that the two antipodal planes {113} and {$\bar{1}\bar{1}\bar{3}$} can be composed entirely of atoms of the opposite type, that is, Si and C or vice versa. The existence of {113} can therefore be attributed to the rearrangement of one type of atom (as similarly observed in Si), whereas the absence of {$\bar{1}\bar{1}\bar{3}$} can be related to the absence of such surface deformation as in diamond. This type of study has been extended to all polytypes of SiC in which good agreement has been found between derived and observed forms (Table 2; Fig. 7).

2.4 The Influence of Adsorption on Surface Structure

The change of crystal morphology by adsorption has been the subject of many investigations in the past (e.g., Buckley[38]). In a more recent investigation on gaseous etching the habit-changing effects of gaseous oxygen, sulfur, chlorine, bromine, and iodine on silicon single crystals have been studied.[9] The adsorption of oxygen and sulfur produces a cuboctahedral crystal morphology by stabilizing the "ideal" surface configuration of tetrahedral coordination in the {001} and related planes containing silicon atoms with two "dangling" bonds, whereas chlorine favors the simple octahedral morphology of the planes containing only atoms with single "dangling" bonds (e.g., {111}). The same study shows that when chlorine is replaced by bromine for steric reasons the latter planes are less favored and that when iodine is substituted the octahedral habit will have been completely suppressed (Fig. 8). Thus the proper application of the morphological theory can supply information about the

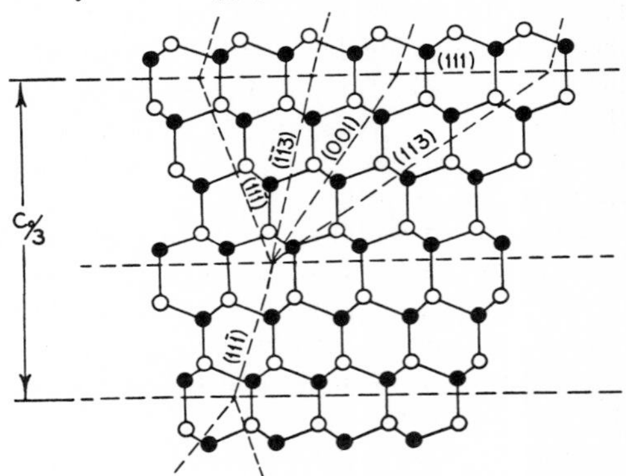

Fig. 7. Geometrical traces of observed and derived planes in 15R SiC polytype (32 32 32 = $(32)_3$ in Zhdanov notation). The 32 stacking is obtained by a 180° rotation about the ⟨111⟩ axis ("twinning") of every three and two double layers of β-SiC (of sphalerite structure), respectively. The planes would be composed of sections of {111}, {001}, and {113} planes as follows, with the first plane index referring to the three-layer stack. (10.1): (11$\bar{1}$)–(11$\bar{1}$); (01.8): (001)–(11$\bar{1}$); (01.5): ($\bar{1}\bar{1}$3)–(11$\bar{1}$); (01.14): (113)–(11$\bar{1}$). The combinations (01.5) and (01.14) form as a result of surface deformation. The actual surface configuration also depends on the ionicity of the surface atoms and on the energy ratio of *c*-directional to *a*-directional bonding in the crystal.

TABLE 2

A Comparison Between Observed and Calculated Planes for SiC Polytypes*

	a	*a*	*b*	*a*	*c*	*a*	*b*	*a*	*d*	*b*	*b*	*b*
	4H	6H	8H	15R	19H	21R	27R	33R	51Ra	75R	84R	87R
0001	2	67	2	22	2	2	2	1		2	2	2
$01 \cdot \frac{1}{3}(2M + N)$	11	56		12	1	4		5	5	3	1	2
$10 \cdot \frac{1}{3}(M + 2N)$	—	—	—	13	2	3	1	6	5	1	2	4
$10 \cdot \frac{1}{3}(M - N)$	5	26		9		3	1	3	5		2	4
$01 \cdot \frac{1}{3}(M/2 + N) + m_0/2$	8	34	1	6	2	1		4	5			2
$10 \cdot \frac{1}{3}(M + N/2) + n_0/2$	—	—	—	7	1	2	1	5	5	3		3
$10 \cdot \frac{1}{3}(M + 5N)$				1								
$01 \cdot \frac{1}{3}(5M + N)$	—	—	—			1						
$10 \cdot \frac{1}{3}(M + 5N) - p$	5	23		2†		1			2			
$01 \cdot \frac{1}{3}(5M + N) - p$	—	—	—	1	3				5			
$10 \cdot \frac{1}{3}(M + 2N) + p$		17	3	2†		1		2	4	1		
$01 \cdot \frac{1}{3}(2M + N) + p$	—	—	—	5		1		2	4	1	1	
$11 \cdot (M + N)$	5	5		5								

* The table includes only those polytypes for which morphology data were available. The numerical values represent the number of times the respective planes were observed. Lower-case letters denote subreferences to Ref. 37. M and N refer to the sum $\Sigma_i\, mi$ and $\Sigma_i\, ni$ of the first and second Zhdanov notations of all $m_1\, n_1$, $m_2\, n_2$ polytypes, respectively. In the 15R polytype, for example, the Zhdanov notation is 32 32 32 = $(32)_3$ with $M = 9$, $N = 6$, $m_0 = 3$, $n_0 = 0$, and $p = 3$; here m_0 is the number of odd m's appearing in Zhdanov's notation; n_0 is the number of odd n's appearing in Zhdanov's notation; p is the number of mn combinations in Zhdanov's notation. All but the first four types of planes in the first four lines appear on the equilibrium form as a result of surface deformation; they are of the "{113} type" of the sphalerite modification of SiC (= 3R or 30 in Zhdanov notation).

† Identical planes ({10.10} for 15R).

actual surface structure and the bonding arrangement.

Another example is etching of germanium by gaseous $GeCl_4$ and Cl_2. Here the markedly different crystallographic preference of both etchants likewise reveals a different chemical mechanism.

3. METHOD OF ANALYSIS

Information on the bonding and structure in crystal surfaces can be obtained by the use of micro- and macromorphology studies of grown, etched, and fractured (cleaved) crystals. Micromorphology, microhabit in particular, is easily obtained and can be investigated by the light reflection method. The cleavage habit, especially the microcleavage habit, is extremely useful in this respect, for it does not depend on a matrix or an etchant as do the growth and etching habits.[13] (Microcleavage is observed when crystals are mechanically etched, e.g., by abrasion in a wet slurry of coarse SiC powder.)

An idealized ("normalized") crystal habit is represented by a polyhedron. It can be shown that the edges of such a polyhedron represent atomic bond chains or arrays of atoms which are held together by strong bonds. Accordingly, the edges of a polyhedron corresponding to a naturally occurring or experimentally obtained crystal habit represent atomic bond chains which are present in the surfaces of such crystals. A comparison of experimental and theoretical crystal habit, therefore, indicates whether the atomic bond chains at the surface are identical or different from the ones which are present in the bulk. The existence or nonexistence of such atomic bond arrays in both theoretical and experimental crystal forms then leads to direct conclusions regarding the nature of the bonding in the crystal under study.

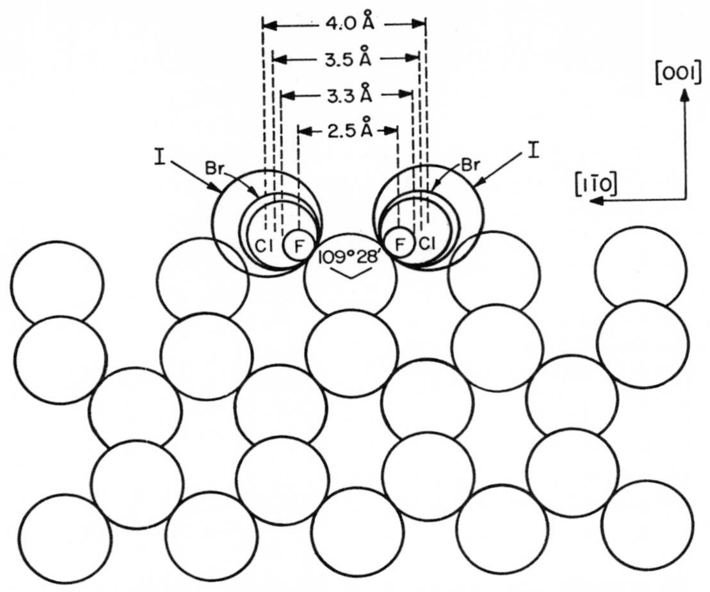

Fig. 8. A schematic picture of silicon in [110] projection showing relative positions and sizes of ambient atoms on {001}. Steric hindrance of the iodine in the preformation of the SiI_2 on {001} provides for a slow attack of iodine on this plane. This process precedes the evaporation of the silicon diiodide.

3.1 Results of Analysis

1. *The theoretical and experimental habits coincide.*

In this case the atomic structure of a crystal surface is most likely identical to its bulk structure. As mentioned before, in a crystal polyhedron adjacent crystal planes are in contact with one another at crystal edges or crystal corners. If one cannot move completely around the whole crystal from plane to plane via *the adjoining edges* (which are all parallel to one another) within a crystal zone, but rather has to move from plane to plane *via adjacent corners*, as, for example, in an octahedron, the atomic bond chains are not straight bond arrays. They are sawtooth or zigzag, as in diamond.

In some structures with this type of atomic bonding array, the crystal habit changes abruptly at certain critical bond strength ratios. In a comparison of experimental and theoretical habits[39] the ratio of bond strengths of, for instance, the nearest- to next-nearest-neighbor interactions can be estimated, as has been done for Se (Fig. 9), Te, As, Sb, Bi,[18] FeS_2 (pyrite, marcasite structure[40]), and many other materials.

2. *The theoretical and experimental habits differ.*

(*a*) If the cleavage or microcleavage habit differs from the theoretical habit, as, for example, in Si, Ge, and other materials, the surface structure differs from the bulk structure. In this case certain surface atoms have repositioned themselves with respect to normal bulk positions, the surface atoms moving to positions of lower energy. These changes can be readily pinpointed.[12,13]

(*b*) If the cleavage or microcleavage does not differ from the theoretical habit but the etching habit does, new atomic bond chains are formed with the help and participation of the adsorbed atoms or molecules. Etching of Si in oxygen gas at elevated temperature is an example.[9]

If, in the calculation of the theoretical habit in compounds, different surface energy contributions are assigned to the various atoms, then different crystal forms result for the various surface energy contribution ratios. From the experimental habit, therefore, conclusions can be drawn regarding the bonding state of the different surface atoms; this is especially applicable to crystals with no

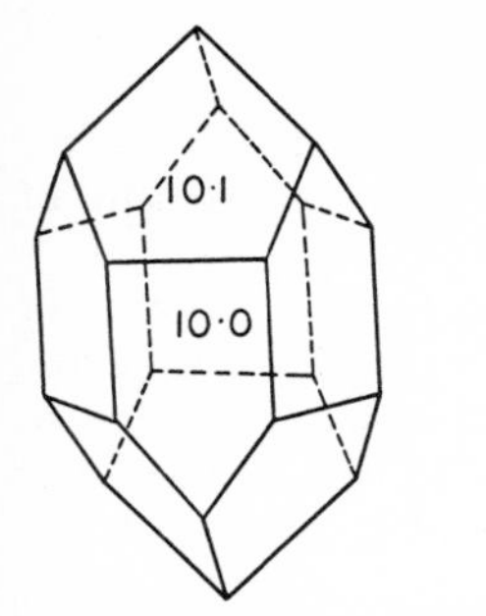

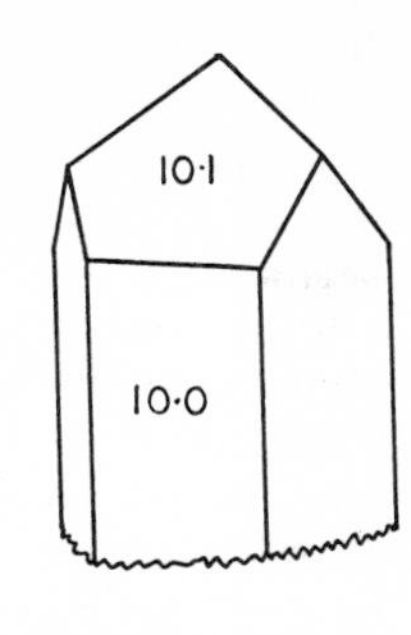

Fig. 9. Calculated and observed crystal form of selenium.[29,39] The former is derived with $\varphi_1 > 2\varphi_2$ where φ_1 and φ_2 represent, respectively, the bond energy within the selenium spiral chains and between these chains.

center of symmetry. In ZnS the Zn atoms contribute at least three times as much to the surface energy as do the S atoms if the observed habit is compared with the theoretically derived crystal morphology.[19] In this way, by the comparison of experimental and theoretical crystal forms—bond energy ratio, changes in the surface structure, the mechanism of certain reactions at the surface, and the properties of covalent surface bonding—have been investigated.[18,19]

3.2 Evaluation of Covalent Surface Bonding

The manner of evaluation may be presented as follows; diamond, silicon, and germanium are used as examples. The cleavage habit of diamond is the octahedron (Fig. 10*a*), as would be expected for an ideal surface structure. The cleavage habit of silicon, germanium, and gray tin, on the other hand, is related to the cuboctahedron of Fig. 10*b*; habit theory explains this change in surface morphology as a rearrangement of atoms within a ⟨$\bar{1}$01⟩ atomic bonding array in such a way that the intrinsic coherence (or "bond chain character") of the array within the crystal surface disappears in all planes between {111} and {110}, (i.e., in {111}, {332}, {221}, {331}), whereas the coherence of the same bonding array is established in the remaining planes of the zone between {111} and {001} (i.e., in {111}, {223}, {112}, {113}).

This can be achieved in two ways, as demonstrated on {001}.

1. *Rearrangement of the surface atoms in the manner shown in Figs. 11a and 11b.* This type of surface structure has also been proposed by R. E. Schlier and H. E. Farnsworth[41] for silicon and germanium based on low-energy electron diffraction data. (See, however, M. Green and R. Seiwatz[42] and D. Haneman.[43])

In this arrangement a fair approximation is to assume that only nearest-neighbor interactions and the energy to deform the bond angle determine stability. The bond energy can be derived from the heat of evaporation, whereas an estimate of the bond angle deformation in diamond can be obtained by interpolation of the respective angular deformation of the cycloalkanes.

A reduction of the surface energy of the deformed {001} plane by 20 to 30 per cent would thus result when compared with the ideal {001} surface. This value differs from results obtained by M. Green and R. Seiwatz[42] who find no reduction of surface energy for {001} in diamond. Although the bond angle deformation energy should be smaller in silicon, germanium, and gray tin, it is felt that in these elements the increased stability resulting from this structural deformation is not sufficient to explain the observed magnitude of the cleavage deviation from the ideal behavior.[12] It is therefore concluded that the actual arrangement is:

2. *The alternate mode of surface deformation on {001}, as shown in Fig. 11c.* In this arrangement the surface atoms would have moved by approximately [$\frac{1}{4}\,\frac{1}{4}\,\frac{1}{4}$] from the ideal tetrahedral s^2p^6 positions to interstitial positions of nearly octahedral p^6 coordination (with some small amount of admixed bonding of other types).[44] In Fig. 11*c* these new positions are marked by the open broken

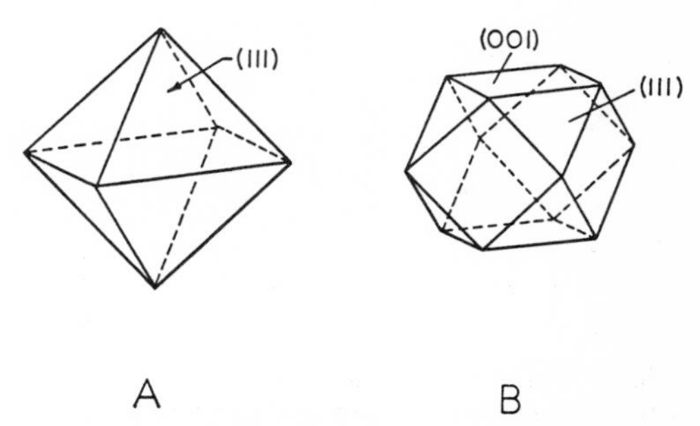

Fig. 10. Idealized cleavage habit for diamond-structure materials: (*a*) diamond, arsenolite (As_4O_6), senarmontite (Sb_4O_6), and (*b*) silicon, germanium, gray tin.

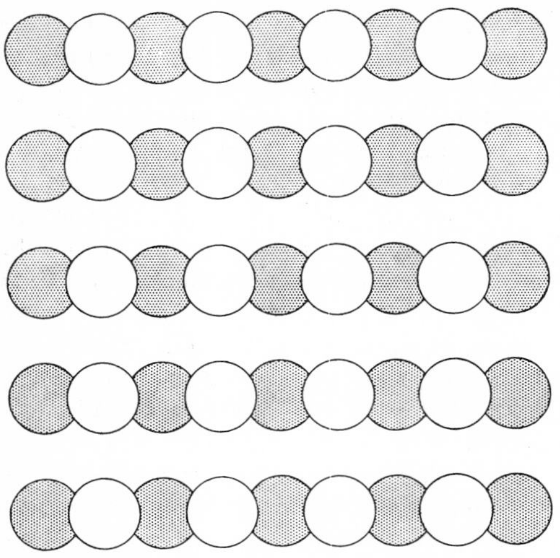

Fig. 11a. Ideal arrangement of atoms in {001} of diamond-structure elements. The open and shaded circles represent the atoms of the upper and lower layers, respectively.

circled positions. Most of the reasons for preference for the assumed octahedral p^6 coordination have been presented in a preceding paragraph. One might also consider this arrangement as related to a two-dimensional simple cubic structure. The related B1 (NaCl) structure ⟨001⟩ bond coherence was

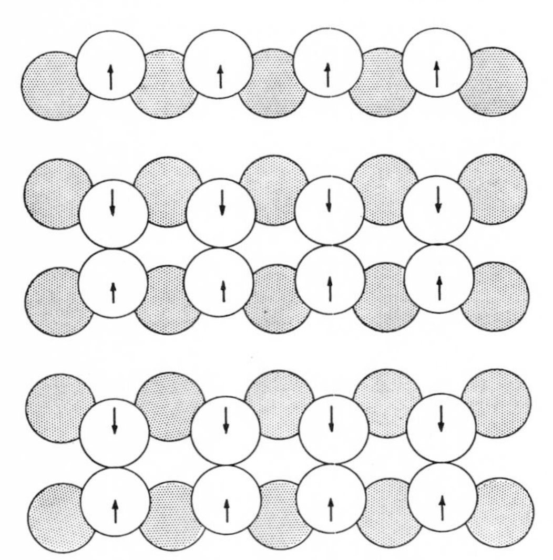

Fig. 11b. The arrangement of the atoms in {001} after the surface structure change when the atoms approach one another (see arrows) and thus partially saturate "dangling bonds." This arrangement has been suggested by R. E. Schlier and H. E. Farnsworth,[41] and M. Green and R. Seiwatz.[42]

also identified by cleavage studies in {001} of HgSe, HgTe, and CdTe.

Two reasons account for the double surface spacing of the open circle arrangement of Fig. 11*c*:

(*a*) According to the general concepts of chemical bonding, the surface atoms of the uppermost layer should most likely bond to four nearest neighbors of the atomic layer below. Since the atoms of this second layer, however, are bonded to two neighbors of the following third layer below and to two atom neighbors of the uppermost first layer (of double spacing), the atoms of this layer must occupy only half of its positions. Of all other

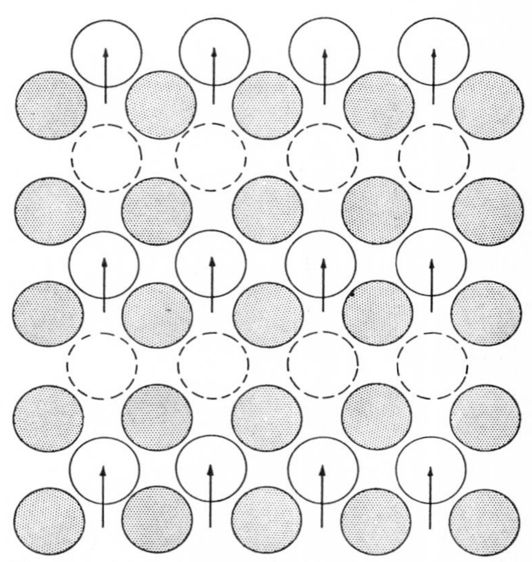

Fig. 11c. The arrangement of atoms in {001} after the surface structure change as described in the text.

structural arrangements which one could possibly expect for this uppermost surface layer, for example, checkerboard type and the like, the alternate occupancy of [110] atomic structure arrays obviously introduces the least stress to the underlying crystal. This arrangement is therefore the surface structure closest to the most preferable arrangement.

(*b*) In silicon, germanium, and gray tin the {001} plane disappears on formation of the {113} plane in the vapor- and gas-phase matrix. In the diamond structure {113} is the only plane formed from single (i.e., unpaired) [1$\bar{1}$0] rows, as are also present in {001}. However, neither pairs of rows nor the arrangement shown in Fig. 11*b* and described in Section 3.2 is possible. Corresponding arguments also hold for silicon carbide of all polytypes.

4. DERIVATION OF THE IDEALIZED EQUILIBRIUM FORM

The theoretical cleavage habit can be derived in the following way[45]: a model of a crystal structure, for example, of graphite structure (Fig. 12), is separated or "cleaved" into two parts parallel to a low index plane (e.g., (10.0) for graphite); the numbers of nearest, next-nearest, and more distant neighbors which are separated in the process are then counted per unit mesh area of the plane $\{hkl\}$:

$$A_{hkl} = ac\left(h^2 + k^2 + hk + \tfrac{3}{4}\frac{a^2}{c^2}l^2\right)^{1/2} \qquad (3)$$

Thus $A_{10.0} = ac$.

If there are several ways of separating the crystal parallel to the selected plane, only the value for which the sum $\Sigma_i\, n_i\varphi_i = 2\sigma_{hkl}$ is a minimum is selected. Here n_1, n_2, n_3 and φ_1, φ_2, φ_3 represent the number of nearest, next-nearest, third-nearest neighbors, etc., and their respective bond energies; σ_{hkl} is the specific surface free energy for which in a fair approximation the specific surface energy can be taken. The proper selection of neighbors is determined by the bond energy ratios φ_2/φ_1, φ_3/φ_1, etc., where for covalent, metallic, and van der Waals bonding $1 > \varphi_2/\varphi_1 > \varphi_3/\varphi_1$, etc. Then two planes are selected for each of two zones which have the (10.0) plane in common. These are (10.1), (00.1) in the [010] zone and (11.0), (01.0) in the [001] zone. Their σ_{hkl} values are determined when, because of the hexagonal symmetry, $\sigma_{01.0} = \sigma_{10.0}$.

Considering the bond energies between nearest and next-nearest neighbors only (namely, φ_1 and φ_2), the σ-values of $\{10.0\}$, $\{10.1\}$, $\{00.1\}$, and $\{11.0\}$ are $2\varphi_1/ac$, $(2\varphi_1 + \varphi_2)/a(c^2 + \tfrac{3}{4}a^2)^{1/2}$, $2\varphi_2/a^2\sqrt{3}$, and $4\varphi_1/ac\sqrt{3}$, respectively. The resulting specific surface energy equation for the planes $h, k, l \geq 0$ is

$$\sigma_{hkl} = \frac{2(h + k)\,\varphi_1 + l\varphi_2}{ac\left(h^2 + k^2 + hk + \tfrac{3}{4}\frac{a^2}{c^2}l^2\right)^{1/2}} \qquad (4)$$

for first- and second-nearest-neighbor interaction. To verify this result the third boundary zone [100] of this spherical polygon has to be checked by comparing $\sigma_{01.1}$ $(= \sigma_{10.1})$ with the formula obtained.

The σ-values for φ_1, and, if necessary, also those for φ_2 and φ_3, are then drawn in spherical coordinates and these in turn are intercepted by the corresponding (hkl) planes normal to these radial σ_{hkl} vectors. The resulting Wulff's construction for the graphite structure is shown in Fig. 1 and its stereographic presentation is given in Fig. 12*b*.

4.1 Crystal Faceting

In the derivation of the Wulff's plot (σ-plot; γ-plot after C. Herring[4]) it is most important to consider only the lowest energy surface configurations in the calculation of the (0°K) σ-values for the various crystal planes. In this case Wulff's plot is always represented by a number of spheres, each of which passes through the crystal origin (pole of the sphere) and through an apex of the crystal equilibrium form (antipole). Generally the derivation of this plot does not pose a problem for structures of exclusively attractive bonding forces,

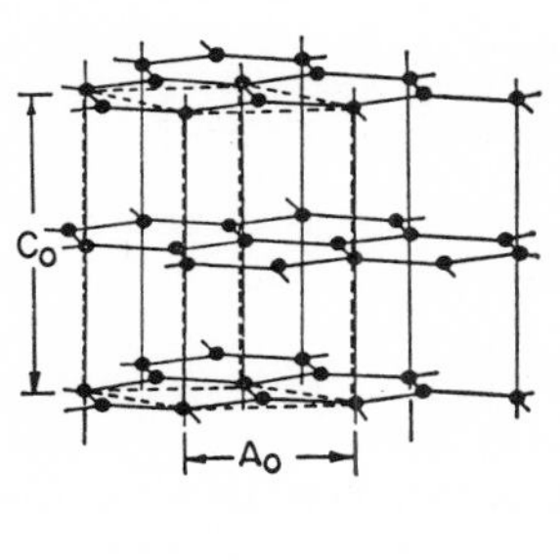
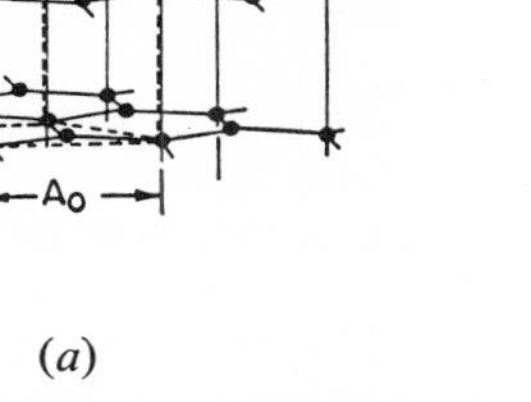

(*a*)

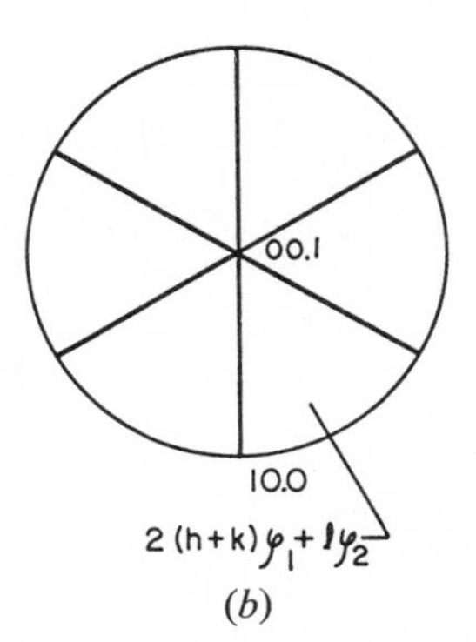

(*b*)

Fig. 12. (*a*) Lattice of normal graphite. (*b*) Stereographic presentation of calculated crystal equilibrium form shown in Fig. 1.

that is of covalent, metallic, and van der Waals bonding. Its derivation for ionic structures may, however, lead to difficulties.[46] The specific surface energy σ_{011} value of an atomically flat {011} plane of NaCl structure[47,48] is much greater than the value $\sigma_{011}\sqrt{2}$ corresponding to a configuration which can best be described as a {011} plane made up of a parallel array of extended rectangular grooves bounded by {001} facets (Fig. 5*a*). If, instead of the correct value $\sigma_{0kl} = \sigma_{001}(k^2 + l^2)^{1/2}$ for (011) and the other (0*kl*) planes of the [100] zone, the incorrect σ_{0kl} value corresponding to the atomically flat planes is taken, the resulting sphere of Wulff's plot no longer passes through either the crystal origin or the apex of the equilibrium form (Fig. 5*b*).

An originally smooth {011} plane would therefore exhibit faceting.[47] The driving force for the faceting process is apparently the repulsive Coulomb interaction φ_2 between the faceted ("grooved") {011} sections in the [10$\bar{1}$] direction. The conditions for faceting have been extensively studied by C. Herring,[4,49] W. W. Mullins,[50] and others. C. Herring states that faceting is expected when the γ-plot of the actual surface free energy at any point passes through the crystal origin and is tangent to the γ-plot at a given orientation. The use of the corresponding plot in spherical coordinates of $1/\sigma$ $(= 1/\gamma)$ instead of the σ-plot $(= \gamma$-plot) makes this test even simpler[51,52]. In this case the tangent sphere passing through the origin inverts into a tangent plane. Wherever the $1/\sigma$ surface is concave, a series of touching "tangent" planes can be made to connect or bridge the adjacent convex sections. This is the criterion that faceting is possible for the orientations bounded by these spherical (coordinate) sections.

Similarly, portions of the crystal surface which are expected to be rounded, are easily recognized by an inspection of this plot. They are represented by sections of the $1/\sigma$ plot which are convexly rounded and cannot be connected by the "tangent" planes. The rounding of a crystal surface would most likely take place at higher temperatures where the surface entropy contribution would affect the *S*- and *K*-planes more than the more densely packed *F*-planes. This would in turn result in a flattening of the spheres in the σ-plot. This satisfies the condition[53]

$$\frac{1}{\sigma}\frac{\partial^2\sigma}{\partial\alpha^2} > -1 \tag{5}$$

A polyhedral $1/\sigma$-plot corresponds to an ideal Wulff's plot constructed of spheres passing through origin and the apices of the inscribed equilibrium form. It should be pointed out, however, that in this case, faceting is also expected for all planes which do not correspond to the cusp planes (*F*-planes) in Wulff's plot, or to the apices planes in the $1/\sigma$-plot. These planes are *K*- or *S*-planes. For these planes there is no, or a very low, nucleation energy barrier; the ever-present fluctuations in matrix concentration and statistical supersaturations should therefore suffice to produce faceting even under equilibrium conditions. Gruber's condition for faceting[53]

$$\frac{1}{\sigma}\frac{\partial^2\sigma}{\partial\alpha^2} < -1 \tag{6}$$

should thus be extended to

$$\frac{1}{\sigma}\frac{\partial^2\sigma}{\partial\alpha^2} \leq -1 \tag{7}$$

ACKNOWLEDGMENTS

This work was supported in part by the Advanced Research Projects Agency under Contract No. AF49(638)-1246 and ARPA Order No. 444.

REFERENCES

1. Hauy R. J., *Essai d'une theorie de la structure des cristaux*. Paris, 1784.
2. Gibbs J. W., *Collected Works*. Longmans, Green & Co., New York 1928.
3. Wulff G., *Z. Krist.* **34** (1901) 449.
4. Herring C., *Phys. Rev.* **75** (1950) 344.
5. Fullman R. L., *Acta Met.* **5** (1957) 638.
6. Stranski I. N., *Z. Krist.* **105** (1943) 91.
7. Hartman P., and W. G. Perdok, *Acta Cryst.* **8** (1955) 49, 521.

8. Volmer M., *Kinetik der Phasenbildung*. Steinkopff, Leipzig, 1939, p. 108.
9. Gualtieri J. G., M. J. Katz, and G. A. Wolff, *Z. Krist.* **114** (1960) 9.
10. Wolff G. A., and J. R. Hietanen, *Proc. Intern. Symp. on Condensation and Evaporation of Solids*, Dayton, 1962. Gordon & Breach, 1964, p. 451.
11. Wolff G. A., and J. R. Hietanen, *J. Electrochem. Soc.* **111** (1964) 22.
12. Wolff G. A., and J. D. Broder, *Acta Cryst.* **10** (1957) 848; **12** (1959) 313.
13. Wolff G. A., and J. D. Broder, *Am. Miner.* **45** (1960) 1230.
14. Bravais A., *Etudes Cristallographiques*. Gauthier-Villars, Paris, 1865.
15. Donnay J. D. H., and D. Harker, *Am. Miner.* **22** (1937) 446.
16. Kossel W., *Leipz. Vortr.*, **1** (1928) 1.
17. Stranski I. N., and R. Suhrmann, *Z. Krist.* (*A*) **105** (1943) 481.
18. Wolff G. A., and J. G. Gualtieri, *Am. Miner.* **47** (1962) 562.
19. Wolff G. A., *Z. Phys. Chem. N.F.* **31** (1962) 1.
20. Jamieson J. C., *Science* **139** (1963) 762.
21. Smith P. L., and J. E. Martin, *Nature* **196** (1962) 762.
22. Banus M. D., R. E. Hanneman, A. N. Mariano, E. P. Warekois, H. C. Gatos, and J. A. Kafalas, *Appl. Phys. Letters* **2** (1963) 35.
23. Jamieson J. C., *Science* **139** (1963) 845.
24. Kasper J. S., and H. Brandhorst, *General Electric Report* 64-RL-3608MC (1964).
25. Minomura S., G. A. Samara, and H. G. Drickamer, *J. Appl. Phys.* **33** (1962) 3196.
26. Mariano A. N., and E. P. Warekois, *Science* **142** (1963) 672.
27. Owen N. B., P. L. Smith, J. E. Martin, and A. J. Wright, *J. Phys. Chem. Solids* **24** (1963) 1519.
28. Corrl J. A., *J. Appl. Phys.* **35** (1964) 3032.
29. Bates C. H., W. B. White, and R. Roy, *Science* **137** (1962) 3534.
30. Jacobs R. B., *Phys. Rev.* **54** (1938) 325.
31. Krebs H., M. Hauke, and H. Weyand, *The Physical Chemistry of Metallic Solutions and Intermetallic Compounds*. New York Chemical Publishing Company, 1960, Vol. II, p. 41.
32. Farnsworth H. E., R. E. Schlier, T. H. George, and R. M. Burger, *J. Appl. Phys.* **29** (1958) 1150.
33. Lander J. J., and J. Morrison, *J. Chem. Phys.* **37** (1962) 729.
34. Farnsworth H. E., private communication.
35. Anderson G. S., and G. K. Wehner, *J. Appl. Phys.* **12** (1960) 2305.
36. Farnsworth H. E., J. B. Marsh, and J. Toots, *Proc. Intern. Conf. on the Physics of Semiconductors*, Exeter, 1962, p. 836.
37. (*a*) Thibault N. W., *Am. Miner.* **29** (1944) 249. (*b*) Ramsdell L. S., and J. A. Kohn, *Acta Cryst.* **5** (1952) 215. (*c*) Ramsdell L. S., and R. S. Mitchell, *Am. Miner.* **38** (1953) 56. (*d*) Thibault N. W., *Am. Miner.* **33** (1948) 588.
38. Buckley H. E., *Crystal Growth*. Wiley, New York, 1958, pp. 339, 529.
39. Palache C., H. Berman, and C. Frondel, *The System of Mineralogy*. Wiley, New York, 1963.
40. Broder J. D., and G. A. Wolff, *Proc. Intern. Conf. on Adsorption and Crystal Growth*, Nancy, France (1965) 171.
41. Schlier R. E., and H. E. Farnsworth, *Semiconductor Surface Physics*. Univ. of Pennsylvania Press, Philadelphia, 1956, p. 114.
42. Green M., and R. Seiwatz, *J. Chem. Phys.* **37** (1962) 458.
43. Haneman D., *Proc. Intern. Conf. on the Physics of Semiconductors*, Exeter 1962, p. 842.
44. Pearson W. B., and G. A. Wolff, *Discussions Faraday Soc.* **28** (1959) 142.
45. Wolff G. A., in *Compound Semiconductors*, Vol. 1, *Preparation of III–V Compounds*, R. K. Willardson and H. L. Goering, eds. Reinhold, New York, 1962, p. 34.
46. Yamada M., *Phys. Z.* **24** (1924) 364; also cited in F. Seitz, *Modern Theory of Solids*, McGraw-Hill, New York, 1940, p. 97.
47. Stranski I. N., *Z. Phys. Chem.* (*A*) **136** (1928) 259.
48. Honigmann B., *Gleichgewichts- und Wachstumsformen von Kristallen*, Steinkopff, Darmstadt, 1958, p. 123.
49. Stranski I. N., *Naturwiss.* **30** (1942) 425.
50. Mullins W. W., *Phil. Mag.* **6** (1961) 1313.
51. Frank F. C., in *Growth and Perfection of Crystals*. R. H. Doremus, B. W. Roberts, and D. Turnbull, eds., Wiley, New York, 1958, p. 3.
52. Meijering J. L., *Acta Met.* **11** (1963) 847.
53. Gruber E., private communication, cited in W. M. Robertson, *Acta Met.* **12** (1964) 241.

chapter 6

Bonding as Interpreted by Spatial Correlation of Electrons

K. Schubert

K. SCHUBERT

Max Planck Institut für Metallforschung
Stuttgart, Germany

1. INTRODUCTION

Since it was first realized that all materials are composed of comparatively few chemical elements, an important source of information on chemical bonding has been the constitution of simple mixtures of elements. Today the term "constitution," as applied to metallic systems, is understood to mean not only the relative amount of the components in an alloy and the statements of the phase diagram but also the positions of the component atoms in the crystal structure. From compilation of constitutional data as well as from other sources (see Chapters 3, 4, 5, and 7), chemists, crystallographers, and physicists have evolved over the the last one hundred years phenomenological theories of bonding from which in turn many concepts have been adapted into the quantum-mechanical bonding theory (see Chapter 2). Because of the computational difficulties of that theory, much experimental work is still being done[1] to improve our empirical knowledge, and such work will certainly be needed for some time in the future. In the hands of materials scientists and engineers the phenomenological bonding theories are therefore a very useful method to help them assess alloying behavior and crystal structure from the properties of the component elements. Additional theories relating structure and bonding with physical properties also help the materials experts in their search for materials with desired properties.

How does constitutional research give information on bonding?

1. The wealth of experimental information on constitution and structure is systematized.
2. From the experimental material, rules are inferred for the occurrence of compounds and for their crystal structure. It can be assumed that for each type of structure, one or several rules exist.
3. Physical reasons and interpretations of the rules and relations are sought. Partial interpretations of structures can be called "structural arguments." The fact of relative

uniformity in structural constitution is an indication of comparative simplicity of the structural arguments.

In this chapter we consider empirical rules and arguments arising from constitutional research. First some well-known viewpoints for assessing the bonding are mentioned and afterward a method of expressing bonding which permits discussion of further features of constitutional observations is described.

2. BASIC CONCEPTS

2.1 Homology and Electron Counts

The periodic system of chemical elements gives interrelations which are a valuable means of orientation in constitutional research. The most important relation is *homology*, that is, the membership of elements in the same column of the periodic system. It is useful to give the columns and the lines the names in Fig. 1. Quasihomology denotes membership in neighboring columns. Not only may elements be homologous, but also alloy systems and compounds. Following the quantum-mechanical theory, the periodic system is connected with the electron occupation of the quantum shells of the various atoms; thus relationships expressing homology may be formulated in terms of the number of electrons. Only the outer electrons of the atoms are of importance in these electron counts, for the inner electrons only weakly influence the chemical bonding. The *outer electrons* are understood to comprise all electrons outside the highest completed noble gas shell. For the transition or T elements the outer electrons are distributed on two different shells, the *outer core electrons* (mostly *d*-electrons) and the *valence electrons* (mostly *s*- and *p*-electrons). Frequently in the chemistry of metal compounds only the average number of valence (or outer) electrons per atom (*valence electron concentration, VEC*) is of importance to the bonding. Some rules for the appearance of phases (or structures) connected with VEC are the following:

1. If the VEC and composition (atomic per cent) of two phases are the same, they are often isotypical (isostructural) or homotypical (have very closely related structures).
2. If VEC does (not) change rapidly with composition, small (broad) ranges of homogeneity of intermetallic phases are found.
3. Isotypy and quasihomology are necessary conditions for continuous solid solubility.
4. Many special rules for the appearance of special structures exist (e.g., those of Hume-Rothery and Westgren for the brass structures or of Grimm and Sommerfeld for diamond structures; for these rules refer to Section 3.
5. Some electron compounds obey the rules of inorganic chemistry (e.g., the appearance of the NaCl structure in AB phases in which the noble gas shell of the B component is completed). Metallic phases in which this octet rule applies are often called Zintl phases, for Zintl first pointed out that B^n elements with $n > 3$ but not those with $n \leq 3$ can be completed to anions with a noble gas shell.
6. The different kinds of bonding are connected with VEC. Following some historical theories of bonding we distinguish ionic bonds, covalent bonds, molecular bonds, and metallic bonds (Chapter 2), all of which may be roughly connected with intervals of VEC: VEC $= 0 \cdots 3$, metallic bond; VEC $= 3 \cdots 5$, ionic and covalent bond; VEC $= 5 \cdots 8$, molecular bond.

2.2 Size and Microelasticity

The ample results in the literature, both of phase-diagram investigations and of structure determinations, show that alloying behavior (compound formation) is also influenced by the size of atoms. From the lattice constants of the elements, together with the lattice constants of certain solid solutions, a system of metallic radii for 12 coordination was derived (see Fig. 1). Although the size of an atom is only a phenomenological concept and has no strict meaning, it is very useful in the determination and systematization of crystal structures. Sometimes it is more convenient to use atomic volumes instead of atomic radii or distances (which require a knowledge of crystal structure). Typical rules concerning atomic radii are the following:

1. In an alloy system with continuous miscibility the lattice constant is an approximately linear function of composition (Vegard's rule). If the function is not linear, it is generally smoothly curved.
2. In substitutional solid solutions which are quasihomologous with Cu(Zn) the limiting

A^1	A^2	T^3	T^4	T^5	T^6	T^7	T^8	T^9	T^{10}	B^1	B^2	B^3	B^4	B^5	B^6	B^7	B^8	
																1 **H** 1.008 0.78	2 **He** 4.003 H^2	*K*
3 **Li** 6.940 H^2, B^1 1.57											4 **Be** 9.02 H^2 1.13	5 **B** 10.82 R^{12}, T^{50}, R^{108} 0.92	6 **C** 12.01 $(F^2), H^4_a$ 0.86	7 **N** 14.01 H^4_b 0.80	8 **O** 16.0000	9 **F** 19.00	10 **Ne** 20.18 F^1	*L*
11 **Na** 22.99 H^2, B^1 1.92											12 **Mg** 24.32 H^2 1.60	13 **Al** 26.98 F^1 1.43	14 **Si** 28.09 F^2 1.34	15 **P** 30.98 Q^4_c 1.3	16 **S** 32.06 S^{32}, u. a. 1.3	17 **Cl** 35.46 Q^4_b	18 **Ar** 39.94 F^1	*M*
19 **K** 39.10 B^1 2.36	20 **Ca** 40.08 F^1, B^1 1.97	21 **Sc** 44.96 $(F^1), H^2$ 1.65	22 **Ti** 47.90 H^2, B^1 1.45	23 **V** 50.95 B^1 1.36	24 **Cr** 52.01 B^1, F^1 1.28	25 **Mn** 54.94 B^{29}, C^{20}, F^1, B^1 1.31	26 **Fe** 55.85 B^1, F^1, B^1 1.27	27 **Co** 58.94 H^2, F^1 1.26	28 **Ni** 58.71 F^1 1.24	29 **Cu** 63.54 F^1 1.28	30 **Zn** 65.38 H^2 1.37	31 **Ga** 69.72 Q^4_a 1.39	32 **Ge** 72.60 F^2 1.39	33 **As** 74.92 R^2 1.48	34 **Se** 78.96 H^3 1.6	35 **Br** 79.92 Q^4_b	36 **Kr** 83.8 F^1	*N*
37 **Rb** 85.48 B^1 2.53	38 **Sr** 87.63 F^1, H^2, B^1 2.16	39 **Y** 88.92 H^2 1.81	40 **Zr** 91.22 H^2, B^1 1.60	**Nb** (Cb) 92.91 B^1 1.47	42 **Mo** 95.95 B^1 1.40	43 **Tc** 99 H^2 1.34	44 **Ru** 101.1 H^2 1.32	45 **Rh** 102.91 F^1 1.34	46 **Pd** 106.4 F^1 1.37	47 **Ag** 107.87 F^1 1.44	48 **Cd** 112.41 H^2 1.52	49 **In** 114.82 U^1_a 1.57	50 **Sn** 118.70 F^2, U^2 1.58	51 **Sb** 121.76 R^2 1.61	52 **Te** 127.60 H^3 1.7	53 **I** 126.90 Q^4_b	54 **Xe** 131.30 F^1	*O*
55 **Cs** 132.91 B^1 2.74	56 **Ba** 137.36 B^1 2.25	57–71 see below	72 **Hf** 178.50 H^2, B^1 1.59	73 **Ta** 180.95 B^1 1.46	74 **W** 183.86 B^1 1.41	75 **Re** 186.22 H^2 1.37	76 **Os** 190.2 H^2 1.34	77 **Ir** 192.2 F^1 1.35	78 **Pt** 195.09 F^1 1.38	79 **Au** 197.0 F^1 1.44	80 **Hg** 200.61 R^1_a 1.55	81 **Tl** 204.39 H^2, B^1 1.71	82 **Pb** 207.21 F^1 1.75	83 **Bi** 209.00 R^2 1.82	84 **Po** 210.0 R^1_b	85 **At** 210	86 **Rn** 222	*P*
87 **Fr** 223	88 **Ra** 226	89–103 see below														Limit of Metallic Property		

	T^3	T^4	T^5	T^6	T^7	T^8	T^9	T^{10}	B^1	B^2	B^3	B^4	B^5	B^6	B^7
Lanthanides	57 **La** 138.92 H^4_c, F^1 1.86	58 **Ce** 140.13 H^2, F^1 1.82	59 **Pr** 140.92 H^4_c, F^1 1.82	60 **Nd** 144.27 H^4_c 1.82	61 **Pm** 147 1.80; 2.00	62 **Sm** 150.35 R^3 1.80; 2.00	63 **Eu** 152.0 B^1 2.04	64 **Gd** 157.26 H^2 1.79	65 **Tb** 158.93 H^2 1.77	66 **Dy** 162.51 H^2 1.77	67 **Ho** 164.94 H^2 1.75; 1.95	68 **Er** 167.27 H^2 1.75	69 **Tm** 168.94 H^2 1.74	70 **Yb** 173.04 F^1 1.93	71 **Lu** 174.99 H^2 1.74
Actinides	89 **Ac** 227 F^1	90 **Th** 232.05 F^1, B^1 1.80	91 **Pa** 231 U^1_c	92 **U** 238.07 Q^2, T^{30}, B^1 1.57	93 **Np** 237 O^8, T^4, B^1	94 **Pu** 242 $M^{16}, N^{17}, S^8, U^1_b, B^1$ 1.60	95 **Am** 243 H^4_c	96 **Cm** 247	97 **Bk** 249	98 **Cf** 249	99 **E** 254	100 **Fm** 253	101 **Mv** 256	102 **No** 253	103 **Lw** 257

Fig. 1. Periodic system of the elements giving atomic number, symbol, atomic weight, crystal structure (low, room, high-temperature phase), and atomic radius for 12 coordination.

VEC = 1.4 is not reached if the difference between the smallest distances in the structure of the component elements exceeds 15 per cent (this 15 per cent rule of Hume-Rothery may be extended to other classes of compounds).

3. Alloy formation usually results in volume contraction. There is a correlation between the relative contraction and the molar mixing enthalpy in the sense that if one is large the other is also (Richards' rule).

4. The radius ratio of the components can frequently serve as a stability criterion for certain structures (e.g., radius ratio of 1.2 as a necessary condition for the appearance of a Laves structure).

Reference is made to such rules in Section 3.

Because of the usefulness of the radius concept, the dependence of atomic radii on other factors has been investigated[2] with the following results:

1. The atomic distance is dependent on the kind of bonding. In ionic compounds it is useful to work with ionic radii which depend on the charge of the ion; in covalent compounds covalent radii are useful. In certain alloy phases (e.g., in systems of transition metals with aluminum surprisingly short distances have been verified[3]).

2. Atomic radii depend on the coordination number (V. M. Goldschmidt).

3. Atomic radii are dependent on the size and kind of the other component (e.g., in Laves phases).

4. Atomic volume is slightly dependent on ordering.

The accommodation of atoms of different sizes in the same lattice can be considered as a problem of elasticity and thus affords a basis for Hume-Rothery's 15 per cent rule.[4,5] If the components display equal outer electronic structure, second-order elasticity theory gives only positive departures from Vegard's value and this is in agreement with experience.[6] If the components have different outer electronic structure, the volume behavior of alloys can be assessed on the basis of the free electron gas.[7]

The electronic factors (§ 2.1) and the size factors are important for phase formation not only in binary alloy systems but also in ternary alloy systems. The homogeneity range in ternary systems often extends in the direction of constant electron concentration, especially in alloys in which the size factor is less important (§ 2.1, rule 2). If, on the other hand, the size factor is of primary importance, we may find a solid solution in which the electron concentration varies in a manner that may be traced in structural or physical properties.

Some work has been done on the connection between the elastic interactions in mixtures and the thermodynamic potentials (Chapter 3). The thermodynamic potential can be used to explain such features in phase diagrams as the appearance, disappearance, or transformation of phases and their dependence on temperature or the difference between the observed composition and the ideal stoichiometric ratio.[8] These questions are preliminaries to the problem of bonding for which we must consider a further influential factor.

2.3 Electrical Charge and Spatial Correlation

It is known from quantum mechanics that the reduced density matrix $P(x_1', x_2'; x_1, x_2)$, in which x_i are the coordinates of the ith electron, contains all the energetic information on a system. The reduced density matrix for temperature T can be called the *bonding relation* for T. If spin is not taken into explicit consideration, the diagonal part of P can be called the *spatial correlation* and the diagonal of $\int P\, \delta(x_2' - x_2)\, dx_2$ the *spatial probability* of the electrons. The assumption can now be made that the kinetic energy of an electron together with the potential energy relative to the cores H_1 can be found to a first approximation by putting the interaction energy of two electrons $V_{12} = 0$. The ground state can then be described by a determinant of one-electron states which means that the electrons are completely independent. This special probability amplitude leads to the theory of the periodic system of chemical elements and to the band model of electron theory (Chapter 2). If V_{12} is now taken into consideration, the amplitude and therefore the density matrix is altered, but the energy arising from H_1 is presumably not seriously changed. In this way crystal structure depends on two further energy contributions:

(1) a low energy arising from V_{12}, that is a favorable spatial correlation, and (2) a spatial probability remaining favorable with respect to the first approximation.

Assumptions concerning the spatial probability have been made since the time of Berzelius' electrostatic bonding theory. These assumptions have been substantiated by experience with electrolytes, and they have led to the electrostatic lattice theory (Chapter 2). Energy contributions arising from spatial probability (Coulomb energies) are not of direct experimental importance in metal chemistry since no dissociation into ions exists. Assumptions for kinetic energy are fairly well determined, for example, from Sommerfeld's model.

Assumptions on spatial correlation in crystals are scarce. Before the zero-point energy was clearly understood, Haber[9] suggested that the electrons are ordered in a space lattice which is inserted into the positively charged lattice of the atomic cores. Hume-Rothery[10] said also: "It is suggested that the beta phases of alloys of copper, silver, and gold (with B metals) are interpenetrating space lattices of atoms and electrons for which the fundamental ratio is three electrons to two atoms with a certain amount of variation on either side." Further short remarks were given by Bradley, Laves and Norbury. Concepts of the covalent bond or resonating covalent bond (Pauling[11]) are also closely related to a spatial electron correlation, but these do not imply explicit models for the spatial correlation in the whole lattice. Even if the V_{12} terms are not taken into consideration, a certain degree of correlation is given in the antisymmetry of a probability amplitude on interchanging electrons, as has been shown by Wigner and Seitz[12]; these authors have proposed a spatial correlation of the A2 type in alkali metals.

If we are not interested in the determination of the first approximation, we may omit discussion of the thermodynamic data and take some average level of energy for granted. With regard to the question of the structure assumed by a given compound, this first approximation may be equivalent in both cases, so that the higher contributions become important. We shall see later that a consideration of the spatial correlation of electrons is vital to an understanding of various phenomena of structural behavior. If this is accepted, it must be expected that for each structure type (and each subtype containing quasi-homologic phases) there exists an electronic rule for its stability. It is an important task of constitutional research to discover these rules and to analyze their physical significance.

2.4 A Simple Method for Treating Spatial Correlation

The spatial correlation of electrons is such that the electrons avoid one another. Because such a (possibly liquid like) structure of the electron gas is developing in a crystal structure, it will itself be more or less similar to a crystal structure if only the outer electrons are considered. A further simplification is given by the assumption that the correlation depends as a first approximation only on the difference of the spatial coordinates of two electrons. In this way the spatial correlation can be represented by a function in three-dimensional space, and it can be called a freely translatable *electron lattice*. Another simplifying assumption is that the electron lattice is not completely free but limited to certain ranges. In order to have a low energy associated with V_{12} one can assume a correlation similar, for example, to A1 (Cu type), A2 (W type), or B1 (NaCl type). Which of these possibilities is chosen depends on different further circumstances. Because of the high kinetic energy of the electrons, the electron lattice is much less well defined than a crystal lattice. The *range* at the end of which periodicity fades out is of the order of 50 Å, and can be assessed from shift structures* (§ 3.2).

The electron lattice need not be completely occupied, we must therefore compare the place number (PN) with the number of electrons supplied by the crystal (ES). This is possible only with certain assumptions of the electron count: A simple case is given by the assumption that the valence electrons of the phase components belong to the same gas, band, or correlation. A more complicated case is the *penetration correlation* in which the valence electrons of the one component are in correlation with the outer-core electrons of the other. A primary aim of this systematization of phases must be to find out which shells of one component are in correlation (or coupled) with which shells of the other.

If the V_{12} energy is now made low by the assumption of a latticelike spatial correlation of the valence electrons (or outer electrons),

* Editor's note: Related structure types produced by stacking variations. See References 14 and 16.

the energy of the electron gas relative to the crystal lattice must be low. This is achieved partly by the assumption that the translation group of the crystal is a subgroup of the translation group of the correlation; in other words, the electron lattice should fit into the crystal lattice by whole-number relations. The physical reason for this *commensurability* requirement lies in the fact that if one electron lies in a minimum of potential energy a whole set of electrons does also. The commensurability requirement is not a strict one; there are many occasions on which it is only partly fulfilled.

As already mentioned, the density matrix and therefore the spatial correlation may also contain parts belonging to excited or degenerate states. Thus electron lattices of different orientation may be contained in the spatial correlation. Because in this way the simple lattice character of the correlation is lost, we prefer to consider only one component of the correlation and to call the phenomenon a *twinned correlation*. This should be distinguished from a *coexisting correlation*, by which is meant the coexistence of a core electron correlation with a valence electron correlation.

Although we shall consider several examples that give strong evidence for the existence and the influence of latticelike spatial electron correlation in metallic phases,[13] we must mention here a curious connection of spatial correlation considerations with band model considerations. Both kinds of consideration are formally very similar: in band model arguments the Fermi sphere in wave-number space must be fitted into the Brillouin zone polyhedron, which depends on crystal structure. In spatial correlation arguments the electron lattice must be fitted into the crystal structure itself. The contact of the Fermi sphere with a Brillouin plane of a large Fourier component of the potential corresponds to a good commensurability of an electron lattice, which also can be formulated with Fourier components. It is probable that spatial correlation as a structure-determining factor is of greater influence than the band structure.

Because at present no experiments have been developed to give direct evidence of spatial correlations, we have to deduce them indirectly from structural knowledge. Examples in the following paragraphs show that this method yields some insight into constitutional phenomena which were not previously understood. A helpful means for finding probable spatial correlation proposals is a system of average electron distances (Fig. 2 which was drawn from several probable proposals[13]). If no use is made of electron distances, the number of valence electrons per cell is distributed over an assumed electron lattice which must be fitted into the crystal cell so that good commensurability results without much deformation of the electron lattice. As simple examples, the correlation proposals for In or Si should be considered (§§ 3.6 and 3.7). The electron distances are even less constant than the atomic radii, but there are rules to show how they are influenced in different phases. It is clear that under similar conditions similar distances should be found (metric coherence of proposals).

In order to have a short expression for a spatial correlation proposal, we give the electron distance d or some equivalent length indexed with the type of the correlation (e.g., d_{A1}) expressed by the lattice constant of the considered phase. From the symmetry of the phase one distance vector may often be extended to an electron net. The number of nets stacked, for example, in the direction of the c-axis is called l_c.

In the following paragraphs examples are given of metallic compounds together with discussion of their bonding. It is well known that the systematization of phases begins with *structural types*. It is often convenient to divide types into different *branches* if the bonding is different and to collect types into structural *families* if the bonding is similar.

3. EXAMPLES OF THE SPATIAL ELECTRON CORRELATION CONCEPT

3.1 Brasslike Alloys

B–B phases showing a valence electron concentration between 1 and 2 may be called brasslike phases. The phase diagrams of brasslike alloys show many phases with broad homogeneity ranges (§ 2.1, rule 2). Three basic types of structure (*basic structures*) have been found.

1. Face-centered cubic close-packed structures of the Cu type.

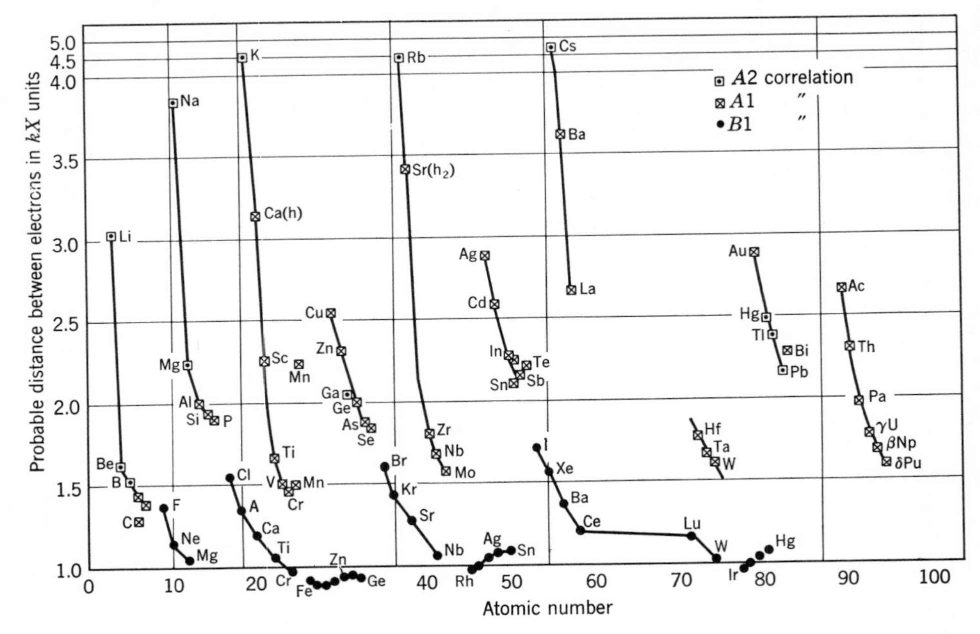

Fig. 2. Probable distances of electrons in elements. The data of this figure have been abstracted from our knowledge of structures. For each structure or class of structures, proposals for the electron correlation are made (some of them are noted in this chapter, others are to be found in Ref. 13), and from these the electron distances are derived.

2. Hexagonal close-packed structures of the Mg type.
3. Body-centered cubic structures of the W type.

These structure types are either the closest packing of spheres of equal size or a comparatively close packing. Necessary conditions for the formation of such densely packed structures are (1) a highly filled shell at the border of the core (*d*-core), which gives an unsaturated attracting force in all directions, and (2) a not too highly filled valence electron gas which does not essentially counteract the attracting forces of the cores. All three basic structure types can be observed with different distortions, superstructures, and sometimes vacancy variants. The appearance of the basic structure is governed by the Hume-Rothery rules which give valence electron concentrations favoring each type—Cu type: 1 . . . 1.4; Mg type: 1.5 or 1.75; W type: 1.5. These rules have been explained in the framework of the band model of electron theory. However, in many different alloy systems, numerous secondary structural features are found in addition to basic structure which also should be explained by bonding properties. An outstanding superstructure type in brasslike alloys is given, for example, by the *shift structures* which were first found for $TiAl_3$[14] and afterward for the brasslike phase CuAu(II). An explanation of this simple structure and its secondary features (e.g., the direction of the shift system and distortions) allows us to understand some basic properties of the spatial correlation of valence electrons in brasslike alloys (§ 3.2). In a fairly straightforward manner the correlation of A1 basic structure phases can be extended to phases with A3 and A2 basic structures (§§ 3.4 and 3.5). On the other hand, this interpretation of A1 shift structures may also lead to an interpretation of T–B phases of the kind $TiAl_3$ (§ 3.9).

3.2 Shift Superstructures of Phases with fcc Basic Structures (Cu Type) in Brasslike Alloys

Structures of this kind (Chapter 9) may be observed on the basis of a Cu_3Au- or a CuAu-type of superstructure.[15,16] They display a valence electron rule (in analogy to the Hume-Rothery rule) that is related to the *shift density* which is the reciprocal of the distance of two shift planes [assumed parallel (001)] times half the basic structure lattice constant in the direction of the normal to the shift plane; in other words, the number of shifts per atom layer parallel to (001). The connection between shift density and valence electron concentration is shown in Fig. 3; it is nearly linear and approximately independent of the individual alloy system,[16] and can be understood in terms of the spatial correlation of valence electrons. In a structure of the Cu type with valence electron concentration $V \neq 1$ the correlation must deviate from congruence; either further electron places must be interstitially inserted in the correlation belonging to $V = 1$ or the correlation must be contracted relative to the crystal lattice in order to give on the average more places per atom. In brasslike alloys the latter possibility is realized, whereas in alloys with $V > 4$ the other possibility will be met with. Because a loss of commensurability is energetically unfavorable, the contraction of the correlation will take place only in one direction. In this direction the crystal will be somewhat strained because of the stiffness of the correlation. As is well known (Chapter 23), the stiffness in Cu-type phases is minimum in the [001] direction, so that a small homogeneous strain will choose the [001] direction. From Fig. 3 we infer that to a first approximation $V \approx D + 1$ ($D =$ shift density), or rewritten, $(V - 1)\,D^{-1} \approx 1$, which means that to each shift plane there belongs one additional correlation plane (Fig. 4). We are thus quite naturally led to the model of two space lattices which are congruent in two directions but interfere in the third.[16,17] Because all the valence electrons lead to dispersion forces (Chapter 2) of fairly long range, the loss of commensurability between crystal and correlation leads to a rearrangement of dispersion forces. The difference between the *eigen*-dispersion force and the average dispersion force acts on the minority component of the shift superstructure. This difference gives rise to an attractive force between minority atoms of one and the same shift domain and a repelling force between minority atoms of different neighboring shift domains. These forces are fluctuating in the crystal lattice before the ordering transformation; however, if the crystal transformation from the disordered

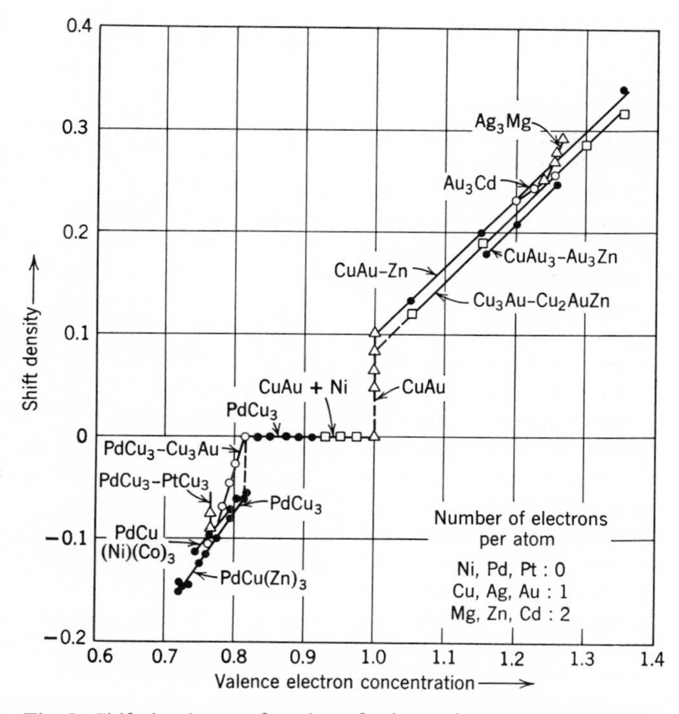

Fig. 3. Shift density as a function of valence electron concentration.

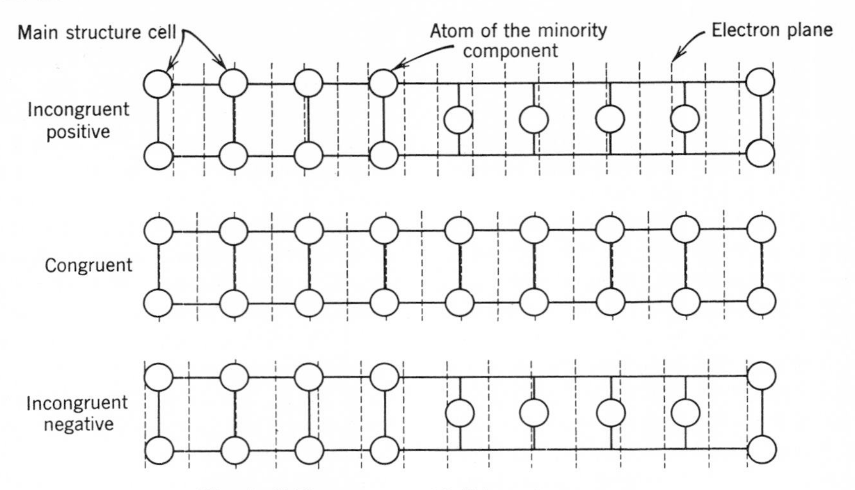

Fig. 4. Shift structure with $L1_2$ main structure.

state into the ordered state with shifts has taken place, the forces settle somewhat in relation to the crystal; that is, the electron lattice loses some freedom in the *c*-direction. If the difference in properties between the minority atom and the majority atom in an alloy is small, the forces producing the shift also become small so that the shift may not be found, as, for example, in Cu–Zn. But the tendency to a shift arrangement will remain and will be observable in the damping properties; in the Cu(Zn) solid solution a special damping capacity has indeed been found.[18] If V diminishes, the shift distance will become greater, and considerations of the model discussed show that the apparent shift distance divided by the lattice constant need not be a whole number. The relative shift distance in the crystal will fluctuate about a mean value which is given by V, and there will be a considerable mixing of the different lengths which is greater than normal. It has been shown that in this case sharp diffraction lines will result.[19] If V approximates 1, the shift distance will eventually equal the range of the correlation. The distance will then become statistically distributed as has been shown with dark-field electron micrographs.[20] If the correlation lattice had ideal cubic symmetry, the increase of the axial ratio of the crystal should exceed the observed one by a factor of 10. This will be understood by the assumption that the stiffness of the core electron correlation exceeds the stiffness of the valence electron correlation by a factor of 10. The structures with more than one shift system[21] seem to be connected with deviations from the ideal 1:3 composition of the basic structure.

3.3 Other Distortions of Phases with fcc Basic Structure (Cu-Type) in Brasslike Alloys

Other distortions of Cu type structures in brasslike phase diagrams have been found which are not primarily connected with the shift mechanism. The best known example is the structure type CuAu(I). Johansson and Linde[22] who first found this structure suggested that the relative atomic radii of the components explained the structure. This can be only partly right, since PtCu has a rhombohedrally distorted superstructure of another type. A further condition is derived from the table of representatives of the CuAu type (Chapter 9): the valence electron concentration must be approximately 1. In this case the valence electron correlation will be congruent to the basic structure so that from this viewpoint the possible tetragonal distortion is not disturbed. It is found that there is no good correlation between the axial ratio c/a and the ratio of the atomic diameters for 12 coordination for the brasslike representatives of this structure type; nevertheless, this does no harm, because the diameters generally refer to a state with another valence electron concentration. We should therefore take a system of *ad hoc* radii related to the core electron correlation. It thus turns out that the primary structural argument for CuAu(I) is the interaction of the core electrons. CuAu(II) shows the CuAu(I) superstructure and distortion, together with a shift system, which develops so that an *a*-axis is normal to the shift planes; this direction has a smaller stiffness than the *c*-direction. The reason for the transformation CuAu(I → II) lies in a temperature-dependent valence electron contribution of the Au atoms, which is also met in other cases.

Another type of distortion, the inhomogeneous or inner distortion, can be described with the example $Au_3Zn_{1+}(r)$[23,24] (Fig. 5). This structure develops in a diffusionless transformation from an undistorted shift structure based on the Cu_3Au type, of shift density $\frac{1}{4}$, the phase $Au_3Zn(h_1)$. In the transformation the lattice constants of the basic structure decrease a little, the symmetry becomes very slightly orthorhombic, and the lattice constant in the *c*-direction increases somewhat; furthermore, the atoms leave their ideal places in the (h_1) structure so that the atomic distances become unequal (the $d_{(Au-Zn)}$ distances become 2.5 per cent smaller than the $d_{(Au-Au)}$ distances) and the repeat distance in the basic structure is multiplied by the factor $\sqrt{2}$. Now it has already been found that in the simple shift structures the spatial correlation lattice loses some freedom in the *c*-direction in order to make the shift transformation possible; we can therefore assume that it becomes more probable to meet an electron plane precisely in the undistorted (Au, Au) plane between two shifted (Au, Zn) planes parallel to the base. Then any distorted (Au, Au) plane is surrounded by two electron planes, one of which with increased probability coincides in its projection with the enclosed (Au, Au) plane and thus makes the position of the Au atoms somewhat unstable so that they are subjected to an inhomogeneous distortion. This distortion also causes the position of the Zn to be shifted in the *c*-direction.

A third type of distortion, which is also inhomogeneous but for a somewhat different reason, can be illustrated by the phase Ir_3Si which is brasslike and has a U_3Si-type structure. The lattice constants of the basic structure are $a = 3.69$, $c/a = 1.07$. The superstructure is fundamentally of the Cu_3Au type, there is no shifting, so that the possibilities discussed above for the bonding relation are not present. Here all valence electrons are supplied by the B component so that there is no even distribution of them over the structure. In this case the transformation of the correlation lattice of Fig. 6 becomes possible if the electron distance is smaller than it would be in the average A1 correlation. The Ir atoms then give place to the valence electrons and, consequently, electron tetrahedra neighboring in the *c*-direction have opposite orientation with the result that the axial ratio increases. Some further types of distortion are also known (e.g., that of Pt_3Si), which are not yet

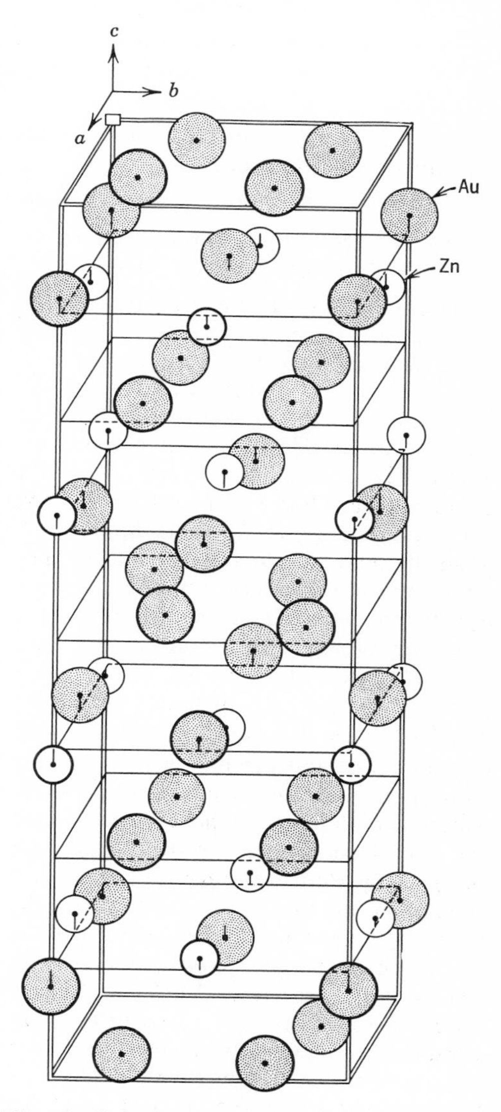

Fig. 5. Crystal structure of $Au_3Zn_{1+}(r)$. The structure is fundamentally of A1 type. (Wilkens and Schubert.[23])

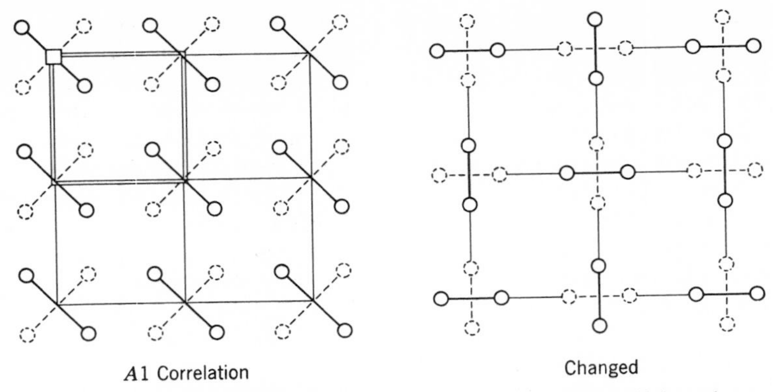

Fig. 6. Possibility of change of an A1 correlation in a $L1_2$ structure if the volume of the T component is not smaller than that of the B component.

understood in terms similar to those used here for the structures of CuAu, $Au_3Zn_{1+}(r)$, and Ir_3Si.

3.4 Phases with hcp Basic Structure (Mg Type) in Brasslike Alloys

Since the close-packed structures with a Cu-type basic structure are fairly well understood in terms of spatial correlation of electrons, we may ask whether spatial correlation arguments may be considered in brasslike phases with Mg structure. If the valence electron concentration increases in an A1 structure from 1.0 to 1.4, the valence electron correlation is, as we found in Section 3.2, gradually contracted to a body-centered translation lattice (A2) because of the core electron correlation which hinders the tetragonal distortion of the crystal structure. At these electron concentrations an A2 correlation is probably energetically preferred to an A1 correlation. At the valence electron concentration 1.4 (more precisely $\sqrt{2}$) the electron correlation must become cubic, provided the distortion of the crystal lattice can be neglected and all places of the electron lattice are occupied by electrons. Now, in the relative orientation of the electron lattice to the crystal lattice assumed in the structures of Cu type much of the good commensurability at VEC = 1 is lost when VEC = 1.4. Thus another orientation might give a better commensurability. If we turn the electron lattice in a Cu-type basic structure with VEC = 1.4, 45° about the c-axis, the (111) plane of the crystal structure becomes parallel to (111) of the A2 electron lattice. This means a gain in commensurability, though the relation between the repeat distances of the crystal and electron lattice in the common (111) plane is not rational. If we idealize the ratio of separation of atoms to separation of electrons in that plane from $\sqrt{2}$ to $\frac{4}{3}$, we get (in the case of the ideal axial ratio 3×0.816 of the crystal lattice and of an ideally cubic body-centered electron lattice) in the [111] direction, 3.0 electron planes per crystal plane of the kind $(111)_{A1}$. Since three planes parallel to the base of a hexagonally described rhombohedral lattice give a repeat in the direction normal to the base, we find that, because of the commensurability phenomenon, a lattice of the A3 type is favored, compared to a lattice of the A1 type. The expected value of the valence electron concentration in the described case is 1.68 in fair agreement with observation.

In brasslike phases of the Mg-type structure we find that the axial ratio c/a decreases with increasing VEC.[25] In terms of the spatial correlation this means that the commensurability in the c-direction remains constant, whereas the orientation of the hexagonal nets in the base may be changed. If a and c are the lattice constants of the crystal cell, then $(-a_1 + 3a_2)/3$ is slightly changed with respect to the $4a/3$ correlation; it has 2.08 places per atom and leads to the axial ratio 1.47. Just as in the shift structures on the A1 basis, the ideal axial ratio is (by a factor of 10) not reached here. But two phases $Cu_{10}Sb_3$ and $Au_{3.5}In(r)$ (Fig. 7) have been found which show a superstructure giving evidence of the proposed basic correlation. As in the A1 shift structures, we assume that the type of

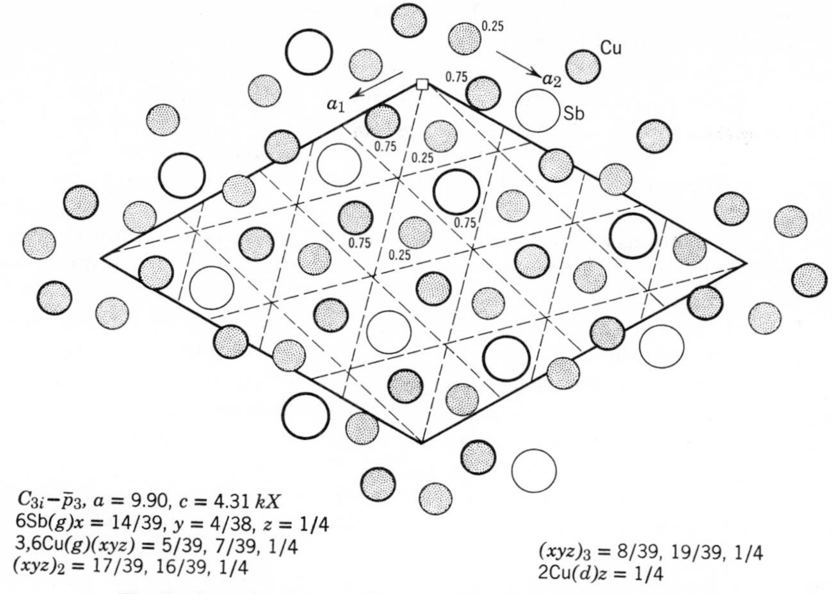

Fig. 7. Crystal structure of $Cu_{3.3}Sb(h)$. (Günzel and Schubert.[38])

superstructure is some sort of evidence for a changed commensurability of the correlation to the crystal cell.

Shift structures have also been observed on the basis of an A3 crystal structure, but they are not yet completely understood in terms of spatial correlation of valence electrons. Distortions of the hexagonal symmetry are uniaxial strains in the basal plane, two types of which have been observed.[26] Furthermore, stacking variants of the basic structure have been found in brasslike alloys. In the $TiNi_3$ type (which also has appeared without superstructure in brasslike alloys) it can be said that it occurs in the transition range between the A1 and A3 basic structures. The fact that the basic structure axial ratio is higher than ideal is accounted for by the spatial correlation to be assumed here.

3.5 Phases with bcc Basic Structure (W Type) in Brasslike Alloys

At VEC 1.5 the bcc structure (the W basic structure) becomes stable in brasslike systems (Chapter 10). Two different views have been presented in regard to bonding in these phases. The earlier one, from Hume-Rothery,[27] suggests an A1 correlation occupied to 75 per cent; a later one is based on the band model of electron theory. The first view is in better agreement with the phenomenon of vacancy formation with increasing VEC (Chapter 10). Although the number of electrons per lattice cell in the band model should remain constant with increasing VEC, in the spatial correlation model an electron-atom substitution process is expected (Norbury[28]), that is a slowly increasing number of valence electrons per cell when VEC increases. In Fig. 8 the broken line refers to a constant number of valence electrons

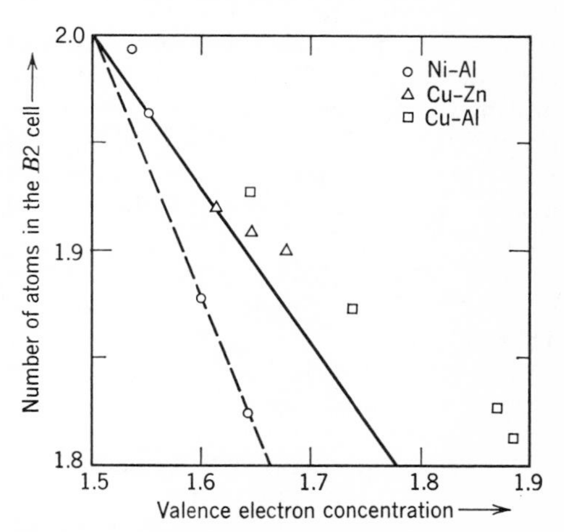

Fig. 8. Number of atoms per cell as a function of electron concentration in structures related to the B2 type. The broken line corresponds to a constant number of electrons per cell. The solid line corresponds to the atom-electron replacement process.

per cell, whereas the full line refers to the substitution process. The points for NiAl are plotted under the assumption of zero-valent Ni which is in contradiction to the probability that there is a positive contribution of valence electrons on the part of Ni. The fact of the vacancy formation process in beta brass phases suggests that the electron lattice in these structures is not freely translatable, that is, not all of the different translatory positions of the electron lattice have equal probability.

Cases of vacancy formation are known for both a statistical distribution of vacancies within the cell, as in NiAl and similar phases, and for ordered distributions; for example, in Cu_5Zn_8 (gamma brass) and Ni_2Al_3, $AuAl_2$ (fluorite type).* Although the Cu_5Zn_8 structure may occur at different compositions, this property is lost in the structure types of Ni_2Al_3 and $AuAl_2$. The latter structure type shows clearly that the electron number can exceed three electrons per basic structure cell; it has even been observed with NiSi and $PtSn_2(h)$. As final members of this class of structures, the diamond type and its derivative structures may be considered. We return to these phases in a later section of this chapter.

There are some phases, even brasslike ones, which can be considered as derivatives of the CsCl structure by vacancy formation but which do not belong to the bonding class of CuZn (CsCl type): The phase $NiHg_4$ has a body-centered cubic structure in which the Hg atoms form a cubic primitive lattice with eight atoms per cell and the Ni atoms are distributed in the hexahedral interstices; isotypes include $MnGa_4$, $CrGa_4$, $PtMg_4$. It is probable that in these phases we will have to assume an A2 correlation of the valence electrons.

Many complicated variants are known, especially of the gamma brass structures. Different vacancy arrangements and correlations of the core electrons play roles in this class of alloy structures.

3.6 Alloys Between B Elements with VEC < 4

The bonding relation of these alloys is relatively easy to identify, for the occupation of the valence shell is unambiguously known in them. There is no distribution of the valence electrons between two different shells as, for example, in T–T alloys. Nevertheless core electrons occupying a filled *d*-shell are of some influence in the structural behavior of these alloys. To a first approximation we can consider the *d*-shells as hard spheres seeking contact with one another. If additional electrons are added to the valence shell by changing the atomic number, the contact distance between atoms will not primarily be strained to give a lower energy to the valence electrons but rather the number of contacts, the coordination number, will be decreased so as to increase the spatial volume for the valence electrons.

Evidence for this is given in Fig. 9 in which the difference of the atomic volume V_{at} and the volume of the cores $4\pi r_0^3/3$ (r_0 = half of the nearest distance in the element structure) is plotted against the atomic number. One conclusion is that the core electrons are of largest influence in B–B alloys with few valence electrons.

The element structure of Zn is an example of core electron influence. In order to understand this, we must make an assumption in regard to the structure of the core electron correlation in brasslike alloys which has until now been considered only in a qualitative manner. A cubic primitive lattice (B1 correlation) of cube edge $a/4$, in which a is the lattice constant of the face-centered structure of Cu, for instance, is suggested for several reasons for the *d*-electron correlation. If we describe the Cu structure in a hexagonal cell (a_h, c_h) with three atoms, the *d*-electron correlation of A1 type would be described by a hexagonal net with lattice constant $d_{B1}\sqrt{2} = a_h/2$ which is stacked with $l_c = 12$ layers per c_h-axis in a rhombohedral sequence. If nine layers per two atomic planes were stacked parallel to the hexagonal base plane, a crystal axial ratio of $c/a = \sqrt{8/3} \cdot 9/8 = 1.83$ similar to the axial ratio found for Zn would result. If this is the real reason for the anomalous axial ratio of Zn, we must understand it as an example of the commensurability phenomenon: the increased density of the valence electrons compared with Cu seeks to strain the core electron correlation, but this is possible for Zn only by snapping into a new commensurability position. Because nine electron planes in rhombohedral stacking repeat in the c_h direction, commensurability

* See Part III, Chapters 8–13.

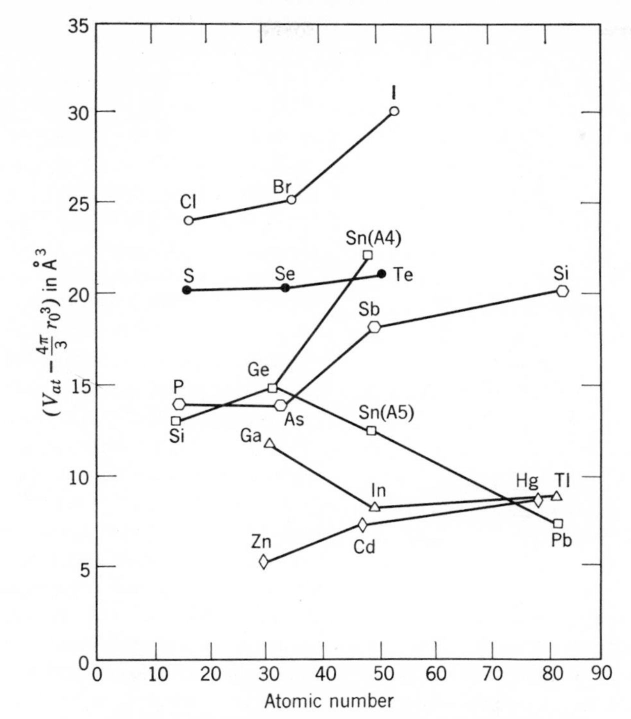

Fig. 9. Volume of valence electrons in B elements.

favors the A3 stacking of the crystal planes. As the axial ratio of Zn at room temperature ($c/a = 1.86$) is somewhat larger than the suggested value 1.83, it is to be assumed that the valence electrons also are in a correlation which favors the high axial ratio and are thus equally responsible[13] for the remarkable structure.

Another B element that shows a distorted closest packing of atoms is indium. Because we shall find later that we can assume an A1 correlation of valence electrons in Ge and gray Sn, we might ask how an A1 correlation of three valence electrons per atom could be fitted into the tetragonal cell of In, which is described in a face-centered tetragonal lattice (a, c) with four atoms per cell. If we build up a square electron lattice with repeat distance $d_{A1} = a/2$ on the basis of In and stack it in an A1 manner so that $l = 3$ planes to correspond to one c-distance, then with an undistorted electron lattice the axial ratio c/a of the crystal cell becomes 1.06, whereas 1.075 is observed. For aluminum we assume that the correlation of In is twinned in all three possible directions. In Ga it should be pointed out that there is a greater influence of the core electrons similar to Zn. The spatial correlation of In is substantiated by the structural behavior of the In-rich phases in the alloy system Mg–In. $MgIn_{2.5}$ has an A1 structure and MgIn, a tetragonal CuAu structure with an axial ratio $c/a = 0.96$. The assumption of a completely occupied A1 electron correlation predicts somewhat lower values of the axial ratio, (see Fig. 10); the observed behavior therefore must be understood in terms of a correlation that is not completely occupied, as we have found in the phases of the CuZn type. It is of special interest that in the neighborhood of $MgIn_{2.5}$ a $Cu_3Au(Ll_2)$-type of structure was found in $MgIn_3$. One would perhaps have expected a $TiAl_3$ type of superstructure because of the dispersion forces discussed with brasslike shift structures. There are indeed such shift structures as $TiAl_3$ and $ZrAl_3$ in this range of electron concentration, but because $MgIn_3$ is cubic the shifting forces average out to zero, for the normal of the shift system can assume any of the three possible directions. The fact that $MgIn_{2.5}$ is completely disordered seems to mean that in this phase forces which tend to shift and so destroy any tendency to the Cu_3Au superstructure are sensible.

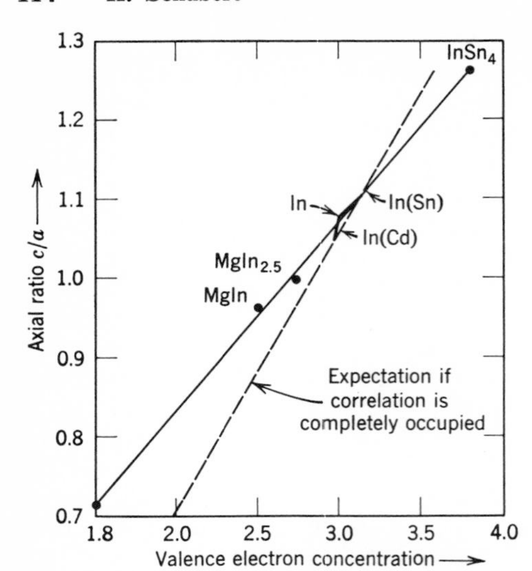

Fig. 10. Axial ratio of some phases related to indium.

Indium solid solutions also are compatible with the above spatial correlation. On the electron-poor side the decrease in axial ratio is larger than expected, since cubic symmetry is approximated, and on the electron-rich side the increase is smaller than expected, since the structure departs from cubic symmetry with an increasing tetragonal compression of the A1 correlation.

A representative of the bonding relation of the In type, rich in electrons, is the structure of $HgSn_6$. The (100) plane of this hexagonal structure is quasitetragonal and to be compared metrically with the In basal plane. If we build up in this plane a correlation net similar to that in In and stack it in the A1 manner, a valence electron concentration of 3.5 results, whereas the observed value is 3.8. We must conclude that the valence electron correlation is tetragonally compressed under the influence of the core electron correlation. Because of the quasisymmetry of the described array, different positions relative to the crystal lattice (i.e., twinning) are possible and in them the crystal lattice assumes the hexagonal symmetry. We should expect that with increasing valence electron concentration the ratio c/a should decrease as has indeed been found.[29] One structure is closely related to the structure of $HgSn_6$ but does not belong to its bonding class: this is the phase CdHg, which has a primitive quasibody-centered structure with two atoms in the cell and an axial ratio that gives nearly hexagonal (110) nets. It is probable that in this phase we will have to assume an A2 correlation of valence electrons with $a\sqrt{2} \approx c = a_{A2}$. The occurrence of an A2 correlation at VEC ≈ 2 can be understood by the assumption that electrons of equal spin avoid one another more than electrons of different spin, especially if they have the same spatial probability amplitude ψ.

3.7 Alloys Between *B* Metals with VEC > 4

The diamond structure of Si, Ge, and gray Sn shows in its cubic basal plane a square array, the primitive mesh of which is metrically related to the centered basal plane of In. If we assume an electron lattice similar to that in In and stack it in the A1 manner, we get the spatial correlation of the diamond lattice: $a/2 = a_{A1}$. We can understand this structure and correlation also as the end point of the vacancy process found in beta brass phases. It is easily realized that there is no simple contradiction between this correlation and the generally accepted covalent bond. The diamond lattice is not so stable as one would be inclined to assume. Half an electron per atom suffices to effect a stacking variation, for example, in the phase SnSb, which has a rhombohedrally strained rocksalt structure. It is known that here the [111] direction shows a minimal elastic modulus so that, as in the shift structures, it is preferred for straining morphotropy. A near relative to the SnSb structure is the structure of As which has one more electron per atom than Ge. Further relatives are contained in Fig. 11. From the material plotted in this figure we can infer that the axial ratio is not linearly related to the valence electron concentration. The reason for this may be seen in the following assumptions: the valence electrons in the range of concentration $3 \cdots 4$ are in A1 correlation. If a B atom bears more than four valence electrons, the first four are in A1 correlation but the fifth, sixth, and following electrons fill up the octahedral interstices of the A1 correlation to build up locally a B1 correlation. This assumption is compatible with Hund's rule of maximal multiplicity of terms in atomic spectroscopy, and it leads to a better understanding of structures with B elements of more than four electrons per atom.

As the beta brass phases show, there is a

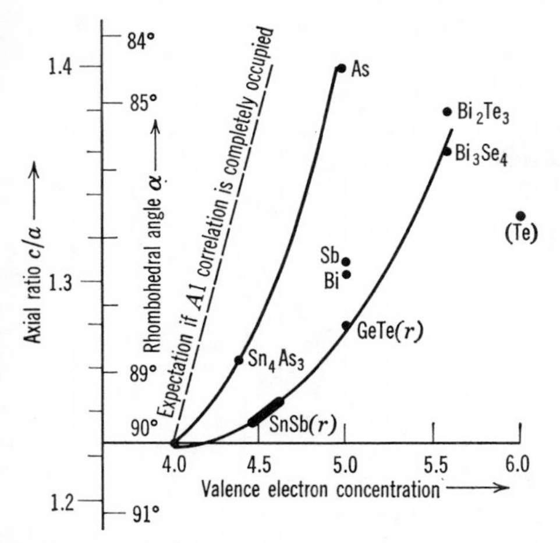

Fig. 11. Trigonally distorted structures related to the B1 structure. α = rhombohedral angle in setting corresponding to B1 structure. c/a = axial ratio of three atom layers parallel to the basal plane.

possibility for the electron lattice to become fixed in relation to the crystal lattice, that is, to lose some probability of certain translatory positions. The same possibility is valid for the inserted electrons. In this way close contacts (e.g., in the diamond structure) may eventually be interrupted by inserted valence electrons. The Hume-Rothery 8-n rule (Chapter 8) may similarly be understood. A particularly simple example is given by the structure of black phosphorus (Fig. 12). The A1 correlation for four valence electrons per atom is shown by broken lines, and areas of higher probability for the inserted fifth electron per atom are shaded. Similar spatial correlation proposals may be given for As, and the difference between the structures of both homologous elements should be sought in the influence of the core electrons.

The above-mentioned filling-in process is not a rigid one. If, for example, the number of valence electrons per atom is six, the A1 correlation contains only two more electrons per atom than the filling in lattice; if one electron per atom goes over to the minority, equilibrium is attained. This might be the reason for the phenomenon in which an A2 correlation is sometimes favored with B^6 atoms. As an example, Fig. 13 shows that the A2 correlation fits excellently in sulfur.

It must be mentioned that compounds between very different B elements, for example, oxides, are not simply to be understood with one correlation lattice. These structures show different kinds of bonding involving penetration of correlations into the core of the metal component. (§ 2.4)

3.8 Alloys Between *T* Elements

It is known spectroscopically that the outer electrons in T atoms, that is, the electrons energetically above the highest closed noble gas shell, distribute among a higher s-shell and a lower d-shell. A similar distribution is often assumed in the crystal. In the s-band there is about one electron per atom, so that in view of the findings in brasslike and B–B alloys we have to assume a strong influence of the

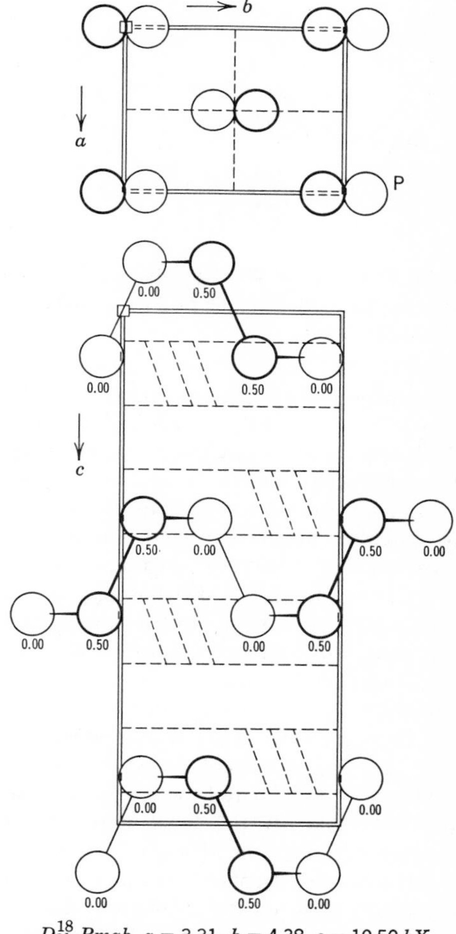

D_{2h}^{18} *Bmab*, $a = 3.31$, $b = 4.38$, $c = 10.50\,kX$
$8P(f)\,y = 0.090$, $z = 0.098$

Fig. 12. Crystal structure of phosphorus (A17SB 3, 6, Hultgren, Gingrich, and Warren.[39])

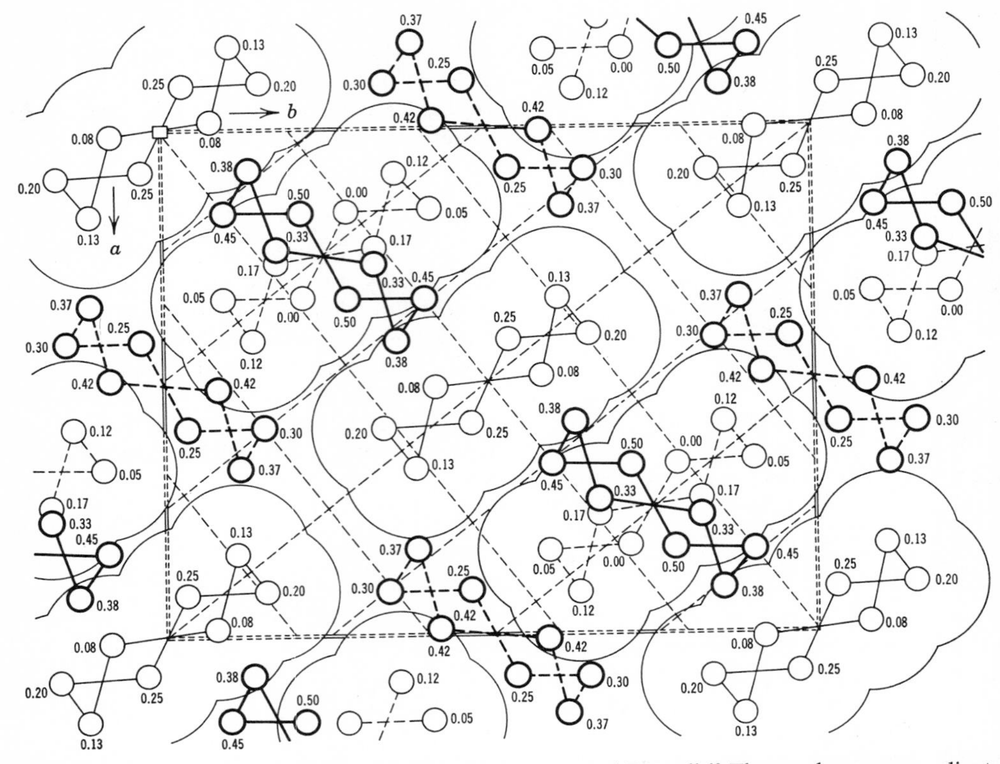

Fig. 13. Projection of the structure of sulfur after Warren and Burwell.[40] The numbers are coordinates perpendicular to the drawing plane. In broken lines is given the A2 correlation $a_{A2} = a/\sqrt{25} \approx b/\sqrt{41}$; $l_c = 24$, PN = ES.

d-electrons on the alloying behavior which results in a dissimilarity between T–T alloys and brasslike alloys. Since the *s*-electrons keep the atoms separated to a certain degree, twinned core electron correlations that are not easy to analyze may be expected. Therefore it is necessary first to state some empirical rules. In Fig. 14 the frequency distributions of the W, Mg, and Cu types are plotted as functions of the outer electron concentration. We find distinct maxima, which for short we call "branches," and which show that these types are favored at certain electron concentrations. In the gap between OEC 6 and 7, that is, between the V branch and the Ru branch, we find "βW" and βU (σ) types and related structures. Although the phases of Fig. 14 are alloyed from components whose outer electron contribution does not differ by more than 5, we can also find representatives of the W, Mg, Cu, βW, and βU types in alloys with components that differ more widely.

In the V branch of A2 structures an A2 correlation of *s*-electrons seems to be probable, and the *d*-electrons will be accommodated in an A1 correlation that could perhaps be considered similar to the correlation of valence electrons in aluminum. In the Ru branch of A3 structure a certain dependence of axial ratio c/a on OEC has been found which is quite similar to the dependence of c/a on VEC in brasslike phases with Mg structure but here observed at an electron concentration about four times larger (Fig. 15). This makes it probable that in the phases of the Ru branch there is a B1 correlation similar to the A2 correlation in the brass branch of the A3 structure. It must be remarked that with these proposals a similar filling-in process of A1 correlation to B1 correlation is assumed as in B–B alloys.

The more involved T–T structures such as βU or αMn are all related to the simple Cr_3Si (or "βW") structure. If we compare Cr_3Si with CrSi (B20), whose bonding relation is discussed in the last section, we are led to

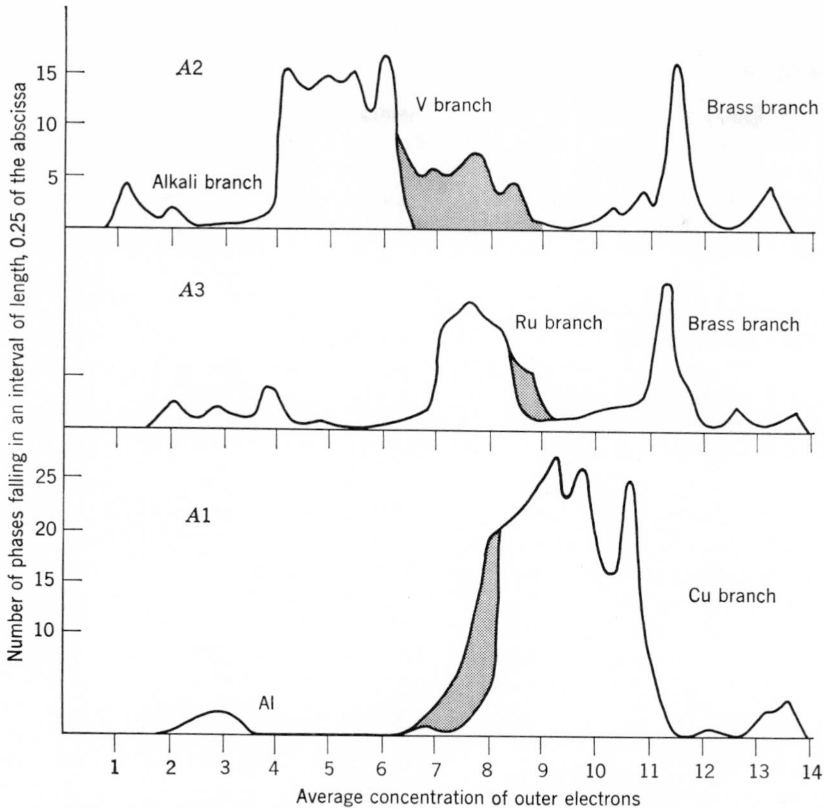

Fig. 14. Frequency distribution of 120 A2 and B2 phases, 85 A3 and D019 phases, and 140 A1, $L1_2$, and $L1_0$ phases as a function of the average concentration of outer electrons. The shaded areas represent phases containing ferromagnetic components. Phases with a difference of electron contribution of the components exceeding five are not counted.

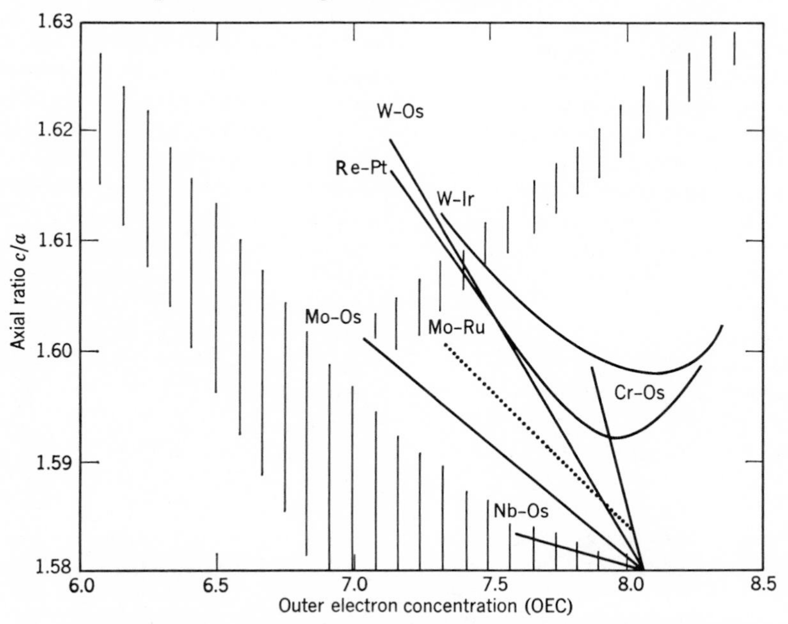

Fig. 15. Axial ratio of A3 phases of the Ru branch. (The shaded area contains the c/a of brasslike A3 phases as a function of 4 VEC = OEC.)

ascribe to the Cr_3Si structure an A1 correlation with $a/2 = a_{A1}$. There are four places per atom, which are not sufficient for the outer electrons of these atoms in the elementary cell. We must therefore assume that some of the outer electrons are contained in a separate valence correlation and that some fill the A1 correlation locally to a B1 correlation. Since for this filling-in process some positions (the 12 coordinated) in the structure are preferred,[30] the number of outer electrons per atom is limited but the electron rule is more complicated than with brasslike alloys.[30] In the βU structure we have $a\sqrt{29} = d_{A1}$, $l_c = 4$, which gives 3.9 places per atom, and in αMn there are perhaps somewhat more than four places. The similarity of place numbers in the *d*-correlation makes it probable that the choice of structure is essentially influenced by the *s*-electron correlation on the one hand and by the correlation of the filling-in core electrons on the other.

There are some actinide structures for which simple proposals are found. Without doubt, the actinide elements have their outer electrons above the Rn noble gas shell. They need not be *d*-electrons but may have a partial *f*-character. This results in the assumption that an A1 correlation is retained at electron concentrations above five and that this should lead to peculiar structures which are not comparable to structures of common T–T alloys. As an example, we may mention (1) Pa with a tetragonally distorted W structure ($a = 3.92$ Å, $c/a = 0.82_5$); if we build up a square net in the basal plane with $a/2 = d$ (= distance in the square net) and stack 2.5 such nets in A1 manner, we get a slightly compressed A1 correlation with PN = 10 (ES = 10). (2) Pu(h_4) which has a similar distorted Cu structure ($a = 4.70$ Å, $c = 4.49$); if we build up a square net in the basal plane with $a/3d$ and stack four such nets in A1 manner, we get PN = 36 and ES = 32.

An important group of structures observed frequently in T–T alloys includes the multiple-replacement structures to which the Laves phases of Cu_2Mg, Zn_2Mg, and Ni_2Mg structure types belong.* We easily find the A1 correlation of valence electrons valid in the Cu_2Mg structure, and a VEC rule[31] governing the stacking sequence can be reduced to the Hume-Rothery rules. The spatial correlation of the Cu_2Mg structure is not valid for all Cu_2Mg representatives; hence the Cu_2Mg type is composed of different branches, but other rules can be given which allow their bonding relation to be assessed.[37]

3.9 *T–B* Alloys

We consider first some phases with Cu-type basic structures, the best known of which are $TiAl_3(D0_{22})$ and $ZrAl_3$. In them we must assume A1 penetration correlation with electron distance $d_{A1} = a/2$ in which a is the basal lattice constant. If there are three electron planes per c_b axis of the basic structure cell, then (following the rules for brasslike shift structures) we have to expect the $D0_{22}$ type. Though there might be some variation of that number of planes we can assume that the electron count is $Ti^3Al_3{}^3$. In $ZrAl_3$ there must be about five electron planes per domain length, that is, 2.5 electron planes per c_b; if we assume at most 2.75, the electron count should be $Zr^2Al_3{}^3$. At any rate, there are more electrons per atom in the relevant correlation of $TiAl_3$ than in that of $ZrAl_3$. This is confirmed by the observations $ZrAl_2Si(D0_{22})$ and $\sim HfZnGa_2(D0_{23})$.[32,33] In the latter case, with higher Zn content, A1 and $L1_2$ phases are found which is in agreement with the assumed spatial correlation and is analogous to the Mg–In alloys discussed. The structure of $ZrSi_2$ belongs to the $TiAl_3$ family and becomes quasitetragonal in $TiAl_{0.3}Si_{1.7}$[34] ($a = c = 3.58$, $b = 13.49kX$). If we compare that compound with the neighboring $TiSi_2$(C54) ($a = 8.24$, $b = 4.77$, $c = 8.52$—pseudohexagonal axis) we are led to a hexagonal net with $b/\sqrt{7} = d_{A1}$ in the pseudohexagonal basal plane which is to be stacked about six times per *c*-axis. For $CrSi_2$(C40) and $MoSi_2$(C11b) belonging to the same family of structures we find more electron planes to be stacked per atom layer, and this must be connected with the crystalline stacking sequence in these types. The bonding relation in the basal plane of these related structure types can be traced to the structure types of WAl_5, WAl_4, Mn_4Al_{11}, Co_2Al_5, and $MnAl_6$ (Fig. 16), though the connection between the spatial correlation and the stacking sequence and atom sites needs further elaboration.

Comparing $CrSi_2$ and CrSi(B20), we find the probability of an A1 correlation $d_{A1}\sqrt{2} = a/2$, $l_c = 4$ which fits excellently into the B20

* See Chapter 12.

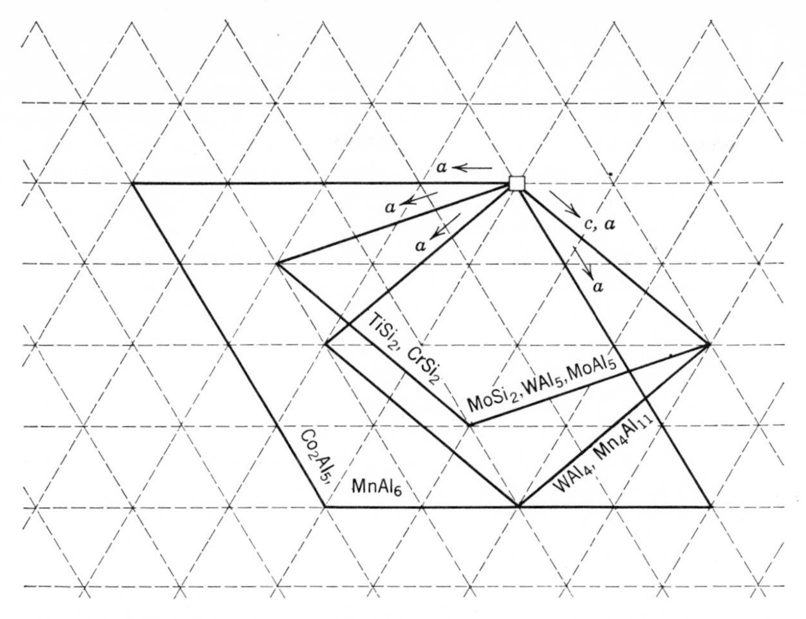

Fig. 16. Electron lattice on the basal plane of some phases with penetration bonding (of the Raynor-Taylor class).

structure. The electron counts in the table of representatives are understandable only if we assume that the A1 correlation in the neighborhood of the T atom is filled to a B1 correlation. This curious effect is the reason why electronic rules in T–B phases have remained hidden for so long a time. It is well known[35,36] that the FeSi(B20) type is linked to the CsCl(B2) type on the one hand and to the MnP(B31) type on the other. We can presume that in CsCl type phases such as FeAl there is primarily a brasslike bonding relation; that is, the valence electron of the B component is in correlation with the *s*-electrons of the T component. The reason for this should be sought in the large electron distance of the B^3 component. On the other hand the bonding relation in MnP-type phases is closely related to that in FeSi. If we consider the FeSi structure along [111] we find broken helices of T atoms about [111]; in the MnP structure similar helices are parallel to the pseudohexagonal axis. The geometric relation between the FeSi and MnP structures is similar to the relation between the Cu and Mg structures. A special feature is the orthorhombic distortion of the quasihexagonal basal plane which must be ascribed to the insertion of valence electrons of the B component into the A1 correlation about these atoms. We can distinguish two different modes of distortion of the basal plane, one "long" (example MnP) and one "short" (example CoP), in arriving at higher electron concentrations. The distortion of the CoP type can be composed of two distortions of MnP type and the nondistortion of the NiAs type can be composed of three.[36]

The structural sequence NiAl–FeSi–MnP–CoP–NiAs is found once more in the sequence $NiSi_2$–FeS_2(Pyrite)–$FeAs_2$(Lollingite)–FeS_2(Marcasite)–CdI_2. Similar structural arguments may be applied to either of them.[13]

4. CONCLUDING REMARKS

In this survey* of bonding in metallic phases we have stressed the spatial electron correlation relative to electron count and size factors, since good accounts of electron count and size factor are already available in the literature. Spatial correlation has seldom been discussed in the past because the probable electron distances were not known. However, as indicated in Section 3, a system of distances can be found in an analysis of present structural knowledge. Considerations from the

* The author wishes to thank Dr. Hornbogen and Dr. Westbrook for stylistic help.

viewpoint of the spatial electron correlation give such a valuable insight into the phenomena of alloy formation and into finer structural features that it is worthwhile to follow them. Any consideration of bonding in alloys has to take into account spatial correlations; whether or not it will be made in the present simplified manner is a question of further development. In any case, the present considerations are concerned with unsolved problems of high interest to the experimentalist; for example, the reason for shift superstructures, tetragonality of In, and orthorhombicity of MnP. The arguments will help him to find (1) a systematization of his collected empirical facts and (2) useful directions for further research. The present state of the theory can be given here in only a cursory manner. On the other hand, there are many open questions and further work will be needed to produce a more comprehensive picture of bonding in metallic phases.

REFERENCES

1. Hansen M., and K. Anderko, *Constitution of Binary Alloys*. McGraw-Hill, New York, 1958.
2. Laves F., in *Theory of Alloy Phases*, ASM Cleveland, 1956, p. 124.
3. Taylor W. H., *Acta Met.* **2** (1954) 684.
4. Friedel J., *Adv. Phys.* **3** (1954) 446.
5. Eshelby J. D., *Solid State Phys.* **3** (1956) 115.
6. Gschneider K. A., and G. H. Vineyard, *J. Appl. Phys.* **33** (1962) 3444.
7. Sarkisow E. S., *Zh. Fiz. Khim.* **34** (1960) 432.
8. Raynor G. V., in *Theory of Alloy Phases*, ASM Cleveland, 1956 p. 321.
9. Haber F., *Verhandl. Deut. Phys. Ges.* **13** (1911) 1117, 1128.
10. Hume-Rothery W., *J. Inst. Met.* **35** (1926) 295.
11. Pauling L., in *Theory of Alloy Phases*, ASM Cleveland, 1956, p. 220.
12. Wigner E., and F. Seitz, *Phys. Rev.* **43** (1933) 804.
13. Further discussion may be found in Schubert K., *Kristallstrukturen zweikomponentiger Phasen.* Springer Verlag, Berlin, 1964. Several figures of this paper are drawn from that monograph.
14. Fink W. L., K. R. van Horn, and P. M. Budge, *Trans. AIME* **93** (1931) 421.
15. Johansson C. H., and J. O. Linde, *Ann. Phys. Leipz.* **25** (1936) 1.
16. Schubert K., B. Kiefer, M. Wilkens, and R. Haufler, *Z. Metallk.* **46** (1955) 692.
17. The arguments discussed by H. Sato and R. S. Toth [*Phys. Rev.* **127** (1962) 469] give energies which seem too low as compared with the observed distortion energy.
18. Zener C., *Trans. AIME* **152** (1943) 122.
19. Fujiwara K., *J. Phys. Soc. Jap.* **12** (1957) 7.
20. Glassop A. B., and D. W. Pashley, *Proc. Roy. Soc.* **A250** (1959) 132.
21. Watanabe D., and S. Ogawa, *J. Phys. Soc. Jap.* **11** (1956) 226.
22. Johansson C. H., and J. O. Linde, *Ann. Phys. Leipz.* **78** (1925) 439.
23. Wilkens M., and K. Schubert, *Z. Metallk.* **49** (1958) 633.
24. Iwasaki H., *J. Phys. Soc. Jap.* **17** (1962) 1620.
25. Löhberg K., *Z. Metallk.* **40** (1949) 68; see also T. B. Massalski, *Met. Rev.* **3** (1958) 45.
26. Burkhardt W., and K. Schubert, *Z. Metallk.* **50** (1959) 442.
27. Hume-Rothery W., *J. Inst. Met.* **35** (1926) 295.
28. Norbury A. L., *J. Inst. Met.* **65** (1939) 355.
29. Raynor G. V., and J. A. Lee, *Acta Met.* **2** (1954) 616.
30. Kasper J. S., in *Theory of Alloy Phases*, ASM Cleveland, 1956, p. 264.
31. Laves F., and H. Witte, *Metallwirtschaft* **15** (1936) 15, 840.
32. Pötzschke M., and K. Schubert, *Z. Metallk.* **53** (1962) 548.
33. Raman A., and K. Schubert, *Z. Metallk.* **56** (1965) 99.
34. Brukl C., H. Nowotny, O. Schob, and F. Benesovsky, *Mh. Chem.* **92** (1961) 781.
35. Zhdanov G. S., and V. P. Glagoleva, *Tr. Inst. Kristallogr. Akad. Nauk SSSR* **9** (1954) 211.
36. Esslinger P., and K. Schubert, *Z. Metallk.* **48** (1957) 126, 193.
37. Schubert K., *Z. Metallk.* **56** (1965) 93.
38. Günzel E., and K. Schubert, *Z. Metallk.* **49** (1958) 124.
39. Hultgren R., N. S. Gingrich, and B. E. Warren, *J. Chem. Phys.* **3** (1935) 351.
40. Warren B. E., and J. T. Burwell, *J. Chem. Phys.* **3** (1935) 6.

chapter 7

Bonding as Revealed by Catalytic Behaviors

G-M. SCHWAB
Institute of Physical Chemistry
University of Munich,
Munich, Germany

1. INTRODUCTION

The term "intermetallic compounds" can be applied in different ways, depending on the point of interest. In this chapter it designates binary systems in the solid state in which a new kind of crystal lattice, different from those of the components, is formed. In the liquid state, however, systems are also considered in which evidence is present for a covalent relationship between neighboring unlike atoms rather than a metallic bond. The catalytic action of intermetallic compounds has, as yet, been investigated to a relatively small extent, and this is why most of the papers known to the author came from his own laboratories. The reason apparently is that there is as yet little commercial interest in the use of these compounds as catalysts; there is no case in which especially favorable catalytic activity has been observed whereas, on the contrary, an activity poorer than that for the components is relatively frequent. The importance of intermetallic compounds for a general theory of catalysis is also somewhat restricted because usually the question of the type of bond in these compounds cannot yet be entirely answered and therefore relationships between electronic bonding and catalysis can be established only in favorable cases. The Hume-Rothery alloys between group IB metals and those of groups II–V are the most useful ones in this respect. Beside these compounds, the intermetallic compounds of the III–V elements can be evaluated on the basis of the theory of semiconductors. However, in favorable cases it is possible to draw conclusions about the electronic structure of the solid on the basis of catalytic observations. This is also the reason why the major portion of the catalytic literature refers to one special reaction in which the electronic conditions of catalytic activity are known: the dehydrogenation of formic acid

$$HCOOH \rightarrow H_2 + CO_2$$

Here, from a rich literature gained on metallic solid solutions, it is known that unoccupied electronic levels in the solid are favorable for catalysis. Hence a low activation energy is an indication of a relatively empty conduction band and a relatively low Fermi energy and vice versa.

2. STATES OF ORDER OR DISORDER

If a rigorous interpretation of the definition of a new phase is accepted, our notion of intermetallic compounds will also include the formation of an ordered solid solution showing a superstructure compared with the statistically disordered phase of the same composition and basic lattice. An interesting example exists in the Au–Cu series of solid solutions. At the composition Cu_3Au the decomposition of formic acid shows an activation energy of 25 kcal/mole for the disordered (quenched) alloy. After annealing, this value decreases to 21 kcal/mole[1] for samples in the ordered state. The same phenomenon occurs also at the composition CuAu; here the difference amounts to 3 kcal/mole in the same direction. Schneider[2] was able to show that during the annealing process two intermediate states are passed through, which differ from the ordered as well as from the disordered state by having very large activation energies (increases by 8 and 5 kcal/mole, respectively). Similar differences between ordered and disordered states are also observed in the systems Cu–Pd and Cu–Pt, but only in diamagnetic compositions. Since the process of ordering usually produces a decrease in the magnetic susceptibility, the reduced activation energy must be due to an influence of the spin compensation of electrons, probably in the 3*d*-bands. The increase in the intermediate states has not yet found an interpretation.[3]

3. HUME-ROTHERY ALLOYS

A clearer insight into the electronic conditions of catalysis is, of course, offered by the Hume-Rothery alloys. In these Mott and Jones have developed certain ideas on the

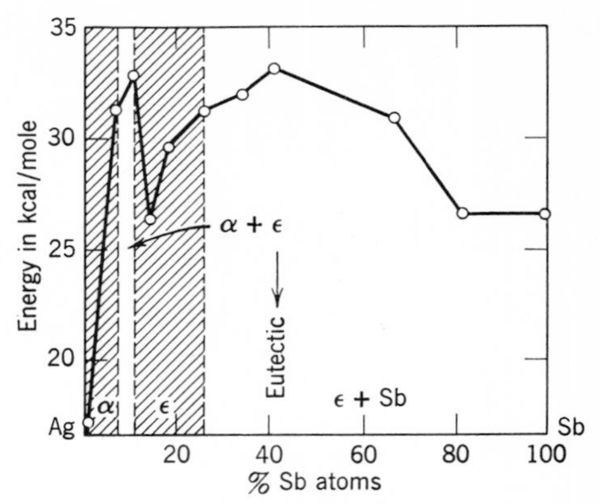

Fig. 1. Activation energy of formic acid dehydrogenation over the system silver-antimony.

occupation of the conductivity band by electrons: within the domain of every intermetallic phase (compound) an increase in the mole fraction of the component with the higher valence signifies an increase in the valence electron concentration (VEC), hence a higher degree of occupation of the first Brillouin zone. According to the principle mentioned, this must result in an increase of the activation energy for reactions of the dehydrogenation or hydrogenation variety and has been amply shown for α terminal solid solutions.[4] Similarly, when the activation energy is followed throughout the whole composition range of a binary system, it is observed that this energy increases with increasing electron concentration within every intermetallic phase. This can be seen within the face-centered cubic α-phase of the system Ag–Sb in Fig. 1, within the α-phase of the system Cu–Sn in Fig. 2, within the hexagonal

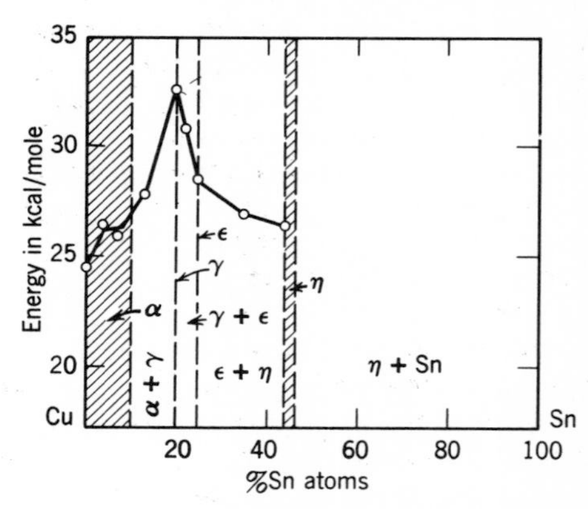

Fig. 2. Activation energy of formic acid dehydrogenation over the system copper-tin.

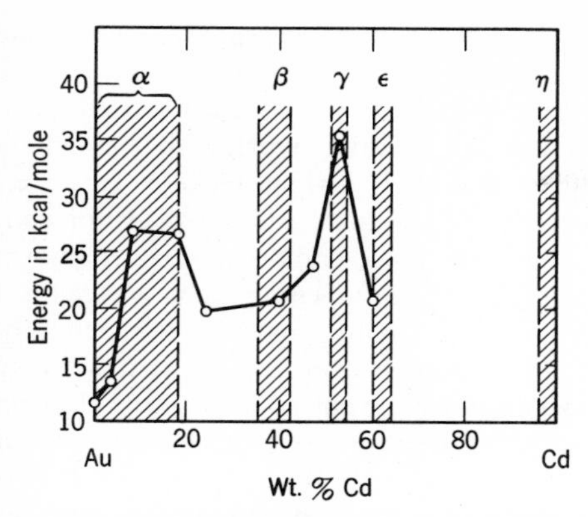

Fig. 3. Activation energy of formic acid dehydrogenation over the system gold-cadmium.

ε-phase of the system Ag–Sb, again in Fig. 1, and within the α-phase of the system Au–Cd in Fig. 3. (Here it must be remarked that in Fig. 1 the third point from the left must be situated in reality within the α-phase.) The increase in activation energy at the left edge of each phase field depends on the square of the increase of the number of electrons.[4] Only qualitative statements can be made with respect to the proportionality factor.

Comparison between different intermetallic phases of the same system also provides some very instructive information: in all three systems investigated it is apparent that a sharp maximum of the activation energy occurs in the γ-phase (a very large cubic cell containing fifty-four atoms), whereas the other phases do not differ essentially from the α-phase. These facts can easily be explained by the zone theory of Mott and Jones: in phases α, β, ε, and probably η the Fermi sphere touches the polyhedron of the first Brillouin zone at only a few points; that is, 30 per cent of the possible electron states are unoccupied. However, in the γ-phase which has numerous lattice planes with intense reflection, the polyhedron is nearly spherical, so that only 11 per cent of the possible states remain unoccupied. These findings constitute strong support for the view that unoccupied conductivity levels are used for substrate chemisorption in hydrogenation and dehydrogenation. It has also been shown[5] that corresponding relationships hold for the dehydrogenation of methanol and ethanol as well as for hydrogenations.

The above-mentioned electron concentration dependence of the activation energy of catalyzed reactions is very characteristically the same as that of the electrical resistivity. For example, in intermetallic phases the same maximum in the γ-phase occurs for resistance as well as for mechanical hardness.[6]

4. HUME-ROTHERY-LIKE PHASES

This term is used to designate intermetallic compounds, the components of which should form Hume-Rothery phases but do not, either because the electron concentration is beyond the limits of Hume-Rothery alloys (η-phases have a VEC = 2.5) or because other structure types are formed (Laves phases). The occupation of the bands in these phases is still essentially unknown, and covalent as well as polar contributions to the binding are probable. In just such cases observations of the catalytic properties permit conclusions about these questions under the supposition that the above-mentioned relationships between conductivity zones and catalysis hold also in these systems. For example, the 32 kcal/mole activation energy for the phase $AuSb_2$[7] does not differ significantly from the value of 31 kcal/mole which has been measured for an alloy containing 3 per cent antimony. The latter apparently represents the saturated α-solution (pure gold shows 12 kcal/mole). Although the formal electron concentration amounts to 3.7 a high maximum value similar to that for a γ-phase is out of the question, and in agreement with the pyrite lattice of this phase it must be concluded that the true conduction electron concentration is relatively low because antimony is bound by covalent or probably even quasiionic bonds. Analogous statements hold for the compound Cu_2Sb[7] with an activation energy of 23 kcal/mole, a value not measurably different from that for the saturated α-phase or for the hexagonal ε-phase. Here also the crystal structure (the antimony atoms are isolated from one another by copper atoms) suggests a quasiionic bonding or at least neutral nature of the antimony atoms.

In the system Cu–Mg two congruently melting compounds Cu_2Mg and $CuMg_2$ are formed. The first one shows an activation energy higher than that of copper (24 instead of 20 kcal/mole[7]). Note that it has been indicated that this fails to be an α-phase (VEC = 1.33) only because of the extreme radius ratio which causes it to adopt a structure of its own; its bonding must be considered as purely metallic. For the other phase $CuMg_2$, a very low activation energy had been found originally.[7] Later, however,[8] it was discovered that in this (and possibly even in the other compound) alloy catalysis did not in fact proceed on the surface of the alloy but rather on finely divided copper in a MgO-carrier formed by the chemical action of the formic acid.

The older findings on Cu_2Mg, however, are in agreement with those on Au_2Pb which has the same structure. Here also, because of the extreme radius ratio, in spite of the formal VEC of 1.33 (lead assumed to be bivalent) an α-phase is not formed; however, a metallic bond can be assumed for the same reasons. Since traces of lead increase the activation energy of gold from 12 to as much as 20 kcal/mole,[9] the proposed interpretations are confirmed by the fact that the activation energy measured for the compound Au_2Pb is as high as 30 kcal/mole.

In the system iron-carbon the compound poorest in carbon is cementite, Fe_3C; however, martensite, the metastable solid solution of carbon in body-centered cubic α-iron, might be considered as an intermetallic compound in the sense that carbon occupies lattice sites different from those of the iron lattice. From this point of view it is interesting that hardened carbon steel containing martensite shows an activation energy of 25 kcal/mole for formic acid decomposition as compared with 21 kcal/mole for the same steel in the annealed state and 22 kcal/mole for pure soft iron. It is suggestive to ascribe this to the fact that carbon contributes four electrons to the electron gas, without, however, increasing the number of occupied atom sites in the iron lattice. This would lead to an increase of VEC by 0.19 electrons/atom per weight per cent of carbon. Considering the parallelism between activation energy and hardness[6] we can thus trace the hardness of martensitic steel back to its increased electron concentration.

5. NICKEL ARSENIDE LATTICES

The hexagonal layer lattice of NiAs is exhibited by the chalcogenides NiS, NiSe, and NiTe and by NiAs, NiSb, and NiBi. Unfortunately, the catalytic action of these compounds cannot be measured by the formic acid method because all of them react with the nascent hydrogen of the decomposed acid, forming the volatile hydride of the added element and finely divided nickel.[10] However, an interesting remark can be made in connection with these compounds: It is known that sulfur or sulfur compounds are catalyst poisons for hydrogenation or dehydrogenation reactions over nickel, and this can be understood by the entrance of a free electron pair from the sulfur atom into the 3*d*-states of the nickel atom.[11,12] With this in mind, it is surprising that nickel sulfide as well as molybdenum sulfide are used as active catalysts for pressure hydrogenation reactions. However, in Fig. 2 we have seen that the η-phase of the system Cu–Sn shows a low activation energy for the dehydrogenation reaction and it must be emphasized that this η-phase has the same lattice as NiAs or NiS. These chalcogenides are thus, in this sense, also η-phases and as such they offer more room in the phase space for conductivity electrons than does poisoned metallic nickel.

6. III-V COMPOUNDS

Intermetallic compounds of elements of the III and V groups of the periodic table crystallize in the diamond lattice with the zincblende structure and are intrinsic semiconductors, capable of being doped as *n*- or *p*-conductors. Among these compounds the catalytic activity of InAs, InSb, and AlSb has been investigated. Systematic differences have been discovered between *n*- and *p*-states of the same compound. Table 1 is a summary of these results,[13] which show unambiguously that for such reactions as the dehydrogenation of formic acid or ethanol or the hydrogenation of ethylene the excess conducting state yields an activation energy 8 to 19 kcal/mole higher than that of the defect-conducting state. This observation is in agreement with the facts observed on metallic conductors which show that unoccupied electronic states or low Fermi energies favor reactions of this type. The opposite type of reaction, that is, the mobilization of oxygen (e.g., in the decomposition of hydrogen peroxide), because of the oxidative properties of the substrate can be observed only with AlSb. This material decomposes in an aqueous solution into aluminum oxide and

TABLE 1

Catalytic Action of III-V Compounds

Catalyst	Conduction Type	Reaction	Activation Energy kcal/mole	Δq kcal/mole
InAs	*n*	$C_2H_4 + H_2$	24	19
	p		5	
	n	HCOOH →	47	8
	p		39	
	n	C_2H_5OH →	48	12
	p		36	
InSb	*n*	$C_2H_4 + H_2$	17	12
	p		5	
	n	HCOOH →	43	11
	p		32	
AlSb	*n*	HCOOH →	38	17
	p		21	

metallic antimony in powder form and therefore continuously offers a fresh pure surface to the solution. Here it has been found that on *n*-type materials hydrogen peroxide is strongly adsorbed and rapidly decomposed, whereas on *p*-type samples the adsorption is weak and the decomposition is slow. Thus, in this case, the chemical activation of the peroxide is favored by the presence of mobile electrons.[14]

7. LIQUID METAL COMPOUNDS

Although in the liquid state intermetallic compounds do not manifest themselves by the occurrence of special phases, there are indications of their existence. One of these is the catalytic behavior of liquid alloys, especially the deviations observed from the regularities found for solid compounds. For example, the activation energy of the formic acid dehydrogenation over the surface of mercury amounts to 45 kcal/mole. By addition of a few per cent per atom of copper it decreases to 26; of silver, to 22; of cadmium, to 21; and of lead, to 24 kcal/mole.[15] Here, for the first time, additives with more electrons than mercury show the same effect as those with fewer electrons, and the only conclusion possible is that in all these cases electrons are not contributed to the electron gas, but, on the contrary, conduction electrons of mercury are used and localized by the formation of covalent bonds in intermetallic molecules. In the original paper arguments which come from different experimental methods are mentioned in favor of this view. Analogous findings obtain for liquid tin: the low activation energy of 17 kcal/mole is scarcely altered by copper; however, antimony lowers it to 10 kcal/mole.[16] With a metallic bond between tin and antimony, an increase of activation energy would have been expected. Here also, one must think of covalent molecules, and this is corroborated by the deviation from Vegard's rule.

8. ADSORPTION BEHAVIOR OF INTERMETALLIC COMPOUNDS

A phenomenon only indirectly connected to our point of view is the following: when, in a solid-state powder metallurgical reaction, an intermetallic compound is formed, it may happen that in an intermediate state an increased specific surface is obtained as manifested by a high gas adsorption for nitrogen. Such a surface increase occurs at about 300°C in the system Co–W during the formation of the compound WCo_3 and vanishes around 500°C.[17] It is probably due to a very fine dispersion of the first nuclei of the compound which in turn would contribute toward an increased catalytic activity.

9. FINAL REMARKS

In this chapter a number of original papers have been treated to show that the catalytic properties of intermetallic compounds, at least for reactions sensitive to electronic effects, can be understood on the basis of the electronic theories of metals on the one hand and of catalysis on the other. Catalytic observations even permit certain conclusions to be drawn regarding the distribution of valence electrons among localized bonding states and conductivity zones. The remaining references are to a number of summarizing reports[18–23]; these papers do not contain facts in addition to those reported.

REFERENCES

1. Rienäcker G., *Z. Anorg. Allgem. Chem.* **228** (1936) 65.
2. Schneider A., *Z. Elektrochem.* **45** (1939) 727.
3. Schneider A., *Z. Elektrochem.* **46** (1940) 321.
4. Schwab G-M., *Trans. Faraday Soc.* **42** (1946) 689; Schwab G-M., and S. Pesmatjoglou, *J. Phys. Chem.* **42** (1948) 1046.
5. For example, Schwab G-M., and E. Schwab-Agallidis, *Ber. Deut. Chem. Ges.* **76** (1943) 1228.
6. Schwab G-M., *Experientia* **2** (1946) 103; *Trans. Faraday Soc.* **45** (1949) 383; Schwab G. M., and M. Tsipuris, *Z. Physik. Chem., Neue Folge* **14** (1958) 65.
7. Schwab G-M., and G. Petroutsos, *J. Phys. Chem.* **54** (1950) 581.
8. Schwab G-M., J. Block, W. Müller, and D. Schultze, *Naturwissenschaften* **44** (1957) 582.
9. Schwab G-M., and G. Holz, *Z. Anorg. Allgem. Chem.* **252** (1944) 205.
10. Schwab G-M., and S. Pesmatjoglou, *Hedvall-Festschrift*, Gothenburg (1948) 533.
11. Schwab G-M., *Second Intern. Symposium on Reactivity of Solids*, Gothenburg (1952) 515.
12. See, for example, Bond G. C. *Catalysis on Metals*. Academic Press, London (1962).
13. Schwab G-M., G. Greger, St. Krawczynski, and J. Penzkofer, *Z. Physik. Chem., Neue Folge* **15** (1958) 363.
14. Schwab G-M., and G. Greger, *Z. Physik. Chem., Neue Folge* **13** (1957) 248.
15. Schwab G-M., and A. Hell, *Z. Elektrochem.* **61** (1957) 6.
16. Schwab G-M., *Dechema Ber.* **38** (1960) 205.
17. Schwab G-M., and R. Ammon, *Z. Metallk.* **52** (1961) 583.
18. Schwab G-M., H. Noller, and J. Block, in Schwab, G-M. *Handbuch der Katalyse V*, Springer, Wien, 1956 p. 330 ff.
19. Natta G., and G. Rigamonti, *ibid*, p. 487 ff.
20. Schwab G-M., and A. Karatzas, *Z. Elektrochem.* **50** (1944) 242.
21. Schwab G-M., *Research* **1** (1948) 717.
22. Schwab G-M., *Z. Elektrochem.* **43** (1949) 274.
23. Schwab G-M. "Catalysts in Theory and Practice," *Chalmers Tekniska Högskolas Handlingar* **81** (1949).

part III

Crystal Structure

chapter 8

Factors Governing Crystal Structure*

F. Laves.

F. LAVES

Institut für Kristallographie und Petrographie
der Eidgenössischen Technischen Hochschule
Zürich, Switzerland

1. INTRODUCTION

As is well known, the chemical compositions of intermetallic compounds usually do not follow the normal valence concepts of chemistry. Take for example the binary system K–Na. Why should the compound KNa_2 exist? Why is there no compound KNa, as both elements are univalent, or even K_7Na?

One of the main reasons for the difficulty in answering such questions is that the attracting forces between the atoms in metallic phases are of a rather general character. Therefore, *the geometrical properties of three-dimensional space* become more significant than in other classes of compounds and they, to a great extent, determine the chemical composition of intermetallic compounds.

2. GEOMETRICAL PRINCIPLES

When metals crystallize, their individual bonding tendencies must adjust themselves to the properties of space. They must meet, as well as possible, the requirements of certain geometrical principles which are recognized, on the basis of many structure determinations, as being primarily responsible for the formation of the structure of metals and their compounds.

A discussion of the structure of the elements will make clear the main principles involved. Let us divide the periodic table by a vertical line (the *Zintl line*) which runs between columns C, Si, Ge, Sn, and Pb on the right side, and B, Al, Ga, In, and Tl on the left, and let us call the elements to the left of this line "metals." Then, looking at the crystal structures of these metals, we note that most structures are cubic or hexagonal close-packed with the coordination number (CN) equal to 12. (This is actually true for 52 out of 91

* Reprinted from *Advanc. X-Ray Anal.*, **6** (1963) with the permission of the author and the publishers: Denver Research Institute, University of Denver, and Plenum Press, New York.

structures.) If we admit some tetragonal or rhombohedral deformation, and if we consider distances differing by not more than 10% as equivalent, we even have 58 close-packed structures. (A survey is given elsewhere.[1])

One may well conceive that some—if not most—metals would prefer arrangements with higher CN's, e.g., 18 or another number greater than 12. This is, however, geometrically impossible if the atoms of a metal are to show indistinguishable behavior. Under this condition, 12 is the highest CN possible.

If we consider the atoms as spheres in contact, space will be filled in the best possible way when the atoms are arranged in cubic or hexagonal close-packed structures with the "ideal" CN = 12. As most of the metals crystallize in these two types, the tendency to good space filling may be called the "space principle."

The next most frequent type of structure—23 of the remaining 33—is the body-centered cubic structure with CN = 8. Thus, there must be factors (e.g., temperature and bond factor, to be discussed later) which counteract the space principle. In this connection, however, it appears remarkable that the CN drops from 12 to 8, whereas sphere packings are available with a CN less than 12 but greater than 8, which would lead to a better space filling than that reached by the body-centred cubic arrangement. However, sphere packings with CN equal to 11, 10, or 9 would not be as symmetrical as the body-centered cubic structure. Thus a tendency to form arrangements of high symmetry becomes apparent. This tendency may be called the "symmetry principle."

A third geometrical tendency which plays an important role in the formation of compound structures may be called the "connection principle." Let us first define this concept of "connection." Imagine a structural arrangement where each atom is connected with all the others. There will be a shortest link. If all the links except the shortest are dropped, those atoms that are still connected with each other form a "connection." If the connection consists of structurally equivalent atoms, it is called *homogeneous*: if it consists of structurally nonequivalent atoms, it is called *heterogeneous*. The connections can be of finite or infinite extent and be one, two, or three-dimensional. Accordingly, we name them islands, chains, and nets or lattices. They are symbolized by the letters *I, C, N,* or *L* (capital letters in the case of homogeneous, small letters in the case of heterogeneous connections). By adding the CN's to these letters, the main geometrical features of structures that are not too complex can be represented by a short symbol. We may now define the *connection principle*: It is the tendency of the atoms to form connections of "high" dimension.

Whereas the importance of the connection principle can best be seen in a discussion of compound structures, it is already observable in the structures of the elements. In these, three-dimensional lattice connections are certainly the most frequent.

A discussion of the hexagonal close-packed metals may be of special interest. As is known, the CN = 12 is only exactly realized if the ratio $c/a = 1.63$. If c/a deviates from this value, and if we retain the exact definition of the CN, the latter drops from 12 to 6. Any deviation of c/a to higher values leads to *net* connections with a CN of 6, whereas any deviation of c/a to smaller values leads to *lattice* connections with a CN also 6.

Table 1

Axial Ratios c/a of Metals with Hexagonal Close-Packed Structure

c/a	Elements[a]
1.56	Li (in the low-temperature state)
1.57	Be, Y, Tm
1.58	Tb, Dy, Er, Ru, Os
1.59	Sc, Gd, Ti, Zr, Hf, Lu
1.60	Tc, Tl
1.61	La, Pr, Nd, Sm, Ho, Re, Am
1.62	Mg, Ce, Co
1.63	Ca, Sr
1.86	Zn
1.89	Cd

[a] Some of the elements listed are polymorphous.

Table 1 shows the actual data. Note that only zinc and cadmium have c/a values above 1.63; the majority has values lower than 1.63. Thus we see that, whereas the space principle is somewhat violated by the deviation from the ideal c/a value, the connection principle is not (except in the cases of zinc and cadmium to be discussed later).

If all these three principles, space principle, symmetry principle, and connection principle, are exactly followed, the cubic close-packed structure results. However, many metals or some of their modifications do not crystallize in the cubic close-packed structure and do not exactly follow the main principles. Thus, there are counteracting factors, and these will now be discussed.

3. FACTORS THAT MAY COUNTERACT THE GEOMETRICAL PRINCIPLES

3.1 Formation of Special Bonds (Bond-Factor)

1. Completion of electron shells where *s* and *p* levels are involved.

(*a*) *Hume-Rothery rule*, which holds for the elements to the right of the Zintl line and leads, e.g., to the CN 4 in the cases of germanium and gray tin.

(*b*) *Hume-Rothery tendency*, which causes white tin to have the CN 6 and influences the structures just to the left of the Zintl line; e.g., gallium has a CN 1 or 7 (depending on the extent to which distances may differ and still be treated as equal); indium has a CN 4, and zinc and cadmium form 6*N* connections, if only the very shortest distances are considered.

2. Completion of electron shells where *d* or *f* levels are involved. Examples are the α and β modifications of manganese, uranium, and neptunium and the 8*L* connections of the modifications of iron and chromium stable at room temperature.

3.2 Effect of Temperature (Temperature-Factor)

1. *Goldschmidt's Rule:* Rising temperature favors lower CN; e.g., titanium, zirconium, hafnium, thorium, thallium have 12*L* connections at low, but 8*L* connections at high temperatures.

2. Sometimes the effect of temperature is reversed: The thermal movement counteracts the bond factor in such a way that the CN increases with rising temperatures; e.g., gray tin and white tin; α and β chromium; α and γ iron; α and γ manganese; γ and δ plutonium.

The same principles and factors can also be recognized in discussing the structure of metallic compounds. However, additional factors play a role due to differences between the components. Before discussing these factors, the space principle must be reconsidered in the light of some very important discoveries published by Kasper.[2]

4. KASPER'S POINT OF VIEW

Kasper observed that several complex structural types in which transition metals play a role (σ-phase, e.g., $Co_{14}Cr_{16}$; μ-phase, e.g., Fe_7Mo_6, and others) have some very interesting characteristics in common which (1) make it relatively easy to describe their geometrical features and (2) open up a very important new line of insight into the problems of CN and space filling.

The main point of the discovery is as follows: if one disregards the chemical difference between the components and considers only the number and positions of the atoms X which surround an atom Y, one finds that compounds of transition elements very frequently show four YX_n groupings with n = 12, 14, 15, or 16, and having the following characteristics: (1) The X atoms form the corners of convex polyhedra bounded by triangular faces only; (2) at least five but not more than six triangles meet in each corner; (3) each atom of the structure can be considered as a Y atom in the above sense. We shall call these four polyhedra "*Kasper Polyhedra.*"

As it is not easy to grasp the geometry of such arrangements in three dimensions, Fig. 1 indicates its essence in two. One sees an *AB* arrangement with small *A* dots and larger *B* dots. Looking at this as an AB compound one would "conventionally" say that it is a structure with a CN 4 as each *A* has four *B* neighbors and vice versa. However, such a description would miss essential points. Following Kasper, one should say: Let us forget about the *A*–*B* difference, but let us look simply at the coordination of the dots regardless of whether they are *A* or *B* dots. Then, we see that *A* has the CN 5 and that *B* has the CN 7, and on the average the CN is 6. Using the X and Y letters in the above sense, we notice that there are YX_5 groupings and YX_7 groupings, and that each dot can be considered as a Y center. Thus, the arrangement can be considered to be the mutual interpenetration

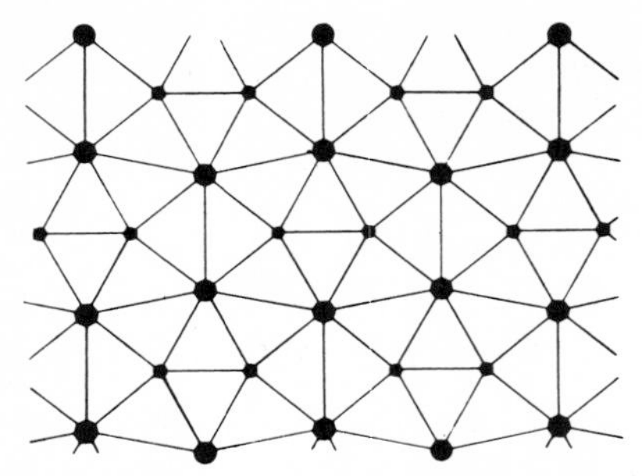

Fig. 1. Two-dimensional *AB* arrangement for demonstrating Kasper's concept of the coordination number. If the small dots represent *A* atoms and the larger ones *B* atoms, this arrangement corresponds to an *AB* compound with the "classical" CN 4, for each *A* has four *B* neighbors and vice versa. If, however, only the number of neighbors is considered but not their chemical difference, and if some variation in distance is admitted, the *A* dots have five neighbors and the *B* dots seven. See in addition Fig. 2.

of polygons with 5 and 7 edges. Figure 2 shows two of them marked more clearly. Such two dimensional polygons correspond to the three-dimensional Kasper polyhedra, and Fig. 3 shows these four polyhedra as originally drawn by Kasper.[2]

Two years later Frank and Kasper[3] showed that these four polyhedra are the only geometrically possible ones which are (1) convex, (2) bounded only by triangles, and (3) have corners in which at least five triangles meet.

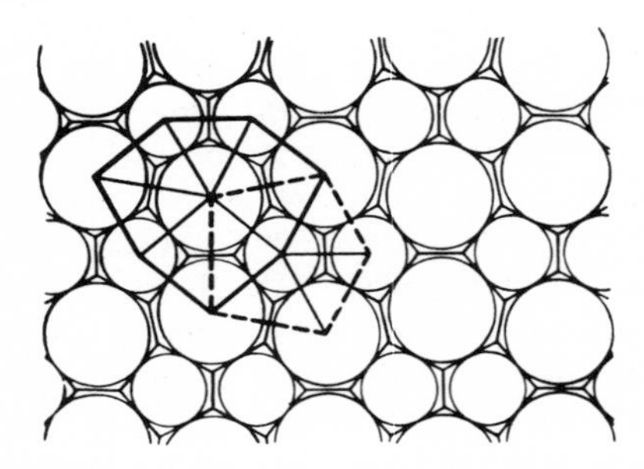

Fig. 2. Two-dimensional analog of Kasper polyhedra, here sketched as polygons, marked by dotted and solid lines. The small and large circles represent *A* and *B* atoms in the positions of Fig. 1. Note that the plane is entirely divided into irregular triangles. They correspond to the irregular tetrahedra into which three-dimensional space can be divided completely and uniquely, if each atom of a structure can be considered to be the center of a Kasper polyhedron.

Table 2 lists their main topological features; e.g., the polyhedron having the CN 16 with respect to its center has 12 corners in which 5 triangles (or 5 edges) meet, and 4 corners in which 6 triangles (or 6 edges) meet, and it is bounded by 28 triangles.

TABLE 2

Geometrical Features of the Kasper Polyhedra

Co-ordination number (CN)	Corners	Number of triangular faces	Edges
12	12 C_5	20	30
14	12 C_5, 2 C_6	24	36
15	12 C_5, 3 C_6	26	39
16	12 C_5, 4 C_6	28	42

The fact that the Kasper polyhedra are bounded by triangles has a very important consequence for the concept of space filling. We consider a grouping YX_n in which the X are at the corners of a Kasper polyhedron, bounded by triangles. If we connect these triangles formed by the X atoms with the Y center, we divide the space of the polyhedron into irregular tetrahedra. Therefore, if a structure is made up of atoms, all of which can be considered as centers of Kasper polyhedra, the whole structure can be divided into irregular tetrahedra. Of course, these tetrahedra *must* be irregular, for space cannot be divided up entirely into regular tetrahedra. However, the important point is that structures in which each atom can be considered as the center of a Kasper polyhedron are made up of a compact packing of somewhat deformed tetrahedra.

If for reasons of merit and convenience we call such structure types "Kasper types," it is interesting to compare these with the "close-packed structures." As can easily be realized, the close-packed structures have tetrahedral and octahedral voids, or, expressed differently, they are packings of tetrahedra and octahedra. In the close-packed structures, the corners of these tetrahedra and octahedra are occupied by the centers of the spheres which touch each other at the middle of the edges. Thus, there are relatively large fluctuations in density within a close-packed structure, a high density in the tetrahedral regions and a low density in the octahedral regions. On the other hand, in the Kasper types, the density distribution is

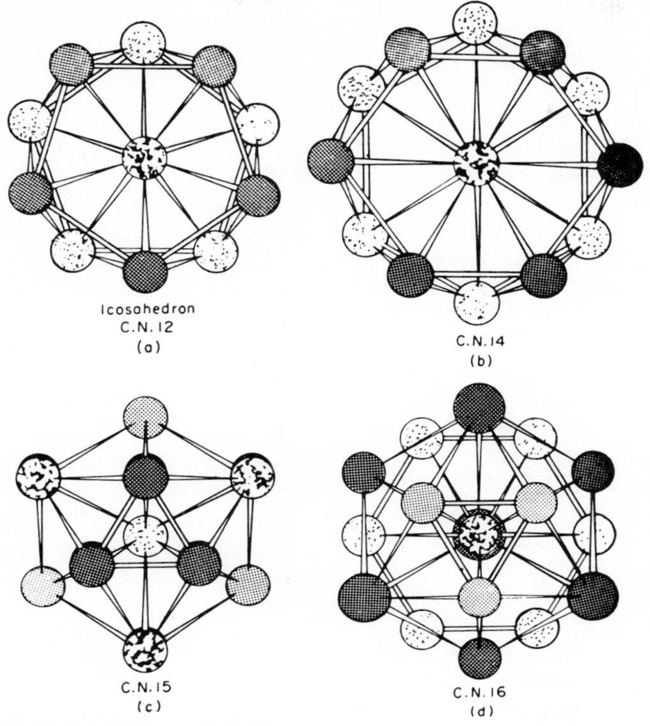

Fig. 3. The four Kasper polyhedra, as first drawn by Kasper. (Reprinted by permission from *Theory of Alloy Phases*, ASM 1956.)

more even, as they are entirely divided into tetrahedra, albeit irregular ones. Thus, the space principle can be satisfied in two ways:

1. Ideal packing of spheres with one CN, namely, 12, but with several types of voids (tetrahedra and octahedra).
2. Ideal packing of tetrahedra with several CN's, namely, two or more of 12, 14, 15, and 16, but with a single type of void (tetrahedra only).

These two subcases of the space principle can be classified as realizations of a long-range space principle and a short-range space principle.

Whereas the metal elements prefer following the space principle by the formation of sphere packings, the metallic compounds follow it rather frequently by the formation of tetrahedral packings, probably for two reasons:

1. The irregularity of the tetrahedra, occasioned by the varying length of their edges, allows for differences in the size of the components.
2. The irregularity of the tetrahedra, occasioned by the varying size of the edge angles, provides some flexibility in approaching special bond angles which may contribute to the stability of a compound, especially in cases where transition elements or elements near to, or to the right, of the Zintl line are involved.

It is rather important to note that both these subdivisions of the space principle are extremes and that several structure types (e.g., α-manganese) are known in which the two subprinciples enter into competition with one another.

5. DEFINITION OF A COMPOUND

For the following discussion of alloy structures, it may be useful to define what I mean when I speak of compounds.

A phase in a q-component system may

exhibit either of two forms: (1) The phase may have a range of homogeneity; i.e., it may form solid solutions by changing its composition without changing its structure type. If the phase has no range of homogeneity or if its range of homogeneity does not extend continuously to a phase in a p-component system which is part of the q-component system (p smaller than q), it is called a compound or more specifically a q-component compound. (The words binary, ternary, and so on are usually used if q is small.) (2) If the range of homogeneity does extend continuously to a phase of the same structure type occurring in a p-component system which is part of the q-component system, it is not a q-component compound, but a q-component solid solution. It may be a solid solution of a p-component compound if the approached p-component phase satisfies the compound definition just given in (1) above.

For some purposes it may be convenient to classify structure types in a similar fashion. A structure type may be called a q-component type if (considering present-day knowledge) q components are needed to form such a type. By the use of this definition we can formulate a rule which is frequently followed by compounds: q compounds tend to crystallize in p-structure types, where p is smaller than q.

As examples of compounds that crystallize in the structure type of elements I mention the binary Hume-Rothery compounds crystallizing in the magnesium, tungsten, and β-manganese types; the ternary compound $ZnCu_2Al$ crystallizing in the tungsten type; the binary compound $Mg_{17}Al_{12}$ and many ternary compounds such as $Fe_{36}Cr_{12}Mo_{10}$ crystallizing in the α-manganese type.

As examples of ternary compounds crystallizing in binary types, I mention MgNiZn crystallizing in the $MgCu_2$ type and $Mg_2Cu_6Al_5$ crystallizing in the Mg_2Zn_{11} type.

In most of such cases the observed isomorphism can be understood on the basis of our present-day knowledge of the influence of the radius ratio of the components, their atomic structure (especially with respect to the number of valence electrons), and their electrochemical differences or similarities. Vice versa, it is frequently possible to make intelligent guesses as to which elements should be alloyed in a q-component system to produce a desired special p-structure type. As an example, I mention the production of the ternary MgLiSb crystallizing (like Mg_2Sn) in the binary CaF_2 type.

6. THE CONNECTION PRINCIPLE IN COMPOUNDS

Proceeding now to a more detailed discussion of compound structure types, I should first like to stress the power of the connection principle in the realm of the structures of intermetallic phases.

Figure 4 shows the structure of the NaTl type, a rather frequent type. At the top right one sees the whole structure and can recognize that it is a superstructure of the tungsten type. We let black dots represent sodium and white ones thallium, at the bottom we see the position of sodium and thallium drawn separately. We can recognize that in both cases the connection is a three-dimensional one, i.e., a lattice connection. Actually each element considered alone forms the lattice connection of the diamond structure. The geometry of the type is summarized in the top left corner in the form of a "connection map." This map shows that sodium and thallium considered together form a heterogeneous lattice connection with the CN 4, whereas considered alone they form homogeneous lattice connections, also with the CN 4. So far, the structure conforms very well to the connection

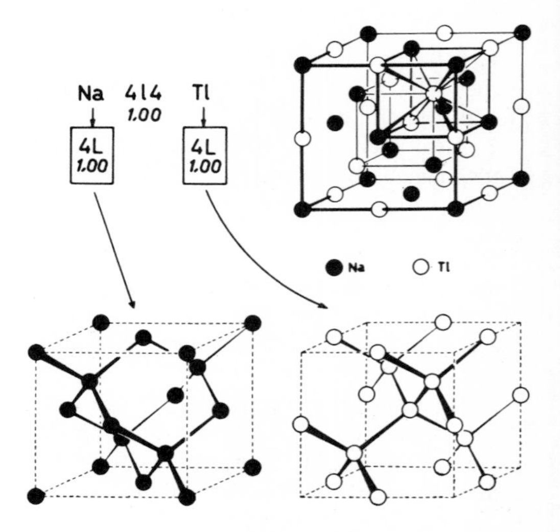

Fig. 4. NaTl-type structure and its lattice connections. (Reprinted by permission from *Theory of Alloy Phases*, ASM, 1956.)

principle. One further important point has to be discussed. Note the figures written below the connection symbols. They represent the relative values of the shortest distances within the respective connections. In this connection map, as in the others to be shown later, the following procedure is adopted: The shortest distance between different elements is taken as unity, as one would generally expect that the *different* components of a compound would attract each other most strongly and, therefore, approach each other as closely as possible, i.e., till the atoms which we consider as spherical touch each other.

The shortest distances within the homogeneous connections are then expressed by figures related to the chosen unit value and written below the respective connection symbol. One may now notice that in this NaTl type the distances in the heterogeneous connection are equal to those in the homogeneous connections. This means that either the A atoms or the B atoms in this structure type do not touch each other, because it would be a chance occurrence if both sorts of atoms were of the same size. In addition, one can conclude from the same kind of reasoning that there cannot be an A–B contact. Following Zintl,[4] who discovered the structure type, the explanation for its existence might be the following: The strongly electropositive sodium gives away its valence electron, which is used by the thallium to develop a sort of covalent bond as in the diamond type of structure. Thus, one may visualize this structure as a somewhat negative lattice connection of the B atoms in which positive A atoms are inserted. The connection principle thus receives a somewhat expanded meaning when we are dealing with compound structure types: *Like atoms may tend to approach each other as closely as possible, as their bonds may add to the stability of the compound.* For this reason the connection principle may compete with the space principle, as can be shown by a comparison of the NaTl and the CsCl types. In Fig. 5 the CsCl type is drawn for different radius ratios. As can readily be understood, A–B contact is possible within a wide range of different values of the radius ratio. Thus, from the simplified point of view that alloy structures may be visualized as sphere packings, the CsCl type would allow a more dense packing than the NaTl type, except for the special radius ratio

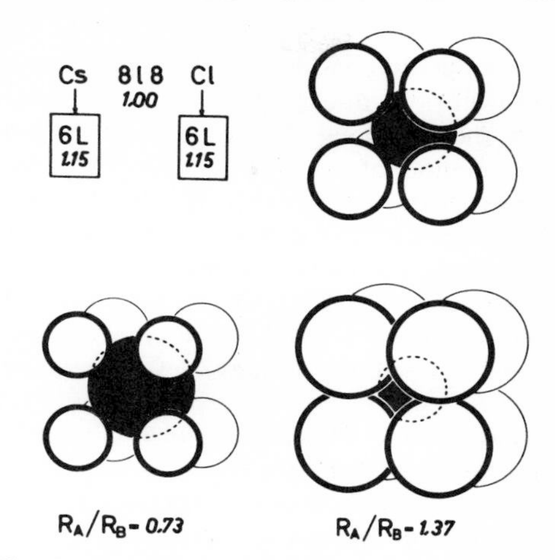

Fig. 5. CsCl-type structure drawn for different radius ratios. (Reprinted by permission from *Theory of Alloy Phases*, ASM, 1956.)

equal to 1, in which case the space filling would be the same in both types.

That the connection principle can also compete with the symmetry principle is dramatically shown by the fact that many AB compounds crystallize in the NiAs type, instead of in the NaCl type, which has the higher symmetry. The reason is that the relative position of the octahedral voids in a hexagonal close-packed arrangement of the "negative" partner provides the Ni "ions" with the opportunity to approach each other more closely than in the cubic NaCl-type structure.

7. ADDITIONAL STRUCTURE-INFLUENCING FACTORS IN COMPOUNDS

In the discussion of element structures I mentioned briefly the influence of factors which counteract structure type expectations based on the main geometrical principles. The same factors apply, of course, to the structures chosen by intermetallic compounds. In addition, however, some new factors come into play due to the fact that compounds consist of atoms differing among themselves. These factors are usually called "size factor" and "electrochemical factor." The special influence of these factors depends largely on the differences the atoms composing an alloy may or may not have with respect to size and degree of electronegativity.

7.1 Size Factor

First let us consider the influence of the size factor. In discussing it, difficulties arise from the beginning. It is true that we have some notion of the size of metallic atoms thanks to the work of V. M. Goldschmidt, W. Hume-Rothery and G. V. Raynor, but the atomic size of an element *A* may vary considerably depending on the partners with which an alloy is formed. A collection of many details pertinent to this question was presented in 1956.[1]

Despite these difficulties, which cannot be discussed here in detail, it can be stated that many compounds owe their existence predominantly to the fact that the radius ratio of their components has a value exactly corresponding to that required to build up a structure satisfying the main geometrical principles outlined above. Looking back to the question put forward at the beginning of this chapter, namely, why does a compound KNa_2 exist, I well remember a discussion with a very good friend of mine about the conditions to be satisfied for a compound to have an MgX_2-type structure. I maintained that the main condition is for the radius ratio of the components to be near the value $1.23 = \sqrt{3}/\sqrt{2}$, which can be derived on purely geometrical lines if one believes in the validity of the main geometrical principles. As the radius of potassium is 2.36 and the radius of sodium is 1.92, the radius ratio $2.36/1.92 = 1.255$, i.e., it is the ideal value for a compound with a MgX_2-type structure. Therefore, I said, "I bet that KNa_2 has an MgX_2-type structure." My friend did not believe me and challenged me by asking: "Why don't you do the structure?" I replied that I did not like handling compounds consisting of alkali metals. So we decided that somebody else would do the job for us. Fortunately, there are chemists in this world, and so two years later Böhm and Klemm[5] proved the existence of a compound KNa_2 by presenting *d* values from an *X*-ray photograph. The issue of the journal arrived just on the evening when I was waiting at the Institute for news of the birth of my first child. To calm myself I took a slide rule and calculated the *d* values to be expected, assuming that KNa_2 has the $MgZn_2$ structure (as it could be seen immediately that the reported *d* values were incompatible with a cubic structure of the $MgCu_2$ type). And the assumption turned out to be correct! So, I won the bet and a child in the same night.

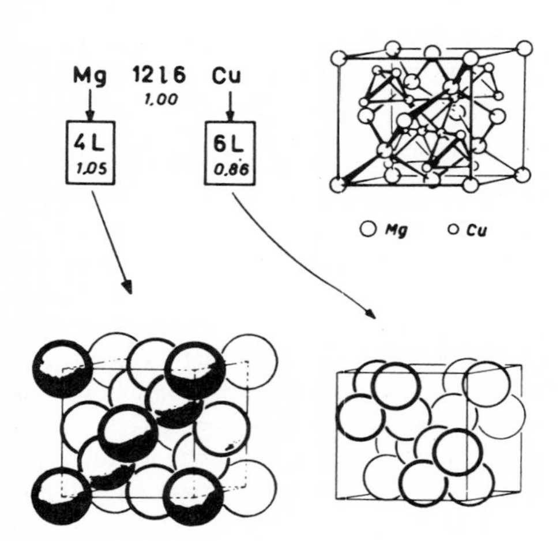

Fig. 6. $MgCu_2$-type structure and its lattice connections. (Reprinted by permission from *Theory of Alloy Phases*, ASM, 1956.)

I said just now that the MgX_2 types best satisfy the requirements of "the main principles" if compounds AB_2 are considered which have a radius ratio approaching 1.23. I would like to discuss this statement using Fig. 6, which represents the cubic version of $MgCu_2$.

Note that the *space principle* is satisfied as indicated by the highest CN's known for AB_2 compounds. Note that the *symmetry principle* is satisfied as the MgX_2 types have either cubic or hexagonal symmetry. Note that the *connection principle* is satisfied as the heterogeneous and both homogeneous connections are three-dimensional ones. In addition, *it is even a Kasper-type structure* as both the magnesium and the *X* atoms are surrounded by Kasper polyhedra. Thus, there is no ground for astonishment that the MgX_2-type structures are met with most frequently both as binary and as polynary compounds. H. J. Goldschmidt[6] points to the fact that if transition metals are involved, MgX_2 types may occur in which the radius ratio of the components becomes rather irrelevant. This is expressed by the fact that large homogeneity regions exist. In such cases the influence of the ratio between valence electrons and the number of atoms ("electron compounds") is considered to play an important role. Earlier,[7] the influence of the electron concentration was considered only with the question of whether the $MgCu_2$ or the $MgNi_2$ or the $MgZn_2$ type is the favored one. Unpublished work[8] of the author shows also that the radius ratio derived from the elementary radii can be quite unimportant for the formation of an MgX_2-type compound if transition metals are involved. For it was found that a compound Mn_2Cu_3Al crystallizes in the $MgCu_2$ type with aluminum (which has the largest radius) in the copper position. This compound forms a series of solid solutions with a compound $MnInCu_4$ in which the chemically analogous indium (because of its very large radius) occupies the magnesium position together with the manganese. Thus, in addition to the geometrical requirements "electron concentrations" and the formation of special bonds within a homogeneous lattice connection appear to play a decisive role in the formation of MgX_2-type compounds and probably in other types of compounds as well.

Before leaving these MgX_2 types I should like to point out one further interesting feature of their structure. If we again choose as unity the shortest distance between unlike atoms—as can be seen in the connection map—and if we measure the shortest distance between like atoms with the same unit, we observe that the average of the distances between like atoms is smaller than the distance between the unlike atoms. This means that if we visualize the type as made up by spheres that touch each other only, the like atoms touch each other and the unlike do not.

This interesting fact can be expressed[9] by a simple quotient Q. Let d_A be the distance between A atoms, d_B the distance between B atoms, and d_{AB} the A–B distance; if we now calculate the quotient

$$Q = (d_A + d_B)/2d_{AB}$$

this Q value differs from structure type to structure type, and is rather indicative of the bond character which may or may not develop in a certain type. Table 3 gives some data concerning structure types. As a rule one can say that *the smaller Q is, the less likely is the occurrence of polar bonds.*

The CN's (in a rigid sense) 12 and 6 are apparently the highest that are geometrically possible for AB_2 compounds and two more types with these high CN's are known: the AlB_2 and the $ThSi_2$ type. (Note: Roman B stands for the element boron; an italic B stands for any element in a compound with the stoichiometrical formula AB_2.) However, these types do not satisfy our main principles as well as the MgX_2 types do and they are, therefore, limited to compounds of a less metallic character, in which Hume-Rothery tendencies may have an essential influence. As far as the radius ratio is concerned, however, these types are comparable to the MgX_2 types.

TABLE 3

Some Q Values of Structure Types

$Q = (d_A + d_B)/2d_{AB}$

Structure Type	Coordination Numbers (CN)	Q
SiO_2	4 and 2	1.82
TiO_2	6 and 3	1.47
CaF_2	8 and 4	1.39
$CuAl_2$	8 and 4	1.06[a]
$MoSi_2$	10 and 5	1.11
$MgCu_2$	12 and 6	0.96
AlB_2	12 and 6	1.01
$ThSi_2$	12 and 6	1.04
ZnS	4 and 4	1.63
NaTl	4 and 4	1.00
NaCl	6 and 6	1.41
NiAs[b]	6 and 6	1.19
CsCl	8 and 8	1.16

[a] Average values of the distances are used.
[b] As c/a varies with different compositions the Q value given refers here to the NiAs compound.

Figure 7 shows the $ThSi_2$ type first described by Brauer and Mitius.[10] Note the dominating $3L$ connection of the silicon atoms which form a sort of three-dimensional graphite structure. Obviously, rather covalent bonds within this $3L$ connection are responsible for the stability of this type, and accordingly the representatives crystallizing in this AB_2 type are those which have chemical compositions with B lying near to the Zintl line. As examples, we can mention: {La, Ce, Pr, Nd, Th, Np, U, Pu}Si_2 and {Pr, Pu}Ge_2.

In a similar way the chemical composition of the representatives of the AlB_2 type (Fig. 8) is characterized by components having (1) the proper radius ratio, and (2) a B component lying near the Zintl line. As examples, we can mention: {Mg, Al, Ti, V, Cr}B_2; $ThAl_2$; {Ca, La}Ga_2; USi_2. It is interesting to compare the structures of the two compounds $CaAl_2$ and $CaGa_2$. In both compounds the radius ratio is virtually the same and suitable for the formation of an MgX_2 type. Thus, for both cases the MgX_2 type was "expected." Whereas $CaAl_2$ was found to have the $MgCu_2$ type, $CaGa_2$ has not. It has the AlB_2 type. Obviously, in the latter case a Hume-Rothery tendency plays a decisive role, as gallium shows this tendency already in the elementary state (CN 1 to 7) whereas aluminum itself crystallizes in a cubic close-packed structure.

Apparently no restrictions exist on the chemical character of the components of MgX_2 type compounds, as the following examples show: KBi_2, KNa_2, $BiAu_2$, $NaAu_2$, and (Au, Be)Be_2. On the other hand AB_2 compounds occur in the $ThSi_2$ or AlB_2-type structures if the B component tends to form bonds of more covalent character (with the small CN 3) than would be possible in a MgX_2-type structure (where the CN is 6). (More data on the influence of the size factor on chemical compositions and structure of metallic compounds have been collected in an earlier paper.[1] A rather recent discussion on the size, and other factors involving the stability of MgX_2-type structures has been given by Schulze.)[11]

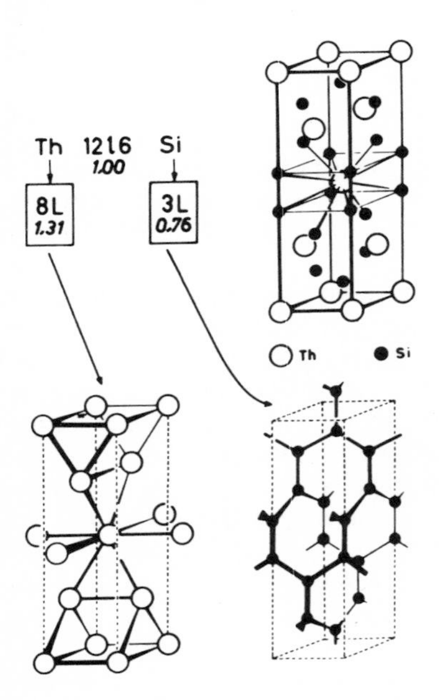

Fig. 7. $ThSi_2$-type structure and its lattice connections. (Reprinted by permission from *Theory of Alloy Phases*, ASM, 1956.)

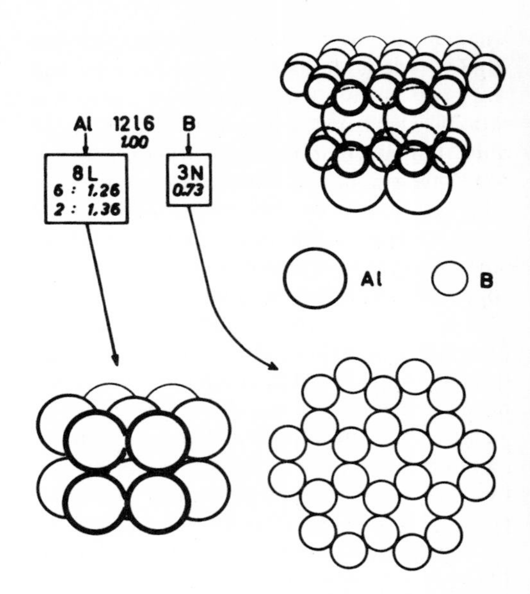

Fig. 8. AlB_2-type structure and its connections. The B atoms form a net connection only.

7.2 Electrochemical Factors

The influence of the "electrochemical factor" can best be demonstrated by discussing the so-called "Zintl-phases."[12] They are compounds in which one component (an electronegative one) lies to the right of the Zintl line and the other (an electropositive one, which as an ion would have a noble gas configuration) to the left, and which can be formally explained on the basis of the classically recognized valences. As an example, I mention Mg_2Sn or $Mg_2{}^{2+}Sn^{4-}$, which crystallizes in the CaF_2-type structure. It is characteristic of the Zintl phases that their structure types have Q values larger than those of other structure types chosen by metallic compounds with analogous stoichiometric and radius ratios, but which are composed by elements which do not follow the restrictions outlined above (Table 3—the CaF_2 and $CuAl_2$ types). This fact indicates that the bonds within the Zintl phases are of a more polar character than in other compounds.

This formal valence concept "explaining" the existence of Zintl phases could be used to predict the existence of many isotopic ternary compounds. As an example I mention the compound $Mg^{2+}Li^{1+}Sb^{3-}$, which crystallizes as $Mg^{2+}Mg^{2+}Sn^{4-}$, and I would not be surprised if these two compounds formed a continuous series of quaternary mixed crystals with the CaF_2-type structure. In the last decade the Zintl phases have gained considerable interest as semiconductors.[13]

As already mentioned, several principles and factors may compete with each other and their ratio of importance may change within a series of compounds which have certain features in common. Table 4 gives such a series. At the top the compound Li_2O is a typical saltlike structure. Going downward we pass the Zintl phases MgLiSb, Mg_2Sn, $MgLi_2Sn$. When the strong electropositive alkali and earth-alkali atoms are exchanged for the less electropositive Cu, Ni, and Mn, the metallic character of the compounds becomes increased and the "holes" in the CaF_2-type structure are filled up to conform with the space principle, which is a leading one for metallic substances. At the bottom of the table, we end up with the iron structure. Thus, a more or less continuous series of compounds exists (in part forming continuous series of solid solutions) in which the bonding character changes from that of a typical salt (as in Li_2O) to that of a typical metal (as in the element iron).

In the Zintl phases the ratio electrons/atoms is clearly defined and can easily be understood on the basis of classical valence considerations. However, there are several further important groups of compounds in which the number of electrons available for bond formation plays a decisive role for the chemical composition of the compounds to be formed, and in which the size factor loses much of its influence.

TABLE 4

CaF_2 or "Antifluorite"-Type Compounds and Structurally Related Substances

Chemical Formula	Lattice Constant a	Element Distribution on Different Positions (only one representative of a face-centered cubic lattice given) 000	$\frac{1}{4}\frac{1}{4}\frac{1}{4}$	$\frac{3}{4}\frac{3}{4}\frac{3}{4}$	$\frac{1}{2}\frac{1}{2}\frac{1}{2}$	
Li_2O	4.61	O	Li	Li		Increasing polar character ↑
Li_2S	5.71	S	Li	Li		
Li_2Se	6.01	Se	Li	Li		
Li_2Te	6.50	Te	Li	Li		
MgLiAs	6.21	As	Mg	Li		
MgLiSb	6.61	Sb	Mg	Li		
MgLiBi	6.75	Bi	Mg	Li		
Mg_2Sn	6.75	Sn	Mg	Mg		
$MgLi_2Sn$[a]	6.69[a]					
	6.75[a]	Sn	Mg, Li	Mg, Li	Mg, Li[b]	
MgCuSb	6.15	Sb	Cu		Mg	
MgCuBi	6.26	Sb	Cu		Mg	
MgCuSn	6.22	Sn	Cu		Mg	
MgNiSb	6.04	Sb	Ni		Mg	
$MgNi_2Sn$	6.05	Sn	Ni	Ni	Mg	
$(Cu, Ni)_3Sb$	5.86	Sb	Cu, Ni	Cu, Ni	Cu, Ni[b]	
Cu_3Sb	6.00	Sb	Cu	Cu	Cu	
Cu_2MnSn	6.17	Sn	Cu	Cu	Mn	
Cu_2MnAl	5.9	Al	Cu	Cu	Mn	
$(Cu, Mn)_3Al$	5.9	Al	Cu, Mn	Cu, Mn	Cu, Mn[b]	↓
Cu_3Al	5.84	Al	Cu	Cu	Cu	Increasing metallic character
Fe_3Al	5.78	Al	Fe	Fe	Fe	
Fe	5.72	Fe	Fe	Fe	Fe	

[a] Two phases of such composition (or near to it) were found to coexist (unpublished work of the author).
[b] No decision has been reached yet on the actual position of Mg and Li, Cu and Ni, Cu and Mn.

I mention the well-known Hume-Rothery phases,[14,15] for example, the CsCl-type (or tungsten-type) compounds in which the electron/atom ratio has a value near to 1.5. (Other well-known Hume-Rothery compounds crystallize in the β-manganese type, in the γ-brass type, and in the ϵ-brass type, which is a disordered hexagonal close-packed structure).

Within the last decade a large number of other structure types have become known, the stability of which also depends mainly on the ratio of valence electrons/atoms or on the ratio of missing *d* electrons/atoms. It is characteristic for the representatives of these types that their composition is dominated by transition elements. They are usually called "electron compounds" and have at first sight rather complex atomic arrangements. However, applying the points of view developed by Kasper[2] and Frank and Kasper,[3] we find we can grasp the main features of the structures relatively easily. Among others the σ-phase structure and the α-manganese-type structure should be mentioned here. P. Beck and his co-workers[16] in Urbana were especially successful in exploring the chemistry of these compounds.

There is still another interesting group of types in which the number of electrons available plays an important role. I am referring to those types in which the presence of boron, carbon, nitrogen, or oxygen helps the formation. I mention the $E9_3$ type: Fe_3W_3C; the $D8_8$ type: Mn_5Si_3 for which the name Nowotny-phases has been proposed,[17] and the $L'1_2$ type: $AlFe_3C$ or BT_3M, where *B* stands for a *B*-metal (e.g., magnesium, zinc, and aluminum) and *T* for a transition element (e.g., iron).

The last type has in recent years been studied especially by Stadelmaier[18] and coworkers. Figure 9 shows its features. In the ideal case it is cubic and its atoms take the

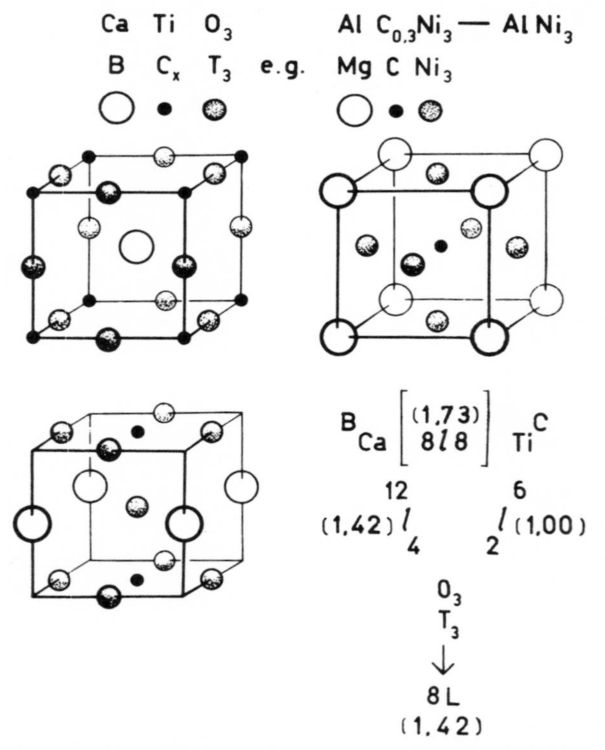

Fig. 9. Different views of the perovskite ($CaTiO_3$–$AlCFe_3$) structure corresponding to different choices of the origin. In the connection map, bottom right, the structurally unimportant Ca–Ti (or B metal-carbon) connection is bracketed.

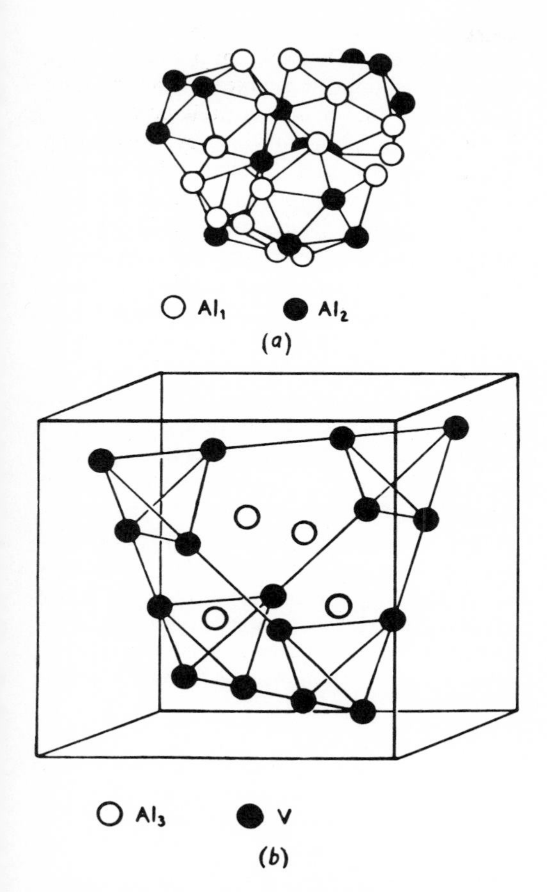

Fig. 10. The structure of α(V–Al) = VAl_{10}, after Brown.

positions of the perovskite structure, i.e., $CaTiO_3$. This already indicates that heteropolar forces are involved in its formation. Figure 9 shows three different views of the structure; depending on which sort of atom is placed in the center of the cell. In addition, it shows the connection map. Let us concentrate on the right top drawing: Note that the shortest distance is the one between carbon and the *T* metal, here taken as nickel. Note the three-dimensional lattice connection made up of carbon and the *T* metal. It will be a negative one and the *B* metal helps to build it up by contributing valence electrons. A further discussion would indicate that the type may best be explained as made up of a negative transition-metal framework in which positive *B* metals and metalloid "ions" are inserted, in much the same way as was discussed at the beginning of this chapter in reference to the NaTl-type structure. I believe Stadelmaier has hitherto found 30 or more representatives of this interesting ternary $AlFe_3C$-type structure.

The last group I propose to discuss, however briefly, is the terrible group of compounds which have the chemical composition T_pB_q in which T means a transition metal and B means a B metal, e.g., aluminum, and in which q is much larger than p. I choose two examples, VAl_{10}, determined by P. J. Brown,[19] and $V_4Al_{23} \approx VAl_6$, determined by J. F. Smith[20] and co-workers.

Figure 10 shows a drawing of the structure as given by Brown. The bottom drawing gives the position of the vanadium (black points). Around these vanadium atoms aluminum atoms are clustered in the form of Kasper polyhedra, as shown in the top drawing.

Another way of looking at the structure would be to search for pertinent connections in the sense discussed before. For this purpose I draw your attention to the vanadium arrangement in the bottom drawing of Fig. 10. It is the same as the copper arrangement in the $MgCu_2$ type. Table 5 contains the atomic distances as given by Brown. There is one remarkably short V–Al distance equal to 2.57.

TABLE 5

Interatomic Distances of (V–Al) in VAl_{10} as Determined by P. J. Brown[19]

Kind of Surrounded Atom	Kind and Number of Surrounding Atoms	Distance
V	6 Al_1	2.826
	6 Al_2	2.572
Al_3	12 Al_1	3.083
Al_1	2 Al_1	2.943
	2 Al_1	2.740
	1 Al_1	2.679
	2 Al_2	2.880
	1 Al_2	2.679
	2 Al_3	3.083
	1 V	2.826
Al_2	4 Al_2	2.882
	2 Al_1	2.679
	4 Al_1	2.880
	2 V	2.572

The space group is *Fd*3*m* with Al_1 in 96(*g*), Al_2 in 48(*f*), Al_3 in 16(*d*) and V in 16(*c*).

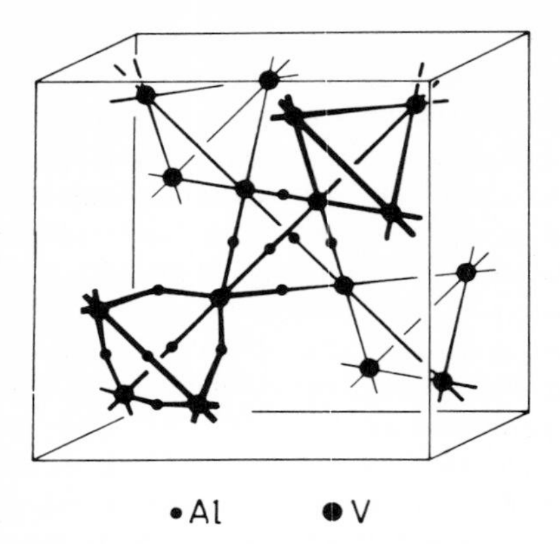

Fig. 11. VAl_3 lattice connection within the VAl_{10} structure. The Al positions are near the centers of the tetrahedra edges. In two of the tetrahedra they are marked by small dots. In one of the tetrahedra the slightly bent edges are sketched. Because the radii of V and Al are 1.36 and 1.43 Å in the elementary state, the actual V–Al distance equal to 2.57 is considerably smaller than the sum of the radii $1.36 + 1.43 = 2.79$. The actual Al–Al distance equal to 2.88, however, is near to the sum $1.43 + 1.43 = 2.86$. For this drawing an origin different from the one used in Fig. 10 was chosen in order to better demonstrate the connection feature here discussed.

Let us now look for the heterogeneous connection which results if we consider this short distance of 2.57 only. The result is drawn in Fig. 11. It shows a VAl_3 framework made up of vanadium tetrahedra, the edges of which are nearly centered by aluminum. Thus, the structure can be visualized as a strongly bonded VAl_3 framework the voids of which are filled by aluminum of normal metallic size.

If we analyze the V_4Al_{23} structure in the same fashion we again find vanadium tetrahedra bonded by aluminum. This time, however, chains only are formed, probably due to the relatively smaller amount of aluminum present (see Fig. 12).

However, in both cases the difference in the electronic structure of the components has a remarkable and specific effect, and the geometrical properties of space finally determine the queer chemical formulas observed.

8. SUMMARY

I hope I have been able to convey the meaning of some geometrical principles, and of some physicochemical factors which may counteract them. Both the principles and the factors play important roles in determining the composition of intermetallic compounds. From another point of view, by considering these principles and factors, rules can be recognized which help in predicting the constitution of alloy systems not yet investigated. The principles are: space principle, symmetry principle, and connection principle.

The factors are: size factor, temperature factor, and electrochemical factor (or bond factors). In principle, every structure type can be discussed by considering its known representatives and by examining them to discover which of the principles and which of the factors appear to be the most determining ones. On the basis of such a discussion, certain rules emerge the value of which can be checked by preparing new mixtures of metals which should conform to such rules. In many cases "predictions" have been verified, thus adding to our confidence in a certain rule. In other cases compounds are formed that do

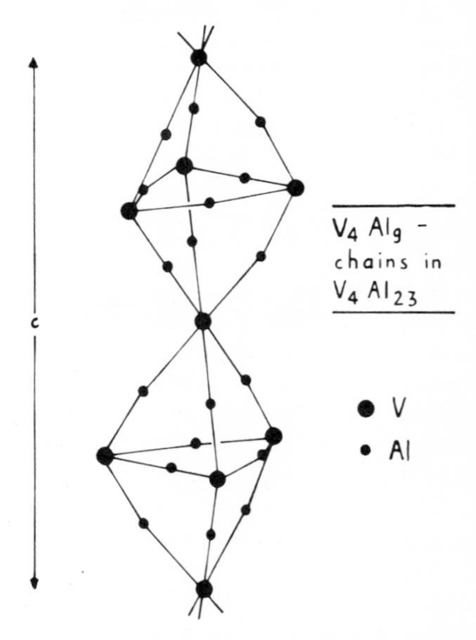

Fig. 12. V_4Al_9 chain connection within the V_4Al_{23} structure. As in the case of Fig. 11, the V–Al distances (2.52, 2.58, and 2.60) are rather small, having the average value 2.57. On the other hand, the Al–Al distances in the drawn connection (2.83, 2.89, and 2.91) are quite normal, having the average value 2.88. Note that the average values of this connection are exactly the same as the corresponding values of the VAl_3 connection depicted in Fig. 11.

not conform to the predicted result. If this is the case a new window is usually opened shedding light on a new direction of knowledge concerning the strange relations between chemical composition and the structure of metallic compounds. Slater[21] (1956) once made a comment as follows:

I don't understand why you metallurgists are so busy in working out experimentally the constitution of a polynary metal system. We know the structure of the atoms, we have the laws of quantum mechanics, and we have electronic calculating machines, which can solve the pertinent equations rather quickly.

However, no structure of a compound has as yet (1962) been predicted by using such higher methods of approach. On the other hand, many unknown compounds and their structures have been predicted and verified by past experience (expressed as "rules" and by the intuition of persons (like Hume-Rothery[14]) who were intrigued by questions like "Why do the compounds Cu_5Zn_8 and Cu_9Al_4 exist, since they have no similarity in chemical composition, but are isomorphous, and have atomic arrangements of a ridiculous complexity?"

REFERENCES

1. Laves F., *Theory of Alloy Phases*. ASM, Cleveland, 1956, pp. 124–198.
2. Kasper J. S., *Theory of Alloy Phases*. ASM, Cleveland, 1956, pp. 264–279.
3. Frank F. C., and J. S. Kasper, *Acta Cryst*. **11** (1958) 184–191; **12** (1959) 483–499.
4. Zintl E., and G. Woltersdorf, *Z. Elektrochem*. **41** (1935) 877–879.
5. Böhm B., and W. Klemm, *Z. Anorg. Allgem. Chem*. **243** (1940) 69–85.
6. Goldschmidt H. J., *J. Less-Common Metals* **2** (1960) 138–153.
7. Laves F., and H. Witte, *Metallwirtschaft* **15** (1936) 840–842.
8. Laves F., *Z. Krist*. to be published.
9. Laves F., and H. J. Wallbaum, *Z. Angew. Mineral*. **4** (1941) 17–46.
10. Brauer G., and A. Mitius, *Z. Anorg. Allgem. Chem*. **249** (1942) 325–335.
11. Schulze G. E. R., *Z. Krist*. **115** (1961) 261–268.
12. Laves F., *Naturwiss*. **29** (1941) 244–254.
13. Mooser E., and W. B. Pearson, *J. Electron*. **1** (1955-6) 629–645; *Progress in Semiconductors, Vol. 5*, Heywood Comp., London, 1960, pp. 103–139.
14. Hume-Rothery W., *J. Inst. Metals* **35** (1926) 295–361.
15. Witte H., *Metallwirtschaft* **16** (1937) 237–245.
16. Beck P., and co-workers (only two of his papers can be quoted here): 1. K. R. Gupta, N. S. Rajan, and P. A. Beck, *Trans. Met. Soc. AIME* **218** (1960) 617; 2. B. N. Das and P. A. Beck, *Trans. Met. Soc. AIME* **218** (1960) 733.
17. Parthé E., *Acta Cryst*. **10** (1957) 768–769; Parthé E., and J. T. Norton, *Acta Cryst*. **11** (1958) 14–17.
18. Stadelmaier H. H., *Z. Metall*. **52** (1961) 758–762; *Acta Met*. **7** (1959) 415–419.
19. Brown P. J., *Acta Cryst*. **10** (1957) 133–135.
20. Smith J. F., and A. E. Ray, *Acta Cryst*. **10** (1957) 169–172; see in addition *Acta Cryst*. **13** (1960) 876–884.
21. Slater J. C., *Theory of Alloy Phases*. ASM, Cleveland, 1956, pp. 1–12.

chapter 9

Close-Packed Structures

Harry J. Beattie, Jr.

HARRY J. BEATTIE, Jr.

General Electric Company
Schenectady, New York

1. GENERAL FEATURES

The lattice types discussed in this chapter are based on "cannonball" packing. This consists of coplanar layers of spheres arranged like hexagonal tiles, as shown in Fig. 1. Successive layers are stacked with centers of spheres falling directly over the centers of triangular interstices of the two-dimensional layer. Each set of four spheres, consisting of three in one layer in mutual contact at the corners of an equilateral triangle and one from the next layer in mutual contact with the three, enclose a "tetrahedral" interstice—the smallest possible three-dimensional interstice formed by spheres—so named because the centers of the four spheres lie at the corners of a tetrahedron. This represents the densest packing of rigid equal spheres.

The general discussion in this section and the one on stacking sequence in the next apply to all of the kinds of structures covered in this chapter and are the only geometric features of a homogeneous collection of atoms. Ordered structures of binary types with atoms of nearly equal size are discussed next, and finally interstitial types with small atoms in the octahedral interstices of metal lattices of this same general class are presented. Not covered are interstitial types with nonoctahedral interstices but still conceivable as close packings of different-size spheres, such as Fe_3C, and topologically close-packed arrangements, which are described in another chapter. The type of structure discussed here is stabilized particularly by filled or nearly filled outermost *d*-shells, which act as almost rigid spheres. In the case of the interstitial compounds, a preference for octahedral covalent bonding is indicated, as will be discussed below. Calculations of X-ray structure factors for the most prevalent ordering scheme and for several stacking sequences are given in the Appendix.

The layers in Fig. 1 have two sets of triangular interstices: one with the apex angle up and the other with the apex angle down; they are designated $\triangle$ and ∇, respectively.*

* This notation was introduced by F. C. Frank.

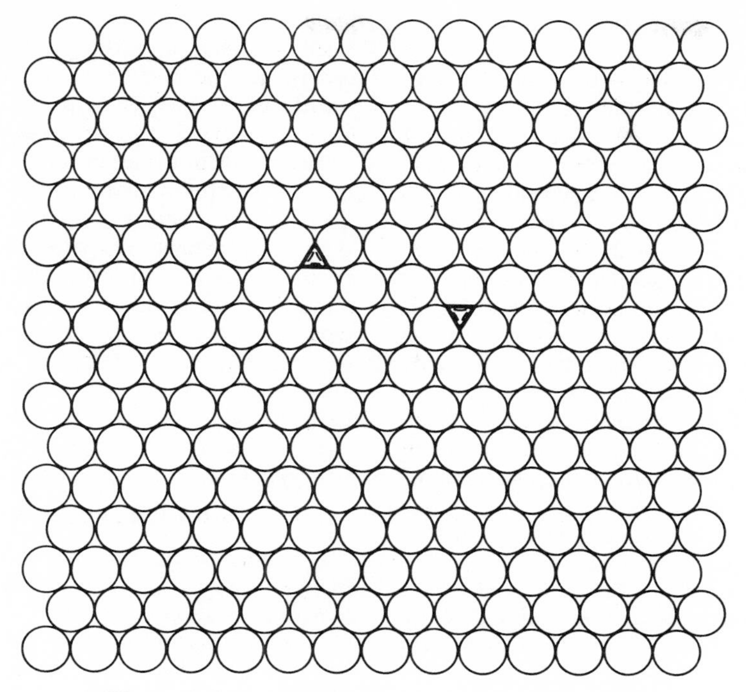

Fig. 1. A close-packed layer of "cannonball" structures.

Each of the two sets forms the same lattice as the spheres themselves, so that the spheres of the next layer will lie on either the $\triangle$ set or the $\triangledown$ set but not both. Let A designate the first layer and B, the next. If the spheres of B fall on the $\triangle$ interstices of A, the $\triangle$ interstices of B will overlie the $\triangledown$ interstices of A. The six spheres that form the triangular antiprism, $\triangledown$ over $\triangle$, enclose an "octahedral" interstice, for a triangular antiprism is also an octahedron. Let $\triangle_A$ stand for "the $\triangle$ interstices of layer A," and so on, and S_A for "the spheres (centers) of layer A"; then between layers A and B there are two sets of tetrahedral interstices, $S_B/\triangle_A$ and $\triangledown_B/S_A$, and one set of octahedral interstices, $\triangle_B/\triangledown_A$. Each of these three sets of interstices lies on the same lattice (translated) as the spheres in a layer; therefore there are just as many octahedral and twice as many tetrahedral interstices contained between layers as there are spheres in a layer. Neglecting surface effects, there is a one-to-one correspondence between the layers and the interlayer regions, so that there is a general rule for all kinds of "cannonball" packings: *there are two tetrahedral and one octahedral interstice per atom.*

2. STACKING SEQUENCE

Let A, B, and C now represent horizontal positions of the close-packed layers. With A at an arbitrary reference position, there is freedom to position the next layer on either the $\triangle_A$ or $\triangledown_A$ interstices. This twofold freedom of choice exists again for each additional layer, that is, the atoms of the *n*th layer may be set upon either the $\triangle_{(n-1)}$ or the $\triangledown_{(n-1)}$ interstices. Thus there are an infinite number of possible stacking sequences. Let the *B* position arbitrarily coincide with $\triangle_A$; then the C position coincides with $\triangledown_A$ (or $\triangle_B$). All successive layers must lie directly over A, B, or C. The various stacking possibilities are generated simply by writing various sequences of the letters A, B, and C, taking care not to have two consecutive letters alike.

The simplest stacking sequence is found in the familiar hexagonal close-packed structure. The layers simply alternate between two of the three horizontal positions in the sequence ABABABAB . . . , or, more briefly, AB,A . . . , where the comma marks the completion of one stacking cycle. Here each basal plane is a plane of symmetry, that is, the entire lattice below it is a mirror reflection of the entire lattice above it.

The simplest stacking sequence that includes all three horizontal positions is ABC,A . . . . Each atom in this case is located at a center of symmetry instead of lying on a plane of symmetry. This structure generally belongs to the rhombohedral class. The primitive rhombohedral cell has a rhombohedral angle close to 60°. It can also be regarded as a face-centered rhombohedron with an angle close to 90°. If this angle is exactly 90°, the lattice loses its uniaxial anisotropy, since all four body diagonals become equal, and the lattice then has the familiar face-centered cubic form. This three-layer stacking sequence is seldom found in a form other than face-centered cubic. The structures of mercury and indium are exceptions, the former being rhombohedral, the latter tetragonal.

The great majority of close-packed phases occur in one of these two stacking sequences, AB,A . . . or ABC,A . . . . In both cases each atom has a coordination of twelve neighbors: six in its own layer and three from each of the two neighboring layers. There is a difference in geometry of the coordinations of the two structures, however, and it is related to the difference in symmetries. The triangular interstices lying above and below an atom in the AB,A . . . sequence are in the planosymmetric prism relationship, for example, △/△, while in the ABC,A . . . sequence they are in the centrosymmetric antiprism relationship, △/▽. Longer stacking sequences involve combinations of these two kinds of coordination.

The possible longer stacking sequences up to six layers consist of one four-layer, one five-layer, and two six-layer cycles. The four-layer cycle alternates the planosymmetric with the centrosymmetric coordinations. We could use an alternative set of symbols to define stacking sequences, writing *p* for the planosymmetric and *c* for the centrosymmetric coordination. The six sequences mentioned are listed in Table 1 in both sets of symbols:

Two-thirds of the atoms in the 6*a* sequence have the centrosymmetric coordination of the *fcc* structure. Its diffraction pattern contains a set of prominent lines appearing to belong to a *fcc* lattice that predominate over the remaining weak lines. Similarly, the 6*b* sequence, which is two-thirds *p*-coordinated, would give a diffraction pattern resembling a mixture of phases predominated by a simple hexagonal close-packed structure.

The missing three-symbol simplified notation, *cpp*, leads to a nine-layer cycle:

ABABCBCAC,A . . . .

The passage of *cpp* through three cycles and of *pcc* through two to complete a unit cell illustrate a general rule that any sequence with an even number of *p*'s passes through one or, more often, three cycles, while any sequence with an odd number of *p*'s must pass through two. This rule results from the fact that each *p* in the sequence reverses the parity of the *c*'s that follow. The element *samarium* has the nine-layer structure.[10] Some other lanthanons and actinons have the four-layer sequence. By and large, however, elements and disordered solid solutions having close-packed lattices are restricted to the two- and three-layer cycles, in which all coordinations are the same. There are a few chemically homogeneous compounds with the four-layer and the six-layer (6*a*) cycles among ordered and interstitial compounds. The ordering pattern of the ordered compounds is defined within each layer, that is, the layers are identical. One of the four-layer compounds, Ni_3Ti, has been determined in several ternary phase diagrams,

TABLE 1

Number of Layers	Conventional Symbols	Coordination Symbols	
		Unit Cell	Simplified
2	AB,A . . .	*pp*	*p*
3	ABC,A . . .	*ccc*	*c*
4	ABAC,A . . .	*cpcp*	*cp*
5	ABCBC,A . . .	*ccppc*	*ccppc*
6*a*	ABCACB,A . . .	*pccpcc*	*pcc*
6*b*	ABCBCB,A . . .	*cpcppp*	*cpcppp*

Fig. 2. Triangular ordering for AB_3 compounds.

and its single-phase field is often a point,* that is, it is not usually found without the ideal composition. A possible explanation is based on the fact that stacking sequences of four or more layers in structures of this kind must be determined by interactions beyond the immediate nearest neighbors. These long-range interactions are probably weak and inoperative unless the lattice is nearly perfect.

3. ORDER

There is a large general class of intermetallic compounds that have ordered structures of cannonball-type packing. These often precipitate coherently with an alloy matrix of similar but disordered lattice structure, and thereby impart high strength. In this case groups of elements, acting as one, substitute as a solid solution at one lattice site or another, creating an "ordered solid solution."

For compounds in which the predominating element lies in the periodic group range from VII–A to III–B, Laves and Wallbaum[8] have shown that this class of structures is preferred to a Laves phase when the outermost *d*-shell is filled or nearly filled. This presumably promotes the tendency to behave as rigid spheres. It now appears that the filled or nearly filled outermost *d*-shell stabilizes the ordered close-packed structures with respect generally to the *bcc* derivative structures and the topologically close-packed phases. (See Chapters 10 and 12.)

In the great majority of cases the ordering is defined within the close-packed layers, which are identical with one another. The most common intralayer ordering pattern is depicted in Fig. 2.[1,2] In this pattern, which has the AB_3 composition, the A atoms themselves form a triangular lattice similar to that of the nonordered close-packed layer and which therefore also has hexagonal symmetry. The unit cell in the basal plane has twice the lattice parameter of the disordered form. Examples of this ordering pattern are known with stacking sequences of the 2-,[3–6] 3-,[2,7–8] 4-,[1,2,8] and 6*a*-layer[9] type. Those having the cubic three-layer sequence far outnumber all the rest combined.[10] In terms of the cubic unit cell, the A atoms are at the cell corners and the B atoms at the face centers; this is the familiar "Cu_3Au structure" (Fig. 4*a*).

The second intralayer ordering pattern is shown in Fig. 3. This also has the AB_3 composition, but in this case the A atoms themselves form a rectangular lattice. The hexagonal symmetry of the close-packed layer decomposes with ordering into a twofold symmetry, which places it in the orthorhombic system. The unit cell face on the basal plane is one of the rectangles outlined by the A atoms.

Both AB_3-type structures are governed by the stacking rule that all A atoms shall be in contact with all-B triangular interstices in neighboring layers, so that each A atom is coordinated entirely by twelve B atoms. This

* Line-like single-phase loci of the $(A,C)_3B$ or $A_3(B,C)$ types have been noted by I. I. Kornilov and others; see Chapter 19.

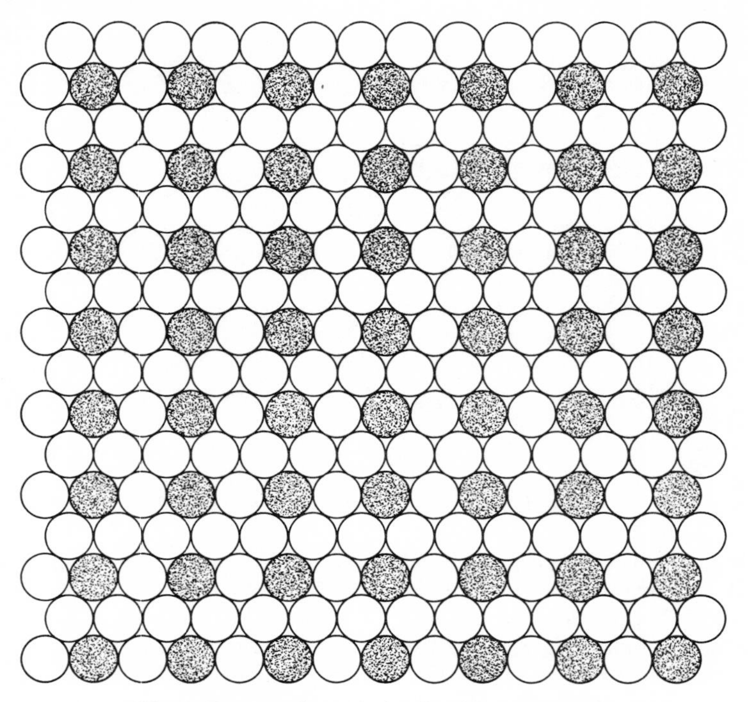

Fig. 3. Rectangular ordering for AB_3 compounds.

principle can generate both the *p*- and *c*-type coordinations with both the triangular and the rectangular orderings. This means that all of the stacking sequences are possible for both orderings; this was intimated above for the triangular ordering where the various stacking sequences that occur were indicated. Two stacking sequences for the rectangular ordering are recognizable among known structures. These represent the all-*p* and all-*c* type coordinations and correspond to the two- and three-layer stacking sequences described in the previous section. The all-*p* two-layer form, designated as the $TiCu_3$ structure, has an orthorhombic unit cell[11–14]; the orthorhombic symmetry is due to the fact that ordering usually distorts the lattice geometrically by a small elongation along the vertical columns of Fig. 3. The all-*c* three-layer form of the rectangular ordering introduces a new contrapuntal polycycle. In order to conform with the aforementioned rule that each A atom be coordinated with B atoms only, the superlattice must span two three-layer stacking cycles. The superlattice stacking sequence may be written ABCA′B′C′,A . . . . These symbols serve to indicate that, while there are only three layer positions as before, there are six superlattice positions.[13]

That this is not so in triangular ordering can be seen as follows: The A atoms themselves form a triangular net with △ and ▽ vertices, as already mentioned. The rule that neighboring layers be placed with A atoms upon all-B interstices amounts to specifying that A atoms of neighboring layers be located over the centers of the △ or ▽ triangles of the A network; therefore, the relationship is identical with that of simple close-packed layers. This equivalence entails the limitation to three positions for the superlattice.

There is another way to illustrate that the cubic stacking cycle of the rectangular superlattice is twice that of the triangular. The A atoms in both nets are centrosymmetric within the layer. The horizontal displacement of A atoms between adjacent layers may be represented by a vector from an A atom to one of its neighboring all-B triangular interstices. The cubic stacking creates for the A atoms a three-dimensional centrosymmetry, as was previously mentioned; therefore, the layer on the other side is displaced by the vector of opposite sign going to the centrosymmetrically opposite all-B triangular interstice. Since the centrosymmetric environment for the A atom holds throughout the lattice, the same horizontal displacement vector occurs between all adjacent layers. Thus, the position of the *n*th layer is found by a translation of *n* vector lengths in the same direction. Applying this to Figs. 2 and 3, the third layer of the triangular ordering lies with the A atoms in register with the A atoms of the initial layer thus completing the cycle, while the third layer of the rectangular ordering lies with A atoms in register with B atoms of the initial layer, and it takes another three layers before the A atoms are in register with those of the initial layer to complete the superlattice cycle.

The superlattice doubling of the *c*-axis entails a doubling of one of the "cubic" axis when the all-*c* coordinated stacking of the rectangular ordering is referred to the cubic system, so that it actually becomes tetragonal with $c/a \sim 2$. Actual compounds that have this structure are tetragonally distorted so that $c/a < 2$; the actual unit cell is tetragonal, for the distortion deviates the "orthorhombic" axes lined along and normal to the close-packed planes (i.e., axes that would be orthorhombic without this tetrahedral distortion and trigonal without the nontrigonal ordering) from right angles.[13] This gives the $D0_{22}$, or Al_3Ti structure.[15] It resembles the Cu_3Au structure in that square-lattice layers normal to the tetragonal axis consisting one-half of A atoms and one-half of B atoms alternate with all-B layers. In the Cu_3Au structure all of the mixed layers are in register, while in the Al_3Ti structure successive mixed layers are displaced by $[\frac{1}{2}, \frac{1}{2}, 0]$, as shown in Figs. 4*a* and 4*b*.

Another ordered AB_3-type structure, namely the $D0_{23}$, or Al_3Zr structure,[15] combines the Cu_3Au and the Al_3Ti structures. It is tetragonal with $c/a \sim 4$. It can be derived by reflecting the Al_3Ti unit cell through one of the all-B tetragonal planes. This produces pairs of successive mixed layers displaced $[\frac{1}{2}, \frac{1}{2}, 0]$ from the next pair, and so on (Fig. 4*c*). The ordering pattern in the close-packed planes is a hybridization of the triangular and rectangular orderings, so that strips of rectangles alternate with strips of △ and ▽ superlattice triangles (Fig. 5). The A atoms lie at the intersections of the semiregular tessellation[16] with the Schläfi symbol $3^3 \cdot 4^2$. The vector method described

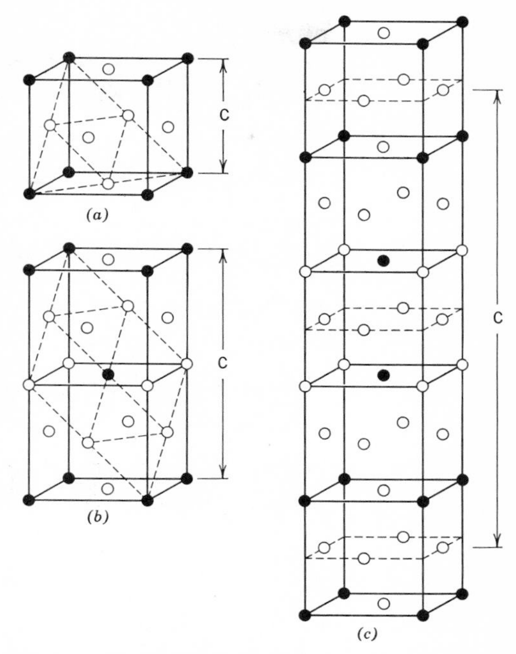

Fig. 4. Tetragonal unit cells of AB_3 compounds in all-*c* coordination stacking. (*a*) Cu_3Au-type (triangular ordering); (*b*) Al_3Ti-type (rectangular ordering); (*c*) Al_3Zr-type (dashed lines indicate reflection planes).

above for determining the superlattice positions in successive layers of the all-*c*-type coordination stacking shows that twelve close-packed layers are necessary to complete a superlattice cycle. There are instances within the twelve layers where some but not all of the A atoms lie above A atoms of the starting layer, so that a vector reaching an A atom no longer necessarily indicates the end of the cycle; the terminal A atom must be equivalent to the initial A atom by translation, not by rotation, reflection, or inversion.

The long superlattice stacking cycles of the Al_3Ti and Al_3Zr structures do not necessarily involve the weak long-range forces that are involved in the positional four- and six-layer stackings of the triangular ordering. The latter result from combinations of *p*- and *c*-type coordination, whereas the former are the unavoidable consequence of the principle that A atoms be centrosymmetrically all-B coordinated.

Other variations of AB_3 intralayer ordering exist along the lines of that shown in Fig. 5, in which panels of blocks (rectangles) alternate

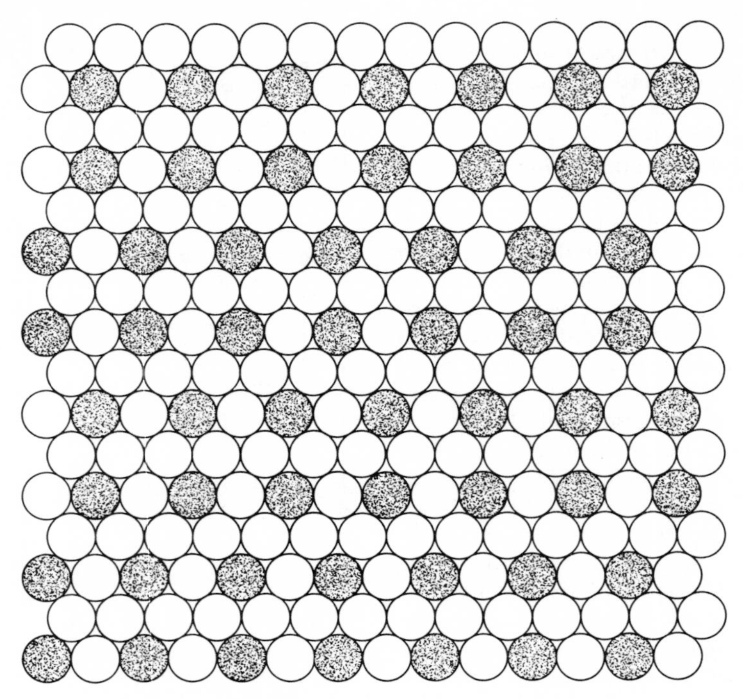

Fig. 5. Tessellated $3^3 \cdot 4^2$ ordering for AB_3 compounds.

with panels of triangles of the *A* atoms. The "one-block" tessellation shown in Fig. 5 exists not only in the pseudocubic form just described but also takes the second (*b*) six-layer and the nine-layer stacking sequences in Au_3(Cd, In) compounds.[18] This same series of compounds also produces the "two-block" tessellation (Fig. 6) in the nine-layer sequence and an eight-layer sequence not previously mentioned. The latter sequence is

ABABACAC,A . . .

or, in simplified coordination symbols, *cppp*.* The Au_3(Cd, In) series also includes a four-block tessellation (Fig. 7) with a two-layer stacking sequence.

This gives an idea of the infinite variety of structures just within the AB_3 intralayer ordering group. Intralayer ordering compounds of AB_4 composition will be considered now.

The Dl*a*, or Ni_4Mo structure,[14,17] is an interestingly complex variation of the centrosymmetric stacking sequence. The close-packed layers are equivalent intralayer orderings with an AB_4 composition. All three close-packed directions have repeat units of one A atom followed by four B atoms: in other words, along any close-packed row of atoms, every fifth one is an A atom. One can construct such a layer (Fig. 8), and, using the vector method described above, establish that a centrosymmetric stacking sequence with all-B coordinated A atoms results in a *fifteen-layer* superlattice cycle. Any vector from an A atom to any of the all-B triangular interstices, near or distant, leads to the same fifteen-layer cycle; hence, no other structures are generated by this principle for this ordering.

The superlattice stacking sequences can be exhibited by the use of relatively simple ideographs that are here proposed. Starting with an A atom a straight line is followed along the normal axis through the close-packed planes; a plane where this line passes through an atom is represented by a hexagon, ⬡, and where it passes through a triangular interstice is represented by a triangle, △ or ▽. Thus the two-layer sequence is ⬡ △ ⬡ △, . . . and the three-layer "cubic" sequence is ⬡ △▽ ⬡△ ▽, ... A dash attached to any of

* The reverse symbol, *pccc*, is also possible in eight layers. Try it. Then try *cpppp*.

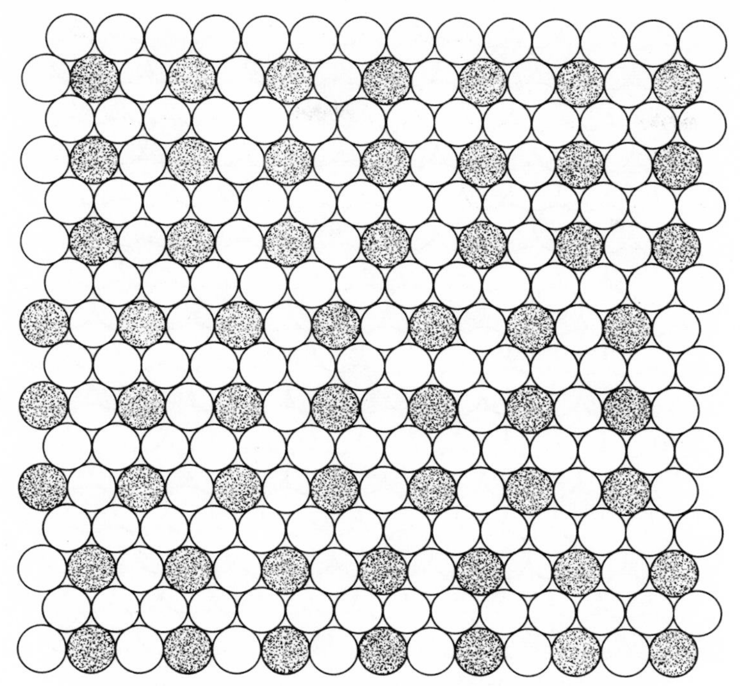

Fig. 6. Two-block tessellated ordering for AB_3 compounds.

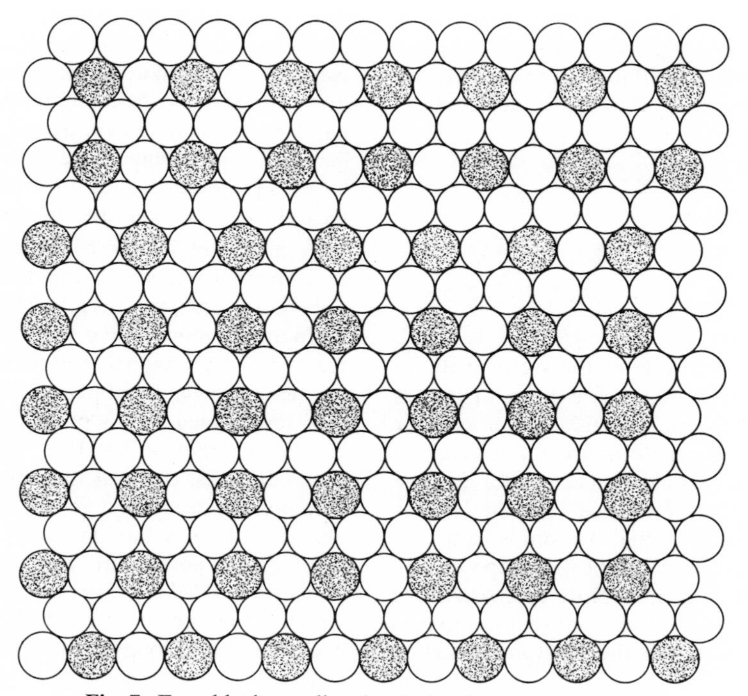

Fig. 7. Four-block tessellated ordering for AB_3 compounds.

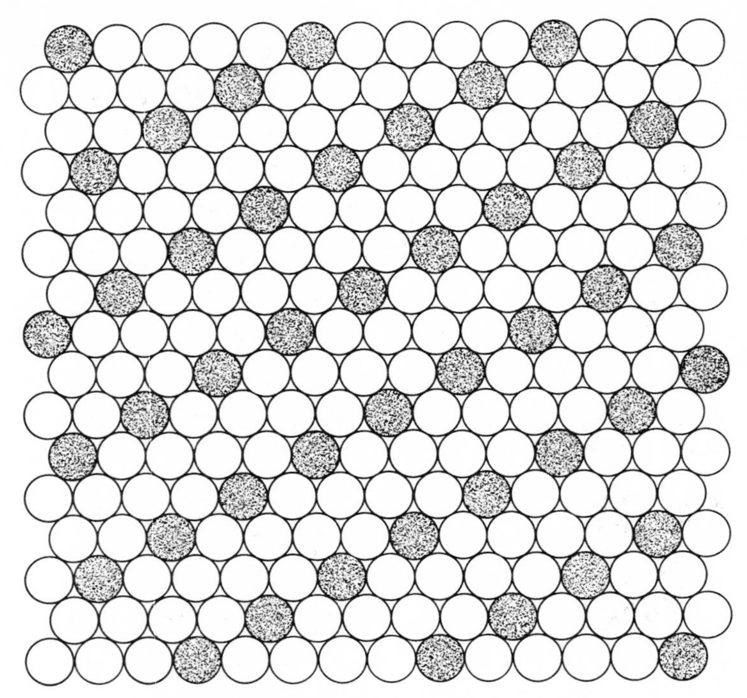

Fig. 8. Oblique ordering for AB_4 compounds.

the corners of the hexagons or triangles or in the center of a hexagon indicates the location of an A atom. The four examples of cubic sequence described above are depicted as follows:

Cu_3Au ⏣▽△,⏣...

Al_3Ti ⏣▽△⬡▽△,⏣...

Al_3Zr ⏣▽△⏣▽△⬡▽△⬡▽△,⏣...

Ni_4Mo ⏣▽△⬡▽△⬡▽△⬡▽△⬡▽△,⏣...

These are all derivable by the vector method, except for a restriction on the choice of vectors in the case of the Al_3Zr. This superlattice layer differs in that the A atoms are not in a centrosymmetric environment. The choice for the all-B triangular interstice as the vector terminus from the initial A atom is restricted to those centrosymmetrically opposed by another all-B interstice. Note that Al_3Zr has the only sequence of the above four that is not inverted through the A atoms (starting points).

Returning to the Ni_4Mo structure, we might expect by analogy with the other structures that its unit cell could be tetragonal with a_0 equal to one fcc cell and the *c*-axis five cells deep to correspond with the fifteen-layer cycle. Actually,[14,17] unlike the case with the AB_3 compositions, the A atoms in the tetragonal basal plane do not come at the corners of a fcc face, but, being more diluted, form a larger square lattice that is twisted from the fcc cell by 18.5° about the tetragonal axis, and with a period longer by a factor of $\sqrt{\frac{5}{2}}$. It is surprising, however, that the *c*-axis spans only one fcc cell, not five. An explanation can be found by representing both the disordered fcc and the tetragonal superlattice unit cells on the tetragonal basal plane (Fig. 9). The fcc cells represented in this projection are joined corner-to-corner along [111] with respect to the cubic axes. Starting with both cells on a common center, they go out of phase in neighboring cells but ultimately get back in step at the coordinate [555] in the cubic system or [425] in the tetragonal system, which is just the length of the fifteen-layer cycle. Thus the long cycle arises as a "beat frequency" between the disordered lattice and the superlattice. Analogies exist in the music of George Gershwin, whose "exploitations of concurrent and apparently irreconcilable

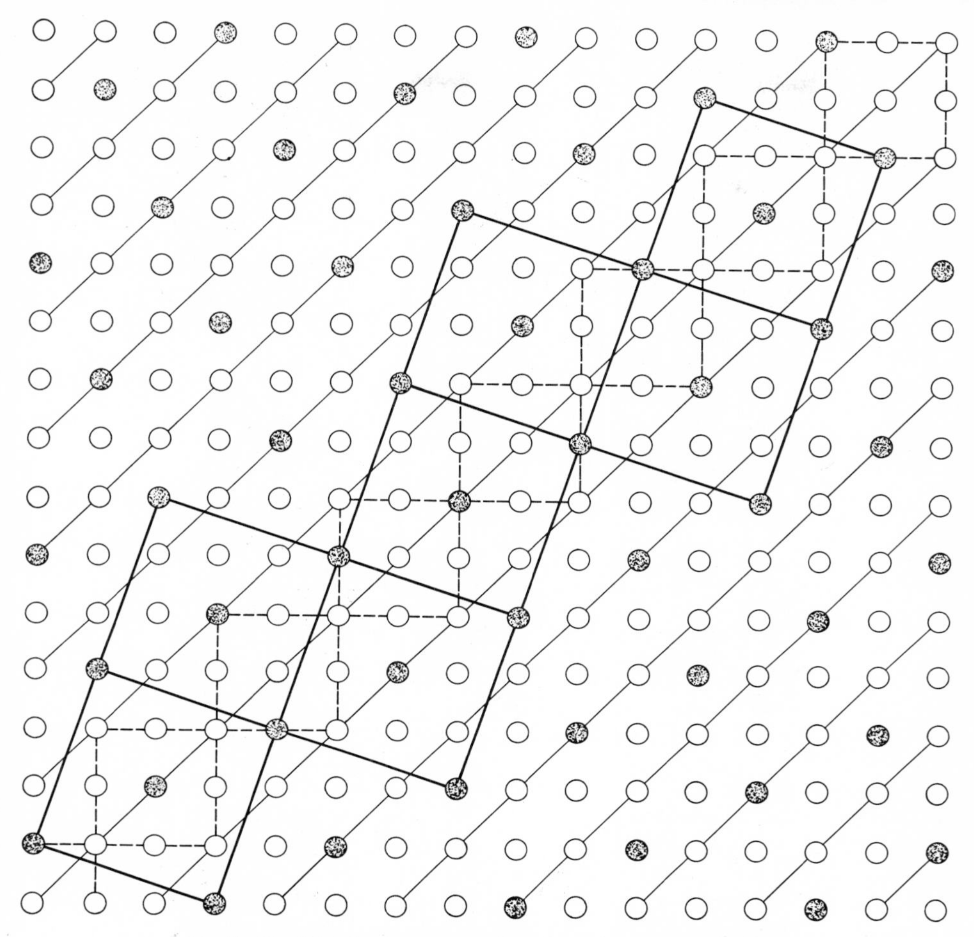

Fig. 9. The Ni_4Mo structure projected on the tetragonal plane.

rhythms are at first exasperating and eventually exciting."[19]

Another intralayer ordering of the AB_4 composition has been found recently in Au_4Zr[14] (see Fig. 10). It occurs in the two layer all-*p*-coordinated stacking sequence and also conforms with the rule that A atoms be all-B coordinated. The A atoms are not centrosymmetric with respect to the superlattice. By vectoring them on an all-*B* triangular interstice with a centrosymmetrically opposite partner, a cubic stacking sequence may be hypothesized to preserve the all-B coordination for A atoms. This gives a six-layer superlattice sequence of the form ⏣▽△̍ -⬡▽̩△, ⏣. The -⬡ figure shows the stacking to be noncentrosymmetric in contrast with the previous six-layer centrosymmetric example, Al_3Ti. If the Au_4Zr ordering occurs in cubic stacking sequence, it has not been recognized to the author's knowledge. It would seem likely that the noncentrosymmetric layer environment of the A atoms would discourage a centrosymmetric stacking sequence, but the Al_3Zr structure shows that this need not be so. The distribution of A atoms in a cubic or tetragonal plane is topologically equivalent to that in Fig. 10; thus either system of axes yields an orthorhombic superlattice.

Figure 11 shows an ordering pattern for the A_3B_{10} composition. It exists in two-layer stacking sequence as the compound $Cu_{10}Sb_3$.[20] As in the preceding compounds, the A atoms are all-B-coordinated, and this is possible only with the all-*p* stacking sequence because there are three sets of A atoms; in centrosymmetric stacking each set would have to be vectored in a different direction.

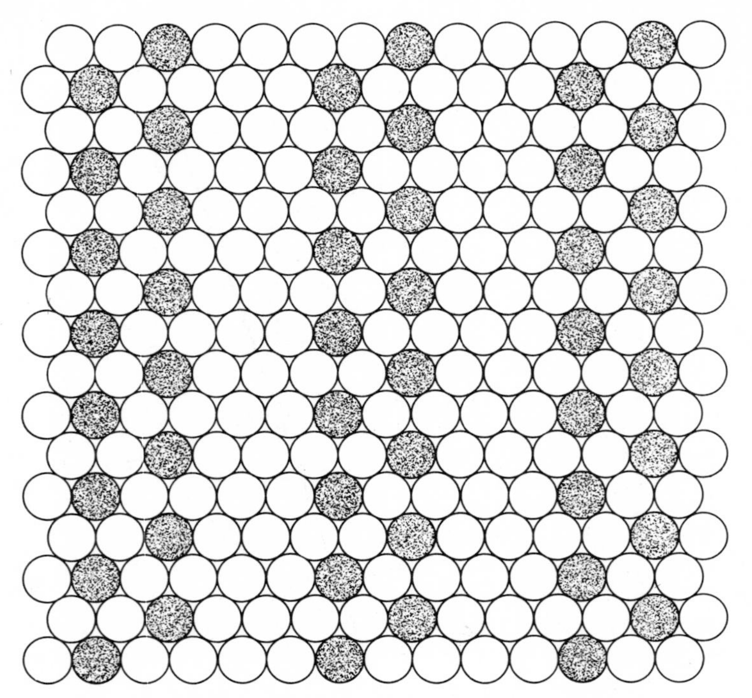

Fig. 10. Triangular-striped ordering for AB_4 compounds.

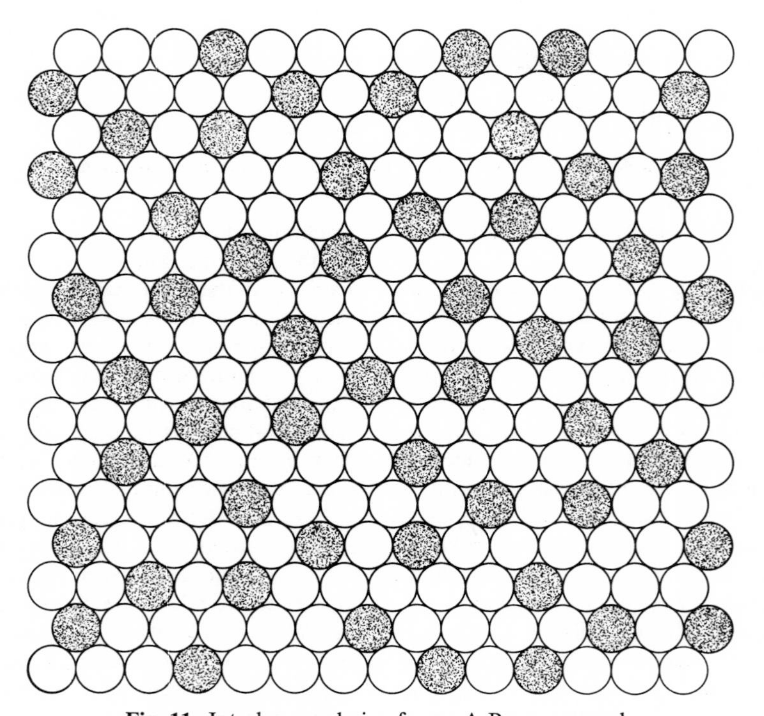

Fig. 11. Intralayer ordering for an A_3B_{10} compound.

The last example of intralayer ordering in "cannonball" structures has the AB composition, where the two kinds of atom are on an equivalent basis and it is impossible to form a coordination entirely of one species. The close-packed layers consist of close-packed rows alternately all-A and all-B (Fig. 12), and with these layers forming the basal planes of the unit cell, the system is orthorhombic. This ordering occurs in the two-layer (AuCd type, B19), the three-layer (CuAu type, $L1_0$) and the six-layer (TaRh type)[49] stacking sequences. The three-layer structure has a tetragonal unit cell, distorted from the fcc unit cell by shortening along the normal to the tetragonal planes, which are alternately all-A and all-B sheets. The six-layer structure is a twinned version of this.

Next to be considered is the question whether interlayer ordering may occur, that is, with nonequivalent close-packed layers. For instance, close-packed sheets of A and B atoms can alternate in a regular manner. The tungsten carbide, WC, structure is sometimes depicted as ordered hexagonal close-packed with alternating sheets of tungsten and carbon atoms, but this is not one of the "cannonball" structures under discussion; the cannonball lattices entail contact of neighboring atoms within the layers, whereas in WC the carbon atoms are enclosed in interstices of the tungsten lattice and hence cannot be regarded as in contact with each other. Even the tungsten lattice does not conform with our definition of close packing, for the successive layers are stacked in direct coincidence instead of following the atoms-upon-triangular-interstices rule that was established in the beginning.

There is a metal-metalloid-interstitial compound that may be regarded as an interlayer ordered stacking of layers of pure metal and metalloid on a six-layer cycle, but its axial *c/a* ratio is some 29 per cent below the ideal ratio for closest packing of equal spheres. This is titanium carbosulfide,[21] earlier known as "Y phase" or "tau phase," which often occurs as an inclusion in commercial alloys with titanium additives.[22,23] Using chemical-symbol subscripts on the stacking symbols, the close-packed lattice is written as:

$$A_{Ti}B_SA_{Ti}B_{Ti}A_SB_{Ti},A_{Ti}\ldots$$

The carbon atoms are in the octahedral interstices between the consecutive titanium layers, and the formula is Ti_2SC.

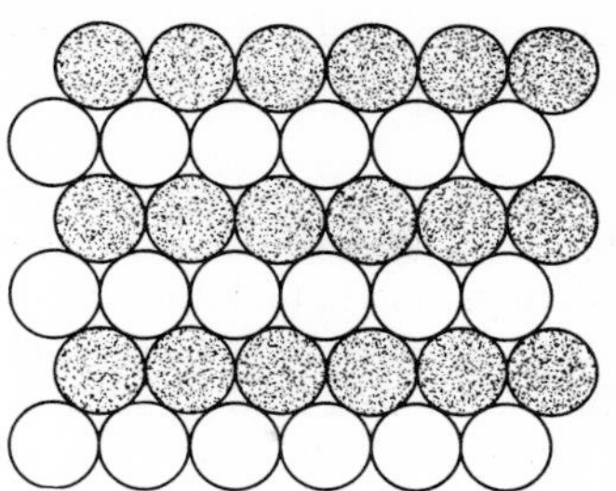

Fig. 12. Ordering for AB compounds.

Interlayer orderings of true cannonball structure are found in the platinum-rich Cu–Pt phases. In the CuPt, or $L1_1$, structure alternate Cu and Pt layers are stacked in cubic sequence to give a six-layer cycle for the superlattice: $A_{Cu}B_{Pt}C_{Cu}A_{Pt}B_{Cu}C_{Pt},A_{Cu}\ldots$. In this case the lattice is rhombohedrally distorted with a slight compression along the normal to the close-packed layers. The rhombohedral unit cell contains eight face-centered pseudocubes and therefore thirty-two atoms. The platinum-rich phases of the Cu–Pt system are also rhombohedrally distorted cubic sequences of six-layer superlattice cycles of two kinds of layers.[24] The sequences are of the same nature as in the CuPt phase, except that the copper layers are replaced with ordered mixtures of copper and platinum. In $CuPt_3$ all-Pt layers alternate with CuPt layers of the same form as in Fig. 12, whereas in $CuPt_7$ all-Pt layers alternate with $CuPt_3$ layers in the triangular ordering form (Fig. 2).

The final example is cited because at first glance it appears to be an interlayer ordering with the hypothetical five-layer stacking sequence. It is the Al_3Ni_2, or $D5_{13}$, structure. Topologically it is equivalent to the five-layer structure, and is represented in our notation by $A_{Al}B_{Ni}C_{Al}B_{Al}C_{Ni},A_{Al}\ldots$. The axial ratio, however, departs very widely from the ideal for close packing. Bradley and Taylor[25] interpret it as a deficit NiAl structure (ordered bcc) with an entire layer of nickel atoms missing out of six layers along the trigonal axis. This interpretation is geometrically sound, for the average "rhombohedral" angle between layers is 106°, which is quite close to the value, 109.5°, for the body-centered cubic lattice and very remote from the value 60°, for ideal cannonball packing. This example is therefore excluded from the class of structures under discussion, and the five-layer stacking sequence remains only hypothetical for this class.

The number and variety of interlayer orderings is much smaller than those of intralayer orderings, and to find them one generally has to turn to compounds in which forces other than those that lead to sphere packing play a significant role. The only exception noted here is the Pt-rich Cu–Pt phases.

4. INTERSTITIAL STRUCTURES

In 1930 Hägg[26] noted that a number of carbides, nitrides, oxides, hydrides, and borides consisted of lattices of metal atoms in cannonball-type packing with the smaller carbon, nitrogen, oxygen, or boron atoms in the octahedral interstices or the still smaller hydrogen atoms in the tetrahedral interstices. These comprise the bulk of what may be called "simple" structures among interstitial compounds. Hägg found empirically that if the ratio of atomic radii taken as the interstitial-to-metal atom exceeds 0.59, the interstitial compound has a complicated structure. The latter structures provide larger interstitial holes, with the surrounding metal atoms usually forming the corners of triangular prisms or, in a few structures, square antiprisms. If the critical atomic radius ratio is interpreted as the limit of accommodation in octahedral interstices, one avoids the difficulties presented by the simple AlB_2 structure for borides with atomic sizes above the critical ratio, and the complicated Fe_3Mo_3C structure, whose carbon-surrounding metal component, Mo, brings the ratio below critical; AlB_2 has triangular-prism, while Fe_3Mo_3C has octahedral interstices for carbon. This interpretation meets with some difficulty in the carbide WC and the diborides ZrB_2 and HfB_2, which have triangular-prism interstices at the ratio 0.55. Notwithstanding this, the Hägg ratio can still be regarded as the limit of accommodation in octahedral interstices, only instead of being a critical value between mutually exclusive regions, it is the upper limit of a region of overlap. Thus, cannonball interstitial compounds are excluded among pairs of elements with radius ratios exceeding the Hägg limit, 0.59.

There are two well-populated types of interstitial compounds, in which the interstitial atoms lie in the octahedral interstices of cannonball structures. They reflect the elemental cannonball structures in that they are either two-layer (hexagonal) or three-layer (cubic) sequences. These consist primarily of carbides and nitrides, which are mutually intersoluble so that a continuous series of "carbonitrides" may be synthesized. Similarly, there is complete mutual solubility between pairs of metal components that form such compounds so long as their atomic radii differ by no more than 15 per cent, in conformity with the Hume-Rothery rule for solid solutions. (A large amount of data and reference to original work on this exists in Reference 10.)

In principle, all of the octahedral interstices are occupied by C or N in the cubic phases. Actually up to 25 per cent of the octahedral interstices may be vacant at random, so that compositions vary from MX to M_4X_3, where M is the metallic and X the interstitial element. In the hexagonal phase, on the other hand, generally one-half or less of the octahedral interstices are occupied, and they are customarily referred to as the M_2X phases. (An exception is δ-NbN. A two-layer stacking with all octahedral interstices filled would have the "NiAs," or $B8_1$ structure, and δ-NbN is the only interstitial compound by our definition that Pearson[10] lists under this structure.) Jack[27] has observed superlattice lines for Fe_2N, and has thus shown that the N atoms occupy half of the interstices not at random but in an ordered pattern. He also found that other Fe:N ratios, notably 3:1, have a similar significance. This ordering of the interstitial element has not been detected in most of the other M_2X phases, as this requires an unusually sensitive technique, but the consistent occurrence of this stoichiometric ratio as an upper limit for the interstitial element serves to indicate that it exists in all M_2X phases.

Andrews and Hughes[28] have noticed a curious regularity in surveying a large number of carbides and nitrides of transition elements. Almost without exception, a hexagonal close-packed transition element forms only cubic interstitial phases, while a face-centered cubic transition element forms only hexagonal interstitial phases; furthermore, the body-centered cubic transition elements without close-packed allotropes (groups V and VI) form both cubic and hexagonal carbides and nitrides. The only exception to this mutual-exclusion rule is cobalt, which has hexagonal form both as an element and as a nitride.

Before Andrews and Hughes discovered this general rule, it was well-known that the hexagonal close-packed group IV elements and the body-centered cubic group V elements form face-centered cubic lattices as carbides and nitrides. Rundle[29] suggested that the fcc metal lattice is stabilized by the octahedral coordination it provides for the interstitial atoms, thus imposing suitable boundary conditions for a stable set of orthogonal octahedrally-directed wave functions expressible as a linear combination of the *s* and the three *p* spherical harmonics. (Orthogonal sets of directed wave functions expressed as linear combinations of the spherical harmonics are called "hybrid orbitals.") Hume-Rothery[30,31] pointed out that this does not account for the preference for the octahedral holes in the fcc lattice over those in the hcp lattice, but noted that the lattice type does make a difference as far as the coordination of nonmetal atoms around the metal atoms is concerned. The nonmetal atoms surround the metal octahedrally in the fcc and prismatically in the hcp lattice. The octahedral coordination fixes suitable boundary conditions for the stable d^2sp^3 octahedral hybrid orbitals of the transition metals. This could explain the great stability of the cubic form at the MX composition. The cubic MX carbides as precipitate particles in steels are generally not replaced by other carbides after extensive aging as the M_2X carbides are.[28]

The cubic MX carbide may be represented by stacking symbols as follows:

$$AXBX'CX'',A\ldots$$

where A, B, and C represent the three positions of metallic layers as before, and the X's represent a complete filling of the octahedral holes between the metallic bilayers. The primes on the X's represent a lateral shift in position from that of any dissimilarly adorned X.

There are two examples of the MX composition with long stacking sequences. The ternary nitride Ta_3MnN_4 has a four-layer interlayer stacking cycle[32] in the form:

$$A_\alpha XB_\beta XA_\alpha X'C_\beta X',A_\alpha\ldots$$

where the α subscript represents a MnTa composition, while β represents an all-Ta layer. Note that the α-layers have centrosymmetric type coordination while the β-layers have planosymmetric type. The other example is one of the several structures reported for MoC. Its low-temperature, highly purified form, first determined by Nowotny et al.,[33] has a six-layer stacking cycle: AXBX′CXAXCX′BX,A . . . . Note that both the four- and six-layer cycles comply with the rule that the interstitial layer X changes position on either side of a *c*-type coordination but is positioned similarly on either side of a *p* type. The six-layer sequence of MoC is the one consisting of two-thirds *c*-type and one-third *p*-type coordination. As mentioned earlier, with a nearly ideal axial ratio for sphere packing, which MoC has, the X-ray diffraction pattern appears to be dominated by a face-centered cubic pattern with a set of weaker extra lines and was so initially reported by Nowotny and Kieffer.[34] This form of MoC has been found only in synthesized specimens carefully prepared for a high degree of purity. It has since been confirmed by others.[35] Small substitutions of other elements for either the metal or the interstitial stabilizes the cubic form,[36] as do combined elevated temperature and pressure.[37] A cubic form of tungsten carbide at very high temperatures has also been reported.[38] Apparently, interstitial impurities in MoC cause it to crystallize in the non-cannonball structure found by Kuo and Hägg.[39,10] The stringent conditions for the formation of the six-layer MoC structure, like those for the four-layer Ni_3Ti structure mentioned earlier, again reflect the operation of weaker long-range forces that reach beyond the nearest-neighbor coordination of atoms.

Monoxides of all of the first-row transition elements except chromium have the cubic MX structure. The monoxide of manganese and the group VIII elements of the first long period undergo rhombohedral (Mn, Fe, Ni) or tetragonal (Co) distortion when cooled below the antiferromagnetic Néel temperatures.[40,41,10] Vanadium monoxide has a range of homogeneity from V_4O_3 to V_5O_6.[42,43] Since no more than 100 per cent of octahedral sites can be occupied, there is an implication that in the region of oxygen deficiency, this phase is a metallic interstitial compound similar to the carbides and nitrides, while in the region of excess oxygen, it is an ionic compound with the V^{2+} cations in the octahedral interstices of the close-packed lattice of O^{2-} anions. Excess oxygen has been observed for most of the other monoxides, and deficient-oxygen phases seem lacking except for that of vanadium.[10] We are now at the edge of the domain of metallic compounds and must withdraw.

Here we are concerned with metallic interstitial compounds, which, to a first approximation, can be viewed as packing of neutral spheres with two classes of size determined by the limiting Hägg ratio. This view is justified by (1) the metallic electrical conductivities, (2) the disregard of usual valences, and (3) the continuously variable compositions. Attempts have been made experimentally to show a partial electron transfer from the one atomic species to the other, but different approaches indicate opposite directions of transfer. Evidence for the atomic neutrality of titanium nitride has been offered by Philipp with electrochemical measurements.[44] At the MX composition, we have seen, there are factors other than sphere packing, such as Rundle's[29] and Hume-Rothery's[30] suggestion of stable octahedral wave functions, which stabilize the cubic form or moderate departures from it.

Turning now to interstitial compounds in which no more than one-half of the octahedral holes are filled, it has been found that the orderings of the interstitial atoms generally follow a mutual-avoidance principle.

Jack[27,31] found two nitrogen orderings for the Fe_2N composition. There is no way to fill one-half of the octahedral interstices in a single bilayer in an ordered arrangement and maintain hexagonal symmetry. The hexagonally symmetric Fe_2N compound alternates bilayers that are two-thirds filled with ones that are one-third filled. The occupied octahedral interstices of the two-thirds filled bilayers are at the nodes of a hexagonal net with the vacant interstices at the centers of the hexagons. The one-third filled bilayer is the reverse of this, the occupied interstices falling directly over the vacant interstices of the neighboring bilayers and lying at the nodes of a trigonal net, which defines the unit cell of the superlattice, this having a parameter $\sqrt{3}$ times the dimension of the metallic layers. The metallic layers, being planosymmetric, have one set of triangular interstices, say $\triangledown$, forming tetrahedral interstices on both sides and the other set, $\triangle$, forming octahedral interstices on both sides. All of the $\triangle$ set have nitrogen on one side and a vacant interstice on the other. If X_h represents the hexagonal network of interstitials filling two-thirds of the octahedral interstices and X_t represents the trigonal net that fills one-third of these holes, the stacking symbol for this M_2X structure is: AX_hBX_t,A . . . .

Two other orderings of interstitials in M_2X phases are represented in the structures of ζ-Fe_2N and Co_2N.[45] Unlike the above structure for ϵ-Fe_2N, these have intralayer ordering with the interstitial atoms filling one-half of the octahedral holes in each bilayer. The superlattice is reduced to orthorhombic symmetry. In the ζ-Fe_2N phase, the nitrogen atoms form zigzag chains topologically similar to those in Fig. 10. In the Co_2N phase the nitrogen atoms form rectangular nets like the A atoms in Fig. 3; however, the metal lattice is quite distorted with the close-packed layers rippled. If X_r represents the rectangular nets and X_c the zigzag chains, the stacking symbols are:

$$Co_2N\text{: } AX_rBX'_r,A \ldots$$

$$\zeta\text{-}Fe_2N\text{: } AX_cBX'_c,A \ldots$$

The ϵ-Fe_3N structure contains only the trigonal nets that fill one-third of the holes, and is represented by:

$$\epsilon\text{-}Fe_3N\text{: } AX'_tBX_t,A \ldots$$

There is a continuous solid solution from ϵ-Fe_2N to ϵ-Fe_3N. The X_h layers in Fe_2N can be regarded as two X_t sublattices: $X_h = X'_t + X''_t$. In the intermediate range between Fe_2N and Fe_3N there must be random vacancies on, for example, the X''_t sublattice of these layers, whereas the X'_t sublattice of these layers and X_t of the alternating layers remain fully occupied.

The γ'-Fe_4N phase consists of a fcc lattice of iron with one nitrogen atom per unit cell occupying the octahedral interstice at the body center, so that the nitrogen atoms themselves lie on a primitive cubic lattice, still apparently governed by a principle of maximum separation. The phase Fe_3AlC has a similar structure, except that the metal atoms are ordered in Cu_3Au ($L1_2$) fashion; with this arrangement the carbon atoms are coordinated entirely by iron atoms, and the interstices they occupy (25 per cent of all that are available) are the only ones so coordinated. Stadelmaier[46]

TABLE 2

Summary of Cannonball-Structure Types

Description	Prototype	Structure Symbol	Population
I. Elemental			
A. AB,A . . .	Mg	A3	~25 elements
B. ABC,A . . .			
1. Cubic	Cu	A1	~25 elements
2. Tetragonal	In	A6	1
3. Rhombohedral	Hg	A10	1
C. ABAC,A . . .	La	—	4
D. ABABCBCAC,A . . .	Sm	—	1
II. Ordered			
A. Intralayer orderings			
1. AB_3 composition, all-B coordination for A.			
a. Triangular superlattice			
(1) AB,A . . .	Ni_3Sn	DO_{19}	15
(2) ABC,A . . .			
(a) Cubic	Cu_3Au	$L1_2$	75[2]
(b) Tetragonal	Ti_3Cu	$L6_0$	3
(3) ABAC,A . . .	Ni_3Ti	DO_{24}	7[2]
(4) ABCACB,A . . .	Co_3V	—	1[9]
b. Rectangular superlattice			
(1) AB,A . . .	Cu_3Ti	—	7[11–13,20]
(2) $(ABC)^2$,A . . .	Al_3Ti	DO_{22}	7
c. $3^3 \cdot 4^2$ tessellated superlattice			
(1) $(ABC)^4$,A . . .	Al_3Zr	DO_{23}	1
(2) ABCBCB,A . . .	$Au_{78.5}Cd_{13.5}In_8$	—	1[18]
(3) ABABCBCAC,A . . .	$Au_{75}Cd_{15}In_{10}$	—	1[18]
d. Two-block tessellation			
(1) ABABCBCAC,A . . .	$Au_{75}Cd_{11}In_{14}$	—	1[18]
(2) ABABACAC,A . . .	$Au_{75}Cd_{3.3}In_{21.7}$	—	1[18]
e. Four-block tessellation			
(1) AB,A . . .	$Au_{75}Cd_{2.5}In_{22.5}$	—	1[18]
2. AB_4 composition, all-B coordination for A			
a. Oblique superlattice			
(1) $(ABC)^5$,A . . .	Ni_4Mo	$D1_a$	7[14]
b. Triangular-striped superlattice			
(1) AB,A . . .	Au_4Zr	—	1[14]
(2) $(ABC)^2$,A . . .	(hypothetical)	—	0
3. A_3B_{10} composition, all-B coordination for A			
a. AB,A . . .	$Cu_{10}Sb_3$	—	1[20]
4. Superlattice stacking schemes of the previous cubic or pseudocubic $(ABC)^n$,A . . . stackings			
a. Triangular (tetragonal $c/a \sim 1$)			
|⦶▽△|⦶. . .	Cu_3Au, Ti_3Au	$L1_2$, $L6_0$	
b. Rectangular (tetragonal, $c/a \sim 2$)			
|⦶▽△|-⬡-▽△|⦶. . .	Al_3Ti	DO_{22}	
c. Tessellated (tetragonal, $c/a \sim 4$)			
|⦶▽△|⦶▽△|-⬡-▽△|-⬡-▽△|⦶. . .	Al_3Zr	DO_{23}	
d. Oblique (fractional enlargement of tetragonal superlattice creates long "beat" cycle)			
|⦶▽△|-⬡▽△|⬡▽△|⬡▽△| ⬡-▽△|⦶. . .	Ni_4Mo	$D1_a$	

TABLE 2 (*Continued.*)

Description	Prototype	Structure Symbol	Population
II. ***A.*** **4** (*contd.*)			
e. Striped (orthorhombic superlattice on both hexagonal and cubic axes)			
|⏀ ▽△|–⬡ ▽△|⏀. . .		(hypothetical)	
5. AB composition by alternating rows			
a. |–⏀– ▽|–⏀–. . . (hexagonal)	AuCd	B19	3[14]
b. |–⏀– ▽△|–⏀–. . . (cubic)	CuAu	$L1_0$	19
c. |–⏀– ▽△–⏀–△▽|–⏀–. . . . (orthorhombic)	TaRh	—	4[49]
B. Interlayer ordering			
1. Layer α–layer β–layer α–interstitial (X) (hexagonal) $A_\alpha B_\beta A_\alpha X\ B_\alpha A_\beta B_\alpha X, A_\alpha$ . . .	Ti_2SC	—	2[21]
2. Layer α–layer β (rhombohedral) $A_\alpha B_\beta C_\alpha A_\beta B_\alpha C_\beta, A_\alpha$ . . .			
a. Pure α and pure β	CuPt	$L1_1$	1
b. Pure α, mixed β			
(1) Alternating rows in β	$CuPt_3$	—	1
(2) Triangularly ordered β	$CuPt_7$	—	1
III. Interstitial compounds*			
A. MX types			
1. Two-layer: $A_MX\ B_MX, A_M$ . . .	δ-NbN	$B8_1$	1 interstitial 48 metalloid
2. Three-layer: $A_MX\ B_MX'C_MX'', A_M$ . . .			
a. Cubic	TiC	B1	50 interstitial 80 metalloid
b. Rhombohedral	NiO(L.T.)	—	3
c. Tetragonal	CoO(L.T.)	—	2
3. Four-layer: $A_\alpha X\ B_\beta X\ A_\alpha X'\ C_\beta X', A_\alpha$ . . .			
a. α = MnTa, β = Ta	Ta_3MnN_4	—	1
4. Six-layer: $A_MX\ B_MX'\ C_MX\ A_MX\ C_MX'\ B_MX, A_M$ . . .	MoC	—	1
B. M_2X to M_3X types			
1. Hexagonal/trigonal interlayer $A\ X_hB\ X_t, A$ . . .*	ε-Fe_2N or W_2C	L_3'	9
2. Orthorhombic intralayer			
a. Rectangular $A\ X_rB\ X'_r, A$ . . .	Co_2N	—	2
b. Chained $A\ X_cB\ X'_c, A$ . . .	ζ-Fe_2N	—	1
3. Trigonal intralayer $A\ X'_tB\ X_t, A$ . . .	ε-Fe_3N	—	4
C. M_4X type, cubic			
1. Interstitial order	γ'-Fe_4N	L_1'	2
2. Metallic and interstitial order	Fe_3AlC	$E2_1$	59[46]

* Subscripts on X symbols:
unadorned: all octahedral holes filled
h—$\frac{2}{3}$ of the octahedral holes filled in hexagonal nets
t—$\frac{1}{3}$ of the octahedral holes filled in trigonal nets
r—$\frac{1}{2}$ of the octahedral holes filled in rectangular nets
c—$\frac{1}{2}$ of the octahedral holes filled in zigzag chains
prime indicates a change in position from a predecessor

has investigated a large number of ternary systems consisting of transition metals, non-transition metals, and interstitial elements. He found this structure in thirty-nine carbide and twenty nitride systems with the interstitial atom coordinated entirely by transition metals.

In some of the newer superalloys a precipitation that occurs within precipitated, ordered γ-$Ni_3(Al, Ti)$ particles is apparently a carbide of this type with the metallic composition of the ordered particle.[47]

5. TABULATION

Table 2 lists all of the structure types discussed in this chapter, together with stacking symbols, prototypes, structure symbols as given by Pearson,[10] and population of each type known to date.

The approach used here is topological; it is considered the proper one for a materials scientist or engineer. Crystallographic details on lattice parameters, space groups, and position assignments are given in Pearson in most cases, as are the lists of examples of the various types. Significant additions to Pearson's listings, or structure types not covered in Pearson, are found in the references designated by superior numbers beside the population figures.

The population figures refer to binary and, in a few cases, ternary examples of the structure types. They do not include substitutional solution series.

The practice of using ionic or covalent compounds as prototypes has been abandoned here. These would be "NaCl" for B1, "NiAs" for $B8_1'$ and "perovskite" or "$CaTiO_3$" for $E2_1$. The equivalence of these structures to their metallic isomorphs (or "anti-isomorphs") is geometric only. It seems more appropriate to have as a prototype a member of the class that is represented.

TABLE 3
Structures with More than Five Examples

Prototype	Structure Symbol	Population
Cu_3Au	$L1_2$	75
Fe_3AlC	$E2_1$	59
TiC	B_1	50
CuAu	$L1_0$	19
Ni_3Sn	$D0_{19}$	15
M_2X	L_3'	9
Ni_3Ti	$D0_{24}$	7
Al_3Ti	$D0_{22}$	7
Ni_4Mo	D1a	7
Cu_3Ti	—	6

Table 3 lists all of the compound structure types that have more than five members.

One feature of the cannonball structures is the ease with which a detailed description of the topology of their structures lends itself to expression in rather simple ideographic symbols. For materials scientists and engineers this is the sort of description that is the most meaningful. Not only do the symbols describe the structures, but they can be used to work out structural relationships with greater ease than can be done even with three-dimensional models of the structures. Perhaps they could also serve as a basis for a uniform nomenclature of alloy phases[48] of this type.

REFERENCES

1. Laves F., and H. J. Wallbaum, *Z. Krist.* (*A*) **101** (1939) 78.
2. Dwight A. E., and P. A. Beck, *Trans. AIME* **215** (1959) 976.
3. Dehlinger U., *Z. Anorg. Allgem. Chem.* **194** (1931) 223.
4. Rahlfs P., *Metallwirtsch.* **16** (1937) 343.
5. Babich M. M., E. N. Kislyakova, and Ya. S. Umanskii, *Zh. Tekh. Fiz.* **8** (1938) 119, 122.
6. Magnéli A., and A. Westgren, *Z. Anorg. Allgem. Chem.* **238** (1938) 268.
7. Bradley A. J., and A. Taylor, *Proc. Roy. Soc.* (*A*) **159** (1937) 56.
8. Wallbaum H. J., *Naturwiss.* **31** (1943) 91.
9. Saito S., *Acta Cryst.* **12** (1959) 500.
10. Pearson W. B., *A Handbook of Lattice Spacings and Structures of Metals and Alloys.* Pergamon Press, New York, 1958.
11. Kurdjumov G., V. Miretskii, and T. Stelletskaya, *J. Phys. USSR* **3** (1940) 297.
12. Karlsson N., *J. Inst. Met.* **79** (1951) 391.
13. Saito S., and P. A. Beck, *Trans. AIME* **215** (1959) 938.

14. Stolz E., and K. Schubert, *Z. Metallk.* **53** (1962) 433.
15. Brauer G., *Z. Anorg. Allgem. Chem.* **242** (1939) 1.
16. Frank F. C., and J. S. Kasper, *Acta Cryst.* **12** (1959) 483.
17. Harker D., *J. Chem. Phys.* **12** (1944) 315.
18. Wegst J., and K. Schubert, *Z. Metallk.* **49** (1958) 533.
19. Armitage Merle, *George Gershwin, Man and Legend.* Duell, Sloan and Pearce, New York, 1958, p. 53.
20. Günzel E., and K. Schubert, *Z. Metallk.* **49** (1958) 533.
21. Kudielka H., and H. Rohde, *Z. Krist.* **114** (1960) 447.
22. Gemill M. G., H. Hughes, J. D. Murray, F. B. Pickering, and K. W. Andrews, *J. Iron Steel Inst.* **184** (1956) 122.
23. Brown J. F., W. D. Clark, and A. Parker, *Metallurgia* **56** (1957) 215.
24. Schneider A., and U. Esch, *Z. Elektrochem.* **50** (1944) 290.
25. Bradley A. J., and A. Taylor, *Phil. Mag.* **23** (1937) 1049.
26. Hägg G., *Z. Phys. Chem.* (*B*) **11** (1930) 152; **12** (1931) 413.
27. Jack K. H., *Acta Cryst.* 3 (1950) 392; **5** (1952) 404.
28. Andrews K. W., and H. Hughes, *J. Iron Steel Inst.* **193** (1959) 304.
29. Rundle R. E., *Acta Cryst.* **1** (1948) 180.
30. Hume-Rothery W., *Phil. Mag.* **44** (1953) 1154.
31. Hume-Rothery W., and G. V. Raynor, *The Structure of Metals and Alloys*, 3rd ed. *The Institute of Metals*, London, 1954, pp. 216–227.
32. Schönberg N., *Acta Chem. Scand.* **8** (1954) 213.
33. Nowotny H., E. Parthé, R. Kieffer, and F. Benesovsky, *Monatsh. Chem.* **85** (1954) 255.
34. Nowotny H., and R. Kieffer, *Z. Anorg. Chem.* **267** (1951–52) 261.
35. Davis A. M., and H. J. Beattie, Jr., unpublished investigation of material furnished by J. G. McMullin and labeled "MoC."
36. Nowotny H., *Electronic Structure and Alloy Chemistry of the Transition Elements*, P. A. Beck, ed. AIME Symposium, Interscience, New York, 1963, pp. 185–186.
37. Clougherty E. V., K. H. Lothrop, and J. A. Kafalis, *Nature* **191** (1961) 1194.
38. Goldschmidt H. J., and J. A. Brand, *ASD-TDR-62-25*, Part I, (March 1962).
39. Kuo K., and G. Hägg, *Nature* **170** (1952) 245.
40. Tombs N. C., and H. P. Rooksby, *Nature* **165** (1950) 442.
41. Willis B. T. M., and H. P. Rooksby, *Acta Cryst.* **6** (1953) 827.
42. Schönberg N., *Acta Chem. Scand.* **8** (1954) 221.
43. Andersson G., *ibid.*, 1599.
44. Philipp W., *Acta Met.* **10** (1962) 583; **12** (1964) 740.
45. Clarke J., and K. H. Jack, *Chem. Ind.* **46** (1951) 1004.
46. Stadelmaier H. H., *Z. Metallk.* **52** (1961) 758.
47. Wlodek S. T., private communication.
48. "What can be Done to Improve Alloy Phase Nomenclature?" *ASTM Bulletin No. 226* (1957) 27.
49. Giessen B. C., H. Ibach, and N. J. Grant, *Trans AIME* **230** (1964) 113.

APPENDIX

X-Ray Structure Factors

For the purpose of calculating X-ray line intensities some discussion is given here on structure factors for the close-packed types that are referable to hexagonal axes.

The principal variable considered here is stacking sequence. Of the intralayer orderings, only the triangular ordering (Fig. 2) can be discussed. The others throw the crystal class into a nonhexagonal system,* so that it is impossible to express the ordering in terms of the hexagonal system in which it is convenient to express stacking sequence. The triangular ordering, however, has been found in at least four stacking sequences, as we have seen. Some of the interstitial compounds also appear to have a tendency to assume a long-cycle stacking sequence. The structure factors calculated for various stacking sequences of close-packed layers will be approximately correct for the interstitial compounds, because the interstitial atoms, being of low atomic number, generally have little influence on the X-ray scattering.

For trigonal ordering, the structure factor can be expressed as the product of a factor due to

* This is not true of the ordering shown in Fig. 7, but its rarity does not warrant the space it would take to present its calculation.

ordering and a factor due to stacking sequence. The triangular ordering factor will now be considered. Let the origin [000] be placed at an A atom. This places the three B atoms at $[0, \frac{1}{2}, 0]$, $[\frac{1}{2}, 0, 0]$, and $[\frac{1}{2}, \frac{1}{2}, 0]$. This is not always strictly correct, for it has been found[6] that the B atoms may be slightly displaced from these positions so that another location for the origin is better suited. However, the structure factors calculated here will still be approximately correct, which is all that is intended.

Applying the structure factor formula to the ordering:

$$F_0 = \sum_n f_n e^{2\pi i(hx_n+ky_n+lz_n)}$$

$$= f_A + f_B(e^{\pi ih} + e^{\pi ik} + e^{\pi i(h+k)})$$

Case 1, h and k both even:

$$F_0 = f_A + 3f_B$$

Case 2, h and k both odd:

$$F_0 = f_A - f_B$$

Case 3, h and k mixed:

$$F_0 = f_A - f_B$$

Thus any reflection with h or k or both odd is weak. This is another way of saying that a_0 in the superlattice is double that of the disordered lattice.

It was demonstrated above that the triangular ordering together with the rule that A atoms lie upon all-B triangular interstices leads to the very same stacking conditions for the superlattice as for nonordered close-packed layers. Thus, the superlattice relations above, combined with the structure factors to be derived due to various stacking sequences, lead to all possibilities up to six layers (the five-layer sequence being omitted) for this kind of ordering.

Two-layer sequence. This is handled as follows: calculations are tidier if the origin is located at a centrosymmetric point. In this all-p-coordinated structure the atoms do not lie at such points; however, the centers of the octahedral interstices (Δ over ∇) do. With the origin at these points, the atomic coordinates are $(\frac{1}{3}, \frac{2}{3}, \frac{1}{4})$ and $(\frac{2}{3}, \frac{1}{3}, \frac{3}{4})$, or, since in a lattice $\frac{2}{3} = -\frac{1}{3}$ and $\frac{3}{4} = -\frac{1}{4}$, the positions may be written $\pm(\frac{1}{3}, -\frac{1}{3}, \frac{1}{4})$. The stacking phase factor (S) is:

$$S = e^{2\pi i(h/3-k/3+l/4)} + e^{-2\pi i(h/3-k/3+l/4)}$$

$$= 2\cos 2\pi\left(\frac{h-k}{3} + \frac{l}{4}\right) = 2\left[\cos\frac{2\pi}{3}(h-k) \times \cos\frac{\pi l}{2} - \sin\frac{2\pi}{3}(h-k)\sin\frac{\pi l}{2}\right]$$

If l is even, the term in sines vanishes, whereas if l is odd, the term in cosines vanishes. Thus there is always only one term, so there is no need to bother with signs since the structure factors are squared to give the intensities. If $(h - k) = 3n$, $\cos 2\pi/3(h - k) = 1$, and $\sin 2\pi/3(h - k) = 0$. If $(h - k) = 3n \pm 1$, $\cos 2\pi/3(h - k) = -\frac{1}{2}$, and $\sin 2\pi/3(h - k) = \pm(\frac{1}{2})\sqrt{3}$. This covers all possibilities, and a table may now be written:

l	$(h-k)$	
	$3n$	$3n+1$
$2n$	± 2	± 1
$2n+1$	0	$\pm\sqrt{3}$

S—factors for AB, A . . .

In all stacking sequences that follow, the lateral positions A, B, and C may be identified with the (X, Y) coordinates $(0, 0)$, $+(\frac{1}{3}, -\frac{1}{3})$, and $-(\frac{1}{3}, -\frac{1}{3})$. The terms of the S-factors due to the A position are independent of h and k, while those due to the B and C positions have the phase angle $\pm(2\pi/3) \times (h - k)$. Hence, only the cases $(h - k) = 3n$ and $(h - k) = 3n \pm 1$ need be considered as far as the first two indices are concerned.

The three-layer sequence is best expressed in the cubic system if the axial ratio is ideal. However, rhombohedral distortions from cubicity are known, and they are conveniently handled in the hexagonal system.

Three-Layer Sequence. ABC, A . . . (rhombohedral)

Positions at (000); $\pm(\frac{1}{3}, -\frac{1}{3}, \frac{1}{3})$

$$S = 1 + 2\cos(2\pi/3)(h - k + l)$$

$$(h - k + l) = 3n, \qquad S = 3$$
$$(h - k + l) = 3n \pm 1, \qquad S = 0$$

This amounts to the extinction rule for rhombohedral lattices in general when referred to hexagonal axes.

Four-Layer Sequence. ABAC, A . . .

Positions at (000), $(0, 0, \frac{1}{2})$, $\pm(\frac{1}{3}, -\frac{1}{3}, \frac{1}{4})$

The last pair of positions is the same as those of the two-layer sequence, so we may write immediately:

$$S = 1 + (-1)^l + 2\,[\cos(2\pi/3)(h - k)\cos \pi/2l + \sin(2\pi/3)(h - k)\sin \pi l/2]$$

and write the following S-factor table:

l	$(h-k)$	
	$3n$	$3n \pm 1$
$4n$	4	1
$4n \pm 1$	0	$\pm\sqrt{3}$
$4n \pm 2$	0	3

S-factors for *ABAC, A . . .*

Six-Layer Sequence a. ABCACB, A . . .

Positions at (000), $(0, 0, \frac{1}{2})$, $(\frac{1}{3}, -\frac{1}{3}, \frac{1}{6})$, $(\frac{1}{3}, -\frac{1}{3}, -\frac{1}{6})$, $(-\frac{1}{3}, \frac{1}{3}, \frac{1}{3})$, $(-\frac{1}{3}, \frac{1}{3}, -\frac{1}{3})$

$$S = 1 + (-1)^l + 2\left[\exp\frac{2\pi i}{3}(h-k)\cos\frac{\pi l}{3} + \exp\left(\frac{-2\pi i}{3}\right)(h-k)\cos\frac{2\pi}{3}l\right]$$

For $h - k = 3n$:

$$S = 1 + (-1)^l + 2\left(\cos\frac{\pi}{3}l + \cos\frac{2\pi}{3}l\right)$$

For $h - k = 3n \pm 1$:

$$S = 1 + (-1)^l - \left(\cos\frac{\pi}{3}l + \cos\frac{2\pi}{3}l\right) \pm i\sqrt{3}\left(\cos\frac{\pi}{3}l - \cos\frac{2\pi}{3}l\right)$$

l	$\left(\cos\frac{\pi}{3}l \pm \cos\frac{2\pi}{3}l\right)$		S	
			$h - k = 3n$	$h - k = 3n \pm 1$
	(+)	(−)		
$6n$	2	0	6	0
$6n \pm 1$	0	1	0	$\pm i\sqrt{3}$
$6n \pm 2$	−1	0	0	3
$6n \pm 3$	0	−2	0	$\pm i2\sqrt{3}$

Six-Layer Sequence b. ABCBCB, A . . .

Positions: same as above except $(\frac{1}{3}, -\frac{1}{3}, \frac{1}{2})$ replaces $(0, 0, \frac{1}{2})$.

$$S = 1 + \left(\exp\frac{2\pi i}{3}(h-k)\right)\left((-1)^l + 2\cos\frac{\pi}{3}l\right) + 2\left(\exp\frac{-2\pi i}{3}(h-k)\right)\cos\frac{2\pi}{3}l$$

For $h - k = 3n$:

$$S = (-1)^l + 2\left(\cos\frac{\pi}{3}l + \cos\frac{2\pi}{3}l\right)$$

For $h - k = 3n \pm 1$:

$$S = 1 - \tfrac{1}{2}(-1)^l - \left(\cos\frac{\pi}{3}l + \cos\frac{2\pi}{3}l\right) \pm i\sqrt{3}\left(\tfrac{1}{2}(-1)^l + \cos\frac{\pi}{3}l - \cos\frac{2\pi}{3}l\right)$$

l	$\left(\cos\frac{\pi}{3}l \pm \cos\frac{2\pi}{3}l\right)$		S	
			$h - k = 3n$	$h - k = 3n \pm 1$
	(+)	(−)		
$6n$	2	0	6	$-\frac{3}{2} \pm i(\frac{1}{2})\sqrt{3}$
$6n \pm 1$	0	1	0	$\frac{3}{2} \pm i(\frac{1}{2})\sqrt{3}$
$6n \pm 2$	−1	0	0	$\frac{3}{2} \pm i(\frac{1}{2})\sqrt{3}$
$6n \pm 3$	0	−2	0	$\frac{3}{2} \pm i(\frac{5}{2})\sqrt{3}$

See Table 4 for S^2 values from the above S-factor tables. In the six-layer, sequence b, case with $(h - k) = 3n \pm 1$ there are complex S-factors of the form $(a + ib)$. The S^2 values are obtained by multiplying by the complex conjugates to give $S^2 = (a^2 + b^2)$.

The X-ray intensities are proportional to pF^2S^2, where p is the multiplicity factor. The table of S^2 values may be used to assist in a rapid calculation of approximate X-ray intensities of various possible close-packed arrangements.

TABLE 4

X-Ray Phase Factors for Various Stacking Sequences

Stacking	l	S^2: $h - k = 3n$	S^2: $h - k = 3n \pm 1$
AB, A	$2n$	4	1
	$2n \pm 1$	0	3
ABC, A	$3n$	9	0 (nonzero only for $h - k + l = 3n$)
	$3n \pm 1$	0	0 (nonzero only for $h - k + l = 3n$)
	$3n \pm 2$	0	9 (nonzero only for $h - k + l = 3n$)
ABAC, A	$4n$	16	1
	$4n \pm 1$	0	3
	$4n \pm 2$	0	9
ABCACB, A	$6n$	36	0
	$6n \pm 1$	0	3
	$6n \pm 2$	0	9
	$6n \pm 3$	0	12
ABCBCB, A	$6n$	36	3
	$6n \pm 1$	0	3
	$6n \pm 2$	0	3
	$6n \pm 3$	0	21

chapter 10

Body-Centered Cubic Derivative Structures

[signature: Austin E. Dwight]

AUSTIN E. DWIGHT

Argonne National Laboratory
Argonne, Illinois

1. DISORDERED BCC PHASES

There is a relatively small family of intermetallic compounds which have a disordered body-centered cubic crystal structure. These compounds are usually referred to as β-phases, although there are also β-phases which exhibit an ordered structure. The β-phases are most numerous in binary systems based on copper or silver. Table 1 lists the disordered bcc phases which have been reported. The β-phases are of historical interest in that they were among the first of the electron compounds to be studied. Hume-Rothery[1] in 1926 discovered that the three compounds CuZn, Cu_3Ga, and Cu_5Sn correspond to a ratio of three valence electrons to two atoms. Later work has shown that the stoichiometric ratios CuZn, Cu_3Ga, and Cu_5Sn are only approximately correct; actually the atomic percentage ranges of the second element are 45–48 Zn, 23.7 Ga and 14.9 Sn at the minimum temperature at which the compound is stable. However, the concept of electron compounds remains useful in understanding many compounds of copper group metals.

Although the 3:2 ratio is common among disordered bcc phases, not all such phases exhibit this ratio. MnNi and Mn_3Zn_2 are among the exceptions. A characteristic feature of the β-phases is that they are stable only at high temperatures. This is true of all the phases listed in Table 1 except Mn_3Si. Nearly all are formed from the melt, and they decompose eutectoidally.

It appears that relatively few β-phases become ordered at lower temperatures. CsCl-type structures exist at low temperatures in the Cu–Zn and Ag–Cd systems, although in the latter another structure intervenes between the bcc and CsCl types. In the Ag–Zn system an unstable CsCl-type phase may be retained by quenching the β-phase. There seems to be a tendency to form a CsCl-type phase from a β-phase only when the two elements come from the Cu and Zn groups of the periodic table. In the Cu–Al system there is a β-phase at high temperature and an ordered $BiF_3(D0_3)$ type at lower temperatures. Of the eighteen

TABLE 1

Disordered BCC Phases

Formula	Composition[a]	Reference	Formula	Composition[a]	Reference
β-Cu_2Be[b]	31 Be	2	β-AgZn[b]	34–39 Zn	2
β-CuZn	45–48 Zn	2	β-AgCd[b]	48.5 Cd	2
δ-$CuZn_3$[b]	74 Zn	2	β-Ag_3Al	25 Al	2
β-Cu_3Al[b]	24 Al	2	β-Ag_3In[b]	25 In	2
β-Cu_3Ga[b]	23.7 Ga	2	β-Au_4Al[b]	19 Al	2
β-Cu_4In[b]	20 In	2	β-Pd_2Cd_3[b]	57 Cd	2, 3
β-Cu_5Si[b]	16 Si	2	ht-MnNi[b]	47 Ni	2
β-Cu_3Ge[b]	27 Ge	2	β-Mn_3Zn_2[b]	40 Zn	2
β-Cu_5Sn[b]	14.9 Sn	3	Mn_3Si	25 Si	2

[a] Atomic per cent at the lowest temperature of stability.
[b] Decomposes eutectoidally.

disordered bcc phases listed in Table 1, four may be considered to be parents of low-temperature ordered structures.

2. ORDERED BCC PHASES FROM DISORDERED TERMINAL SOLID SOLUTIONS

Ordered bcc phases are far more numerous than disordered ones, and, accordingly, most ordered phases are presumed to form from the melt rather than from a disordered phase. There is a scarcity of experimental data on the existence of ordering in many systems at high temperature. It has been found that in the V–Fe[2] and V–Mn[4] systems an ordered structure exists at low temperatures below a bcc terminal solid solution. In these systems the A and B elements are separated by only one and two groups of the periodic table, and it is assumed that the tendency to form ordered structures is marginal. Fe_3Al is another compound known to form from a disordered bcc terminal solid solution. A fairly rare mode of formation is exhibited by the Cu–Pd system in which the CsCl-type CuPd is formed from a face-centered cubic solid solution.

3. BINARY CsCl-TYPE PHASES

In a recent paper Nevitt[5] has complied 144 binary CsCl-type phases, with data on lattice parameters and contractions. Table 2 lists only those binary CsCl-type phases which were not included in Nevitt's paper.

The large number of CsCl-type binary phases, 169, makes this the second largest family of intermetallic compounds, exceeded only by the 220 known Laves phases. In order to analyze such a large group of phases, it is convenient to describe them in terms of A and B elements. An A element is one lying to the left of the Cr group in the periodic table, and a B element is one lying to the right of the Cr group. This division of the periodic table follows from the observed fact that neither Cr, Mo nor W will take part in the formation of a CsCl-type phase. It was noted earlier by Dwight[7] that an A element in a group relatively far to the left of the Cr group tends to form a CsCl-type structure with B elements relatively far to the right. Conversely, VMn illustrates the tendency of an A element close to the Cr group to select a B partner which also is close to the Cr group.

The division of the periodic table into A and B elements is shown in Fig. 1. The radial form of the periodic table is most useful in this work because the lanthanide and thoride series are included rather than appended. The lanthanides as A elements form a large number of CsCl-type phases with B elements from the Cu and Zn groups. In general, the lanthanide elements behave as might be expected from their position in the Sc group. The elements Be and Mg are placed in group IIb rather than in IIa; that is, Mg lies above Zn, not Ca. Mg and Be thus become B instead of A elements. Placing Mg and Be in group IIb has been done by other writers, most recently by Makarov.[8] In the case of CsCl-type phases, SrMg, YMg, and REMg fall into place as AB compounds, rather than as A′A″ if Mg were a group IIa element. Mg

TABLE 2

Some Recently Reported CsCl-Type Phases

AB	a_0, Å	$D_{AB} - d_{AB}$	R_A/R_B	Reference
$Zr_{46}Rh_{54}$	3.260	+0.044	1.19	6
$Hf_{48}Rh_{52}$	3.248	+0.033	1.17	6
TmRh	3.358	+0.097	1.297	6
DyRh	3.403	+0.086	1.317	6
ErIr	3.367	+0.113	1.294	6
TbAg	3.625	+0.001	1.232	47
TmAg	3.562	+0.019	1.208	47
TbAu	3.576	+0.039	1.235	47
DyAu	3.555	+0.048	1.229	47
TmAu	3.516	+0.056	1.210	47
TbCu	3.480	−0.037	1.393	48
HoCu	3.445	−0.022	1.380	48
TmCu	3.414	−0.015	1.365	48
YZn	3.577	+0.010	1.291	48
NdZn	3.667	−0.049	1.305	48
SmZn	3.627	−0.033	1.292	48
GdZn	3.602	−0.011	1.292	48
TbZn	3.576	−0.008	1.277	48
DyZn	3.563	−0.006	1.271	48
HoZn	3.547	+0.002	1.266	48
ErZn	3.532	+0.006	1.259	48
TmZn	3.516	+0.009	1.252	48
GdMg	3.787	+0.030	1.125	49
GdCd	3.748	+0.031	1.149	49
GdHg	3.719	+0.061	1.146	49

RADIAL PERIODIC TABLE

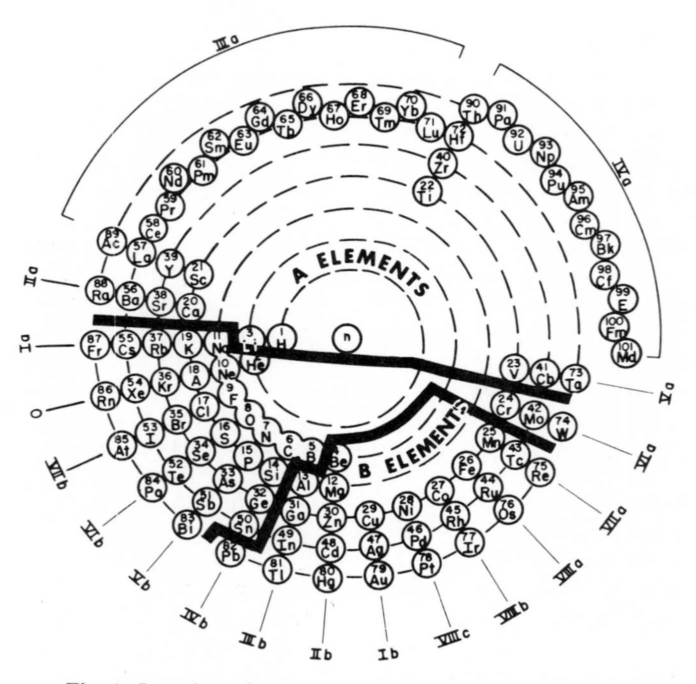

Fig. 1. Location of A and B elements in the periodic table.

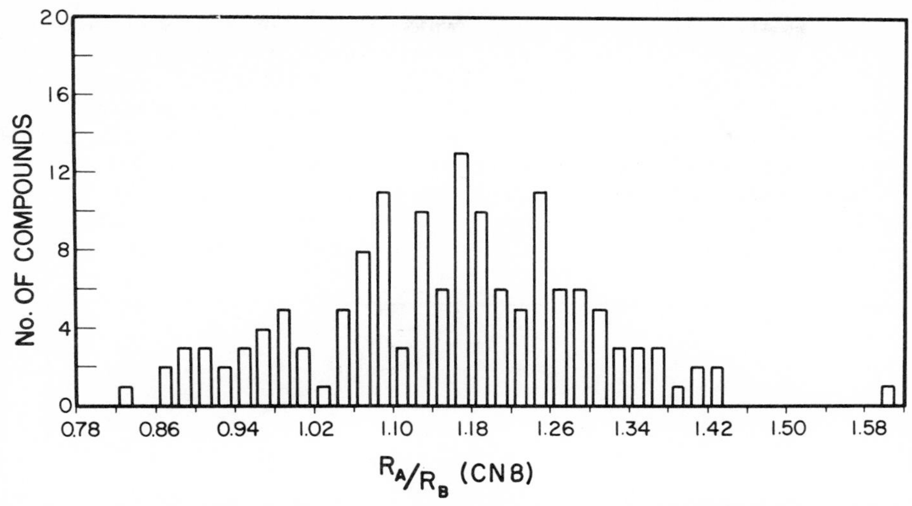

Fig. 2. Distribution of CsCl-type compounds with respect to Goldschmidt radius ratio.

sometimes acts as an A element, as in the series MgHg, CaHg, SrHg, BaHg. It is uncertain whether Be is an A or B element in BeCu, BeNi, and BeCo. All CsCl-type phases are combinations of either A and B or of two B elements.

The existence of certain intermetallic compounds is controlled by relative atomic size. In order to evaluate the influence of relative atomic size on the stability of CsCl-type phases, the distribution plot shown in Fig. 2 was made. There is a peak in occurrence from a radius ratio of approximately 1.06 to 1.26. Probably of more importance is the broad span from 0.82 to 1.60 over which the CsCl structure has been found. It is concluded that relative atomic size is not a major factor in controlling the occurrence of CsCl-type phases.

An investigation was made of the relation between the radius ratio and the interatomic contraction. The latter term is defined by $D_{AB} - d_{AB}$, in which D_{AB} is the sum of the two CN8 radii R_A and R_B, and d_{AB} is half the cube diagonal $(a_0\sqrt{3}/2)$ in the unit cell. The radius ratio, R_A/R_B, was plotted against the interatomic contraction, and lines were drawn through points representing alloy families having a common B element. From the contraction plot, certain regularities became apparent. These are summarized as follows:

1. For a given B element and for A elements from one group of the periodic table, a decreasing interatomic contraction generally accompanies an increasing radius ratio.
2. For a given B element, the contraction moves toward a lower positive (or higher negative) value with increasing group number (valence) of the A element.
3. When the A element is held constant and the B element varied within one long period (LP), an orderly shift in contraction occurs.
4. When we compare contraction curves in which the B element varies within one group, for example, from Cu to Ag to Au, the curves for Cu and Au are nearly continuous, but the curve for Ag lies distinctly lower.

The above four observations are discussed in more detail, with reference to Figs. 3 through 5.

1. Figure 3 shows the measured contractions for the families of alloys in which the B elements are Ag and Rh, respectively, and the A elements are members of the Sc group. A smooth curve can be fitted to all points except those for Y, Ce, and Gd. With increasing radius ratio, the interatomic contraction decreases, and for the Ag series the contraction becomes negative, which is actually an expansion. For families of alloys in which the B element is Ir, Pd, Cu, Au, Zn, Cd, and Hg, respectively, the contraction also decreases with increasing radius ratio.

2. The influence of the group number, or valence, of the A element on the interatomic contraction may be seen by a comparison of

data for ScCo and ZrCo, whose radius ratios do not differ greatly. ScCo (Sc valence +3, contraction +0.091) and ZrCo (Zr valence +4, contraction +0.008) illustrate that the higher valence of Zr is accompanied by a lower contraction. A comparison of contractions for BaZn (Ba valence +2) and LaZn (La valence +3) confirms the effect of the valence of the A element on contractions.

is visible at Gd in a curve of radius versus atomic number. It is believed that the deviation of GdAg from the contraction curve in Fig. 3 is a manifestation of the same phenomenon that causes the cusp at Gd in other plots. No explanation is available for the abnormally high contraction in YAg.

3. When the A element is held constant and the B element varied within one LP, an

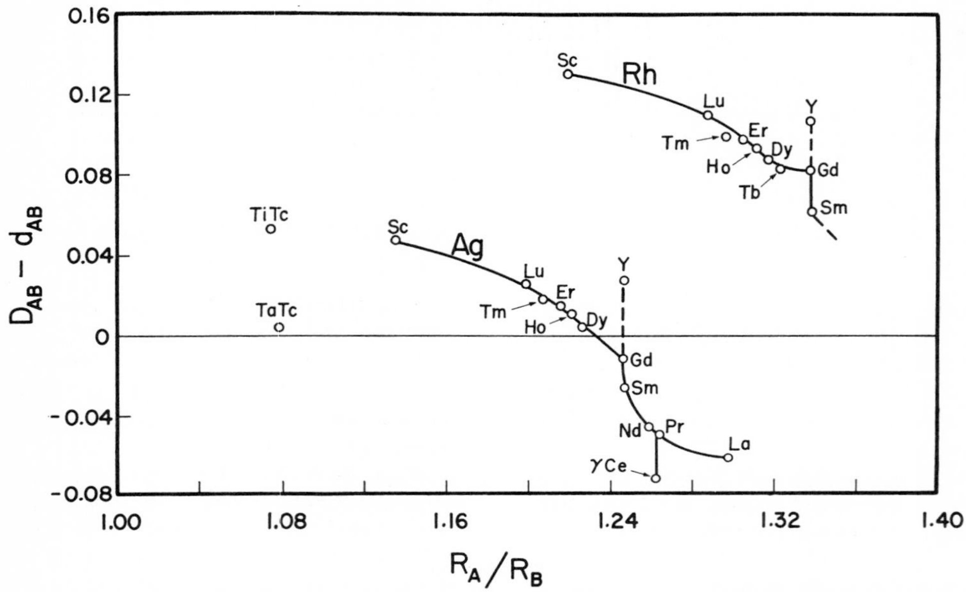

Fig. 3. Interatomic contraction in CsCl-type compounds of the RE–Ag and RE–Rh families.

The deviation of CeAg from the Ag curve in Fig. 3 may be explained in terms of the valence of Ce. The radius of γCe was assumed in computing D_{AB} and according to Teatum, Gschneidner, and Waber[9] this corresponds to a valence of 3.1 as compared to 3 for Pr, Nd, Tb, etc. If the radius of αCe (valence 3.6) had been used, the deviation from the contraction curve would be even greater. It is implied that the deviation shown by CeAg is due to the higher than normal valence of Ce.

The contraction GdAg deviates from the contraction curve in the opposite direction from that of CeAg. This deviation is not explainable in terms of valence, since Gd is generally considered to have a valence of +3. It has been pointed out by Gschneidner[45] that for many families of compounds, Gd compounds exhibit a larger than normal lattice parameter. For the pure lanthanide elements and for the bcc equiatomic compounds, a cusp

orderly shift in contraction occurs. It was noted by Aldred[46] that for compounds in which the A element is Sc, the contraction increases when the B element is taken from the Cu, Ni, and Co groups, but falls when B is Ru of the Fe group. A similar trend exists for the other A elements of the Sc group. In Fig. 4 are shown curves for alloy families in which the A elements are Sc, Lu, and Tm, respectively. In all these families the contraction rises as the B element is shifted from Ag to Pd to Rh. At Rh an inflection point in the curve is noted, with a decrease in contraction at Ru when A is Sc and Tm. Lack of an equiatomic compound in the Sc–Tc and Tm–Tc systems prevents the extension of the Sc and Tm curves beyond Ru. However, there is a pair of cubic equiatomic compounds in the Ti–Ru and Ti–Tc systems. The contractions of TiRu and TiTc, illustrated in Fig. 4, show a decrease in going from Ru to Tc. The effect of varying the

B element from Ag to Tc is to reach a maximum contraction at Rh followed by a decline to Tc.

The contractions of TiNi, TiCo, and TiFe are also shown in Fig. 4. This family differs from those discussed earlier in that an increasing contraction accompanies a decreasing radius ratio. However, contraction increases on shifting the B element from Ni to Fe, as in the other families. It is noted that in all the examples shown in Fig. 4, increasing lattice contraction accompanies decreasing separation in the periodic table, which is counter to that expected on the basis of the electronegative valence effect. Beck[10] has previously discussed this trend for the sequence TiFe, TiCo, and TiNi.

4. When contraction curves are compared in which the B element varies within one group of the periodic table, it is always the curve for the second LP element which lies below the curves for the first and third LP elements. To illustrate this point, the Ag curve in Fig. 5 lies below the Cu and Au curves. Furthermore, the Pd curve lies below the Ni and Pt curves. This means that lattice contraction is less for an alloy containing a B element from the second LP than for its counterpart from the first or third LP.

If one considers the curves for Cu and Au to be continuous (Ni and Pt to be continuous also) the sequence of curves from left to right is Ag, Cu–Au, Pd, Ni–Pt. The Pd curve then is about midway between the Cu–Au and Ni–Pt curves. This can be rather crudely expressed by the observation that Pd behaves as if it were midway between Au and Pt, rather than directly over Pt in the periodic table. Insofar as lattice contractions in CsCl-type phases are concerned, the second LP elements Rh, Pd, Ag, and Cd behave as if they should be displaced half a step to the right in the periodic table in comparison with their counterparts in the first and third long periods.

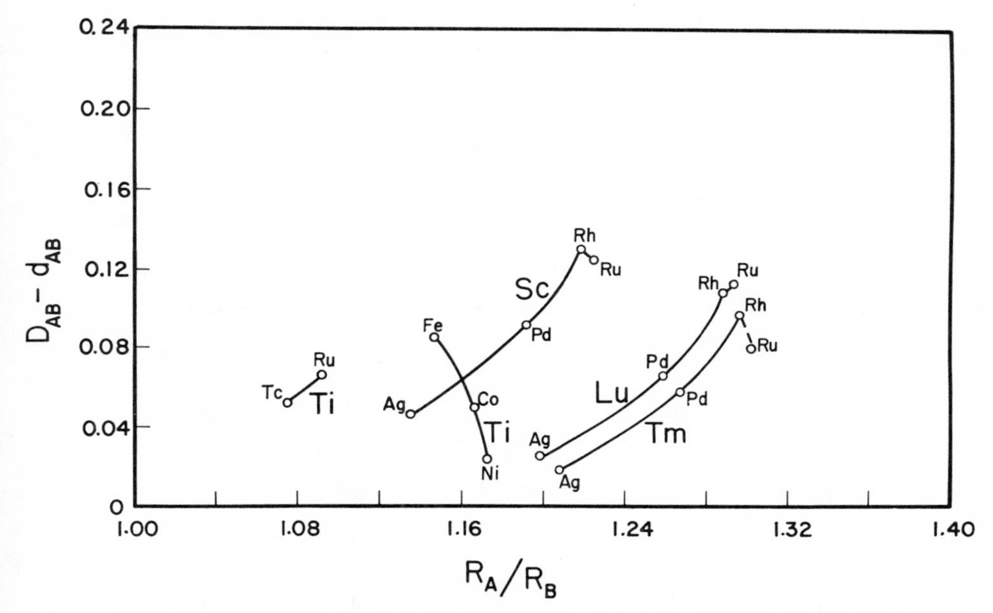

Fig. 4. Interatomic contraction in CsCl-type compounds of the Sc–X, Tm–X, Lu–X, and Ti–X families.

4. TERNARY CsCl-, Cu_2MnAl-, Fe_3Al-, and NaTl-TYPE COMPOUNDS

These four families are crystallographically related to binary CsCl-type and other bcc structures, in that all may be described in terms of the space model shown in Fig. 6. There are four atomic positions, designated *a*, *b*, *c*, and *d*. The location of elements in these positions in either a random or ordered manner distinguishes six crystal structure types, which are listed in Table 3.

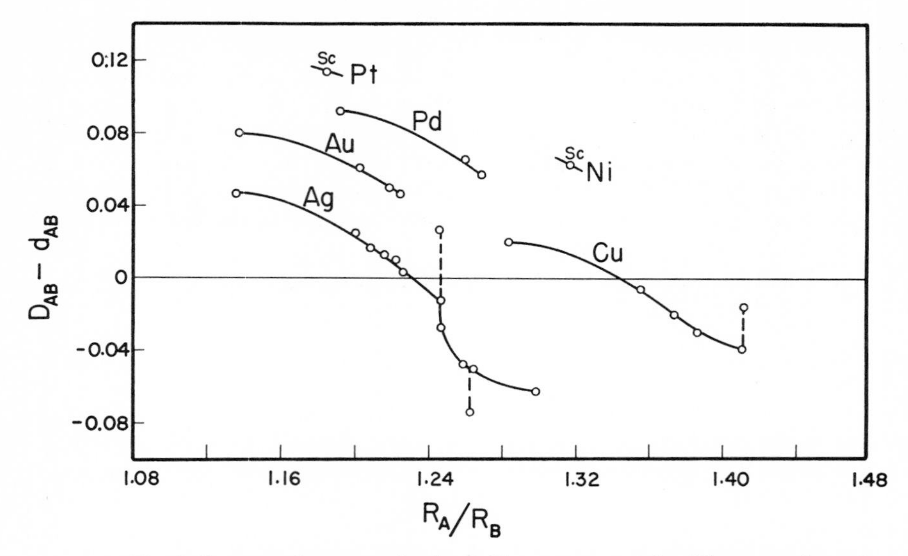

Fig. 5. Interatomic contraction in CsCl-type compounds of Sc-group elements with Cu, Ag, Au, Ni, Pd, and Pt.

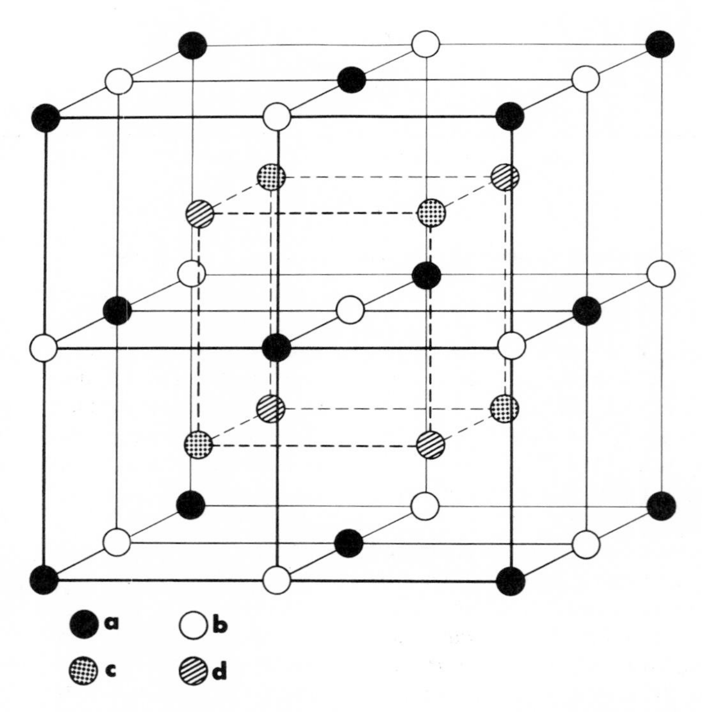

Fig. 6. The four atomic positions obtained by stacking eight small body-centered cubes.

TABLE 3

Some Crystallographically Related BCC Structures

Structure Type	Name	Example	Element in Position			
			a	*b*	*c*	*d*
A2	bcc	VRe	V or Re	V or Re	V or Re	V or Re
B2	CsCl-binary	VOs	V	V	Os	Os
B2	CsCl-ternary	Ag_2TiAl	Ag	Ag	Al or Ti	Al or Ti
$L2_1$	Cu_2MnAl	Cu_2MnAl	Cu	Cu	Mn	Al
$D0_3$	Fe_3Al	Fe_3Al	Fe	Fe	Fe	Al
B32	NaTl	LiAl	Li	Al	Al	Li

It is of interest to consider the distinguishing features between a ternary CsCl-type and a Cu_2MnAl-type compound. In each structure the *a* and *b* sites are fully occupied by a single species. In the ternary CsCl-type structure the *c* and *d* sites are randomly occupied by the other two species. The relative quantity of the two species may vary widely, as in an alloy of $Ti_{50}Re_{47}Pt_3$, in which the *c* and *d* positions are probably occupied by a random mixture of Re and Pt atoms because there are too few Pt atoms present to occupy either the *c* or *d* position completely and there is no evidence of secondary ordering. The alloy Ag_2TiAl has the proper stoichiometric ratio so that Ti and Al atoms could selectively occupy *c* or *d* positions; but since there is no evidence that they do, it is assumed that here also *c* and *d* positions are occupied by a random mixture of Ti and Al atoms. The known ternary CsCl-type compounds are listed in Table 4.

TABLE 4

Ternary CsCl-Type Compounds

Composition	a_0, Å	Reference
Li_2HgTl	3.361	2
Ti_2MoAl		11
Ti_2MoGa		11
$Ti_5Mo_3Ir_2$	3.124	6
$Ti_5Re_2Ir_3$	3.096	6
Ti_5Re_4Pd	3.109	6
$Ti_{50}Re_{47}Pt_3$	3.113	6
Ti_5Re_4Au	3.113	6
Ti_5Rh_4Re	3.093	6
$Ti_5Rh_3Nb_2$	3.136	6
Ti_5Rh_4Cr	3.083	6
Ti_5Rh_4Mo	3.100	6
Ti_5Rh_4W	3.166	6
Hf_2FeNi	3.172	6
Hf_2ReIr	3.259	6
$V_{10}Re_9Rh$	3.034	6
V_5Re_4Ir	3.031	6
$Co_{13}V_3Si_5$	2.807	12
Ni_2LuHf	3.244	6
Cu_2TiAl	2.936	13
Cu_2TiIn	3.016	13
Cu_2TiSn	2.956	13
Cu_2ZrAl	3.106	13
Ag_2TiAl	3.206	14
Al_5MnFe	?	15
$Al_5Mn_3Co_2$	2.920–2.974	18
$Al_5Mn_3Ni_2$	2.92–2.97	17
$Al_5Mn_3Cu_2$	2.984	16

4.1 Cu_2MnAl-Type Compounds

The Cu_2MnAl-type compounds are sometimes called Heusler alloys, after the discoverer of the original group. The unique feature of the Heusler alloys is their ferromagnetism, since the constituent elements (of the prototype at least) are not ferromagnetic. In this review the term Heusler alloy is reserved for those compounds which are ferromagnetic. In other words, all Heusler alloys have the Cu_2MnAl-type structure, but not all compounds with the Cu_2MnAl-type structure are Heusler alloys because some are not ferromagnetic. All known examples of Cu_2MnAl-type alloys are listed in Table 5. It is worth noting that Cu_2MnAl becomes disordered above about 625°C.[2]

The Cu_2MnAl-type compounds occur only at a 2:1:1 atomic ratio. All four atomic positions in Fig. 6 are occupied by a definite species. The essential difference between a ternary CsCl type and a Cu_2MnAl type is whether the distribution of atoms in positions *c* and *d* is random or ordered.

There is no completely satisfactory explanation as to why an alloy of 2:1:1 atomic ratio

TABLE 5

CuMnAl-Type Compounds

Composition	a_0, Å	Reference
Li_2MgSn		44
Li_2MgPb	6.781	21
Co_2TiAl	5.847	50
Co_2ZrAl	6.081	53
Co_2HfAl	6.009	53
Co_2NbAl	5.946	53
Co_2TaAl	5.927	53
Co_2NbGa	5.954	53
Co_2TaGa	5.923	53
Co_2MnGa	5.77	54
Co_2MnSi	5.670	12
Co_2MnGe	5.745	19
Co_2MnSn	6.003	19
Ni_2TiAl	5.872	2
Ni_2ZrAl	6.123	53
Ni_2HfAl	6.081	53
Ni_2VAl		55
Ni_2NbAl	5.974	53
Ni_2TaAl	5.949	53
Ni_2HfGa	5.945	53
Ni_2NbGa	5.958	53
Ni_2TaGa	5.933	53
Ni_2MnGa	5.86	54
Ni_2TiIn	6.099	50
Ni_2MnIn	6.07	54
Ni_2MgIn	6.167	50
Ni_2MnGe	5.701	19
Ni_2MnSn	6.057	2
Ni_2MgSn	6.109	2
Ni_2MnSb	6.013	2
Ni_2MgSb	6.062	2
Pd_2MnSb	6.38	54
Cu_2ZrAl	6.215	53
Cu_2HfAl	6.172	53
Cu_2MnAl	5.949	2
Cu_2MnGa		2
Cu_2TiIn	6.222	50
Cu_2MnIn	6.187	2
Cu_2MnSn ht	6.173	2
Cu_2FeSn ht	5.932	2
$Cu_2(CrNi)Sn$ ht	5.982	2
Cu_2CoSn	5.982	2
Cu_2NiSn	5.958	2
Cu_2MnSb	6.096	20
$Cu_{45}Ni_{29}Sb_{26}$	5.869	2
Au_2MnAl	6.36	20
Zn_2CuAu	6.112	22
Zn_2AgAu		56
Cd_2AgAu		56

should choose either the ternary CsCl-type or Cu_2MnAl-type structure in preference to the other. For example, Cu_2TiSn is a ternary CsCl type, but Cu_2MnSn is a Cu_2MnAl type. A trend, to which there are exceptions, is that the Cu_2MnAl-type structure is preferred when the two elements that supply 75 at.%. of the composition are small atoms from the first long period.

4.2 Fe_3Al-Type Compounds

The known Fe_3Al-type compounds are listed in Table 6. The elements which take part in this structure come from many different regions of the periodic table. The first eight

TABLE 6

Fe_3Al-Type Compounds ($D0_3$-Type)

Composition	a_0, Å	Reference
Li_3Bi	6.722	2
Li_3Hg	6.561	2
Li_3Sb	6.572	2
Fe_3Al	5.792	2
Fe_3Si	5.655	2
Cu_3Al	5.832	2
Cu_3Sn	6.117	2
Cu_3Sb(ht)	6.012	2
[a]Mg_3La	7.509	23
[a]Mg_3Ce	7.428	23
[a]Mg_3Pr	7.430	23
[a]Mg_3Nd	7.410	23
[a]Mg_3Sm	7.327	23
[a]Cd_3Ce	7.228	23
[a]Cd_3Pr	7.200	23
[a]Cd_3Nd	7.182	23
[a]Cd_3Sm	7.233	23

[a] Reported as B32 structure, but probably $D0_3$.

Fe_3Al-type compounds in Table 6 do show a marked similarity to Cu_2MnAl-type compounds in that they form from related systems. For example, Cu_3Sn has a Fe_3Al-type structure and Cu_2MnSn(ht) has a Cu_2MnAl-type. From Fig. 6 it is apparent that replacement of a Cu atom from a *c* position with a Mn atom converts the structure from Fe_3Al type to Cu_2MnAl type.

4.3 NaTl-Type Compounds

The NaTl type, or B32, is a cubic structure whose known representatives are listed in Table 7. The structure may be shown by means of the space model in Fig. 6, if Na atoms are in *a* and *d* positions and Tl atoms in *b* and *c* positions. There appears to be a strong electronegativity effect, since the seven examples all are composed of elements widely separated

TABLE 7

NaTl-Type Compounds

Composition	a_0, Å	Reference
LiZn	6.221	2
LiGa	6.207	2
LiAl	6.37	2
LiCd	6.700	2
LiIn	6.80	2
NaIn	7.312	2
NaTl	7.488	2

in the periodic table. Laves[24] has discussed the interrelation of NaTl- and CsCl-type compounds.

It has occasionally proved difficult to identify positively the structure of certain alloys. For example, LiCd was reported both as NaTl type and CsCl type, and $PrMg_3$ was reported both as Fe_3Al type and NaTl type. Tables 4 through 7 list the structures which the author considers most probable.

4.4 Related Cubic Structures

There are other cubic crystal structures which can be derived from Fig. 6 by leaving certain positions vacant and specifying ordered or random distribution in the remaining positions. Four such structures have as their prototypes CaF_2, MgAgAs, ZnS, and NaCl. Table 8 illustrates the location of atoms.

is pure fcc. In the case of Fe_3Si and Cu_2MnSb the mixed bc and fc characteristics are obvious on a powder diffraction pattern.

4.5 Distorted CsCl-Type Compounds

The vast majority of CsCl-type compounds are cubic, but a few exist in distorted cubic forms. The more common variant is a tetragonally distorted CsCl type. Table 9 lists the known representatives.

TABLE 9

Tetragonally Distorted CsCl-Type Compounds

Composition at. % A	Composition at. % B	Lattice Parameter, Å c_0	a_0	c/a	Reference
V_{50}	Ru_{50}	3·09	2.96	1.04	2
Nb_{60}	Ru_{40}	3.263	3.134	1.04	6
Ta_{60}	Ru_{40}	3.27	3.13	1.05	6
V_{50}	Ir_{50}	3.68	2.74	1.34	6
Ti_{50}	Rh_{50}	3.36	2.95	1.14	6
Ti_{50}	Ir_{50}	3.468	2.915	1.19	6
Mn_{50}	Au_{50}		3.27	0.95	2
Zn_{50}	Ni_{50}	3.178	2.747	1.157	2

Another distorted CsCl-type structure exists in UCo. The diffraction pattern appears cubic, with $a_0 = 6.356$, which is nearly double the parameter of the undistorted CsCl-type compounds. A description of the structure has been published by Makarov.[8] Essentially, planes of atoms which are flat in ordinary

TABLE 8

Other Related Cubic Structures

Structure Type	Name	Example	Atomic Positions (From Fig. 6) *a*	*b*	*c*	*d*
Cl	CaF_2	$AuIn_2$	Au	vacant	In	In
Cl_b	MgAgAs	CuMgSb	Cu	vacant	Mg	Sb
B3	ZnS	AsIn	As	vacant	In	vacant
B1	NaCl	UTe	U	Te	vacant	vacant

The four structures listed above are fcc rather than bcc. However, when they are considered as part of a series (six from Table 3 followed by four from Table 8), it is noted that the series exhibits a gradual transition from bcc to fcc. VRe, VOs, and Ag_2TiAl have a pure body-centered structure; Cu_2MnAl, Fe_3Si, NaTl, $AuIn_2$, CuMgSb, and ZnS have a mixture of bc and fc characteristics; and UTe

CsCl-type compounds have become rippled in UCo. A grouping of eight small cubes is necessary to construct the large unit cell.

5. OTHER BCC STRUCTURES

5.1 Gamma Brass

There are three structure types, $D8_1$, $D8_2$, and $D8_3$, which are closely related. Each is

TABLE 10

Type $D8_a$ Structures

Composition	a_0, Å	Reference
Th_6Mn_{23}-Type Phases		
Th_6Mn_{23}	12.523	2
Th_6Mg_{23}	14.27	26
Sr_6Mg_{23}	14.91	26
Ba_6Mg_{23}	15.21	26
G or $Mg_6Cu_{16}Si_7$-Type Phases		
$Mg_6Cu_{16}Si_7$	11.65	2
$Sc_6Ni_{16}Si_7$	11.429	30
$Ti_6Ni_{16}Si_7$	11.187	27
$Zr_6Ni_{16}Si_7$	11.423	27
$Hf_6Ni_{16}Si_7$	11.39	26
$V_6Ni_{16}Si_7$	11.153	27
$Nb_6Ni_{16}Si_7$	11.249	27
$Ta_6Ni_{16}Si_7$	11.215	27
$Mn_6Ni_{16}Si_7$	11.15	26
$Sc_6Co_{16}Si_7$	11.438	30
$Ti_6Co_{16}Si_7$	11.201	27
$Zr_6Co_{16}Si_7$	11.417	27
$Nb_6Co_{16}Si_7$	11.235	27
$Ta_6Co_{16}Si_7$	11.198	27
$Sc_6Ni_{16}Ge_7$	11.663	30
$Ti_6Ni_{16}Ge_7$	11.421	27
$Zr_6Ni_{16}Ge_7$	11.689	27
$Nb_6Ni_{16}Ge_7$	11.504	27
$Mn_6Ni_{16}Ge_7$	11.41	26
$Zr_6Co_{16}Ge_7$	11.625	27
$Mg_6Ni_{16}Ge_7$	11.532	42

cubic and contains fifty-two atoms per unit cell. Each belongs to a different space group, with different conditions limiting possible reflections from lattice planes. However, the difference is slight between the $D8_2$ and $D8_3$ types.

Hume-Rothery and Raynor[1] have explained the structure of Cu_5Zn_8 (type $D8_2$) as a cubic cell formed by stacking twenty-seven unit cells of bcc structure, then removing two of the fifty-four atoms. After some readjustments, the remaining fifty-two atoms form the gamma-brass structure. Since the known examples of gamma-brass structures have been recently tabulated,[1,2] the tabulation is not repeated here.

Prototypes of the three gamma-brass structures are Fe_3Zn_{10}, Cu_5Zn_8, and Cu_9Al_4. All three prototypes are 21:13 electron compounds, meaning that twenty-one valence electrons are present for every thirteen atoms, although the first one has to be considered as Fe_5Zn_{21}, with zero valence for Fe.

One example of the $D8_2$-type structure is V_5Al_8, discovered by Carlson et al.[(2)] and confirmed by the present writer. The most prominent lines on the powder diffraction pattern are those of a bc cube with a_0 = 3.074 Å. The weaker lines may be indexed by tripling the a_0 to 9.222 Å. The indices selected on the basis of the tripled unit cell edge satisfy the condition $h + k + l = 2n$, hence the structure is bcc.

5.2 Th_6Mn_{23} and $Mg_6Cu_{16}Si_7$-Type Phases

The structure type $D8_a$ occurs in Th_6Mn_{23} and $Mg_6Cu_{16}Si_7$-type phases. These compounds are related in that Cu and Si together replace Mn. Gladyshevskii et al.[26] refer to $Mg_6Cu_{16}Si_7$ as the superlattice type of Th_6Mn_{23}. The prototype, $Mg_6Cu_{16}Si_7$, was discovered by Witte.[2] Somewhat later Beattie and VerSnyder[25] discovered $Ti_6Ni_{16}Si_7$ and described it as the G phase. The crystal structure may be derived from a stack of small bc cubes. The list of known examples is given in Table 10. Spiegel, Bardos, and Beck[27] have recently discussed the G phases.

5.3 Ir_3Sn_7-Type Compounds, $D8_f$ Type

This family contains the eight representatives listed in Table 11.

This family is one whose components are closely restricted to certain groups of the periodic table. The element which occupies twelve of the forty sites in the unit cell comes only from group VIII. Examples in which Rh, Fe, or Os participate have not yet been reported, but Rh in particular may be expected to participate. The other element, occupying

TABLE 11

Ir_3Sn_7-Type Compounds, $D8_f$ Type

Composition	a_0, Å	R_A/R_B	Reference
$Co_3Al_3Si_4$	?	0.915	28
$Ni_3(Al, Si)_7$	?	0.911	28
Ru_3Sn_7	9.351	0.868	2
Pd_3Ga_7	?	0.974	2
Ir_3Ge_7	8.735	0.99	2
Ir_3Sn_7	9.360	0.878	2
Pt_3Ga_7	8.799	0.982	2
Pt_3In_7	9.435	0.834	2

TABLE 12

CrB-Type Compounds, B_f Type

Composition	Lattice Parameters, Å			R_A/R_B	Reference
	a_0	b_0	c_0		
ThRu	3.878	11.29	4.071	1.343	31
ThCo	3.74	10.88	4.16	1.436	2
ThIr	3.894	11.13	4.266	1.325	31
ThRh	3.866	11.24	4.220	1.337	31
ThPt	3.900	11.09	4.454	1.296	31
CeNi	3.77	10.46	4.37	1.376	32
ZrNi	3.268	9.937	4.101	1.286	33
HfNi	3.220	9.820	4.12	1.268	33
PuNi	3.59	10.21	4.22	1.277	34
CaAg	4.08	11.48	4.65	1.366	52
YAl	3.884	11.522	4.385	1.258	51
ZrAl	3.359	10.887	4.274	1.119	35
HfAl	3.25	10.83	4.28	1.103	36
[a]ThAl	4.42	11.45	4.19	1.256	2
PrGa	4.452	11.331	4.195	1.295	23
[a]GdGa	4.341	11.02	4.066	1.277	37
[a]DyGa	4.300	10.89	4.067	1.257	37
CaSi	4.59	10.795	3.91	1 497	2
YSi	4 251	10.526	3.826	1.365	38
CaGe	4.575	10.845	4.001	1.442	2
PrGe	4.474	11.098	4.064	1.335	23
[a]GdGe	4.175	10.61	3.960	1.316	37
[a]DyGe	4.112	10.81	3.924	1.295	37
CaSn	4.821	11.52	4.349	1.216	2

[a] Crystallographic axes translated from those originally reported.

TABLE 13

FeB-Type Structures, B27 Type

Composition	Lattice Parameters, Å			R_A/R_B	Reference
	a_0	b_0	c_0		
[a]YNi	7.12	4.10	5.51	1.445	39
[a]GdNi	6.931	4.353	5.428	1.446	37
[a]DyNi	6.895	4.319	5.353	1.423	37
ThNi	14.51	4.31	5.73	1.443	2
[a]GdPt	7.164	4.458	5.574	1.299	37
[a]DyPt	7.118	4.453	5.466	1.278	37
CeCu	7.30	4.30	6.36	1.342	40
ZrSi	6.982	3.786	5.302	1.215	2
[a]ThSi	7.88	4.15	5.89	1.363	41
USi	7.665	3.908	5.661	1.183	2
PuSi	7.933	3.847	5.727	1.246	2

[a] Crystallographic axes translated from those originally reported.

twenty-eight of the forty sites, is a member of either the Al or Si group, with Tl and Pb not yet reported.

5.4 Ru_3Be_{17}-Type Compounds

Sands et al.[43] have reported the compounds Ru_3Be_{17} and Os_3Be_{17} which have a bcc crystal structure and belong to space group Im3, No. 204. Somewhat earlier, Y_3Mg_{17} was reported by Gibson and Carlson[29] who did not determine the structure. It seems probable from the similarity of Mg and Be that these compounds are isostructural. Parameters are given below:

Y_3Mg_{17}	$a_0 = 11.26$ Å
Ru_3Be_{17}	$a_0 = 11.337$
Os_3Be_{17}	$a_0 = 11.342$

6. ORTHORHOMBIC COMPOUNDS

6.1 CrB or B_f Type

The CrB-type compounds are not derived from a bcc structure, but since they occur in alloy systems where a bcc compound might be expected they are discussed in this chapter. Two alloy series in which both CsCl-type and CrB-type compounds appear are those between group IVB elements and Co and Ru. TiCo, ZrCo, and HfCo have the CsCl-type structure, but ThCo has the CrB type. Likewise, TiRu, ZrRu, and HfRu are CsCl type, but ThRu is CrB type. An apparent reason for the change in crystal structure is that CrB-type compounds are associated with large radius ratios, whereas CsCl-type compounds, as shown in Fig. 2, are not highly selective. The published examples of CrB-type compounds (excluding borides) are listed in Table 12. Several investigators who have reported CrB-type compounds have chosen different crystallographic axes. For consistency, the compounds listed in Table 12 have had their axes translated when necessary in order to conform with space group Cmcm.

6.2 FeB or B27 Type

The FeB compounds, like the CrB type, sometimes occur in alloy systems in which we might expect a CsCl type. For example, CeCu has the FeB-type structure although four other members of the RE–Cu series have the CsCl-type structure. More often FeB-type compounds are found in systems in which the B element has too high a group number to favor the CsCl type. A large radius ratio favors the FeB type, as may be seen from Table 13. In this table, crystallographic axes are translated where necessary to make them increase in the order *b*, *c*, *a*.

The factors which control the choice between a CrB- and an FeB-type structure are not clear. The behavior of Y and Zr is unexplainable. As shown in the following tabulation

CrB	FeB
YSi	ZrSi
ZrNi	YNi
HfNi	DyNi
ThCo	ThNi

Y favors one structure when the B element is Si and another structure when the B element is Ni.

ACKNOWLEDGMENTS

The author wishes to acknowledge many helpful discussions with Dr. M. V. Nevitt and other members of the Alloy Properties Group, Metallurgy Division, Argonne National Laboratory. This work was performed under the auspices of the U.S. Atomic Energy Commission.

REFERENCES

1. Hume-Rothery W., and G. V. Raynor, *The Structure of Metals and Alloys*. The Institute of Metals, London, 1962.
2. Pearson W. B., *Handbook of Lattice Spacings and Structures of Metals and Alloys*. Pergamon Press, New York, 1958.
3. Hansen M., *Constitution of Binary Alloys*. McGraw-Hill, New York, 1958.
4. Darby J. B., Jr., *Trans. AIME* **227** (1963) 1460.
5. Nevitt M. V., in *Electronic Structure and Alloy Chemistry of the Transition Elements*. Wiley-Interscience, New York, 1963, p. 101.
6. Dwight A. E., unpublished data.
7. Dwight A. E., *Trans. AIME* **215** (1959) 283.

8. Makarov E. S., *Crystal Chemistry of Simple Compounds*, Consultants Bureau, New York, 1959.
9. Teatum E., K. Gschneidner, Jr., and J. Waber, Los Alamos Report LA-2345, 1960.
10. Beck P. A., OOR Conference on Basic Research in Metallurgy, Durham, N.C., 1958.
11. Böhm H., and K. Löhberg, *Z. Metallk.* **49** (1958) 173.
12. Gladyshevskii E. I., et al., *Soviet Phys.-Cryst.* **6** (September–October 1961) 207.
13. Heine W., and U. Zwicker, *Z. Metallk.* **49** (1962) 391.
14. Dwight A. E., in *Annual Report for 1961*. Metallurgy Division, Argonne National Laboratory, ANL-6516, 259 (1962).
15. Tsuboya L., and M. Sugihara, *J. Phys. Soc. Japan* **15** (1960) 1534.
16. Tsuboya I., and M. Sugihara, *J. Phys. Soc. Japan* **16** (1961) 571.
17. Tsuboya I., and M. Sugihara, *J. Phys. Soc. Japan* **16** (1961) 1257.
18. Tsuboya I., and M. Sugihara, *J. Phys. Soc. Japan* **17** (1962) 410.
19. Cherkashin Ye. Ye., et al., *Zh. Neorgan. Khim.* **3** (1958) 166.
20. Oxley D. P., R. S. Tebble, and K. C. Williams, *J. Appl. Phys.* **34** (1963) 1362.
21. Ramsey W. J., *Acta Cryst.* **14** (1961) 1091.
22. Schubert K., et al., *Naturwiss.* **43** (1956) 248.
23. Iandelli A., "Intermetallic Compounds of the Rare Earth Metals," in *Physical Chemistry of Metallic Solutions and Intermetallic Compounds*. H.M. Stationery Office, London, 1959.
24. Laves F., in *Theory of Alloy Phases*. American Society for Metals, 1956 p. 124.
25. Beattie H. J., Jr., and F. L. VerSnyder, *Nature* **178** (1956) 208.
26. Gladyshevskii E. I., et al., *Soviet Phys.-Cryst.* **6** (1961) 615.
27. Spiegel F. X., D. Bardos, and Paul A. Beck, *Trans. AIME* **227** (1963) 575.
28. Nowotny H., in *Electronic Structure and Alloy Chemistry of the Transition Elements*. Wiley-Interscience, New York, 1963, p. 179.
29. Gibson E. D., and O. N. Carlson, *Trans. ASM* **52** (1960) 1084.
30. Dwight A. E., R. A. Conner, Jr., and J. W. Downey, *Nature* **197** (1963) 587.
31. Thomson J. R., *Acta Cryst.* **15** (1962) 1308.
32. Finney J. J., and A. Rosenzweig, *Acta Cryst.* **14** (1961) 69.
33. Kirkpatrick, M. E., D. M. Bailey, and J. F. Smith, *Acta Cryst.* **15** (1962) 252.
34. Cromer D. T., and R. B. Roof, Jr., *Acta Cryst.* **12** (1959) 942.
35. Spooner F. J., and C. G. Wilson, *Acta Cryst.* **15** (1962) 621.
36. Edshammar L., *Acta Chem. Scand,* **15** (1961) 403.
37. Baenziger N. C., and J. L. Moriarty, Jr., *Acta Cryst.* **14** (1961) 946).
38. Parthé E., *Acta Cryst.* **12** (1959) 559.
39. Beaudry B. J., and A. H. Daane, *Trans. AIME* **218** (1960) 854.
40. Larson A. C., and D. T. Cromer, *Acta Cryst.* **14** (1961) 545.
41. Jacobson E. L., et al., *J. Am. Chem. Soc.* **78** (1956) 4850.
42. Teslyuk M. Yu., and V. Ya. Markiv, *Soviet Phys.-Cryst.* **7** (1962) 103.
43. Sands D. E., Q. C. Johnson, O. H. Krikorian, and K. L. Kromholtz, *Acta Cryst.* **15** (1962) 1205.
44. Mooser A., and W. Pearson, *J. Chem. Phys.* **26** (1957) 893.
45. Gschneidner, K. A., Jr., *Rare Earth Alloys*, Van Nostrand, New York, 1961.
46. Aldred A. T., *Trans. AIME* **224** (1962) 1082.
47. Chao C. C., H. L. Luo, and P. Duwez, *J. Appl. Phy's.* **34** (1963) 1971.
48. Chao C. C., H. L. Luo, and P. Duwez, *J. Appl. Phys.* **35** (1964) 257.
49. Iandelli A., *Rend. Sci. Fis. Mat. Nat.* **29** (1960) 62.
50. Markiv V. Ya., and M. Yu. Teslyuk, *Dopovidi Akad. Nauk Ukr. RSR* (1962) No. 12, 1607–9.
51. Dagenham T., *Acta Chem. Scand.* **17** (1963) 267.
52. Calvert L. D., H. S. Dunsmore, L. V. Kuhi, and R. S. Tse, *Acta Cryst.* **10** (1957) 775.
53. Markiv, V. Ya., Yu. V. Voroshilov, P. I. Kripyakevich, and E. E. Cherkashin, *Kristallografiya* **9** (1964) 737.
54. Hames F. A., *J. Appl. Phys.* **31** Supplement (1960) 370S.
55. Giessen B. C., J. S. Benjamin, and N. J. Grant, *J. Metals* **16** (1964) 773,
56. Muldawer L., *A Study of the IB-IIB β-Phase Alloys*. TID-18632, Annual Progress Report, May 1, 1962 to April 1, 1963.

chapter 11

Wurtzite and Zinc Blende Type Structures

ERWIN PARTHÉ

School of Metallurgical Engineering
and
Laboratory for Research on the Structure of Matter
University of Pennsylvania
Philadelphia, Pennsylvania

1. INTRODUCTION

The current interest in the semiconducting properties of materials has prompted an extensive search for new compounds with wurtzite, zinc blende or related structures. These structures belong to the so-called *adamantine* structures* which form a unique group.

Their compositions and structural features can be predicted by using two electron rules and one empirical relationship. This chapter is concerned for the most part with the two electron rules, the first having as the important parameter the *electron*-to-*atom* ratio, $N_{Electrons}/N_{Atoms}$ (also known as valence electron concentration), the second using the *electron* to *anion* ratio, $N_{Electrons}/N_{Anions}$. It is also shown that for an adamantine structure to occur both rules must be satisfied because adamantine structures are tetrahedral structures and are obtained only with normal valence compounds.

* Adamantine in its original meaning is: "like diamond in hardness and luster." Here it is to be understood as being similar in structure to diamond or a stacking variation of it.

2. THE GEOMETRICAL FEATURES OF ADAMANTINE STRUCTURES

The adamantine structures show the geometrical features of wurtzite, zinc blende, or a stacking variation of them. Every atom on the Zn sites or cation sites is tetrahedrally surrounded by four anions.* Conversely, every atom of the S sites or anion sites is tetrahedrally surrounded by four or fewer cations. There are two kinds of adamantine structures: *normal* and *defect* adamantine structures. In normal adamantine structures every anion has four nearest cation neighbors. However, in defect adamantine structures cation vacancies occur, and some anions then have fewer than four cation neighbors, but these cations are still located at corners of a surrounding tetrahedron. Such vacancies never occur on the anion sites. If we disregard for the moment all cations and study only the arrangement of the anions (the so-called anion partial structure), we find that the anions are close-packed in all adamantine structures; close-packing can thus be considered a characteristic feature of these structures. The anion partial structure of wurtzite is hexagonal close-packed, whereas that of zinc blende is cubic close-packed. Two compounds are known which have adamantine structures with more complicated stacking variations of the close-packed anion partial structure. One of these is silicon carbide which occurs with as many as seventy-five different adamantine structure modifications;† the other is zinc sulfide with eleven modifications.‡ Seven of the latter are shown in Fig. 1. Except for these two compounds, all presently known adamantine structure compounds have hexagonal or cubic close-packed anion partial structures and, therefore, are directly related to wurtzite or to zinc blende. The different adamantine structures to be discussed below differ only in the way different atoms and vacancies can be distributed over the cation sites of zinc blende and wurtzite. In some cases there are also different anions on the corresponding S sites of wurtzite and zinc blende. The adamantine structures occur with binary, ternary, quaternary, and multicomponent compounds. We include for completeness also the structure of diamond itself and the structure of the recently observed wurtzite-silicon,[1] although the cation-anion concept has no meaning in this case.

3. ADAMANTINE STRUCTURES, A SUBDIVISION OF THE TETRAHEDRAL STRUCTURES

The adamantine structures form a subgroup of the so-called *tetrahedral structures*. The term tetrahedral structures applies to a separate group of structures which are characterized by tetrahedral bonds. Every atom in a tetrahedral structure has as next-neighbors four other atoms (in special cases, fewer), which are located at the vertices of a surrounding tetrahedron. We distinguish two kinds of tetrahedral structures: *normal* and *defect*. In normal tetrahedral structures every atom without exception is tetrahedrally surrounded by four tetrahedrally coordinated atoms. Every atom in such a structure has sp^3 hybridized orbitals, for which one needs an average of four valence electrons per atom. In defect tetrahedral structures every atom again has four tetrahedral orbitals, but not every orbital is used for bonding. Thus in defect tetrahedral structures some atoms have fewer than four neighbors. All these orbitals, which are not used for bonding, obtain one extra electron each and become nonbonding orbitals.

The geometrical features of tetrahedral structures and their electron rules have been studied recently.[2,3] It was shown that a correlation exists between the electron-to-atom ratio, $N_{Electrons}/N_{Atoms}$, and the observable structural features. This correlation was expressed mathematically by the *general tetrahedral structure equation*.

$$\mathrm{VEC} = 4\left(1 + \frac{1}{y}\right) \tag{1}$$

VEC is the electron-to-atom ratio, $N_{Electrons}/N_{Atoms}$, commonly known among metallurgists as valence electron concentration. The term has been formulated by Hume-Rothery

* It is customary to use the term "anion" even if ionic bonding is not the prevailing bonding mechanism. The term applies then to the more electronegative atom.

† Forty-six SiC modifications are given in Appendix A of reference 2, and twenty-nine new ones are listed by Kuo Chang-Lin, "X-Ray Studies of Polytypism of Silicon Carbide II," *Sci. Sinica* **13** (1964) 1773–1784. Recently, a new structure determination was published by A. H. Gomes de Mesquita, "The Structure of a Silicon Carbide Polytype 24R," *Acta Cryst.* **18** (1965) 128.

‡ Ten ZnS modifications are listed in Appendix B of Reference 2. A new twenty-four layer polytype was recently reported by M. Farkas-Jahnke, "A New Polytype of Zinc Sulfide Crystals," *Acta Cryst.* **18** (1965) 571–572.

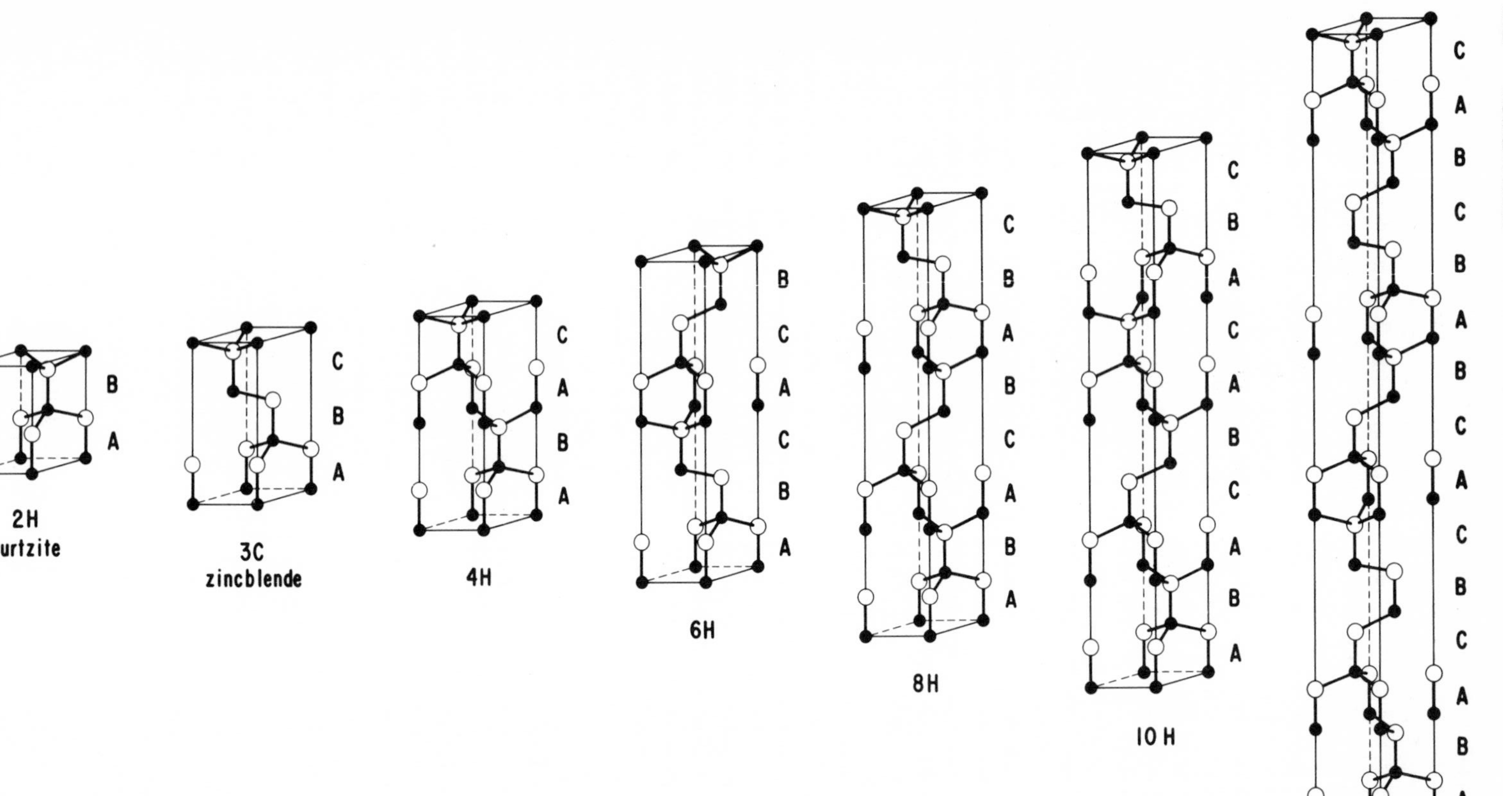

Fig. 1. Seven of the zinc sulfide modifications.

to account for the occurrence of some intermetallic phases. However, up to now the valence electron concentration has found no application to covalent or ionic inorganic compounds. If the VEC value is four, we have a normal tetrahedral structure; if it is larger than four, a defect tetrahedral structure. Compounds with a VEC value smaller than four cannot crystallize with a tetrahedral structure.

The y value is a quantity which denotes the tetrahedral bond defects. For its evaluation we must determine for a structure the average number of nonbonding orbitals per atom, usually obtained by counting bonds and atoms. The y value is defined as four times the reciprocal of the average number of nonbonding orbitals per atom. In normal tetrahedral structures there are no bond defects; thus the y value is infinite. In defect tetrahedral structures, however, the y value is a positive quantity, but not infinity.

The different solutions of the general tetrahedral structure equation are summarized in Table 1. We notice that adamantine structures are listed only with $4.0 \leq \text{VEC} \leq 4.8$. Adamantine structures with a VEC value of four are normal adamantine structures; all others are defect adamantine structures. For VEC values larger than 4.8, adamantine structures cannot occur for there are now so few bonding orbitals that it becomes impossible to construct a three-dimensionally netted array having a close-packed anion partial structure.

The composition formulas of tetrahedral structures are customarily written by using integers, which denote the numbers of the valence electrons of the different atoms. In the case of defect tetrahedral structures these formulas contain also the symbol 0 (= zero); for example, $3_2 0 6_3$ for Ga_2S_3. By convention the 0 (= zero) is used for every quadruplet of nonbonding orbitals per formula unit. For a defect tetrahedral structure compound with composition $X_m Y_n Z_o 0_s$, where X, Y, and Z are different atoms and 0 denotes quadruplets of nonbonding orbitals the general tetrahedral structure equation is conveniently rewritten in the form[2]

$$\frac{me_X + ne_Y + oe_Z}{m + n + o + s} = 4 \tag{2}$$

where e_X, e_Y, and e_Z represent the numbers of valence electrons of elements X, Y, and Z. This form of the general tetrahedral structure equation will be used for later calculations. The y value can now be expressed in a different way. It is the ratio of the total number of atoms divided by the number of quartets of

TABLE 1

Different Solutions of the General Tetrahedral Structure Equation[a]

$$\text{VEC} = 4\left(1 + \frac{1}{y}\right)$$

VEC	y	Type of Structures
<4	negative	No tetrahedral structures
4	∞	Normal tetrahedral structures, some of them normal adamantine structures, no vacancies
$4 < \text{VEC} \leq 4.80$	$\infty > y \geq 5$	Defect tetrahedral structures, some of them defect adamantine structures
>4.80	<5	Other defect tetrahedral structures, but no adamantine structures

[a] VEC—valence electron concentration or electron-to-atom ratio; y—number of atoms per unit cell divided by the number of quadruplets of nonbonding orbitals or four times the reciprocal of the average number of nonbonding orbitals per atom.

nonbonding orbitals per formula unit.

$$y = \frac{m + n + o}{s} \tag{3}$$

For example, the y value of Ga_2S_3 is 5.

In defect *tetrahedral* structures the nonbonding orbitals can be arranged in any arbitrary way as long as their number per formula unit is correct. Thus with the defect tetrahedral structures we find molecular, fiber, layer, and three-dimensionally netted structures. In defect *adamantine* structures, however, only three-dimensionally netted structures with fully occupied, close-packed anion partial structures occur. Atoms are missing on the cation sites, which are always tetrahedrally surrounded by anions. Each of the four anion neighbors surrounding a vacancy has one nonbonding orbital directed toward the vacancy. A quadruplet of nonbonding orbitals in a defect adamantine structure thus corresponds to a structural vacancy.

For defect adamantine structures the previously used symbol 0 (= zero) can now be interpreted as structural vacancy. It has been suggested that in this case a □ (square) be used instead of a 0 (zero). The formula of the adamantine structure compound Ga_2S_3 can then be written as $3_2\square 6_3$. Correspondingly, for the special case of defect adamantine structures the y value can be formulated as the number of atoms divided by the number of cation vacancies per formula unit.

It is peculiar to defect adamantine structure compounds that a temperature treatment does not change the number of vacancies in the structure; rather its only influence is possibly to change their mutual arrangement. This is a clear indication that these vacancies are an integral and required part of the structural features.

The elements that have been found to participate in tetrahedral and also adamantine structures are given below, grouped according to their valence electron contribution to the bonding mechanism. They include all elements of the second and third period excepting Li, Na, F, and inert gases, and include further all B-group elements* with the exception of some very heavy ones like Au, Pb, and Bi. Elements with empty or partially filled *d*-shells, alkali, and alkaline earth metals have not been found in adamantine structures. Exceptions are a very small number of Mn- and Fe-containing compounds.

One-electron elements: Cu and Ag
Two-electron elements: Be, Mg, Zn, Cd, Hg, and Mn
Three-electron elements: B, Al, Ga, In, Tl, and Fe
Four-electron elements: C, Si, Ge, and Sn
Five-electron elements: N, P, As, and Sb
Six-electron elements: O, S, Se, Te, and Po
Seven-electron elements: Cl, Br, and I

4. ADAMANTINE STRUCTURE COMPOUNDS AS NORMAL VALENCE COMPOUNDS

The second electron rule for adamantine structure compounds follows from the observation that all adamantine structure compounds are normal valence compounds. Compounds, which are composed of B elements* and which have at least one component from the silicon, phosphorus, sulfur, or chlorine group, have an electron distribution that obeys the octet principle. The mathematical formulation of this situation is given by the *general equation for valence compounds*. For a compound C_mA_n it can be shown that

$$m(e_C - CC) = n(8 - e_A - AA) \tag{4}$$

where m and n are composition parameters, e_C and e_A the outer valence electrons of cation C and anion A, CC the average number of cation-cation bonds per cation (or, in a different interpretation, the average number of electrons per cation which are not involved in cation-anion bonds), and AA the average number of anion-anion bonds. An incomplete version of this formula was first published by Mooser and Pearson[4] in 1956; it has been improved over the years by number of authors.[5–9]

The general equation for valence compounds is conveniently rearranged in such manner that the electron to anion ratio

$$\frac{N_{Electrons}}{N_{Anions}} = \frac{me_C + ne_A}{n} \tag{5}$$

is separated from the other parts. We obtain thus

$$\frac{N_{Electrons}}{N_{Anions}} = 8 + \frac{mCC - nAA}{n} \tag{6}$$

* B elements include elements of the second and third period, except alkali and alkaline earth elements, and those from higher periods with filled *d*-shells.

All compounds that obey this rule are called *valence compounds*. Depending on the value of the electron to anion ratio N_E/N_A we distinguish between three groups of valence compounds: *polycationic valence compounds* with $N_E/N_A > 8$, *normal valence compounds* with $N_E/N_A = 8$, and *polyanionic valence compounds* with $N_E/N_A < 8$.

Table 2 gives the different solutions of the general equation for valence compounds. In the most simple cases polycationic valence compounds have no anion-anion bonds (AA = 0), normal valence compounds only cation-anion bonds (AA = CC = 0), and finally polyanionic valence compounds no cation-cation bonds (CC = 0). For polyanionic valence compounds the number of anion-anion bonds is then given by

$$AA = 8 - \frac{N_E}{N_A} \qquad (7)$$

We may now compare with this result the kind of atom netting in the structures of elements, as described by the Hume-Rothery 8-N rule. Applying the general equation for valence compounds (4) to elements we obtain

$$0 = 8 - e_A - AA \qquad (8)$$

or

$$AA = 8 - e_A = 8 - \frac{N_E}{N_A} \qquad (9)$$

Equations (7) and (9) are identical. Thus the anion-anion netting in polyanionic valence compounds is similar to the netting of atoms in the structure of the elements with equal N_E/N_A value. The different solutions of Eq. (7) are given in Table 3. The last row of Table 3 gives the results for $N_E/N_A = 8$; in this case the compound is called a normal valence compound.

The structural resemblance of the anion array in polyanionic valence compounds and the atom array in element structures was first formulated by Busmann[10] and has since been described by other authors.[6,9]

Goodman[11] observed that wurtzite- and zinc-blende-related compounds are normal valence compounds since the valence electrons provided by all elements on the cation sites have just the right number to complete the octet shells of the atoms on the anion sites. These ideas have been extended to defect structure compounds by Suchet[12] and have since been covered in a number of papers.[2,13,14]

TABLE 2

Solutions of the General Equation for Valence Compounds^a

$$\frac{N_{Electrons}}{N_{Anions}} = 8 + \frac{mCC - nAA}{n}$$

Polycationic Valence Compounds		Normal Valence Compounds		Polyanionic Valence Compounds	
$\frac{N_{Electrons}}{N_{Anions}} > 8$		$\frac{N_{Electrons}}{N_{Anions}} = 8$		$\frac{N_{Electrons}}{N_{Anions}} < 8$	
$CC > \frac{n}{m} AA$		$mCC = nAA$		$\frac{m}{n} CC < AA$	
CC	AA	CC	AA	CC	AA
$\frac{n}{m}\left(\frac{N_E}{N_A} - 8\right)$	0	0	0	0	$8 - \frac{N_E}{N_A}$
$\frac{n}{m}\left(\frac{N_E}{N_A} - 7\right)$	1	n	m	1	$8 + \frac{m}{n} - \frac{N_E}{N_A}$
Example: GaSe		ZnS, NaCl		KGe, ZnP_2	

^{a} m, n—composition parameters, CC—average number of cation-cation bonds per cation, AA—average number of anion-anion bonds per anion, N_E—total number of valence electrons, N_A—total number of anions.

TABLE 3

The Anion Array in Polyanionic Valence Compounds[a]

N_E/N_A[b]	AA[c]	Type of Anion-Anion Netting
4	4	Three-dimensional netting as in diamond
5	3	Two-dimensional netting as in arsenic, or isolated tetrahedrons as in white phosphorus
6	2	One-dimensional netting as in selenium
7	1	Dumbbells as in chlorine
8	0	Isolated atoms as in inert gases; compound is called a normal valence compound

[a] The recognizable anion array in polyanionic valence compounds resembles the array of atoms in the structure of the elements provided that (1) no electrons are involved in forming cation-cation bonds and there are no unshared valence electrons on the cations, or in other words CC = 0; (2) anions occupy only one set of equivalent sites in the structure or, if they occupy different sites, they have the same number of anion nearest-neighbors.
[b] N_E—total number of valence electrons; N_A—total number of anions.
[c] AA—average number of anion-anion bonds per anion.

Normal valence compounds occur for $N_E/N_A = 8$, and according to the last entry in Table 3 we expect to find in the structures of these compounds isolated anions. This agrees well with the wurtzite- and zinc-blende-related structures, in which only cation-anion bonds occur.

A difficulty arises with some ternary and multicomponent adamantine structure components because on first sight it seems difficult to decide which elements are cations and which elements anions. For example, in the adamantine structure compound Cu_3AsS_4($1_3 5 6_4$) both Cu and As have to be considered as cations, whereas in the defect adamantine structure compound Zn_3AsI_3 ($2_3\square 5 7_3$) the As atoms are counted as anions. However, we can overcome this difficulty if we observe that the anion sites of wurtzite or zinc blende must be always completely occupied with electronegative atoms and, further, that the number of electrons provided by the atoms on the Zn or cation sites must be just right to complete the octet shell of the atoms on the S or anion sites.

5. THE CALCULATION OF ADAMANTINE STRUCTURE COMPOSITIONS

For the calculation of possible adamantine structure compositions two equations will be used: the general tetrahedral structure equation and the equation for normal valence compounds.

Normal adamantine structure compositions $X_mY_nZ_o$ are obtained if we apply

$$\frac{me_X + ne_Y + oe_Z}{m + n + o} = 4 \quad \text{and} \quad \frac{N_E}{N_A} = 8$$

Defect adamantine structure compositions $X_mY_n\square_sZ_o$ can be derived from

$$\frac{me_X + ne_Y + oe_Z}{m + n + o + s} = 4 \quad \text{and} \quad \frac{N_E}{N_A} = 8$$

where

$$4 < \frac{me_X + ne_Y + o_Z}{m + n + o} \leq 4.8$$

On the following pages some examples are given, which show how the actual compositions of adamantine structures can be calculated.

5.1 Normal Adamantine Structures

Binary Compounds. If we assume a general binary compound of composition C_mA_n, the equation for normal valence compounds $N_E/N_A = 8$ can be written as

$$me_C = n(8 - e_A) \qquad (10)$$

which, combined with the equation for normal tetrahedral structures

$$me_C + ne_A = 4(m + n) \qquad (11)$$

gives as a result

$$m = n \tag{12}$$

Thus the only possible solutions are equiatomic compounds with compositions:

17, 26, 35, and 44

These compounds are the well-known *Grimm-Sommerfeld* compounds, which are listed at the top of Table 5.

Ternary Compounds. Goryunova[13] has shown that for ternary normal adamantine structure compounds, we must distinguish two separate cases: "two-cation" phases and "two-anion" phases.

1. "*Two-cation*" phases $(C_mD_n)A_o$: Two different kinds of cations, C and D, occupy the Zn sites, whereas the S sites are filled with the third component A. Combining the normal tetrahedral structure equation

$$\frac{me_C + ne_D + oe_A}{m + n + o} = 4 \tag{13}$$

with the equation for normal valence compounds

$$me_C + ne_D = o(8 - e_A) \tag{14}$$

we obtain

$$o = m + n \tag{15}$$

It is advantageous to introduce now a new parameter R which is the ratio of m over n

$$R = \frac{m}{n} \tag{16}$$

Inserting (15) and (16) in (13) gives

$$R = \frac{e_D + e_A - 8}{8 - e_C - e_A} \tag{17}$$

To find possible solutions for R, we may construct a table as shown in Table 4. Only five solutions exist which are neither zero, infinite, nor negative. They correspond to compositions

136_2, 14_25_3, 1_246_3, 1_356_4, and 245_2

A study of Table 5 reveals that examples are known for each of these compositions.

2. "*Two-anion*" phases $C_m(A_oB_p)$: The C atoms occupy the cation sites, whereas A and B atoms are distributed over the anion sites of wurtzite or zinc blende.

The equation for normal valence compounds can be written for this case

$$me_C = o(8 - e_A) + p(8 - e_B) \tag{18}$$

which, combined with the normal tetrahedral structure equation, gives

$$m = o + p \tag{19}$$

Introducing again a parameter $R = \frac{m}{o}$ leads to

$$R = \frac{e_B - e_A}{e_C + e_B - 8} \tag{20}$$

which has five solutions for $R > 1$ corresponding to compositions:

2_257, 3_34_27, 2_347_2, 2_437_3, and 3_246

No examples for these five compositions can be found in Table 5. The only known representative is a 3_246 compound, Al_2CO, which has a structure based on wurtzite, showing, however, some statistical occupancy of octahedral sites.[15]

Quaternary and Other Multicomponent Compounds with Normal Adamantine Structure. The same electron equations can be used to calculate the compositions of possible quaternary and multicomponent compounds. Goryunova[13] has obtained thirty-seven different quaternary compositions. The number of possible compositions increases accordingly if quinary phases are considered. Some representatives of quaternary and quinary adamantine structure compounds are listed in the middle part of Table 5. More quaternary and quinary alloys with adamantine structure have been synthesized in the course of studies to determine the mutual solubility of adamantine structure compounds. However, these compositions have mostly been left out of Table 5.

Structure Types for Normal Adamantine Structures. The arrangement of the different atoms on the zinc blende or wurtzite structure sites may be ordered or not ordered, but the "cations" are always on the zinc sites and the "anions" on the sulfur sites.* Theoretical

* It is characteristic of these compounds, and also another proof for the correctness of the electronic rules that the phase extent of these compounds is very small. Particularly accurate data are available for the binary compound GaAs which has a phase extent of 0.01 atomic percent at the most, according to M. E. Straumanis and C. D. Kim, "Phase Extent of Gallium Arsenide determined by the Lattice Constant and Density Method," *Acta Cryst.* **19** (1965) 256–259.

TABLE 4

Solutions of Equation (17) for Ternary Compounds of the "Two-cation" Type

$$R = \frac{e_D + e_A - 8}{8 - e_C - e_A} \quad \text{for } C_m D_n A_o \quad \text{with} \quad R = \frac{m}{n}$$

	$e_C = 1$					$e_C = 2$				$e_C = 3$			$e_C = 4$		$e_C = 5$
	$e_A = 3$	$e_A = 4$	$e_A = 5$	$e_A = 6$	$e_A = 7$	$e_A = 4$	$e_A = 5$	$e_A = 6$	$e_A = 7$	$e_A = 5$	$e_A = 6$	$e_A = 7$	$e_A = 6$	$e_A = 7$	$e_A = 7$
$e_D = 2$	$-\frac{3}{4}$	$-\frac{2}{3}$	$-\frac{1}{2}$	0	∞										
$e_D = 3$		$-\frac{1}{3}$	0	$\boxed{1}$	∞	$-\frac{1}{2}$	0	∞	−2						
$e_D = 4$			$\boxed{\frac{1}{2}}$	$\boxed{2}$	∞		$\boxed{1}$	∞	−3	∞	−2	−3			
$e_D = 5$				$\boxed{3}$	∞			∞	−4		−3	−2	$-\frac{3}{2}$	$-\frac{4}{3}$	
$e_D = 6$					∞				−5			$-\frac{5}{2}$		$-\frac{5}{3}$	$-\frac{5}{4}$

calculations have been made to determine what kind of supercells should occur if all atoms arranged themselves in orderly fashion. However, only five supercells of normal adamantine structures have been found in nature. These superstructures are shown in Fig. 2 together with their call name. Three are related to zinc blende: *chalcopyrite* observed with 136_2 and 245_2 compounds, *famatinite* with 1_3564 compounds, and *stannite*, a superstructure type for quaternary adamantine structure compounds. Two superstructures can be derived from wurtzite: the $BeSiN_2$ type, which has not yet been observed with other compounds, and *enargite*, another structure type for 1_356_4 compounds. There is some indication that a wurtzite analog of stannite exists; however a definite proof has not yet been given.

5.2 Defect Adamantine Structures

Binary Compounds. To find the possible compositions of binary defect adamantine structures with formula $C_m\square_sA_o$, we must combine the general tetrahedral structure equation

$$\frac{me_C + oe_A}{m + o + s} = 4 \tag{21}$$

with the equation for normal valence compounds

$$me_C = o(8 - e_A) \tag{22}$$

We obtain

$$o = m + s \tag{23}$$

which, inserted in (21) together with $R = m/o$, gives:

$$R = \frac{8 - e_A}{e_C} \tag{24}$$

Because $s > o$ and therefore $m < o$, only solutions for $R < 1$ are of interest. There are nine of these which correspond to compositions 4_305_4, 3_206_3, 406_2, $5_20_36_5$, 207_2, 30_27_3, 40_37_4, 50_47_5, and 60_57_6.

It was stated above that the valence electron concentration should not be larger than 4.8, or the corresponding y value not smaller than 5. With this restriction, only two compositions remain

$$\boxed{4_3\square 5_4 \text{ and } 3_2\square 6_3}$$

$4_3\square 5_4$ compounds with adamantine structure have not yet been found but eight phases with composition $3_2\square 6_3$ are listed in the last row of Table 5. Their structures can be seen in the right hand part of Fig. 3.

Ternary Compounds with Defect Adamantine Structure. In a ternary phase diagram, these composition series appear as lines. For example, the alloys of the series $(3_2\square 6_3)_x(26)_{1-x}$ and $(3_2\square 6_3)_x(35)_{1-x}$ are the possible ternary adamantine structure compositions in the ternary systems 2–3–6 and 3–5–6. Not all of the possible compositions may have an adamantine structure; it happens, however, in the ternary system Ga–As–Se, for example, that a "line" compound with adamantine structure is formed which starts at GaAs and extends all the way to Ga_2Se_3. A thorough analysis of these "line compounds" is given in *Crystal Chemistry of Tetrahedral Structures*.[2]

The analytical approach again becomes useful if it is intended to find possible compositions for structures with a given y value. We may, for example, consider the possible compositions of ternary compounds with $y = 7$, having two kinds of "cations" on three of every four former Zn sites. The composition of such a ternary compound is $C_mD_n\square_sA_o$, and its defect tetrahedral structure equation is

$$\frac{me_C + ne_D + oe_A}{m + n + o + s} = 4 \tag{25}$$

which, combined with

$$me_C + ne_D = o(8 - e_A) \tag{26}$$

gives

$$o = m + n + s \tag{27}$$

Using again $R = m/n$, we can transform (25) into

$$R = \frac{e_D + e_A - 8 - \frac{s}{n}(8 - e_A)}{8 - e_C - e_A} \tag{28}$$

The parameter s/n appears now in the formula for R. However, since it was specified that

$$y = \frac{m + n + o}{s} = 7 \tag{29}$$

the equation for R can be simplified to

$$R = \frac{3e_D + 4e_A - 32}{32 - 3e_C - 4e_A} \tag{30}$$

A table similar to Table 4 will show that this equation has eleven different solutions, which correspond to eleven possible compositions for a structure with $y = 7$.

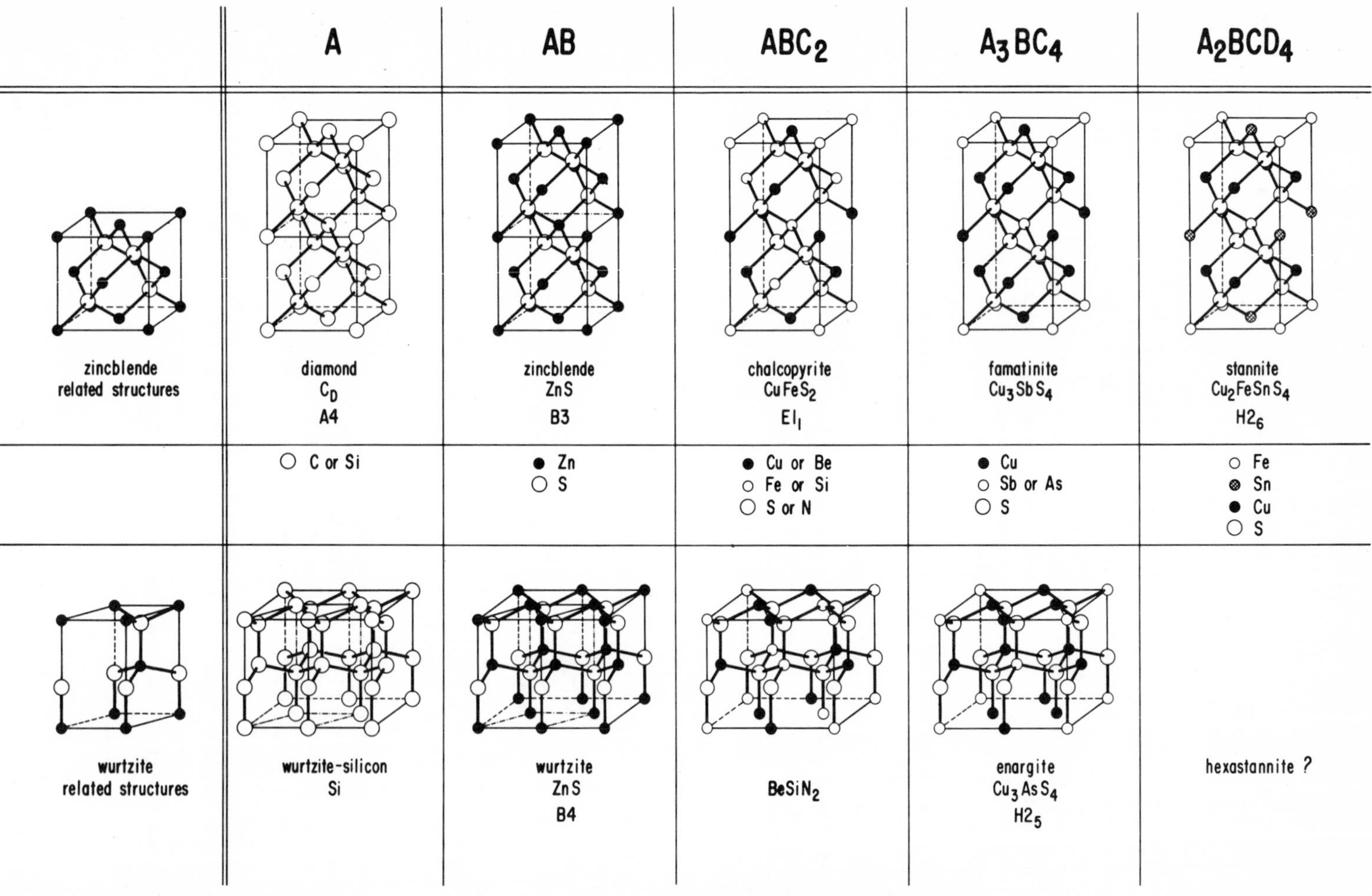

Fig. 2. Structure types of normal adamantine structures.

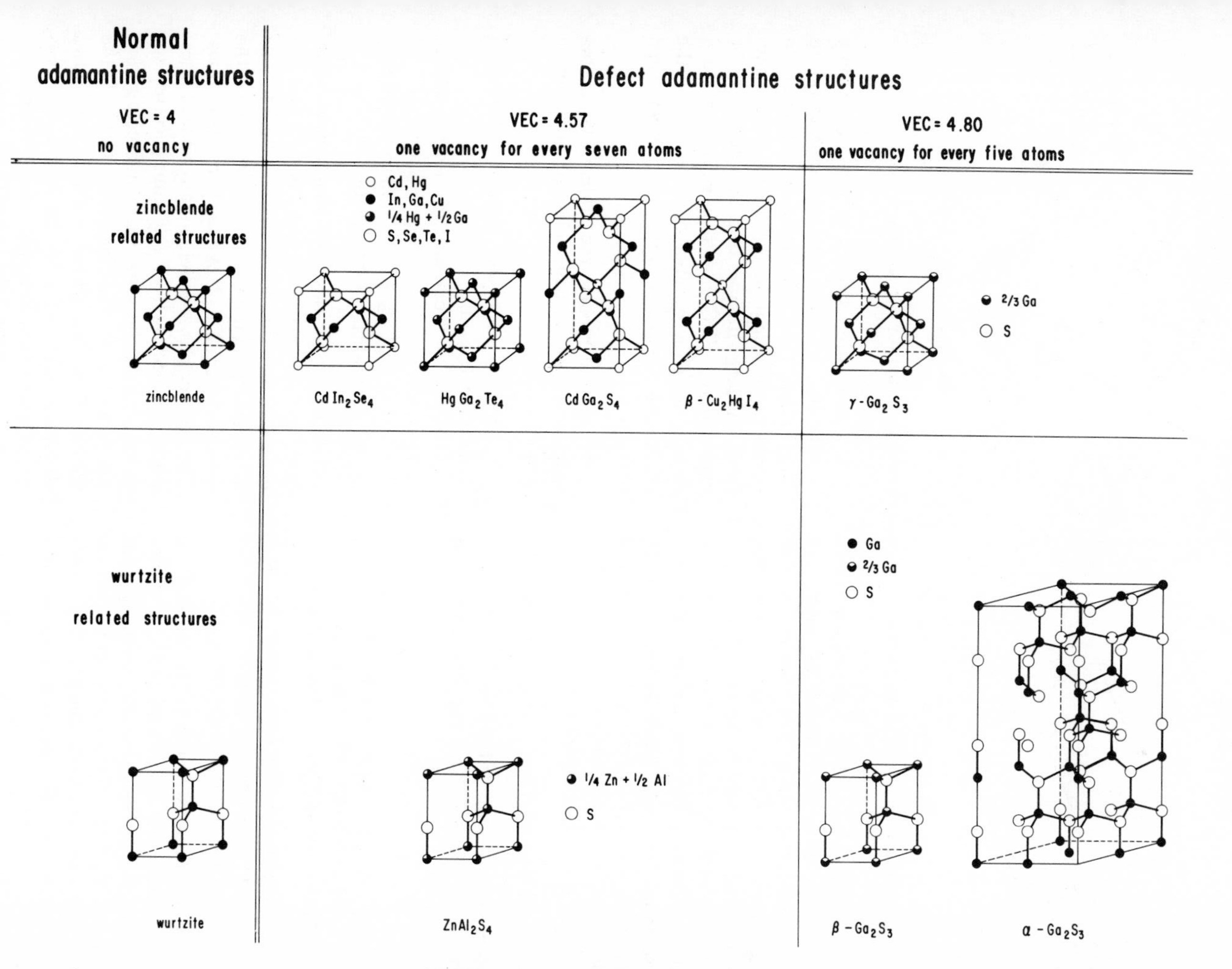

Fig. 3. Defect adamantine structures.

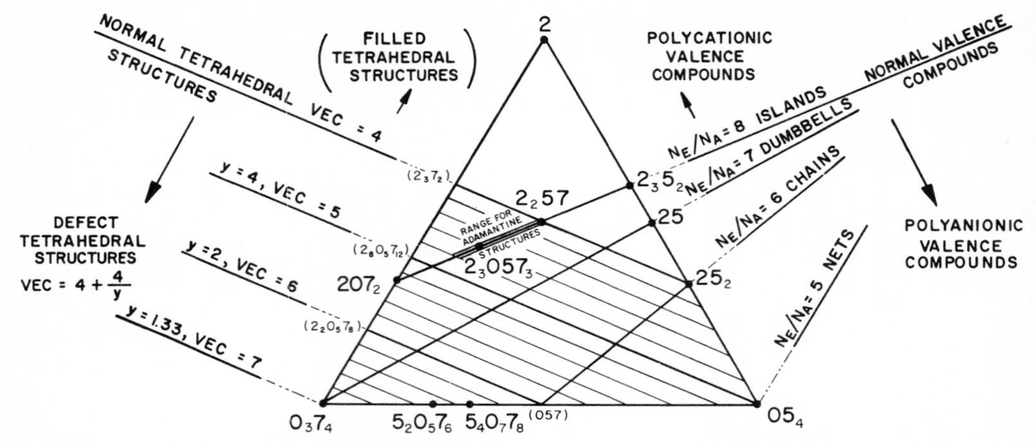

Fig. 4. The electron rules for covalent compounds demonstrated for the ternary system 2–5–7.

The middle part of Fig. 3 shows five structure types for $y = 7$. Three of them are ordered structure types. The "anion" partial structure is the same as in zinc blende, but two kinds of "cations" in the ratio 2:1 occupy three out of every four former zinc sites in orderly fashion. To find compositions for these ordered structure types, we must consider only those solutions of R in (30) which are either $\frac{1}{2}$ or 2. Three solutions exist and they correspond to the compositions

$$1_2 2\square 7_4, \quad 23_2\square 6_4, \quad \text{and} \quad 2_2 4\square 6_4$$

Representatives of all three composition formulas are known. They can be found in the proper rows of Table 5.

6. APPLICATION OF THE ELECTRON RULES FOR IONIC-COVALENT COMPOUNDS

The adamantine structures and their geometric and electronic properties are conveniently defined with three statements:

1. Adamantine structures are normal or defect tetrahedral structures with a close-packed, completely occupied anion partial structure. Vacancies may occur on the cation sites. The number of vacancies is then determined by the valence electron concentration value.

2. Compounds with adamantine structure are normal valence compounds. The structures can thus occur only if both the general tetrahedral structure equation and the equation for normal valence compounds are satisfied.

3. Adamantine structures can occur only within the valence electron concentration range from 4 to 4.80. Because cation-anion bonds alone occur in the adamantine structures, the large number of cation defects to be expected for VEC > 4.80 is equivalent to a reduction of the number of bonds below the minimum necessary for three-dimensional netting.

We can formulate the statements mathematically. The electron rules for adamantine structures are:

1. $\text{VEC} = 4 + 4/y$, general tetrahedral structure equation.
2. $N_E/N_A = 8$, equation for normal valence compounds.
3. $4.0 \leq \text{VEC} \leq 4.8$, boundary condition for tetrahedral structure equation.

The electronic rules for adamantine structures and the way in which they fit into the overall picture for ionic-covalent compounds in general is demonstrated in Fig. 4 for the ternary system 2–5–7. We see two sets of lines. The first set has parallel, equally spaced lines, which connect loci of the same electron-to-atom ratio or the same valence electron concentration value. The line for VEC = 4 is the composition line where normal tetrahedral structures may possibly be found, whereas defect tetrahedral structures are to be expected in the field below. The second set consists of four, nonparallel lines of opposite slope. These lines connect points of equal electron-to-anion ratio, $N_{\text{Electron}}/N_{\text{Anion}}$. Normal valence compounds occur with compositions which are on the line for $N_E/N_A = 8$. Polyanionic valence compounds are below that line, and polycationic valence compounds are above. The compositions of adamantine structures are found on the line for $N_E/N_A = 8$ with a valence electron concentration value between 4.0 and 4.8. The application of the electron rules is conveniently demonstrated with the examples marked by filled circles.

- 2_35_2: Zn_3P_2 is a normal valence compound with isolated anions. It does not crystallize with a tetrahedral structure.
- 25: ZnSb is a polyanionic valence compound with Sb–Sb pairs. It does not have a tetrahedral structure.
- 25_2: ZnP_2 has a normal tetrahedral structure with P–P spiral chains.
- 2_257: Composition for a normal adamantine structure compound.
- 2_37_2: An unknown polycationic valence compound that would have a normal tetrahedral structure.
- $2_3\square57_3$: Zn_3PI_3 has a defect adamantine structure with one vacancy for every three cations.
- 207_2: ZnI_2 has a layer structure, which belongs to the defect tetrahedral structures but not to adamantine structures. The I anions are isolated; they are bonded only to Zn cations. (The 0 or zero—here and in the following examples—no longer means a vacancy but a quadruplet of nonbonding orbitals.)
- 0_37_4: I_2 has a defect tetrahedral structure characterized by I_2 pairs.
- 05_4: Sb has a defect tetrahedral structure characterized by buckled nets.
- $5_20_57_6$ and $5_40_77_8$: PI_3 and P_2I_4 belong to the defect tetrahedral structures. They crystallize with molecules having four and six atoms, respectively.
- 057: An unknown compound for which we should expect a chainlike molecule, as observed with S or Se.

7. THE THIRD RULE FOR ADAMANTINE STRUCTURE COMPOUNDS

A survey of all known adamantine structure compounds in Table 5 shows that all compositions in which adamantine structures are observed can be calculated by using the given electron rules.* However, these electron rules are necessary but not sufficient conditions. Compounds are known in which the electron rules are satisfied but the structures occurring are not adamantine structures. To predict whether a compound (with a permitted composition) will actually crystallize with an adamantine structure, we can use an empirical rule developed from Mooser-Pearson plots.

Mooser and Pearson[16] have studied the occurrence of the different structure types among the normal valence compounds and have found that, if they plotted different compounds of one composition in a diagram with average principal quantum number $\bar{n}$ as ordinate and electronegativity difference of the components Δx as abscissa, all compounds with a particular structure type have their plots in a special area of the diagram. For example, in the $\bar{n}$ versus Δx diagram for equiatomic compounds there is such an area where only plots of Grimm-Sommerfeld compounds are found. Similarly, in $\bar{n}$ versus Δx diagrams for other compositions the plots for adamantine structure compounds (they may be more complicated normal adamantine or defect adamantine structure compounds) are always closely gathered in one part of the diagram, sharply separated from the plots of compounds having other structures.[2] These results can be formulated analytically. To a first approximation, adamantine structures will occur when

$$\bar{n} + 2\Delta x \leq 6 \qquad (31)$$

Goryunova[13] has suggested a different empirical formula which uses as parameters the "specific electron affinities" of the different atoms.† Goryunova has plotted all equiatomic binary compounds in a diagram with specific electron affinity of the cation E_C as ordinate and specific electron affinity of the anion E_A as abscissa. She found that the plots

* Exceptions are CuH and CuD, which supposedly crystallize in the wurtzite structure. If these data are correct, hydrogen must be able to form tetrahedral orbitals using a not yet understood bonding mechanism.

† The specific electron affinity of an atom with x valence electrons is equal to the xth ionization potential (for reaction $M^{(x)+} + e^- \leftrightarrow M^{(x-1)+}$) divided by x.

of compounds having adamantine structures are sharply separated from the plots of compounds crystallizing with other structures. To a first approximation, adamantine structures will occur when

$$E_C - 0.1\, E_A \geq 6 \qquad (32)$$

There are obviously two different influences which may prevent the occurrence of adamantine structures. One is a strong tendency to draw the bonding electrons closer to one atom than the other. This condition is characterized by a large Δx value and leads to the formation of ions and the occurrence of typically ionic structures. Another disturbance is found when $\bar{n}$ becomes large. The ionization energies of all elements decrease when the principal quantum number increases. Compounds with large values of $\bar{n}$ will show metallic properties and will crystallize with metallic crystal structures. The formulas given above show that both influences have to be

TABLE 5

List of Adamantine Structure Compounds[a]

Compounds with normal adamantine structures

Composition Formula	VEC	Wurtzite or Structure Related to It	Zinc Blende or Structure Related to It
4	4.0	Si (wurtzite silicon)	C (diamond), Si, Ge, α-Sn (grey tin)
44	4.0	SiC (2H)	β SiC, (Si-Ge)
35	4.0	BN, AlN, GaN, InN	BN (borazon), BP, BAs, AlP, AlAs, AlSb, GaP, GaAs, GaSb, InP, InAs, InSb
26	4.0	BeO (bromellite), MgTe, ZnO (zincite), α-ZnS (wurtzite), ZnSe, ZnTe,[g] α-CdS (greenockite) αCdSe, CdTe,[i] βMnS, γ-MnSe	BeS, BeSe, BeTe, BePo, ZnO ?, β-ZnS (sphalerite or zinc blende), ZnSe, ZnTe, ZnPo, βCdS (hawleyite), βCdSe, CdTe, CdPo, HgS (metacinnabarite), HgSe (tiemannite), HgTe (coloradorite), βMnS, βMnSe
17	4.0	CuCl, βCuBr, βCuI, βAgI (iodyrite)	CuCl, γCuBr, γCuI, γAgI
136_2	4.0	$AgInS_2$	$CuBSe_2$, $CuAlS_2$, $CuAlSe_2$, $CuAlTe_2$, $CuGaS_2$ (gallite), $CuGaSe_2$, $CuGaTe_2$, $CuInS_2$ (roquesite), $CuInSe_2$, $CuInTe_2$, $CuTlS_2$, $CuTlSe_2$, $CuFeS_2$ (chalcopyrite), $CuFeSe_2$, $AgAlS_2$, $AgAlSe_2$, $AgAlTe_2$, $AgGaS_2$, $AgGaSe_2$, $AgGaTe_2$, $AgInS_2$, $AgInSe_2$, $AgInTe_2$, $AgFeS_2$ (argentopyrite)
245_2	4.0	$BeSiN_2$	$MgGeP_2$, $ZnSiP_2$, $ZnSiAs_2$, $ZnGeP_2$, $ZnGeAs_2$, $ZnSnP_2$,[h] $ZnSnAs_2$, $CdSiP_2$,[b] $CdSiAs_2$,[h] $CdGeP_2$, $CdGeAs_2$, $CdSnP_2$,[h] $CdSnAs_2$
1_356_4	4.0	Cu_3PS_4, Cu_3AsS_4 (enargite)	Cu_3AsS_4 (luzonite), Cu_3AsSe_4, Cu_3SbS_4 (famatinite), Cu_3SbSe_4
14_25_3	4.0		$CuSi_2P_3$, $CuGe_2P_3$
1_246_3	4.0	Cu_2SiS_3[c]	Cu_2SiS_3[c], Cu_2SiTe_3, Cu_2GeS_3, Cu_2GeSe_3, Cu_2GeTe_3, Cu_2SnS_3, Cu_2SnSe_3, Cu_2SnTe_3
1_2246_4	4.0	Ag_2CdGeS_4,[e] Ag_2CdSnS_4,[e] $Ag_2CdSnSe_4$[e]	Cu_2ZnGeS_4, $Cu_2ZnGeSe_4$,[d] $Cu_2CdGeSe_4$,[e] $Cu_2HgGeSe_4$,[d] Cu_2FeGeS_4, $Cu_2(Fe,Zn)GeS_4$[j] (briartite), $Cu_2FeGeSe_4$, Cu_2ZnSnS_4,[d] $Cu_2ZnSnSe_4$,[d] $Cu_2ZnSnTe_4$,[d] Cu_2CdSnS_4,[d] $Cu_2CdSnSe$,[d] Cu_2HgSnS_4,[d] $Cu_2HgSnSe_4$,[d] $Cu_2HgSnTe_4$,[d] Cu_2FeSnS_4 (stannite), $Cu_2FeSnSe_4$,[d] $Ag_2ZnGeSe_4$[e]
134_25_4	4.0		$CuGaGe_2P_4$
$1_x2_y3_x6_{2x+y}$	4.0		Complete solid solution between 26 and 136_2 found with $Cu_xZn_yGa_xS_{2x+y}$,[h] $Cu_xCd_yIn_xTe_{2x+y}$,[h] $Ag_xCd_yIn_xTe_{2x+y}$, $Ag_xHg_yIn_xTe_{2x+y}$
$1_x3_{x+y}5_y6_{2x}$	4.0		Complete solid solution between 35 and 136_2 found with $Cu_xIn_{x+y}As_yTe_{2x}$[h]
$2_x3_y4_x5_{2x+y}$	4.0		Complete solid solution between 35 and 245_2 found with $Zn_xIn_yGe_xAs_{2x+y}$, $Zn_xIn_ySn_xAs_{2x+y}$, $Cd_xIn_ySn_xAs_{2x+y}$[h]
$2_x3_y5_y6_x$	4.0		Complete solid solution between 26 and 35 found with $Cd_xAl_ySb_yTe_x$,[h] $Zn_xGa_yAs_ySe_x$,[h] $Hg_xIn_yAs_yTe_x$[h]
$1_23_345_36_3$	4.0		$Ag_2In_3SnSb_3Te_3$

Table 5 (*Continued*)

	Composition Formula	VEC	Wurtzite or Structure Related to It	Zinc Blende or Structure Related to It
Defect adamantine structure compounds	$2_5 3_2 \square 6_8$	4.27		$Hg_5Ga_2Te_8$, $Hg_5In_2Te_8$
	$12_3 3_3 \square 6_8$	4.27		$AgHg_3In_3Te_8$
	$1_2 2_2 \square 6 7_4$	4.44	$Ag_2Hg_2SI_4$	
	$1_2 2 \square 7_4$	4.57		Cu_2HgI_4, Ag_2HgI_4
	$2 3_2 \square 6_4$	4.57	$ZnAl_2S_4$	$ZnAl_2Se_4$, $ZnAl_2Te_4$, $ZnGa_2S_4$, $ZnGa_2Se_4$, $ZnGa_2Te_4$, $ZnIn_2Se_4$, $ZnIn_2Te_4$, $CdAl_2S_4$, $CdAl_2Se_4$, $CdAl_2Te_4$, $CdGa_2S_4$, $CdGa_2Se_4$, $CdGa_2Te_4$, $CdIn_2Se_4$, $CdIn_2Te_4$, $HgAl_2S_4$, $HgAl_2Se_4$, $HgAl_2Te_4$, $HgGa_2S_4$, $HgGa_2Se_4$, $HgGa_2Te_4$, $HgIn_2Se_4$, $HgIn_2Te_4$
	$2_2 4 \square 6_4$	4.57		Zn_2GeS_4,[f] Zn_2GeSe_4,[f] Hg_2GeSe_4[f]
	$2_3 \square 5 7_3$	4.57		Zn_3PI_3, Zn_3AsI_3
	$1 3_2 \square 6_3 7$	4.57		$CuIn_2Se_3Br$, $CuIn_2Se_3I$, $AgIn_2Se_3I$
	$1_2 2_3 \square_2 6 7_6$	4.66		$Ag_2Hg_3SI_6$
	$3_2 \square 6_3$	4.80	Al_2S_3, Al_2Se_3, Ga_2S_3, β-In_2Se_3	γGa_2S_3, Ga_2Se_3, Ga_2Te_3, In_2Te_3
	$3_{2y+x} \square_y 5_x 6_{3y}$	4.0–4.80		Complete solid solution between 35 and $3_2 \square 6_3$ found with $Ga_{2y+x}P_xS_{3y}$,[h] $Ga_{2y+x}As_xSe_{3y}$,[h] $In_{2y+x}As_xTe_{3y}$[h]

[a] If not specially marked the literature references to all compounds can be found in Reference 2.
[b] Vaipolin A. A., N. A. Goryunova, E. O. Osmanov, Yu. V. Rud, and D. N. Tretyakov, "Investigation of Crystals of $ZnSiP_2$, $CdSiP_2$ and $ZnSiAs_2$," *Dokl. Chem.* **154** (1964) 146–149.
[c] Rivet J., J. Flahaut, and P. Laurelle, "Sur un groupe de composés ternaires à structure tétraèdrique," *Compt. Rend. Acad. Fr.* **257** (1963) 161–164.
[d] Hahn H., and H. Schulze, "Über quaternäre Chalkogenide des Germaniums und Zinns," *Naturwissenschaften* **52** (1965) 426.
[e] Deitch R. H., and E. Parthé, to be submitted to *Acta Cryst.*
[f] Hahn H., and A. de Lorent: "Untersuchungen über ternäre Chalkogenide. Über ternäre Sulfide und Selenide des Germaniums mit Zink, Cadmium und Quecksilber," *Naturwissenschaften* **24** (1958) 621–622.
[g] Shalimova K. V., A. F. Andrushko, I. Spynulesku-Kanaru, and B. P. Seredinskij, "Crystal structure of films of zinc telluride with various deviations from the stoichiometric composition," (in Russian), *Kristallografija* **9** (1964) 741–743.
[h] Goryunova N. A., *The Chemistry of Diamond-like Semiconductors*, Chapman and Hall Ltd., London, 1965.
[i] Weinstein M., G. A. Wolff, and B. N. Das, "Growth of Wurtzite CdTe and Sphalerite Type CdS Single-Crystal Films," *Bull. Amer. Phys. Soc.* **10** (1965) 84.
[j] Francotte J., J. Moreau, R. Ottenburgs, and C. Levy: "La briartite, $Cu_2(Fe,Zn)GeS_4$, une nouvelle espèce minerale," *Bull. Soc. Franc. Miner. Crist.* **88** (1965) 432–437.

considered simultaneously. The empirical rules still need an exact physical derivation; however, to a first approximation they describe the known data very well.

The two electron equations and the empirical rule are certainly very useful for the prediction of new adamantine structure compounds and for the classification of all those phases which are already known. However, the structural details of adamantine structure—such as choice of wurtzite or zincblende type, degree of ordering or amount of distortion—are not yet accessible from theoretical considerations. More theoretical and experimental work has to be done before these problems can be solved.

REFERENCES

1. Wentorf R. H., Jr., and J. S. Kasper, *Science* **139** (1963) 338–339.
2. Parthé E., *Crystal Chemistry of Tetrahedral Structures*, Gordon & Breach Science Publishers, New York, 1964.
3. Parthé E., *Z. Krist.* **119** (1963) 204–225.
4. Mooser E., and W. B. Pearson, *J. Electron.* **1** (1956) 629–645.
5. Hulliger F., and E. Mooser, *J. Phys. Chem. Solids* **24** (1963) 283–295.
6. Pearson W. B., *Acta Cryst.* **17** (1964) 1–15.
7. Kjekshus A., *Acta Chem. Scand.* **18** (1964) 2379–2384.
8. Hulliger F., *J. Phys. Chem. Solids* **26** (1965) 639–645.

9. Hulliger F., and E. Mooser, *Progress in Solid State Chemistry*, Vol. 2 Pergamon Press, New York, 1965 pp. 330–377.
10. Busmann E., *Z. Anorg. Allgem. Chem.* **313** (1961) 90–106.
11. Goodman C. H. L., *J. Phys. Chem. Solids* **6** (1958) 305–314.
12. Suchet J. P., *J. Phys. Chem. Solids* **12** (1959) 74–88.
    Suchet J. P., *Chemical Physics of Semiconductors*, Van Nostrand Company, New York, 1965. A translation of *Chimie physique des Semiconducteurs*, Dunod, Paris, 1962.
13. Goryunova N. A., *Vestnik Leningrad University* 1961, No. 10, *Ser. Fiz. i Khim.* (1961) 112–124.
    Goryunova N. A., *Chemistry of Diamond-Like Semiconductors*, Chapman and Hall Ltd., London, 1965. A translation of a Russian book with the same title, Leningrad University Press, 1963.
14. Pamplin B. R., *J. Phys. Chem. Solids* **25** (1964) 675–684.
15. Amma E. L., and G. A. Jeffrey, *J. Chem. Phys.* **34** (1961) 252–259.
16. Mooser E., and W. B. Pearson, *Acta Cryst.* **12** (1959) 1015–1022.

chapter 12

Topologically Close-Packed Structures

Jack Harry Wernick

JACK HARRY WERNICK

Bell Telephone Laboratories, Inc.
Murray Hill, New Jersey

1. INTRODUCTION

Of the three main factors controlling the formation and structure of intermediate phases—atom size, relative valency, and relative electronegativities of the atoms (often referred to as the electrochemical factor)—the size factor appears to some to be the most important in the discussion of close-packed intermediate phases. The size factor is generally discussed in terms of the radius ratio of the atoms, and it has been found that whole families of intermediate phases (for example, AB_2 Laves phases)[1] form because of the ease with which atoms of a given radius ratio can most efficiently fill space. Nevertheless, the relative valence factor, that is, the difference in the valences of the component atoms, is also an important consideration in the discussion of close-packed intermediate phases and in some cases appears to predominate over the size factor. The relative valence factor is generally discussed in terms of the changing electron concentration within the Brillouin zone of a host lattice that changes structure or in terms of the electron concentration of an intermediate phase that corresponds to a full or nearly full Brillouin zone. The relative electronegativities of the atoms indicate qualitatively whether a new phase is likely to form and the nature of the bonding in the phase; they have qualitative significance only in discussions of non-close-packed semi-metallic and semiconducting intermediate phases. The intermediate phases to be discussed in this chapter are the Laves phases, the sigma phases, and related topologically close-packed structures.

Kasper[2] and Frank and Kasper[3,4] have described the structures of a number of complex close-packed phases in terms of sphere packing. The basic stacking units are 12-, 14-, 15-, and 16-coordinated polyhedra (Fig. 1).

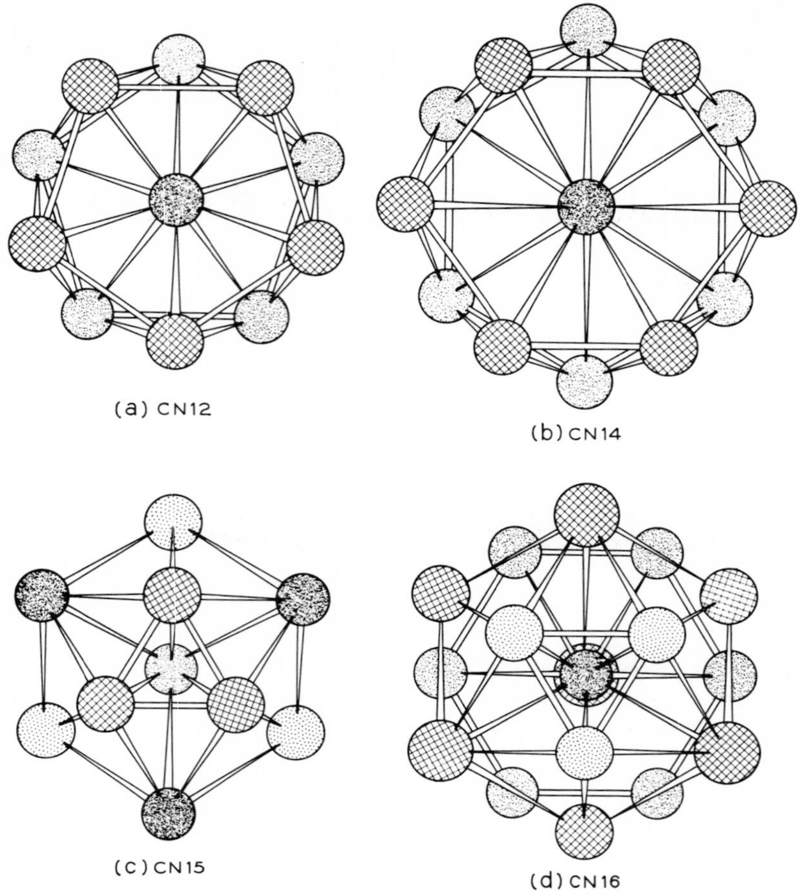

Fig. 1. Projections of the 12-, 14-, 15-, and 16-coordination polyhedra. (*a*) For CN-12, the fivefold axis is normal to the plane of the paper and passes through the central sphere. A sphere lies above and below this sphere along this axis. For equal-size spheres, the CN-12 polyhedron corresponds to a regular icosahedron, and the surface spheres are in contact only with the central sphere. (*b*) The projection of the CN-14 polyhedron is normal to a sixfold axis. As for (*a*), there are two spheres along the sixfold axis. The twelve coordinated spheres can be equal in size, but they are different from the central one, and the two spheres lying along the sixfold axis differ in size from the twelve. (*c*) Although not shown for CN-15, the spheres coordinated to the central one are not equivalent in size. (*d*) For CN-16, there is a tetrahedral arrangement of the four larger spheres about the large central sphere. In addition, there are twelve equidistant smaller spheres coordinated to the larger central sphere, and they are in contact with one another. This arrangement is met when the radius ratio of the two sphere sizes is 1.225. (After Kasper.[2])

Although it is pointed out that the importance of packing considerations relative to the electronic factors cannot be readily ascertained in most cases, it is suggested that packing considerations are the more important ones.

2. THE LAVES PHASES

The Laves phases[5–8] denote a large group of related intermetallic compounds, AB_2, having one of the three following structure types: (*a*) $MgCu_2$, (*b*) $MgZn_2$, and (*c*) $MgNi_2$. The $MgCu_2$ structure is cubic, and the $MgZn_2$ and $MgNi_2$ structures are hexagonal. A large number of intermetallic compounds are isostructural to the above Mg compounds (Tables 1, 2, and 3).*

* Tables 1, 2, and 3 were compiled mainly from References 9, 24, 28–31, 36–44, 111–113, 115 and 124.

TABLE 1

Intermediate Phases with the $MgCu_2$ Structure

$AgBe_2$	$ErNi_2$	$LaNi_2$	$PuCo_2$	$TiBe_2$
$BaPd_2$	$EuAl_2$	$LaOs_2$	$PuFe_2$	$TiCo_2$
$BaPt_2$	$EuPt_2$	$LaPt_2$	$PuMn_2$	$TiCr_2$(lt)[a]
$BaRh_2$	$EuRh_2$	$LaRh_2$	$PuNi_2$	$TmAl_2$
$BiAu_2$	$EuRu_2$	$LaRu_2$	$PuRu_2$	$TmCo_2$
$CaAl_2$	$FeBe_5$(E)	$LuAl_2$	$PuZn_2$	$TmFe_2$
$CaIr_2$	$GdAl_2$	$MgCu_2$	$RbBi_2$	$TmNi_2$
$CaPd_2$	$GdCo_2$	$Mg_3Cu_4Be_2$	$ScCo_2$	UAl_2
$CaPt_2$	$GdFe_2$	MgNiZn	$ScIr_2$	UCo_2
$CaRh_2$	$GdIr_2$	$MnSnCu_4$	$SmAl_2$	UFe_2
CdCuZn	$GdMg_2$	$NaAu_2$	$SmCo_2$	UFeNi
$CeAl_2$	$GdMn_2$	$NbBe_2$	$SmFe_2$	UIr_2
$CeCo_2$	$GdNi_2$	$NbCo_2$	$SmNi_2$	UOs_2
$CeFe_2$	$GdPt_2$	$NbCr_2$	$SrIr_2$	UMn_2
$CeIr_2$	$GdRh_2$	$NdAl_2$	$SrPd_2$	WOs_2
$CeMg_2$	$GdRu_2$	$NdCo_2$	$SrPt_2$	YAl_2
$CeNi_2$	$HfCo_2$	$NdIr_2$	$SrRh_2$	YCo_2
$CeOs_2$	$HfCr_2$(ht)[a]	$NdNi_2$	$TaCo_2$	YIr_2
$CePt_2$	$HfFe_2$(lt)[a]	$NdPt_2$	$TaCr_2$(lt)[a]	YPt_2
$CeRh_2$	$HfMo_2$	$NdRh_2$	TaFeNi	YRh_2
$CeRu_2$	HfV_2	$NdRu_2$	$TaV_{1.5}Mn_{0.5}$	$YbAl_2$
$CsBi_2$	HfW_2	$NpAl_2$	$TbAl_2$	$YbNi_2$
$CuBe_2$	$HoAl_2$	$PbAu_2$	$TbCo_2$	$ZrCo_2$
CuTaV	$HoCo_2$	$PrAl_2$	$TbFe_2$	$ZrCr_2$(ht)[a]
$DyAl_2$	$HoFe_2$	$PrIr_2$	$TbMn_2$	$ZrFe_2$
$DyCo_2$	$HoMn_2$	$PrNi_2$	$TbNi_2$	$ZrIr_2$
$DyFe_2$	$HoNi_2$	$PrOs_2$	$ThIr_2$	ZrMnNi
$DyMn_2$	KBi_2	$PrPt_2$	$ThMg_2$	$ZrV_{0.5}Ni_{1.5}$
$DyNi_2$	$LaAl_2$	$PrRh_2$	ThMnAl	$ZrMo_2$
$ErAl_2$	$LaIr_2$	$PrRu_2$	$ThOs_2$	ZrV_2
$ErCo_2$	$LaMg_2$	$PuAl_2$	$ThRu_2$	ZrW_2
$ErFe_2$				$ZrZn_2$

[a] ht = high-temperature form; lt = low-temperature form.

2.1 The Three Structure Types

The $MgCu_2$ Structure Type. The $MgCu_2$ structure type is cubic with eight formula units per cell (Fig. 2). It belongs to space group $Fd3m$–O_h^7. There are eight A atoms in position (a): (000, $0\frac{1}{2}\frac{1}{2}$; ↻) + 000, $\frac{1}{4}\frac{1}{4}\frac{1}{4}$, Sixteen B atoms are in positions (d): (000, $0\frac{1}{2}\frac{1}{2}$; ↻) + $\frac{5}{8}\frac{5}{8}\frac{5}{8}$; $\frac{5}{8}\frac{7}{8}\frac{7}{8}$; ↻. The structure can be regarded as being made up of two interpenetrating lattices of A and B atoms. The B atoms lie at the corners of tetrahedra, and the tetrahedra are joined at the points (Fig. 3a). The "holes" enclosed by the tetrahedra accommodate the larger A atoms. The cubic array of A atoms (Fig. 2) is identical to that of Si or Ge (diamond cubic). Although the coordination to an A atom is four other equidistant A atoms at a distance $a\sqrt{3}/4$, there are twelve B atoms at a somewhat smaller distance $a\sqrt{11}/8$. The coordination then is quite high, effectively sixteen, and corresponds to the CN16 polyhedron of Fig. 1. Each B atom is surrounded by 6B at a distance $a\sqrt{2}/4$ and 6A atoms at a distance $a\sqrt{11}/8$.

The $MgZn_2$ Structure Type. The $MgZn_2$ structure type contains four formula units per cell and belongs to space group $P6_3/mmc$-D_{6h}^4. The four A atoms are in position (f): $\frac{1}{3}$, $\frac{2}{3}$, z; $\frac{2}{3}$, $\frac{1}{3}$, z; $\frac{2}{3}$, $\frac{1}{3}$, $\frac{1}{2}+z$; $\frac{1}{3}$, $\frac{2}{3}$, $\frac{1}{2}-z$. Two of the eight B atoms are in positions (a): 000; $00\frac{1}{2}$. The remaining six B atoms are in positions (h): x, $2x$, $\frac{1}{4}$; $2\bar{x}$, $\bar{x}$, $\frac{1}{4}$; x, $\bar{x}$, $\frac{1}{4}$; $\bar{x}$, $2\bar{x}$, $\frac{3}{4}$; $2x$, x, $\frac{3}{4}$; $\bar{x}$, x, $\frac{3}{4}$. For the ideal structure, the parameters $x = -\frac{1}{6}$ and $z = \frac{1}{16}$.

In this structure the smaller B atoms are also arranged in tetrahedra, but the tetrahedra

TABLE 2
Intermediate Phases with the $MgZn_2$ Structure

$BaMg_2$	Mg_2Cu_3Si(lt)	TaNiV
$CaCd_2$	$MgZn_2$	$ThMn_2$
$CaLi_2$	$MnBe_2$	$ThRe_2$
$CaMg_2$	$MoBe_2$	$TiCr_2$(ht)
$CrBe_2$	$MoFe_2$	$TiFe_2$
$CdCu_2$	$NbFe_2$	$TiMn_2$
DyCoAl	$NbMn_2$	$TiZn_2$
DyFeAl	$NdOs_2$	$TmMn_2$
$DyOs_2$	$PrOs_2$	UNi_2
$ErMn_2$	$PuOs_2$	URe_2(ht)
$ErOs_2$	$PuRe_2$	VBe_2
$ErRu_2$	$ReBe_2$	WBe_2
$FeBe_2$	$ScOs_2$	WFe_2
$GdOs_2$	$ScRe_2$	YOs_2
$GdRu_2$	$ScRu_2$	YRe_2
$HfCr_2$(lt)	$ScTc_2$	YRu_2
$HfFe_2$	$SmOs_2$	$ZrAl_2$
$HfMn_2$	$SrMg_2$	$ZrCr_2$(lt)
$HfOs_2$	$TaCr_2$(ht)	$ZrIr_2$
$HfRe_2$	$TaFe_2$	$ZrRu_2$
$HfTe_2$	$TaMn_2$	$ZrMn_2$
KNa_2	TaCoCr	$ZrOs_2$
KPb_2	Ta_2Cr_3Cu	$ZrRe_2$
$LuRe_2$	TaCoTi	$ZrRu_2$
$LuRu_2$	TaCoV	$ZrTc_2$
$LuOs_2$	TaCrNi	Zr_2VCo_3
MgCuAl(lt)	$TaCu_{0.5}V_{1.5}$	

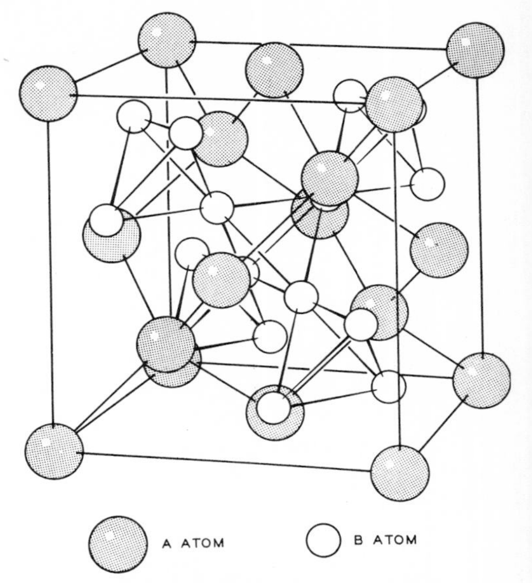

Fig. 2. The $MgCu_2$ structure (AB_2).

are joined alternately base-to-base and point-to-point (Fig. 3*b*). The larger A atoms fit into the holes enclosed by the tetrahedra, and these holes have the same shape as in the $MgCu_2$ structure. Each A atom is also tetrahedrally coordinated to four other A atoms, but the arrangement is identical to the hexagonal Wurtzite structure.

The $MgNi_2$ Structure Type. The $MgNi_2$ structure type contains eight formula units per cell and belongs to space group $P6_3/mmc$-D_{6h}^4. The distribution of the twenty-four atoms among the allowable atom positions are as follows:

Four A atoms in (*e*): $0, 0, z; 0, 0, \bar{z}; 0, 0, \frac{1}{2} + z; 0, 0, \frac{1}{2} - z$.
Four A atoms in (*f*): $\frac{1}{3}, \frac{2}{3}, z; \frac{2}{3}, \frac{1}{3}, \bar{z}; \frac{2}{3}, \frac{1}{3}, \frac{1}{2} + z; \frac{1}{3}, \frac{2}{3}, \frac{1}{2} - z$.
Six B atoms in (*g*): $\frac{1}{2}, 0, 0; 0, \frac{1}{2}, 0; \frac{1}{2}, \frac{1}{2}, 0; \frac{1}{2}, 0, \frac{1}{2}; 0, \frac{1}{2}, \frac{1}{2}; \frac{1}{2}, \frac{1}{2}, \frac{1}{2}$.
Six B atoms in (*h*): $x, 2x, \frac{1}{4}; 2\bar{x}; \bar{x}, \frac{1}{4}; x, \bar{x}, \frac{1}{4}; \bar{x}, 2\bar{x}, \frac{3}{4}; 2x, x, \frac{3}{4}; \bar{x}, x, \frac{3}{4}$.
Four B atoms in (*f*): as above.

For the ideal structure, the parameter z is equal to $\frac{3}{32}$ and $\frac{27}{32}$ for the (*e*) and (*f*) positions, respectively, for the A atoms. For the B atoms, occupying the (*f*) positions, $z = \frac{1}{8}$. The parameter x is equal to $\frac{1}{6}$ for the (*h*) positions.

The arrangement of the B-atom tetrahedra in this structure is a mixture of the other two (Fig. 3*c*). The A atoms are tetrahedrally coordinated to four other A atoms, but now there is a mixture of the diamond cubic and hexagonal Wurtzite-type structures. The A-atom structures can also be considered in terms of stacking of three double layers (A′, B′, and C′) of hexagonally arranged A atoms (Fig. 4). The stacking sequence for the $MgZn_2$, $MgCu_2$, and $MgNi_2$ structures is A′B′A′B′A′B′, A′B′C′A′B′C′, and A′B′A′C′A′B′A′C′, respectively.

TABLE 3
Intermediate Phases with the $MgNi_2$ Structure

$HfMo_2(\alpha)$	$ThMg_2$
MgCuAl(ht)	$TiCo_2(\beta)$
Mg_2Cu_3Si(ht)	$U(Co, Ni)_2$
Mg_2CuZn_3	$U(Fe, Ni)_2$
$MgNi_2$	$U(Mn, Ni)_2$
$NbZn_2$	UPt_2
$ScFe_2$	

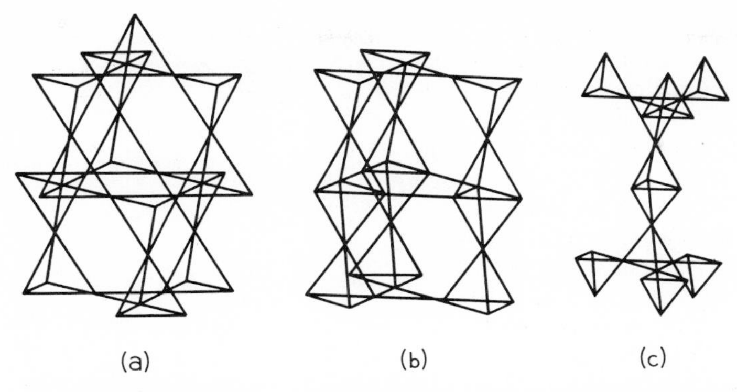

Fig. 3. The arrangement of the tetrahedra of B atoms in (*a*) the $MgCu_2$ structure, (*b*) the $MgZn_2$ structure, and (*c*) the $MgNi_2$ structure. [After Raynor (1949).]

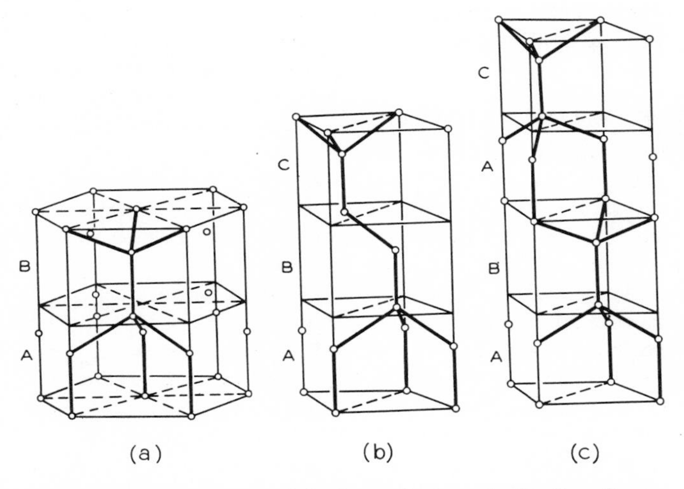

Fig. 4. The arrangement of the larger A atoms in the three Laves-phase structure types. (*a*) $MgZn_2$, (*b*) $MgCu_2$, and (*c*) $MgNi_2$. [After Raynor (1949).]

2.2 The Atomic Size Factor

An examination of the large number of Laves phases shows that a determining factor in the formation of these phases is the relative atomic size of the constituent atoms. On the basis of a hard sphere-packing model, A atoms touching A atoms and B atoms touching B atoms, the ideal ratio of the atom radii, r_A/r_B, for the formation of the Laves phases is 1.225. In practice, the ratio of the Goldschmidt radii of the pure elements, R_A/R_B, for the known Laves phases varies from 1.05 to 1.68, whereas the *effective* radius ratio is close to 1.225.[9,10] The A and B atoms, having a wide range of Goldschmidt radii, contract or expand in order to achieve the effective ratio.* Berry and Raynor[15] suggest that R_A/R_B affects the type of Laves structure formed. However, Dwight[9] concludes that the Goldschmidt radii of the pure elements is not an important factor in dictating which of the

* The UNi_5-type structure is closely related to the cubic Laves phase,[11,12] and the hexagonal $CaCu_5$-type structure is related to the hexagonal $MgZn_2$ phase.[13] The (rare-earth) Ni_5 and (rare-earth) Co_5 compounds[14] have the latter structure. The radius-ratio range characteristic of the UNi_5 and $CaCu_5$-type compounds and a further discussion of the relation between the AB_2 and AB_5 phases have been given recently by Dwight[9] and Raynor.[120] The Goldschmidt radius ratio for the $CaCu_5$-type compounds lies between 1.29 and 1.61 and this ratio is less than 1.29 for the UNi_5 type.

three structures form; it is only necessary that the A atom be larger than the B atom and that they be able to contract or expand to achieve nearly the ideal ratio of 1.225. In this connection, we note that Th forms the AlB_2 structure type with Al and Ni, even though the radius ratio of the Goldschmidt radii is favorable for Laves-phase formation. Al in $ThAl_2$ has an abnormally small radius (1.27 Å) and Th an abnormally large radius.[123] The addition of Mn to $ThAl_2$ stabilizes the $MgCu_2$ structure type.[111] $ThMn_2$ has the $MgZn_2$ structure. In the cubic ternary Th–Al–Mn alloys, Al does not have the abnormally small radius. This then suggests that electronic factors (as indicated qualitatively in some instances by positions in the periodic table) are very important and will be dealt with next.

2.3 Influence of the Electronic Factor

The role of the electronic factor in the formation and stability of many of the Laves phases is well known.[8,15–21] For example, Laves and Witte have shown that the electron concentration determines which of the three structures form in the pseudobinary systems of $MgCu_2$ and $MgZn_2$ with Al, Ag, and Si. One or both of the hexagonal Laves phases form as intermediate ternary phases along the pseudobinary joins with increasing electron concentration, and valency considerations successfully dictate the structure formed. This is illustrated in Fig. 5 and Table 4.[8,18] The homogeneity ranges, expressed as valence electrons per atom, are shown for the MgC_2u, $MgNi_2$, and $MgZn_2$-type structures in ternary Mg alloys.

It is to be noted that the maximum concentration of electrons in each structure type

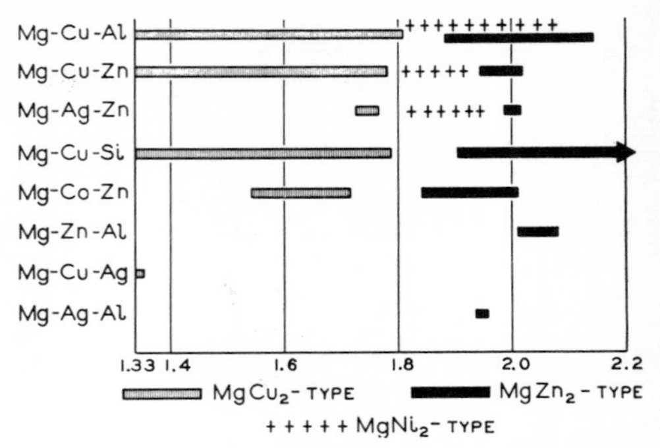

Fig. 5. Homogeneity ranges in terms of valence electron-to-atom ratios for several ternary Mg alloys. [Laves and Witte (1936).]

in all of the systems is nearly the same, suggesting that only a certain number of allowed electron states are available for filling. This would be directly connected to a certain-size Brillouin zone, and it suggests that a new structure will form in order to accommodate additional electrons. Indeed, Klee and Witte[20] have performed experiments that strongly support the Brillouin-zone effect. They measured the magnetic susceptibility of a number of alloys along the $MgCu_2$–$MgZn_2$, $MgCu_2$–$MgAl_2$, and $MgCu_2$–$MgSi_2$ pseudobinary joins. Since the susceptibility is related to the density of electron states at the Fermi surface, information regarding the form of the $N(E)$ versus E curve inside the zone can be obtained. States near zone boundaries often have small effective masses, and crystals in which the Fermi surface is near or at zone boundaries may have high diamagnetic susceptibilities from the conduction electrons and holes.[22] The susceptibility results of Klee and Witte, together with the phase diagram, are illustrated in Fig. 6 for the $MgCu_2$–$MgZn_2$ system. Note

TABLE 4

Homogeneity Ranges of Phases with the $MgCu_2$-, $MgNi_2$-, and $MgZn_2$-Type Structures in Ternary Magnesium Alloys[a]

System	$MgCu_2$ Type	$MgNi_2$ Type	$MgZn_2$ Type
$MgCu_2$–$MgZn_2$	1.33–1.75	1.83–1.90	1.98–2.00
$MgCu_2$–$MgAl_2$	1.33–1.73	1.84–1.95	2.03–2.05
$MgCu_2$–$MgSi_2$	1.33–1.71	1.81–1.89	1.81–2.01
$MgAg_2$–$MgZn_2$	1.72–1.75	1.78–1.90	1.98–2.00

[a] The concentration ranges are expressed in terms of valence electrons/atom; Cu, Ag = 1; Mg, Zn = 2; Al = 3; Si = 4.

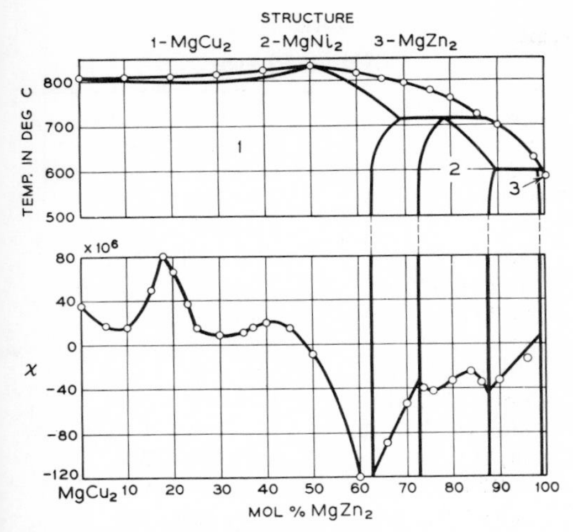

Fig. 6. Phase diagram and variation of magnetic susceptibility in the $MgCu_2$–$MgZn_2$ system. [After Klee and Witte (1954).]

the large diamagnetic contribution to the susceptibility as e/a approaches 1.75 and the changes in the susceptibility at the other phase boundaries. The $MgNi_2$ structure-type phase has a largely diamagnetic susceptibility. The proposed Brillouin zone for the $MgCu_2$ structure can accommodate 1.83 electrons per atom, and it is suggested that the zone is about 94 per cent full at the phase boundary. Brillouin zones for the two hexagonal Laves phases have also been proposed by Klee and Witte, and qualitative agreement is obtained between observed maximum electron concentrations and those corresponding to a full zone.

The electronic factor is also important in considering the formation and stability of the Laves phases containing transition elements. For example, in the $MgNi_2$–$MgZn_2$ system, the ternary phase MgNiZn forms and has the $MgCu_2$ structure. If the valency of Ni is taken as zero, then $e/a = \frac{4}{3}$, in agreement with the data on Fig. 5. Yet, from Table 4 we would expect the effective valency of Ni to be between 1 and 2. The compound $Mg_3Zn_2Cu_2Ni_2$ is also isostructural to $MgCu_2$, and it also has $e/a = \frac{4}{3}$ if Ni is assumed to contribute no valence electrons. The addition of Co to $MgZn_2$ gives the $MgCu_2$ structure, and this is also consistent if Co is also assumed to contribute no valence electrons. According to Raynor,[23] it appears that Ni has effectively a zero valence and that the $MgNi_2$ structure type observed in the ternary alloys of Table 4 is just a "structural intermediate" between the $MgCu_2$ and the $MgZn_2$ structures.

When the A element is Ta, Ti, Zr, Hf, or Nb and the B element is Cr, Mn, Fe, Co, Ni, or V, the crystal structure of the Laves phase goes from $MgCu_2$ to $MgZn_2$ to $MgCu_2$.[8,19,24,121] Similarly, this sequence of structures occurs for the Laves phases of Zr (A atom) with W, Re, Os, and Ir.[9] Dwight[9] has discussed the crystal-structure variation among the Laves phases composed of elements of the first, second and third long periods as B elements. He shows that the crystal structure of the Laves phase is related to the positions of the partner elements in the periodic table and that these relationships can be utilized to predict the structural relations in unknown binary, ternary and some quaternary systems. In connection with this, we note that Fe, Co, and Ni (B atoms) form the $MgCu_2$ structure with Er and Tm, whereas Mn forms the $MgZn_2$ structure.[30]

Gupta, Rajan, and Beck[25] have shown that the addition of Si to σ-phases of transition elements stabilize the σ-phase at electron concentrations higher than those at which the phase normally occurs, a result they attributed to a tendency for Si to act as an acceptor of electrons. This finding prompted their search for ternary Laves phases of the form $A_2(B_3Si)$ in transition metal systems, in which AB_2 binary Laves phases do not exist.[26] Indeed, a number of new ternary phases with the $MgZn_2$ structure type did form (Table 5), which suggests that Si appears to decrease the effective electron concentration, thereby lowering the Fermi energy and free energy. The temperature range of stability of the binary Laves phase WFe_2 is extended by the addition

TABLE 5

Ternary Laves Phases ($MgZn_2$ type) Containing Si

Ti_2Ni_3Si	$Ti_{30}Mn_{50}Si_{14}$
$V_4Ni_5Si_3$	$Mo_{25}Mn_{47.5}Si_{27.5}$
Nb_2Ni_3Si	W_2Mn_3Si
Ta_2Ni_3Si	MoFeSi
Mo_2Ni_3Si	MoCoSi
W_2Ni_3Si	MoNiSi
$V_4Co_5Si_3$	WFeSi
Mo_2Co_3Si	WCoSi
W_2Co_3Si	WNiSi
$W_{40}Fe_{50}Si_{10}$	

of Si. Also shown in Table 5 are the Laves phases of stoichiometry ABSi reported by Gladyshevskii and Kuzma[27] which indicates the presence of broad homogeneity ranges in some of these ternary systems.

The structural and magnetic properties of Laves phases of Y, La, and the $4f$ transition elements (A atoms), and Al, Mn, Fe, Co, and Ni (B atoms) have recently been investigated.[28–32] The magnetic properties are, of course, related to the electronic states of the constituent atoms, which in turn is manifested in the structural relationships.* A few of the salient features of this work are mentioned here. The free atom electron configuration of the rare-earth atoms is $4f^x5d^16s^2 (x = 1 \cdots 14)$; we shall use the word "valence" to denote the degree to which the s, d, and f-electron wave functions, respectively, overlap in the solid to form a band. Since the atom size is determined by the state of ionization, the structural data indicate that in the compounds CeB_2, where B = Fe, Co and Ni, $Ce(4f^15d^26s^2)$ appears to have a valence substantially greater than three, suggesting that the one $4f$-electron is not completely localized. However, Ce in $CeAl_2$ appears to be trivalent. Ce in $CePt_2$ appears to be more nearly trivalent in character, but has a valence greater than three in $CeRu_2$ and $CeRh_2$. The divalent character of $Eu(4f^65d^16s^2)$ and $Yb(4f^{13}5d^16s^2)$ metal is apparent in $EuAl_2$ and $YbAl_2$. This is a manifestation of the fact that the $4f^65d^1$ and $4f^{13}5d^1$ configurations effectively correspond to a half-full and a full $4f$-shell, respectively, and represent rather stable electron configurations. However, Yb appears to be trivalent in $YbNi_2$; this is probably related to the ease with which the Ni $3d$-orbitals can be filled.

Jaccarino et al.[33] have determined the magnitude and sign of the conduction electron polarization in the $(RE)Al_2$ compounds by nuclear magnetic resonance and electron paramagnetic resonance measurements. The rare-earth spin moment, S, of the unpaired $4f$-electrons polarize the conduction electron spins, s, in such a manner that the spins of the rare-earth ion and the conduction electrons are antiferromagnetically aligned, if a uniform polarization of the conduction electrons is assumed. This is the first determination of the sign of the conduction electron polarization in a magnetic metal, and it enables us to understand the magnetic properties of the rare-earth alloys, intermetallic compounds, and solid solutions among them.

* Similar investigations of these properties have been carried out for the (rare-earth) Ni_5 and (rare-earth) Co_5 compounds.[14,34,35] The results are discussed in Chapter 27 which is concerned with magnetic properties.

TABLE 6

Sigma Phases[a]

VMn (13.4–24.5% V)	CrCo (58–63% Cr)
VFe (33.5–56% V)	CrRu (63.5–66% Cr)
VCo (43.5–68.0% V)	CrRe (37.5–36.8% Cr)
VNi (54.5–71%)	CrOs (66% Cr)
Nb_2Al	MoMn (40% Mo)
NbNi	MoFe (50% Mo)
NbRh (60% Nb)	MoCo (60.5% Mo)
NbPd (60% Nb)	MoRu (60% Mo)
NbRe (60% Nb)	MoRe (33–52% Mo)
NbOs (60% Nb)	MoOs (65% Mo)
NbIr (65% Nb)	MoIr (70% Mo)
NbPt (62.5% Nb)	MoTc (25–30% Mo)
Ta_2Al	WFe (60% W)
TaRh (60% Ta)	WCo (50% W)
TaRe (50% Re)	WRu (60% W)
TaOs (50%–75% Ta)	WRe (37–58% W)
TaIr (65–85% Ta)	WOs (50–75% W)
TaPt (68–82% Ta)	WIr (70% W)
CrMn (16–24% Cr)	Cr_8Ni_5W
CrFe (43–49% Cr)	

[a] The percentages are atomic and only approximate. The table was complied mainly from References 36, 52, 55, 67, 68, 69, 76, 77, and 114. The known temperature ranges of stability are not given, but References 52 and 67 give this information.

3. THE SIGMA PHASE

3.1 Occurrence and Structure

The sigma phase has a complex tetragonal structure with thirty atoms per cell,[45–51] and it is generally found in systems involving the transition elements.[52–55] Significant features of this phase are broad homogeneity ranges, and the composition and temperature range of stability are not the same in the different alloy systems. These facts are illustrated in Table 6 and Fig. 7. In Fig. 7a the σ-phase forms from the primary α-solid solution and is stable to room temperature; in Fig. 7b it forms by peritectic reaction and is stable to room

* Recently binary sigma phases containing a non-transition element have been discovered.[56–60]

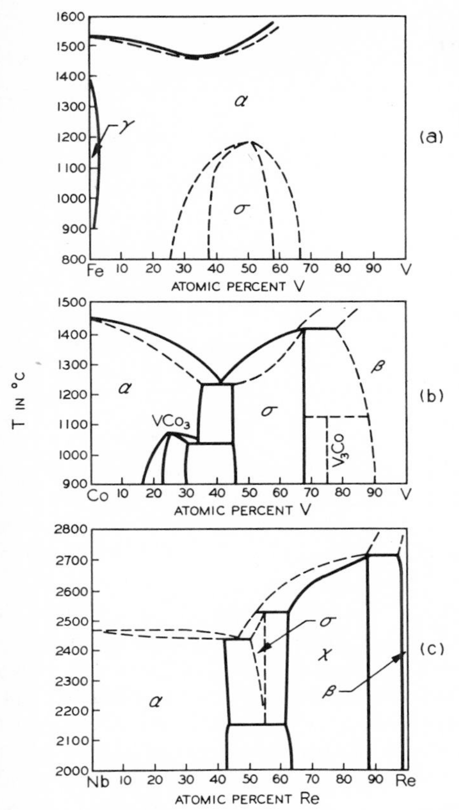

Fig. 7. σ-phases in the Fe–V, Co–V, and Nb–Re systems.

temperature; in the Nb–Re system (Fig. 7*c*), it forms by peritectic reaction, but it undergoes a eutectoid decomposition to the primary α-terminal solid solution and to the χ-phase (α-Mn type).[66]

Sigma phase belongs to space group $P4_2/mnm$-D_{4h}^{14}. There are five kinds of nonequivalent crystallographic positions for the thirty atoms in the unit cell: (*a*), (*f*), (*ic*), (*id*), and (*j*). If we designate the atoms occupying the sites as A, B, C, D, and E, there are:

2A atoms in positions (*a*): 000; $\frac{1}{2}\,\frac{1}{2}\,\frac{1}{2}$
4B atoms in positions (*f*): $x\,x\,0$; $\bar{x}\,\bar{x}\,0$; $\frac{1}{2}+x, \frac{1}{2}-x, \frac{1}{2}$; $\frac{1}{2}-x, \frac{1}{2}+x, \frac{1}{2}$.
8C atoms in positions (*ic*): $x, y, 0$; $\bar{x}, \bar{y}, 0$; $\frac{1}{2}+x, \frac{1}{2}-y, \frac{1}{2}$; $\frac{1}{2}-x, \frac{1}{2}+y, \frac{1}{2}$; $y, x, 0$; $\bar{y}, \bar{x}, 0$; $\frac{1}{2}+y, \frac{1}{2}-x, \frac{1}{2}$; $\frac{1}{2}-y, \frac{1}{2}+x, \frac{1}{2}$.
8D atoms in positions (*id*): same as above.
8E atoms in positions (*j*): x, x, z; $\bar{x}, \bar{x}, z$; $\frac{1}{2}+x, \frac{1}{2}-x, \frac{1}{2}+z$; $\frac{1}{2}-x, \frac{1}{2}+x, \frac{1}{2}+z$; $x, x, \bar{z}$; $\bar{x}, \bar{x}, \bar{z}$; $\frac{1}{2}+x, \frac{1}{2}-x, \frac{1}{2}-z$; $\frac{1}{2}-x, \frac{1}{2}+x, \frac{1}{2}-z$.

The atoms are arranged in layers, and the structure bears a strong resemblance to a hexagonal closed-packed structure (Fig. 8).* Certain atoms, however, are displaced from the basal plane (in the hcp consideration) to positions intermediate between the layers leading to improved overall packing.[47] The crystallographic site parameters for a number of sigma phases are approximately as follows: for (*f*), $x \simeq 0.4$; for (*ic*), $x \simeq 0.54$, $y \simeq 0.13$; for (*id*), $x \simeq 0.73$, $y \simeq 0.065$; for (*j*), $x \simeq 0.2$, $z \simeq 0.25$.

The fact that large homogeneity ranges are exhibited by the σ-phase suggests that some of the atoms are statistically distributed and the atomic size factor is important in the formation of these compounds. Evidence for ordering in some of the σ-phases has been obtained by detailed X-ray and neutron diffraction work.[47,51,61–63] More precise information has been obtained by neutron diffraction because of the larger difference in nuclear scattering factors of the atoms that have only slightly different X-ray scattering factors. In Table 7 we show the distribution of the atoms among the five different sites for a number of σ-phases.[63]

Kasper and Waterstrat[62] have shown that a generalization regarding the positions occupied by the atoms can be made based on the ordering occurring in a number of binary σ-phases that contain the first long row

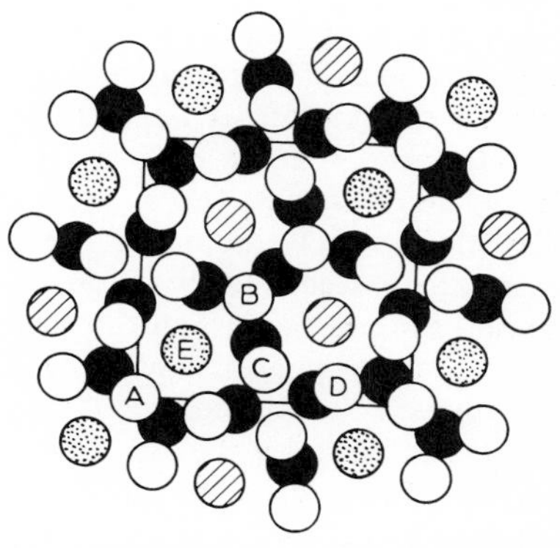

Fig. 8. The (001) projection of the σ-phase. Open circles: atoms in the planes $z = 0, 1, \ldots$; filled circles: atoms in the planes $z = \frac{1}{2}, \ldots$; shaded circles: each circle represents two atoms per unit cell with z coordinates equal to approximately $\frac{1}{4}$ and $\frac{3}{4}$. [After Bergman and Shoemaker (1954).]

* The structure of β-U is similar to the σ-phase structure (see, for example, References 47 and 51).

TABLE 7

Atomic Ordering in Some σ-Phases

Kind of Atom	Ni–V	Fe–V	Fe–Cr	Co–Cr	Mn–Cr	Mn–Mo	Fe–Mo
A	Ni	Fe	Fe	Co	Mn	Mn	Mo
D	Ni	Fe	Fe	Co	Mn	Mn	Fe
B	V	V	Cr	Cr	Mixed	Mo	Mo
C	Mixed	Mixed	Mixed	Mixed	Mixed	Mixed	Fe
E	V	Mixed	Cr	Mixed	Mixed	Mixed	Mo

elements and Mo. If we designate the elements left of Mn (V, Cr, and Mo) as A atoms and those to the right of Mn (Fe, Co, and Ni) as B atoms (Mn itself is considered in both groups), the order of occupancy of the crystallographic sites is:

Position	Coordination Number	Occupancy
(2*b*)	12	B
(4*f*)	15	A
(8*ic*)	14	Mixed
(8*id*)	12	B
(8*j*)	14	Mixed

The A and B atoms together can occupy the sites of intermediate coordination. In all these structures, the A atom is larger than the B atom. However, for the σ-phases, (CrRe)σ, (CrRu)σ, and (CrOs)σ, the A atom, Cr, is smaller than the B atom. In these three σ-phases there seems to be a random distribution of atoms because of the smaller Cr atom.[55] In the σ-phases containing Si (see below), the smaller Si occupies the low CN sites. Al also occupies the low CN sites in Nb_2Al and Ta_2Al, even though the metallic radius for Nb, Ta, and Al is 1.43 Å.[59,60,64,65]

3.2 The Electronic Factor

It appears that both the atomic size factor and the electronic factor contribute to the formation and stability of the σ-phases, whereby the first factor predominates in one group and the second in another. The σ-phase, as an electron compound, has been discussed by a number of authors.[47,52,70–72] Bloom and Grant[72] assume that the σ-phase compositions are determined by an electron concentration of 210 per unit cell or an average of seven valence electrons per atom, and Greenfield and Beck[52] have shown that 6.93 electrons per atom can be obtained from the *mean* composition of the known σ-phases.* The second Brillouin zone[47] can contain 6.97 electrons per atom, which is consistent with the above treatment. However, Bergman and Shoemaker[47] doubt that the average valence of the metal atoms can be as high as seven and feel that it is more likely to be closer to 5.76. This latter value is based on the observed interatomic distances.

Gupta, Rajan, and Beck[65] have determined the effect of Si on the stability of the following binary σ-phases: (V, Fe)σ, (V, Co)σ, (V, Ni)σ, and (Cr, Co)σ at 1175°C, and (Cr, Mn)σ at 1000°C. They have also confirmed the existence of a ternary σ-phase in the Cr–Ni–Si system, first reported by Aronsson and Lundstrom,[64] and have determined the composition range of stability of this phase at 1175°C. The pertinent portions of the isothermal sections of the phase diagrams for the above six systems are shown on Fig. 9. It is to be noted that the composition range of stability is increased by the addition of Si, particularly in the Cr–Mn–Si and V–Fe–Si systems. The existence of the ternary σ-phase in the Cr–Ni–Si system, even though a binary (Cr, Ni)σ does not exist, is

* Note that Nb_2Al and Ta_2Al do not fit into this scheme.

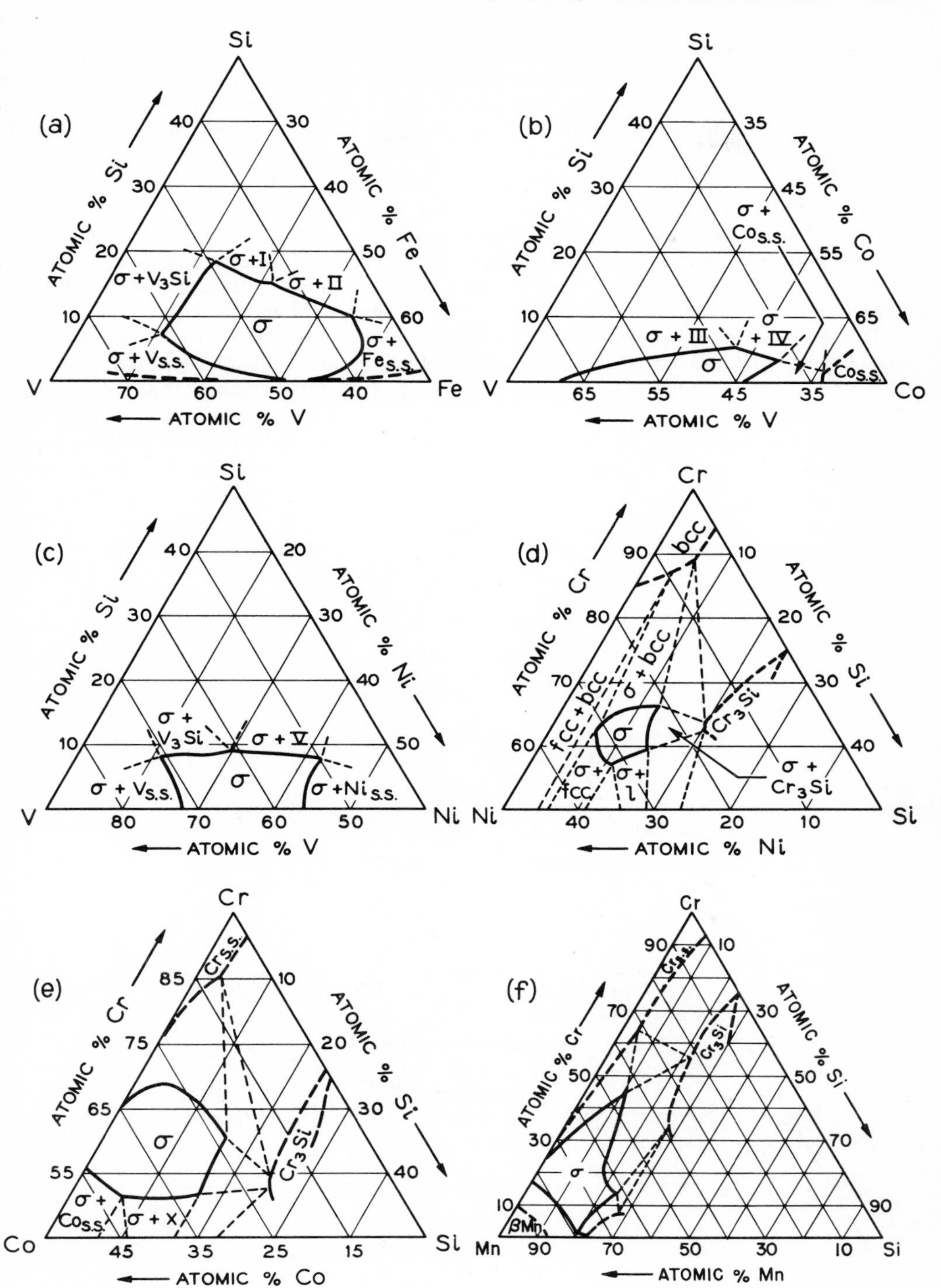

Fig. 9. Sigma-phase homogeneity ranges: (*a*) at 1175°C in the V–Fe–Si system; (*b*) at 1175°C in the V–Co–Si system; (*c*) at 1175°C in the V–Ni–Si system; (*d*) at 1000°C in the Cr–Mn–Si system; (*e*) at 1175°C in the Cr–Co–Si system; (*f*) at 1175°C in the Cr–Mn–Si system.

further evidence for the stabilizing effect of Si. Not only is the binary composition range of stability extended, but evidence for the extension of the temperature range of stability has been obtained in the Cr–Fe–Si system. The (Cr, Fe, Si)σ phase[73] is stable[64] up to 1200°C, although the (Cr, Fe)σ is stable only up to 825°C. Si tends to occupy the lower coordination sites, *D*, in the ternary σ-phases, substituting for the smaller transition element, for example, Fe and Mn.

The addition of Al to the (V, Fe)σ and (Cr, Co)σ phases has an effect opposite of that of Si, that is, it tends to suppress the binary σ-phases.[65]

The effect of Si is interpreted on the basis of the fact that the binary σ-phase has a full Brillouin zone and that Si can accept electrons in a manner to reduce the free electron concentration. This then will allow for increased solubility of the transition elements, because there are now allowed electron states available for filling. σ-phase ternary solid solutions form from the binary σ-phase in the Cr–Co–Ni, Cr–Co–Fe, Cr–Co–Mo, and Cr–Fe–Mo systems.[70,74,75] In the Cr–Ni–Mo system a ternary σ-phase forms, although no binary σ-phases exist in this system (Fig. 10).[70] A ternary P-phase also forms in this system. In the Cr–Co–Fe system, the (Cr, Co)σ and (Cr, Fe)σ form a complete series of σ-phase solid solutions at 800°C, but at 1200°C the extent of the ternary solid solution is limited and it is confined to the Cr–Co-rich alloys.[70]

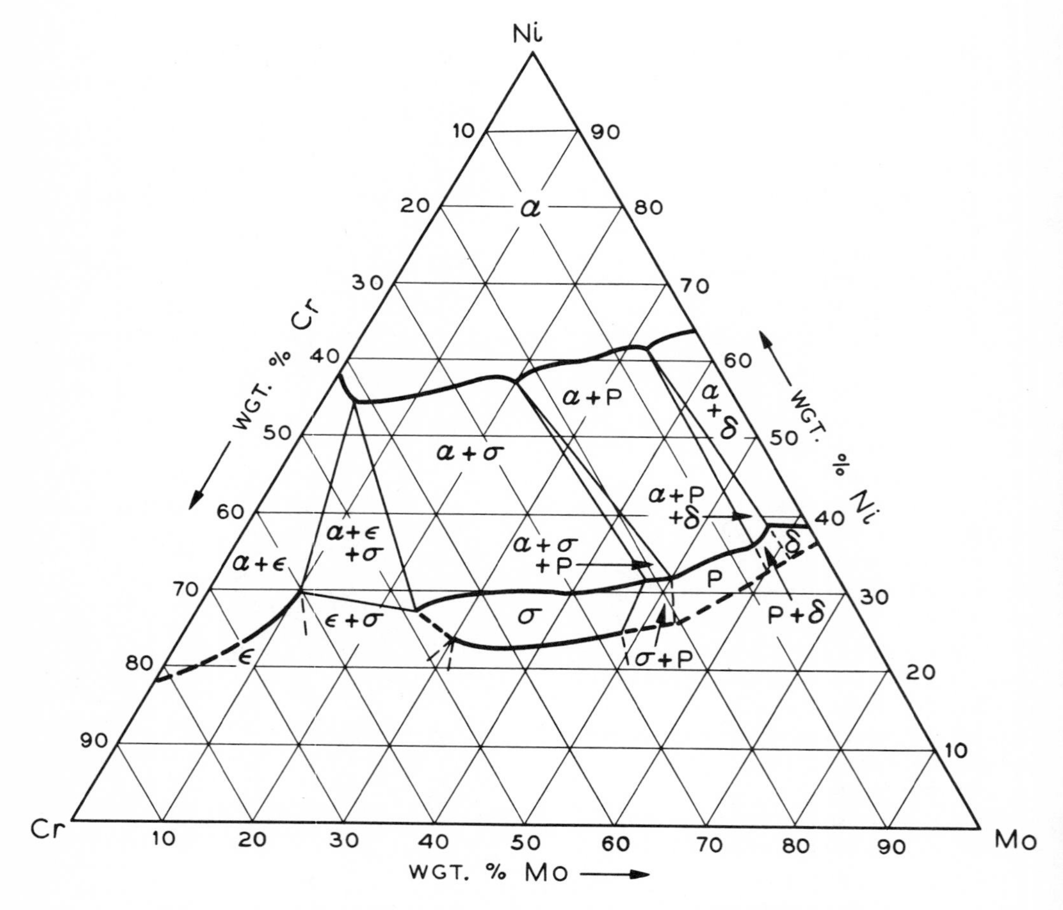

Fig. 10. The 1200°C isothermal section of the Cr–Ni–Mo phase diagram, showing the ternary σ- and P-phases. (After Rideout et al.[70])

TABLE 8

Intermediate Phases with the α-Mn Structure*

Mn–Ti(9.1 % Ti)	TaOs	$Cr_{12}Fe_{36}Mo_{10}$ (χ-phase)
Re_6Zr	Tc–Sc(12.5 % Sc)	Cr–Ti–Fe
Re–Hf(14–50 % Hf)	Tc–Ti	$V_3Ni_5Ge_2$
Re–Nb(18–40 % Nb)	Tc_6Zr	(V–Fe–Si)
ReTi	Tc–Hf(12.5 % Hf)	(V–Co–Si)
Re–Ta(25–37 % Ta)	Tc_3Nb	(V–Ni–Si)
Re–Mo(<21 % Mo)	Tc–Ta(16.7 % Ta)	$NbOs_2$
Re–W(<27 % W)	$Al_{12}Mg_{17}$	Nb–Pd(40 % Pd)

* The percentages are atomic. The table was compiled mainly from References 36, 52, 65, 67, 81–84, 88–95. These references should be consulted for details of preparation, which indicate to some degree the temperature region of stability.

A ternary R-phase exists in the Cr–Co–Mo system.[70]

In the Cr–Co–Ni and Cr–Co–Fe systems the σ-phase extends in a direction which corresponds to approximately constant Cr content, indicating that Ni and Fe are capable of partially replacing Co in forming the σ-phase. In the Cr–Co–Mo system the σ-phase extends in a direction which corresponds to approximately constant Co content, indicating that Mo is replacing the Cr in the binary σ-phase.

Although Nb does not form binary σ-phases with Co and Ru, extensive ternary solid solutions can be obtained with these elements and Ir, Os, and Pd.[76]

Raub[77] has discussed the σ-phase formation of Ru and Os with the group 6A elements (Cr, Mo, and W). The σ-phase compositions are not the same in the various binary systems, and they tend to follow the shift in the solid solubility limit of the ε-terminal solid solution. This suggests that these σ-phases are electron compounds. Blaugher and Hulm[78] have determined the occurrence of superconductivity in the binary σ-phases formed by Nb, Mo, Ta, and W with Ru, Rh, Pd, Re, Os, Ir, and Pt (as well as α-Mn structure type compounds). The superconducting critical temperatures, when plotted against the valence electron-to-atom ratio, show a maximum at 6.5* (similar to that observed for the β-tungsten compounds); it is suggested that this is connected with the existence of a peak at this electron concentration in the density of states at the Fermi surface.

Rocher and Friedel[79] have proposed a model of the electronic structure for some complex crystalline substances, including the β-tungsten, sigma, and α–Mn-type phases (see next section). Their model predicts a high electron density at the Fermi level for a *d*-band filled up to more than one-third. They suggest that the superconductivity of a number of compounds with these structures is due to a high density of electron states, similar to that proposed by Clogston and Jaccarino.[80]

4. PHASES CLOSELY RELATED TO THE SIGMA PHASE (α-Mn TYPE, P, R, μ, AND Δ)

The structure of α-Mn is common to a number of transition metal intermediate phases. In Table 8 we list the intermediate phases known to-date, which have this structure. No intermediate phases of Tc with V, Cr, Mo, and W, which have the α-Mn structure, have been reported up to the time of this writing.† Recently ternary phases involving Si have been reported to have the α-Mn

* The Fe group elements are assigned a valence of 8; Co group, 9; and the Ni group, 10.

† The superconducting transition temperatures of the Re and some Tc α-Mn structure-type phases have been reported recently.[78,83]

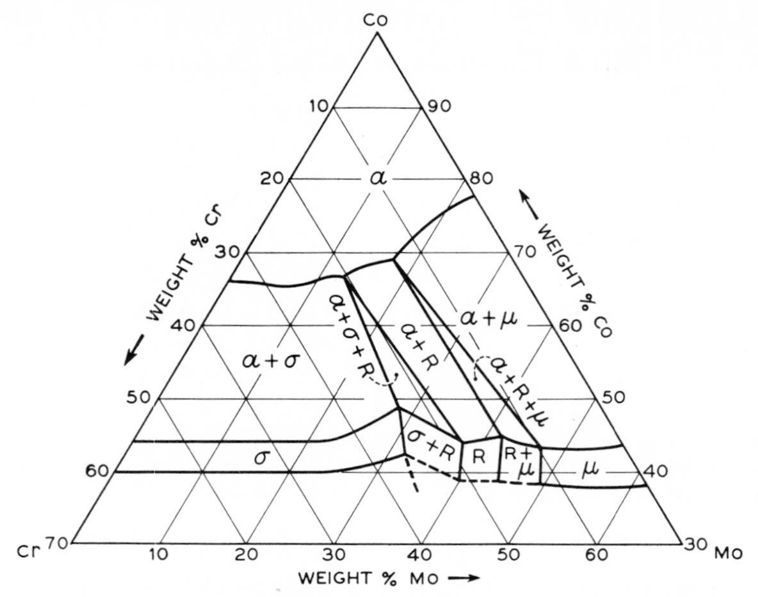

Fig. 11. A portion of the 1200°C isothermal section of the Cr–Co–Mo system. (After Rideout et al.[70])

structure.[88] The ternary systems, in which this phase forms, are V–Fe–Si, V–Co–Si, and V–Ni–Si.

The α-Mn structure type is cubic with fifty-eight atoms per unit cell.[85] It belongs to space group $I43m\text{-}T_d^3$. There are four non-equivalent crystallographic positions for the fifty-eight atoms in the unit cell. If we designate the atoms occupying these sites as A_1, A_2, A_3 and A_4, the atomic distribution is as follows:

equivalent positions (000; $\frac{1}{2}\,\frac{1}{2}\,\frac{1}{2}$)*
$2A_1$ in (*a*): 000.
$8A_2$ in (*c*): xxx; $x\bar{x}\bar{x}$; $\bar{x}x\bar{x}$; $\bar{x}\bar{x}x$.
$24A_3$ in (*g*): xxz; zxx; xzx; $\bar{x}x\bar{z}$; $\bar{z}x\bar{x}$; $\bar{x}z\bar{x}$; $x\bar{x}\bar{z}$; $z\bar{x}\bar{x}$; $x\bar{z}\bar{x}$; $\bar{x}\bar{x}z$; $\bar{y}\bar{x}x$; $\bar{x}\bar{z}x$.
$24A_4$ in (*g*): as above

In α-Mn itself the interatomic distances vary from 2.24 to 2.96 Å. Even within each of the four crystallographically different sites, the interatomic distances are not the same. We can assign a total coordination number to each site and an average interatomic distance. Thus atoms A_1 and A_2 have CN 16, whereas A_3 and A_4 have CN 13 and 12, respectively. This structure, in terms of basic coordination polyhedra, has been discussed by Kasper.[2] This, together with the average interatomic distances, suggests only three physically distinguishable kinds of atoms, atom A_1 and A_2 being one kind. Pauling has suggested that there are two kinds of Mn atoms present, A_1, A_2, and A_3 being the same.

Studies of the occurrence of order in the intermediate phases with the α-Mn structure have been limited. For the cases studied ($Fe_{36}Cr_{12}Mo_{10}$, $Mg_{17}Al_{12}$, and 30 at.% Fe in α-Mn), the smaller atoms prefer sites occupied by atoms A_3 and A_4.[2]

The P- and R-phases were first discovered by Rideout et al.[70] in the Cr–Ni–Mo and Cr–Co–Mo systems (Figs. 10 and 11). Subsequently, Das and Beck[96] found that the P- and R-phases existed in the Mn–Mo–Co system and that the R-phase existed in the Mn–Mo–Fe system (Fig. 12). Further work by Bardos, Gupta, and Beck[88] led to the discovery of a number of ternary R-phases containing Si (Table 9), which show once again the stabilizing effect of Si (and the importance of free electron concentration) previously found for some Laves and sigma phases.[25,26] The binary R-phase, $Ti_{0.17}Mn_{0.83}$, was shown to exist by Waterstrat.[89]

The binary μ-phases are Fe_7W_6, Fe_7Mo_6, Co_7W_6, and Co_7Mo_6; they are known to exist over a range of composition.[97–100] In addition,

* Add the coordinates of the equivalent positions to the coordinates of the *a*, *c*, and *g* sites to obtain the coordinates of all of the atoms.

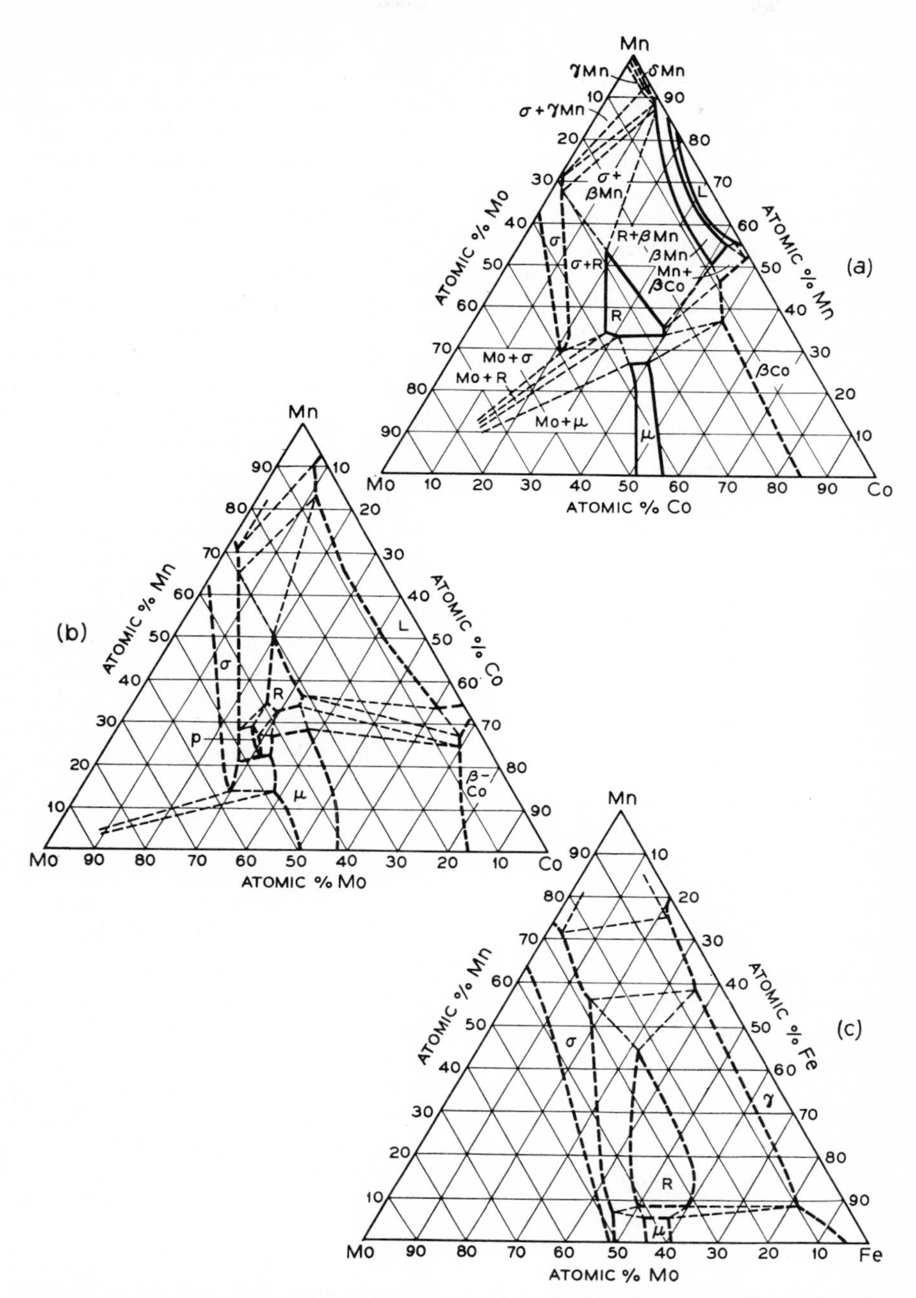

Fig. 12. (*a*) The 1175°C isothermal section of the Mo–Mn–Co system illustrating the relation between the μ-, R-, σ-, and β-Mn phases. (*b*) The 1240°C isothermal section of the Mo–Mn–Co system illustrating the relation between the P-, R-, μ-, and σ-phases. (*c*) The 1240°C isothermal section of the Mo–Mn–Fe system illustrating the relation between the R-, μ-, and σ-phases. (After Das and Beck.[96])

ternary μ-phase solid solutions are present in the Fe–Mo–Ni and Co–Mo–Cr systems.[70,102]

The structure of the P-phase is orthorhombic, belonging to space group D_{2h}^{16}-$Pbnm$, and contains fifty-six atoms per cell.[103,104] It is closely related to the σ-phase, but there is a kind of 16-fold coordination in the P-phase which is absent in the σ-phase. There are some features of the P-phase common to the α-Mn structure, namely, a resemblance in the layers and the occurrence of 16-fold coordination. A possible ordering scheme for the Mo–Ni–Cr P-phase has been discussed by Kasper.[2]

TABLE 9

Ternary R-Phases Containing Silicon

Composition (Approximate)	Annealing Temperature (°C)	c (Å)	a (Å)	c/a
$Ti_2Mn_{79}Si_{19}$	1000	19.23	10.87	1.77
$Nb_5Mn_{79}Si_{16}$	1000	19.28	10.89	1.77
$Ta_5Mn_{79}Si_{16}$	1000	19.19	10.86	1.77
$Mo_3Mn_{78}Si_{19}$	1000	19.18	10.85	1.77
$W_3Mn_{78}Si_{19}$	1000	19.22	10.86	1.77
$V_{37}Fe_{41}Si_{22}$	1100	19.23	10.79	1.78
$V_{45}Co_{40}Si_{15}$	1100	19.14	10.78	1.78
$V_{45}Ni_{40}Si_{15}$	1100	19.10	10.82	1.77

The crystal structure of the R-phase in the Mo–Co–Cr system was recently determined by Komura et al.[105] It is rhombohedral (space group C_{3i}^{2}-$R\bar{3}$) with fifty-three atoms per cell. It exhibits 12-, 14-, 15-, and 16-fold atomic coordination. The more highly coordinated sites are occupied predominately by Mo, a situation similar to that which probably occurs in the Mo–Ni–Cr P-phase[2] and in the σ-phases.

The crystal structure of the μ-phase is rhombohedral (space group D_{3d}^{5}-$R\bar{3}m$) with thirteen atoms per cell. The ideal stoichiometry is A_7B_6. There are five crystallographically different kinds of atoms distributed among the sites as follows:

1 A_1 atom in position (a): 000.
6 A_2 atoms in position (h): xxz; xzx; zxx; $\bar{x}\bar{x}\bar{z}$; $\bar{x}\bar{z}\bar{x}$; $\bar{z}\bar{x}\bar{x}$.
2 B_1 atoms in position (c): xxx; $\bar{x}\bar{x}\bar{x}$.
2 B_2 atoms in positions (c): as above.
2 B_3 atoms in positions (c): as above.

The larger atoms, W or Mo, occupy the sites of highest coordination (c).[2]

The importance of the electronic factor in the formation and stability of the σ-phases (as well as some of the Laves phases) was noted above. The α-Mn, P-, R-, μ-, and Δ-phases* are related structurally to the σ-phase. In addition to this close structural relation, there is strong evidence that these phases can also be considered as "electron compounds" since they all tend to occur in relatively narrow electron concentration ranges.[52,70,96,102] The common occurrence of these phases in transition metal systems, and in the order σ, P, R, μ, α-Mn, presumably corresponding to increasing electron concentration, is suggested as additional support for the importance of electron concentration in determining the formation and stability of these phases.[96]

Haworth and Hume-Rothery[106] have shown that in binary transition alloy systems there is a definite sequence in the appearance of intermediate phases. In a binary system, in which the terminal members are bcc and fcc, the sequence is bcc → β-tungsten → σ → α-Mn → hcp → fcc. It is not necessary that all phases occur, but the order is maintained. Dwight,[76] as a result of his studies, has modified this sequence further as follows: bcc → β-tungsten → σ → Laves → α-Mn → hexagonal variant → fcc.

5. INTERMEDIATE PHASES WITH THE AlB_2 AND $ThSi_2$ STRUCTURE

It is appropriate finally to consider briefly the intermediate phases of stoichiometry AB_2 having the hexagonal AlB_2 (Table 10) and the tetragonal α-$ThSi_2$ structure (Table 11).[122] Although the number of phases having the α-$ThSi_2$ structure is small in comparison to the number having the AlB_2 structure, it is quite likely that Si and Ge will form this structure with nearly all of the $4f$ rare earths.

The AlB_2 structure belongs to space group $P6/mmm$-D_{6h}^{1}, with one Al in position (a), 000; and two B atoms in positions (d), $\frac{1}{3}\,\frac{2}{3}\,\frac{1}{2}$, $\frac{2}{3}\,\frac{1}{3}\,\frac{1}{2}$ (Fig. 13).† The larger A atoms lie between planes of hexagonal arrays of B

* The Laves and β-tungsten phases are also related to these phases, but are less complex.

† α-$ThSi_2$ belongs to space group $I4_9/amd$, with three formula units per cell.

TABLE 10

Intermediate Phases with the Hexagonal AlB_2 Structure[a]

AlB_2	$LaHg_2$	$ThCu_2$
$BaGa_2$	MgB_2	$ThNi_2$
$CaGa_2$	MoB_2	β-$ThSi_2$
$CeGa_2$	$NaHg_2$	TiB_2
CrB_2	NbB_2	TiU_2
$DyGa_2$	$NdGa_2$	UB_2
$ErGa_2$	$PrGa_2$	UHg_2
$GdGa_2$	Pu_2Si_3	β-USi_2
Ge_3Pu_2	$SmGa_2$	VB_2
HfB_2	$SrGa_2$	ZrB_2
$HoGa_2$	TaB_2	$ZrBe_2$
$LaCu_2$	$TbGa_2$	YGa_2
$LaGa_2$	$ThAl_2$	

[a] Compiled from References 36, 108, and 109.

TABLE 11

Intermediate Phases with the Tetragonal α-$ThSi_2$ Structure[a]

$CeSi_2$	$PuGe_2$
$LaSi_2$	$PuSi_2$
$NdSi_2$	α-$ThSi_2$
$NpSi_2$	α-USi_2
$PrSi_2$	YGe_2

[a] Compiled from References 36 and 110.

atoms. It is positioned in the open central region between the hexagonal arrays. For the case of packing of touching spheres, the ideal $c/a = 1.07$.[10] Laves[10] has discussed the relation of the c/a ratio on the A–A, A–B, and B–B contacts in the AlB_2-type phases as well as the atomic radii determined from some of these phases. In addition, he has discussed the speculation regarding the bonding in some of these intermediate phases.

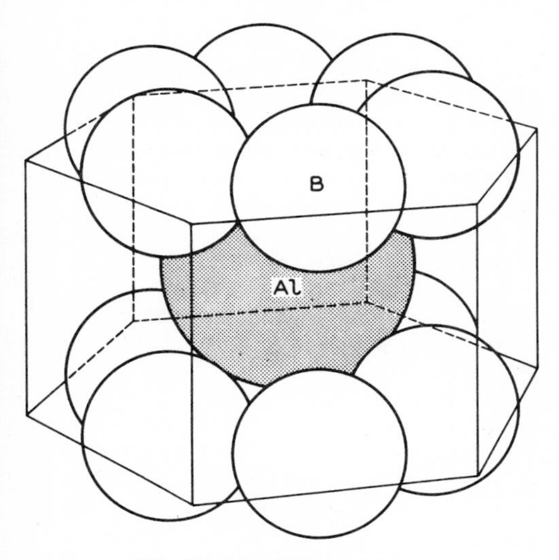

Fig. 13. The AlB_2 structure.

It is interesting to note that, whereas UAl_2, $PuAl_2$, and $NpAl_2$ have the cubic $MgCu_2$ structure, $ThAl_2$ has the AlB_2 structure. However, $ThAl_2$ and UAl_2 exhibit large mutual solid solubility and U exhibits local magnetic moments in $ThAl_2$.[107]*

6. SUMMARY

Some of the important close-packed intermediate phases have been discussed generally in terms of the factors that contribute to their formation and stability. There is no doubt that the electronic factor is very important, and more detailed knowledge of the electronic state in these materials will become available as a result of magnetic susceptibility, low-temperature heat capacity, magnetic resonance, magnetoresistance and magnetoacoustic, de Haas-van Alphen, and Hall effect measurements.

Recent literature indicates increased activity in the application of the above experimental techniques to metallic intermediate phases (as opposed to semiconductors).† For some of these measurements, single-crystal specimens having high-resistance ratios are necessary in order to obtain meaningful information. High-resistance ratios (low resistivity at liquid He temperatures) imply high chemical purity and low-defect content so that the electrons have a large mean-free path. A number of intermetallic compounds can be prepared in the form of large, single, quite pure crystals, and because they exhibit rather complete long-range order they have sufficiently high resistance ratios. The use of high magnetic fields for these measurements decreases the resistance ratio requirement.

In addition to heat capacity data, determination of heats of formation of intermediate phases are being made at a number of laboratories (see Chapter 3 on thermodynamic measurements and reference 119). This information is a necessary complement to the transport data.

* See also the discussion regarding the $ThAl_2$–$ThMn_2$ system in Section 2.2.

† See, for example, References 116–118.

REFERENCES

1. Laves F., and H. Witte, *Metallwirtschaft* **14** (1935) 645.
2. Kasper J. S., in *Theory of Alloy Phases*. ASM Cleveland, Ohio, 1956.
3. Frank F. C., and J. S. Kasper, *Acta Cryst.* **11** (1958) 184.
4. Frank F. C., and J. S. Kasper, *Acta Cryst.* **12** (1959) 483.
5. Friauf J. B., *J. Am. Chem. Soc.* **49** (1927) 3107; *Phys. Rev.* **29** (1927) 34
6. Laves F., and K. Lohberg, *Nachr. Akad. Wiss. Göttingen, Math. Physik. Kl. IV*, Neue Folge 1, **6** (1932) 59.
7. Laves F., and H. Witte, *Metallwirtschaft* **14** (1935) 645.
8. Laves F., and H. Witte, *Metallwirtschaft* **15** (1936) 840.
9. Dwight A. E., *Trans. ASM* **53** (1961) 479.
10. Laves F. in *Theory of Alloy Phases*. ASM Cleveland, Ohio, 1956.
11. Misch L., *Metallwirtschaft* **14** (1935) 897.
12. Baenziger N. C., R. E. Rundle, A. E. Snow, and A. S. Wilson, *Acta Cryst.* **3** (1950) 34.
13. Nowotny H., *Z. Metallk.* **34** (1942) 247.
14. See, for example, Wernick J. H., and S. Geller, *Acta Cryst.* **12** (1959) 662; Haszko S. E., *Trans. Met. Soc. AIME* **218** (1960) 763.
15. Berry R. L., and G. V. Raynor, *Acta Cryst.* **6** (1953) 178.
16. Lieser K. H., and H. Witte, *Z. Metallk.* **43** (1952) 396.
17. Lieser K. H., and H. Witte, *Z. Physik. Chem.* **202** (1954) 321.
18. Witte H., *Zur Strucktur und Materie der Festkörper*, Springer Verlag, Berlin, 1952.
19. Kuo K., *Acta Met.* **1** (1953) 720.
20. Klee H., and H. Witte, *Z. Physik. Chem.* **202** (1954) 352.
21. Brook G. B., G. I. Williams, and E. M. Smith, *J. Inst. Metals* **83** (1955) 271.
22. Kittel C., *Introduction to Solid State Physics*, 2nd Edition. Wiley, New York, 1960.
23. Raynor G. V., *Progress in Metal Physics*, B. Chalmers, ed., Butterworths Scientific Publications, London, 1949.
24. Elliot R. P., and W. Rostoker, *Trans. ASM* **50** (1958) 617.
25. Gupta K. P., N. S. Rajan, and P. A. Beck, *Trans. Met. Soc. AIME* **218** (1960) 617.
26. Bardos D. I., K. P. Gupta, and P. A. Beck, *Trans. Met. Soc. AIME* **221** (1961) 1087.
27. Gladyshevskii E. I., and Yu. B. Kuzma, *Zh. Strukt. Khim.* **1** (1960) 66.
28. Wernick J. H., and S. Geller, *Trans. Met. Soc. AIME* **218** (1960) 866.
29. Haszko S. E., *Trans. Met. Soc. AIME* **218** (1960) 958.
30. Wernick J. H., and S. E. Haszko, *J. Phys. Chem. Solids* **18** (1961) 207.
31. Wernick, J. H., S. E. Haszko, and D. Dorsi, *J. Phys. Chem. Solids* **23** (1962) 567.
32. Williams H. J., J. H. Wernick, E. A. Nesbitt, and R. C. Sherwood, "Proceedings of the International Conference on Magnetism and Crystallography," *J. Phys. Soc. Japan* **17** SB-1 (1962) 91.
33. Jaccarino V., B. T. Matthias, M. Peter, H. Suhl, and J. H. Wernick, *Phys. Rev. Letters* **5** (1960) 6.
34. Nesbitt E. A., H. J. Williams, J. H. Wernick, and R. C. Sherwood, *J. Appl. Phys.* **32** (1961) 342s.
35. Nesbitt E. A., H. J. Williams, J. H. Wernick, and R. C. Sherwood, *J. Appl. Phys.* **33** (1962) 1674.
36. Pearson W. B., *Lattice Spacings and Structure of Metals and Alloys*. Pergamon Press, New York, 1958.
37. Wood E. A., and V. B. Compton, *Acta Cryst.* **11** (1958) 429.
38. Compton V. B., and B. T. Matthias, *Acta Cryst.* **12** (1959) 651.
39. Elliot, R. P., *Trans. ASM* **53** (1961) 321.
40. Matthias B. T., V. B. Compton, and E. Corenzwit, *J. Phys. Chem. Solids* **14** (1961) 130.
41. Wilson C. G., *Acta Cryst.* **12** (1959) 660.
42. Zhuravlev N. N., T. A. Mingazin, and G. S. Zhdanov, *Soviet Phys. JETP* (*English Transl.*) **7** (1958) 566.
43. Cherkaskin E. E., E. I. Gladyshevskii et al., *Zh. Neorgan. Khim.* **3** (1958) 650.
44. Zhuravlev N. N., *Soviet Phys. JEPT* (*English Transl.*) **7** (1958) 571.
45. Shoemaker D. P., and B. G. Bergman, *J. Am. Chem. Soc.* **72** (1950) 5793.
46. Bergman B. G., and D. P. Shoemaker, *J. Chem. Phys.* **19** (1951) 515.
47. Bergman B. G., and D. P. Shoemaker, *Acta. Cryst.* **7** (1954) 857.
48. Dickens G. J., A. M. B. Douglas, and W. H. Taylor, *J. Iron. Steel Inst.* **167** (1951) 27.
49. Dickens G. J., A. M. B. Douglas, and W. H. Taylor, *Nature*, (London) **167** (1951) 192.
50. Kasper J. S., B. F. Decker, and J. R. Belanger, *J. Appl. Phys.* **22** (1951) 361.
51. Dickens G. J., A. M. B. Douglas, and W. H. Taylor, *Acta Cryst.* **9** (1956) 297.

52. Greenfield P., and P. A. Beck, *Trans. AIME* **200** (1954) 253; **206** (1956) 265.
53. Nevitt M. V. and P. A. Beck, *Trans. AIME* **203** (1955) 669.
54. Nevitt M. V., and J. W. Downey, *Trans. AIME* **203** (1957) 669; **209** (1957) 1072.
55. Waterstrat R. M., and J. S. Kasper, *Trans. AIME* **209** (1957) 872.
56. Corenzwit E., *J. Phys. Chem. Solids* **9** (1959) 93.
57. McKinsey C. R., and G. M. Faulring, *Acta Cryst.* **12** (1959) 701.
58. Edshammer L. E., and B. Holmberg, *Acta Chem. Scand.* **14** (1960) 1219.
59. Brown P. J., and J. B. Forsyth, *Acta Cryst.* **14** (1961) 362.
60. Gupta K. P., *Trans. Met. Soc. AIME* **221** (1961) 1047.
61. Decker B. F., R. M. Waterstrat, and J. S. Kasper, *Trans. AIME* **200** (1954) 1406.
62. Kasper J. S., and R. M. Waterstrat, *Acta Cryst.* **9** (1956) 289.
63. Kasper J. S., in *Theory of Alloy Phases*, ASM Cleveland, Ohio, 1956.
64. Aronsson A., and T. Lundstrom, *Acta Chem. Scand.* **11** (1957) 365.
65. Gupta K. P., N. S. Rajan, and P. A. Beck, *Trans. Met. Soc. AIME* **218** (1960) 617.
66. Grant N. J., and B. C. Giessen, *Refractory Metal Constitution Diagrams*, Air Force Contract 33 (616)-6023, WADD Tr-60-132, 1960.
67. Knapton A. G., *J. Inst. Metals* **87** (1958) 28.
68. Raub E., *Z. Metallk.* **45** (1954) 23.
69. Raub E., and P. A. Walter, *Hereus Festschrift* **124** (1951).
70. Rideout S., W. D. Manly, E. L. Kamen, B. S. Lement, and P. A. Beck, *Trans. AIME* **191** (1951) 852.
71. Sully A. H., *J. Inst. Metals* **80** (1952) 173.
72. Bloom D. S., and N. J. Grant, *Trans. AIME* **197** (1952) 88.
73. Anderson A. G. H., and E. R. Jette, *Trans. ASM* **24** (1936) 375.
74. Beck P. A., and W. D. Manly, *Trans. AIME* **185** (1949) 354.
75. Baen S. R., and P. Duwez, *Trans. AIME* **191** (1951) 331.
76. Dwight A. E., in *Columbium Metallurgy*, D. L. Douglas and F. W. Kunz, ed., Met. Soc. AIME, Interscience, 1961.
77. Raub E., *J. Less-Common Metals* **1** (1959) 3.
78. Blaugher R. D., and J. K. Hulm, *J. Phys. Chem. Solids* **19** (1961) 134.
79. Rocher Y. A., and J. Friedel, *J. Phys. Chem. Solids* **21** (1961) 287.
80. Clogston A. M., and V. Jaccarino, *Phys. Rev.* **121** (1961) 1357.
81. Lam D. J., J. B. Darby, Jr., J. W. Downey, and L. J. Norton, *Nature* **192** (1961) 744.
82. Savitskii E. V., M. A. Tylkina, and I. A. Tsyganova, *At. Energ.* **7** (1959) 231; *Chem. Abstracts* **54** (1960) 58h.
83. Compton V. B., E. Corenzwit, J. P. Maita, B. T. Matthias, and F. J. Morin, *Phys. Rev.* **123** (1961) 1567.
84. Niemiec J., and W. Trzebiatowski, *Bull. Acad. Polon Sci.* **4** (1956) 601; *Chem. Abstracts* **51** (1957) 7280a.
85. Bradley A. J., and J. Thewlis, *Proc. Roy. Soc.* **A115** (1927) 456.
86. Kasper J. S., *Acta Met.* **2** (1954) 456.
87. Laves F., K. Lohberg, and P. Rahlfs, *Nachr. Gese. Wisse. Göttingen*, **1** (1934) 67.
88. Bardos D. I., K. P. Gupta, and P. A. Beck, *Nature* **192** (1961) 744.
89. Waterstrat R. M., *Trans. Met. Soc. AIME* **221** (1961) 687.
90. Boriskina N. G., and I. I. Kornilov, *Russ. J. Inorg. Chem.* **4** (1959) 986.
91. Matthias B. T., V. B. Compton, and E. Corenzwit, *J. Phys. Chem. Solids* **19** (1961) 130.
92. Bucher E., F. Heiniger, and J. Muller, *Helv. Phys. Acta* **34** (1961) 843.
93. Gladyshevskii E. I., P. I. Kripyakevich, M. Yu. Teshyuk, et al., *Soviet Phys. Cryst.* (*English Transl.*) **6** (1961) 207.
94. Blaugher R. D., A. Taylor, and J. K. Hulm, *IBM J. Res. Develop.* **6** (1962) 116.
95. Matthias B. T., T. H. Geballe, and V. B. Compton, *Rev. Mod. Phys.* **35** (1963) 1.
96. Das B. N., and P. A. Beck, *Trans. AIME* **218** (1960) 733.
97. Arnfelt H., and A. Westgren, *Jernkontorets Ann.* **119** (1935) 185.
98. Magnéli A., and A. Westgren *Z. Anorg. Chem.* **238** (1938) 268.
99. Babiche M. M., E. M. Kisljakova, and J. S. Umanskii, *Zh. Tekhn. Fiz. SSSR* **9** (1939) 533.
100. Henglein E., and H. Kohsok, *Rev. Met.* **46** (1949) 569.
101. Ellinger F. H., *Trans. ASM* **30** (1942) 607.
102. Das D. K., S. P. Rideout, and P. A. Beck, *Trans. AIME* **194** (1952) 1071.
103. Brink C., and D. P. Shoemaker, *Acta Cryst.* **8** (1955) 734.
104. Shoemaker, D. P., C. Brink Shoemaker, and F. C. Wilson, *Acta Cryst.* **10** (1957) 1.

105. Komura Y., W. G. Sly, and D. P. Shoemaker, *Acta Cryst.* **13** (1960) 575.
106. Haworth C. H., and W. Hume-Rothery, *Phil. Mag.* **3** (1958) 1013.
107. Jaccarino V., J. H. Wernick, and H. J. Williams, *Bull. Am. Phys. Soc.*, Series 2, **7** (1962) 556.
108. Haszko, S. E., *Trans. Met. Soc. AIME* **221** (1961) 201
109. Storm A. R., and K. E. Benson, *Acta Cryst.* **16** (1963) 701.
110. Matthias B. T., E. Corenzwit, and H. Zacchariasen, *Phys. Rev.* **112** (1958) 89.
111. Wernick J. H., and R. R. Soden, *J. Phys. Chem. Solids.* **25** (1964) 449.
112. Hatt B. A., *Acta Cryst.* **14** (1961) 119.
113. Kripyakevich P. I., V. F. Terekhova, O. S. Zarechnyuk, I. V. Burov, *Soviet Phys. Cryst.* (*English Transl.*) **8,** 2 (1963) p. 203.
114. Darby J. B., Jr., and S. T. Zegler, *J. Phys. Chem. Solids* **23** (1962) 1825.
115. Nevitt M. V., in *Electronic Structure and Alloy Chemistry of the Transition Elements*, P. A. Beck, ed. Interscience, 1963.
116. Jaccarino V., W. E. Blumberg, and J. H. Wernick, *Bull. Am. Phys. Soc.* **6** (1961) 104.
117. Rayne J. A., *Phys. Letters* **7** (1963) 114.
118. Beck A., J.-P. Jan, W. B. Pearson, and I. M. Templeton *Phil. Mag.* **8** (1963) 351.
119. *The Physical Chemistry of Metallic Solutions and Intermetallic Compounds Symposium*, Vol. 1, 1960. Chemical Publishing Company, New York.
120. Raynor G. V., *ibid.*
121. Elliot R. P., AFOSR-TR-58-49, ASTIA Document 154-199.
122. See Nowotny H., in Reference 115 for further discussion of phases related to AlB_2 and $ThSi_2$ structure types.
123. The Al-Al distances are unusually short in Zr_2Al and Hf_2Al having the $CuAl_2$ structure type. Nowotny, H., Reference 115.
124. Wernick J. H., to be published.

chapter 13

Miscellaneous Structures of Fixed Stoichiometry*

M. V. NEVITT

Argonne National Laboratory
Argonne, Illinois

A substantial number of structure types are represented by families of intermetallic compounds, which occur at simple stoichiometric ratios. Most of them show close adherence to their characteristic stoichiometries, a behavior which is in contrast to that of the sigma phase and related phases. The latter have compositions that shift with the periodic-table positions of their components. The compounds of fixed stoichiometry constitute a rather heterogeneous grouping, in which structures and compositions are not amenable to a simple unifying treatment. However, many of the factors that influence the occurrence of intermetallic compounds are brought into focus by a consideration of them.

1. COMPOUNDS AT A_3B

Cr_3Si-Type Phases

The cubic Cr_3Si-type structure, (Strukturbericht type A15), which belongs to the space group O_h^3–$Pm3n$, is the principal occupant of the A_3B composition in systems where the component A is an element of the titanium, the vanadium, or the chromium group, and B is from the manganese, iron, cobalt, nickel, copper, aluminum, silicon, or phosphorous group. Table 1 lists the compounds having this structure that are presently known. In this table and in subsequent ones the compilation of Pearson[2] is used as the primary reference. Original papers are cited for information that is not reported by Pearson or is believed to be more accurate than what is reported. A more recent compilation is embodied in Reference 69.

The structure has also been called the beta-tungsten type and the Cr_3O type, but both of these terms now seem inappropriate. There is no evidence of polymorphism in tungsten, and there is much uncertainty about the composition and the atomic-ordering arrangement in "W_3O", which was reported by Hägg and Schönberg[14] to have the A15 structure and was thought to be the substance that earlier workers had mistakenly identified

* Work performed under the auspices of the U.S. Atomic Energy Commission.

TABLE 1

Cr_3Si-Type Compounds

(Strukturbericht Type *A*15)

A_3B	*a*, Å	R_A/R_B	Ω_A/Ω_B	$\Sigma\Omega$, Å^3	*V*, Å^3	$\frac{\Sigma\Omega - V}{V} \times 100$	Reference
Ti_3Sb	5.217	0.920	0.584	166.3	142.0	17.1	1
Ti_3Ir	5.007	1.077	1.246	134.2	125.5	6.9	2, p. 705
Ti_3Pt	5.033	1.054	1.168	136.1	127.5	6.7	2, p. 825
Ti_3Au	5.096	1.014	1.040	139.8	132.3	5.6	2, p. 444
Ti_3Hg	5.189	0.929	0.754	152.6	139.7	9.2	2, p. 690
V_3Si	4.722	1.020	0.693	123.3	105.3	17.1	2, p. 857
V_3Co	4.681	1.075	1.250	105.5	102.2	3.2	2, p. 527, 3[b]
V_3Ni	4.712	1.080	1.268	105.1	104.5	0.6	2, p. 794, 3[b]
V_3Ga	4.815	0.954	0.708	122.5	112.7	8.7	2, p. 672, 3[b]
V_3Ge	4.769	0.983	0.613	128.5	108.5	18.5	2, p. 680
V_3As	4.75	0.968	0.645	126.3	107.2	17.8	2, p. 405
V_3Rh	4.786	1.001	1.008	110.8	108.3	2.3	2, p. 835, 3[b]
V_3Pd	4.816	0.978	0.942	112.7	111.7	0.9	4
V_3Cd	4.943	0.858	0.643	126.4	120.8	4.6	5
V_3Sn	4.94	0.871	0.661	125.3	120.6	3.9	2, p. 864
V_3Sb	4.92	0.846	0.459	143.7	119.1	20.7	2, p. 846
V_3Ir	4.785	0.992	0.980	111.6	109.6	1.8	2, p. 705
V_3Pt	4.808	0.970	0.919	113.5	111.1	2.1	2, p. 825
V_3Au	4.88	0.933	0.818	117.2	116.2	0.8	2, p. 445
V_3Pb	4.937	0.769	0.458	143.9	120.3	19.6	5
Cr_3Si	4.550	0.972	0.599	112.1	94.2	19.0	2, p. 562
Cr_3Ga	4.645	0.909	0.612	111.2	100.2	11.0	6
Cr_3Ge	4.623	0.936	0.530	117.3	98.8	18.7	2, p. 539
Cr_3Ru	4.683	0.957	0.880	99.3	102.7	−3.3	2, p. 558
Cr_3Rh	4.656	0.953	0.872	99.5	100.9	−1.4	2, p. 558
Cr_3Os	4.677	0.948	0.858	100.0	102.3	−2.2	7
Cr_3Ir	4.668	0.945	0.848	100.3	103.0	−2.6	2, p. 539
Cr_3Pt	4.706	0.924	0.794	102.2	104.2	−1.9	2, p. 556
Zr_4Sn[a]	5.65	1.037	1.108	182.6	180.4	1.2	8
Zr_3Au	5.482	1.111	1.372	173.6	164.8	5.4	9
Zr_3Hg	5.558	1.018	0.995	186.4	171.7	8.6	2, p. 691
Nb_3Al	5.187	1.025	1.083	141.1	139.6	1.1	6
Nb_3Ga	5.171	1.040	0.917	147.1	138.3	6.4	6
Nb_3Ge	5.174	1.072	0.794	153.2	138.5	10.6	9
Nb_3Rh	5.132	1.091	1.306	135.4	133.8	1.2	2, p. 771, 3[b]
Nb_3In	5.303	0.883	0.668	160.1	149.1	7.4	10
Nb_3Sn	5.289	0.950	0.856	149.9	148.0	1.3	2, p. 773, 9[b]
Nb_3Sb	5.26	0.923	0.595	168.3	145.5	15.7	2, p. 772
Nb_3Os	5.136	1.085	1.285	135.9	134.3	1.2	2, p. 770, 3[b]
Nb_3Ir	5.136	1.082	1.270	136.2	135.1	0.8	2, p. 703, 3[b]
Nb_3Pt	5.153	1.058	1.190	138.1	136.8	0.9	2, p. 770
Nb_3Au	5.203	1.018	1.060	141.8	141.4	0.3	2, p. 438
Nb_3Bi	5.320	0.864	0.508	178.7	150.6	18.6	11
Nb_3Pb	5.270	0.839	0.593	168.5	146.4	15.1	5
Mo_3Al	4.950	0.978	0.938	126.6	121.3	4.4	6
Mo_3Si	4.888	1.061	0.777	133.5	116.8	14.3	2, p. 760
Mo_3Ga	4.94	0.992	0.794	132.6	120.6	10.0	2, p. 668
Mo_3Ge	4.933	1.023	0.688	138.7	120.0	15.5	2, p. 677
MoTc[a]	4.934	1.029	1.095	119.1	120.1	−0.8	12
Mo_3Os	4.973	1.035	1.112	121.4	123.0	−1.3	2, p. 754
Mo_3Ir	4.973	1.032	1.100	121.7	123.0	−1.0	2, p. 703
Ta_3Sn	5.276	0.950	0.854	151.6	146.9	3.2	2, p. 862
Ta_3Sb	5.260	0.923	0.593	168.0	145.5	15.5	9
Ta_3Au	5.222	1.017	1.057	141.5	142.4	−0.6	13

[a] Note deviations from A_3B stoichiometry. Values of $\Sigma\Omega$ calculated for $Zr_6(Sn_{1.6}Zr_{0.4})$ and $(Mo_4Tc_2)Tc_2$.
[b] Lattice parameter from this reference.

as beta tungsten. Tungsten and oxygen atoms were believed to occupy the lattice sites randomly. Millner[15] was unable to find the A15 structure in a compound having the formula W_3O, although he did find it in tungsten containing about 0.5 at.% oxygen. He, therefore, suggested that the phase is stabilized in tungsten by a small concentration of oxygen. The occurrence of the structure in metallic films deposited on glass in vacuo[16] seems consistent with this view. There are similar discrepancies in the reports concerning Cr_3O. Schönberg[17] stated that it could be prepared by several techniques and that it has the A15 structure, but Kihlborg[18] suggested that there was no definite proof of its existence, after he had tried unsuccessfully to produce it by one of Schönberg's methods. On the basis of present evidence it seems clear that the intermetallic compounds, having A_3B stoichiometry and an ordered arrangement of the atoms, have as their prototype Cr_3Si, the first compound definitely shown to have the structure.[19]

In the Cr_3Si-type structure the A atoms in positions 2(*a*) form three mutually perpendicular chains parallel to the cubic crystal axes, and the B atoms in 6(*c*) are in body-centered cubic positions. A model of the structure is shown in a review paper by Laves.[20] The coordination shell around the A atom is somewhat unique in the extent to which it deviates from sphericity. The nearest neighbors of an A atom are two A atoms, in the same chain, at a distance $\frac{1}{2}\,a_0$. Four B atoms are situated at a distance $\frac{1}{4}\sqrt{5}\,a_0$, and eight other A atoms are separated from the central A atom by $\frac{1}{4}\sqrt{6}\,a_0$. The latter distances are greater than the nearest-neighbor distance by 12 per cent and 22 per cent, respectively. There are no B–B nearest neighbors. Each B atom is surrounded by twelve equidistant A atoms arranged in a distorted icosahedron.

Considerable attention has been given to the apparent sizes of the atoms in Cr_3Si-type phases. A scheme has been proposed for reproducing the lattice parameters by adding the empirically derived radii of the atoms involved and multiplying by the appropriate geometric factor.[13,21] It appears to depend on the fact that each atomic species has a "size" in the A–B direction, which is not influenced by the identity of its partner. However, since they are derived from A–B distances, the radii cannot be uniquely determined from the measured lattice parameters and the single assumption of A–B contacts.[9] Furthermore, it has been shown that there is an increasing contraction of the A–A distance with a decrease in the size of the B atom.[22] Thus the A atom exhibits a dual behavior as regards its effective size that is clearly inconsistent with a spherical shape. In the A–B direction it appears to have a characteristic size that does not change significantly in the various Cr_3Si-type phases, whereas in the A–A direction it displays an effective size that varies with the size of the other component. Radius is, therefore, not a meaningful term for describing atomic size in this instance.

An alternative approach involves the use of atomic volumes. The concept of atomic volume has been discussed with reference to the alloy phases of the noble metals.[23] Some clarification of the role of atomic size can be derived from Table 1. All of the known Cr_3Si-type phases are tabulated with the ratios of their atomic radii, R_A/R_B and the ratios of the atomic volumes derived from the volumes of the unit cells of the solid elements Ω_A/Ω_B. The values of the ratios have been obtained from a recent compilation.[24] Also given are the volumes of the unit cells of the A_3B phases, V; the sums of the atomic volumes of the six A atoms and the two B atoms which occupy the cell, $\Sigma\Omega$; and the differences between V and $\Sigma\Omega$, expressed as a fraction of V in per cent. Despite the limitations of the atomic radius concept, the occurrence of the phase can be associated more definitively with a critical radius ratio than with a ratio of atomic volumes. The distribution of R_A/R_B peaks sharply at unity, and all but eight of the phases have values that are within 10 per cent of unity. However, the values for the ratios of atomic volumes lie between 0.5 and 1.5 in a distribution without a well-defined peak.

Although the ratio of the atomic volumes is not a sensitive indicator of the extent to which size governs the occurrence of the structure, an interesting relationship can be observed between the volumes of the unit cells and the sums of the volumes of the atoms in the cells. When A is a vanadium-group or chromium-group atom and the B atom is from the iron, cobalt, nickel, or copper group, there is agreement between the V and $\Sigma\Omega$ to within about 3 per cent. It may be concluded that the

Cr_3Si-type structure has about the same space-filling efficiency as the structures of the pure metals and that the atoms of the aforementioned groups have essentially the same effective volumes in Cr_3Si-type phases as they have in the structures of the pure metals. The volumes of the vanadium and the niobium atoms are perhaps slightly smaller in the Cr_3Si-type phases than in their own metallic lattices, whereas for the chromium and the molybdenum atoms a reverse relationship exists between the atomic volumes in the two environments. For the compounds involving titanium and zirconium, $\Sigma\Omega$ is significantly larger than $\bar{V}$. Since the same transition-metal B components are involved, it is logical to attribute the volume discrepancy to the titanium and zirconium atoms, that is, their volumes in the Cr_3Si-type phases appear to be about 8 per cent smaller than Ω_{Ti} and Ω_{Zr}. This trend is probably related to a decreasing atomic compressibility, which characterizes the shift from the titanium to the vanadium to the chromium group.

The other volume discrepancies in Table 1 are observed when the B component is a nontransition metal from the aluminum group (with the exception of aluminum itself), the silicon group, or the phosphorous group. Because these elements do not crystallize in close-packed structures, their values of Ω are disproportionately large, and it is not surprising that this effect shows up in the present analysis. The effective atomic volumes of silicon and germanium appear to be about 40 per cent smaller in Cr_3Si-type phases than in their own structures, whereas the volume of the tin atom is about 5 to 10 per cent smaller.

Cr_3Si-type phases adhere closely to A_3B stoichiometry, with the possible exceptions of MoTc and Zr_4Sn. In neither of the latter phases has the atomic ordering arrangement been determined. In the case of Cr_3Os there is an indication that the phase may exist over a range of compositions on the chromium-deficient side of stoichiometry and that partial disordering may also be a characteristic.

Some of the selectivity involved in the formation of the structure is due to the influence of the relative sizes of the atoms. It has been proposed* that a necessary condition for the occurrence of the phase is that the CN 12 atomic radii of the components do not differ by more than about 15 per cent.[9] However, other factors must also enter, since many combinations having favorable radius ratios do not form the compound. With respect to the A component, in the first long period elements of the fourth, fifth, and sixth columns have a particular affinity for forming the structure, and there is a decreasing tendency for the formation upon shifting the A component to the second and to the third long periods. Tantalum is the only A component from the third long period, if the oxygen-stabilized phase involving tungsten is not considered.

Some data are available for the mutual solubility of various Cr_3Si-type phases. The following pairs of compounds appear to be soluble in all proportions:

V_3Co with V_3Ni[25] and V_3Rh[25]
V_3Si with V_3Ga[25], Cr_3Si, and Mo_3Si[26]
V_3Ir with Nb_3Ir[27]
Cr_3Si with Mo_3Si[26]
Nb_3Sn with Mo_3Si, Mo_3Ge, and Nb_3Ge[28]

Complete miscibility has also been reported in the ternary system V_3Sn–Nb_3Sn–Ta_3Sn.[29] Several systems have been found to have partial solubility. Iron can replace 0.7 or more of the nickel in V_3Ni, but is able to replace virtually none of the cobalt in V_3Co.[25] Partial miscibility in the systems V_3Co–V_3Ir and V_3Co–V_3Si has also been observed.[25]

2. COMPOUNDS AT A_2B

2.1 Ti_2Ni-Type Phases

Among the crystal structures that occur at A_2B stoichiometry, the Ti_2Ni-type or η-carbide type (Strukturbericht type $E9_3$) is the one found most frequently in binary systems, in which a titanium-group element is the A component and the B component is from the cobalt or the nickel group. Table 2 lists the binary phases.

The structure of Ti_2Ni is that of a large face-centered cubic unit cell containing ninety-six metal atoms; the space group is O_h^7–$Fd3m$. The atomic ordering in Ti_2Ni has been determined by X-ray diffraction[35] and by neutron diffraction.[31] The nickel atoms are in positions 32(e), for which there are twelve nearest-neighbors, three nickel and nine titanium,

* Since this proposal was made, the phase V_3Pb has been reported; its components have radii differing by 24 per cent.

TABLE 2

Binary Ti_2Ni-Type Phases

(Strukturbericht Type $E9_3$)

A_2B	a, Å	R_A/R_B	Ω_A/Ω_B	$\Sigma\Omega$, Å^3	V, Å^3	$\frac{\Sigma\Omega - V}{V} \times 100$	Reference
Sc_2Ni	12.120	1.317	2.285	1950.0	1780.4	9.5	30
Sc_2Pd	12.427	1.193	1.698	2071.1	1919.1	7.9	30
Ti_2Co	11.30	1.168	1.589	1631.3	1442.9	13.0	2, p. 526
Ti_2Ni	11.319	1.173	1.613	1626.0	1451.0	12.1	31
Nb_3Fe_2	11.262	1.152	1.528	1487.5	1428.4	4.1	32
HfMn	11.812	1.212	1.586	1747.4	1648.0	6.0	33
Hf_2Fe	12.055	1.240	1.897	1805.6	1751.9	3.1	33
Hf_2Co	12.104	1.262	2.010	1784.3	1773.3	0.6	33
Hf_2Rh	12.326	1.175	1.622	1869.3	1872.7	−0.2	34
Hf_2Ir	12.352	1.164	1.577	1882.0	1884.6	−0.1	34
Hf_2Pt	12.461	1.139	1.478	1912.3	1934.9	−1.2	34

arranged in a distorted icosahedron. Titanium atoms are assigned to positions 16(*c*), which are CN 12 sites having an equal number of titanium and nickel nearest-neighbors, and to positions 48(*f*), the nearest-neighbors to which are ten titanium and four nickel atoms.

The large face-centered cubic cell of Ti_2Ni can be visualized as composed of eight cubic subcells having the two alternating patterns that are shown in Figure 1.[31] The tetrahedra and octahedra represent tetrahedral and octahedral clusters of nickel and titanium atoms, respectively, which have been collapsed about their centers so that the internal structure of the subcells is revealed. In Ti_2Ni the oxygen positions in subcell (*b*) are empty.

The relative sizes of the atoms apparently affect the stability of the structure since, for the binary phases, the average CN 12 radius ratio, R_A/R_B, is about 1.20, and individual values do not deviate from the average by more than 10 per cent; the atomic volume ratio is about 1.72 ± 30 per cent. However, certain aspects of the selectivity in the formation of the phase do not seem to be attributable to an effect of atomic size. Binary phases involving zirconium are completely absent although R_{Zr}/R_B is within the favorable range

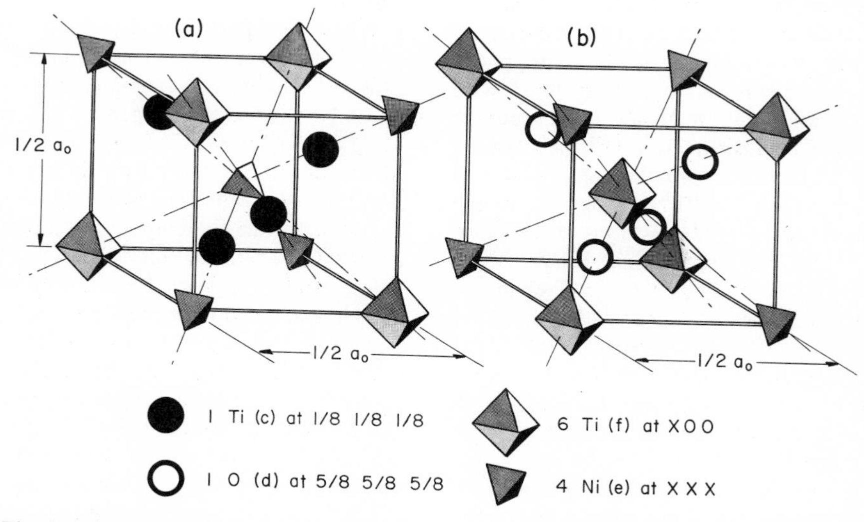

Fig. 1. Schematic diagram showing crystallographic subcells found in Ti_2Ni and Ti_4Ni_2O. (After Mueller and Knott.[28])

for many of the combinations. There are also restrictions with respect to the B components, which do not have a basis in atomic size. The frequency of occurrence of the phase varies with the group number of the B component, rising to a maximum at the cobalt group. The present author has suggested that this is an electron concentration effect, basing his conclusion on the occurrence of ternary phases in the systems Hf–Ni–Ru, Hf–Ni–Re, and Hf–Ni–Os.[33] These three ternary systems pair hafnium with two B components, one of which is nickel lying to the right of the cobalt group and the other is a second- or third-long-period element lying to the left of the cobalt group. None of the limiting binary systems has a Ti_2Ni-type phase. The occurrence of the phases in the ternary systems suggests that a necessary condition is the occurrence of an electron-atom ratio roughly corresponding to that of Hf_2Co.

With several notable exceptions, the phases have unit cell volumes which differ by less than 10 per cent from the sums of the atomic volumes of the component atoms. Scandium and titanium atoms appear to have smaller volumes here than they have in their elemental forms; this situation was noted previously for titanium in Cr_3Si-type phases.

An important characteristic of Ti_2Ni and many other compounds listed in Table 2 is the ability to dissolve oxygen. Oxygen atoms are accommodated in the structure by the filling of positions that are interstitial with respect to the metal atoms, as has been demonstrated by density measurements, by the shapes of the single-phase fields in ternary systems with oxygen,[36] and by the X-ray[37] and neutron diffraction[31] results. In Ti_4Ni_2O[31] the titanium and nickel atoms are in the same positions that they occupy in Ti_2Ni, whereas oxygen atoms occupy the 16(*d*) positions of the space group, which are empty in the binary phase (see Fig. 1). In Ti_4Cu_2O[31] the copper atoms substitute for the nickel atoms in the 32(*e*) positions. In Ti_2Ni, and in virtually all of the other isostructural phases in which oxygen solubilities have been studied, the upper limit of the oxygen concentration is about 14 at.%, which coincides with the filling of the sixteen holes in the unit cell containing ninety-six metal atoms. In Fig. 2 are shown the Ti_2Ni-type phase fields in the Ti–Ni–O, the Ti–Co–O, the Ti–Fe–O, and the Ti–Mn–O systems; they

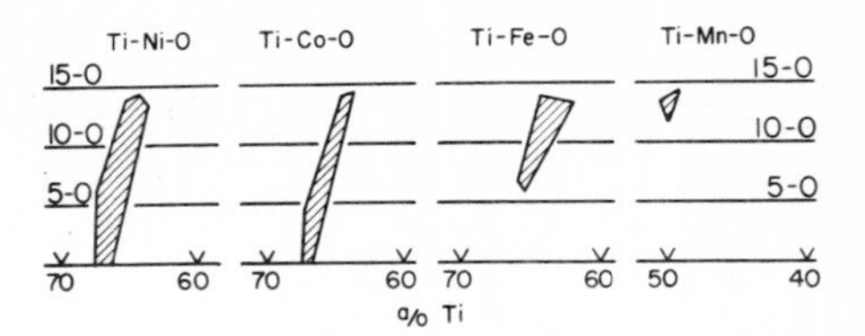

Fig. 2. Ranges of concentration of oxygen in Ti_2Ni-type phases in the ternary systems Ti–Ni–O, Ti–Co–O, Ti–Fe–O, and Ti–Mn–O.

are characteristically narrow with respect to the ranges of concentration of the metallic components. The figure illustrates another aspect of the alloy chemistry of these phases, namely, that oxygen has the ability to stabilize ternary phases. The overall effect of oxygen is to enlarge by more than a factor of three the total number of combinations that form Ti_2Ni-type phases. Figure 2 also shows a systematic trend in the minimum oxygen concentration. In the nickel and cobalt systems the phase fields extend to zero oxygen concentration, while in the iron and manganese systems the minimum oxygen concentrations are respectively 6 and 12 at.%. It has been proposed that this is another manifestation of an electron concentration effect, with oxygen assuming the role of an electron acceptor.[36] In Ti_4Ni_2O and Ti_4Cu_2O the apparent radius of the oxygen atom in the direction of Ti–O contacts is 0.65 to 0.67 Å, in close agreement with Pauling's[38] single-bond covalent radius, 0.66 Å.

When the B partner is chromium or manganese, there is a shift in stoichiometry from A_2B to AB, as, for example, in the binary phase HfMn and the oxygen-stabilized phases Ti_3Cr_3O, Ti_3Mn_3O, Zr_3Cr_3O, and Zr_3Mn_3O. Minor deviations from stoichiometry, in the direction of a deficiency in the B component, are observed in the compounds occurring in Zr–Rh–O, Zr–Ir–O, and Zr–Pt–O. Density measurements indicate that the nonstoichiometry results from a replacement of the zirconium atoms by the atoms of the B component, rather than by the occurrence of vacant lattice sites.[33] The isostructural compound Nb_3Fe_2 also shows a deviation from normal stoichiometry and is the only one of the binary compounds whose A component is not an element of the titanium group or scandium. There is a need for additional studies of atomic ordering in these phases, so

TABLE 3

Ternary Ti_2Ni-Type Phases
(Strukturbericht Type $E9_3$)

A. Oxygen-Stabilized Phases

Comp. at Max. O Conc.	R_A/R_B	Reference	Comp. at Max. O Conc.	R_A/R_B	Reference
Ti_3Cr_3O	1.140	39, 40	$Zr_{4.2}Ir_{1.8}O$	1.180	33
Ti_3Mn_3O	1.121	36	$Zr_{4.4}Pt_{1.6}O$	1.155	33
Ti_4Fe_2O	1.148	2, p. 662	$Nb_3Zn_3O_{0.4}$		43
Ti_4Cu_2O	1.144	2, p. 618	$Mo_3Mn_3O^a$	1.074	44
Ti_4Rh_2O	1.087	33	$Mo_3Fe_3O^a$	1.099	44
Ti_4Ir_2O	1.077	33	$Mo_3Co_3O^a$	1.118	44
Ti_4Pt_2O	1.054	33	$Mo_3Ni_3O^a$	1.124	44
$Ti_4Pd_2O^a$	1.062	41	Hf_4Ni_2O	1.268	33
Zr_3V_3O	1.190	42	$Hf_4Pd_2O^a$	1.148	34
Zr_3Cr_3O	1.250	42	$Ta_3Mn_3O^a$	1.125	45
Zr_3Mn_3O	1.228	42	$W_3Mn_3O^a$	1.080	44
Zr_6Fe_3O	1.258	33	$W_3Fe_3O^a$	1.105	44
$Zr_{4.2}Rh_{1.8}O$	1.191	33	$W_3Co_3O^a$	1.125	44
Zr_4Pd_2O	1.164	33	$W_3Ni_3O^a$	1.130	44

B. Nitrogen-Stabilized Phases

A_2BN_x	R_A/R_B	Reference
Ti_2ZnN_x	1.049	43
Zr_2ZnN_x	1.149	43
Hf_2ZnN_x	1.133	43

C. Ternary Phases Without Oxygen or Nitrogen

A_2B	R_A/R_B	Reference
$Hf_2Ni_{0.67}Ru_{0.33}$	1.237	33
$Hf_2Ni_{0.67}Re_{0.33}$	1.226	33
$Hf_2Ni_{0.67}Os_{0.33}$	1.232	33

[a] Composition not firmly established.

that the significance of deviations from the normal stoichiometry can be clarified.

Table 3 lists the oxygen- and nitrogen-stabilized isostructural phases and other ternary phases. Except as noted, the compositions given for the oxygen-stabilized phases correspond to the maximum oxygen concentration observed experimentally.

The cubic η_1- and η_2-carbides constitute another group of isostructural compounds, having the ideal formulas A_3B_3C and A_4B_2C, respectively. There are ninety-six metal atoms in the unit cell, with an additional complement of sixteen carbon atoms. A is zirconium, niobium, molybdenum, or tungsten, and B is vanadium, chromium, manganese, iron, cobalt, or nickel. Tantalum may also serve as a component in a group of "triple" carbides, in which there is a composite B component involving chromium or vanadium with iron, cobalt, nickel, or copper. The atomic ordering scheme reported for η_2-carbide W_4Co_2C[46] is equivalent to the arrangement in Ti_2Ni.[43] There is a need for detailed information on the composition boundaries of the carbide phase fields. It has been suggested that a favorable electron-atom ratio plays a role in the stability of the carbides.[42] Table 4 gives the carbides

TABLE 4

Carbides of the η_1 and η_2 Types[2]

(Strukturbericht Type $E9_3$)

η_1 Carbides		η_2 Carbides
Zr_3V_3C	$Ta_3(Cr, Fe)_3C$	$(Ti, Nb)_4Ni_2C$
Nb_3Cr_3C	$Ta_3(Cr, Co)_3C$	$(Ti, Ta)_4Co_2C$
$Nb_3(V, Ni)_3C$	$Ta_3(Cr, Ni)_3C$	$(Ti, Ta)_4Ni_2C$
Mo_3Mn_3C	$Ta_3(Cr, Cu)_3C$	$(V, W)_4Ni_2C$
Mo_3Fe_3C	W_3Mn_3C	Mo_4Fe_2C
Mo_3Ni_3C	W_3Fe_3C to W_2Fe_4C	Mo_4Co_2C
$Ta_3(V, Fe)_3C$	W_3Co_3C to W_2Co_4C	Mo_4Ni_2C
$Ta_3(V, Co)_3C$	W_6Co_6C	W_4Co_2C
$Ta_3(V, Ni)_3C$	W_3Ni_3C	

listed by Pearson.[2] Lattice parameters vary from 10.65 to 12.12 Å for the η_1 carbides and from 11.21 to 11.58 Å for the η_2 carbides.

2.2 $MoSi_2$-Type Phases

Another family of compounds at the same stoichiometry has the $C11_b$ structure, whose prototype is $MoSi_2$. The space group is D_{4h}^{17}–I/mmm. Over one-half of the compounds involve an element of the titanium group or a lanthanide as the A component, and palladium or a member of the copper group as the B component.

The compounds can be classified in two groups on the basis of their axial ratios. One group is characterized by c/a values lying between 2.29 and 2.60. In $HfAu_2$, a typical member of this group, a hafnium atom lying in the plane $Z \sim \frac{1}{2}$ has ten gold next-neighbors, eight at an identical distance in the planes $Z \sim \frac{1}{3}$ and $Z \sim \frac{2}{3}$, and two at essentially the same distance along the c-direction in the planes $Z \sim \frac{1}{6}$ and $Z \sim \frac{5}{6}$. The coordination shell around the gold atom has the same form, but five atoms are of like kind and five are of unlike. In the other group, for which the axial ratio varies between 3.188 and 4.684, only the atoms lying in the planes $Z \sim \frac{1}{3}$ and $Z \sim \frac{2}{3}$ are equidistant from the central atom. The distance to the two atoms in the c-direction is 25 to 30 per cent greater, and four more atoms in the a-direction are at a distance about 10 per cent greater. Figure 3 shows the structure of Ti_2Cu, a member of the latter group. Only the compounds Cr_2Al and U_2Mo fail to fit this classification; they have axial ratios that lie midway between the ranges of the two groups.

All of the compounds are listed in Table 5. In the systems Mn–Au, Zr–Pd, Zr–Au, and Hf–Au the structure is known to occur at both A_2B and AB_2.

From the geometry of the structure and the assumption that there are simultaneous contacts between spherical atoms in the A–A and A–B directions it may be determined that the ideal radius ratio for $MoSi_2$-type phases is $r_A/r_B = 1$.[20] For the most part the values of the ratios of the CN 12 atomic radii R_A/R_B lie above unity. It has been shown that in forming the compounds the A and B atoms make cooperative changes in their apparent radii as large as 10 per cent in the nearest-neighbor directions in order to fulfill the geometric requirements.[22]

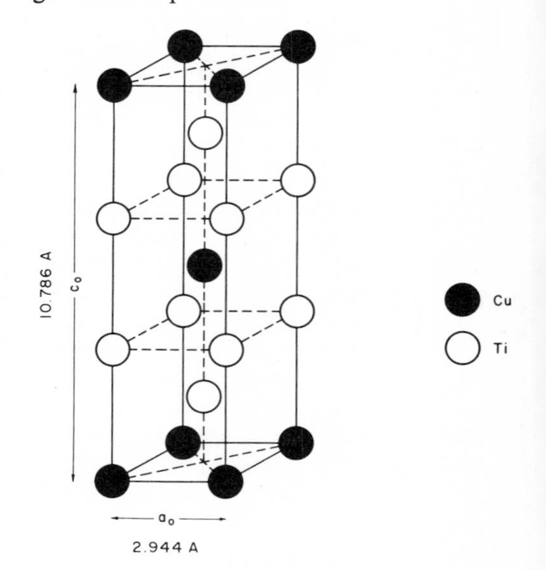

Fig. 3. The crystal structure of Ti_2Cu.

TABLE 5

$MoSi_2$-Type Phases

(Strukturbericht Type $C11_b$)

Compound (A_2B or AB_2)	a, Å	c, Å	c/a	R_A/R_B	Ω_A/Ω_B	$\Sigma\Omega$, A^3	V, A^3	$\frac{\Sigma\Omega - V}{V} \times 100$	Reference
$MgHg_2$	3.838	8.799	2.293	1.018	0.994	140.04	129.6	8.0	2, p. 685
$ScAg_2$	3.519	8.992	2.535	1.136	1.466	118.19	111.4	6.1	48
$ScAu_2$	3.508	8.725	2.487	1.138	1.474	117.86	107.4	9.8	48
Ti_2Cu	2.944	10.786	3.664	1.144	1.495	94.19	93.5	0.8	31
Ti_2Zn	3.04	10.70	3.52	1.05	1.160	102.99	98.9	4.1	49a
Ti_2Pd	3.090	10.054	3.254	1.062	1.198	100.03	96.0	4.2	50
Ti_2Cd	2.865	13.42	4.684	0.932	0.818	113.74	110.2	3.2	51
$TiAu_2$	3.419	8.514	2.490	1.014	1.040	103.15	99.5	3.6	52
Cr_2Al	3.004	8.647	2.878	0.895	0.723	81.21	78.0	4.1	2, p. 323
Mn_2Au	3.34	8.58	2.57	0.904	0.830	90.2	95.7	−5.2	53
$MnAu_2$	3.37	8.78	2.60	0.904	0.830	96.0	99.7	−3.7	54
YAg_2	3.701	9.201	2.486	1.246	1.936	134.19	126.0	6.5	48
YAu_2	3.691	8.959	2.427	1.249	1.945	133.86	122.0	9.7	48
Zr_2Cu	3.220	11.183	3.473	1.254	1.972	116.70	115.9	0.6	50
Zr_2Zn	3.30	11.26	3.41	1.15	1.530	123.50	122.6	0.7	49b
Zr_2Pd	3.306	10.894	3.295	1.164	1.581	122.54	119.1	2.9	50[a], 53
$ZrPd_2$	3.426	8.644	2.523	1.164	1.581	105.44	101.7	3.6	49c, 20[a]
Zr_2Ag	3.246	12.004	3.698	1.109	1.365	127.19	126.5	0.5	50
Zr_2Au	3.28	11.6	3.536	1.111	1.372	127.0	124.8	1.8	50
$ZrAu_2$	3.544	8.732	2.464	1.111	1.372	114.41	109.7	4.3	22
$MoSi_2$	3.203	7.887	2.462	1.061	0.777	111.29	80.9	37.5	2, p. 760
$MoGe_2$	3.313	8.195	2.494	1.023	0.688	121.71	89.9	35.3	2, p. 676
Tc_2Al	2.98	9.50	3.19	0.950	0.856	90.0	84.4	6.7	55
$GdAg_2$	3.728	9.296	2.494	1.247	1.942	134.39	129.2	4.0	56
$GdAu_2$	3.732	9.014	2.415	1.250	1.951	134.06	125.5	6.8	56
$TbAg_2$	3.710	9.247	2.492	1.233	1.875	132.14	127.3	3.8	48
$TbAu_2$	3.707	8.987	2.424	1.236	1.884	131.80	123.5	6.7	48
$DyAg_2$	3.6957	9.213	2.493	1.227	1.849	131.24	125.8	4.2	56
$DyAu_2$	3.6940	8.956	2.424	1.230	1.858	130.91	122.2	7.1	56
$HoAg_2$	3.681	9.181	2.494	1.222	1.826	130.44	124.4	4.8	48
$HoAu_2$	3.677	8.940	2.431	1.225	1.835	130.11	120.9	7.6	48
$ErAg_2$	3.668	9.159	2.497	1.216	1.797	129.48	123.2	5.1	48
$ErAu_2$	3.662	8.920	2.436	1.218	1.806	129.15	119.6	8.0	48
$TmAg_2$	3.642	9.140	2.503	1.208	1.765	128.38	121.2	5.9	48
$TmAu_2$	3.648	8.900	2.440	1.211	1.774	128.05	118.4	8.1	48
$LuAg_2$	3.628	9.113	2.512	1.200	1.730	127.19	119.9	6.0	48
$LuAu_2$	3.623	8.875	2.450	1.202	1.739	126.86	116.5	8.9	48
Hf_2Cu	3.170	11.133	3.512	1.236	1.892	112.91	111.9	0.9	50
Hf_2Zn	3.25	11.21	3.45	1.13	1.468	119.72	118.4	1.1	49a
Hf_2Pd	3.251	11.061	3.402	1.148	1.516	118.76	116.9	1.6	50
$HfPd_2$	3.399	8.658	2.547	1.148	1.516	103.55	100.0	3.5	49c, 22[a]
Hf_2Cd	3.27	11.88	3.63	1.008	1.035	132.47	127.0	4.3	49b
Hf_2Au	3.231	11.606	3.592	1.096	1.316	123.24	121.2	1.7	50
$HfAu_2$	3.527	8.663	2.456	1.096	1.316	112.51	107.8	4.4	6, 22[a]
WSi_2	3.211	7.868	2.450	1.068	0.791	111.85	81.1	37.9	2, p. 857
$ReSi_2$	3.129	7.674	2.452	1.042	0.734	109.56	75.1	45.8	2, p. 832
U_2Mo	3.427	9.834	2.870	1.114	1.403	118.52	115.5	2.6	57

[a] Lattice parameters from this reference.

TABLE 6

$CuAl_2$-Type Phases

(Strukturbericht Type C16)

A_2B	a, Å	c, Å	c/a	R_A/R_B	Ω_A/Ω_B	$\Sigma\Omega$, A^3	V, A^3	$\frac{\Sigma\Omega - V}{V} \times 100$	Reference
Na_2Au	7.417	5.522	0.7445	1.325	2.328	383.79	303.8	26.3	2, p. 438
Al_2Cu	6.066	4.874	0.8035	1.120	1.406	180.01	179.3	0.4	2, p. 329
Sc_2Co	6.374	5.616	0.8811	1.311	2.251	244.42	228.2	7.1	30
Ti_2B	6.11	4.56	0.7463	1.492	3.271	162.75	170.2	−4.4	20
Cr_2B	5.18	4.32	0.8340	1.308	2.225	117.59	115.9	1.4	2, p. 898
Mn_2B	5.148	4.208	0.8174	1.331	2.609	122.51	111.5	9.8	2, p. 900
Fe_2B	5.109	4.248	0.8315	1.300	2.182	115.74	110.9	4.4	2, p. 899
Co_2B	5.016	4.220	0.8413	1.278	2.058	110.42	106.2	4.0	2, p. 897
Ni_2B	4.990	4.244	0.8505	1.271	2.028	109.10	105.7	3.2	2, p. 903
Ge_2Fe	5.911	4.951	0.8376	1.075	1.924	228.22	173.0	31.9	2, p. 628
Zr_2Al[a]	—	—	—	1.119	1.402	—	—	—	58
Zr_2Si	6.6120	5.2943	0.8007	1.215	1.162	289.60	231.5	25.1	2, p. 858
Zr_2Co	6.367	5.513	0.8659	1.280	2.096	230.61	223.5	3.2	59
Zr_2Ni	6.490	5.270	0.8120	1.286	2.127	229.94	222.0	3.6	58[b], 60, 61
Zr_2Ga	6.712	5.443	0.8109	1.135	1.187	264.60	245.2	7.9	62
Mo_2B	5.543	4.735	0.8542	1.429	2.886	146.15	145.5	0.5	2, p. 901
In_2Ag	6.883	5.615	0.8158	1.151	1.532	277.09	266.0	4.2	2, p. 289
Sn_2Mn	6.660	5.445	0.8176	1.185	1.492	224.30	241.5	−7.1	2, p. 747
Sn_2Fe	6.533	5.323	0.8148	1.213	1.784	215.07	227.2	−5.3	2, p. 659
Sn_2Co	6.361	5.452	0.8571	1.234	1.891	212.41	220.6	−3.7	2, p. 523
Sn_2Rh	6.411	5.654	0.8819	1.149	1.526	223.04	232.4	−4.0	2, p. 834
Sb_2Ti	6.66	5.81	0.8724	1.088	1.713	312.41	257.7	21.2	2, p. 845
Sb_2V	6.55	5.63	0.8595	1.181	2.178	297.34	241.5	23.1	2, p. 846
Hf_2Al	—	—	—	1.103	1.345	—	—	—	58
Hf_2Si	6.48	5.21	0.808	1.198	1.114	258.8	218.8	18.3	63
Hf_2Ni	6.74	5.58	0.8279	1.268	2.041	222.37	253.5	−12.3	61
Hf_2Ga	6.686	5.295	0.7920	1.120	1.139	257.04	236.7	8.6	62
Hf_2Ge	6.58	5.37	0.815	1.154	0.986	269.2	232.5	15.8	64
Ta_2B	5.785	4.867	0.8413	1.497	3.323	165.00	162.9	1.3	2, p. 905
Ta_2Si	6.157	5.04	0.818	1.112	0.895	223.57	191.1	17.0	2, p. 853
W_2B	5.564	4.740	0.8519	1.437	2.938	148.40	146.7	1.1	2, p. 910
Pb_2Rh	6.664	5.865	0.8801	1.301	2.204	297.67	260.5	14.3	2, p. 806
Pb_2Pd	6.849	5.833	0.8516	1.272	2.060	301.52	273.6	10.2	2, p. 804
Pb_2Au	7.325	5.655	0.7720	1.214	1.788	310.49	303.4	2.3	2, p. 439
Th_2Al	7.62	5.86	0.769	1.256	1.979	329.21	340.3	−3.2	2, p. 384
Th_2Cu	7.28	5.74	0.788	1.407	2.783	310.02	304.2	1.9	2, p. 616
Th_2Zn	7.60	5.64	0.742	1.290	2.160	323.63	325.8	−0.6	2, p. 871
Th_2Ag	7.56	5.84	0.772	1.244	1.927	331.00	333.8	−0.8	2, p. 306
Th_2Au	7.42	5.95	0.802	1.247	1.936	330.67	327.6	0.9	2, p. 443
Th_2In	7.787	6.113	0.785	1.081	1.258	367.2	370.7	−0.9	65

[a] Zr_2Al also occurs with the Ni_2In-type ($B8_2$) structure.[66]

[b] Lattice parameter from this reference.

It is a curious fact that many $MoSi_2$-type compounds have values of R_A/R_B, which are closer to the ideal ratio for Laves phases (1.225) then for the $MoSi_2$-type structure. This anomaly, which is part of the general pattern that appears to exclude the elements of the copper group from forming Laves phases with A components from the long periods, has no obvious explanation at the present time, but it indicates clearly that atomic-size effects are not solely responsible for the stability of these intermediate phases. The only indication of a restrictive atomic-size effect in $MoSi_2$-type phases is the fact that there are apparently upper limits for the ranges of R_A/R_B and Ω_A/Ω_B, 1.25 and 1.97, respectively.

2.3 $CuAl_2$-Type Phases

A third structure type occurring at A_2B is the tetragonal $CuAl_2$ type (Strukturbericht type C16), belonging to the space group D_{4h}^{18}–I_4/mcm. It often forms in preference to the two other structure types, and this may result from its less restrictive atomic-size requirements. It has been concluded that the $CuAl_2$-type structure imposes only the limitation that the relative sizes of the component atoms be such as to produce only A–B contacts.[20] When R_A/R_B exceeds the critical limits for the Ti_2Ni type or the $MoSi_2$ type, as, for example, in Zr_2Co, Zr_2Ni, Hf_2Ni, Th_2Cu, Th_2Ag, and Th_2Au, the $CuAl_2$-type structure is favored. The phases having this structure are presented in Table 6.

3. COMPOUNDS AT AB_{13}

3.1 The $NaZn_{13}$-Type Phases

The structure of $NaZn_{13}$ (Strukturbericht type $D2_3$) is face-centered cubic with 112 atoms per unit cell. The space group is O_h^6–$Fm3c$.

TABLE 7

$NaZn_{13}$-Type Phases

(Strukturbericht Type $D2_3$)

AB_{13}	a, Å	R_A/R_B	Ω_A/Ω_B	$\Sigma\Omega$, A^3	V, A^3	$\frac{\Sigma\Omega - V}{V} \times 100$	Reference
$NaZn_{13}$	12.2836	1.371	2.597	1902.5	1853.4	2.6	2, p. 768
$MgBe_{13}$	10.166	1.420	2.869	1028.4	1050.6	−2.1	2, p. 455
KZn_{13}	12.360	1.704	4.979	2192.3	1888.2	16.1	2, p. 708
KCd_{13}	13.803	1.515	3.508	2850.0	2633.8	8.2	2, p. 483
$CaBe_{13}$	10.312	1.750	5.366	1190.3	1096.6	8.5	2, p. 452
$CaZn_{13}$	12.15	1.416	2.859	1934.4	1794	7.8	2, p. 477
$ScBe_{13}$	10.10	1.456	3.086	1042.5	1030	1.2	67
$RbCd_{13}$	13.91	1.624	4.313	2988.9	2691	11.0	2, p. 490
$SrZn_{13}$	12.23	1.543	3.704	2037.2	1829	11.4	2, p. 866
YBe_{13}	10.2408	1.597	4.074	1106.5	1074.0	3.0	68
$ZrBe_{13}$	10.047	1.420	2.873	1028.7	1014.2	1.4	2, p. 459
$CsCd_{13}$	13.92	1.742	5.322	3163.1	2697	17.3	2, p. 479
$LaBe_{13}$	10.44	1.664	4.619	1141.8	1138	0.33	70
$LaZn_{13}$	12.079[a]	1.346	2.461	1780.7	1762.4	1.0	71
$CeBe_{13}$	10.375	1.520	3.489	1068.7	1116.8	−4.3	2, p. 452
$PrBe_{13}$		1.621	4.266				69
$NdBe_{13}$		1.614	4.219				69
$SmBe_{13}$	10.28	1.598	4.088	1107.4	1086	2.0	73
$EuBe_{13}$	10.288	1.595	4.102	1108.3	1088.9	1.8	73
$YbBe_{13}$	10.239	1.543	3.672	1080.4	1073.4	0.65	73
$HfBe_{13}$	10.010	1.401	2.756	1021.1	1003.0	1.8	72
$ThBe_{13}$	10.395	1.594	4.055	1105.3	1123.2	−1.6	2, p. 458
UBe_{13}	10.256	1.383	2.696	1017.3	1078.8	−5.7	2, p. 458
$NpBe_{13}$	10.266	1.378	2.686	1016.6	1081.9	−6.0	2, p. 456
$PuBe_{13}$	10.282	1.401	2.471	1002.7	1087.0	−7.8	2, p. 457
$AmBe_{13}$	10.283	1.605	4.168	1112.6	1087.3	2.3	2, p. 392

[a] Lattice parameter is for La-rich phase. Same workers report a_0 = 12.096 Å for Zn-rich phase.

The compounds having this structure, listed in Table 7, involve a divalent B component, beryllium from group IIA or zinc or cadmium from group IIB.

The A atoms are in positions 8(*a*) surrounded by a nearly regular snub cube of twenty-four B_I atoms, the latter occupying positions 96(*i*). Atoms B_{II} are in positions 8(*b*), surrounded by a nearly regular icosahedron of B_I atoms.

It may be inferred from the distribution of R_A/R_B given in Table 7, and the fact that compounds such as $NbBe_{13}$ ($R_A/R_B = 1.301$), $TaBe_{13}$ ($R_A/R_B = 1.300$), $CaBe_{13}$ ($R_A/R_B = 1.750$), and $RbZn_{13}$ ($R_A/R_B = 1.826$) are absent, that radius ratios deviating by more than about 15 per cent from the mean value, 1.542, are unfavorable for the occurrence of the structure. On this basis it seems likely that other members of the lanthanide group will also form the compound with beryllium. On the other hand, there is small probability for an extensive series involving the lanthanides with zinc or cadmium.

REFERENCES

1. Kjekshus A., *Scientific Paper No.* 418, 10th Inter-Scandinavian Chem. Conf., Stockholm, 1959.
2. Pearson W. B., *Handbook of Lattice Spacings and Structures of Metals and Alloys.* Pergamon Press, New York, 1958.
3. Zegler S. T., and J. W. Downey, *Trans. AIME* **227** (1963) 1407. Also Zegler S. T., *Phys. Rev.*, in press.
4. Köster W., and W. Haehl, *Z. Metallk.* **49** (1958) 647.
5. Holleck H., H. Nowotny, and F. Benesovsky, *Mh. Chem.* **94** (1963) 473.
6. Wood E. A., V. B. Compton, B. T. Matthias, and E. Corenzwit, *Acta Cryst.* **11** (1958) 604.
7. Waterstrat R. M., and J. S. Kasper, *Trans. AIME* **209** (1957) 872.
8. Schubert K., T. R. Anantharaman, H. O. K. Ata, H. G. Meissner, N. Potzschke, W. Rossteutscher, and E. Stolz, *Naturwissenschaften*, **47** (1960) 512.
9. Nevitt M. V., *Trans. AIME* **212** (1958) 349.
10. Branus M. D., T. B. Reed, H. C. Gatos, M. C. Lavine, and J. A. Kafalas, *J. Phys. Chem. Solids* **23** (1962) 971.
11. Killpatrick D. H., *J. Metals* **16** (1964) 98.
12. Darby J. B., Jr., D. J. Lam, and J. W. Downey, *J. Less-Common Metals* **4** (1962) 558.
13. Geller S., *Acta Cryst.* **9** (1956) 885.
14. Hägg G., and N. Schönberg, *Acta Cryst.* **7** (1954) 351.
15. Millner T., *J. Inorg. Chem. U.S.S.R.* **3** (1958) 946.
16. Moss R. L., and I. Woodward, *Acta Cryst.* **12** (1959) 255.
17. Schönberg N., *Acta Chem. Scand.* **8** (1954) 221.
18. Kihlborg L., *Acta Chem. Scand.* **16** (1962) 2458.
19. Borén B., *Arkiv Kemi Mineral. Geol.* **11A** (1933) No. 10.
20. Laves F., in *Theory of Alloy Phases*, ASM Cleveland, 1956.
21. Geller S., *Acta Cryst.* **10** (1957) 380.
22. Nevitt M. V., in *Electronic Structure and Alloy Chemistry of the Transition Elements.* Wiley-Interscience, New York, 1963.
23. Massalski T. B., and H. W. King, *Progr. Mater. Sci.* **10** Pergamon Press, New York (1962).
24. Teatum E., K. Gschneidner, Jr., and J. Waber, *Los Alamos Scientific Laboratory Report No. LA*-2345 1960.
25. Zegler S. T., and J. W. Downey, *Trans. AIME* **227** (1963) 1407.
26. Kotel'nikov R. B., *J. Inorg. Chem. U.S.S.R.* **3** (1958) 841.
27. Nevitt M. V., unpublished work.
28. Holleck H., F. Benesovsky, and H. Nowotny, *Mh. Chem.* **93** (1962) 996.
29. Gody G. D., J. J. Hanak, G. T. McConville, and F. D. Rosi, in *Proc. VIIth Int. Conf. on Low Temp. Phys.* Univ. Toronto Press, Toronto, 1961.
30. Aldred A. T., *Trans. AIME* **224** (1962) 1082.
31. Mueller M. H., and H. W. Knott, *Trans. AIME* **227** (1963) 674.
32. Goldschmidt H. J., *Research* **10** (1957) 289.
33. Nevitt M. V., J. W. Downey, and R. A. Morris, *Trans. AIME* **218** (1960) 1019.
34. Nevitt M. V., and L. A. Schwartz, *Trans. AIME* **212** (1958) 700.
35. Yurko G. A., J. W. Barton, and J. G. Parr, *Acta Cryst.* **12** (1959) 909; *Acta Cryst.* **15** (1962) 1309.
36. Nevitt M. V., *Trans. AIME* **218** (1960) 327.

37. Karlsson N., *Nature* **168** (1951) 588.
38. Pauling L., *The Nature of the Chemical Bond.* Cornell Univ. Press, Ithaca, 1960.
39. Rostoker W., *Trans. AIME* **203** (1955) 113.
40. Wang C. C., and N. J. Grant, *Trans. AIME* **200** (1954) 200.
41. McQuillan A. D., and M. K. McQuillan, *Titanium.* Butterworths, London, 1956.
42. Kuo K., *Acta Met.* **1** (1953) 301 (reporting unpublished results of N. Karlsson).
43. Stadelmaier H. H., and R. A. Meissner, private communication.
44. Schönberg N., *Acta Chem. Scand.* **8** (1954) 932.
45. Schönberg N., *Acta Met.* **3** (1955) 14.
46. Westgren A., *Jernkontorets Ann.* **116** (1933) 1.
47. Jeitschko W., H. Nowotny, and F. Benesovsky, *Mh. Chem.* **95** (1964) 156.
48. Dwight A. E., private communication.
49*a*. Schubert K., K. Frank, R. Gohle, A. Maldonado, H. G. Meissner, A. Raman, and W. Rossteutscher, *Naturwissenschaften* **50** (1963) 41.
49*b*. Schubert K., H. G. Meissner, A. Raman, and W. Rossteutscher, *Naturwissenschaften* **51** (1964) 287.
49*c*. Schubert K., S. Bahn, W. Burkhardt, R. Gohle, H. G. Meissner, M. Pötzschke, and E. Stolz, *Naturwissenschaften* **47** (1960) 303.
50. Nevitt M. V., and J. W. Downey, *Trans. AIME* **224** (1962) 195.
51. Schablaske R. V., B. S. Tani, and M. G. Chasanov, *Trans. AIME* **224** (1962) 867.
52. Pietrokowsky P., *J. Inst. Met.* **90** (1962) 434.
53. Stolz E., and K. Schubert, *Z. Metallk.* **53** (1962) 433.
54. Hall E. O., and J. Royan, *Acta Cryst.* **12** (1959) 607.
55. Darby J. B., Jr., J. W. Downey, and L. J. Norton, *J. Less-Common Metals* **8** (1965) 15.
56. Baenziger N. C., and J. L. Moriarty, Jr., *Acta Cryst.* **14** (1961) 948.
57. Rough F. A., and A. A. Bauer, *Constitution Diagrams of Uranium and Thorium.* Addison-Wesley, Reading, Mass., 1958.
58. Nowotny H., O. Schob, and F. Benesovsky, *Mh. Chem.* **92** (1961) 1300.
59. Nevitt M. V., and J. W. Downey, *Trans. AIME* **221** (1961) 1014.
60. Kirkpatrick M. E., and W. L. Larsen, *Trans. ASM* **54** (1961) 580.
61. Kirkpatrick M. E., D. M. Bailey, and J. F. Smith, *Acta Cryst.* **15** (1962) 252.
62. Pötzschke M., and K. Schubert, *Z. Metallk.* **53** (1962) 474.
63. Nowotny H., E. Laube, R. Kieffer, and F. Benesovsky, *Mh. Chem.* **89** (1958) 701.
64. Nowotny H., F. Benesovsky, and O. Schob, *Mh. Chem.* **91** (1960) 270.
65. Murray J. R., *J. Less-Common Metals* **1** (1959) 314.
66. Pötzschke M., and K. Schubert, *Z. Metallk.* **53** (1962) 548.
67. Laube E., and H. Nowotny, *Mh. Chem.* **93** (1962) 681.
68. Lundin C., in *The Rare Earths.* Wiley, New York, 1961.
69. Schubert K., *Kristallstrukturen zweikomponentiger Phasen.* Springer-Verlag, Berlin, 1964.
70. Paine R. M., and J. A. Canabine, *Acta Cryst.* **13** (1960) 680.
71. Schablaske R. V., and E. Veleckis, private communication.
72. Rudy E., F. Benesovsky, H. Nowotny, and L. E. Toth, *Mh. Chem.* **92** (1961) 692.
73. Matyushenko N., V. Karev, and L. Verkhorobin, *J. Inorg. Chem. U.S.S.R.* **8** (1963) 1788.

part IV

Microstructure and Substructure

chapter 14

Antiphase Boundaries and Domains

Joe R. Beeler, Jr.

JOE R. BEELER, Jr.
Nuclear Materials and Propulsion Operation
General Electric Company
Cincinnati, Ohio

1. INTRODUCTION

This chapter is concerned with two particular types of microscopic inhomogeneities peculiar to orderable alloys, namely, *antiphase boundaries* (APB's) and *antiphase domains* (APD's). By definition, all atoms of a given type tend to position themselves on a particular subset of lattice sites in an orderable alloy. It is observed, however, that the subset of positions occupied by atoms of a given type can vary as a function of position in an alloy crystal. This variation rise to APB's and APD's, which exert an important influence on the alloy's mechanical, electrical, and magnetic properties. Mechanical strength is affected because these inhomogeneities influence dislocation structure, generation, and movement. Electrical and magnetic properties are influenced because the electronic energy and Brillouin zone structure are altered by the presence of APB's and APD's.

A two-dimensional example for an AB alloy will serve to illustrate the concept of antiphase configurations, that is, the essential nature of APB's and APD's. (An AB alloy is one that contains equal numbers of A and B atoms.) This example is based on a model for an orderable alloy, which assumes that the crystal energy is minimized by maximizing the number of first-neighbor AB atom pairs, that is, that each B atom tends to surround itself with A atoms and vice versa. The limitations of this AB pair model are discussed in Section 5. Figure 1 shows parts of three perfectly ordered antiphase domains, APD1, APD2, and APD3. Inside each of these domains any two adjacent sites are occupied by different atom types, but note that adjacent sites positioned on opposite sides of the antiphase boundaries (APB1 and APB2) are occupied by atoms of the same type. In this example, the ordering sequence in a given APD, along either a column or a row of lattice sites, is the mirror image of that in any adjacent domain. The names antiphase domain and antiphase boundary were suggested by this reversal in the ordering sequence across an APB. If we denote atoms arranged as in APD1 and APD3 by "o", and those arranged as in APD2 by "x", Fig. 1*a* goes over into Fig. 1*b*. The latter representation is perhaps a more vivid illustration of the antiphase concept. The APB's in Fig. 1 are defined by a reversal

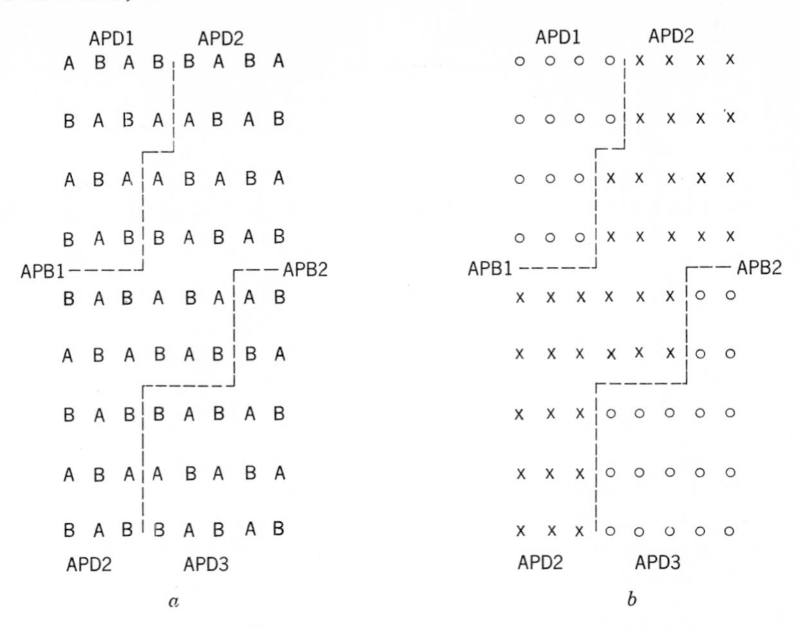

Fig. 1. Antiphase boundaries (APB) and antiphase domains (APD) in a two-dimensional AB alloy.

in the ordering sequence of *first neighbors* at an APB. In A_3B alloys an APB can also arise from a reversal in the ordering sequence of second neighbors without that of first neighbors being disturbed.

A perfectly ordered single crystal often turns out to be a single ordered domain. In terms of our overly simplified example, all atoms in such a crystal would be arranged in alternating sequences, ABAB . . . , along both the x- and y-directions. Observe that in this state all A atoms are located on one sublattice of sites (a-sublattice), and all B atoms on a second but equivalent sublattice (b-sublattice). This type of structure is called a superlattice. By definition, the long-range order (LRO) of a crystal is expressed in terms of the fractions of A and B atoms, which occupy sites on the a-sublattice and b-sublattice, respectively, which comprise the superlattice. Hence if a perfectly ordered crystal is perturbed by interchanging several A and B atoms, the degree of LRO is reduced. The LRO of a crystal is measured by the parameter[1]

$$S = \left[\frac{F(\mathrm{A}, a) - F(\mathrm{A})}{1 - F(\mathrm{A})}\right] = \left[\frac{F(\mathrm{B}, b) - F(\mathrm{B})}{1 - F(\mathrm{B})}\right] \tag{1}$$

where $F(\mathrm{A}, a)$ is the fraction of A atoms on the a-sublattice and $F(\mathrm{A})$ is the fraction of all atoms which are A atoms. $F(\mathrm{B}, b)$ and $F(\mathrm{B})$ are similarly defined. S decreases monotonically with increasing temperature; $S = 1$ for a perfectly ordered crystal but vanishes at and above a critical temperature T_c, peculiar to each alloy. Reviews by Nix and Shockley,[2] Lipson,[3] Muto and Takagi,[1] and Guttman[4] describe how S is determined experimentally and the theory of LRO. Rudman discusses the physical interpretation of order parameters in Chapter 21 of this book.

Many important considerations regarding the behavior of APD's and APB's deal directly with the *local correlation* between the occupancy of a given site and that of its neighboring sites. This local correlation is called short-range order (SRO). The SRO parameter, σ, for first neighbors, as defined by Bethe,[1,5] is

$$\sigma = \frac{q - q_r}{q_m - q_r} \tag{2}$$

where q is the number of AB pairs and q_m and q_r are, respectively, the maximum possible number of AB pairs and the average number for a random arrangement. Although σ

decreases with temperature, it does not vanish at $T \geq T_c$. Two general relations between σ and S are: (1) $\sigma = 1$ if and only if $S = 1$, (2) $S^2 \leq \sigma$. SRO parameters, which measure the correlation between the occupancy of higher order neighbor sites, have been defined by Warren.[6] They are commonly referred to as Cowley parameters. It is important to note that SRO is not defined in terms of sublattice occupancy.

An initially disordered crystal ($S = 0$) will develop LRO, that is, a superlattice, if its temperature is gradually lowered to a temperature $T < T_c$. Experimental evidence supports the idea that this ordering process is not a continuous change of atomic arrangement within a single-phase system, as was first supposed. In particular, the equilibrium-ordered state is thought to evolve from the disordered state via the formation and coalescence of APD's. Because of this, the physics of domain behavior is fundamentally important. A simple model for the transition from a disordered to an ordered state was first formulated by Sykes and Jones[7,8] for Cu_3Au. Their two-stage description for the ordering of Cu_3Au is a good first approximation to the general ordering process provided the lattice structure does not change during ordering and there is no strong magnetic ordering interference. In the first stage, localized ordered regions (nuclei) grow at a rapid rate and independently of one another until their boundaries touch. If two such nuclei exhibit the same A and B atom sublattices, their juncture results in a larger ordered domain. However, if the A atom sublattice for one is the B atom sublattice for the other, they form an APB at contact, as in Fig. 1, and become contiguous APD.

After boundary contact a second and slower ordering stage begins, wherein these nuclei and/or APD coalesce. During coalescence the disordered remnants between nuclei become ordered, and certain large nuclei grow at the expense of smaller nuclei. Ultimately each grain becomes a single ordered domain. A more detailed account of the ordering process in Cu_3Au is given by Quimby et al.[9–11] and Nagy et al.[12–14] In general, the lattice constant decreases slightly during the ordering process, and in some instances the crystal structure also changes. Each of these distortions gives rise to a strain energy at the boundary of an ordered nucleus or APD, which tends to limit its size and influences the mode of coalescence.

The stable-equilibrium ordered state need not be that wherein the entire crystal is a single ordered domain. In important instances the equilibrium state structure is a periodic array of regularly shaped APD. These periodic arrangements of APD are called long-period superlattices. It is not possible to explain the existence of the long-period superlattice solely on the basis of a preference for AB pairs over AA and BB pairs. It is necessary to consider the electronic energy in this instance.

At present there is no general quantitative theory of the ordering process. Considerable progress has been made for those alloys, in which ordering is associated with a marked increase in the number of AB pairs relative to the number of AB pairs for a random distribution of atoms. However, orderable alloys exist in which this increase is either small or zero. Briefly, there appear to be two principal effects which promote ordering. If the size of A and B atoms is significantly different, the lattice-strain energy relief afforded by the formation of AB pairs is usually the dominant mode for lowering the crystal energy. This mechanism is associated with the repulsive interaction between ion cores. However, the crystal energy of an ordered arrangement may also be lowered by a decrease in the electronic energy as well. The electronic contribution is believed to be important in the formation of long-period superlattices and in alloys that order without there being an increase in the number of AB pairs over that for a random arrangement.

In general, an APD need not be perfectly ordered, as are those in Fig. 1, and an APB exhibits a finite width which increases with temperature.[15–17,61] A more complete idea of what constitutes an APB and an APD will develop from the simple descriptions given above as the pertinent research work is subsequently described, particularly the work on the direct observation of APB's and the theory of long-period superlattices. A formal description of antiphase configurations is given in Section 2.

The APB is perhaps a more fundamental inhomogeneity than is the APD in the sense that it can be observed directly by means of transmission electron microscopy. In addition,

APB's can exist independently of APD's as the result of dislocation motion. Antiphase boundaries created by dislocations can have a different structure and energy than those which bound APD's.[61] In general, an APB is a high-energy boundary.

2. TYPES OF ANTIPHASE BOUNDARIES AND DOMAINS

This section describes the types of thermal APB and the types of APD, which can exist in binary alloys with cubic structure and composition close to either AB or A_3B. In general, we can regard an APB as having been formed as the result of translating one part of a perfectly ordered alloy crystal while holding the other part fixed. Figure 1 affords a partial illustration of this principle. When an APB is formed by dislocation movement, this translation actually occurs; however, it is a purely conceptual operation in the case of thermal APB's. A thermal APB is one formed when differently ordered nuclei make contact during growth. Marcinkowski's[18] notation will be used.

An APB is completely characterized by stating the associated translation vector (**p**-vector) and the plane in which the APB lies. The **p**-vector is written as

$$\mathbf{p} = \frac{a_0}{n}\,[uvw] \tag{3}$$

where a_0 is the width of the superlattice unit cell, n is an integer and $[uvw]$ is the conventional notation for a crystallographic direction. A particular APB is designated by appending the Miller indices (hkl) of its plane to its **p**-vector, that is, by writing

$$\frac{a_0}{n}\,[uvw]\,(hkl) \tag{4}$$

Two general types of APB can arise: Type 1 corresponds to the case wherein **p** lies in the APB plane and Type 2 to the case wherein **p** does not lie in this plane. These types will be illustrated in the discussion of the B2-type superlattice.

2.1 B2-Type Superlattice

AB alloys with a bcc (A2) structure in the disordered state can form a B2 superlattice in the ordered state. The unit cell for this superlattice is shown in Fig. 2. This structure can be thought of as two interpenetrating simple cubic sublattices I and II, with I occupied by B atoms and II by A atoms. As indicated in Fig. 2, an APB can be established by holding one part of the crystal fixed and bodily translating the *atoms* of the other part along $\langle 111\rangle$ in such a way that translated B atoms go onto the II sublattice and the translated A atoms onto the I sublattice. The appropriate class of **p**-vectors for this translation is $(a_0/2)\langle 111\rangle$. A particular translation, $\mathbf{p} = (a_0/2)[\bar{1}\,\bar{1}\,1]$ is illustrated in Fig. 2.

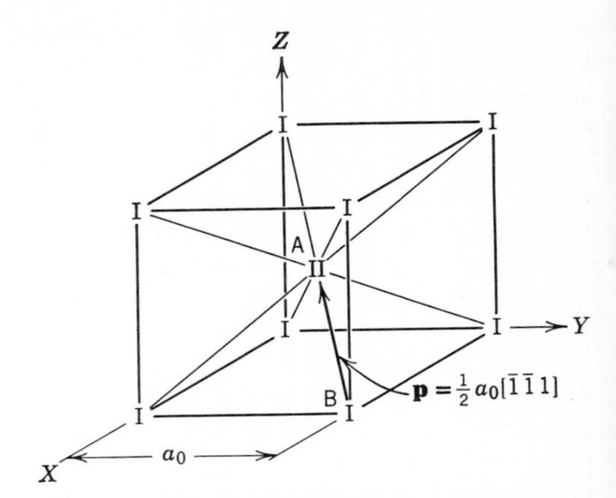

Fig. 2. Unit cell and APB **p**-vector for the B2 superlattice. (Figure 1 of Reference 18.)

We will illustrate the difference between Type 1 and Type 2 APB's by using examples from the B2 superlattice. It should be emphasized that these are general types that also occur in other superlattices.

Type 1 APB: If the operation **p** corresponds to simple shear in a $\{110\}$ plane a **p** $\{110\}$ APB is formed in the B2 superlattice. A member of this class, $(a_0/2)\,[\bar{1}\,11]\,(110)$, is described in Fig. 3. A Type 1 APB is formed when **p** lies in (hkl) and is such that the number of AB pairs it disrupts is divided into equal numbers of AA and BB pairs. This circumstance is important in APB energy computations for long-period superlattices and dislocation-APB interactions.

Type 2 APB: A Type 2 APB is formed when **p** does not lie in (hkl). The $\{100\}$ APB formed in a B2 superlattice via $\mathbf{p} = (a_0/2)\langle 111\rangle$ and the removal of a plane of either A or B atoms is a Type 2 APB. A particular case in which an A atom plane is removed and the translation $\mathbf{p} = (a_0/2)\;[\bar{1}\,\bar{1}\,\bar{1}]$ is effected to produce an

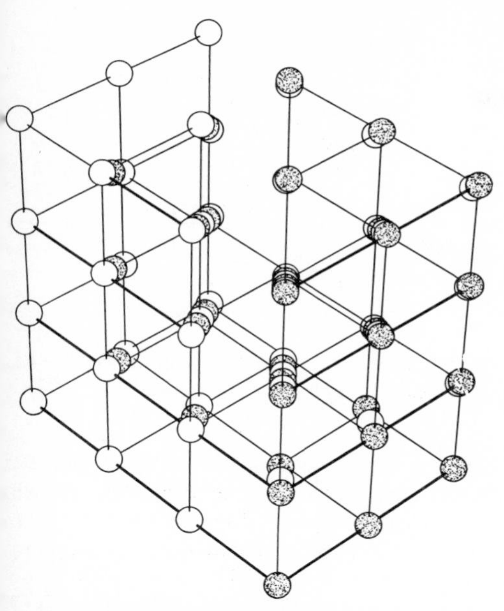

Fig. 3. Type 1 APB in the B2 superlattice.

$(a_0/2)$ $[\bar{1}\,\bar{1}\,\bar{1}]$ (010) antiphase boundary is shown in Fig. 4. As a result of removing the A atom plane, only BB pairs exist at this APB, that is, all AB pairs are converted to BB pairs. This creates a *local* excess of B atoms at the boundary, an effect which always occurs when **p** is not in the boundary plane. Cahn and Kikuchi[16,17] have shown how this local excess of B atoms can develop and treat its contribution to the energy of a Type 2 APB.

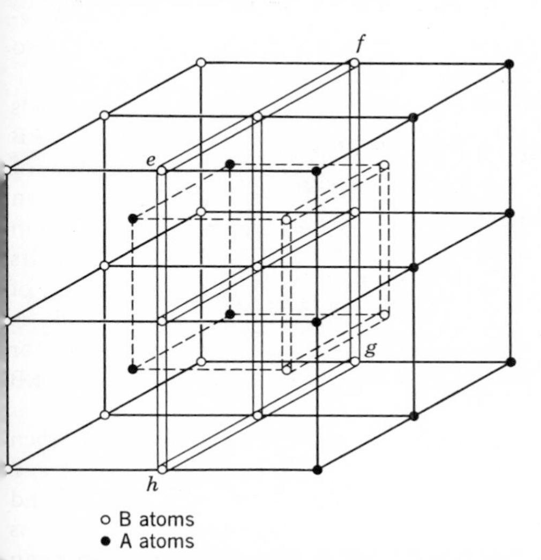

Fig. 4. Type 2 APB in the B2 superlattice. (Figure 2 of Reference 18.)

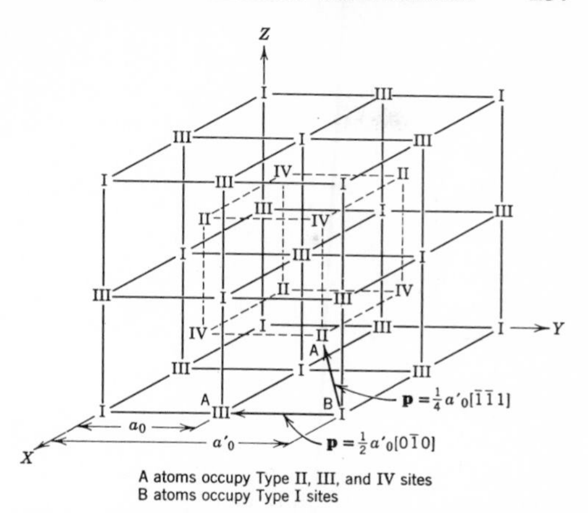

Fig. 5. Unit cell and APB **p**-vectors for the $D0_3$ superlattice. (Figure 3 of Reference 18.)

2.2 $D0_3$-Type Superlattice

Alloys with A_3B composition and a bcc structure in the disordered state can form a $D0_3$ superlattice in the ordered state. As shown in Fig. 5, the $D0_3$ unit cell is defined in terms of four interpenetrating fcc sublattices. In the perfectly ordered state, B atoms occupy lattice I and A atoms occupy sublattices II, III, and IV. The unit cell width, a_0', for the superlattice is twice that of (a_0) for the disordered bcc structure. In the $D0_3$ superlattice, APB's can be formed either by moving B atoms from I to II ($\mathbf{p} = (a_0'/4)\langle 111\rangle$) or from I to III ($\mathbf{p} = (a_0'/2)\langle 100\rangle$), as shown in Fig. 5. Boundaries defined by $\mathbf{p} = (a_0'/4)\langle 111\rangle$ include only "wrong" first-neighbor pairs, that is, BB nearest-neighbors. In contrast, an APB defined by $(a_0'/2)\langle 100\rangle$ does not alter the number of AB nearest-neighbors but exists because it leads to wrong *second-neighbor* BB pairs.

2.3 $L1_2$- and $L1_0$-Type Superlattices

The unit cell and sublattices for the $L1_2$ superlattice are defined by Fig. 6. Alloys with A_3B composition and a fcc (A1) structure in the disordered state can adopt this superlattice. Following the same line of argument used above we find the **p**-vector for this superlattice to be of the class $(a_0/2)\langle 110\rangle$. The vector $(a_0/6)$ $[\bar{2}11]$, shown in Figs. 6 and 7, is associated with both stacking fault and APB formation. Although pertinent to the treatment of dislocation behavior in superlattices (see §6), it is not associated with thermal APB's.

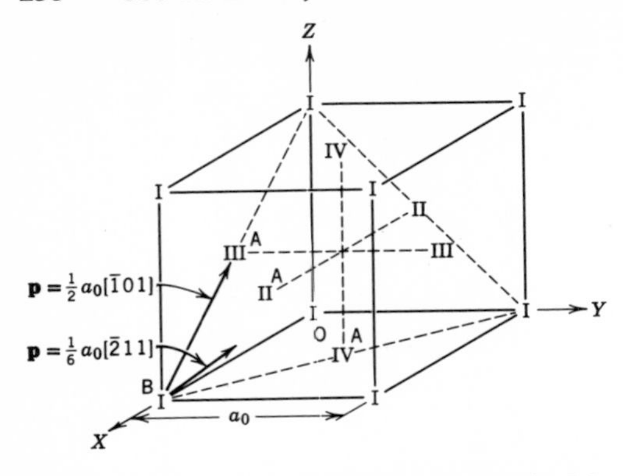

Fig. 6. Unit cell and APB **p**-vectors for the $L1_2$ superlattice. (Figure 4 of reference 18.)

The sublattices and **p**-vector for the $L1_0$ superlattice are given by Fig. 7. This structure is tetragonal. Because the *c*-axis for a domain may be along any of the three principal directions, and B atoms may chose either of the two sublattices shown in Fig. 7, there exist six possible APD ordering schemes.

2.4 Antiphase Domains

In the case of binary cubic alloys, the number of different APD which can exist is equal to the number of sublattices required to define the superlattice unit cell. This is true because the minority atom type (always B atoms in this chapter) can choose any one of the sublattices as its "proper" sublattice when an ordered nucleus is formed. If four or more domain types are possible, as in DO_3 and $L1_2$ superlattices, for example, a stable

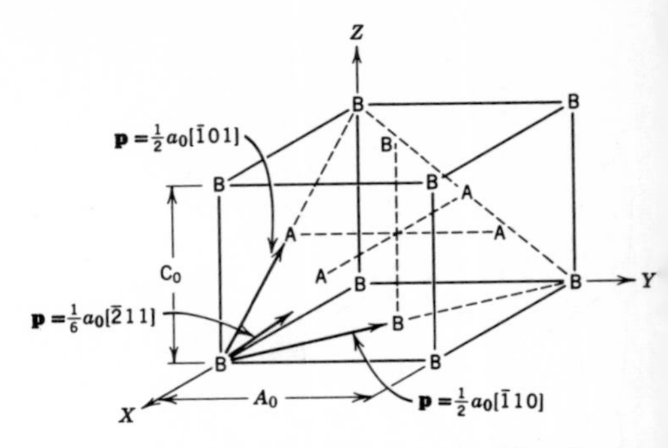

Fig. 7. Unit cell and APB **p**-vectors for the $L1_0$ superlattice. (Figure 6 of Reference 18.)

domain structure can exist. This is called a foam structure. In these instances, domain coalescence ceases after the stable foam structure has been attained at a given temperature. A detailed discussion of the foam structure is given by Fisher and Marcinkowski,[33] and examples of this structure appear in Section 4. The foam structure is not a periodic array of APD's and the individual domains exhibit different sizes. This distinguishes the foam structure from the long-period superlattice.

2.5 Summary

Table 1 summarizes the essential geometrical characteristics of APB's and APD's in cubic structure superlattices. Guttman[4] gives an extensive list of realizations for all superlattice structures studied up to 1956 and a bibliography for each realization.

TABLE 1

APB p-Vectors and the Possible Number of Different APD Types for Common Cubic Superlattices

Superlattice Type	APB p-vector	Number of APD Types	Disordered Structure
B2	$(1/2)a_0\langle 111\rangle$[a]	2	bcc (AB)[b]
DO_3	$(1/2)a_0\langle 111\rangle$ $a_0\langle 100\rangle$	4	bcc (A_3B)
$L1_2$	$(1/2)a_0\langle 110\rangle$ $(1/6)a_0\langle 112\rangle$	4	fcc (A_3B)
$L1_1$	$(1/2)a_0\langle 110\rangle$ $(1/6)a_0\langle 112\rangle$	6	fcc (AB)

[a] a_o is the width of the unit cell for the disordered structure.
[b] A_nB indicates composition.

3. INDIRECT OBSERVATION OF ANTIPHASE BOUNDARIES AND DOMAINS

A number of indirect techniques can be used to detect the presence of APD's. All of these methods measure the average of some characteristic over a relatively large volume of the crystal compared with the average size of an APD. The most quantitative indirect methods for studying APD's are X-ray and electron diffraction. In the case of X-ray diffraction, the presence of a superlattice is indicated by the appearance of the superlattice lines which occur in addition to the fundamental lines of the disordered alloy after a considerable portion of the ordering process has been accomplished. The width of the superlattice lines is inversely proportional to the size of the APD's in the ordering material. As the APD's grow in size, the superlattice lines become progressively sharper. Superlattice lines do not appear until the average size of the ordered nuclei is $\sim 5.5 \times 10^{-7}$ cm or greater, that is, about fifteen lattice constants in most systems. The apparent size ϵ of ordered nuclei is given by[8]

$$\epsilon = \frac{\lambda \cos x/2}{\beta} \qquad (5)$$

where λ is the X-ray wavelength, x twice the Bragg angle, and β the true angular breadth of the line. The true width, β, is obtained from the observed width by making corrections for finite slit width, size of the specimen, and doublet separation of the incident X-ray beam.[19] If APD's are bounded by parallel, planar boundaries and are regularly shaped, diffraction techniques can be used to determine their shape as well as to measure their size. When the APD shape is irregular, X-ray diffraction data, per se, give no information on the APD structure aside from the average domain size. However, recent developments[24] indicate that the structure of irregularly shaped APD's may possibly be obtained from X-ray data via an electronic computer. The techniques of X-ray diffraction measurements and the interpretation of X-ray diffraction pictures as they pertain to orderable systems are described in references.[20–23]

Electron diffraction has been used by investigators to determine APD structure in films and foils. The present theory of long-period superlattices, for example, is based on electron diffraction measurements. References[25–30] and [35,55–59] describe the use of electron diffraction methods to determine APD structure.

4. DIRECT OBSERVATION OF ANTIPHASE BOUNDARIES

Direct observation of APB has been accomplished by using electron transmission microscopy. The dynamical theory of electron diffraction as applied to the detection of stacking faults is used to interpret the micrographs. This theory was developed by Whelan and Hirsch.[31,32] Marcinkowski et al.[33–35] have done an extensive analysis of diffraction contrast pertinent to the direct observation of APD's and APB's. We shall, therefore, only sketch out the main points involved.

APB's can be seen, that is, they show contrast, only if the specimen is so oriented that it is diffracting into one of the superlattice reflections. There are three principal types of contrast. These are illustrated schematically in Fig. 8: (*a*) Normal boundary contrast is believed to result because, for a superlattice reflection, the diffracted beam from one side of the boundary is antiphase with the diffracted beam from the other side. (*b*) Tilted-boundary contrast is effectively the same as

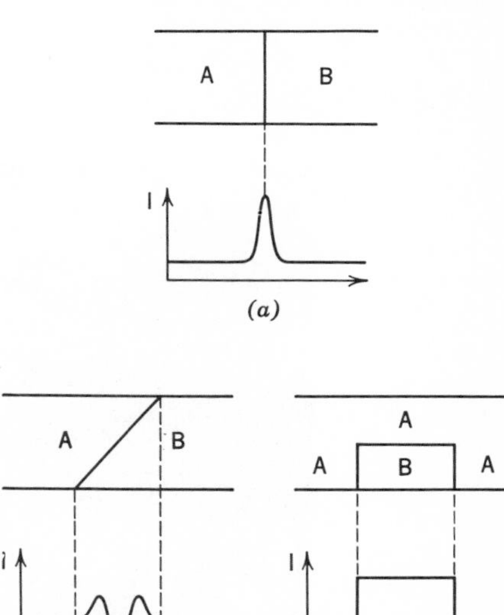

Fig. 8. Contrast effects: (*a*) normal boundary contrast; (*b*) tilted boundary contrast; (*c*) domain contrast. (Figure 4 of Reference 38.)

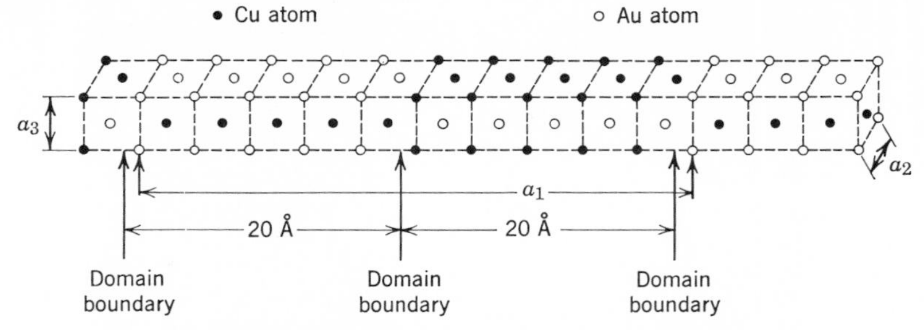

Fig. 9. Unit cell of CuAu(II). (Figure 1 of Reference 38.)

stacking-fault contrast and shows up as a number of black and white fringes. The number of fringes depends (inversely) upon the thickness of the film and upon the extinction distance. (*c*) Domain contrast occurs when the APB does not extend through the film. In this instance, superlattice reflections from the top half of the boundary are out of phase with those from the bottom half, and destructive interference occurs.

4.1 Copper-Gold Alloys

The CuAu System. The alloy CuAu forms two distinguishable superlattices. Below 380°C it exhibits the $L1_0$ structure ($c/a \sim 0.92$) defined by Fig. 7. This superlattice is called CuAu(I). Between 380°C and the critical temperature of 410°C, the (002) planes are again alternately all gold and all copper, but there is a periodic arrangement of APB perpendicular to [100] which are separated a distance equal to the width of five unit cells of the CuAu(I) superlattice (~20 Å). This long-period superlattice (LPS) is called CuAu(II). The CuAu(II) unit cell is shown in Fig. 9 and was first determined by Johannson and Linde[36] using X-ray diffraction.

Ogawa et al.,[25] in 1957, were the first to observe an APD structure directly. Using transmission electron microscopy, they made photographs of APB's in evaporated thin films (single crystal) of ordered CuAu(II) at a magnification of 720,000 X. The APB were observed as regularly spaced parallel lines separated by a distance of 20 Å. Shortly after the publication of Ogawa's results, Glossop and Pashley[37] reported direct observation photographs of APB in CuAu(II), which exhibited a larger area of resolved structure and gave more detail of the arrangement of APB and APD. Pashley and Presland[38] also studied the transition from CuAu(I) to

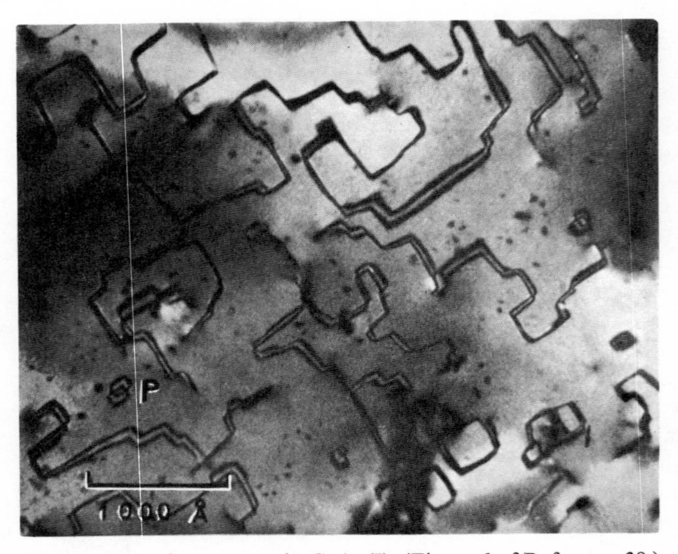

Fig. 10. Domain structure in CuAu(I). (Figure 6 of Reference 38.)

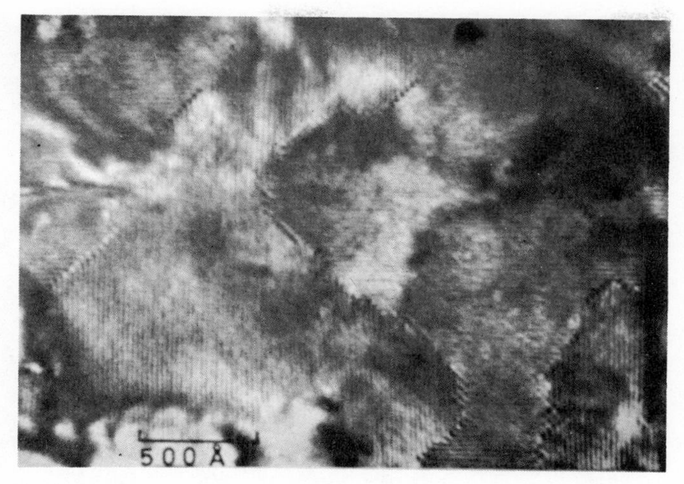

Fig. 11. Domain structure of CuAu(II). (Figure 5 of Reference 38.)

CuAu(II), using transmission electron microscopy and again working with evaporated films. Recently Hirabayashi and Weissmann[39] investigated the ordering of CuAu(I) in foils prepared from bulk material.

Figures 10 and 11 show the domain structure observed in CuAu(I) and CuAu(II), respectively, by Pashley and Presland. The cube planes, on which the copper and gold atoms separate, are parallel to the plane of the film, so that the *c*-axis of CuAu(I) or the *a*-axis of CuAu(II) (see Fig. 9) always lies perpendicular to the film plane. This is believed to occur because, when ordering takes place, the *c*-axis of the CuAu(I) or the *a*-axis of the CuAu(II) is contracted relative to the original *a*-axis of the disordered CuAu, and the strain energy is minimized if this contraction is normal to the film plane. The work of Hirabayashi and Weissmann on CuAu(I) foils, thinned from bulk material, gives a thorough account of strain considerations in CuAu(I).

The CuAu(I) APB's in Fig. 10 are tilted slightly with respect to the film plane and hence, in accord with Fig. 8, give rise to a pair of black fringes instead of the single fringe we would observe if they were perpendicular to the film plane (normal contrast). In general, the APB's are of two geometrical types: (*a*) they form closed loops; (*b*) they terminate within the specimen. The majority of the APB's in Fig. 10 are of the latter type. In a two-domain system, the APB's should all be of type (*a*), in the absence of dislocations. It can be readily shown that type (*b*) APB's are introduced by dislocations. If a dislocation is introduced into the sample by slip, then at the slip plane the (002) planes of copper atoms join (002) planes of gold atoms to give an APB that terminates on a dislocation line. In some instances it is possible to detect the terminal dislocation, but, since the diffraction conditions for obtaining dislocation contrast are not the same as those for APB contrast, it is not easy to show both on the same micrograph.

The structure of CuAu(II), shown in Fig. 11, is divided into two kinds of area, referred to as unidirectional zones. Within each zone the periodic domain boundaries are parallel to one cube plane of the original disordered structure. Each unidirectional zone represents one particular orientation of the cell in Fig. 9, the *a*-axis of the cell being along one of the two cube directions in the film plane. The spacing between APB's is 20 Å, and within one unidirectional zone the structure is that deduced by Johannson and Linde.[36] The domains usually extend through the thickness of the film at a fairly early stage in the ordering process, so that subsequent growth is in a direction parallel to the film plane. Figure 12 illustrates the growth of CuAu(II) from CuAu(I) at a late stage in the transition.

When the tetragonal ordered CuAu(I) structure is formed from the fcc (a_0 = 3.88 Å) disordered state, a considerable amount of strain is introduced. In order to reduce this strain, Harker[40] proposed that a lamellar

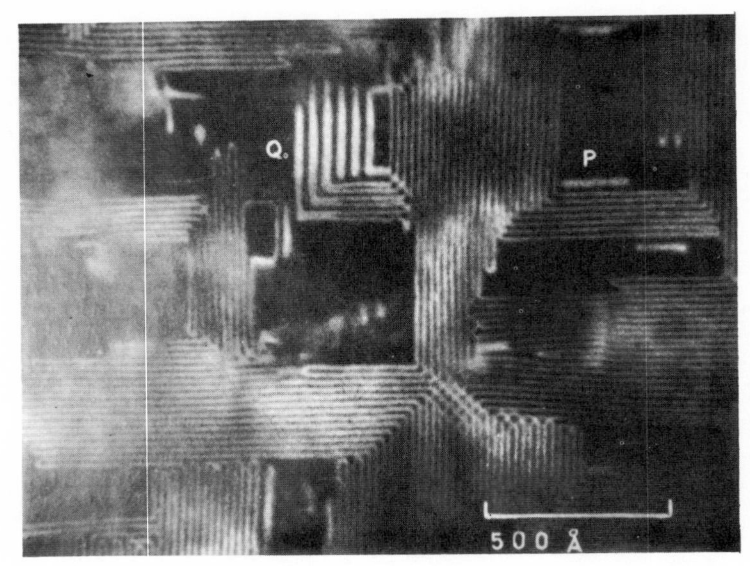

Fig. 12. Growth of CuAu(II) from CuAu(I). Dark areas are CuAu(I). (Figure 9 of Reference 38.)

structure was formed parallel to the {111} planes in such a way that the *c*-axis direction of the tetragonal CuAu(I) alternate in three mutually perpendicular directions, as shown in Fig. 13*a*. Kuczynski et al.[41,42] found that the formation of CuAu(I) was accompanied by twinning with the twin planes parallel to the {101} planes of the disordered structure. They proposed that these twins appeared simultaneously with the inception of ordering at elevated ordering temperatures, but that ordering began before the twinlike structure

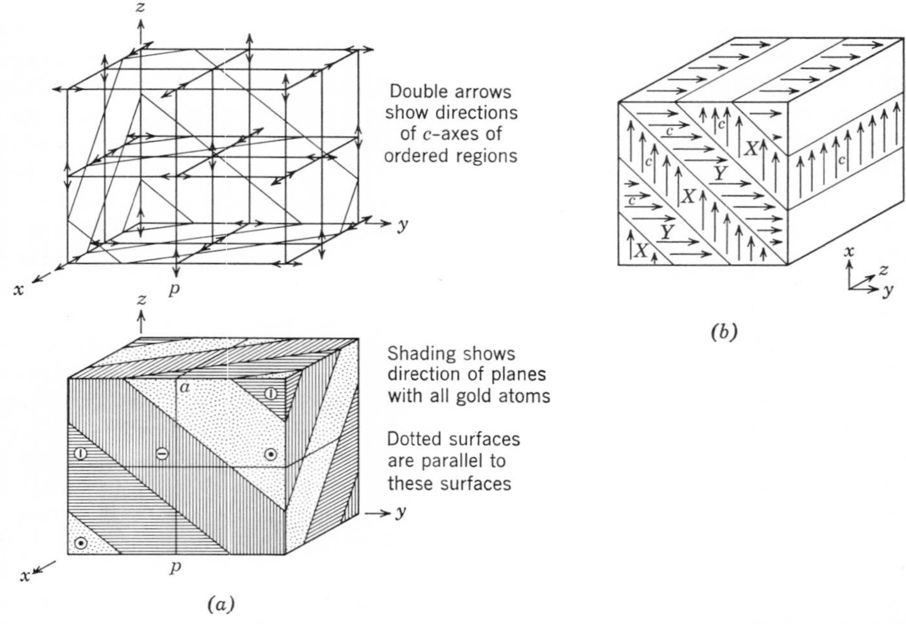

Fig. 13. (*a*) Lamellar structure for CuAu(II) proposed by Harker;[40] (*b*) lamellar structure observed by Hirabayashi and Weissman (Figure 11 of Reference 39).

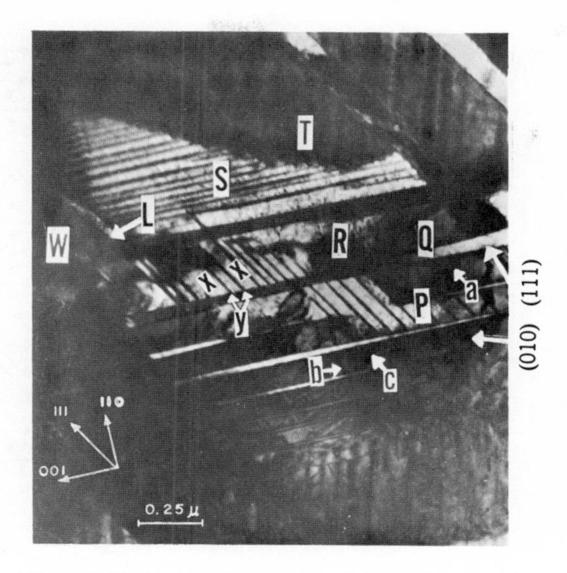

Fig. 14. Twin lamellae and antiphase domains in CuAu. (Figure 5 of Reference 39.)

appeared at low ordering temperatures. It was assumed that, at low temperatures, the critical size of twinning is larger than that for ordering and, therefore, the ordered lattice domains grow coherently within the disordered matrix. The reverse case was assumed for elevated temperatures. Hirabayashi and Weissmann obtained the micrograph of Fig. 14 for CuAu foils annealed at 300°C for 114 hours. These foils were prepared by thinning bulk material. Microtwins are denoted by x and y, and coarse lamellae by P, Q, R, S, T, L, and W. Within the large lamellae P and R, the directions of the microtwins x and y are identical, except for a parallel displacement of the y microtwins by the interferring lamella Q. During the early stage of ordering, P and R comprised a single region in which microtwins x and y alternated. This region was subsequently traversed at a more advanced stage of ordering by the large lamella Q. The boundaries of the coarse lamellae usually followed the crystallographic directions associated with twinning. However, when the ordering had not progressed far enough, the interfaces of the large lamellae followed irregular, noncrystallographic directions, such as the boundary between S and T. The irregular boundary between S and W was in the process of becoming strictly crystallographic by the interposition of the large twin lamella L. Hirabayashi and Weissmann visualize the ordering process in CuAu as follows: Small ordered regions in the disordered matrix form first and function as nuclei for the ordered structure. These nuclei are coherent with the disordered matrix and hence give rise to elastic strains, which are subsequently reduced by microtwinning. In contrast to Harker's proposal, however, the twin interface is parallel to (110) rather than to (111). The alternation of the c-axis follows the sequence X, Y, X, Y, . . . , as in Fig. 13*b*, rather than the sequence X, Y, Z, X, Y, Z, . . . of Fig. 13*a*. In Fig. 15 we see that a compressed region of the disordered matrix (C in Fig. 15) becomes a tension region (T in Fig. 15) due to alternation of the c-axis orientation. This configuration will minimize or even cancel the tetragonality strains.

In order to achieve an effective reduction of tetragonality strains by twinning, one governing principle has to be obeyed: Twinning has to occur in such a manner that the twin planes do not contain the unique c-axis of the ordered tetragonal CuAu(I) structure. This restricts the number of possible twinning modes. Since any one of the cube axes of the disordered structure may become the unique c-axis of the ordered tetragonal structure, the c-axis orientation will differ among the various nucleation centers. With increased ordering, the twins

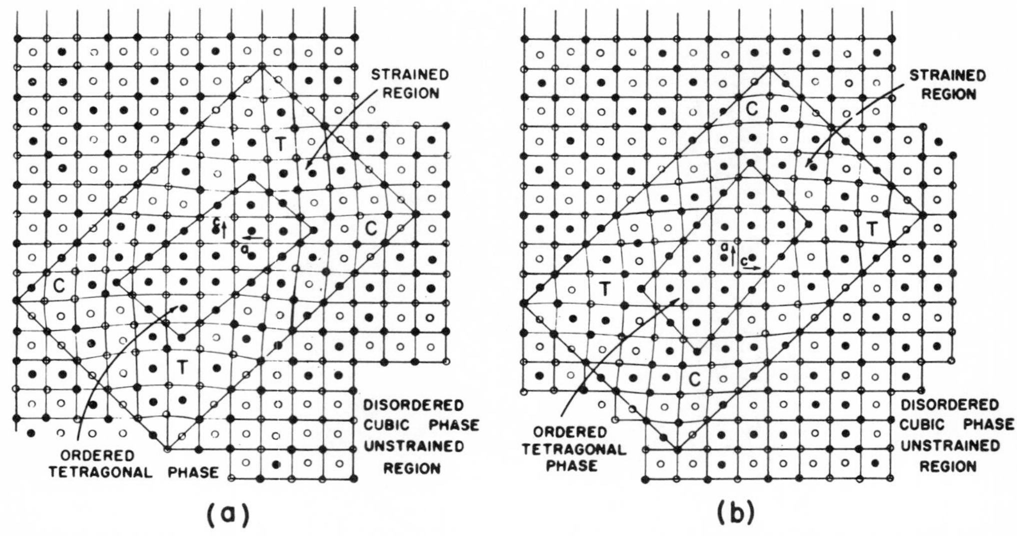

Fig. 15. Lattice distortions due to ordering of CuAu. Change in *c*-axis direction causes the tension (*T*) and compression (*C*) regions to become reversed. *C*-axis directions are noted at the centers of (*a*) and (*b*). (Figure 10 of Reference 39.)

emanating from various nucleation centers will eventually meet.

Long-range strains arise at the juncture of coarse lamellae growing from nucleation centers with different *c*-axis orientation. These strains are relieved by the interposition of another coarse twin lamella having a *c*-axis perpendicular to that of the other lamellae forming the juncture (see Fig. 14). There are, therefore, two mutually related twinning operations: (1) microtwinning at an early ordering stage which relieves short-range strains created by ordering, (2) formation of coarse twin lamellae, during a more advanced ordering stage, which result from the accommodation of long range strains. Advancing coarse twin lamellae may destroy existing order at their interfaces, but the order remains completely preserved within these lamellae. From an energetic viewpoint, the relief of lattice strains by twinning appears to be more important for the progress of ordering than the local destruction of existing order by advancing twins. In this regard, an increase in mechanical strength was correlated with the coherency strains between ordered nuclei and the disordered matrix and a decline in mechanical strength to the elimination of coherency strains by microtwinning.

The Cu_3Au System. Until 1960 it was believed that Cu_3Au exhibited only an ordinary superlattice. During this year Scott[43] published

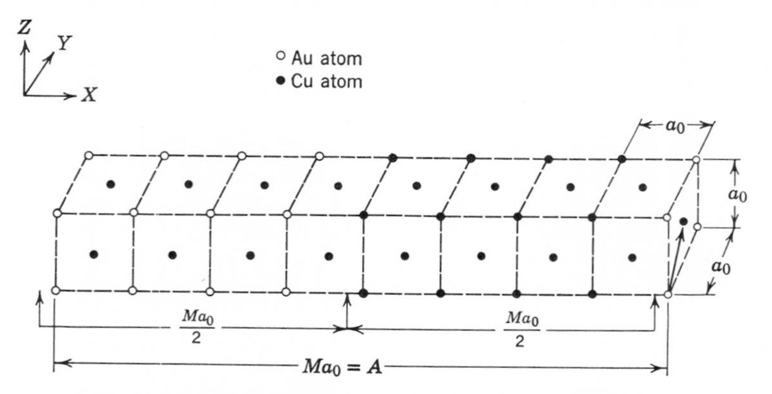

Fig. 16. Unit cell for $Cu_3Au(II)$. (Figure 3 of Reference 35.)

his discovery of a LPS structure in Cu_3Au. As in the case of CuAu, the ordinary superlattice, which exists up to 380°C, will be denoted as $Cu_3Au(I)$ and the LPS, which exists over the range 380 to 408°C, as $Cu_3Au(II)$. The unit cell for $Cu_3Au(II)$ is defined in Fig. 16.

Fisher and Marcinkowski[33] first observed APB's in $Cu_3Au(I)$ specimens electrothinned from bulk material. In this system there are four types of APD's and two types of APB's, as defined in Section 2. Figure 17 is one of their photographs of APB's in fully ordered $Cu_3Au(I)$. The APB's lie essentially on cube planes, and the APD structure consists of a network of rectangular blocks, each face of which is surrounded by a domain of different type. No indication of any two-phase regions was found in partially ordered $Cu_3Au(I)$, and they conclude that the degree of order within the domains is homogeneous but imperfect.

They showed both theoretically and experimentally that the specimen must be oriented for a strong superlattice reflection in order to obtain APB contrast by using the electron microscope. Because only two thirds of the APB's in $Cu_3Au(I)$ will show contrast for any given particular strong superlattice reflection, the maze pattern (Fig. 17) cannot be directly interpreted as indicating a rectangular block network of APD's. The existence of this block structure was inferred from an analysis, which showed that the APB that gave contrast enclosed superdomains within which all of the boundaries that did not show contrast were of identical type. In the case of CuAu all APB are of a single type and hence show contrast simultaneously.

Marcinkowski and Zwell[35] have obtained micrographs of $Cu_3Au(II)$, one of which is shown in Fig. 18. Regions immediately adjacent to B show the $Cu_3Au(II)$ structure with APB's lying in (010) planes. The average spacing between APB's was ~35 Å. Domains which appear black, such as A, correspond to

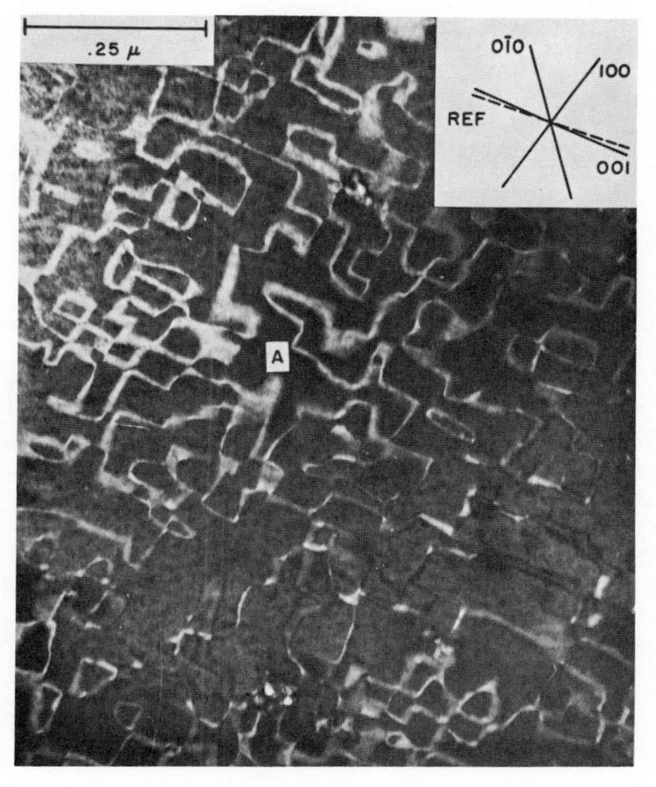

Fig. 17. APB's in $Cu_3Au(I)$. This is a picture of a stable foam structure. The rectangular APD block structure does not appear complete because only $\frac{2}{3}$ of the APB's show contrast for a given superlattice reflection.

Fig. 18. Domain structure in Cu_3Au(II). (Figure 10 of Reference 35.)

areas of the foil which were not diffracting. The light areas B correspond to domains in which the APB's lie on (001) planes.

4.2 Iron-Aluminum Alloys

All of the micrographs discussed for copper-gold systems indicated that APB's lie preferentially on cube planes. In Fe_3Al this is not the case. Following the approach of Fisher and Marcinkowski for $AuCu_3$, Marcinkowski and Brown[34] made the first micrographs of APB's in Fe_3Al and analyzed the energy of these APB's. A typical APB photograph is shown in Fig. 19. Two types of ordered arrangements were observed. The first type, observed in cooling from above the critical temperature, was based on an imperfect B2-type structure which is stable between 560° and 800°C. In this structure iron atoms occupy one sublattice, whereas the remaining equal numbers of iron and aluminum atoms are randomly distributed on the second sublattice. Marcinkowski and Brown showed that the APD's that form in this superlattice are arranged so that one of the two possible domain types is completely surrounded by domains of the second type. The APB's are curved. This is explained on the basis of the relatively small energy difference between APB's lying on different planes in the alloy. The B2-type ordering transformation occurs so rapidly that it is not possible to quench-in the disordered state. Below 560°C, the B2-type structure transforms to a $D0_3$-type structure, in which those iron and aluminum atoms randomly distributed in the B2-type structure rearrange themselves on one of two sublattices so that each aluminum atom has only iron atoms as second nearest-neighbors. The $D0_3$ APD's form inside the larger existing B2

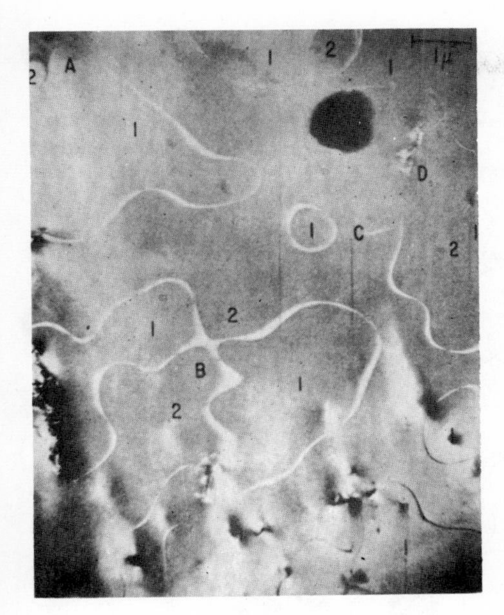

Fig. 19. B2-type domain structure in Fe_3Al. (Figure 4 of Reference 34.)

APD's and are, therefore, small in comparison. Again the APB's show no preference for crystallographic planes.

4.3 Computer Experiments

Theoretical "direct" observation studies on the growth of APD's and APD shapes have been performed using a high-speed electronic computer. This has been done in two different ways. In one instance, APD formation was studied by simulating atom movements in cubic AB alloys and in the other instance by using a computer to construct APD's from experimentally obtained SRO parameters.

In the first instance, it was assumed that atom movement proceeds via the vacancy mechanism and that the ordering process is governed primarily by a tendency to maximize the number of first-neighbor AB pairs. On this basis, a computer was used to trace out individual vacancy migration histories in AB alloys and thereby generate the rearrangement of atoms as the alloy ordered. This was done using the Monte Carlo method devised by Flinn and McManus[44] for approximating the effect of vacancy movement in an orderable alloy.

In a monatomic crystal, a migrating vacancy executes a symmetric random walk on the crystal lattice, that is, the jump probability at a given site is the same for all possible jump directions and is independent of position. However, in an orderable alloy, the jump probability is, in general, different for each possible jump direction and is also position-dependent. This is true because jumps that result in an increase in the number of AB pairs are energetically preferred to those that decrease the number of AB pairs. Flinn and McManus devised a Monte Carlo sampling scheme for approximating the vacancy jump probability in an orderable system. They used this method to compute equilibrium properties of a B2 superlattice and to approximate order-disorder kinetics. Their calculations were performed in a crystallite of 2000 atoms and used, in effect, periodic boundary conditions. Because periodic boundary conditions were adopted, the growth of individual APD could not be studied.

Beeler and Delaney[45] applied the Flinn-McManus method in a study of APD growth by increasing the crystallite volume affected by vacancy migration to include 2.3×10^5 atom sites and dispensing with periodic boundary conditions. They treated AB alloys with square planar and simple cubic (sc) lattices. Using the same method, Beeler[46] studied APD growth in AB alloys with sc, bcc and fcc structures. He found that a single vacancy would construct contiguous APD's with a diameter of $\sim$20 Å, during the course of 3×10^4 jumps, in an initially disordered alloy with either sc or bcc structure. However, only disjoint-ordered nuclei were formed in 3×10^4 jumps, when the disordered phase had a fcc structure. An example of an APD, from the contiguous structure given by his calculations for a bcc system, is shown in Fig. 20. The APD boundary pattern on a planar cross section through this contiguous APD structure is shown in Fig. 21. There is a striking similarity between this "computer" pattern and that observed by Marcinkowski and Brown[34] for the Fe_3Al "imperfect" B2 superlattice. In addition, Beeler found that vacancies tended to become trapped at APD boundaries, a result which is consistent with the findings of Brown and Cupshalk[47] and Nagy et al.[12–14]

Gehlen and Cohen[24] have used a computer to estimate APD structure from measured values of S and the Warren-Cowley SRO parameters α_1, α_2, and α_3. In their computations the computer selects two different atoms,

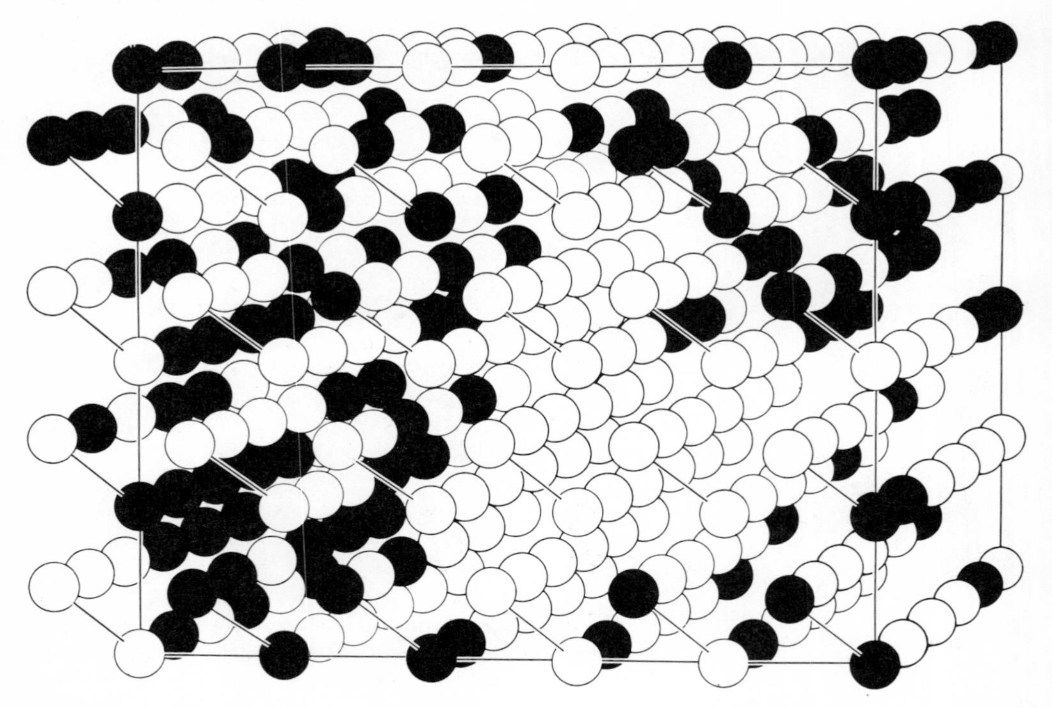

Fig. 20. APD in B2 superlattice given by Beeler's computer experiments[46] during initial stage of ordering.

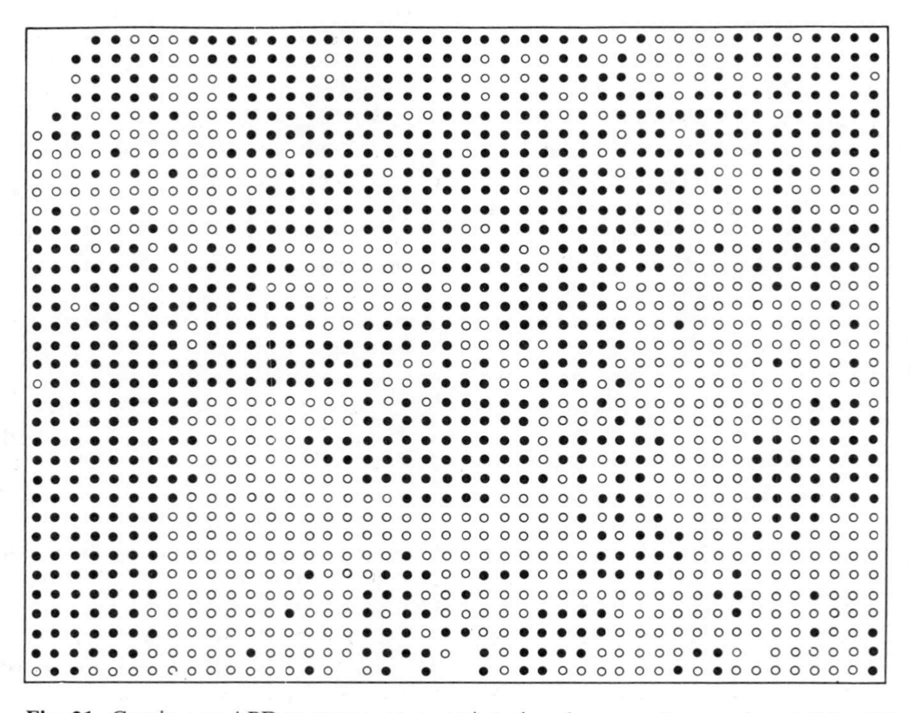

Fig. 21. Contiguous APD structure cross section given by computer experiments[46] for B2 superlattice. Compare with Fig. 19.

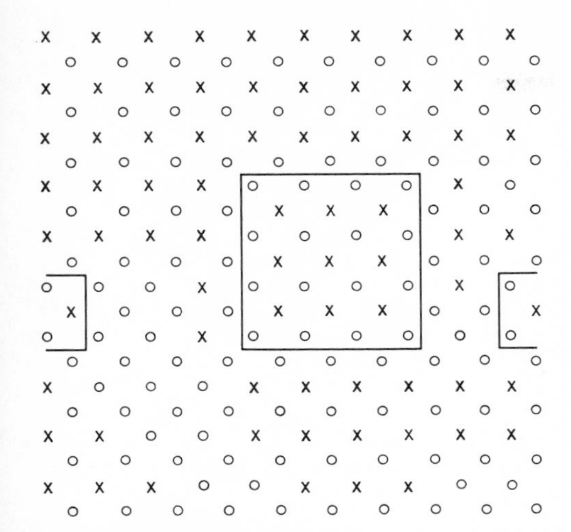

Fig. 22. APD structure in Cu_3Au ($S = 0.8$), computed by Gehlen and Cohen[24] from measured Warren-Cowley parameters.

essentially at random, and computes the change in the SRO parameters which would result if the positions of these atoms were interchanged. If the associated change is toward the measured parameter values, the interchange is executed; otherwise it is rejected and another atom pair selected, etc. When a configuration is attained that exhibits order parameters equal to those observed experimentally, the computation is halted, and a map of atom positions on all {100} planes is printed. An example of such a map is shown in Fig. 22 for Cu_3Au with LRO parameter $S = 0.8$. The program is run from both perfectly ordered and completely disordered initial states to check the uniqueness of the final configuration. This new approach appears to be a very promising technique for analyzing SRO data. By combining (100) maps, such as that from Fig. 22, APD volumes similar to those computed by Beeler[46] (Fig. 20) can be obtained.

4.4 Field-Ion Microscopy

Until recently, the electron microscope provided the only method for direct observation of very fine scale phenomena. However, as amply illustrated in the preceding sections, many features of importance in the study of APB's and APD's are characterized by a dimensional scale below the limit of resolution (~20A) of transmission electron microscopy. In this regard, a resolution of the order of 2–3A is possible in the case of field-ion microscopy and the application of this technique to the study of order-disorder processes has been under development at Cambridge University since 1962. At this writing, the most complete study of the order-disorder process via field-ion microscopy is that done for the PtCo system by Southworth and Ralph.[119] Their paper is particularly informative because it discusses not only the order-disorder process per se, but also those aspects of field-ion microscope use and of image interpretation peculiar to order-disorder systems. Because the use of field-ion microscopy in the study of APD's and APB's is still in the development stage an extensive discussion in this chapter is not yet appropriate. References 119–122 pertain to the use of field-ion microscopy in the study of APD's and APB's.

5. THEORY OF THE LONG-PERIOD SUPERLATTICE

A long-period superlattice (LPS) is a periodic array of regularly shaped APD's of a definite size. It is a *stable equilibrium* structure. CuAu(II) and Cu_3Au(II), discussed in Section 4, are two realizations of the LPS structure. At present, LPS theory is based on the relative magnitudes of the decrease in the electronic contribution to the crystal energy, associated with the existence of a periodic APD structure, and the increase in crystal energy associated with the existence of the APB's that separate these domains. The fundamental ideas in LPS theory are those used to explain the origin of the general ordering process. We will, therefore, outline the present theory for the origin of ordering and that for the APB energy, prior to describing LPS theory. Although the qualitative picture given by LPS theory is in accord with experimental data, it is not yet possible to compute the electronic and APB energy terms involved in a rigorous manner.

5.1 Origin of the Ordering Process

Hume-Rothery and Powell[48] explained the ordering tendency on the basis of lattice-strain relaxation. They suggested that in those cases where the sizes of A and B atoms were significantly different, the preference of one atom type for nearest-neighbors of the other type arose because the strain associated with size differences was minimized by such an arrangement. Although this model is in accord with a significant proportion of the experimental findings, it cannot be the general basis for the ordering process. This is true by virtue of at least two counter examples. In particular, the alloy CuPt, although orderable, possesses the same number of AB bonds per atom in the perfectly ordered state as in the disordered state. Its ordered structure consists of alternate (111) planes being occupied exclusively by Cu and Pt atoms. A more general counterexample is furnished by the existence of the LPS. This equilibrium structure cannot be explained by the AB-pair model, because this model predicts the total absence of APB's in a stable equilibrium-ordered structure.

The total energy of a crystal is made up of two contributions; one from the ion-core interactions and the other from the valence electrons. Only the ion-core interaction is affected in the model of Hume-Rothery and Powell. It has been suggested that perhaps the core interaction energy always swamps the order-sensitive electronic contribution in those alloys for which the AB pair model gives a qualitative description of the ordering process, Cu_3Au being cited as an example. However, this is not generally true, as was subsequently shown by Scott's discovery of the Cu_3Au(II) LPS. The extra periodicity of the ordered state relative to that for the disordered state, introduces additional Brillouin zones. The additional zones serve to depress the electronic energy. Jones et al.[49–51] explored the consequences of this effect in their theory for the stability of alloy phases. Subsequently, Slater[52] and Lipson[3] pointed out how the existence of additional Brillouin zones could lead to an ordering tendency. The first explicit three-dimensional treatment was given by Nicholas[53] in 1953 for CuPt, CuAu, and Ag_3Mg. Using the perturbed free-electron approximation, Flinn[54] demonstrated that a second-order, SRO-dependent term appears in the expression for the crystal energy. This result indicates that an ordering effect on the electronic energy exists even in the absence of a superlattice. In the disordered state the electron energy density of states per unit volume, $N(E)$, has the general shape of curve (*a*) in Fig. 23. The extra periodicity of the ordered state, which gives rise to superlattice lines,

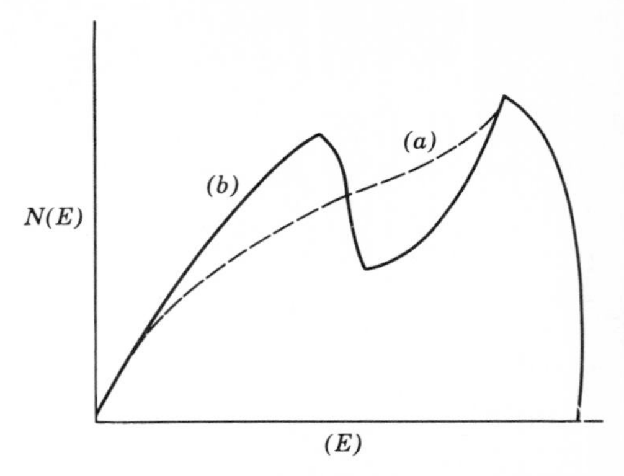

Fig. 23. Electron energy density of states $N(E)$: (*a*) disordered state and (*b*) ordered state. (Figure 1 of Reference 53.)

splits the Brillouin zone, that is, establishes an inner zone contained within that for the disordered state. As a consequence, $N(E)$ changes to the form of curve (*b*) in an ordered alloy. Nicholas[53] demonstrated that this occurrence should always lower the electronic energy contribution in the ordered state and alter the Fermi surface relative to the disordered state. In their theory of the long-period superlattice, Sato and Toth[55–59] extended this idea to include the additional Brillouin zone splitting introduced by a periodic APD structure.

5.2 Antiphase Boundary Energy

All APB energy calculations known to the author are based on the Ising model, because there are still no accurate quantum-mechanical calculations. According to the Ising model[60] the configurational potential energy of a binary alloy can be written as

$$V = (N_{AA}v_{AA} + N_{BB}v_{BB} + N_{AB}v_{AB}) \tag{6}$$

where v_{AA}, v_{BB}, and v_{AB} are the interaction energies for AA, BB, and AB pairs, respectively, and N_{AA}, N_{BB}, and N_{AB} are the numbers of such pairs. If the ordering energy, v, defined as

$$v = v_{AB} - \frac{v_{AA} + v_{BB}}{2} \tag{7}$$

is negative, a binary alloy will order at a sufficiently low temperature. An ordering energy v_n for nth-neighbor pairs can be defined from Eq. 7. All APB energy calculations to-date consider only either first- or first- and

TABLE 2

Partial List of Thermal APB Energies for S = 1. If S < 1, Multiply Energy Given by S^2

Superlattice Type	APB Type	APB Energy[a]	Reference
B2	$(1/2)a_0\langle 111\rangle$	$4v_1h/a_0^2\sqrt{N}$ $h \geq k \geq l$	62
AB_3-B2	$(1/2)a_0\langle 111\rangle$	$[v_1h + v_2(h+k+l)]/a_0^2\sqrt{N}$ $h \geq k \geq l$	34
DO_3	$a_0\langle 100\rangle$	$v_2(h+k+l)/a_0^2\sqrt{N}$ $h \geq k \geq l$	34
$L1_2$	$(1/2)a_0\langle 110\rangle$	$2v_1h/a_0^2\sqrt{N}$ $h \geq k$	62

[a] (hkl) are the Miller indices of the APB plane. $N = h^2 + k^2 + l^2$. a_0 is the width of the unit cell for the disordered structure.

second-neighbor interactions, that is, v_1 and v_2. These calculations have been performed by Brown,[61] Cahn and Kikuchi,[16–17] Flinn,[62] Marcinkowski and Brown,[34,63] and Marcinkowski, Brown, and Fisher.[64] The results of this work are listed in Table 2.

Figures 24 and 25 are the energy surfaces for an $(a_0/2)$ [111] APB in the B2 superlattice and an $(a_0/2)$ [001] APB in the $L1_2$ superlattice, as computed by Flinn[62] on the basis of a first-neighbor interaction model. The indices shown in these figures are the Miller indices (hkl) of the APB plane. Since there are some wrong first-neighbor bonds present for all APB orientations in a B2 superlattice, the APB energy is never zero. The maximum energy occurs for {100} APB planes, the minimum for {111} APB planes. The energies for APB on {110} planes are saddle points on the energy surface.

The energy surface for APB's in the $L1_2$ superlattice goes to zero at (001), because Flinn did not consider second-neighbor interactions. As he points out, second-neighbor interactions will produce a nonzero (001) APB energy. The energy maxima occur at (±1, 0, 0) and (0 ± 1, 0), and saddle points at (±1, ± 1, 0).

Cahn and Kikuchi and Brown computed the free energy of an APB. The calculations of Cahn and Kikuchi are quite extensive. They treated the free energy as a function of orientation and composition at absolute zero temperature and at nonzero temperatures.

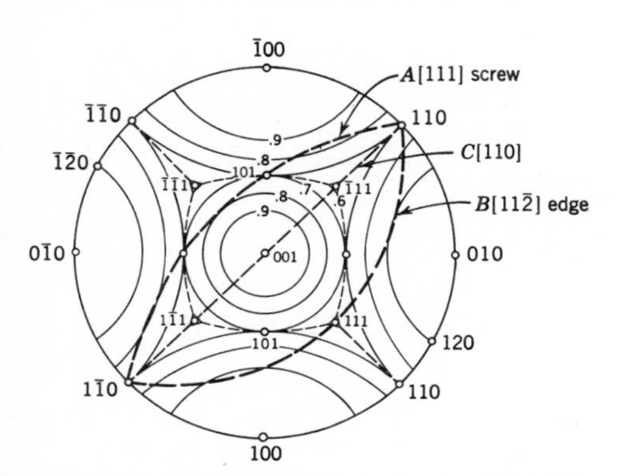

Fig. 24. Energy surfaces for an $(a_0/2)$ [111] APB in a B2 superlattice with plane (hkl). (Figure 7 of Reference 62.)

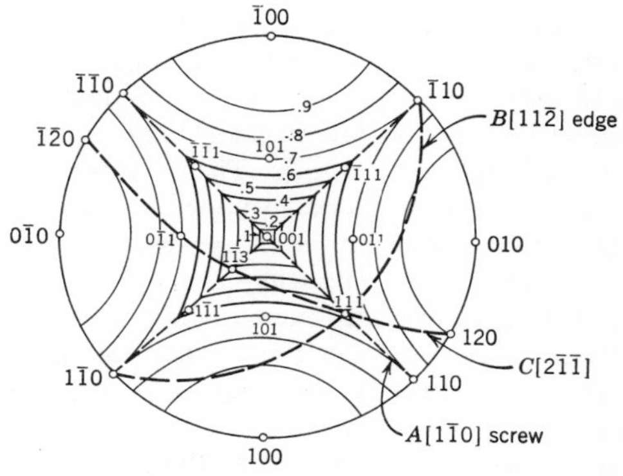

Fig. 25. Energy surfaces for an $(a_0/2)$ [001] APB in a $L1_2$ superlattice with plane (hkl). (Figure 6 of Reference 62.)

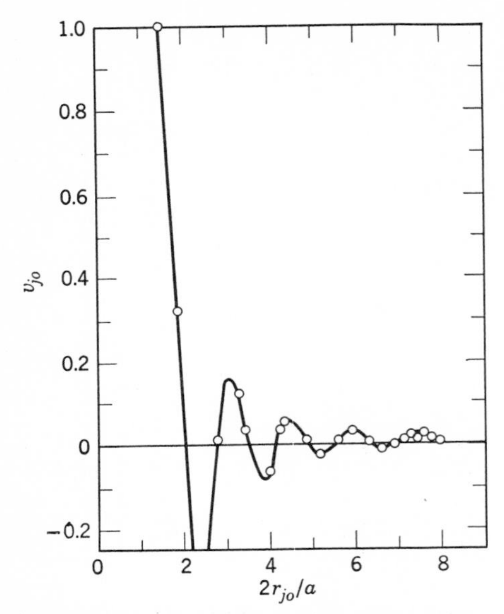

Fig. 26. Relative ordering energy v_{j_0} as a function of neighbor distance r_{j_0}, the distance to the jth site for β-brass. The lattice parameter is a. All values referred to $v_{10} = v_1$. (Figure 1 of Reference 68.)

The energy contribution from the excess of B (or A) atoms at a Type 2 boundary was computed by introducing material reservoirs in equilibrium with the superlattice. The energy change of these reservoirs, associated with Type 2 boundary formation, was then included in the APB formation energy.

It is not possible to fully utilize even the first approximation expressions for the APB energy listed in Table 2 because the magnitudes of v_1 and v_2 are not known. This is unfortunate because these energies are important in a detailed consideration of mechanical properties as well as in consideration of the LPS. Harrison and Paskin[65] have computed the ordering energy for first-and higher-order neighbor pairs in β-brass. Their calculations are based on Mott's polar model for an alloy, and recent results on the charge density variation about an impurity atom.[66,67] They obtained an ordering energy which is a long-range, oscillating function of the atom-atom separation distance, as shown in Fig. 26. Using this long-range function and Cowley's[6] statistical theory of order, Paskin[68] found about the same average behavior for the order parameters as that given by a first-neighbor interaction model. In this regard his results are consistent with the neutron diffuse scattering results of Walker and Keating[69] for β-brass. The magnitudes of v_1 and v_2 obtained by Harrison and Paskin stand in the ratio $v_2/v_1 \simeq 0.3$. In the case of Fe_2Al, Rudman[70] obtains the value 0.5 and Matsuda[72] the value 0.67 for this ratio. Fosdick[71] found that the ratio 0.25 gave the best fit between his computer studies for Cu_3Au and experimental data, whereas Cowley[6] suggests a value of 0.1. Fosdick also observed that T_c varied linearly with v_2/v_1 in his computer experiments.

5.3 Long-Period Superlattice Theory

Figure 27 defines one- and two-dimensional LPS's in an fcc A_3B alloy. The one-dimensional LPS for an AB fct alloy was defined in Section 3 for CuAu(II). In the one-dimensional antiphase structure, the superperiod exists only in one direction in any particular part of the crystal. This structure can be described by using a long unit cell composed of a row of fundamental unit cells for the disordered fcc lattice. This long cell lies along the direction of the period and contains two APD's of size M_3a_3. Here, a_3 is the dimension of the unit cell for the disordered alloy in the $z = x_3$ direction, and M_3 is the number of these cells per APD in this direction. For CuAu(II), $M_3 = 5$. The distance $2M_3$ is called the period of the one-dimensional LPS.

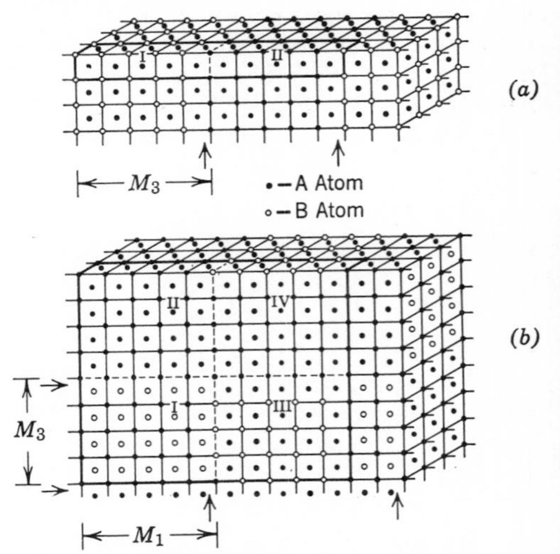

Fig. 27. Unit cells and domain types for (*a*) one-dimensional A_3B LPS and (*b*) two-dimensional A_3B LPS. Arrows indicate APB's and Roman numerals domain types. M_1, and M_3 are the domain size in each direction. (Figure 1 of Reference 56.)

The two-dimensional LPS is defined in terms of an orthorhombic unit cell that contains four APD's of size $M_1a_1xM_3a_3$. The half-periods M_1 and M_3 are assumed to be integers. The two-dimensional periodic structure is not a simple superposition of a one-dimensional periodic structure. Rather, the APD's in the two directions are connected by a definite relation, that is, the nature of the antiphase relations between APD's I and II, in Fig. 27, and that between III and IV should be the same. This is also true for I and III and II and IV. A three-dimensional structure would be described by an orthorhombic unit cell containing eight APD's; however, this structure has never been observed, and Sato and Toth show that its existence is improbable.

The basic ideas of the Sato-Toth theory are as follows: (1) The stabilization of a LPS is principally determined with a minimization of the internal energy; that is, entropy considerations are unimportant. (2) The existence of a LPS depends upon the relative magnitudes of the decrease in the free electron kinetic energy, due to the splitting of the Brillouin zone by the introduction of an extra period, and the increase in the energy required to create APB's. Electronic energy considerations are the most important in the case of small periods. The theory predicts that the half-period, M, for one-dimensional APD's is given by

$$e/a = \frac{\pi}{12t^3}\left(2 \pm \frac{1}{M} + \frac{1}{4M^2}\right)^{3/2} \tag{8}$$

where e/a is the electron per atom ratio, t is a truncation factor used to represent the non-sphericity of the Fermi surface, and M is the stable half-period. The positive sign in Eq. 8 is used if e/a exceeds the critical value $(e/a)_c$, defined in the next paragraph; if $e/a < (e/a)_c$, the negative sign is used.

The relation between e/a and $1/M$ was found by experiment to be linear, as shown in Fig. 28. This was determined by introducing other elements into the alloy of interest and observing the change in M for a known e/a. The value of e/a for an infinite period, that is, $1/M = 0$, is called the critical electron to atom ratio, $(e/a)_c$. It was found by experiment that if $e/a > (e/a)_c$ in the unperturbed alloy, M would decrease when e/a was increased, and that the converse was true if $e/a < (e/a)_c$. Sato and Toth's theoretical

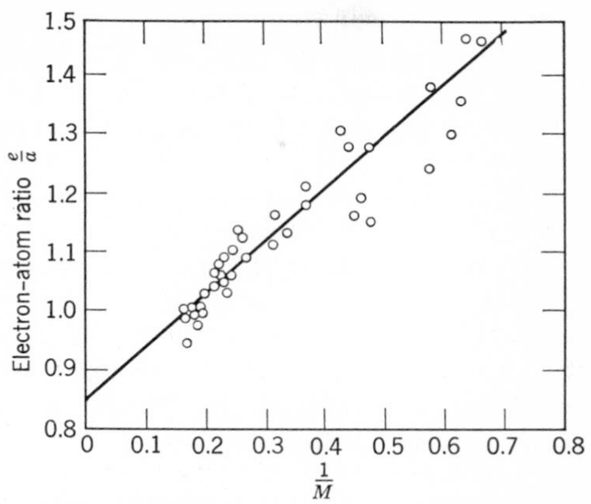

Fig. 28. Electron-atom ratio as a function of reciprocal domain size ($1/M$), observed when additional elements were added in CuAu(II). (Figure 16 of Reference 55.)

approach was based on these fundamental observations. An outline of the theoretical argument follows.

The first Brillouin zone boundary for the disordered CuAu alloy is represented by the thin lines in Fig. 29. It is bounded by {111} and {200} planes. The volume of this zone (in k-space) is $4/a^3$ and corresponds to two electrons per atom. The volume of the inscribed sphere for this zone is $0.681 \times 4/a^3$, which corresponds to 1.362 electrons per atom.[56] If we assume that the Fermi surface is spherical and ascribe one electron each to Cu and Au, it follows that the Fermi surface for the disordered alloy lies well within the first

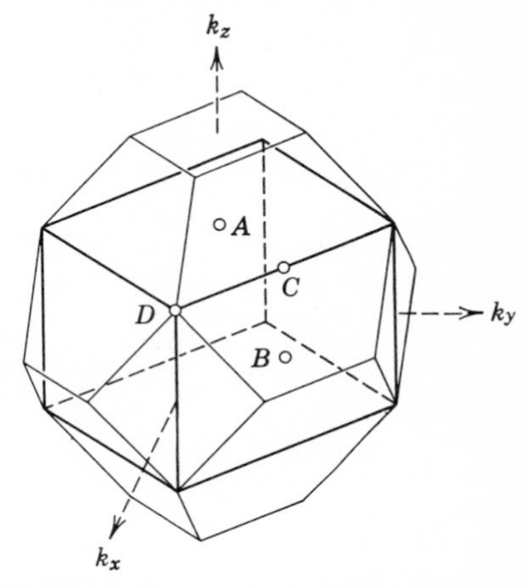

Fig. 29. Brillouin zone structure for CuAu. Outer thin lines define the s zone for the disordered phase; thick lines define the added zone for ordered CuAu(I). (Figure 17 of Reference 55.)

Brillouin zone. However, when the alloy orders to CuAu(I), the Brillouin zone splits due to the added periodicity, and a new Brillouin zone, defined by heavy lines, is formed. This new zone is bounded by {001} and {110} planes. Its volume is $2/a^2c$, and it accommodates one electron per atom. The associated inscribed sphere now accommodates only 0.26 electrons per atom. Hence one can expect overlap of the Fermi surface into the second zone. Free electron energies at points A, B, C, and D, on the new zone boundaries, were estimated to be 2.4, 4.8, 7.1, and 9.5 eV, respectively, and the energy at the surface of the inscribed sphere was estimated to be 6.5 eV. This indicates that overlap should occur, at least, in the c-direction, that is, the direction associated with point A and k_z in Fig. 29. Up to this point the argument of Sato and Toth parallels that of Nicholas.[53]

When the alloy orders to CuAu(II), there is an additional periodicity in the b-direction (that of k_y) which causes a further splitting of the Brillouin zone, as illustrated in Fig. 30. The section of the zone shown in this figure is the one cut by a (001) plane through the origin. Zone changes for directions associated with large overlap are not considered, because they cannot affect the behavior of electrons in the stabilization process. The separation of zones is measured by the separation of superlattice spots in an electron diffraction photograph, as shown by the two radial lines drawn from the origin to the spots in the upper right part of the figure. As the APD period becomes smaller these spots separate more widely, and hence so does the zone separation. The degree

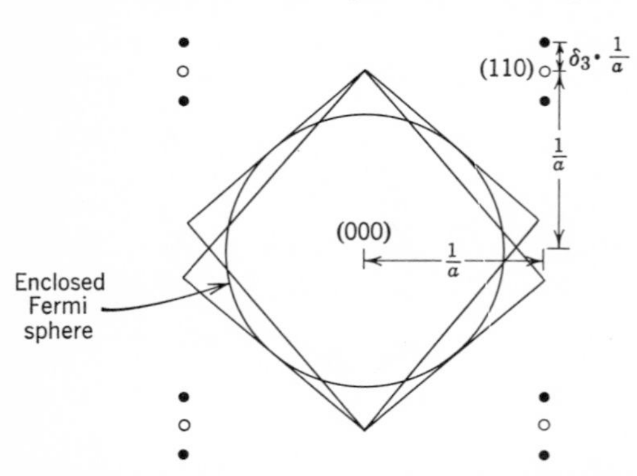

Fig. 30. Structure of the Brillouin zone CuAu(II) and enclosed Fermi sphere in a plane through the origin of reciprocal space and parallel to (001). (Figure 4 of Reference 56.)

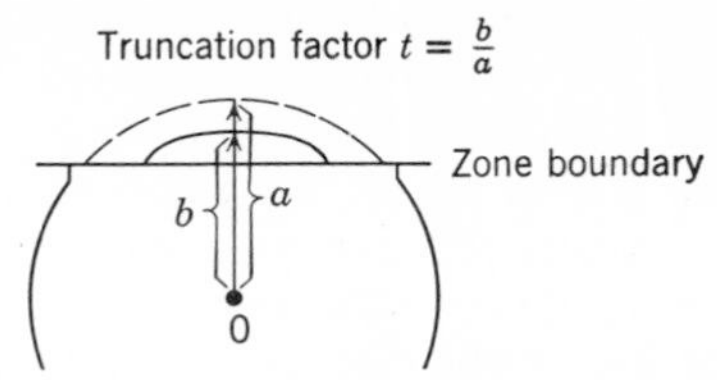

Fig. 31. Schematic drawing of Fermi surface showing the overlapping of the inner zone boundary.

of overlapping, as well as the degree of energy stabilization at a zone boundary, depends sensitively on the energy gap at the boundary. A reasonable magnitude of the energy gap is of the order of a fraction of an electron volt. Because the Fermi surface comes close to the {110} plane boundaries, the energy gap across this boundary is very important in determining electron behavior. The Fermi surface is quite close to the *outer* boundary of the split zone in view of the energy values stated for the points A–D. As the period of the APD structure decreases, the separation of the boundaries increases to provide more electrons per atom inside the outer boundary.

The volume V of the inscribed Fermi surface to the outer zone can be calculated from the separation, $2\delta/a$, of superlattice spots, which is directly related to the APD period by $\delta = 1/2M$ for a primary reflection. The result is

$$V = \frac{\pi}{6a^3}\left(2 + \frac{1}{M} + \frac{4}{4M^2}\right)^{3/2} \tag{9}$$

Taking into account the nonsphericity of the Fermi surface through a truncation factor, t, defined by Fig. 31, and the fact that there are four atoms per unit cell, the number of electrons per atom is

$$e/a = \frac{\pi}{12t^3}\left(2 + \frac{1}{M} + \frac{1}{4M^2}\right)^{3/2} \tag{10}$$

The value, $t = 0.95$, was obtained for CuAu(II) by taking $M = 5.0$ and $e/a = 1$ for stoichiometric CuAu(II).

If we carry through a similar analysis for the case in which $e/a < (e/a)_c$, it follows that stabilization is determined by the relation between the inscribed Fermi surface and the *inner* zone boundary. Fig. 32 shows the variation of e/a versus M for each of these two cases. Experimental data on the Cu–Pd system confirm the result for $e/a < (e/a)_c$. Whenever the theory predicts a large but finite period, an infinite period is observed. It

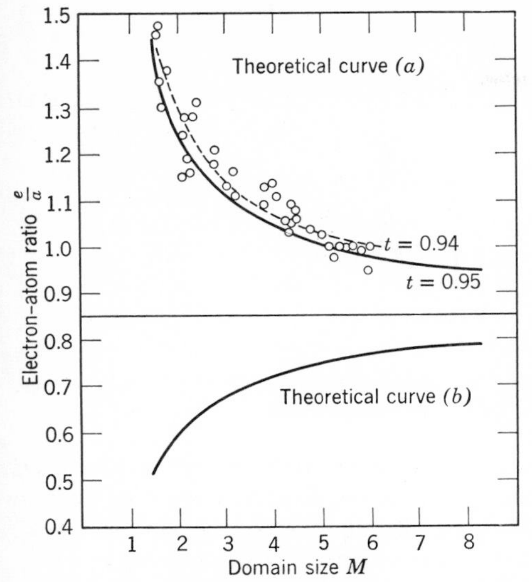

Fig. 32. Theoretical curve of electron-atom ratio versus domain size M: (a) stabilization by outer Brillouin zone; (b) stabilization by inner Brillouin zone. Points O represent experimental data for CuAu(II). (Figure 20 of Reference 56.)

is reasoned, in this case, that, because a large period is associated with only a small splitting of the Brillouin zone, the associated electron kinetic energy decrease is too small to compensate for the energy increase associated with the associated APB formation for a LPS.

Calculations for one-dimensional long-period superlattices in a fcc A_3B alloy give exactly the same conditions for long-period stability found for CuAu(II). In the case of two-dimensional APD's for a fcc A_3B alloy, there is no way of inscribing one sphere to all the separated Brillouin zone boundaries. To find the condition for a maximum reduction of the internal energy in this case, detailed information about energy contours and the energy gap at the boundary is required. It was possible to avoid this difficulty in the one-dimensional calculations because the boundary separations were equal. The application of this theory to two-dimensional LPS's gives qualitative agreement with experiment, but it does not give a unique solution, as it does in the case of one-dimensional APD's. This difficulty arises because a detailed knowledge of the alloy electronic structure is required in the two-dimensional case.

Toth and Sato[55] have subjected their theory to a number of tests. A very dramatic test was one in which a LPS structure was induced in Au_3Cu, as predicted by their theory, by adding aluminum to increase its e/a ratio. Normally, Au_3Cu does not form a LPS.

6. MECHANICAL EFFECTS

Westbrook[73] has reviewed the literature up to 1959 on the general effects of order upon the mechanical properties of intermetallic compounds, and Lawley (Chapter 24) covers more recent studies. The purpose of this section is to describe those effects primarily associated with the presence of APD's, that is, domain strengthening. A general discussion of APB-dislocation interactions is given by Brown in Chapter 15 of this book. Excellent summaries of the present knowledge and theory related to domain strengthening are given in recent articles by Marcinkowski and Fisher[74] and Stoloff and Davies.[75] These discussions treat alloys that exhibit B2, $D0_3$, $L1_2$, and $D0_{19}$ superlattices in the ordered state. Descriptions of electron microscope studies on order strengthening, up to July 1961, are included in a recent book edited by Thomas and Washburn.[77]

The influence of the state of order on mechanical properties, that is, dislocation behavior, changes its character as a function of LRO and SRO. This circumstance requires that we touch upon the salient features of APB-dislocation interactions and the current theories of order strengthening in order to define the effect of APD's per se. An ordinary dislocation leaves an APB in its wake as it moves through an ordered alloy, as shown schematically in Fig. 33. Because the energy required to form an APB is generally nonzero (see § 5), it is harder to move an ordinary

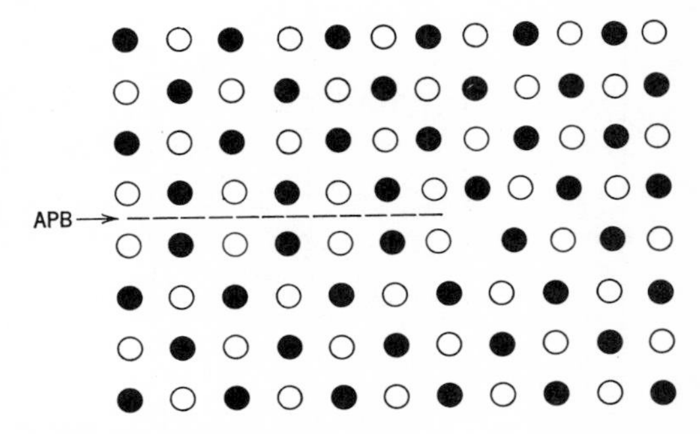

Fig. 33. APB created by passage of an ordinary dislocation (unit dislocation).

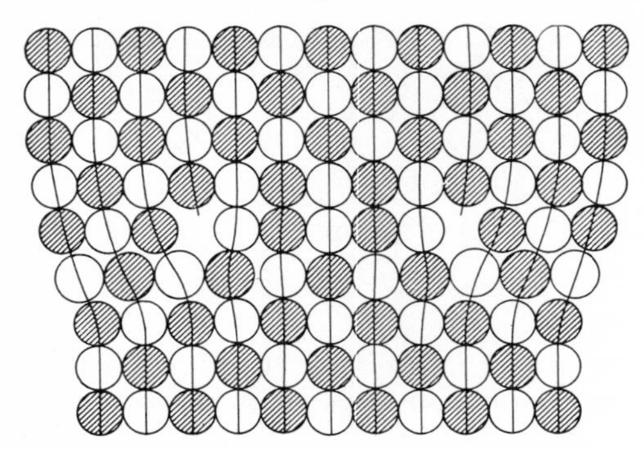

Fig. 34. Superdislocation in a B2 superlattice. (Figure 8 of Reference 62.)

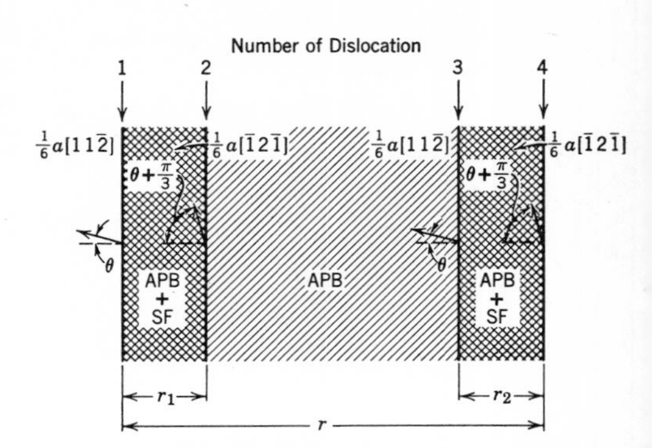

Fig. 36. Superdislocation in a $L1_2$ superlattice. (Figure 2 of Reference 64.)

dislocation through an ordered region than through a disordered region. Koehler and Seitz[76] first suggested that a pair of dislocations, bound together by a strip of APB, could move through an ordered region more easily than an ordinary dislocation. As the dislocation pair moves, disorder created by the leading dislocation is erased by the trailing dislocation. This configuration is indeed observed[64,74,78,79,81] and is called a superdislocation or superlattice dislocation. The ways in which either ordinary or superdislocations move in a particular alloy are governed by its SRO, LRO, crystal structure, and composition. For example, Cu_3Au has such a low critical temperature that deformation mechanisms that depend upon atomic diffusion cannot be operative in ordered Cu_3Au. Also, dislocation motion in Cu_3Au is different from that in Fe_3Al, because APB follow crystallographic planes in Cu_3Au but not in Fe_3Al.

In the B2 superlattice[61] a superdislocation is composed of two ordinary dislocations joined by a strip of APB, as shown in Fig. 34. This is the configuration Koehler and Seitz used to illustrate the superdislocation concept; however, more complicated superdislocation configurations can exist. Superdislocations in the $D0_3$ superlattice[63] are expected to be qualitatively the same as those shown in Fig. 35. In this figure NN stands for nearest-neighbor and NNN for next-nearest-neighbor. The superdislocation in a $L1_2$ superlattice[64] is composed of two pairs of partial dislocations, that is, two extended dislocations, as shown in Fig. 36. Both symmetric and asymmetric superdislocation configurations are possible in the $L1_0$ superlattice.[64] The asymmetric structure appears in Fig. 37. Again four partials are involved. In the symmetric configuration, the two extended dislocations are not bound, and the structure degenerates into two independent partial dislocations. The separation distance between the components of a superdislocation increases rapidly as the APD size decreases or LRO increases. This effect is described by Figs. 38 and 39.

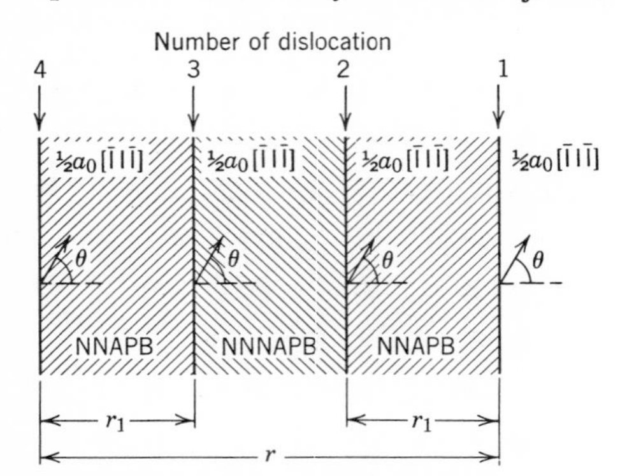

Fig. 35. Superdislocation in a $D0_3$ superlattice. (Figure 3 of Reference 63.)

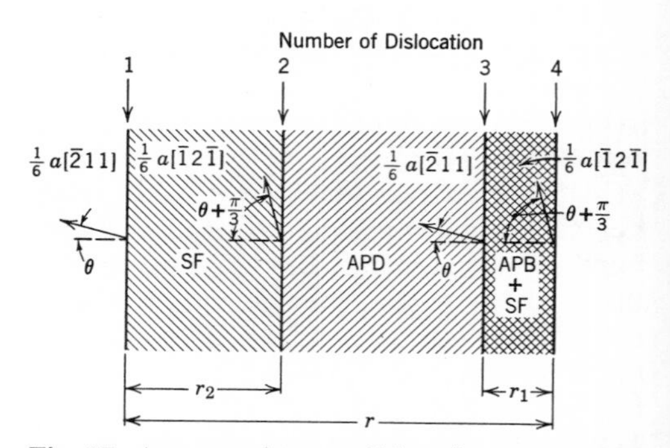

Fig. 37. Asymmetric superdislocation structure in a $L1_0$ superlattice. (Figure 5 of Reference 64.)

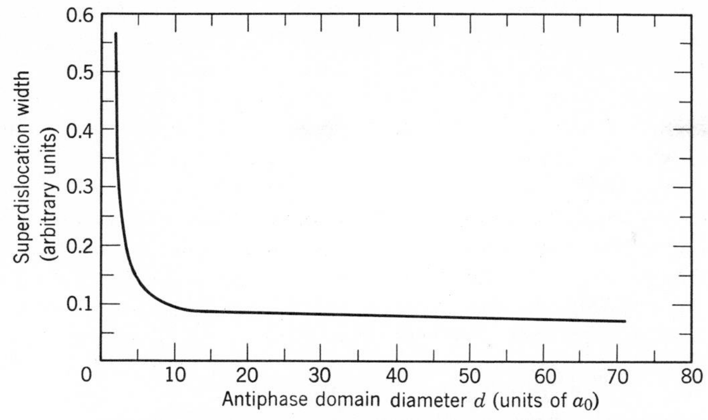

Fig. 38. Effect of APD size (d) on the width of a superdislocation in a B2 superlattice. (Figure 3 of Reference 74.)

6.1 Theoretical Models for Order Strengthening

Each model listed below (with the possible exception of the models due to Fisher and Cottrell) was devised to explain the behavior of a particular alloy. Consequently, no one model should be expected to be generally applicable. There appears to be one general feature of order strengthening, which has been observed in all cases studied thus far. This feature is the appearance of a maximum in the flow stress-quenching temperature curve,[81] which occurs either at or below the critical temperature. It has been observed for CuZn,[61] FeCo–V,[75] Cu_3Au, CuAu, Cu_3Pd, Cu_3Pt[82] Ni_3V, Pd_3V, Co_3V,[83] and Fe_3Al.[84] Any general model must explain this maximum. In addition, the strength of an alloy always depends upon the degree of LRO and, when they exist, upon the size of APD's.

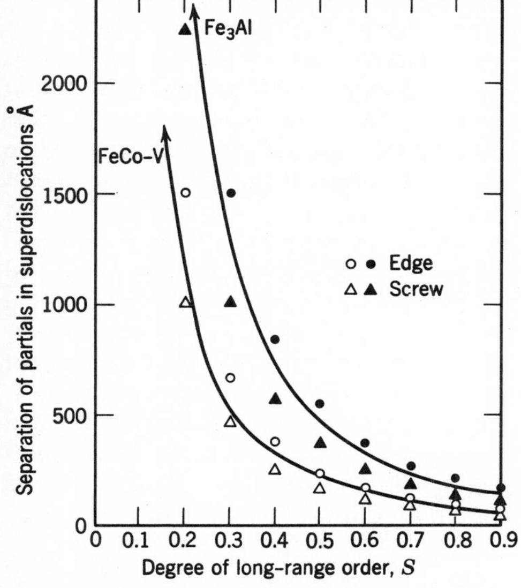

Fig. 39. Width of a superdislocation as a function LRO. (Figure 11 of Reference 75.)

According to Fisher[85] the SRO stress increment, τ(SRO), seen by a dislocation is

$$\tau(\text{SRO}) = \frac{\gamma}{b} \tag{11}$$

where γ is the APB energy and b the magnitude of the Burgers vector. τ(SRO) should be nearly temperature-independent at low temperatures because (in the absence of diffusion) the stress increment required to maintain the displacement of a dislocation in SRO material is a measure of the SRO strengthening effect. This increment is not determined by a thermal activation process. Fisher's idea also can be applied to the case of the departure from perfect LRO, as well as to the case of the departure from a completely random arrangement.[86] Finally, in Fisher's model a larger stress is required to start deformation than to maintain it because the first dislocations to pass across the slip plane will produce more disorder than those that follow.[87]

Cottrell[87] showed that the strength of an alloy can depend explicitly on the APD diameter, ϵ. His equation for the yield stress, τ, is

$$\tau = \frac{\gamma}{\epsilon}\left(1 - \frac{\alpha t}{\epsilon}\right) \tag{12}$$

where ϵ is the APD diameter, t the APB thickness, and γ the APB energy. $\alpha \simeq 6$ is a domain shape factor. Eq. 12 predicts a maximum in the yield stress for $\epsilon \simeq 20$ Å. Brown[61] suggested that Eq. 12 is more closely related to flow stress than yield stress for large ϵ, because it was derived by assuming (implicitly) that yielding is associated with a shear displacement of $\epsilon/2$ along the slip plane.

Sumino[88] considered both the SRO effect and the interaction of the dislocation stress field with the change in nearest-neighbor distance caused by ordering. In the latter instance, stress-induced alignment produces oriented SRO. The tendency to form oriented SRO in a stress field is pronounced in iron-aluminum alloys,[89,90] hence dislocation locking due to oriented SRO is quite feasible in these systems. LRO will interfere with the development of oriented SRO in dislocation stress fields. Because of this, Sumino's model predicts a maximum in the dislocation-locking force at the critical temperature.

Brown[61] showed that a stress is required to move a superdislocation and that it should attain a maximum at some value of intermediate degree of LRO. Assuming first-neighbor interactions and Bragg-Williams theory,[1] he obtained Eq. 13 for the stress τ to move a superlattice dislocation in a B2 superlattice.

$$\tau = \frac{NkTS^2}{4b} \tag{13}$$

S is the equilibrium LRO in the perfect crystal, N is the density of atoms on the (110) plane. b the magnitude of the Burgers vector, k is Boltzmann's constant, and T the absolute temperature in °K. Brown's model gives a sharp peak in the yield point for β-brass, as a function of temperature, at 515°K. This peak is observed experimentally at 493°K.

None of the models described above considered the effects of diffusion. This aspect is included in Flinn's[62] theory of deformation in superlattices. If diffusion occurs, dislocations may climb, and order-produced pinning may be altered. At temperatures high enough for diffusion to occur during deformation, portions of the dislocations rings produced during deformation may leave the slip plane when a lower energy orientation for the common plane of the superdislocation pair exists. The component dislocations would then be more difficult to move, because each would produce APB if it moved. This suggests that the yield stress should, in some appropriate temperature range, increase with temperature. This effect is observed in alloys based on Ni_3Al,[62] for example, and in β-brass single crystals.[38] According to Flinn's theory, the dominant resistance to dislocation motion at elevated temperatures would be viscous drag.

Marcinkowski and Miller[81] proposed another strengthening model to account for their data on the maximum in the flow stress-temperature curve for Ni_3Mn. They proposed that superdislocations, even though little resisted by LRO, are resisted by an impediment due to SRO. This resistance should decrease near the critical temperature where the SRO falls off sharply. When APD's are large, $\epsilon \sim 500$ Å, superdislocations meet with little LRO resistance. However, because the APB energy varies as S^2, it must decrease rapidly with decreasing LRO as the temperature is raised. As this occurs, the component dislocations in a superdislocation will tend to move more and more independently of one another as their separation distance increases. In this event, the partially independent imperfect dislocations must do work in the destruction of SRO. This argument suggests that partially ordered alloys should be stronger than either perfectly ordered or completely disordered alloys.

6.2 Experimental Results

Ardley[86] carried out a systematic experimental study on order strengthening in single crystal Cu_3Au, which is regarded as one of the classic investigations performed in this field. In brief, he found that as the order within APD's increases, the room-temperature strength decreases. Furthermore, as the size of the APD's increases, the strength at first increases with respect to that for the disordered state, attains a sharp maximum, and subsequently decreases to the equilibrium value for perfect LRO. At the critical temperature there is a considerable weakening of the alloy as it passes into the disordered state. Above the critical temperature, the strength once again rises and then falls. The temperature, at which the latter peak occurs, increases with strain rate; it was attributed by Ardley to strain aging.

Figure 40 is Ardley's plot of the critical resolved shear stress versus APD size. The peak occurs at $\epsilon \sim 30$ Å. That part of the

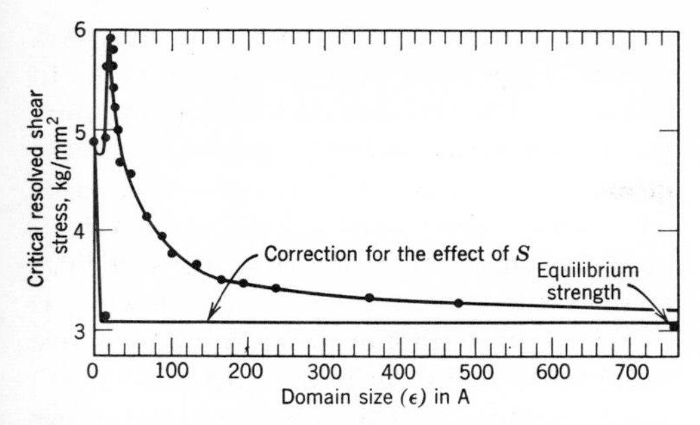

Fig. 40. Effect of APD size (ϵ) on flow stress in Cu_3Au. (Figure 2 of Reference 86.)

strengthening contributed by the presence of APD's is the difference between the peaked curve and the line *BC*. The latter corresponds to the LRO contribution to strength. The experimental points obtained by Ardley are fitted by the equation

$$\tau_S = \frac{\gamma}{\epsilon}\left(\frac{\epsilon - t}{\epsilon}\right)^3 \tag{14}$$

where τ_s is the critical resolved shear stress, ϵ the APD diameter, and t the APD boundary thickness. This equation reduces to Cottrell's equation (Eq. 12) for $\epsilon \gg t$. Therefore, it is felt that the behavior of Cu_3Au is largely due to the interaction of dislocations with APD boundaries.

Brown[61] found no essential difference in the temperature dependences of the yield point for single-crystal and polycrystalline β–CuZn. Both exhibited a sharp drop near and above the critical temperature, which Brown attributed to the Sumino mechanism. His order-strengthening mechanism was developed to explain the peak observed below the critical temperature.

Lawley et al.[84] used polycrystalline samples of Fe_3Al in a systematic order-strengthening study. Fe_3Al does not undergo a discontinuous change in LRO at the critical temperature (as does Cu_3Au, for example), and hence it was possible to vary LRO from zero to about 0.8 in this study. The major purpose of this study was the detection of variations in strength caused purely by changes in atomic configuration. It was found that the variation in flow stress at room temperature was not large, but, nevertheless, a peak occurred at the critical temperature. They suggest that the flow stress was determined by Fisher's mechanism for quenching temperatures $T_q > T_c$ and by Sumino's mechanism for $T_q < T_c$, with the combination of the two mechanisms giving a peak at the critical temperature. This point has also been discussed by Rudman.[91] In one alloy composition (27.8 per cent Al) there was a sudden rise in flow stress above the critical temperature at 800°C. They remark that it is reasonable to interpret this as being due to a sudden decrease in APD size in view of Taylor and Jones's[92] observation of a sudden broadening of superlattice lines at this temperature. Isothermal annealing and hardness results were in general accord with Cottrell's model. There was no sign of a continued variation of flow stress with domain size as proposed by Flinn.[62] They concluded that Brown's mechanism did not apply because at the time it was thought, on the basis of foil studies, that superdislocations did not exist in Fe_3Al.[63] However, more recent work indicates that superdislocations do exist in bulk Fe_3Al.[74]

Direct observation of dislocations in Ni_3Mn by Marcinkowski and Miller,[81] coupled with flow stress measurements, revealed behavior that could not be explained by the mechanisms described in Section 6.1. Plastic strains up to 10 per cent were considered. The rate of work hardening observed in the ordered alloy was about twice as large as that in the disordered alloy. Similar behavior was observed by Flinn[62] in an alloy based on the Ni_3Al superlattice and by Biggs and Broom[93] in Cu_3Au. Flinn suggested that this high rate of work hardening was caused by a progressive decrease in APD size which accompanies increasing amounts of deformation. Cohen and Bever[94] argue that this is the process by which cold work destroys order on the basis of superlattice line broadening with cold work.

A peak in the flow stress versus quenching temperature existed even at 10 per cent strain. This fact appears to eliminate Brown's mechanism as the principal cause of the peak in Ni_3Mn, because most of the initially pinned dislocations should be freed from their equilibrium APB configurations, at such a large strain, with the result that their capability to cause a maximum should disappear. The Sumino mechanism was excluded because it had been established previously[95,96] that no detectable change in lattice constant occurs with ordering in Ni_3Mn. It had also been established that the APD size was at least 500 Å at the temperature at which the peak occurred, which rules out Cottrell's mechanism as a principal causative factor. In view of these difficulties, they suggested that superlattice dislocations experience a resistance in Ni_3Mn due to SRO. This is plausible because APD formation ($\epsilon \simeq 500$ Å) is well established in Ni_3Mn not far below the critical temperature. It was further suggested that increasing ease of cross slip might contribute to order strengthening by forcing a given number of dislocations to travel over a greater number of slip planes, resulting in the destruction of large amounts of SRO.

6.3 Recapitulation

Marcinkowski and Fisher[74] and Stoloff and Davies[75] have critically examined the data and theory on order strengthening. Their conclusions appear to furnish a general basis for explaining the difficulties mentioned in Section 6.2. These authors base their discussion on a transition from deformation resulting predominantly from superdislocation motion, in the perfectly ordered state, to that resulting from unit dislocation motion in the disordered state. The principal ideas involved are those introduced by Fisher,[85] Brown,[61] and Marcinkowski and Miller.[81] They also discuss the competition between slip and twinning, as a function of the state of order, and the influence of order on the slip mode. Marcinkowski and Fisher treat systems containing APD's and therefore cast their description in terms of APD size. Stoloff and Davies use LRO as a reference in order to cover systems without an APD structure. In each case, the essential idea is that, starting from a perfectly ordered state, the resistance to superdislocation motion increases with temperature as superdislocations tend to dissociate into independent dislocations. This occurs because the superdislocation component separation increases with decreasing APD size (decreasing LRO). This allows the components to move more and more independently and as a consequence to produce progressively more APB. At the same time the APB energy which is proportional to S^2, falls off rapidly as the critical temperature is approached from below. The composite effect of increasing APB production and diminishing APB energy leads to a peak in the flow stress at or below the critical temperature.

If the alloy does not exhibit a discontinuity in LRO at the critical temperature, this peak occurs below the critical temperature. In alloys that do exhibit such a discontinuity superdislocations suddenly dissociate at the critical temperature. This circumstance leads to a flow stress peak at the critical temperature. In the final analysis, the particular manifestation of this general behavior is governed by both the absolute and relative magnitudes of the ordering energies for nth order neighbors. The absolute magnitudes determine the resistance to ordinary dislocations, whereas the relative magnitudes play the dominant role in determining the stability of superdislocation configurations and the stress required to move them.[74]

Figure 41 illustrates the effect of APD size on the stress required to move a dislocation, as determined by Marcinkowski and Fisher for the B2 superlattice. In this figure, d is the APD diameter, a_0 the lattice constant, and v the first neighbor ordering energy. The maximum resistance to a superdislocation (superlattice dislocation) occurs for a domain size of $4a_0$ A schematic description of the variation in flow stress with temperature, given by Stoloff and Davies, appears in Fig. 42 for three particular alloys. The sharp drop for Cu_3Au at $T = T_c$ is the result of its LRO discontinuity at the critical temperature.

The presence of order is known to suppress mechanical twinning.[80] Therefore, in addition to its effect upon ductility, via an influence upon dislocation motion, it can also affect ductility by restricting the number of deformation modes. Furthermore, the presence of APD's can also influence the mode of slip.[74] For example, in the B2 superlattice, the minimum APB energy is that for {110} orientation. Hence multiple and wavy slip, often associated with disordered bcc alloys, should decrease upon ordering because of a tendency for {110} planes to be preferred. This tendency is enhanced by a large ordering energy and/or a small APD size.

7. ELECTRICAL RESISTANCE AND GALVANOMAGNETIC COEFFICIENTS

It has been known for some time that both the electrical resistivity (1938) and possibly the Hall coefficient (1941) vary with APD size. The first systematic exploration of the effect of APD size on electrical resistance was made by

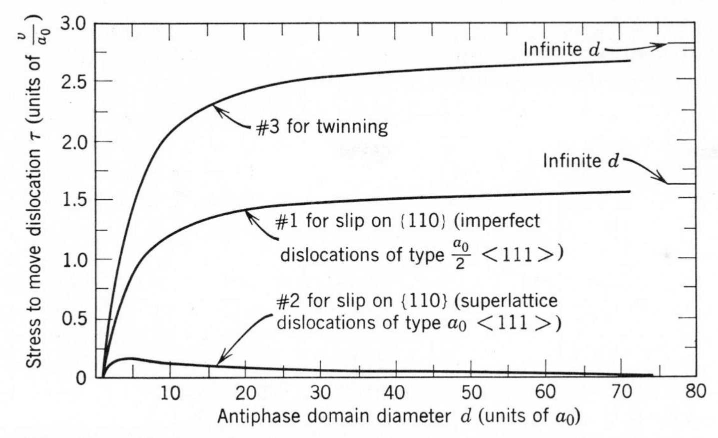

Fig. 41. Variation of critical stress for slip and twinning with APD size (d) for perfect and imperfect dislocations in a B2 superlattice. (Figure 2 of Reference 74.)

Sykes et al.[8] for Cu_3Au. They found that the electrical resistance decreased as the APD size increased, the resistivity, ρ, being given by

$$\rho = \frac{h}{e^2}\left(\frac{3}{\pi N^2}\right)^{1/3}\left(\frac{1}{\lambda_0} + \frac{r}{\epsilon}\right) \tag{15}$$

where h is Planck's constant, e the electronic charge, N the number of conduction electrons per unit volume, r the probability of a conduction electron being scattered at an APD boundary, λ the electron mean free path, and ϵ the average APD diameter. This study pertained to temperatures below the critical temperature. However, prior to 1955, nearly all attempts to investigate the influence of APD's on electronic behavior were based solely on electrical resistivity measurements. However, since that time a considerable amount of information on the relationship between APB's and APD's and the electronic structure of an alloy, and their essential role during the evolution of the ordered state, has been obtained by simultaneous measurements of electrical resistance and galvanomagnetic coefficients. Most of this work has been done on CuAu and Cu_3Au. After we outline the reasons for employing this type of experimental approach, the principal results for the Cu–Au system will be summarized.

In most orderable alloys, the electrical resistance decreases monotonically as the LRO increases. At one time it also was assumed that the electrical resistivity was a monotonically decreasing function of SRO above the critical temperature. It is now known that no general correlation exists between the direction of the electrical resistivity change and the increase of either LRO or SRO. In a first approximation the electrical resistance, ρ, is given by

$$\rho = \frac{m}{e^2 n}\langle\tau\rangle \tag{16}$$

where n is the effective number of free electrons per unit volume, $\langle\tau\rangle$ is the average relaxation time for electron scattering, taken over the Fermi surface (FS), and e and m are, respectively, the electronic charge and mass. According to Eq. 16, the major effects of LRO and/or SRO upon ρ must arise via their influence on n and $\langle\tau\rangle$. The experiments of Sykes et al.[7,8] showed that the existence of APD boundary perturbs $\langle\tau\rangle$. Damask[97] demonstrated that ρ

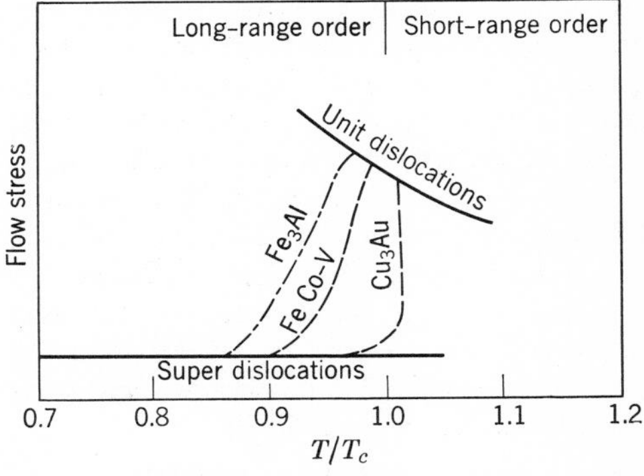

Fig. 42. Schematic variation of flow stress with quenched-in order for Fe_3Al, FeCo–V, and Cu_3Au. (Figure 13 of Reference 75.)

may either increase or decrease with SRO above the critical temperature, that is, in the absence of LRO. This result indicated that a more subtle effect than additional APB scattering was operative. Gibson,[98] therefore, theoretically considered the effect of SRO on ρ, on the basis of the Nordheim approximation, and concluded the following: SRO may be expected to decrease the resistivity of a binary solid solution whose constituents have different valence when the Fermi surface (FS) passes close to a Brillouin zone boundary for the random lattice. An increase should occur when the FS is close to a zone boundary for the superlattice. Beal[99] has treated the problem for temperatures both above and below the critical temperature. Beal's calculations are consistent with Damask's experimental results and indicate a decrease in resistance with increasing LRO below the critical temperature. The resistance anomaly has also been treated by Pick for both ordinary superlattices and LPS's[100] and by Hall.[101] Phillips,[102,103] however, has found the residual resistance of ordered Cu_2NiZn to be larger than that of the disordered state down to 4.2°K. He suggests that this is due to a decrease in the effective number of electrons associated with a splitting of the Brilluoin zone during ordering and to the presence of APB's or to the latter phenomena alone. In addition to Phillips' work, it is also known that ρ is larger in the ordered state of CoPt than in the disordered state.

Recognition that the use of electrical resistance measurements alone could not be used to monitor all changes in the ordered state stimulated investigations on the behavior of the galvanomagnetic coefficients. As mentioned in Section 5, electron states near the FS are changed because of Brillouin zone splitting during ordering. Such changes generally cause a considerable alteration in the values of the galvanomagnetic coefficients. A sensitive measure of changes in states near the FS is given by simultaneous measurements of ρ and the magnetoresistance coefficient, B. This coefficient is determined by the ratio $f(\mathbf{k}) = m/m^*(\mathbf{k})$, where m^* is the effective electron mass, and the electron scattering relaxation time τ. In particular

$$B = \frac{e}{mc^2}\left(\frac{\langle f^2\tau^3\rangle\langle\tau\rangle - \langle f\tau^2\rangle}{\langle\tau\rangle^2}\right) \tag{17}$$

All averages $\langle\rangle$ in Eq. 17 are taken over the FS. The magnetoresistance change, $\Delta\rho$, produced by an external magnetic field $\mathbf{H}$ is

$$\frac{\Delta\rho}{\rho} = BH^2 \tag{18}$$

The coefficient B measures fluctuations in f, that is, m^*, at the FS and the product

$$q^2 = \rho^2 B \tag{19}$$

measures distortion of the FS. The truth of this statement is clearly shown by examination of Eq. 77.2 on page 282 of Mott and Jones.[104] This feature allows the detection and measurement of FS distortion, produced as a consequence of ordering, by measuring ρ and B.

The thermoelectric power (Q) is sensitive to the detailed nature of the electron scattering process responsible for the electrical resistivity. Q is determined by $N(E)$, τ, and the electron velocity, v, at the Fermi surface through the function

$$S = \left(\frac{\pi^2k^2T}{3}\right)\left[\frac{\partial}{\partial E}\log\int\frac{\tau v^2 d\sigma}{|\nabla E_k|}\right]_{E=E_F} \tag{20}$$

The thermoelectric power Q_{ab} of metal a with respect to metal b is given by

$$Q_{ab} = -\left(\frac{S_a - S_b}{e}\right) \tag{21}$$

Combined measurements of Q, q^2 and ρ provide a nearly complete description of the relation between electron state changes and the state of order. Hall coefficient (R) measurements allow one to detect changes in the relative amounts of hole and electron conduction associated with ordering. R is given by

$$R = -\left|\frac{1}{ec}\right|\left[\frac{n_e\mu_e^2 - n_h\mu_n^2}{n_e\mu_e + \mu_n n_n^2}\right] \tag{22}$$

where n and μ are the carrier concentration and mobility, respectively. The subscript e denotes electrons and h denotes holes.

The Hall coefficient for Cu_3Au has been studied by Komar et al.,[105,106] Von Neida and Gordon,[107] and Elkholy and Nagy.[13] Von Neida and Gordon investigated the dependence of R on LRO, SRO, and APD size. The work of Elkholy and Nagy was principally concerned with characterizing different stages in the evolution of the ordered state. In Cu_3Au, R is negative when the alloy is disordered and positive when it is ordered. The fact that R is always negative in CuAu shows that the change of sign observed in Cu_3Au is not a general characteristic. However, a positive change in R with increasing LRO does appear to be a general characteristic.

The data of Von Neida and Gordon, together with those of Flanagan and Averbach,[108] give R values more positive than those given by the free electron approximation over the entire range of composition for Cu–Au alloys. This indicates the presence of hole conduction in both the pure metals and the alloys, a behavior which is consistent with Pippard's[109] construction of the FS for copper. The observed variation of R with SRO was small. It was therefore concluded that the FS is not much affected by the diffuse Brillouin zone splitting due to SRO proposed by Averbach et al.[110] Von Neida and Gordon suggest that the observed changes of R with SRO could also be attributed to changes in the electron scattering relaxation time.

In order to study the effect of APD size (ϵ) on R, independently of the degree of order in the APD, the normalized parameter

$$R_n \equiv \frac{R - R_0}{R_\infty - R_0} \tag{23}$$

was plotted against ϵ. In Eq. 23, R_0 is the value of R for the disordered state and R_∞ that for the largest domain size attainable at a given temperature, that is, equilibrium S value. This plot is shown in Fig. 43. At the two temperatures well removed from the critical temperature (390°C), R_n becomes negative at $\epsilon \simeq 35$ Å. It is thought that an interaction of conduction electrons with APB's produces the ϵ-dependence of R. If this were true, then the larger is the electron mean free path, the larger should be the ϵ-value at which R becomes negative. Measurements at liquid nitrogen temperature, where the electron mean free path is longer than that at room temperature, supported this hypothesis. The observed shift to a larger APD size is shown in Fig. 44.

When the APD size is much larger than the electron mean free path, only a small fraction of the conduction electrons will interact with APB's, and the isolated influence of LRO on R can be estimated. A plot of R versus temperature appears in Fig. 45 for two APD sizes. In the case of the larger APD size, the variation of R with temperature was very similar to that of the LRO; however, R and S were not linearly related.

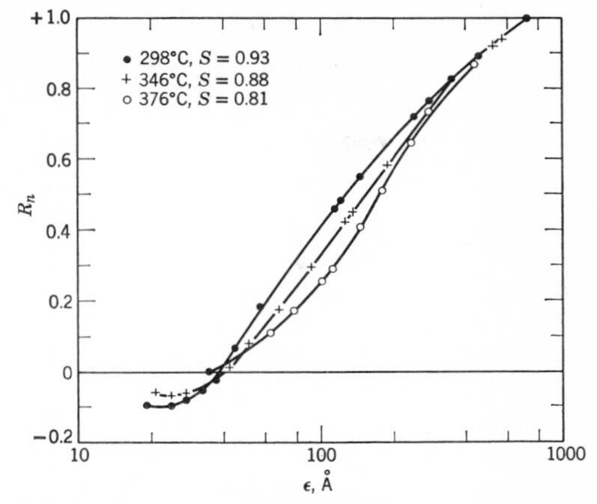

Fig. 43. Variation of normalized Hall coefficient, R_n, with domain size (ϵ) for three degrees of LRO. (Figure 4 of Reference 107.)

Having separated the effects of APB scattering and LRO on the Hall coefficient, Von Neida and Gordon discussed its change in sign at the critical temperature in terms of Brillouin zone splitting. In Figs. 46 and 47 the Brillouin zone for the disordered state and the first two additional zones induced by order splitting are shown together with the free electron FS and the Pippard-like FS of the disordered alloy. There is a nonzero energy gap at both the additional zone boundaries induced by ordering. The principal effect on R must come from the outer {110} added zone because the inner added zone should lie almost entirely within the FS for the ordered alloy. Hence the major cause for the observed

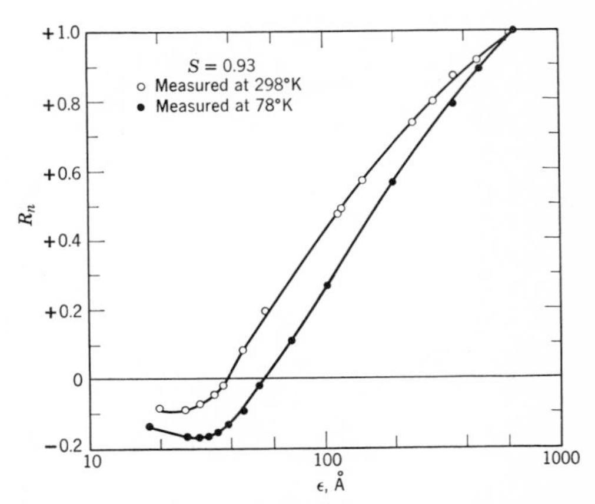

Fig. 44. Variation of R_n with ϵ at room temperature and 78°K. $S = 0.93$ is constant. (Figure 6 of Reference 107.)

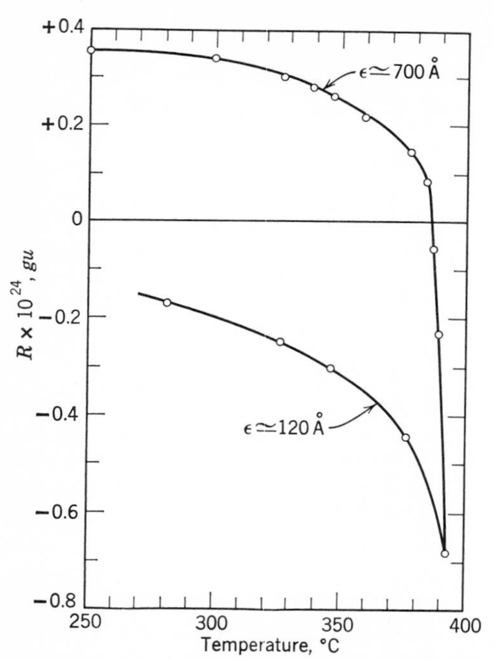

Fig. 45. Hall coefficient of Cu_3Au as a function of temperature at which the alloy was equilibrated for a small and large domain size. Measurements made at room temperature. (Figure 7 of Reference 107.)

sign change is due to the alteration of the FS, with respect to that for the disordered state, resulting from the appearence of an energy gap across the outer {110} zone faces. In addition, as the energy gap across the inner zone boundary increases with increasing LRO, the spherical caps of the FS in the ⟨110⟩ direction will shrink whereas those in the ⟨100⟩ direction will expand into the corners of the outer boundary. (See Fig. 47). The number of electron states is diminished by the ⟨110⟩ shrinkage and the number of hole states increased by the ⟨100⟩ expansion.

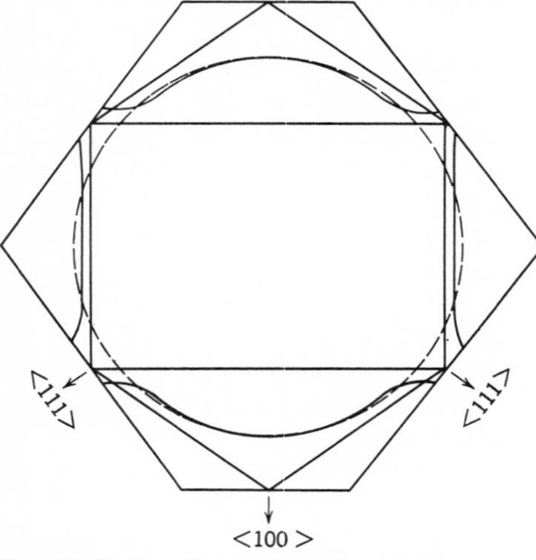

Fig. 46. Unit cell of CuAu(II). (Figure 9 of Reference 107.)

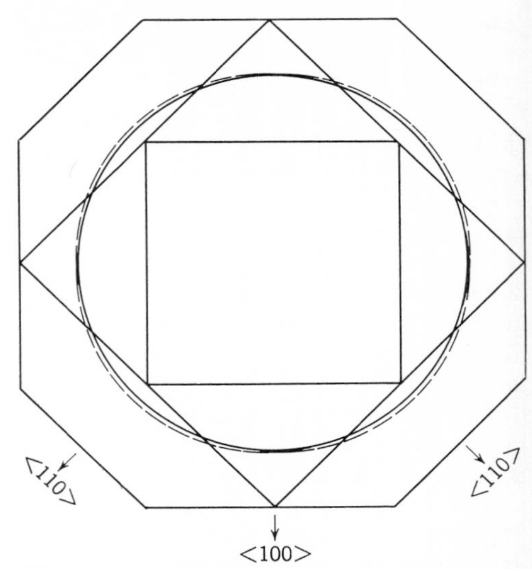

Fig. 47. Domain structure in CuAu(I). (Figure 10 of Reference 107)

A series of related papers by Nagy and Nagy,[12] Elkholy and Nagy,[13] and Nagy and Toth[14] treats the evolution of order in Cu_3Au as deduced from resistivity, Hall coefficient, and thermoelectric power measurements. From the observed behaviors of ρ, R, and Q, as functions of annealing time after quenching from the disordered state, they concluded that APB movement proceeds via a different diffusion mechanism than that operative during the initial growth of ordered nuclei. It was assumed that ordered nuclei grow as a consequence of atomic rearrangement during bulk diffusion. They obtained activation energies of 0.84 eV for APB movement and 1.8 eV for atomic movement during the growth of ordered nuclei. The description of APD evolution thus obtained by Nagy et al. meshes nicely with that obtained by Weisberg and Quimby[9] on the basis of Young's modulus measurements.

Wiener et al.[111] and Cooper et al[112] measured the magnetoresistance coefficient (B) and thermoelectric power (Q) of CuAu as a function of the ordered state. This work

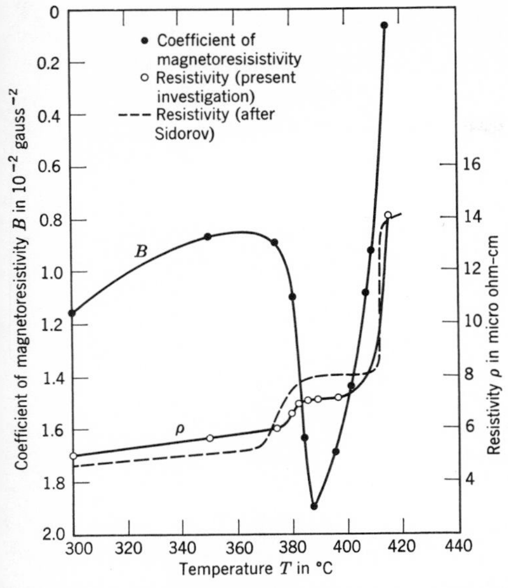

Fig. 48. Room temperature values of ρ and B of CuAu as a function of the temperature at which equilibrium was established. (Figure 1 of Reference 111.)

clearly illustrates that measurements of ρ, B, and Q characterize different features of the electronic state structure as a function of antiphase domain configuration. The behavior of ρ and B in CuAu are shown in Fig. 48. In reading Fig. 48, recall that B furnishes a measure of FS distortion (through $q^2 = \rho^2 B$) and that ρ is primarily determined by APB scattering in CuAu by virtue of the fact that the electron concentration in the ordered and disordered states of this alloy are nearly the same. In this context we see immediately that the results of Wiener et al. strongly confirm the fundamental ideas of Brillouin zone splitting in an ordinary superlattice and a LPS. The ordinary CuAu(I) superlattice transforms into CuAu(II) LPS at 380°C. At 408°C the LPS transforms into the disordered fcc structure. The formation of APB, in the transformation from CuAu(I) to CuAu(II), is in accord with the sharp rise in ρ over the 10°C interval centered on 380°C. After this, ρ is constant up to the critical temperature. An intense distortion of the FS is indicated by the sharp maximum in B that occurs within the temperature range 380 to 408°C, for which the LPS structure exists as an equilibrium structure. Moreover, B drops off extremely rapidly at the critical temperature when the LPS decomposes. From these data it is clear that the electron states near the FS are altered considerably during the order-disorder transformation. These data furnish strong additional support for the theories of Nicholas[53] and Sato and Toth.[55–59] Cooper et al. combined their data on the time variation of Q during constant temperature annealing of quenched specimens with those of Wiener et al. to show that ρ, B, and Q measurements are needed to specify the state of CuAu.

8. MAGNETIC EFFECTS

There is an extensive literature on experimental studies, which correlate changes in magnetic properties with a variation in order. The existing theoretical tools used for a qualitative interpretation of the data from these experiments are based on either band theory[113,114] or a nearest-neighbor spin interaction approach.[92,115–117] Neither point of view considers the influence of APD size or structure explicitly. Theoretical difficulties in this area are especially difficult because both magnetic and atomic ordering must be considered simultaneously. References to the work in this area can be found, for example, in articles by Pugh et al.[118] Marcinkowski and Brown,[96] and Muto and Takagi,[1] as well as in Chapter 27 of this book.

9. SUMMARY

Direct observation of antiphase domain structure in the electron microscope has confirmed earlier interpretations of X-ray data on the average domain size and the long-period superlattice structure. In addition, direct observation has furnished a detailed description of the role of strain processes during order transitions. Coupled with measurements of the galvanomagnetic coefficients, these observations outline the principal features of the competition between the ion core and electronic contributions to the crystal energy in the formation of an antiphase domain array, be it a foam structure or a long-period superlattice. The study of antiphase domains too small to be resolved in the electron microscope has been approached via the computer experiment. The field ion microscope provides a means for studying small domains in alloys with a high melting temperature.

A much greater effort has been applied to studying the effects of antiphase domains on mechanical properties than to electrical and magnetic properties. However, it would seem that the latter type of investigation is more fundamental because of the subtle but important effects of electronic behavior on antiphase domain stabilization. A refined treatment of the antiphase boundary energy awaits a thorough treatment of the spatial variation of the ordering energy.

REFERENCES

1. Muto T., and Y. Takagi, *Solid State Phys.* **1** (1955) 193.
2. Nix F. C., and W. Shockley, *Rev. Mod. Phys.* **10** (1938) 1.
3. Lipson H., *Progr. Metal Phys.* **2** (1950) 1.
4. Guttman L., *Solid State Phys.* **3** (1956) 145.
5. Bethe H., *Proc. Roy. Soc.* **A150** (1935) 552.
6. Cowley J. M., *Phys. Rev.* **120** (1960) 1648; **77** (1950) 669.
7. Sykes C., and F. W. Jones, *Proc. Roy. Soc.* **A157** (1936) 213.
8. Sykes C., and F. W. Jones, *Proc. Roy. Soc.* **A166** (1938) 376.
9. Weisberg L. R., and S. L. Quimby, *J. Phys. Chem. Solids* **24** (1963) 1251.
10. Weisberg L. R., and S. L. Quimby, *Phys. Rev.* **110** (1958) 338.
11. Burns F., and S. L. Quimby, *Phys. Rev.* **97** (1955) 1567.
12. Nagy E., and J. Tóth, *J. Phys. Chem. Solids* **24** (1963) 1043.
13. Elkholy H., and E. Nagy, *J. Phys. Chem. Solids* **23** (1962) 1613.
14. Nagy E., and I. Nagy, *J. Phys. Chem. Solids* **23** (1962) 1605.
15. Cahn J. W., and J. E. Hilliard, *J. Chem. Phys.* **28** (1958) 258.
16. Kikuchi R., and J. W. Cahn, *J. Phys. Chem. Solids* **23** (1962) 137.
17. Cahn J. W., and R. Kikuchi, *J. Phys. Chem. Solids* **20** (1961) 94.
18. Marcinkowski M. J., in *Electron Microscopy and Strength of Crystals*, G. Thomas and J. Washburn, ed. Interscience Publishers, 1963, p. 333.
19. Jones F. W., *Proc. Roy. Soc.* **A166** (1938) 16.
20. Suoninen E., and B. E. Warren, *Acta Met.* **6** (1958) 172.
21. Rudman P. S., and B. L. Averbach, *Acta Met.* **5** (1957) 65.
22. Keating D. T., and B. E. Warren, *J. Appl. Phys.* **22** (1951) 286.
23. Wilson A. J. C., *X-ray Optics*. Methuen, London, 1949.
24. Gehlan P. C., and J. B. Cohen, *Phys. Rev.* **139** (1965) A844.
25. Ogawa S., D. Watanabe, H. Watanabe, and T. Komoda, *Acta Cryst.* **11** (1958) 872.
26. Watanabe D., M. Hirabayashi, and S. Ogawa, *Acta Cryst.* **8** (1955) 510.
27. Watanabe D., and S. Ogawa, *J. Phys. Soc. Japan* **11** (1956) 226.
28. Watanabe D., *J. Phys. Soc. Japan* **13** (1958) 535.
29. Fujiwara K., M. Hirabayashi, D. Watanabe, and S. Ogawa, *J. Phys. Soc. Japan* **13** (1958) 167.
30. Raether H., *Z. Angew. Physik* **4** (1952) 53.
31. Whelan M. J., and P. B. Hirsch, *Phil. Mag.* **2** (1957) 1121.
32. Whelan M. J., *J. Inst. Metals* **87** (1958–59) 392.
33. Fisher R. M., and M. J. Marcinkowski, *Phil. Mag.* **6** (1961) 1385.
34. Marcinkowski M. J., and N. Brown, *J. Appl. Phys.* **33** (1962) 537.
35. Marcinkowski M. J., and L. Zwell, *Acta Met.* **11** (1963) 373.
36. Johansson C. H., and J. O. Linde, *Ann. Physik.* (*Lpz.*) **78** (1925) 439; **21** (1936) 1.
37. Glossop A. B., and D. W. Pashley, *Proc. Roy. Soc.* **A250** (1959) 132.
38. Pashley D. W., and A. E. B. Presland, *J. Inst. Metals* **87** (1958–59) 419.
39. Hirabayashi M., and S. Weissmann, *Acta Met.* **10** (1962) 25.
40. Harker D., *Trans. Amer. Soc. Metals* **32** (1944) 210.
41. Kuczynski G. C., R. F. Hochman, and M. Doyama, *J. Appl. Phys.* **26** (1955) 871.
42. O'Brien J. L., and G. C. Kuczynski, *Acta Met.* **7** (1959) 803.
43. Scott R. E., *J. Appl. Phys.* **31** (1960) 2112.
44. Flinn P. A., and G. M. McManus, *Phys. Rev.* **124** (1961) 54.
45. Beeler J. R., Jr., and J. A. Delaney, *Phys. Rev.* **130** (1963) 962; *Bull. Am. Phys. Soc.* **8** (1963) 339.
46. Beeler J. R., Jr., *Phys. Rev.* **138** (1965) A1259.
47. Brown N., and S. Cupshalk, *Bull. Am. Phys. Soc.* **8** (1963) 217.

48. Hume-Rothery W., and H. M. Powell, *Z. Krist.* **91** (1935) 23.
49. Jones H., and N. F. Mott, *Proc. Roy. Soc.* **A162** (1937) 49.
50. Jones H., *Proc. Roy. Soc.* **A147** (1934) 396.
51. Jones H., *Proc. Roy. Soc.* **A144** (1934) 225.
52. Slater J. C., *Phys. Rev.* **82** (1951) 538; **87** (1952) 807.
53. Nicholas J. F., *Proc. Phys. Soc.* **A66** (1953) 20.
54. Flinn P. A., *Phys. Rev.* **104** (1956) 350.
55. Toth R. S., and H. Sato, *J. Appl. Phys.* **35** (1964) 698.
56. Sato H., and R. S. Toth, *Phys. Rev.* **124** (1961) 1833.
57. Sato H., and R. S. Toth, *Phys. Rev.* **127** (1962) 469.
58. Sato H., and R. S. Toth, *Phys. Rev. Letters* **8** (1962) 239.
59. Sato H., and R. S. Toth, *J. Appl. Phys.* **33** (1962) 3250.
60. Hill T. L., *Statistical Mechanics*, McGraw-Hill, New York, 1956, p. 286.
61. Brown N., *Phil. Mag.* **4** (1959) 693.
62. Flinn P. A., *AIME Trans.* **218** (1960) 145.
63. Marcinkowski M. J., and N. Brown, *Acta Met.* **9** (1961) 764.
64. Marcinkowski M. J., N. Brown, and R. M. Fisher, *Acta Met.* **9** (1961) 129.
65. Harrison R. J., and A. Paskin, *J. Phys. Radium* **23** (1962) 613.
66. Kohn W., and S. H. Vosko, *Phys. Rev.* **119** (1960) 912.
67. Silverman B. D., and P. R. Weiss, *Phys. Rev.* **114** (1959) 989.
68. Paskin A., *Phys. Rev.* **134** (1964) A246.
69. Walker C. B., and D. T. Keating, *Phys. Rev.* **130** (1963) 1726.
70. Rudman P. S., *Acta Met.* **8** (1960) 321.
71. Fosdick L. D., *Phys. Rev.* **116** (1959) 565.
72. Matsuda S., *J. Phys. Soc. Japan.* **6** (1951) 131.
73. Westbrook J. H., *Mechanical Properties of Intermetallic Compounds*. Wiley, New York 1960.
74. Marcinkowski M. J., and R. M. Fisher, *J. Appl. Phys.* **34** (1963) 2135.
75. Stoloff N. S., and R. G. Davies, *Acta Met.* **12** (1964) 473.
76. Koehler J. S., and F. Seitz, *J. Appl. Mech.* **14** (1947) A217.
77. Thomas G., and J. Washburn, *Electron Microscopy and Strength of Crystals*. Interscience Publishers, New York, 1963.
78. Marcinkowski M. J., *Acta Met.* **12** (1964) 473, Reference 30.
79. Hull D., in *Electron Microscopy and Strength of Crystals*. Interscience Publishers, New York, 1963. p. 439.
80. Cahn R. W., and J. A. Coll, *Acta Met.* **9** (1961) 138.
81. Marcinkowski M. J., and D. S. Miller, *Phil. Mag.* **6** (1961) 871.
82. Köster W., *Z. Metallk.* **32** (1940) 145.
83. Köster W., and W. Gmöhling, *Z. Metallk.* **51** (1960) 385.
84. Lawley A., E. A. Vidoz, and R. W. Cahn, *Acta Met.* **9** (1961) 287.
85. Fisher J. C., *Acta Met.* **2** (1954) 9.
86. Ardley G. W., *Acta Met.* **3** (1955) 525.
87. Cottrell A. H., *Properties and Microstructure*. ASM, Cleveland, 1954.
88. Sumino K., *Sci. Rept. Res. Inst. Tohoku Univ.* (1958) A-10, No. 4.
89. Shyne J. C., and M. J. Sinnoti, *Trans. AIME* **218** (1960) 861.
90. Birkenbeil H. J., and R. W. Cahn, *J. Appl. Phys.* **32** (1961) 362(S).
91. Rudman P. S., *Acta Met.* **10** (1962) 253.
92. Taylor A., and R. M. Jones, *J. Phys. Chem. Solids* **6** (1958) 16.
93. Biggs W. D., and T. Broom, *Phil. Mag.* **45** (1954) 246.
94. Cohen J. B., and M. B. Bever, *Trans. AIME* **218** (1960) 155.
95. Hahn R., and E. Kneller, *Z. Metallk.* **49** (1958) 426.
96. Marcinkowski M. J., and N. Broom, *J. Appl. Phys.* **32** (1961) 375.
97. Damask A. C., *J. Phys. Chem. Solids* **1** (1956) 23.
98. Gibson J. B., *J. Phys. Chem. Solids* **1** (1956) 27.
99. Beal M. T., *J. Phys. Chem. Solids* **15** (1960) 72; **18** (1961) 156.
100. Pick R., *J. Phys. Chem. Solids* **24** (1963) 741.
101. Asch A. E., and G. L. Hall, *Phys. Rev.* **132** (1963) 1047.
102. Phillips V. A., *Acta Met.* **9** (1961) 976.
103. Phillips V. A., and R. B. Jones, *Trans. ASM* **53** (1961) 775.
104. Mott N. F., and H. Jones, *The Theory of the Properties of Metals and Alloys*. Dover, 1958.

105. Komar A., and S. Siderov, *J. Tech. Phys.* (*U.S.S.R.*) **11** (1941) 711.
106. Komar A. P., N. V. Volkenshtein, and G. V. Fedorov, *Soviet Phys.* "*Doklady*" (*English Transl.*) **4** (1959–60) 359.
107. VonNeida A. R., and R. B. Gordon, *Phil. Mag.* **7** (1962) 1129.
108. Flanagan W. F., and B. L. Averbach, *Phys. Rev.* **101** (1956) 1441.
109. Pippard A. B., *Phil. Trans.* **250** (1957) 325.
110. Averbach B. L., P. A. Flinn, and M. Cohen, *Acta Met.* **2** (1954) 92.
111. Wiener B., P. Schwed, and G. Groetzinger, *J. Appl. Phys.* **26** (1955) 609.
112. Cooper H., P. Schwed, and R. W. Webeler, *J. Appl. Phys.* **27** (1956) 516.
113. Goldman J. E., *Rev. Mod. Phys.* **25** (1953) 108.
114. Mattuck R. D., *Phys. Rev.* **127** (1962) 738.
115. Piercy G. R., and E. R. Morgan, *Can. J. Phys.* **31** (1953) 529.
116. Carr W. J., Jr., *Phys. Rev.* **85** (1952) 590.
117. Nathans R., M. T. Pigott, and C. G. Shull, *J. Phys. Chem. Solids* **6** (1958) 38.
118. Dreesen J. A., and E. M. Pugh, *Phys. Rev.* **120** (1960) 1218.
119. Southworth H. N., and Ralph, B., to be published (1966).
120. Ralph B., Ph.D. Thesis, Cambridge University (1964).
121. Ralph B., and Brandon, D. B., *J. Roy. Microscop. Soc.* **82** (1964) 185; *Journées Int. des Applic. du Cobalt* (June 1964).
122. Müller E. W., *Imperfections in Crystals*, Interscience Publishers, New York (1961); *Platinum Metals Rev.* **9** (1965) 84.

chapter 15

Lattice Defects

Norman Brown

NORMAN BROWN

School of Metallurgical Engineering
and
Laboratory for Research on the Structure of Matter
University of Pennsylvania
Philadelphia, Pennsylvania

1. INTRODUCTION

Because the class of materials that are called intermetallic compounds range from a kinship with terminal solid solutions on the one hand to that with ionic crystals on the other, their defects exhibit a similar kinship. The literature on defects in pure cubic metals, in their dilute solid solutions, and in ionic crystals is quite large.[1–5] This chapter, therefore, emphasizes those defects that are peculiar to intermetallic compounds and merely indicate those that are common to the other well-known classes of materials. All defects distort the lattice and consequently interact with one another. Thus in a real crystal the isolated defect is not the usual state; consequently, the interaction among defects is discussed. Details concerning antiphase boundaries and surfaces are omitted because they are covered in other chapters of this book.

Because direct evidence of a defect in a particular material is usually lacking, it is useful to be able to infer the probable nature of the defect from what is known about other materials. The following aspects of structure have been found to be the most useful for inferring the nature of the defects in a compound: (1) crystal structure, (2) state of aggregation, and (3) strength and type of the interatomic bond.

The advantage of knowing the crystal structure is obvious. If, for example, we know that an intermetallic compound has a tetrahedral structure, its defect structure can be inferred to a large extent from the extensive work on germanium and silicon. To determine whether a particular intermetallic compound is likely to exist with a tetrahedral structure, we should refer to the extensive work by Parthé who has also written Chapter 11 of this book. Parthé[6] has also pointed out the importance of the equilibrium vacant lattice defect in the tetrahedral structure.

The state of aggregation of intermetallic compounds may be described by a spectrum

Spectrum of Aggregation

← Randomness or Order →

Random Solid Solution	Clustering or Short-Range Order	Disorder and Long-Range Order	Dilute Disorder	Complete Order to Melting Point
RSS	C or SRO	LRO	DD	CO

of aggregation which ranges from that of the random concentrated terminal solid solution to that of the perfect order of the ionic crystal. Although the spectrum may be conceived primarily in terms of the degree of randomness or order of the atomic species, the randomness or order of the electrons follows a roughly parallel course; in the random solid solution the electrons are free to move about the lattice, whereas the electrons are tightly bound to the lattice site in the perfect ionic crystal. The following diagram may be useful for visualizing the spectrum of aggregation.

A particular intermetallic compound does not necessarily occupy a unique position in the above spectrum, because variations in cooling rate from elevated temperature, plastic deformation, irradiation, and small changes in composition may change the state of aggregation. In fact, if we think in terms of the state of lowest energy, probably all single-phase materials consist of ordered multiphases or complete order, because any of the randomness that exists is merely a frozen-in nonequilibrium state according to the third law of thermodynamics. A similar remark has been made by Matthias[7] concerning the order among electrons; it leads to the conclusion that all metals should be ferromagnetic, ferrimagnetic, or superconducting at 0°K.

The state of aggregation of a particular intermetallic compound depends upon (1) how well it is ordered at its melting point, (2) how rapidly it is cooled from an elevated temperature, and (3) the rate of diffusion with which it approaches its equilibrium state of order. It is quite possible that some intermetallic alloys that are classified as being completely ordered up to their melting point may have an appreciable amount of disorder at elevated temperatures. It is rather strange that ordered materials only fall into the following three categories: (1) those that are completely ordered up to the melting point, (2) those whose critical temperature for LRO is below the melting point and directly measurable, and (3) those whose critical temperature for ordering is at too low a temperature to be measured. There should be a fourth category for those materials which are in a state of partial LRO at their melting point. It seems that the experimental difficulties at elevated temperatures have excluded this fourth category from our consideration. In a similar vein many materials that are classified as being completely ordered may only be ordered to within 97 per cent, but, again, X-rays have a limited sensitivity in determining the last few per cent of LRO. However, a few per cent of disorder, DD, may have a large effect on the electrical resistivity.[8]

The strength of the ordering energy between atoms is indicated by T_c in the case of crystals that undergo LRO. This strength of the interatomic bonds often determines the slip plane and whether dislocations are extended as in the case of tetrahedral structures. The melting point is the simplest indication of the strength of the bond in completely ordered crystals.

2. POINT DEFECTS

2.1 Nonstoichiometry

Point defects are assumed to include nonstoichiometry, impurity atoms, vacancies, interstitials, and combinations of them. If the intermetallic compound has an extended phase field, the most common defect is nonstoichiometry; in such cases it is very difficult to prepare the compound with the exact stoichiometric composition or to determine whether a compound is within 0.1 per cent of stoichiometry. In many ways the amount by which the compound is off stoichiometry is equivalent to an equal concentration of foreign atoms. The compounds with the strong covalent bond do not suffer from the nonstoichiometric defect. The effects of the nonstoichiometric defect would tend to be greater the greater the order of the state of aggregation. In the RSS slight amounts of nonstoichiometry would tend to have a small effect on properties because stoichiometry has no special significance in this state.

2.2 Impurity Atoms

A given small concentration of impurity atoms may not by themselves produce as large an effect in a RSS as they do in a pure metal. However, in an ordered structure the advent of impurity atoms might be concommitant with the introduction of a vacancy similar to the case of a divalent impurity in an ionic crystal being accompanied by a vacancy. Thus the impurity-vacancy combination could have a far greater effect than that of the isolated impurity atom. For example, the strengthening of an impurity-vacancy combination should be very large according to Fleischer,[9] because the lattice would be tetragonally distorted.

2.3 Vacancies

Vacancies may be divided into three types: (1) the extrinsic vacancy, (2) the intrinsic or thermal vacancy, and (3) the crystal structure vacancy. The extrinsic vacancy corresponds to the example in the previous paragraph. The number of extrinsic vacancies is determined by the change in energy of the crystal with the electron-atom ratio. For example, in NiAl the nonstoichiometric aluminum atoms are accompanied by vacancies (Fig. 1). Similarly, it would be expected that trivalent and tetravalent impurity atoms would produce extrinsic vacancies in NiAl. In order to determine whether a system is truly a defect structure in that it contains extrinsic vacancies, it is necessary to measure both changes in density along with changes in lattice parameter. It was recently pointed out that some structures, which have been called defect structures, actually contain microporosity instead of vacancies.[10] However, systems that are definitely known to be defect structures include NiAl, CoAl, Cu–Al, Cu–Ga, NiAs, NiSb and NiGa. Presumably trivalent and tetravalent substitutional impurity atoms in stoichiometric compositions of the above compounds would produce extrinsic vacancies. Recent work indicates that there is no experimental evidence for extrinsic defects in any primary solid solution.[11]

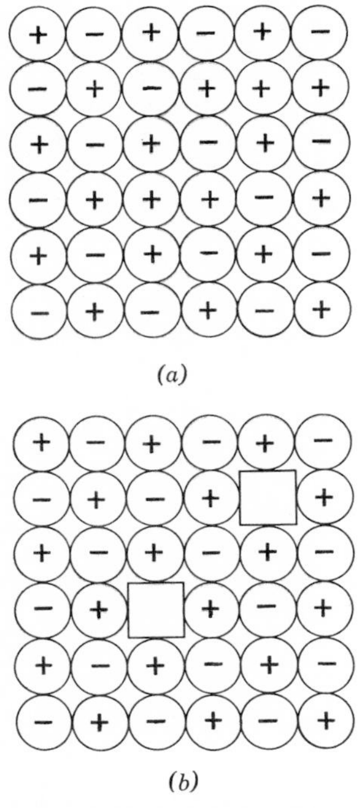

Fig. 1. Nonstoichiometric defects: (*a*) substitutional, excess positive; (*b*) vacancy defect structure, excess positive.

Intrinsic or thermal vacancies vary in concentration according to the equation

$$C = e^{-(U_F/kT)}$$

where U_F is the energy of formation. The usual experiment to determine U_F consists in rapidly quenching from a given temperature and in measuring the electrical resistivity of the quenched-in vacancies as a function of the quenching temperature. The rate of decay of disappearance of the excess vacancies follows the equation

$$\text{rate} \sim e^{-(U_M/kT)}$$

where U_M is the energy of motion. Self-diffusion experiments in pure fcc metals indicate that

$$D = D_0 e^{-(U_D/kT)}$$

where $U_D = U_F + U_M$. Experiments indicate that U_F is only slightly larger than U_M in pure fcc metals. Although it has been suggested that U_F is less than U_M in fcc concentrated solid solutions, the paucity of data leaves this point unanswered. Quenching[12] and diffusion experiments[13] on AuCd, which is a CsCl structure, show that U_F is about 0.4 eV and U_D is about 1.2 eV. Recent work on β Cu–Zn also indicates that it is easy to quench in high concentrations of vacancies in intermetallic compounds that order with the CsCl structure.

The following rules of thumb are suggested for determining U_D, U_F, and U_M for fcc- and bcc-type intermetallic solid solutions if only the melting point is known:

$$U_D \approx 38T_M \qquad \text{for RSS}$$

and

$$U_D \approx U_{D(RSS)} + \frac{ZRT_c}{4} \qquad \text{for LRO}$$

with U_D in calories/mole, T_M in °K, Z is the coordination number, and R is the gas constant. Assume that $U_D = U_F + U_M$. For an fcc-type structure take $U_F \approx \frac{1}{2}U_D$; for bcc or CsCl structure assume $U_F \approx \frac{1}{3}U_D$. In the absence of experimental data, the above rules may be useful in estimating the intrinsic vacancy concentration, the mobility of vacancies, and the rates of diffusion in a particular intermetallic compound.

Vacancies in tetrahedral structures such as a III–V compound occupy the site for which U_F is least. Thus in self-diffusion experiments it is expected that the component will diffuse fastest whose site has the highest vacancy concentration. It has also been suggested[14] that when a vacancy forms in a tetrahedral structure there is a large relaxation of the surrounding atoms which acts as a barrier to recombination with interstitials.

The final class of vacancies is called a crystal structure vacancy. Parthé[6] has pointed out that in order to form a tetrahedral structure the average electron/atom ratio must be four. Therefore, certain tetrahedral crystal structures occur only if enough vacancies are present to produce the necessary e/a ratio. These vacancies form an ordered arrangement on the lattice just like any other species in the alloy. The important point is that the very existence of the tetrahedral structure may be contingent upon the existence of the vacancies. In such a system high vacancy concentrations are to be expected.

Vacancies in intermetallic compounds do not usually exist as random isolated defects. If, as in Fig. 1*b*, the positive atoms are in excess, the vacancy is expected to be a nearest-neighbor to the atom which is in excess. Such a configuration reduces the ordering energy by minimizing the number of like nearest-neighbors.

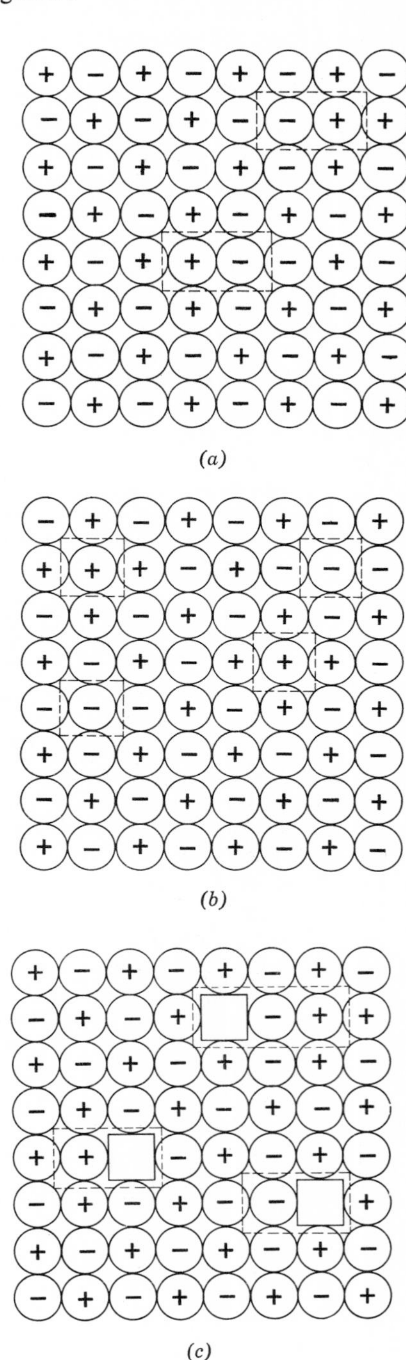

Fig. 2. (*a*) Bound wrong pairs; (*b*) unbound wrong pairs; (*c*) vacancies bound to wrong atoms.

2.4 Dilute Disorder

Dilute disorder[8] is a defect in the ordered structure which consists of a pair of atoms in wrong sites. Figure 2*a* shows that the misplaced atoms may be nearest-neighbors bound, or unbound as in Fig. 2*b*. The unbound state is more likely the higher the temperatures. Figure 2*c* shows how vacancies are likely to bind with a wrong atom in order to minimize the energy associated with order. Little work has been done on the effect of dilute disorder on properties.

2.5 The Interstitial

There is practically no information about interstitial atoms in intermetallic compounds. The occurrence of an extrinsic interstitial defect to minimize the energy associated with a change in the electron/atom ratio has not been reported. The high energy of formation of an interstitial atom in closely packed structures most likely renders its existence doubtful in measurable concentration after quenching. The formation of interstitials is most prominent after irradiation. Vook[14] has studied the lengthening associated with electron bombardment of III–V compounds. He attributes the lengthening to the formation of interstitial vacancy pairs. Because the lengthening for a given dose of radiation was greater in the III–V compounds than in Si and Ge, Vook[14] suggested that the interstitial in the III–V compounds expands the lattice to a greater extent for there is a coulombic repulsion between the interstitial and its like nearest-neighbors.

3. PLANAR DEFECTS

This section is concerned with stacking faults and antiphase domain boundaries (APDB). The nature of APDB is not discussed in detail because it is covered by Beeler in Chapter 14. In this section the APDB are separated into three cateogries: (1) the thermal APDB, (2) the slip-produced APDB, and (3) the periodic APDB, or the long-period boundary. The thermal APDB is produced by the random nucleation of LRO. The slip-produced APDB's occur when single dislocations move in a superlattice. The periodic APDB has a period which varies with the electron/atom ratio (*e*/*a*) (Fig. 3).

3.1 Thermal APDB

As is the case with all APDB's, they tend to form on planes of lowest energy. As pointed out by Brown,[15] the width of the APDB is a function of temperature. Close to T_c the thickness may be appreciably wider than a single atomic plane. Since the APDB contains an excess of like nearest-neighbors, vacancies will tend to form on an APDB in order to minimize the number of like neighbors. Recent experiments have shown that this is the case in quenched βCu–Zn; Fig. 4 shows dislocation loops at an APDB where the dislocation loops are a result of a much higher vacancy concentration of thermal vacancies at the APDB than within the domain. The attraction between APDB's and vacancies is similar to the suggestion by Rudman[16] that SRO enhances the vacancy concentration in the region between disorder and SRO. Figure 5 shows thermal and slip-produced APDB's.

3.2 The Slip-Produced APDB

The slip-produced APDB lies on the slip plane of the dislocation which produces it. The phase of the LRO parameter changes abruptly across the boundary, and therefore the number of like nearest-neighbors is greater

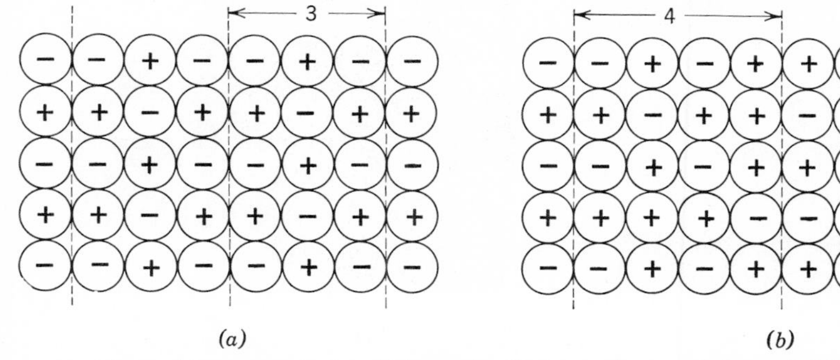

Fig. 3. (*a*) Long-period stoichiometric; (*b*) longer long-period nonstoichiometric.

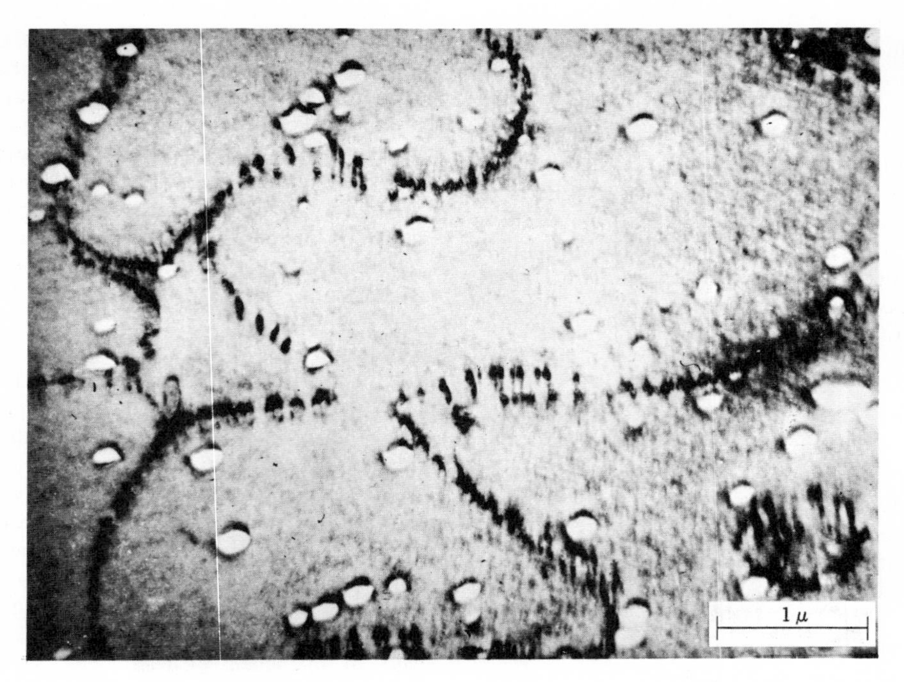

Fig. 4. Dislocation loops at antiphase domain boundaries of quenched β CuZn. (Courtesy of S. Cupschalk.)

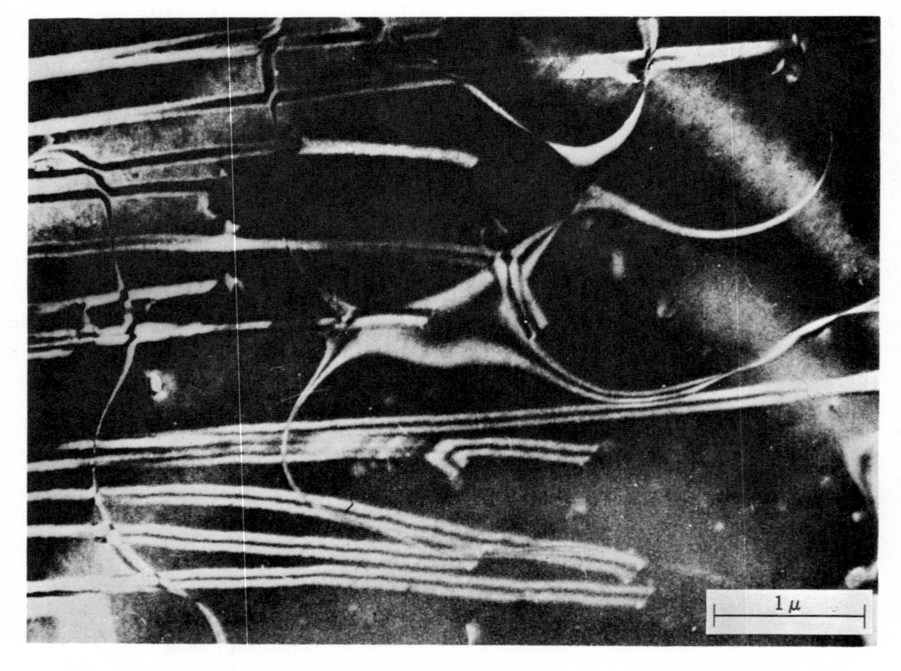

Fig. 5. Slip-produced and thermal antiphase domain boundaries. (Reference 39.)

than is typically the case for the thermal APDB. As a result, vacancies and impurity atoms may be more strongly attracted to a slip-produced APDB than to a thermal APDB.

3.3 Periodic APDB

The periodic APDB is associated with a long period over and beyond the repeat distance associated with LRO. For example, in CuAu the long period is ten times the superlattice repeat distance. The long period has been discussed extensively in a review paper by Sato and Toth.[17] The length of the long period is based on the *e*/*a* ratio. Since a nonstoichiometric defect changes the *e*/*a* ratio, it can also change the length of the long period as illustrated in Fig. 3. It is difficult to guess whether vacancies would or would not tend to segregate at the APDB associated with the long period, because the long period APDB is an equilibrium defect, whereas the random APDB is a nonequilibrium interface. Little is known about the effect of the periodic APDB on properties.

3.4 Stacking Faults

Stacking faults in close-packed structures (fcc and hcp) occur on the close-packed planes. As in the case of terminal solid solutions, it is expected that the stacking fault energy would depend on the *e*/*a* ratio.[18] However, in the presence of LRO the stacking fault coincides with an APDB in most cases. All faults on {111} planes of Cu_3Au-type structures also produce an APDB, but faults on only certain {111} planes in CuAu produce an APDB. If a stacking fault coincides with an APDB, its energy changes, but the direction and magnitude of the change have still to be well determined.[19]

The composition along a stacking fault may be different from the average composition in accordance with the "Suzuki effect." Impurity segregation or depletion is expected at stacking faults. As in the case of APDB's, vacancies should segregate at faults.

As pointed out by Parthé, there exist other sets of crystal structures which are like the fcc and hcp in that within the set one structure may be converted into another by changing the stacking sequence. Parthé calls these sets homeotect structures. The zinc blende and wurtzite structures are homeotect. The stacking unit consists of a double layer of atoms. The faulting occurs between the double layers and not within them as dictated by the energy to break bonds. Growth faults are most likely in those systems which have the lowest stacking fault energy. It is expected that those systems which exhibit both the zinc blende and the wurtzite modifications would have the lowest stacking fault energy. Parthé points out that the greater the difference in electronegativity of the atoms, the greater the tendency to go from the zinc blende to the wurtzite structure. Just as the *e*/*a* ratio decreases the stacking fault energy in a fcc structure, it may be reasoned that alloying, which increases the electronegativity of the atoms within the double layer of zinc blende, decreases its stacking fault energy.

4. DISLOCATIONS

Since the nature and behavior of dislocations depend primarily upon the crystal structure, this section is so divided. The emphasis is on those aspects of dislocation behavior that distinguish the intermetallic compound from the terminal solid solution and the pure metal. Many of the well-known properties of dislocations are omitted. The sections on LRO will be abbreviated because Marcinkowski[19] has recently written an excellent review article on dislocations in superlattices. Amelinckx and Delavignette[20] have written a review article concerning extended dislocations in layer structures of graphite, molybedum sulfide, talc, aluminum nitride (wurtzite), bismuth telluride, tin sulfide, and nickel bromide. It is expected that intermetallic compounds, which have the same crystal structure as any of the above materials, would be expected to slip in a similar way.

4.1 Close-Packed Structures

As is well known, the dislocation in the disordered fcc lattice glides on the {111} planes and has a $\frac{1}{2}\langle 110\rangle$ Burgers vector. The dislocation is extended so that it consists of two $\frac{1}{6}\langle 112\rangle$ partial dislocations which are connected by a ribbon of stacking fault. The stacking fault energy determines how much the dislocation is extended and also how difficult it is for the dislocation to leave its slip plane by cross-slip. The presence of LRO further extends the dislocation by having it

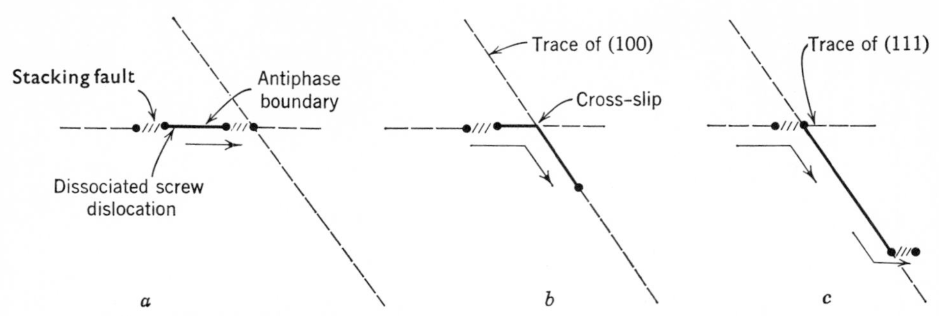

Fig. 6. Trapping of a superlattice dislocation on the (100) plane of Cu_3Au.[22]

consist of a pair with $\frac{1}{2}\langle 110\rangle$ Burgers vectors connected by an APDB. The unusual aspects in the slip behavior of the superlattice dislocation stem from the fact that the APDB energy may not be a minimum on the glide plane so that the superlattice dislocation may not be entirely in one plane. As pointed out by Flinn,[21] the partial dislocations may lie on parallel slip planes and be joined by an APDB; when each partial dislocation moves, the dislocations generate APDB. Recently Kear[22] has shown that in Cu_3Au the superlattice dislocation, which may be extended on a single plane, may be trapped on a $\{100\}$ plane because the APDB has its lowest energy on that plane (Fig. 6). On the other hand, Kear shows that the disordered Cu_3Au behaves like a terminal solid solution. Whereas jogs on dislocations in RSS may produce point defects as the dislocation moves, jogs in superlattice dislocations may, in addition, produce tubular trails of APDB, as pointed out by Vidoz and Brown.[23]

All experimental data[24] to date show that a critical state of partial LRO is stronger than the completely ordered or the completely disordered state. Such behavior is expected from general considerations based on the fact that the superlattice dislocation becomes more extended as the degree of LRO decreases. The more extended the dislocation becomes, the more likely it is that its component $\frac{1}{2}\langle 110\rangle$ dislocations will *not* move in tandem at a constant separation; therefore, more APDB will be generated by the type of processes discussed above. However, when the degree of LRO becomes too low, the energy per unit area of the associated APDB decreases to the point where there is a net decrease in energy, although the total area of APDB may be increasing. The temperature of maximum strength does not occur at the same critical degree of LRO for all systems. Still no generally accepted quantitative theory has been produced.

The ordered CuAu structure is unusual in that it is tetragonal and thus produces asymmetric superlattice dislocations. The tetragonality is expected to give an extra amount of strengthening which would be absent in the Cu_3Au type of superlattice. Marcinkowski[19] has tabulated the c/a ratio for CuAu intermetallic compounds and has suggested that their strength would be enhanced by an amount which depends on their degree of tetragonality.

The hcp structure, such as Mg_3Cd, undergoes an order-disorder transition. In the disordered state the slip system is $\frac{1}{2}\langle 11\bar{2}0\rangle$ (0001); and, analogously to the Cu_3Au fcc structure, the $\frac{1}{2}\langle 11\bar{2}0\rangle$ is extended with two $\frac{1}{2}\langle 10\bar{1}0\rangle$ partials connected by a ribbon of stacking fault. In the state of LRO the superlattice dislocation is again like the Cu_3Au system in that it consists of four $\frac{1}{2}\langle 10\bar{1}0\rangle$ partials connected by APDB's and stacking faults.[19] However, on the $\{1010\}$ planes the ordinary $\frac{1}{2}\langle 11\bar{2}0\rangle$ dislocation does not produce a disordering of nearest-neighbors during slip; therefore, a superlattice dislocation is not expected on the $\{10\bar{1}0\}$, and slip should be more likely relative to $\{0001\}$ slip in the state of LRO than it is in the state of disorder. Davies and Stoloff[25] have shown that basal plane slip occurs in Mg_3Cd in both the LRO and disordered state, but that $\{10\bar{1}0\}$ slip occurs only for LRO.

4.2 Body-Centered Cubic and CsCl-Type Structures

Much of what was said for the close-packed system applies to the bcc type. It is important to note that cross-slip is easier in bcc-type structures. In CsCl structures, as pointed out by Rachinger and Cottrell,[26] the ⟨100⟩ may become the slip direction instead of the ⟨111⟩ depending on the ordering energy. It was estimated that if $kT_c/4$ were greater than 0.06 eV, the ⟨100⟩ slip directions would be preferred over the ⟨111⟩ directions. If the critical temperature for ordering T_c is the same as the melting point, the criterion for ⟨100⟩ slip is that APDB energy on a (100) plane be greater than 250 erg/cm^2. Punching experiments showed that βCu–Zn and βAg–Mg slip along ⟨111⟩, and AuCd, AuZn, LiTl, and MgTl glide along ⟨100⟩.

With respect to the ductility of cubic types of intermetallic compounds, it should be expected that they should slip readily in the pure state, because there is no reason to expect an inordinately high innate lattice resistance (Peierls-Nabarro force). It is expected that impurity segregation and the effects of nonstoichiometry may pin dislocations more strongly by interacting with their APDB ribbon. Just as impurities segregate to APDB, it is expected that grain boundaries, which are also regions of disorder, will be a stronger attraction for impurities and vacancies in ordered systems than disordered systems. Stoloff and Davies[24] have discussed the possible effects of LRO on ductility.

4.3 Tetrahedral Structures

Parthé[6] has discussed in detail the various types of tetrahedral structures which fall in the class of layerlike crystal structures. In this section differences between some of the tetrahedral type structures are discussed from the dislocation viewpoint.

The zinc blende structure is the most common and is associated with most III–V and II–VI type binary compounds. The stacking of {111} is aαbβcγaα, etc., where the Roman letters are a layer of III-type atoms and the Greek letters the V-type (or vice versa). Between the α and b, the β and c, and the γ and a layers there are three nearest-neighbor bonds per atom; between the α and a, the β and b, and the γ and c layers there is only one bond. Thus slip usually occurs between the latter pairs of planes and not the former. The dislocation would be extended into partial dislocations joined by a ribbon of stacking faults similar to the case of fcc metals. The stacking-fault ribbon would be a layer of wurtzite structure. The wurtzite structure is the same as the zinc blende structure, except that it has the stacking sequence of aαbβaα, etc. Thus the extended dislocation in wurtzite also consists of partials joined by a ribbon of stacking faults consisting of a layer of zinc blende structure. One of the unique features of the edge-type dislocations in zinc blende and wurtzite structures is their polarity. Because one glide interface consists of one type of atom (e.g., III) and the other glide interface consists of the other type (e.g., V), the end atom of the extra half plane will be either III or V. Therefore, the so-called plus and minus edge dislocations will not be simply upside down relative to one another, but the atoms along the edge of their extra half plane will be different.[27] The difference in polarity of the dislocations has been associated,[28,29] with a difference in the way that pits develop. Also, if these zinc blende or wurtzite type crystal structures are cleaved along their slip plane, each surface has a different surface layer of atoms.

The details of the core structure of dislocations in tetrahedral type structures are similar to the pure tetrahedral structures, silicon and germanium. In Van Bueren's book[2] there is a detailed discussion of the core of the dislocations in the diamond cubic structure. (One of the important questions concerning the core of the dislocation is whether a dangling or distorted bond is more favorable.) Because there is a strong directional bond between nearest-neighbors, the width of the dislocation is narrow and the innate lattice resistance (Peierls-Nabarro force) should play a larger part in resisting the motion on dislocations than is the case for the cubic and hcp systems.

It has been suggested that the point defects produced by plastic deformation would occur in pairs as a double vacancy or a double interstitial consisting of a III- and a V-type atom. The double defect occurs because the jog height is equal to the double layer of atoms separating adjoining slip planes. Thus the single vacancies produced by quenching would anneal out at a different rate than the complex vacancies produced by plastic deformation as pointed out by Van Bueren.[2]

Another tetrahedral layer structure of interest is that typified by GaSe. The stacking is A$\beta\beta$AB$\gamma\gamma$BC$\alpha\alpha$C, etc., where the Roman letters are planes of Se and the Greek are Ga. Because there are weak bonds between the A and B, B and C, and C and A planes, the slip plane is between these planes. When the bond is weak, the stacking fault energy would be low, and the dislocations should be greatly extended as shown by Basinski, Dove, and Mooser.[30] By analogy with the zinc blende structure, where the jog in a dislocation is two layers high, it would be expected that in the GaSe-type structure jogs in dislocations might be four layers high. Thus point defects that are produced by the nonconservative motion of jogs should be rather complex in this crystal structure.

4.4 Sigma Phases

The crystal structure of the sigma phase is the same as β-uranium. Many of the intermetallic compounds that consist entirely of transition elements have the β–U structure. The structure is simple tetragonal and consists of thirty atoms per unit cell. The lattice may be described in terms of a stacking of Kagomé tiling. Holden[31] has deformed β-uranium at room temperature and has shown that the system is $\{110\}\langle 001\rangle$, which are perpendicular to the layers stacked as Kagomé tiling. There is, therefore, reason to believe that sigma phases may be deformed plastically if they are prepared with the proper purity. Kronberg[32] has presented a detailed model for the atom movement in β-uranium during slip based on zonal dislocations. The Kronberg model is very unique in that a rather complex coordination of atom movements is necessary in order to make the least disturbance in the hard-sphere model. The important point, however, is that complex crystal structures may deform plastically, but the atom movements may be somewhat more complex than is the case for the simpler crystal structures.

Marcinkowski and Miller[33] have observed stacking faults in the Fe–Cr sigma phase, which is ordered. The faults were parallel to the layers with sessile dislocations terminating the faults. They attributed the brittleness and corresponding lack of zonal dislocations to the high energy required to form APDB's in an ordered sigma phase. Marcone and Coll[34] have made a preliminary investigation of the order-disorder transformation in the Co–V sigma phase. A sharp increase in hardness with decreasing quenching temperatures indicated along with resistivity and dilatometric results that T_c was about 550°C.

4.5 Laves Phases

The Laves phases have the composition AB_2. and the crystal structure may be built by stacking layers of composition AB_2. The stacking may be ABAB, etc., analogous to wurtzite or ABCABC, etc., analogous to zinc blende. There is no information on the plastic deformation of Laves phases. A cursory examination of the crystal structure indicates that slip might occur parallel or perpendicular to the stacked layers. From the behavior of other intermetallic compounds there is some expectation that the Laves phases will exhibit plasticity if prepared sufficiently pure and if deformed under the proper conditions.

4.6 Miscellaneous Intermetallic Structures

Recently Lewis[35] has reviewed the properties of beryllides whose stoichiometric compositions take the forms, MBe_2, M_2Be_{17}, MBe_{13}. The MBe_2 is the hexagonal form of the Laves phase. The M_2Be_{17} is a complex rhombohedral structure of which little is known. The MBe_{12} is a body-centered tetragonal structure which gives some indication of slip. The MBe_{13} structure is a cubic structure with the Be atoms grouped as an icosahedron. The MBe_{13} has been hot-extruded, and the crystal structure indicates that slip on $\{001\}\langle 110\rangle$ is a distinct possibility.

The disilicide structure, such as $MoSi_2$, is formed by stacking planes of hexagonal packing so that Mo is completely surrounded by Si atoms. Instead of stacking the layers so that atoms in adjacent planes occupy tetrahedral sites, as in the case of close-packed structures, they sit on the saddle points. Thus slip might be expected to occur on $\{10\bar{1}0\}\langle 11\bar{2}0\rangle$; since only Si–Si bonds occur across these planes, this slip does not produce disorder.

5. EXPERIMENTAL METHODS

Most of the experimental methods that have been used to obtain information about defects will be mentioned. Some methods are direct, such as the observation of dislocations, stacking faults, and APDB's with the electron microscope. The indirect methods vary in their degree of directness depending upon the type of theory and of assumptions that are required to interpret the experimental data. Thus, the less direct the method, the greater is the number of methods that must be used to arrive at what would be considered a useful, and not simply a speculative, picture of the particular defects. In most cases, even when all possible experimental methods have been applied, the nature of the defect is not known unambiguously, but we may eliminate certain possibilities with varying degrees of confidence. Therefore, many experimental papers that are concerned with defects will often end with a discussion based on trying to guess what is the most likely form of the defect that produced the observed effects. To this end, rigorous theories on the structure of defects play a vital part in deciding the nature of defects. Even the observation of dislocations by X-ray or electron microscopy is called "direct" only because the image contrast can be predicted by an exact theory based on diffraction. The experimental methods for determining the existence and nature of defects will be presented in the form of a running commentary.

The degree of nonstoichiometry is primarily determined by a chemical analysis. However, it is difficult to determine the amount of stoichiometry within 0.1 per cent. Thus the effect of nonstoichiometry is determined with a series of alloys which straddle the stoichiometric composition. In semiconducting intermetallic compounds, the degree and direction of nonstoichiometry will determine the type and concentration of the current carrier. In PbSe an excess of Pb gives an n-type material, and an excess of Se produces a p-type. Thus electrical resistivity should be a very sensitive measure of nonstoichiometry.

Extrinsic vacancies are directly determined by measuring both the change in lattice parameter and the density with composition.[10] Care must be taken in density measurements to avoid microporosity.[11] The existence of intrinsic vacancies may be determined by measuring concurrently changes in length and the lattice parameter as a function of temperature[36] where the vacancy concentration is given by $C = 3(\Delta L/L - \Delta a/a)$. By this method concentrations as small as 10^{-4} per cent can be measured.

Internal friction has been used to determine the nature of point defects. The effect of solute atoms, the so-called "Zener relaxation,"[37] is readily observable, but the details of how the solute atoms produce the peak are still not well established.[38] In general, a relaxation effect will be produced by point defects if the distortion from the point defect lowers the symmetry of the crystal. The internal friction method has been widely used to show the effect of pinning of dislocations by point defects.[39]

The use of electrical resistivity measurements has been widespread in the case of metals. The full interpretation of electrical resisitivity data is often based on a theory or experiments[36] which determines how much the resistivity changes per unit concentration of defect. This information is not necessary in order to determine the activation energy of formation and of motion of the defect. Since changes in electrical resistivity as low as 10^{-3} per cent can be measured, then point defect concentrations of the same order of magnitude can be detected by this method. In semiconductors the electrical resistivity has an added significance depending on whether the defect acts like a donor or acceptor.

Self-diffusion experiments have played a large part in determining the nature of point defects. If one component diffuses with a lower activation energy, it might be assumed that the defect responsible for diffusion is more closely tied to one component than the other. In III–V compounds the group III elements have a lower activation energy for diffusion than the group V; as a result, it has been assumed that the III element jumps into vacancies which occupy second-nearest neighbor III–type sites.

Point defects may be observed directly with the field ion microscope.[40] Vacancies, interstitials, grain boundaries, and impurity dislocations have been observed on the atomic scale. Long-range order can be distinguished from disorder. To date, the technique has been limited to metals of melting point $> \sim 1000°C$. The Mössbauer[41] effect has been used to study defects. The Mössbauer spectrum is sensitive to the immediate surroundings of the emitting and absorbing atoms; the effect of point defects, short-range order, long-range order, and clustering should be observed. Electron paramagnetic resonance[42] has been used to study defects in high purity nonconductors like silicon. It may be used for intermetallic compounds if they have a high purity in order to increase the mean free path of the electron and if their surface-to-volume ratio is increased by using a fine dispersion. Metals keep the magnetic field from penetrating deeply; therefore, the increased surface would increase the effective volume of sample and thereby increase the sensitivity. Electron paramagnetic resonance measures the electron spin associated with the defect and compares it with the spin from an atom surrounded by perfect crystal. Point defects and dislocations have been studied by measuring thermal conductivity[43] at low temperatures. Defects in the lattice have an effect on superconductivity and especially on the critical current; presumably, superconductivity may be used as a tool to study the defects themselves.[44]

Measurements of yield point,[8] work hardening,[46] and hardness[47] are also used to infer the nature of defects. Irradiation[48] by different particles is used to study defects by determining the critical energies for producing a defect and then studying their subsequent annealing kinetics. Length changes, which are produced by irradiation or which occur during annealing, have been observed. Some of the techniques for measuring changes in length are as follows: (1) direct measurement of length by means of a traveling microscope which measures changes of 10^{-3} per cent, (2) capacitance gage[49] to 10^{-5} per cent, (3) dilatometer with optical lever[50] to 10^{-5} per cent, and (4) deflection of suspended wire[51] to 10^{-3} per cent. Density changes,[52] using the displacement method, may be measured within 10^{-4} per cent.

Electron microscopy not only detects dislocations, but makes it possible to determine the sign of the Burgers vector.[53] Therefore, dislocation loops which are produced by condensation of interstitial atoms may be distinguished from loops produced by vacancies. Stacking faults and antiphase boundaries, both periodic[54] and random,[55] may be observed. The periodic boundaries produce extra diffraction spots associated with the long period. Although single vacancies are not resolvable, vacancy clusters of about 50 Å may be seen.[56]

Stacking faults are observed directly with the electron microscope. The stacking-fault probability on bulk specimens may be obtained from an analysis of the diffuse X-ray scattering.[57] Stacking faults may also be seen by observing the etched surface optically, as has been done with silicon.[58] APDB's are observed with the electron microscope, and the antiphase domains may be observed optically if the APDB can be etched selectively, as in the case of Fe_3Al.[59] Analysis of diffuse scattering from X-rays[60] and neutrons[61] have been used to measure short-range order.

Dislocations may also be observed directly by X-rays using the Lang[62] method if the density is less than about $10^6/cm^2$. Dislocations may be observed in transparent crystals after they have been decorated.[63] In the case of silicon, infrared light has been used to make the crystal transparent.[64] Etch pitting has been used to study dislocations.[65] Thermal changes in etch pits[66] associated with dislocations have been used to study the migration of point defects along dislocations.

In summary, there are many experimental methods for studying only the concentration and distribution of point defects. Yet, it is usually difficult to specify exactly the structure of the defect and the mechanism by which it produces the observed behavior of the material in which we happen to be interested. A most recent book by Damask and Dienes[67] is an excellent presentation of how theory and experiment are combined for the purpose of determining the nature of point defects.

ACKNOWLEDGMENTS

The discussions with Dr. J. H. Westbrook and Dr. E. Parthé were most helpful. Support was received from the Atomic Energy Commission.

REFERENCES

1. *Symposium on Vacancies and Other Defects in Metals and Alloys*, 1957. The Institute of Metals, London.
2. Van Bueren H. G., *Imperfections in Crystals*. North-Holland Publ. Co., Amsterdam, 1960.
3. *International Conference on Crystal Lattice Defects*, 1962. Physical Society of Japan. In three volumes; I, Tokyo, on Mechanical Aspects; II and III, Kyoto, on Point Defects.
4. Kröger F. A. and Vink, H. J. *Solid State Physics* **3** (1956) 307.

5. Kröger, F. A. *Chemistry of Imperfect Crystals*, North-Holland Publ. Co., Amsterdam.
6. Parthé E., *Z. Krist.* (*English Transl.*) **119** 3/4, (1963) 204.
7. Matthias B. T., *Science* **144** (1964) 373.
8. Clark J. S., and N. Brown, *J. Phys. Chem. Solids*, **19** (1961) 291.
9. Fleischer R. L., *Acta Met.* **10** (1962) 835.
10. Helfrich W. J., and R. A. Dodd, *Acta Met.* **11** (1963) 982.
11. *Ibid.*, **12** (1964) 667.
12. Wayman C. M., K. M. Thein, and N. Nakanishi, private communications.
13. Huntington H. B., N. C. Miller, and V. Nernes, *Acta Met.* **9** (1961) 749.
14. Vook F. L., Reference 3, Vol. II, p. 190.
15. Brown N., *Phil. Mag.* **4** (1959) 693.
16. Rudman P. S., *Acta Met.* **10** (1962) 116.
17. Toth R. S., and H. Sato, *J. Appl. Phys.* **35** (1964) 698 (this article contains references to their earlier papers).
18. Howie A., and P. R. Swann, *Phil. Mag.* **6** (1961) 1215.
19. Marcinkowski M. J., Electron Microscopy and Strength of Crystals, G. Thomas and J. Washburn ed. Interscience, New York, 1963.
20. Amelinckx S., and P. Delavinette, *ibid.*
21. Flinn P. S., *Trans. AIME* **218** (1960) 145.
22. Kear B. H., *Acta Met.* **12** (1964) 555.
23. Vidoz A. E., and L. M. Brown, *Phil. Mag.* **7** (1962) 1167.
24. Stoloff N. S., and R. G. Davies, *Acta Met.* **12** (1964) 473.
25. Davies R. G., and N. S. Stoloff, *Trans. AIME* **230** (1964) 390.
26. Rachinger W. A., and A. H. Cottrell, *Acta Met.* **4** (1956) 109.
27. Haasen P., *Acta Met.* **5** (1957) 598.
28. Venables J. D., and R. M. Broudy, *J. Appl. Phys.* **29** (1958) 1025.
29. Lavine M. C., H. C. Gatos, and M. C. Flinn, *J. Electrochem. Soc.* **108** (1962) 974.
30. Basinski Z. S., D. B. Dove, and E. Mosser, *J. Appl. Phys.* **34** (1963) 469.
31. Holden A. N., *Acta Cryst.* **5** (1952) 182.
32. Kronberg M. L., *J. Nucl. Mater.* **1** (1959) 85.
33. Marcinkowski M. J., and D. S. Miller, *Phil. Mag.* **7** (1962) 1025.
34. Macrone N. J., and J. A. Coll, *Acta Met.* **12** (1964) 742.
35. Lewis J. R., *J. Metals* **13** (1962) 357.
36. Simmons R. O., and R. W. Baluffi, *J. Appl. Phys.* **31** (1960) 2284.
37. Zener C., *Elasticity and Anelasticity*. The Chicago Univ. Press, 1948.
38. Nowick A. S., and B. S. Berry, *Acta. Met.* (Internal Friction Conf.) **10** (1962) 312.
39. Gordon R. B., *ibid.*, 339.
40. Muller E. W., Reference 3, Vol. II, p. 1.
41. Gonser U., and H. Wiedersich, Reference 3, Vol. II, p. 47.
42. Watkins, G. D., Reference 3, Vol. II, p. 22.
43. Mendelssohn K., Reference 3, Vol. II, p. 17.
44. Livingston J. D., *Rev. Mod. Phys.* **36** (1964) 54.
45. Marcinkowski M. S., and N. Brown, *J. Appl. Phys.* **33** (1962) 537.
46. Peiffer H. R., *J. Appl. Phys.* **34** (1963) 298.
47. Seybolt A. U., and J. H. Westbrook, *Acta Met.* **12** (1964) 449.
48. Lucasson P. G., and R. M. Walker, *Phys. Rev.* **127** (1962) 485, 1130.
49. Abramson E., and M. E. Caspari, *Phys. Rev.* **129** (1963) 536.
50. Takamura J., *Acta Met.* **9** (1961) 547.
51. Bauerle J. E., and J. K. Koehler, *Phys. Rev.* **107** (1957) 1493.
52. Kuhlmann-Wilsdorf D., and K. Sezaki, Reference 3, Vol. III, p. 57.
53. Mazey D. J., R. S. Barnes, and A. Howie, *Phil. Mag.* **7** (1962) 1861.
54. Ogawa S., D. Watanabe, H. Watanabe, and T. Komoda, *Acta Cryst.* **11** (1958) 872.
55. Fisher R. M., and M. J. Marcinkowski, *Phil. Mag.* **6** (1961) 1385.
56. Thomas G., and J. Washburn, *Univ. Calif. Rept. Lawrence Rad. Lab.* UCRL 10674 (1963).
57. Warren B. E., and E. P. Warekois, *Acta Met.* **3** (1955) 473.
58. Mendelson S., *J. Appl. Phys.* **35** (1964) 1570.
59. Stoloff N. S., private communication.
60. Warren B. E., and B. L. Averbach, *Modern Research Techniques*. ASM, 1952, p. 95.
61. Walker C. B., and D. T. Keating, *Phys. Rev.* **130** (1963) 1726.

62. Lang A. R., *J. Appl. Phys.* **30** (1959) 1748.
63. Hedges J. M., and J. W. Mitchell, *Phil. Mag.* **44** (1953) 223.
64. Dash W. C., *J. Appl. Phys.* **27** (1956) 1153.
65. Gilman J. J., and W. P. Johnston, *J. Appl. Phys.* **30** (1958) 129.
66. Doherty P. E., and R. S. Davis, *Acta Met.* **7** (1959) 118.
67. Damask A. C., and G. J. Dienes, *Point Defects in Metals.* Gordon and Breach, New York, 1963.

chapter 16

Internal and External Interfaces

ROBERT BAKISH
Bakish Materials Corporation
Englewood, New Jersey

1. INTRODUCTION

Although substantial progress has certainly been made in the understanding of interfaces in pure metals, the concepts of the true nature of the interfaces in intermetallic compounds are still in the process of being worked out.

Webster's Dictionary defines an interface as "a surface forming a common boundary of two bodies, spaces or phases." Interfaces in intermetallic compounds can be either external, that is, surface, or internal, that is, grain boundary or antiphase boundary. In subsequent sections an attempt is made to review the limited studies available on the geometry, structure, chemistry, and properties of both external surfaces and internal interfaces or grain boundaries.

2. EXTERNAL INTERFACES OR FREE SURFACES

The exact atomistic structure and behavior of a surface of a pure metal is a difficult problem. Although substantial theoretical contributions have been made, more progress has been achieved experimentally. The experimental approaches include direct measurement of surface tension, measurement of the effects of surface-to-volume ratio on properties, and measurement of effects of environment on properties with particular attention to the influence of crystalline anisotropy. In the case of intermetallics, in which the details of interfaces are complicated by the bonding parameters of the constituent metals (which can vary within wide limits for the different compounds), there has been considerably less effort to clarify the situation and theoretically to devise a quantitative picture. Experimentally, efforts on intermetallics have been confined to the last of the three aforementioned approaches. Progress has been hindered by the difficulties in preparing suitable specimens. A large part of the available literature is therefore confined to semiconductors, in which a highly advanced technology of single-crystal synthesis has developed.

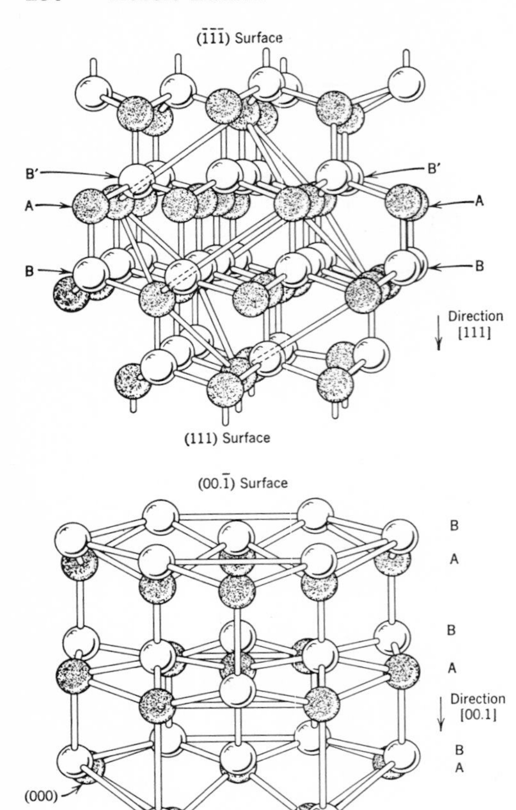

Fig. 1. Structure symmetry in zinc blende-ABCABC and wurtzite-ABAB layers. (Reference 7.)

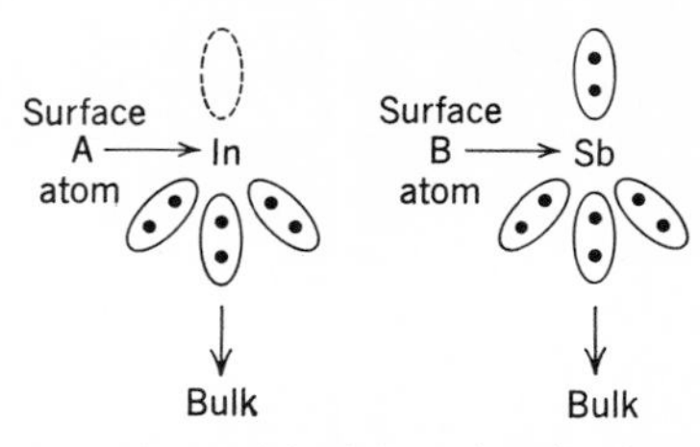

Fig. 2. Atomic model of A and B {111} surfaces. (Reference 4.)

Recently Gatos and his associates at the Lincoln Laboratory have done a considerable amount of work on the III–V[1–9] and II–VI[10] compounds and in particular have examined some low index faces, that is, {111}, of these compounds. Inasmuch as the second group of compounds are not truly intermetallics but are metal-metalloid compounds, reference to them could be considered outside the scope of this book. Yet, discussion of these data should help amplify the picture of the structure of surfaces in intermetallics. As a consequence of these studies, several ideas about the structure of some surfaces in these compounds have been advanced. Information pertaining to their chemical, electrical, and mechanical properties has also been generated.

The III–V compounds—the so-called polar compounds—can be characterized by the zinc blende structure; we can consider InSb a typical example. Mercuric telluride can be taken as a typical example of the II–VI compounds,[10] although to some CdSe, with a structure closely related to wurtzite, might perhaps seem a better example. The basic difference between the two types of structures is the stacking of the different atomic layers. Reference to this will also be made later. Figure 1 shows the characteristic ABCABC (cubic) and ABAB (hexagonal) layer symmetries, respectively. The most important aspect of materials with these structures is their noncentrosymmetric character showing polarity in the ⟨111⟩ directions and thus leading in turn to two types of {111} surfaces. They terminate in groups of III and V atoms, respectively, and because of their atomic structure we can envisage atomic arrangements in the ⟨111⟩ directions as shown in Fig. 2. We see here that the A surface has all electrons employing in bonding, whereas the B surface has two dangling electrons in the dangling bonds. If we reflect on the partially ionic character of these bonds, we can envisage the electrical dipoles oriented as shown in Fig. 3.

Because known surface chemical properties of intermetallics are limited almost exclusively to them, it is worthwhile to show the behavior of the {111} surfaces under chemical attack and to note the imperfection

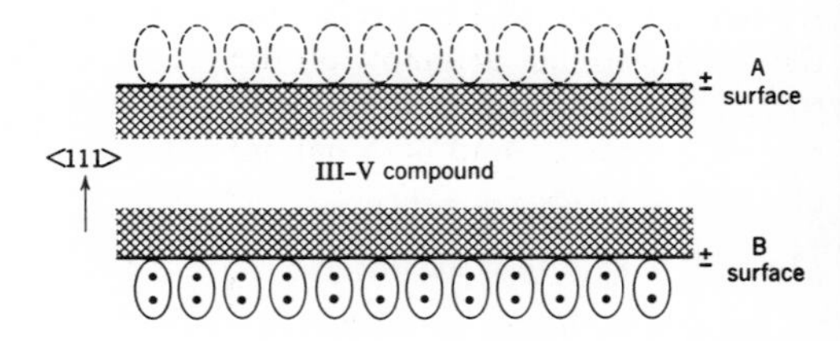

Fig. 3. Schematic representation of the dangling bonds on the A and B surfaces of III–V compounds and the associated surface dipole movements. (Reference 10.)

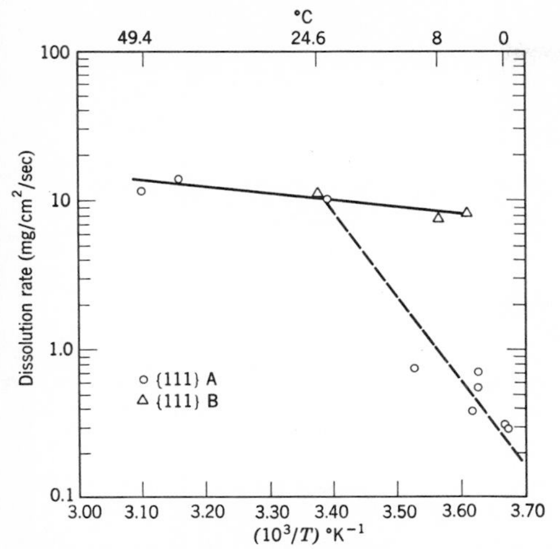

Fig. 4. Dissolution rate of InSb as a function of temperature in modified CP-4 [70% HNO_3 ((2 parts by volume)), 48% HF (1 part), and glacial CH_3COOH (1 part)]. (Reference 4.)

structures that are seen on them. Figure 4 shows the rate of dissolution of the two types of surfaces in indium antimonide, and Fig. 5 shows the respective etch morphologies of them. Figures 6 and 7 display imperfection structures as delineated on the surfaces. A schematic of the atomic arrangements believed responsible for this behavior are shown in Figs. 8 and 9. Similar differences in chemical reactivity, X-ray diffraction behavior, and morphology of such polarized surfaces are observed in II–VI compounds. Figure 10 shows the different X-ray reflections of the (0001) and $(000\bar{1})$ and the (111) and $(\bar{1}\bar{1}\bar{1})$ surfaces in CdSe and HgSe. The difference in etching behavior of equivalent and opposite

TABLE 1

Radii of Curvature and the Corresponding Elastic Strain Energy (ESE) of {111} InSb Wafers[11]

Thickness of Wafer μ (±0.3)	Measured Radius cm (±10%)	Total ESE Energy ergs × 10^4	ESE per Surface Gram-Atom (ergs/gm.at.) × 10^{-6}
7.8	104	3.0	0.8
7.9	81	5.2	1.4
11.1	205	2.3	0.6
13.1	194	4.1	1.1

faces in this group of crystals is shown in Figs. 11 to 14 for HgTe, ZnS, ZnTe, CdS, CdSe, CdTe, HgSe, and HgTe. The bonding here is believed to be analogous to that of the III–V compounds already discussed.

More recently the same group of investigators have proposed that considerable elastic stresses are associated with the surfaces of the III–V compounds,[11] and they have shown their presence in InSb. Figure 15 shows the extent of bending resulting from the structural differences in opposing {111} surfaces in a 7 to 8 μ InSb[11] wafer. Table 1 shows additional data from this study and the level of the strain energies observed.

Students of chemical bonding have made contributions to the understanding of surfaces in intermetallics, although their interests in these studies is considerably broader. Among

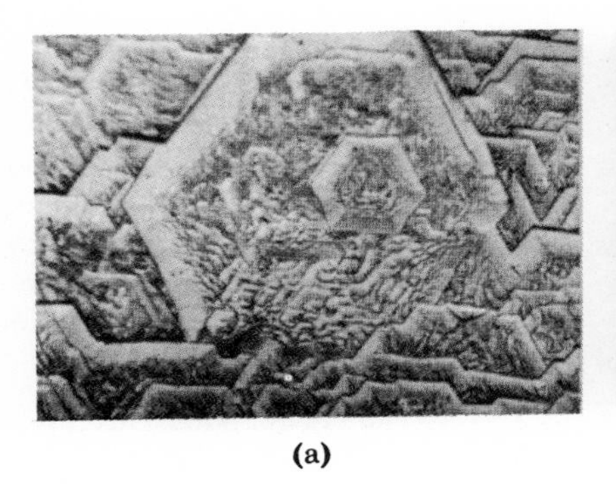

(a)

(b)

Fig. 5. Etch morphology of the A (*a*) and B (*b*) surfaces of InSb wafer etched in Fe^{3+} etchant for 30 min at 87°C × 315. (Reference 4.)

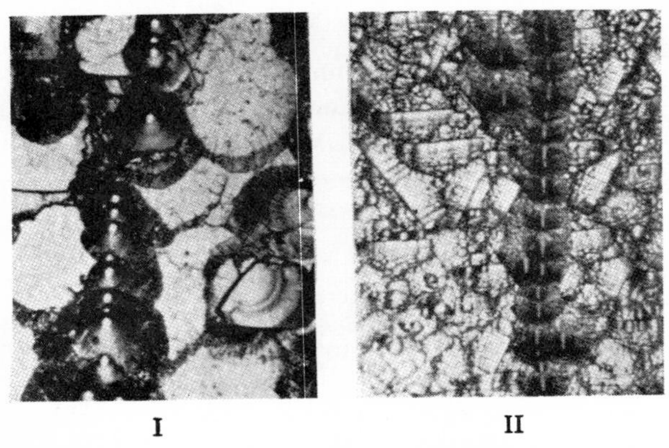

Fig. 6. Dislocation etch pits on the A (I) and B (II) surfaces of an {111} wafer of InSb; etchant, 0.2N Fe^{3+} in 6N HCl. Temperature 95°C. 50 X. The identical row of dislocations is running on both sides. (Reference 8.)

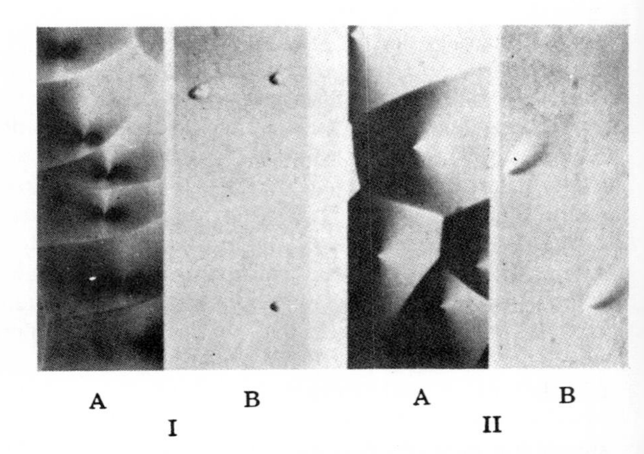

Fig. 7. Dislocation etch pits on two {111} wafers of InSb containing approximately 0.4% Pb (I) and 0.4% Ge (II). A and B designate the A and B surfaces, respectively. Etchant, parts by volume, 1 of 30% H_2O_2; 1 concentrated HF; 2 H_2O. 157.5 X. (Reference 8.)

the different criteria utilized by them is the crystallography of cleavage so as to obtain information on surface atom configurations and in turn bonding parameters.* We are here particularly interested in information that has been generated to broaden concepts of the nature of surfaces.

It has been shown[12,13,14] that crystals of sphalerite structure with electronic ratios $q/e \neq 0$ for the surface atoms are expected to cleave on the (011), whereas those with neutral $q/e = 0$ are expected to cleave on (111). It has further been demonstrated[12,13,14] that from the ratios of the (011) to (111) cleavage we obtain information on the bonding character of compounds and, in turn, on the structure of respective cleavage faces. According to Wolff, the following principles determine the cleavage plane:

1. The number of bonds (nearest-neighbors) to be separated per unit area of the plane is a minimum relative to all other crystal planes.

* See Chapter 5 by Wolff.

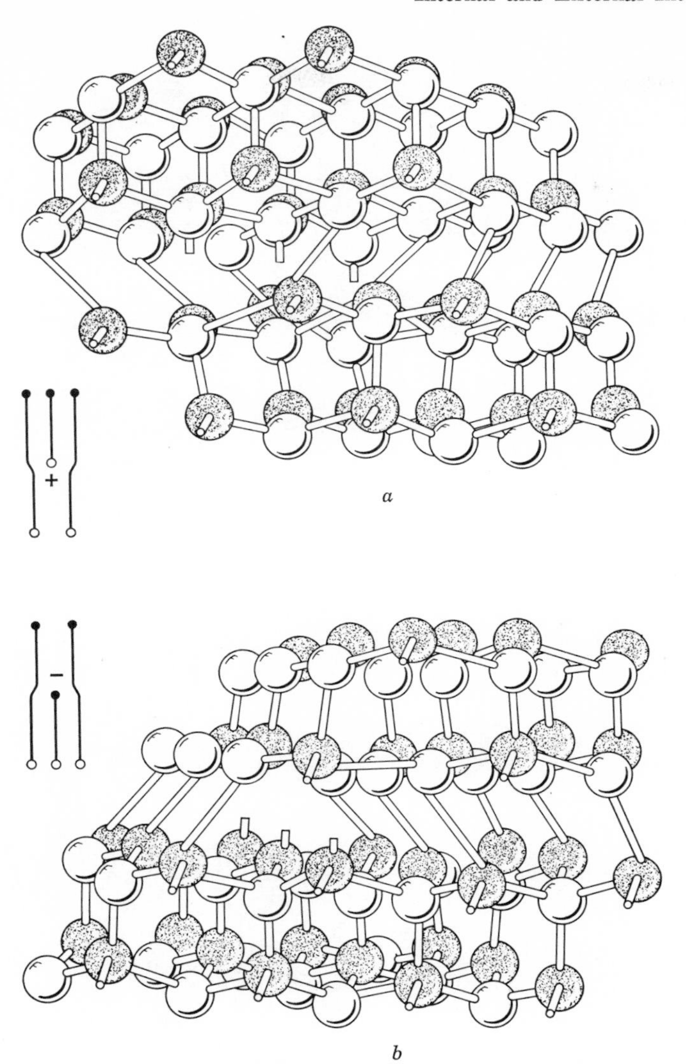

Fig. 8. Atom model showing a 60° edge dislocation intersecting the A surface of a III–V compound. The surface of intersection is defined by the atoms with protruding bonds: (*a*) dislocation with row of B atoms; (*b*) dislocation with a row of A atoms. (Reference 5.)

2. The plane is electrically neutral with alternate arrays of positive and negative structure elements (checkerboard arrangement), which permits the two separating surface layers to repel when shifted with respect to one another.

Winkler[12] showed that, for materials with diamond and sphalerite structure, conditions 1 and 2 require (111) and/or (011) cleavage; (011) cleavage was found in all antimonides, phosphides, and arsenides of In, Ga, and Al. The (111) cleavage was found[13] in InSb, InAs, ZnP, and GaSb, verifying the presence of covalent bonding in the later compounds. Wolff[15] further established that AlP, AlAs, AlSb, GaP, GaAs, GaSb, InP, InAs, InSb, ZnS, ZnTe, CdSe, HgSe, and HgTe, exhibit (011) or so-called type 1 cleavage. AlSb, GaAs, InP, and ZnTe, and to greater extent GaSb, InAs, InSb, and CdTe also show cleavage in

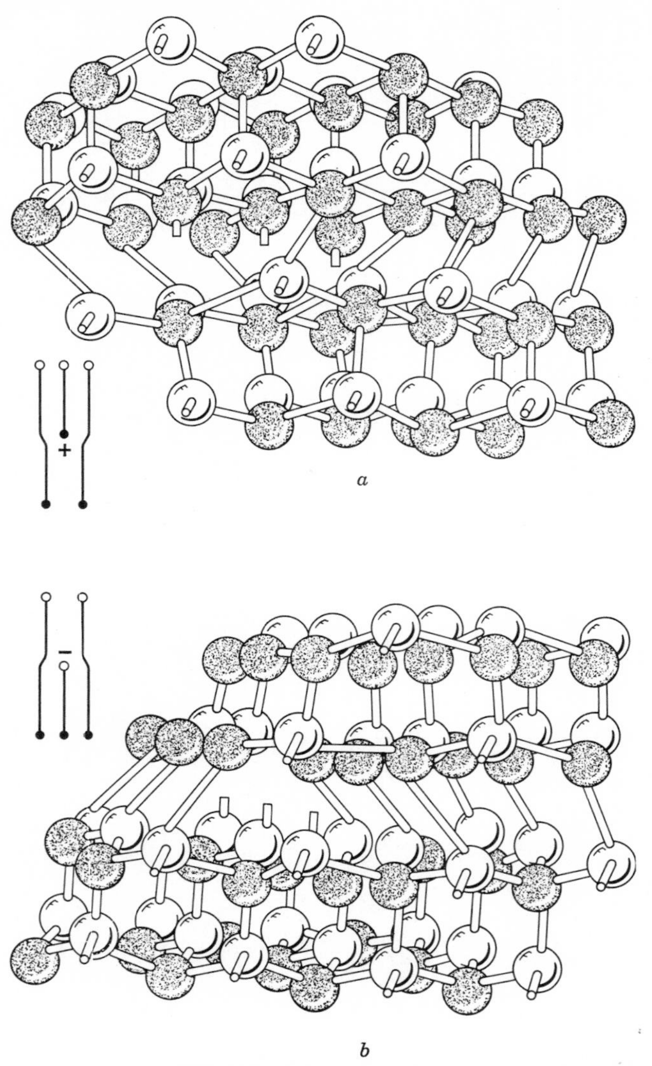

Fig. 9. Atom model showing a 60° edge dislocation intersecting the B surface of a III–V compound. The surface of intersection is defined by the atoms with protruding bonds: (*a*) dislocation with a row of A atoms; (*b*) dislocation with a row of B atoms. (Reference 5.)

(*hkl*) ($h \geq l$) planes; this is referred to as type 2. A third-type cleavage is also observed taking place in planes of the ⟨001⟩ zone. CdTe, for example, is a material where all three cleavages have been observed.

Fluorite-type intermetallics, that is, those with structures analogous to that of CaF_2, as, for example, Mg_2Sn, $AuGa_2$, and $AuIn_2$, show (111) and (110) microcleavage as well as some type 3 cleavage.

Complex intermetallics containing structural vacancies, as, for example, Ga_2Te_3, In_2Te_3, Cu_3AsS_3, $HgIn_2$, and $NiIn_2$, seldom show cleavage, although (111) cleavage has been observed in Ga_2Te_3 and In_2Te_3.

Now that we have reviewed the type of cleavages observed, it is important to indicate the type of structural information that we can acquire on the surfaces of these intermetallics. In materials with ionic and covalent bonding, cleavage informs us of the absolute q/e value of the surface atoms, which may be different

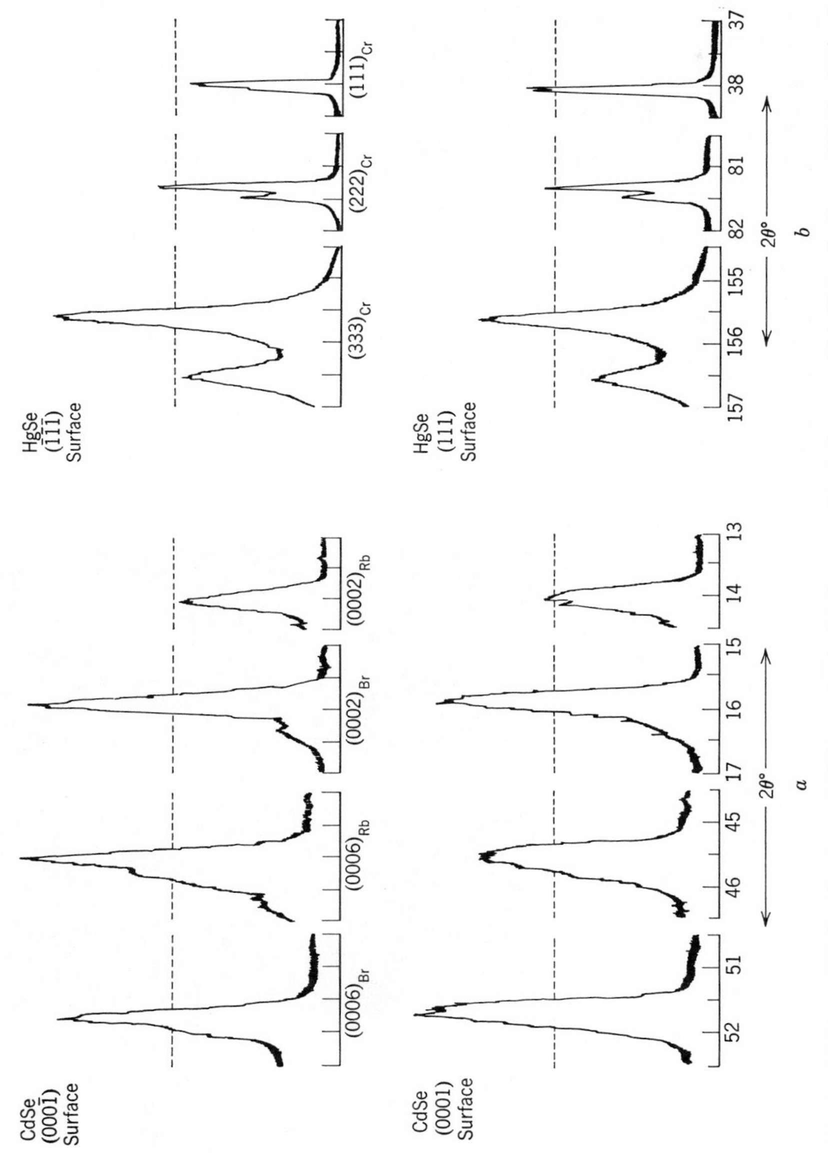

Fig. 10. (*a*) X-ray (0001) reflections of CdSe obtained with fluorescent techniques and (*b*) (*hkl*) reflections of HgSe obtained with sealed-off X-ray tube. (Reference 10.)

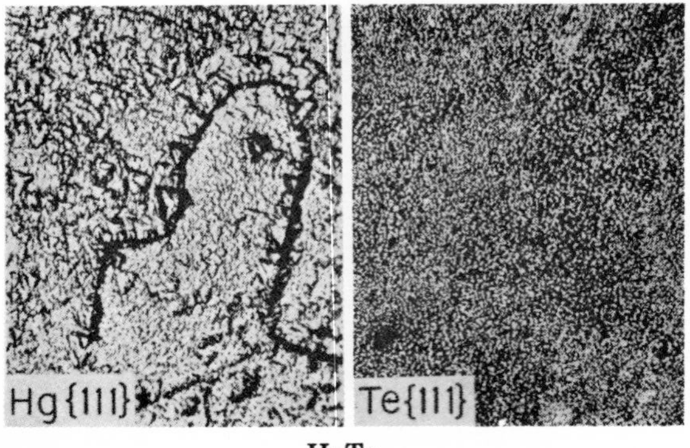

Fig. 11. The {111} surfaces of HgTe etched in 1 (part by volume) concentrated HCl + 1 concentrated HNO_3 at room temperature for three 1 min intervals at 25°C after chemical polishing. The interrupted etching results in more clearly defined surfaces. The mercury surface develops triangular etch pits in an overall background of triangular etch figures. In the photomicrograph, the pits can be seen defining a boundary. The tellurium surface has a flat uniform "grainy" appearance. Two crystals of this material were available and gave identical results. 200 X. (Reference 10.)

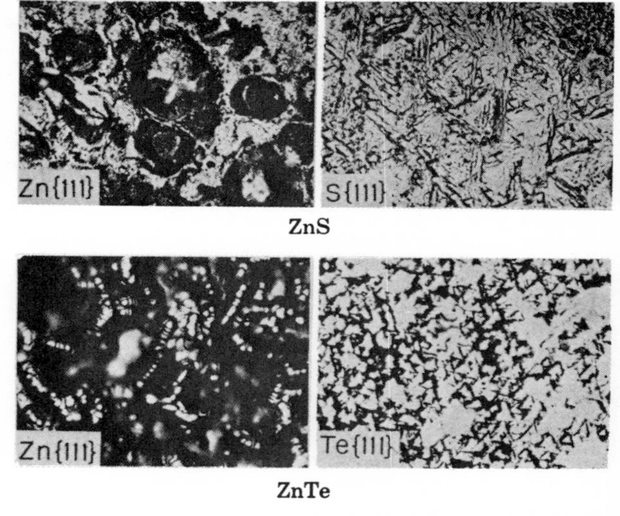

Fig. 12. The {111} surfaces of natural ZnS and ZnTe etched in 1 (part by volume) concentrated HCl + 1 concentrated HNO_3. ZnS: The sulfur surface develops triangular etch pits in a background of flat triangular etch figures. The zinc surface has a rough nonstructural appearance. Both surfaces develop a discontinuous sulfur film. ZnTe: The zinc surface develops triangular etch pits, which can be seen to be in arrays. The pits are small in size and quite deep. The tellurium surface develops flat triangular etch figures and, with continued etching, tends to become somewhat polished. Magnification: ZnTe, Zn {111}—640 X; all others—160 X. (Reference 10.)

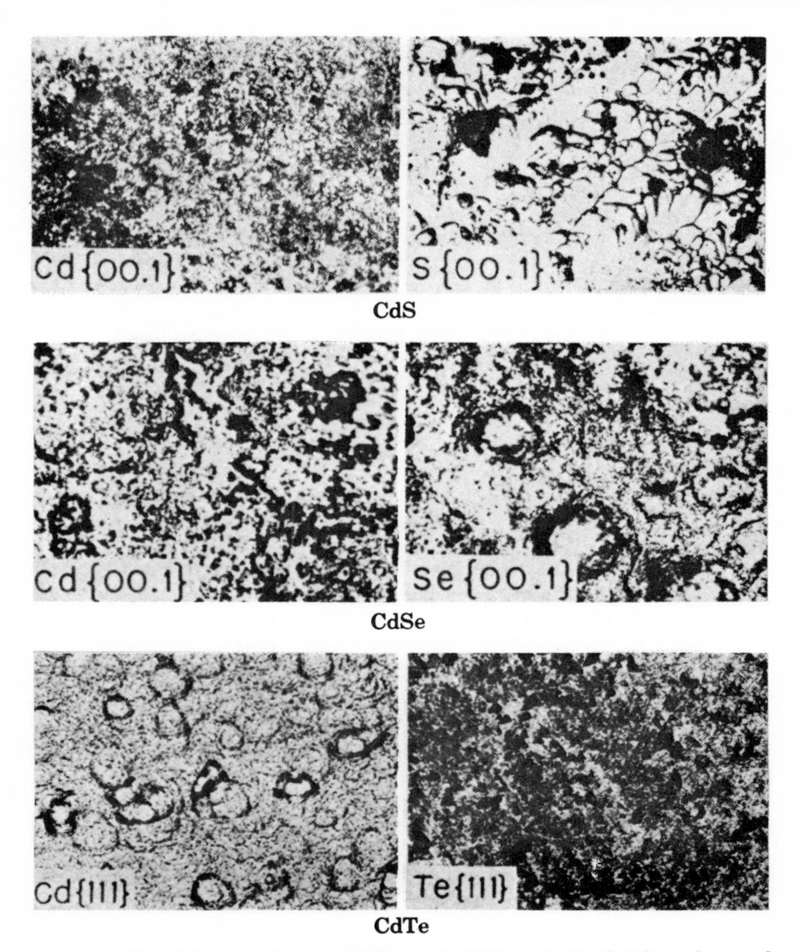

Fig. 13. The {00.1} surfaces of CdSe and CdS, and the {111} surfaces of CdTe etched in 1 (part by volume) concentrated HCl + 1 concentrated HNO_3. CdS: The sulfur surface develops very clearly defined conical etch pits with a suggestion of hexagonal symmetry. The cadmium surface is flat and grainy, somewhat obscured by the sulfur deposit. CdSe: Both surfaces appear to etch equivalently. The surfaces are nondescript and obscured by a strongly adherent selenium film. CdTe: The cadmium surface develops triangular etch figures; the tellurium surface is covered with a film and has a random distribution of very flat triangular figures, which may lie in the film. 200 X. (Reference 10.)

from the q/e value of the atoms of the bulk, but we are here interested only in the surface parameters. The surface value of q/e is expressed as $(q/e)_{hkl}$ to differentiate it from (q/e) bulk.

If $(q/e)_{hkl}$ is negligible in all planes in the bulk, (111) cleavage predominates; however, if $(q/e)_{hkl}$ is appreciably large, (011) cleavage occurs in addition or exclusively.

In both (111) and (011) cleavages the constituents A and B must separate from one of their four original neighbors in the bulk of the crystal. According to Wolff[15] the atoms retain their effective charge when (*a*) an "unshared" or "ionic" binding electron pair remains with B, or (*b*) the two electrons of an originally "shared" or covalent electron pair separate and stay with A and B.

In these terms, the condition $(q/e_{\text{bulk}} = 0$ obtains for 25, 50, and 75 per cent ionic

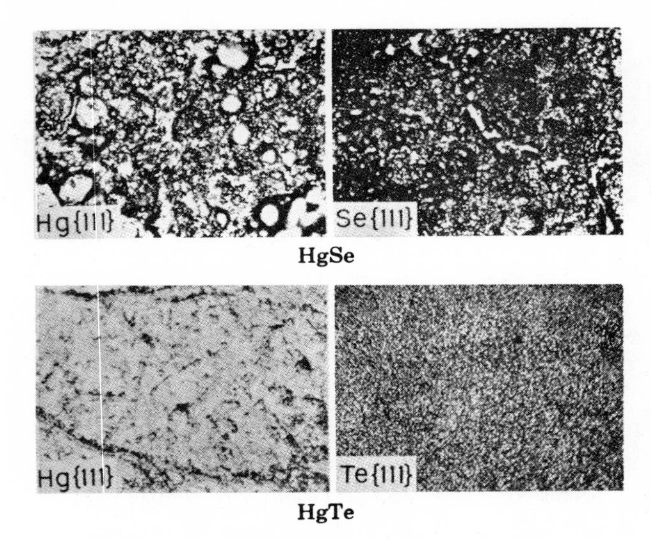

Fig. 14. The {111} surfaces of HgSe and HgTe etched in 1 (part by volume) concentrated HCl + 1 concentrated HNO_3. HgSe: Both surfaces etch somewhat equivalently; however, in the initial period of the etch, some tendency is observed of the mercury surface toward the development of pitlike structure. Continued etching failed to enhance this tendency. Both surfaces developed an adherent selenium film, which was partially dissolved in concentrated H_2SO_4. HgTe: The mercury surface develops triangular etch pits in a background of flat triangular etch figures. In the photomicrograph the etch pits can be seen in arrays defining boundaries. The tellurium surface has a flat grainy character with no resolvable structure. Magnification: HgSe, 160 X; HgTe, 176 X. (Reference 10.)

bonding character in A(III)B(V), A(II)B(VI), and A(I)B(VII), even though not all compounds of these groups are truly applicable to the scope of materials being considered here.

We should also mention the state of affairs on surfaces of materials that show the third-type cleavage. Again, according to Wolff[15] examination of the details of cleavage in HgTe, HgSe, and CdTe indicate that the surfaces of these intermetallics are deformed, so that atoms from the first, and possibly to a minor extent the second, surface layers shift to new equilibrium positions. This is especially true for the (001) plane.

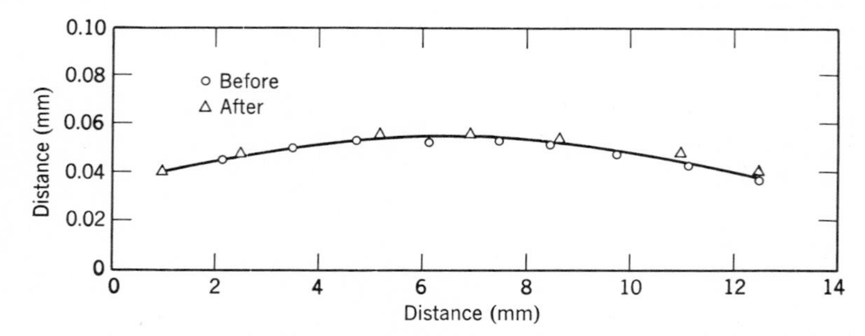

Fig. 15. Curvature exhibited by an InSb {111} wafer 7.8 ± 0.3 μ thick before and after annealing at 450°C for 7 hours. (Reference 11.)

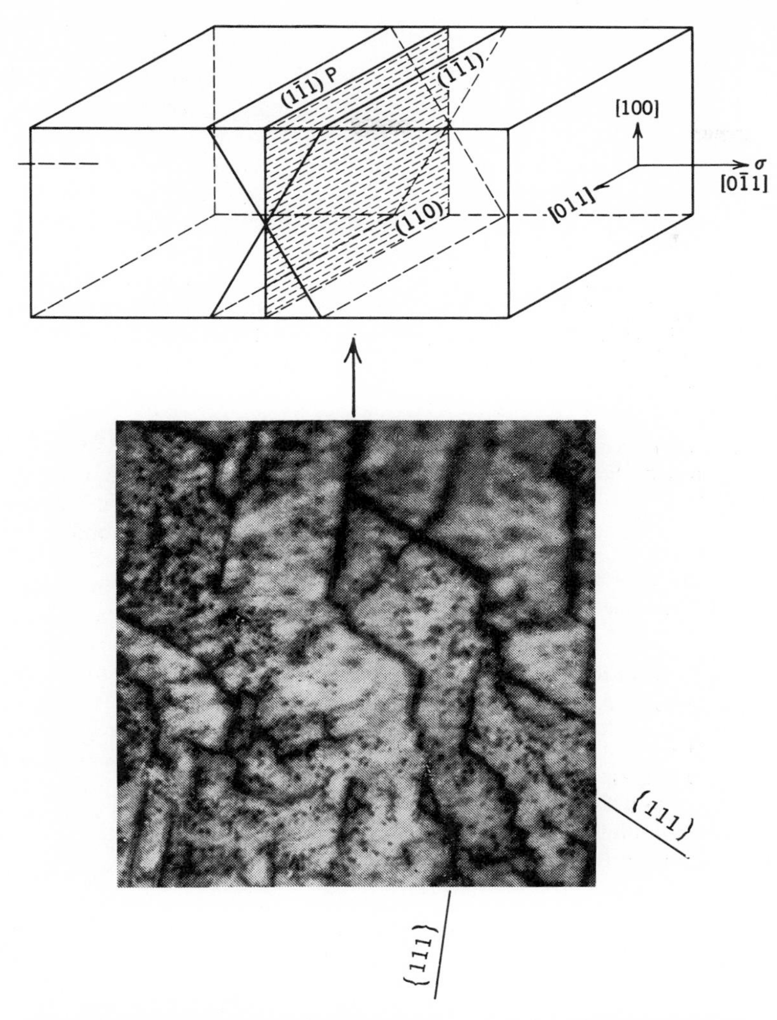

Fig. 16. Fine steps in a chemically [$FeCl_3$(2% aqueous)] embrittled (110) face of Cu_3Au. 2000 X. (Reference 16.)

Before completing the discussion of cleavage and surfaces, the writer wishes to refer to some of his early work[16] in which he reported unexpected stress-corrosion-induced cleavage in Cu_3Au. Figure 16 shows a cleaved surface and the steps associated with the (111) glide plane in this material. It is also of interest to observe the surface of the (111) glide plane itself (Fig. 17[17]) and the regions of corrosive attack associated with it. They were believed to be dislocations in this material. Results of a recent investigation[18] show very similar findings (Fig. 18) and confirm the nature of the attacked region.

Another case of cleavage associated with corrosion processes has been reported for Mg_2Sn.[19] In the two foregoing cases the atomistics of the surfaces were not investigated.

The above discussion pertains to specific surfaces of very specific intermetallics and is not sufficient to give a true picture of external surfaces of intermetallics; yet, no more can be stated with concreteness and accuracy about surfaces of intermetallics. We can make a number of analogies from knowledge of surfaces of pure metals, but this will not substantially change the picture. It is certain that a considerable number of studies on surfaces of intermetallics and their behavior

Fig. 17. Imperfections observed on a (111) face of a Cu_3Au crystal, corroded by $FeCl_3$(2% aqueous) 200 X. (Reference 17.)

will have to be performed before we can have a complete picture of this subject.

3. INTERNAL INTERFACES—GRAIN BOUNDARIES

The internal interfaces in intermetallic compounds are comprised of grain boundaries and antiphase domains. Grain boundaries, like those in all crystalline solids, according to today's concepts differ from the bulk material in several respects:

(*a*) The grain boundary as an internal surface is a likely site for nucleation of additional phases formed either by precipitation or other solid state transformations.

(*b*) The grain boundary is by definition an interface and as such it is a demarcation of a geometric transition from one crystallographic orientation to another.

(*c*) As consequence of its structure the grain boundary is a high energy region.

(*d*) As a consequence of both its structure and its high energy state, the grain boundary region is a site at which chemical segregation may occur in systems containing more than one component.

3.1 Geometry

It is believed that the three-dimensional grain-boundary distribution in an intermetallic can be assumed to correspond to that of pure metals. Smith in his detailed study[20] treats the topic of grain shapes in metals without reference to intermetallics. He shows that grain boundaries in metals form a network closely analogous in shape to soap bubbles in a froth, in which surface energy and space filling alone determine the structure. Limited observations made by the author and other investigators on Cu_3Au, CuSn, β-brass, and Mg_2Sn[19] indicate that the ideas presented by Smith for metals and alloys are a good representation, if not of all, at least of a large number of the intermetallics. In the cases referred to above, the writer did not use the stereomicroscopic technique,[20] but rather

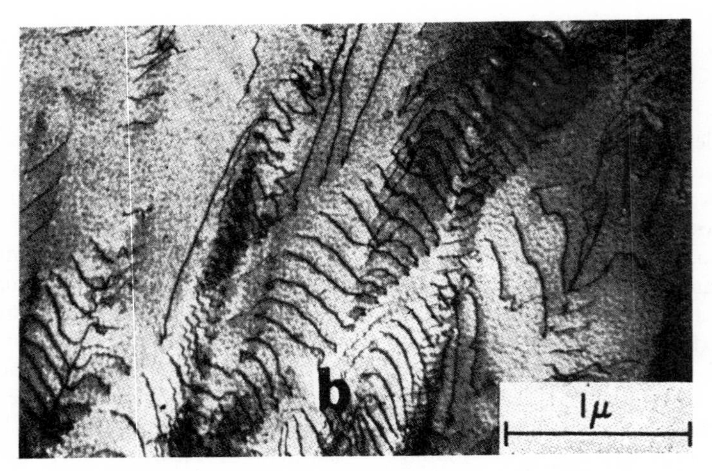

Fig. 18. Electron micrograph showing advanced $FeCl_3$(2% aqueous) attack in disordered Cu_3Au along imperfections. 25,000 X. (Reference 18.)

cursorily examined individual grains, which he obtained by virtue of the high incidence of intergranular fracture, promoted by a variety of methods from plain stressing to a combination of stress-corrosion cracking treatments.

In Smith's[20] study average grains were found to have ~14 faces with the average face $5\frac{1}{7}$ edges. It was also indicated that for a significant fraction (4 per cent) of the corners, the number of meeting edges differs from that of the typical tetrahedral arrays that result from equal surface tensions. It is believed that such behavior can be ascribed to the presence of impurity at the interfaces and to variation of the interfacial energy with variation of grain boundary orientation. It would be particularly desirable to investigate these factors in detail in intermetallic compounds. This problem can be significantly simplified by selection of brittle intermetallics, which virtually disintegrate along their interfaces and permit evaluation of interface networks as well as possibly reveal some of the parameters leading to their formation. In such studies we should be especially careful with the purity of the intermetallics, so as to be able to separate interface embrittlement effects due to impurity contamination from those due to the interface properties themselves. This problem is discussed again later in this chapter.

3.2 Chemistry

The chemistry of interfaces can be deduced in principle either on the basis of direct chemical analysis or on the basis of interfacial property manifestations and changes as a consequence of chemical changes at interfaces.

In intermetallics, the possible role of impurities in the interpretation of property data is apparently often ignored. Although there have been only limited investigations into the direct analytical chemistry of grain boundaries in intermetallics, it can be stated with certainty that, because of their structure, grain boundaries are the sites of impurity segregations. As a result, we can make the rather sweeping statement that no studies of inherently pure grain boundaries in intermetallics have been made. The great desirability of such studies can hardly be underestimated.

When we consider the problem of gaseous impurities more data on their distribution appear to be available. Recently data have been obtained on grain boundary properties of intermetallics in which a slight excess of the more electropositive one of the constituents is present. Reference is made to recent work by Westbrook and Wood[21] in which AgMg, NiAl, NiTi, CoAl, NiAl, Ir_3Cr, AgTi, Zr_2Be_{17}, $TaBe_{12}$, $ZrBe_{13}$, MgAl, and $MoSi_2$ were investigated. It was shown that changes in interface chemistry, namely the increase in local concentration of oxygen and/or nitrogen, affect the properties of the boundary, specifically its mechanical properties. These changes are ascribed to the interaction of the excess component with O_2 or N_2, although this behavior at this stage is more hypothesized than proved; data are shown in Table 2. Additional composition factors investigated in this study are indicated by Table 3, in which results obtained as a consequence of ternary additions to Mg-rich and Ag-rich MgAg are seen.

An indirect insight into the chemistry of interfaces can be obtained by referring to early work by the writer,[22] in which electrochemical potentials in Cu_3Au and CuAu were measured; Figs. 19 and 20 show the type of behavior observed. It can be seen that in the case of Cu_3Au there appears to be a constant potential difference between the interface and the matrix. In the case of CuAu, on the other hand, no such behavior is observed, and the potential difference between interface and matrix is similar to the type of potentials obtained for pure copper. It is believed that even though no claim for absolute purity of these compounds can be made here, the observed facts are due to the properties of the interfaces themselves.

These electrochemical measurements help to interpret the measured properties of these materials, their behavior as structural materials, and corrosion-cracking susceptibility. A certain similarity between these findings and the results of the studies[21,23,24] by Westbrook and associates, aimed at understanding the occurrence of the so-called "pest" phenomena in these intermetallics, can be noted. Both cases will be discussed subsequently.

It is difficult to understand why greater effort has not been exerted in the study of chemistry of the grain boundary in intermetallics, now that microprobe analysis is becoming a relatively common tool. The

TABLE 2

Approximate Grain-Boundary Hardening ΔH at Room Temperature in Several Intermetallic Compounds

$$\Delta H\,(\%) = \left(\frac{H_{\text{boundary}} - H_{\text{bulk}}}{H_{\text{bulk}}}\right) \times 100$$

<table>
<tr><th rowspan="2">Compound (A_mB_n)</th><th rowspan="2">Crystal Structure</th><th colspan="2">$\Delta H\,(\%)$</th></tr>
<tr><th>Excess A</th><th>Excess B</th></tr>
<tr><td>AgMg</td><td>CsCl</td><td>0</td><td>20–50</td></tr>
<tr><td>NiAl</td><td>CsCl</td><td>0</td><td>25–60</td></tr>
<tr><td>NiTi</td><td>CsCl</td><td>0</td><td>~25</td></tr>
<tr><td>CoAl</td><td>CsCl</td><td>0</td><td>~20</td></tr>
<tr><td>Ni_3Al</td><td>CuAu</td><td>0</td><td>~15</td></tr>
<tr><td>Ir_3Cr</td><td>Cu_3Au</td><td colspan="2">~20[a]</td></tr>
<tr><td>AgTi</td><td>CuAu</td><td>0</td><td>~20</td></tr>
<tr><td>TiAl</td><td>CuAu</td><td colspan="2">~20[a]</td></tr>
<tr><td>Zr_2Be_{17}</td><td>Complex hexagonal</td><td colspan="2">~35[a]</td></tr>
<tr><td>$TaBe_{12}$</td><td>Complex tetragonal</td><td colspan="2">~60[a]</td></tr>
<tr><td>$ZrBe_{13}$</td><td>$NaZn_{13}$ (fcc)</td><td colspan="2">25–50[a]</td></tr>
<tr><td>γ-MgAl</td><td>α-Mn (bcc)</td><td colspan="2">~15[a]</td></tr>
<tr><td>$MoSi_2$</td><td>Body-centered tetragonal</td><td colspan="2">~33[a]</td></tr>
</table>

[a] Exact composition not known.

TABLE 3

Effect of Ternary Solute Additions to Both Ag-Rich and Mg-Rich AgMg on Grain-Boundary Hardening

Mg/Ag	A/o Mg in Ternary Compound	A/o Ternary Solute	Grain Boundary Hardening
1	49.5	1.0 Sn	Present
<1	48.9	1.0 Sn	Present
<1	48.5	2.3 Sn	Present
<1	49.4	1.0 Zn	Not present
>1		1.0 Si	Present
>1	50.2	1.0 Zn	Not present
>1	50.0	2.0 Zn	Not Present
>1	50.1	0.3 Cu	Not present
>1	50.0	1.0 Cu	Not present
>1	50.0	0.2 Al	Present
>1	50.0	1.1 Al	Present
>1	50.6	0.03 La	Not present
>1	50.6	0.05 Ce	Not present
>1		0.5 Au	Not present

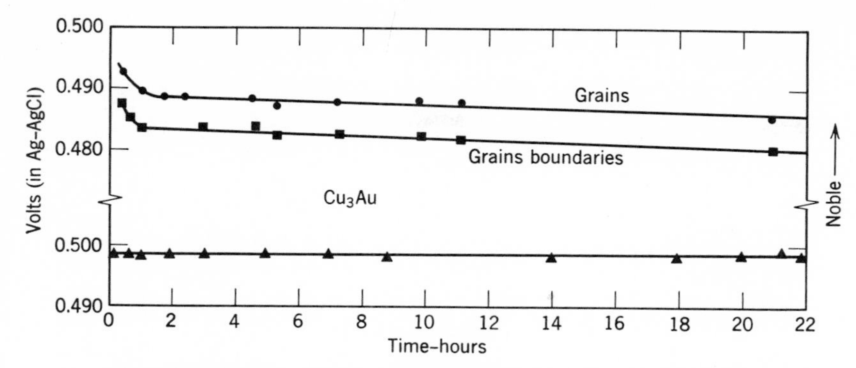

Fig. 19. Galvanic potentials of Cu_3Au grains, grain boundaries, and polycrystalline aggregates in 2% ferric chloride (aqueous). (Reference 22.)

writer feels that substantial benefits could be derived through better understanding of the chemistry of intermetallic grain boundaries, whether by direct or indirect measurements.

3.3 Properties

Perhaps one of the easiest and most productive approaches to obtaining data on the nature of intermetallic grain boundaries is to study their mechanical properties and, in particular, plastic deformation. Unfortunately, it appears that again there have been no studies of plastic flow along grain boundaries in intermetallics, whether of random or selected and predetermined orientation. Such study conducted on judiciously selected compounds should be capable of generating worthwhile information.

With the substantial advances in crystal-growing techniques and the improved supply situation on ultrahigh purity metals, it appears that studies of the mechanical properties—and particularly the plastic deformation—at interfaces in intermetallics should be reexamined. They should be of greatest interest as a function of orientation for different binary intermetallics; they should help understand structural behavior and, specifically, clarify the question of whether differences exist between intermetallic interfaces and those in

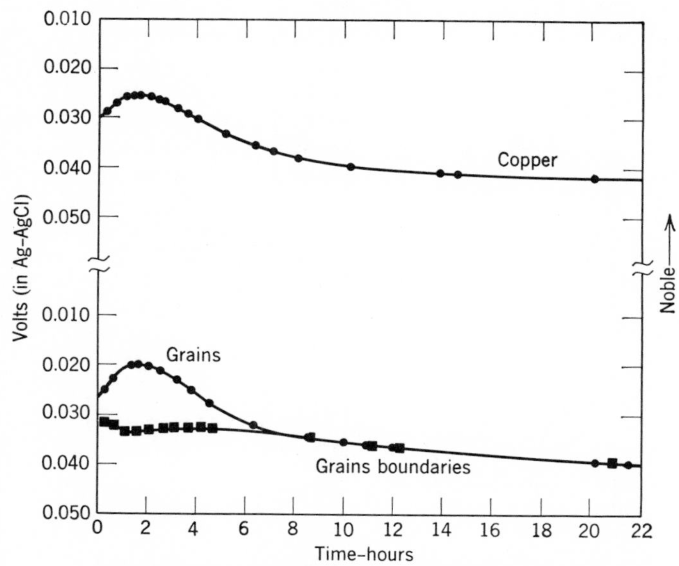

Fig. 20. Galvanic potentials of copper grain boundaries and polycrystalline copper. (Reference 22.)

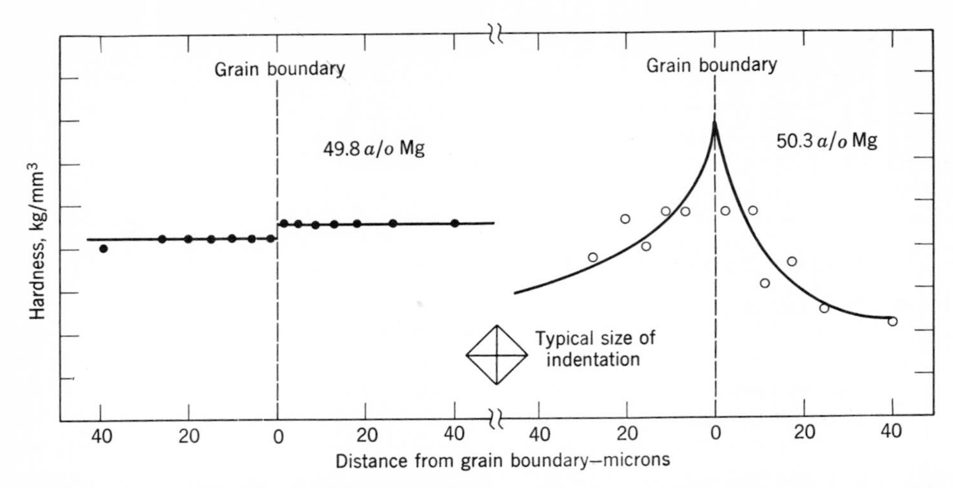

Fig. 21. Microhardness traverses using a 10 gm load across boundaries in Ag-rich and Mg-rich AgMg compounds. (Reference 21.)

pure metals. Similarly, the possible effect of order and disorder on the interface itself could be revaluated. From electrochemical measurements, for example, it is implied that some intermetallics behave like pure metals, whereas others behave in a fashion of their own. The Cu–Au system is one which should be studied in particular because of the great deal of information already accumulated for this system. Such a study would not only clarify some of the points made above, but it would also throw light on the type of interfaces that we should find in ordered and disordered intermetallic compounds. Full control of the variables of importance may not be very easy, however.

The only significant accumulation of property data on interfaces appears to be on hardness. Here reference is made to the work of Westbrook and Wood[21,23] and Seybolt and Westbrook,[24] who measured hardness across interfaces on a variety of materials. They indicate that at least in some cases substantial local increase of hardness is to be observed. (See Tables 2 and 3.) The hardness profiles of two AgMg compounds are given in Figs. 21 and 22. The effect of composition on this hardness is shown in Fig. 23. It appears that, in all instances given in Table 2, compounds having an excess of the electropositive element over that of the stoichiometric composition showed grain-boundary hardening, whereas those with concentrations of that component

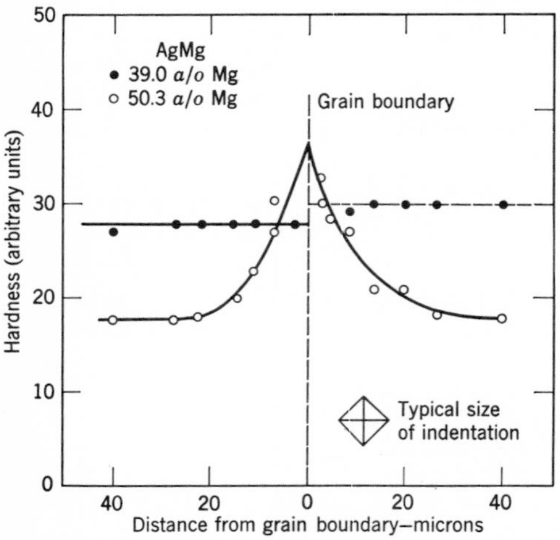

Fig. 22. Microhardness traverses using a 1 gm load, across grain boundaries in Ag-rich and Mg-rich AgMg compounds. (Reference 21.)

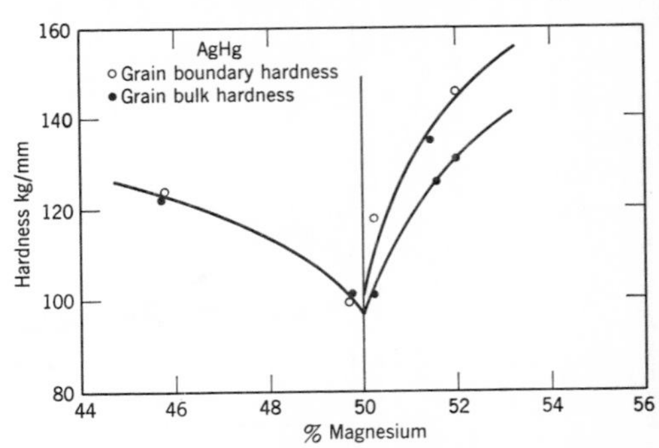

Fig. 23. Bulk and grain boundary hardness of AgMg as a function of composition. (Reference 21.)

below the stoichiometric composition did not show this effect. Westbrook and Wood[21] also investigated the effects of small ternary additions to the intermetallics given in Table 3, in which the changes in hardness as consequence are also given. This study showed that hardness increase could be brought about by some of these ternary additions and obliterated by others. These studies also show a strong correlation between segregation of the gaseous elements (O_2 and N_2) and grain-boundary hardening. The suggested explanations for intergranular fracture based on internal stresses imparted by the segregating elements or changes in surface energy brought about by them these would not account for the hardness effects observed by Westbrook and Wood. Instead, it is believed that the hardening may be produced by a mechanism similar to that proposed by Ainslie[25] et al. for the case of sulphur segregation in α-iron, namely a diffusion generated dislocation network associated with the grain boundary. The width of this network, as determined by Ainslie, is similar to the hardness profile noted for AgMg, that is, of the order of 10 to 15 μ.

Although not exactly parallel to this, the observations made in Cu_3Au in which very rapid attack on relatively broad grain-boundary bands by Fe_2Cl (2 per cent aqueous solution) appear analogous to the heavy density of dislocations in the sulphur segregated ion. Heavily deformed Cu_3Au (in which, we know, heavy concentration of dislocations exists) is attacked at rates analogous to those of the grain boundaries.[26] Although it is possible by analogy to Ainslie's case to propose the existence of wide grain-boundary regions, the studies of Pickering and Swann[18] would not support such speculations. The high energy state of the grain boundary in the Cu_3Au, however, is strong enough to lead to the observed potential differences. Pickering and Swann's study[18] also reveals some very interesting information on the nature of antiphase domains and their role in corrosion processes.

Hopkin,[27] in his studies of intergranular brittleness in CuSb, showed that it was associated with equilibrium segregation of antimony in compositions within the solid solution field. He also found that equilibrium segregation is associated with two types of embrittlement, a high- and low-temperature process, respectively. He ascribes the high-temperature process to an excess vacancy content and the low-temperature one to dislocation pileup on slip planes that terminate against the hardened boundaries.

In an attempt to complete the discussion of properties of intermetallic grain boundaries, we might find it of interest to mention briefly the so-called "pest" phenomenon that takes place in beryllides, silicides, and aluminides. There is an intermediate temperature range in which these materials, which show good oxidation resistance at both high and low temperatures, virtually disintegrate at their grain boundaries. It has been shown[28] that this is a consequence of gaseous contaminants, and the medium temperature range coincides with the temperature in which diffusion rates of O_2 and N_2 along the grain boundaries exceed bulk diffusion rates.

The great influence of the grain boundary on the properties of many different intermetallic compounds cannot be overemphasized. From brittle fracture through stress-corrosion cracking and the "pest" phenomenon, we have an array of failure mechanisms that cause considerable difficulties today. The fact that we have done so very little in studying the state of affairs at the grain boundary in intermetallics should be only a glowing attraction for new studies.

4. INTERNAL INTERFACES—ANTIPHASE DOMAIN BOUNDARIES

The antiphase domain boundary constitutes another type of internal interface. Because this topic has been treated in detail by Beeler in Chapter 14 and more briefly by Brown in Chapter 15, it will not be reviewed in detail here. Suffice here to point out that the antiphase domain boundary exerts effects on properties not only directly through its structure, but also because it is a likely site for the segregation of solute elements or vacancies.*

5. SUMMARY

The writer has attempted in the preceding sections to present highlights of available

* Editor's Note: This possibility has recently been the subject of a theoretical study by Popov et al. *Phys. Stat. Solidi* **13** (1966) 569.

information pertinent to interfaces in intermetallics. The presentation was initiated with discussion of external interfaces or surfaces and the known concepts pertaining to them. It was followed by discussion of internal interfaces or grain boundaries, their nature, the type of information that their properties reveal, and the importance of grain boundaries in certain intermetallics relative to their useful physical properties.

Although a considerable amount of information has been generated on the various aspects of intermetallics, the writer can only say that no clear picture of the true and detailed nature of their surfaces or internal interfaces has emerged.

A number of areas of research endeavor, which are certain to bring forth valuable information, are suggested. First and foremost, there is need to work with ultrapure intermetallics, so that the intrinsic properties of their grain boundaries can be established. In the quest for new data, studies of the plastic deformation and electrochemical properties of ultrapure intermetallics should yield worthwhile results. These studies should be conducted systematically in the different groups of intermetallics, so that we can eventually separate the chemistry of the components from the properties of the intermetallics themselves. These basic studies should lead to an accumulation of background information, which should substantially assist in the interpretation of interface problems leading to failure in structural intermetallic materials.

REFERENCES

1. Gatos H. C., *J. Appl. Phys.* **32** (1961) 1232.
2. Gatos H. C., M. C. Finn, and M. C. Lavine, *J. Appl. Phys.* **32** (1961) 1174.
3. Gatos H. C., P. L. Moody, and M. C. Lavine, *J. Appl. Phys.* **31** (1961) 212.
4. Gatos H. C., and M. C. Lavine, *J. Phys. Chem. Solids* **14** (1960) 169.
5. Lavine M. C., H. C. Gatos, and M. C. Finn, *J. Electrochemical Soc.* **108** (1961) 645.
6. Gatos H. C., M. C. Lavine, E. P. Warekois, *J. Electrochem. Soc.* **108** (1961) 974.
7. Gatos H. C., *Science* **137** (1962) 311.
8. Gatos H. C., and M. C. Lavine, *J. Appl. Phys.* **31** (1961) 743.
9. Gatos H. C., and M. C. Lavine, *J. Electrochem. Soc.* **107** (1960) 427.
10. Warekois E. D., M. C. Lavine, A. N. Mariano, and H. C. Gatos, *J. Appl. Phys.* **33** (1962) 690.
11. Hanneman R. E., M. C. Finn, and H. C. Gatos, *J. Phys. Chem. Solids* **23** (1962) 1553.
12. Winkler H. G. F., *Structure and Properties of Crystals*. Springer, Berlin, 1950, p. 219.
13. Wolff G. A., *Electrochemical Soc. Meeting*, *Electr. Div.*, *Enlarged Abstracts*, 1954, p. 101.
14. Pfister H., *Z. Naturforsch* **10a** (1955) 79.
15. Wolff G. A., and J. D. Brode, *Acta Cryst.* **12** (1959) 313.
16. Bakish R., *Trans. AIME* **209** (1957) 494.
17. Bakish R., *Acta Met.* **3** (1955) 513.
18. Pickering H. W., and P. R. Swann, *Corrosion* **19** (1963) 373T.
19. Uhlig H. H., and W. D. Robertson, *J. Appl. Phys.* **19** (1948) 864.
20. Smith C. S., *Metal Interfaces*. ASM, Cleveland, Ohio, 1951, p. 65.
21. Westbrook J. H., and D. L. Wood, *J. Inst. Metals* **91** (1963) 174.
22. Bakish R., and W. D. Robertson, *J. Electrochem. Soc.* **103** (1956) 320.
23. Westbrook J. H., and D. L. Wood, *Nature* **192** (1961) 1280.
24. Seybolt A. U., and J. H. Westbrook, *Acta Met.* **12** (1964) 449.
25. Ainslie N. G. et al., *Acta Met.* **8** (1960) 528.
26. Bakish R., and W. D. Robertson, *Trans. AIME* **206** (1956) 1277.
27. Hopkin L. M. T., *J. Inst. Metals* **84** (1956) 102.
28. Westbrook J. H., and D. L. Wood, *J. Nuclear Mat.* **12** (1964) 208.

part V

Formation, Stability, and Constitution

chapter 17

Formation Techniques

ALLAN BROWN
University of Uppsala
Uppsala, Sweden

J. H. WESTBROOK
General Electric Research and Development Center
Schenectady, New York

1. INTRODUCTION

Preparative inorganic chemistry depends largely on the solubility of acids, bases, salts, and complexes in a range of liquids at temperatures that rarely exceed 100°C. Reactions are readily carried out in solution, and phase mixtures are separated into their components by the selective control of individual solubility limits. The very different character of intermetallic compounds prevents the application of these techniques, and in the absence of suitable solvents the separation of phase mixtures becomes very difficult or impracticable. Accordingly, preparation of intermetallics is often limited to the direct union of elements at temperatures between 200°C and 3,000°C and, as such, traditionally belongs together with alloy manufacture to the sphere of metallurgy. Formation methods for intermetallic synthesis have been brought to their highest level of sophistication in the synthesis of the semiconducting III–V and II–VI compounds. For this reason, much of the work reviewed here has been developed for this class of compounds, although the principles are much more broadly applicable.

From a thermodynamic viewpoint, the formation of a compound or a mixture of compounds occurs only if the free energy of the compound or mixture is lower than that of competing compounds or mixtures. Such

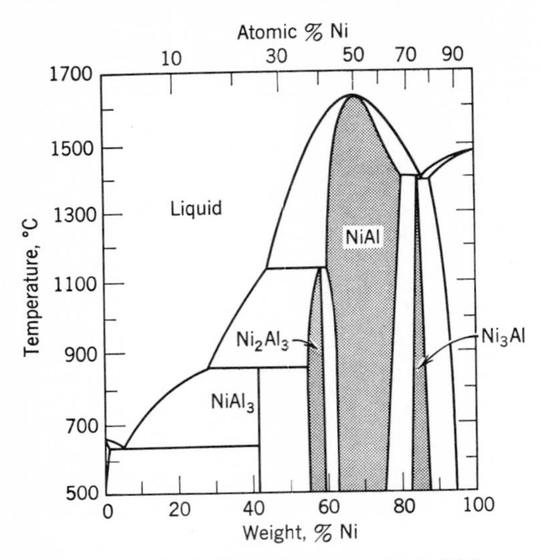

Fig. 1. The Ni–Al phase diagram; a typical binary system with intermetallic compound formation.

equilibrium relationships are customarily represented by the constitutional or phase-equilibrium diagram, which shows the phase condition in a system of components as a function of composition and temperature (and/or pressure). A typical binary phase diagram is that of the Ni–Al system shown in Fig. 1. Here are illustrated a number of characteristics of intermetallic compounds that are the basis for some of the difficulties in their synthesis.

We may first note that intermetallic compounds often occur over a range of compositions; that is, they exhibit a range of solid solubility for the combining elements. The phases Ni_2Al_3, NiAl, and Ni_3Al of Fig. 1 are examples of such behavior and contrast with so-called "line" compounds, such as $NiAl_3$. This circumstance not only illustrates the difficulty that is often encountered in preparing certain intermetallics in a "phase-pure" condition, but also suggests that, in other cases, control of composition within single-phase regions may be critical with respect to properties.

Secondly, it is obvious that before a compound can be prepared under optimum conditions it is necessary to have some idea of these equilibrium conditions and the range of temperature and composition over which it exists as a single entity. For purposes of discussion it is convenient to make a distinction between compounds that can be formed from the melt, for example, the aluminides of Fig. 1, and those, such as peritectoid phases, superlattices, and low temperature polymorphs, that can only be obtained by reactions or changes occurring in the solid state, for the kinetic influences are different in the two cases.

We also may distinguish among compounds formed from the melt according to the congruency of this reaction. So-called congruently melting materials are comprised of a single-phase solid, which at some composition passes directly into a single-phase melt, as with NiAl. Incongruently melting compounds have their archetype in peritectic compounds formed by a reaction:

$$A_{\text{solid}} + B_{\text{liquid}} \rightarrow C_{\text{solid}}$$

Thus, in all cases, the solidifying compound C has a composition different from that of the reacting melt, and the formation temperature always lies below the liquidus curve. Examples of peritectic formation in Fig. 1 include $NiAl_3$, Ni_2Al_3, and Ni_3Al. Obviously, the incongruently melting class presents much more difficult problems of synthesis and homogenization.

In the case of a range of homogeneity, congruent melting is limited to one, two, or three compositions, depending upon the free energy relationships between the liquid and solid phases in the system. At the remaining compositions the solid solutions melt incongruently, and each can be regarded as forming by a sequence of solid-liquid reactions in which the compositions and proportions of the solid and liquid phases are progressively altered as the temperature falls. If the range of solubility is particularly narrow, it may be difficult to resolve the points of congruent and incongruent melting. In instances when it is essential to prepare a phase in a highly pure and homogeneous condition, it must be recognized that a range of solubility, although narrow, does exist and that over this small range compositions corresponding to congruent and incongruent melting must occur. The significance of this fact is brought out in a later section dealing with formation from the melt.

Mention should also be made of a further point that is often overlooked in dealing with compounds formed from the melt, namely, the

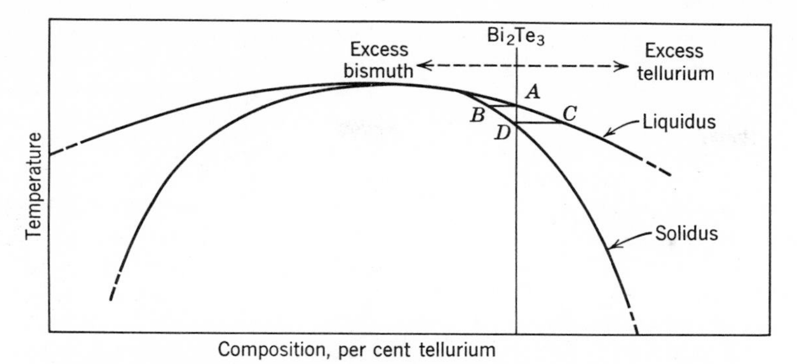

Fig. 2. Noncoincidence of stoichiometry and maximum melting point in Bi_2Te_3. (From Cadoff.[14])

frequent lack of coincidence, at one atmosphere, between the compositions of congruent melting and stoichiometry.[13] The composition for congruent melting is determined by the relationship between the free energy curves for the solid and liquid phases in a system and not by simple atomic ratios. It seems probable that in most intermetallic solid solutions such melting occurs at nonstoichiometric compositions, although the departure from stoichiometry in many of these materials is doubtless very small. An example is shown in Fig. 2 for the case of Bi_2Te_3, one of the semiconducting compounds in which this situation is of special importance.

A frequent difficulty in the preparation of intermetallic compounds is the vapor pressure of a volatile constituent over the melt. For example, the vapor pressure of phosphorus over gallium phosphide at its melting point of about 1500°C is approximately 20 atm of phosphorus. Not only does such a circumstance dictate the use of pressurized reaction systems, but it introduces an ever-present explosion hazard while working with these compounds.

It has already been pointed out that intermetallic synthesis requires a temperature regime far above that of ordinary inorganic preparation. Beyond this, special difficulties are often encountered, because the melting point of the compound may well exceed that of either of the constituents, such as with NiAl in Fig. 1. More extreme cases are by no means rare, for example, Li_3Bi, mp 1145°C; cf. Li mp 180°C, Bi mp 271°C.

Instances of rapid reactivity of the component elements with the environment may also pose special problems. In addition to introducing undesirable impurities, such reactions, if not prevented, may even inhibit compound formation completely. However, the inverse case is also known when relatively inert elements, or those forming self-protective films, exhibit an extraordinary susceptibility to rapid atmospheric degradation when united to form a compound.[15] Thus after synthesis protection is required for the compound, whereas it was not necessary for the component elements.

A final source of difficulty is the expansion on freezing that occurs with many compounds containing elements such as bismuth or antimony. Unless precautions are taken, the containing apparatus may be broken, or undesirable deformation damage may be imparted to the compound.

2. SOME GENERAL PRINCIPLES

A reaction or change of state may or may not occur immediately when the thermodynamic factors favor the formation of a new phase. There is, instead, an induction stage associated with the formation of suitable nuclei, and, following this, growth of the new phase occurs at a rate dependent upon the kinetics of the change. Nucleation requires an initial supply of energy associated with the creation of a new interface, and the length of the nucleation period depends on the ease with which this energy can be obtained. In cases in which a phase is favored by increasing the temperature above the equilibrium value, nucleation is expected to occur smoothly, for energy is readily available. For phases favored by decreasing temperatures, the energy requirement may be more difficult to satisfy, and some degree of supercooling may occur before the free energy difference becomes sufficiently great to trigger the nucleation of the stable phase. This effect is encountered in the freezing of congruently melting materials of high purity, in which the conditions for nucleation are uniform throughout the liquid. In such cases of homogeneous nucleation, supercooling may occur to the extent of several tens or hundreds of degrees. Generally, however, nucleation occurs heterogeneously on traces of impurity phases, gas bubbles, or at the walls of a mold or crucible, and in the case of solid state processes suitable sites are provided by lattice vacancies, crystal boundaries, and dislocations, all of which offer less stringent energy requirements for the formation of a new interface.

Once the reaction product is nucleated, the rate at which equilibrium is attained depends ultimately on the thermal energy of the system. An aggregate of atoms experiencing a reaction or change of state must pass through an acitvated state in which the energy is higher than in the initial or final states. This energy difference, the activation energy E, is derived from the thermal energies of the atoms, and only those atoms capable of attaining activation can take part in the process. At an absolute temperature T, the proportion p of activated atoms in an aggregate is given by the expression $p = Ae^{-E/RT}$, where A is a constant characteristic of the process. The rate for the process is proportional to p and therefore increases with T.

For a thermodynamically reversible process, transfer of atoms from a phase of high free energy to one of low free energy occurs more readily than the reverse, because activation is more easily achieved in this direction. For the exclusive formation of one phase, therefore, the free energy difference between it and the competing phase must be made appreciable, so that only a small proportion of atoms can attain activation for the reverse transfer. Equilibrium in the kinetic sense, is reached when the numbers of atoms are the same in the competing processes.

When the energy difference favors the high-temperature phase, upon increasing the temperature the quantitative formation of the more stable phase occurs smoothly and rapidly. Thus there is little difficulty in melting a solid, given adequate temperatures, for with increasing temperature there is an increase in the probability of an atom attaining activation and passing into the liquid phase. A different situation arises when formation is favored by lowering the temperature. Although the free energy difference between the stable and unstable phases increases as the temperature falls below the equilibrium level, the proportion of atoms in the unstable phase capable of attaining activation decreases. Under certain circumstances a sufficiently low temperature may be reached at which the proportion of activated atoms is negligible, and the higher energy phase can be regarded as existing in metastable equilibrium with the phase or phases more favored on purely thermodynamic grounds. The extent to which the unstable phase is retained must depend on the rate at which cooling through the equilibrium temperature occurs, and in the case of solid state transformations when this temperature is low, it is often possible to quench-in such phases quantitatively.*

Reactions that occur in the solid state are, with the exception of martensitic transformations, dependent on mass transfer along a gradient of chemical potential, which for most purposes approximates to a gradient of chemical composition. This process of diffusion occurs by activation and is temperature-controlled, the rate being given by $D_0e^{-E/RT}$, where D_0 is the diffusion coefficient appropriate to the system. Rates of diffusion are considerably less than those of reaction, and diffusion therefore constitutes the rate-determining step in many solid state processes. Diffusion may occur along grain boundaries, across free crystal surfaces, or through crystal lattices. The diffusion rates associated with these paths, designated D_B, D_S, and D_L, respectively, make different contributions to the overall diffusion rate depending on the different energy conditions in these regions. In general, $D_S > D_B > D_L$, and at low to moderate temperatures compared with the melting point, diffusion in polycrystalline materials occurs principally along grain boundaries. Table 1 relates D_B/D_L and the grain size at which the two processes contribute equally to overall diffusion.[16] For $D_B/D_L = 10^6$ the boundary contribution falls from 90 per cent at a grain size of 2.6×10^{-2} cm

* See, in particular, Chapter 18 by Duwez.

TABLE 1

Effect of Grain Size on the Relative Contributions of Grain Boundary (D_B) and Lattice (D_L) Diffusion

D_B/D_L	10^6	10^5	10^4	10^3
Grain size in cm for which boundary- and lattice-diffusion contributions are equal	2.6×10^{-1}	1.4×10^{-3}	2.15×10^{-5}	10^{-6}

to 10 per cent at a grain size of 2.6 cm. As grain sizes increase, the number of diffusion channels and hence the boundary contribution are reduced.

In a solid state reaction A + B → C, the product forms at the A–B interface, and reaction continues by diffusion of the reactants through the accumulating layers of C. Accumulation of product leads to a reduction of composition gradient, and the growth of the product layer of thickness x has a parabolic dependence on time, $x^2 = Kt$, where K is a constant incorporating diffusion coefficients appropriate to the system. If x attains 10^{-4} cm in four hours, the thickness of the layer will be doubled in twenty hours, assuming constant diffusion rates. Equilibrium will therefore be reached after four hours, if the grain size of the reactants is 10^{-4} cm, but at least one hundred hours are required for the consumption of reactants having a grain size of 10^{-3} cm.

For a general treatment of the topics of this section the reader is referred to the text by Darken and Gurry.[17] The special features of the subject pertinent to intermetallic compounds are treated elsewhere in this book as follows: diffusion by Hagel in Chapter 20, ordering by Rudman in Chapter 21, and crystallographic transformation by Christian et al. in Chapter 22.

3. CONTAINER MATERIALS AND ATMOSPHERES

Because of the high reactivity of most metals at elevated temperatures, the preparation of an intermetallic compound is best regarded as a multicomponent reaction that is likely to involve container materials and atmospheres in contact with the elements. The choice of container materials and formation conditions must therefore be made with a view to suppressing all possible side reactions. Kinetic considerations are again important, for the extent to which such reactions occur must depend on the duration of, and the temperatures reached in, the heating cycle and the possible presence of melts in appreciable quantities.

The requirements for crucible materials are corrosion compatibility towards the reactants and a high resistance to thermal shock in addition to a softening point in excess of the chosen reaction temperature. Corrosion resistance under equilibrium conditions can be judged from values of ΔG_T, the free energies of formation of the various possible compounds, assuming that the appropriate thermodynamic constants are known. Departure from equilibrium conditions occurs, if one of the reactants is capable of forming a volatile compound as a result of a side reaction with the container. Typical examples are the formation of volatile TiO and TaO by reaction between titanium- or tantalum-bearing melts and refractory oxides. Less obvious instances are the partial reduction of silica containers to SiO effected by many intermetallic compounds or the reaction of silica containers with borides leading to the formation of silicides or silicoborides.[18] However, when thermodynamic data predict corrosion, the formation of a thin product layer at the melt-crucible interface can effectively inhibit further reaction if diffusion rates are sufficiently slow. This often occurs with graphite crucibles, which have found wide use in the past as containers for liquid metals, although stable carbides are formed by most metallic elements. Nevertheless, it is clear that unless all the parameters are known there is considerable risk of contamination in such cases.

The most commonly used container materials are the refractory oxides, of which those listed below find the widest application. The sequence follows the order of thermodynamic stability at 25°C; the melting point in degrees centigrade and some indication of the resistance to thermal shock (*g* = good, *m* = moderate, *p* = poor) are given after each compound.

CaO, 2600, *p*
ThO_2, 3300, *m*
BeO, 2500, *g*
MgO, 2800, *m*
Al_2O_3, 2015, *g*
ZrO_2, 2677, *p* but *m* for stabilized material
CeO_2, 2600, *p*
SiO_2, 1728, *m* to *g* depending on temperature range

Free energies of formation for many metal oxides are given in Reference 19, but complete appraisal of resistance to attack cannot be made in the absence of data for the formation of all possible intermetallic compounds. Thus molten magnesium does not normally attack thoria, but in the presence of bismuth, attack occurs at 500°C because of the favorable energy change resulting from the formation of $ThBi_2$.[20] An experimental method for investigating compatibility under appropriate conditions is described in Reference 21. More detailed discussions of refractory oxides and of more specialized carbides, nitrides, and sulphides appear in References 2 to 4. Table 2 summarizes suitable containers for various liquid metals. A preliminary selection for an intermetallic can be made by determining mutual suitability with each of the component elements. The reader is also warned that significant differences in reactivity exist among the various commercially available forms of the same compound.

A frequently useful construction is to coat the inside surface of a refractory container with a film of another material. Thus pyrolytic carbon coatings can be applied to silica to control wetting characteristics or to reduce the hazard of silicon contamination, and alumina slips may be utilized to form fragile but convenient liners.

Metallic containers find some application. Copper and silver-plated copper are often used for crucibles or hearths for induction or arc melting. The noble metals, gold and especially platinum, have found some service. Even other intermetallic compounds are sometimes employed[24] because of their frequently low solubility for impurities and because it is sometimes possible to choose for the container a compound whose components would not be deleterious to the desired product compound.

Because of the ease with which most metals and their compounds react when heated in the atmosphere, it is often expedient to conduct preparations in vacua. Apparatus design is critical, however, since discontinuities such as metal-metal junctions represent possible points of leakage. In extreme cases it may be necessary to evacuate to more than 10^{-8} mm Hg to avoid oxidation during prolonged heat treatment. In such circumstances small leaks, which are difficult to eliminate, may prove damaging particularly where the speed of pumping and the gettering action of the reactants combine to hide the true state of affairs. The best use of high vacuum is, therefore, made when containers in the form of small capsules of steel, silica, or glass can be sealed after evacuation.

The use of an inert atmosphere provides an alternative solution to the problem, because a positive pressure or a low vacuum can be employed, thereby eliminating or minimizing the risk of leakage. Where some form of vacuum is desirable, leakage into the reaction vessel can be made up virtually entirely of inert gas by enclosing the vessel in a metal, glass, or flexible plastic envelope filled with gas under a positive pressure. The most commonly used inert gases are helium and argon which may be gettered by passage over turnings of titanium or titanium-40%-zirconium alloy heated to 700°C.

The principal source of atmosphere contamination is likely to arise during heating by the desorption of gases from the walls of crucibles and reaction vessels. In this respect ceramic materials and graphite can be the most troublesome, because of their marked powers of absorption, and, where traces of oxygen are likely to prove damaging, use of these materials should be avoided as far as possible. Such contamination can be minimized by thorough cleaning and by a preliminary "bake-out" of the affected areas at 200 to 400°C under a vacuum of more than 10^{-4} mm Hg for an hour or so. The use of reactants in massive rather than powdered form is also advantageous, for finely divided metals are capable of either adsorbing oxygen or acquiring an appreciable oxide layer.

Use of a vacuum or inert atmosphere implies enclosure in a vessel, for which the chief requirement is gas tightness. One important factor here is porosity, which is likely to increase with increasing temperature. Double enclosure, with one evacuated container lying within a second, can be used to limit the effects of porosity, but, where the source of heating is external to the container, this arrangement must lead to a loss of efficiency.

With externally heated systems the limiting temperature of the sample is 100 to 500°C below the softening point of the container material, depending on the pressure gradient. As an example, experience shows that a single furnace tube of alumina, 4 cm in diameter and with a wall thickness of 3 mm, can be expected to collapse after two twelve-hour periods at 1500°C and 10^{-5} mm Hg. Double-enclosure systems, in which the intercontainer region is evacuated to about 1 mm Hg, appear to have a better life expectancy.

Container materials commonly used for high-temperature work are, in order of porosity, vitreous silica (to 1100°C under vacuum), mullite (to 1400°C under vacuum), and alumina (to 1600°C under vacuum). Although silica has very low porosity at 1000°C, it suffers from a tendency to devitrify as a result of contamination by small traces of many metallic elements. Attack by metal vapors at high temperatures is also possible in initially evacuated systems, as is shown by the formation of $ThSi_2$ at 1000°C during the preparation of Th–Zn alloys in evacuated silica capsules. Conventional silica tubing is good for pressures up to about 25 atm, and special thick-walled tubing, $\frac{1}{2}$ in. i.d., $\frac{1}{8}$ in. wall, can withstand 100 atm.

When the source of heating lies within the container, the walls of the vessel and connections to vacuum pumps and gauges can be cooled by water circulation or by a directed blast of cold air. Cold-wall systems may be constructed of steel, which has negligible porosity.

Finally, mention must be made of the several "crucibleless" melting processes that include floating zone melting, levitation melting, skull melting, and the Verneuil process. These methods usually (but not always) are applied, when the compound has already been synthesized and the object is further purification or conversion to single-crystal form. Of those cited, only the floating-zone technique has much versatility for intermetallics and hence is the process most frequently used in the crucibleless class.

4. PREPARATION OF THE CHARGE

It is usually desirable to utilize materials as pure as possible in the synthesis of compounds. Investigators concerned with synthesis of semiconducting compounds naturally had an early appreciation of this necessity, but ultrapurity is becoming important also to others concerned with other types of properties. Although many elements are now available commercially in purity levels that would have been considered fantastic even twenty years ago, is still frequently necessary for the experimenter to undertake a preliminary purification of his component elements. For this purpose he may use zone refining, distillation, chemical treatment, or the formation of an intermediate compound species and purification thereof by one of the other processes.

It is usually necessary to rely on physical property measurements, radiochemistry, or X-ray fluorescence rather than conventional wet analytical chemistry to assess the success of purification efforts. This topic will be discussed more extensively in subsequent sections.

Favorite intermediate types in compound synthesis are the halides and hydrides, because of the ease with which they are formed and then purified and transported in the vapor state. Use of an intermediate species may aid the synthesis itself, especially when component fusion is not feasible and solid state reactions prove too sluggish. For example, the monophosphides of a series of transition elements have been prepared by reacting calcium phosphide with transition metal powders or chlorides.[26] Excess calcium phosphide and calcium are removed by leaching with 0.1 N HCl; good yields of fairly pure materials have been obtained. It is claimed that, for the preparation of substantial amounts of phosphide, this method is preferable to the prolonged heating of mixtures of the elements. Again,

TABLE 2

Atmospheres and Containers for the Melting of Metals and Alloys

Class	Metal	Refractory for Crucible*	Protective Atmosphere or Flux
Alkali metals	Li, Na, K, Rb, Cs	Pyrex or silica at low temperatures; steel at higher temperatures	Argon preferable; all except Li can be melted under heavy oil or paraffin
Alkaline earth metals	Be	BeO or ThO_2	Argon
	Mg	Al_2O_3, MgO, graphite or steel	Argon or sulphur vapor; under KCl, $CaCl_2$, CaF_2 flux; avoid nitrogen
	Ca, Sr, Ba	Steel, graphite for Ba	Argon
	Al	Al_2O_3, MgO, BeO, CeS graphite; avoid siliceous material	In argon under charcoal, can be melted in air without flux; N_2 or H_2 in graphite.
Refractory metals	Ti, Zr, Hf	Mo_3Al for Ti; arc melting on copper mold	In argon or vacuum
	V	ThO_2	In argon or vacuum
	Nb, Ta	See Ti, Zr, Hf; for Nb:ThO_2 or BeO	Hydrogen
	Cr	ThO_2, ThO_2 lined Al_2O_3, BeO; reacts slightly with Al_2O_3	Hydrogen or argon
	Mo, W	Arc melting on water-cooled copper mold	Hydrogen or argon
Th + U	Th	ThO_2 results in some ThO_2 pick up; does not react with CaO or BaO	In argon or vacuum
	U	BeO, ThO_2, MgO up to 1500°C, CeS to 1900°	In argon or vacuum
Mn + Re	Mn	Al_2O_3, spinel	Hydrogen or argon; too volatile for vacuum
	Re	Arc melting on water-cooled copper mold	In argon, vacuum, or nitrogen
Iron group	Fe	Al_2O_3, BeO, CaO, ZrO_2, ThO_2	Hydrogen
	Co	Al_2O_3, ZrO_2	In hydrogen, argon, or vacuum
	Ni	Al_2O_3, BeO, ZrO_2 ThO_2	In argon or vacuum, H_2 or He for BeO, ZrO_2, ThO_2
Group VIII precious metals	Ru, Os	ThO_2, or ZrO_2	In nitrogen, hydrogen, or vacuum
	Rh, Ir	ThO_2 or ZrO_2	Nitrogen or argon
	Pd, Pt	ZrS, ZrO_2, Al_2O_3, CaO	Carbon monoxide, nitrogen, or argon.
Copper, silver, and gold	Cu, Ag, Au	Graphite, Al_2O_3, MgO, spinel	Under charcoal, nitrogen, argon, or carbon monoxide; hydrogen in graphite

(*Continued*)

TABLE 2 *(Continued)*

Class	Metal	Refractory for Crucible*	Protective Atmosphere or Flux
Zinc, cadmium, and mercury	Zn, Cd	Pyrex at low temperatures; Al_2O_3, graphite, mullite	Under charcoal or halide fluxes, argon
	Hg	Pyrex	In argon or vacuum
Group III	Ga	Pyrex at low temperatures; graphite and Al_2O_3 at high temperatures; avoid siliceous material at high temperatures	Under charcoal or halide fluxes Under potassium chloride or charcoal; argon or vacuum
	In	Porcelain, pyrex, mullite	Hydrogen flame or argon
	Tl	Porcelain, pyrex	Hydrogen
Group IV	Si	Silica, TiO_2, ZrO_2, ThO_2	In argon, vacuum, or helium
	Ge	Graphite, silica, Al_2O_3	Nitrogen or vacuum
	Sn, Pb	Pyrex, porcelain, graphite, spinel, mullite	Under charcoal; hydrogen
Pnictogens	As	Pyrex, silica	Low As under halide flux, high As in sealed tubes
	Sb	Porcelain, graphite, silica	Under charcoal; hydrogen
	Bi	Pyrex, porcelain, silica, graphite, CeS	Hydrogen
Chalcogens	Se, Te	Silica or graphite	In argon or vacuum
Sc, Y, and rare-earth metals	Ce	CeS	
	Sc, Y, La, Pr, etc.	Graphite jacketed tantalum or CaO or BeO	Argon or vacuum in vacuum

* Note that arc or induction melting on water-cooled metallic hearths can be practiced advantageously more widely than indicated in the table. Note further that for semiconductor grades of purity, many of the recommended containers will not provide sufficient freedom from contamination. For guidance in such cases the reader is directed to the literature on specific semiconducting compounds and to References 9 and 11.

niobium-rich niobium pnictides (group V_A compounds) are not readily prepared by direct reaction from the elements, but can be obtained by thermal decomposition of $NbAs_2$ and $NbSb_2$, both of which are easily made. NbAs is obtained after complete degradation of $NbAs_2$ at 1100°C; Nb_5Sb_4 and Nb_3Sb are obtained by degradation of $NbSb_2$ at 830° and 1000°C, respectively.[27]

Special handling and cutting techniques may have to be devised to prevent recontamination of the purified elements. Pieces of component metals that form the charge may require a series of surface-cleaning operations to minimize, insofar as possible, ultimate contamination of the product compound. Mechanical methods, organic solvents, acid cleaning, etc., are frequently used in combination, the exact nature and sequence of which will be specific to the compound being synthesized. Compounds such as the aluminides or titanides, containing an oxygen-avid element, pose a particularly difficult problem, because of the virtually inevitable formation of a tenacious oxide film. Even if removed by mechanical or chemical means, it will immediately reform. The presence of oxide particles within the compound is deleterious to the mechanical properties and, by acting as nucleation centers, they can effectively prevent

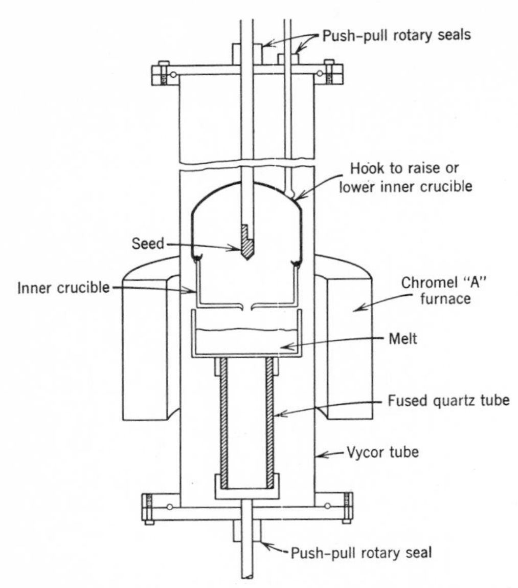

Fig. 3. GaSb crystal grower using the double crucible method to eliminate oxides. (After Allred.[29])

the growth of single crystals. Seybolt and Westbrook[28] have also found that oxygen in solution in a variety of compounds can produce grain-boundary hardening and catastrophic disintegration in the so-called "pest" reaction. The total oxygen concentration required is apparently less than a few parts per million—below the limit of ordinary analytical techniques and less than that required for a perceptible change in the lattice parameter. In addition to the conventional techniques alluded to, other methods of oxygen removal that have been practiced include prolonged exposure of the melt to a very high vacuum, rapid rotation of the crucible to drive oxide particles to the walls, skimming the surface of the melt, freezing the top layer of the melt to a paddle which is then removed before proceeding with the compound synthesis, and double crucible methods. One version of the latter, adapted for the Czochralski method of crystal growth, is illustrated in Fig. 3. Melting is initially carried out in the lower crucible; the upper crucible is then immersed beneath the surface of the melt, so that oxide-free melt is forced through the aperture in the bottom of the upper crucible to provide a clean pool in the upper container for growth onto the seed crystal. For compound synthesis alone an alternative arrangement provides for initial melting in the upper crucible; the molten liquid flows through the bottom hole into the lower crucible leaving the oxide dross behind.

5. METHODS OF COMPOUND SYNTHESIS

5.1 Direct Fusion of the Components

Direct fusion of the component elements is the simplest method of synthesis and therefore the one usually employed in the absence of complicating problems.

Low-Temperature Techniques. For low-melting-point materials, glass or fused silica containers in combination with resistance furnace heating is the usual practice. A particular embodiment of this technique, devised by Lawson[30] to minimize oxide contamination, is the *Y*-tube arrangement shown in Fig. 4. Components are inserted into the open branches of the *Y*, which are then sealed off. Following outgassing of the crucible, the components are melted under flowing dry hydrogen, and they run down and fill the crucible. Oxide dross is left behind in the side arms. Finally, the hydrogen is pumped off and the reaction chamber sealed off.

Another technique was practiced very early in the synthesis of alkali metal compounds. The two component elements are separately

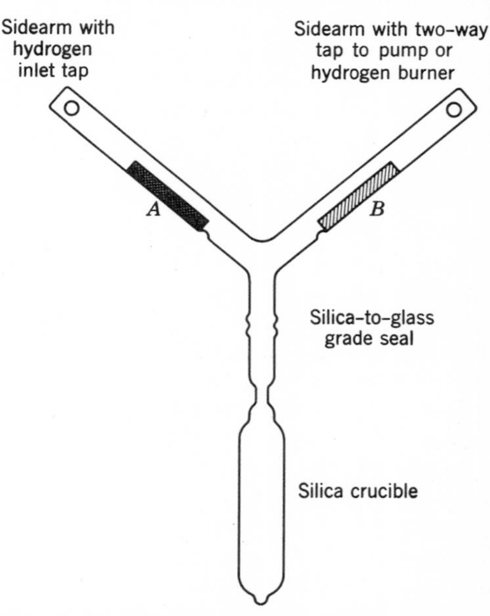

Fig. 4. *Y*-tube apparatus for direct synthesis. (After Lawson.[30])

fused under liquid paraffin and then mixed at a temperature below the melting point of the compound. If mixed in stoichiometric proportions, the compound crystals can frequently be readily separated by mechanical means. If an excess of either component is present, it must be removed—perhaps initially by mechanical means and then by heating with liquid ammonia or other solvents. Such reactive compounds must be dried under nitrogen or inert gas, or otherwise suitably protected from atmospheric attack.

High-Temperature Techniques. High-melting-point materials usually require arc melting or induction heating. Arc melting is conducted in a cold-wall system using an inert atmosphere, generally argon, at reduced pressure (20 to 40 cm Hg). The cathode may either be inert (tungsten) or consumable, in which it is made up either of the product compound or a composite of the component metals. The arc is produced by a stream of electrons emitted from the cathode and drawn through a plasma of ionized gas to a water-cooled copper anode in the form of a crucible or hearth. The arc length is of the order 1 to 2 cm, and the potential drop is 18 to 25 V for an arc current of 200 to 300 A. Temperatures up to 3000°C are easily attained, and all intermetallics can therefore be melted by this method.

Reaction under these conditions is virtually instantaneous, and contamination is minimized by using a cooled hearth and an inert atmosphere, initially gettered by melting zirconium. Reactants may be used in the form of lumps or as compacted powders. The chief limitation of the method is the volatility of the reactants at the temperature of the melt. In general, elements that develop less than 10 mm Hg at 1700°C can be melted with inappreciable loss. One of the authors, Westbrook, has successfully synthesized the iron-group aluminides[31] as well as a group of transition metal compounds[32], using a multiple-hearth arc furnace designed by Rossin.[33] Elements that develop 10 mm Hg between 1500 and 1700°C occupy a marginal position, and preparation may be successful if the temperature of the melt does not exceed 1500°C. Thus one of the authors, Brown, has melted Ce–Sn alloys with only inappreciable weight losses, whereas the corresponding Th–Sn alloys showed large weight losses, the products being highly porous because of the boiling of tin during melting. A voltage controller at the d.c. source serves as a means of limiting the temperature to the level necessary for maintaining melting, and it is essential when relatively volatile elements are reacted.

The water cooling of the anode results in a sharp temperature gradient across the sample, which is therefore unable to attain true thermodynamic equilibrium. Depending on sample size and other factors, it may be difficult to melt the entire original charge. Even under the best circumstances, a homogeneous distribution of the elements can only be approached by turning and remelting the button several times. Subsequent cooling of the sample is rapid and uncontrolled; segregation is therefore to be expected on freezing, and many peritectic and most peritectoid reactions do not go to completion during the short cooling period allowed. In addition to these inhomogeneities, the products are frequently porous, and densities may be as low as 80 per cent of theoretical. Judicious control of the voltage prior to extinguishing the arc may lead to the development of crystals sufficiently large to allow examination by X-ray diffraction techniques. A more detailed discussion of fundamentals is given in Reference 34, and some aspects of arc-melting alloys of refractory metals can be found in Reference 35.

Electron-beam melting uses a focused beam emitted by a tungsten filament or similar source and accelerated in high vacuum through a hollow anode at ground potential. Volatility of the reactants is the principal limitation of this technique, and usually only the more refractory alloys can be melted satisfactorily. Thus ZrB_2 was observed to vaporize in the electron beam when the surface of the sample became molten,[36] whereas alloys such as those of the tungsten-tantalum solid solution series are readily prepared. However, Barton et al.[37] successfully synthesized CoTi by electron-beam melting, although its melting point is below 1500°C.

Induction heating depends on the circulating currents induced in a piece of metal placed in an alternating magnetic field. The reactance of the metal susceptor to these currents leads to a power loss which appears in the form of heat—an effect supplemented in magnetic materials by hysteresis losses.

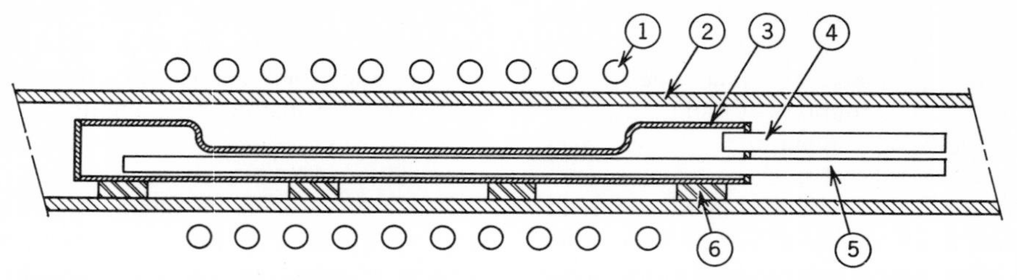

Fig. 5. Apparatus for induction melting on water-cooled metallic hearth. (1) induction coil, (2) quartz tube, (3) silver boat, (4) water inlet, (5) water outlet, and (6) supports. (After Mayer and Mlavsky.[38])

In its simplest form the furnace comprises an inductor, made from a few turns of water-cooled copper tubing, connected to a suitable generator, and the susceptor, a graphite or refractory metal sleeve or the alloy charge itself supported at the axis of the inductor coil. At kilocycle frequencies the circulating currents are restricted to the surface layers of the susceptor, where the magnetic flux is greatest, and heating in this region becomes intense. With the induction of sufficiently powerful currents the temperature of the susceptor can be raised to and beyond its melting point.

Induction furnaces operate at fixed frequencies, the power transfer to the susceptor being controlled by a coupling circuit whose characteristics can be adjusted to provide different temperatures. Tuning of the inductor to give a specified temperature depends on the physical nature of the charge and the different elements that may be present. Alternatively, induction heaters may be controlled by varying the plate voltage on the oscillator tubes. Temperatures may be measured with thermocouples or, more usually, by pyrometer, in which case some knowledge of the emissivity of the charge is required. Frequencies that are used range from a few kilocycles to several megacycles depending on the heating requirements and the materials involved. An all-purpose system may operate at 800 kc/sec with an inductor current of 100 A and a power consumption of 5 to 10 kW.

Cold-wall systems can easily be employed, the inductor lying either outside a vitreous silica tube or inside a metal chamber. The furnace may operate in a vacuum or inert atmosphere at any pressure provided that the conditions do not lead to a gaseous discharge that reduces the efficiency of power transfer to the susceptor. The molten charge may be supported in a crucible, but this involves the search for suitable container materials. As an alternative to the use of refractory materials, a water-cooled crucible manufactured from copper, silver, or aluminum tubing has been employed, in which the crucible acts as an intermediate susceptor.[38–40] Water-cooled fused silica may also be employed.[41] Mayer and Mlavský[38] have used the arrangement shown in Fig. 5 for the preparation of silicides. Disadvantages of the cold crucible method include the high thermal stresses developed in the solidified ingot, the difficulties of single-crystal growth, and the restriction to non-wetting melts.

With poorly conducting or semiconducting materials, megacycle frequencies may be required to promote melting. Alternatively, the charge may be heated by radiation from an intermediate susceptor, and cyclinders of molybdenum or tantalum are useful for this purpose. Graphite also couples readily and can perform two functions simultaneously, if used as a crucible material.

Homogeneity in a molten charge is ensured by electromagnetic stirring, and, because the variable coupling permits some degree of temperature control, the worst forms of segregation can be avoided during freezing. Arrests can also be made at specified temperatures to promote solid-liquid and solid-solid reactions and transformations, although it should be noted that the tuning of the inductor may require frequent adjustment to maintain a constant temperature because of the changing character of the susceptor. It should also be noted that skin-heating effects become more pronounced with increasing frequency and that in the megacycle range a temperature gradient may develop between the middle and the surface of solid charges.

High-frequency techniques are readily adapted for the melting of all intermetallic compounds and, if high vacua are not prohibited by vaporization of the charge, a good approach to the theoretical density can be achieved. The only limitations of the method lie in the somewhat greater difficulties entailed in controlling and measuring temperature. A more detailed discussion of the high-frequency principle is given in Reference 42.

Very useful general accounts of methods for the generation, measurement, and control of high temperatures may be found in References 1–4.

Extraction Methods. Sometimes preparation of an intermetallic compound in a state approaching the phase-pure state proves to be essentially impossible either by direct fusion of the components or by any of the other techniques to be described in subsequent paragraphs. In such cases it may be feasible to extract crystals of the compound from a multiphase alloy, particularly if conditions have been controlled so as to permit substantial growth of the crystals.[43] Compounds of limited solubility for the components can be grown even from melts whose composition is far removed from the stoichiometric ratio.

If the separation from either the partially or completely solidified alloy can be successfully carried out, it may be possible to consolidate the compound phase by other means.[44] In special cases, if the crystals are of adequate size, structure or physical-property determinations may be carried out directly on the separated crystals.

Compounds formed in systems of low-melting metals may be separated by mechanical means by using sieves, centrifuges, heavy liquid flotation, etc. Acid treatment or liquid-ammonia leaching may be used either for extraction or for further perfecting of a prior mechanical separation. In the case of systems with a volatile element that has low volatility over the compound sought, distillation can be used. In many systems there will be sufficient electrochemical differences between phases in a solidified alloy, so that electrolytic extraction, as illustrated in Fig. 6, is an attractive process. Raynor et al.[45,46] and Clare[47] have had good success with this technique in extracting compounds such as $CrAl_7$ and $MnAl_6$ from aluminim-rich alloys.

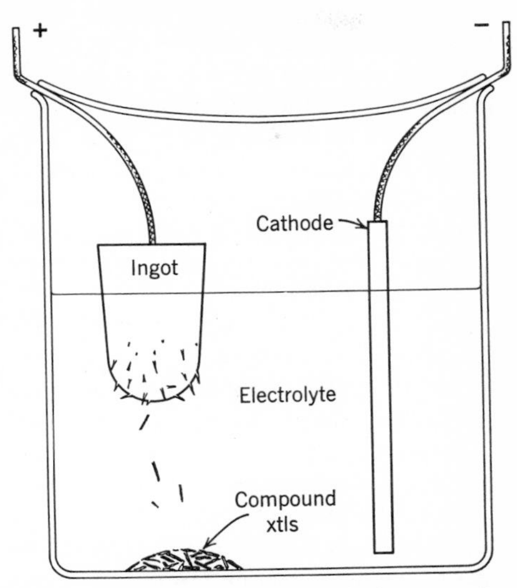

Fig. 6. Apparatus for electrolytic extraction of compounds from alloys. (After Raynor and Wakeman.[45])

All of these separation techniques should be regarded with suspicion, if a 100 per cent phase separation is of importance to subsequent measurements. The difficulties are great: the reagent attack may not be as specific as desired, entrainment of excess constituent may occur, films of peritectic- or peritectoid-forming compounds may remain at crystal surfaces, and inclusions or precipitates may be present internally. Nonetheless, the method has its uses and, with care, can be very valuable.

5.2 Solid-Liquid Reactions

Solid-liquid reactions constitute a common class of intermetallic synthesis because of the frequent incidence of peritectic formation of intermetallic compounds. Although difficult to prepare by conventional fusion technique,[43] peritectically formed compounds can frequently be prepared with good success by zone melting,[48–50] and Barber[51] has prepared several peritectic-type aluminides by a modified Czochralski technique.

It is clear that solid-liquid reactions are assisted by the removal of product layers from the diffusion interface, the reactants being subsequently reunited. This can be accomplished if the reactants are shaken togethe vigorously at temperature. One of the authors, Brown, has used this approach for the preparation of a number of thorium and uranium compounds with the low melting point elements Zn, Cd, Hg, In, Sn, Pb, and Bi.[52,53] Thus UPb was obtained after continuously shaking a mixture of the elements for 24 h at 500°C whereas an earlier investigation indicated that heating to 1220°C was necessary to promote reaction.[54]

Shaking equipment is simple, consisting of a reciprocating arm mounted on bearings and connected through a variable-speed gear box to an electric motor. Reactants are sealed under vacuum into glass or silica capsules and loaded into a stainless-steel or Nimonic-alloy bomb attached to the free end of the arm. The heating source is a conventional resistance furnace. Standard shaking conditions are 200 rpm through an amplitude of 10 cm for powdered materials; when there is more than 50 g of dense liquid, slower rates are used to prevent fracture of the capsule. In most cases molten phases represent transient stages in preparation and make little contact with the capsule. There is, therefore, no serious corrosion problem at low temperatures.

In addition to providing a means of reacting elements with a disparity of melting point, this technique has also been used to prepare a range of Bi–Te and Sb–Te solid solutions, in which the tendency towards segregation is otherwise marked.[55] In each system it has been possible to prepare single phases that could not be obtained even over periods of months by using conventional homogenizing techniques. Volatile components can also be handled by using silica capsules to withstand the high pressures initially generated. Thus, in the preparation of Ti_3P, the calculated amount of phosphorus was fixed in the surface layers of titanium turnings at 800°C without noticeable corrosion of the silica capsule. Ti_3P was then formed by arc-melting this inhomogeneous but involatile product.[56]

Vibrational mixing has also been used for the synthesis of some high-melting semiconductor materials that are homogenized only with difficulty by using conventional techniques.[57] The system utilizes a quartz rod sealed at one end to a silica capsule containing the reactants and connected at the other end to an electromagnetic vibrator. Vibrational frequencies of 100 cps have been used in the temperature range 1000 to 1400°C, reducing homogenization periods by factors of ten or twenty.

It seems probable that the vibrational method functions largely by maintaining a high concentration of rapid diffusion paths in materials that exhibit low rates of diffusion at low temperatures and macrocrystal growth at high temperatures.* The obvious disadvantage is that high temperatures are still necessary for diffusion, and corrosion of the container becomes a serious problem if components of a more metallic character are employed.

The application of ultrasonic frequencies to vibrational mixing is clearly indicated, for it might be expected to assist diffusion at low temperatures. Effective coupling of the transducer to the charge may be difficult, however, particularly because changes in the charge during reaction are likely to entail repeated adjustment of the tuning circuit.

One clever alternative to the continuous shaking procedure has been described. An arrangement devised for intermittent ball milling of the product-reactant mixture while contained within the evacuated reaction vessel was reported by Gross et al.[59] for the synthesis of uranium-bismuth compounds.

5.3 Vapor-Liquid Reactions

Vapor-liquid reactions are commonly employed when one component is a low-melting-point, low-volatility metal and the other is a highly volatile element. The arsenides are a typical compound family often synthesized by this method. The main features of the experimental arrangement are shown in Fig. 7. The furnace arrangement is such that all parts of the reaction chamber are kept above the temperature of the arsenic supply in order to prevent undesired condensation of the arsenic. The metal to be reacted is heated above its melting point, and by slow heating of the more volatile species (arsenic in this instance) the vapor is made to diffuse along the tube and to react with the liquid metal. Although reaction occurs when the maximum temperature is between the melting point of the metal and that of the compound, completion is necessarily slow by virtue of the slow diffusion of the

---

* However, it should be noted that Burghard and Brotzen[58] have found that vibration decreased the ordering rate of CuAu at all frequencies and all ordering temperatures investigated.

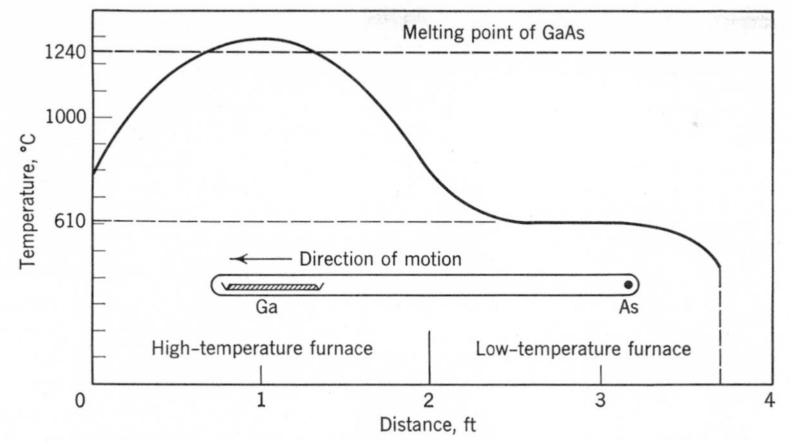

Fig. 7. Vapor-liquid synthesis exemplified by GaAs formation. (After Cunnel et al.[60])

volatile species through the solid skin of compound formed on the liquid. It is therefore preferable to exceed the melting point of the compound. In any event, when the reaction is complete, the liquid compound is caused to freeze by moving the reaction tube through a temperature gradient in the direction indicated.

Although control of stoichiometry can be effected by arsenic pressure control (i.e., furnace temperature) transient variations in pressure can cause void formation in the ingot when the arsenic pressure falls below the dissociation pressure of the molten compound for that composition and temperature. Excess vapor pressure of the volatile species poses an explosion hazard. Thus the rate of reaction and the quantity of the volatile constituent must be carefully controlled. Details of these techniques have been presented for the general case by Gaus et al.[61] and specifically applied to GaAs by Cunnell[62] and to AlAs by Stambaugh.[63]

5.4 Vapor-Solid Reactions

The direct reaction of one solid component with the other in the vapor form is seldom practiced, although the first synthesis of Cs_3Sb which yielded crystals suitable for X-ray structure determination was by this technique.[64] The slow diffusion through the reacted material means that this class of reaction is usually limited to a powder product. The vapor-phase reactions of greater significance are discussed under halide reduction in Section 5.6.

5.5 Liquid-Liquid Reactions

The direct reaction of the two constituent elements in the molten state to form the compound has already been discussed in Section 5.1.

Reactions exemplified by

$$MX_2 + N \rightarrow MN + MX_3$$

where MX_2 is a suitable low-melting compound, usually a halide, may be carried out in the liquid state to form the compound MN as described by Antell and Effer.[65,66] A complex product may result, which must subsequently be decomposed by heating to yield the desired product.

Reactions in liquid ammonia are of interest when metal-ammonia solutions can be prepared.[67,68] Thus solutions of the alkali metals and calcium have been used to reduce halides or cyanides of Zn, Cd, Hg, Tl, Pb, and Bi. Products include compounds such as $NaZn_4$, Ca_7Zn, NaCd, K_4Pb_9, Ca_2Pb_3, Ca_3Hg_2, Na_3Bi_5, and $NaTl_2$. Some alkali metal arsenides, antimonides, selenides, and tellurides have also been obtained by this technique.

Mixed acidic solutions of the component metals may be coreduced to form an intermetallic product in powder form at temperatures $\leq 100°C$ and at atmospheric pressure. Kulifay[69] used hypophosphorous acid, acidified ammonium sulfite, and particularly ammoniacal hydrazine as reducing agents to obtain mercurides, bismuthides, antimonides, arsenides, tellurides, Au_3Pt, and Cu_3Pd, and later extended the same technique to selenides.[70]

Benzing[71] earlier had obtained selenides by hydrazine reduction of selenites.

Reaction and crystallization in liquid metals also offer some attractive possibilities. Although compound formation by mixing of amalgams was the formation technique practiced most frequently by the pioneer investigators of intermetallics over one hundred years ago, this procedure has been little used of late. It was briefly revived by Russell, Kennedy et al. in the 1930's, and Fitzer and Gerasimov[73] have more recently applied it to the synthesis of NiAl. Crystals of α-$ThSi_2$ have been prepared by reacting thorium and silicon in molten aluminum at 1150°C.[74] The use of bismuth as a molten medium permits formation at lower temperatures, and the following compounds have been crystallized: β-$ThSi_2$, USi_2 and USi_3, $ThGe_2$ and $Th_{0\cdot9}Ge_2$, and $ThAu_3$.[75–77] In each case the compound was isolated by dissolving the bismuth matrix in nitric acid, but filtration on a pyrex frit at 300°C under argon may be employed to isolate compounds less resistant to acid attack. The preparation of USi_2 and $ThGe_2$ is of particular interest, for they are peritectoid phases that have not been obtained in the binary system. Analysis shows that they are not stabilized by bismuth, whereas heat treatment experiments show that they have relatively low decomposition temperatures.

The scope of the liquid metal technique might be enlarged by the use of other low-melting elements such as Sn, Pb, Ga, In, Tl, and Cd. The possibility of contamination or phase stabilization by the liquid element is strong, however, and the need for the careful analysis of any product not obtained by direct union is evident. Other difficulties in extraction methods are outlined in Section 5.1.

5.6 Halide Reduction

The possibilities of compound synthesis by halide reduction have already been alluded to in the section on liquid-liquid reactions. The more usual form of halide reduction, however, is via the vapor phase. The general form of the reaction is

$$3AX + 2B \rightleftarrows AX_3 + 2AB$$

where AX is the halide and AB the intermetallic sought. This reaction has been especially employed in the synthesis of certain III–V compounds, such as InAs, InP, GaAs, and GaP, in which all participants in the above reactions are in the vapor state except the product compound. The method is particularly advantageous in these cases, because the synthesis can be carried out at a satisfactory rate well below the melting point of the compound, in which the equilibrium pressure of the group V element is quite large. Antell[78] has given further details.

Silicides are frequently prepared as coatings by hydrogen reduction of silicon tetrachloride on a refractory metal substrate. The technique was first practiced by Fitzer[79] for molybdenum silicide, using temperatures in the range of 1000 to 1800°C. Below 1420°C (approximately the melting point of silicon) the coating consists of Si, $MoSi_2$, and MoSi, but at higher temperatures free Si is not usually present. If formed, free silicon can be converted to the silicide by a diffusion anneal. Deposition rates of up to $\frac{1}{2}$ mil/min have been achieved. Applications of this concept to the formation of other compounds are given by Powell et al.[8]

5.7 Powder Metallurgy Techniques

By heating them to appropriate temperatures, powder mixtures of elements may be induced to react in the solid state. Such reactions depend on diffusion, as described in Section 2, and the phase condition of the products is determined by the amounts of the elements present and in mutual contact, the possible phases in the system, and whether or not equilibrium is attained. Rapid attainment of equilibrium is assisted by very fine particle sizes, by a thorough mixing of the reactants, and by close contact between individual particles to develop a large diffusion interface. Powders may be compacted and subsequently sintered, or they may be hot-pressed or hot-extruded to synthesize the compound in bulk form.

Disc or block compacts can be supported in refractory crucibles with the minimum of contact with the container, and, because reaction is limited by the range of solid diffusion, corrosion of the refractory is insignificant. Accordingly, this technique offers an excellent means of handling elements or compounds which, in the molten condition, are too corrosive for easy containment. If, moreover, reactivity decreases on compound formation, the products may subsequently be melted in a suitable container. The technique is also useful for preparing incongruently melting materials that segregate easily during freezing, for inhomogeneities are on a microscale and may be eliminated by a sufficiently long period of heating.

The method has several limitations. Some elements are not easily obtained in powder form or, if powders are obtainable, they are sensitive to contamination by oxygen. Seybolt[80] has circumvented the difficulties of making powder from the relatively tough NiAl by preparing instead powder of the brittle compound Ni_2Al_3 and admixing it with the appropriate amount of nickel powder. Upon subsequent hot extrusion, NiAl of the desired composition was formed as a dense polycrystalline rod. Contamination of charges may be restricted by using freshly prepared powders (obtained, whenever possible, by hydride decomposition), by limiting the minimum particle size used (possibly at the expense of a rapid attainment of equilibrium), and by carrying out all mixing and pressing operations under inert atmospheres.

A more serious limitation occurs if the reactants have a marked disparity of melting points, so that grains of the low-melting component coalesce and grow or melt before reacting, thus reducing the diffusion interface. In such cases product layers rapidly accumulate at the interface, and equilibrium is attained only after prolonged heating. Raising the temperature is likely to be of little assistance, because the layer probably experience crystal growth as a result, thereby reducing the effectiveness of grain boundary diffusion. Other difficulties are the frequent inability to achieve theoretical density and the impossibility of forming single crystals.

Sinel'nikova[81] has compared the production of the refractory aluminides by powder metallurgy methods to arc melting and other alternative processes.

More detailed discussions of procedures in the applications of powder-metallurgy techniques are given in References 1, 2, and 3.

5.8 Electrodeposition

Electrodeposition techniques sometimes offer advantage in purity, ease of formation, or the ability to produce coatings. Dense continuous coatings are of special interest both for practical applications and because in this form they are convenient for studies of oxidation resistance, surface potential, and various other electrical properties. The formation of borides and phosphides by electrolysis in borate and phosphate melts is reviewed in Reference 82, and the specific preparation of VP, NbP, and TaP is described in Reference 83. In the case of vanadium the metal pentoxide was dissolved in molten $Na_4P_2O_7$, and VP was separated at the cathode on electrolysis with a current density of 1.2 A cm^{-2} and a voltage of 3.6. Cook[84] has successfully deposited borides, silicides, and beryllides from fused salt melts as coatings on refractory metals. The compound AuSn was formed at a gold electrode in $2M$ boiling HCl containing $0.001M$ Sn^{II} by Leidheiser and Carver,[85] who suggested extension of the method to other noble metal systems containing Pb, Sn, or Cd. Uhlig et al:[86] readily formed the compound PtZn by electrolysis in zinc sulfate baths, although its formation by other techniques is difficult because of the volatility of the zinc and the extreme disparity in melting points. The electrodeposition and passivity of NiSn, Cu_6Sn_5, and $FeSn_2$ have been studied by Clarke and Britton.[87]

5.9 Order-Disorder Reactions

Some special problems in synthesis may develop when the desired compound is an ordered phase that forms from a disordered solid solution. The critical temperature T_c, below which ordering begins, is highest at stoichiometric compositions, and, in general, occurs at an approximately constant fraction (0.6) of the melting point.[88] Clearly, with each successive decrease of sample temperature, the attainable degree of order increases until in favorable instances a fully ordered structure becomes possible. In some compounds, in which lattice energies and bonding cooperate, ordering may be complete on cooling from the melt. An example of such a compound is provided by Bi_2Te_3, which occurs in a solid solution range extending from ~30 to ~61 at% Te. In compounds with more metallic character, the ordering process is slower and heat treatment at temperatures below T_c is required to promote rearrangement of the atoms on ordered sites. Compounds of this type are exemplified by CuAu and Cu_3Au.

The development of an ordered phase follows the pattern of nucleation and growth.

The growth process occurs by short-range diffusion of atoms over Å distances via the vacancy mechanism. The effect of thermal energy is critical, because an increase in temperature aimed at accelerating lattice diffusion sets a limit to the degree of order attainable. Rates of ordering therefore have a time-temperature dependence expressed in the form of *TTT* curves. Factors favoring rapid ordering at temperatures below T_c are a high defect concentration and dislocation density produced either by quenching from a higher temperature or by deformation. If the ordering temperature lies below the recrystallization temperature, however, residual cold work may retard the ordering process.

Studies of compositions close to Fe_3Al show that quenching from successively higher temperatures assists the low-temperature transition from FeAl order to Fe_3Al order, presumably by the freezing-in of increasing numbers of vacancies. Cold work remaining after recrystallization reduces the ordering rate, probably by shortening the vacancy lifetime. The grain size of the material appears to have no effect on ordering in this instance, but it may be of significance in other solid solutions.

Ordering may also be accelerated by exposure to radiation when the operative mechanism is apparently a radiation-enhanced diffusion. Several experiments have been performed that demonstrate the effect in both real and model systems. A particularly striking example is the formation under irradiation of FeNi, a compound that could not be obtained by heat treatment alone.[90]

Another method of accelerating the ordering process and promoting grain growth is to cycle thermally the sample in an appropriate temperature range.[91,92] The thermal cycling acts as a periodic quenching and tempering of the samples, and the acceleration of ordering is thought to be brought about in part by diffusion of excess vacancies to sinks.

Rudman has discussed order-disorder as well as radiation effects in detail in Chapter 21.

5.10 Effects of Pressure

Under high pressure many solids undergo a phase transformation to another crystal structure—although these new structures cannot in general be retained upon return to atmospheric pressure. The possible retention of a high-pressure modification must depend upon the free-energy gain and the height of the activation barrier involved in the transformation to the modification found at barometric pressures. Clearly, the best possible chance of retaining such a phase in a metastable condition is provided by first removing surplus thermal energy before releasing the pressure. It is to be expected that if it were at all possible to retain a high pressure modification, the most favorable conditions would be provided by the development, during the formation period, of large crystals with a minimum dislocation density. The growth of such crystals, however, invariably requires a heat input which by promoting thermal agitation must oppose the influence of pressure, thereby favoring a return to a less densely packed configuration. Increased pressures are therefore needed, and the degree of retention in these materials may exhibit a time-temperature-pressure dependence.

Pressure-induced transformations in intermetallics have been observed particularly in III–V and II–VI compounds, in which the change to a more densely packed structure is found to cause a change from the semiconducting to the metallic state.[93–97] Superconductivity has been observed in some cases in which it has been possible to preserve metastably the metallic form at low pressures and temperatures.[98,99] Application of high pressure to the semiconductor Mg_2Si,[100] however, causes a polymorphic transformation but no change in conductivity type. High pressure may also be used to shift the transition temperature of a compound that exhibits a temperature-dependent polymorphism at atmospheric pressure.[101] Finally, we may note the influence of high pressure on the nucleation and growth of certain compounds during the diffusive formation of intermetallics.[102]

6. HOMOGENIZATION

Unless special procedures are adopted during or subsequent to the synthesis process itself, the product compound will have inadequate homogeneity for the intended purpose.

6.1 Sources of Segregation

We have to consider nonuniform distributions of several chemical species—the component elements themselves, impurities, and intentional solute additions. The heterogeneous distributions may arise from the nature of the formation reaction, for example, peritectics; from solidification, either dendritic or so-called "inverse" segregation of incongruently melting compositions; from anisotropic segregation associated with crystal faceting; or from solute segregation at free surfaces and grain boundaries. The first case has been discussed in Section 5.2, and the second includes well-known and extensively studied metallurgical phenomena, the underlying theory of which has been thoroughly reviewed in a recent monograph by Chalmers.[5] Anisotropic segregation assumes particular importance in semiconducting compounds, in which it has been extensively studied.[103] It is found in InSb, for example, that solidification from the melt occurs most rapidly parallel to {111} facets that can form at the solid/liquid interface thus trapping impurities in the solid. Crystals pulled in a ⟨111⟩ direction tend to concentrate Se and Te along the central axis instead of the edges. Solute segregation at free surfaces and internal surfaces, such as grain boundaries and antiphase domains was once thought to be strictly a matter of preferential Gibbsian adsorption. It now seems clear that the effects observed are larger and more extensive than can be accounted for in this way. Recent speculations have been in terms of a solute dumping process as a result of annihilation of vacancies when solute-vacancy complexes reach appropriate sinks.[104]

6.2 Means of Combating Segregation

Mechanical agitation by stirring, rocking, vibrating, electromagnetic stirring, etc., while the material is in the molten or semimolten state, is often helpful, as has been discussed in Section 5.1.

Cases of solidification segregation are most severe when the segregation coefficient k, the ratio of the solid to the liquid solubility, departs appreciably from unity. It seems preferable in such cases to cool rapidly to develop a highly cored but fine-grained material, thus limiting the distance over which composition differences occur. Prolonged heat treatment at temperatures below the solidus should then develop a homogeneous product by solid state diffusion. Ordinarily, annealing is carried out under normal pressures and protective atmospheres. In the case of InAs–In_2Se_3 alloys, however, Goryunuova et al.[105] found that homogenization could be achieved more quickly by annealing under about 800 atm. of hydrogen at the usual temperature (600 to 700°C).

It may sometimes be possible to utilize pronounced grain-boundary segregation of an impurity to effect a purification of the material. Either we may extract from a coarse grained polycrystal single crystal specimens for subsequent use, or a complete separation of grains can be effected, their surfaces removed by chemical treatment, and the assemblage reconsolidated.

Although neither normal freezing nor zone melting yield a chemically homogeneous product, except in special cases when the segregation coefficient of all relevant solutes approximates unity, repeated reverse passage of a molten zone through an ingot can produce a very effective homogenization in all but the last zone to freeze. The reader is referred to Pfann's book[6] for further details on this zone-leveling technique.

As with ordinary alloys, severe deformation combined with heat treatment is the most satisfactory means of homogenizing an ingot that is initially of nonuniform composition. This technique obviously is not applicable to any single crystal, if it is to be maintained in that form, and is of limited applicability to polycrystalline intermetallics because of their poor deformability under ordinary conditions. However, by successive hot extrusion—a process favoring deformation of brittle materials—Wood and Westbrook[106,107] have been able to prepare adequately homogeneous specimens of such compounds as Bi_2Tl_3, AgMg, and NiAl. The hot-deformation processing has also the advantage of improving low-temperature ductility and accomplishing size and shape changes. See also Section 9.5.

6.3 Tests for Homogeneity

As with several other aspects of intermetallic compound synthesis, the highest level of development of techniques has been achieved with the semiconducting compounds, and most of our examples must necessarily be chosen from this class of materials. Although electronic parameters, such as resistivity, Hall coefficient, or carrier lifetime, are extremely sensitive to small variations in composition, they must be amenable to probe techniques;

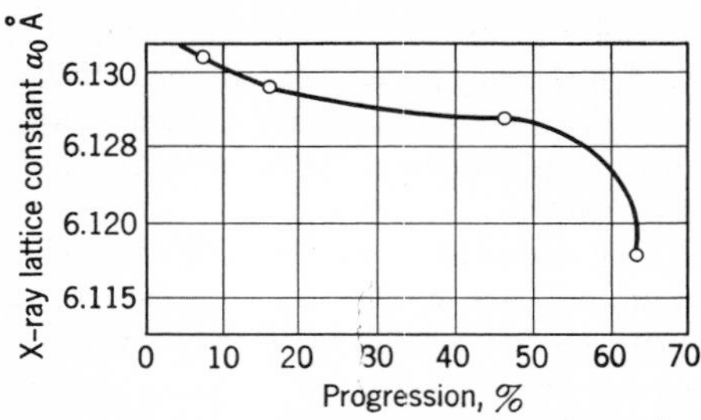

Fig. 8. Lattice parameter as a function of progression along zone cast $(Ga_{0.5}Al_{0.5})Sb$; casting rate 0.05 in./h. (After Miller et al.[108])

otherwise the survey of homogeneity will necessarily lack resolution or will be destructive of the sample itself. Therefore, other nondestructive probe techniques, such as emission spectroscopy, the electron microprobe, X-ray lattice parameter, or hardness surveys, are often preferred. Figure 8 shows an example of a lattice-parameter survey of a zone-refined (Ga, Al)Sb sample. It is seen that the central portion of the ingot is relatively homogeneous. Microhardness measurements are very convenient and sometimes give information when all other tests fail. Oxygen segregation, which frequently occurs in intermetallics near free surfaces and grain boundaries, can be readily detected even when the overall content of oxygen is only a few parts per million. Figure 9 shows a result obtained on a NiAl bicrystal that has picked up oxygen from the furnace atmosphere. Prior to this exposure, the hardness value was independent of position.

Most of the techniques cited so far give only qualitative indications of homogeneity and do not disclose the solute species responsible for

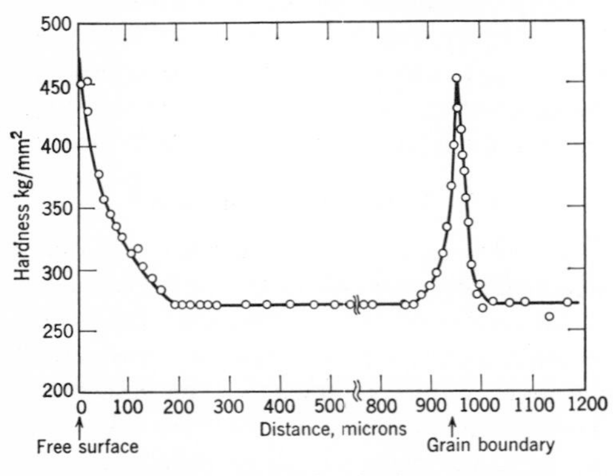

Fig. 9. Oxygen-induced hardening near free surfaces and grain boundaries in NiAl. (Unpublished data, but from work reported in Reference 28.)

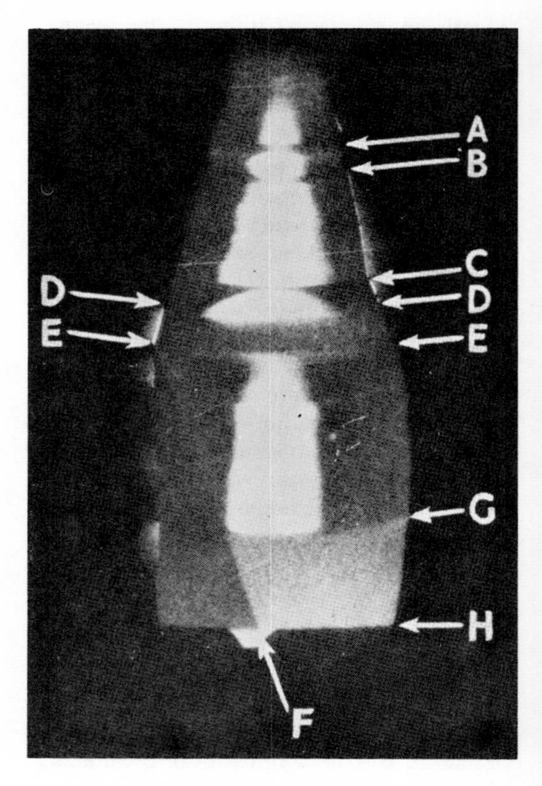

Fig. 10. Autoradiography of a vertical {110} section of ⟨111⟩ InSb crystal grown from a melt containing initially 160 ppm of radio-Te. (*A*, *B*, and *G* show changes in stirring conditions; meltback and regrowth occurred at *C*, and twin formation at *D*, *E*, and *F*.) (After Mullin.[103])

the property variation observed. Techniques such as the emission spectrograph or electron microprobe,[109] which are specific to a particular solute, are more informative and, are therefore, preferred whenever they can be conveniently applied. Another specific method that can be used in favorable cases is autoradiography. An example of this technique, showing the distribution of radio-Te on a {110} slice of a ⟨111⟩ melt-grown InSb crystal from the work of Mullin,[103] is reproduced in Fig. 10. The variations in distribution of tellurium are due to the faceting type of segregation as well as to changes in stirring conditions and growth accidents. As a final method of monitoring heterogeneity, we may consider another case of solid solution formation in quasibinary systems of compounds. It has been found that equilibrium is often extremely

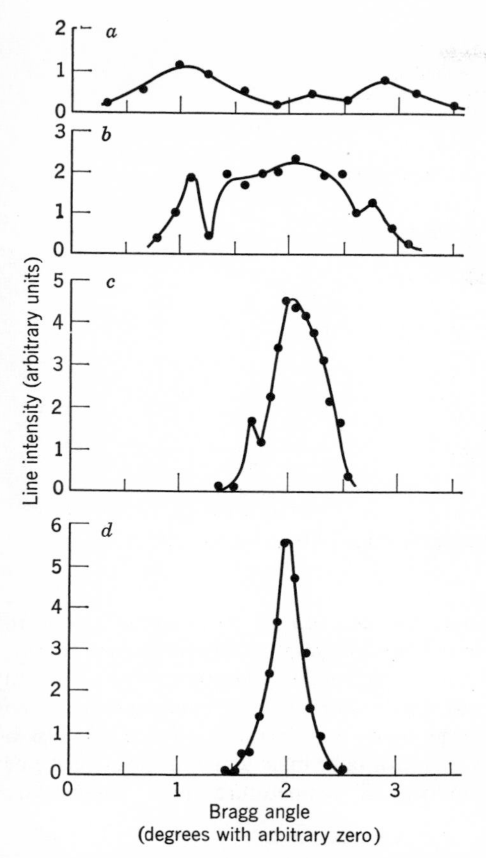

Fig. 11. Profile of 422-line for 50 mole % GaSb–50 mole % InSb alloy after various periods of annealing: (*a*) after 18 h at 480°C, (*b*) after 214 h at 480°C and 120 h at 520°C, (*c*) after 214 h at 480°C, 170 h at 520°C, and 260 h at 540°C, and (*d*) after 214 h at 480°C, 170 h at 520°C, and 910 h at 540°C. (After Woolley, Smith, and Lees.[110])

difficult to obtain in such cases even with prolonged annealing. Figure 11 shows the use of X-ray line width as a monitoring technique for the approach to equilibrium in (Ga,In)Sb.[110] Taylor[111] has illustrated the evaluation of treatments for homogenization of coring in NiAl by the resolution of end-doublets in the diffraction lines.

7. SINGLE-CRYSTAL GROWTH

The availablity of several good books,[9–11] devoted specifically to methods of single-crystal growth, obviates the need for a detailed review in the present context. Here we will simply outline the various classes of method and cite a few examples of their application to growth of intermetallic crystals. Illustrations of some typical products are shown in Fig. 12.

7.1 Bridgman-Stockbarger Method

This method depends on the relative movement of the specimen and the furnace in such a way that a temperature gradient, bracketing the melting point of the compound, is caused to pass through the specimen. Designs have been worked out using both horizontal and vertical configurations. Although the natural thermal gradient of a singly wound furnace can be used, or two windings on a single-furnace tube, a preferred arrangement is that of Fig. 13 with two furnaces separated by a baffle plate to give a sharp gradient. Typical gradients used are 25°C/cm with a growth rate of 1 cm/h. For compounds with a volatile component a sealed silica capsule is used as shown in the figure; in other cases an open crucible of graphite or other suitable refractory

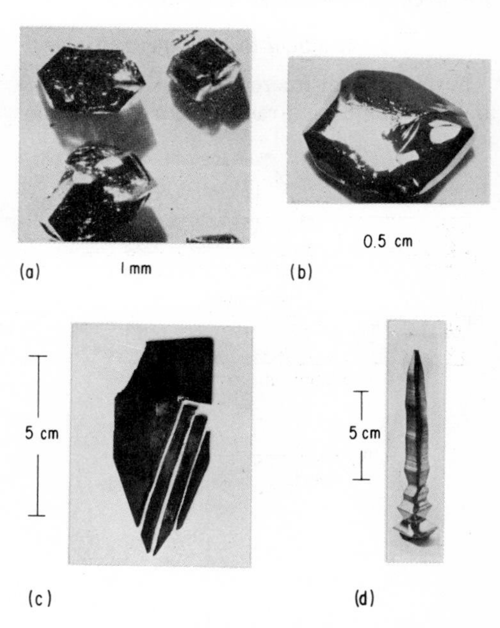

Fig. 12. Some examples of intermetallic crystals. (*a*) $MoAl_{12}$ crystals extracted from slowly cooled alloy (after Clare[47]); (*b*) vapor-grown PbSe crystal of typical quality (after Prior[113]); (*c*) AgMg single-crystal specimens cut from vertically zone-melted strip ingot (after Wood and Westbrook[106]); (*d*) GaAs crystal grown by the Czochralski technique, ⟨111⟩ orientation (after Weisberg et al.[112]).

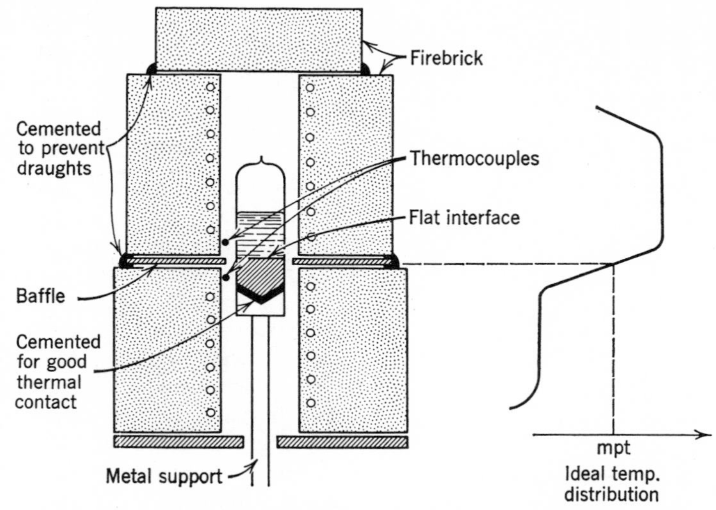

Fig. 13. Crystal growth by the Stockbarger method. (After Lawson and Neilson.[114])

can be employed. Other examples of intermetallic crystal growth by this method are given in References 115 and 116.

7.2 Gradient-Freeze Method

In the gradient-freeze technique, the furnace and growing crystal remain in fixed position, and the temperature gradient is caused to move by control of the furnace power supply. It is of particular applicability in the growth of crystals with a volatile component or when large highly pure crystals are desired. Disadvantages are the high temperature and extreme control of temperature and temperature

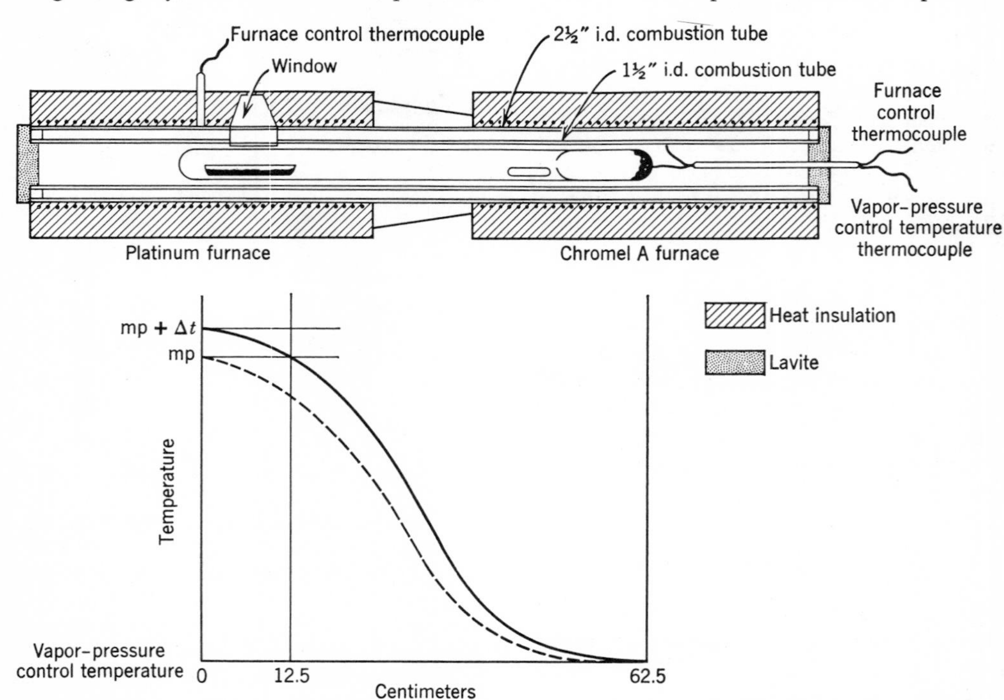

Fig. 14. Furnace setup for gradient-freeze growth of crystals and the associated temperature profile. (After Miller.[117])

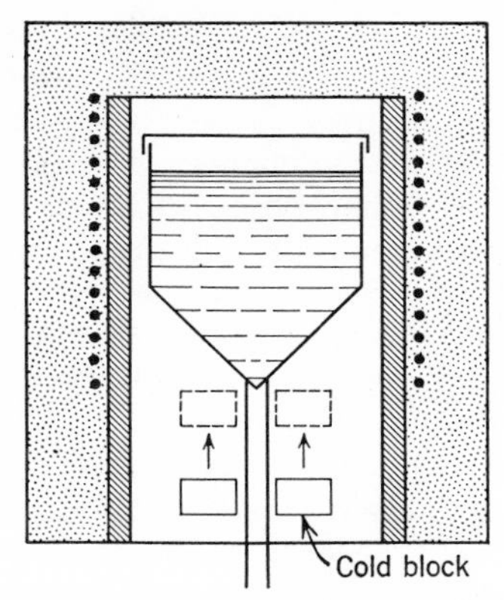

Fig. 15. Crystal growth by controlled solidification by Strong's method[118] with a moving heat sink.

gradient required. The principle and arrangement of apparatus utilized in the synthesis of crystals of a compound, such as GaAs, are shown in Fig. 14. The right-hand furnace controls the vapor pressure of the volatile component (As), which is transferred in the vapor state to react with the other molten component in the left-hand furnace. The initial temperature gradient is adjusted in the manner of the solid line in the temperature profiles. Programming of the furnace controls is devised in such a way as to bring the gradient smoothly but slowly to that of the dashed profile, thus causing the freezing point to sweep through the boat. Gradients of 2.5°/cm with growth rates up to 2.5 cm/h are typical. In another (and historically earlier) variant of the moving-gradient concept, movement of an external heat sink can be employed, as illustrated schematically in Fig. 15.

7.3 Czochralski Pulling

The Czochralski technique of slowly raising a seed crystal from a molten bath held near the melting point finds its major advantage in its

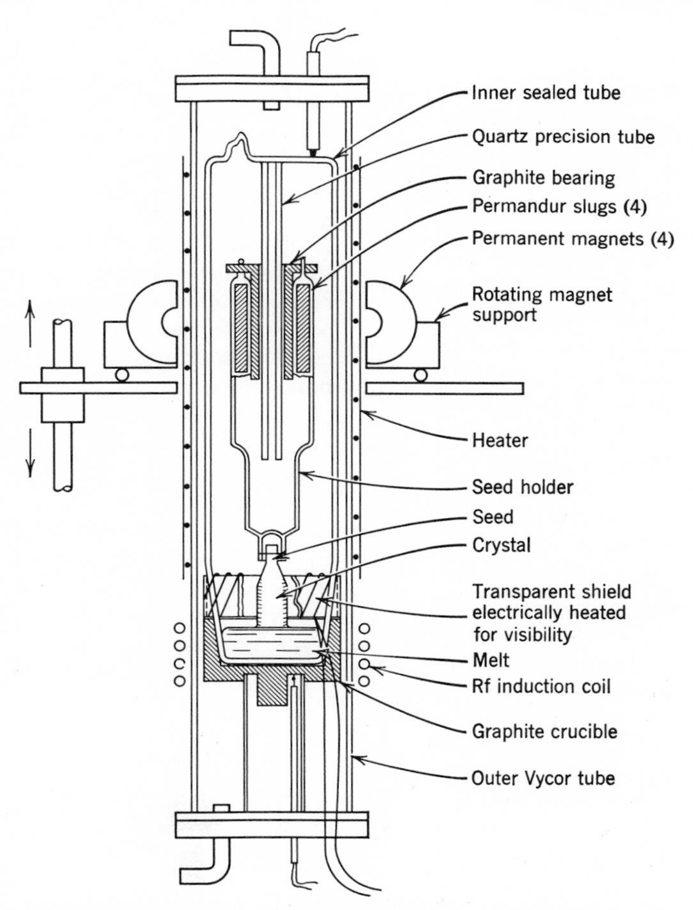

Fig. 16. Magnetically coupled Czochralski growth apparatus for GaAs, designed by J. R. Woolston. (After Weisberg et al.[112])

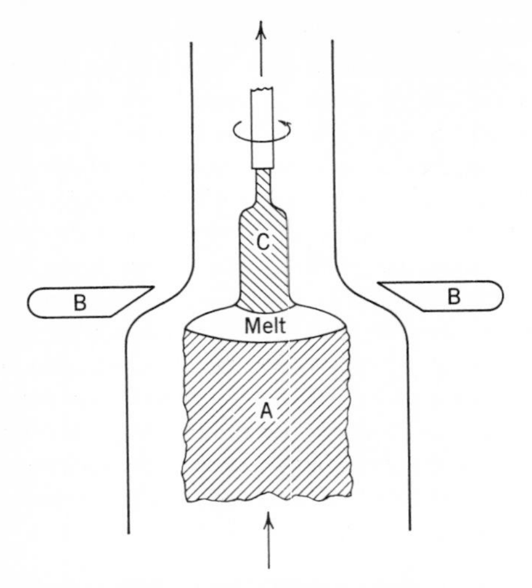

Fig. 17. A crucible-free adaptation of the Czochralski technique of crystal growth. As the single crystal C is drawn from the melt, polycrystalline boule A is moved up into the heater B to replenish the melt. (After Tannenbaum.[120])

simplicity, which makes it readily adaptable to a wide variety of configurations. A particular problem, to which it has been adapted in the case of III–V semiconducting compounds, is growth in a sealed system to prevent loss of the volatile component. The requisite motion of the growing crystal, as the seed is withdrawn from the melt, can be accomplished by magnetic coupling, as shown in Fig. 16, designed by Woolston[112] after an original design by Gremmelmaier.[119] Liquid seals and a piston-type syringe arrangement have been employed as alternate schemes for transmitting motion in a sealed system. Note that the growing crystal is rotated as it is withdrawn; in some designs the crucible is also rotated, but in the opposite direction. The principal difficulties with the Czochralski technique arise from the crucible. The crucible wall may serve to nucleate undesired crystals or may contaminate the melt. In an effort to avoid this problem, some work has been done combining the Czochralski method with a self-supported floating zone, as shown in Fig. 17. Both the rate of growth and the crystal perfection are found to be sensitive to crystal orientation. The noncentrosymmetric compounds even exhibit differences for the different polarities of a given crystallographic direction.

7.4 Zone Melting

Zone melting was originally developed by Pfann[121] as a refining process taking advantage of the different solubilities of solutes in the solid and liquid states. Because it is by definition a controlled freezing technique, it also constitutes a means for isolating congruently melting phases from incongruently melting mixtures or solid solutions, or a means for the growth of single crystals. It may also be adapted to crucible-less melting as already described. In zone refining, a narrow molten zone is passed slowly along the length of an ingot of the material to be purified, as shown schematically in Fig. 18. Considering only values of k less than unity, the solute is rejected into the liquid until a concentration of C_0/k is reached. Up to this point the refrozen ingot behind the molten zone has a solute content of less than C_0; beyond this point passage of the zone leaves the concentration unchanged at C_0, with the exception of the last zone to freeze in which the solute content rises again. The value of k is a measure of the ease of refinement; as k departs from unity, a greater length of ingot can be traversed by the molten zone before a steady state is reached. The ultimate distribution of solute is attained by repeatedly passing the zone along the ingot; relatively few passes are sufficient for $k < 0.5$, but for $k = 0.9$ many hundreds of passes may be necessary. In addition to the k value determined by the phase-diagram relationships, the actual compositional gradients achieved depend on the effectiveness of stirring of the liquid and rate of motion of the zone.

Zone-fusion equipment generally utilizes vitreous silica envelopes, and owing to the ease with which narrow zones can be formed by high-frequency techniques, this method is frequently used to accomplish melting, the envelope being water- or air-cooled as required. Apparatus that use arc[122,123] or electron-beam[124] heat sources have also been described

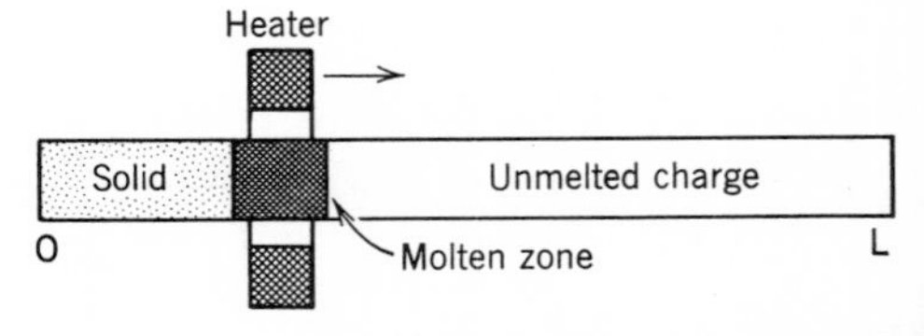

Fig. 18. Schematic arrangement for zone melting.

for refractory materials. Resistance heating is common for low-melting-point materials.

In addition to the conventional crucibles, there are at least two alternative means of supporting the molten zone. In the floating zone method, the ingot is mounted vertically and the melt is contained by surface-tension forces. Critical factors in addition to the surface tension are the diameter of the ingot and the density of the material. Figure 19 illustrates a typical floating zone apparatus for III–V compounds. In the magnetic suspension method, a direct current is passed through a horizontal ingot, and the gravitational field in the molten zone is counterbalanced by a vertical force produced by the interaction of the current with a constant magnetic field. This field is generated by a horizontal source normal to the ingot axis.

Although crystal-seeding techniques can be used in single-crystal growth by zone melting, this is not usually necessary. Growth-rate anisotropies are such that repeated passage of the zone usually results in a single crystal (so-called self-seeding) or at least in a very coarsely polycrystalline sample. Composition control is important, particularly in cases

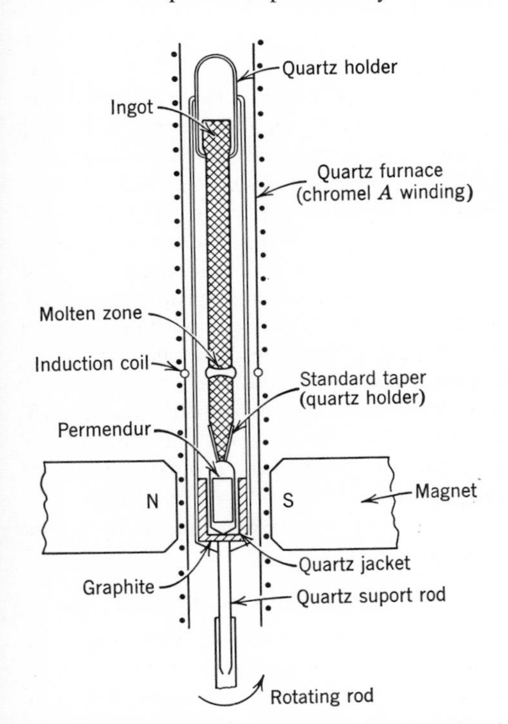

Fig. 19. Floating-zone apparatus for growth of III–V crystals. (After Allred.[125])

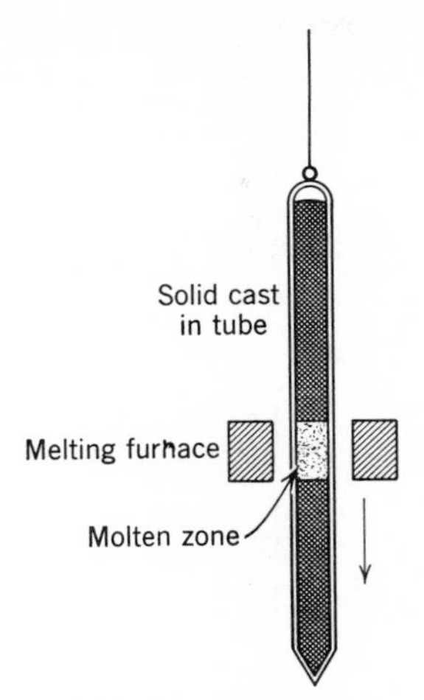

Fig. 20. Sealed-solid method for zone-melting growth of crystals. (After Heumann.[129])

involving volatile components, for variations in composition can lead to twinning or to the nucleation of new crystals. Volatile impurities may also be transferred in an undesired fashion.[126–128] Another problem of compositional variations in the floating-zone method is the development of zone instability.

Heumann[129] has devised a zone-melting technique known as the "sealed solid" method, which minimizes the equipment required, yet results in product of improved homogeneity by elimination of the possibility of vapor transfer. An ingot of the compound is sealed into a suitable tube, the principal requirements being that the ingot be free to move in the tube prior to zoning and that the material not react with or stick to the tube. Lowering the sealed tube through a single narrow-zone furnace, as in Fig. 20—an adaptation by Lorenz and Halstead[130]—causes the zone to traverse the ingot. The solid material above and below the melt acts as a vapor seal and prevents transfer of the volatile species. The fact that the major part of the ingot is cold, minimizes undesired back diffusion of impurities in the solid state.

Reviews of zone-fusion techniques are given by Pfann[6] and by Parr.[7]

7.5 Vapor Growth

Single-crystal growth by vapor processes is based simply on mass transfer in the vapor state through a temperature gradient. In addition to the use of elemental vapors, hydrogen compounds, such as H_2Se and H_2Te,

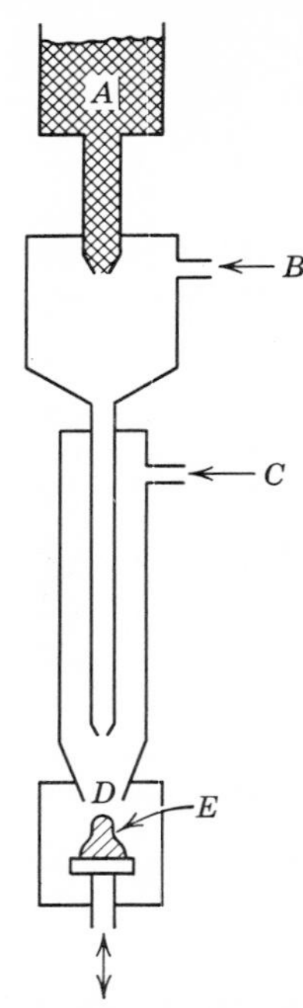

Fig. 21. Schematic diagram of Verneuil apparatus for growing crystals from a melt without a crucible. *A* is the powder feed, *B* the combustion gas, *C* the oxygen, *D* the burner tip, and *E* the growing boule. (After Tannenbaum.[120])

or various halides may be used. Apart from providing a seed for single-crystal growth, no new principles or techniques are involved beyond those discussed in Section 5.3 on vapor-liquid synthesis and in Section 5.6 on halide reduction.

7.6 Verneuil Process

The Verneuil process for single-crystal growth has been very popular for growth of oxides and other highly refractory compounds. It consists essentially of dropping particles through a heated gas, so that they strike a seed crystal in the molten state and there freeze and grow epitaxially. A schematic of the apparatus is shown in Fig. 21. Adapted for heating with an inductively coupled plasma rather than a gas flame, it has found some use for synthesis of silicide and beryllide crystals.

7.7 Strain Anneal

This favorite method for single-crystal growth of metals can rarely be practiced for intermetallics because of their restricted low-temperature ductility. Wang et al.[92] have been able to use this method, however, for NiTi, which exhibits an anomalous ductility near room temperature. Cyclic variation in temperature during the subsequent anneal was found to aid in single-crystal growth. Wood[131] was also able to grow AgMg single crystals by the strain-anneal method on wires prepared by double-extrusion.

8. SPECIAL PROBLEMS

8.1 Control of Stoichiometry

Stoichiometric control is one of the frequently encountered problems in intermetallic synthesis or growth.[13] In any of the processes involving a component in the vapor state, compositional control, as reviewed in preceding sections, is usually effected by vapor-pressure (i.e., temperature) control of the source of that species in the apparatus. In order to use these techniques to their fullest extent, a rather complete knowledge of the pressure-temperature-composition relationships in the system is required. The best example of such study is the one on ionic compound, PbS, shown in Fig. 22. Similar results in intermetallic systems are those for GaAs[133] and InAs,[133] PbTe,[134] and CdTe.[135] Sometimes, particularly for those compounds exhibiting a broad range of homogeneity, preparation of a particular composition is no more difficult than achieving a desired composition in any other alloy. However preparing a compound of exactly ideal stoichiometric composition, or indeed even knowing exactly how far deviant the resulting composition is from ideal stoichiometry, can be very difficult. Not only may problems develop because of differences in composition between the maximum melting point and the ideal stoichiometry, but also the compositional sensitivity is usually beyond ordinary analytical procedures. Resort must then be made to physical or mechanical property measurements, many of which have been found to be extraordinarily sensitive to small deviations from stoichiometry.

8.2 Vapor-Pressure Influences

Many of the metallic elements exert an appreciable vapor pressure at moderate to high temperatures, and the consequences of volatilization must be taken into account, particularly in evacuated systems, during the

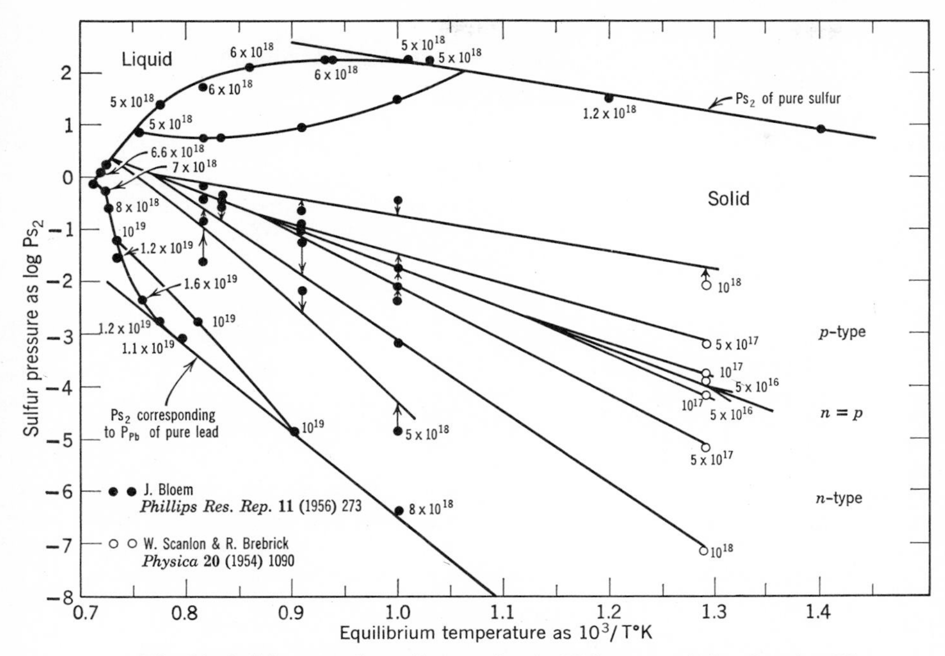

Fig. 22. Solid-vapor phase diagram for the PbS system. (After Scanlon.[132])

formation and melting of intermetallic compounds. Although this matter has been discussed briefly in some of the preceding sections, the topic is sufficiently important to warrant some repetition and amplification.

Some impression of the volatility of the metals and semimetals can be gained from an inspection of Tables 3 and 4. Thus at temperatures in the region of the *melting point* a pressure of at least 1 mm Hg is exerted by the elements As, Ca, Cr, Er, Eu, Mg, Mn, P, Sm, Sr, Te, Yb, and Zr. At temperatures of the order of 1000°C, a moderate temperature for melting many intermetallic compounds, a pressure of at least 1 mm Hg is exerted by As, Ba, Bi, Ca, Cd, Cs, Eu, Hg, K, Li, Mg, Na, P, Pb, Po, Rb, Sb, Se, Sr, Te, Tl, Yb, and Zn.

The kinetics of the reaction play a leading part in determining the influence of vaporization in the attainment of equilibrium because, following combination, the vapor pressure of an element can be expected to fall below the value for the free element. The physical condition of the compound is important in this connection, however, for in many molten intermetallics the vapor pressures of the elements are only a little lower than those for the free elements at the same temperature.

In instances where volatility is suspected prolonged heating in continuously evacuated systems is clearly to be avoided for, at the very least, it is likely to lead to inhomogeneous products as a result of diffusion-controlled depletion of the surface layers.

Procedures for preparations in sealed containers depend on whether vapor pressures over the final product are, or are not, negligible. If the ultimate vapor pressure is negligible, the reactants may simply be sealed into a container under vacuum. It is necessary, however, to control the distribution of temperature over the container to prevent subsequent deposition of elements away from the reaction site. For this purpose the more remote parts of the container are generally maintained at a temperature of not less than 100°C below the melting point of the compound.

Alternatively, the compounds may be prepared by two-stage techniques. A master alloy rich in the volatile element (no free element present) is first obtained in a homogeneous condition. The exact composition of this material is found by analysis, and compounds may be prepared by mixing in appropriate amounts of the less volatile components and then melting. A further alternative is provided by heating calculated amounts of reactants in a sealed container until the

TABLE 3

Melting and Boiling Points of Metals and Semimetals[a]

(Temperatures in °C)

1a	2a	3a	4a	5a	6a	7a	8	8	8	1b	2b	3b	4b	5b	6b		
1347 Li 180	2477 Be 1283											3675 B 2300	[b] C 3700				
883 Na 98	1108 Mg 650											2060 Al 660	3310 Si 1412				
765 K 63	1440 Ca 850	2480 Sc 1539	3270 Ti 1677	3000 V 1910	2480 Cr 1903	2041 Mn 1244	2735 Fe 1537	2870 Co 1492	2732 Ni 1453	2580 Cu 1083	907 Zn 420	1980 Ga 30	2700 Ge 958	[b] As 613	685 Se 220		
680 Rb 39	1370 Sr 770	3200 Y 1509	4370 Zr 1850	3300 Nb 2468	4800 Mo 2625	4600 Tc 2100	3700 Ru 2400	3700 Rh 1960	3100 Pd 1552	2180 Ag 961	765 Cd 320	2050 In 156	2270 Sn 232	1440 Sb 630	990 Te 450		
685 Cs 29	1500 Ba 710	La-	5400 Hf 2225	5400 Ta 2980	5500 W 3380	5500 Re 3180	4200 Os 3045	5300 Ir 2443	3800 Pt 1769	2660 Au 1063	357 Hg −40	1470 Tl 304	1750 Pb 327	1420 Bi 271	960 Po 254		
680 Fr 30	1500 Ra 700	Ac-															
Lanthanides			3470 La 920	2927 Ce 795	3020 Pr 935	3100 Nd 1024	2700 Pm 1000	1590 Sm 1052	1430 Eu 830	2700 Gd 1312	2500 Tb 1356	2300 Dy 1500	2300 Ho 1461	2600 Er 1500	1700 Tm 1545	1500 Yb 824	1900 Lu 1730
Actinides			3300 Ac 1230	4200 Th 1842	4000 Pa 1200	3930 U 1132	— Np 637	— Pu 640									

[a] Values taken from the *International Tables of X-ray Crystallography*, Vol. III, pp. 46–56 (1962).
[b] Sublimes.

volatile component has been incorporated in the form of stable intermetallics into the surface of the other components. This material may then be melted to form a homogeneous product. These techniques are particularly useful when phases rich in a high-melting component are desired.

If the ultimate vapor pressure is appreciable, evaporation of a volatile component can be prevented by connecting the reaction vessel to a reservoir containing this element, heated to a temperature sufficient to maintain an atmosphere at the equilibrium vapor pressure. Alternatively, the reaction may be carried out in an inert atmosphere at a pressure sufficiently high to minimize the partial pressure of the volatile element.

8.3 Structural Imperfections

Cooling conditions—by affecting the gross cooling rate, the shape of the solid-liquid interface, and nucleation events—determine grain size, shape, and compositional heterogeneities, as already discussed. Over and above these influences, cooling conditions can affect the thermal stresses built into the product as well as the dislocation substructure. Dislocation densities in cast materials range typically from 10^4 to $10^7/cm^2$, although dislocation-free single crystals (silicon) have sometimes been grown. Czochralski pulling of crystals tends to give much lower dislocation densities than floating zone methods. In general, the dislocation distributions with either technique are not homogeneous; marked gradients in dislocation density usually are present along both radial and axial directions of grown crystals. Dislocation densities can be reduced by climb and mutual annihilation during anneals close to the melting point. Dislocation densities and distribution have most often been examined by etch pit studies.[138]

8.4 Polarity Effects

The crystallographic polarity of noncentrosymmetric crystals gives rise to a number of effects pertinent to the synthesis of intermetallic compounds. Not only will growth rates and consequent crystal morphology be affected, but due to preferential adsorption, impurity contamination may also be greatly different on antipodal faces of a crystal. Polarity of grown crystals may be determined by etching, X-ray, or hardness techniques. See Chapter 16 for further details.

9. SPECIAL GROWTH FORMS

Most of the foregoing sections have treated the synthesis of massive polycrystalline or single-crystal bodies in which external form was of little consequence. For many experimental property studies or for the direct commercial application of intermetallics it is both convenient and desirable to have the compound in a form in which it can be used without further processing. The latter preference is motivated by the unfavorable consequences of contamination and/or structural damage that may result from cutting and final surface preparation.

9.1 Dendritic Growth

Dendritic growth has been perfected especially for growing ribbons of semiconducting or ferroelectric materials. A review of dendritic growth of III–V compounds has been provided by Lindburg and Faust.[139] Dendritic growth is really a special case of Czochralski pulling with certain differences: a dendritic seed, a supercooled melt because growth is crystallographically (rather than thermally or physically) limited, and much higher growth rates and pulling speeds (3 to 50 cm/min). To accomplish the latter it is usually necessary to provide special equipment for long and fast pulling; take-up reels are sometimes provided. The dendrites referred to in this context consist of long thin platelets comprised of two broad faces of unusual smoothness and constant width with essentially no side arms and slightly faceted edges. Twin planes occur parallel to the growth faces and indeed are requisite to this type of growth. Dendritic growth of III–V compounds has been studied particularly in InSb[140] and GaAs and differs from that for the elemental metals largely in the effects of stoichiometry and polarity. Although the polar character of the $\{111\}$ planes has little or no effect in InSb, it is thought to account for the differences in the width of the ribbon faces in GaAs.

9.2 Fibers and Whiskers

Fibers may be defined as slender solid filaments, and whiskers in turn as fibers of such size and perfection that theoretical strengths are obtained. Interest in this growth form stems from this latter circumstance, which both holds theoretical interest and offers the hope of a practical composite material fashioned from a multiplicity of such fibers embedded in an appropriate matrix. Many intermetallics grow naturally in filamentary form, for example, in the formation of Mn_5Si_3 by halide reduction,[141] or can be induced to do so by unidirectional solidification of an appropriate eutectic. Hertzberg and associates[142,143] have so formed whiskers of $NiAl_3$ that exhibited whisker strengths, and Davies[144] has studied similar fibers of Cu_6Sn_5. In the latter case a strength-size effect was found which was associated with the stress concentrations produced by growth steps on the fiber surfaces. Fibrous growth of intermetallics is a challenging area in which more work should be done.

9.3 Thin Films

The motivation for thin-film production of intermetallics, like that of other materials, has come from the demands of electronic technology for high-component density in circuits and for minimization of the consumption of high-cost materials. Because most electronic

TABLE 4

Vapor Pressures of Metals and Semimetals[a]

(Temperatures in °K for Vapor Pressures in mm Hg)

	10^{-8}	10^{-6}	10^{-4}	10^{-2}	1	10^{2}
Ac	1001	1150	1348	1625	2050	2770
Ag	852	961	1105	1305	1610	2120
Al	950	1080	1245	1480	1820	2350
As_4	377	423	477	550	645	790
Au	1045	1180	1355	1605	1980	2580
B	1650	1850	2100	2430	2930	3730
Ba	560	640	740	900	1140	1570
Be	972	1100	1260	1485	1840	2370
Bi	590	672	781	934	1165	1555
Ca	555	630	725	865	1090	1480
Cd	346	393	455	540	665	883
Ce	1256	1431	1662	1986	2430	3219
Co	1200	1345	1535	1790	2180	2770
Cr	1125	1250	1435	1665	2010	2550
Cs	256	295	348	425	550	786
Cu	1005	1135	1305	1545	1895	2460
Dy	900	1030	1180	1390	1680	2140
(Er)	947	1080	1260	1503	1872	2480
Eu	556	634	739	884	1100	1455
Fe	1150	1290	1480	1740	2120	2710
Ga	845	961	1115	1330	1645	2170
(Gd)	1000	1138	1322	1580	1960	2580
Ge	1085	1225	1415	1680	2070	2710
(Hf)	2010	2270	2620	3090	3780	4830
Hg	199	228	265	318	398	534
(Ho)	823	941	1100	1320	1655	2215
In	770	877	1020	1220	1515	2010
Ir	1720	1950	2220	2580	3100	3900
K	294	337	396	481	614	857
La	1260	1430	1650	1970	2420	3180
Li	505	577	672	806	1010	1355
(Lu)	1290	1455	1670	1960	2380	3010
Mg	462	524	603	715	885	1175
Mn	807	910	1040	1220	1500	1975
Mo	1855	2110	2440	2900	3570	—
Na	350	400	468	563	710	967
Nb	2020	2280	2610	3050	3680	4610
Nd	1005	1150	1335	1615	2050	2800
Ni	1185	1330	1520	1770	2150	2750
(Os)	—	—	2537	2940	3494	—
P_4[b]	320	355	403	460	535	640
Pa	1382	1560	1737	2092	2525	3185
Pb	617	705	824	992	1250	1695
Pd	1035	1180	1375	1640	2040	2690
Po	387	432	493	587	745	1025
Pr	1075	1225	1430	1710	2120	2820
Pt	1560	1755	2015	2350	2860	3630
Pu	1106	1266	1479	1777	2228	2983
Ra	—	—	—	—	—	—
Rb	270	312	368	449	573	800
Re	2200	2480	2830	3330	4070	5200
Rh	1550	1745	1980	2300	2800	3530

(*Continued*)

TABLE 4 *(Continued)*

	10^{-8}	10^{-6}	10^{-4}	10^{-2}	1	10^{2}
(Ru)	—	—	2331	2704	3219	—
Sb	550	615	700	815	1030	1570
Sc	1052	1200	1396	1667	2070	2725
Se	357	394	440	505	620	820
Si	1200	1355	1555	1820	2200	2740
Sm	645	732	845	1000	1225	1580
Sn	937	1070	1250	1500	1885	2540
Sr	499	571	667	804	1015	1380
Ta	2230	2510	2860	3340	4010	5070
Tb	1190	1350	1555	1850	2290	2940
(Tc)	—	—	—	2650	3200	—
Te_2	451	503	569	656	793	1065
Th	1510	1703	1952	2285	2760	3480
Ti	1330	1500	1715	2000	2450	3130
Tl	558	637	741	888	1110	1480
Tm	731	827	953	1123	1368	1749
U	1405	1600	1855	2200	2720	3540
V	1428	1600	1824	2120	2560	3220
W	2340	2640	3030	3570	4200	5250
Y	1190	1350	1560	1850	2290	2940
(Yb)	526	605	712	863	1095	1500
Zn	396	449	519	615	758	1005
Zr	1745	1975	2275	2670	3250	4040

[a] Values quoted by Margrave.[136] Temperatures above 2000°K are subject to uncertainties of ±15°K. Larger uncertainties exist for elements in parentheses.
[b] The vapor pressure over condensed phosphorus depends on the modification present. For details see von Wazer.[137]

devices only make use of a thin layer near the surfaces of the active materials, the remaining bulk is both unnecessary and undesirable, and equivalent or superior performance should theoretically be achievable in thin films. General reviews of the science and technology of thin films have been provided by Holland[145] and by Neugebauer et al.[146] The specific class of III–V compounds has been covered in three chapters of Reference 11.

Intermetallic thin films have been produced by pyrolysis, electrodeposition, casting, vacuum evaporation and squeezing of droplets. Only the latter two methods have been practiced with any large degree of success for intermetallics, and thus discussion here is confined to them. Some comments on the other techniques may be found in Section 5. Films prepared by evaporation techniques range in thickness from 0.05 to 5 μ, with the most interesting properties being obtained in the lower end of this range; droplet squeezing has produced films about 10 μ thick.

Vacuum evaporation of intermetallic thin films may be accomplished either by direct evaporation of the compound or by multisource evaporation of the component elements, which then combine on the substrate to form the compound film. Success with the former method has been confined, among the semiconducting compounds, largely to InSb, because it presents the favorable conditions of low-melting point, ready availability of high-purity source material, and small differences in volatility of components. Attainment of a film of desired stoichiometry can be achieved either by a timewise selection of a fraction of the vaporized product or by complete evaporation of the source followed by a homogenization anneal.

The various techniques of multisource evaporation of compound thin films may be understood in terms of the schematic drawing of Fig. 23.[147] Three classes of method are

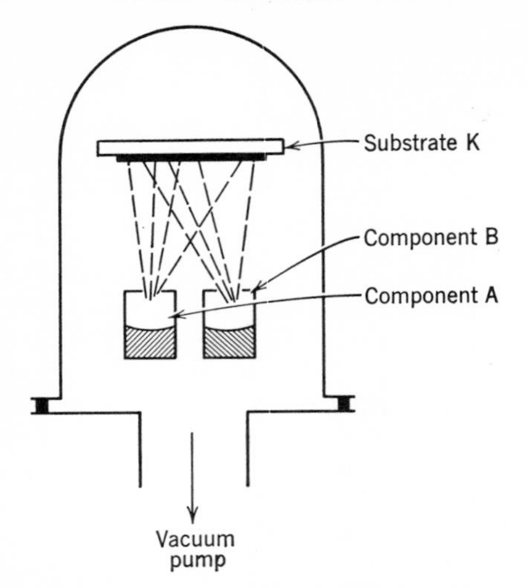

Fig. 23. Schematic of the apparatus for the preparation of compound thin films by evaporation of the components. (After Gunther.[147])

employed. In the first, complete evaporation of both component sources is effected either simultaneously or alternately. Adequate refinement of successive layers of the deposit is usually achieved by some means of continuous quantized addition of source elements to the individual evaporators. In another type of process, A and B are simultaneously evaporated and composition-controlled through control of the temperatures T_A and T_B and/or by lateral displacement of the sources T_A and T_B, so that a spatial variation in stoichiometry is obtained on the substrate. The third or so-called three-temperature process, in which T_A, T_B, and T_K are all controlled, is the preferred method. It has been found that it is not difficult to adjust the excess of the more volatile species, so that T_K can be high enough to obtain good homogeneity in the deposit by diffusion without excessive reevaporation. Most films as deposited are polycrystalline, but single-crystal formation is favored by high values of T_K and the use of single-crystal substrates. Postdeposition anneals may also be used to increase grain size. It is extremely important in all vacuum evaporation preparations of thin films to minimize impurity contamination by use of ultrahigh vacua, clean apparatus, and high impingement rates.

Bate and Taylor[148] as well as Pietrokowsky[149] have prepared thin films by squeezing molten drops between flat platens. A drawing of the equipment used by the former authors is shown in Fig. 24. Variations of the technique consist of alternative ways of mechanically loading the piston and programming the heating cycle. Freezing of the droplet is accomplished by shutting off the heater power to the lower platen and by the rapid extraction of heat through the cold upper platen. Bate and Taylor were interested only in equilibrium phases and in films of moderate thickness (10 to 20 μ). Duwez, on the other hand, by the use of much higher cooling rates, has obtained much thinner films and metastable structures. For further details of his work, see Chapter 18.

The electrical properties of semiconducting thin films approach those of bulk material, but in general are not equivalent. Even when the best matching of film to bulk properties is achieved, the carrier mobilities in the films tend to be low because of scattering effects at grain boundaries.

9.4 Powders

Powders can, of course, be obtained by comminution of most intermetallics. Because this involves the hazard of contamination and because it is inapplicable in cases of even moderate toughness, direct methods of powder production have been of some interest. Amalgam reaction, electrodeposition, and

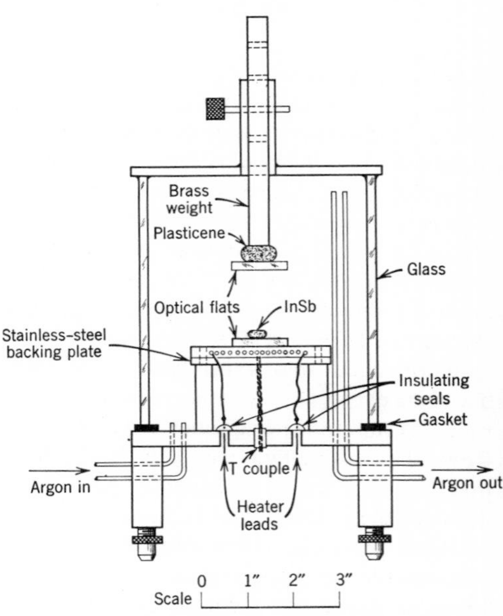

Fig. 24. Apparatus for the preparation of compound thin films by rapid squeezing of molten drops. (After Bate and Taylor.[148])

vapor state reaction are among the techniques which can be adapted to direct production of powders. Coreduction from solution is also possible.[69,70,71,150] The reader is directed to Section 5 for further details of these methods.

9.5 Mechanical Fabrication

Although brittle at ordinary temperatures, most intermetallic compounds become sufficiently deformable at high homologous temperatures to be fabricable by mechanical means. Slow-deformation rates and high confining pressures favor deformability. The extrusion process, which permits achievement and ready control of all these conditions, has been a favored technique.[151–154] By this method Wood and Westbrook[106] were able to produce 40 mil wire and 10 mil sheet of AgMg by extrusion. Even the relatively high-melting NiAl could be prepared in 100 mil rod by extrusion through a multiple-hole die.[155] Rolling of intermetallics as slabs or in multilayer packs has also been practiced, with the billet either bare or clad. The cladding is used not so much for oxidation protection or confinement but to prevent chilling by the rolls. Rolling in general is more difficult than extrusion because conventional equipment does not allow adequate control of temperature or deformation speed. More work is needed in this area of mechanical fabrication, because it is the only method of producing small dimensioned wire, sheet, and tubing on a truly large scale for both practical and experimental purposes. Lawley has given further details on plastic deformation in Chapter 24.

10. DETECTION OF IMPURITIES

Analysis of intermetallic compounds demands the application of the most powerful and sophisticated techniques at our command, for the effects of stoichiometric deviation and free-surface and grain-boundary segregation have been found to be so great. In addition the semiconducting properties, as with all semiconductors, are extraordinarily sensitive to the presence of small amounts of impurities. Resistivity ratios, optical absorption, mass spectrometry, radiochemistry, activation analysis, and semiconducting properties have all been applied as analytical tools. Space does not permit review of these topics here. The reader is referred to the analytical literature, particularly that for semiconductors, for details.

11. REVIEW OF SOME USEFUL REFERENCE WORKS

In recent years a number of reference books have appeared which offer "how-to-do-it" information on material synthesis and crystal growing as well as valuable reviews of underlying principles. Although several of these have been cited in the text, it appears warranted to review briefly the features of each that are relevant to intermetallic synthesis.

Burke and Seybolt[1] is the most general, but contains no specific information on intermetallics. Their treatments of temperature generation, measurement, and control, refractories, controlled atmospheres, and vacua are useful, and the chapter on powder metallurgical procedures is especially to be commended. Campbell and Sherwood[2] is concerned primarily with materials and techniques associated with temperatures of 1500°C and above. One chapter is devoted to intermetallics, but little is said about their synthesis. Other chapters deal with other refractory classes, melting techniques, etc. Kingery[3] and Bockris, White and Mackenzie[4] both emphasize property measurements at high temperatures, but contain useful data on refractories, furnaces, temperature controls, etc. Kingery particularly treats ceramic aspects, but neither consider intermetallics explicitly.

The general subject of solidification is dealt with by Chalmers[5] and contains much of fundamental pertinence to such problems of intermetallic synthesis as zone melting, peritectic formation, and incongruent melting. Zone melting and allied topics are covered in detail by Pfann[6] and by Parr,[7] but with little specific treatment of intermetallics. Vapor-deposition techniques reviewed by Powell et al.[8] include those relating to silicides.

Three books in English are concerned with single-crystal growth. Lawson and Nielson[9] was the first to appear and is the most concise; it contains a single chapter on the preparation of compounds. Gilman[10] is the most general, covers, as the title implies, the art and science equally well, and includes two chapters on the preparation of semiconducting compounds, most of which are intermetallics. The book by

Willardson and Goering[11] is restricted to semiconducting compounds, but covers this subject in much detail. Sections of the book relevant to intermetallic synthesis include purification of the elemental components, detection of impurities, single-crystal growth, thin films, segregation and surfaces, as well as three chapters on the preparation of specific semiconducting compounds.

The German text by Smakula[12] covers all aspects of single-crystal work. Although its treatment of crystal-growth techniques is not as detailed as that of the English works just cited, it is valuable for its coverage of underlying fundamentals, such as structure, bonding, nucleation, and defects in crystals, as well as for its discussion of treatments and applications of grown crystals. However, little specific discussion of intermetallics is included.

REFERENCES

General References

1. Seybolt A. U., and J. E. Burke, *Procedures in Experimental Metallurgy*, Wiley, New York, 2nd ed. 1953.
2. Campbell I. E., and E. M. Sherwood, eds., 2nd ed. *High Temperature Technology*, Wiley, New York, in press.
3. Kingery W. D., *Property Measurements at High Temperatures*. Wiley, New York, 1959.
4. Bockris J. O'M., J. R. White, and J. D. Mackenzie, eds., *Physicochemical Measurements at High Temperatures*. Butterworths Scientific Publications, London, 1959.
5. Chalmers B., *Principles of Solidification*. Wiley, New York, 1964.
6. Pfann W. G., *Zone Melting*. Wiley, New York, 1958.
7. Parr N. L., *Zone Refining and Allied Techniques*. George Newnes Ltd., London, 1960.
8. Powell C. F., I. E. Campbell, and B. W. Gonser, *Vapor-Plating*. Wiley, New York, 1955.
9. Lawson W. D., and S. Nielsen, *Preparation of Single Crystals*. Butterworths Scientific Publications, London, 1958.
10. Gilman J. J., ed., *The Art and Science of Growing Crystals*. Wiley, New York, 1963.
11. Willardson R. K., and H. L. Goering, eds., *Compound Semiconductors, Vol. I Preparation of III-V Compounds*. Reinhold Publ. Corp., New York, 1962.
12. Smakula A., *Einkristalle* (*Wachstum, Herstellung und Anwendung*). Springer Verlag, Berlin 1962.

Specific References

13. Hodgkinson R. J., *J. Electronics* (1955–56) 612.
14. Cadoff I., Chapter 26 of this book.
15. Westbrook J. H., and D. L. Wood, *J. Nucl. Mat.* **12** (1964) 208.
16. Turnbull D., in *Atom Movements*. ASM (1951) 129.
17. Darken L. S., and R. W. Gurry, *Physical Chemistry of Metals*. McGraw-Hill, 1953.
18. Aronson B., E. Stenberg, and J. Åselius, *Nature* **183** (1959) 1318.
19. Margrave J. L., Reference 4, p. 353.
20. Brown A., and A. Chitty, *J. Nucl. Energy B* **1** (1960) 145.
21. Mackenzie J. D., Reference 4, p. 240.
22. Hume-Rothery W., J. W. Christian, and W. B. Pearson, *Metallurgical Equilibrium Diagrams*. The Institute of Physics, London, 1953, p. 72.
23. Mackenzie J. D., Reference 4, p. 337.
24. Crandall W. B., C. H. McMurtry, and D. D. Button, *Research on Intermetallic Containers for Melting Titanium*, WADC TR56-633 (Feb. 1957).
25. Makarov E. S., and R. S. Gudkov, *Kristalografiya* **1** (1956) 650.
26. Ripley R. L., *J. Less-Common Metals* **4** (1962) 496.
27. Furuseth S., and A. Kjekshus, *Acta Chem. Scand.* **18** (1964) 1180.
28. Seybolt A. U., and J. H. Westbrook, *Plansee Proceedings* (1964) 845.
29. Allred W. P., Reference 11, p. 187.
30. Lawson W. D., *J. Appl. Phys.* **23** (1952) 495.
31. Westbrook J. H., *J. Electrochem. Soc.* **103** (1956) 54.
32. Westbrook J. H., *Trans. AIME* **209** (1957) 898.

33. Rossin P. C., *General Electric Research Lab Memo* **7** (1953) 2.
34. Kuhn W. E., Reference 2, p. 288.
35. Kuhn W. E., ed., *Arcs in Inert Atmospheres and Vacuum*. Wiley, New York, 1956.
36. Seagle S. R., R. L. Martin, and O. Bertea, *J. Metals* **14** (1962) 812.
37. Barton P. W., E. J. Hughes, and A. A. Johnson, *Proc. Int. Conf. on Electron and Ion Beam Tech.*, R. Bakish, ed. (1965).
38. Mayer S. E., and A. T. Mlavsky, "Thermal and Electrical Properties of Some Silicides," in *Properties of Elemental and Compound Semiconductors*, H. C. Gatos, ed. AIME-Interscience, New York 1960, p. 261.
39. Sterling H. F., and R. W. Warren, *Metallurgia* **67** (1963) 301.
40. Ware R. M., "The Preparation and Purification of Transition Metal Silicides," in *Ultra-Purification of Semiconductor Materials*, M. S. Brooks and J. K. Kennedy, eds. Macmillan Co., 1962.
41. Lamb D. M., and J. L. Porter, "Horizontal Zone Purification of Silicon in Non-Reactive Crucibles," in *Ultra-Purification of Semiconductor Materials*, M. S. Brooks and J. K. Kennedy, eds. Macmillan Co., 1962, p. 469.
42. Curtis F. W., *High Frequency Inductive Heating*, 2nd ed. McGraw-Hill, New York, 1950.
43. Barton P. J., and G. W. Greenwood, *J. Inst. Metals* **86** (1957/58) 504.
44. Petty E. R., *J. Inst. Metals* **89** (1960/61) 343.
45. Raynor G. V., and D. W. Wakeman, *Proc. Roy. Soc. A* **190** (1947) 82.
46. Raynor G. V., and B. J. Ward, *J. Inst. Metals* **86** (1957/58) 182.
47. Clare J. W. H., *J. Inst. Metals* **89** (1961) 232.
48. Mason D., and J. Cook, *J. Appl. Phys.* **32** (1961) 475.
49. Eisner R. L., R. Mazelsky, and W. A. Tiller, *J. Appl. Phys.* **32** (1961) 1833.
50. Nester J. F., and J. B. Schroeder, *Trans. AIME* **233** (1965) 249.
51. Barber B. J., *J. Appl. Phys.* **35** (1964) 398.
52. Brown A., and J. J. Norreys, *J. Inst. Metals* **89** (1961) 238.
53. Brown A., and J. J. Norreys, *J. Less-Common Metals* **5** (1963) 302.
54. Teitel R. J., *J. Inst. Metals* **85** (1956/57) 409.
55. Brown A., and B. Lewis, *J. Phys. Solids* **23** (1962) 1597.
56. Lundstrom T., *Acta Chem. Scand.* (in press).
57. Broshchevskii A. S., and D. N. Tretyakov, *Soviet Phys.-Solid State* **1** (1960) 1360.
58. Burghard H. C., and F. R. Brotzen, *Trans. AIME* **215** (1959) 863.
59. Gross P., D. L. Levi, and R. H. Leiven, "Activities in U-Bi Alloys and the Free Energies of Uranium Bismuth Compounds," in *Physical Chemistry of Metallic Solutions and Intermetallic Compounds I*, HMSO, London, 1959, p. 362.
60. Cunnell F. A., J. T. Edmond, and W. R. Harding, *Solid-State Electron.* **1** (1960) 97.
61. Gaus F., J. Lagrenaudie, and P. Seguin, *Compt. Rend.* **237** (1953) 310.
62. Cunnell F. A., Reference 11, p. 207.
63. Stambaugh E. P., Reference 11, p. 184.
64. Jack K. H., and M. M. Wachtel, *Proc. Roy. Soc.* **239A** (1957) 48–60.
65. Antell G. R., and D. Effer, *J. Electrochem. Soc.* **106** (1959) 509.
66. Effer D., and G. R. Antell, *J. Electrochem. Soc.* **107** (1960) 252.
67. Audrieth L. F., and J. Kleinberg, *Non-Aqueous Solvents*, Wiley, New York, 1953 p. 101.
68. Pray A. R., *Comprehensive Inorganic Chemistry V*, Sneed, M. C., and R. C. Brasted, eds. Van Nostrand, Princeton, N.J., 1956, pp. 164–166.
69. Kulifay, S. M., *J. Am. Chem. Soc.* **83** (1961) 4916.
70. Kulifay S. M., *J. Inorg. Nucl. Chem.* **25** (1963) 75.
71. Benzing W. C., I. B. Conn, J. V. Magee, and E. J. Sheehan, *J. Am. Chem. Soc.* **80** (1958) 2657.
72. Russell A. S., T. R. Kennedy et al., *J. Chem. Soc.* (1932) 841, 852, 857, 2340; *ibid.* (1934) 1750.
73. Fitzer E., and P. Gerasimoff, *Z. Metallk.* **50** (1959) 187.
74. Brown A., unpublished research.
75. Brown A., and J. J. Norreys, *Nature* **183** (1961) 673.
76. Brown A., and J. J. Norreys, *Nature* **191** (1962) 61.
77. Brown A., *Acta Cryst.* **15** (1962) 652.
78. Antell G. R., Reference 11, p. 288.
79. Fitzer E., *Berg- Huettenmaenn. Monatsh.* **97** (1952) 81.
80. Seybolt A. U., unpublished research.
81. Sinel'nikova V. S., *Tsvetn. Metal.* **10** (1964) 69.
82. Andrieux J. L., *J. Four Elec.* **57** (1948) 54.

83. Hartmann H., and W. Massing, *Z. Anorg. Chem.* **266** (1951) 98.
84. Cook N. C., *Ceramic Bonding to Diffusion Layers of Silicides and Borides on Metals*, Am. Ceram. Soc. Mtg. Chicago, 1964.
85. Leidheiser Jr., H., and T. G. Carver, *J. Electrochem. Soc.* **109** (1962) 68–69.
86. Uhlig H. H., J. S. MacNairn, and D. A. Vaughn, *Acta Met.* **3** (1955) 302.
87. Clarke M., and S. C. Brittan, *Corrosion Sci.* **3** (1963) 207.
88. Oriani R. A., *Acta Met.* **2** (1954) 343.
89. Bransky J., and P. S. Rudman, *Trans. ASM* **55** (1962) 335.
90. Dautreppe D., J. Langler, J. Pauleve, and L. Néel, *Radiation Damage in Solids II.* Int. Atomic Energy Agency, Vienna, 1962, p. 21.
91. Shalvo S. T., *Fiz. Metal. i Metalloved* **17** (1964) 633.
92. Wang F. E., A. M. Syeles, W. L. Clark, and W. J. Buehler, *J. Appl Phys.* **35** (1964) 3620.
93. Jayaraman A., R. C. Newton, and G. C. Kennedy, *Nature* **191** (1961) 1288.
94. Musgrove M. J. P., and J. A. Raple, *Phys. Chem. Solids* **23** (1962) 321.
95. Minomura S., and H. G. Drickamer, *Phys. Chem. Solids* **23** (1962) 451–456.
96. Samara G. A., and H. G. Drickamer, *Phys. Chem. Solids* **23** (1962) 457–461.
97. Poniatovsky E. H., and H. I. Peresada, *Dokl. Akad. Nauk SSSR* **144,** No. 1 (1962) 129–131.
98. Tittmann B. R., A. J. Darnell, H. E. Bommel and W. F. Libby, *Phys. Review* **135,** No. 5A (1964) A1460–A1462.
99. McWhan D. B., G. W. Hull, T. R. R. McDonald, and E. Gregory, *Science* **147** (1965) 1441.
100. Cannon P., and E. T. Conlin, *Science* **145,** No. 3631 (1964) 487–489.
101. Bames M., *Science* **147** (1965) 732.
102. Adda Y., M. Beyeler, A. Kirpianenko, and B. Pennot, "Influence of Pressure on the Diffusive Formation of Intermetallic Compounds," *Proc. 4th Int. Symp. on Reactivity of Solids*, Amsterdam (1960). Elsevier, 1961, p. 512.
103. Mullin J. B., Reference 11, p. 365.
104. Westbrook J. H., *Metallurgical Rev.* **9** (1964) 415.
105. Goryunova N. A., S. I. Radautsan, and V. I. Deryabina, *Fiz. Tverd. Tela* **1** (1959) 512; *Soviet Phys.-Solid State* **1** (1959) 460.
106. Wood D. L., and J. H. Westbrook, U.S. Air Force Contract WADD Tech. Report 60-184, Part I (Apr. 1960).
107. Wood D. L., and J. H. Westbrook, *Trans. AIME* **224** (1962) 1024.
108. Miller J. F., A. L. Goering, and R. C. Himes, *J. Electrochem. Soc.* **107** (1960) 527.
109. Wittry D. B., J. M. Axelrod, and J. O. McCaldin, in *Properties of Elemental and Compound Semi-Conductors*, H. C. Gatos, ed. AIME-Interscience, New York, 1960.
110. Wooley J. C., B. A. Smith, and D. G. Lees, *Proc. Phys. Soc. (London)* **B69** (1956) 1339.
111. Taylor A., *X-Ray Metallography*, Wiley, New York, 1961, p. 368.
112. Weisberg L. R., F. D. Rosi, and P. G. Herkart, "Materials Research on GaAs and InP," in *Properties of Elemental and Compound Semiconductors*, H. C. Gatos, ed. AIME-Interscience, New York, 1960, p. 25.
113. Prior W. M., *J. Electrochem. Soc.* **108** (1961) 82.
114. Lawson W. D., and S. Nielson, Reference 10, p. 365.
115. Lawson W. D., *J. Appl. Phys.* **22** (1951) 1444.
116. Kroger F. A., and D. deNobel, *J. Electron.* **1** (1955) 190.
117. Miller S. E., Reference 11, p. 274.
118. Strong J., *Phys. Rev.* **36** (1930) 1663.
119. Gremmelmaier R., *Z. Naturforsch.* **119** (1956) 511.
120. Tannenbaum M., in *Semiconductors*, N. B. Hannay, ed. Reinhold Publ. Corp. 1960, p. 87.
121. Pfann W. G., *Trans. AIME* **194** (1952) 747.
122. Geach G. A., and F. O. Jones, *J. Less-Common Metals* **1** (1959) 56.
123. Carlson O. N., F. A. Schmidt, and W. M. Paulson, *Trans. ASM* **57** (1964) 356.
124. Calverley A., M. Davis, and R. F. Lever, *J. Sci. Instr.* **34** (1957) 142.
125. Allred W. P., Reference 11, p. 266.
126. Van den Boomgaard, J., *Phillips Res. Rept.* **10** (1955) 319.
127. Van den Boomgaard J., *Philips Res. Rept.* **11** (1956) 27.
128. Van den Boomgaard J., *Phillips Res. Rept.* **11** (1956) 91.
129. Heumann F. K., *J. Electrochem. Soc.* **109** (1962) 345.
130. Lorenz M. R., and R. E. Halsted, *J. Electrochem. Soc.* **110** (1963) 343.
131. Wood D. L., unpublished research.

132. Scanlon W. W., "Stoichiometry in Compound Semiconductors," in *Properties of Elemental and Compound Semiconductors*, H. C. Gatos, ed. AIME-Interscience, New York, 1960, p. 185.
133. Van den Boomgaard J., and K. Schol, *Philips Res. Rept.* **12** (1957) 127.
134. Brebrick R. F., *J. Chem. Phys.* **36** (1962) 183.
135. Lorenz M. R., *J. Phys. Chem. Solids* **23** (1962) 939.
136. Margrave J. L., Reference 4, p. 369.
137. Von Wazer, *Phosphorous and Its Compounds I.* Interscience, New York, 1958, pp. 106–117.
138. Abrahams M. S., and L. Ekstrom, in *Properties of Elemental and Compound Semiconductors*, H. C. Gatos, ed. Interscience, New York, 1960, p. 225.
139. Lindburg O., and J. W. Faust, Reference 11, p. 294.
140. Faust J. W., H. Nicholson, and R. Moss, *Electrochem. Soc.*, *Electronics Division Abstracts* **9** (1960) 165.
141. Riebling E. F., and W. W. Webb, *Science* **126** (1957) 309.
142. Hertzberg R. W., and F. D. Lemkey, *Mechanical Behavior of Lamellar (Al–Cu) and Rod Type (Al–Ni) Unidirectionally Solidified Eutectic Alloys*, ASM Fall Meeting (1963).
143. Ford J. A., and R. W. Hertzberg, *United Aircraft Corp. Rept.* B-9160068-1, 2 (1963).
144. Davies G. J., *Phil. Mag.* **9** (1964) 953.
145. Holland B., *Vacuum Deposition of Thin Films.* Chapman and Hall Ltd., London, 1956.
146. Neugebauer C. A., J. B. Newkirk, and D. A. Vermilyea, eds., *Structure and Properties of Thin Films.* Wiley, New York, 1959.
147. Gunther K. G., Reference 11, p. 313.
148. Bate G., and K. N. R. Taylor, Reference 11, p. 337.
149. Pietrokowsky P., *J. Sci. Instr.* **34** (1963) 445.
150. Bando Yoshichika, *J. Phys. Soc. Japan* **16** (1961) 2342–2343.
151. Savitskii E. M., and V. V. Baron, *Doklady Akad. Nauk SSSR* **64** (1949) 649.
152. Wood D. L., in *Mechanical Properties of Intermetallic Compounds*, J. H. Westbrook, ed. Wiley, New York, 1959.
153. Tanner L. E. et al. *Man Labs. Report ASD TDR* 62-1087, Parts I-III (1962–64).
154. Mannas D. A., and J. P. Smith, *J. Metals* **14,** No. 8 (1962).
155. Wood D. L., and J. H. Westbrook, *U.S. Air Force Contract WADD Tech. Rpt.* 60-184, Part II (May 1961).

chapter 18

Metastable Phases

POL DUWEZ
W. M. Keck Laboratory of Engineering Materials
California Institute of Technology
Pasadena, California

1. INTRODUCTION

Metastable phases in alloys are very familiar to metallurgists. In fact, most of the alloys used every day in practical applications are metastable. Quenched steels and precipitation-hardened aluminum alloys are striking illustrations of this statement. Until recently, however, metastable structures in alloys have always been obtained by quenching from the solid state, and little effort has been spent on what could be done by retaining some of the features of the structure of liquid alloys by extreme rates of cooling through the liquidus-solidus transition. Because diffusion rates are very high in the liquid state, the traditional rates of quenching are probably several orders of magnitude too small to achieve any spectacular changes in the transition from the liquid to the solid state. Recent developments in techniques for rapid cooling from the liquid state have demonstrated that unusual metastable structure could indeed be obtained. The main object of Section 2 of this chapter is to summarize the results obtained so far.

Another rather recent development is the application of high-pressure techniques to alloy systems. Most of the work in this field has been concerned with the effect of pressure on phase changes (effect of pressure on equilibrium diagrams). In some cases, however, polymorphic transformations have been found in intermediate phases. This question is treated in Section 3.

2. METASTABLE PHASES OBTAINED BY QUENCHING FROM THE MELT

2.1 Experimental Techniques

The traditional techniques used in metallurgy for retaining metastable phases in solid alloys by quenching from high temperature make use of the high rates of heat transfer from a solid to a fluid, either gas or liquid. In this case, the heat is transferred to the cooling medium by convection. The rate of heat transfer can be increased by increasing the velocity of the quenching fluid in contact with the solid. The extension of this method to liquid alloys is difficult because, beyond a rather limited velocity of the cooling fluid, the liquid globule disintegrates into a fine powder. Recently developed techniques for rapidly cooling liquid alloys are based on the principle of conduction rather than convection heat transfer. The basic principle is to force a thin layer of the liquid to solidify when suddenly placed in contact with a metallic surface having a high thermal conductivity. A simple apparatus based on this principle was used for obtaining a complete series of solid solutions in copper-silver alloys which, under normal conditions, solidify in two-phase alloys.[1] The technique consists of propelling a small globule of liquid against the inside surface of a copper-lined cylinder rotating at high speed. The globule is accelerated by means of a shock wave and strikes the cylindrical surface at an angle of about 15°. The rotation of the cylinder (in a direction opposite to that of the globule) provides for additional spreading of the liquid alloy, and also introduces a centrifugal force that results in a good thermal contact between the molten metal and the cooling target. The foils obtained by this technique are in general very thin (ranging from a few microns down to 0.2 microns) and are very uneven and discontinuous.

An improved version of this fast-cooling apparatus involving a vertical "gun" and a nonrotating target has been described in Reference 2. The layer of metal deposited by the "gun" technique is about 1 cm wide and 2 or 3 cm long. In general, the foil does not adhere strongly to the target and can be easily removed for structure analysis by X-ray diffraction methods. Physical properties (electrical, magnetic, etc.) are difficult to measure because of the lack of thickness uniformity, and mechanical properties are practically impossible to study.

Another approach to the rapid-cooling technique was described recently in Reference 3. This technique is also based on conduction cooling, but, instead of accelerating the liquid globule on a fixed copper target, a free-falling globule is caught between a fixed anvil and a piston moving at a high speed. The result is a foil of solidified material of uniform thickness (varying from 25 to 100 microns) and as large as one inch in diameter. This new technique will make possible detailed studies of both physical and mechanical properties of metastable phases obtained by rapid cooling from the liquid state.

The actual rate of cooling achieved with either the "gun" or the "piston and anvil" techniques is still unknown, and its measurement presents a challenge to the imagination of experimental physicists. Based on the fact that the entire solidification process takes place in less than a few milliseconds in either technique, the order of magnitude of the average rate of cooling from the liquid state to room temperature is approximately two million degrees C per second. The rate at which the liquidus-solidus boundary is crossed may be even higher. In any case, whatever the actual rate of cooling is, it is high enough to result in very unusual structures, which, in many cases, do not have any obvious relationships with the equilibrium phases.

2.2 Metastable Crystalline Intermediate Phases

Metastable intermediate phases have been found under conventional rates of cooling. In cadmium-antimony alloys, for example,[4] it is well known that two phase diagrams may be established depending on the rate of cooling. The intermediate phase CdSb forms under very slow cooling rates, and Cd_3Sb_2 results at somewhat faster rates. When the new techniques for achieving extreme rates of cooling are used, metastable intermediate phases appear to be the rule rather than the exception.

The first new phase* obtained after quenching from the liquid state was found in silver-germanium alloys.[5] In this system (of the eutectic type under normal conditions) a hexagonal close-packed structure can be obtained between approximately 15 and 26 at. % germanium. In addition, the terminal silver-rich solid solution was extended from 8 to 13.5 at. % germanium. The lattice parameters of the hexagonal phase vary with composition, and a still unexplained rapid change in the c/a ratio was found around 22 at %. germanium.[6] The hexagonal phase appears to be stable up to about 150°C, where decomposition into the equilibrium phase takes place by a process of nucleation and growth.

A similar hexagonal phase has been found in gold-germanium alloys,[7] which, under

* A summary of metastable phases obtained so far is given in Table 1.

TABLE 1

Metastable Phases Obtained by Rapid Cooling from the Liquid State

Alloy Systems	Equilibrium Phases	Metastable Phases	Reference
Ag-Ge	Eutectic	hcp phase between 15 and 26 at.% Ge	5, 6
Au-Ge	Eutectic	hcp intermediate phase around 25 at.% Ge plus second phase having complex crystal structure	7
Te-Au	Two eutectics with Te_2Au phase	Simple cubic phase between 10 and 45 at.% Ag	8
Te-Ag	Two eutectics with $TeAg_2$	Simple cubic phase between 10 and 35 at.% Ag	8
Te with Ge, Al, Ga, and In	Two eutectics with intermediate phases	Amorphous structures between 10 and 25 at.% Ge, around 8 at.% Al, 10 to 30 at.% Ga, and 10 to 30 at.% In	12, 13
Pd-Si	Eutectic between Pd and Pd_2Si with Pd_3Si by peritectic reaction	Amorphous structure between 15 and 20 at.% Si	To be published

normal conditions, are also of the eutectic type. In this case, however, the alloys are not single-phase, and the X-ray diffraction patterns indicate the presence of another phase that has a very complex diffraction pattern. Because this pattern contains low-angle reflections corresponding to spacings as high as 8 Å, the unit cell must be large, and no indexing has been possible so far.

Intermediate phases with a most unusual crystal structure were found in alloys of tellurium with gold and with silver.[8] Under equilibrium conditions these alloy systems consist of two eutectics separated by the intermetallic compounds Te_2Au or $TeAg_2$. After rapid cooling from the melt, a simple cubic phase with one atom per unit cell was found between 10 and 45 at.% gold and 10 and 40 at.% silver. The only element crystallizing in such a structure is polonium, and no alloy phases have been reported with this crystal-structure type. A detailed study was made of the relative intensities of the diffraction peaks in the Te–Au alloy containing 37.5 at.% Au, and the agreement between observed and calculated intensities was satisfactory. The effect of gold and silver concentration on the lattice parameter of the simple cubic phases is shown in Figs. 1 and 2. It is interesting to note the difference in slopes of lattice parameters versus either gold or silver concentration. The usually accepted atomic sizes of gold and silver are almost equal; however, this is not true

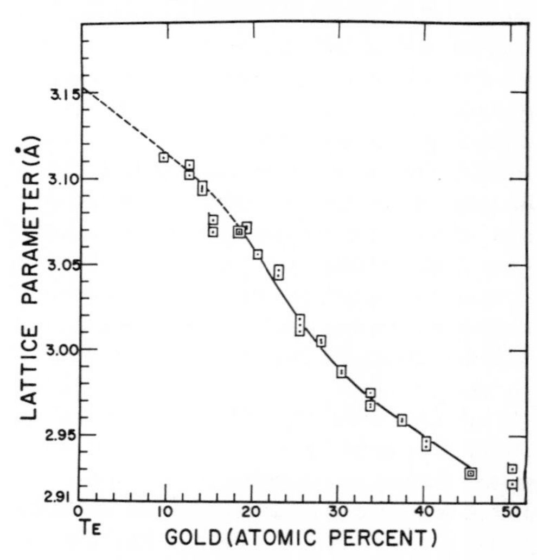

Fig. 1. Lattice parameter versus composition of metastable simple cubic phases in tellurium-gold alloys.

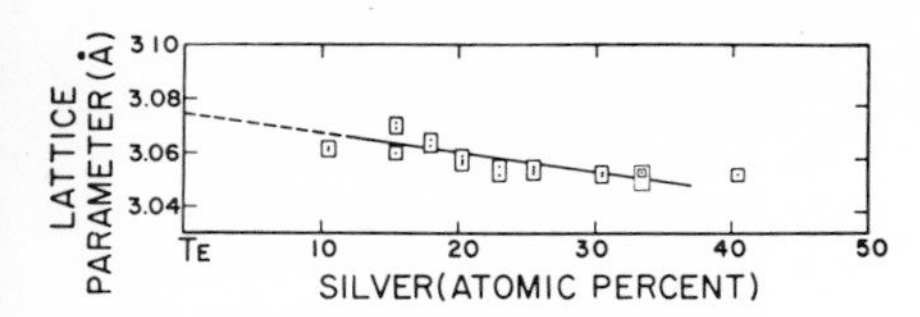

Fig. 2. Lattice parameter versus composition of metastable simple cubic phases in tellurium-silver alloys.

when they are alloyed with tellurium in the simple cubic structure.

Decomposition of the simple cubic structures occurs at temperatures as low as 110°C, but the alloy containing 37.5 at.% gold was stable for at least 10 minutes as high as 165°C. The occurrence of a simple cubic phase in tellurium-gold and tellurium-silver alloys is tentatively explained in Reference 13, based on the fact that the crystal structure of pure tellurium can be considered a distorted simple cubic structure, as proposed by Von Hippel.[9]

An interesting aspect of the application of rapid-cooling techniques from the melt to the study of alloy phases is that it has already led to the synthesis of phases which, although expected to be present under equilibrium conditions, were obtained only after fast cooling. In alloys involving silver and the rare-earth metals, many intermediate phases with the CsCl crystal structures exist at equiatomic percentages. The search for similar phases was recently extended to gold and rare-earths alloys, and ten CsCl phases were discovered in these alloys.[10] However, five of these phases (Au with Y, Pr, Nd, Sm, and Gd) could only be obtained in the alloys quenched from the liquid state and were not present under equilibrium conditions. The number of CsCl-type phases in silver and gold alloys with the rare-earth elements is now about twenty-three and, as shown in Fig. 3, a correlation exists between the lattice parameter of these phases and the trivalent ionic radius of the rare-earth metals. This correlation is discussed in Reference 10.

2.3 Noncrystalline Intermediate Phases

Amorphous solids, such as glass, can be obtained by relatively slow cooling from the liquid state. When very high rates of cooling are used, it appears that some combinations of elements, which normally would not be expected to result in glassy structures, can actually be forced to solidify in an amorphous state. It should perhaps be mentioned here that amorphous solids, including metals and alloys, have been obtained in thin-film form by vacuum deposition. When the substrate is cooled to liquid helium temperature during deposition, the chances of obtaining amorphous structures are enhanced, and several binary metallic alloys have been obtained in vitreous form by this technique.[11] These thin films are extremely unstable and crystallize at temperatures as low as 25°K. The techniques for rapid cooling from the melt have been successfully used for preparing amorphous binary alloys, and, contrary to what has been observed in vapor-deposited alloys, the rapidly quenched foils are relatively stable. In several cases crystallization does not start below about 125°C, and in some cases amorphous structures are stable up to about 300°C.

A large number of amorphous phases have been found in binary alloys involving tellurium with Ge, Ga, and In.[12] The concentration limits for these various tellurium-base alloys are as follows: Ge, from 10 to 25 at,%; Ga and In, from 10 to 20 at.%. The amorphous tellurium-germanium alloys have been the most extensively studied ones so far. The X-ray diffraction patterns of the amorphous phases are very similar to those of the liquid state, and a typical microphotometer trace of a diffraction pattern for an alloy containing 20 at.% germanium is shown in Fig. 4. This pattern was obtained with molybdenum K_α radiation monochromatized by the (200) planes of a

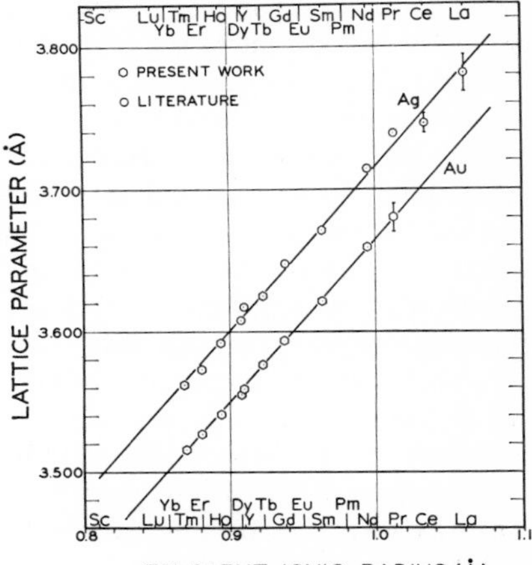

Fig. 3. Correlation between lattice parameter and trivalent ionic radius in CsCl structures in alloys of gold or silver with rare-earth elements.

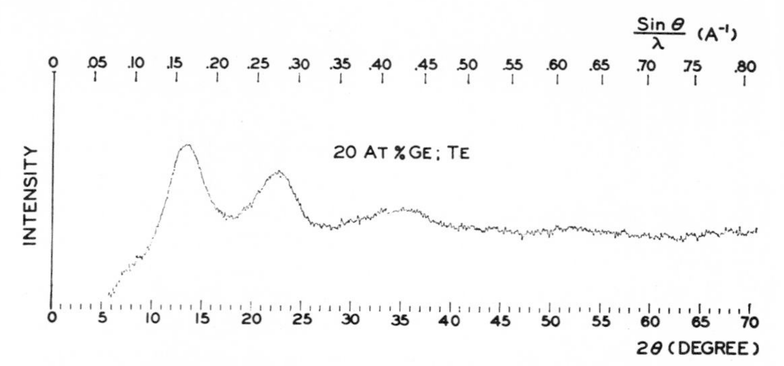

Fig. 4. Microphotometer trace of an X-ray diffraction pattern of a tellurium–20 at.% germanium amorphous alloy (Monochromatic Mo K_α radiation).

lithium fluoride crystal. The radial distribution function for this particular alloy has been analyzed, and results indicate that the amorphous structure can be correlated with that of solid tellurium.[13] In hexagonal tellurium, there are two nearest-neighbors at a distance of 2.86 Å, four atoms at 3.47 Å, and eight atoms at about 4.46 Å. The radial distribution function of the amorphous Te–Ge alloys indicates that 1.6 atoms are at a distance between 2.65 and 2.75 Å, and 12 atoms are at distances between 4.05 and 4.20 Å. These next-to-nearest neighbors are at approximately the average distance of the 12 next-to-nearest neighbors in solid tellurium.

The mechanism of crystallization of some of the tellurium-germanium alloys was studied by Willens.[14] Because the thickness of amorphous foils rapidly quenched from the melt varies within a given foil from a few microns down to less than 0.1 micron, it is possible to find regions that are thin enough for direct transmission electron microscopy without subjecting the foil to any mechanical or chemical thinning. As expected, no structure was detected in the foils as quenched. A specimen was then heated within the microscope and, as soon as diffraction spots appeared, the heating was stopped. At this stage, microscopy revealed the presence of dendrites, as shown in Fig. 5. This is very similar to the crystallization process in a supercooled liquid or in amorphous solids such as glass.

The possibility of quenching tellurium-base alloys into amorphous solids might have been anticipated because tellurium is very similar to selenium, and pure selenium can be obtained in an amorphous state by relatively fast cooling. Less predictable results, however, were obtained recently when amorphous structures were discovered in palladium-silicon binary alloys. Under equilibrium conditions, the palladium-rich side of the palladium silicon system consists of a eutectic between palladium and Pd_2Si with Pd_3Si present in-between as a result of a peritectic reaction. The eutectic occurs at about 15.5 at.% Si at a relatively low temperature (780°C) compared to that of palladium (1552°C). Within the range of concentrations from 15 to about 20 at.% Si, amorphous structures can be retained by rapid cooling from the melt. Preliminary results indicate that these amorphous alloys are metallic in character. Their electrical resistivity is only two or three times that of purest palladium, and decreases with decreasing temperature down to 2.3°K. The fast-quenching technique from the liquid state has, therefore, made possible the synthesis of amorphous solids having a metallic character.

3. METASTABLE PHASES OBTAINED UNDER HIGH PRESSURE

The pioneering work of Bridgman in the field of high pressures clearly demonstrated that polymorphic transitions can be obtained at room temperature, provided that the pressure was sufficiently high. The phases existing at high pressure are actually stable phases, but in most cases they revert to the equilibrium atmospheric phase when the pressure is released. If they can be retained at atmospheric pressure, they are indeed metastable.

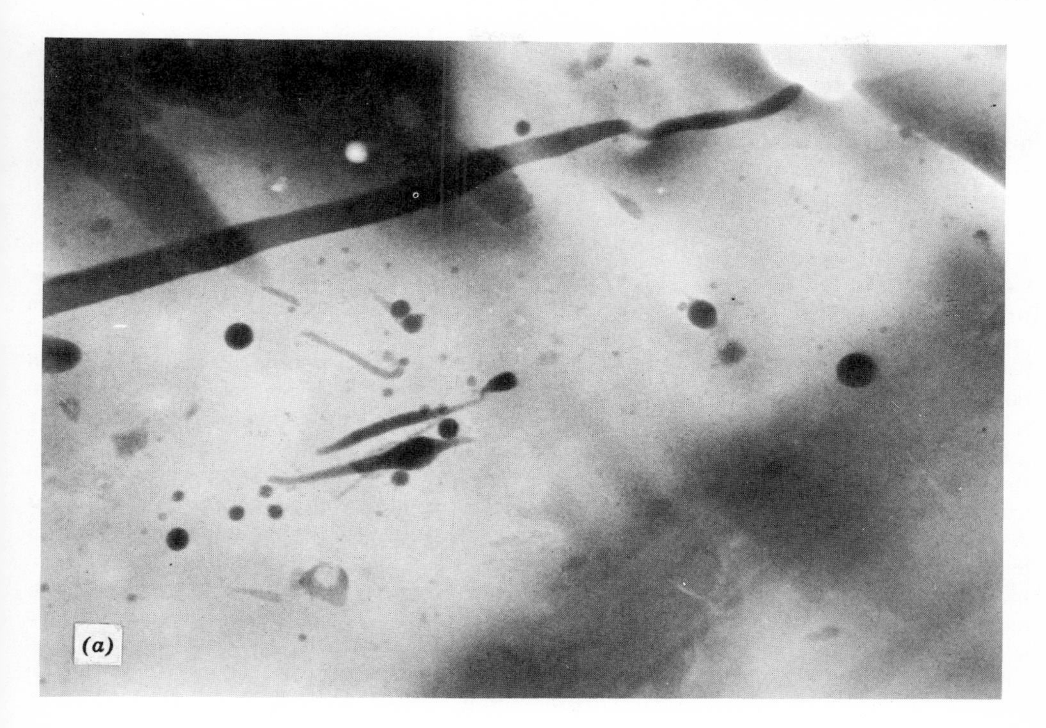

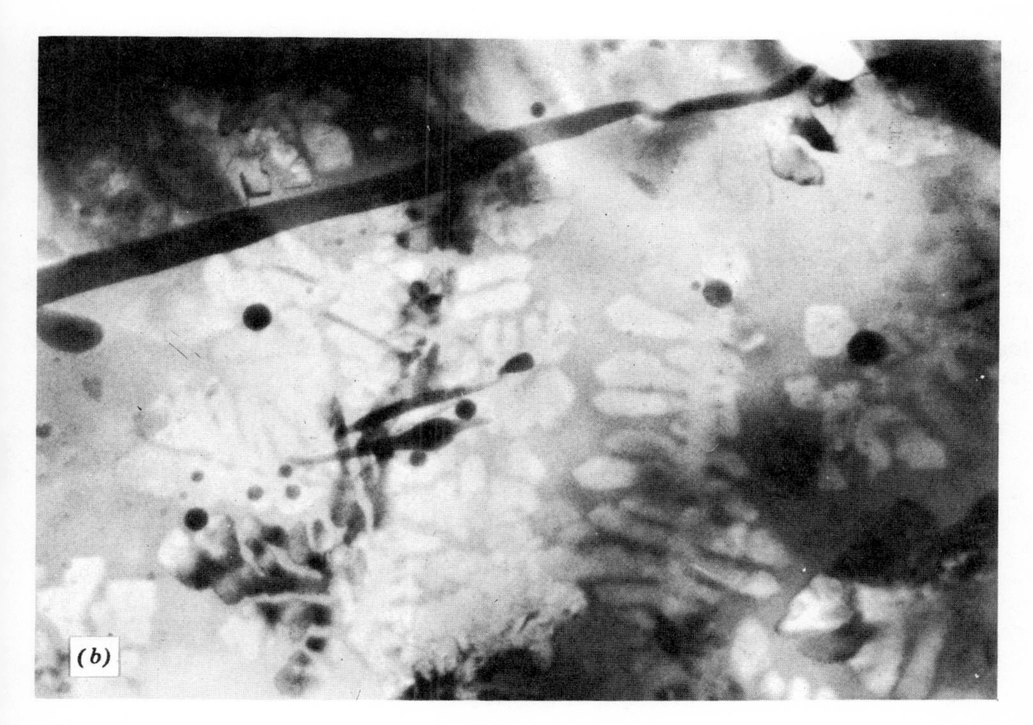

Fig. 5. Electron micrograph of a tellurium–15 at.% germanium alloy: (*a*) amorphous state, (*b*) after heating, showing evidence for dendritic growth. (Magnification about 80,000 ×.)

The retention of phases stable at high temperature is made possible by very rapid cooling either from the solid or the liquid state. The corresponding approach to phases stable at high pressure would require a very rapid pressure release, which might be called "pressure quenching." Suitable technique for achieving this goal might be difficult, but there are no physical principles against this possibility, and it is hoped that some "invention" will eventually achieve this goal. At present, only a few phases found under high pressure have been retained at atmospheric pressure, but this field of research is still unexplored. It is indeed significant that most of the high pressure phases have been reported in a steady stream of short papers published within the last three years.

A relatively small number of intermediate phases have been found to undergo a polymorphic change under high pressure. In fact, most of the reported interesting phase changes involve elements of the second and third column of the periodic table, combined with those of the sixth and fifth respectively (so-called III–V and II–VI compounds). These compounds and a few others are discussed in the following sections.

3.1 High-Pressure Phases in III-V and II-IV Compounds

Most of the III–V compounds crystallize into the ZnS (zinc blende) cubic structure, which is similar to that of diamond but contains two different atoms instead of one. The lattice of diamond does not represent a close packing of atoms, and it is intuitively logical to expect that high pressure would favor a different crystal structure bringing the atoms closer together. The possibility of transforming gray tin (diamond structure) into the white-tin structure with an increase in density of about 21 per cent was established by Bridgman.[15] More recently, transition from a diamond to a white-tin structure has been reported for germanium and silicon.[16,17]

By analogy, it was anticipated that the III–V and II–VI compounds might transform into a more densely packed structure under high pressure. This has been found to be true for a number of compounds,[18–23] and a partial list of the results published recently is given in Tables 2 and 3. By taking X-ray diffraction patterns while the solid is under pressure, the high-pressure phase was found to be either that of white tin or sodium chloride. In all cases a large volume change was found, as shown in Table 2. In addition (as in the case of (tin the change in electrical resistivity between the low-pressure and the high-pressure phases indicate a shift from semiconductivity to metallic conductivity.

The possibility of retaining the white-tin polymorphic form of InSb at atmospheric pressure has been reported.[20] When the compound was cooled under a pressure of 23 Kb down to liquid nitrogen temperature and the pressure then released, the metallic form of InSb was stable for weeks at −63°C. The lattice parameters of the metallic form were measured at room temperature and were $a = 5.72$ and $c = 3.18$ Å. These values are in poor agreement with those listed in Table 2, which were measured under pressure. Further work will probably remove these inconsistencies.

One of the most striking examples of what can be expected when the type of atomic

TABLE 2

High-Pressure Phases in III-V Compounds

Compound	Crystal Structure (1 atm)	Crystal Structure (High Pressure)	Transition Pressure (Kb)	Lattice Parameters (Å) High Pressure Phase		Volume Change V_p/V_o
				(a)	(c)	
AlSb	ZnS blende	White tin	125	5.375	2.892	0.867
GaSb	ZnS blende	White tin	90	5.348	2.937	0.903
InSb	ZnS blende	White tin	22	5.357	2.970	—
GaAs	ZnS blende	—	240?	—	—	—
InAs	ZnS blende	NaCl	102	5.514	—	0.926
InP	ZnS blende	NaCl	133	5.710	0	0.920

TABLE 3

High-Pressure Phases in II-VI Compounds[21,22]

Compound	Crystal Structure (1 atm)	Crystal Structure (High Pressure)	Transition Pressure (Kb)	Lattice Parameters (Å) High-Pressure Phase	
				(*a*)	(*c*)
HgS	ZnS blende	HgS cinnabar	—	4.15	9.50
HgSe	ZnS blende	HgS cinnabar	7	4.32	9.68
HgTe	ZnS blende	HgS cinnabar	14	4.46	9.17
CdS	ZnS wurtzite	NaCl	20	5.32	—
CdSe	ZnS wurtzite	NaCl	23	5.49	—
CdTe	ZnS blende	NaCl	33	5.81	—

bonding is changed by the application of high pressure is the discovery of superconducting properties in the high-pressure metallic form of InSb. The superconducting transition of this compound is about 2.1°K, according to Reference 20.

Most of the II–VI compounds crystallize in the ZnS cubic (blende) or ZnS hexagonal (wurtzite) structures. Many of these compounds undergo a polymorphic transformation at high pressure to a form which is either that of HgS (cinnabar type) or NaCl.[21,22] In all cases there is a definite increase in density and a sharp increase in electrical conductivity.

3.2 Other Compounds

Another interesting and quite unusual compound has been described recently.[23,24] Under a pressure of about 32 kb, indium telluride (InTe) undergoes a phase change from a tetragonal TlSe structure to a cubic NaCl structure; the temperature-pressure diagram is given in Reference 23. The most interesting feature about this high-pressure phase is that it can be retained as a metastable phase at atmospheric pressure, as described in Reference 23. According to Reference 24, the crystal structure of the high-pressure phase would be simple cubic with an edge side of 3.056 Å. This conclusion was refuted in Reference 25, in which a cubic NaCl-type crystal structure, with a parameter equal to 6.154 Å, is proposed. In the two studies, the observed spacings from powder patterns are essentially the same, and it is obvious that the unit cell of the simple cubic structure proposed in Reference 24 is exactly half that of the NaCl structure. It is most probable that the NaCl structure is the correct one.

The transition from ordinary InTe to the high-pressure phase is accompanied by a sharp decrease in electrical resistivity, and the high-pressure form is metallic with a resistivity of the order of 6×10^{-5} ohm-cm. In addition, this metastable phase is superconducting with a transition temperature that is not yet well-established.

According to Reference 26, the zero-field transition temperature is 2.18°K, with a very sharp transition width of 0.01°K. According to Reference 25, the transition temperature would be about 3.5°K. Obviously, additional studies are required to reconcile these differences, but it is quite certain that metastable InTe is a superconductor.

The stability of metallic InTe at room temperature is also a still unsolved problem. According to Reference 26, this compound is stable up to about 125°C. In Reference 25, however, it is stated that after four months about 20 per cent of a specimen was transformed at room temperature. The transformation was 75 per cent complete after one hour at 100°C, and no appreciable stability was found above 125°C.

REFERENCES

1. Duwez Pol, R. H. Willens, and W. Klement, Jr., *J. Appl. Phys.* **31** (1960) 1136.
2. Duwez Pol, and R. H. Willens, *Trans. AIME* **277** (1963) 362.
3. Pietrokowsky P., *J. Sci. Instr.* **34** (1963) 445.

4. Hansen M., and K. Anderko, *Constitution of Binary Alloys*. McGraw-Hill Book Company, New York, 1958.
5. Duwez Pol, R. H. Willens, and W. Klement, Jr., *J. Appl. Phys.* **31** (1960) 1137.
6. Klement W., Jr., *J. Inst. Metals.* **90** (1961) 27–30.
7. Luo H. L., and W. Klement, Jr., U.S. At. Energy Comm., TID 20489 (1964).
8. Luo H. L., and W. Klement, Jr., *J. Chem. Phys.* **36** (1962) 1870–1874.
9. von Hippel A., *J. Chem. Phys.* **16** (1948) 372.
10. Chao C. C., H. L. Luo, and Pol Duwez, *J. Appl. Phys.* **35** (1964) 257–258.
11. Buckel W., *Z. Physik* **138** (1954) 136.
12. Luo H. L., and Pol Duwez, *Appl. Phys. Letters* **2** (1963) 21.
13. Luo H. L., Ph.D. Thesis, California Institute of Technology, 1961.
14. Willens R. H., *J. Appl. Phys.* **33** (1962) 3269–3270.
15. Bridgman P., *Physics of High Pressure*. Bell and Sons, 1949.
16. Jamieson J. C., *Science* **139** (1963) 762.
17. Minomura S., and H. G. Drickamer, *J. Phys. Chem. Solids* **23** (1962) 451.
18. Jayaraman A., R. C. Newton, and G. C. Kennedy, *Nature* **191** (1961) 1288.
19. Jamieson J. C., *Science* **139** (1963) 845.
20. Bommel H. E., A. J. Darnell, W. F. Libby, and B. R. Tittman, *Science* **139** (1963) 1301.
21. Mariano A. N., and E. P. Warekois, *Science* **142** (1963) 672.
22. Jayaraman A., W. Klement, Jr., and G. C. Kennedy, *Phys. Rev.* **130** (1963) 2277–2283.
23. Darnell A. J., and W. F. Libby, *Science* **139** (1963) 1301.
24. Darnell A. J., A. J. Yencha, and W. F. Libby, *Science* **141** (1963) 713.
25. Banus M. D., R. E. Hanneman, M. Strongin and K. Gooen, *Science* **142** (1963) 662.
26. Bommel H. E., A. J. Darnell, W. F. Libby, B. R. Tittman, and A. J. Yencha, *Science* **141** (1963) 714.

ADDENDUM

The subject field of this chapter has been moving very rapidly. Since the original preparation of this manuscript, a number of valuable papers have appeared which are not discussed. The most important of these are listed below as References 27–38.

27. Kane R. H., B. C. Giessen, and N. J. Grant, *Acta Met.* **14** (1966) 605.
28. Klement W., Jr., *Trans. Met. Soc. AIME* **233** (1965) 1182.
29. Willens R. H., and E. Buehler, *Appl. Phys. Letters* **7** (1965) 25.
30. Anantharaman T. R., *Physica Status Solidi* **10** (1965) K3.
31. Linde R. K., *Trans. Met. Soc.* **236** (1966) 58.
32. Willens R. H., and E. Buehler, *Trans. Met. Soc. AIME* **236** (1966) 171.
33. Duwez Pol, R. H. Willens, and R. C. Crewdson, *J. Appl. Phys.* **36** (1965) 2267.
34. Predecki P., B. C. Giessen, and N. J. Grant, *Trans. Met. Soc. AIME* **233** (1965) 1438.
35. Tsuei C. C., and Pol Duwez, *J. Appl. Phys.* **37** (1966) 435.
36. Moss M., D. L. Smith, and R. A. Lafever, *Appl. Phys. Letters* **5** (1964) 120.
37. Anantharaman T. R., H. L. Luo, and W. Klement, Jr., *Trans. Met. Soc. AIME* **233** (1965) 2014.
38. Klement W., Jr., *Trans. Met. Soc. AIME* **233** (1965) 1180.

chapter 19

Constitution of Metallide Systems

И. Корнилов

I. I. KORNILOV

Baikov Institute of Metallurgy
Moscow, U.S.S.R.

1. METALLOCHEMISTRY AND METALLIDES

1.1 Definition

Metallic alloys, as far as their structure is concerned, consist mainly of solid solutions and metallic compounds. These structural constituents are found in alloys either as a homogeneous single phase or as part of a heterogeneous mixture in different combinations. In systematic investigations of the structure and properties of alloys metallic compounds are of special interest; they are also the basis for the creation of materials with special physical properties. We encounter among them materials with high hardness and with semiconductor, superconductor, magnetic and other properties.

Scientific investigations of equilibrium systems of metallic alloys began with the discovery (1876–1878) of J. Gibbs' phase rule[1] and with the foundation of the theory of physicochemical analysis of N. S. Kurnakov[2,3] at the beginning of the twentieth century. D. I. Mendeleyev's periodic law[4,5] plays a significant part in investigations of the general character of chemical interaction among metals. The periodic system of chemical elements, created by him on the basis of this law, serves as the basis of development of scientific chemistry. It also helps to establish the general rules for the formation of solid solutions and compounds in metallic systems. Further application of the ideas of the periodic law of chemical elements and of physicochemical analysis to the study of the chemical interaction of metals has led to the creation of a special scientific discipline—chemistry of metallic alloys[3,6] or metallochemistry.[7,8] It develops at the juncture of inorganic and physical chemistry, physical metallurgy, and solid state physics.

The general problems of metallochemistry have been formulated as follows: "Investigation of the chemical interaction of metallic elements of the periodic system with each other or with metalloids when the latter form solutions and compounds with interatomic bonds of a metallic character, and the establishment of general regularities of the formation, structure and properties of a broad class of solid solutions and intermetallic compounds."[7] These regularities refer to the interaction of the more than eighty metallic elements of the periodic system. Many modern problems of metallochemical reactions were considered in reports at a special symposium on the physical chemistry of metallic solutions and compounds in London in 1959.[9]

The subject of metallochemistry is an investigation of the following main reactions:

a. formation of solid solutions of metals
b. formation of metallic compounds of both constant and variable composition
c. interactions of metallic compounds with each other
d. crystallochemical reactions taking place in metallic systems in the solid state.

In this chapter, problems of the formation of intermetallic compounds and equilibrium systems composed of these compounds are considered. We prefer to use the term *metallides*, first proposed by N. S. Kurnakov in 1905[10,11] instead of the expression *intermetallic compounds*, which means only compounds between metals. The term "metallides" more correctly reflects the nature of such a class of inorganic compounds, and from this general term other derivative terms become understandable—beryllides, borides, aluminides, stannides, silicides, carbides, nitrides, and other metal-metal or metal-metalloid compounds.[12]

The problem of the formation of metallides and of their physicochemical nature and properties, is discussed in the works of many scientists.[6–19] A detailed historical review of the investigations of metallic compounds, their classification and characteristics, and the conditions for the formation of compounds and their solid solutions has been given.[18]

The classification of inorganic chemical compounds suggested by N. S. Kurnakov[2,3] reduces the great variety of metallides to two types:

a. compounds of variable composition (berthollides)
b. compounds of constant composition (daltonides)

Interstitial phases, such as the carbides, nitrides, and borides, could be also referred to these types of compounds. We can also single out a group of metallides that are formed as a result of transformation of terminal solid solutions or of disordered intermediate phases. These compounds are characterized by superlattice structures.

Many metallic compounds are characterized by a metallic type of interatomic bond. As a rule, they do not obey valence theory and have individual types of crystal lattices. Another large group of metallides belong to semiconducting compounds and with their covalent bonding represent compounds intermediate between metallic and ionic compounds.

In equilibrium systems, metallides act as independent components, and on composition-property diagrams, with the exception of certain compounds of variable composition, they are revealed by singular points. The existence of these compounds in metallic systems is expressed by the three schemes of equilibrium diagrams and composition-property diagrams shown in Fig. 1.

1.2 Kurnakov Compounds

The first schematic form of the equilibrium diagram (Fig. 1*a*) corresponds to the formation of metallides from primary solid solutions. As is now well-known, upon slow cooling many solid solutions form compounds that have ordered structures. Such compounds were originally discovered in 1914 in the Au–Cu system by N. S. Kurnakov,[11,20] who showed that the definite compounds, Cu_3Au and CuAu, are formed by transformation from a continuous solid solution series. The latter play (relative to these compounds) the same role as do liquid solutions in the generation of the

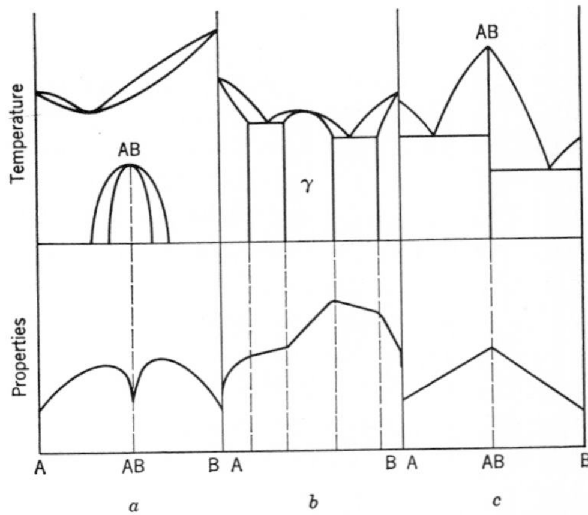

Fig. 1. Typical phase diagrams with metallic compounds. (*a*) Kurnakov compounds, (*b*) compounds of the berthollide type, and (*c*) compounds of the daltonide type.

majority of individual solids. Subsequently, such compounds with ordered structures were discovered in many solid solutions and were called Kurnakov* compounds.[7]

In a number of recent works[21–23] previously expressed opinions [2,6,7,11] were confirmed to the effect that the formation of compounds from solid solutions obeys the phase rule and that they can be regarded as independent phases with definite phase boundaries between ordered and disordered phases, as shown in Fig. 1*a*. It must be recognized that compounds of such type exist only over a limited temperature range (see Fig. 1*a*). Kurnakov compounds and phases based on them are formed not only in continuous solid solution series, but also in systems with limited regions of solid solubility. In Hansen's monograph[24] on binary systems, we can find many examples of such compounds: $MnAu_3$, Fe_3Al, VNi_3, VCo_3, $MgAg_3$, Au_3Zn, $AuZn_3$, etc. Two new compounds Ti_6Al and Ti_3Al, which have been reported[25] to form from α-solid solutions of titanium with aluminum, can also serve as examples.

The existence and thermal stability of metallides depend on the degree of difference of the electrochemical properties of the interacting components. An increased difference in electrochemical potential leads to increased stability, if the thermodynamic heat of formation is taken as characteristic of relative stability. From this point of view, Kurnakov compounds are the least stable, because, as a rule, they are formed from solid solutions of isomorphous metals and thus exhibit small thermal effects. As has been shown,[23] in accordance with the increase of the difference in electronegativity between nickel and the more electropositive metals of groups VIII–IV (Fe, Mn, Cr, V, and Ti), the stability of $MeNi_3$-type compounds increases in that order. The heats of formation and the temperatures of dissociation of these compounds increase in the following order:

$$FeNi_3 \rightarrow MnNi_3 \rightarrow CrNi_3 \rightarrow VNi_3 \rightarrow TiNi_3$$

Some violation of the succession of the change of properties in the case of $CrNi_3$ requires additional studies. It is characteristic that the first four compounds are formed from solid solutions and belong to Kurnakov compounds, whereas $TiNi_3$, which is the most stable of all these compounds, is formed by congruent solidification with the largest heat of formation.[23]

1.3 Compounds of the Berthollide Type

Metallides of variable composition (berthollides) formed during crystallization are characterized in the phase diagram by the presence of homogeneity regions and by the absence of clearly expressed singular points on composition-property diagrams (see Fig. 1*b*). Such behavior is typical of two special cases: (1) where the stoichiometric ratio lies outside the region of phase stability, or (2) where the compound is completely disordered at the test temperature. In terms of their relative stability, these compounds are situated between Kurnakov compounds and compounds of the daltonide type. Compositions and crystal structures of berthollides are often determined by electron-to-atom ratios. The quantum theory of the formation of compounds of such type has been described in several papers.[14–17,26,27] These compounds are formed chiefly by the interaction of metals of subgroup B situated close to each other in the periodic system and with small differences in electronegativity.

1.4 Compounds of the Daltonide Type

With increasing differences in electronegativity, ionization potential, and the number of valence electrons, the tendency toward the formation of metallides of the daltonide type (see Fig. 1*c*) is increased. This type is especially characteristic of the interaction of the most electropositive metals of groups I and II with the most electronegative metals of groups II–VI (subgroup B). In this respect it is remarkable that Kurnakov's ideas have come true to the effect that the alkali and alkaline earth metals, having the strongest metallic character, form very stable compounds with the heavy (most electronegative) metals—Hg, Cd, Zn, Pb, Sn, Bi.[10,11] These compounds are stable up to their melting points and give rise to singular points on composition-property diagrams, that is, intersections of property curves occur at the ordinates corresponding to stoichiometric compositions of these compounds.

* Editor's note: Observe that superlattices fall within this definition, but all Kurnakov compounds are not superlattices. See, for example, the discussion of the Fe–Cr–V system in Section 3.1.

The types of intermetallic compounds considered above are genetically connected with each other. In a number of papers, transitions have been shown of compounds of variable composition into daltonides and vice versa. Kurnakov compounds are formed from solid solutions or phases of variable composition.

In addition to the classification of compounds following Kurnakov, classifications of metallic compounds according to crystal-structure type[28–31] and according to the character of the chemical bond[32–34] are also well-developed. A most complete systematization of intermetallic compounds has appeared in papers[18,29–32] where classifications are made according to types of structures and chemical bond. In this chapter we cannot consider these classifications in detail.

2. THE MAIN CONDITIONS OF INTERACTION OF METALLIDES

Metallic compounds in equilibrium systems should be considered as individual components. In the interaction of metallic compounds it is possible to form

(*a*) continuous solid solutions
(*b*) limited solid solutions
(*c*) peritectic and eutectic mixtures

These main types of interaction between metallic compounds correspond to three types of equilibrium diagrams of metallide systems that represent quasibinary cross sections in ternary and more complex systems.

In Fig. 2, schematic equilibrium diagrams are given to show the formation (during interaction of metallides) of continuous solid solutions (Fig. 2*a*), limited solid solutions (Fig. 2*b*), and simple eutectic mixtures (Fig. 2*c*). In analogy with solid solutions of metals, we have suggested that such solid solutions based upon metallic compounds be called "metallide solid solutions."[7,18,35–37]

What physicochemical factors determine that metallide solid solutions (continuous and limited) occur in some systems and mechanical mixtures in other systems? It is possible to draw the general conclusion that the main determining factors for the formation of solid solutions between metallides are the position of the component elements in the periodic system, the electronic structure of the constituent atoms, and the crystal structures of the compounds. Geometric factors and hence atom size also play an important role in the formation of metallide solid solutions. The application of this factor for metallic solid solutions is discussed by many authors.[14–17]

2.1 Conditions for the Formation of Continuous Solid Solutions

Under certain conditions, continuous solid solutions of metallides are formed in systems

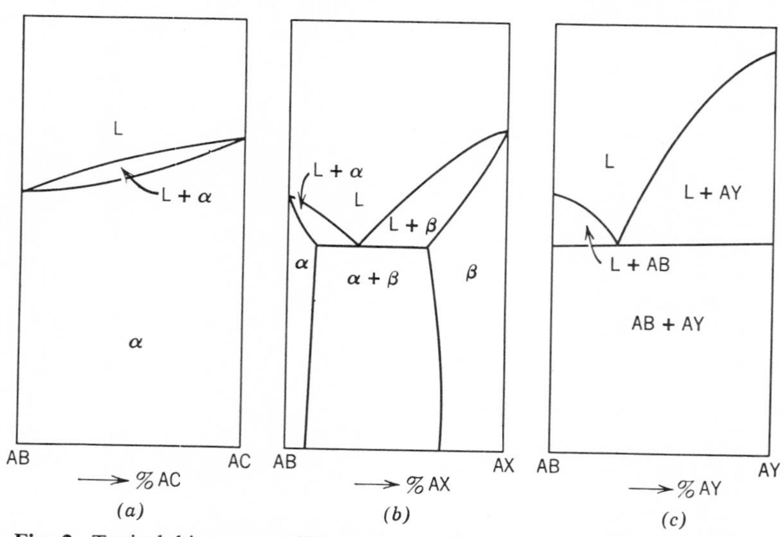

Fig. 2. Typical binary metallide systems: (*a*) continuous solid solutions, (*b*) limited solid solutions, and (*c*) eutectic mixtures.

consisting of

(*a*) Kurnakov compounds
(*b*) compounds of the berthollide type
(*c*) compounds of the berthollide and daltonide types
(*d*) compounds of the daltonide type
(*e*) compounds based upon interstitial phases.

The problems of formation of such solid solutions have been considered by the author in several papers.[35–37] The ideas expressed in these works can be more precisely described by specifying the following conditions, which favor the formation of continuous solid solutions between metallides:

1. Compounds should have the same type of crystal structure with close values of the lattice parameters.
2. Compounds should have the same type of chemical bond.
3. Compounds should be comprised of atoms of analogous elements.
4. Compounds should have the same stoichiometric formula.

The first two of these factors are necessary and other two merely favorable for the formation of continuous metallide solid solutions. Thus in the system shown in Fig. 2*a* compounds AB and AC should have an isomorphous structure and should contain atoms of elements which by their analogous chemical properties are capable of substituting for one another in that structure.

It should be pointed out that polymorphic modifications of different compounds and the formation of ternary compounds can influence the character of the interaction of metallides and, therefore, the equilibrium diagrams. From the above rules it also follows that compounds with different types of crystal lattice cannot form continuous solid solutions due to the impossibility of continuous transition from one type of lattice to another. We can also say that compounds with different types of chemical bond, for instance, with metallic and covalent bonds, or with metallic and partly or completely ionic bonds, cannot form continuous solid solutions.

The third favorable factor for the formation of continuous solid solutions of metallides is based on the fact that the two interacting compounds should contain analogous elements from the same group of the periodic system, for instance, metals of group VIII—Fe, Co, Ni, or group V—V, Nb, Ta, etc., which in binary systems form continuous solid solutions with one another. In addition to this last requirement, in a number of cases it is necessary to have atoms of the same element. This will favor mutual substitution of atoms as the composition of the metallide system is varied between the two compounds. A number of systems may be cited in which the existence of the same kind of atoms in different compounds promotes the formation of continuous solid solutions between these compounds, for instance, in aluminides of metals of the transition groups. Thus the third condition for the formation of continuous solid solutions between metallides is connected with the atomic size and requires the presence in the system of identical, or structurally closely related, atom in the compositions of two compounds. The last condition—the presence of compounds of the same formula type—follows from the fact that compounds with different atomic ratio, as a rule, have different types of crystal structure. In many cases they also have different types of chemical bond, which prevents the formation of continuous solid solutions. Continuous solid solutions can exist when the requirements of the same type of lattice and interatomic bond among these compounds are satisfied.

Our ideas on the formation of continuous metallide solid solutions have been confirmed by subsequent papers by a number of authors.[38–50] In one of these papers[38] the problems of the formation of continuous solid solutions of metallides were discussed with special reference to compounds with a covalent bond. In other papers[39–41] the same problems are considered for interstitial phases. In the paper[38] devoted to isomorphism of compounds with a covalent bond, the possibility of the formation of solid solutions between cadmium and zinc tellurides (CdTe–ZnTe) and cadmium and mercury tellurides (CdTe–HgTe) was proved. These compounds with less marked covalent bonding are capable of forming continuous substitutional solid solutions, whereas compounds of the type of indium antimonide and arsenide (InSb and InAs), which are characterized by a more strongly covalent bond, have not been proved to form continuous solid solutions. It follows that metallides with a covalent bond have more rigid conditions for the formation of continuous solid solutions than compounds with a purely metallic bond. Developing these ideas, we can expect the absence of continuous solid solutions between compounds with covalent and metallic bonds.

In Reference 39 problems of the formation of continuous solid solutions among carbides, borides, silicides, and nitrides of metals of transition groups have been considered. On the basis of our ideas for the conditions for the formation of continuous solid solutions among metallides, the author of the paper[39] deems it possible to retain the following conditions for compounds of the interstitial type:

1. Compounds should have the same type of crystal lattice.
2. Compounds should be comprised of metals which form continuous solid solutions between themselves.
3. Compounds should have chemical bonds which are of the same type.

From this point of view, the author has considered the available material on the interaction of diborides of the transition-group metals, and in general has confirmed the conditions necessary for continuity of solid solutions among interstitial phases. In Reference 41 the possibility of the formation of continuous solid solutions in the TiB_2–CrB_2 system was experimentally established as well as the absence of such continuity in the system ZrB_2–CrB_2. The results of these investigations are in conformity with the rules for the formation of continuous solid solutions between metallides. The paper[40] on borides of transition metals gives the structural characteristics of many hexagonal borides of different stoichiometric composition, as well as the character of the bond between the metallic atoms and boron in such compounds. In such systems, atoms of metals with an unfilled *d*-shell interact with boron atoms. The tendency to form continuous solid solutions in such systems, according to Reference 40, is limited to cases where the difference of the atomic radii of the metals is not more than 15 per cent. In the case of greater difference, only limited solid solutions are formed. Similar conclusions were obtained earlier in a paper,[51] which considered the interaction of phases in systems of nitrides and carbides of titanium, zirconium, vanadium, and niobium.

2.2 Conditions of Formation of Limited Solid Solutions

There are no explicit studies of limited solid solutions between metallides. Limited solid solutions are formed in the absence of isomorphism of the compounds and/or of identity of chemical bond type. The formation of metallide solid solutions with limited solubility also should be expected when the atoms of the compounds differ in their chemical properties and atomic radii (more than 15 per cent). Under such conditions two compounds AB and AX form limited solid solutions, as shown in Fig. 2*b*.

From this point of view, some ideas expressed in the paper[52] on the problem of the formation of limited solid solutions based on the compound Ni_3Al are of interest. In this paper it is pointed out that:

1. Elements close in atomic dimensions to nickel or aluminum have the highest solubility.
2. Elements having an electronic structure similar to that of nickel substitute for nickel atoms in the compound, and atoms similar to aluminum substitute for aluminum atoms.
3. Solubility of elements in Ni_3Al depends in part on their positions in the periodic system.

In the limiting case, when differences in the character of the chemical bond, and sizes and properties of the atoms constituting the compounds, reach maximum values, conditions are absent for the formation of even limited solid solutions between metallides. Then metallide systems, as shown in Fig. 2*c* (AB–AY), become simple peritectic or eutectic types, without any significant solubility for the components.

2.3 Classification of Phases Based on Metallides

The problem of classification of ternary phases based on metallic compounds has been adequately dealt with only recently, although there have been many previous investigations. In References 36 and 37, some regularities of the formation of continuous ternary solid solutions based upon binary metallic compounds have been established. The chemical nature of a number of ternary berthollide and daltonide phases has been considered.[53] Problems of the classification of some ternary phases from the viewpoint of similarities of electron concentration have also been discussed.[54,55]

In References 56 and 57, a more complete analysis has been made of the physicochemical nature of ternary metallic phases, and an attempt is made to classify such ternary phases by the extent and character of their homogeneity-regions within their equilibrium diagrams. A review[58] of crystal structures of ternary intermetallic compounds and their classification by crystallochemical indices represented a further development.

Vulf[56] in his classification of ternary phases proceeds from more general principles of their formation on the basis of metals and binary and ternary metallides. These principles are based on N. S. Kurnakov's ideas on compounds of the berthollide and daltonide types.[2,3] From this point of view, the ternary phases are divided into three groups:

(*a*) ternary metallic phases of variable composition—solid solutions based on metals

(*b*) ternary solid solutions based on binary metallic compounds

(*c*) ternary metallic compounds of variable or constant composition (berthollides and daltonides) and phases based on them.

Of these three groups of ternary metallic phases, the last two, which comprise compounds and phases based on binary and ternary metallides, have bearing on the problem considered in this chapter. Therefore, out of the eight general types of ternary phases given in Reference 56 we consider four main types which are based on metallic compounds. These types are indicated in Fig. 3.

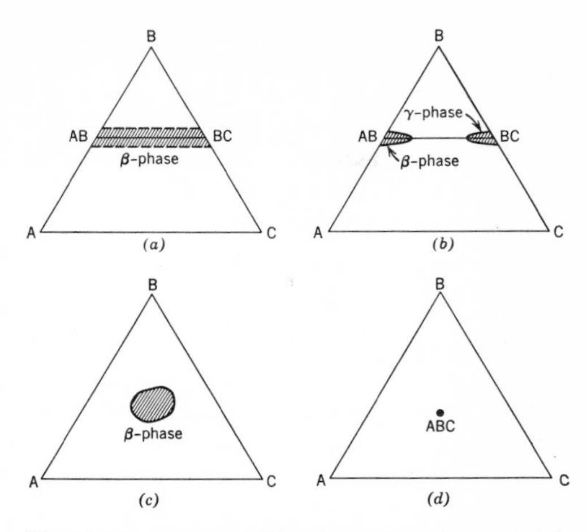

Fig. 3. Ternary metallic phases based on metallides: (*a*) continuous solid solutions between two compounds, (*b*) limited solid solutions between two compounds, (*c*) a phase based on a ternary compound of the berthollide type, and (*d*) a ternary compound ABC of the daltonide type without solid solution formation with the components.

The first type (Fig. 3*a*) refers to the ternary system, in which two compounds are formed in the binary systems at the compositions AB and BC. Having crystallochemical similarities, these compounds interact to form continuous solid solutions. Such continuous solid solutions can be also imagined between a binary compound and one of the pure metal components of the system. If the properties and the different types of crystal lattices for both compounds differ greatly, only limited solid solutions will form, as shown in Fig. 3*b*. Two ternary phases (β and γ) are then formed on the basis of the two binary compounds AB and BC with a limited region of extension into the ternary system.

The third type (Fig. 3*c*) embraces phases based on a ternary compound of variable composition with a wide region of homogeneity in the ternary system. Such ternary compounds of the berthollide type are often encountered in the corresponding ternary systems of aluminum, magnesium, etc. By way of examples we can cite the ternary phase T (MgCuZn) in the Mg–Zn–Cu system, phase U (AlMgCu) in the ternary system Al–Mg–Cu, etc.

Finally, the fourth type of ternary phase (Fig. 3*d*) is represented by ternary metallic compounds of constant composition—daltonides—which do not form solid solutions with their components. As is evident from Fig. 3*d*, the position of this compound in the corresponding ternary system is characterized by a point. As will be shown later, these types of ternary phases based upon binary and ternary metallic compounds have been confirmed in a number of experimentally investigated systems.

The main principles of phase classification in ternary metallic systems can be extended to quaternary and more complex systems. By triangulation and tetrahedronization methods it is possible to single out quasibinary and quasiternary systems which consist of metallic compounds.

3. EXPERIMENTAL INVESTIGATIONS OF EQUILIBRIA AMONG METALLIDES

3.1 Binary Systems with Continuous Solid Solutions

Investigations of equilibria of metallide systems naturally began in connection with studies of the equilibrium diagrams of multicomponent metallic systems. There are at present extensive experimental results in the literature from surveys of a large group of binary, ternary, and more complex systems composed of metallides. Brief information about these systems is given below.

Let us consider systems with continuous solid solutions composed of metallides by turning our attention to Kurnakov compounds, that is, those formed by ordering of a terminal metallic solid solution, and to compounds crystallized from a liquid phase.

One of the first systems of Kurnakov compounds that we investigated was the ternary Fe–Cr–V system,[43] containing the metallides FeCr and FeV. As is known,[24] in the binary systems Fe–Cr and Fe–V, the compounds FeCr and FeV are formed from continuous α-solid solutions. These two compounds satisfy the main conditions for the formation of continuous solid solutions between them, according to both chemical and structural factors. An experimental investigation of a quasibinary cross section of the ternary Fe–Cr–V system, passing through the compositions of the two compounds FeCr and FeV, has shown that ternary alloys along this cross section crystallize as continuous α-solid solutions.[43] Under conditions of slow cooling or prolonged heat treatment at temperatures below those critical for the formation of the compounds, α-solid solutions are transformed into metallide solid solutions of FeCr–FeV compounds, called σ-phase. This $\alpha \rightleftarrows \sigma$ phase transition along the cross section investigated takes place on a continuous curve from the compound FeCr to the compound FeV.

The equilibrium diagram of alloys on the cross section FeCr–FeV (see Fig. 4) shows that the alloys crystallize as continuous ternary α-solid solutions with a body-centered cubic

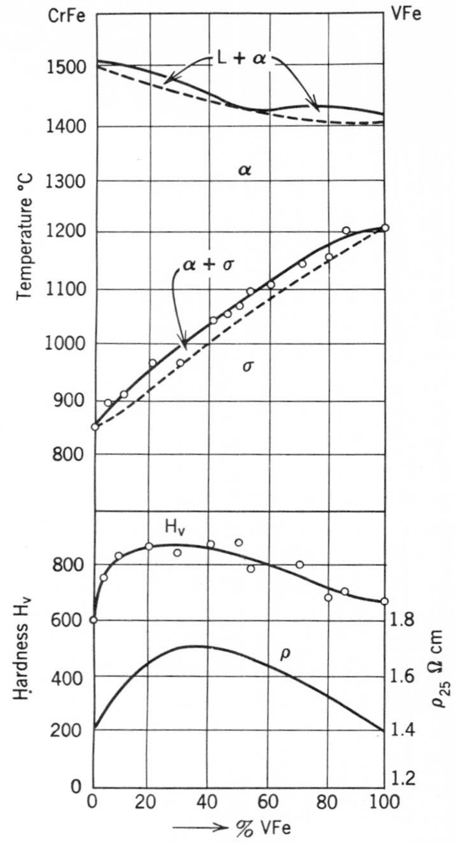

Fig. 4. A quasibinary system of the cross section FeCr–FeV. Upper part—thermal analysis; lower part—properties of annealed alloys.

lattice, and over a definite interval they are transformed into σ-solid solutions. Sigma phase has the tetragonal β-uranium structure. The continuous character of the metallide solid solutions (σ-phase) has been confirmed by X-ray diffraction and by an investigation of hardness and electric resistivity. The investigated properties change continuously over the region of continuous solid solutions of σ-phase (see the lower part of Fig. 4). As was shown in Reference 43, the two kinds of solid solutions α and σ, which in principle differ in their physical-chemical nature, also differ sharply in their properties. The hardness of σ-solid solutions reaches 900 H_v, but that of the α-solid solution of the same composition (in the quenched state) is only 270 to 300 H_v.

It is possible to cite many cases of the formation of continuous solid solutions among compounds of the Kurnakov type. In an investigation[60] of the ternary system VNi_3–VCo_3–VPd_3 it was shown that between the compounds VNi_3–VCo_3 and VNi_3–VPd_3 there are continuous solid solutions. In these systems the phase transformation from a ternary terminal solid solution into a metallide solution also takes place along continuous curves. In VCo_3–VPd_3, the third binary system studied by the author, limited mutual solubility of the two metallides and a two-phase region have been established. This is accounted for by the difference in lattice type of the subject compounds, by a discontinuity of the solid solutions in the binary Pd–Co system, and by the formation of intermediate phases in it. Such continuous solid solutions of the Kurnakov-type compounds can be assumed in many systems not yet investigated experimentally. Some of them are: AuCu–PtCu, FePt–FePd, CrCo–VFe, $AuCu_3$–$PtCu_3$, $PdCu_3$–$PtCu_3$, $FeNi_3$–$MnNi_3$, $MnNi_3$–$CrNi_3$, $CrNi_3$–$FeNi_3$.

We may cite as one interesting example of solid-solution formation between a Kurnakov compound and a daltonide that is crystallized from a liquid phase the system Ti_3Al–Ti_3Sn. We have established[61] that the Ti_3Al compound is formed from solid solutions of the Ti–Al system; the Ti_3Sn compound, according to Reference 24, is crystallized from a liquid phase in the system Ti–Sn. Both compounds are isomorphous and have a hexagonal lattice. From study of the quasi-binary cross section, Ti_3Al–Ti_3Sn, it was shown[62] that solid-solution formation occurs over almost the entire system. The formation of a metallide solid solution of these two compounds is shown in Fig. 5 by a dashed line (we do not have sufficient experimental data). The continuous character of the solid solutions between these metallides is confirmed by the hardness and electric resistivity curves of long-time annealed alloys along the cross section Ti_3Al–Ti_3Sn. These properties change continuously with composition and pass through a smooth maximum at a Ti_3Sn concentration of about 50 per cent (see curves in Fig. 5). However, the chemical difference of the Al and Sn atoms makes it likely that a solubility gap may exist in this system; additional study is therefore required.

Systems of metallides that form continuous solid solutions upon crystallization are described in the literature.[42,43–50,63] Such a

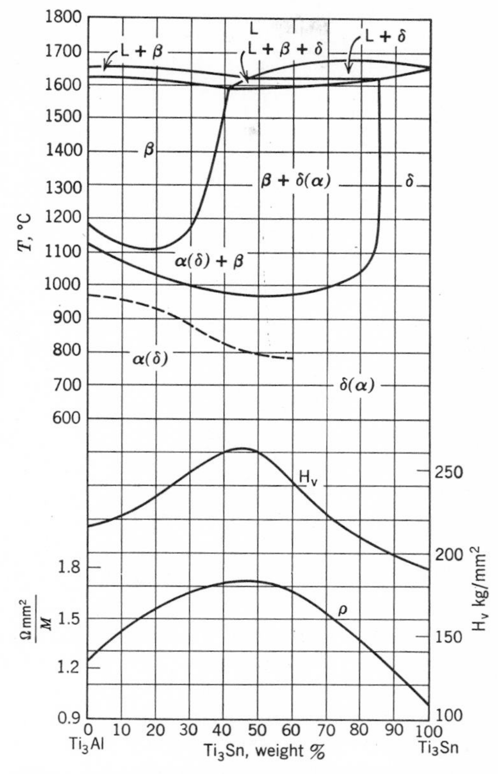

Fig. 5. A quasibinary system along the cross section Ti_3Al–Ti_3Sn. Upper part—thermal analysis; lower part—properties of annealed alloys.

system of compounds, $NbNi_3$–$TaNi_3$, has been investigated by the author.[42] Both compounds are classed as daltonides. They occur in corresponding binary systems,[24] are isomorphous, and form continuous solid solutions. Subsequent investigation has experimentally proven the existence of analogous continuous solid solutions in the related binary metallide systems: $TiNi_3$–$NbNi_3$ and $TiNi_3$–$TaNi_3$.[63] It has recently been demonstrated[64] that the compounds $NbNi_3$ and $TaNi_3$ possess orthorhombic symmetry and belong to the structural type β-Cu_3Ti.

Because there are some indications of the possible existence of a rhombic modification of the lattice in $TiNi_3$,* we can assume that in these two binary systems the main condition of isomorphism for the formation of continuous solid solutions between metallides may be fulfilled. The continuous character of the solid solutions in three binary systems of these

* According to unpublished work by the author.

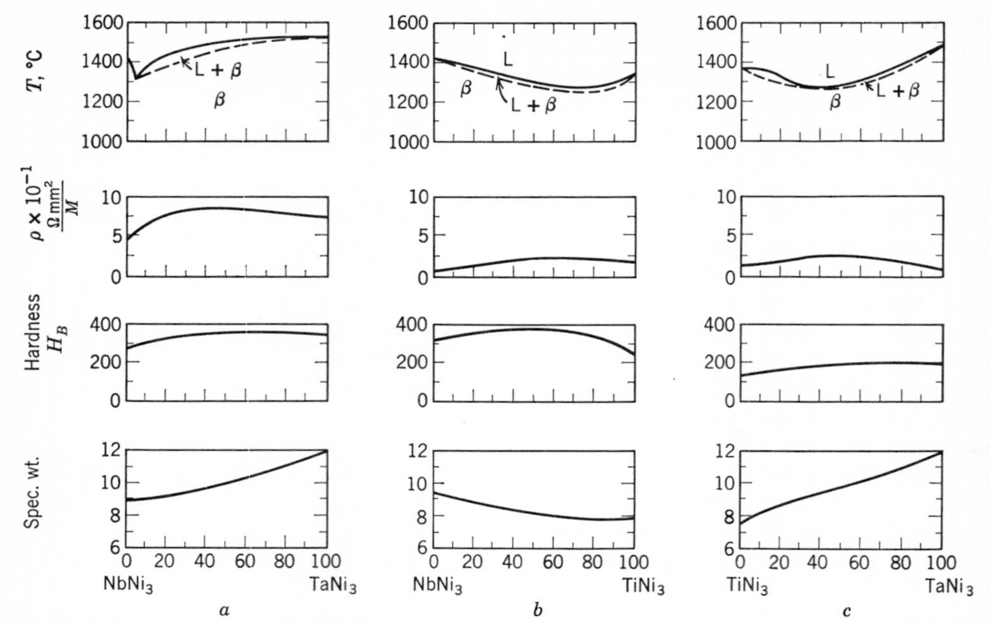

Fig. 6. Binary metallide systems and corresponding properties of annealed alloys: (*a*) system $NbNi_3$–$TaNi_3$, (*b*) system $TiNi_3$–$NbNi_3$, and (*c*) system $TiNi_3$–$TaNi_3$.

metallides is confirmed by data from thermal analysis and by investigation of properties as a function of composition. Figure 6, gives thermal diagrams with properties such as hardness, specific electrical resistivity, and specific weight as a function of composition, according to data in References[42,63]. These diagrams confirm the continuity of solid solutions in the systems $NbNi_3$–$TaNi_3$, $TiNi_3$–$NbNi_3$, and $TiNi_3$–$TaNi_3$. The formation of a hexagonal modification of $TiNi_3$ at low temperatures, in contrast to the orthorhombic lattice of the compounds $NbNi_3$ and $TaNi_3$ which exists over the entire temperature range, can be expected to cause some discontinuity in solid solubility at low temperatures in the systems $TiNi_3$–$NbNi_3$ and $TiNi_3$–$TaNi_3$.

The possible existence of a discontinuity of solid solubility of metallides as a result of the existence of polymorphic modifications has been experimentally shown[65] in an investigation of the quasibinary $TiFe_2$–$TiCr_2$ from the system Ti–Fe–Cr. These two Laves phases, $TiFe_2$ and $TiCr_2$, are formed in the corresponding binary systems.[24] $TiFe_2$ is formed upon solidification and $TiCr_2$ from β-solid solutions of the Ti–Cr system. Investigations made by various authors[24] have shown that $TiFe_2$ has a hexagonal modification of the $MgZn_2$ type without polymorphism and $TiCr_2$ has two modifications—the hexagonal $MgZn_2$ structure for temperatures from the melting point down to $\sim$1000°C and the face-centered cubic $MgCu_2$ type below this temperature.[66] Taking into account these observations, the authors of the paper[65] have established (from a survey of the lattice parameters) that in the system $TiFe_2$–$TiCr_2$ (high-temperature modification) continuous solid solutions exist. A small discontinuity on the lattice parameter curve near the $TiCr_2$ side of the $TiFe_2$–$TiCr_2$ system was explained by the existence of a two-phase region: $TiCr_2$ (with a face-centered cubic lattice) and $Ti(FeCr)_2$ (with a hexagonal lattice).

Continuous solid solutions between metallides exist in many other systems too. We can mention continuous solid solutions in systems consisting of cobalt and nickel, iron and nickel monaluminides and perhaps iron and cobalt monaluminides: CoAl–NiAl, NiAl–FeAl, and CoAl–FeAl.[67–69] Of these three isomorphous compounds, CoAl and NiAl have congruent melting points and cusped maxima and minima in their property-composition curves; FeAl, however, does not show a singular melting point, and it belongs to the berthollide type of compound. Therefore, among these three binary systems CoAl–NiAl represents a system of daltonide-type compounds, and the two other systems CoAl–FeAl and NiAl–FeAl are combinations of the daltonide compounds CoAl and NiAl with compounds of the berthollide-type FeAl.

A transition of a compound of the daltonide type into a berthollide through a continuous series of solid solutions has been shown in a study[6,70] of alloys along the cross section NiSb–FeSb in the ternary system Fe–Ni–Sb. Both compounds are isomorphous in structure (of the NiAs type), but one of them (NiSb) is a daltonide and the other (FeSb) a berthollide. As shown in Fig. 7, there are continuous solid solutions between these two compounds.

The possible formation (at high temperatures) of continuous solid solutions between compounds of the daltonide AuZn with a singular melting point and disordered intermediate-phase berthollide CuZn types has been predicted.[6] Such phases of variable composition are called Kurnakov phases[6] in honor of the scientist who first pointed out the possibility of the existence of these phases in ternary metallic systems. If the conditions for the formation of continuous solid solutions between metallides (see above) are satisfied, we can expect the formation of such solutions in many systems consisting of purely metallic compounds, in compounds based on interstitial phases, and among compounds of the semiconductor and superconductor types.

As an illustration we can point out the

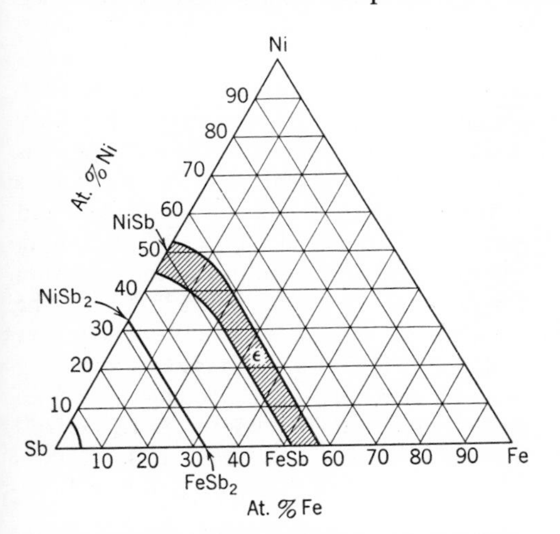

Fig. 7. Continuous solid solutions of metallides: $NiSb_2$–$FeSb_2$ and NiSb (a daltonide)—FeSb (a berthollide).

formation of extensive solid solutions in a system of Laves phases: $MgCu_2$–$MgNi_2$. As a result of an investigation of the thermal diagram of the ternary system Mg–Cu–Ni, it has been established[45] that alloys on the quasibinary cross section $MgCu_2$–$MgNi_2$ form solid solutions over virtually the entire concentration interval. The "backbone" line of crystallization of these solutions connects compositions of binary compounds, and the liquidus and solidus curves correspond to systems of solid solutions of the compounds. However, these two Laves phases have a different crystal structure (C15 and C36, respectively), and therefore we conjecture some interruption of solid solubility in this system.

Continuous solid solution formation between compound analogs has also been established in the ternary system Ag–Ce–La.[46] Here cerium and lanthanum are analogs; they form continuous solid solutions with each other. Silver forms a number of compounds with these analogs: $CeAg_3$ and $LaAg_3$; $CeAg_2$ and $LaAg_2$; CeAg and LaAg. The first and third pairs of compounds solidify with an open maximum, whereas the second pair of compounds form by a peritectic reaction. The $CeAg_3$–$LaAg_3$ and CeAg–LaAg systems, according to the data given by the authors, solidify as continuous solid solutions.

Among metallides there are a great number of systems that form continuous solid solutions involving compounds formed from a metal and a metalloid. There are many examples in the literature[47,71–76] of the formation of such solutions among silicides: Fe_3Si–Mn_3Si; Cr_3Si–Mo_3Si; V_3Si–Mo_3Si; V_3Si–Cr_3Si; V_5Si_3–Mo_5Si_3; V_5Si_3–Nb_5Si_3; $MoSi_2$–WSi_2 and $NbSi_2$–VSi_2, and others, produced by the substitution of atoms of analogous metals.

Compounds based on interstitial phases—borides, carbides, nitrides, and even, in some individual cases, oxides (suboxides of metals of transition groups)—when the necessary conditions obtain, also form solid solutions with one another. Many of them have a metallic type of bond, some have covalent bonding, and in some a ionic type of bond appears. Many papers have been devoted to an investigation of these problems, and there are many examples indicating the formation of continuous solid solutions between compounds in this class.[39–40,49,51,77–79]

We have at the present time only some

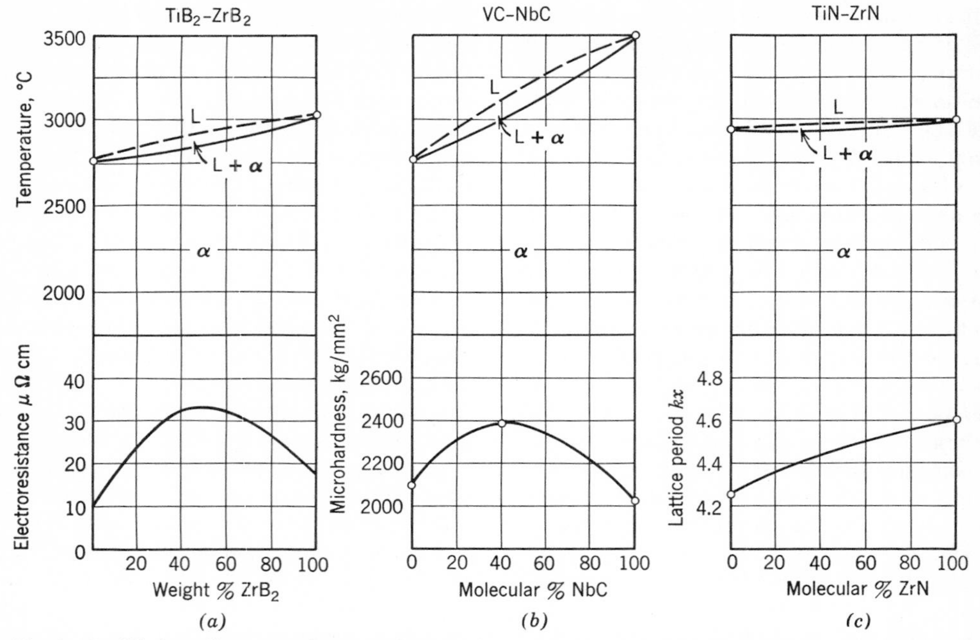

Fig. 8. Equilibrium diagrams of systems from borides, carbides, and nitrides: (*a*) the system ZrB_2–TiB_2 and composition dependence of specific electric resistivity; (*b*) the system VC–NbC and composition dependence of microhardness; and (*c*) the system TiN–ZrN and composition dependence of lattice parameter.

illustrations rather than a detailed review of these results. In Fig. 8, three equilibrium diagrams are given for the systems of titanium and zirconium borides (Fig. 8*a*), vanadium and niobium carbides (Fig. 8*b*), and titanium and zirconium nitrides (Fig. 8*c*), all of which form continuous solid solutions. Due to the absence of data for their thermal diagrams, the systems are given schematically. In the lower part of the figure curves are shown for the change in the lattice parameter and hardness of the systems, which demonstrate the continuity of the solid solutions between these compounds.

Among interstitial phases (if certain conditions are satisfied) continuous solid solutions frequently exist between nitrides and oxides (e.g., TiN–TiO) as was considered in the paper.[80] There are also cases of the formation of solid solutions between carbides and nitrides TiN–TiC, NbN–TiC, and others.[51] Apparently, in such phases the elements C, N, and O with small atomic radii can substitute mutually for one another and give interstitial phases of a variable composition. With a greater difference in atomic dimensions between metal and metalloid atom, for instance, with boron and silicon atoms, no interstitial phases are formed, and, therefore, it is impossible to expect the existence of continuous solid solutions of borides and silicides with carbides and nitrides.

Continuous solid solutions have been found in dozens of cases among systems composed of compounds of the semiconductor type, despite more rigid conditions for solid-solution formation among covalent compounds.[38] It should be pointed out that such solutions are formed among these compounds with different combination of metallic atoms with semimetals and metalloids—bismuth and antimony, arsenic and phosphorus, selenium, tellurium, etc. Investigations in this direction are described in many papers.[38,48,81–89] The existence of continuous solid solutions has been proven in the following systems consisting of metallides of the semiconductor type: GaSb–InSb[48], PbTe–PbSe,[81] SnTe–GeTe,[82] InP–InAs,[84] GaP–GaAs,[84] InSb–InBi,[87] CdSb–InSb,[88] HgTe–HgSe,[89] and many others.*

* In accordance with the latest paper of R. I. La Botz et al. [*J. Electrochem. Soc.* **110** (1963) 127], the system Mg_2Si–Mg_2Ge also exhibits continuous solid solubility.

It is of interest to note from this brief list that the compounds considered are formed by various elements, whereby an element A in a compound is taken from groups II (Hg), III (Ga, In), IV (Pb, Ge) and the element B, from elements of groups V (P, As, Sb) and VI (Se and Te). Despite such a variety in element combinations in compounds of this type, all of them are characterized by a covalent bonding, are isomorphous in their structures, and have the same stoichiometric composition. This explains the fact that with chemical similarity of one group of elements A–II, A–III, A–IV, and concurrent similarity of the other group of elements B–V, B–VI, systems of such compounds form continuous solid solutions with one another.

It is also interesting that, although elements such as Sn and Ge do not themselves form continuous solid solutions, in the system SnTe–GeTe such continuous solutions do exist. Apparently, such a deviation from the general rule of the formation of continuous solid solutions of intermetallic compounds[35–37] testifies to the presence of additional conditions for semiconductor compounds having a covalent type of bond. They require additional consideration.

As an example, Fig. 9 gives two typical systems, GaSb–InSb[48] and PbSe–PbTe,[81] in which the formation of continuous solid solutions is shown. Although in the early paper on the system GaSb–InSb[86] the absence of continuity of solid solutions in this system was demonstrated, additional investigations[48] have proved, by methods of thermal and

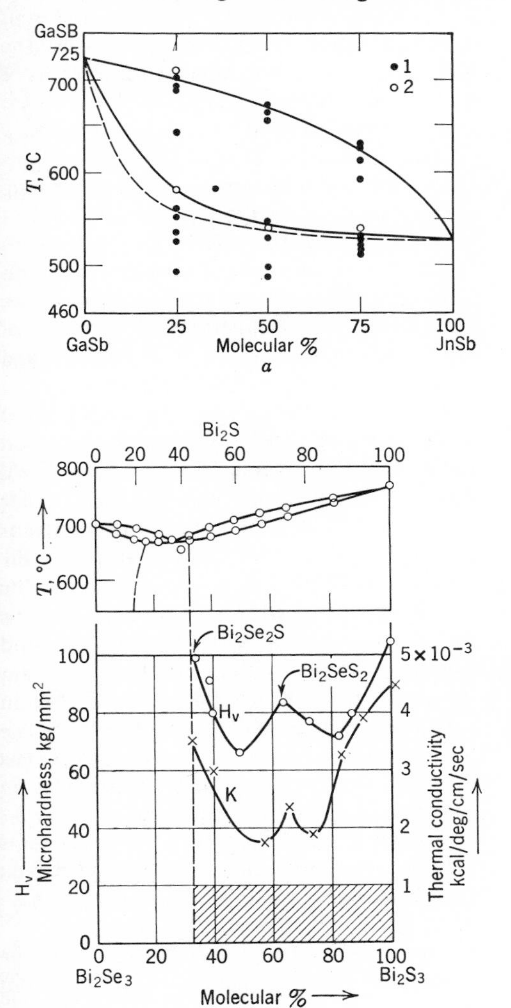

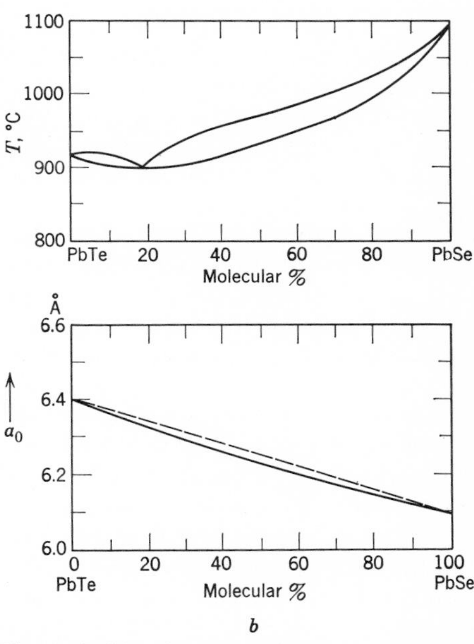

Fig. 9. Solid solutions of systems of semiconductor compounds: (*a*) system GaSb–InSb, (*b*) system PbTe–PbSe, and (*c*) system Bi_2Se_3–Bi_2S_3.

structural analyses of long-time annealed alloys, the existence of continuous solubility in the system of these two compounds.†

An intermediate type of transition to limited solid solutions is represented by the system Bi_2Se_3–Bi_2S_3. Compounds of this system are not isomorphous. According to Reference 83, despite the chemical resemblance of S and Se and the common atom Bi, this system exhibits a small discontinuity of solubility and an eutectic reaction, as is shown in Fig. 9*c*. In the region of the solid solutions the composition dependence of microhardness and thermal conductivity indicated[83] the formation of ternary compounds of the compositions Bi_2Se_2S and Bi_2SeS_2 (see Fig. 9*c*); these compounds may be regarded as Kurnakov compounds formed from solid solutions.

3.2 Binary Systems with Limited Solid Solutions

When conditions are unfavorable to the formation of continuous solid solutions between metallides, either limited solid solutions or mechanical mixtures without solid solutions are formed. It can be stated that mutual solubility of metallides in the solid state decreases with an increase of the difference in chemical properties of elements (metals and metalloids) which form these compounds or, in other words, with increased separation of these elements in the periodic system.

This general conclusion was illustrated in Reference 44 with the example of the different solubility of the metals Zn, Al, Sn, and Sb in the compound $MgCu_2$, and in Reference 52 with the example of the limited solubility of components in the compound Ni_3Al. The solubility limit of Zn, which is situated close to Mg and Cu in the periodic system, is maximal in the compound $MgCu_2$. This limit decreases from Zn to Al to Sn to Sb, in which an extremely low solubility is found because antimony is the most electronegative element and the farthest removed in the periodic table from magnesium and copper. The same regularities can be shown with examples of the change in mutual solubility in systems between metallides with an increase in the difference of chemical properties of elements that comprise these compounds. Solubility gaps in solid solution binary compounds can appear also due to the presence (in the corresponding systems) of polymorphic modifications and due to the formation of ternary compounds. In the References 56 and 57 attention was drawn to these important factors. They, in a number of cases, may even upset the supposed quasibinary character of systems between two compounds in a ternary system and make impossible the existence of continuous solid solutions between compounds, even when the aforementioned conditions for the formation of continuous solid solutions are satisfied.

Let us consider some examples of the formation of limited solid solutions between different classes of compounds. Examples from the Kurnakov-type compounds are the systems of the compounds FeCo–$FePd_3$ and FePd–FeCo formed from solid solutions in the ternary system Fe–Co–Pd.[90] Components of this system are metal analogs capable of forming continuous solid solutions; discontinuities of solubility are due to different iron and cobalt modifications and the formation of nonisomorphous compounds from solid solutions. As a result of investigations of alloy transformations of this system in the solid state, the authors[90] give phase diagrams for the cross sections FeCo–$FePd_3$ and FePd–FeCo. The compound FeCo has a body-centered lattice, $FePd_3$ has a face-centered lattice, and FePd possesses a tetragonal lattice. Due to the superposition of the $\alpha \rightleftarrows \gamma$-phase transformation ranges with the formation of the compounds FePd, FeCo, and $FePd_3$, and the different types of lattices of these phases, we would assume that discontinuities of solid solubility would exist in the cross sections investigated. This has been proven by an experimental investigation.[90] The corresponding equilibrium diagrams are given in Fig. 10*a* for the cross section FeCo–$FePd_3$ and in Fig. 10*b* for the cross section FeCo–FePd. As is evident from the figure, in both systems there are discontinuities of solubility between compounds and the appearance of two-phase and three-phase regions.

The system $NbCr_2$–$NbNi_3$[91] is one of the examples of the formation of limited solid solutions between compounds of the daltonide

† For a discussion of nonequilibrium and equilibrium states in this system, see the papers of N. I. Ivanov-Omsky and B. T. Koloniets, *Fiz. Tverd. Tela* **1** (1959) 913, and Wooley and Smith, *Proc. Phys. Soc.* (*London*) **22** (1958) 214.

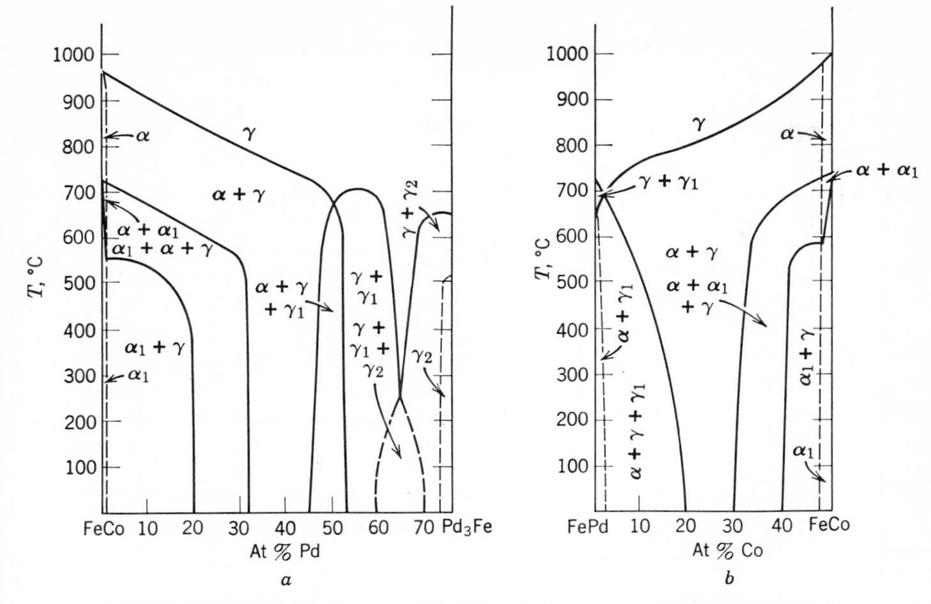

Fig. 10. Equilibrium diagrams of alloys of the ternary system Fe–Co–Pd: (*a*) cross section FeCo–$FePd_3$ and (*b*) cross section FeCo–FePd.

type. As is seen from the different compositions of these compounds, the one common metal (niobium) and the lack of analogy between chromium and nickel would lead us to expect only the formation of limited solid solutions. Figure 11 shows the equilibrium diagram of this system according to the data of the paper[91] illustrating limited solubility with the formation of a eutectic mixture. The presence of two polymorphic modifications of the compound $NbCr_2$—a high-temperature ϵ (of the $MgZn_2$ type) and a low-temperature β (of the $MgCu_2$ type)—causes the appearance of an intermediate, two-phase ($\beta + \epsilon$) region (see Fig. 11).

A considerable contrast in limited mutual solubility can be shown by two systems, wherein component element pairs that are closely similar and sharply different in their chemical properties are compared. Figure 12 presents the systems UAl_2–$ZrAl_2$ and UAl_2–U_3Si_2.[92–93] The first system of compounds (with the common element aluminum) contains the elements uranium and zirconium that form between themselves continuous solid solutions at high temperatures, whereas in the second system the compounds do not possess the same stoichiometric atom ratio, and, although uranium is common to both compounds,

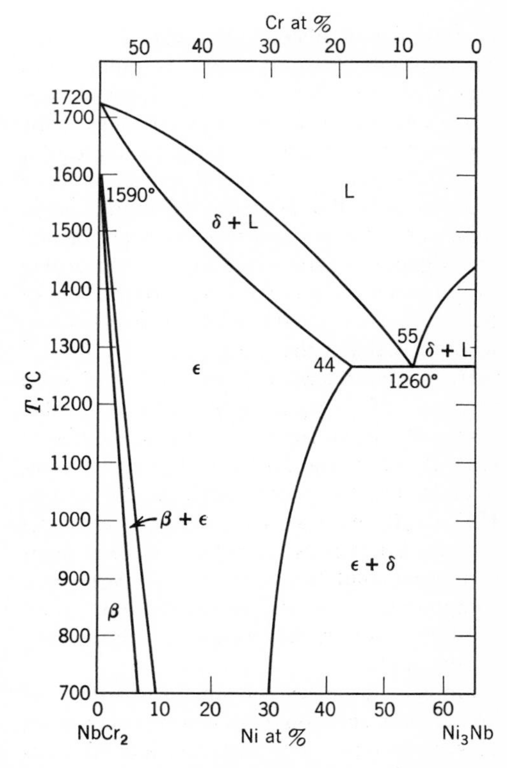

Fig. 11. An equilibrium diagram of the system $NbCr_2$–$NbNi_3$.

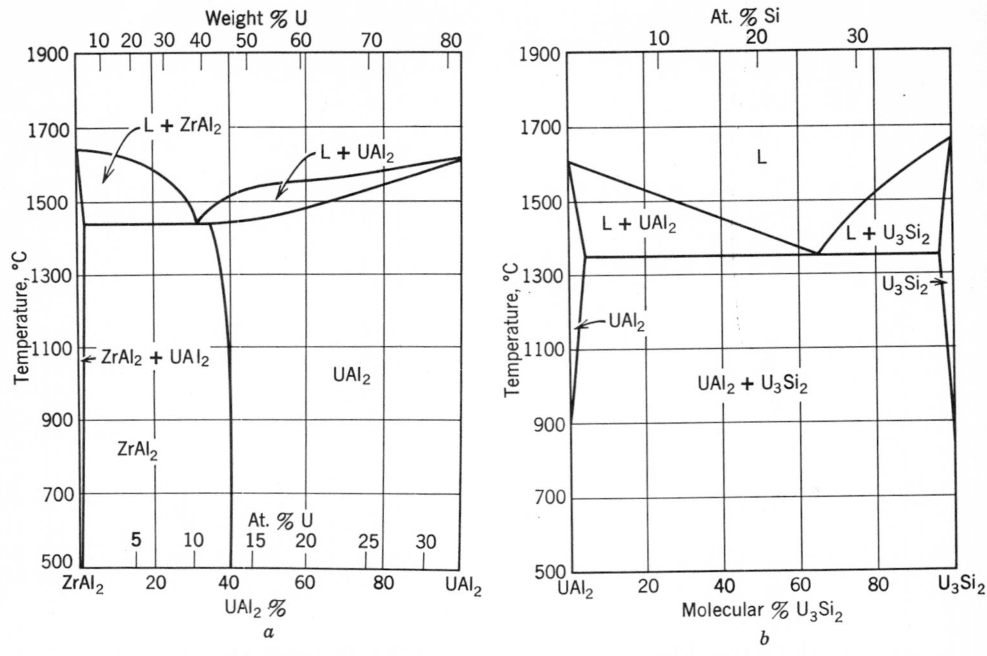

Fig. 12. Equilibrium diagrams of compounds with limited solubility: (*a*) system UAl_2–$ZrAl_2$ and (*b*) system UAl_2–U_3Si_2.

aluminum and silicon differ sharply in their chemical properties. It should be pointed out that UAl_2 and $ZrAl_2$ are both Laves phases but are not isomorphous; UAl_2 belongs to the $MgCu_2$ type, but $ZrAl_2$ belongs to the $MgZn_2$ type.[92] Therefore, we could expect in both systems only the existence of limited solid solutions. This is confirmed by experimental investigations of them (see Fig. 12), according to References 92 and 93. Upon comparison of the equilibrium diagrams of the two systems we observe that in the first system UAl_2–$ZrAl_2$[92] there is a considerable region of limited solid solutions of $ZrAl_2$ in UAl_2, and a possible continuity is violated only by the different type of the crystal lattice of these two compounds. On the other hand, in the system UAl_2–U_3Si_2 with sharply different elements (Al and Si) there is almost complete absence of mutual solubility between the investigated compounds.[93] In its extremely low solid solubility this system resembles the binary system Al–Si.[24]

Similar instances of the formation of limited solid solutions exist among compounds of various types. For instance, limited solubility has been established in the Ni_3Al–Ni_3Nb system[94]; in the $FeSi_2$–$CrSi_2$ system where the compounds are not isomorphous[95]; among compounds of the interstitial type: carbides, borides, nitrides, etc.,[75–79] semiconductor compounds,[96–100] and other types. We cannot consider all these cases and shall confine ourselves to expressing some opinions on limited solid solutions among systems of semiconductor compounds. First, as was pointed out above, limited solid solutions are formed in systems in which the compounds are not isomorphous, for instance, Bi_2Se_3–Bi_2S_3 (see Fig. 9*c*). Second they are formed when the components fail to have the same formula type, for instance, in the systems AlSb–Al_2Te_3,[98] PbTe–Bi_2Te_3 and SnTe–Sb_2Te_3,[99] and PbSe–Bi_2Se_3.[100] Third they are formed when a nonisomorphous ternary compound occurs by the interaction of two binary compounds, as has been shown for the system Bi_2Te_3–Bi_2S_3.[97] The latter system is characterized by limited solubility of bismuth sulfide in bismuth selenide and by the formation of a ternary compound Bi_2Te_2S, which corresponds to the natural mineral tetradymite and possesses a rhombohedral structure. Analogous ternary compounds were revealed upon investigation of the system PbSe–Bi_2Se_3[100] where, in addition to limited solubility in the system, the formation of three discrete phases,

3PbSe·2Bi_2Se_3, PbSe·Bi_2Se_3, and PbSe·2Bi_2Se_3, was established.[100] All these phases possess semiconductor properties.

In conclusion to illustrate the above statement, Fig. 13 shows three systems of semiconductor compounds of which the first one, AlSb–Al_2Te_3 (Fig. 13*a*), is characterized by limited solubility without ternary compounds; the second one, Bi_2Te_3–Bi_2S_3 (Fig. 13*b*), is characterized by formation of a single ternary compound Bi_2Te_2S; and the third one, PbSe–Bi_2Se_3 (Fig. 13*c*), shows the three ternary compounds: 3PbSe·2Bi_2Se_3, PbSe·Bi_2Se_3, and PbSe·2Bi_2Se_3.

3.3 Ternary and More Complicated Metallide Systems

Experimental investigations of ternary and more complex systems composed of metallides are still very limited. Here, as in the case of

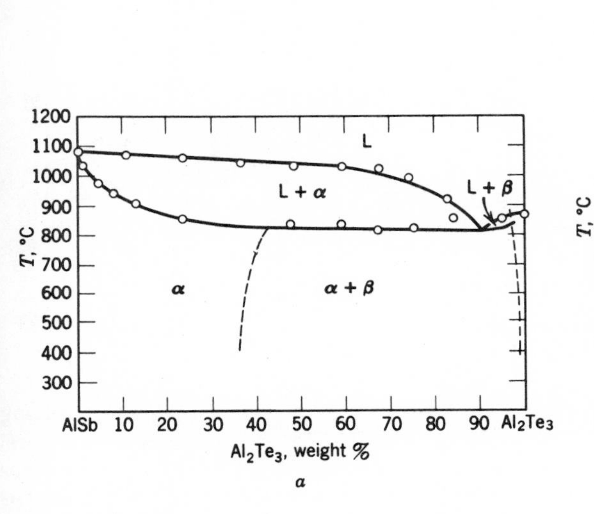

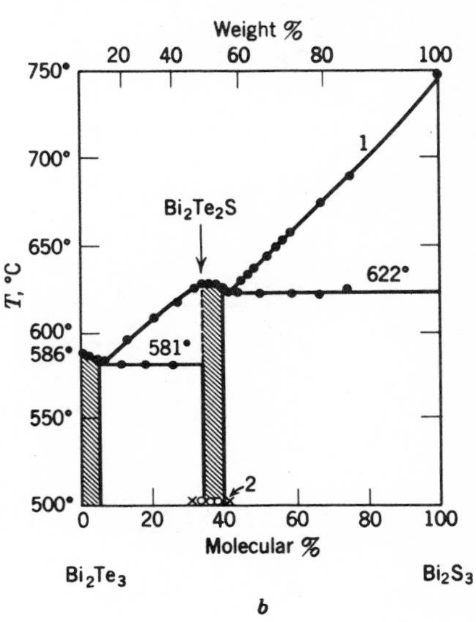

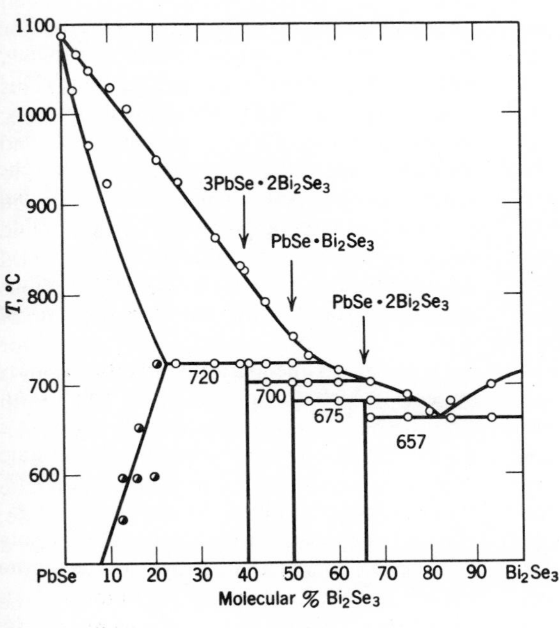

Fig. 13. Equilibrium diagrams of systems of semiconductor compounds with limited solubility: (*a*) system AlSb–Al_2Te_3, (*b*) system Bi_2Te_3–Bi_2S_3, and (*c*) system PbSe–Bi_2Se_3.

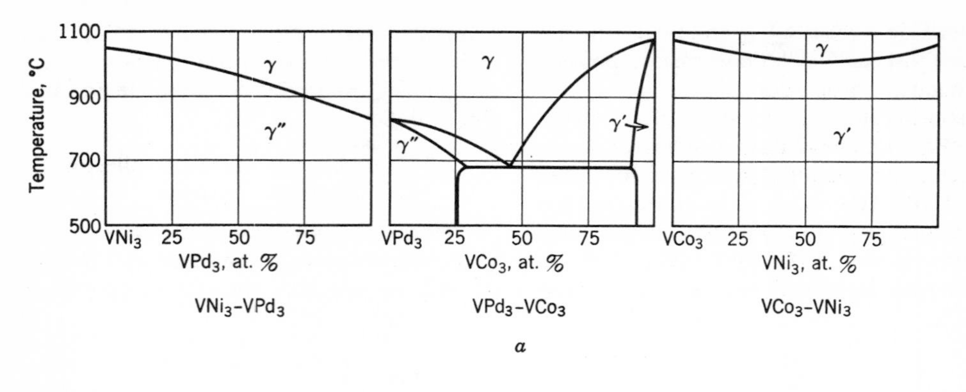

a

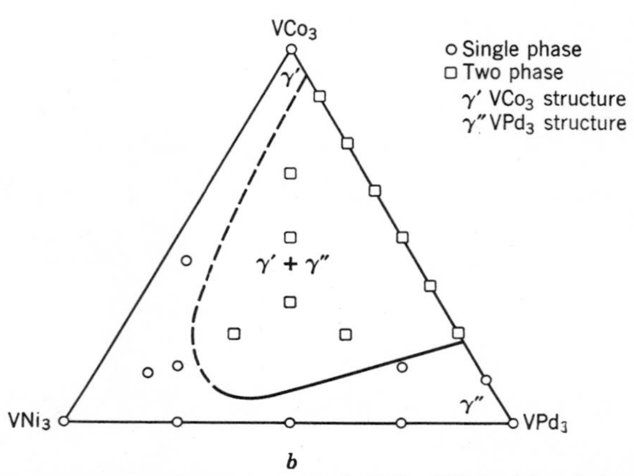

b

Fig. 14. An equilibrium diagram of the system VNi_3–VCo_3–VPd_3: (*a*) polythermal cross sections VNi_3–VCo_3, VCo_3–VPd_3, and VNi_3–VPd_3; (*b*) VNi_3–VCo_3–VPd_3, the isotherm at 600°C.

binary metallide systems, we can consider ternary solid solutions both of Kurnakov compounds, which are formed from solid solutions, and similar compounds, which crystallize from a liquid phase.

In References 36 and 37 many examples are cited of possible ternary systems composed of Kurnakov compounds in which, we suppose, continuous solid solutions should exist. Among these the following systems have been noted:

$FeNi_3$–$CrNi_3$–$MnNi_3$, CuAu–CuPd–CuPt, CrFe–VFe–CrCo, etc.

In the literature there are no such systems investigated experimentally.

Of the ternary systems with limited solid solutions of Kurnakov compounds we can mention the system VNi_3–VCo_3–VPd_3.[60] They are formed from the γ-solid solutions of the corresponding binary systems. As already mentioned, two of the binaries of this ternary system, VNi_3–VCo_3 and VNi_3–VPd_3, form continuous solid solutions, and the system VCo_3–VPd_3 shows limited solubility. In Fig. 14*a* vertical sections of the quasibinary systems VNi_3–VCo_3, VCo_3–VPd_3, and VNi_3–VPd_3 are given. Figure 14*b* presents an equilibrium diagram of the ternary system VNi_3–VCo_3–VPd_3 at 600°C. The diagrams show the regions of the ternary solid solutions γ' and γ'' based on the compounds VCo_3 and VPd_3, respectively, and a two-phase region $\gamma' + \gamma''$ in the binary system VCo_3–VPd_3, which extends partially into the ternary system VNi_3–VCo_3–VPd_3.

Of the experimentally investigated ternary systems, which form continuous solid solutions between compounds crystallizing from a liquid phase, we can mention the system $NbNi_3$–$TaNi_3$–$TiNi_3$.[101] In earlier papers[36,37] it has been shown on the basis of an investigation of one composition of an alloy (from three compounds) that alloys of the entire system should correspond to homogeneous metallide solid solutions. Subsequently, more detailed analysis of the ternary system $NbNi_3$–$TaNi_3$–$TiNi_3$ established the actual existence of continuous solid solutions in this system.[101] The investigated ternary system of these three metallides represents a triangular cross section of the quaternary system Ni–Nb–Ta–Ti, as is shown in the cross section of the tetrahedral figure in Fig. 15*a*. This quasi-ternary system, $NbNi_3$–$TaNi_3$–$TiNi_3$, was studied by methods of thermal and structural analysis and hardness measurements. As we show in Fig. 15*b* by projections of the liquidus surface, alloys of these cross sections solidify as continuous solid solutions. In the annealed state alloys of this ternary system have a homogeneous structure and a continuous character of the composition dependence of hardness. The possible existence of a small solubility discontinuity caused by the hexagonal modification of $TiNi_3$ probably will be

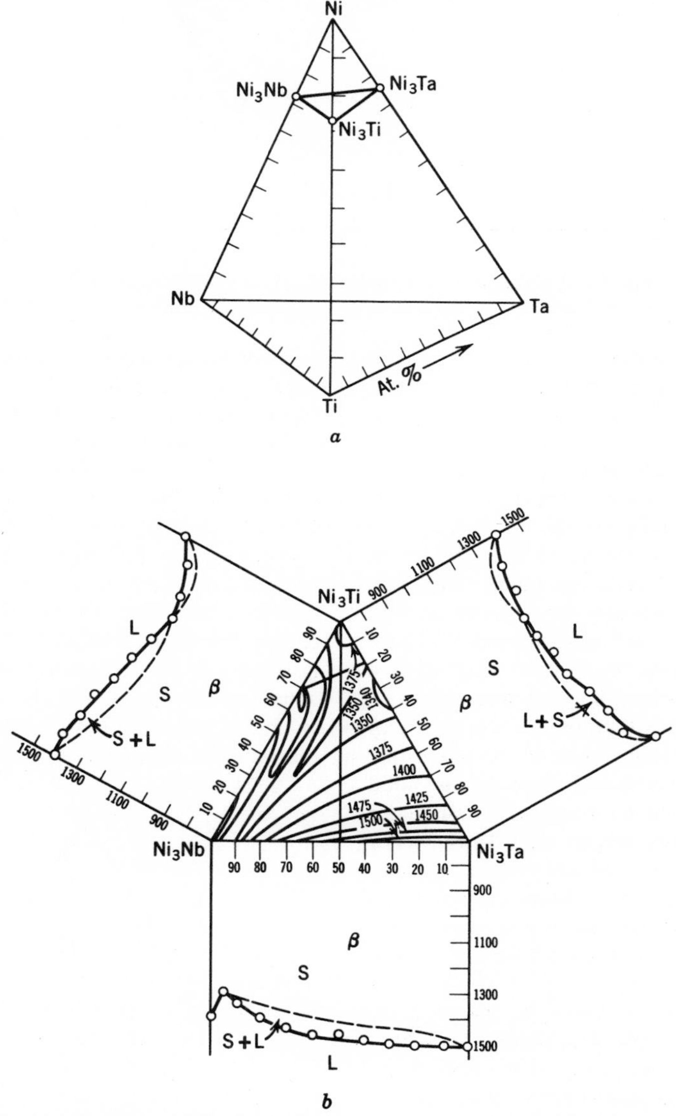

Fig. 15. (*a*) Tetrahedrality of the quaternary system Ni–Nb–Ta–Ti with the separation of the ternary system $NbNi_3$–$TaNi_3$–$TiNi_3$; (*b*) cross sections of investigated ternary alloys and projections of the surface of liquidus of the system $NbNi_3$–$TaNi_3$–$TiNi_3$.

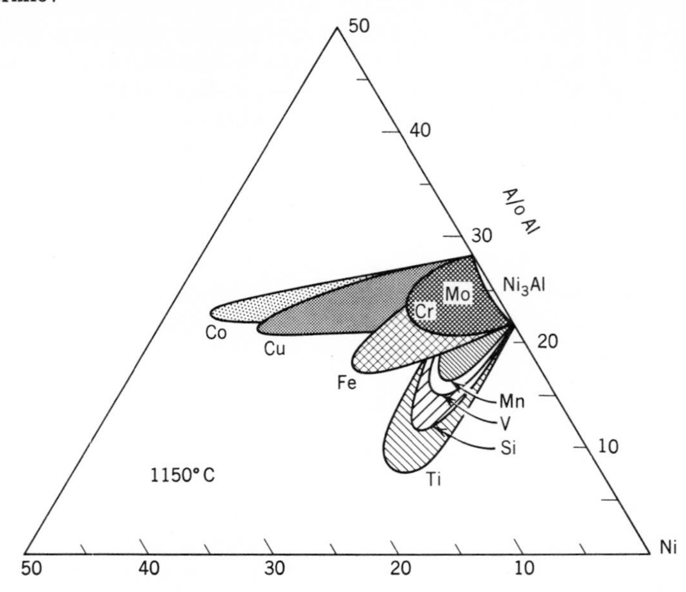

Fig. 16. Regions of limited solid solutions on the basis of Ni_3Al in different ternary systems at 1150°C (semischematic).

reflected in the appearance of a two-phase region in the corner of the ternary system $TiNi_3$–$NbNi_3$–$TaNi_3$ which is adjacent to this compound.

Similarly, we can assume the formation of such solutions in the ternary system of compounds $TiNi_3$–$ZrNi_3$–$HfNi_3$. In this system with the common metal, nickel, the other elements in the three compounds:—titanium, zirconium, and hafnium—are analogs, and all three compounds are isomorphous.[24] These factors give reason to believe that such a ternary system will belong to the group of continuous solid solutions. In the work of other authors[102] we have one of the interesting examples of the formation of the ternary solid solution among the following superconductor compounds: V_3Sn, Nb_3Sn, and Ta_3Sn. They are all isomorphous with the β-W structure.

The existence of continuous ternary solid solutions can also be expected in many other families of compounds—interstitial phases, semiconductor types, etc. As examples, such solutions can be supposed in the systems TiB_2–ZrB_2–HfB_2, TiB_2–VB_2–NbB_2, TiC–ZrC–HfC, TiC–NbC–TaC, and many other compounds with elements similar in their chemical properties.

There are also examples of the existence of limited solid solutions in ternary metallide systems. In Reference 52 the authors have investigated the solubility limit of the elements Cr, Co, Fe, V, Mo, Ti, and Si in the compound Ni_3Al as shown in Fig. 16. Another example of limited solid solutions based on metallides is the quasiternary system of compounds $CrNi_3$–$TiNi_3$–Ni_3Al.[103] As a result of a detailed investigation of this system by X-ray methods and microstructural analyses, phase diagrams of this system were established at 1000° and 750°C. In Fig. 17, according to data in References 103 and 104, an isotherm is given at 750°C; it is seen then that there are considerable regions of limited solid solutions (γ' and γ) based on Ni_3Al and $CrNi_3$ and that the compound $TiNi_3$ is almost void of

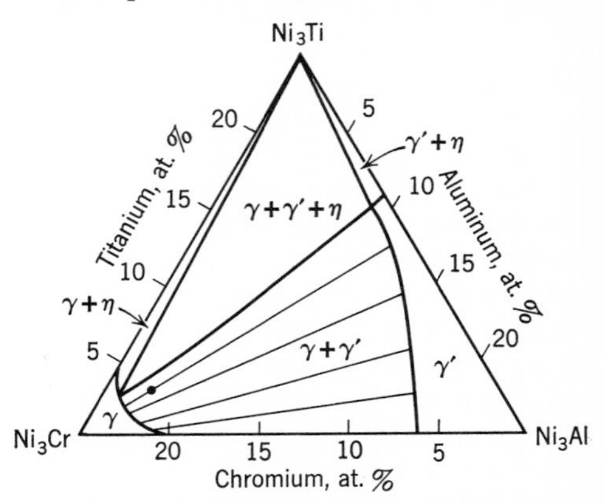

Fig. 17. A phase diagram of the quasiternary system $CrNi_3$–$TiNi_3$–Ni_3Al at 750°C.

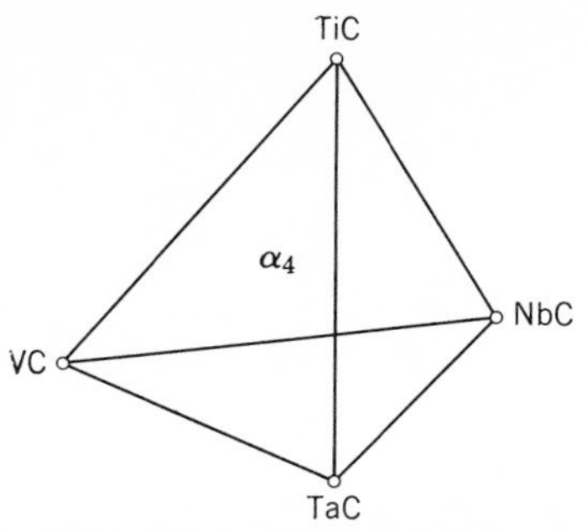

Fig. 18. Continuous quaternary solid solutions of monocarbides TiC–VC–NbC–TaC (schematic).

solubility for the components. The extent of the limited solid solutions in these compounds follows from the general rules for the formation of solutions among metallides (see above).

By successive development of the principles of phase equilibria between metallides in systems more complex than the ternary ones, the possibilities of the existence of solid solutions based on compounds in quaternary and more complex systems of metallides may be shown. By considering the main factors which determine the formation of solid solutions among metallides, we can obtain an idea of the formation of continuous and limited solid solutions in such systems. Compounds with purely metallic bonds, interstitial phases, and semiconductor compounds can take part as components.

Figure 18 shows schematically in a tetrahedral model a quaternary system of monocarbides TiC–VC–NbC–TaC. All these carbides are isomorphous and have a lattice of the NaCl type. In these compounds there is a common carbon atom, and the metal atoms Ti, V, Nb, Ta are close in atomic structure and form continuous solid solutions with one another.[24] Thus the entire quaternary system of these carbides will correspond to homogeneous solid solutions, as is shown in Fig. 18. Identical suppositions concerning the formation of complex solid solutions among multi-component systems involving titanium, vanadium, niobium, tantalum, zirconium, hafnium, and uranium monocarbides have been expressed in Reference 105. We can expect also the existence of such solutions in quaternary systems of borides, nitrides etc.

The system $TiNi_3$–$CrNi_3$–Ni_3Fe–Ni_3Al can serve as an example of an experimental investigation of quaternary systems of metallides with limited solid solutions. This system was studied in a paper[106] in connection with an investigation of alloys of a five-component system Ni–Cr–Fe–Ti–Al. In Fig. 19, according to the data in References 104 and 106, an equilibrium diagram of a quasiquaternary system $TiNi_3$–$CrNi_3$–Ni_3Fe–Ni_3Al is presented, from which it is seen that between the compounds $CrNi_3$–$FeNi_3$ there is a continuous series of solid solutions. The compound Ni_3Al has limited solubility, and $TiNi_3$ shows complete insolubility for the components.

The formation of solid solutions (continuous or limited) can be imagined not only among multicomponent systems composed of binary metallic compounds, but examples of such solutions among multicomponent systems composed of ternary compounds can also be shown. If ternary compounds contain elements that are analogs and if these elements obey the rules of formation of solid solutions, such compounds, despite the complexity of their compositions, should form mutual solutions. One of the interesting examples of such a case is the experimentally studied quasi-quaternary system:

$AgSbSe_2$–$AgSbTe_2$–$AgBiSe_2$–$AgBiTe_2$.[107]

As is seen, the system consists of four ternary compounds whose compositions are characterized by the existence of one common metal-silver and a pair of element analogs—Sb and Bi, Se and Te. The latter form between themselves continuous solid solutions.[24] At

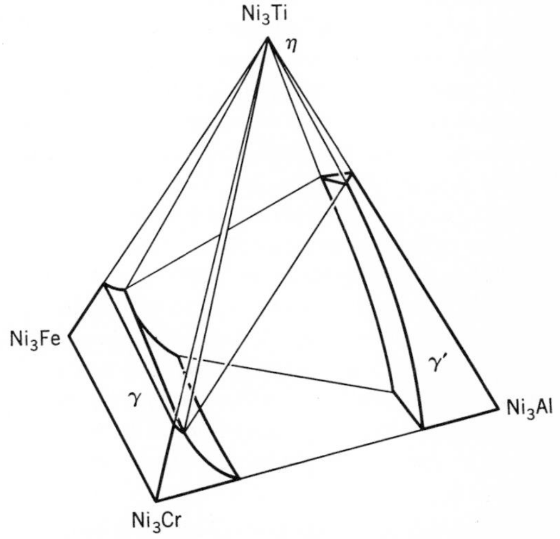

Fig. 19. The quasiquaternary system, $CrNi_3$–$TiNi_3$–Ni_3Fe–Ni_3Al with limited solid solutions.

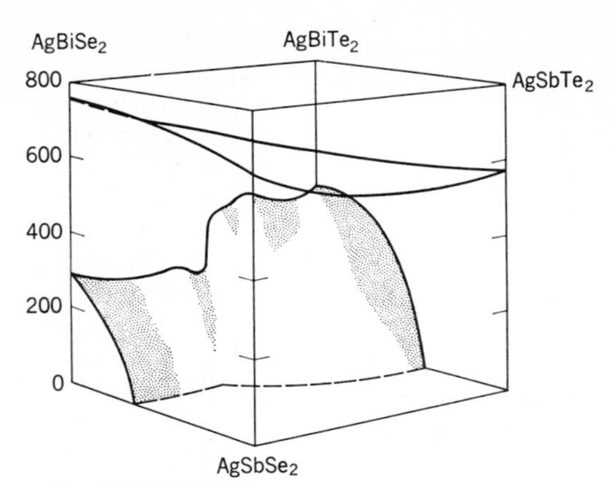

Fig. 20. A spatial model of the quaternary system $AgSbSe_2$–$AgSbTe_2$–$AgBiSe_2$–$AgBiTe_2$.

high temperatures the compounds $AgBiSe_2$, $AgBiTe_2$, $AgSbSe_2$, and $AgSbTe_2$ are found[107] to have a cubic lattice of the NaCl type. The compound $AgBiSe_2$ has a polymorphic rhombohedral modification at a temperature lower than 287°C, and at temperatures lower than 428°C the compound $AgBiTe_2$ becomes thermodynamically unstable and decomposes.

By a preliminary investigation of the six quasibinary systems formed from the ternary compounds $AgSbSe_2$, $AgSbTe_2$, $AgBiSe_2$, and $AgBiTe_2$, the authors[107] have shown that at high temperatures in all six quasibinary systems there are continuous solid solutions. In the quasibinary systems with the compounds $AgBiSe_2$ and $AgBiTe_2$ there is a solubility gap resulting from the polymorphic modification of $AgBiSe_2$ and from the solid state decomposition of the compound $AgBiTe_2$. As a result of a generalization of the data on the investigated quasibinary systems, the authors presented a quasiquaternary system of these ternary compounds in a spatial model (see Fig. 20). As is evident from the figure, at high temperatures in the system $AgSbSe_2$–$AgSbTe_2$–$AgBiSe_2$–$AgBiTe_2$ there are continuous quaternary solid solutions of compounds, and at low temperatures a discontinuity is observed.

Thus equilibrium investigations have shown that among simple and multicomponent metallide systems, comprised of a variety of metallic compounds with different types of chemical bonds, as well as among metallic systems in general, there are both continuous and limited solid solutions. The formation of such solutions is determined by the physicochemical factors enumerated in the Section 2 of this chapter. Studies of such systems can be made by the conventional methods of physicochemical analysis.

4. CONCLUSIONS

Metallic compounds or metallides are one of the main constituents in alloys. They play an important role in the complex of physicochemical and mechanical properties of metallic alloys. Such compounds are of independent value as a basis for many materials with specific physical properties. Properties of metallides can be considerably changed by alloying them with one another or with metals. During such alloying there may be formed: metallide solid solutions, eutectic mixtures, and more complex compositions of ternary and multielement compounds and phases based on them. These statements show the importance of investigations of regularities of metallide interaction, of the establishment of the nature of the equilibria between them, as well as of the functional relationships between various properties and the change of the composition and structure of these phases.

Of the many different systems of classification of metallic compounds, the one suggested by N. S. Kurnakov is used in this chapter, as well as a division of the great number of compounds into berthollides and daltonides. In addition to these two types of compounds, we have singled out Kurnakov compounds, which are formed with an ordered structure as a result of transformation of terminal solid solutions and of disordered intermediate phases. Depending on their physical and chemical nature, they can also be referred to as berthollides and daltonides.

In this chapter the main factors considered were those that determine the character of the interaction of metallides and the conditions for the formation of solid solutions (continuous and limited) and mechanical mixtures. For the formation of continuous solid solutions of metallides the following factors are important: (1) the identical type of crystal lattice, (2) the identical type of chemical bond, (3) the existence of analogous or identical elements in the compounds, and (4) the identical formula type. Violation of these conditions prevents complete solid solubility and leads to the formation of limited solid solutions or to no detectable solubility. These conclusions have been confirmed by many authors with examples of the formation of continuous and limited solid solutions among Kurnakov compounds, berthollides, and daltonides.

In addition to the consideration of experimental results from investigations of equilibrium systems among such metallides, the generalization of these data for use in the development of the theory of metallic-compound interaction is also of interest. These conclusions are also useful for predicting the interaction character of a number of unexplored systems based on metallides. By analogy with metallic systems, investigation of metallide systems can be carried out for binary, ternary, and multicomponent systems by taking as components binary, ternary, and more complex compositions of compounds.

ACKNOWLEDGMENTS

The author expresses gratitude to Corresponding Member of the U.S.S.R. Academy of Sciences N. V. Ageev, to Dr. B. K. Vulf, and to Dr. J. H. Westbrook for their review of the manuscript and their valuable comments and remarks on the contents of this chapter.

REFERENCES

1. Gibbs J. W., *Collected Works*. Longmans, Green and Company, New York, 1928.
2. Kurnakov N. S., *Introduction to Physico-Chemical Analysis*, The Publishing House of the USSR Academy of Sciences, 1940.
3. Anosov V. Ya., and S. A. Pogodin, *Foundations of Physico-Chemical Analysis*. The Publishing House of the USSR Academy of Sciences, 1947.
4. Mendeleyev D. I., *Scientific Archives*, Vol. 1. *The Periodic Law*, A. V. Topchiyev, ed. The Publishing House of the USSR Academy of Sciences, 1953.
5. Mendeleyev D. I., *Foundations of Chemistry*, Vols. I and II. Khimizdat, USSR, 1954.
6. Ageyev N. V., *The Chemistry of Metallic Alloys*. The Publishing House of the USSR Academy of Sciences, 1941.
7. Kornilov I. I., *Izv. Akad. Nauk SSSR, Otd. Khim. Nauk*, No. 4, (1957) 397.
8. Kornilov I. I., "Metallochemistry and Some of Its Problems, a Report at the International Colloquium 'The Impact of Science on Material Technology,'" *Argentina*, April 2–7, 1962.
9. *Proceedings of the National Physical Laboratory Symposium, The Physical Chemistry of Metallic Solutions and Intermetallic Compounds*, Vols. I, II. Her Majesty's Stationery Office, London, 1959.
10. Kurnakov N. S., and N. I. Stepanov, *J. Russ. Phys.-Chem. Soc.* **37** (1905) 568.
11. Kurnakov N. S., *Collection of Selected Works*, Vol. 2. GONTI, 1939, p. 24. Moscow.
12. Commission on Nomenclature of Chemical Compounds of the USSR Academy of Sciences, *Draft Nomenclature of Inorganic Compounds*. The Publishing House of the USSR Academy of Sciences, 1959.
13. Bochvar A. A., *Science of Metals*, Metallurgizdat, 1956. Moscow.
14. Umansky Ya. S., B. N. Finkelshtein, M. N. Blanter, S. T. Kishkin, N. S. Fastov, and S. S. Gorelik, *Physical Foundations of Science of Metals*. Metallurgizdat, 1955. Moscow.
15. Hume-Rothery W., *The Structure of Metals and Alloys*. 3rd ed. The Institute of Metals, London, 1954.
16. Darken L. S., and R. W. Gurry, *Physical Chemistry of Metals*. McGraw-Hill Book Co., New York, 1953.
17. Pauling L., *The Nature of Chemical Bond*, 3rd ed. The Cornell University Press, Ithaca, 1960.
18. Kornilov I. I., and B. K. Vulf, *Usp. Khim.* (Advances in Chemistry) **28,** 9 (1959).
19. Westbrook J. H., *Mechanical Properties of Intermetallic Compounds*. Wiley, New York, 1959.
20. Kurnakov N. S., *Izv. St. Petersburg Poly. Inst.* **22** (1914) 487; *J. Inst. Metals* **15** (1916) 305.

21. Newkirk J. B., R. Smoluchowski, A. H. Geisler, and D. L. Martin, *J. Appl. Phys.* **22** (1951) 290.
22. Rhines F. N., and J. B. Newkirk, *Trans. ASM* **45** (1953) 1029.
23. Kornilov I. I., and N. M. Matveyeva, *Dokl. Akad. Nauk SSSR* **139,** 4 (1961) 880.
24. Hansen M., and K. Anderko, *Constitution of Binary Alloys.* McGraw-Hill Book Co., New York, 1958.
25. Grum-Grzhimailo, N. V., I. I. Kornilov, E. N. Pylayeva, and M. A. Volkova, *Dokl. Acad. Sci. SSSR* **137,** 3 (1961).
26. Mott N., and H. Jones, *The Theory of the Properties of Metals and Alloys,* 1936.
27. Konobeyevsky S. T., *Izv. Sektora Fiz.-Khim. Analiza, Akad. Nauk SSSR* **16,** 19 (1943).
28. Boky G. B., *Introduction to Crystallo-Chemistry.* The Publishing House of the Moscow State University, Moscow, 1954.
29. Kripyakevich P. I., and E. E. Cherkashin, Systematization of Binary Intermetallic Phases, *Izv. Sektora Fiz.-Khim. Analiza, Akad. Nauk SSSR* **24** (1954) 59.
30. Makarov E. S., *Zh. Neorgan. Khim.* **1** (1956) 1583.
31. Dehlinger U., and G. E. Schulz, *Z. Krist.* **102** (1940) 377.
32. Ageyev N. V., *The Nature of Chemical Bond in Metallic Alloys.* USSR Academy of Sciences, 1947.
33. Samsonov G. V., *Poroshkovaya Met., Akad. Nauk SSSR* (Powder Metallurgy), 2 (1962) 3.
34. Zhuze V. P., (ed.), *Semiconductor Compounds-Problems of Chemical Bond, Collection of Articles.* Foreign Literature Publishing House, Moscow, 1960.
35. Kornilov I. I., *Dokl. Akad. Nauk SSSR* **81,** 4 (1951) 597.
36. Kornilov I. I., *Dokl. Akad. Nauk. SSSR* **106,** 3 (1956) 476.
37. Kornilov I. I., "Solid Solutions of Metallic Compounds," *Proc. NPL Symposium No.* 9 *Physical Chemistry of Metals and Alloys* (1959), V. I, Paper 3J.
38. Goryunova N. A., and N. N. Fedorova, *Dokl. Akad. Nauk SSSR* **90,** 6 (1953) 1039.
39. Kotelnikov R. B., *Zh. Neorgan. Khim.* **3,** 4 (1958) 841.
40. Post B., F. Glaser, and D. Moskowitz, *Acta Met.* **2** (1954) 20.
41. Meyerson G. A., G. V. Samsonov, R. B. Kotelnikov, M. S. Voikova, I. P. Yevteyeva, and S. D. Krasnenkova, *Zh. Neorgan. Khim.* **3,** 4 (1958) 898.
42. Kornilov I. I., and E. N. Pylayeva, *Dokl. Akad. Nauk SSSR* **97,** 3 (1954) 455.
43. Kornilov I. I., and N. M. Matveyeva, *Dokl. Akad. Nauk SSSR* **98,** 5 (1954) 787.
44. Gladyshevsky E. I., Dissertation for the Scientific Degree: Cand. Chem. Sci. Lvov State University, Lvov 1953.
45. Mikheyeva V. I., and G. G. Babayan, *Dokl. Akad. Nauk SSSR* **108,** 6 (1956) 1086.
46. Vogel R., and H. Klose, *Z. Metallk.* **45,** 12 (1954) 670.
47. Cherkashin E. E., E. I. Gladyshevsky, P. I. Kripyakevich, Yu. B. and Kuzma, *Zh. Neorgan. Khim.* **3** (1958) 650.
48. Gorshkov I. E., and N. A. Goryunova, *ibid.* **3** (1958) 668.
49. Samsonov G. V., and Ya. S. Umansky, *Solid Compounds of High-Melting Metals.* Metallurgizdat, 1957.
50. Nowotny H., R. Kieffer, F. Benesovsky, C. Brukl, and E. Rudy, *Monatsh. Chem.* **90,** 5 (1959) 669.
51. Duwez P., and F. Odell, *J. Electrochem. Soc.* **97** (1950) 299.
52. Westbrook J. H., and R. W. Guard, *Trans. Met. Soc. AIME* **215** (1959) 807.
53. Mikheyeva V. I., *Izv. Sektora Fiz.-Khim. Analiza, Akad. Nauk SSSR* **17** (1949) 174.
54. Pratt J., and G. Raynor, *Proc. Roy. Soc. A* **205** (1951) 103.
55. Pratt J., and G. Raynor, *J. Inst. Metals* **79** (1951) 211.
56. Vulf B. K., *Fiz. Metal. i Metalloved.* **3,** 1 (1956) 97.
57. Vulf B. K., *Usp. Khim.* **29,** 6 (1960) 774.
58. Boky G. B., B. K. Vulf, and N. L. Smirnova, *Zh. Strukt. Khim.* **2,** 7 (1961) 74.
59. Rhines F. N., *Phase Diagrams in Metallurgy.* McGraw-Hill Book Co., New York, 1956.
60. Köster W., and W. Gmohling, *Z. Metallk.* **51,** 7 (1960) 385.
61. Kornilov I. I., and T. T. Nartova, *Dokl. Akad. Nauk SSSR* **131,** 4 (1960) 837.
62. Kornilov I. I., and T. T. Nartova, *Dokl. Akad. Nauk SSSR* **140,** 4 (1961) 829.
63. Kornilov I. I., and E. N. Pylayeva, *Zh. Neorgan. Khim.* **3** (1958) 673.
64. Pylayeva E. N., E. N. Gladyshevsky, and P. I. Kripyakevich, *Zh. Neorgan. Khim.* **3,** 7 (1958) 1626.
65. Van Thyne R. J., H. D. Kessler, and M. Hansen, *Trans. AIME* **197** (1953) 1209.
66. Levinger B. W., *Trans AIME* **197** (1953) 196.
67. Bradley A., and A. Taylor, *J. Inst. Metals* **66,** 2 (1940) 53.
68. Ivanov O. S., *Izv. Sektora Fiz.-Khim. Analiza, Akad. Nauk SSSR* **19** (1950) 503.
69. Ivanov O. S., *Dokl. Akad. Nauk SSSR* **78,** 6 (1957) 1157.

70. Ageyev N. V., and E. S. Makarov, *Compt. Rend. Acad. Sci. URSS* **38** (1943) 20.
71. Gladyshevsky E. I., P. I. Kripyakevich, and Yu. B. Kuzma, *Fiz. Metall. i Metalloved.* **2** (1956) 454.
72. Kieffer R., F. Benesovsky, and H. Schnoth, *Z. Metallk.* **44** (1953) 437.
73. Nowotny H., *Monatsh. Chem.* **85** (1954) 242.
74. Parthé E., and H. Nowotny, *Monatsh. Chem.* **86** (1955) 385.
75. Samsonov G. V., *Silicides and Their Use in Engineering*. Publishing House of the Ukrainian Academy of Sciences, 1959.
76. An-Si-Yun, *An Investigation of the Phase Equilibrium Diagram of Molybdenum Alloys with Chromium and Silicon.* Moscow Institute of Steel, Autoreferat, 1962.
77. Hagg G., and R. Kiessling, *J. Inst. Metals* **81** (1952) 58.
78. Nowotny H., R. Kieffer, F. Benesovsky, and C. Brukl, *Monatsh. Chem.* **90,** 2 (1959) 86.
79. Samsonov G. V., L. Ya. Markovsky, A. F. Zhigach, and M. G. Valyaskho, *Boron, Its Compounds and Alloys*. The Publishing House of the Ukrainian Academy of Sciences, 1960.
80. Samsonov G. V., T. S. Verkhoglyadova, S. N. Lvov, and V. F. Nemchenko, *Dokl. Akad. Nauk SSSR* **142,** 4 (1962) 862.
81. Yelagina E. I., and N. Kh. Abrikosov, *Dokl. Akad. Nauk SSSR* **111,** 2 (1956) 353.
82. Abrikosov N. Kh., A. M. Vasserman, and L. V. Poretskaya, *Dokl. Akad. Nauk SSSR* **123,** 2 (1958) 273.
83. Beglaryan M. L., and N. Kh. Abrikosov, *Dokl. Akad. Nauk SSSR* **128,** 2 (1959) 345.
84. Folberth O. G., *Z. Naturforsch.* **10a** (1955) 502.
85. Baruch P., and M. Desse, *Compt. Rend. Acad. Sci.* **241** (1955) 1040.
86. Köster W., and B. Thoma, *Z. Metallk.* **46** (1955) 293.
87. Peretti E. A., *Trans. Met. Soc. AIME* **1** (1959) 79.
88. Strauss A. J., and Farnell Lyne, *Quatern. Progress Rept. Solid State Research*, October 28, 1959.
89. Strauss A. J., and Farnell Lyne, *Quatern. Progress Rept. Solid State Research*, April 24, 1960.
90. Grigoryev A. T., and V. V. Kuprina, *Zh. Neorgan. Khim.* **6,** 8 (1961) 1897.
91. Pan V. M., *Dopovidi Akad. Nauk Ukr. SSR* (in Ukrainian), 3 (1961) 332.
92. Petzow G., S. Steeb, and J. Ellinghaus, *J. Nucl. Mater.* **4,** 3 (1961) 316.
93. Petzow G., and J. Kvernes, *Z. Metallk.* **52,** 10 (1961) 603.
94. Mints R. S., G. F. Belyayeva, and Yu. S. Malkov, *Dokl. Akad. Nauk SSSR* **143,** 4 (1962) 871.
95. Dubrovskia, L. B., and P. V. Geld, *Zh. Neorgan. Khim.* **7** (1962) 145.
96. Abrikosov, N. Kh., K. A. Duldina, and T. A. Danilyan, *Zh. Neorgan. Khim.* **3,** 7 (1958) 1632.
97. Beglaryan M. L., and N. Kh. Abrikosov, *Dokl. Akad. Nauk SSSR* **129,** 1 (1959) 135.
98. Mirgalovskaya M. S., and E. B. Skudnova, *Zh. Neorgan. Khim.* **4,** 5 (1959) 1111.
99. Yelagina E. I., and N. Kh. Abrikosov, *Zh. Neorgan. Khim.* **4,** 7 (1959) 1638.
100. Yelagina E. I., *Collection of Proc. IV Conf. Problems of Metallurgy and Physics of Semiconductors.* Publishing House of the USSR Academy of Sciences, 1961, p. 153.
101. Kornilov I. I., and E. N. Pylayeva, *Izv. Akad. Nauk SSSR Otd. Khim. Nauk*, **2** (1961) 197.
102. Lody G. D., and I. I. Tank, *Low Temperature Physics*, University of Toronto, Proceed. of the VII Conference, October 1960.
103. Taylor A., *J. Metals* (1956) 1356.
104. Betteridge W., *The Nimonic Alloys.* Edward Arnold Ltd., London, 1959.
105. Nowotny H., R. Kieffer, F. Benesovsky, and E. Laube, *Monatsh. Chem.* **88** (1957) 336.
106. Taylor A., *J. Metals* **9** (1957) 72.
107. Wernick J. H., S. Geller, and K. E. Benson, *Phys. Chem. Solids*, **2/3** (1958) 240.

part VI

Kinetics and Transformations

chapter 20

Diffusion

WILLIAM C. HAGEL
Engineering Recruiting Service
General Electric Company
Schenectady, New York

1. INTRODUCTION

During a 1950 seminar on atom movements, Mehl[1] noted that, "Knowledge of diffusion rates in intermediate solid phases (e.g., beta, gamma, delta, etc., brass) is nonexistent .... No one has attempted measurements seriously." Aided by an abundance of radioactive isotopes from nuclear reactors, such studies did begin, and toward the end of 1962 some seventy references could be found treating diffusion in intermetallic compounds.

A large impetus was provided by the commercial possibilities of compound semiconductors, and Russian investigators have been rather prolific in data gathering. Other primary reasons for measuring diffusion coefficients are the need to uncover mechanisms by which atoms or ions change position with one another and to establish a basis of comparison for the related rate processes of precipitation, creep, oxidation, grain growth, or sintering. For the research scientist, intermetallic compounds thus present a wide variety of crystal structures, bonding types, and lattice defects; in certain cases, it is possible to produce single crystals whose purity and perfection exceed those of common metals by orders of magnitude. For the materials engineer, who must have tomorrow's components and devices today, much guesswork can be eliminated.

There are numerous excellent texts and review articles on the overall field of diffusion, but none possess more than a few paragraphs concerning intermetallic compounds. Barrer,[2] Jost,[3] Crank,[4] and LeClaire[5] are classics, Birchenall[6,7] ponders mechanisms; Tomizuka[8] covers experimental techniques; Lazarus[9] ably simplifies theory. Familiarization with diffusion in semiconductors, mainly germanium and silicon, can be obtained from Hobstetter,[10] Smits,[11] and Reiss and Fuller.[12]

It has long been recognized that diffusion in crystalline solids should bear some relationship to absolute melting temperature T_m. The law of corresponding states suggests replacement of the usual Arrhenius expression

$$D = D_0 \exp(-Q/RT)$$

with

$$D = D_0 \exp(-b\, T_m/T),$$

where D is the diffusion coefficient at absolute temperature T, D_0 is the frequency factor, Q is the activation energy, and R is the gas constant. Sherby and Simnad[13] have found good empirical correlation for substitutional self-diffusion in metals and dilute alloys, where b is the sum of a crystal-structure factor K_0 and valence V. Although the comparisons should not be forced, they appear correct in form and permit the rough prediction of diffusion coefficients whenever experimental data are not available.

All aspects considered, the approach used in this chapter is to group determinations according to compound crystal structure and to plot them as a function of T_m/T. When a solidification range is present, T_m is taken as the halfway point; Hansen's[14] binary diagrams are the major references. Anomalous results and areas requiring further study become readily apparent. Let us start historically with the brasses and conclude with a prognostication of where the next dozen years may lead.

2. BODY-CENTERED CUBIC STRUCTURES

2.1 β-CuZn, γ-CuZn, and β-AgZn

While a large quantity of data on diffusion in the fcc terminal solid solution of α-CuZn was being accumulated, compositions were occasionally extended into the bcc β-CuZn-phase field. As early as 1928, Köhler[15] ran couples of α- and β-CuZn at 350°C to find interdiffusion coefficients ($\tilde{D}$) of 5.8×10^{-11} and 1.3×10^{-9} cm^2/sec, respectively. Using Dunn's[16] vaporization method, Seith and Krauss,[17] almost concurrently with Petrenko and Rubinstein,[18] compared interdiffusion in the α- and β-phases of CuZn, AgZn, and AgCd. These investigators reported that diffusion in the β-phases is approximately two orders of magnitude faster than in the α-phases and that the β-activation energies are half those of the α. Such differences were attributed to the less dense atomic packing and lower melting temperatures of the β-phases. Although the vaporization method has since declined in repute as a means of providing precision measurements, the primitive $\tilde{D}$'s are qualitatively correct.

Using thin, neutron-irradiated brass foil welded between polycrystalline β-CuZn compacts, Inman et al.[19] determined the self-diffusion of ^{64}Cu and ^{65}Zn by a sectioning technique, in which the short and long half-lives of the isotopes provided counting discrimination. Their frequency factors and activation energies are listed in Table 1 with other determinations pertinent to this section. The general term *intrinsic diffusion* in the table is used to denote the self-diffusion or impurity diffusion of one elemental species rather than the interdiffusion of more than one.

Landergren et al.[20] measured interdiffusion coefficients and marker displacements in β-CuZn. For the composition range of 43.5 to 49.0 a/o Zn, they found D_0 varying from 1.8×10^{-2} to 1.3×10^{-2} cm^2/sec with Q changing from 19.9 to 18.2 kcal/mole. At 750°C, the compositional dependence of $\tilde{D}$ was sigmoidal in shape. Tungsten wire markers moved to the high-zinc side of each couple for distances consistent with Inman's self-diffusion data substituted in Darken's relations.[21] A ring-type diffusion mechanism seemed highly improbable. In fact, the system has yet to be discovered in which marker movement (known as the Kirkendall effect) does not occur. The work of Resnick and Balluffi[22] coincided with Landergren's. Here zinc was diffused from the vapor phase into marker-covered β-CuZn samples, and concentration-penetration curves were obtained by chemical analysis of parallel slices. Since surface concentration varied in an unknown manner, it was possible only to calculate the ratio D_{Zn}/D_{Cu}. Their ratios at 600°, 700°, and 800°C were about 30 per cent higher than those of Inman et al. and did not show any marked variation with temperature.

Kuper et al.[23] measured the self-diffusion of ^{64}Cu and ^{65}Zn, and the impurity diffusion of ^{124}Sb, in β-CuZn single crystals from 264° to 817°C. Therefore, the change from nearly complete ordering of the CsCl type to complete randomness could be followed for the same bcc crystal structure. Above 500°C, their data varied linearly with $1/T$ in fair agreement with Inman et al., if we remember that the earlier measurements were performed on samples of lower zinc content. Below 468°C, diffusion fell off rapidly as ordering increased, and some curvature in the Arrhenius plot occurred. D_{Zn} was approximately a

TABLE 1

Intrinsic Diffusion in β-CuZn, γ-CuZn, and β-AgZn

Letter	Compound	Composition	Melting Temperature, °C	Crystal Structure	Element	Melting Point, °C	Reference	Temperature Range, °C	D_0, cm²/sec	Q, kcal/mole
a	β-CuZn	45 a/o Zn	888	A2-cubic	^{65}Zn	419.5	19	636–870	3.1×10^{-2}	23.3
b					^{64}Cu	1083		636–870	3.8×10^{-2}	25.0
c	β-CuZn	47–48 a/o Zn	880	A2-cubic	^{65}Zn	419.5	23	499–718	3.5×10^{-3}	18.8
d					^{64}Cu	1083		497–817	1.1×10^{-2}	22.0
e					^{124}Sb	630.5		498–594	8.0×10^{-2}	23.5
f	β'-CuZn	47–48 a/o Zn	880	B2-cubic	^{65}Zn	419.5	23	264–480	163–78,000	36.3–44.2
g					^{64}Cu	1083		292–480	80–180	36.0–37.1
h					^{124}Sb	630.5		351–459	163–78,000	36.3–44.2
i	γ-CuZn	65.3–65.7 a/o Zn	790	$D8_2$-cubic	Zn	419.5	27	$D(375°C) = 1.4 \times 10^{-8}$ cm²/sec		
								$D(475°C) = 1.5 \times 10^{-7}$ cm²/sec		
j					Cu	1083		$D(375°C) = 1.4 \times 10^{-9}$ cm²/sec		
								$D(475°C) = 1.6 \times 10^{-8}$ cm²/sec		
k	β-AgZn	50 a/o Zn	690	A2-cubic	^{65}Zn	419.5	28	350–600	55	27.0
l					^{110}Ag	960.5		350–600	2.18×10^{-3}	16.6

factor of two greater than D_{Cu}, although an equality seems possible at still lower temperatures. D_{Sb} was much the same as D_{Zn}, except for a small increase at 600°C. The investigation sought to distinguish between the interchange, vacancy, and interstitialcy mechanisms.[24] None was clearly evident, and almost by default interstitialcies were postulated. Shortly thereafter, Slifkin and Tomizuka[25] reverted to suggesting a vacancy mechanism via nearest-neighbor jumps; then Lidiard[26] stated that their argument was inconsistent with the condition of thermodynamic equilibrium. After more experimental results are documented, we shall proceed to theoretical considerations in Section 6.

Interdiffusion in γ-CuZn, a more complicated cubic phase possessing a defect superlattice rather than random defects, has been studied by Mehl and Lutz.[27] They used vapor-solid couples and applied the Boltzmann-Matano analysis to zinc composition gradients obtained with an electron microprobe. $\tilde{D}$ values were reported for the temperature range of 375° to 650°C and compositions of 59 to 68 a/o Zn. Activation energies decreased continuously from about 23 to 15 kcal/mole with increasing zinc content; no maximum was observed at the stoichiometric composition of 61.5 a/o Zn. Isothermal $\tilde{D}$'s increased with increasing zinc content in a nonmonotonic manner; no minimum was observed at the stoichiometric composition. Comparison between the β- and γ-phases shows that $\tilde{D}$ values for the latter are one to ten times larger than the former. From marker-movement data and the Darken relations, self-diffusion coefficients can be approximated, but direct measurements using ^{64}Cu and ^{65}Zn are still required.

A few data are starting to accumulate for self-diffusion in the α- and β-phases of the silver-zinc system. Kuzmenko et al.[28] observed an increase in both D_{Zn} and D_{Ag} values on going from 30 to 50 a/o Zn between 350° and 600°C. Gertsriken et al.[28] report that D_{Zn} is approximately one order of magnitude faster than D_{Ag} in β-AgZn between 500° to 700°C; for example, at 500°C $D_{Zn} = 10^{-7}$ cm²/sec and $D_{Ag} = 10^{-8}$ cm²/sec. The difference decreases at lower temperatures. From all appearances, the situation is the same as with α- and β-CuZn.

In Fig. 1 the dashed lines show the experimental limits found by Sherby and Simnad[13] for self-diffusion in pure metals possessing the bcc, hcp, or fcc crystal structures. Solid lines are identified by a letter assigned each determination listed in Table 1; they were drawn from the accompanying information. Many correlations were tried to aid in predicting which component of a binary compound would be the faster moving species, and almost without exception it is the element possessing the lower melting point. When reduced-temperature comparisons are made, notably high diffusion rates are characteristic of this particular group of compounds containing zinc.

2.2 β-NiAl, β-CoAl, and β-FeAl

On following the variations in lattice parameter and density for β-NiAl, Bradley and Taylor[30] uncovered defects in the form of unfilled lattice sites on the aluminum-rich side of stoichiometry. At 50 a/o Al, there are two interpenetrating simple cubic lattices of nickel and aluminum; each atom has eight nearest-neighbors of the opposite kind and six next-nearest neighbors of the same kind. Below 50 a/o Al, some of the aluminum atoms are replaced at random by nickel atoms. Because the latter are smaller and heavier, lattice parameter decreases and density increases, as would be expected. Above 50 a/o Al, however, the converse does not apply; the lattice parameter decreases rather than increases and the density decreases more than would be calculated from normal substitution. The results can be explained by postulating that the aluminum cubic lattice remains filled and nickel atoms are subtracted from its cubic lattice, leaving an excess of vacancies.

To study the effect of a large vacancy concentration on diffusion, Smoluchowski and Burgess[31] measured the impurity diffusion of ^{60}Co in seven β-NiAl alloys after 18 hours at 1150°C. They detected little change in D values between 48 and 53 a/o Al and raised doubt as to the ability of excess vacancies to increase diffusion rates in a highly ordered structure, in which the nickel-vacancy size is smaller than that of an aluminum atom. Nix and Jaumot[32] investigated the self-diffusion of ^{60}Co in five β-CoAl alloys (which also possess the bcc CsCl-type structure) at 1050°, 1150°, and 1250°C. At each temperature, there was a

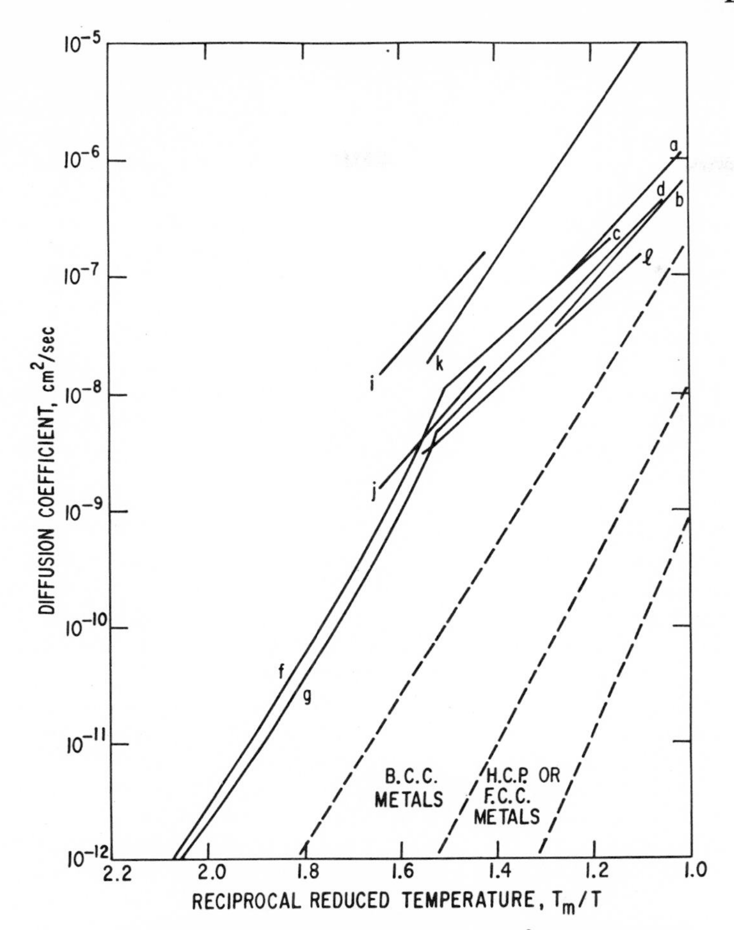

Fig. 1. Intrinsic diffusion coefficients for β-CuZn, γ-CuZn, and β-AgZn calculated from the data of Table 1 and plotted as a function T_m/T.

minimum in D values at the stoichiometric composition with an indisputable increase on the aluminum-rich side; here, the activation energy dropped off sharply. The excess vacancies as next-nearest neighbors to the cobalt atoms appeared to be of much consequence for their rapid self-diffusion.

Berkowitz et al.[33] reconsidered in further detail the diffusion of ^{60}Co in six β-NiAl alloys. Their results for three of these compositions can be compared in Fig. 2 with those of Smoluchowski and Burgess. A definite inverse cusping converged on the stoichiometric composition. At 49.3 a/o Al, a maximum Q of 80.6 kcal/mole was determined; at higher aluminum contents, Q dropped to an erratic low level of about 56 kcal/mole. If we treat the stoichiometric Q as consisting of a ΔH_f (for vacancy formation) and a ΔH_m (for vacancy motion), ΔH_f could be around 25 kcal/mole—a reasonable value. Gertsriken and Dekhtyar[34] also reexamined the self-diffusion of ^{60}Co in β-CoAl containing 42, 49, and 50.7 a/o Al, with results similar to those of Nix and Jaumont. Gertsriken et al.[35] then explored the iron-aluminum system from 3.47 to 52.2 a/o Al using ^{60}Co and ^{59}Fe, and a sharp decrease in Q above 50 a/o Al was likewise noted. The work bears repeating on alloys prepared in smaller increments and closer to the stoichiometric β-FeAl composition. Use of aluminum isotopes for all compounds involves procedural difficulties owing to their low specific activity, high expense, and the short half-lives of some.

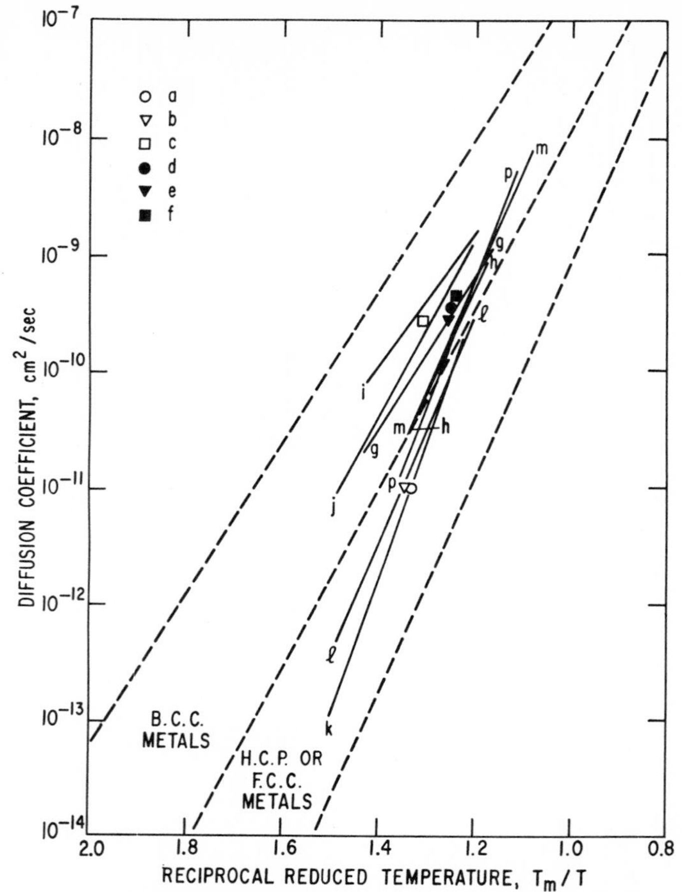

Fig. 2. Intrinsic diffusion coefficients for β-NiAl, β-CoAl, and β-FeAl calculated from the data of Table 2 and plotted as a function of T_m/T.

Again, the data available for listing in Table 2 have been converted to reduced-temperature points and lines in Fig. 2. The Nix-Jaumot and Berkowitz D values conform well to those measured for pure bcc metals; the high vacancy concentration in the aluminum-rich alloys does not cause any unusual displacements. Diffusion rates in the near-stoichiometric compounds do, on the other hand, tend to lie along the low side of the scatter band. The Gertsriken-Dekhtyar β-CoAl investigation gave the lowest results for cobalt-rich alloys. All too frequently diffusionists make only Q- or D_0-value arguments when actually quite large errors can reside in these numbers, yet it takes woefully poor technique to shift D-value positions by an order of magnitude.

2.3 β-AgMg and β-AuCd

The hardness[36] and tensile behavior[37] of β-AgMg suggested the possibility that the greater strengthening and lower flow stress Q's observed in magnesium-rich alloys of β-AgMg, compared to silver-rich alloys, result from a large concentration of vacancies similar to those appearing in β-NiAl. Although a part of this effect was found later to result from grain-boundary hardening,[38] Hagel and Westbrook[39] measured the self-diffusion of ^{110}Ag in three β-AgMg alloys containing 45.8, 49.8, and 52.0 a/o Mg; lattice parameters and densities were followed on a more numerous assortment and wider range of compositions. The conclusion was that only substitutional defects are present on both sides of stoichiometry.

TABLE 2

Intrinsic Diffusion in β-NiAl, β-CoAl, and β-FeAl

Letter	Compound	Composition	Melting Temperature °C	Crystal Structure	Element	Melting Point, °C	Reference	Temperature Range, °C	D_0, cm²/sec	Q, kcal/mol
a	β-NiAl	52.5 a/o Al	1620	B2-cubic	^{60}Co	1495	31	1150	$D = 1 \times 10^{-11}$ cm²/sec	
b		49.0	1640						$D = 1 \times 10^{-11}$ cm²/sec	
c		42.0	1590						$D = 2.6 \times 10^{-10}$ cm²/sec	
d	β-CoAl	51.0 a/o Al	1630	B2-cubic	^{60}Co	1495	32	1250	$D = 3.4 \times 10^{-10}$ cm²/sec	
e		49.7	1640						$D = 2.8 \times 10^{-10}$ cm²/sec	
f		43.0	1620						$D = 4.2 \times 10^{-10}$ cm²/sec	
g	β-NiAl	52.7 a/o Al	1620	B2-cubic	^{60}Co	1495	33	1050–1350	4.7×10^{-2}	56.6
h		49.3	1640					1150–1350	5.77×10^{1}	80.6
i		44.5	1615					1050–1350	7.2×10^{-3}	47.1
j	β-CoAl	50.7 a/o Al	1620	B2-cubic	^{60}Co	1495	34	1000–1300	1.3	65.0
k		49.0	1645						3.33×10^{4}	102
l		42.0	1620						1.84×10^{2}	85.0
m	β-FeAl	52.2 a/o Al	1270	B2-cubic	^{60}Co	1495	35	900–1150	1.48×10^{2}	67.0
n		52.2	1270		^{59}Fe	1534			6.0×10^{1}	66.0
p		47.3	1320		^{60}Co	1495			6.3×10^{3}	79.0

Domian and Aaronson[40] next provided ^{110}Ag self-diffusion data for six β-AgMg alloys containing on the average 41.1, 43.6, 48.5, 48.7, 52.8, and 57.1 a/o Mg. Accepting a six-jump cycle as the mechanism of diffusion in CsCl-type compounds, they went through an involved many-parameter calculation of Q values showing fair agreement with experimental Q's. The development required knowledge of quantities for the sums of ΔH_f and ΔH_m for both magnesium and silver in *disordered* β-AgMg. Because β-AgMg is ordered up to its melting temperature, no real data can be obtained, but the maneuver was used of going back to the case of β-CuZn containing 47–48 a/o Zn. Then it was assumed that ratios for $Q_{Zn}/T_m(\beta\text{-CuZn})$ and $Q_{Cu}/T_m(\beta\text{-CuZn})$ equalled the disordered $(\Delta H_f + \Delta H_m)_{Mg}/T_m(\beta\text{-AgMg})$ and disordered $(\Delta H_f + \Delta H_m)_{Ag}/T_m(\beta\text{-AgMg})$, respectively. As seen in Fig. 1, diffusion in zinc-containing compounds is unusually fast and provides abnormally low Q values, which should not be used as standards for other systems. Choice of any mechanistic model for β-AgMg still requires the difficult experimental determination of magnesium self-diffusion.

Question was raised as to the proper dependence of activation energy with composition on the silver-rich side of stoichiometric β-AgMg. Diffusion rates in three large-grained cast samples containing 45.0, 43.8 and 43.6 a/o Mg were measured[41] to find that Q does decrease slightly, but not to the extent reported by Domian and Aaronson. Within experimental error, the compositional dependence is directly proportional to the slowly falling T_m. The wrought samples used by Domian and Aaronson possessed a smaller grain size than is customarily acceptable for accurate volume-diffusion determinations, and it is suspected that some bias from grain-boundary diffusion was present. On the magnesium-rich side of stoichiometry, however, both groups of investigators found that Q drops off more rapidly than Hansen's T_m's would allow; a redetermination[40] of solidus temperatures may provide better agreement. The magnesium-rich alloys also have a higher affinity for oxygen, and it is possible that a small amount of oxygen contamination would act to open up the compound lattice, making diffusion easier and decreasing Q.

Using three single crystals containing 50 a/o Cd, Huntington et al.[42] measured the self-diffusion of ^{198}Au and ^{115}Cd in β-AuCd. The D_{Cd}/D_{Au} ratio was consistently 4:3. Their activation energy of ~28.0 kcal/mole for both elements was higher than would be expected from Wechsler's[43] resistivity results of 8.8 kcal/mole and 13.5 kcal/mole for ΔH_f and ΔH_m, respectively. Clark and Brown[44] found an even larger discrepancy between Kuper's Q of 36 kcal/mole at 364 to 381°C and their resistivity results of 7.2 to 8.0 kcal/mole and 14 kcal/mole for ΔH_f and ΔH_m, respectively. It seems that small equilibrium amounts of short-range order can be quenched-in along with mobile vacancies, while conducting the resistivity experiments, to give lower total activation energies for resistivity studies. Some upward curvature of data points near the melting temperature of β-AuCd was detected; this also occurs in β-CuZn near the critical temperature for ordering.

In a graduate-student investigation,[45] D. Gupta determined the simultaneous self-diffusion of ^{195}Au and ^{109}Cd in single-crystal β-AuCd samples containing 47.5, 49.0, and 50.5 a/o Cd. Owing to the presence of diffusionless transformations below 60°C, all processing was done at about 75°C. Density measurements were performed which indicated excess lattice vacancies in the 50.5 a/o Cd alloy, but no evidence for these was found on the gold-rich side of stoichiometry. Because D_{Cd}/D_{Au} varied from 1.6 to 0.6 on the cadmium-rich side, an analysis based on the six-jump cyclic mechanism was conducted. For 49.0 a/o Cd, D_{Cd} was only slightly higher than D_{Au}; for 47.5 a/o Cd, D_{Au} was higher than D_{Cd} at lower temperatures with a crossover at 560°C. Above that temperature, both self-diffusion rates curved up rapidly.

Many of the lines calculable from the information listed in Table 3 closely overlap, and therefore only a few were plotted in Fig. 3 to illustrate representative features. Again, these data group on the low side of the range for bcc metals, with ^{110}Ag in near-stoichiometric β-AgMg being the lowest. The presence of a large concentration of vacancy defects (line *o*) coincides with the fastest diffusion rates, but these are not as high as we might expect. The difference in lines *h* and *l* is the amount of experimental variance between the two groups of investigators who studied ^{110}Ag self-diffusion in β-AgMg containing 43.6 a/o

TABLE 3

Intrinsic Diffusion in β-AgMg and β-AuCd

Letter	Compound	Composition	Melting Temperature °C	Crystal Structure	Element	Melting Point, °C	Reference	Temperature Range, °C	D_0, cm²/sec	Q, kcal/mole
a	β-AgMg	52.0 a/o Mg	815	B2-cubic	^{110}Ag	960	39	500–700	0.134	38.0
b		49.8	820						0.280	40.6
c		45.8	815						1.53	41.3
d	β-AgMg	57.1	780	B2-cubic	^{110}Ag	960	40	500–600	0.05	28.7
e		52.8	812					500–700	0.31	36.6
f		48.7	818						0.38	39.7
g		48.5	817						0.36	39.5
h		43.6	809						0.15	35.3
i		41.1	790						0.09	33.2
j	β-AgMg	45.0	815	B2-cubic	^{110}Ag	960	41	500–700	0.90	40.1
k		43.8	809					550–700	0.65	38.3
l		43.6	809					550–700	0.65	38.3
m	β-AuCd	50.0 a/o Cd	627	B2-cubic	^{115}Cd	321	42	300–590	0.23	28.0
n		50.0			^{198}Au	1063			0.17	27.9
o	β-AuCd	50.5	626	B2-cubic	^{109}Cd	321	45	350–600	0.78	29.2
p		50.5			^{195}Au	1063			0.12	27.0
q		49.0	626		^{109}Cd	321			1.50	31.2
r		49.0			^{195}Au	1063			0.61	30.0
s		47.5	625		^{109}Cd	321			1.36	31.0
t		47.5			^{195}Au	1063			0.23	28.1

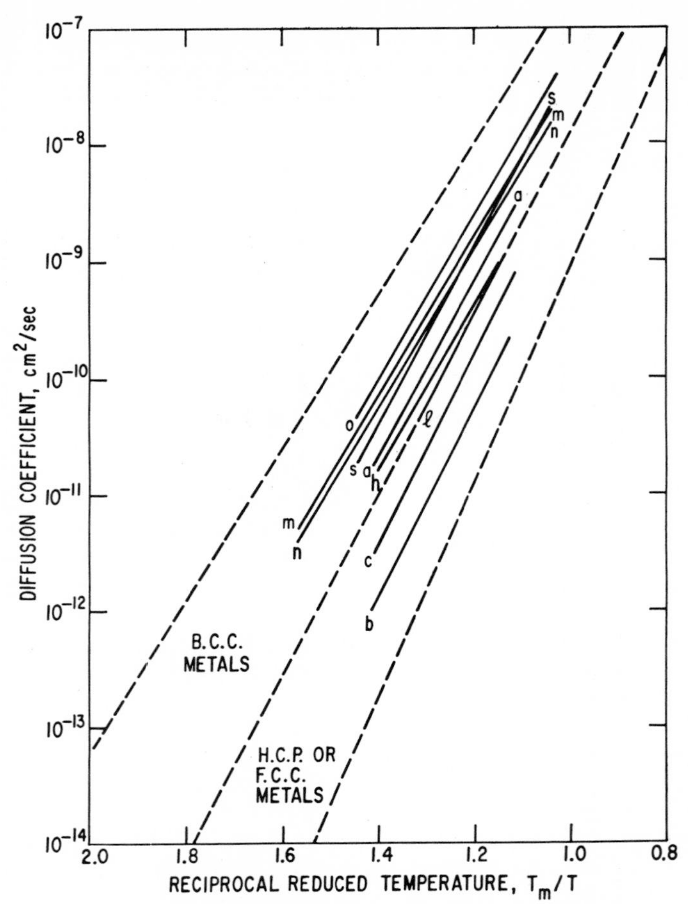

Fig. 3. Intrinsic diffusion coefficients for β-AgMg and β-AuCd calculated from the data of Table 3 and plotted as a function of T_m/T.

Mg. A greater variance exists between the data of Huntington et al. and Gupta. The latter's best line for ^{109}Cd in β-AuCd containing 49.0 a/o Cd falls closely below line s as compared with the formers' determination (line m) on single crystals, whose composition is given as stoichiometric β-AuCd. All might be more consistent if it were assumed that closer to 50.2 a/o Cd were actually present in the samples of Huntington et al. This points up one of the main difficulties of conducting measurements on intermetallic compounds—the problem of maintaining constant composition during diffusion anneals within the experimental accuracy of chemical analyses. A variety of undetected factors affecting the sample could exercise large effects on data derived from surface sectioning.

3. ZINC BLENDE STRUCTURES

3.1 InSb, GaSb, and AlSb

An appreciation of diffusion phenomena in these III–V semiconducting compounds comes from the realization that the zinc blende structure is quite open with large, accessible interstitial holes analogous to the diamond structure of germanium and silicon. Each atomic species is located on one of the two mutually displaced fcc lattices; each atom has four nearest neighbors of the opposite kind and twelve next-nearest neighbors of the same kind. It is thought that the tetrahedral arrangement around the central atom is a sign of covalent bonding, because ionic bonding tends to favor higher coordination numbers. Both substitutional and interstitial types of diffusion have been recognized and studied.

Using large-grained ingots of InSb, Boltaks and Kulikov[46] determined the self-diffusion of ^{114}In and ^{124}Sb and the impurity diffusion of ^{127}Te, by sectioning. Little detailed information was given, but the anneals were between 300° and 500°C for time periods of 50 to 180 hours. Resultant D_0 and Q values are listed in Table 4. It is customary for semiconductor technologists to think in terms of electron volts and Boltzmann's constant k; to maintain consistency throughout this chapter, these units have been changed to kcal/mole and the gas constant R by multiplying by 23.06.

Eisen and Birchenall[47] utilized a ring-and-piston grinding technique for removing thin sections (of the order of a few microns in thickness) from single-crystal or slightly polycrystalline InSb and GaSb. Residual activity was counted after each section was taken. With diffusion anneals running from two weeks to a few months, their D values are among the lowest yet measured by sectioning. Both indium and gallium diffused at significantly faster rates than antimony in each compound. A product of grain-boundary diffusivity D_B and boundary width δ equal to 3.8×10^{-16} cm^3/sec was found for polycrystalline InSb at 502°C; a product of dislocation diffusivity and dislocation radius squared between 10^{-21} and 10^{-22} cm^4/sec was observed in GaSb near 700°C. Their self-diffusion results all seem to indicate a vacancy mechanism, in which each species is moving along its own fcc sublattice. On plotting their least-squares lines versus reciprocal reduced temperature, as in Fig. 4, we note a significant displacement for these ordered compounds below that of substitutional self-diffusion in pure germanium.[48] In subsequent figures, the line for pure germanium will be taken as the lower reference line for diffusion in diamond-structure metals.

The self-diffusion of ^{124}Sb in GaSb was also measured by Boltaks and Gutorov,[49] along with the impurity diffusion of ^{114}In, ^{113}Sn, ^{75}Se, and ^{127}Te. The indium, tin, and antimony were diffused from thin layers deposited on the surface of GaSb single-crystal or large polycrystalline samples; selenium and tellurium were deposited from the vapor phase. All samples were annealed in evacuated and sealed quartz ampules. Their interpretational interest seems centered on activation energies, and some correlation of decreasing Q with increasing covalent radius or a decreasing number of valence electrons was made. Each best straight line contained only three or four points.

Goldstein[50] studied the impurity diffusion of electroplated ^{65}Zn in single-crystal wafers of InSb from 362° to 508°C. Zinc is an acceptor in InSb and is thought to enter the lattice substitutionally by replacement of indium. As with indium, it then seems logical that zinc diffuses via vacancy migration to next-nearest neighbors. Hulme and Kemp[51] used the *p-n* junction method developed by Fuller[52] to measure the self-diffusion of zinc from the vapor phase into tellurium-doped *n*-type single crystals of InSb. The *p-n* junctions were located by thermolectric probing at −196°C, and their positions were marked with the same sharp tungsten point. Sze and Wei[53] continued experiments on the impurity diffusion of ^{65}Zn and ^{115}Sn in single-crystal and polycrystalline InSb. The volume diffusion of ^{115}Sn proceeded at a rate about twice as high as ^{65}Zn, and the boundary diffusion of ^{115}Sn was observed to be much greater than that of ^{65}Zn. This caused the authors to rationalize that a tin atom would look alike upon indium and antimony vacancies, whereas a zinc atom would favor taking only an indium vacancy for pure sublattice diffusion. Then, if the grain boundaries were rich in antimony vacancies from antimony vaporization, they would serve as open channels to tin but not to zinc.

By means of an X-ray method, which followed the growth kinetics of a AlSb layer between layers of aluminum and antimony, Piñes and Chaikovskii[54] made a rough determination of the self-diffusion of aluminum and antimony in this compound at only 570° and 620°C. Their Q values were 41 kcal/mole for aluminum and 35 kcal/mole for antimony. Wieber et al.[55] measured the fast interstitial diffusion of ^{64}Cu in single-crystal samples of AlSb from 150° to 500°C with results that preclude the use of copper as a dopant in forming *p-n* junctions for high-temperature operation and warn against situations in which copper could become a contaminant.

Of all these determinations, those provided by Eisen and Birchenall seem the most reliable. Diffusion in ordered zinc blende structures

TABLE 4

Intrinsic Diffusion in InSb, GaSb, and AlSb

Letter	Compound	Melting Temperature °C	Crystal Structure	Element	Melting Point, °C	Reference	Temperature Range, °C	D_0, cm²/sec	Q, kcal/mole
a	InSb	530	B3-cubic	^{114}In	156	46	300–500	1.8×10^{-9}	6.46
b				^{124}Sb	630			1.4×10^{-6}	17.3
c				^{127}Te	450			1.7×10^{-7}	13.2
d	InSb	530	B3-cubic	^{114}In	156	47	478–520	5.0×10^{-2}	41.8
e				^{124}Sb	630			5.0×10^{-2}	44.6
f	GaSb	706	B3-cubic	^{72}Ga	30		658–700	3.2×10^{3}	72.6
g				^{124}Sb	630			3.4×10^{4}	79.4
h	Pure Ge	958	A4-cubic	^{71}Ge	958	48	766–928	7.8	68.5
i	GaSb	706	B3-cubic	^{114}In	156	49	320–650	1.2×10^{-7}	12.2
j				^{113}Sn	232			2.4×10^{-5}	18.5
k				^{124}Sb	630			8.7×10^{-3}	26.1
l				^{75}Se	217		$D(400°) = 2.4 \times 10^{-13}$; $D(650°) = 1.37 \times 10^{-10}$		
m				^{127}Te	450			3.8×10^{-4}	27.7
n	InSb	530	B3-cubic	^{65}Zn	419.5	50	362–508	0.5	31.1
o	InSb	530	B3-cubic	Zn	419.5	51	350–500	1.6×10^{6}	53.0
p	InSb	530	B3-cubic	^{65}Zn	419.5	53	400–512	1.4×10^{-7}	19.8
q				^{115}Sn	232			5.5×10^{-8}	17.3
r				^{65}Zn	419.5	(In polycrystalline InSb)		1.1×10^{-6}	19.6
s				^{115}Sn	232		$D_B(390°) = 4.3 \times 10^{-8}$; $D_B(512°) = 1.27 \times 10^{-7}$		
t	AlSb	1065	B3-cubic	Al	660	54	$D(570°) = 1.2 \times 10^{-11}$; $D(620°) = 5.1 \times 10^{-11}$		
u				Sb	630		$D(570°) = 7.0 \times 10^{-11}$; $D(620°) = 2.6 \times 10^{-10}$		
v	AlSb	1065	B3-cubic	^{64}Cu	1083	55	150–500	3.5×10^{-3}	8.3

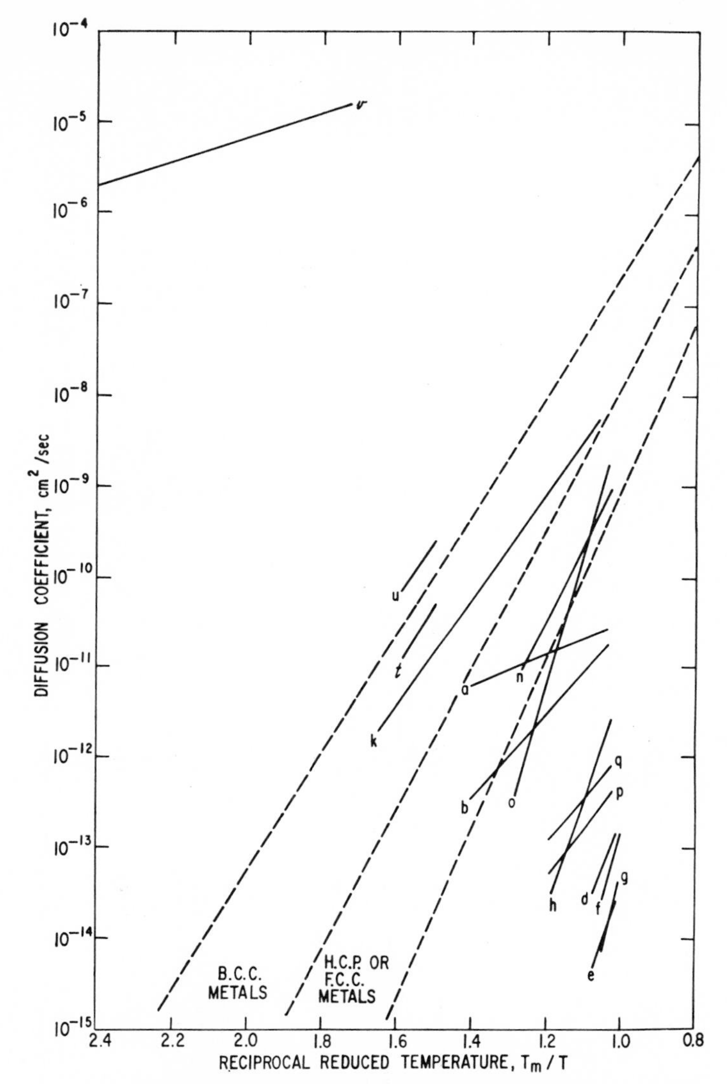

Fig. 4. Intrinsic diffusion coefficients for InSb, GaSb, and AlSb calculated from the data of Table 4 and plotted as a function T_m/T.

should proceed at a somewhat slower rate than in the elemental diamond structures. Comparison with Boltaks' data for ^{114}In and ^{124}Sb in InSb and for ^{124}Sb in GaSb, discloses an unusually large discrepancy to the extent that one is prone to dismiss results from that group. Figure 4 also shows that Pines' indirect measurements for self-diffusion in AlSb give unreasonably high values, although the faster diffusion rate of antimony would be expected from its lower melting point. Both the Goldstein and the Hulme-Kemp lines for zinc in InSb appear high compared to the more recent and probably more accurate Sze-Wei line. However, based on the relatively minor shift evidenced by tin in InSb, the Sze-Wei proposal of tin diffusion along both sublattices does not seem well-founded. The D values for ^{64}Cu in AlSb are clearly above the level at which any lattice substitution might occur.

3.2 InAs, InP, and GaAs

Following Dunlap's[56] use of the p-n junction method for studying the diffusion of impurities in germanium, Schillman[57] similarly

TABLE 5

Intrinsic Diffusion in InAs, InP, GaAs, ZnSb, and CdSb

Letter	Compound	Melting Temperature, °C	Crystal Structure	Element	Melting Point, °C	Reference	Temperature Range, °C	D_0, cm²/sec	Q, kcal/mole
a	InAs	943	B3-cubic	Mg	649	57	600–900	1.98×10^{-6}	27.0
b				Zn	419.5			3.11×10^{-3}	27.0
c				Cd	321			4.35×10^{-4}	27.0
d				Ge	937			3.74×10^{-6}	27.0
e				Sn	232			1.49×10^{-6}	27.0
f				S	119			6.78	50.7
g				Se	217			1.25×10^{1}	50.7
h				Te	450			3.43×10^{-5}	29.5
i	InP	1062	B3-cubic	^{114}In	156	61	830–990	1.0×10^{5}	88.8
j				^{32}P	44		900–1000	7.0×10^{10}	130.3
k	GaAs	1237	B3-cubic	^{72}Ga	30		1125–1230	1.0×10^{7}	129.
l				^{76}As	817		1200–1230	4.0×10^{21}	235.
m				^{115}Cd	321	60	868–1149	5.0×10^{-2}	56.0
n				^{65}Zn	419.5		660–980	1.5×10^{1}	56.0
o				^{33}S	119	61	1000–1200	4.0×10^{3}	93.2
p				^{75}Se	217		1025–1200	3.0×10^{3}	95.9
q	GaAs	1237	B3-cubic	^{64}Cu	1083	67	100–600	3.0×10^{-2}	12.0
r	GaAs	1237	B3-cubic	Li	180	68	250–500	0.53	23.1
s	ZnSb	546	B3-cubic	^{124}Sb	630	69	315–400	4×10^{-11}	4.62
							400–483	3×10^{1}	41.6
t				^{113}Sn	232		315–400	3.2×10^{-9}	8.24
							400–483	2.3	36.0
u	ZnSb	546	Deformed Diamond	^{59}Fe	1534	70	270–385	2.23×10^{-9}	6.08
								2.80×10^{-10}	3.89
v				^{114}Cd	321			4.67×10^{-9}	12.7
w	CdSb	456	Deformed Diamond	^{75}Se	217	71	100–200		

measured the diffusion of Mg, Zn, Cd, Ge, Sn, S, Se, and Te in InAs. Advantages of this method over tracer techniques are that it is quite rapid and can be used for elements whose isotopes are difficult to obtain or work with; disadvantages are that the elements must be electrically active and the base compounds must be semiconducting. Schillman's results are listed in Table 5, and two of these are plotted in Fig. 5. At 900°C, zinc shows the highest D value of 2.8×10^{-8} cm²/sec, and tin shows the lowest of 1.5×10^{-11} cm²/sec. To the author's knowledge, tracer diffusion in InAs has not been reported.

Goldstein designed a precision lapping device[58] for sectioning single-crystal wafers; from the findings of Boomgaard and Schol[59] he could include appropriate amounts of phosphorus and arsenic to prevent the dissociation of InP and GaAs in their annealing ampules. Subsequently, self-diffusion of all components in InP and GaAs was measured, along with the impurity diffusion of cadium, zinc, sulfur, and selenium in GaAs.[60,61]

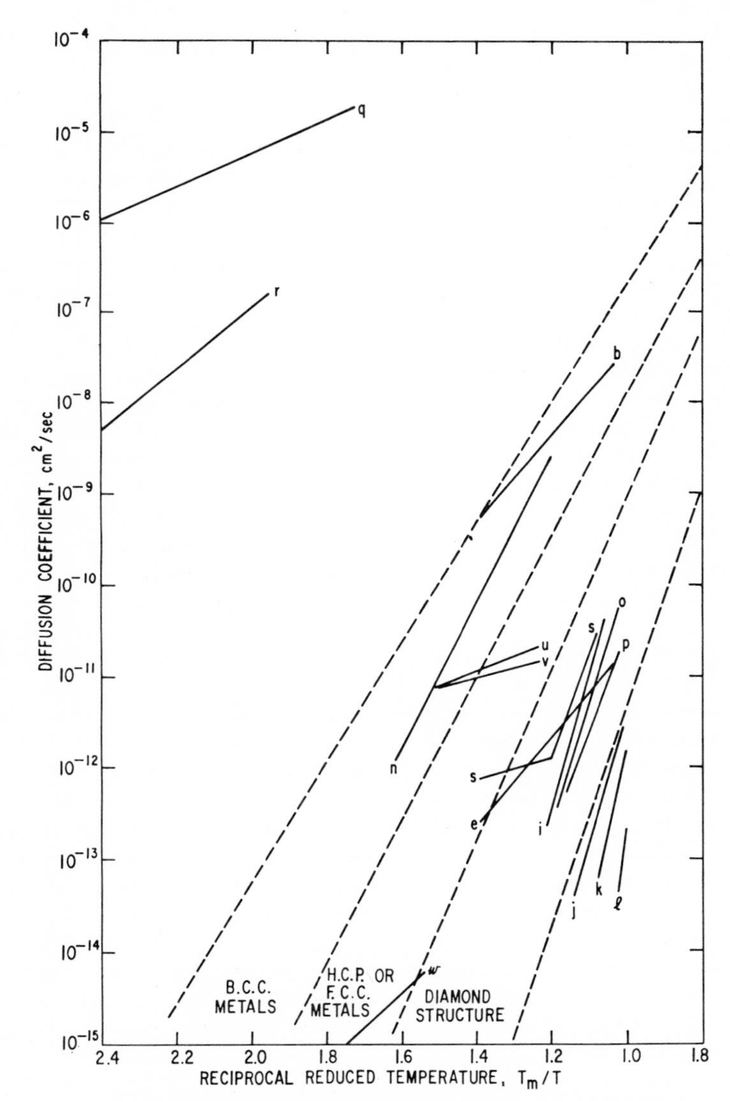

Fig. 5. Intrinsic diffusion coefficients for InAs, InP, GaAs, ZnSb, and CdSb calculated from the data of Table 5 and plotted as a function of T_m/T.

Isotopes of cadmium, phosphorus, arsenic, sulfur, and selenium were diffused from the vapor phase, and the others were electroplated as thin films. Strong evidence was presented to support atomic migration along individual sublattices as the basic mechanisms of substitutional diffusion. The acceptors cadmium and zinc seem to replace gallium and the donors sulfur and selenium replace arsenic. Unusually large D_0 values were reported for phosphorus in InP and arsenic in GaAs.

Allen and Cunnell[62] recognized that zinc atoms may diffuse at different rates in GaAs and attributed this behavior to their being ionized or un-ionized. Cunnell and Gooch[63] diffused ^{65}Zn from the vapor phase into GaAs as a function of zinc and arsenic pressure over the sample. Their penetration curves displayed sharp knees, giving evidence for the rapid diffusion of un-ionized zinc at the zinc-rich surface and the much slower diffusion of ionized zinc further inside. Allen[69] rationalized this anomalous behavior and discussed how the *p-n* junction method would then fall in error from fictitiously high values for surface concentration. No unequivocal diffusion coefficients could be given. Using somewhat different terminology, Longini[65] proposed (when zinc doping is very heavy) that zinc goes into donor interstitial positions rather than acting as an acceptor in its normal gallium-substitutional positions. With GaAs tunnel diodes, deterioration from interstitital diffusion should be faster when a forward bias is applied.

From autoradiographic evidence, Fuller and Whelan[66] found ^{64}Cu to diffuse in GaAs single crystals at rates comparable to those for copper in germanium and silicon. Again, a sharp dropoff in penetration was observed, allowing the possibility of an interstitial-substitutional mechanism. Hall and Racette[67] provide Fickian D values for ^{64}Cu in GaAs from 100 to 600°C; their interpretation is that interstitial copper behaves as a single donor and substitutional copper as a double acceptor. Using flame analyses and conductivity changes, Fuller and Wolfstirn[68] found that lithium in GaAs shows a strong disposition to be substitutional. Because the kinetics do not strictly follow Fick's second law, they could only present an effective diffusion coefficient characteristic of interstitial lithium.

Inspection of Fig. 5 discloses that Goldstein's results for self-diffusion in InP and GaAs (lines *i*, *j*, *k*, and *l*) are displaced somewhat lower than those for self-diffusion in diamond structures. Considering that phosphorus possesses a lower melting point than indium, the level for indium self-diffusion in InP seems high.

Lines for zinc in both GaAs (*n*) and InAs (*b*) are intermediate to those for lithium (*r*) and interstitial copper (*q*) in GaAs. Explaining this large spread in D values is a problem requiring more attention. Even the data for sulfur (*o*) and selenium (*p*) in GaAs are higher than what we would expect for purely substitutional diffusion along the arsenic sublattice.

3.3 ZnSb and CdSb

Some attention has been paid these II–V compounds. Boltaks[69] diffused ^{124}Sb and ^{113}Sn into ZnSb from 315° to 483°C; his D_0 and Q values are listed in Table 5. At about 400°C, a change in slope occurred for the temperature dependence of each of the tracers. As might be expected from its lower melting point, the diffusion rate of tin was always higher than that of antimony. Also using a sectioning technique, Kuliev[70] determined the impurity diffusion of ^{59}Fe and ^{114}Cd in ZnSb from 270° to 385°C. Here the iron diffused faster than cadmium, and the D_0 and Q values were unusually low. A change in slope similar to that found for antimony in ZnSb would be anticipated, if the measurements were extended to higher temperatures. On dissolving ^{75}Se in carbon-bisulfide solution for coating the surface of cast CdSb crystals, followed by evaporating the carbon bisulfide, Kuliev and Abdullaev[71] measured very low diffusion rates for this impurity from 100° to 200°C. Almin[72] has shown that both ZnSb and CdSb possess a rather unique crystal structure at room temperature. Each atom is surrounded by one atom of the same kind and three of the second kind to form a deformed tetrahedron. Although classified as orthorhombic with sixteen atoms per unit cell, the structure may be regarded as a strongly deformed diamond structure.

The large change in slope experienced by ^{124}Sb in ZnSb (line *s* of Fig. 5) could result from some low-temperature short-circuiting mechanism. Lines *u*, *v*, and *w* all follow the same pattern. Careful, accurate self-diffusion determinations in ZnSb, CdSb, and other II–V compounds are still required.

4. HEXAGONAL STRUCTURES

4.1 Bi_2Se_3, Bi_2Te_3, Tl_2Se, and $NbNi_3$

Using a sectioning technique, Boltaks[73] measured the impurity diffusion of ^{113}Sn and ^{124}Sb in Bi_2Se_3 and Bi_2Te_3. Both thermoelectric compounds are rhombohedral-hexagonal and were presumably stoichiometric, although the phase field of Bi_2Te_3 can vary from 53 to 63 a/o Te. At higher temperatures, antimony diffuses faster than tin; the D_0 and Q values for tin seem unrealistically low. Kuliev and Abdullaev[71] determined the self-diffusion of ^{75}Se and the impurity diffusion of electroplated ^{65}Zn in Bi_2Se_3. Their diffusion anneals were conducted in a vacuum of about 10^{-3} mm Hg. From an activation-energy analysis, they assume that zinc diffuses mainly along octahedral sites. Selenium may then diffuse substitutionally at a much slower rate along its own sublattice. Nevertheless, the plotted datum points do not support their Arrhenius expressions, and there is much to doubt about the whole study. Kuliev[70] also reported four points apiece for the impurity diffusion of ^{59}Fe and ^{114}Cd in Bi_2Se_3; iron was found to diffuse more slowly with a lower activation energy.

Stoichiometric Bi_2Te_3 is *p*-type; of the various additives, which can convert it to an *n*-type semiconductor, copper is a suitable donor impurity. However, copper will on standing diffuse to the surface and render the doped compound less *n*-type. There is also a marked anisotropy to many of the properties, for example, Hall coefficient, electrical resistivity, cleavage, etc., of Bi_2Te_3. Carlson[74] therefore set out to measure the impurity diffusion of ^{64}Cu parallel to and perpendicular to the cleavage planes in Bi_2Te_3. At room temperature, he found that $D_{\parallel}$ is $\sim 10^{-6}$ cm²/sec and that an extrapolated $D_{\perp}$ is $\sim 3 \times 10^{-15}$ cm²/sec. Because it is known that Bi_2Te_3 possesses the layer structure (Te—Te-Bi-Te-Bi-Te—Te-Bi · · ·), Carlson suggests that ^{64}Cu will move rapidly between adjacent Te—Te layers owing to weak electrostatic bonding forces and a large spacing distance. Because it is perpendicular to the easy cleavage plane (or parallel to the hexagonal *c* axis), copper may diffuse interstitially; the covalent and ionic bonding between bismuth and tellurium atoms should make penetration more difficult.

Using ^{110}Ag as the tracer, Keys[75] sectioned single crystals of Bi_2Te_3 by peeling off layers of <1 to 50 μ thickness parallel to the cleavage plane. This provided penetration data for the impurity diffusion of silver perpendicular to the cleavage plane. The results are listed in Table 6 and plotted as line *k* in Fig. 6. Comparison with Carlson's data for ^{64}Cu diffusing in the same direction (line *j*) discloses a considerable shift downwards for ^{110}Ag. Boltaks and Fedorovich[76] concurrently conducted the same measurements as Keys and included diffusion rates parallel to the cleavage plane. Fair agreement exists with Keys' data (comparing lines *k* and *m*) for diffusion perpendicular to the cleavage plane. Although diffusion rates parallel to the cleavage plane (line *l*) are 3 to 4 orders of magnitude higher, they are still 2 to 3 orders of magnitude lower than what Carlson found for ^{64}Cu in the same direction (line *i*).

Because the thickness and composition of diffusion-grown Tl_2Se determines the characteristics of thallium-selenium rectifying devices, Akhundov and Abdullaev[77] measured the self-diffusion of both components in Tl_2Se. Some question exists as to the crystal structure of this compound. Hahn and Klingler[78] found it nonisotypic with Tl_2S, a hexagonal structure of the C6 type, but the difference should not be very great. In contrast to most cases, the lower melting element ^{75}Se diffuses at a slower rate (line *o*) than ^{204}Tl (line *n*). As ionic character increases, ionic size may become more important. The high-temperature portion of line *o*, however, is within the range for self-diffusion in hexagonal pure metals.

Shinyaev[79] studied the self-diffusion of ^{63}Ni in alloys of the $NbNi_3$–$TaNi_3$ system in which concentrations of $NbNi_3$ were 100, 96, 80,40 and 20 weight per cent. Diffusion coefficients for only pure $NbNi_3$ are listed in Table 6. Figure 6 shows that the respective data (line *p*) are displaced reasonably lower than the normal range for hcp pure metals. Shinyaev maintained that the lower *D* values and higher activation energies of these materials are responsible for their hot strength.

The marked anisotropy of diffusion rates for copper and silver in Bi_2Te_3 leads us to suspect that the same should occur for other elements in more compounds of this group. Too little attention was paid to this possibility by earlier investigators. The low D_0's and Q's of lines *a*, *c*, *f*, and *g* could easily result from

TABLE 6

Intrinsic Diffusion in Bi_2Se_3 Bi_2Te_3, Ti_2Se, and $NbNi_3$

Letter	Compound	Melting Temperature, °C	Crystal Structure	Element	Melting Point, °C	Reference	Temperature Range, °C	D_0, cm²/sec	Q, kcal/mole
a	Bi_2Se_3	706	C33-rhombic	^{113}Sn	232	73	360–540	4.0×10^{-9}	9.50
b				^{124}Sb	630.5		360–500	1.8×10^{-3}	29.5
c	Bi_2Te_3	585	C33-rhombic	^{113}Sn	232	73	260–470	3.0×10^{-8}	11.5
d				^{124}Sb	630.5		270–500	4.3×10^{-4}	24.0
e	Bi_2Se_3	706	C33-rhombic	^{75}Se	217	71	100–200	8.5×10^{-9}	50.1
f				^{65}Zn	419.5		300–400	5.0×10^{-9}	17.6
g	Bi_2Se_3	706	C33-rhombic	^{59}Fe	1534	70	300–400	1.25×10^{-7}	11.0
h				^{114}Cd	321			1.39×10^{-3}	21.3
i	Bi_2Te_3 ‖ [a]	585	C33-rhombic	^{64}Cu	1083	74	25–300	3.4×10^{-3}	4.84
j	Bi_2Te_3 ⊥						200–500	7.1×10^{-2}	18.45
k	Bi_2Te_3 ⊥			^{110}Ag	960.5	75	270–540	5.1×10^{-3}	21.2
l	Bi_2Te_3 ‖			^{110}Ag	960.5	76	100–500	2.2×10^{-3}	9.68
m	Bi_2Te_3 ⊥						300–500	2.3×10^{-1}	27.0
n	Tl_2Se	390	pseudo-hexagonal	^{204}Tl	303	77	150–300	1.17×10^{-3}	14.1
o				^{75}Se	217			2.25×10^{-5}	13.4
p	$NbNi_3$	1403	pseudo-hexagonal	^{63}Ni	1453	78	1132–1252	3×10^{2}	88.7

[a] ‖ parallel to cleavage plane; ⊥ perpendicular to cleavage plane.

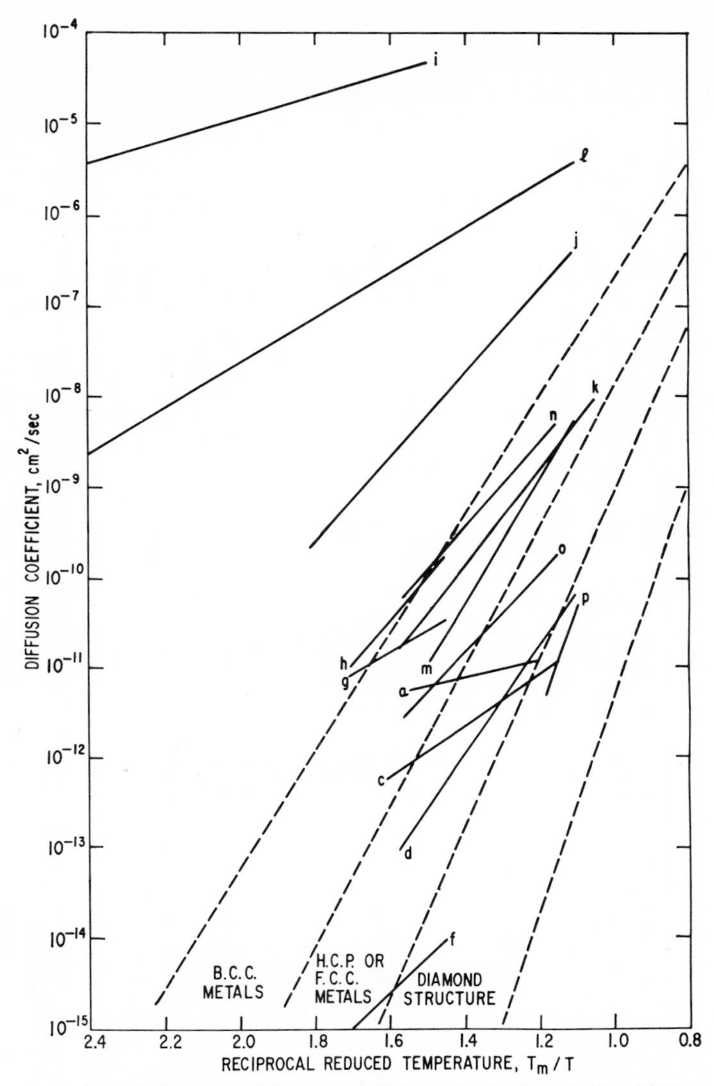

Fig. 6. Intrinsic diffusion coefficients for Bi_2Se_3, Bi_2Te_3, Tl_2Se, and $NbNi_3$ calculated from the data of Table 6 and plotted as a function of T_m/T.

using mixed or disoriented crystals. Perhaps, the wide scatter of results in Fig. 6 will become more meaningful as work continues.

Further understanding can now be obtained from investigations on representative metal-metalloid compounds, which are not intermetallic in the strict sense.

4.2 CdS, CoTe, NiS, and FeS

A barrier layer of copper is generally deposited on the surface of cadmium-sulfide single crystals used for photovoltaic devices. To determine how much penetration occurs as a function of time and temperature, Clarke[80] measured the impurity diffusion of ^{64}Cu both parallel and perpendicular to the cleavage plane of CdS single crystals. With this wurtzite-type compound, in which each atom is surrounded by an irregular tetrahedron of four different atoms, little difference was found between diffusion in the two directions. Curved concentration versus distance-squared plots indicated the presence of volume-diffusion short circuiting either along the

TABLE 7

Intrinsic Diffusion in Wurtzite-, NiAs-, and NaCl-type Compounds

Letter	Compound	Melting Temperature °C	Crystal Structure	Element	Melting Point, °C	Reference	Temperature Range, °C	D_0, cm²/sec	Q, kcal/mole
a	CdS	~1000	B4-hexagonal	^{64}Cu	1083	80	450–750	1.5×10^{-3}	17.6
b	CoTe	968	B8-hexagonal	^{60}Co	1495	81	$D_V(410°) = 1.6 \times 10^{-9}$ & $\delta D_B = 2.2 \times 10^{-11}$;		
							$D_V(805°) = 1.5 \times 10^{-7}$ & $\delta D_B = 2.5 \times 10^{-8}$		
c	NiS	920	B8-hexagonal	^{63}Ni	1453		$D_V(570°) = 5.10^{-10}$ & $\delta D_B = 1.1 \times 10^{-11}$;		
							$D_V(883°) = 10^{-7}$ & $\delta D_B = 2.3 \times 10^{-9}$		
d	FeS ∥ [a]	1190	B8-hexagonal	^{55}Fe	1534	83	350–697	2.9×10^{-4}	19.2
e	FeS ⊥							2.5×10^{-3}	21.2
f	$Fe_{0.85}S$ ∥	1150						3.3×10^{-3}	22.4
g	$Fe_{0.85}S$ ⊥							7.6×10^{-3}	23.0
h	FeS	1190		^{35}S	119		895–1057	10^{33}	240.
i	PbTe	917	B1-cubic	Pb	327	84	250–500	2.9×10^{-5}	13.8
j				^{127}Te	450	85	500–800	2.6×10^{-6}	17.3
k				^{124}Sb	630.5			4.9×10^{-2}	35.5
l				^{113}Sn	232			3.1×10^{-2}	36.0
m	PbSe	1076	B1-cubic	^{75}Se	217		650–850	2.1×10^{-5}	27.7
n				^{124}Sb	630.5			0.34	46.2
o	BiSe	605	B1-cubic	^{75}Se	217	71	100–200	2.5×10^{-9}	15.5
p				^{65}Zn	907		300–400	9.3×10^{-8}	33.5
q	PbSe	1076	B1-cubic	^{210}Pb	327	86	400–800	4.98×10^{-6}	19.1
r	PbSe + 0.5 m/o Bi_2Se_3							4.28×10^{-2}	37.1
s	PbSe + 0.5 m/o Ag_2Se							4.41×10^{-7}	12.7
t	PbS	1119	B1-cubic	^{210}Pb	327	87	460–770	1.4	42.0
u	PbS	1119	B1-cubic	Cu	1083	89	100–500	5.0×10^{-3}	7.13
v	PbS	1119	B1-cubic	Ni	1453	90	200–500	17.8	22.0
w	TiC	3250	B1-cubic	^{95}Nb	2468	91*a*	1935–2170	2.4	84.0
x	TiC + 50 m/o NbC	~3500	B1-cubic	^{95}Nb	2468		2120–2280	4.7×10^{2}	120.0

[a] ∥ Parallel to cleavage plane; ⊥ perpendicular to cleavage plane.

the crystal surfaces or into imperfections; suitable corrections had to be made. Arkharov et al.[81] reported a few volume- and boundary-diffusion rates for ^{60}Co in CoTe and for ^{63}Ni in NiS. Both compounds possess the NiAs-type hexagonal structures. The isotope ^{125}Te plainly outlined the cast structure of their CoTe in an autoradiograph. At only 880°C, Klotsman et al.[82] found the self-diffusion of ^{63}Ni and ^{35}S perpendicular to the cleavage plane of NiS single crystals to be 7.2×10^{-8} and 3.5×10^{-11} cm^2/sec, respectively.

Condit[83] measured the self-diffusion of ^{55}Fe and ^{35}S in FeS as a function of crystal orientation and deviation from stoichiometry. His overall relationship for iron self-diffusion from 350° to 697°C is $D_{Fe} = D^{\circ}_{\circ a,c}\ [\delta + 10^{-2} \exp(-4/RT)]^{1/2} \exp -(19 + 20\delta)/RT$, where δ is the iron deficit in the formula $Fe_{1-\delta}S$, $D^{\circ}_{\circ a}$ equals 0.16 cm^2/sec parallel to the cleavage plane and $D^{\circ}_{\circ c}$ equals 0.27 cm^2/sec perpendicular to the cleavage plane. The exponential numbers 4, 19, and 20 are in units of kcal/mole. Between 895–1057°C, sulfur was found to diffuse at much slower rates; they were not strongly anisotropic or dependent on deviations from stoichiometry. Some of the actual results are listed in Table 7 and plotted in Fig. 7. Condit believed that iron migrates between lattice interstices via a vacancy mechanism, although interstitialcies may play a role.

5. NaCl-TYPE STRUCTURES

Prompted by the rising interest in PbTe as a semiconducting compound for photoelectric and thermoelectric devices, Boltaks and Mokhov[84] first measured the self-diffusion of lead by the *p-n* junction method. From the low Q (listed in Table 7) and high D values (shown in Fig. 7), they implied that an interstitial mechanism is involved. Work was then continued by these two investigators on determining the self-diffusion of ^{127}Te in PnTe, the impurity diffusion of ^{124}Sb and $^{113+123}Sn$ in PbTe, the self-diffusion of ^{75}Se in PbSe, and the impurity diffusion of ^{124}Sb in PbSe.[85] Although both compounds possess the NaCl-type structure, the bonding is more covalent than ionic in nature. Only single crystals were used; diffusion anneals were carried out in sealed and evacuated quartz ampules. Kuliev and Abdullaev[71] have also measured the self-diffusion of ^{75}Se and the impurity diffusion of ^{65}Zn in BiSe, but their values are too low to be plotted in Fig. 7.

The self-diffusion of ^{210}Pb in undoped, bismuth-doped, and silver-doped PbSe single crystals has been reported recently by Seltzer and Wagner.[86] Comparison with the Boltaks-Mokhov data for ^{75}Se in PbSe discloses that lead self-diffusion is more than one order of magnitude greater. Silver doping decreases activation energy, and bismuth-doping increases it; the authors have correlated increased diffusion rates with the presence of a higher concentration of interstitial lead ions caused by silver-doping. Hence an interstitial mechanism is thought more likely than diffusion via vacancies, although these cannot be ignored completely. It was concluded that Frenkel disorder prevails on the cation sublattice.

Before many tracers became available, Anderson and Richards[87] followed the diffusion of radioactive lead sulfide in various compositions of compressed PbS from 460° to 770°C. They commented on the possibility of D values changing with deviations from stoichiometry and inferred that Schottky disorder predominates in PbS. Brebrick and Scanlon[88] exposed natural PbS crystals to vaporous sulfur; calculations based on the *p-n* junction method gave an interdiffusion coefficient of 2×10^{-6} cm^2/sec at 550°C. Bloem and Kröger electroplated both copper[89] and nickel[90] onto single crystals of PbS and heated them in H_2, H_2S, argon, or mixtures of these gases between 100° and 500°C. Impurity diffusion rates were determined by the *p-n* junction method. These two elements are thought to penetrate the lattice interstitially and to give rise to *n*-type conductivity. Under sulfurizing conditions, neither copper nor nickel will enter the crystal; if already present, they can be drawn out again. Nickel diffuses somewhat more slowly than copper. Personal communication from J. B. Wagner, Jr., informed the author of new determinations of the self-diffusion of ^{210}Pb and ^{35}S in PbS single crystals. These determinations, and a study of the impurity diffusion of ^{63}Ni in PbS, should appear shortly.

Very little has been done to determine self-diffusion in carbides, borides, aluminides, or any of the high-temperature intermetallic

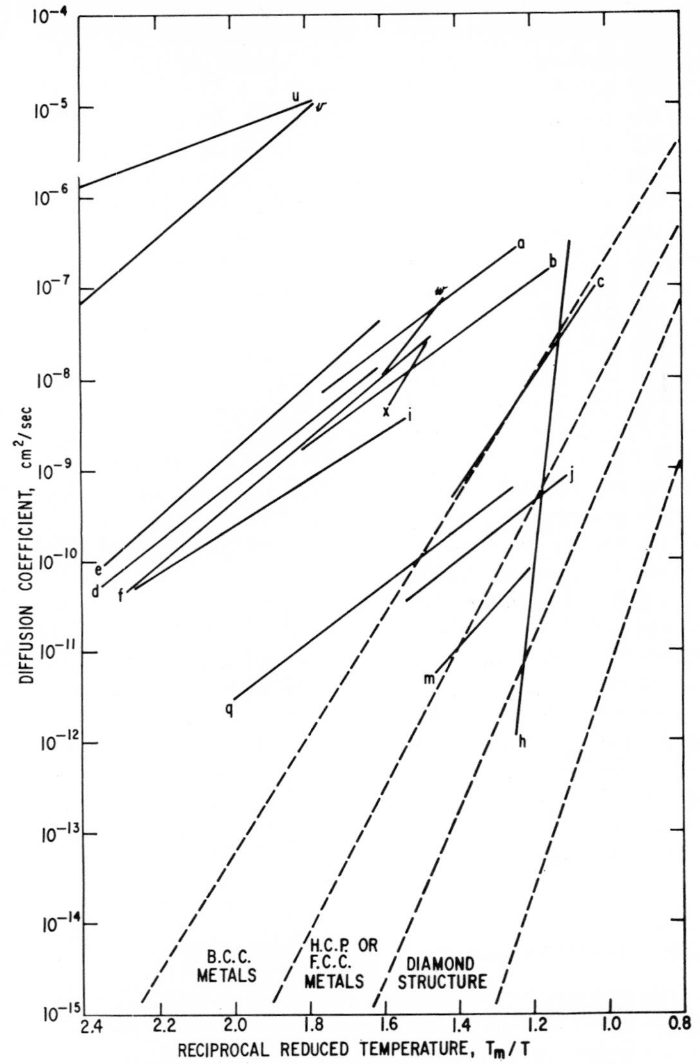

Fig. 7. Intrinsic diffusion coefficients in wurtzite-, NiAs-, and NaCl-type compounds calculated from the data of Table 7 and plotted as a function T_m/T.

compounds. Preliminary steps in this direction have been taken by Baskin et al[91a]; they measured the diffusion of ^{95}Nb in TiC from 1935° to 2170°C and in an equimolar solid solution of TiC and NbC from 2130° to 2280°C. This was followed by measurements[91b] of the self-diffusion of ^{185}W in WC, TaC, TiC, and various combinations of WC–TaC–TiC from 2130° to 2300°C. D_0 values for the first three pure carbides are, respectively, 7.33, 19.2, and 7.7 cm²/sec; the related Q values are 138, 142, and 115 kcal/mole. Gel'd and Lyubimov[92] have presented talks on the diffusion of ^{95}Nb and ^{14}C in NbC and Nb_2C from 1600° to 2000°C, but no D_0 and Q values are yet available.

6. THEORY

After Lidiard[26] pointed out that vacancies cannot move in the manner described by Slifkin and Tomizuka and still satisfy their equilibrium requirements, Huntington's personal communication to Slifkin consolidated thinking about nearest-neighbor ring mechanisms that do not create disorder. Elcock and McCombie[93] then suggested a correlated series of six vacancy jumps around the sides of a square consisting of two B(A) atoms, an A(B) atom, and an a(b) vacancy. It results, as far as diffusion is concerned in the two B(A) atoms exchanging places across a diagonal and the single A(B) atom moving a similar distance to replace the a(b) vacancy. This would give a maximum ratio of $D_{A(B)}/D_{B(A)}$ equal to the two experimentally determined by Kuper et al.[23] for D_{Zn}/D_{Cu}. A year later Elcock[94] presented the same idea in considerably more detail. For the low-temperature range $0 \leq T/T_c \leq 0.5$ where T_c is the critical temperature of ordering, he proposed that $\frac{1}{2} \leq D_{A(B)}/D_{B(A)} \leq 2$ for a simple cubic lattice. He acknowledged that at higher temperatures, more complicated unit processes come into play and correlations between successive vacancy jumps become relatively unimportant. For bcc alloys, the six-jump unit process proceeds around the sides of a rhombus and the limits $\frac{2}{3} \leq D_{A(B)}/D_{B(A)} \leq \frac{3}{2}$ apply. Elcock could see no reason why $D_{A(B)}/D_{B(A)}$ should be restricted by ordering effects in fcc alloys. Unfortunately, no data are available for self-diffusion in any of the fcc Cu_3Au-type alloys to determine how markedly different their diffusion behavior may be. The work of Huntington et al.[42] and Gupta[45] does seem to support Elcock's reasoning, but at present it should be considered an attractive possibility rather than a fact of nature. Except for the presence of an a(b) vacancy, which permits the Kirkendall effect, Elcock's rings are geometrically the same as Zener's[95] four-atom rings; no incontrovertible evidence for the latter has yet been found. Of course, the kinetics of Elcock rings are completely different from the Zener rings.

Recognizing that the simple random-walk approach[96] may not hold for ordered compounds, Kikuchi[97] has derived self-diffusion coefficients for β-CuZn, which decrease more rapidly below T_c than above. His analysis is based on a more general path-probability method for irreversible cooperative phenomena and a cluster-variation technique, assuming the vacancy mechanism. Flinn and McManus[98] studied order-disorder transformations in β-CuZn and β-AgZn by using the Monte Carlo method for a 2000-atom crystal. Agreement between calculated and experimental long-range order, as a function of reduced temperature T/T_c, was very good. They reason that the rate-limiting process for diffusion in these ordered compounds is the jumping of vacancies to next-nearest neighbors. Proceeding with further Monte Carlo computational experiments, Beeler and Delany[99] followed vacancy migration in AB alloys possessing square planar, simple cubic, and bcc lattices. Ordering was found to contract the extent of vacancy migration relative to that for a symmetric random walk; the degree of contraction was greater in two dimensions than in three. They found that three types of jump sequences would maintain order during a net transport of atoms. One was a ten-jump double loop, the second a six-jump Elcock loop, and the third an inefficiently negotiated Elcock loop with retracted spurs; the first occurred half as frequently as Elcock loops and the third occurred 3.5 times as frequently.

A thought-provoking way of considering how crystal structure affects diffusion has been presented by Azaroff.[100] His interest lay in the coordination polyhedra of voids in cubic and hexagonal close packings. With the latter, octahedral voids form continuous chains, whereas tetrahedral voids form isolated pairs; with cubic close packings, each kind of void shares faces with unlike voids. The diffusion path and mechanism for any structure type thus depends on how voids are filled; voidal diffusion takes place when one-half or less of one kind of voids is occupied. In the case of cubic zinc blende structures, the unoccupied set of tetrahedral voids and the empty octahedral voids form veritable tunnels for easy interstitial diffusion. With hexagonal wurtzite structures, voidal diffusion should again occur. In NiAs- or NaCl-type structures, where all octahedral voids are filled, diffusion should take place only by vacancy or interstitial mechanisms. The effect of available diffusion paths on activation energies was considered for β-AgI, ZnO, BiSe, and Bi_2Se_3.

Accepting the presence of Schottky disorder in such binary ionic compounds as PbSe, PbTe, and PbS, Brebrick[101] developed phenomenological expressions for interdiffusion coefficients as a function of compositional deviations from stoichiometry. Whether the basic assumptions are valid can be tested experimentally.

7. DISCUSSION AND SUMMARY

Nearly all the aforementioned determinations were provided by counting of radioactive isotopes on sample faces or in separated residues. As has been established with pure metals, the various sectioning techniques can give calculated results reproducible from one laboratory to another within ± 5 to 10 per cent. Even with brittle intermetallic compounds, suitable methods for removing thin layers have been devised, and similar precision seems feasible. Why then do certain independent investigations, for example, Boltaks and Kulikov[46] versus Eisen and Birchenall,[47] report D values for the same isotope in the same material which are orders of magnitude apart? One frequent source of error is carelessness in maintaining and measuring equilibrium annealing temperatures; another is the failure to keep binary samples at constant, known compositions. More collaboration with skilled analytical chemists may help to show that compounds are truly what they are thought to be before and *after* each anneal. Diffusion coefficients are primarily indices of the binding and structure of a substance at any given temperature; if the sample itself remains an undefined variable, D values lose their significance.

Techniques that follow *p-n* junctions or scan chemical gradients with an electron-beam microanalyzer may at times be preferable to sectioning, but, again, the emphasis should be placed on the condition of the sample rather than on gaining readings to the fourth significant figure. Of course, there are special diffusion problems requiring the utmost of precision, such as correlation effects, the influence of strain, or small concentrations of dilute impurities. Such tasks are being faced by those who study pure metals; they would seem premature in the case of intermetallic compounds where so much general understanding is still to be gained. With the intermetallics, only a few attempts have been directed beyond volume diffusion toward the effects of pressure, grain boundaries, dislocations, and surfaces.

Reviewing briefly what has been done, we see that intrinsic diffusion in Hume-Rothery intermediate phases containing zinc proceeds more rapidly than expected. The interstitialcy mechanism has not been laid to rest. Zinc is the faster moving species, and its presence could somehow aid the motion of other atoms. One advantage of these alloys is that diffusion rates can be determined above and below T_c, and increasing order is found to decrease D values. Many other CsCl-type compounds remain ordered very close to their congruent melting points; intrinsic diffusion in near-stoichiometric compositions of these is somewhat depressed as compared to what occurs in pure bcc metals. At constant temperature, D values on both sides of stoichiometry rise sharply upwards in a nonlinear manner from an inverted cusp. Large vacancy concentrations (a rather rare occurrence) increase diffusion rates and decrease Q, but not to the extent seen in zinc-containing alloys. When the self-diffusion of both species is measured, the element possessing the lowest melting point usually proceeds the fastest, as though it were passing along its own sublattice. Ratios corresponding to Elcock-loop predictions are sometimes found, but the data are too sparse to permit general acceptance of that mechanism.

Although atoms in simple metals exchange with vacancies by near-neighbor random walk and ions in certain ionic compounds diffuse only along their respective sublattices via next-nearest neighbor exchange, it may be that many of the intermetallic compounds are literally intermediate (depending on degree of order) in following both courses of action. Experimental control of near-neighbor and next-nearest neighbor diffusion paths can be obtained in intermetallic compounds and perhaps will contribute greatly to fundamental theory.

With semiconducting compounds possessing the zinc blende structure, self-diffusion can proceed at a slower rate than in pure germanium. A wide range of intrinsic diffusion rates is observed, depending on whether an element migrates substitutionally along its own sublattice, goes into both, or moves through large interstitial holes. Copper and lithium diffuse rapidly in the latter manner. Zinc may diffuse interstitially if its surface concentration is sufficiently high, or it may migrate substitutionally if deposited as a very thin layer.

There is a large anisotropy effect in the

thermoelectric compound Bi_2Te_3, in which copper penetrates parallel to the cleavage plane at a rate many orders of magnitude faster than in the other direction; silver shows the same effect at lower levels. Diffusion anisotropies in intermetallic compounds are much more pronounced than those appearing in hexagonal metals. Crystal orientation should, therefore, be carefully specified.

Self-diffusion in some of the more ionic NiAs- and NaCl-type metalloid compounds has been studied in considerable detail as a function of deviations from stoichiometry and doping additions. Both lattice vacancy and interstitial mechanisms have been identified under different conditions; current emphasis seems to be placed on determining whether Schottky or Frenkel defects are present.

There are many structural types and systems for which little or no information has been acquired. In time, proprietary interests may complete some of the gaps. Work is just beginning on refractory-metal carbides, and all other high-temperature compounds remain unexplored because of the difficulty of attaining stable annealing temperatures over the range of possible application. Virtually virgin from a diffusional point of view is an understanding of what to expect in cubic Cu_3Au-, cubic $MgCu_2$-, hexagonal $MgNi_2$-, and hexagonal $MgZn_2$-type intermetallic compounds. Such phases possess a simple structure easily amenable to unambiguous interpretation, and there are many examples of each, from some thirty systems, whose melting temperatures vary over a wide range.

In decreasing order of importance and difficulty (my own opinion), future experimental studies should be conducted on one simple system in depth, on those compounds not yet considered, and on data refinement of work done to date. To quote John Dryden, "Errors like straws, upon the surface flow; he who would search for pearls must dive below." Theoretically, we desire detailed models of how different atoms might move in various geometric arrays and, eventually, an elucidation of the forces holding two or more chemical entities together. The ideal binary system for experimental purposes should possess at least one ordered compound of broad compositional limits which can be disordered 50 to 1000°C below its solidus; satisfactory isotopes should be available for each species; ease of processing into samples is desired. Preliminary studies must be made to establish limits of thermodynamic equilibrium and the nature of defect structure on both sides of stoichiometry. Each self-diffusion determination for differing compositions should be conducted from the liquidus, through T_c, to as low a temperature as counting techniques permit, so that the importance of the state of order can be correlated with other properties. These qualifications soon reduce consideration of available systems to a small number, of which silver-palladium may be a good possibility. From resistivity measurements, Savitskii and Pravoverov[102] propose that the compounds AgPd and Ag_2Pd_3 exist up to about 1200°C (respectively, 100 and 150°C below the liquidus), but their presence requires independent verification.

Another study whose significance demands more attention is that of Laurent and Bénard.[103] They measured the self-diffusion of Na^+ and Cl^- in single crystal and polycrystalline NaCl) Here, only the slower-moving species (Cl^-. was accelerated by grain boundary short-circuiting, and D values for the faster-moving species (Na^+) were independent of boundary surface area. If this holds true in oxides and other ionic and intermetallic compounds, implications for the fields of oxidation, grain growth, and sintering are widespread. A proper corroboration would consist of determining self-diffusion of both species in any of the CsCl- or NaCl-type compounds as a function of grain size over a broad temperature range. For example, the two groups of investigators reporting silver diffusion in AgMg find differences that may result from the samples of one having more boundary surface area than the other. Because magnesium possesses the lower melting point and is probably the faster moving species, its self-diffusivity is expected from the investigation of Laurent and Bénard to be independent of grain size, whereas silver diffusion would be grain-size dependent. More along these lines might be done.

Now gaining momentum, this literature on diffusion in intermetallic compounds should keep accumulating at an exponential rate; the service of a review will be required over shorter time intervals. As with most worthwhile endeavors, there are numerous pitfalls to trap the unwary before we actually learn *what is* happening.

ACKNOWLEDGMENTS

C. E. Birchenall, H. B. Huntington, and J. H. Westbrook were very helpful in reviewing the manuscript critically and offering suggestions as to its improvement.

REFERENCES

1. Mehl R. F., *Atom Movements*, ASM Seminar, Cleveland, Ohio, 1951, p. 239.
2. Barrer R. M., *Diffusion In and Through Solids*. Cambridge University Press, Cambridge, 1951.
3. Jost W., *Diffusion in Solids, Liquids, and Gases*. Academic Press, New York, 1952.
4. Crank J., *Mathematics of Diffusion*, Clarendon Press, Oxford, 1956.
5. LeClaire A. D., *Prog. Metal Phys.* **1** (1950) 306; *ibid.* **4** (1953) 265.
6. Birchenall C. E., *Met. Rev.* **3** (1958) 225.
7. Birchenall C. E., "The Mechanisms of Diffusion in Solids," *Proc. 4th Intern. Symp. on Reactivity of Solids*, 1961 p. 24.
8. Tomizuka C. T., *Methods in Exp. Phys.* **6A** (1959) 364.
9. Lazarus D., *Solid State Phys.* **10** (1960) 72.
10. Hobstetter J. N., *Prog. Metal Phys.* **7** (1958) 1.
11. Smits F. M., *Ergeb. Exakt. Naturw.* **31** (1959) 167.
12. Reiss H., and C. S. Fuller, *Semiconductors*, Reinhold, New York, 1959, p. 222.
13. Sherby O. D., and M. T. Simnad, *Trans. ASM* **54** (1961) 227.
14. Hansen M., *Constitution of Binary Alloys*. McGraw-Hill, New York, 1958.
15. Köhler W., *Zentr. Hütten Walzwerke* **31** (1928) 650.
16. Dunn J. S., *J. Chem. Soc.* (London) **2** (1926) 2973.
17. Seith W., and W. Krauss, *Z. Electrochem.* **44** (1938) 98.
18. Petrenko B. G., and R. E. Rubinstein, *Zh. Fiz. Khim.* **13** (1939) 508.
19. Inman M. C., D. Johnston, W. L. Mercer, and R. Shuttleworth, "The Measurement of Self-Diffusion Coefficients in Binary Alloys," *Proc. of Radioactive Conf.*, Oxford, II. Butterworths Scientific Publications, London, 1954, p. 85; Inman M. C., *J. Inst. Metals* **81** (1953) 744.
20. Landergren U. S., C. E. Birchenall, and R. F. Mehl, *Trans. AIME* **206** (1956) 73.
21. Darken L. S., *Trans. AIME* **175** (1948) 184.
22. Resnick R., and R. W. Balluffi, *Trans. AIME* **203** (1955) 1004.
23. Kuper A. B., D. Lazarus, J. R. Manning, and C. T. Tomizuka, *Phys. Rev.* **104** (1956) 1536.
24. Slifkin L., and C. T. Tomizuka, *Phys. Rev.* **97** (1955) 836.
25. Slifkin L., and C. T. Tomizuka, *Phys. Rev.* **104** (1956) 1803.
26. Lidiard A. B., *Phys. Rev.* **106** (1957) 823.
27. Mehl R. F., and C. F. Lutz, *Trans. Met. Soc. AIME* **221** (1961) 561.
28. Kuzmenko P. P., E. I. Kharkov, and G. P. Grinevich, *Ukrain. Fiz. Zh.* **5** (1960) 683.
29. Gertsriken S. D., T. K. Yatsenko, and L. F. Slastnikova, *Sb. Nauchn. Rabot Inst. Metallofiz, Akad. Nauk Ukr. SSSR* **13** (1961) 93.
30. Bradley A. J., and A. Taylor, *Proc. Roy. Soc.* **159** (1937) 56.
31. Smoluchowski R., and H. Burgess, *Phys. Rev.* **76** (1949) 309.
32. Nix F. C., and F. E. Jaumot, *Phys. Rev.* **83** (1951) 1275.
33. Berkowitz A. E., F. E. Jaumot, and F. C. Nix, *Phys. Rev.* **95** (1954) 1185.
34. Gertsriken S. D., and I. Ya. Dekhtyar, *Fiz. Metal. i Metalloved., Akad. Nauk SSSR* **3** (1956) 242.
35. Gertsriken S. D., I. Ya. Dekhtyar, N. P. Plotnikova, L. F. Slastnikova, and T. K. Yatsenko, *Issled. po Zharoproch. Splavam, Akad. Nauk SSSR* **3** (1958) 68.
36. Westbrook J. H., *J. Electrochem. Soc.* **104** (1957) 369.
37. Wood D. L., and J. H. Westbrook, *Trans. Met. Soc. AIME* **224** (1962) 1024.
38. Westbrook J. H., and D. L. Wood, *J. Inst. Metals* **91** (1963) 174.
39. Hagel W. C., and J. H. Westbrook, *Trans. Met. Soc. AIME* **221** (1961) 951.
40. Domian H. A., and H. I. Aaronson, *Trans. Met. Soc. AIME*, **230** (1964) 44.
41. Hagel W. C., and J. H. Westbrook, in *Diffusion in Body-Centered Cubic Materials*, ASM Conference, Metals Park, Ohio, 1965, 197. Westbrook J. H., and W. C. Hagel, *Trans. Met. Soc. AIME*, **221** (1963) 193.
42. Huntington H. B., N. C. Miller, and V. Nerses, *Acta Met.* **9** (1961) 749.
43. Wechsler M. S., *Acta Met.* **5** (1957) 150.

44. Clark J. S., and N. Brown, *J. Phys. Chem. Solids* **19** (1961) 291.
45. Gupta D., *Self-Diffusion Studies in β-Phase Ordered Alloys of Au-Cd by Radioactive Tracer Technique.* Ph.D. Thesis, Univ. of Illinois, 1961.
46. Boltaks B. I., and G. S. Kulikov, *Soviet Phys.—Tech. Phys.* **2** (1957) 67.
47. F. H. Eisen, and C. E. Birchenall, *Acta Met.* **5** (1957) 265.
48. Portnoy W. M., H. Letaw, and L. Slifkin, *Phys. Rev.* **98** (1955) 1536.
49. Boltaks B. I., and Yu. A. Gutorov, *Soviet Phys.—Solid State* **1** (1960) 930.
50. Goldstein B., *Properties of Elemental and Compound Semiconductors*, AIME Met. Soc. Conf. **5** (1960) 155.
51. Hulme K. F., and J. E. Kemp, *J. Phys. Chem. Solids* **10** (1959) 335.
52. Fuller C. S., *Phys. Rev.* **86** (1952) 136.
53. Sze S. M., and L. Y. Wei, *Phys. Rev.* **124** (1961) 85.
54. Pines B. Ya., and E. F. Chaikovskii, *Soviet Phys.—Solid State* **1** (1959) 864.
55. Wieber R. H., H. C. Gorton, and C. S. Peet, *J. Appl. Phys.* **31** (1960) 608.
56. Dunlap W. C., Jr., *Phys. Rev.* **94** (1954) 1531.
57. Schillman A., *Z. Naturforsch.* **11A** (1956) 472.
58. Goldstein B., *Rev. Sci. Inst.* **28** (1957) 289.
59. Van den Boomgaard J., and K. Schol, *Philips Res. Rept.* **12** (1957) 127.
60. Goldstein B., *Phys. Rev.* **118** (1960) 1024.
61. Goldstein B., *Phys. Rev.* **121** (1961) 1305.
62. Allen J. W., and F. A. Cunnell, *Nature* **182** (1958) 1158.
63. Cunnell F. A., and C. H. Gooch, *J. Phys. Chem. Solids* **15** (1960) 134.
64. Allen J. W., *J. Phys. Chem. Solids* **15** (1960) 134.
65. Longini R. L., *Solid-State Electron.* **5** (1962) 127.
66. Fuller C. S., and J. M. Whelan, *J. Phys. Chem. Solids* **6** (1958) 173.
67. Hall R. N., and J. H. Racette, *Bull. Am. Phys. Soc.* **7** (1962) 234.
68. Fuller C. S., and K. B. Wolfstirn, *J. Appl. Phys.* **33** (1962) 2507.
69. Boltaks B. I., *Dokl. Akad. Nauk SSSR* **100** (1955) 901.
70. Kuliev A. A., *Fiz. Tverd. Tela* **1** (1959) 1176
71. Kuliev A. A., and G. B. Abdullaev, *Soviet Phys.—Solid State* **1** (1959) 545.
72. Almin K. E., *Acta Chem. Scand.* **2** (1948) 400.
73. Boltaks B. I., *Zh. Tekh. Fiz.* **25** (1955) 767.
74. Carlson R. O., *J. Phys. Chem. Solids* **13** (1960) 65.
75. Keys J. D., *J. Phys. Chem. Solids* **23** (1962) 820.
76. Boltaks B. I., and N. A. Fedorovich, *Soviet Phys. —Solid State* **4** (1962) 400.
77. Akhundov G. A., and G. B. Abdullaev, *Soviet Phys.* "*Doklady*" **3** (1958) 390.
78. Hahn H., and W. Klingler, *Z. Anorg. Chem.* **260** (1949) 110.
79. Shinyaev A. Ya., *Issled. po Zharoproch. Splavam, Akad. Nauk SSSR* **9** (1962) 19.
80. Clarke R. L., *J. Appl. Phys.* **30** (1959) 957.
81. Arkharov V. I., S. M. Klotsman, A. N. Timofeev, and I. Sh. Trakhtenberg, *Phys. Metals Metallogr.* **14** (1962) 68.
82. Klotsman S. M., A. N. Timofeev, and I. Sh. Trakhtenberg, *Phys. Metals Metallogr.* **12** (1961) 150.
83. Condit R. H., *Self-Diffusion of Iron and Sulfur in Ferrous Sulfide.* Ph.D. Thesis, Princeton Univ., 1960.
84. Boltaks B. I., and Yu. N. Mokhov, *Zh. Tekh. Fiz.* **26** (1956) 2448.
85. Boltaks B. I., and Yu. N. Mokhov, *Zh. Tekh. Fiz.* **28** (1958) 1046.
86. Seltzer M. S., and J. B. Wagner Jr., *J. Chem. Phys.* **36** (1962) 130.
87. Anderson J. S., and J. R. Richards, *J. Chem. Soc.* Part 1 (1946) 537.
88. Brebrick R. F., and W. W. Scanlon, *Phys. Rev.* **96** (1954) 598.
89. Bloem J., and F. A. Kröger, *Philips Res. Rept.* **12** (1957) 281.
90. Bloem J., and F. A. Kröger, *Philips Res. Rept.* **12** (1957) 303.
91*a*. Baskin M. L., V. I. Tret'yakov, and I. I. Chaporova, *Phys. Metals Metallogr.* **12** (1961) 72.
91*b*. Baskin M. L., V. I. Tret'yakov, and I. I. Chaporova, *Phys. Metals Metallogr.* **14** (1962) 422.
92. Gel'd, P. V., and V. D. Lyubimov, *Self-Diffusion of Niobium and Carbon in Niobium Oxides and Carbides.* Seminar—Acad. of Sciences, Ukrain. SSR, Inst. of Powder Metallurgy and Special Alloys, Kiev, 1961.
93. Elcock E. W., and C. W. McCombie, *Phys. Rev.* **109** (1958) 605.
94. Elcock E. W., *Proc. Phys. Soc.* **73** (1959) 250.
95. Zener C., *Acta Cryst.* **3** (1950) 346.
96. Zener C., *J. Appl. Phys.* **22** (1951) 372.

97. Kikuchi R., *J. Phys. Chem. Solids* **20** (1961) 35.
98. Flinn P. A., and G. M. McManus, *Phys. Rev.* **124** (1961) 54.
99. Beeler J. R., Jr., and J. A. Delany, "Order-Disorder Events Produced by Single Vacancy Migration," TM 62-11-10, N.M.P.O., General Electric (1962).
100. Azaroff L. V., *J. Appl. Phys.* **32** (1961) 1658; *ibid.* **32** (1961) 1663.
101. Brebrick R. F., *J. Appl. Phys.* **30** (1959) 811.
102. Savitskii E. M., and N. L. Pravoverov, *Zh. Neorg. Khim. Akad. Nauk SSSR* **6** (1961) 499.
103. Laurent J. F., and J. Bénard, *Compt. Rend.* **241** (1955) 1204.

chapter 21

Order-Disorder and Radiation Damage

PETER S. RUDMAN
Battelle Memorial Institute
Columbus, Ohio

1. INTRODUCTION

As the name implies, order-disorder (O-D) deals with the transition states between the asymptotic limits of perfect order and perfect randomness in multicomponent systems. These imperfectly ordered states may be obtained by thermal disordering, by alloying to a nonstoichiometric composition, or by mechanical disordering, such as high-energy radiation or plastic deformation. O-D is treated in most texts on physical metallurgy, solid state physics, and statistical thermodynamics; it has also been discussed in several recent reviews. Elementary introductions to O-D concepts are given by Hume-Rothery and Raynor,[1] Lipson,[2] and Elcock.[3] The statistical-thermodynamic treatments have been reviewed by Muto and Takagi[4]; Guggenheim[5] has summarized the quasichemical approach; Domb[6] has reviewed recent advances in calculations of lattice statistics. Physical-property changes accompanying the O-D transformation have been reviewed by Muto and Takagi[4] and Elcock,[3] and mechanical properties have been covered by Westbrook.[7] The definition and the determination of order by X-ray diffraction has been reviewed by Lipson,[2] Elcock,[3] Muto and Takagi,[4] Guttman,[8] Warren,[9] Guinier[10] and Wooster.[11] The review by Guttman[8] and a review by Cohen and Fine[12] also cover a variety of other aspects.

It does not appear to be necessary to cover all of this ground again, and this chapter is limited to a few aspects of O-D in which, it is felt, adequate treatments of problems of current or potential importance are lacking.

We shall first consider long-range order (LRO). In the past, attention has been directed mostly at some structurally simple, stoichiometric binary alloys, and in these cases the definition of LRO and its determinability by X-ray diffraction intensity measurements are generally understood. However, in cases of nonstoichiometry or of multicomponent systems, these problems lack discussion. We shall develop here a general approach to LRO definition and its diffraction determination for arbitrary systems.

We shall next consider the definition of short-range order (SRO) and its determination by X-ray diffraction. We shall compare the approach of A. J. C. Wilson with that of B. E. Warren. A new derivation of the SRO X-ray diffuse scattering equation will be given. Although generally reproducing the results of the Warren approach, the new derivation leads to the prediction that from small-angle scattering, $\theta \lesssim \frac{1}{2}°$, a mean square radius of SRO propagation can be obtained.

Finally, we shall review the somewhat related problems of O-D transformation kinetics and high-energy radiation disordering. They are complex, combining the problems of mobile defect identification and enumeration with the problem of the mode of propagation of order (or disorder). This is now an area of rapidly expanding study, but a review of the current state of prejudice should be of interest.

1.1 Long-Range Order Diffraction Effects

Any crystal structure can be defined by specifying a system of interpenetrating sublattices and the occupation of each sublattice. To give a simple example: NaCl can be described as two face-centered cubic (fcc) sublattices, with Na occupying one sublattice and Cl the other. More complicated structures can be defined by employing more sublattices and more components. Let us assume that a given crystal structure can be described by l sublattices, numbering them $1 \cdots j \cdots l$. Let the crystal contain n atomic species, numbering them $1 \cdots i \cdots n$, and let the fraction of sublattice j occupied by species i be x_{ij}. There are thus nl variables x_{ij}, and the first question we ask is how many independent variables are there? The x_{ij}'s are related by l equations

$$\sum_{i=1}^{n} x_{ij} = 1 \tag{1}$$

and by n equations

$$\sum_{j=1}^{l} X_j x_{ij} = x_i \tag{2}$$

where X_j = fraction of lattice sites of type j, and x_i = average overall fraction of species i. The n equations (Eq. 2) are not independent, but are related by

$$\sum_{i=1}^{n} x_i = 1 \tag{3}$$

so that there are $n + l - 1$ independent relations among the nl variables. Hence the number of independent variables, ϕ, is

$$\phi = nl - (n + l - 1) = (n - 1)(l - 1) \tag{4}$$

This result has been obtained previously by Wojciechowski[13] and Men.[14] The next problem is to decide which of the x_{ij}'s, or combinations of them, are the most convenient to be selected as LRO parameters, and how many of the ϕ independent LRO parameters are experimentally determinable.

The integrated intensity of a Bragg reflection is proportional to $F^2(hkl)$, where $F(hkl)$ is the structure factor. The experimental details of the intensity measurement and the specification of the proportionality constants have been reviewed recently[8–11] and will not be considered here. Rather, it will be assumed that $|F(hkl)|$ can be obtained, and we shall only discuss the relationship of the $|F(hkl)|$ to the state of LRO.

If the scattering factor of atom i is f_i, the average scattering factor of sublattice j is

$$\langle f_j \rangle = \sum_{i=1}^{n} x_{ij} f_i \tag{5a}$$

and the average scattering factor of the crystal is

$$\langle f \rangle = \sum_{i=1}^{n} x_i f_i \tag{5b}$$

The structure factor of the crystal can be written as

$$F(hkl) = \sum_{j=1}^{l} \mathscr{F}_j \langle f_j \rangle \, e^{-2\pi i(hu_j + kv_j + lw_j)} \tag{6}$$

where $u_j \, v_j \, w_j$ define the origin of sublattice j, and $\mathscr{F}_j$ is the geometric structure factor of sublattice j (e.g., for a fcc sublattice, $\mathscr{F}_j = 0$ for hkl mixed, $\mathscr{F}_j = 4$ for hkl unmixed, etc.). Although it does not appear to be easily proved in general, it seems to follow from Eq. 6 that we can obtain only $(l - 1)$ independent types of reflections that can be used to distinguish between the sublattice occupations. We shall later justify this conclusion by means of some specific examples. Assuming this conclusion we thus infer that there are R LRO parameters, not determinable by means of a diffraction intensity measurement alone, in which

$$R = \phi - (l - 1) = (n - 2)(l - 1) \quad (7)$$

The most important consequence of this result is the fact that only in a binary system can an arbitrary state of LRO be determined unambiquously by a diffraction intensity measurement alone. Let us now consider the above results for some specific examples.

Example 1: *Subdivision of a body-centered cubic (bcc) into two simple cubic (sc) sublattices:* $\alpha(000)$ *and* $\beta(\frac{1}{2}\,\frac{1}{2}\,\frac{1}{2})$.

We obtain for the structure factors

$$F_f = f_\alpha + f_\beta = 2\langle f\rangle \quad (8a)$$
$$h + k + l = 2n$$
$$F_S = f_\alpha - f_\beta = \sum_{i=1}^{n} (x_{i\alpha} - x_{i\beta})f_i \quad (8b)$$
$$h + k + l = 2n + 1$$

Reflections with structure factors of the form of Eq. 8a are independent of the degree of LRO and are called fundamental reflections. There is only one type of reflection that depends on the relative sublattice occupations. These reflections are called superlattice reflections. We can generalize from the form of Eq. 8b to define LRO: a structure exhibits LRO if at least two sublattices differ in composition. We can also see from Eq. 8b that the sublattice composition differences, the $(x_{i\alpha} - x_{i\beta})$, are natural choices for LRO parameters. For the special case in which the degree of LRO is determinant, $n = 2$; if we let the two species be A and B, the superlattice structure factor can be written as

$$F_S = S(f_A - f_B) \quad (8c)$$

where $S = (x_{A\alpha} - x_{A\beta}) = (x_{B\beta} - x_{B\alpha})$ is the Bragg-Williams LRO parameter. For the case of perfect LRO (where, e.g., only A is on α and only B on β) $S = 1$; for complete disorder $S = 0$.

Example 2: *Subdivision of a fcc lattice into two sublattices: a sc* $\alpha(000)$ *and a* $\beta(\frac{1}{2}\,\frac{1}{2}\,0, \frown)$, *made up of three sc lattices.*

For the structure factors we obtain

$$F_f = f_\alpha + 3f_\beta = 4\langle f\rangle \qquad hkl \text{ unmixed}$$
$$F_S = f_\alpha - f_\beta = \sum_{i=1}^{n} (x_{i\alpha} - x_{i\beta})f_i \qquad hkl \text{ mixed}$$

Thus the superlattice structure factor is exactly the same as in the preceding case; it requires no further elaboration, except to illustrate one additional property of the Bragg-Williams LRO parameters as defined here—they can obtain the value of unity only at certain stoichiometric ratios. In addition to this convenient limit, at least when there is only one independent LRO parameter, the maximum value attainable by the Bragg-Williams LRO parameter is simply a linear function of composition. We thus have a very simple means of estimating the maximum degree of order that can be obtained at arbitrary composition. As an illustration, let us consider the process of placing all the possible A atoms on the α-sublattice and letting the β-sublattice take the overflow. We then obtain for the bcc lattice of Ex. 1

$$S = x_{A\alpha} - x_{A\beta} = 2(x_{A\alpha} - x_A) \quad (9)$$

so that

$$S_{max} = 2x_A, \; x_B \geq \tfrac{1}{2}$$
$$= 2x_B, \; x_B \leq \tfrac{1}{2}$$

For the fcc lattice of Ex. 2, with AB_3 as the stoichiometric composition, we obtain

$$S = x_{A\alpha} - x_{A\beta} = \tfrac{4}{3}(x_{A\alpha} - x_A) \quad (10)$$

so that

$$S_{max} = 4x_A, \; x_B \geq \tfrac{3}{4}$$
$$= \tfrac{4}{3}x_B, \; x_B \leq \tfrac{3}{4}$$

Example 3: *Subdivision of a bcc lattice into three sublattices: a fcc* $\beta(000)$, *a fcc* $\gamma(\frac{1}{2}\,\frac{1}{2}\,\frac{1}{2})$, *and a* $\alpha(\frac{1}{4}\,\frac{1}{4}\,\frac{1}{4},\ \frac{3}{4}\,\frac{3}{4}\,\frac{3}{4})$, *made up of two fcc lattices.*

The origin coordinates, *u v w* here refer to a cubic unit cell comprising eight bcc unit cells, as illustrated in Fig. 1. For this case, the structure factor is nonzero only for *hkl* unmixed and is

$$F_f = 4(f_\alpha + 2f_\beta + f_\gamma) = 16\langle f\rangle$$
$$h + k + l = 4n$$
$$F_S = 4(f_\beta + f_\gamma - 2f_\alpha)$$
$$= 4\sum_{i=1}^{n} (x_{i\beta} + x_{i\gamma} - 2x_{i\alpha})f_i$$
$$h + k + l = 2(2n + 1)$$
$$F_S' = 4(f_\beta - f_\gamma)$$
$$= 4\sum_{i=1}^{n} (x_{i\beta} - x_{i\gamma})f_i$$
$$h + k + l = 2n + 1$$

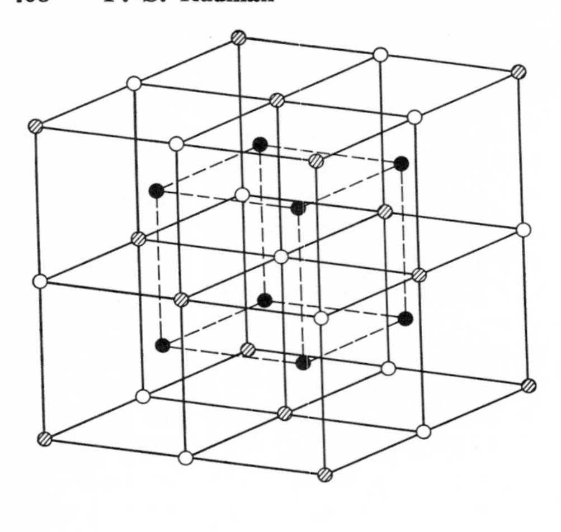

● α-sites

⊘ β-sites

○ γ-sites

Fig. 1. Long-range order structure: body-centered cubic lattice, three sublattices.

For the determinate case, where $n = 2$, we obtain for the two independent superlattice structure factors

$$F_S = 8S(f_A - f_B)$$
$$F_S' = 4S'(f_A - f_B)$$

where $S = x_{A\alpha} - \frac{1}{2}(x_{A\beta} + x_{A\gamma})$, and $S' = x_{A\gamma} - x_{A\beta}$. This type of two simultaneous orderings is found in Fe-rich Fe–Al alloys and has been discussed among others, by Rudman.[15]

Let us now return to the rather surprising conclusion that arbitrary LRO is not completely determinate in systems other than binary. Again, let us consider a specific, simple example—a ternary alloy of composition x_A, x_B, x_C, ordering in a bcc lattice with two sc sublattices $\alpha(000)$ and $\beta(\frac{1}{2}\,\frac{1}{2}\,\frac{1}{2})$. We can define three LRO parameters: $S_1 = (x_{A\alpha} - x_{A\beta})$, $S_2 = (x_{B\beta} - x_{B\alpha})$, and $S_3 = (x_{C\beta} - x_{C\alpha})$, they are related as

$$S_1 - S_2 - S_3 = 0$$

If we choose S_1 and S_3 as the two independent variables, we can write the superlattice structure factor as

$$F_S = S_1(f_A - f_B) + S_3(f_B - f_C)$$

F_S is thus a plane in $S_1 - S_3$ space, and in Fig. 2 a possible form of this plane is illustrated. It can be seen that, except for F_S(max), every value of F_S can be attained by a continuous, although limited, range of S_1 and S_3. However, $F_S = 0$ is not a singular point, and hence the absence of a superlattice intensity in this case does not necessarily eliminate LRO. We can see by a generalization of this example that, although it is possible to determine the complete structure of a perfectly ordered

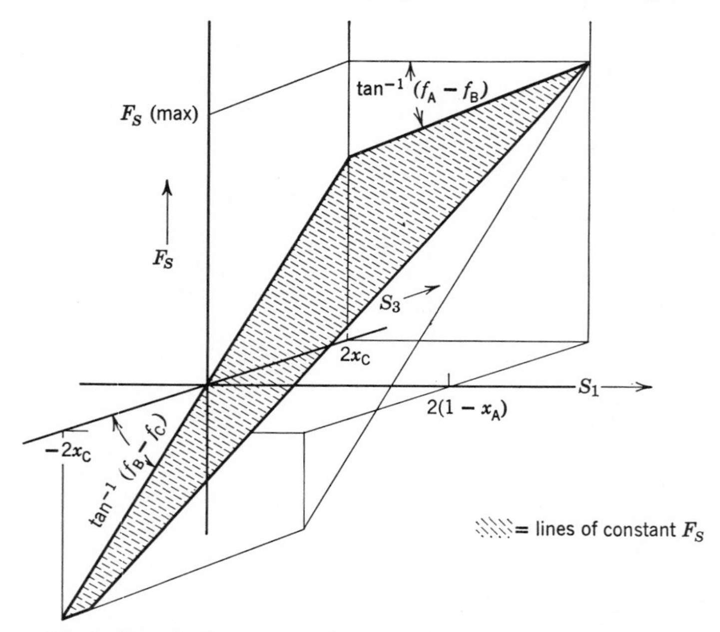

Fig. 2. Superlattice structure factor limits in a ternary alloy.

multispecies crystal, we are unable to determine completely an arbitrary degree of LRO in the same crystal.

1.2 Short-Range Order Diffraction Effects

The term short-range order is conventionally employed in two different senses. It describes the general near-neighbor identity correlations in a solid solution and the specific case of the near-neighbor correlations that is the antecedent of LRO. Because it is now recognized that another type of near-neighbor correlation also exists—clustering,[16] the antecedent of precipitation—it follows logically that one of the definitions of short-range order should be changed. However, as is often the case, logic yields to convention; we shall retain this ambiguous nomenclature, but shall employ the abbreviations SRO for the general case and sro for the specific.

As in the previous discussion of LRO, we shall adopt natural SRO parameters based on X-ray (or neutron or electron) diffraction information. A complete set of LRO parameters (Eq. 4) uniquely defines the state of LRO in a homogeneous crystal, and hence all complete sets of LRO parameters are equivalent, regardless of how they are obtained or defined. On the other hand, the SRO parameters that can be obtained by diffraction do not furnish a unique description of the state of SRO, and hence the diffraction-defined SRO parameters are not necessarily equivalent to those obtained by some other technique. Because it has been found that some diffraction-defined SRO parameters have the virtue of being relatable to parameters employed in most statistical-thermodynamic theories of ordering, and because of the present dominance of diffraction determinations of SRO, there has been little need to consider other parameters.

Let us briefly note some relatively new techniques that can contribute SRO information in terms of other parameters. SRO nuclear quadrupole effects in NMR experiments on alloys have been summarized by Cohen and Reif.[17] Mössbauer-effect line shapes have been discussed briefly by Flinn and Ruby[18] in terms of SRO. In solid solutions of mixed salts, the color-center absorption spectra may be related to anion SRO in systems such as NaCl–KCl[19] or to cation SRO in systems such as KCl–KBr.[20] The ultimate technique for determining SRO would be by means of a microscope capable of resolving individual atoms, and the preliminary field-ion microscope results reviewed by Cottrell[21] indicate that the enjoyment of this technique may not be far off. An artificial approximation to this ultimate technique, the electronic-computer Monte Carlo "experiments," has already made a significant contribution[22–26] to the O-D problem in general, and it is the non-diffraction technique likely to be of most importance in the immediate future.

In view of the difficulty that we have already encountered in the determination of LRO in systems other than binary, we can limit our present discussion of SRO to binary systems only, with little loss in practicality. The first X-ray diffraction analysis of SRO appears to have been made by Wilchinsky,[27] and most of the more recent treatments closely follow this analysis.

Wilchinsky and many of the others who developed this treatment did so at the suggestion and with the help of B. E. Warren; Therefore, we shall hereafter refer to it as the Warren approach. Even though the Warren formulation of the SRO diffraction problem has been reviewed several times,[8–11] some aspects of the approach have not had adequate discussion. Accordingly, we shall reconsider this treatment in some detail.

We start with the general expression for scattered X-ray intensity in absolute units

$$I = \sum_m \sum_n f_m f_n \, e^{ik\cdot r_{mn}} \tag{11}$$

where

$$k = \frac{2\pi}{\lambda}(s - s_0)$$

s, s_0 = unit vectors in the directions of the scattered and incident beams, respectively; λ = scattered wavelength; r_{mn} = vector between atoms at sites m and n. We divide this into Bragg reflections with nonzero values only at reciprocal lattice points (rlp's) and diffuse scattering with no singularities at rlp's The average value of $f_m f_n$ will vary with r_{mn}, but if the order is only short-range, then for any solution and for any particular state of order there will be a limiting constant, $(f_m f_n)_{\lim}$, for large r_{mn}. Thus if we set

$$f_m f_n = (f_m f_n)_{\lim} + [f_m f_n - (f_m f_n)_{\lim}] \tag{12}$$

the first term is independent of r_{mn} and the

second has nonzero values only for small r_{mn}. When we incorporate Eq. 12 into Eq. 11, the first term gives rise to sharp Bragg reflections, and the second term produces diffuse scattering with no singularities at rlp's. Assuming rapid convergence of the second term in Eq. 12, we can replace the double summation in Eq. 11 by N times a sum over neighbors of an average atom, so that

$$I_{\text{SRO}} = N \sum_{i=0} [f_m f_n - (f_m f_n)_{\text{lim}}] \, e^{ik \cdot r_i} \quad (13)$$

Let us now consider the evaluation of $f_m f_n$ in terms of pair probabilities in a solution not exhibiting LRO. We have

$$f_m f_n = x_A f_A (q_i f_A + p_i f_B) + x_B f_B (q_i' f_B + p_i' f_A) \quad (14)$$

where q_i = probability that the atom at r_i from an A is an A, p_i = probability that the atom at r_i from an A is a B, q_i' = probability that the atom at r_i from a B is a B, and p_i' = probability that the atom at r_i from a B is an A. However, the four probabilities are related by three relations

$$\begin{aligned} p_i + q_i &= 1 \\ p_i' + q_i' &= 1 \\ x_A p_i &= x_B p_i' \end{aligned} \quad (15)$$

so that there is only one independent probability, which we shall take as p_i. We thus obtain

$$f_m f_n = x_A f_A^2 + x_B f_B^2 - x_A p_i (f_A - f_B)^2 \quad (16)$$

If we let p_{lim} be the value for p_i in the limit of large r_i

$$f_m f_n - (f_m f_n)_{\text{lim}} = x_A (p_{\text{lim}} - p_i)(f_A - f_B)^2 \quad (17)$$

Now, if we let

$$\alpha_i = \frac{p_{\text{lim}} - p_i}{x_B} \quad (18)$$

which is equal to the SRO coefficient, we obtain

$$I_{\text{SRO}} = L \sum_{i=0} \alpha_i \, e^{ik \cdot r_i} \quad (19)$$

where

$$L = N x_A x_B (f_A - f_B)^2 \quad (20)$$

which is the Laue monotonic scattering.

Equation 19 for the diffuse scattering intensity is the same formal result as obtained by the Warren formulation.[9] However, the short-range order coefficients, the α_i's, defined by Eq. 18, are not exactly equivalent. The Warren definition *assumes* that $p_{\text{lim}} = x_B$. That this is only an approximation can be seen by considering I_{SRO} as k approaches an rlp; then from Eq. 19

$$I_{\text{SRO}}(k \to \text{rlp}) \to L \sum_{i=0} \alpha_i \quad (21)$$

If we note

$$\begin{aligned} \sum_{i=0} 1 &= N \\ \sum_{i=0} p_i &= x_B N \end{aligned} \quad (22)$$

we see easily that $\sum_{i=0} \alpha_i = 0$. Thus the Warren formulation yields $I_{\text{SRO}}(k \to \text{rlp}) \to 0$; it does so independently of the system or state of order. This is not a reasonable result. In a perfectly random solution, it can be rigorously shown that $I_{\text{SRO}} = L$ everywhere, including rlp's. In solutions that are just beginning to decompose, as in Guinier-Preston zone formation, a strong small-angle scattering is observed, and there is no indication that this diffuse intensity would diminish to zero as $k \to 0$ more closely.

The difficulty lies in the "obvious" but wrong assumption that $p_{\text{lim}} = x_B$. Unfortunately perhaps, p_{lim} cannot be given any a priori, calculable value. Because the discrepancy in $I_{\text{SRO}}(k \to \text{rlp})$ is of the order of $\pm L$, the discrepancy in the definition of α_i must be of the order of $\pm N^{-1}$. Thus for analyzing diffuse scattering away from rlp's, the Warren definition of α_i is adequate.

Let us now consider another consequence of the separation that we effected in Eq. 12. The Bragg reflection term is

$$I_{\text{Bragg}} = (f_m f_n)_{\text{lim}} \sum_m \sum_n e^{ik \cdot r_{mn}} \quad (23)$$

When LRO exists, this term yields both the fundamental and the superstructure reflections. This expression can be readily shown to reduce to the consideration of the structure factors that we have discussed previously. It can thus be seen that LRO and SRO are, diffractometrically speaking at least, independent effects. With the onset of LRO, superlattice lines appear superposed on the diffuse sro scattering rather than the sro scattering peaking up to form superlattice lines. Actually, there is a competition for diffracted intensity between superlattice lines and diffuse scattering.

Another point of view,[2] however, considers sro formation through a continuous process of successive partitioning of a LRO system into antiphase domains. In this case, the sro diffuse scattering is essentially a particle-size broadening of superlattice lines. The diffraction theory for this process has been given by Wilson.[28] During early nonequilibrium stages of LRO formation, or as a result of plastic deformation of a LRO structure, sufficiently small antiphase domains may be obtained for the superlattice particle-size broadening to yield an apparent diffuse scattering. As the domain size grows, the diffuse scattering will appear to peak up into superlattice lines. Although the Warren approach to sro with its essentially infinite domain size may seem to be a more useful approximation, the true description of sro must actually be a mixture of both of these descriptions. The degree of SRO certainly exhibits fluctuations throughout the crystal, so that there is a nebulous domain structure. On the one hand, the Warren approach averages over these fluctuations and thus yields appropriately averaged SRO parameters. On the other hand, there apparently is an antiphase juxtapositioning of the fluctuations in accord with Wilson's model. Thus Wilson's prediction of a lens-shaped distribution of intensity centered on superlattice reciprocal lattice points has been observed in disordered Cu_3Au by Cowley[29] and in disordered CuAu by Borie[30] from data of Roberts.[31] The presence of antiphase juxtapositions in the absence of LRO is most strikingly demonstrated by the electron diffraction work of Watanabe and his coworkers.[32] From this work it is clear that even in the disordered solution there are fluctuations exhibiting a high degree of order and extending over sufficient dimensions to include many antiphase boundaries.

Cohen and Fine[12] have reviewed other evidence for SRO fluctuations.

The formulation for SRO in the presence of LRO has been given by Roberts and Vineyard.[33] Although not so stated, their result appears to be valid only for stoichiometry, and the general result rigorously derived from Eq. 13 has not been given.

Equation 19 is the basis for single-crystal diffuse scattering determinations of the SRO parameters. The general solution involves a Fourier transformation of the scattered intensity throughout reciprocal space; Cowley's[29] study in Cu_3Au is a classic on this topic. Similar studies on complete single-crystal solutions have been made by Sutcliffe and Jaumot[36] on an off-stoichiometry Cu_3Au, by Roberts[31] on CuAu, and by Batterman[37] on $CuAu_3$. This technique is tedious, and resort has also been made to single-crystal partial solutions, by which the intensity distribution on a reciprocal lattice plane only is determined, as in the study of Norman and Warren[38] in Ag–Au alloys, or along a given direction only in reciprocal space, as in the studies of Roberts et al. in CuAu[39] and Cu_3Au.[33]

Unfortunately, SRO is not the only source of diffuse scattering, and it is a formidable problem to separate the SRO contribution from the superimposed Compton-modified scattering plus temperature diffuse scattering plus size-effect diffuse scattering. The aforementioned single-crystal studies have employed a variety of techniques for performing this separation and should be consulted for details. However, two more recent suggestions for corrections to the observed diffuse scattering should also be noted: Borie[30] on the size-effect separation and Walker and Keating[40] on a temperature effect.

However, by far most SRO diffuse-scattering studies have employed powder-pattern analysis which is a much simpler technique. By averaging Eq. 19 over all orientations of $\mathbf{r}_n$, we obtain the SRO diffuse-scattering powder-pattern expression

$$I_{\mathrm{SRO}} = L \sum_{i=0} c_i \alpha_i \frac{\sin kr_i}{kr_i} \tag{24}$$

where c_i = coordination number of ith distant shell from origin atom center, r_i = distance of ith shell from origin atom center, and $k = \mathbf{k} = 4\pi \sin\theta/\lambda$.

In general the α_1 term in Eq. 24 dominates the diffuse scattering modulations, thus a good idea of the general form of the SRO diffuse scattering in a powder pattern can be obtained by considering the function

$$I_{\mathrm{SRO}}/L = 1 + c_1\alpha_1 \frac{\sin kr_1}{kr_1} \tag{25}$$

For the clustering case, $\alpha_1 > 0$; for the sro case, $\alpha_1 < 0$. In Fig. 3, this function is plotted, taking $c_1\alpha_1 = 1$. Thus we see that clustering is characterized by an increasing intensity with

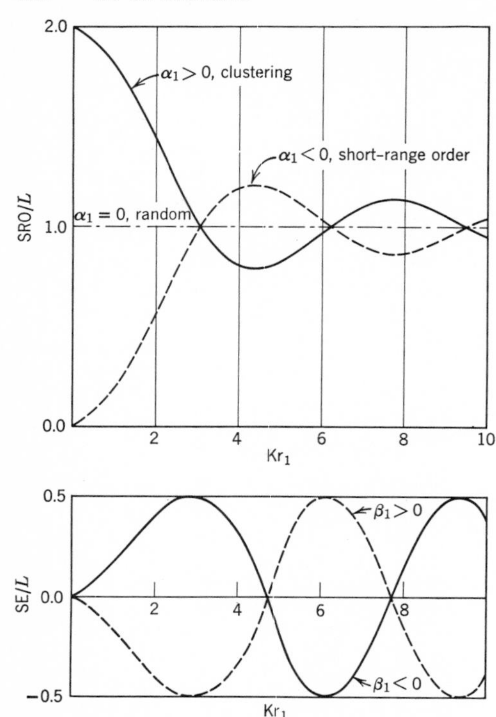

Fig. 3. Short-range order and size-effect diffuse scattering modulation functions.

decreasing angle in the range $kr_1 < 4$, whereas for sro the diffuse intensity peaks at about $kr_1 = 4$ and then decreases with decreasing angle.

The diffuse scattering contribution most disturbing to the analysis for SRO is the size-effect modulation.[39] The dominant size-effect (SE) term is

$$I_{\mathrm{SE}}/L = -c_1\beta_1\left(\frac{\sin kr_1}{kr_1} - \cos kr_1\right) \quad (26)$$

where

$$\beta_1 = \left(\frac{1}{\eta - 1}\right)\left[-\left(\frac{x_{\mathrm{A}}}{x_{\mathrm{B}}} + \alpha_1\right)\epsilon_{\mathrm{AA},1} + \left(\frac{x_{\mathrm{B}}}{x_{\mathrm{A}}} - \alpha_1\right)\eta\epsilon_{\mathrm{BB},1}\right] \quad (27)$$

is the first size-effect parameter

$$\eta = \frac{f_{\mathrm{B}}}{f_{\mathrm{A}}}$$

$$\epsilon_{\mathrm{AA},1} = \frac{r_{\mathrm{AA},1} - r_i}{r_i} \quad (28)$$

$$\epsilon_{\mathrm{BB},1} = \frac{r_{\mathrm{BB},1} - r_i}{r_i}$$

If the larger atom, say A, also has the larger scattering factor, we can expect $\epsilon_{\mathrm{AA},1} > 0$, and $\eta < 1$ generally to yield $\beta_1 > 0$; however, if the larger atom has the smaller scattering factor, generally we would expect $\beta_1 < 0$. In Fig. 3, the size effect modulation function of Eq. 26 is plotted for $|c_1\beta_1| = 0.5$. If $|\alpha_1| \gg |\beta_1|$, the size effect can be considered as shifting the positions of SRO modulation maxima and minima. On the other hand, if $\beta_1 \simeq \alpha_1$, then, as emphasized by Munster and Sagel,[41] the general form of the diffuse scattering may be so distorted it may not even be possible to say qualitatively whether clustering or sro exists. In Fig. 4 representative diffuse-scattering powder data are plotted: Al–Zn[16] is fcc, illustrating clustering, no size effect; Au–Ni[42] is also fcc, illustrating a dominant-size effect, $\beta_1 > 0$, and even after formal analysis of these data the question of whether sro or clustering exists is still not clear;[41] Mo–Ti[43] is bcc,

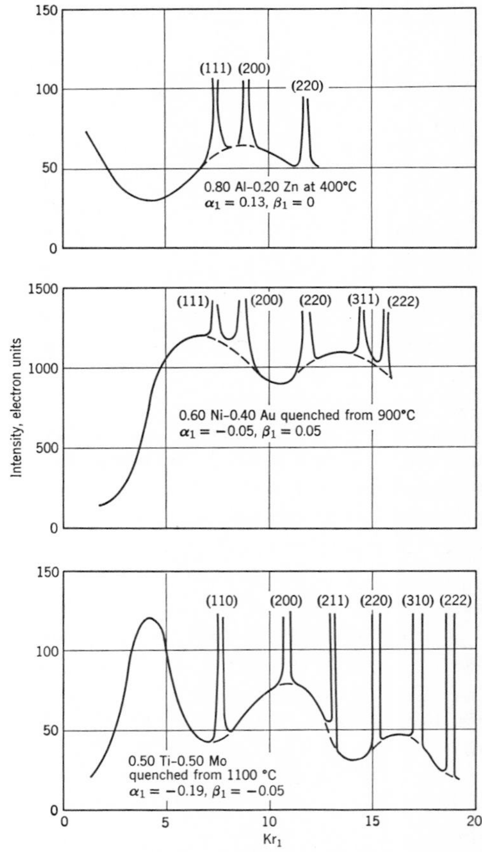

Fig. 4. Representative powder patterns showing diffuse scattering.

illustrating sro with a large-size effect, but $\beta_1 < 0$. A relatively up-to-date tabulation of SRO determinations can be found in Guinier's review.[10]

The Fourier analysis of powder patterns for SRO and size-effect parameters has been given by Flinn et al.[44] Although powder-pattern analysis cannot yield potentially as much information as single-crystal analysis, the amount and accuracy of usable information are, in fact, approximately the same for both methods. One reason for this is simply the fact that a major quantitative use of SRO data is in statistical-thermodynamic treatments of solid-solution energetics,[16] where usually only α_1 is exploited, which generally can be obtained just as well from powder data. Single-crystal data usually provide better higher-order α_i's (generally $i < 10$), but at the moment there does not appear to be any exploitation of such data. Presumably, the succession of the α_i's reflects the distribution functions of the degree and range of order propagation, that is, the fluctuations in the solid solution. Many physical and mechanical properties of alloys are sensitive to the fluctuation distribution, and hence their determination is an important problem and an interesting challenge in SRO analysis.

Toward this end, a consequence of the SRO diffuse-scattering derivation given here seems to be the capacity of small-angle scattering to be analyzed in terms of such order fluctuations. To see this, we perform a series expansion of Eq. 24, and obtain for small k

$$I_{\mathrm{SRO}}/L = \sum_{i=0} c_i \alpha_i \left[1 - \frac{(kr_i)^2}{6}\right] \tag{29}$$

Thus we see that from the intercept and the initial slope we can obtain a measure of the range of order propagation

$$\langle r^2 \rangle = \frac{\sum_{i=0} c_i \alpha_i r_i^2}{\sum_{i=0} c_i \alpha_i} \tag{30}$$

which we can consider as a mean-square radius of SRO domains.

2. ORDER-DISORDER KINETICS

The reasons for interest in O-D kinetics are indicative of the times. About thirty years ago, when the initial studies were made, interest was essentially academic. Subsequently, it was found that many O-D systems exhibit very useful physical and mechanical structure-sensitive properties, and, hence an understanding of O-D kinetics has been of continuing interest in developing optimum heat-treatment procedures. Since about 1950, however, the interest has shifted from O-D kinetics per se to its employment to elucidate the identity and the energetics of the lattice defects that are now generally accepted as required for diffusive atomic movement. This interest stems from the fact that lattice defects are generally difficult to study directly and are more easily studied through their effects on diffusive kinetics of LRO, which is more easily measured. However, in the last few years, various techniques have been developed,[45] which are sensitive enough to study the defects directly in pure metals and without the complications that O-D introduces. Thus the signal amplification role of O-D in the study of lattice defects is no longer so important. However, now that O-D has been relieved of this role, and defect identification and energetics can be accomplished for pure metals, order-disorder kinetics can assume the role of elucidating what is different about defects and their motions in alloys relative to pure metals. There are still relatively few results on this latter aspect.

We shall review here various theoretical models that have been proposed for order-disorder kinetics, attempting to define their areas of validity. We shall consider in particular the problem of determining vacancy energetics from order-disorder kinetics.

2.1 Microstructural Modes of Long-Range Ordering

As in decomposition transformation kinetics, it is profitable to consider order-disorder kinetics in relation to microstructure morphology. Although there are many analogous features in both transformations and, in fact, in equilibrium the ordering and the decomposition of a regular solution are energetically equivalent,[5] it is important to note a fundamental difference when considering transformation kinetics. In Fig. 5, the characteristic binary phase diagrams of the respective transformations are sketched. The decomposition transformation results in macroscopic composition differences, and the transformation

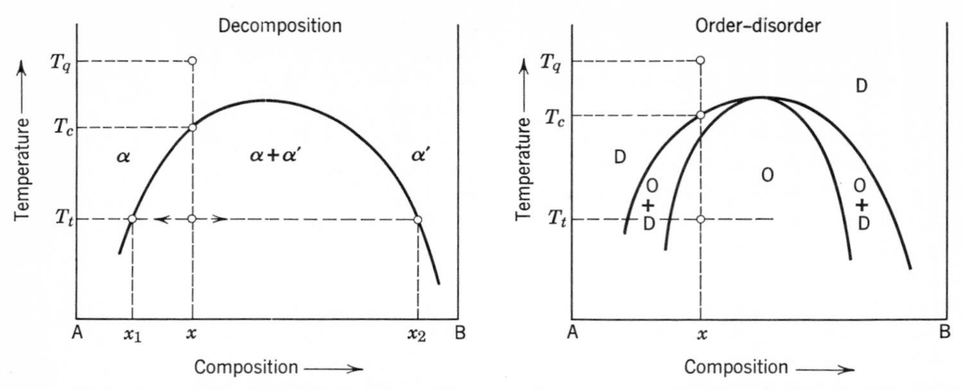

Fig. 5. Characteristic binary phase diagrams for the decomposition and order-disorder transformations.

kinetics are microstructurally determined through the imposition of boundary conditions on the mass transport diffusion equation. In contrast, the LRO transformation is essentially isocompositional, and hence no analogy is possible with the diffusion-gradient-controlling role of microstructure. Although it is true that decomposition can occur also during the order-disorder transformation, as illustrated in Fig. 5, this is not an essential feature, and is presumably simply the superposition of decomposition and ordering transformations. We shall see that it is difficult enough to consider order-disorder kinetics without this added complication, and we shall exclude decomposition from further consideration in the order-disorder problem.

Referring to Fig. 5, let us consider an alloy of arbitrary composition, x, and hold it at some temperature, T_q, in the disordered phase, D, and then instantaneously quench it to some temperature, T_t, in the ordered single-phase region, O, below the critical temperature for ordering, T_c. Let us consider a microstructural description of the ensuing isothermal ordering in a structure which is defined by two equivaelnt sublattices, k and l.* In some regions of the lattice, LRO will start (nucleate) with A atoms preferentially occupying the k sublattice, which will thus also by definition become an α-sublattice of the ordered structure. However, because the k and l sublattices are equivalent, in other regions of the lattice LRO will start with A atoms preferentially occupying the l sublattice, which will thus also by definition become an α-sublattice. Thus initially there will be ordered nuclei (domains), some of them with antiphase juxtaposition (k sublattice is α versus l sublattice is α). There will thus be present O domains, D domains, and O-domain—D-domain interfaces. This is illustrated schematically in Fig. 6 as stage (*a*). At a further stage in the transformation, the domains will grow until they meet, and then there will also be antiphase-domain interfaces, as illustrated in Fig. 6, stage (*b*). Eventually, there will not remain any disordered material, and only antiphase-domain interfaces will be present, as illustrated in Fig. 6, stage (*c*). Simultaneously with the growth of ordered domains, the degree of LRO within the domains may be increasing. Let us now formulate a general treatment of order-disorder kinetics embracing both of these aspects.

The observed degree of LRO, as properly defined by a crystal structure determining measurement, such as an X-ray diffraction superlattice intensity measurement, is a volume average, so that

$$S_{\text{obs}} = \langle S \rangle_{\text{vol}} = \frac{1}{V}\int SV(S)dV \tag{31}$$

The observed order-disorder kinetics thus become

$$\frac{dS_{\text{obs}}}{dt} = \frac{1}{V}\int S\frac{dV(S)}{dt}dV + \frac{1}{V}\int \frac{dS}{dt}V(S)dV \tag{32}$$

The first term on the right-hand side of Eq. 32 describes the O domain growth aspect, and

* See also Chapter 14 by Beeler for a description of antiphase domain structures and their development.

the second term describes what we shall refer to as the homogeneous ordering rate. Either term may be dominant in any given ordering transformation. There is no a priori reason to expect both rate processes to follow the same kinetics, and therefore successful analysis of order-disorder kinetics first requires identification of the mode of transformation. Let us now identify three idealized modes of the O-D transformation, which can be treated quantitatively and which singly, in combination, or in succession appear adequate to account for most of the experimental results: (1) homogeneous ordering, (2) O-domain growth, and (3) antiphase-domain coalescence.

We would expect homogeneous ordering to be rate-controlling in the case of coarse nucleation and rapid advancement of the O-domain—D-domain interface and for a change in the degree of order between two LRO states. In this case, Eq. 32 takes the form

$$\frac{dS_{\text{obs}}}{dt} = \frac{dS}{dt} \tag{33}$$

The X-ray diffraction manifestation of this mode is the gradual shifting of the position of fundamental lines (assuming that the lattice parameter is a function of the degree of LRO)

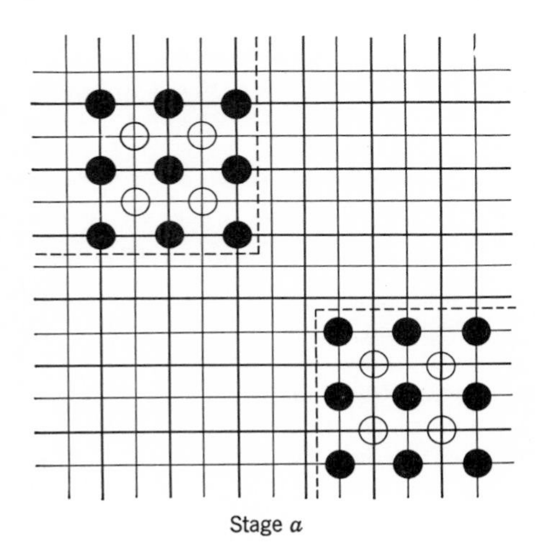

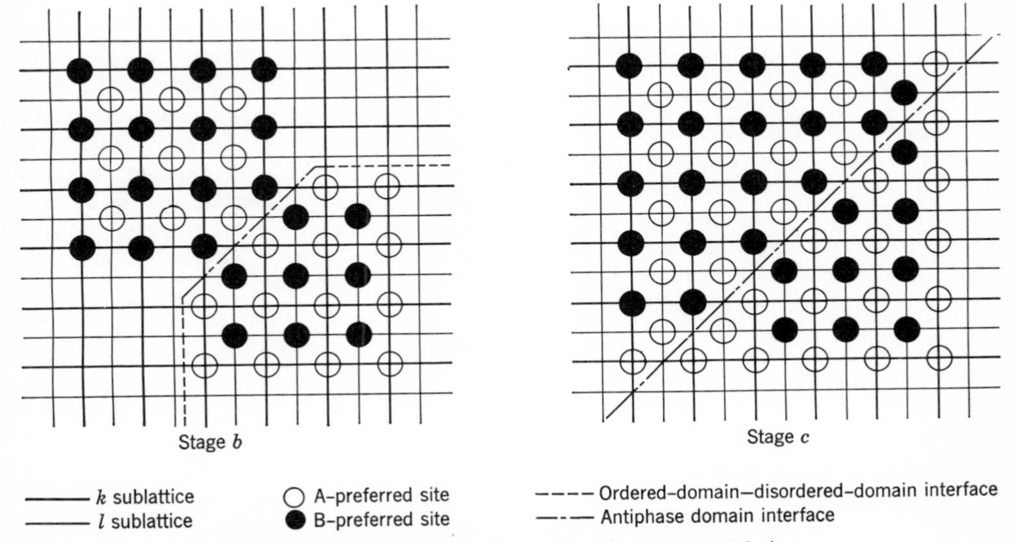

Fig. 6. Microstructural developments in long-range ordering.

and a gradual shifting of the position plus an increase in intensity of sharp superlattice lines.

We would expect O-domain growth to be rate-controlling in the case of coarse nucleation with rapid homogeneous ordering. In this case, Eq. 32 takes the form

$$\frac{dS_{\text{obs}}}{dt} = S\frac{1}{V}\frac{dV(S)}{dt} \tag{34}$$

which can be simply integrated, yielding

$$S_{\text{obs}} = \frac{S_e V(S_e)}{V} \tag{35}$$

where S_e is the effective equilibrium degree of LRO at T_t. The X-ray diffraction manifestation of this mode is the simultaneous appearance of discrete fundamental lines characteristic of the disordered and the ordered phases, with those of the latter increasing in intensity at the expense of the former. The superlattice lines increase in intensity at a fixed position with perhaps some sharpening.

We would expect the antiphase-domain coalescence mode in the case of fine nucleation and rapid homogeneous ordering and rapid O-domain—D-domain interface advancement. The antiphase-domain size is apparently never small enough to contain a significant fraction of the volume that is disordered, so that by the time this stage is reached the ordering proceeds essentially at constant $S_{\text{obs}} = S_e$. The X-ray manifestation of this mode is nonalteration in fundamental lines and a sharpening of superlattice lines at constant integrated intensity.[46] Let us now consider the expected kinetics for each of these modes.

2.2 Homogeneous Ordering Kinetics

Dienes'[47] chemical-rate-theory treatment appears to be the correct approach to this problem, and we shall first consider it by transcribing it into the notation that we introduced in the preceding section. After Dienes, the order-disorder transformation is describable as a "chemical" reaction

$$A_\alpha + B_\beta \underset{K_O}{\overset{K_D}{\rightleftarrows}} A_\beta + B_\alpha \tag{36}$$

where A_α is an A atom on an α site, etc. The chemical rate equation is thus

$$\frac{d[A_\alpha]}{dt} = K_O\,[A_\beta][B_\alpha] - K_D[A_\alpha][B_\beta] \tag{37}$$

which, in terms of the sublattice occupation probabilities, becomes

$$\frac{d(x_{A\alpha}X_\alpha)}{dt} = K_O(x_{A\beta}X_\beta)(x_{B\alpha}X_\alpha) - K_D(x_{A\alpha}X_\alpha)(x_{B\beta}X_\beta) \tag{38}$$

As we have already noted, for the binary, two-sublattice systems there is only one independent variable, and if we take this as S, we obtain

$$\frac{dS}{dt} = K_O\left(\frac{x_A}{X_\beta} - S\right)\left(\frac{x_B}{X_\beta} - S\right) - K_D\left(\frac{x_A}{X_\beta} + S\right)\left(\frac{x_B}{X_\alpha} + S\right) \tag{39}$$

as the fundamental O-D kinetics equation. Dienes intuitively evaluated the rate constants K_O and K_D as

$$\left.\begin{aligned} K_O &= \nu_O\, e^{-Q/kT} \\ K_D &= \nu_D\, e^{-(Q+V)/kT} \end{aligned}\right\} \tag{40}$$

where Q = activation energy for diffusion; V = ordering energy difference accompanying a disordering interchange; and ν_O, ν_D = atomic vibration frequencies. In equilibrium, $\frac{dS}{dt} = 0$, so that

$$\frac{K_O}{K_D} = \frac{\nu_O}{\nu_D}\exp\left[\frac{V}{kT}\right] = \frac{(x_A + X_\beta S)\,(x_\beta + X_\alpha S)}{(x_A - X_\alpha S)\,(x_B - X_\beta S)} \tag{41}$$

Dienes noted that by taking $V = V_0 S$ and $\nu_O = \nu_D = \nu$ he obtained the Bragg-Williams equilibrium solution.[48] We can thus expect that to the approximation that the Bragg-Williams solution agrees with the observed equilibrium solution, Dienes' chemical rate theory will agree with experimental order-disorder kinetics, that is, only as a rough approximation. With these assumptions, we thus obtain

$$\frac{dS}{dt} = \nu\, e^{[-Q/kT]f(S,\,T)} \tag{42}$$

where

$$f(S,T) = \left(\frac{x_A}{X_\alpha} - S\right)\left(\frac{x_B}{X_\beta} - S\right) - \exp\left[\frac{V_0 S}{kT}\right]\left(\frac{x_A}{X_\beta} + S\right)\left(\frac{x_B}{X_\alpha} + S\right) \tag{43}$$

Dienes investigated the kinetic behavior prediction of this theory for ordering of the stoichiometric bcc AB (β-brass) type and the fcc AB_3 ($AuCu_3$) type by taking $\nu = 10^{13}\ \text{sec}^{-1}$, $Q/k = 5000°\text{K}$ ($\equiv 0.43$ eV), and $V_0/k = 1000°\text{K}$ ($\equiv 0.082$ eV). Because the critical temperature according to Bragg-Williams theory is given as

$$T_c = \frac{cV_0}{k} \tag{44}$$

where $c = 0.250$ for bcc AB and $c = 0.205$ for fcc AB_3, Dienes' assumed that V_0/k corresponds to critical temperatures of $T_c = 250°\text{K}$ and 205°K for the respective transformations. These assumed critical temperatures are not very realistic, and we would also expect $Q \simeq 2$ eV as more reasonable; however, because the assumed activation energy for diffusion and the assumed critical temperatures for ordering are too low by about a factor of three, Dienes' calculations should still be qualitatively valid for realistic ordering temperatures about a factor of three greater than those given in his calculation. In Fig. 7, Dienes' calculated ordering rate curves for AB LRO are presented, and, mostly following Dienes, the important points to be noted with respect to these results are:

(*a*) The rate at $S = 0$ is always zero so that ordering must start by fluctuations. However,

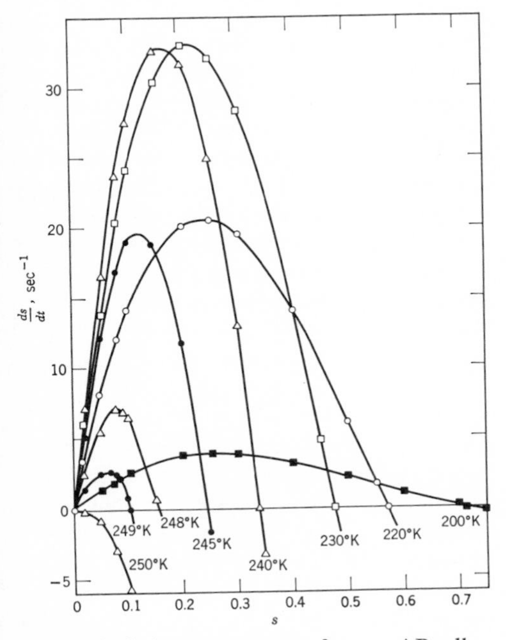

Fig. 7. Ordering rate curves for an AB alloy. $T_c = 250°\text{K}$. (From Dienes.[47])

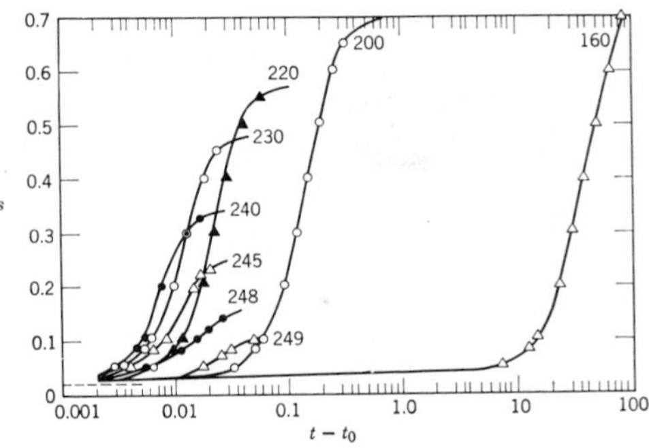

Fig. 8. Isothermal ordering curves for an AB alloy. $T_c = 250°\text{K}$. (From Dienes.[47])

because dS/dt as a function of S has a positive initial slope at all subcritical temperatures, large fluctuations are not required, and long induction periods may not occur. The rate theory says nothing about the time scale or temperature dependence of the required fluctuations.

(*b*) An isothermal S versus time curve can be constructed by numerical integration of Eq. 42, yielding

$$\int_{S_0}^{S} \frac{dS}{F(S)} = t - t_0 \tag{45}$$

where t_0 is the time required to reach the initial degree of order, S_0. Typical sigmoid curves are obtained as shown in Fig. 8, as taken from Dienes. If we assume that the induction period is not relatively large, these curves would describe the complete ordering kinetics. Although they do have the sigmoid shape that generally characterizes nucleation and growth processes, the shapes have been obtained entirely from chemical rate theory.[5]

It can also be seen from both Figs. 7 and 8 that the maximum ordering rate occurs at about 240°K; it can be generalized, from what we have already said, to a maximum ordering rate in the range $T/T_c = 240/250 = 0.96$, that is, only slightly under the critical temperature.

In Fig. 9 Dienes' calculated ordering rate curves for AB_3 LRO are given. The AB_3 ordering transformation is first-order with a discontinuous jump in S from zero to a finite value S_c at T_c. This is manifested in the ordering rate curves by the fact that over a large temperature range below T_c, virtual disordering exists for small values of S for an alloy starting at $S = 0$. This means that S must change discontinuously from $S = 0$ to a

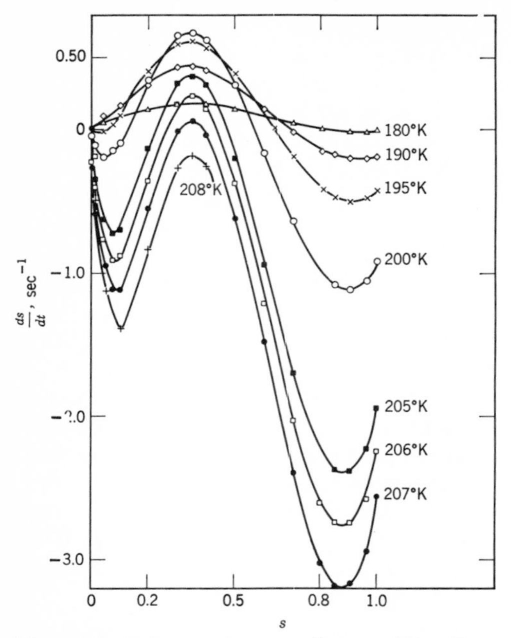

Fig. 9. Ordering rate curves for an AB_3 alloy. $T_c = 205°K$. (From Dienes.[47])

finite value before ordering can take place. Thus long induction periods may be required, for large fluctuations are needed to start the ordering reaction. Otherwise, the ordering kinetics are very similar to the AB case, although it should be noted that the AB ordering rates are of the order of 100 times the AB_3 ordering rates at comparable temperatures. This fact, coupled with the possibility of much longer induction times in AB_3 ordering, would predict much slower ordering rates for the AB_3 than for the AB ordering, which is in general agreement with observations.

Actually, this solution, which we attributed to Dienes, was given in very nearly this form in the classic Bragg-Williams paper of 1934.[48] Bragg and Williams, however, rightly concluded that by far the dominant temperature-dependent factor is $\exp[-Q/kT]$ and simply did not pay detailed attention to $f(S, T)$, although they did note that it goes to zero as $S \rightarrow 0$ and hence the O-D transformation should be "sluggish" in the neighborhood of the critical temperature.

Dienes' treatment is for a two-atom direct-interchange mechanism. Presumably a vacancy interchange mechanism would be more generally applicable. What amounts to the extension of the chemical-rate-theory approach, taking the vacancy mechanism into account, has been given by Vineyard.[49] Actually, Vineyard starts with an apparently rigorous treatment of the equations of motion of the complete set of probability functions; however, by the time that he has simplified his results to a manipulatable analytic form, his solution amounts to the kinetic counterpart of the zeroth order quasichemical approximation, which has been shown to be equivalent to the Bragg-Williams approximation.[4] Vineyard's kinetics solution is qualitatively the same as Dienes', that is, it approximately yields Eq. 42, with $f(S, T)$ behaving approximately as in Eq. 43. However, for purposes of employing O-D kinetics to determine vacancy energetics, Vineyard's result is essentially different and yields

$$\frac{dS}{dt} = x_V F(S, T) \tag{46}$$

where x_V = atomic fraction of vacancies, and $F(S, T)$ is a complex function of the degree of LRO and the temperature, which contains as constants: E_{mA}, E_{mB}, the migration activation energies for an A-atom-vacancy interchange, and a B-atom-vacancy interchange, respectively, and $(V_{AA} - V_{AB})$ and $(V_{BB} - V_{AB})$, where the $V_{ij} = i\text{-}j$ pair bond energy. Although we can with some confidence accept the functional form of Eq. 46, the calculation of $F(S, T)$ is crude. Therefore, one consequence of Vineyard's result is that, although the number of vacancies, and hence their formation energy, is separable from the S dependence, the vacancy migration energies are inextricably involved in the S dependence.

Let us now consider some analyses of experimental results, which have been based on the homogeneous ordering theory reviewed here.

Nowick et al.[50,51] found that for relatively small changes in order, the integration of Eq. 45, essentially employing Eqs. 42 and 43, resulted in a very good fit between the experimental and calculated S versus t curves. From their analysis they obtained an activation energy $Q = 2.03$ eV for ordering in Cu_3Au. Although this is a very reasonable sort of number, in the light of Vineyard's analysis we can conclude that it is not a very useful number for its relation to the activation energy for diffusion is obscure.

Bransky and Rudman[52] used Vineyard's result, Eq. 46, to determine the activation energy for vacancy formation, E_f, in Fe_3Al. They assumed that at the start of the ordering transformation, following a rapid quench from T_q, the equilibrium number of vacancies at T_q are present. Thus

$$x_V \propto \exp\left[-\frac{E_f}{kT_q}\right] \qquad (47)$$

and transformation at a constant temperature T_t and to a constant small change in LRO, so that $\Delta F(S, T)$ is a constant, is proportional to $x_V(T_q)$. By measuring $dS/dt_{t\to 0, T_t=\text{const}}$ versus T_q, they obtained $E_f(Fe_3Al) = 0.75$. Bransky and Rudman also attempted to analyze the kinetics at later stages in the transformation by allowing for the decay of the excess vacancies, but with only limited success. They did, however, observe, as have many others, that residual cold work retarded ordering kinetics, and they attributed this to a more rapid decay to dislocation sinks of the excess, quenched-in vacancy concentration.

It has been rather generally observed[52,53] that quenching from higher temperatures accelerates the ordering kinetics. This finds a ready explanation in the increased number of quenched-in vacancies, whereas if the rate were dominantly dependent on the number of nuclei we would expect an opposite effect (despite the reasoning of Kuczynski et al.[53] to the contrary).

2.3 Ordered-Domain Growth Kinetics

The kinetics of this mode of ordering have been documented for CoPt by Newkirk et al.[54] and for CuAu by Kuczynski et al.,[53,55] yet there does not appear to be any attempt at theoretical analysis. As pointed out by Newkirk et al.,[54] this mode microstructurally resembles precipitation. However, as we noted before, the kinetics probably are not analogous, because the development of long-range concentration gradients presumably does not occur in the O-D transformation. The growth kinetics might thus be expected to resemble those of recrystallization. Let us attempt to develop here the O-D kinetics behavior for this mode, using the applicable features of the precipitation and recrystallization transformations.

The theory of recrystallization kinetics has been reviewed by Burke and Turnbull,[56] and Turnbull[57] has reviewed phase change kinetics; we shall employ the methods summarized in these reviews.

If we assume that the number of nuclei that grow is constant, as is generally observed in precipitation,[57] and that they are the supercritical sized-order fluctuations inherited from the disordered solution, we obtain for the volume fraction ordered at T_t after a time t

$$X = 1 - e^{-fND_1D_2D_3} \qquad (48)$$

where X = volume fraction ordered, f = ordered domain shape factor, N = number of nuclei per unit volume, and D_i = orthogonal dimensions of the growing domains, neglecting impingement. In Table 1 we summarize the values of the shape factor, f, and the dimensions D_i, for various domain shapes.

We obtain the linear growth rate from absolute reaction rate theory as

$$G = a(R_{D\to O} - R_{O\to D}) \qquad (49)$$

where $R_{D\to O}$ is the rate of ordered configuration production at the O-domain—D-domain interface, and $R_{O\to D}$ is the analogous rate for disordered configuration production, $a \simeq$ lattice parameter, so that

$$G = \left(\frac{ea}{h}\right) \exp\left[\frac{S_D}{k}\right] \Delta\mathscr{F}(T_t) x_V \exp\left[-\frac{E_m}{kT_t}\right] \qquad (50)$$

where e = base natural logarithms, S_D = activation entropy for diffusion, and $\Delta\mathscr{F}(T_t)$ = free energy change for disordering at T_t. The growth rate goes to zero at $T_t \to T_c$ because of the function $\Delta\mathscr{F}(T_t)$ and at lower temperatures because of the usual freeze due to the term $\exp -E_m/kT_t$. For n-dimensional domain growth we thus obtain for the transformation-time relation

$$ln\,(1 - X) \propto N(T_q, T_t)G^n(x_V, T_t)t^n \qquad (51)$$

TABLE 1

Ordered-Domain Growth Rate Parameters for Different Domain Shapes

(G = linear growth rate)

Domain Shape	f	D_1	D_2	D_3
Cube	1	Gt	Gt	Gt
Sphere	$\pi/6$	Gt	Gt	Gt
Disk	$\pi/4$	Gt	Gt	Gt
Cylinder	$\pi/4$	Gt	d	d

Let us consider what sort of information about vacancy energetics could be obtained from analysis of the kinetics of this mode of ordering. If we examine a series of experiments for example, of measuring the time to transform a given fraction as a function of T_t, holding T_q constant, or as a function of T_q holding T_t constant, we see that the temperature dependence always contains the product $N(T_q, T_t)G^n(T_q, T_t)$; hence the vacancy energetics, which are contained in the G factor, cannot be obtained without independent knowledge of the temperature dependence of the number of nuclei. Burns and Quimby[58] have discussed this point, and. although their calculation is questionable (their surface energy calculation appears to be merely a redundant calculation of the volume energy), their conclusion that N is a strong function of T_t is probably valid. N probably has a relatively weak dependence on T_q, at least for $T_q > T_c$. However, N is, in principle at least, an experimentally determinable quantity. Thus a T_t constant, T_q variable experiment should yield the vacancy formation energy, E_f. It should be noted, however, that the apparent activation energy that will be obtained will be nE_f, and thus the domain morphology must also be determined. Platelike growth was observed in CoPt;[54] this is probably the most general mode of precipitation in solid solution morphology, and therefore an apparent activation energy of $2E_f$ would generally be expected. A T_q constant, T_t variable experiment, however, would not easily yield the vacancy migration energy, E_m, because of the strong nuclei number dependence on T_t and also because of a strong free energy of ordering, $\Delta\mathscr{F}$, dependence on T_t.

Although there is actually a large literature on the kinetics of this mode, there are apparently no data that can be employed to determine unambiguous vacancy activation energies. The general qualitative kinetics behavior[4,53,54] is as predicted by the equations that we have derived.

2.4 Antiphase-Domain Coalescence Kinetics

As we have already noted, this transformation mode occurs at essentially a constant degree of LRO and perhaps should not properly be considered a mode of ordering. However, in general, LRO kinetics are not followed by a superlattice intensity measurement, but by some more convenient physical property measurement, the most popular being electrical resistivity, and these measure ments do "see" the domain size. Thus, for probably the great majority of O-D kinetics studies in the literature, it is not clear whether it is a true change in the amount of order or simply domain coalescence that has been followed. Because it is certainly present in all ordering transformations, we shall also consider this mode.

Just as we found domain growth to be closely analogous to recrystallization, we might expect antiphase-domain coalescence to be closely analogous to grain growth. We shall, in fact, pursue an approach closely related to the usual grain-growth treatment, as reviewed by Burke and Turnbull.[56] Although the analogy is actually very straightforward, we must first divest the domain-coalescence problem of an ingenious suggestion made by Bragg[59] in 1940, which has dominated thinking on this subject for twenty-odd years, but is incorrect. Bragg noted that domain coalescence in bcc β-brass was orders of magnitude faster than in fcc $AuCu_3$. Today we attribute this simply to the much greater diffusivities in bcc lattices relative to fcc lattices. Bragg, however, proposed that it was due to the fact that metastable-ordered domain configurations occurred in the $AuCu_3(AB_3)$ structure but were not possible in the β-brass (AB) structure. We have noted previously that this AB_3 structure consists of four sublattices. As ordering commences in different regions of the disordered structure, the A atoms, defining the α-sublattices, may prefer any of the four sublattices defining four different antiphase domains: A, B, C, and D. In the AB lattice, there are only two sublattices, leading to the possibility of only two domains, A and B. In Fig. 10 the models for Bragg's domain stability suggestion and also a model of a stable configuration for grain size are sketched. For the AB LRO (Fig. 10*a*), the A domain can consume the B domain with decreasing interfacial energy; the interfacial area increase that occurs in developing boundary curvature is more than compensated for by disappearance of other interfacial areas. In the AB_3 configuration (Fig. 10*b*), however, there is no such interface disappearance concurrent with increased area due to curvature; therefore, the

A domain can consume the B domain only with increasing interfacial energy, and thus it is in a metastable configuration. The stable grain-size configuration, (Fig. 10*c*) is stable for the same reason as the AB_3 domain structure. The establishment of the latter structure and hence the cessation of grain growth at an early stage generally do not occur, because, to quote Burke and Turnbull,[56] "The introduction of even one grain with a non-equilibrium shape upsets the balance of the whole system, and it cannot be restored in general, because, as boundaries migrate in a direction to restore the equilibrium configuration, grains disappear so the non-equilibrium configuration is self-perpetuating. Thus occurs grain growth." Analogously, domain coalescence in LRO thus occurs. Figure 11 depicts an idealized but more realistic array of four-domain configurations, just at the point of growth to impingement (dashed boundaries). (The array was generated by assigning each cell a domain identity, according to the fall of two coins: h = head coin #1, H = head coin #2, etc.; h + H = A, t + T = B, h + T = C, and H + t = D.) From this illustration, it can be seen that in any randomly

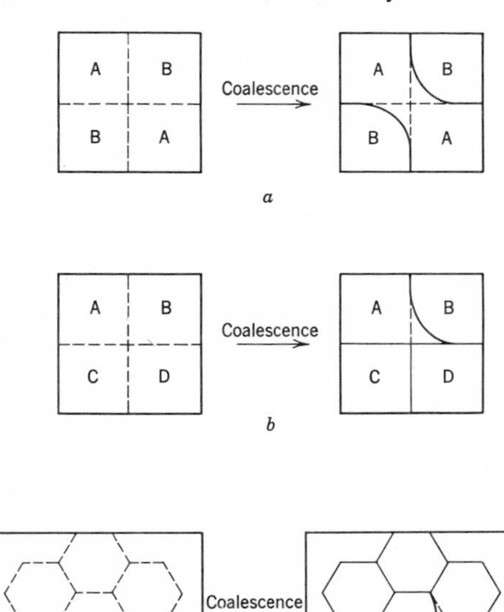

Fig. 10. Bragg's model for domain stability: (*a*) unstable domain size in bcc LRO, (*b*) stable domain size in fcc AB_3 LRO, and (*c*) stable grain size analog.

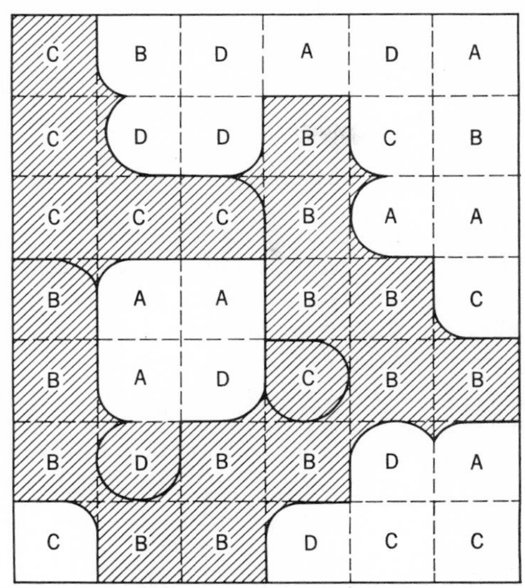

Fig. 11. Domain-size instability in a randomly nucleated domain array in fcc AB_3, LRO.

generated array of domains, there will always be a sufficient number of unstable configurations for extensive domain growth. The growth of domains C and B by virtue of such unstable configurations is also illustrated in Fig. 11. We can now pursue our development of domain coalescence kinetics by analogy with grain growth without having to make any distinction among the different LRO crystal structures.

Applying the same absolute-rate-theory approach as for the domain-growth case, we must merely obtain a new expression for the free energy driving force, $\Delta\mathscr{F}$, in Eq. 50. For an effective domain diameter, D, ($D \simeq$ average domain diameter), there will be $1/fD^3$ domains/unit volume, and an interfacial area of $f'D^2$ per domain (where f' is a shape factor). There will thus be an interfacial area per unit volume of f'/fD and an interfacial free energy of σ times this, where σ = surface tension. The free energy driving force in the absolute rate theory formulation thus becomes

$$\Delta\mathscr{F} \simeq a\left[\frac{\partial}{\partial D}\left(\frac{f'}{fD}\right)\right]\left(\frac{a^2}{f'/fD}\right) \simeq v\sigma/D \qquad (52)$$

where v = atomic volume.

Because the growth rate $G = dD/dt$, by combining Eqs. 50 and 52 we obtain

$$\frac{dD}{dt} = \frac{K}{D} \qquad (53)$$

where

$$K = \left(\frac{ea}{h}\right) \exp\left[\frac{S_D}{k}\right] x_V \exp\left[\frac{-E_m}{kT}\right] v\sigma$$

Integration of Eq. 53 yields

$$D^2 - D_0^2 = 2K(t - t_0) \quad (54)$$

Eq. 54 has also been derived by Lifshitz.[60]

The employment of domain-coalescence kinetics to determine vacancy energetics follows the same considerations that we have given previously concerning domain growth. Although apparently there are no data from which to determine vacancy energetics, the classical domain-growth study by Jones and Sykes[61] does provide some data with which to check the time-dependence prediction of Eq. 54. In Fig. 12, is plotted D^2 versus t for Cu_3Au for the conditions indicated. For domain diameters less than approximately 500 Å, Eq. 54 is obeyed within the limit of experimental error. For longer times, the growth rate drops off. This longer-time drop-off is also observed in grain growth[56] and has been attributed to a number of factors, among which the analog of attainment of a relatively stable domain structure, as proposed by Bragg, is a good possibility. Inclusion pinning of boundaries is also a possibility. The electron-transmission microscopic observation of Brown and Cupschalk[62] that the excess quenched-in vacancies appear to be swept through the crystal by the growing domains and then deposited (precipitated) at the anti-phase domains, may provide one mechanism for forming such inclusions.

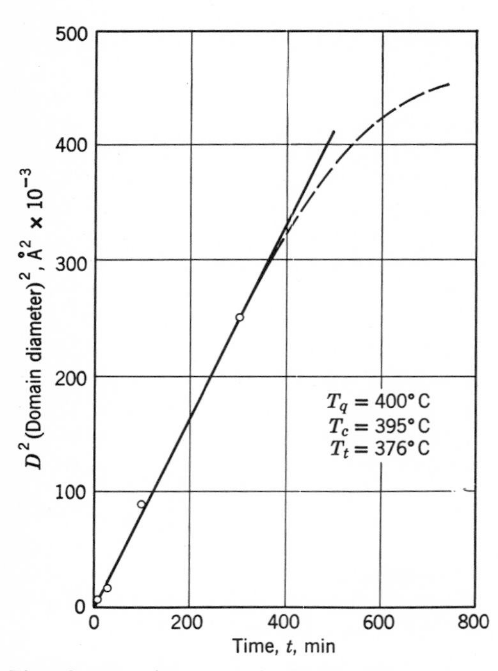

Fig. 12. Domain growth in Cu_3Au. (From Jones and Sykes.[60])

With the perfection in recent years of electron-transmission microscopy wherein anti-phase domain sizes and configurations can be directly seen, as, for example, in the work of Fisher and Marcinkowski,[63] the detailed study of domain coalescence kinetics is now possible, but such a study does not appear to have been made yet.

2.5 Short-Range Ordering Kinetics

Theories of short-range ordering kinetics have been given by Iida[64] and by Kidin and Shtremel[65] for the vacancy mechanism of interchange. Both treatments give essentially the same result that the kinetic behavior of the first-neighbor SRO parameter can be approximated as

$$\frac{d\alpha_1}{dt} = \frac{(\alpha_1{}^e - \alpha_1)}{\tau_1} \quad (55)$$

where α_1 = Warren first-neighbor SRO parameter and α_1^e = equilibrium value.

$$\tau_1^{-1} = \nu_1 x_V \exp\left[-\frac{E_m}{kT}\right] \quad (56)$$

Because of the tediousness of an X-ray diffuse-scattering determination of SRO, most SRO kinetics studies have employed some physical property as a measure of the degree of order. With the assumption that the electrical resistivity is proportional to $|\alpha_1|$, Damask[66] observed the approximate correctness of Eq. 55 for α-brass, and Iida,[67] employing heat content, made a similar observation for Ni_3Fe.

At first sight SRO kinetics appear very promising for studying vacancy energetics, but complications become evident on closer study. Although the literature on the subject of defect kinetics in alloys is vast (unfortunately, there do not appear to be any comprehensive reviews on the subject since 1954[68]), there are actually few results that are completely unambiguous. We shall here only attempt to illustrate some of the difficulties in SRO kinetics analysis.

The physical-property changes due to changes in the degree of SRO are of the same order of magnitude as those due to changes in the vacancy concentration or configuration. For example, Fig. 13 gives the excess resistivity due to some thermally activated defect observed in an alloy Cu-15 a/o Al by Wechsler and Kernohan.[69] They analyzed these data in terms of vacancy formation. They attributed the high-temperature resistivity drop-off in the quenching experiments to nonretention of the equilibrium vacancy concentration. However, they found it difficult to rationalize an apparent energy of vacancy formation of $E_f \simeq 0.2$ eV. A subsequent X-ray diffuse scattering study by Davies and Cahn[70] shows that sro retention behavior provides a more consistent explanation of these results, if we assume that the resistivity decreases with increasing degree of sro. Furthermore, because according to the quasichemical theory, the temperature dependence of the degree of SRO is given as[71]

$$\alpha_1 \simeq x_A x_B \left\{ \exp\left[-\frac{2v}{kT} \right] - 1 \right\} \qquad (57)$$

we see that $v \simeq 0.1$ eV, which is the expected order of magnitude, provides a ready explanation for the observed apparent defect formation energy of $\simeq 0.2$ eV.

In the preceding discussion, we had to assume that the resistivity decreased with increasing degree of sro. Unlike LRO, in which the resistivity always decreases with increasing degree of LRO, the SRO dependence of the resistivity apparently can be either increasing or decreasing. For example, the resistivity in Cu–Au alloys apparently increases with increasing degree of sro,[72] whereas in α-brass the resistivity apparently decreases.[66] Theoretical treatments[73] of the dependence of resistivity on SRO provide a posteriori justification of this perhaps surprising behavior, but they are not of much predictive help.

It is also not easy to distinguish sro from a sluggishly formed LRO. Thus the question of whether the low-temperature, $T \lesssim 200°$C, annealing behavior of α-brass is due to sro or due to LRO is, after a disproportionate effort, still unanswered.[74]

A controversy has persisted during the past several years over the observations of Kuczynski et al.[75] of apparent phase transi-

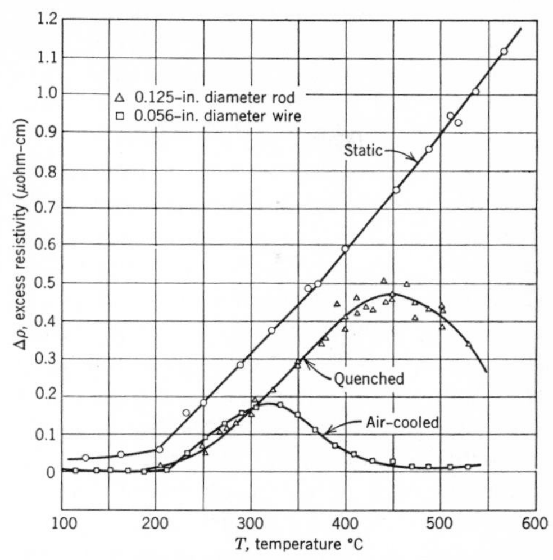

Fig. 13. Excess resistivity versus temperature for Cu-15 *a/o* Al. (From Wechsler and Kernohan.[69])

tions in what was hitherto considered a continuous disordered Cu–Au solution. This is illustrative of either the unexpected complexity of the SRO state, or of the importance of unknown effects due to vacancies, dislocations, and their interactions with SRO. Although Borie and Warren[76] have shown by an X-ray diffuse-scattering study that part of the explanation of the Kuczynski et al. results may lie in failure to quench- in the equilibrium degree of sro, this cannot be the complete explanation, because most of the anomalous results were obtained at temperature. Despite a very detailed study by Damask et al.[77] there is no clear explanation of these anomalies. In fact, Damask et al. added observations of a few more anomalies to the list.

3. RADIATION EFFECTS IN ORDER-DISORDER ALLOYS

Radiation effects in solids due to high energy particles (fission fragments, alphas, gammas, neutrons, protons, and electrons) are well documented. Recent general reviews have been given in the books by Dienes and Vineyard[78] and Billington and Crawford,[79] in articles by Kinchin and Pease,[80] Seitz and Koehler,[81] and Seitz,[82] and in the proceedings of symposia.[83,84] In view of the extensively reviewed nature of this field, we shall give only a brief discussion of radiation effects in O-D alloys. Paradoxically, it is found that radiation can both order a disordered alloy and disorder an ordered one, and we shall consider each of these effects separately.

3.1 Radiation Disordering

In a pure metal following irradiation, there may remain a number of improperly positioned atoms. They are primarily Frenkel pairs: a dissociated vacancy and interstitial. Through various physical-property measurements, the number of the remaining defect *displacements*, D, can be estimated. Following irradiation of an ordered alloy, it is generally found that the degree of LRO is decreased. By the definition of LRO, this means that a certain number of unlike atoms have interchanged places, and they are called *replacements*. R. Siegel[85] in 1949 was apparently one of the first to observe that under fast neutron irradiation $R \simeq 10D$. Several mechanisms have been proposed to explain this ratio, but to date no experiment has been performed to adequately discriminate among the various suggestions.

Seitz[86] has suggested a *thermal-spike* mechanism. Each primary neutron produces secondary knocks-ons which, in passing through the lattice, heat their "wakes," perhaps even above the melting point. The wakes are rapidly quenched and hence may have the atomic distribution of the disordered phase. However, after more detailed consideration, Seitz and Koehler[81] conclude that the amount of disordering available by this mechanism is insufficient to account for the observations. The kinetics of thermal-spike disordering after Cook and Cushing[87] is simply

$$\frac{dX_D}{dn} = (1 - X_D)\gamma \tag{58}$$

where X_D = disordered volume fraction, n = number of irradiating particles, and γ = volume fraction disordered per particle. Thus the LRO-dose dependence is (see Eq. 35)

$$S = S_0 \, e^{-\gamma n} \tag{59}$$

as observed.[87]

Kinchin and Pease[88] proposed a *replacement-collision* mechanism. They proposed that a knock-on atom may have sufficient energy to knock another atom from its lattice site, but may itself be of insufficient energy to escape from the created vacancy and hence be captured, thus creating a replacement without another displacement. The replacement collision was used by Kinchin and Pease to account for Aronin's[89] Ni_3Mn data. Aronin also showed that a replacement-collision disordering mechanism leads to the same disordering kinetics as for the thermal spike mechanism (see Eq. 59). Dienes and Vineyard,[78] however, conclude that this mechanism is improbable in Cu_3Au.

Seitz[81] proposed a *thermal spike-plastic flow* radiation disordering mechanism. Seitz concluded that, due to thermal stresses, dislocations would be generated and moved by thermal spikes with an ensuing destruction of order. However, Billington and Crawford[79] point out that total disordering of Cu_3Au by strain has not been observed in the absence of radiation, and hence they would discount the importance of this mechanism.

Seitz[82] has also noted that *focusing collision replacements* may also contribute to radiation disordering. In this mechanism, the knock-on energy is transferred down a close-packed row, creating a vacancy at the point of initial knock-on, an interstitial at the end of the row, and a one-atom distance shift for each intermediate atom in the row. This type of motion rapidly damps out as the difference in the masses of the components increases and hence would be more efficient in an alloy such as Ni_3Mn than in Cu_3Au.

Thus, despite a mass of experimental data and several ingenious suggestions, a dominant radiation disordering mechanism has not been established, and it may be that all of these mechanisms play a significant role, as suggested by Billington and Crawford.[79]

3.2 Radiation Ordering

Radiation ordering is apparently due to only one generally accepted mechanism: *radiation enhanced diffusion*. The quantitative accounting for this behavior is a very complex problem, because it combines the O-D kinetics that we have reviewed here with vacancy recombination kinetics.

After Dienes and Vineyard,[78] we can consider radiation-enhanced diffusion in terms of the diffusion coefficient, D

$$D = A \exp\left[-\frac{E_m}{kT}\right]\left\{\exp\left[-\frac{E_f}{kT}\right] + \Delta x_V\right\} \tag{60}$$

where Δx_V = excess radiation-produced vacancies. The vacancy excess is a function of the irradiation flux and the rate of decay, the latter being a function of the vacancy concentration, sink concentration (dislocations, grain boundaries, etc.), and temperature. We can identify three temperature intervals in discussing radiation-diffusion effects.

(*a*) *At high temperatures*, the equilibrium vacancy concentration greatly exceeds the excess vacancy concentration, because the excess vacancies anneal out so rapidly. Thus irradiation has not been found to have any measurable effect in the usual diffusion experiment.

(*b*) *At low temperatures*, $\Delta x_V \gg \exp(E_f/kT)$, and the diffusion coefficient can be enhanced enormously. For example, in α-brass Dienes and Damask[90] estimated a 10^{10}-fold increase at room temperature; in the Fe–Ni system Dautreppe et al.[91] have formed under irradiation a new ordered phase, FeNi, which cannot be obtained by heat treatment because of a low critical temperature.

(*c*) *At intermediate temperatures*, D will be increased, but the behavior will be complex.

The low-temperature region is probably of greatest interest and has been studied extensively. In general, at low temperatures following a thermal ordering treatment, the degree of order, LRO or SRO, will be less than the equilibrium value and hence under radiation-enhanced diffusion the degree of order will increase. It can be seen from Eq. 60 that, provided the irradiation is carried out at a sufficiently low temperature for all the excess vacancies to be frozen-in, the temperature dependence of the diffusion coefficient is given by $D \propto \exp - (E_m/kT)$. Dugdale[92] has studied low-temperature ordering in irradiated Cu_3Au, and in this manner he obtained $E_m \simeq 0.9$ eV. However, our previous discussion of LRO kinetics should be recalled, where we noted that the physical significance of this number is still obscure for it includes unevaluated contributions of the binding energies, the V_{ij}'s. Dugdale,[92] however, observed another interesting ordering effect in these experiments: after obtaining partial order at 230°C, he raised the temperature and then observed an initial disordering followed by further ordering. Dugdale attempted to explain this effect in terms of "soft wrong pairs," nearest neighbors, and "hard wrong pairs," non-nearest neighbors. It was later noted by Rudman[93] that Dugdale's effect could also be explained by the fact that the excess-vacancy concentration is not homogeneous, but is less in the neighborhood of sinks.

Thus although there is good general understanding of O-D kinetics and the effects of irradiation, the detailed picture involving order dependence, defect concentration, and defect distribution is very complex, and, in view of the already vast amount of experimental data without unambiguous resolution of these problems, no complete solution can be expected in the near future, barring a breakthrough somewhere. If there is any new technique that promises such a breakthrough, it is the electronic-computer "experiments." The Monte Carlo experiments in O-D systems of Flinn and McManus[25] and Beeler and Delaney[26] have already provided much insight into diffusive atomic motions in ordering alloys. A computer experiment of radiation damage events in a pure metal by Vineyard[94] and co-workers has also proved very valuable. A computer experiment of radiation damage in an O-D system has apparently not been performed yet, but would probably go far in resolving several of the controversial points of view that we have noted.

REFERENCES

1. Hume-Rothery W., and G. V. Raynor, *The Structure of Metals and Alloys*. The Institute of Metals, London, 1954.
2. Lipson H., *Progr. Metal Phys.* **2** (1950) 1.
3. Elcock E. W., *Order-Disorder Phenomena*. Methuen, London, 1956.
4. Muto T., and Y. Takagi, *Solid State Phys.* **1** (1955) 194.
5. Guggenheim E. A., *Mixtures*. Clarendon Press, Oxford, 1952.
6. Domb C., *Advan. in Phys.* **9** (1960) 149.
7. Westbrook J. H., "Mechanical Properties of Intermetallic Compounds—A Review of the Literature," in *Mechanical Properties of Intermetallic Compounds*. Wiley, New York, 1960.

8. Guttman L., *Solid State Phys.* **3** (1956) 146.
9. Warren B. E., and B. L. Averbach, "The Diffuse Scattering of X-Rays," in *Modern Research Techniques in Physical Metallurgy*. ASM, Cleveland, 1953.
10. Guinier A., *Solid State Phys.* **9** (1959) 294.
11. Wooster W. A., *Diffuse X-ray Reflections From Crystals*. Clarendon Press, Oxford, 1962.
12. Cohen J. B., and M. E. Fine, *J. Phys. Radium*, **23** (1962) 749.
13. Wojciechowski K. F., *Acta Met.*, **6** (1958) 396.
14. Men A. N., *Fiz. Metal. i Metalloved.* **7** (1959) 633.
15. Rudman P. S., *Acta Met.* **8** (1960) 321.
16. Rudman P. S., and B. L. Averbach, *Acta Met.* **2** (1954) 576.
17. Cohen M. H., and F. Reif, *Solid State Phys.* **5** (1957) 322.
18. Flinn P. A., and S. L. Ruby, *Phys. Rev.* **124** (1961) 34.
19. Zeedijk H. B., E. Ottens, and W. G. Burgers, *Proc. Konikl. Ned. Akad. Westenschap.* Amsterdam **B64** (1961) 231.
20. Smakula A., N. Maynard, and A. Repucci, *J. Appl. Phys.* (Suppl.) **33** (1962) 453.
21. Cottrell A. H., *J. Inst. Metals* **90** (1962) 449.
22. Fosdick L. D., *Phys. Rev.* **116** (1959) 565.
23. Ehrman J. R., L. D. Fosdick, and D. C. Handscomb, *J. Math. Phys.* **1** (1960) 547.
24. Guttman L., *J. Chem. Phys.* **34** (1961) 1024.
25. Flinn P. A., and G. M. McManus, *Phys. Rev.* **124** (1961) 54.
26. Beeler J. R., Jr., and J. A. Delaney, *Phys. Rev.* **130** (1963) 962.
27. Wilchinsky Z. W., *J. Appl. Phys.* **15** (1944) 806.
28. Wilson A. J. C., *X-Ray Optics*. Methuen, London, 1949.
29. Cowley J. M., *J. Appl. Phys.* **21** (1950) 24.
30. Borie B., *Acta. Cryst.* **14** (1961) 472.
31. Roberts B. W., *Acta Met.* **2** (1954) 597.
32. Sato K., D. Watanabe, and S. Ogawa, *J. Phys. Soc. Japan* **17** (1962) 1647.
33. Roberts B. W., and G. H. Vineyard, *J. Appl. Phys.* **27** (1956) 203.
34. Cowley J. M., *Phys. Rev.* **120** (1960) 1648.
35. Warren B. E., *Private Communication.*
36. Sutcliffe C. H., and F. E. Jaumot, *Acta Met.* **1** (1953) 725.
37. Batterman B. W., *J. Appl. Phys.* **28** (1957) 556.
38. Norman N., and B. E. Warren, *J. Appl. Phys.* **22** (1951) 483.
39. Warren B. E., B. L. Averbach, and B. W. Roberts, *J. Appl. Phys.* **22** (1951) 1493.
40. Walker C. B., and D. T. Keating, *Acta Cryst.* **14** (1961) 1170.
41. Munster A., and K. Sagel, "Short Range Order and Thermodynamic Properties of Metallic Solutions," in *Phys. Chem. Metallic Sol. Intermetallic Comp.*, Her Majesty's Stationery Office, London, 1959.
42. Flinn P. A., B. L. Averbach, and M. Cohen, *Acta Met.* **1** (1953) 664.
43. Dupouy J. M., and B. L. Averbach, *Acta Met.* **9** (1961) 755.
44. Flinn P. A., B. L. Averbach, and P. S. Rudman, *Acta Cryst.* **7** (1954) 153.
45. Seitz F., *Rev. Mod. Phys.* **34** (1962) 656.
46. Chipman D., and B. E. Warren *J. Appl. Phys.* **21** (1950) 696.
47. Dienes G. J., *Acta Met.* **3** (1955) 549.
48. Bragg W. L., and E. J. Williams, *Proc. Roy. Soc.* **145A** (1934) 699.
49. Vineyard G. H., *Phys. Rev.* **102** (1956) 981.
50. Nowick A. S., and L. R. Weisberg, *Acta Met.* **6** (1958) 260.
51. Feder R., M. Mooney, and A. S. Nowick, *Acta Met.* **6** (1958) 266.
52. Bransky J., and P. S. Rudman, *Trans. ASM* **55** (1962) 335.
53. Kuczynski G. C., R. F. Hochman, and M. Doyama, *J. Appl. Phys.* **26** (1955) 871.
54. Newkirk J. B., A. H. Geisler, D. L. Martin, and R. Smoluchowski, *Trans. AIME* **188** (1950) 1249.
55. O'Brien J. L., and G. C. Kuczynski, *Acta Met.* **7** (1959) 803.
56. Burke J. E., and D. Turnbull, *Progr. in Metal Phys.* **3** (1952) 220.
57. Turnbull D., *Solid State Phys.* **3** (1956) 226.
58. Burns F. P., and S. L. Quimby, *Phys. Rev.* **97** (1955) 1567.
59. Bragg W. L., *Proc. Phys. Soc.* **52** (1940) 105.
60. Lifshitz I. M., *Soviet Phys. JETP* **15** (1962) 939.
61. Jones F. W., and C. Sykes, *Proc. Roy. Soc.* **166A** (1938) 376.
62. Brown N., and S. Cupschalk, *Bull. Am. Phys. Soc.* **8** (1963) 217.
63. Fisher R. M., and M. J. Marcinkowski, *Phil. Mag.* **6** (1961) 1385.

64. Iida S., *J. Phys. Soc. Japan* **10** (1955) 769.
65. Kidin I. I., and M. A. Shtremel, *Fiz. Metal. i Metalloved* **11** (1961) 641.
66. Damask A., *J. Appl. Phys.* **27** (1956) 610.
67. Iida S., *J. Phys. Soc. Japan* **10** (1955) 9.
68. Broom T., *Phil. Mag. Suppl.* **3** (1954) 26.
69. Wechsler M. S., and R. H. Kernohan, *Acta Met.* **7** (1959) 599.
70. Davies R. G., and R. W. Cahn, *Acta Met.* **10** (1962) 170.
71. Flinn P. A., *Phys. Rev.* **104** (1956) 350.
72. Damask A., *J. Phys. Chem. Solids* **1** (1956) 23.
73. Gibson J., *J. Phys. Chem. Solids* **1** (1956) 27.
74. Clarebrough L., M. Loretto, and M. Hargreaves, *Proc. Roy. Soc.* **257A** (1960) 326.
75. Kuczynski G. C., M. Doyama, and M. E. Fine, *J. Appl. Phys.* **27** (1956) 651.
76. Borie B., and B. E. Warren, *J. Appl. Phys.* **27** (1956) 1562.
77. Damask A. C., Z. A. Fuhrman, and E. Germagnoli, *J. Phys. Chem. Solids* **19** (1961) 265.
78. Dienes G. J., and G. H. Vineyard, *Radiation Effects in Solids*. Interscience Publishers, New York, 1957.
79. Billington D. S., and J. H. Crawford, *Radiation Damage in Solids*. Princeton University Press, Princeton, 1961.
80. Kinchin G. H., and R. S. Pease, *Repts. Progr. in Phys.* **18** (1955) 1.
81. Seitz F., and J. S. Koehler, *Solid State Phys.* **2** (1956) 307.
82. Seitz F., *Rev. Mod. Phys.* **34** (1962) 656.
83. *Properties of Reactor Materials and the Effects of Radiation Damage*, Proc. International Conference at Berkeley Castle, Gloucestershire, England (1962). Butterworths, London.
84. *Radiation Damage in Solids*, Proc. Symposium at S. Giorgio Maggiore, Venice (1962). International Atomic Energy Agency, Vienna.
85. Siegel S., *Phys. Rev.* **75** (1949) 1823.
86. Seitz F., *Discussions Faraday Soc.* **5** (1949) 271.
87. Cook L. G., and R. L. Cushing, *Acta Met.* **1** (1953) 539.
88. Kinchin G. H., and R. S. Pease, *J. Nuclear Energy* **1** (1955) 200.
89. Aronin L. R., *J. Appl. Phys.* **25** (1954) 344.
90. Dienes G. J., and A. C. Damask, *J. Appl. Phys.* **29** (1958) 1713.
91. Dautreppe D., J. Langler, J. Paulevé, and L. Néel, "Effects of Neutron Irradiation on the 50-50 Fe-Ni Alloy," *Radiation Damage in Solids*, II. International Atomic Energy Agency, Vienna, 1962, p. 21.
92. Dugdale R. A., *Phil. Mag.* **1** (1956) 537.
93. Rudman P. S., *Acta Met.* **10** (1962) 195.
94. Gibson J. B., A. N. Goland, M. Milgram, and G. H. Vineyard, *Phys. Rev.* **120** (1960) 1229.

chapter 22

Crystallographic Transformations

J. W. CHRISTIAN*
Oxford University
Oxford, England

THOMAS A. READ
University of Illinois
Urbana, Illinois

C. M. WAYMAN
University of Illinois
Urbana, Illinois

1. INTRODUCTION

This chapter is concerned primarily with the class of solid state changes, in which the initial phase undergoes a change in crystal structure but none in chemical composition. The term *diffusionless transformation* is often used to characterize this kind of reaction and includes both the *martensitic* and *massive* types of transformation.

Martensitic-reaction phase changes exhibit certain characteristic features, which have been discussed by Bilby and Christian.[1] The most significant feature, and the one proposed by Bilby and Christian as a criterion for these reactions, is a characteristic shape change, which occurs as a volume of the crystal transforms. Other important crystallographic features of this type of transformation include the orientation relationship and the habit plane. The first of these refers to the fact that, after partial transformation, the crystal axes in the parent and product crystals have a characteristic relative orientation. The habit plane, which is the interface between the new and old structures, also has a characteristic orientation

* On sabbatical leave at the University of Illinois during the preparation of this manuscript.

relative to the crystal axes. These two crystallographic features are always associated with martensite transformations, but may also be exhibited by other types of phase changes, such as precipitation from solid solution. Thus they cannot serve as criteria for martensitic transformations in the way that the characteristic shape change does.

Massive transformations, on the other hand, do not exhibit surface-relief effects, and in extreme cases a massively formed phase bears no definite crystallographic-orientation relationship to its parent. However, massive transformations are related to martensitic transformations in that they involve no long-range diffusion.

Diffusionless transformations have been extensively studied in intermetallic compounds not only because these compounds frequently serve as useful model materials for experimental studies of such transformations in commercially important terminal solid solutions, but also because the many and varied special features which they exhibit present intriguing problems. In the following sections of this chapter we first review the crystallographic theory of martensitic transformations, which, in its present form, is equally applicable to both terminal solid solutions and intermediate phases. Next are summarized some of the experimental studies made of martensitic transformations in intermetallics for which crystallographic results can be compared with theory. Finally, the subject of massive transformations in intermetallic compounds is briefly reviewed.

2. CRYSTALLOGRAPHIC THEORY OF MARTENSITIC TRANSFORMATIONS

Substantial progress has been made in accounting for the crystallographic features of martensitic transformations on the basis of a phenomenological theory developed independently by Bowles and Mackenzie[2–4] and Wechsler, Lieberman, and Read.[5] The main ideas of this theory are very simple and are described here, but the mathematical formulation is rather complex. More detailed descriptions than are possible here are given in the review articles by Bilby and Christian[1,6] and in books by Christian[7] and Wayman.[8]

The phenomenological theory originated with a famous paper by Greninger and Troiano,[9] in which it was shown that the shape change in a martensitic transformation is macroscopically an invariant-plane strain and, furthermore, that application of this deformation to the parent lattice does not produce the product lattice. They therefore proposed that, in addition to the observable shape change, the transformation of lattices involves a further deformation which is macroscopically invisible, because it is cancelled on a fine scale by another deformation which does not affect the lattice. Greninger and Troiano considered both the components of the total lattice deformation to be simple shears, but in fact this is not possible for there is usually a volume change associated with the transformation. Later phenomenological theories, however, retain the assumption that the invisible component of the lattice deformation is a simple shear.

The physical significance of Greninger and Troiano's observation arises from the difficulty of fitting together two phases of different structures and fixed lattice parameters. Martensitic transformations involve highly coordinated atomic movements, as is clear from the shape change, and these take place without thermal activation, for growth at very high velocities has been observed near 0°K. Such growth is clearly possible only if the parent and product phase remain coherent during growth, and the crystallographic theory of martensites investigates the conditions under which this is possible. If the structures are fully coherent, there exists a plane (not necessarily a low-index plane) which is common to both lattices; this plane is then an invariant plane in a deformation which carries the points of one lattice into those of the other. Some transformations involving changes from fcc to hcp structures are of this type, and the invariant plane is then the habit plane of the martensite plates.

In most transformations, the two structures cannot be fully coherent; this follows because the mathematical condition for full coherence is rather restrictive and will generally be satisfied only coincidentally. Appreciable coherence is still possible, however, if the accumulating misfit of the structures is periodically corrected in some way, so that there is an average or macroscopic matching at the interface. Various ways of accomplishing this result have been suggested and are described below; the resultant interface is sometimes described as semicoherent.

The phenomenological theories do not consider directly the structure of the habit-plane interface, although they are clearly related to it in the manner just described. They involve the basic assumption that the habit plane is an invariant plane of the macroscopic shape deformation or a close approximation to it. The starting point of the theories is a knowledge of the crystal structures and lattice parameters of the parent and product phases. From these structures a lattice correspondence is deduced, in most cases by inspection. The lattice correspondence relates planes and directions in the product structure to the corresponding planes and directions from which they are derived in the parent structure. This permits the specification of a homogeneous deformation which, when applied to a unit cell of the parent phase, yields a product unit cell of the proper dimensions. In some cases the pure lattice deformation thus obtained does not relate all the atoms within these unit cells; additional rearrangements of the atoms are then necessary and are generally called "shuffles."

As already explained, the lattice deformation is usually such that there is no invariant plane. The procedure used by Wechsler, Lieberman, and Read is to combine the pure lattice deformation with a slip or twinning deformation, which modifies the total shape change without changing the unit cell. The slip or twinning elements have to be specified at this stage, and the magnitude of this "lattice invariant deformation" is then adjusted so that the combined deformation contains two macroscopically undistorted planes. Finally, a rotation, which returns one of these planes to its original orientation—thus making it an invariant plane—is added. The habit plane and the magnitude of the shape deformation are specified by the combined effects of the lattice and lattice invariant deformations and by the rotation, whereas the pure lattice deformation and the rotation give the total lattice deformation and specify the orientation relations between the phases. (There are naturally two such relations when twinning is involved.) Once the correspondence and the elements of the lattice invariant deformation have been specified, the theory yields unique predictions of the shape change, orientation relation, habit-plane orientation, and magnitude of the lattice invariant strain. In the general case, there are four nonequivalent solutions for any one set of input data, but in many transformations these solutions are degenerate and yield only one or two crystallographically distinguishable cases.

The Bowles-Mackenzie theory is completely equivalent to the Wechsler-Lieberman-Read theory, although formulated in a slightly different way. The total lattice deformation is factorized into two invariant-plane strains, one of which is identified with the shape deformation, and the other (assumed to be a simple shear) is invisible in the sense used by Greninger and Troiano. This invisible part of the lattice deformation is just the inverse of the lattice invariant deformation of the Wechsler-Lieberman-Read theory.

A novel feature introduced into the theory by Bowles and Mackenzie is the idea that there may be some small macroscopic misfit at the habit plane, which thus becomes only approximately an invariant plane of the shape deformation. The distortion they envisaged is a uniform dilatation, so that macroscopic vectors remain unchanged in direction but may change in length by up to $\sim$1 per cent. This introduces an arbitrary parameter, and has enabled many experimental results which were otherwise inconsistent to be fitted to the theory. More recently, the possibility of nonuniform deviations from the exact invariant-plane strain condition has been considered.[8]

The other disposable parameters used to vary the predictions are the elements of the lattice invariant deformation. This deformation must produce zero change of volume, and it has usually been assumed to be a simple shear with rational elements. When these are slip elements, the interface may be pictured as an array of parallel dislocation lines, the motion of which produces the shear. The slip planes of the dislocations are then corresponding planes in the two lattices, and they meet edge-to-edge in the habit-plane interface. The line of the dislocations is common to two invariant planes and must thus be an invariant line of the lattice deformation. The Burgers vector is inclined to the habit plane (except for pure screws), and for a given Burgers vector the magnitude of the lattice invariant shear determines dislocation spacing. This model was first used by Frank[10] for the limiting case of {225} martensite in steels, where the dislocations are pure screws. If the dislocations are taken to be normal slip dislocations in one or the other lattice, the Burgers vectors and slip planes are rational, and there are only one or two possibilities to consider. It is theoretically possible, however, to have glissile arrays containing more than one kind of dislocation, so that the resultant simple shear has either an irrational direction or an irrational plane.

The twin model for martensite is very similar, the product now consisting of a stack of fine parallel twins. The twin boundaries meet the habit-plane interface in the invariant line of the lattice deformation, and the twin direction is either inclined to the interface or parallel to this line in the limiting case. The magnitude of the lattice invariant deformation is now specified by the relative amounts of the twin orientations. In recent years, thin-film electron microscopy has shown that many martensite plates do consist of stacks of fine twins, and some of the predictions of this model have been confirmed.

Returning to the dislocation model, it is clear that surfaces of discontinuity will be left in the product if the interface dislocations have Burgers vectors which are not lattice vectors. In general, such surfaces would have high energy and, therefore, are not to be expected. In some close-packed structures, however, stacking faults of low energy could be left by suitable partial dislocations in the interface, and this situation might be energetically preferred because of the smaller Burgers vectors. In fact, martensite products containing arrays of stacking faults have also been observed, as discussed in more detail in the next section.

3. MARTENSITIC TRANSFORMATIONS IN SPECIFIC INTERMETALLIC PHASES

3.1 Copper-Aluminum

The β-phase in the Cu–Al system has been widely investigated. This phase undergoes a eutectoid decomposition at 565°C, but at temperatures near the congruent melting temperature (1048°C) it extends from approximately 10 to 15 wt.% Al.[11] The β-phase transforms martensitically rather than decomposing into its equilibrium phases if the cooling rate is 1°C/sec or greater.[12] Although the composition range of the β-phase is rather limited, occurring over a range of only 5 per cent, within these limits at least three different martensites form, in addition to order-disorder effects. At present, there is still confusion and disagreement concerning the crystal structures of the martensitic phases.

A great deal of early work on the Cu–Al system was done in Europe, particularly by Kurdjumov and colleagues, who showed that the temperature at which the martensite transformation began in the β Cu–Al alloys (i.e., M_s temperature) was composition-dependent, much like the martensite transformation in carbon steels.[13]

In equilibrium, the β-phase is disordered bcc, but alloys containing 11 to 15 per cent Al will order (forming β_1, DO_3 structure) during cooling below the eutectoid temperature; the ordering temperature increases with increasing Al content. Greninger[12] studied both hypo- and hypereutectoid β-alloys and was able to determine the martensite habit-plane and orientation relationship. More recently, Swann and Warlimont[14] studied the morphology and crystallography of Cu–Al martensites by transmission-electron microscopy. Figure 1,

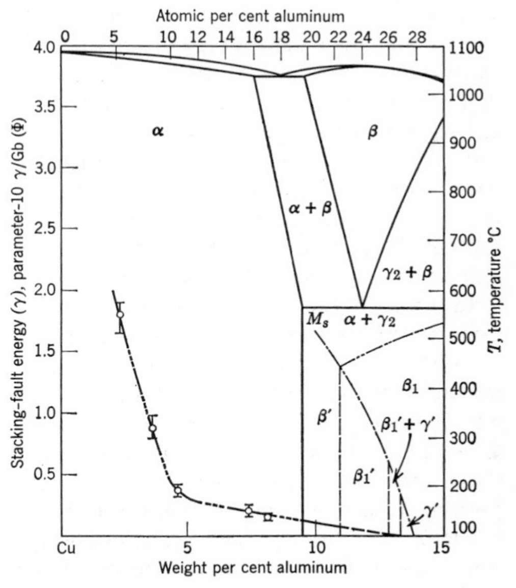

Fig. 1. Partial copper-aluminum equilibrium diagram, showing M_s temperatures of β', β_1', and γ' martensites, $\beta \rightarrow \beta_1$ ordering temperatures, and stacking-fault energies of α-phase alloys. (Swann and Warlimont.[14])

taken from their work, shows the relevant portion of the Cu–Al phase diagram and the (metastable) phases involved. It will be observed that alloys having more than 11 per cent Al become ordered during quenching from the β-field to form β_1 and then transform martensitically into either β_1' or γ'. Alloys containing less than 11 per cent Al transform into β', a disordered martensite. (Both β_1' and γ' are ordered, although earlier work failed to distinguish between β' and β_1'.) An alloy containing 9.3 per cent Al undergoes a massive transformation upon quenching from 1020°C, according to Greninger.[12]

The work of Greninger showed that both β' and γ' martensites were platelike in form. The plate size of the β' was larger with a milder quench. The γ' martensite plates exhibited fine parallel internal striations, observed previously by Gridnew and Kurdjumov,[15] which were suggestive of mechanical twinning; the β' plates also exhibited parallel striations, but these were attributed to "strain-transformation markings." Isaitschev, Kaminsky, and Kurdjumov[16,17] determined the crystal structures of the β' and γ' martensites; γ' was reported to have a hcp structure with an ordered arrangement of atoms, whereas their description of the β' structure corresponds to a distorted hcp structure ([00.1] 2° from (00.1) normal and angle between (10.0) and (01.0) 1° from 120°). Nakanishi[18] also found the β' crystal structure to be distorted hcp after extensive consideration of both monoclinic and triclinic models. He observed that the [0001] direction makes an angle of about three degrees from the normal to the (0001) plane. Greninger confirmed the crystal-structure determined of γ' made by Isaitschev et al.[16,17]

In studies of the deformation of martensite, Greninger[12] reported that slight deformation of either β' or β_1 resulted in γ'. There was no structure change upon deformation of γ', although superlattice reflections disappeared. Careful determinations of the habit planes of the β' and γ'-martensite plates relative to the β-phase were made, and the results are shown in Fig. 2. In addition, orientation relationships were determined by back-reflection Laue analysis. In the 13.1 per cent Al alloy, it was observed that every Laue photograph of γ' showed two twin-related hcp lattices, denoted as A_1 and A_2, from each plate. The twin relation was about a (10.1) plane, and it was

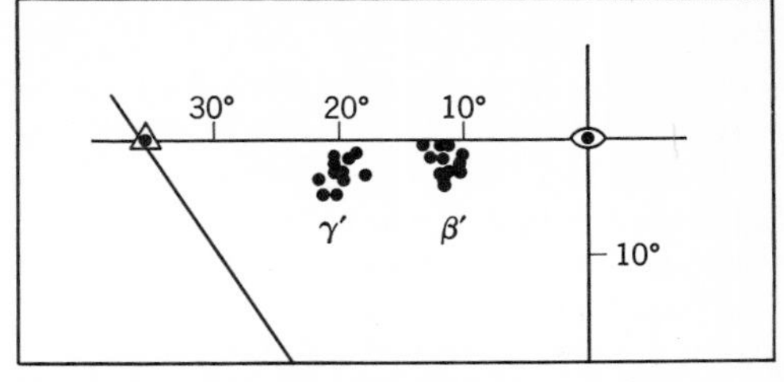

Fig. 2. Habit planes for β' (Cu-12.5% Al) and γ' (Cu-14.7% Al) martensites. (Greninger.[12])

suggested that twinning was an inherent part of the transformation mechanism for (10.1) is the usual deformation twinning plane in hcp metals. It was also found that adjacent parallel plates of γ' exhibited twin orientations $\mathbf{B}_1$ and $\mathbf{B}_2$ in contrast to the orientations A_1 and A_2. The X-ray results did not suggest the presence of twins in β'-plates.

As shown in Fig. 2, the habit-plane determinations by Greninger were (approximately):

$$2° \text{ from } \{133\}\beta_1 \text{ for } \beta'$$
$$3° \text{ from } \{122\}\beta_1 \text{ for } \gamma'$$

The γ'-β_1 orientation relationship was given as

$$\left.\begin{array}{l}(00.1)_{\gamma'} \parallel (110)_{\beta_1}\\ [01.0]_{\gamma'} \parallel [1\bar{1}1]_{\beta_1}\end{array}\right\} A_1 \text{ (or } \mathbf{B}_1)$$
$$\left.\begin{array}{l}(00.1)_{\gamma'}\ 4° \text{ from } (011)_{\beta_1}\\ [01.0]_{\gamma'} \parallel [1\bar{1}1]_{\beta_1}\end{array}\right\} A_2 \text{ (or } \mathbf{B}_2)$$

The uncertainty in the β' crystal structure and other difficulties made the β'-β_1 orientation relationship difficult to establish, but the inconclusive results obtained agreed with the relationship given above between γ' and β_1.

The recent electron-microscope investigation of Swann and Warlimont[14] showed that Cu–Al alloys, having between 10 and 13 per cent Al quenched from 900°C, exhibited lenticular martensite plates of β' and β_1' (recall the distinction in that β' is disordered and β_1' is ordered) with fine internal striations. Under the proper contrast conditions, partial dislocations were observed, and the fine striations were described as stacking faults. The electron-diffraction patterns from selected areas showed superlattice reflections in quenched β-alloys containing more than 10.5 per cent Al; no superlattice reflections were found in the Cu–10 per cent Al alloy. Antiphase-domain boundaries were also observed to extend "indiscriminately" across martensite plate boundaries. The previous observations that deformation of the β_1' martensite resulted in γ' (Nakanishi,[18] Greninger[12]) were questioned, because the transmission microscopy of deformed β' and γ' induced by quenching showed the structures to be different. No twin symmetry in the electron-diffraction patterns of β'- and β_1'-plates could be detected, confirming the suggestion of Greninger, based upon X-ray results. The displacement of reflections (streaked) due to the relaxation of the Laue diffraction conditions was explained by the high density of stacking faults in the β'- and β_1'-plates. The faults were said to be intrinsic, but owing to their high density the presence of some growth faults could not be excluded. On the basis of the electron-diffraction patterns, it was suggested that β' was a heavily faulted fcc structure. The β_1' structure was thought to be produced from the $D0_3$ structure of β_1, that is, basically a bcc to fcc transformation with a correspondence of atomic positions persisting throughout the transformation. The β_1' unit cell supposedly consists of two fcc cells forming a tetragonal unit cell with Al atoms

Fig. 3. Internally twinned γ' martensite (Cu-14.1% Al), also showing fine cross-striations within twins. Transmission-electron micrograph. (Swann and Warlimont.[14]) (Courtesy of *Acta Met.* and P. R. Swann.)

at positions of the type $\frac{1}{2}\frac{1}{2}0$, $00\frac{1}{2}$, and Cu atoms at 000, $0\frac{1}{2}\frac{1}{4}$, $\frac{1}{2}0\frac{1}{4}$, $0\frac{1}{2}\frac{3}{4}$, $\frac{1}{2}0\frac{3}{4}$, $\frac{1}{2}\frac{1}{2}\frac{1}{2}$.

The structure of the deformed β_1' martensite was found to be hexagonal in nature but different from the hexagonal γ', and it was concluded that β_1' did not transform into γ' due to deformation, as suggested previously.

The γ' martensite forms in quenched alloys containing more than 13 per cent Al, and is readily distinguished from β_1' in the electron microscope, although these martensites are indistinguishable under the optical microscope.[14] The γ'-plates exhibited parallel bands 100 to 500 Å thick, which were twin-related (twin-matrix ratio varied from 1.3:1 to 3.3:1). Also, finer cross striations were seen, which were ascribed to twins or stacking faults within the twins (see Fig. 3). Antiphase-domain boundaries were found to be discontinuous across the twin boundaries, showing that the martensite transformation preceded the ordering reaction. It was also reported that twinning of the γ' martensite disturbs the order to a greater extent than the faulting of the β_1' martensite. Deformation of the γ' by cold-rolling 40 per cent gave rise to new heavily faulted twins in addition to heavy faulting of the γ'-transformation twins.

As was the case for β_1' martensite, the transmission electron diffraction patterns of γ' were difficult to interpret. In the latter case, there was reciprocal lattice streaking in two different directions in a given plate of γ'. Nonetheless, the observations of Isaitschev et al.[16,17] that the structure was hcp was confirmed (disregarding the superlattice). The superlattice of γ' is orthorhombic, and the γ'-plates were internally twinned with a twinning plane $\{201\}$, $\{121\}$ orthorhombic ($\{1\bar{1}01\}$ hcp). No reflections in the twinned-diffraction pattern could be ascribed to the fine cross-striations (trace analysis), but streaks perpendicular to the twinning plane were observed, due to relaxation of the Laue conditions in this direction. It was concluded that the β-phase transforms into β' or β_1' (both basically fcc) and γ', a cph martensite. However, due to the heavy density of stacking faults in β' and β_1', certain diffraction maxima were shifted to hexagonal positions. The stacking fault density in both β' and β_1' was (within the limits of the analysis, which assumed a random fault distribution) independent of Al content, and it was concluded that the lattice invariant deformation of the β-to-β' and β-to-β_1' transformations was faulting. The lattice invariant deformation for the β-to-γ' transformation is twinning.

Swann and Warlimont[14] further suggested that the lattice correspondence for the β_1-toβ_1' transformation was the usual fcc-bcc correspondence proposed by Bain[19] on the basis of the relationship between the parent-martensite superstructures. Using the Bain correspondence and the WLR analysis[5] (lattice invariant shear on $\{110\}_{\beta_1}$ $\langle 110\rangle_{\beta_1}$) the magnitude of the lattice invariant shear was calculated to be $g = 0.251$. (This is usually defined such that the tangent of the shear angle is $g/2$.) This value corresponds to a stacking fault probability (density) $\alpha = 0.355$, which is not in favorable agreement with the experimentally observed value $\alpha_{\text{exp}} = 0.44$, but the discrepancy was rationalized on the basis that in calculating the value $\alpha = 0.355$, it was assumed that the faults extended completely across the martensite plates. In fact, the partial dislocations observed within the martensite plates showed that such an assumption is unwarranted.

The Wechsler-Lieberman-Read theory was also used to calculate the habit plane of the β_1' martensite (bcc-to-fcc transformation) for the lattice invariant shear mentioned previously, and the predicted habit plane turned out to be in good agreement with the results obtained by Greninger. Previously, calculations for the β_1' habit plane were not carried out because of the uncertainty of the β_1' crystal structure.

Using their phenomenological theory, Bowles and Mackenzie[2–4,20] analyzed the β'-to-γ' transformation for a Cu-13.6% Al alloy by employing Greninger's X-ray results and the observation that the γ'-plates were twinned on $\{10.1\}$ (this corresponds to $\{201\}$ bcc or $\{121\}$ relative to an orthorhombic basis). Their habit-plane prediction was in good agreement with the experimental poles determined by Greninger; agreement between the theoretical and experimental orientation relationship was also good. The magnitude of the lattice invariant shear required, however, predicted the relative volumes of the two twin orientations to be 3.5 :1. This is not in striking agreement with Greninger's observation that the relative twin volumes were about 2 :1 and with the transmission-microscopy results of Swann and Warlimont which yielded ratios from 1.3 :1 to 3.3 :1. In the latter case, these workers explained the discrepancy on the basis that the twins do not extend completely across the martensite plate. It is difficult to reconcile this point with the requirement of the crystallographic theories of martensite formation that the inhomogeneities extend completely to the interface.

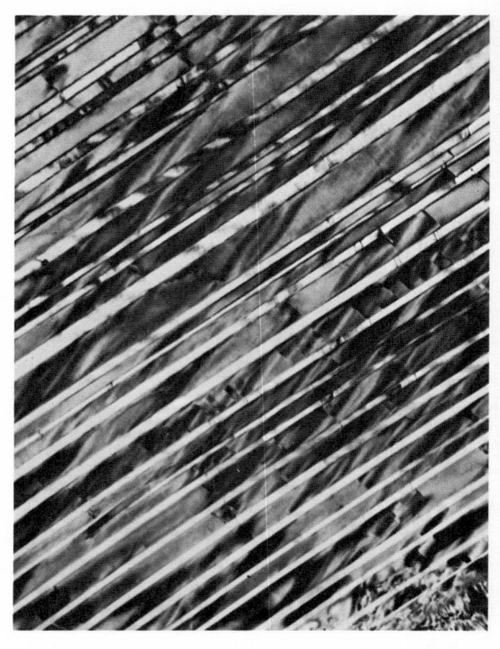

Fig. 4. Transmission-electron micrograph of β'-martensite plates after quenching Cu-10.0% Al from 1000°C. Note the presence of the (lighter) regions of preferential dissolution in the larger plates. (Swann and Warlimont.[14]) (Courtesy of *Acta Met.* and P. R. Swann.)

Swann and Warlimont argued that in the β-to-β' transformation there are twenty-four habit-plane variants, as would be expected from the theory, but in the case of the β_1-to-β_1' transformation consideration of nearest-neighbor violations reduces the number of variants from twenty-four to eight. These workers also found some convincing evidence which seems to indicate that martensite plates first form as long, thin plates, which subsequently thicken by sidewise growth. Due to the rapidity of the growth of martensite plates in general (the time of formation of a martensite plate in steels is less than one microsecond), a more or less "natural" means was necessary to reveal this point. Figure 4 (from Swann and

Warlimont) shows martensite plates in a Cu-10% Al alloy quenched from 1000°C to 100°C. Because the M_s temperature is above the ordering temperature in this alloy, the superlattice is not detectable. However, these authors suggested that the β' martensite contains regions of short-range order, corresponding to the darker regions of the plates in the figure. The central, lighter regions were said to be preferentially dissolved during electropolishing (thinning) and presumably to correspond to martensite formed from disordered β, whereas the darker areas evidently formed from β which underwent short-range order at lower temperatures (during the quench) before the martensite interface advanced into this region. It would be expected that the (smaller) martensite plates, which nucleated at lower temperatures, would not contain the midrib region inherited from a disordered parent. Observations supported this expectation.

Finally, the work of Swann and Warlimont concerning the fine structure of the martensite emphasized that the internal twinning and faulting did not exert an appreciable strengthening effect (i.e., barrier to dislocation movement). The superior hardness of the β_1' martensite over that of β' was attributed to ordering.

Recently, Wilkens and Warlimont[21] made further studies of the crystal structure of the β_1' martensite because of the "additional" reflections which entered into the crystal-structure analysis of the β' and β_1' martensite by Swann and Warlimont by means of electron diffraction. Because the phenomenological theory predicted a stacking-fault probability of $\alpha = 0.355$, Wilkens and Warlimont assumed a model for β' in which there were four stacking faults (in regular distribution) per eleven close-packed planes, that is, $\alpha = 0.36$. Designating with subscripts I and II the two possible positions of Al atoms in any given layer, they arrived at the following stacking sequence: $B_IA_IB_IC_IB_IC_IA_{II}C_IA_{II}B_{II}A_{II}B_{II}$ $A_{II}B_{II}C_{II}B_{II}C_{II}A_IC_{II}A_IB_IA_I$. This sequence was confirmed by diffractometer analysis, and it appears that the structure proposed by Nakanishi[18] is essentially correct; moreover, rather than the interpretation of Swann and Warlimont,[14] an *ordered* array of stacking faults can account for the uncertain reflections. If this is the case, it is not surprising that the phenomenological theory for the bcc-to-fcc with the usual Bain correspondence predicts the correct habit plane, because this results when the lattice deformation *and* lattice invariant deformation occur. It turns out that the magnitude of the inhomogeneous shear is such that the fcc structure generated by the lattice deformation is "converted" into the imperfect or "distorted" hexagonal structure by the (amount of) lattice invariant shear, which insures that the habit plane is undistorted. Hence the results for the β-to-β' (or β_1') transformation are in good agreement with the crystallographic theory.

In retrospect, the crystallographic features of the β-to-γ' transformation are also in agreement with the theory. As was pointed out, the Bowles-Mackenzie analysis (using a lattice invariant shear system, which was later verified experimentally by Swann and Warlimont[14]) gave good agreement between the theoretical and experimental habit planes and orientation relationships. The theory predicts the relative twin thicknesses to be 3.5 :1, in contrast to the ratio 2:1 given by Greninger.[12] However, Greninger's deduction of this twin ratio was based on relative Laue spot intensities, which are not entirely quantitative. The refined values of the relative twin ratios varying from 1.3 to 3.3 :1 based on transmission-microscopy studies rationalize any discrepancy on the magnitude of the lattice invariant shear, if we accept the explanation given by Swann and Warlimont.

Hull and Garwood[22] investigated the martensitic transformation of the β-phase in a ternary Cu–Al–Ni alloy containing 12.8% Al and 7.7% Cu, and found that the parent phase (of given composition) would transform into two different types of martensite, depending upon the prior thermal history of the specimen. In both cases, the transformation occurred isothermally, as well as during continuous cooling, and the martensites were identified as γ' and β'. By preparing wedge-shaped specimens and transforming in a temperature gradient, a single interface β-to-γ' transformation was made to occur as in Au–Cd β-phase alloys, but no β-to-β' single-interface transformation was observed.

Hull and Garwood[23] used a two-surface analysis to determine the martensite habit planes and the angle of shape deformation. The β' habit plane was found to be $\sim\{155\}_\beta$, and the γ' martensite exhibited a habit plane between $\{133\}_\beta$ and $\{144\}_\beta$. The shape deformations were analyzed as a habit-plane shear, and the directions of displacement (β' and γ') were essentially in the habit plane, that is, near $\langle 011\rangle_\beta$. The average shear angle for the shape deformation for $\beta \rightarrow \beta'$ (9.5°) was larger than that for $\beta \rightarrow \gamma'$ (5.2°). It was suggested that the larger shear angle for β-to-β' prevented the single-interface transformation from occurring. The β-phase could be made to transform into the γ'-phase upon tempering, and this "conditioning" treatment was ascribed to growth of ordered domains of the β_1' (Cu_3Al) superstructure. A similar effect has been observed in Au–Cd alloys containing approximately 47.5 at.% Cd. Under (slow) continuous cooling conditions, the transformation is from the β_1 (CsCl) parent phase into the β' (orthorhombic) martensite. However, holding at a temperature (>80°C) immediately above the β_1-to-β' transformation temperature[24] results in a transformation upon subsequent cooling to a different martensite (β'').

3.2 Copper-Tin

The β-phase in the Cu–Sn system is analogous to that in Cu–Al, having a disordered bcc structure and decomposing eutectoidally upon slow cooling (Hansen[11]). However, upon quenching β alloys having a composition near Cu-25% Sn, a martensitic transformation occurs. Smith[25] in 1933 pointed out that the parallel bands found in quenched and deformed Cu-25.08% Sn were parallel to the {133} plane of the β-parent and were definitely not {112} deformation twins. Greninger and Mooradian[26] found that these markings also occurred in a Cu-25.6% Sn alloy quenched from 700°C. Parallel, but shorter, markings could be generated by plastic deformation. A careful analysis of the habit planes of the parallel markings showed that the poles were close to $\{133\}_\beta$, with deviations ranging from 0 to 2°. It was further reported that the direction of deviation from $\{133\}_\beta$ was consistently in the direction of the adjacent $\{110\}_\beta$ plane.

The transformation induced by filing this alloy gave rise to diffuse X-ray lines, which were not analyzed, but were reported to be "completely different" from the patterns of deformed β-brass. It is probable that this transformation is basically the same as the β-to-β' transformation in Cu–Al alloys, in which a $\{133\}_\beta$ habit plane was observed.

3.3 Copper-Zinc

There have been numerous investigations of the martensitic transformations in quenched β-phase copper-zinc alloys, and in many ways these transformations appear to be quite similar to those in other β-phase copper alloys, such as copper-aluminium (§ 3.1) and copper-tin (§ 3.2). Noteworthy features are the variety of product structures that have been reported, the formation of stress-induced martensites in composition ranges where spontaneous transformation does not occur, and the phenomenon of thermoelastic martensite.

The β-phase extends from ~37 to ~56 wt.% zinc at temperatures close to the solidus, where it has a disordered body-centered cubic structure, but the equilibrium phase field is more restricted at lower temperatures and lies on the copper-rich side of the equiatomic alloy. Alloys with up to 39 wt.% zinc generally undergo a massive transformation to the face-centered cubic α-phase on quenching to room temperature, and this change is thought to be noncrystallographic in character, although the crystals of massive α often show planar facets.[22] The remaining alloys may be quenched to room temperature without transforming, except for the spontaneous ordering to the CsCl (B2) type structure which is sometimes called β_1. Martensitic transformation of the quenched β_1 alloys is then obtained by cooling to subzero temperatures, or by cold-working the quenched alloys at room temperature or below.

The early work on the spontaneous transformation is largely due to Kurdjumov and his co-workers, and to Greninger. Isaitchev and Miretskii[27] reported that alloys containing 39 to 41.8 wt.% zinc transform wholly or partially to an unknown structure of lower symmetry on cooling to −160°C, and Isaitchev, Kaminsky, and Kurdjumov[16,17] found the product structure to be face-centered tetragonal in an alloy containing 39% zinc and 1% lead, iron, and tin as impurities. This observation does not appear to have been repeated, and most workers are agreed that the X-ray diffraction patterns from the martensite phase (β') indicate a more complex structure in the binary alloys.[26] Garwood and Hull[18] believe the structure to be related to a close-packed structure, but with deviations caused by the ordering of the atoms in the

β_1-phase, and by the presence of stacking faults. This work, together with selected-area electron diffraction data,[29] leads to the conclusion that the true structure is close to face-centered orthorhombic. The transformation is then crystallographically similar to that in gold-cadmium alloys, and the lattice correspondence is presumably the same. This correspondence is essentially the inverse of the Bain strain for transformations in steels, and it is highly probable that it applies even if the true structure of the product is distorted from orthorhombic. It is noteworthy that simpler structures appear to result from stress-induced transformation (see below), in which some of the effects caused by ordering of the atoms in the initial β_1-phase may not be present.

The main difficulty in finding the exact structure of the product is that of preparing single crystals, and this has been partially solved by Kunze.[30] He obtained a pseudo-monocrystal of martensite by cooling a single crystal wire of the β_1-phase, the axis of this crystal being close to [110]. From oscillation photographs, Kunze concluded that there are two product phases, namely, a transition structure β_1' with a monoclinic unit cell, and a superlattice structure β'' with a triclinic cell. The structure of β'' is said to be built up of antiphase domains limited by {11, 1, 12}-type planes which are also the habit planes between β_1' and β''. The transformation $\beta_1' \rightarrow \beta''$ is not usually completed, so that the two product structures may coexist. Masson and Govila,[31] however, have published X-ray data on copper-zinc and copper-zinc-gallium alloys which are not in agreement with Kunze's structures, and they point out that his own data also show rather large divergencies from these structures. It is evident that this difficult problem is still not resolved.

The M_s temperature for the transformation to the β_1'-phase decreases from -18°C for an alloy containing 38.54% zinc to -131°C for an alloy containing 40.04% zinc, the variation being approximately linear.[32] A linear extrapolation of this curve, although not entirely consistent with the results of Isaitchev and Miretskii[27] quoted above, suggests that spontaneous transformation is not possible in alloys with more than ~42 wt.% zinc, and this seems to be in agreement with the observations. Direct evidence that the M_s temperature is determined largely by the constraints of the surrounding matrix which oppose the growth of a martensite plate has been obtained by Hull.[33] He found that an alloy containing 39 wt.% zinc, which normally transforms at -60°C, transformed spontaneously at 20°C to a face-centered cubic structure when the material was electrolytically thinned to foil of less than 1500 Å thick. The product of the transformation in such thin foils does not necessarily have the same structure as that of the reaction in bulk materials, as is shown by this particular example, and the crystallographic conditions are also quite different. Because the restraints are only two-dimensional, there is no habit plane in the usual sense, and the assumptions of the crystallographic theories mentioned in the introduction are not valid. Hull found two types of transformation, one leading to a single-crystal product and the other to a finely twinned product. In the first case, the product structure is produced directly by a homogeneous pure strain, which is essentially the inverse of the Bain strain used to describe the lattice correspondence in the martensite reaction in steels. This was observed only in the very thin regions at the edges of the foils, presumably because of the very high accommodation stresses. The other type of transformation minimized the misfit in the plane of the foil; the product twins were about 25 Å wide.

The habit plane for the transformation in bulk specimens has been determined by Hull and Garwood,[22] and the shape deformation has been investigated by Garwood and Hull.[28] The plates form in characteristic zigzag colonies, as shown in Fig. 5, each group of four plates being associated with a {110} plane of the parent structure. The actual habit plane is of the form {2, 11, 12}, and the crystallography appears to be degenerate in the sense that one individual habit may be associated with two oppositely directed macroscopic deformations. Thus the two plates associated with a particular habit, for example, (2, 11, 12), are partially self-accomodating and are also at a small angle to the plates of another habit, ($\bar{2}$, 11, 12). These four plates meet along the (011) plane, which often appears as a pseudo-habit for the transformation. Hull and Garwood state that only two habit planes are associated with each {110} plane—in other words, ($\bar{2}$, 11, 12) and (2, 11, 12) would not appear in the group above. An alternative

Fig. 5. Surface-relief structures produced by a typical colony of plates with zigzag formation in 61.0 per cent Cu, 39.0 per cent Zn alloy, cooled to −40°C after quenching from 850°C to retain the β-phase in a metastable condition. ×285 X. (Garwood and Hull.[28]) (Courtesy of *Acta Met.* and D. Hull.)

possibility might be that oppositely directed tilts correspond to $(\bar{2}, 11, 12)$ and $(2, 11, 12)$, so that there are really four habits in each group. The formation of such pseudohabits is also discussed in connection with the gold-manganese and gold-copper transformations. By analyzing the tilts observed in a group of four plates of the type just described, Garwood and Hull found that the macroscopic shape deformation may be described as a simple shear on the habit plane in a direction close to $[0\bar{1}1]$ and of magnitude 10.9°.

Stress-induced martensite forms in quenched copper-rich β-phase alloys when they are cold-worked above the M_s temperature, and also in alloys which do not transform spontaneously at any temperature. In some alloys, the martensite is thermoelastic, that is, it is in reversible equilibrium with the matrix under the combined mechanical and chemical driving forces. Reynolds and Bever[34] found that alloys containing 60 per cent copper formed martensite plates when compressed at room temperature, but most of these plates disappeared again when the stress was removed. The plates could similarly be induced to grow or shrink by lowering or raising the temperature.

Investigations of the structures obtained by deforming alloys of various compositions have been made by Massalski and Barrett[35] and Hornbogen, Segmuller, and Wassermann.[36] Some differences in the results may be correlated with the severity of the deformation and with the composition ranges investigated. Hornbogen et al. used single-crystal wires in the composition range 39.5 to 39.8 wt.% zinc, and they found that these transformed under tensile stress at room temperature to give a tetragonal structure (α_1) of the AuCu ($L1_0$) type. Because this structure is an ordered form of the face-centered cubic α-brass phase, the transformation may be regarded as body-centered cubic to face-centered cubic, apart from the ordering. In the transformation from β_1 to α_1, another less symmetrical structure was observed; it is obviously possible that this may correspond to the β' structure obtained by spontaneous transformation. It is noteworthy that the α_1-phase probably corresponds to the face-centered cubic structure which Hull found in spontaneously transformed thin foils in an alloy of similar composition. Polycrystalline wires gave similar results; but, as higher stresses were then attained, it was found that the ordered α_1 structure could be further transformed into a disordered face-centered cubic structure.

Massalski and Barrett worked with polycrystalline alloys containing 39.73 to 51.89 wt.% zinc, which were given severe cold deformation by filing at temperatures down to liquid nitrogen temperature, or by scraping at liquid helium temperature. Presumably for this reason, the $L1_0$-type structure was not observed. The product structure was face-centered cubic at the lower zinc contents, but contained stacking faults. The fault density increased with increasing zinc content, and the authors describe the product as changing gradually from cubic to hexagonal close-packing, the hexagonal packing predominating in alloys with more than 50 wt.% zinc. The temperature at which stress-induced transformation can first be detected (M_d) is ~100°C at 48.35 wt.% zinc, and decreases by 47°C for each additional 1.0 wt.% zinc; this curve is nearly parallel to the corresponding curve for spontaneous transformation (M_s) as a function of temperature. In the alloys in which the product approximates to a hexagonal close-packed structure, the axial ratio is close to the ideal value of 1.633. Estimates of the stacking-fault probability for the cubic product fit on to the extrapolated curve of this parameter against composition for the equilibrium α-phase. Because the stacking-fault parameter is already quite large, and the stacking fault energy correspondingly low, at the phase limit of the α-phase, the occurrence of heavily faulted and intermediate structures in transformed β-phase alloys is scarcely surprising. Massalski and Barrett also obtained metallographic evidence of the martensitic nature of the transformation by the room-temperature examination of polished and etched sections.

Similar results were obtained for the stress-induced transformation in ternary copper-zinc-gallium alloys. Four alloys, chosen to lie along a line of constant electron/atom ratio, gave generally similar results, except that the product phase was almost completely hexagonal in the alloy with least zinc, and changed to an apparent mixture of cubic and hexagonal phases in the alloy with least gallium.

No detailed test of the crystallographic theories has been possible for those copper-zinc transformations in which the products are reported to have simple structures, for there is insufficient experimental information on the habit planes, orientation relations, and shape deformation. The habit plane has been measured accurately for the spontaneous transformation, however, and a recent investigation of the transformation in a 39% zinc alloy using thin film electron microscopy[29] has enabled a comparison to be made between theory and experiment. As already noted, the product structure was deduced to be face-centered orthorhombic from electron diffraction and X-ray studies, although this leaves some low angle X-ray lines unexplained. The formal theory is thus identical with that of the gold-cadmium transformation, and there is quite good agreement between calculated and measured habit planes and orientation relations. The experimental errors in the copper-zinc measurements do not allow a very critical comparison, but the deviations (~4° for the habit plane, 9° for the shear direction, 5° for the twin plane normal, and up to 3° in the orientation relation) are generally not much larger than the estimated errors. The martensite plates contained a high density of "faults," spaced ~25 Å apart, and it is believed they are fine twins, although the possibility of stacking faults (as in copper-aluminum) is not eliminated. The anomalous low-angle X-ray diffraction lines are now not thought to arise from these faults, and Jolley and Hull suggested instead that they may represent some additional stress-induced martensite of different structure.

3.4 Gold-Cadmium

The β-phase of the gold-cadmium system exists at elevated temperatures over a composition range of approximately 10 at.% cadmium, centered about the stoichiometric composition AuCd. It possesses the ordered CsCl structure, as originally determined by Olander[37] and Bystrom and Almin.[38] At lower temperatures, this phase remains stable over a slightly reduced composition range, which has not yet been determined with precision. At temperatures somewhat above room temperature, the crystal structure changes to structures of lower symmetry, which vary with composition and previous thermal history of the sample.

The transformation crystallography of slowly cooled alloys with compositions near 47.5 at.% cadmium has been studied experimentally and compared with the predictions of the phenomenological theory by Lieberman, Wechsler, and Read.[39] The low-temperature structure in this case is orthorhombic and forms on cooling at about 60°C. According to the lattice parameters determined by Chang,[40] the pure lattice deformation for this transition involves a contraction of 5.1 per cent along one cube edge of the cesium chloride structure and expansions of 1.4 and 3.5 per cent along the two face-diagonal directions perpendicular to that cube edge.

Direct observation of a lattice invariant deformation was reported by Lieberman, Wechsler, and Read.[39] They observed that the immediate product of transformation was a regular array of plate-shaped twins of the orthorhombic structure, in which the orthorhombic axis obtained by the 5.1 per cent contraction along a cube edge alternated in successive twins between two of the three possible choices. Additional experimental observations yielded the orientation of the interface plane between the cubic and orthorhombic structures, as well as the macroscopic shape change of the twinned orthorhombic product. A determination of the orientation relation between the cubic and orthorhombic crystal axes on either side of the interface between the two structures had been previously made by Chang and Read.[41] Comparison of observed and calculated values of the habit-plane orientation, macroscopic shape change, and orientation relation yielded agreement within experimental error.

The beta-phase alloys with cadmium contents in the range 49 to 51 at.% transform on cooling at about 30°C. The low-temperature structure represents only a small distortion of the high-temperature cubic structure; its symmetry has been reported to be tetragonal by Köster and Schneider[42] and Chen,[43] whereas according to Wilkens[44] the symmetry is rhombohedral. Chen also measured the orientation of the interface between the phases and showed that it agreed well with the orientation calculated from the phenomenological theory with his lattice parameter data. This low-temperature phase may be readily distinguished from the orthorhombic structure discussed above on the basis of electrical resistivity measurements. The former has an electrical resistivity approximately 15 per cent *greater* than the cubic phase at the transition temperature, whereas the latter transformation is accompanied by a *decrease* of approximately 15 per cent on cooling. This difference in electrical resistivity has been used in studies of the effect of prior thermal history on the transformation behavior of gold-cadmium alloys. It was reported by Wechsler and Read[45] that a sample containing 47.5 at.% cadmium, when quenched from 410°C, exhibited an increase of electrical resistivity on transformation. Further studies by Wechsler[46] and by Sturm and Wechsler[47] led them to the conclusion that the immediate effect of quenching is to retain a large number of vacant sites in the lattice, presumably the number present in the crystal in equilibrium at the temperature from which it was quenched.

Additional results of quenching on the transformation behavior of gold-cadmium alloys have been reported by Subramanya, Baker, Lieberman, and Read.[24] They quenched a 47.5 at.% cadmium crystal from a series of temperatures. For the lowest temperature, 212°C, there was little, if any, effect on electrical resistivity or transformation behavior. When quenched from 310°C and 350°C, the specimen still appeared to transform to the orthorhombic structure on cooling, for it still exhibited the characteristic resistivity decrease on transformation. The transformation temperature was depressed, however, to approximately 55°C and 45°C, respectively, and resistance increases were found for both low- and high-temperature phases. Samples quenched from 375°C showed a further increase of resistivity, but the transformation seemed to be a mixture of the two types. If the temperature of quenching was finally raised to 400°C or above, the transformation was observed to be completely of the type which involves an electrical resistivity increase on cooling.

Quenching 47.5 at.% cadmium crystals affects their transformation behavior in still another way. In the annealed condition, these crystals exhibit only athermal transformation, that is, in order for the transformation to proceed, the temperature must be continually lowered. Transformation stops if the temperature is kept constant. The quenched crystals of the same composition exhibited, on the other hand, the phenomenon of isothermal transformation, as reported by Subramanya, Baker, Lieberman and Read.[24] The reason for this change from athermal to isothermal transformation is not known. It is, however, consistent with the general observation that the transformation characteristics of the quenched 47.5 at.% cadmium alloy closely resemble those of the annealed alloys with higher cadmium contents, that is, in the range of 49 to 51 at.% cadmium. Isothermal transformation of these latter alloys has been observed by Mullendore.[48]

The effect of a tensile stress on the transformation behavior of annealed 47.5 at % cadmium alloys has been reported by Intrater, Chang, and Read.[49] They found parallel upward shifts of the transformation temperatures for heating and cooling. In both cases the magnitude of the shift was proportional to the applied stress. These results were in part a consequence of the fact that the length increase during transformation on cooling and the length decrease on heating were equal and independent of the magnitude of the applied tensile stress. Application of the appropriate form of the Clausius-Clapeyron equation permitted calculation of the latent heat of transformation; they reported a value of 62 calories per mole.

3.5 Gold-Copper

Although ordering reactions are discussed in Chapter 21, it is appropriate to consider here the particular case of the formation of the orthorhombic ordered phase in AuCu, because the crystallographic features of the reaction are closely related to the effects observed in martensitic transformations. Ordering of the alloy AuCu below about 380°C leads to the formation of a superstructure of the $L1_0$ type, in which the gold and copper atoms occupy alternate planes normal to the tetragonal c-axis. When the disordered face-centered cubic phase is annealed between 410° and 380°C, however, a more complex orthorhombic phase, AuCu(II) is formed. The AuCu(II) structure may be regarded as being derived by introducing antiphase domains in a regular manner into the $L1_0$ structure; it has a pseudocell which corresponds to the cell of the $L1_0$ structure, but which is slightly orthorhombic with $b/a = 0.997$–0.998. The true unit cell is obtained by stacking ten of the pseudocells in the direction of the a-axis, the occupancy of the (001) planes by alternate gold and copper atoms reversing after five pseudocells. This structure was originally deduced from the X-ray diffraction patterns, but has since been confirmed by electron diffraction and by direct observations in the electron microscope.

Although the ordering reactions must involve atomic interchanges, and must have kinetic features which agree with this conclusion, both transformations in AuCu give rise to characteristic relief effects on a polished surface. In the case of the cubic-to-tetragonal ($L1_0$) reaction, the relief effects are apparently caused by mechanical twinning subsequent to the ordering, but in the cubic-to-AuCu(II) reaction, the surface tilts as an ordered region grows from the cubic phase. There is thus some justification for applying the formal theory of martensite crystallography, and a detailed theoretical and experimental study was made by Smith and Bowles.[50] In view of the close relation between the two ordered structures, it is rather surprising that the mechanism of ordering should apparently be quite different in the two cases.

The experimental results show that ordered plates form in groups of four, appearing on the surface as roughly diamond-shaped figures formed from four nonparallel variants of the same habit plane. The habit plane of each plate is irrational, but is within 3° of a $\{110\}$ plane of the original cubic crystal. As the transformation proceeds, the growth of the plates leads to the development of long bands parallel to a pseudohabit plane of $\{110\}$ type; a similar situation is encountered in copper-zinc alloys (see § 3.3). Each ordered plate contains two twin-related orientations, the twinning plane being an orthorhombic plane of type $\{101\}$. The eight orientations obtained from one group of four plates, and the corresponding surface tilts, were carefully measured by Smith and Bowles. The orthorhombic axes are all within a few degrees of the original cubic $\langle 100 \rangle$ axes, and the shape deformation is relatively small, the angle of tilt for the particular surface investigated being about 2°45′.

The lattice correspondence between the structures may be defined by the unit matrix when axes are selected to correspond to the conventional cubic cell of the disordered structure and the pseudocell of the orthorhombic structure. The experimental observation that the product is twinned on $\{101\}$ indicates that the lattice invariant shear is on a $\{101\}$ plane of the cubic structure, and all variables needed to apply the crystallographic theory have been fixed. The predictions for the habit plane, orientation relations, and tilt of the surface are all in excellent agreement with the experimental measurements, but, in order to obtain this good agreement, it is necessary to assume a value of the Bowles-Mackenzie dilation parameter $\delta = 0.9986$. The correspondence between measured and predicted values of the various measured quantities seems to provide good evidence of the validity of the concept of some dilation in the interface in this particular case and is perhaps more convincing than any of the other examples in which a value of δ other than unity has been assumed.

From a general point of view, it is clear that the growth of an ordered phase involves atomic migration, but not long-range diffusion. The glissile nature of an interface containing a single array of parallel edge dislocations, or equivalently of parallel twin boundaries, is presumably very important in martensitic reactions which take place at low temperatures, but the important factor in an ordering reaction must simply be the matching condition at the interface. Isothermal transformation curves show clearly that the growth of the ordered phase is thermally activated, but the lattice correspondence and associated shape change are presumably retained, because individual atoms move very short distances during the transformation. From this point of view, the theory of martensite should apply to many transformations, in which the phases are coherent, and individual atoms move only a few interatomic distances in transferring from one phase to another; in true martensites, the atoms are usually considered to move less than one interatomic distance relative to their neighbors. Later work by Bowles and Malin[51] indicates that ordering reactions with mechanisms similar to that in AuCu occur in cobalt-platinum and magnesium-cadmium alloys. A martensitic ordering reaction has also been reported for Fe_3Al.[52]

3.6 Gold-Manganese

The β-phase of this system has an ordered CsCl (B2) type structure at high temperatures, and the alloy of equiatomic composition melts congruently at 1260°C. On cooling to room temperature, alloys near the composition AuMn transform to an ordered tetragonal structure, which is simply related to the B2 structure. A remarkable feature in the transformation is that the axial ratio of the tetragonal cell is smaller than unity for alloys containing more than 50 at.% gold and larger than unity for alloys containing more than 50 at.% manganese.[53,54]

The β-phase has a rather wide composition range (~35 to 65 at.% manganese), and

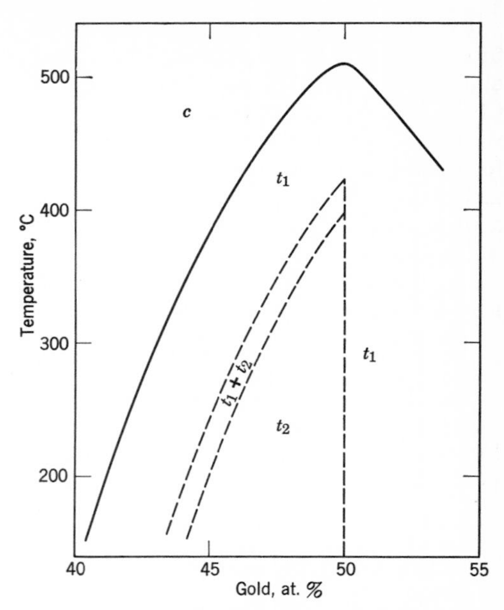

Fig. 6. Upper and lower transformation temperatures for alloys near the composition Au–Mn. c: cubic; t_1: tetragonal with $c/a < 1$; t_2: tetragonal with $c/a > 1$.

strong superlattice lines are observed in X-ray diffraction patterns from alloys with composition differing from AuMn by as much as 10 at.%. X-ray studies by Smith and Gaunt[55] have established that for all alloys within the composition range 50 to 53 at.% gold the initial product formed on cooling the cubic phase is a tetragonal structure (t_1) with $c/a < 1$, and this transformation is associated with the formation of an antiferromagnetic superlattice of the manganese ions.[54] Alloys with less than 50 at.% gold then undergo a further transformation at lower temperatures to form a second tetragonal structure (t_2) with $c/a > 1$. Figure 6 shows the upper and lower transformation temperatures for these reactions, as determined by measurements of Young's modulus and electrical resistivity. Detailed X-ray measurements on an alloy containing 47 at.% gold showed that there is no measurable volume change accompanying the transformations, but appreciable hysteresis was found in the $t_1 \rightarrow t_2$ transformation, small amounts of the t_1 structure remaining well below the temperature at which most of the phase had transformed.

Smith and Gaunt confirmed that the transformations are martensitic by studying the relief markings produced when polished surfaces were heated and cooled through the transformation range, and also by etching specimens at room temperature to reveal the transformation patterns. These patterns are strikingly similar to those obtained after the martensitic transformation from face-centered cubic to face-centered tetragonal phases in indium-thallium alloys[56] and γ-phase manganese alloys.[57] The markings consist of a number of "main bands" or martensite plates approximately parallel to $\{110\}$ planes of the cubic structure, and of sets of subbands within the main bands. The subbands are parallel to traces of $\{110\}$ planes at 60° to the main bands, and different subbands within the same main band represent $\{110\}$ planes which are normal to each other. These subbands are almost certainly sets of tetragonal twins, although this was not proved directly for gold-manganese. The habit plane for the $t_1 \rightarrow t_2$ transformation is the same as that for the cubic-to-t_1 transformation, and Smith and Gaunt were able to show experimentally that the surface tilt is opposite in sign. Orientation relations between the various phases have not been determined.

The crystallography of this transformation appears to be very simple. The lattice correspondence is the unit matrix when the conventional unit cells are taken in the two structures, and the pure lattice deformation for the cubic-to-tetragonal (t_1) change consists in equal expansions of ϵ_1 along two of the cubic axes and a contraction of $2\epsilon_1$ along the third axis. This gives an axial ratio of $1 - 3\epsilon_1 : 1$ and a volume change of zero, apart from terms in ϵ_1^2 and higher orders, which are very small. Twin-related products corresponding to different variants of the lattice correspondence will then give a shape deformation with an invariant plane close to a $\{110\}$ plane of the parent structure. The two twin orientations have to be present in a volume ratio of 2:1, and the exact solution of the crystallographic problem shows that the true habit plane should be about 20′ away from $\{110\}$ for the alloy investigated by Smith and Gaunt. The lattice parameters of the t_1 phase vary relatively rapidly with temperature, and, if the transformation took place directly at a lower temperature, the angle between $\{110\}$ and the predicted habit would increase to 1°7′.

The finely twinned nature of the t_1 phase would be expected to govern the transformation from t_1 to t_2. A hypothetical cubic-to-t_2 transformation could take place on the same habit plane as the cubic-to-t_1 transformation; the twin volume ratio would be 1 : 2 instead of 2:1, and the direction of the macroscopic shear in the shape deformation would be reversed. It follows from this that a t_1-to-t_2 transformation can also take place in the same way, the total shape deformation being the sum of the two cubic-to-tetragonal deformations. The habit plane for the $t_1 \rightarrow t_2$ reaction is then a planar boundary separating two regions, each of which is finely twinned. Boundaries between differently twinned regions of the same structure have been shown to be mobile in indium-thallium alloys,[57,58] and the observation should be equally applicable to twinned regions of different structures which have a common meeting plane. The cinéphotographs of the relief effects obtained by hot-stage microscopy support the conclusion that the habit plane is the same for both transformations, but the sign of the shear is reversed.

As already indicated, this transformation appears to be crystallographically identical with the face-centered cubic to face-centered tetragonal changes in indium-thallium alloys ($c/a > 1$) and in manganese-copper alloys ($c/a < 1$). One unusual feature of all these reactions is that the sets of fine parallel twins, which form in the product in accordance with the formal crystallographic theories, are large enough to be visible by optical microscopy. This observation, which is also made in the transformations from cubic to orthorhombic phases in gold-cadmium and gold-copper alloys (see § 3.4 and § 3.5), seems to be linked to the relatively small pure lattice deformations and shape deformations. There are some indications in some of the alloys investigated that the twins are very fine where they are in contact with the phase interface, but become wider immediately behind this interface, presumably in order to reduce the energy of the twin-boundary interfaces.

Cold-working of the β-phase does not promote the transformation to the low-temperature tetragonal phases, at least not if the deformation is heavy, but rather results in the formation of a new disordered face-centered cubic phase. This transformation was reported by Bumm and Dehlinger,[59] and Smith and

Gaunt[55] confirmed that powders filed from either the cubic or tetragonal ordered phases are almost 100 per cent face-centered cubic, the transformation being accompanied by ~1 per cent reduction in atomic volume.

3.7 Manganese Arsenide and Bismuthide

The transformations of the compounds MnAs and MnBi will be discussed together, because the crystal structure changes involved are similar. Both compounds have the nickel arsenide structure, and their transitions involve no change in macroscopic symmetry. All that occurs are discontinuous changes in the lattice parameters. On the other hand, the two transformations exhibit rather striking differences in other respects.

The transition in manganese arsenide has been studied by Basinski and Pearson.[60] It occurs at about 40°C on cooling with an expansion of the *a*-axis of approximately 1 per cent. There is no measurable change in the *c*-axis dimension. Basinski and Pearson followed the progress of the transformation by observation of a previously polished surface under the microscope. They observed that transformation was accompanied by a breaking-up of the crystal into pencillike subgrains, whose long dimensions were parallel to the *c*-axis.

It is thus apparent that the transformation of manganese arsenide does not exhibit the characteristics of a martensite reaction, although it is diffusionless in character. Evidently, a macroscopically unstrained interface between the new and old phases, which plays a central role in the phenomenological theory of martensite reactions, is not achieved in this case. Instead, the interface stresses lead to cracking. The general course of the cracks parallel to the *c*-axis is consistent with the fact that the dimensional difference between the two structures is maximum in the plane perpendicular to the *c*-axis.

As pointed out by Basinski and Pearson, the transformation behavior they observed is related to the fact that at the relatively low temperature at which the transition occurs manganese arsenide is a brittle material. In addition to this brittleness, however, there is another aspect of the structure change, which has a bearing on its nonmartensitic character. This is the fact that the lattice correspondence is unique, whereas for most phase changes there is a certain multiplicity. For example, in a cubic-to-tetragonal transition the fourfold *c*-axis of the tetragonal structure may correspond to any of the three original fourfold axes of the cubic structure. In such a transition it is frequently observed that the tetragonal product consists of a stack of fine twins, in which the *c*-axis alternates in correspondence between two of the original cube edge directions. If the relative thicknesses of the two twin orientations have appropriate values, the macroscopic shape change will be an invariant-plane strain, corresponding to a martensitic transition. It is to be noted that in this case the lattice invariant deformation does not involve plasticity in the usual sense of slip or mechanical twinning and thus could occur in a material which behaves in a brittle manner in the usual mechanical tests. The uniqueness of the lattice correspondence rules out this possibility in the manganese arsenide case, however.

Manganese bismuthide also has the nickel arsenide structure and also exhibits a transition without change in macroscopic symmetry, at about 350°C. Willis and Rooksby[61] found that on heating there is a discontinuous decrease in lattice spacing of about 3 per cent along the *c*-axis. Simultaneously, the *a*-axis spacing decreases by half this amount, and in consequence only a very small volume change occurs.

A dilatometric study of this transformation has been carried out by Patterson[62] on single-crystal specimens. He observed that the single crystalline character of his specimens was substantially preserved as they were cycled back and forth through the transition temperature. On the other hand, the crystals did undergo a progressive degradation of quality which was observed on Laue patterns; successive cycles of transformation resulted in increasingly blurred spots.

Patterson found a reproducible macroscopic contraction along the *c*-axis direction on heating through the transition, but it amounted to only 0.2 per cent. This is less than one-tenth the lattice spacing contraction along the same direction. As already discussed, it is generally true of martensitic transformations that they exhibit characteristic and reproducible macroscopic shape changes of transformed volumes and that these differ from the pure lattice deformation. Thus the possibility that the manganese bismuthide transition is martensitic in character is not contradicted by the available data. On the other hand, this question cannot be settled definitely until data are available on the habit-plane orientation and the lattice invariant deformation for comparison with theory.

3.8 Silver-Cadmium

Rather complex phase relations are exhibited in this alloy system at and near the equiatomic composition. At elevated temperatures, alloys in this composition range are disordered body-centered cubic solid solutions. Between 220°C and 440°C the stable phase is hexagonal, and below 220°C the ordered cesium chloride structure appears.

Masson and Barrett[63] have shown that this ordered structure undergoes further transformation if either cooled or cold-worked. They found the structure produced by cooling to be orthorhombic, related to the cubic unit cell by a distortion of a few per cent plus some rearrangement of the atoms within the unit cell. Masson[64] investigated the dependence of the cubic-orthorhombic transformation temperature on composition. He found that, as the silver content decreases from 53.4 to 50.9 at.%, the M_s decreases from 138° to 87°K.

Masson and Barrett[63] found that transformation of the cubic phase can be induced by cold work at even room temperature. In this case the product has a close-packed hexagonal structure. This latter structure is unstable and reverts to the cubic phase on aging at room temperature.

Masson[65] studied the rate of the cph-to-cubic reversion as a function of temperature and composition. He concluded that the process is thermally activated with an activation energy of about 23 kcal/mole. The rate is a maximum at the equiatomic composition: a particularly marked decrease in rate was found as the silver content was increased.

3.9 Silver-Zinc

The beta-phase region of the silver-zinc system resembles that of the copper-zinc system in that long-range order does not exist at elevated temperatures, but appears on cooling through a critical temperature. In the case of beta brass, long-range order of the cesium chloride type is found below about 460°C (the temperature varies slightly with composition). At room temperature, the degree of long-range order is very high even for very high rates of cooling. In other words, the ordering process on cooling cannot be suppressed even by a severe quench.

In the case of the silver-zinc alloy the ordered cesium chloride structure is obtained at room temperature *only* if the specimen is cooled rapidly from a temperature above 275°C. Slow cooling to room temperature produces what is referred to as the zeta phase.

It has been reported by Kitchingman and Buckley[66] that the equilibrium temperature for zeta-phase formation is 273°C, but Muldawer[67] found the critical temperature for the cesium-chloride type of ordering to be 272°C. These figures indicate that the bonding energies of the two phases are very nearly equal; this has been shown directly by Orr and Rovel[68] by liquid tin solution calorimetry.

The zeta phase has been determined by Edmunds and Qurashi[69] to have a centrosymmetrical hexagonal unit cell containing nine atoms. The lattice parameters for the 50 at.% alloy are: $a = 7.636$ Å and $c = 2.8197$ Å. They also pointed out that this structure is very closely related to that of the beta phase, except for a peculiar type of ordering of the silver and zinc atoms on the lattice sites. Thus the beta phase can also be described in terms of a hexagonal unit cell containing nine atoms; for it the lattice parameters are $a = 7.7299$ Å and $c = 2.7329$ Å. These values are, respectively, only 1.2 per cent greater and 3.1 per cent less than those for the zeta phase.

This close resemblance of the beta and zeta structures had actually been reported much earlier by Weerts.[70] He determined by X-ray diffraction that on transformation one of the four close-packed ⟨111⟩ directions of the beta phase becomes the c-axis of the zeta phase. A beta single crystal can thus transform to any one of four differently oriented zeta crystals.

The type of order found by Edmunds and Qurashi can be described as follows. One-third of the lines of atoms parallel to the c-axis consist solely of zinc atoms. The silver atoms and the rest of the zinc atoms are randomly distributed on the remaining atom sites. The relationship between the silver and zinc positions in the two structures has been described by Kitchingman[71] in terms of the distribution of these atoms among sites on (112) planes.

In some respects this transformation resembles that for the alloy CuAu discussed by Smith and Bowles.[50] Both involve a change in order and a change of lattice symmetry such that the product phase can consist of differently oriented domains whose crystal axes exhibit a twin relationship. We might, therefore, anticipate in this case also a transformation behavior from an initial single crystal to stacks of fine twins, in accordance with the phenomenological theory discussed in the introduction to this chapter. Experimental studies by Merriman[72] have not, however, borne out this expectation. It is found instead that after partial transformation the zeta-phase regions are single-crystalline rather than twinned in character. Moreover, the shapes of the transformed regions do not correspond to the shape expected for a macroscopically invariant plane strain. It, therefore, appears that this transformation does not fit into the framework of the phenomenological theory of martensite reactions.

3.10 Titanium-Nickel

The existence of three intermetallic compounds in this alloy system with the compositions Ti_2Ni, TiNi, and $TiNi_3$ has been reported by a series of investigators. In the present connection the compound with the equiatomic composition is of particular interest, because experimental results have been published which suggest that it undergoes a martensitic transformation on cooling to room temperature. The most striking evidence for the existence of a transformation of this type is given by Buehler, Gilfrich, and Wiley.[73] They reported that, if an initially straight wire of this composition is deformed at room temperature, as by coiling into a helix, and is then heated, the wire immediately reverts to its original shape at about 65°C. The process can apparently be repeated an indefinite number of times. This phenomenon had been previously observed in gold-cadmium alloys, discussed in Section 3.4. In the latter case, the spontaneous reversion to the original shape on heating could be accounted for in terms of the crystallography of the transformation. On cooling the gold-cadmium alloy transforms from a cubic to an orthorhombic structure. A cubic single crystal thereby changes not into an orthorhombic single crystal but into a polycrystalline structure, which contains, in general, all of the six possible variants of the orientation of the orthorhombic crystal axes. Application of stress to the sample can then produce a shape change by causing motion of the twin interfaces and a change in the relative amounts of the six possible variants. Reheating the sample through the transformation temperature, however, causes all six variants of the orthorhombic structure to revert to the original cubic single crystal. Thus the original specimen shape is regained.

It is not yet known whether the uncoiling of the TiNi helix can be accounted for in a similar manner, for in this case the nature of the transformation has not yet been established. All investigators suggest that above its transformation temperature TiNi has the cesium chloride structure. Divergent conclusions have been published for the low-temperature structure. Purdy and Parr[74] analyzed powder patterns and concluded that this structure (which they call the "pi phase") is hexagonal with the parameters $a = 4.572$ Å and $c = 4.66$ Å. Buehler, Gilfrich, and Wiley, on the other hand, conclude from their X-ray diffraction studies that the transformation on cooling consists of the appearance of a mixture of the TiNi and Ti_2Ni phases, if a tensile stress is present. If the sample is under a compressive stress, the product is a mixture of TiNi and $TiNi_3$. It seems clear that further experimental studies of the nature of this transformation and of the crystal structure and microstructure of the low-temperature phase are needed.

4. MASSIVE TRANSFORMATIONS

The description "mass transformation" was coined by Greninger[12] when he noted that "an alloy (Cu–Al) of 9.3 per cent Al undergoes a "mass transformation" to α (when rapidly quenched from a high temperature." However, Phillips[75] in 1930 reported that "conversion (by quenching) from beta to alpha in brass takes place with very great rapidity if there is no change in composition." The resulting alpha phase in a Cu-37.78% Zn alloy observed by Phillips consisted of large irregular masses. More recent studies by Hull and Garwood[22] of Cu–Zn alloys, and by Massalski[76] of Cu–Ga, Cu–Zn, Cu–Zn–Ga, and Cu–Ga–Ge alloys have somewhat clarified the general picture concerning the nature and characteristics of massive transformations.

While studying a Cu-38.70% Zn alloy, Hull and Garwood[22] observed that a slow quench from 850°C (all β at this temperature) resulted in parallel needles of α, which grew inwardly from β-grain boundaries. With faster quenches, massive alpha formed. According to them, no massive phase formed during the slower quenches because the beta matrix was enriched in Zn content to 39.0 per cent (the limit of solid solubility of Zn in Cu at ~450°C) by the formation of α-needles. This would imply that both the parent and product phases must exist (at different temperatures, of course) at the same composition, or at overlapping compositions, for a massive transformation to occur. However, Massalski's work does not bear this out for Cu–Ga alloys; the β-phase in Cu–Ga alloys will massively transform into the ξ-phase which is only stable, in equilibrium, at lower solute contents. In fact, Massalski's lattice parameter measurements show that the massively formed ξ-phase corresponds *in structure* to the equilibrium ξ_1-phase, but the massive phase is supersaturated to compositions which lie outside the normal homogeneity limits. It would, therefore, appear that a composition overlap is not necessary for a massive transformation. Because the faster quenching rates employed by Hull and Garwood resulted in a decrease in the amount of alpha formed (mostly β was retained), it is clear that massive transformations are suppressible.

Hull and Garwood found that the boundaries between massive α and the parent β consisted of irregular edges made up of "perfectly straight" sections. Although their crystallographic observations were limited to single-surface trace analysis, some of the traces corresponded to low-index (high-density) planes of the parent β, and some others were consistent with $\sim\{155\}_\beta$, the irrational habit plane of the β-brass martensite which forms at a lower temperature. Because of this latter observation, it was suggested that the invariant plane-strain criterion for martensitic transformations should also apply to the formation of massive α. Surface upheavals were not observed for massive α, but their absence was attributed by Hull and Garwood to the strain associated with the coherent interface being rapidly relieved by thermally activated processes. However, it is the atomic movements themselves which define the shape (invariant-plane) strain, that is, surface relief, so that this explanation does not seem to be applicable. Moreover, the linear regions of the massive α-phase were reported parallel to those of the martensite plates, that is, the same habit plane, despite the fact that the crystal structures of the massive α and martensite are different. Because the crystal structures are different, a different lattice deformation would be expected, and the observation of the same apparent habit plane seems more coincidental than meaningful. In addition, as Hull and Garwood pointed out, the same habit plane is also involved when the β-phase is *isothermally* transformed to an fcc "precipitate" at temperatures as high as 350°C. The significance of the straight interfaces for the massive α-blocks is evidently still unresolved.

Massalski[76] worked on Cu–Ga and Cu–Zn alloys with ternary additions and concluded that the ξ-phase, which forms massively from hypereutectoid β alloys during quenching, bears no definite lattice-orientation relationship to its parent. By adjusting the quenching rate, it was possible to precipitate the equilibrium γ-phase within the β-grains prior to the transformation of β into massive ξ. Because the γ-phase precipitated on $\{110\}$ planes of the β, the distribution of the γ-"rosettes" was used to determine the orientation of the original β-grains. When the orientation of the massive ξ was determined by back-reflection X-ray analysis, its orientation relative to the β could be determined. In this way, it was found that ξ bears no specific orientation to β. Massalski also showed that the massive ξ-phase grows readily across β-grain boundaries without changing its orientation.

As determined for massive α in Cu–Zn alloys,[22] the massively formed ξ-phase in Cu–Ga alloys showed no surface relief effects.

When the quench is rapid enough, the β Cu–Ga alloys transform martensitically. From metallographic analyses after intermediate quenching rates, Massalski suggested that the martensitic and massive processes appear to merge into one another. He could not detect a demarcation line between the two phases (in a near-eutectoid alloy), which apparently formed simultaneously.

Because of the rapidity of growth of the massive ξ-phase it is probable that the β-to-ξ (massive) transformation occurs at a relatively high temperature. The isothermal transformation studies conducted by Massalski showed that this is indeed so. By quenching a near-eutectoid (23.86 at.% Ga) alloy from 750°C (all β) to 600°C, holding at 600°C for 5 minutes (eutectoid temperature 616°C), and then quenching in iced brine, it was found that γ precipitated in the massive ξ. The absence of the usual γ-rosettes indicated that no precipitation of γ occurred during cooling through the two-phase ($\beta + \gamma$) region, and that the β-to-ξ massive transformation was completed *during* the transformation of the specimens into the bath at 600°C.

Other findings by Massalski led to the following conclusions:

1. Massive transformations are diffusionless in the sense that no change in bulk composition occurs on quenching; the movement of individual atoms is likely less than a few interatomic distances.
2. Massive transformations experience a nucleation difficulty, but, once nucleated, proceed by the rapid advancement of incoherent boundaries. In certain cases, the nucleation difficulty can be overcome by partially coherent nucleation (hypoeutectoid Cu–Ga alloys) resembling martensitic nucleation, followed by incoherent growth.

Finally, Massalski's observations on a Cu-37.95 wt.% Zn alloy show that the lattice parameter of the massively formed α-phase is the same as that which was obtained by extrapolation of the room-temperature lattice parameters (versus composition) of the equilibrium α-phase. Although the Cu-37.95% Zn alloy overlaps the α-phase region, there is retrograde solubility of Zn in Cu below ~450°C.

Subsequent cinéfilm observations on Cu–Ga alloys by Massalski and coworkers[77] demonstrate clearly that the massive-phase interfaces propagate discontinuously with time and leave shearlike traces behind. The linear interfaces reported by Garwood and Hull[28] for Cu–Zn massive α are clearly seen for the massive ξ-phase in this film.

REFERENCES

1. Bilby B. A., and J. W. Christian, *The Mechanism of Phase Transformations in Metals.* The Institute of Metals, London, 1956, p. 121.
2. Bowles J. S., and J. K. Mackenzie, *Acta Met.* **2** (1954) 129.
3. Bowles J. S., and J. K. MacKenzie, *ibid.*, 138.
4. Bowles J. S., and J. K. Mackenzie, *ibid*,. 224.
5. Wechsler M. S., D. S. Lieberman, and T. A. Read, *Trans. AIME* **197** (1953) 1503.
6. Bilby B. A., and J. W. Christian, *J. Iron. Steel Inst.* **197** (1961) 122.
7. Christian J. W., *The Theory of Transformations in Metals and Alloys.* Pergamon Press, Oxford, 1965.
8. Wayman C. M., *Introduction to the Crystallography of Martensitic Transformations.* The Macmillan Company, New York, 1964.
9. Greninger A. B., and A. R. Troiano *Trans. AIME*, **185** (1949) 590.
10. Frank F. C., *Acta Met.* **1** (1953) 15.
11. Hansen M., *Constitution of Binary Alloys.* McGraw-Hill Book Co., New York, 1958.
12. Greninger A. B., *Trans. AIME* **133** (1939) 204.
13. Gawranek V., E. Kaminsky, and G. Kurdjumov, *Metallwirtschaft* **15** (1936) 370.
14. Swann P. R., and H. Warlimont, *Acta Met.* **11** (1963) 511.
15. Gridnew G., and G. V. Kurdjumov, *Metallwirtschaft*, **15** (1936) 437.
16. Isaitchev I., E. Kaminsky, and G. Kurdjumov, *Trans. AIME* **128** (1938) 337.
17. Isaitchev I., E. Kaminsky, and G. Kurdjumov, *Trans. AIME* **128** (1938) 361.
18. Nakanishi N., *Trans. Japan Inst. Metals*, **2** (1961) 85.
19. Bain E. C., *Trans. AIME* **70** (1924) 25.
20. Mackenzie J. K., and J. S. Bowles, *Acta Met.* **5** (1957) 137.
21. Wilkens M., and H. Warlimont, *Acta Met.* **11** (1963) 1099.
22. Hull D., and R. D. Garwood, *The Mechanism of Phase Transformations in Metals.* The Institute of Metals, London, 1965, p. 219.
23. Hull D., and R. D. Garwood, *J. Inst. Met.* **86** (1957) 485.
24. Subramanya B. S., G. S. Baker, D. S. Lieberman and T. A. Read, *J. Australian Inst. Metals* **6** (1961) 3.
25. Smith D. W., *Trans. AIME* **104** (1933) 48.
26. Greninger A. B., and V. G. Mooradian, *Trans. AIME* **128** (1938) 337.

27. Isaitchev I., and V. Miretskii, *Zh. Tekhn. Fiz.* **8** (1938) 1333.
28. Garwood R. D., and D. Hull, *Acta Met.* **6** (1958) 98.
29. Jolley W., and D. Hull *J. Inst. Met.* **92** (1963–64) 129.
30. Kunze G., *Z. Metallk.* **53** (1962) 329, 396, 565.
31. Masson D. B., and R. K. Govila *Z. Metallk.* **54** (1963) 293.
32. Titchener A. L., and M. B. Bever, *Trans. AIME* **200** (1954) 303.
33. Hull D., *Phil. Mag.* **7** (1962) 537.
34. Reynolds J. E., and M. B. Bever, *Trans. AIME* **194** (1952) 1965.
35. Massalski T. B., and C. S. Barrett, *Trans. AIME* **209** (1957) 455.
36. Hornbogen E., A. Segmuller, and G. Wassermann, *Z. Metallk.* **48** (1957) 379.
37. Olander A., *Z. Krist.* **A83** (1932) 145.
38. Bystrom A., and K. E. Almin, *Acta. Chem. Scand.* **1** (1947) 73.
39. Lieberman D. S., M. S. Wechsler, and T. A. Read, *J. Appl. Phys.* **26** (1955) 473.
40. Chang L. C., *Acta Cryst.* **4** (1951) 320.
41. Chang L. C., and T. A. Read, *Trans. AIME* **191** (1951) 47.
42. Köster W., and A. Schneider, *Z. Metallk.* **32** (1940) 156.
43. Chen Chih-Wen, 1954. Ph.D. Thesis, Columbia University.
44. Wilkens M., *Naturwissenschaften* **44** (1957) 1.
45. Wechsler M. S., and T. A. Read, *J. Appl. Phys.* **27** (1956) 194.
46. Wechsler M. S., *Acta Met.* **5** (1957) 150.
47. Sturm W. J., and M. S. Wechsler, *J. Appl. Phys.* **28** (1957) 1509.
48. Mullendore J. M., 1961. Ph.D. Thesis, University of Illinois.
49. Intrater J., L. C. Chang, and T. A. Read, *Phys. Rev.* **86** (1952) 598.
50. Smith R., and J. S. Bowles, *Acta Met.* **8** (1960) 405.
51. Bowles J. S., and A. S. Malin, *J. Australian Inst. Metals.* **5** (1960) 131.
52. Marcinkowski M. J., and N. Brown, *Phil. Mag.* **8** (1963) 891.
53. Raub E., H. Zwicker, and H. Baur, *Z. Metallk.* **44**·(1953) 312.
54. Bacon G. E., and R. Street, *Proc. Phys. Soc.* **72** (1958) 470.
55. Smith J. H., and P. Gaunt, *Acta Met.* **9** (1961) 819.
56. Bowles J. S., C. S. Barrett, and L. Guttman, *Trans. AIME* **188** (1950) 1478.
57. Basinski Z. S., and J. W. Christian, *J. Inst. Metals.* **80** (1951) 659; *Acta Met.* **2** (1954) 148.
58. Burkhart M. W., and T. A. Read, *Trans. AIME* **197** (1953) 1516.
59. Bumm H., and U. Dehlinger, *Metallwirtschaft* **13** (1934) 23.
60. Basinski Z. S., and W. B. Pearson, *Can. J. Phys.* **36** (1958) 1017.
61. Willis B. T. M., and H. P. Rooksby, *Proc. Phys. Soc.* **B67** (1954) 290.
62. Patterson R. L., 1961. M.S. Thesis, University of Illinois.
63. Masson D. B., and C. S. Barrett, *Trans. AIME* **212** (1958) 260.
64. Masson D. B., *Trans. AIME* **218** (1960) 94.
65. Masson D. B., *Acta Met.* **8** (1960) 71.
66. Kitchingman W. J., and J. I. Buckley, *Acta Met.* **8** (1960) 373.
67. Muldawer L., *J. Appl. Phys.* **22** (1951) 663.
68. Orr R. L., and J. Rovel, *Acta Met.* **10** (1962) 935.
69. Edmunds I. G., and M. M. Qurashi, *Acta Cryst.* **4** (1951) 417.
70. Weerts J., *Z. Metallk.* **24** (1932) 265.
71. Kitchingman W. J., *Acta Met.* **10** (1962) 799.
72. Merriman E. A., 1964. M.S. Thesis, University of Illinois.
73. Buehler W. J., J. V. Gilfrich, and R. C. Wiley, *J. Appl. Phys.* **34** (1963) 1475.
74. Purdy G. R., and J. Gordon Parr, *Trans. AIME* **221** (1961) 636.
75. Phillips A. H., *Trans. AIME* **89** (1930) 194.
76. Massalski T. B., *Acta Met.* **6** (1958) 243.
77. Massalski T. B. et al., presented at AIME Meeting, New York, February 1964.

part VII

Properties and Applications

chapter 23

Elastic Behavior*

LÉON GUILLET
École Centrale des Arts et Manufactures
Paris, France

R. Le ROUX
Conservatoire National des Arts et Métiers
Paris, France

1. INTRODUCTION

The mathematical theory of elasticity of crystals is well known. We, therefore, will not review the definitions of "strain components" or "stress components." Voight, in generalizing Hooke's law, concluded that the stress components are linear functions of the strain components and so defined the "elastic coefficients" C_{11}, C_{12}, etc., of the crystal. In the case of crystals of low symmetry these coefficients are quite numerous, but in cubic crystals they are reduced to three: C_{11}, C_{12}, and C_{44}. The following is a review of the interpretation of these three coefficients.

C_{44} is a measure of the deformation resistance of the crystal, while shear is applied in the [010] direction of the (100) plane.

C_{11} and C_{12} do not have a direct interpretation, but linear combinations of these coefficients have a simple interpretation, namely, $(C_{11} + 2C_{12})/3$ is equal to the modulus of compressibility and is a measure of the resistance of the crystal to deformation by hydrostatic compression.

$(C_{11} - C_{12})/2$ represents the resistance to deformation by shear across the (110) plane in the $[1\bar{1}0]$ direction. Postulating that the strain components are linear functions of the stress components, we may also define elastic constants S_{11}, S_{12}, and S_{44}, which are related

* Translated from the French by A. J. Peat.

to the elastic coefficients by the simple formulas

$$S_{11} - S_{12} = \frac{1}{C_{11} - C_{12}}$$

$$S_{11} + 2S_{12} = \frac{1}{C_{11} + 2C_{12}}$$

$$S_{44} = \frac{1}{C_{44}}$$

2. VOIGHT ELASTIC COEFFICIENTS

There are few intermetallic compounds for which the elastic coefficients are actually known. To determine these experimentally, measurements are made of the rate of propagation of elastic waves in different simple crystallographic directions. Due to the fact that the frequencies used are generally high, the coefficients so determined correspond to adiabatic conditions.

In ionic compounds, the more important interatomic forces derive from the electrostatic interaction of the ions and proceed along the right angles which join the atom centers. We have shown that under these conditions the two coefficients C_{12} and C_{44} should become equal; experiments verify this equality in the case of alkali halides (interstitial fcc lattice), as shown in the values of Table 1.

Some semiconducting compounds, in which the lattice is diamond cubic (ZnS type), have

TABLE 1

Compound	C_{12} (10¹² dynes/cm²)	C_{44} (10¹² dynes/cm²)
NaCl[1,2]	0.123	0.126
NaBr[3]	0.131	0.133
KCl[2]	0.060	0.063
KBr[1,2]	0.054	0.051

TABLE 2

Compound	C_{11} (10¹² dynes/cm²)	C_{12} (10¹² dynes/cm²)	C_{44} (10¹² dynes/cm²)	Ionicity λ
GaAs[4]	1.192	0.598	0.538	0.37
GaSb[5]	0.885	0.404	0.432	0.12
InSb[5]	0.672	0.367	0.302	0.25
AlSb[6]	0.894	0.442	0.413	0

interatomic bonds of covalent nature but also exhibit a certain "ionicity." For crystals of these compounds, the coefficients C_{12} and C_{44} are no longer equal. Table 2 shows the results of recent experiments as well as the degree of ionicity, λ, as defined by Suchet.[7]

This parameter, λ, is the sum of two terms: the atomic ionicity, λ_0, which depends on the crystallographic structure, and the charge displaced by the bond, q/c, (where q designates the effective charge carried by each atom and c the number of electron pairs which would be necessary to establish purely covalent bonds). For a pair of atoms MX, q may be estimated by the formula

$$q_i = n_i - 0.01185 n_i \left\{ \frac{(Z_M)}{(r_X)} + \frac{(Z_X)}{(r_M)} \right\}$$

where n_i, Z_i, r_i are the charge, atomic number, and radius of ion i. The degree of ionicity may also be estimated from the speed of sound (which is a function of the elastic constants and which undoubtedly varies with λ according to an exponential function).[8]

The predominant factors for metallic cohesion are not derived from central forces. No simple relationship between these Voight coefficients could be established for crystals of metallic character. Table 3 gives the values of these constants for several compounds, in which the structure is ordered and which involve in most cases a high value of elastic

TABLE 3

Compound	C_{11} (10¹² dynes/cm²)	C_{12} (10¹² dynes/cm²)	C_{44} (10¹² dynes/cm²)	$\frac{2C_{44}}{C_{11} - C_{12}}$	Lattice
AuCd[10]	0.903	0.829	0.439	11.8	bcc
AuZn[11]	0.50	0.375	0.370	2.8	bcc
CuZn[4]	1.291	1.097	0.824	8.4	bcc
Cu_3Au[13]	1.907	1.383	0.663	2.5	fcc

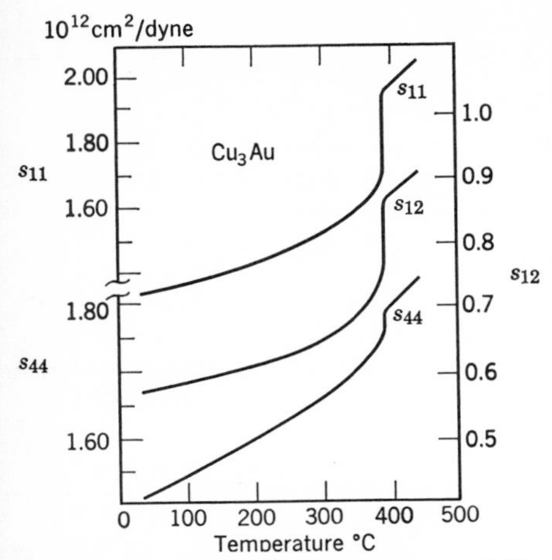

Fig. 1. Variation of elastic constants of Cu_3Au with temperature. (After Siegel.)

anisotropy.* Because for an isotropic body we have the relationship $C_{11} - C_{12} = 2C_{44}$, the elastic anisotropy of a crystal may be defined by the ratio $2C_{44}/(C_{11} - C_{12})$. A strong anisotropy implies a relatively small modulus of rigidity when the crystal is sheared along the (110) plane in the $[1\bar{1}0]$ direction.

The degree of order in these crystals decreases as the temperature is raised and falls considerably in the neighborhood of the critical temperature. The variations of the elastic coefficients with temperature are parallel. The

* There is none for $CaMg_2$ (see Reference 9).

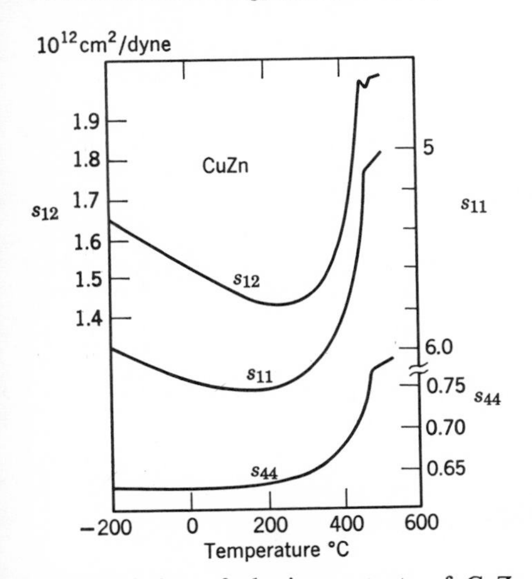

Fig. 2. Variation of elastic constants of CuZn with temperature. (After Siegel, Rinehart, and Good.)

elastic constants show an inverse behavior (Figs. 1 and 2). At the critical temperature the elastic anisotropy is diminished.[13–16]

3. YOUNG'S ELASTIC MODULUS

For an elastically isotropic body the constants which correlate strain with stress are Young's modulus (in pure tension) and Coulomb's modulus (in pure shear). The same is true if the body is pseudoisotropic by compensation. This is the case for a polycrystalline aggregate, in which the grains are small relative to the total volume and are randomly oriented. These conditions are particularly necessary, if the elastic anisotropy of the crystal is high.[16,17]

Young's modulus is easily determined, because it is related to the characteristic vibration frequency of a cylindrical specimen by simple formulas. Resonance techniques are generally used to measure this frequency. Investigation of this modulus in electron compounds and in Laves phases confirmed the atomic character of this value.

The metallic phases are not defined on isothermal modulus-concentration curves by singular points. So it is with β-phases, such as CuZn and Cu_3Al, which have electron-to-atom ratios equal to $\frac{3}{2}$ according to Hume-Rothery; it is also the case for compact Laves phases, such as Al_2Cu, Cu_2Mg, and Zn_2Mg. Table 4 shows that their moduli may be approximately calculated in the same way as for metallic compositions by rule of mixtures.†[,18]

The γ-phases (large cubic cell) of a less metallic character, which correspond to an electron-to-atom ratio of 21/13, have exceptionally high moduli, as shown in Figs. 3, 4, and 5 for Cu_5Zn_8, Cu_9Al_4, and $Cu_{31}Sn_8$.[19] It would be the same for Ag_5Zn_8 and $Cu_{31}Si_8$. Cubic ionic crystals with large unit cells and weak coordination (CaF_2 type), such as Mg_2Sn and Mg_2Pb, are also shown by very acute maxima on isothermal modulus-composition curves (Fig. 6).[20–22]

Finally, the moduli of the semiconducting compounds previously noted have not been determined, but they may be calculated from Voight's elastic coefficients by a formula, which gives valid results in the case of phases with

† For $CaMg_2$, a hexagonal Laves phase, we have shown that compressibility conspicuously obeys the rule of mixtures.[9]

TABLE 4

Composition	Young's Modulus, Kg/mm²		Deviation Per Cent	Lattice
	Experimental Value	Calculated Value		
CuZn	10,000	10,700	7	bcc
Cu_3Al	8,300 (at 600°C)	7,900 (at 600°C)	5	bcc
Al_2Cu	9,640	9,200	4	body-centered tetragonal
Zn_2Mg	7,000	7,400	5	hexagonal Laves (ababab)
Cu_3Au	11,700	11,400	2.5	fcc
CuAu	10,500	10,250	2.4	fcc

weak elastic anisotropy

$$E = \frac{(C_{11} + 2C_{12})(C_{11} - C_{12} + 3C_{44})}{2C_{11} + 3C_{12} + C_{44}}$$

or by another given by Laurent and Eudier,[23] which is a little more complicated and for which the results are close to those that we show. Table 5 permits a comparison of the values of the moduli of these compositions, as calculated by the preceding formula, with those which would correspond to the rule of mixtures. It may be seen that the deviation is considerable as foreseen, according to the nature of the interatomic bonds of these compositions.

The elastic moduli of intermetallic compounds diminish in a continuous fashion when the temperature is raised in the absence of a crystallographic transformation. The low-temperature coefficient of the modulus becomes lower the more refractory the compound. The beryllides are particularly remarkable in this respect. Their high-oxidation resistance and their significant mechanical strength at high

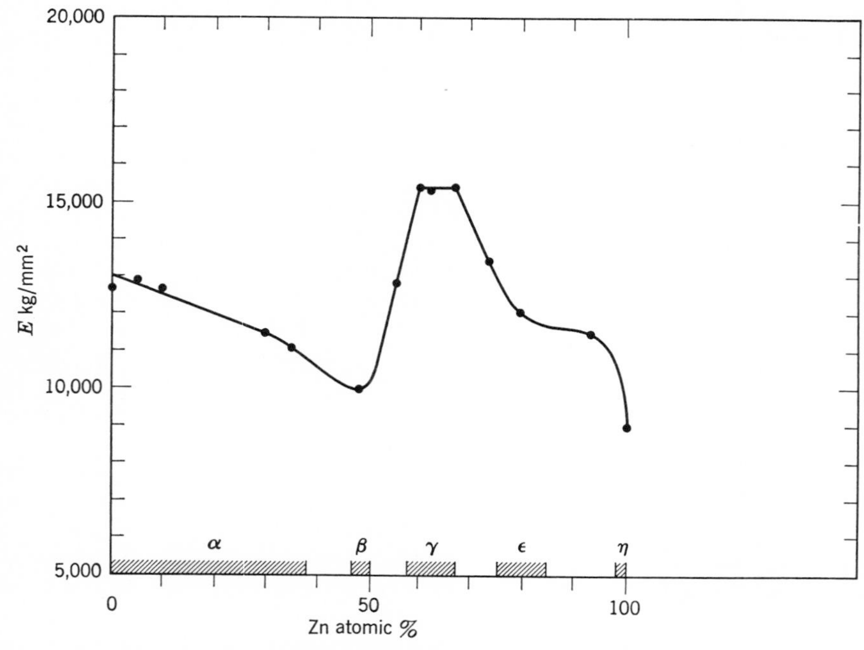

Fig. 3. Variation of elastic modulus with composition for Cu–Zn alloys. (After R. Cabarat, L. Guillet, and R. Le Roux.)

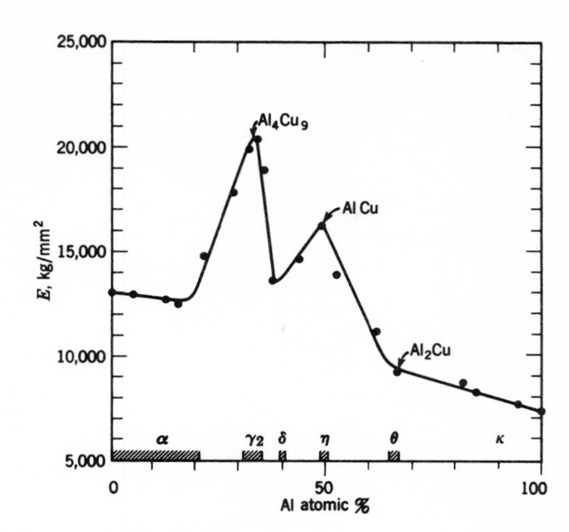

Fig. 4. Variation of elastic modulus with composition for Cu–Al alloys. (After R. Cabarat, L. Guillet, and R. Le Roux.)

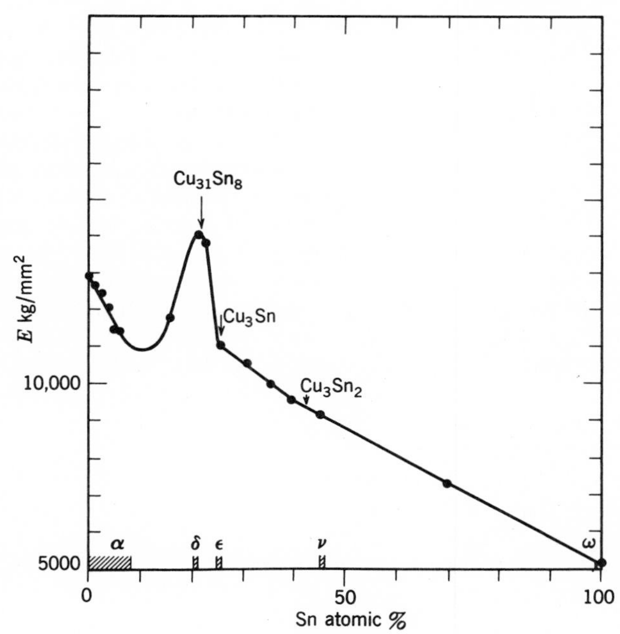

Fig. 5. Variation of elastic modulus with composition for Cu–Sn alloys. (After R. Cabarat, L. Guillet, and R. Le Roux.)

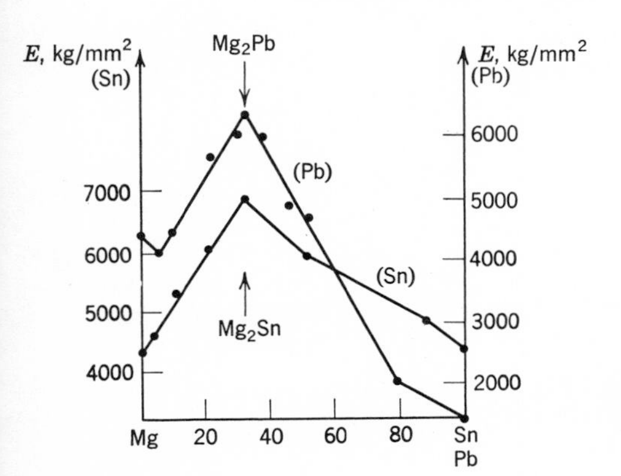

Fig. 6. Variations of elastic modulus of Mg–Pb and Mg–Sn alloys with composition. (After Köster.)

TABLE 5

Composition	E (Voight)	E (rule of mixtures)	Deviation Per Cent
	kg/mm²		
GaAs	12,250	5,000	145
GaSb	8,990	3,300	170
InSb	6,300	3,370	70
AlSb	8,720	6,550	33

TABLE 6

Composition	Young's Modulus, 10^6 psi at Temperature °F: 20	1600	2300	2500	2750
$TaBe_{12}$	45	40	24	13.5	9.5
Ta_2Be_{17}	45	40	26	13	
$ZrBe_{13}$	47	40	40	18.5	10
Zr_2Be_{17}	45	40	25	14.5	10

temperature have attracted the attention of metallurgists in recent years. Their rigidity is not affected by temperatures up to 1600°F. Table 6 gives moduli values of some of these at various temperatures following Stonehouse, Paine, and Beaver.[24]

Transformations in the solid states, in particular, order-disorder and magnetic transformations, are often accompanied by variations in Young's modulus. The ordered states or ferromagnetic states have, in general, the higher moduli.[25] Figure 7 shows the anomalies of the modulus-temperature curves for

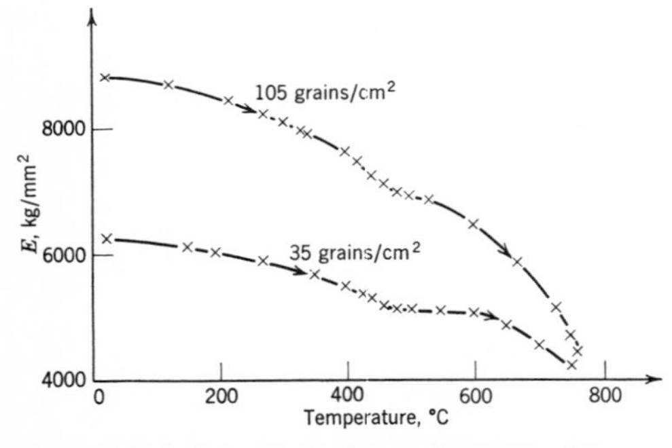

Fig. 7. Variation of elastic modulus for β-CuZn with temperature for two alloys of different grain size. The disappearance of the ordered state above 470°C is manifested by a decline in elastic anisotropy, and the curves rejoin. (After R. Le Roux.)

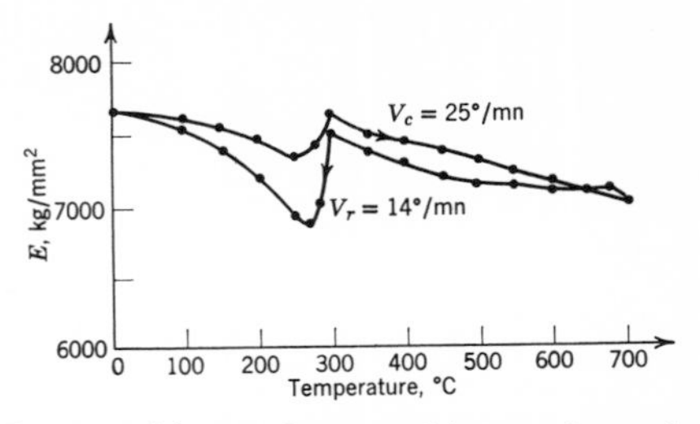

Fig. 8. Evidence of martensitic transformation for Cu_3Al compound by thermoelastic analysis. (After R. Le Roux.)

the compound CuZn after R. Le Roux. Shibuya[26] obtained analogous results for Cu_3Au, confirmed later by Flinn, McManus, and Rayne.[27] Köster also studied Cu_3Pd, and Cu_3Pt[28] as well as Au_3Cd, AuCd, and $AuCd_3$.[29] Investigations on Fe_3Al,[30] Fe_3Pt,[31] and $FeNi_3$[32] resulted in the same conclusions.

Reversible allotropic transformations, such as the martensitic transformations, are also apparent on the modulus-temperature curves (Fig. 8 for Cu_3Al). When transformation causes the appearance (by cooling) and disappearance (by heating) of a γ-phase with high specific modulus, the isothermal variations of the modulus are very marked.[33] Figures 9 and 10 show the effects of the eutectoid reactions:

$$Cu_3Al \rightleftarrows \alpha + Cu_9Al_4 \text{ at } 565°C$$
$$Cu_5Sn \rightleftarrows \alpha + Cu_{31}Sn_8 \text{ at } 525°C$$

W. Köster[25] found analogous curves for the compound AgZn.

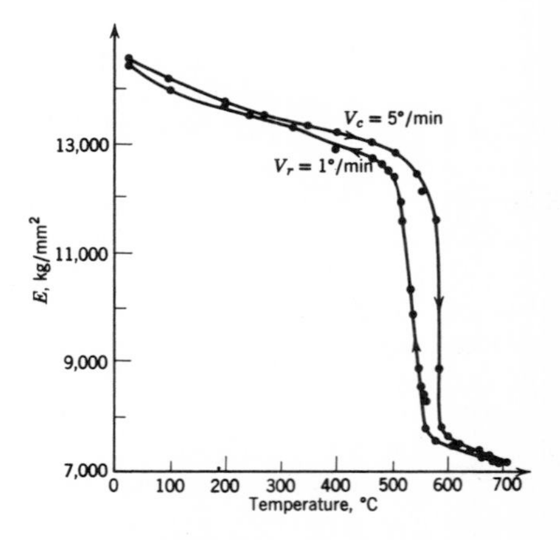

Fig. 9. Thermoelastic analysis of $Cu_3Al \rightleftarrows \alpha + Cu_9Al_4$.

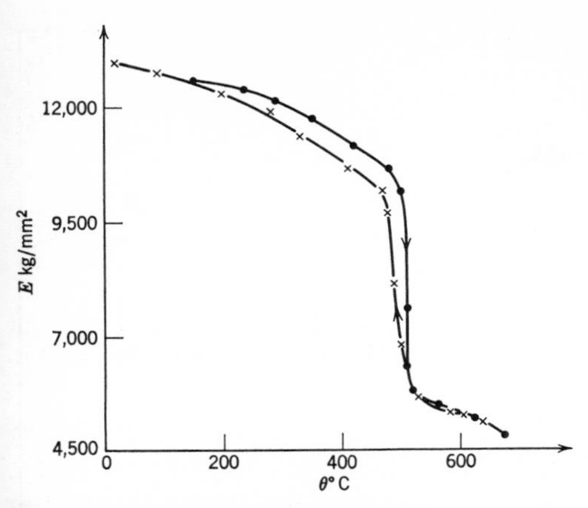

Fig. 10. Thermoelastic analysis of $Cu_5Sn \rightleftarrows \alpha + Cu_{31}Sn_8$. (After R. Cabarat, L. Guillet, and R. Le Roux.)

4. ANELASTICITY

The theory of elasticity is based upon the supposition of a completely instantaneous deformation when stress is applied and a completely instantaneous recovery when stress is relieved. Time factors and relaxation phenomena are not taken into account. In order to obtain a theory which corresponds better to experimental facts, it must be assumed that stresses and their derivatives are linear functions of strains and their derivatives. The mathematical theory of anelasticity[34] is based on the most general type of linear differential equations and homogeneity of the first order between stress and strain. From these equations it may be shown that a periodic stress produces a periodic strain of the same period but with an out-of-phase condition between the two parameters. The difference in phase is a function of the frequency of the excitation and goes through a maximum when the vibration period (or the pulse ω) is of the same order as the "time of relaxation." The stress-to-strain ratio, which is in phase agreement, is the "dynamic" modulus, M_ω, itself a function of the frequency of the oscillations. For high frequencies, it approaches "the instantaneous elastic modulus"; for the very low frequencies, it approaches "the relaxed elastic modulus," the stress-to-strain ratio following total relaxation (Fig. 11).

To measure the phase difference between stress and strain, which characterizes the internal friction, we may observe the progressive decrease of the amplitude of vibrations in free oscillation. This follows an exponential law at a constant frequency. The logarithmic decrement, σ, is the logarithmic ratio of two successive amplitudes. It may be shown that, if the angle φ of the phase difference is small, we have

$$\tan \varphi = \frac{\sigma}{\pi} = Q^{-1}$$

In practice, to determine σ, we note the time τ_n necessary for the amplitude to be reduced to an nth of its initial value; if the vibration frequency is F_0

$$Q^{-1} = \frac{\log n}{\pi F_0 \tau_n}$$

The specimen may also be subjected to forced vibrations and from the resonance curve, the frequency band ΔF, corresponding to the amplitude $A_m/\sqrt{2}$ (A_m being the resonance amplitude) may be determined.

$$Q^{-1} = \frac{\Delta F}{F_0}$$

The two methods are complementary. The first is more convenient when the internal friction is quite weak, the second when it is quite high. The physical interpretation of anelasticity (relaxation by thermal diffusion, relaxation by atomic diffusion, etc.) has been expounded by Zener.[34] The internal friction of intermetallic compounds has been little investigated. A few relative determinations for the Hume-Rothery phases may be found. Methods of measurement and the frequencies used vary among authors.

In polycrystalline aggregates, a source of energy dissipation is the thermoelastic effect and the elastic anisotropy of the crystal. The

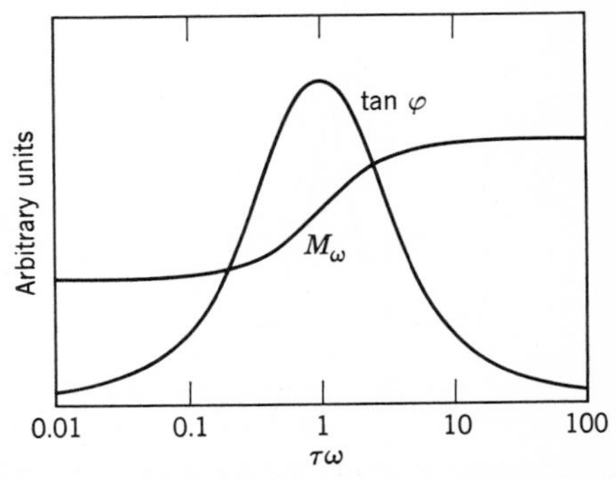

Fig. 11. Effect of frequency on internal friction and dynamic modulus in general. (After Zener.)

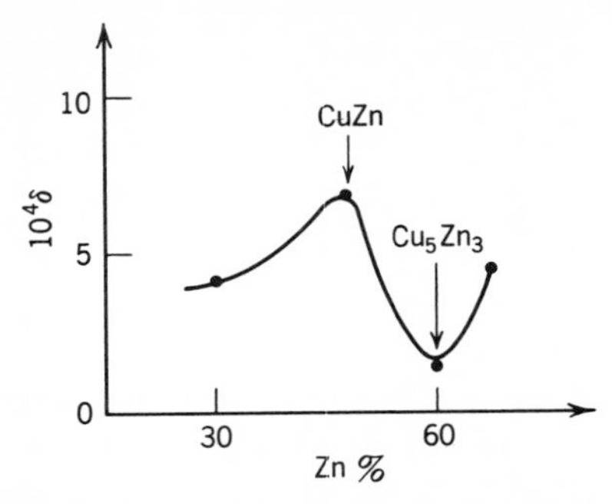

Fig. 12. Variation of internal friction (12,000 cps) with composition for Cu–Zn alloys. (After R. Cabarat, L. Guillet, and R. Le Roux.)

different orientations of the grains have as a result differences in temperature, which are then removed by thermal diffusion. The relaxation time is equal to d^2/D where d is the minimum size of the grains and D the coefficient of thermal diffusivity. The maximum of the internal friction is larger the larger the elastic anisotropy of the crystal. In this way we explain the existence of a maximum[35] of the logarithmic decrement, δ, on the isothermal curves of Figs. 12 and 13 for ordered CuZn and Cu_3Sn compounds[36,37] and a minimum for the Cu_5Zn_8 and Cu_3Sn_8 phases with large cubic cells, in which there is a certain degree of disorder in the distribution of the atoms.

However, when the crystal is subjected to a constraint, there is no heat which diffuses. An elastic stress can, by modifying the equilibrium state between each constitutent and its surroundings, produce a chemical diffusion. In spite of the very weak intensity of the phenomenon, the effect is readily detectable because of

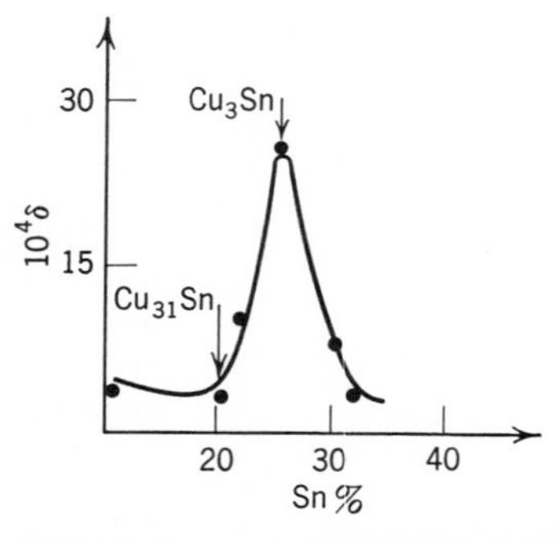

Fig. 13. Variation of internal friction (12,000 cps) with composition for Cu–Sn alloys. (After R. Cabarat, L. Guillet, and R. Le Roux.)

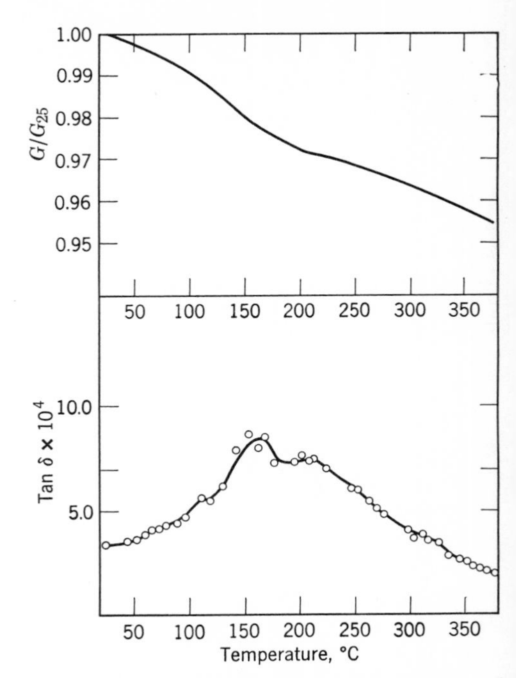

Fig. 14. Variations of elastic modulus and internal friction for InSb compound as a function of temperature. (After R. F. Potter, J. H. Wasilik, and R. B. Flippen.[35]) (Note that the phase difference between stress and strain is designated here by δ rather than φ).

the very large activation energies for diffusion, which can lead to relaxation times of the order of a second. Relaxation by atomic diffusion explains the anomalies which are shown on the internal friction-temperature and modulus-temperature curves when operating at a constant frequency. The time of the relaxation phenomenon varies in effect as exp Q/RT (Q being the activation energy), so that in raising the temperature a decrease in frequency is noted. The curves tan φ versus $\tau\omega$ or tan φ versus T have analogous form (Fig. 14).

To show this type of internal friction, the metal must be subjected to low-frequency oscillations, of the order of 1 cps, in such a way that the period of these oscillations is of the same order of magnitude as that of the relaxation time of the phenomenon.

J. Lulay and C. Wert[38] have shown that the rearrangement of the atoms under the effect of applied stresses is very much influenced by the degree of order that appears naturally in the crystals. If the degree of order is very high, the weak external stresses are not sufficient to

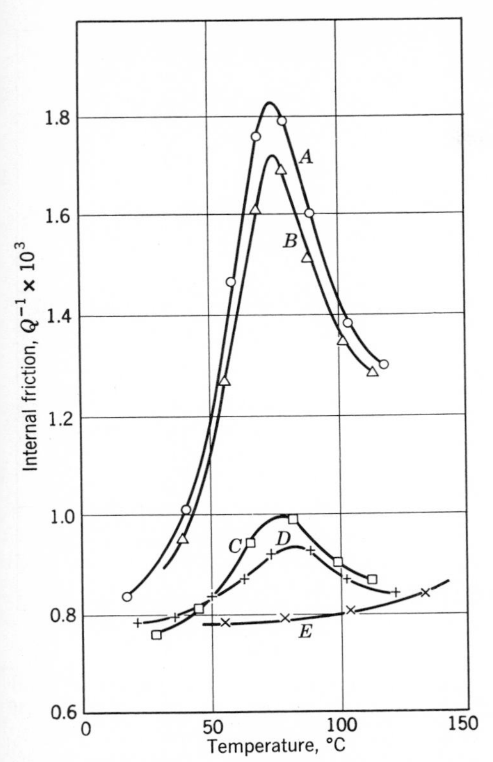

Fig. 15. Influence of composition and heat treatment on the relaxation peak in β-brass at 70°C. ○, polycrystal of batch 4 (45.15 at.% Zn) quenched from 400°C; △, single crystal of batch 4 quenched from 400°C; □, polycrystal of batch 6 (47.64 at.% Zn) quenched from 400°C; +, polycrystal of batch 4 quenched from 600°C; ×, polycrystal of batch 4 slowly cooled from 400°C. Frequency of oscillation 0.9 cps. (After L. M. Clarebrough.)

modify the order perceptibly. Therefore, in magnesium-cadmium alloys, we can show at a temperature near 20°C a peak on the internal-friction curve, the amplitude of which depends very much on the chemical composition of the alloy. This peak becomes negligible at the composition that corresponds to perfectly ordered hexagonal MgCd.

Clarebrough,[39] by subjecting CuZn crystals* to low-frequency oscillations, showed that internal friction increased regularly with temperature if the crystal had been previously slowly cooled from 400°, but showed a maximum at 70° if the crystal had been previously quenched (Fig. 15). He supposed that quenching introduced some disorder in the distribution of the atoms, the zinc atoms being able to occupy two neighboring sites in the [111] directions (Fig. 16), thus causing an elastic distortion of the lattice for they are somewhat larger than the copper atoms. In the absence of applied stress, such pairs of atoms appear at random on the four diagonals of the cube. Under the effect of stress, however, these pairs tend to align themselves along the diagonal that forms the minimum angle with the direction of the stress. Because this change in distribution requires a certain time, the strain lags the stress and thence the existence of internal friction. In order for a pair of atoms to reorient themselves without separation, vacancies must exist in the lattice. It is this fact which explains the presence of an anomaly on the curve obtained with a quenched crystal and its absence on that for a slowly cooled crystal.

Clarebrough also showed another maximum on the internal friction-temperature curve situated at 177°C with the frequency used (Fig. 17). It could also be due to relaxation of external effects through the preferential distribution of the atoms, but here pairs of copper atoms would be involved. In an ordered crystal containing 50 at.% zinc, pairs of neighboring atoms of the same kind would not exist, but they would become possible in a crystal in which the composition deviates a little from stoichiometry. The mechanism would be analogous to that just described, but the preexistence of vacancies would not be necessary. The anomaly is insensitive to heat treatment, because the alloy rapidly reorders following quenching and all disorder in the distribution of the atoms disappears below 177°C (for the rate of heating used). The activation energy measured from the curve is equal to 31 kcal/mole; it is close to that calculated by Kuper

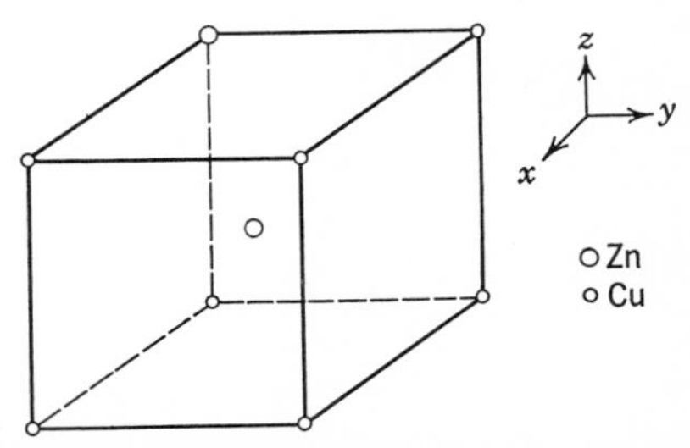

Fig. 16. Unit cell of β-brass with two zinc atoms as nearest neighbors.

* These crystals actually contained somewhat less than 50 a/o Zn.

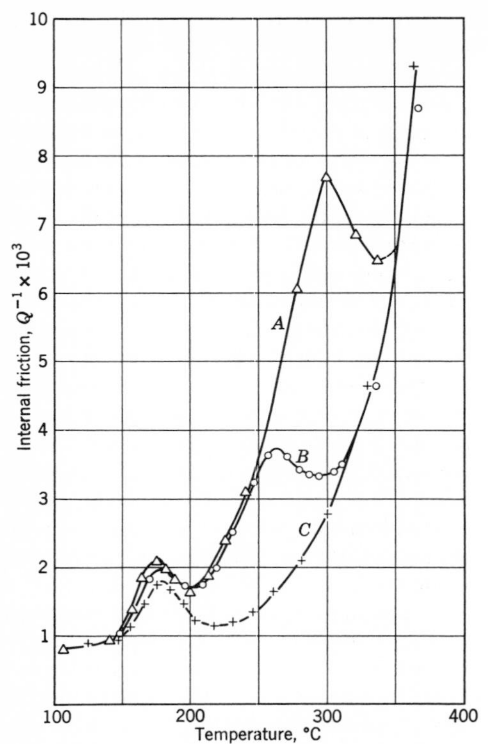

Fig. 17. Internal friction (Q^{-1}) as a function of temperature for a polycrystal and a single crystal of batch 4 (45.15 at.% Zn). △, polycrystal quenched from 600°C; ○, the same specimen slowly cooled from 400°C; +, single crystal slowly cooled from 400°C. Frequency of oscillation approximately 1 cps. The small peaks between 150 and 200°C are due to the phenomenon giving the relaxation peak at 177°C. (After L. M. Clarebrough.)

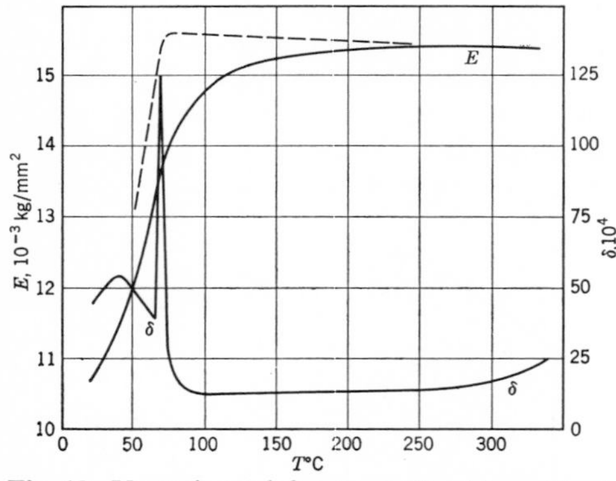

Fig. 18. Young's modulus curve E and logarithmic decrement δ for an alloy Pt 58 wt.%, Fe 42 wt.%. ——— in a nonmagnetic state, as a function of T°C. – – – calculated Young's modulus. (After Kataev and Sirota.)

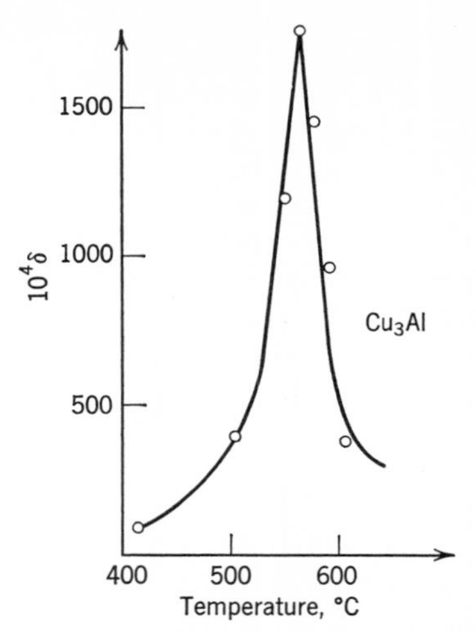

Fig. 19. Variation of internal friction in the neighborhood of the transformation $Cu_3Al \rightleftarrows \alpha + Cu_9Al_4$. (After R. Cabarat, P. Gance, L. Guillet, and R. Le Roux.)

and Tomizuka[40] for the diffusion of copper in β-brass at 350°C.

Finally, on the curve of Fig. 17 we note a maximum at 300°C, which exists only for polycrystalline specimens. This maximum is related to the viscous relaxation of grain boundaries, a phenomenon which was much studied by Kê for the case of metals but which presents no particular interest here because it is not peculiar to intermetallic compounds. The rapid increase in internal friction when

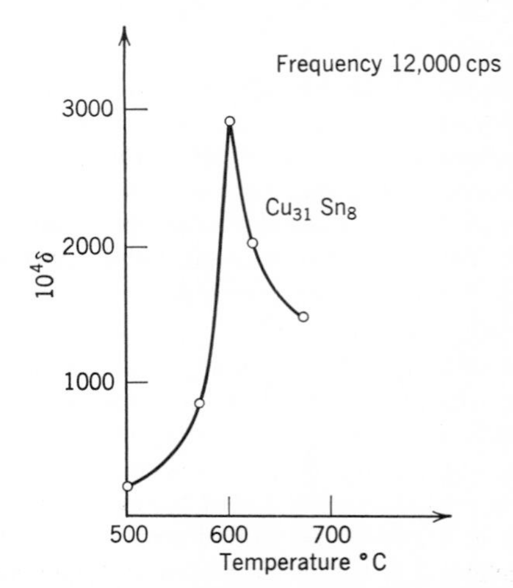

Fig. 20. Variation of internal friction in the neighborhood of the transformation $Cu_5Sn \rightleftarrows \alpha + Cu_{31}Sn_8$.

approaching the critical temperature, that is, that temperature at which long-range order disappears, is in agreement with the results obtained by Köster[25,29,42] and with the theoretical prediction of Zener.[29]

Magnetic transformations are also accompanied by a considerable increase in internal friction. Kataev and Sirota[31] have shown this in studying the temperature dependence of the internal friction of the disordered alloy Fe_3Pt (Curie point 71°C) (Fig. 18).

There are also allotropic transformations which might show anelastic effects. However, this question has not yet been studied theoretically; the structural alteration could, at higher temperatures, cause dislocation motion and hence new sources for dissipation of energy (Figs. 19 and 20).[43,25]

REFERENCES

1. Huntington H. B., *Phys. Rev.* **72** (1947) 321.
2. Galt J. K., *Phys. Rev.* **73** (1948) 1460.
3. Bridgman P. W., *Proc. Am. Acad.* **64** (1929) 19.
4. Bateman T. B., H. J. McSkimin, and J. M. Whelan, *J. Appl. Phys.* **30** (1959) 544.
5. McSkimin H. J., W. L. Bond, G. L. Pearson, and H. J. Hrostowski, *Bull. Am. Phys. Soc.* **1** (1956) 111.
6. Bolef D. I., and Menes, *J. Appl. Phys.* **8** (1960) 1426.
7. Suchet J., Thesis, Paris, 1961. Série A No 3748 n° d'ordre 4599.
8. Le Roux-Hygon P., M. Radot, and J. Suchet, *Compt. Rend.* **254** (1962) 1250.
9. Sumer A., and J. R. Smith, *J. Appl. Phys.* **33** (1962) 2283.
10. Zirinski S., *Acta Met.* **4** (1956) 164.
11. Schwartz M., and L. Muldawer, *J. Appl. Phys.* **29** (1958) 1561.
12. Lazarus D., *Phys. Rev.* **76** (1949) 545.
13. Siegel S., *Phys. Rev.* **57** (1940) 605.
14. Rinehart J. S., *Phys. Rev.* **58** (1940) 365.
15. Rinehart J. S., *Phys. Rev.* **59** (1941) 308.
16. Good W. A., *Phys. Rev.* **60** (1941) 605.
17. Druyvesteyn M. J., and J. L. Meyering, *Physica* **8** (1941) 1059.
18. Portevin A., and L. Guillet, *Compt. Rend.* **203** (1936) 237.
19. Cabarat R., L. Guillet, and R. Le Roux, *Compt. Rend.* **226** (1948) 1374; **227** (1948) 681.
20. Köster W., *Z. Metallk.* **32** (1940) 160.
21. Köster W., and K. Rosenthal, *Z. Metallk.* **32** (1940) 163.
22. Köster W., and W. Rauscher, *Z. Metallk.* **39** (1948) 111.
23. Laurent P., and M. Eudier, *Rev. Met.* **47** (1950) 582.
24. Stonehouse A. J., R. M. Paine, and W. W. Beaver, Chapter 13 in *Mechanical Properties of Intermetallic Compounds*, J. H. Westbrook, ed. Wiley, New York, 1960.
25. Köster W., *Z. Metallk.* **32** (1940) 145, 151.
26. Shibuya Y., *Tohuku Imp. Univ. Sci. Repts. Res. Inst.* **A1** (1949) 161.
27. Flinn P. A., G. M. McManus, and J. A. Rayne, *Phys. Chem. Solids* **15** (1960) 189.
28. Köster W., *Z. Metallk.* **32** (1946) 145.
29. Köster W., and A. Schneider, *Z. Metall.* **32** (1940) 156.
30. Numakura Kenichi., *J. Phys. Soc. Japan* **16** (1961) 2344.
31. Kataev G. I., and Z. D. Sirota, *Zh. Eksperim. i Theor. Fiz.* **38** (1960) 1037.
32. Köster W., *Z. Metallk.* **35** (1943) 194.
33. Cabarat R., L. Guillet, and R. Le Roux, *Compt. Rend.* **231** (1950) 1373.
34. Zener C., *Elasticity and Anelasticity of Metals.* University of Chicago Press, Chicago, 1948.
35. Cabarat R., L. Guillet, and R. Le Roux, *J. Inst. Metals* **75** (1949) 491.
36. Linde J. O., *Ann. Phys.* **5,** 8 (1931) 124.
37. Carlsson O. O., and G. Hägg, *Z. Krist.* **83** (1932) 308.
38. Lulay J., and C. Wert, *Acta Met.* **4** (1956) 629.
39. Clarebrough I. M., *Acta Met.* **5** (1957) 413.
40. Kuper A. B., and C. T. Tomizuka, *Phys. Rev.* **98** (1955) 244.
41. Potter R. F., J. H. Wasilik, and R. B. Flippen, *Mechanical Properties of Intermetallic Compounds*, J. H. Westbrook, ed. Wiley, New York, 1959.
42. Köster W., *Z. Elektrochem.* **43** (1939) 31.
43. Cabarat R., P. Gence, L. Guillet, and R. Le Roux, *J. Inst. Metals* **80** (1951) 151.

chapter 24

Mechanical Properties—Plastic Behavior

ALAN LAWLEY
The Franklin Institute Research Laboratories,
Philadelphia, Pennsylvania

1. INTRODUCTION

In general, intermetallic compounds are characterized by high hardness and limited ductility—properties associated with harmful second-phase embrittlement in most alloy systems. However, for several years there has been a continuing interest in the use of these compounds in bulk form, the necessary stimulus being provided by the constantly increasing demand for improved high-temperature engineering materials. Numerous development projects have been carried out, and are currently in progress, concerning the scope and usage of intermetallic compounds, carbides, borides, silicides, oxides, and phosphides. Parallel studies, aimed at providing a basic understanding of plastic deformation, are also in various stages of completion. In this chapter an attempt is made to provide an up-to-date survey of the plastic properties and to give, wherever possible, a phenomenological description of the deformation process as affected by the common metallurgical variables.

As a means for maintaining clarity of presentation while attempting a comprehensive coverage, the chapter content is broken down into the following sections: an attempted systematization of mechanical properties, the phenomenology of plastic deformation, strengthening mechanisms, crystallography of the deformation process, and general engineering aspects of intermetallic compounds. The latter are defined as intermediate phases in binary and higher order metal-metal systems in both ordered and disordered states. However, metal-metalloid compounds are included, when the deformation behavior is closely related to that of a metal-metal compound. Concurrent with this chapter on the plastic behavior of intermetallic compounds, Stoloff and Davies[1] have made a detailed review of the mechanical properties of ordered alloys. It is suggested that the reader be familiar with both articles.

2. ATTEMPTED SYSTEMATIZATION OF MECHANICAL PROPERTIES

At the 1959 Electrochemical Society Symposium on The Mechanical Properties of Intermetallic Compounds[2] (Philadelphia), Westbrook compiled an extensive review of the literature. Consequently, only a brief outline is given of the systematization attempted. Investigations dating back as far as 1908 established that the hardness of many intermetallic compounds exceeded that of the unalloyed component metals. Hardness minima also occurred in the hardness-composition curve at the stoichiometric composition.

Early attempts at systematization were made in terms of (1) the extent of the ductile-to-brittle transition range and (2) the variation of tensile strength with temperature. Lowrie[3] observed that for a wide range of compounds, the onset of ductility occurred in the temperature range $T/T_m \simeq 0.61 - 0.68$.* Beyond this characteristic, however, no simple correlation could be found between tensile strength and heat of formation, density, or relative decrease in volume on compound formation. The semiconductor compounds InSb and GaSb, having the zinc blende structure, show a change in the mode of deformation at a specific T/T_m value, as indicated by hardness measurements.[4] Similarly, for a wide range of zinc blende or wurtzite structures, Borshchevsky et al.[5] noted a decrease in hardness as the nature of the atomic bonding changed from covalent to ionic. In a somewhat different approach, Schwab[6–8] considered the hardness parameter in relation to the saturation of the Brillouin zone. In accordance with the observed hardness variations in Hume-Rothery compounds, the theory predicts hardness maxima in the γ-phases, and a hardness in the β-, ϵ-, and η-phases similar to that of the saturated terminal solid solutions. Although the zone approach is basically sound, structural effects may be expected to play an important role in determining the mode of deformation. Thus Massalski[9] considers that variations in hardness with a change in composition in the α-solid solutions or β-phases are primarily due to variations in short-range order and stacking-fault energy. In the case of γ-, ϵ-, and η-brasses, large variations in hardness occurring over narrow composition ranges are considered in terms of a complex deformation process rather than electronic (electron:atom ratio) effects.

It is to some extent unfortunate that many of the attempts at systematization have been made in terms of hardness, thereby sacrificing to some extent the possible fundamental significance of the data. Thus, although hardness is known to correlate well with tensile strength, the former is a complex function of work-hardening beyond the yield point. In the light of these complications, and considering the wide range of structural and bonding characteristics of intermetallic compounds, it is not surprising that a concise systematization is lacking. In summary, intermetallic compounds are generally extremely hard and lacking in ductility at ambient temperatures, but show enhanced ductility at sufficiently high temperatures.

* T_m is the melting point in degrees Kelvin.

3. THE PHENOMENOLOGY OF PLASTIC DEFORMATION—EFFECTS OF METALLURGICAL VARIABLES

3.1 Temperature

In attempting to understand the deformation process(es) in intermetallic compounds, temperature has been the variable most frequently studied. Single-crystal work includes yielding and flow behavior of Ag–Al,[10] Cu–Ge,[11–13] InSb,[14] AgMg,[15] Fe_3Be,[16] Ni_3Al,[17] and Cu_3Au.[17] In polycrystalline material, attention has been focused on AgMg,[18,19] FeCo,[20,21] Fe_3Be,[16] Fe_3Al[1] Cu–Al,[22] Cu–Zn,[22] Ni_3Al,[23–25] NiAl,[25] and Mg_3Cd.[26]

The hexagonal Ag-33 at.% Al compound is free from long-range ordering† complexities,‡ and because the c/a ratio (1.61) is close to ideal, short-range order† may be expected to be spherically symmetrical.[10] The effect of temperature on critical resolved shear stress for both basal and prismatic slip may be seen from Fig. 1. The constancy in the stress for basal slip over the range 78–450°K and the appearance of a well-developed upper and lower yield point are shown to be consistent with Suzuki locking of partial dislocations[27,28]

† The abbreviations LRO and SRO are introduced.
‡ The influences of long-range order, defect structure, and compositional change are discussed in detail in the following section.

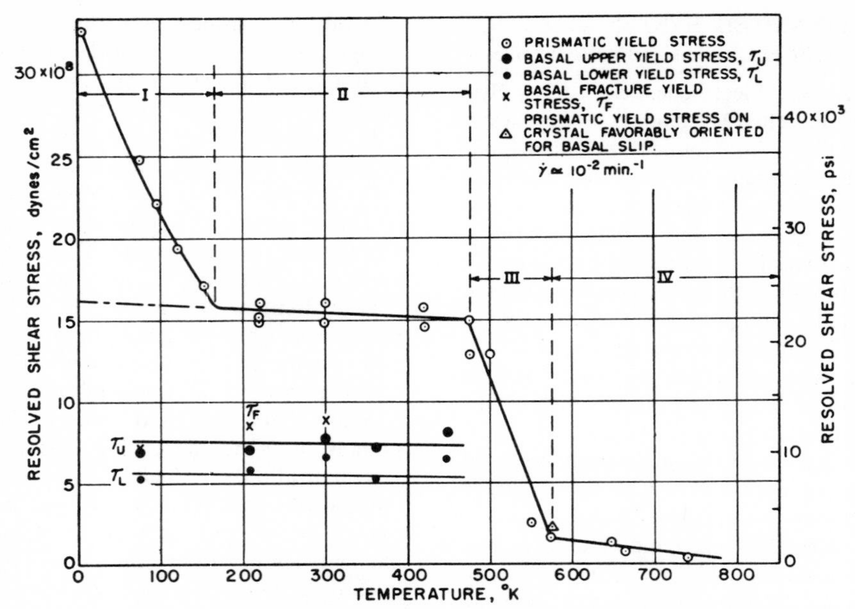

Fig. 1. The effect of temperature on the critical resolved shear stress for prismatic and basal slip in Ag-33 at.% Al. (After Mote et al.[9])

on the basal plane. From the results of strain-rate change experiments, it is concluded that the decrease in shear stress for prismatic slip over the temperature range 4.3–170°K is controlled by the Peierls-Nabarro force, that is, the nucleation and growth of kinks along dislocations lying in potential energy wells. The high yield stress of the alloy (22,500 psi) and its constancy with temperature from 170–450°K are a consequence of SRO hardening superimposed upon the weaker Suzuki hardening. At temperatures $\gtrsim$ 475°K, an exact interpretation of the shear stress behavior is lacking. The decrease in stress in the range 475–575°K is undoubtedly related to the decrease in the degree of SRO with increasing temperature, however, other complicating factors are present. Above ~550°K, basal slip is rarely observed, and crystals oriented for basal slip actually deform by

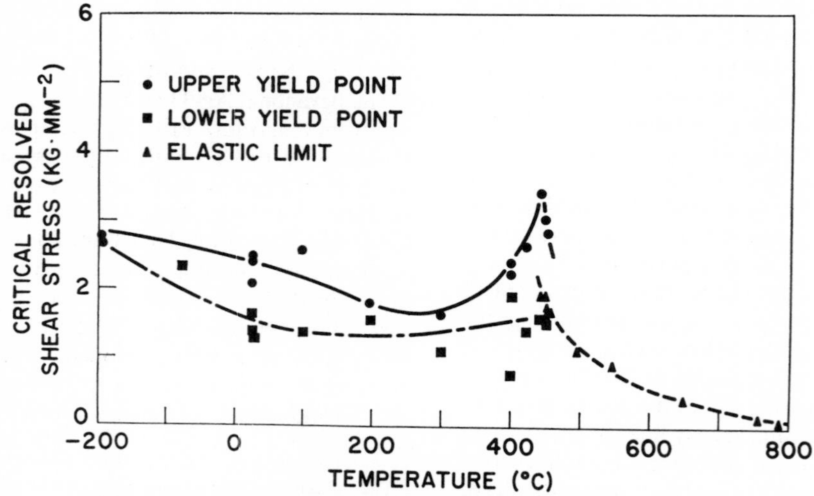

Fig. 2. Temperature dependence of shear stress for basal slip in Cu-13.5 at.% Ge. (After Thornton.[10])

duplex prismatic slip resulting in necking of the specimen.

Thornton[11–13] has examined the deformation behavior of single crystals of the hcp ζ Cu–Ge intermediate phase of compositions 13.5 and 16.2 at.% Ge over the temperature range 4.2–1000°K. In the 13.5 per cent Ge compound, only basal slip was found in tensile deformation over the entire temperature range; crystals oriented for nonbasal slip cleaved on $(10\bar{1}1)$ with no apparent basal slip. Basal slip occurred with a very large single yield point below ~420°K, whereas from ~420°K to 720°K, jerky flow was evident. Above 720°K, only smooth plastic flow occurred. The temperature dependence of the critical shear stress resolved for basal slip is illustrated in Fig. 2. By contrast, crystals of the 16.5 at.% Ge compound showed extensive $\{10\bar{1}0\}\langle\bar{1}\bar{1}20\rangle$ slip, with no yield point. The large yield point in basal slip is believed to be due to pinning by the Suzuki[28] mechanism, whereas the difference in behavior from cleavage in the 13.5 at.% Ge alloy to prismatic slip in the 16.2 at.% Ge alloy may be a consequence of the decrease in axial ratio from 1.635 to 1.625, respectively.

It is of interest to compare the behavior of the hexagonal phases with that of InSb, a typical covalent semiconductor compound having the zinc blende structure. Single-crystal stress-strain curves from 300°C to the melting point (525°C) are characterized by the presence of a double yield point, but with *no* strain aging effects.[14] The proportional limit and the upper and lower yield stress vary linearly with temperature and vanish at the melting point. From a consideration of the dislocation geometry in the zinc blende structure, and from the magnitude and temperature dependence of the energy band gap in InSb, it is believed that the major factor controlling plastic flow is the breaking of the electronic bonds. The appearance of an upper and lower yield point is not related to impurity-dislocation interactions, but is a consequence of the plastic strain rate in the crystal matching the applied strain rate (Johnston[29]). This equivalence defines the upper yield stress. The stress then decreases as the continued generation of dislocations increases the plastic strain in the crystal beyond the required level.

Mukherjee and Dorn[15] have studied the temperature and strain-rate dependence of the flow stress of single crystals of AgMg(B2) over the range 4–350°K. Yielding and flow is a thermally activated process for temperatures below ~250°K, with the critical resolved shear stress increasing from ~3.8×10^8 dynes cm^{-2} at 250°K to ~13×10^8 dynes cm^{-2} at 4°K. It is concluded that the rate-controlling mechanism is that of "kink pair" nucleation on the dislocations.

Bolling and Richman[16] have obtained an increase in critical resolved shear stress of both Fe_3Be (DO_3) and the disordered alloy, with increasing temperature over the range 78–500°K. A theory of twinning is advanced to account for this effect, and for the associated negative strain-rate sensitivity. The increase in strength with increasing temperature ($T/T_m < 0.2$) observed in Ni_3Al and Cu_3Al single crystals[17] is due to a lattice effect, tentatively identified as arising from a difference in vibrational modes among similar and dissimilar metal bonds[30] as a function of temperature.

Turning to studies on polycrystalline material, extensive data are available on the effect of temperature on the flow behavior of AgMg[18] (ordered CsCl structure). Three distinct modes of deformation are apparent: (*a*) at low temperatures ⩽150°C and moderate strain rates (~ 0.005 min^{-1}), deformation occurs by slip as in pure metals; (*b*) for intermediate temperatures (150–350°C) and strain rates ⩽ 0.005 min^{-1}, dislocation interactions with solute atoms (DISA) control the deformation process; (*c*) above ~400°C, and at low strain rates, diffusion processes become important. The form of the load-elongation curve for each of these groupings is illustrated in Fig. 3. DISA phenomena give rise to maxima in the temperature dependence of strain-hardening, and a deviation from the monotonic stress-temperature relationship (Fig. 4). A complete plot of the stress-temperature dependence is illustrated in Fig. 5.

The results of an independent study on AgMg (Terry and Smallman[19]), although in general agreement with those of Westbrook and Wood, do differ in detail. Thus only a small dependence of flow stress on temperature is observed below ~200°C. However, at a slightly higher temperature (the value of which is dependent on strain rate), strength decreases rapidly. This corresponds to the intermediate temperature range described by Westbrook

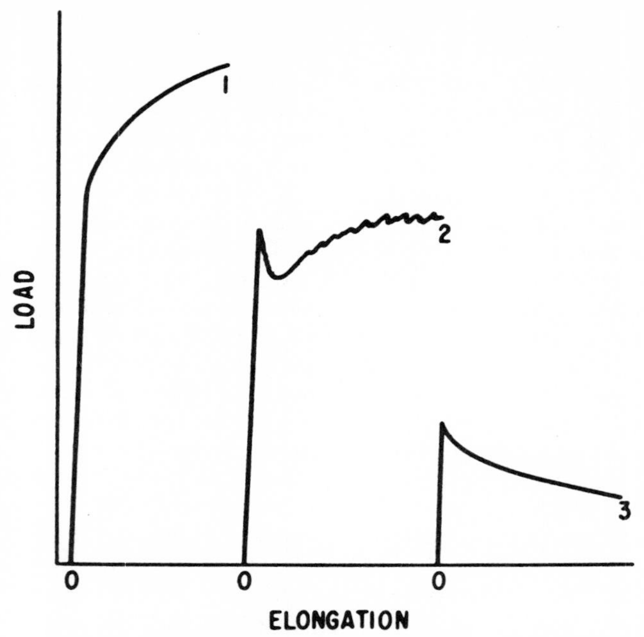

Fig. 3. Load-elongation curves for AgMg: (1) low temperatures, (2) intermediate temperatures, and (3) high temperatures. (After Westbrook and Wood.[12])

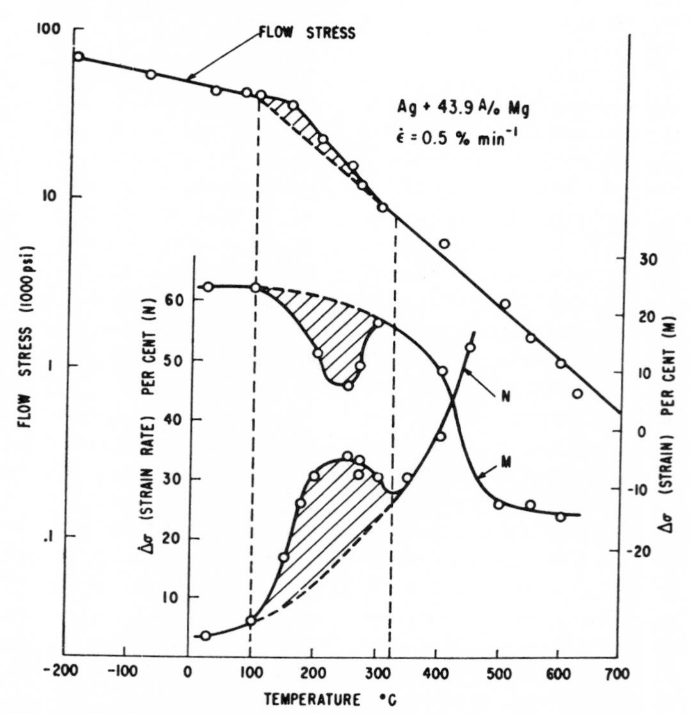

Fig. 4. The temperature dependence of the flow stress; increase in flow stress results from a tenfold increase in strain rate (N); and increase in flow stress after 8 per cent elongation (M) for Ag-43.9 at.% Mg. (After Westbrook and Wood.[12])

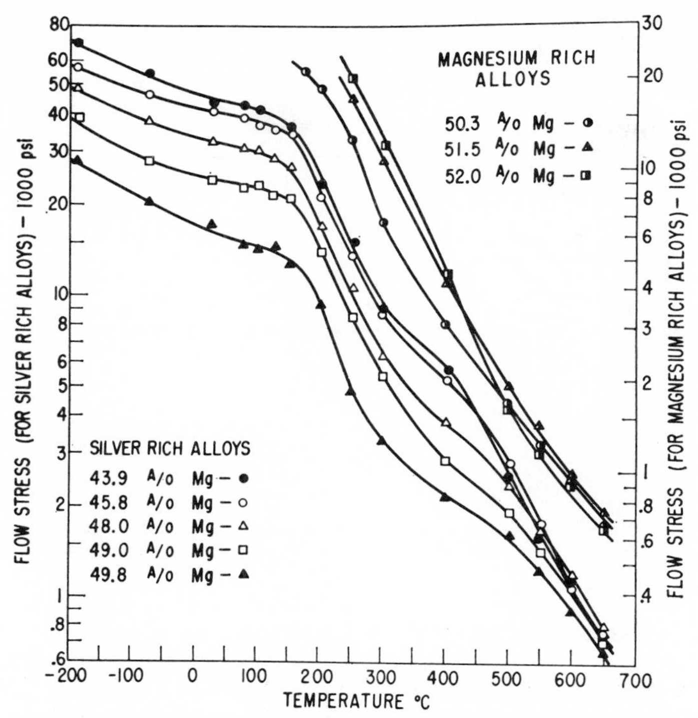

Fig. 5. The flow stress of Ag-rich and Mg-rich alloys as a function of temperature. $\dot{\epsilon} = 0.005\ \text{min}^{-1}$. (After Westbrook and Wood.[12])

and Wood. There is no evidence for a DISA type of interaction in the range 100–300°C, but a small discontinuity in the flow stress-temperature curve at ~500°C is apparent. Without attempting to resolve these discrepancies, we see that factors such as the purity of the starting material and possible contamination, particularly at grain boundaries, must be taken into account.[31]

The Cu–Al and Cu–Zn gamma phases show a brittle-to-ductile transition at $T/T_m \simeq 0.5 - 0.6$,[22] in agreement with the tensile data of Lowrie[3] and the earlier hot hardness and compression data of Mack, Birkle, and Krubsack.[22] A decrease in the transition temperature and in the stress-rupture level with increasing aluminum content of the gamma phase is tentatively attributed to a order-disorder reaction and the consequent variation in antiphase domain size.

3.2 Strain-Rate

The strain-rate dependence of InSb single crystals[14] is illustrated in Fig. 6, in which the measured yield stress refers to the proportional limit. For strain rates less than a critical value, $d\tau/dT$ is strain-rate-dependent with the flow process thermally activated. For greater strain rates, the value of $d\tau/dT$ remains constant, implying that in this region the mechanical energy due to the applied stress supplies the required activation energy for flow. In terms of the actual stress-strain curves, for a given temperature, the proportional limit and the upper yield stress increase with increasing strain rate until a saturation stress is reached. The lower yield stress is independent of strain rate.

Polycrystalline AgMg has been subjected to tensile tests involving a change of strain rate.[18] Below ~150°C, the effect on flow stress is extremely small; however, by using rate-change tests, Westbrook and Wood obtained a rate sensitivity n (defined as Δ log flow stress/Δ log strain rate) ~0.01 for silver-rich alloys. For a coarser grain size or at temperatures close to 150°C, the value of n increased, which is characteristic of pure metals. In the intermediate and high-temperature ranges, stress and strain rate are related exponentially. Activation energies for the deformation process range from ~28 kcal for

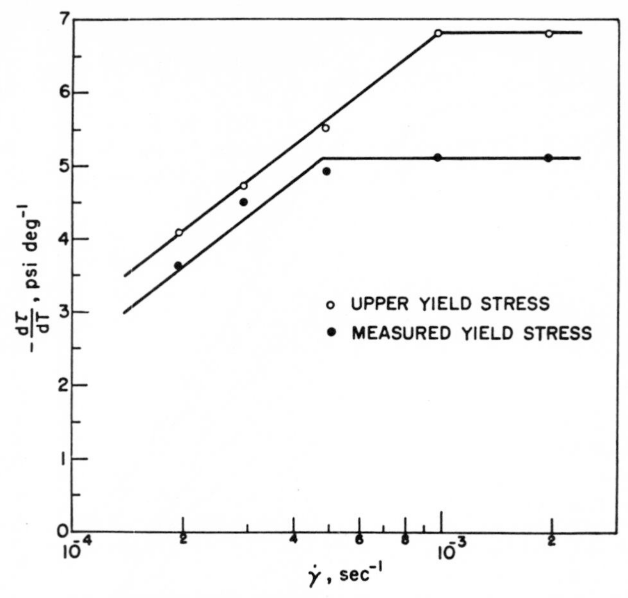

Fig. 6. Effect of strain rate on yield stress in single crystals of InSb. (After Abrahams and Liebmann.[11])

Ag-rich alloys to ~40 kcal for Mg-rich alloys at intermediate temperatures, and from 47 kcal for a 43.9 at.% Mg alloy to 35 kcal for a 52.0 at.% Mg alloy in the high-temperature region. In terms of ducitlity, AgMg is particularly sensitive to strain rate. The transition temperature (bend-test data) is plotted against strain rate for several Mg-rich alloys in Fig. 7. The linear relation between log strain rate and reciprocal transition temperature is in accord with similar observations on pure metals. In an earlier study on hexagonal Bi_2Ti, it was found that the room-temperature elongation decreased from ~70 per cent to ~2 per cent when strain rate was increased from 0.002 min^{-1} to 0.05 min^{-1}. The corresponding yield stress increased from 5000 to ~8500 psi.

3.3 Grain Size

As in the case of strain rate, only limited systematic data are available on grain size. Transition temperature studies on AgMg[31,32] show that there is a linear relation between the inverse of the absolute transition temperature, and the reciprocal of the square root of the grain diameter. In actual magnitude, an increase in grain size from 0.0025 to 0.01 cm dia. raised the transition temperature from ~50°C to ~85°C. It must be stressed, however, that the increase is not completely a function of the grain size because, at the higher annealing temperatures, grain-boundary hardening takes place by impurity segregation.

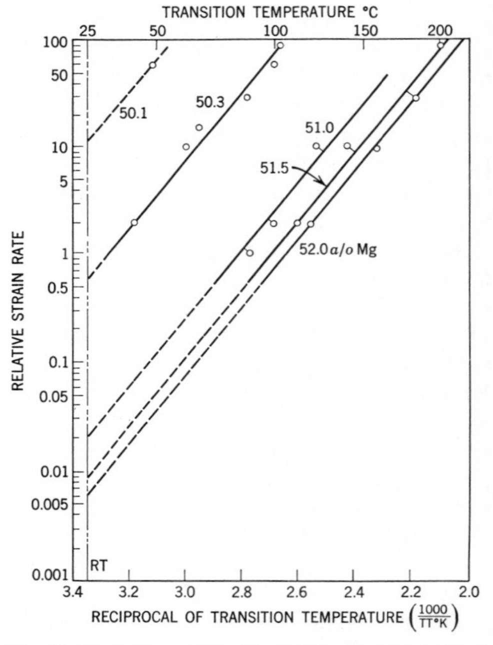

Fig. 7. Variation of ductile-brittle transition temperature with strain rate for polycrystalline AgMg. (After Westbrook and Wood.[19])

Anomalous grain-size effects in the intermediate temperature range (i.e., stress decreases with decreasing grain size), originally attributed to DISA phenomena, are now known to be due to oxygen-induced grain-boundary hardening. The equicohesive temperature range coincides with the lower end of the high-temperature region so that grain size effects are extremely small. Grain-size effects cannot be determined in the high-temperature region, as grain growth is extremely rapid and leads to a coarsened structure.

Grain-size studies on FeCo,[33] FeCo–V,[1] and Ni_3Mn[20] show that the Hall-Petch relationship for yielding and flow is obeyed

$$\sigma_\epsilon = \sigma_{0\epsilon} + k_\epsilon l^{-1/2} \qquad (1)$$

where σ_ϵ is the flow stress for a given strain ϵ, $\sigma_{0\epsilon}$ and k_ϵ are constants for that strain, and l is one half the average grain diameter. The magnitude of k is found to increase with LRO in both Ni_3Mn and FeCo. This increase is related to the change in the number of slip systems with order, because this controls the ease of spreading of slip across grain boundaries.[34]

Grala[25] discusses the tensile strength of Ni_3Al (fcc) and NiAl (bcc) in terms of the grain size obtained following annealing at increasing temperatures. In the light of the behavior of AgMg, it would appear that strength variations are a complex function of the level and distribution of impurities and the subgrain structure, as well as the absolute grain size. The occurrence and importance of grain-boundary hardening and "pest" degradation in these and related compounds is discussed in a later section.

4. STRENGTHENING MECHANISMS

As a general rule, intermetallic compounds are susceptible to strengthening by the mechanisms applicable to pure metals and conventional alloys, namely solid solution strengthening, strain hardening, and strain aging. In addition, the presence of long- or short-range order, a domain structure, and/or a defect lattice can provide a further means of strengthening. This section constitutes a résumé of the current knowledge in each of these areas.

4.1 Solid-Solution Alloying

Conditions governing the formation of solid solutions in intermetallic compounds, either for elements other than their own components or for isomorphous compounds, follow closely those applicable to pure metals. Thus the spatial, crystallographic, and electronic factors are all important, and, as a general rule, the relative atomic size factor difference must not exceed ~15 per cent for extensive solid solubility. These rules are reviewed in some detail in Chapter 19 by Kornilov, who has been responsible for much of the original research in this area.[35–37] Various forms of substitutional alloying may occur, according to whether the substitutional atoms are arranged perferentially in place of one component or the other or are distributed at random on the sublattices of the intermetallic compound. In addition, it is possible for a superlattice to form by ordering on one of the sublattices.

Guard and Westbrook[38] considered the effect of three groups of elements alloyed substitutionally in fcc Ni_3Al, namely (1) elements substituting for nickel—Co and Cu, (2) elements substituting for aluminium—Si, Ti, Mn, and V, and (3) elements substituting for both constituents—Fe, Cr, and Mo. The solid-solution hardening results are complicated relative to binary terminal solid solutions, because of the possibility of virtually independent substitution and of a concomitant strain aging effect. However, a significant increase in hardening occurs at room and elevated temperatures due to silicon or titanium. As in the case of terminal solid solutions, lattice strain plays a major role in hardening. Ideally, the change in hardness (ΔH) should be a single-valued function of the lattice parameter change (Δa_0), provided that only the size factor and the nature of the substitution are important. The observed correlation is illustrated in Fig. 8. The anomalously high $\Delta H/\Delta a_0$ values for Cr and Mo are attributed to electronic and chemical effects.

Recent data on AgMg[31,32] show that solid-solution hardening by ternary solute additions is complex. Thus for additions of Cu, Zn, Sn, Al, and Ce regular increases in flow stress are observed at all temperatures if the base compound has silver in excess of stoichiometry.

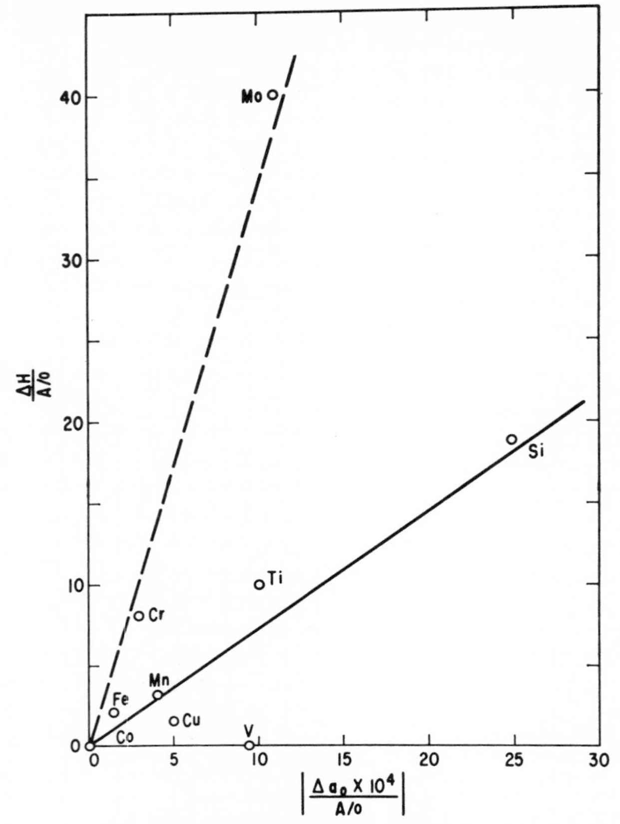

Fig. 8. Relationship between hardening and lattice distortion for various solutes in Ni_3Al. (After Guard and Westbrook.[23])

On the other hand, with an excess of magnesium, the flow stress may increase, decrease, or remain constant, depending on the temperature and specific ternary additions. Significant strengthening, without impairment of ductility can occur in the silver-rich compounds with only small additions. For example, 1 at.% Zn in AgMg at room temperature raises the flow stress by 75 per cent. In addition, there is a marked retention of strength at elevated temperatures as a result of ternary alloying; for example, AgMg + 2.5 at.% Al is ~270 per cent stronger than the binary alloy at 250°C, ($T/T_m \simeq 0.48$). In analyzing the effects of ternary alloying on flow stress and transition temperature, full account has to be taken of any accompanying grain-boundary hardening. In general, alloying additions giving rise to grain-boundary hardening cause an increase in the transition temperature. From a practical standpoint, Zn, Cu, or Al additions (~1 at.%) to stoichiometric AgMg are particularly beneficial because the transition temperature is lowered by ~400°C.

The iron-group aluminides FeAl, CoAl, and NiAl form a series of solid solutions of the form (FeCoNi)Al. Westbrook[2] obtained significant increases in strength especially at high temperature by solid-solution alloying of FeAl and CoAl. For example, at 800°C the hardness levels of CoAl and FeAl are ~28 and 70 kg/mm^2, whereas the (CoFe)Al (25 at.% Co, 25 at.% Fe) alloy has a hardness ~110 kg/mm^2 The effects of Nb, Ti, and Zr in solid solution in NiAl have been discussed by Grinthal,[39,40] Herz[41] and Steinitz.[42] The smaller Nb atom causes only minor lattice strains, so that strength falls off with increasing temperature, as in unalloyed NiAl. Ti and Zr cause severe lattice distortion with a consequent improvement in strength and stress to rupture characteristics. The NiAl + 4 per cent Zr alloy retains a transverse rupture strength ~160,000 psi up to 1100°C, whereas unalloyed NiAl ruptures at ~65,000 psi under these conditions. Grala[25] observed that additions of Mo to stoichiometric NiAl had a beneficial effect on room temperature and elevated-temperature tensile strength. Unfortunately, side effects due to grain refinement and to the scavenging of interstitial impurities (oxygen, nitrogen) make it impossible to assess the true extent of solid-solution hardening. Solid-solution alloying with Mo (~4 per cent), in the intermetallic Cr_2Ti appears to improve stress to rupture life at temperatures ~1000°C.[25] Alloying is limited to ~10 at.% Mo because beyond this amount the Cr_2Ti structure disappears and is replaced by a ternary solid solution. The complex tetragonal σ-phase FeCr shows considerable improvement in high-temperature strength when alloyed with Mo,[43] an effect believed to be due to an increased complexity or ordering in $\sigma Fe_{36}Cr_{12}Mo_{10}$.

4.2 Defect Hardening

Deviations from stoichiometry in intermetallic compounds frequently lead to relatively high equilibrium concentrations of point defects. The defects may take the form of vacancies, substitutional atoms destroying the ordered arrangement, interstitials, or complexes consisting of combinations by pairs of these defects (Brown, Chapter 15). That these defects have a pronounced effect on mechanical properties may be seen from hardness-composition, or flow-stress-composition curves. The defects may either strengthen or weaken the intermetallic compound depending on the temperature;[44] in general, at low homologous temperatures a minimum in strength is observed at the stoichiometric composition, whereas at higher homologous temperatures the reverse holds true. Many examples of the former may be cited including AgMg,[18,45] NiAl,[25,46] Ni_3Al,[38] Ni_3Fe,[47] CoAl,[46] CuAu,[48,49] Cu_3Au,[48] FeCo,[50] and MgCd.[51,52] Strength maxima at high T/T_m levels for the stoichiometric composition are found in AgMg,[45] Ni_3Al,[38] and Bi_5Ti_3.[53] The flow-stress behavior of AgMg is illustrated in Fig. 9.

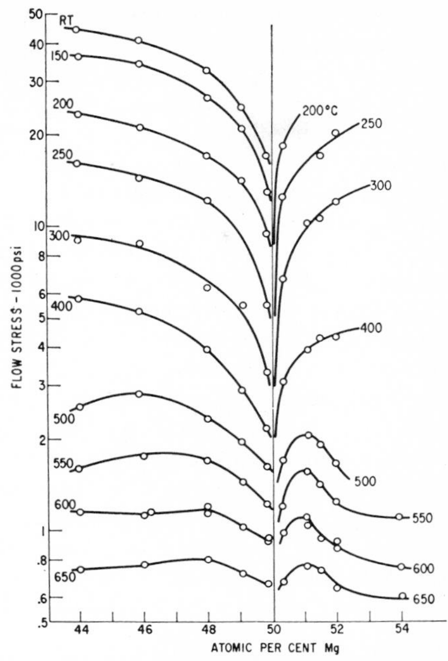

Fig. 9. Flow stress of AgMg as a function of the deviation from stoichiometry. (After Wood and Westbrook.[31])

The effect of temperature on the role of the defects may be understood in a qualitative manner from a consideration of the deformation process. At low temperatures, at which deformation occurs primarily by slip, the relatively immobile point defects perturb the regularity of the crystal structure, and the associated dislocation-point defect interactions lead to an increase in the stress required to initiate slip. In the high homologous temperature region, diffusion-controlled processes predominate[54,55] so that the enhanced diffusion rates afforded by the substitutional and vacancy defects[56] lead to a lowering of strength for nonstoichiometric compositions.

The majority of intermetallic compounds possess substitutional defects for offstoichiometric compositions. However, for NiAl and CoAl, the detailed X-ray studies of Bradley and Taylor[57] and Bradley and Seager[58] show that the defect structure is of the vacancy type on the high aluminium side, and substitutional on the cobalt- or nickel-rich side. Tests of the temperature dependence of indentation hardness of NiAl and CoAl show that for low homologous temperatures the slope of the hardness-composition curve is significantly higher for aluminium-rich compositions than for cobalt- or nickel-rich compositions (Fig. 10). This observation suggests strongly that vacancies have a more potent effect on dislocation motion than do substitutional atoms. In terms of lattice distortion, as measured by lattice parameter changes, the values of ΔH/at.% for Al-rich NiAl and Ni_3Al, when plotted against Δa_0/at.%, fall on the lower curve of Fig. 8.

Apart from strengthening due to equilibrium defect structures, it is possible to increase room-temperature tensile strength by quenching in an excess vacancy concentration. Perhaps the best known example is that of β-brass, in which the strength increase following quenching gradually decays at room temperature.[59] From a theoretical standpoint, Rudman[60] has considered the role of vacancies in the strengthening of ordered alloys, and in particular of β-brass. Following quenching from an elevated temperature, the vacancy-decay rate is inhomogeneous, being faster near vacancy sinks than at areas remote from sinks. Because the rate of ordering is proportional to the vacancy concentration,[61] a nonuniform scheme of long-range order is produced with the degree of order away from vacancy sinks exceeding that in areas close to the sinks. The magnitude of the strengthening effect has been calculated for β-brass assuming quasichemical binding, and a value comparable to the experimental yield stress[59,62] has been obtained.

As a further example, Weiss[63] studied the room-temperature deformation behavior of Fe_3Al as a function of quenching temperature. Yield stress maxima were obtained following quenching from temperatures in the range 500–550°C, 700–750°C, and from ~900°C, respectively. Whereas the 500–550°C peak is attributed to the combined effects of dislocation stress field-order interactions[64] and to short range ordering,[65] the strength maxima for the two higher quenching temperatures are considered to be due to quenched-in vacancies. High concentrations of point defects are often present in metallic materials following radiation damage (Chapter 21), and this is reflected

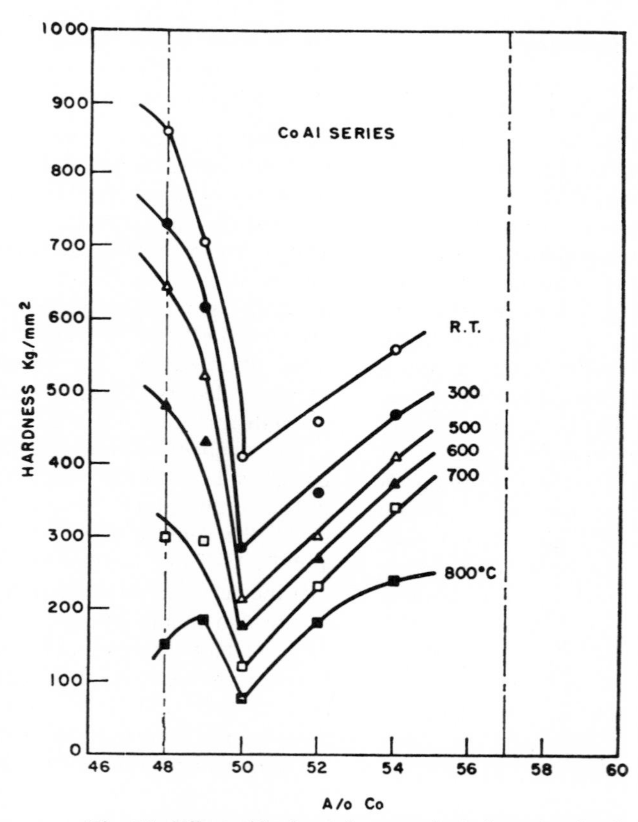

Fig. 10. Effect of lattice defects on the indentation hardness of CoAl at various temperatures. Defect structure is of the vacancy type on the high aluminium side and substitutional on the cobalt-rich side. (After Westbrook.[32])

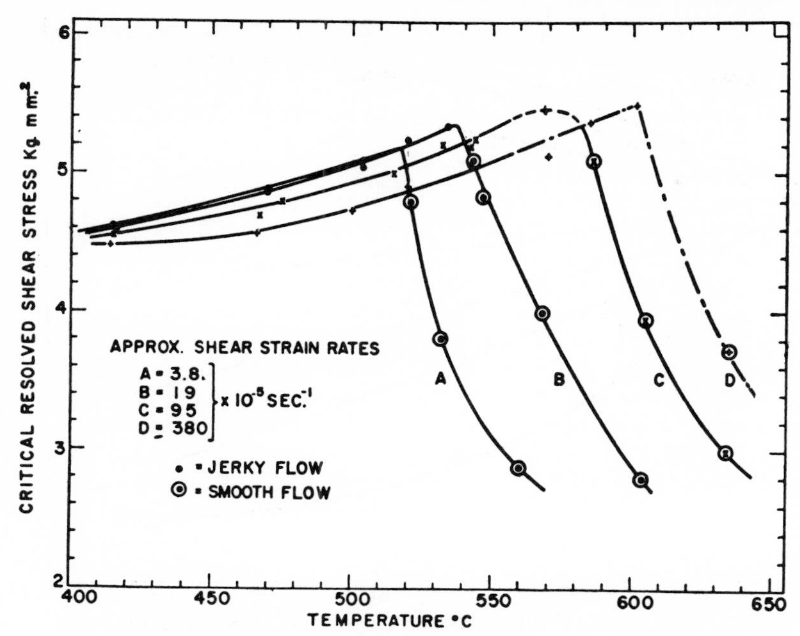

Fig. 11. Variation of the critical resolved shear stress of Cu_3Au with temperature for various applied strain rates. (After Ardley.[51])

in a change in mechanical properties. Typically, TiAl[2] exhibits a 10 per cent increase in hardness and a 48 per cent decrease in impact strength after neutron irradiation exposures $\sim 10^{20}$ nvt.

4.3 Strain Aging

Strain aging and relaxation effects are known to occur in several intermetallic compounds including Cu_3Au,[66] β-brass,[67–69] Ni_3Al,[38] AgMg,[18] Ni_3Fe,[70] and also in InSb.[71,72] The hardness peaks in Ni_3Al are ascribed to carbon, oxygen, and nitrogen; however, from the available data, it is not possible to obtain a clear correlation between the various peaks and a particular impurity or combination of impurities. In the case of β-brass, yield-stress maxima occur at $\sim$200°C, with nitrogen responsible for part of the strengthening; however, the situation is complicated due to a substantial increase in the elastic constants of β-brass at $\sim$200°C.[73] The stress-strain curves are characterized by jerky flow for temperatures below the peak stress whereas above that temperature deformation is smooth. Strain aging in AgMg[18] has been considered in a previous section. It is sufficient to restate here that at intermediate temperatures (150°C–$\sim$350°C) DISA interactions give rise to deviations from monotonic stress-temperature behavior, maxima in the change of strain-rate sensitivity with temperature, minima in the temperature dependence of the strain-hardening coefficient, yield points, serrated stress-strain curves, and relaxation phenomena. (Figures 3–5.)

Ardley[66] observed yield-stress maxima in single crystals of Cu_3Au at temperatures well above the long-range ordering limit ($T_c^* \simeq$ 390°C).* Consistent with the concepts of strain aging, low temperature strength decreased with increasing strain rate, and the stress-strain curves exhibited a pronounced yield drop. The variation of yield stress with temperature for various applied strain rates is given in Fig. 11. In this case, specific impurities were not examined. More recently Kuczynski et al.[74] have measured specific heat, temperature coefficient of expansion, Young's modulus, and yield strength as a function of quenching temperature. On the basis of these measurements it would appear that at $\sim$580°C a phase transition occurs, possibly a sudden decrease in short-range order. Undoubtedly, the deformation behavior of Cu_3Au above T_c is extremely complex and sensitively dependent

* T_c is the critical temperature for long-range order.

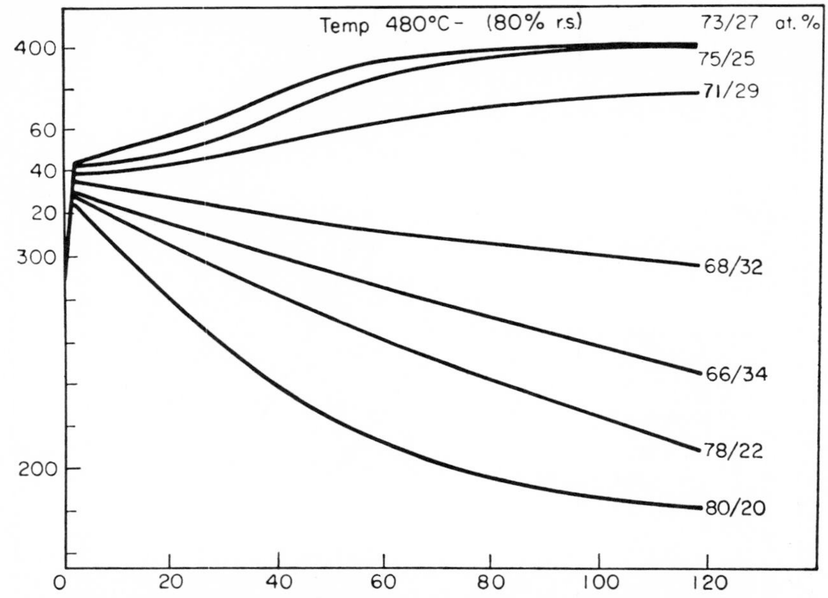

Fig. 12. The hardening kinetics of Ni–Fe alloys at 480°C following cold rolling ~80 per cent reduction. (After Vidoz et al.[55])

on prior heat treatment. It should be pointed out that Ardley's measurements were made actually at temperature, whereas those of Kuczynski et al. refer to measurements at room temperature on quenched Cu_3Au.

Certain nickel-iron alloy compositions are susceptible to hardening by strain aging (Vidoz et al.[70]). In the range 20–34 at.% Fe, strain aging just below T_c ($\equiv$500°C) produces a rapid hardening in the first few minutes. Further hardening during prolonged annealing is restricted to alloys developing a high degree of LRO (23–30 at.% Fe) (Fig. 12). From tensile data, the short-term hardening is shown to be due to an increase in flow stress, whereas the long-term hardening is caused by a greatly enhanced rate of work hardening, both forms of hardening becoming more pronounced with increasing prestrain. (Figures 13*a* and 13*b*.) Ternary Ni–Fe–Al alloys harden by strain aging more rapidly than the binary alloys—a consequence of the enhanced ordering kinetics. Tentatively, the short-term hardening is ascribed to SRO, and the long term hardening is explained on the basis of jog theory of work hardening in ordered material.[75] Stoloff and Davies[1] point out that this jog mechanism cannot explain all aspects of the strain-hardening behavior. They propose that on aging of prestrained material, anti-phase boundaries joining the superlattice dislocations lie along {100} planes, the lowest energy planes in the Ll_2 ordered structure. These dislocations are sessile and act as barriers to the motion of newly created glissile superlattice dislocations on the {111} plane; there are many more such barriers after strain aging than are formed in a continuous tensile test, so that the strain per dislocation source prior to obstruction is less in the strain-aged condition, which results in a higher rate of work hardening through the activation of new sources at a higher stress level.

4.4 Order and Domain Hardening

The mechanical properties of many alloys are altered when ordering takes place. In order to fully understand the behavior of these alloys, it is necessary to consider the separate effects due to LRO, SRO, domain size, and to changes in crystal structure on ordering. Strengthening effects are most marked in systems in which there is a change in the shape of the unit cell on ordering, for example, CuAu, CuPt, CuPd, and CoPt. The severe internal strains set up by the ordered phase in the disordered matrix result in a strengthening over and above that due to atomic rearrangement. Newkirk et al.[76] obtained a peak

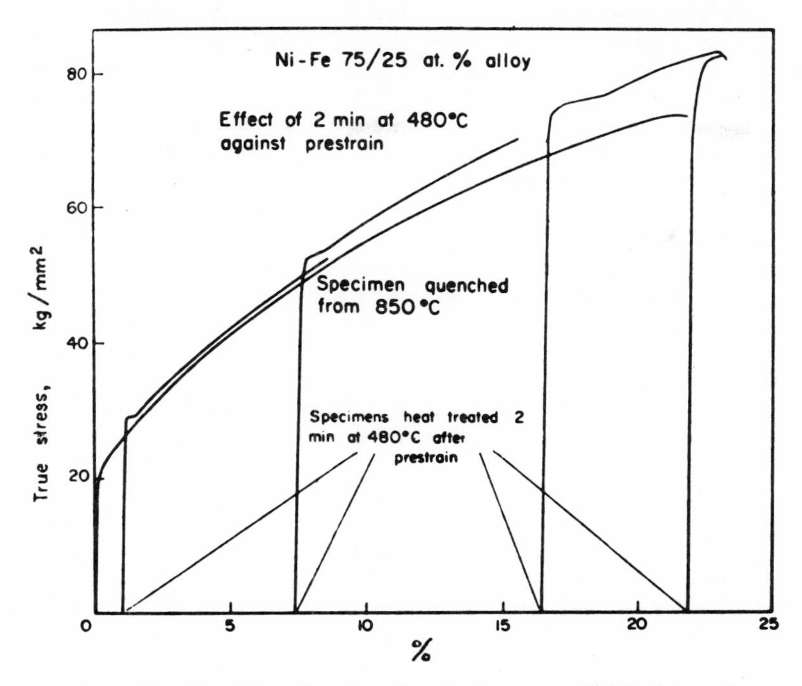

Fig. 13*a*. The effect of strain aging 2 minutes at 480°C after various tensile prestrains.

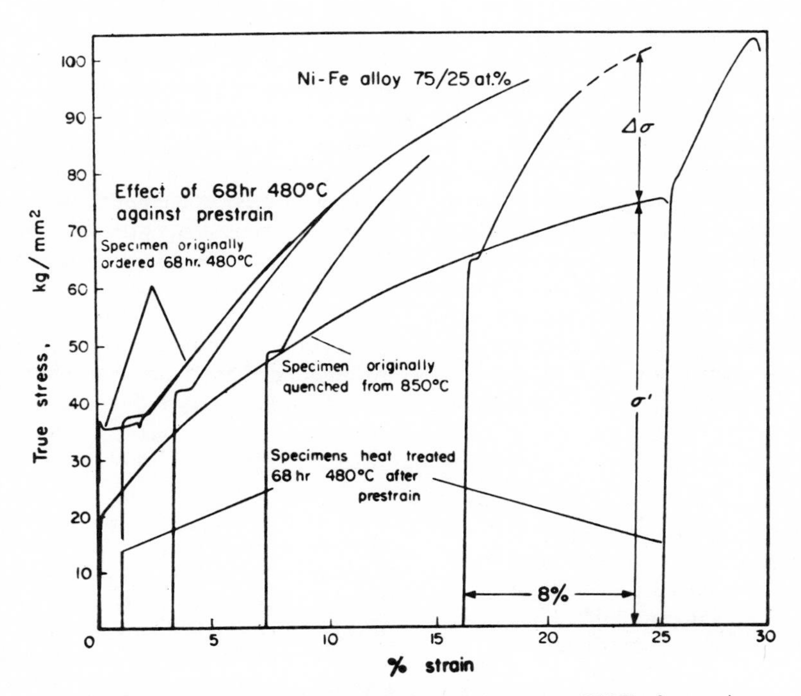

Fig. 13*b*. The effect of strain aging 68 hours at 480°C after various tensile prestrains. (After Vidoz et al.[55])

hardness in CoPt (fcc disordered → fc tetragonal ordered) as a function of the isothermal annealing time. Maximum hardness occurred at the stoichiometric composition, and the magnitude of the hardness peak decreased with increasing temperature. Similar behavior has been reported by Nowack[77] for for CuPt (fcc disordered → trigonal ordered), Köster[78] for CuAu (fcc disordered → fc tetragonal ordered), and by Jaumot and Sawatzky[79] for CuPd (fcc disordered → fc tetragonal ordered).

The intermetallic compounds of Ni and Co with V transform on ordering; Ni_3V: fcc disordered → fc tetragonal ordered; Ni_2V: fcc disordered → ordered orthorhombic; Co_3V fcc disordered → hcp ordered. The associated substructures in these phases have been characterized by transmission electron microscopy.[80] Stacking faults, twins, and/or microtwins are observed, and the maximum strength is associated with the presence of fine platelets of the ordered structure. These observations are similar to those of Hirabayashi and Weissman[21] on the transformation from disordered fcc to ordered fc tetragonal in CuAu. The plate-like ordered CuAu nuclei, which are coherent with the disordered matrix, lie along (101) planes and give rise to lattice strains.

Considering alloy systems, in which there is no change in crystal structure on ordering (CuZn, Cu_3Au, Ni_3Mn, Fe_3Al, Ni_3Fe), it is apparent from the experimental data that a maximum in flow stress occurs at some intermediate degree of order and/or domain size. Detailed analyses of antiphase domains and of line and planar defects are contained in Chapters 14 and 15, respectively. In this chapter, it is appropriate to consider the relationship between theory and experiment. According to Fisher,[65,82] SRO hardening is given by:

$$\sigma = \gamma b^{-1} \tag{2}$$

where γ is the mean energy increase per cm^2 of interface and b is the magnitude of the Burgers vector of the unit single dislocation.

Cottrell[83] has extended the Fisher analysis in order to take into account domain-size effects on yield strength; accordingly

$$\sigma = \gamma l^{-1} (1 - \alpha a l^{-1}) \tag{3}$$

where σ is the resolved shear stress, γ the domain surface energy, α is a shape factor, l the domain size, and a is the thickness of the domain boundary. Experimentally, Broom and Biggs[84] have shown that for polycrystalline Cu_3Au the yield strength is a maximum for a domain size ~50 Å. This is in very good agreement with the value given by the Cottrell theory if a domain width ~2 atomic distances is used (Logie[85]). Ardley's data for single-crystal Cu_3Au[66] show clearly the separate effects due to LRO and antiphase domains. As the degree of LRO within the domains increases, the room-temperature strength decreases (Fig. 14), whereas with an increase in domain size the room-temperature strength first increases and then decreases, with a maximum value at ~25 Å domain diameter.

Three additional mechanisms have been advanced as determining the flow stress of cubic ordering alloys: (1) ordering in the stress field of a dislocation (Sumino[64]), (2) dislocation climb in an ordered lattice (Flinn[86]), and (3) variations in the thickness of domain-boundary ribbons attached to dislocations (Brown[87]). The Sumino theory predicts (*a*) that screw dislocations move more easily than edge dislocations, because the latter have a dilatational component in the strain tensor, and (*b*) that the yield strength reaches a maximum at T_c. CuZn does in fact show an abrupt increase in yield strength at T_c, although there is no maximum.[59,87] It is possible, however, that the latter may be obscured by other effects at T_c. Flinn predicts a maximum in flow stress at temperatures somewhat below T_c, consistent with observations on NiAlFe. Thus maxima on the yield stress-temperature curves occur at ~600°C, with $T_c > 1000$°C. In Brown's theory the thickness of the domain-boundary ribbon depends on the amount of atomic rearrangement possible during the slip process. For a β-brass type of superlattice

$$\tau = \frac{NKTS_0^{\,2}}{4b} \tag{4}$$

where N is the density of atoms on the antiphase boundary, K is Boltzmann's constant, T the absolute temperature, and S_0 the degree of LRO. The essential agreement between theory and experiment for β-brass rests on the presence of a pronounced maximum in yield point of 54×10^7 dynes cm^{-2} at 242°C according to the theory, and 36×10^7 dynes cm^{-2} at 220°C according to the experiment.

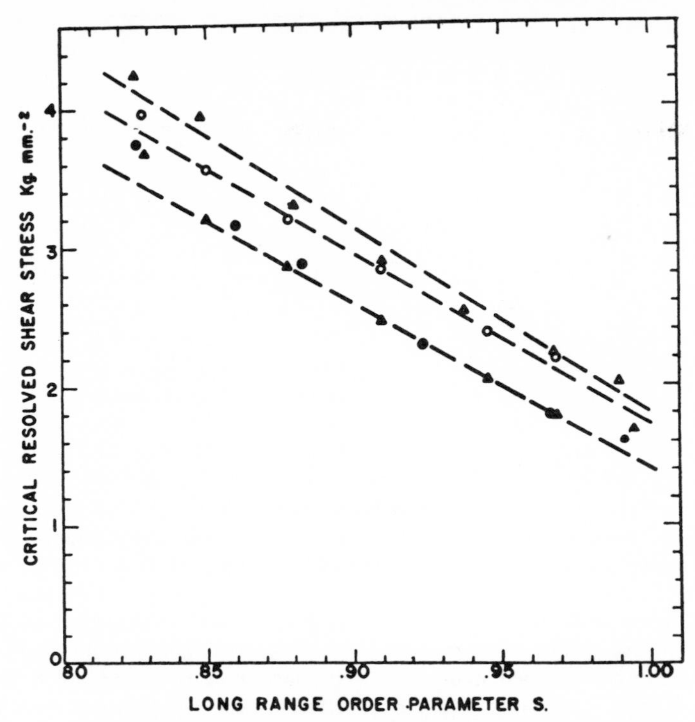

Fig. 14. The dependence of the critical resolved shear stress of Cu_3Au on the degree of long-range order. (After Ardley.[51])

Measurement of the variation of the flow stress of Ni_3Mn with temperature (Marcinkowski and Miller[88]) shows that there is a peak stress at ~15°C below T_c (480°C), which is not explained on the basis of any of the above theories. Instead, Marcinkowski and Miller consider two other possible mechanisms: (1) order strengthening due to destruction of the component of SRO coexisting with LRO (Rudman[89]), and (2) order strengthening associated with the presence of a periodic array of subdomains within the regular larger domains. Evidence for a subdomain structure has been found in Cu_3Au.[90]

Corresponding to the behavior of fcc Cu_3Au-type superlattices, increases in the flow stress of bcc Fe_3Al have been observed at intermediate degrees of order.[91–93] In Fig. 15 the room-temperature proof stress is plotted as a function of the quenching temperature for a 24.8 at.% Al alloy. The peak in flow stress for $T_{\text{quench}} = 540°\text{C}$ ($\equiv T_c$) would indicate a maximum in strength for zero long-range order. However, the X-ray diffraction measurements of Lawley and Cahn[94] show that after quenching in oil, a certain amount of ordering takes place, and in fact for $T_{\text{quench}} = 540°\text{C}$, the degree of long-range order S_0 is ~0.45 (Fe_3Al disordered bcc → Fe_3Al ordered bcc) In a second series of tests, on the 24.8 at.% Al alloy, the flow stress was measured as a function of isothermal annealing time at 440°C for the initially disordered alloy, and a peak flow stress found for an annealing time ~5 minutes, as in Fig. 16. X-ray measurements on the kinetics of the establishment of long-range order show that these maxima are associated with a critical combination of domain size and transient order. Lawley et al. consider the flow stress to be controlled by a combination of the Sumino and Fisher mechanisms, with the latter being of more importance for $T_{\text{quench}} > T_c$. Recently Hren[95] has observed a peak in the elastic modulus of Fe_3Al (measured at temperature) at 540°C. The maximum is believed to be a result of long-range stress fields associated with coexisting ordered and disordered phases of differing lattice parameter.

In view of the possible use of intermetallic

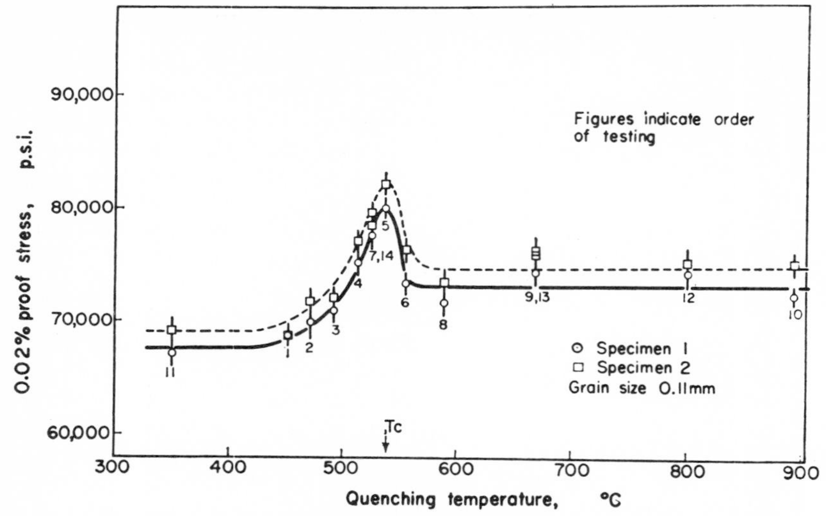

Fig. 15. Flow stress of Fe_3Al (24.8 at.% Al) alloy as a function of quenching temperature. Oil quenches. (After Lawley et al.[74])

compounds as high-temperature materials, creep properties are of prime importance. Published investigations concerned explicitly with the effect of order on creep refer to β-brass (Herman and Brown[96] and Hren[97]), Ni–Fe alloys (Kornilov and Panasyuk[98] and Suzuki and Yamamoto[99]), and Fe–Al alloys (Lawley, Coll, and Cahn[100]). The resistance to deformation of β-brass decreases sharply over a range of a few degrees just above the critical ordering temperature. These observations are especially noteworthy, because in β-brass the degree of order diminishes steadily to zero as the temperature approaches T_c. It is, therefore, the last trace of order which has the largest effect on the resistance of the alloy to plastic deformation. According to Hren, this abrupt change in steady-state creep rate at the critical temperature is primarily due to a difference in the elastic moduli of ordered and disordered phases. A similar situation exists in FeCo at T_c. In the Ni–Fe alloys, resistance to creep at a given temperature and stress is maximum at the stoichiometric composition both below and

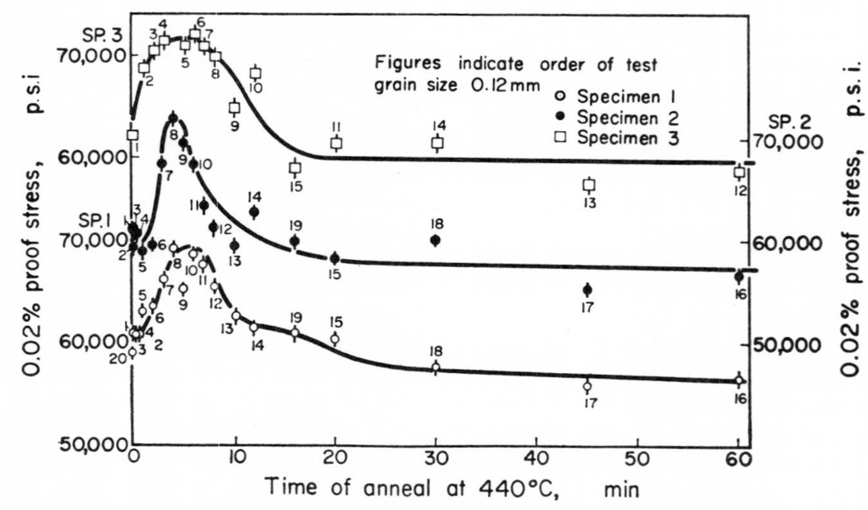

Fig. 16. Flow stress of Fe_3Al (24.8 at.% Al) alloy as a function of annealing time at 440°C. Crystals initially quenched in oil. (After Lawley et al.[74])

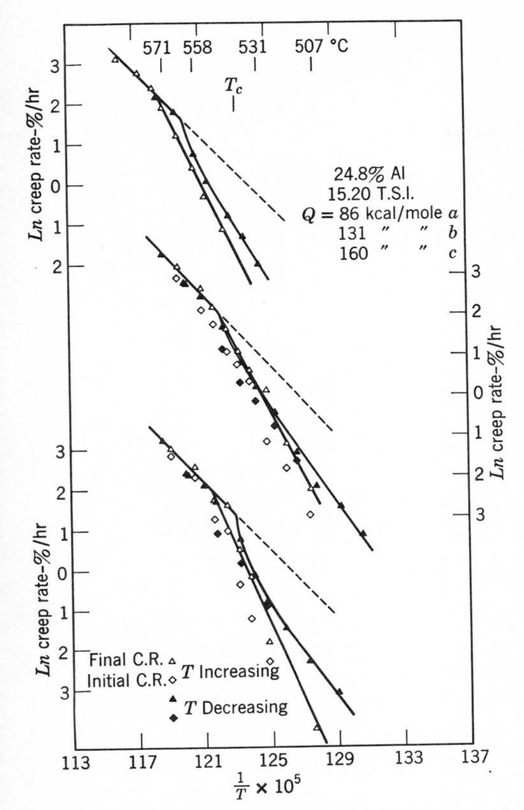

Fig. 17. Steady-state creep rate as a function of temperature for Fe_3Al (24.8 at.% Al) alloy. T_c = 540°C. (After Lawley et al.[81])

above T_c. Near T_c, the creep resistance of the ordered alloy is much higher than that of the same alloy in the disordered condition.[99]

The creep resistance of Fe–Al alloys changes very rapidly close to the critical temperature,[100] and it is concluded that the state of order determines the creep resistance. The alloys exhibit initially accelerating transient creep, which is attributed to a stress-induced change in order; in some instances this effect resulted in intermittent negative creep. Typical plots of log creep rate against the reciprocal of the absolute temperature are given in Fig. 17. The sharp decrease in resistance to deformation at T_c (≡540°C) corresponds to the transformation Fe_3Al ordered → FeAl ordered.* There is a second sharp decrease in creep resistance at ~740°C due to the transformation FeAl ordered → FeAl disordered.

* There is no change in composition, only a rearrangement of atoms.

4.5 Strain Hardening

Strain-hardening characteristics of intermetallic compounds are dependent on the state of order. In general, ordered alloys work-harden more rapidly than do equivalent alloys in the disordered state. Examples of this are found in Cu_3Au,[84,101–104,17] Ni_3Mn,[88] Ni_3Fe,[70] Fe_3Al,[91,43] and in alloys based on Ni_3Al,[86] (Fig. 18). Several theories[86,75,105,106] have been advanced to account for this difference. Flinn[86] argues that, as deformation increases, the domain size decreases, with each increment of strain producing an increasing increment of stress additional to that due to normal work hardening. The mechanism predicts work-hardening characteristics dependent on the domain size in the ordered alloy. However, from experiment, it is found that there is little change in the rate of work hardening as domain size varies. This has prompted Vidoz and Brown[75] to formulate a theory of work hardening based on jog production in the super-dislocation. The theory predicts a relationship between $\Delta\tau$ (the difference in flow stress between ordered and disordered alloy) and strain ϵ of the form: $\Delta\tau = K\epsilon^{1/2}$. The relationship is well obeyed by Ni_3Fe, Cu_3Au, Ni_3Mn, and Au_3Cu. In addition, the theory is able to account for the long-term anneal-hardening in strain-aged Ni_3Fe.[70] From electron microscope observations on Cu_3Au, Kear and Wilsdorf[102,107] have proposed that work hardening in the ordered alloy is controlled by the cross slip of screw dislocations from {111} to {100} cube planes, and also by dislocation reactions of the Lomer type. A more recent theory, also based on cross-slip, has been advanced by Davies and Stoloff.[17,104]

Although β-brass and AgMg both have the ordered CsCl structure, the former work-hardens at a much higher rate. In fact the work-hardening behavior of AgMg is very similar to that of unalloyed silver (Figure 18). Ordered Ni_3Mn, Ni_3Fe, and Ni_3Al, all having the Cu_3Au (Ll_2) structure, show high work-hardening rates.

5. THE CRYSTALLOGRAPHY OF DEFORMATION

The crystallography of deformation in many intermetallic compounds may be studied using the standard two-surface trace techniques

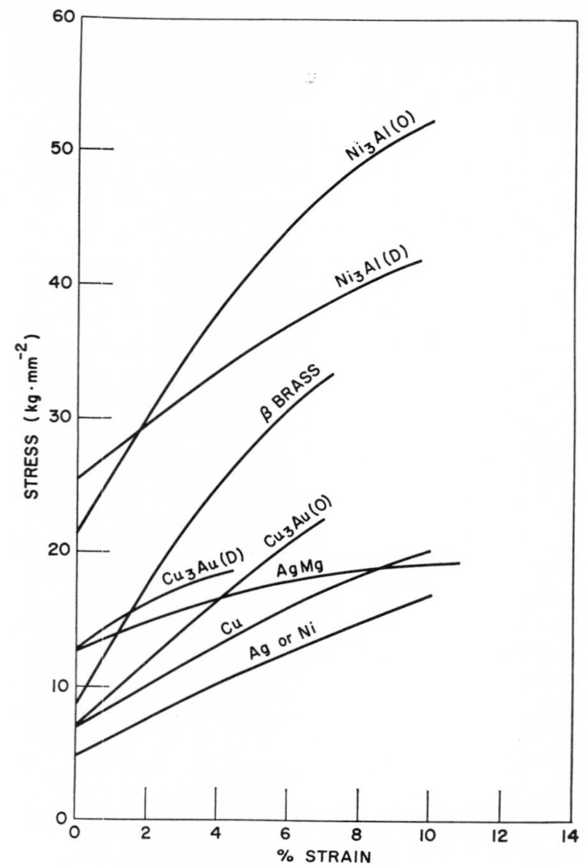

Fig. 18. Typical stress-strain curves for various ordered and disordered alloys.

common to pure metals and alloys.[108] In the case of the more brittle materials, prismatic punching[109,110] and microhardness indentation[111] methods have been used in order to promote deformation. For a limited number of compounds, etch-pit methods reveal the location of dislocations.[112,113] Electron microscopy of surface replicas provides a further means for studying the fine detail of the surface-deformation markings.

5.1 Slip

As in the case of pure fcc metals and alloys, intermetallic compounds having the Cu_3Au type structure deform by slip on the planes of highest atomic density, and in the direction of closest packing, that is, $\{111\}$ $\langle 110 \rangle$. The appearance of the slip markings in the ordered and disordered states is vastly different.[114–116] Disordered alloys show long, sharp, randomly spaced slip lines with single, double, and cross-slip, and with step heights up to 6000 Å. By comparison, in the ordered condition, a fine slip structure is present with an average spacing between slip lines of less than ~300 Å, and with an average step height ~100 Å. Figure 19 shows typical electron micrographs of surface replicas of deformed Ni_3Fe.[116] These differences in slip behavior between ordered and disordered material are comparable with those observed by Wilsdorf and Kuhlmann-Wilsdorf[117] in pure metals and alloys. Pure metals show the elementary ("homogeneous") slip structure and alloys such as α-brass exhibit the sharp ("inhomogeneous") slip.

The slip behavior of β-brass and AgMg (both having the CsCl structure) is similar to that of the pure bcc metals. The slip direction is unambiguously $\langle 111 \rangle$, whereas (110), (211), and (321) have been reported as the active slip planes, depending upon temperature. (See Table 1.) AuZn, AuCd, MgTl, and LiTl slip in the ⟨100⟩ cube direction. This observation has been interpreted in terms of partial dislocations in the simple cubic lattice of the CsCl structure (Rachinger and Cottrell[111]). When the ordering forces are large, partial dislocations are pulled together in pairs by the tension of the superlattice stacking faults joining the pairs. The perfect single dislocations so created can slip along ⟨100⟩. However, when the ordering forces are small, the partials remain dissociated, and slip occurs along ⟨111⟩. CuZn has a low ordering energy and disorders well below the melting point, whereas AgMg is a borderline case. AuZn and AuCd probably have ordering forces just large enough to cause ⟨100⟩ slip.

A similar analysis has been made for NaCl[118] structures. Thus PbTe has a ⟨100⟩ glide direction, but NaCl glides in the ⟨110⟩ direction. At low bonding energies, slip occurs by the movement of coupled partial dislocations of the type $\frac{1}{2}$ a [001]. These partials do not exist in ionic salts, however, and perfect dislocations of the type $\frac{1}{2}$ a [110] are responsible for slip. It should be pointed out that the dislocations in many intermetallic compounds are expected to be quite complex (Brown, Chapter 15). As an example, the zonal dislocations first proposed by Kronberg[121] in connection with β-uranium are expected to be operative in many of the σ-phases, which possess similar structures.

Few systematic studies have been made concerning the formation of deformation bands in intermetallic compounds. In the case of β-brass, Barrett[122] has suggested that a band boundary originates from the pileup of dislocations at an obstacle. The more recent observations of Bassi and Hugo[123] on β-brass, using an etch-pit technique, tend to confirm this idea. Because there is a tendency for dislocation pileup in β-brass, it is to be expected that plastic deformation by glide will quickly cease, and that further plastic deformation will occur by deformation band formation. The amount of lattice tilting necessary for band accommodation is limited, so that beyond this amount brittle fracture should occur, and in order to increase ductility deformation band formation must be delayed. It would appear that the situation is somewhat different in InSb, in which secondary {111} slip occurs within the deformation bands at strains ~40 per cent.[14] Birnbaum and Class[119] find that in β'-AuCd (orthorhombic) deformation bands are formed in regions where slip is predominantly on one-glide system, that is, (110) [001], with the slip traces terminating at the deformation bands. For a 5 per cent deformation, the lattice orientation is spread over ~6°.

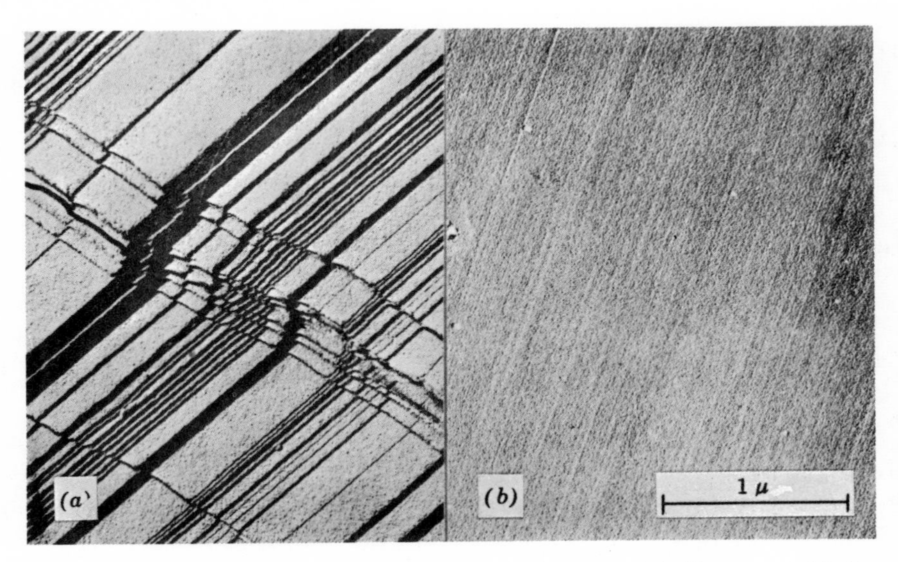

Fig. 19. Electron-micrographs of slip patterns of Ni_3Fe extended 10 per cent. (*a*) disordered, (*b*) ordered. (After Taoka et al.[93])

TABLE 1

Slip Data for Intermetallic Compounds

Intermetallic Compound	Crystal Structure	Slip Elements	Remarks	Reference
β-brass	CsCl B2 superlattice	(112) [11$\bar{1}$] at −196°C	...	
		(110) [$\bar{1}$11] 50°C to 200°C	Slip bands form in clusters	68
		(110) and (123) planes > 250°C	Wavy appearance of slip lines	
		(110) slip plane −196°C to 28°C	...	69
		(110) [111] at 25°C	...	110
AgMg	CsCl B2 superlattice	(321) [$\bar{1}$11] at 25°C	...	110
		(110), (211), and (321) slip planes, −196°C to 400°C	Tendency toward pencil glide with increasing temperature	18
AuZn	CsCl B2 superlattice	(110) [001] at 25°C	...	110
AuCd	CsCl B2 superlattice	(110) [001] at 25°C	...	110
MgTl	CsCl B2 superlattice	(110) [001] at 25°C	...	110
LiTl	CsCl B2 superlattice	(110) [001] at 25°C	...	110
CsBr	CsCl B2 superlattice	{110} ⟨100⟩ slip	Crystal has to be oriented for slip Kink bands form with a ⟨100⟩ compressive load	120
Fe_3Al	Body-centered $D0_3$ superlattice	(110), (211), and (321) slip planes at 25°C	(110) slip predominates Straight slip traces	63
PbTe	NaCl	(110) [001] at 25°C	...	
		(110) [001] at 300°C	...	118
Ag_2Al	Hexagonal	{0001} ⟨11$\bar{2}$0⟩ basal slip and {$\bar{1}$100} ⟨11$\bar{2}$0⟩ prismatic slip −196°C to 177°C	Kinking due to basal slip Fine slip structure	10
ζ-Cu-Ge	Hexagonal	{0001} ⟨11$\bar{2}$0⟩ basal slip	...	13
		{10$\bar{1}$0} ⟨11$\bar{2}$0⟩ 2 −1000°K	...	
AuCd	β′ orthorhombic	(110) [001] at 25°C	Straight slip traces	119
InSb	Zinc blende	(111) [$\bar{1}$10] 200°C	...	
		(110) at 25°C	Dislocation cracks-impact test	71
InSb	Zinc blende	(111) [$\bar{1}$10] 300°C to 515°C	Straight glide traces	14
InSb	Zinc blende	(111) [$\bar{1}$10] at 450°C	Etch pit observations	113

Recent work by Kear[102] on Cu_3Au has revealed the clustering of slip bands in different systems for both ordered and disordered material. Certain features of the slip band patterns are explained in terms of reactions between dislocations with Burgers vectors at 120°. These reactions may form diffuse tilt boundaries between the differently slipped regions of the crystal.

5.2 Twinning

Deformation twinning under a condition of tensile stress, has been observed in β-AuCd (orthorhombic, (111) twin plane), β''-AuCd (tetragonal),[124] and in Ag_2Al (hexagonal, $\{10\bar{1}2\}$ twin plane[10]). Single crystals of Cu-16.2 at.% Ge exhibit pyramidal deformation twinning[13] with indices: $K_1\{10\bar{1}1\}$, $\eta_1\langle 21\bar{2}\rangle$, $K_2\{\bar{1}013\}$, $\eta_2\langle 632\rangle$. In the case of InSb and GaSb, Churchman et al.[4] found twinning on $\{111\}$ and $\{123\}$ planes for $T/T_m \geqslant 0.44$. In the A4 structure the only true coherent twinning shear is $(1\bar{1}0)[11\bar{2}]$ with $S(111) = 0.4084a$, where a is the lattice parameter. For semicoherent boundaries, $\{123\}$ gives the best fit with $(1\bar{2}1)\,[41\bar{2}]$ shear and $S(123) = 0.6552a$.

According to Laves,[125] the B2-superlattice cannot go into the twin orientation by simple shear without disrupting the order. Experimentally, it is found that β-brass does not twin when subjected to various forms of deformation[68,69,122,126] at temperatures down to −196°C. It is to be expected that there will be a critical degree of order, for which the applied stress will just be able to force twinning dislocations through the lattice against the viscous resistance resulting from the destruction of bonds between unlike atoms. Unfortunately, in the case of β-brass, it is not possible to retain the alloy in various states of order at room temperature or subzero temperatures. However, with Fe_3Al ($D0_3$ superlattice), this is possible,[94] and Coll and Cahn[127] have verified that twins are formed at low temperatures only if the alloy is completely disordered or in a partially ordered condition ($S \leqslant 0.5$) (Fig. 20). In a number of intermetallic compounds, for example β-brass, NiAl, and AuCd, deformation takes place by a martensitic transformation, rather than by slip or twinning (see Chapter 22). This in general does not involve any net change in bond type.

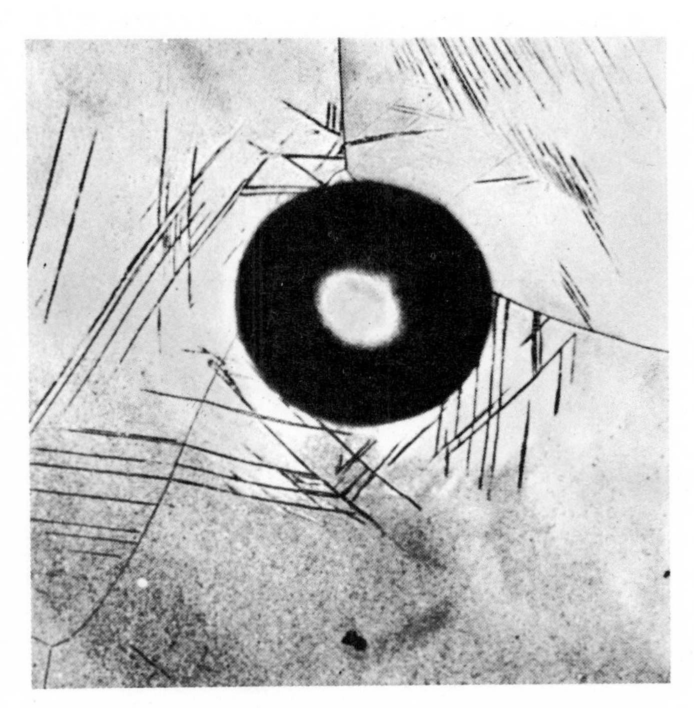

Fig. 20. Deformation twins in disordered Fe_3Al. Magnification 80 X. (After Coll and Cahn.[104])

6. ENGINEERING AND DEVELOPMENT ASPECTS

Many intermetallic compounds combine a high melting point with good oxidation resistance, a favorable strength-to-weight ratio, and a resistance to softening at elevated temperatures. Over the last ten years detailed development studies have been carried out on several intermetallic base alloys in an attempt to arrive at usable high-temperature structural components. The materials investigated include Ni_3Al,[25] $NiAl$,[25,41,128] $TiAl$,[40,41,129] Cr_2Ti,[39] Ni_4Zr,[130] Co–Cr,[131] and Hf–B[132] base alloys. Of necessity, the mode of preparation and processing varied widely for each particular system, and this, combined with the many unavoidable metallurigical variables, make an overall interpretation of the data extremely difficult.

Although several compounds exist possessing the required high-temperature rupture properties, their usefulness is somewhat limited due to brittleness and low-impact strength at room temperature. This drawback appears to have been overcome in a series of nonmagnetic TiNi (nitinol) compounds. Buehler and Wiley[133,134] obtained excellent room-temperature ductility (7–10 per cent) and impact strength (28 ft-lb) combined with a tensile strength ~120,000 psi. In addition, the TiNi and Ni-rich TiNi alloys were shown to be quench-hardenable and capable of maintaining their hardness at elevated temperatures. Apart from the possible use of the nitinols in nonmagnetic structural applications, an *improvement* in impact strength with decreasing temperature denotes a potential use in low-temperature areas. Preliminary data on intermetallic Hf–B compounds[132] show that for many compositions, the alloys are extremely hard without being excessively brittle; for example, HfB_4 has a Vickers hardness > 2000. Several studies are in progress concerning the mechanical behavior of compounds containing the rare earths.

Intermetallic compounds based on beryllium, silicon, and the refractory metals are characterized by having a maximum service temperature ~1600°C. The alloys have a relatively low density, reasonable oxidation resistance, and a high-temperature strength almost ten times greater than that of ceramics. As in the case of the more conventional compounds, however, room-temperature tensile strength is low and ductility practically nonexistent. Compounds of interest that show particular promise include $NbBe_{12}$, Nb_2Be_{17}, Nb_2Be_{19}, $TaBe_{12}$, Ta_2Be_{17}, $ZrBe_{12}$, Zr_2Be_{17}, and WSi_2. As an indication of the high-temperature strength of these alloys, for a test temperature of 1370°C breaking stresses fall in the range 37,000–65,000 psi. The refractory metals tungsten and molybdenum, and the alloy Nb–Zr–Mo, have, corresponding strength level of 50.,000 13,000, and 15,000 psi, respectively[135,136]

Grain-boundary hardening plays a dominant role in the plastic behavior of polycrystalline intermetallics, yet this effect has frequently been overlooked. The phenomenon

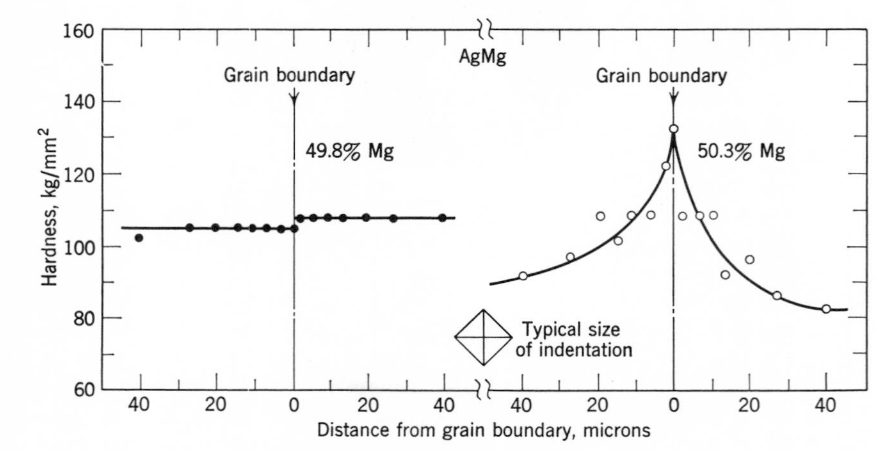

Fig. 21. Microhardness across grain boundaries in Ag-rich and Mg-rich AgMg compounds. (After Westbrook and Wood.[114])

TABLE 2

Approximate Grain-Boundary Hardening, $\Delta H/H$, at Room Temperature in Several Intermetallic Compounds

$$\frac{\Delta H}{H}(\%) = \left(\frac{H_{\text{boundary}} - H_{\text{bulk}}}{H_{\text{bulk}}}\right) \times 100$$

Compound (A_mB_n)	Crystal Structure	$\sim\Delta H$, %		
		Excess A		Excess B
AgMg	CsCl	0		20–50
NiAl	CsCl	0		25–60
NiTi	CsCl	0		~25
CoAl	CsCl	0		~20
CuZn		0		
CuZn[a] + 1 wt. % Al	CsCl		~11	
CuZn[a] + 3 wt. % Al			~13	
Ni_3Al	Cu_3Au	0		~15
Ir_3Cr	Cu_3Au		~20[b]	
AgTi	CuAu	0		~20
TiAl	CuAu		~20[b]	
γ-MgAl	α-Mn (bcc)		~15[b]	

[a] Copper-rich.
[b] Exact composition not known.

occurs quite generally in compounds having a stoichiometric excess of active metal component (Westbrook and Wood[137]), as illustrated for AgMg in Fig. 21, and can result in hardening differences from matrix to grain boundary of as much as 60 per cent (Table 2). The grain-boundary hardening is directly related to the usefulness or inapplicability of the compound in that the ductile-to-brittle transition temperature corresponds to the temperature at which the grain-boundary hardness is equal to that of the matrix. Hardening is a result of a concentration of oxygen and/or nitrogen at the grain boundary, although the specific mechanism is not clear. The grain-boundary hardening of AgMg is extremely sensitive to ternary solute additions. Instances are found of grain-boundary hardening imparted to a base originally free from this effect, as well as elimination of such hardening from magnesium rich bases. The effect can also be modified to some extent by specific annealing treatments. Grain-boundary hardening is closely related to the "pest" phenomenon frequently observed in such compounds as the beryllides, silicides, borides and aluminides,[138] that is, the degeneration to a powder at some intermediate temperature. The grain-boundary hardening, a forerunner of "pest" degradation, has a temperature dependence that may be used to predict the temperature range of "pest" susceptibility in a given compound.

7. SUMMARY

From the preceding sections, it is clear that intermetallic compounds encompass a wide range of crystal structures and bonding types. It is, therefore, not surprising that a comprehensive systematization of plastic behavior, in terms of the common metallurgical variables, is virtually impossible. The continued and increasing demand for high-temperature structural materials insures a wealth and diversity of developmental research. At the same time, investigations of the type carried out by Westbrook and Wood on AgMg, by Mote et al. on Ag_2Al, and current research on V–Ni and V–Co compounds[80] are providing a better understanding of the deformation process. The sensitivity of the mechanical behavior to small amounts of ternary additions or impurities, and to grain-boundary hardening, and the susceptibility to "pest" degradation require that meticulous care be taken in the purity, preparation, and heat treatment of the alloys. As in the study of primary metallic solid solutions, the use of single-crystal material is preferred whenever possible. The techniques of transmission electron microscopy and possibily field-ion microscopy offer potential in the understanding of the mechanical behavior of these materials.

ACKNOWLEDGMENTS

Acknowledgment is made to the several authors and journals for permission to reproduce figures and photographs.

REFERENCES

1. Stoloff N. S., and R. G. Davies, to be published in *Prog. Mater. Sci.* (1966).
2. *Mechanical Properties of Intermetallic Compounds*, J. H. Westbrook, ed. Wiley, New York, 1960.
3. Lowrie R., *Trans. AIME* **194** (1952) 1093.
4. Churchman A. T., G. A. Geach, and J. Winton, *Proc. Roy. Soc.* **238A** (1956) 194.
5. Borshchevsky A. S., N. A. Goryunova, and N. K. Takhtareva, *Zh. Tekhn. Fiz.* **27** (1957) 1408.
6. Schwab G-M., *Experimentia* **2** (1946) 103.
7. Schwab G-M., *Trans. Faraday Soc.* **45** (1949) 385.
8. Schwab G-M., in *Mechanical Properties of Intermetallic Compounds*, T. H. Westbrook, ed. Wiley, New York, 1960, p. 71.
9. Massalski T. B., Private communication, Mellon Institute, Pittsburgh, Pa.
10. Mote J. D., K. Tanaka, and J. E. Dorn, *Trans. AIME* **221** (1961) 858.
11. Thornton P. H., *Phil. Mag.* **8** (1963) 2013.
12. Thornton P. H., *Phil. Mag.* **11** (1965) 71.
13. Thornton P. H., *Acta Met.* **13** (1965) 611.
14. Abrahams M. S., and W. K. Liebmann, *Acta Met.* **10** (1962) 941.
15. Mukherjee A. K., and J. E. Dorn, *Trans. AIME* **230** (1964) 1065.
16. Bolling G. F., and R. H. Richman, *Acta Met.* **13** (1965) 709.
17. Davies R. G., and N. S. Stoloff, *Phil. Mag.* **12** (1965) 297.
18. Westbrook J. H., and D. L. Wood, *Trans. AIME* **224** (1962) 1024.
19. Terry J. C., and R. E. Smallman, *Phil. Mag.* **8** (1963) 1827.
20. Johnston T. L., R. G. Davies, and N. S. Stoloff, *Phil. Mag.* **12** (1965) 305.
21. Marcinkowski M. J., and H. Chessin, *Phil. Mag.* **10** (1964) 837.
22. Mack D. J., A. J. Birkle, and W. L. Krubsack, in *Mechanical Properties of Intermetallic Compounds*, J. H. Westbrook, ed. Wiley, New York, 1960, p. 320.
23. Flinn P. A., *Trans. AIME* **218** (1960) 145.
24. Davies R. G., and N. S. Stoloff, *Trans. AIME* **233** (1965) 714.
25. Grala E. M., in *Mechanical Properties of Intermetallic Compounds*, J. H. Westbrook, ed. Wiley, New York, 1960, p. 358.
26. Stoloff N. S., and R. G. Davies, *Trans. AIME* **230** (1964) 390.
27. Suzuki H., *Sci. Rept. Res. Inst. Tohoku Univ., Ser. A* **4** (1952) 455.
28. Suzuki H. in *Dislocations and Mechanical Properties of Crystals*, J. C. Fisher, W. G. Johnston, R. Thompson, and T. Vreeland, Jr., eds. Wiley, New York, 1957, p. 361.
29. Johnston W. G., *J. Appl. Phys.* **33** (1962) 2716.
30. Johnston T. L., A. J. McEvily, and A. S. Tetelman, *High Strength Materials*, Wiley, New York, 1965, 360.
31. Wood D. L., and J. H. Westbrook, *Trans. AIME* **227** (1963) 771.
32. Westbrook J. H., *Met. Rev.* **9** (1965) 145.
33. Marcinkowski M. J., and R. M. Fisher, *Trans. AIME* **233** (1965) 293.
34. Armstrong R., I. Codd, R. M. Douthwaite, and N. J. Petch, *Phil. Mag.* **7** (1962) 45.
35. Kornilov I. I., *Dokl. Akad. Nauk SSSR* **81** (1951) 597.
36. Kornilov I. I., *Dokl. Akad. Nauk SSSR* **106** (1956) 476.
37. Kornilov I. I., *Izv. Akad. Nauk SSSR* **397** (1957).
38. Guard R. W., and J. H. Westbrook, *Trans. AIME* **215** (1959) 807.
39. Grinthal R. D., "New High Temperature Intermetallic Materials," *WADC Tech. Rep.* 53-190, Parts 5 (Nov. 1956) and 6 (May 1958).
40. Grinthal R. D., in *Mechanical Properties of Intermetallic Compounds*, J. H. Westbrook, ed. Wiley, New York, 1960, p. 337.
41. Herz W. H., "An Investigation of Various Properties of NiAl," *WADC Tech. Rep.* 52-291, Part 3 (April 1955).
42. Steinitz R., in *High Temperature Materials*, R. F. Hehemann and G. M. Ault, eds. Wiley, New York, 1959, p. 182.

43. Westbrook J. H., *Trans. AIME* **209** (1957) 898.
44. Kornilov I. I., *Dokl. Akad. Nauk SSSR* **86** (1952) 721.
45. Westbrook J. H., *J. Electrochemical Soc.*, **104** (1957) 369.
46. Westbrook J. H., *J. Electrochem. Soc.* **103** (1956) 54.
47. Kornilov I. I., *Mechanical Properties of Intermetallic Compounds*, J. H. Westbrook, ed. Wiley, New York, 1960, p. 344.
48. Kurnakov N., S. Zhemchuzhny, and N. Zasedatelev, *J. Inst. Metals* **40** (1916) 305.
49. Syutkina V. I., and E. S. Yakovleva, *Soviet Phys.–Solid State* **4** (1963) 2125.
50. Yokoyama T., *Nippon. Kinzoku Gahkai-shi*, **B14** (1950) 1 [*Chem. Abstracts* **47** (1953) 7412].
51. Urasov G. G., *Z. Anorg. Chem.* **73** (1912) 31.
52. Savitskii E. M., Ref. 2, p. 87.
53. Kurnakov N., S. Zhemchuzhny, and V. Takarin, *Z. Anorg. Chem.* **83** (1913) 200.
54. Mott N. F., *Proc. Phys. Soc.* **64B** (1951) 729.
55. Schoeck G., *Creep and Recovery*. ASM, Cleveland, 1957, p. 199.
56. Berkowitz A. E., A. E. Jaumot, and F. C. Nix, *Phys. Rev.*, **95** (1954) 1185.
57. Bradley A. J., and A. Taylor, *Proc. Roy. Soc.* **A159** (1937) 56.
58. Bradley A. J., and G. D. Seager, *J. Inst. Metals* **64** (1939) 81.
59. Brown N., *Acta Met.* **7** (1959) 210.
60. Rudman P. S., *Acta Met.* **10** (1962) 195.
61. Vineyard G., *Phys. Rev.* **102** (1956) 981.
62. Green H., and N. Brown, *Trans. AIME* **197** (1953) 1240.
63. Weiss B. Z., *Bull. Res. Council Israel* **10F** (1961) 81.
64. Sumino K., *Sci. Rept. Res. Inst. Tohoku Univ., Ser. A* **10** (1958) 283.
65. Fisher J. C., *Acta Met.* **2** (1954) 9.
66. Ardley G. W., *Acta Met.* **3** (1955) 525.
67. Clarebrough L. M., *Acta Met.* **5** (1957) 413.
68. Ardley G. W., and A. H. Cottrell, *Proc. Roy. Soc.* **A219** (1953) 328.
69. Kramer I., and R. Maddin, *Trans. AIME* **194** (1952) 197.
70. Vidoz A. E., D. P. Lazarevic, and R. W. Cahn, *Acta Met.* **11** (1963) 17.
71. Allen J. W., *Phil. Mag.* **2** (1957) 1475.
72. Allen J. W., *Phil. Mag.* **3** (1958) 1297.
73. Good W. A., *Phys. Rev.* **60** (1941) 605.
74. Kuczynski G. C., M. Doyama, and M. E. Fine, *J. Appl. Phys.* **27** (1956) 651.
75. Vidoz A. E., and L. M. Brown, *Phil. Mag.*, **7** (1962) 1167.
76. Newkirk J. B., A. H. Geisler, D. L. Martin, and R. Smoluchowski, *Trans. AIME* **188** (1950) 1249.
77. Nowack L., *Z. Metallk.* **22** (1930) 94.
78. Köster W., *Z. Metallk.* **32** (1940) 145 and 277.
79. Jaumot F. E., and A. Sawatzky, *Acta Met.* **4** (1956) 127.
80. Peters E. T., J. J. Ryan, I. Vilks, and L. E. Tanner, *ASD TDR* 62-1087 Part III (Dec. 1964).
81. Hirabayashi M., and S. Weissman, *Acta Met.* **10** (1962) 25.
82. Fisher J. C., *Phys. Rev.* **91** (1953) 232.
83. Cottrell A. H., *Properties and Microstructure*, ASM, Cleveland, 1954, p. 151.
84. Biggs W. D., and T. Broom, *Phil. Mag.* **45** (1954) 246.
85. Logie H. J., *Acta Met.* **5** (1957) 106.
86. Flinn P. A., *Trans. AIME* **218** (1960) 145.
87. Brown N., *Phil. Mag.* **4** (1959) 693.
88. Marcinkowski M. J., and D. S. Miller, *Phil. Mag.* **6** (1961) 871.
89. Rudman P. S., *Acta Met.* **10** (1962) 253.
90. Raether H., *Z. Angew. Phys.* **4** (1952) 53.
91. Lawley A., E. A. Vidoz, and R. W. Cahn, *Acta Met.* **9** (1961) 287.
92. Kayser F. X., *WAAD Tech. Rep.* 57-298, Part I (1957).
93. Davies R. G., and N. S. Stoloff, *Acta Met.* **11** (1963) 1187.
94. Lawley A., and R. W. Cahn, *J. Phys. Chem. Solids* **20** (1961) 204.
95. Hren J. A., *Phys. Stat. Sol.* **3** (1963) 1603.
96. Herman M., and N. Brown, *Trans. AIME* **206** (1956) 604.
97. Hren J. A., Ph.D. Thesis, Stanford University, April 1962.
98. Kornilov I. I., and I. O. Panasyuk, *Izv. Sektora Fiz. Khim. Analiza, Akad. Nauk SSSR* **27** (1956) 164; *Proc. N.P.L. Symposium on Creep*, H.M.S.O., London, 1956, p. 215.
99. Suzuki T., and M. Yamamoto, *J. Phys. Soc. Japan* **14** (1959) 463.

100. Lawley A., J. A. Coll, and R. W. Cahn, *Trans. AIME* **218** (1960) 166.
101. Sachs G., and J. Weerts, *Z. Phys.* **67** (1931) 507.
102. Kear B. H., *Trans. AIME* **224** (1962) 669.
103. Kear B. H., *Acta Met.* **12** (1964) 555.
104. Davies R. G., and N. S. Stoloff, *Phil. Mag.* **9** (1964) 349.
105. Vasilyev L. I., and A. N. Orlov, *Phys. Met. Metallog.* **15** (1963) 1.
106. Kozlov E. V., and L. E. Popov, *Soviet Phys.* "*Doklady*" **8** (1964) 928.
107. Kear B. H., and H. G. F. Wilsdorf, *Trans. AIME* **224** (1962) 382.
108. Barrett C. S., *Structure of Metals*. McGraw-Hill, New York, 1953, p. 39.
109. Smakula A., and M. W. Klein, *Phys. Rev.* **84** (1951) 1013.
110. Rachinger W. A., and A. H. Cottrell, *Acta Met.* **4** (1956) 109.
111. Westbrook J. H., *ASTM Bull.*, No. 246 (1960) 53.
112. Bakish R., and W. D. Robertson, *Acta Met.* **3** (1955) 513.
113. Venables J. D., and R. M. Broudy, *J. Appl. Phys.* **29** (1958) 1025.
114. Taoka T., and S. Sakata, *Acta Met.* **5** (1957) 61.
115. Taoka T., and R. Honda, *J. Electronmicroscopy* (*Japan*) **5** (1957) 19.
116. Taoka T., K. Yasukochi, and R. Honda, in *Mechanical Properties of Intermetallic Compounds*, J. H. Westbrook, ed. Wiley, New York, 1960, p. 192.
117. Kuhlmann-Wilsdorf D., and H. G. F. Wilsdorf, *Acta Met.* **1** (1953) 349.
118. Rachinger W. A., *Acta Met.* **4** (1956) 647.
119. Birnbaum H., and W. Class, *Acta Met.* **6** (1958) 609.
120. Johnson L. D., *Mechanical Behavior of Single and Polycrystalline CsCl*. Univ. of California, Radiation Lab., Rep. Sept. 1962.
121. Kronberg M. L., *J. Nucl. Mater.* **1** (1959) 85.
122. Barrett C. S., *Trans. AIME* **200** (1954) 1003.
123. Bassi G., and J. P. Hugo, *J. Inst. Metals* **87** (1958) 155.
124. Birnbaum H. K., and T. A. Read, *Trans. AIME* **218** (1960) 381.
125. Laves F., *Naturwissenschaften* **39** (1952) 546.
126. Elam C. F., *Proc. Roy. Soc.* **A153** (1936) 273.
127. Coll J. A., and R. W. Cahn, *Acta Met.* **9** (1961) 138.
128. Imai Y., and M. Kumazawa, *Sci. Rept. Res. Inst. Tohoku Univ.* **A11 210** (1959) 312.
129. McAndrew J. B., and H. D. Kessler, *Trans. AIME* **206** (1956) 1348.
130. Silverman R., "New High Temperature Intermetallic Materials," *WADC Tech. Rep.* 35-190, Part 4 (March 1956).
131. Silverman R., W. Arbiter, and F. Hodi, *Trans. ASM* **49** (1957) 805.
132. Kirkpatrick C. G., "Refractory Intermetallic Compounds," *ASTIA Rep.* AD 289, 364 (Feb. 1962).
133. Buehler W. J., and R. C. Wiley, *Trans. ASM* **55** (1962) 269.
134. Buehler W. J., and R. C. Wiley, *Mater. Design Eng.* **55** (1962) 82.
135. Stonehouse A. J., *Mater. Design Eng.* **55** (1962) 82.
136. Stonehouse A. J., J. Booker, and R. M. Paine, *WADD Tech. Rep.* 60-889, Part II (July 1962).
137. Westbrook J. H., and D. L. Wood, *J. Inst. Metals* **91** (1962–63) 174.
138. Westbrook J. H., and D. L. Wood, *J. Nucl. Mater.* **12** (1964) 208.

chapter 25

Corrosion Behavior

EDWARD A. AITKEN*

Nuclear Materials and Propulsion Operation
General Electric Company
Cincinnati, Ohio

1. INTRODUCTION

Many intermetallic compounds have high melting points and high strengths. In addition, some are oxidation-resistant and are, therefore, a source of materials for high-temperature refractory applications. The metal-metalloid compounds (carbides, nitrides, and borides) have received special attention for high-temperature application, but they (except SiC) do not possess oxidation resistance for temperatures above 1000°C. The silicide, aluminide, and beryllide intermetallic compounds offer promise for use at high temperatures, but they have been rarely used in load-carrying functions because of their lack of ducitility; instead, most of these compounds have been used as coatings over a ductile basis metal. In recent years there has been an increasing amount of publication on the oxidation behavior of these high-melting intermetallic compounds as single-phase materials. Other types of intermetallic compounds, namely group III-V and II-VI compounds, because of their unique electronic properties, have received much attention from the viewpoint of corrosion. Surface contamination of semiconductors by oxides, hydrated oxides, and organic compounds can change both the surface and bulk electronic properties. Because crystal cleavage is usually done in an ambient atmosphere, the surface is readily contaminated by oxygen and water. Etching agents are employed to remove the surface contamination. Thus the technology of surface preparation of semiconductor materials is a source of corrosion data for the semiconductor type of intermetallic compounds. In this review oxidation reactions are emphasized; however, other forms of corrosion are covered briefly for completeness.

Precise kinetic data on many intermetallic compounds are not abundant, because satisfactory control of the purity and composition of the starting material has not been achieved and the density and microstructure have not been completely determined. Therefore, the results are mainly qualitative and represent evaluations as engineering materials. When possible, the discussion here emphasizes description of the basic mechanism rather than quantitative data on corrosion rates because the absolute rates are subject to purity and the

* Present address: Vallecitos Atomic Laboratory, General Electric Company, Pleasanton, California.

process used. Summary data have been provided for comparison purposes, but the reader is cautioned in extrapolating the data beyond the conditions specified. In general, the discussion is confined to intermetallic systems whose oxide films are protective in a useful temperature region. The oxidation characteristics of carbides, nitrides, and borides are omitted because of their general lack of oxidation resistance. Similar data have been omitted on cermets, which have been discussed elsewhere.[1] The silicide, aluminide, and beryllide compounds are arranged in separate sections because of their general interest for high-temperature applications. Silicon carbide has been included because of the similarity of its mechanism with other silicides. The A(III)-B(V) compounds are discussed as a group. Because of the wealth of fundamental knowledge and control of purity and microstructure available with A(III)-B(V) compounds, they provide the most detailed source of data on the corrosion behavior of any intermetallic system. Some compounds have been omitted from discussion mainly because of insufficient data. A comprehensive survey of the properties, including oxidation resistance, of intermetallic compounds with melting points above 1400°C was compiled by Paine, Stonehouse, and Beaver (PSB).[2] This reference includes thirty-five binary systems and is an excellent source for qualitatively indexing the oxidation behavior of intermetallic compounds for the high-temperature field.

2. OXIDATION THEORY AND MECHANISMS

Metal oxidation theories are sufficiently general so that they are pertinent to oxidation of intermetallic compounds as well. The oxidation process involves the absorption, dissociation, and ionization of oxygen molecules, and subsequent incorporation of the ions to form the oxide layer first as a thin film, then as a thick film or scale. The later stages of oxidation can lead to further changes, such as breakaway, spalling, or cracking. Because the standard free energy change for nearly all metals is negative, oxidation should proceed spontaneously on exposure to oxygen. The reaction beyond the monolayer stage encounters an activation barrier, and continued growth becomes kinetically controlled. The controlling processes can be diffusion of cations, anions, vacancies, interstitial ions, electrons, or some combination of them.

The first step of oxidation, chemisorption, is somewhat complex, and the mechanism is not fully understood. This step involves the adsorption of one or more layers of oxygen, probably in the form of O_2^-, O^-, or O^{2-} ions, and occurs very rapidly even at cryogenic temperatures. The work function of the metal is altered as a result of surface coverage, and surface potentials at the gas-metal interface lead to very strong local fields across the adsorbed layer. These fields are sufficiently strong to influence the transport of electrons and ions in subsequent oxidation and effect the kinetics of oxidation during the growth of thin films (<100 Å). For larger thicknesses the influence of the electric field diminishes, and the driving force is provided by the concentration gradient of the rate-controlling defect in the oxide film.[20] If the oxide film is adherent and dense, the oxidation behavior will follow one of the well-known kinetic laws. If the oxide film is not adherent, porous, or is adherent but forms multiple layers, the oxidation behavior is a linear function of time.

The rate expressions for oxidation may be written in generalized form as

$$\frac{dx}{dt} = k_1 x^{-\epsilon} \tag{1}$$

$$\frac{dx}{dt} = k_2\, e^{-\eta x} \tag{2}$$

where x is the extent of reaction at time, t, and ϵ and η are appropriate constants. Rosenberg[18,19] used the differential form of (1) and (2) to describe the oxidation behavior of the A(III)-B(V) compounds, as shown by Equations 3 and 4.

$$-\frac{\partial \log\left(\dfrac{dx}{dt}\right)}{\partial \log x} = \epsilon \tag{3}$$

$$-\frac{\partial \log\left(\dfrac{dx}{dt}\right)}{\partial x} = \eta = \frac{\epsilon}{x} \tag{4}$$

When ϵ and η are plotted as a function of scale thickness, effects of temperature and pressure on the mechanism are readily apparent. The parameter, ϵ, takes on values of 0, 1, 2, respectively, for linear, parabolic, and cubic oxidation. Positive values of ϵ or η indicate that a protective film is forming.

Many mechanisms have been proposed to explain the oxidation process. The derivations are based on a particular set of conditions, which control the transport of cations or oxide ions through the film. These mechanisms have been comprehensively treated by Kubaschewski and Hopkins,[3] and by Hauffe[4] and Evans.[23] A recent review by Roberts[5] describes qualitatively the general state of oxidation theory.

3. NATURE OF THE CORROSION PRODUCT

3.1 Oxidation

The oxidation behavior of intermetallic compounds is in many ways similar to that of the metallic elements or alloys from which they are prepared. The intermetallic compound forms an oxide on exposure to oxygen which grows by one of the common rate laws. The oxide product, formed on the intermetallic compound, is often the same as that on metallic solid solutions containing the same constituents. The oxidation of intermetallic compounds, like many solid-solution alloys, is usually selective, preferentially forming the oxide of the least noble constituent. For the silicide, aluminide, and beryllide compounds, the oxide product, formed at high temperature, is SiO_2,[50] Al_2O_3,[11] and BeO,[16] respectively. For InSb, the oxide of indium is formed preferentially. The oxide can form on the surface or internally by precipitation of the oxide out of the matrix. Clear evidence of internal oxidation, however, has not been reported as yet in intermetallic compounds, but strong evidence of localization of oxygen or nitrogen in the vicinity of the grain boundaries after mild heating has been obtained.

There are cases where the oxide product is a mixture of oxides or a mixed oxide compound. In the oxidation of selenide compounds, the metal selenates[6–9] have been observed. Oxidation of refractory intermetallic compounds at low temperatures has shown a tendency to form more than one oxide. Both TiO_2 and Al_2O_3 have been observed[10] in scales of Ti–Al compounds. The mixed oxide compounds of $Nb_2O_5 \cdot Al_2O_3$[11] have been observed in the oxidation of $NbAl_3$ at low temperatures.

The tendency to form more than one oxide at low temperatures may result from the low mobility of the metal atoms, which prevents the preferential formation of a single oxide that would normally occur at high temperatures. A scale containing the oxides of all of the constituents may produce a structure which is metastable or amorphous and certainly more complex. The oxide composition and structure may vary with atmosphere as in the case of SiC, in which Jorgensen et al.[12] have observed that the tridymite structure is obtained by oxidation in water vapor instead of cristobalite, which is obtained in dry oxygen. Atmospheres may also influence the grain texture and growth morphology of the scale.

The oxidation resistance of the intermetallic compound is achieved by a protective oxide film. If the film cracks or is nonadherent, the oxidation rate will be limited by the permeation rate of oxygen through the oxide layer. The ability of an oxide to protect the intermetallic compound depends on its tenacity with the matrix. Because both the intermetallic compound and the oxide characteristically are difficult to deform, particularly at low temperatures, protective films quite often fracture or spall on thermal cycling. For protective action of the oxide layers, the unyielding character of the intermetallic compound requires that the differential expansion be small and that the molar volume ratio between the oxide scale and the intermetallic compound be near unity.

The oxide products, which are formed on a number of intermetallic compounds, are listed in Table 1. Some properties of the oxides are included which help to characterize the oxide formed in the oxidation process. Additional thermochemical and structure data have been summarized by Kubaschewski and Hopkins.[3]

3.2 Aqueous Corrosion

The presence of moisture enables the corrosion product to deviate from the type of oxide films formed in oxygen. Since the nascent hydrogen is highly reactive it creates the opportunity for formation of hydrates, hydroxides, or gaseous hydrides. The corrosion of antimonides[27] is usually accompanied by the evolution of gaseous SbH_3, leaving behind the oxide or hydroxide of the other component. Aqueous solutions such as etchants leave no film in most cases, but sometimes the substrate can be made passive by formation of an impervious oxide film, which does not dissolve in oxidizing media.

TABLE 1

Summary of Properties of Protective Oxide Scales Formed on Intermetallic Compounds During Oxidation

Oxide Compound	Typical Substrate	Melting Point °C[67]	Structure[a] Low Temperature	Structure[a] High Temperature	Mean Coefficient of Thermal Expansion[67] $(°C)^{-1} \times 10^6$	Density[67] g/cc
SiO_2	$MoSi_2$[43],	1728	amorphous	β-cristobalite	3.0 (300–1100°C) crystalline	2.32
	SiC[12]			tridymite	5 (20–1250°C) vitreous	2.26
BeO	$ZrBe_{13}$[57,58] $NbBe_{12}$[58]	2550	wurtzite and oxide of other metal	wurtzite	9.3 (20–1000°C)	3.01
Al_2O_3	$NbAl_3$[11,14]	2040	γ-Al_2O_3 + $NbAlO_4$	α-Al_2O_3	8.0 (20–1000°C)	3.97
	TiAl[10,63]		TiO_2 + Al_2O_3			
In_2O_3	InSb[19]	volatilizes >900°C	cubic Tl_2O_3[68]		...	7.14[68]
Ga_2O_3	GaAs[b]	1740	hexagonal[67]	monoclinic[67]	...	5.88
TiO_2	TiAl[10,63]	1835	rutile + Al_2O_3	rutile + Al_2O_3	7–8.0 (20–600°C)	4.24
$NiO \cdot Al_2O_3$	NiAl[77]	2030	cubic spinel	cubic spinel	0.82[88] (25–1000°C)	4.52

[a] The type of structure observed is divided into a high- and low-temperature region under which the oxidation was carried out. The dividing point varies with the material; the reader should consult the specific reference for details.
[b] This has not been confirmed, but may be inferred from studies in Reference 18.

4. LOW-TEMPERATURE DISINTEGRATION

Some intermetallic compounds which form protective films at high temperatures virtually disintegrate when heated in a lower-temperature region. The disintegration process takes place without any appreciable bulk oxidation. The phenomenon was observed with $MoSi_2$ by Fitzer,[13] who inappropriately termed it a "pest" reaction because of a similar occurrence with tin. The "tin pest" is the result of a crystallographic transformation, but no such transformation occurs for $MoSi_2$. The "pest" reaction or disintegration has been observed for a broad range of intermetallic compounds regardless of crystal structure. Table 2 lists the compounds that are known to undergo disintegration.[69] Its occurrence seriously hampers the utility of some intermetallic compounds, which otherwise have obvious potential for high-temperature application. The solution of this problem is of prime importance, but satisfactory solutions are lacking because of the surprisingly little attention given to delineating the mechanism for the disintegration process.

The major features of the "pest" reaction are the following:

1. Disintegration occurs over a finite temperature range, which is specific for each compound.
2. Some exposure to a reactive gas (i.e., oxygen, nitrogen, water) is necessary, for disintegration does not occur in inert atmospheres.
3. Disintegration is preceded by an incubation period, in which little or no weight change is observed.
4. The size of fragments formed during disintegration tends toward uniform particles of successively smaller size.
5. The effect has been observed in single crystals, although it is diminished somewhat in comparison to polycrystalline materials.

Low-temperature disintegration is very dramatic in certain compounds, but it appears to be absent in others. Rausch[14] reported that "pest behavior was not present in transition-metal silicides of group IV and in the first members of groups V and VI. Stonehouse[15] has reported that $TaBe_{12}$ is free of the "pest," but it is present in $NbBe_{12}$ and $ZrBe_{13}$. Westbrook and Wood[69] have demonstrated that disintegration occurred in AgMg and NiAl only when a slight excess of Mg or Al was present. Others have indicated that the effect can be aggravated by impurities[14] and moderated[11] by certain alloying.

Grain-boundary hardening has been found by Westbrook and Wood[69] to occur in a variety of intermetallic compounds. (See Table 3.) The hardening is particularly noticeable when the more electropositive element is present in slight excess, as shown in Fig. 1, yet within the single-phase region of the compound. Furthermore, the degree of hardening was decreased by quenching from successively higher temperatures. Oxygen or nitrogen and possibly other reactive gases were found to be a necessary part of the hardening mechanism, because local hardening diminished along grain-boundary paths perpendicular to the exposed metal surface. No second phase was visible at the grain boundary after heat treatment, as is the usual case even under examination by electron microscopy.

The kinetics of grain-boundary hardening have been explored by Seybolt and Westbrook[70] for NiGa, which exhibits the "pest"

TABLE 2

Known Incidence of the Pest Effect Among Intermetallic Compounds

	Other Active Metal-Rich Compounds	MX_3	M_2X_5	MX_2	M_2X_3	MX	M_3X_2	M_2X	M_3X	Other Active Metal-lean Compounds
Aluminides	Co_2Al_9	$CbAl_3$ $TaAl_3$ $(Ta, Cb, V) Al_3$ $MoAl_3$ $FeAl_3$ $MnAl_3$ $NiAl_3$	Fe_2Al_5 Co_2Al_5		Ni_2Al_3	$FeAl$ $NiAl$ $SbAl$			Cr_3Al	
Beryllides	$ZrBe_{13}$ $NbBe_{12}$ Nb_2Be_{17}									
Borides				TiB_2						
Calcides				$PbCa_2$ $AgCa_2$		$PbCa$ $AgCa$ $TlCa$			Pb_3Ca Tl_3Ca	Tl_4Ca_3 Cu_5Ca
Cerides		$BiCe_3$		$SnCe_2$		$AlCe$ $BiCe$	Sn_3Ce_2	Sn_2Ce Al_2Ce Bi_2Ce		Al_4Ce Bi_3Ce_4 Mg_9Ce
Gallides						$NiGa$				
Magnesides			Tl_2Mg_5	$PbMg_2$ $SnMg_2$ $TlMg_2$	Sb_2Mg_3 Bi_2Mg_3	$CdMg$ $AgMg$				
Silicides				$MoSi_2$ WSi_2 $CbSi_2$ $FeSi_2$ $MnSi_2$			Mo_3Si_2		Cu_3Si	Cb_5Si_3

TABLE 3

Incidence of Grain-Boundary Hardening In Some Selected Intermetallic Compounds

Compound	Relative Grain-Boundary Hardening ($\Delta H/H$ per cent)
$MoSi_2$	35
TiB_2	18
$ZrBe_{13}$	20–50
Zr_2Be_{17}	30
$TaBe_{12}$	55
AgMg(Mg-rich)	20–50
NiAl(Al-rich)	25–60

reaction. The degree of hardening was found to be sensitive to the stoichiometry. When microhardness measurements were used the hardening increased in proportion to the square root of time between the temperatures 400 to 800°C. The temperature dependence of the hardening rates showed three branched curves, indicating that more than one process was operating over this temperature region. The grain-boundary hardness increased over a much deeper distance from the gas-metal interface relative to the bulk hardness, and the penetration depth was shown to vary with the oxygen available on the surface, indicating that the hardness gradients are due to oxygen diffusion. The oxygen located in the grain boundary was proposed to be in the form of clusters around the gallium atoms or as precipitates of oxide or suboxide. Attempts to find a precipitate in the grain boundary were unsuccessful.

Noting that many of the compounds which exhibited local hardening also underwent disintegration at low temperatures, Westbrook and Wood[69] have proposed a model based on preferential grain-boundary diffusion of a reactive gaseous element and segregation of the element in the lattice structure near the grain boundary. Although the mechanism of hardening is not definitely established, several observations support a close correlation between the upper-temperature limit for grain-boundary hardening disappearance and the temperature limit for the disappearance of the disintegration effect in the compounds NiAl, $MoSi_2$, $ZrBe_{13}$ (see Fig. 2). The model is shown schematically in Fig. 3. At low temperatures, the oxygen diffusion is too low to penetrate the grain boundaries. At intermediate temperatures the oxygen segregated in regions near the boundary leads to hardening and embrittlement through internal stresses which produce fragmentation. At high temperatures, hardening and embrittlement are relieved and the bulk diffusion and grain boundary diffusion rates for oxygen become comparable.

The grain-boundary hardening thus appears to be a necessary condition for a compound to exhibit the disintegration effect, but it may not be a sufficient condition. For example, the local hardening has been detected in $TaBe_{12}$, but this compound is reported not to undergo low-temperature disintegration. Pure metals and solid solutions do show some grain-boundary hardening; however, the extent of

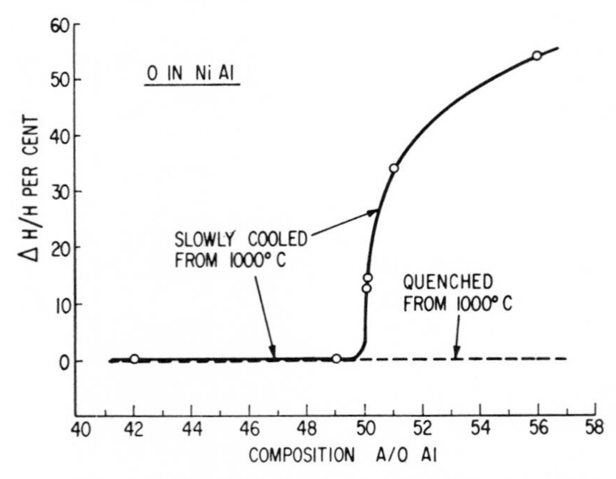

Fig. 1. Effects of heat treatment and stoichiometry on grain-boundary hardening in NiAl.

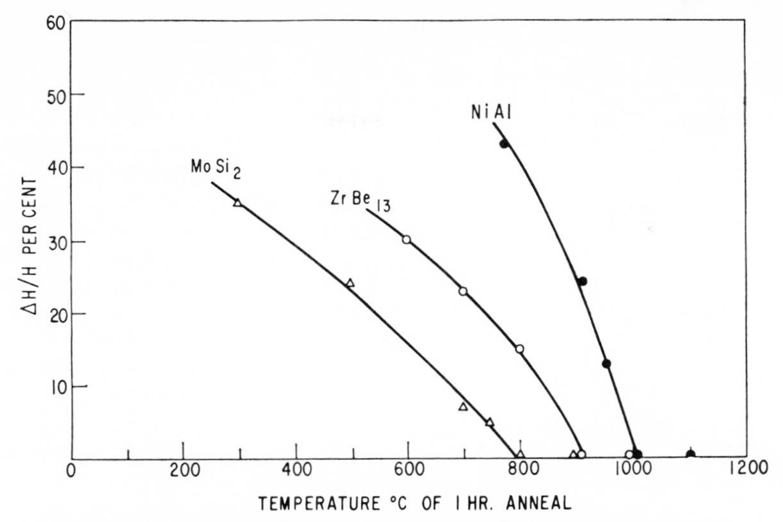

Fig. 2. Grain-boundary hardening as a function of quenching temperature in $MoSi_2$, NiAl, and $ZrBe_{13}$.

the increase over the bulk is small. The question of why such a large increase in local hardening occurs for intermetallic compounds at a specific stoichiometry and temperature lacks a satisfactory solution at the moment.

5. OXIDATION OF SPECIFIC COMPOUNDS

5.1 Silicide Systems

Silicon forms silicide compounds with a large number of metals, some of which have high melting points. SiC and $MoSi_2$ are the best known and are used as heating elements. Two surveys have been conducted to characterize the oxidation resistance of some of the higher melting silicide compounds. One survey by Kieffer, Benesovsky, and co-workers[34] reported that maximum oxidation resistance in the Mo–Si,[35] Ta–Si,[36] W–Si,[37] Zr–Si,[38] and Cr–Si[39] systems in air at temperatures between 1100°C and 1500°C occurred near the disilicide composition. In the other survey Paine, Stonehouse, and Beaver (PSB)[2] have measured the weight gain of some high melting silicides. These data were obtained on sintered plates, which were heated in dry air at 1260°C for 100 hours. The results are summarized in Table 4. Of the group tested by PSB, only Cr_3Si and the titanium silicides showed acceptable oxidation resistance. PSB found the Cr_3Si compound was more oxidation-resistant than compositions with higher silicon, Cr_5Si_3 and Cr_6Si_4; however, Arbiter[40] obtained better oxidation resistance with the compositions

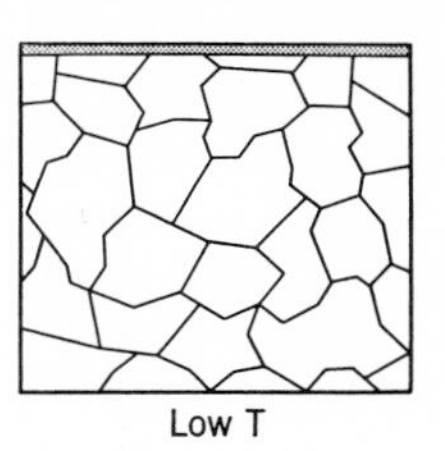

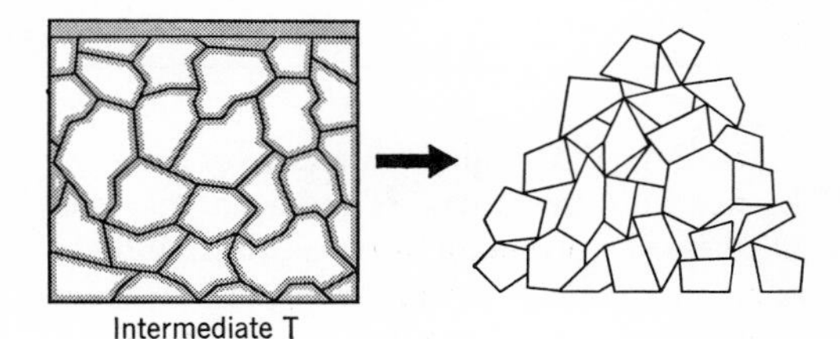

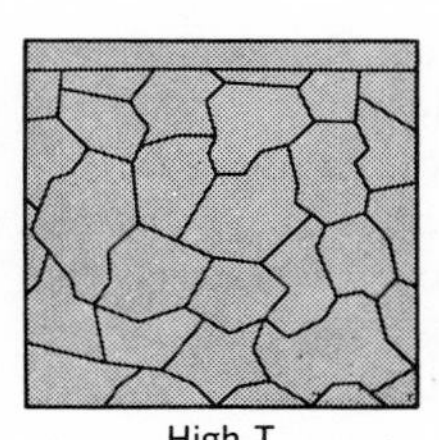

Fig. 3. A grain-boundary segregation model for the pest phenomenon, schematic.

TABLE 4

Summary of Oxidation Data of Various Silicide Compounds

Compound	Density	Test Temperature, °C	Time, hours	Weight Gain mg/cm²	Reference
SiC	3.21	1300	37	+0.64	49
		1135	120	+0.55	50
		1300	100	+0.95	50
		1600	33	+0.95	50
$TiSi_2$	4.02	1200	4	+0.3	34
	3.96	1260	100	+2.2	2
		1370	100	+4.4	2
$ZrSi_2$	4.73	1200	4	+42	34
$ThSi_2$		1200	4	+73	34
VSi_2	4.71	1200	4	4.9	34
$NbSi_2$	5.29	1200	4	−54	34
		1260	100	Oxidation complete	2
$TaSi_2$	8.8	1200	4	−51	34
	8.3	1260	100	Oxidation complete	2
Cr^3Si	5.92	1000	100	+0.5	40
	5.92	1260	100	+7.3	2
		1370	100	+15.0	2
Cr_5Si^3	5.49	1260	100	+57	2
$CrSi_2$	4.73	1200	4	+51	34
		1300	100	7	40
$MoSi_2$	6.12	1200	4	+0.3	34
		1400	2000	+3.7	42
		1500	4	+9	34
		1700	3000	+12.3	42
WSi_2	9.3	1200	4	−17	34
		1500	4	−23	34

containing the higher amount of silicon. Kieffer and Benesovsky[39] reported that the silicides of chromium form porous, loosely adherent scales, which may result from volatilization of chromium oxide. With the likelihood of vaporization, weight-gain measurements would be suspect.

The results of PSB on niobium silicides and tantalum silicides showed low-oxidation resistance, which is in qualitative agreement with the findings of Kieffer, Benesovsky, and coworkers.[34,36] Other silicides of zirconium,[38] cerium,[2] lanthanum,[2] and vanadium[41] have been reported to exhibit poor oxidation resistance above 1000°C and to form scales that spalled on temperature cycling.

PSB[2] found that the titanium silicides exhibited good oxidation resistance, even though the single-phase compositions could not be prepared.

Rausch[14] observed low-temperature disintegration between 500°C and 1000°C in transition-metal silicides of M_5Si_3, M_3Si_2, and MSi_2, in which M was niobium, tantalum, molybdenum, and tungsten. Not all of the silicides were studied, but disintegration was not found in the group IV metal silicides nor in the first member of groups V and VI.

Molybdenum Disilicide. Molybdenum disilicide, $MoSi_2$, is well-known for its oxidation resistance. Fitzer and Schwab[42] obtained film thicknesses of 0.02–0.03 mm after 2000 hours in oxygen at 1400°C and of 0.05–0.1 mm after 3000 hours at 1700°C. The films were self-healing and well-adhering. Maxwell[43] concluded that a protective surface coating of SiO_2, cristobalite, formed on oxidation at elevated temperatures, and the molybdic oxide volatilized away. His conclusions were derived from X-ray diffraction measurements of freshly ground powder of $MoSi_2$. The ignition temperatures were found to vary with the quantity of silicide powder, indicating that the reaction was sufficiently exothermic to raise the control temperature of the packed powder. The residue consisted of needlelike crystals of MoO_3 and a white powder, probably SiO_2. The oxidation kinetics of freshly ground powder exposed to air at room temperature indicated that a protective film was forming. On subsequent heating to 100°C, the oxidation rate increased markedly, and oxidation kinetics appeared to have changed to a first-order time dependence, indicative of a nonprotective film.

Maxwell's data on the presence of a protective silica film were obtained by low-temperature oxidation (<300°C). The extrapolation of his conclusions to oxidation at high temperatures may not be valid. The exothermic reaction of powders around 300°C and the linear oxidation curve at 100°C indicate a possible breakaway phenomenon, which may account for the molybdenum disilicide "pest" reaction reported by Maxwell.

Berkowitz[44] measured the oxidation kinetics of hot-pressed samples of $MoSi_2$ and WSi_2 in the temperature range of 1600°C to 1700°C. The $MoSi_2$ exhibited wide variations in oxidation rate, but the general shape of the oxidation curve was the same. The variation was attributed to variations in sample density and microstructure. At 1700°C a high initial rate was observed, which leveled off after about one half hour to a parabolic rate law. In all cases a glassy film formed on the surface. The oxidation process was described as formation of MoO_3 gas, SiO_2 solid, and Mo_5Si_3 solid. The latter compound was considered to have an important effect on the parabolic stage of oxidation. For WSi_2, a parabolic relationship was observed throughout the entire oxidation process. The oxide film in this case was also glassy, indicative of SiO_2. Treatment of the WSi_2 in water vapor at 1700°C did not produce a glassy film. Free tungsten was observed, however, but no W_5Si_3 compound was found. Under these conditions the silicon was presumed to volatilize as a hydrated oxide.

Confirming Fitzer's[13] earlier observation, Rausch[14] reported that the low-temperature disintegration of $MoSi_2$ occurs at 500°C and 600°C, with an incubation period of 10 to 15 hours for both polycrystalline and single-crystal material. The rate of disintegration was slower in single crystals than polycrystalline material, and the average particle size of the disintegrated product was quite large compared to polycrystalline material. The polycrystalline material developed particles that were widely varying in size during the initial stages of disintegration, but tended toward a single size as disintegration continued.

Silicon Carbide. Silicon carbide is widely used as the basis for high-temperature heating elements. Numerous oxidation studies[45–49] have been carried out on this material. Recently Jorgensen, Wadsworth, and Cutler (JWC)[12,50,51] made a detailed study of the mechanism of oxidation. This work represents the most definitive study on the mechanism of oxidation of a silicide compound.

Silicon carbide formed a silica film, which was protective and followed a diffusion-controlled parabolic oxidation kinetic relationship. JWC[50] found that the product of oxidation was amorphous silica or cristobalite depending on the temperature at which oxidation took place. Both types of silica form protective films; however, the oxidation rates were slightly different for each type. At relatively low temperatures (900°C to 1200°C) the reaction product was found by X-ray diffraction methods to be amorphous silica. Above 1200°C, the amorphous silica film was transformed into cristobalite after an induction period. The transformation corresponded to a change in slope in the parabolic rate equation. The activation energies of the parabolic rate constants were found to be 20.2 and 15.6 kcal/mole for amorphous silica and cristobalite, respectively. The oxidation rate constant of silicon carbide was found to vary with the logarithm of the partial pressure of oxygen.[51] The models of Engel and Hauffe[52] and of Cook and Oblad[53] were used by JWC to explain the observed pressure dependence.

The Engel-Hauffe mechanism involves the transport of cations, in which the driving force is provided by an electric field and a concentration gradient. The field is created by the rapid migration of electrons through the film, which are captured by adsorbed oxygen atoms. The field strength and, therefore, the oxidation rate is proportional to the logarithm of the oxygen pressure, assuming that the chemisorbed oxygen is in equilibrium with the gas. Using this model, JWC obtained activation energies for diffusion of 45.5 kcal/mole. This activation energy was similar to the activation energy calculated by Law[54] for the oxidation of thin films of silicon using the same mechanism.

A second model could account for the observed pressure dependence as well, assuming that the free energy of adsorption varied directly with the degree of surface coverage. Cook and Oblad[53] and also Higuchi, Ree, and Eyring[55] have derived a relationship, which gives a logarithmic dependence of the pressure similar to Engel-Hauffe mechanism. Both mechanisms yield the same activation energy. The latter mechanism yields a heat and entropy of adsorption. The calculated heat of adsorption was −10 kcal/mole, but the calculated entropy values were considered unusually low for a typical adsorption process.

When silicon carbide was oxidized in water vapor, the kinetics and the pressure dependence were the same as with oxygen.[12] However, for water vapor, the calculated energy of adsorption was positive, which tends to discredit the model based on changing free energy of adsorption with surface coverage. The activation energy for diffusion was only 21.1 kcal/mole. JWC showed that in oxidation by water vapor, the oxide film had the tridymite structure above 1218°C instead of cristobalite. The tridymite structure may account for the difference in the activation energies that were observed between water vapor and dry oxygen. The oxidation rate was found to be unchanged, if the oxidation was first started in water vapor and continued in dry oxygen of the same pressure, indicating that the diffusing species was the same in both types of oxidation. The diffusing species could not be identified, but oxygen or silicon ions appeared to be the most likely ones.

5.2 Beryllide Systems

High-Temperature Behavior. Beryllium reacts with most of the transition metals to form intermetallic compounds. Compounds with the compositions M_2Be_{17}, MBe_{12}, and MBe_{13}[2] have particularly good oxidation resistance with the exception of UBe_{13}. Compounds with lower beryllium content, such as MBe_5 and MBe_2, are not generally oxidation-resistant above 1000°C. $CrBe_2$, however, appears to be an exception. A comprehensive study of the oxidation characteristics of beryllide compounds has been carried out by Paine, Stonehouse, and Beaver (PSB)[2] over the temperature range of 1260°C to 1600°C. Weight changes were measured on samples which were initially prepared by sintering prereacted powders. Later samples that were prepared by hot pressing showed improved oxidation resistance. The oxidation data are summarized in Table 5. $ZrBe_{13}$ appears to have the best oxidation resistance; however, several compounds and mixed compounds are nearly as good.

A protective layer of BeO forms on oxidation of most beryllide compounds at temperatures above 900°C. The oxide of the other component is not found in the scale.[56,71] If the oxide scale thickness on $ZrBe_{13}$ is too large, breakaway occurs at temperatures above 1500°C or on cooling, because of the large oxide-to-metal-volume ratio and the large differential thermal expansion between the beryllide compound and BeO.[56,71] Oxidation tests carried out in water vapor produce gaseous beryllium hydroxide and cause the oxidation rate to be increased by the volatilization of the scale.[71] Oxidation tests in flowing moist atmospheres are particularly severe. During oxidation, an underlayer of Zr_2Be_{17} is formed, which appears to grow parabolically with time.[56]

Ervin and Nakata[71] measured the oxidation rates of $ZrBe_{13}$ and $NbBe_{12}$ between 700 and 1500°C using granular samples of known surface areas. The oxidation-rate constants were two to five times higher than with solid bars, but this was attributed to differences in surface roughness. The BeO layer formed during oxidation of $ZrBe_{13}$ was protective between 900 and 1500°C. Parabolic oxidation rates were observed over this temperature range for both $ZrBe_{13}$ and $NbBe_{12}$. The activation energy for $ZrBe_{13}$ above 1200°C was similar to the activation energy for oxidation of Be metal.[72] This activation energy (51.3 kcal/mole) was below the 100 kcal/mole activation energy obtained for Be self-diffusion in BeO reported by Austerman.[73] Recent measurements have indicated that activation energies for oxygen self-diffusion are between 43[74] and 60[75] kcal/mole in reasonable agreement with the values

TABLE 5
Summary of Oxidation[a] Data on Various Beryllide Compounds

Compound	Density[b]	Test Temperature, °C	Time, hours	Weight Gain mg/cm^2	Reference
$TiBe_{12}$	2.11	1260	100	+3.0	2
	2.18	1370	100	9.4, 11.2	2
$ZrBe_{13}$	2.54	1260	100	+7.8	2
	2.64	1370	100	+17.9	2
	2.77	1480	100	+10.4	2
	2.80	1590	100	+46.0	2
	2.70	1370	48	+2.1–2.9	56
	2.70	1500	48	+2.1–5.3	56
	2.70	1370	100	+2.8	71
	2.70	1540	100	+6.0	71
Zr_2Be_{17}	3.01	1260	100	+9.1	2
$NbBe_{12}$	2.88	1260	100	+1.7	2
	2.91	1370	100	+10.7	2
	2.87	1480	100	+59.0	2
	2.85	1370	48	+11.5	56
	2.85	1370	100	+2.5	71
Nb_2Be_{17}	3.07	1260	100	+2.6	2
	3.14	1370	100	6.9, 12.7	2
$TaBe_{12}$	3.76	1260	100	+18.0	2
	4.02	1370	100	+16.5	2
	4.14	1480	100	+7.5	2
$CrBe_2$	4.04	1260	25	+2.0	2
	4.21–4.30	1370	100	8.7–18.9	2
	4.26	1425	48	+2.1	56
$MoBe_{12}$	2.83	1260	100	+2.6	2
	3.02	1370	100	+3.7	2
$ReBe_{16}$	3.55	1370	48	+4.8	56

[a] Tests run in dry air.
[b] Measured density. Some densities may be slightly higher than theoretical for the compound due to the presence of second phases.

found by Ervin and Nakata. These activation energies do not clearly identify which ion is controlling the oxidation process. The preexponential term was small and did not suggest bulk diffusion. Preferential diffusion of oxygen or beryllium along the grain boundaries would not be unlikely in view of the measurements on oxygen self-diffusion in polycrystalline Al_2O_3.[76] A smaller activation energy was obtained for oxidation of $ZrBe_{13}$ in the temperature region between 900–1200°C and for oxidation of $NbBe_{12}$, which was attributed to the influence of the other oxides on the BeO film. The preexponential terms were exceptionally low, again suggesting that bulk diffusion is not involved. The oxidation-rate constant for $ZrBe_{13}$ at 1300°C showed a one-fourth power dependence on the oxygen pressure, instead of a one-sixth dependence calculated for a diffusion process controlled by the beryllium ion.

$NbBe_{12}$ oxidation showed peculiarities in the

TABLE 6

Oxidation of $ZrBe_{13}$ Specimens in Dry Air and Wet Air (dp 18°C)

Temperature, °C	Atmosphere	Total Time, min	Total Weight Change mg/cm²	Specimen Description
650	Wet air	1516	34.8	Powdered completely.
650	Dry air	4487	1.2	Minor metallic powder spalled from surface, leaving lustrous metallic surface
760	Wet air	368	123.0	Powdered with metallic lumps.
760	Dry air	4262	63.1	Gray-white oxide scaling and powdering. No metallic spalling.
870	Wet air	296	38.4	Thick, adherent gray-white oxide coating.
870	Dry air	4071	3.7	Black film with friable, white oxide loosely adhering.

kinetics between 1300 and 1500°C, but its oxide film was protective at all temperatures contrary to the case for $ZrBe_{13}$, which disintegrates at low temperatures.

The effect of impurities on the oxidation of beryllide compounds has been examined only slightly because high-purity beryllium is not readily available. Commercial beryllium contains between 0.5–1.5 per cent BeO. The carbon, iron, and aluminum contents are between 100–1000 ppm. PSB[2] reported improvement in oxidation resistance if Al_2O_3 contamination, which was introduced in the grinding process, was eliminated.

Low-Temperature Disintegration. Lewis[17] showed that $ZrBe_{13}$ readily spalled in laboratory air when heated at 700°C. Aitken and Smith[57] showed that the presence of water vapor in the atmosphere caused complete failure in ten to twenty hours at 700°C, but no visible effects were observed in dry air after 100 hours at the same temperature. When $ZrBe_{13}$ was heated in moist argon (15°C dp), the sample fractured badly one half hour after the moisture was introduced, even though there appeared to be only a very thin oxide layer on the surface. Heating the sample in dry argon showed no effect, but when moisture was introduced, the surface temperature increased 40°C within five to ten minutes, indicating that an exothermic reaction was occurring.

Perkins[58] reported that low-temperature disintegration of $ZrBe_{13}$ occurred in dry oxygen at 650°C as well as in moist oxygen, but the presence of moisture greatly increased the rate. At 870°C the difference in the rate of attack between moist air and dry air was less pronounced. Perkins' results for $ZrBe_{13}$ are summarized in Table 6. The oxidation kinetics were either linear or accelerating, indicating that no protection was provided by the oxide. Perkins observed disintegration of Zr_2Be_{17} similar to $ZrBe_{13}$, but at slightly higher rates.

Aitken and Smith[56] showed that preoxidation at 1370°C had no appreciable effect on retarding the low-temperature disintegration of $ZrBe_{13}$. Some samples showed five per cent linear growth prior to failure. Perkins[58] detected BeO and face-centered cubic or tetragonal ZrO_2 after fifteen minutes at 870°C. Monoclinic ZrO_2 appeared after 115 minutes. ZrO_2 was not observed in the oxide scale after high-temperature oxidation ($>$1000°C). The formation of cubic ZrO_2, followed by subsequent transformation to monoclinic ZrO_2 on cooling, may be responsible for the spallation of the oxide.

Perkins examined the internal structure of $ZrBe_{13}$ metallographically and showed that a very fine grain-boundary precipitate appeared on annealing in dry argon at 870°C for 100 hours, but no matrix spalling or disintegration was evident. The precipitate, which was predominately at the grain boundaries between the $ZrBe_{13}$ phase and the minor Zr_2Be_{17} phase, disappeared after heating at 1150°C in dry oxygen for 145 hours. Some precipitate was observed in the as-received material, which probably appeared because of slow cooling through the supersaturated temperature region during fabrication. When material of the nominal composition Zr_2Be_{17} was annealed in the same temperature region, no precipitate was observed, which indicated that the precipitate was coming from the $ZrBe_{13}$. There were indications that in higher-temperature annealing, large inclusions, commonly present in fabricated $ZrBe_{13}$, also showed a tendency to enter into solution.

When $ZrBe_{13}$ was annealed at 700°C for 100 hours in dry argon after solution treatment at 1150°C, no precipitate appeared. However, on heating briefly at 700°C in air for 7.5 hours to produce spalling, the fine precipitate appeared in grain boundaries near the edge of the specimen. This suggested that oxygen (or H_2O vapor) created a supersaturated condition at a lower temperature than was observed in dry argon. The nature of the impurity precipitate was not identified, but the amount that was present may be responsible for the variability of various $ZrBe_{13}$ samples toward disintegration. The lack of an observed precipitate in the Zr_2Be_{17} material does not support the conjecture that the precipitate is responsible for disintegration, because this composition exhibited as serious spalling as $ZrBe_{13}$ compositions.

Preliminary evaluation of $NbBe_{12}$ and Nb_2Be_{17} by Perkins[58] showed that these compositions oxidized readily in wet and dry air with maximum rates occurring around 1000°C. The oxidation was observed to begin initially at microscopic spots where a minor phase was located and to spread over the surface. Both α- and β-forms of Nb_2O_5 were observed in the oxide scale as well as BeO. A phase transformation from the α- to the β-form occurred between 800°C and 900°C, which may be responsible for the spalling on cooling. A fine precipitate was observed in the Nb_2Be_{17} material at the grain boundaries near the surface after 48 hours in wet air at 900°C.

5.3 Aluminide Systems

Aluminum forms a large number of high melting intermetallic compounds with various transition metals. Several of these compounds have been used as intermetallic coatings of refractory metals, such as niobium and molybdenum.[59] Other aluminides have been studied as mixtures with oxides to form cermets and as dispersion-hardening agents in metals Paine, Stonehouse, and Beaver (PSB)[2] obtained comparative oxidation data at 1200°C and 1370°C for compounds of the type MAl_3, in which M is niobium, tantalum, and molybdenum or zirconium. These data are summarized in Table 7. The zirconium compound showed no resistance at 1260°C, but the $NbAl_3$ and $TaAl_3$ compounds formed tenacious, thin films at this temperature. At 1370°C the oxidation resistance of $NbAl_3$ was considered marginal, and wide variations in the weight gains were noted from sample to sample. The $TaAl_3$ compound showed better oxidation resistance at 1370°C, but it was subject to flaking on cooling from this temperature.

High-Temperature Behavior of Niobium Trialuminide. The oxidation characteristics of $NbAl_3$ compound as a coating for niobium alloys has been reported by Lever[60] and by Wukusick.[11] The coating was formed by dipping niobium into a molten aluminum bath containing small additions of chromium and silicon, which formed successive layers of $NbAl_3$, Nb_2Al, and Nb_3Al by diffusion into the substrate. The chromium and silicon additions did not form reaction compounds, but their presence was important for oxidation resistance. The outer $NbAl_3$ layer was converted to Al_2O_3 and Nb_2Al after about 100 to 200 hours at 1370°C. The Nb_2Al layer disappeared after 500 to 700 hours, and the Nb_3Al layer in turn was eliminated in 800 to 1000 hours by oxidation and aluminum diffusion into the niobium. Internal oxidation in the niobium was noted in the later stages of oxidation. The Nb_2Al and the Nb_3Al compounds were not particularly oxidation-resistant at high temperature, the protection being provided by the Al_2O_3 film.

TABLE 7

Summary of Oxidation Data of Various Aluminide Compounds

Compound	Density	Test Temperature, °C	Time, hours	Weight Gain mg/cm^2	Reference
TiAl	..	800	>40	1.0	81
	..	1150	40	7.0	81
	..	1000	100	12 mils	10
	..	1200	96	35 mils	10
$ZrAl_3$	4.05	1260	1	Oxidation complete	2
Nb_3Al	6.50	1260	1	20 mils scale	2
$NbAl_3$	4.14–4.62	1260	100	2.1–2.8	2
	4.43	1370	100	34.8	2
	4.49	1425	100	176	2
$TaAl_3$	6.33	1260	100	7.1	2
	6.64	1425	100	6.9	2
Mo_3Al	822	1260	1	Oxidation complete	2
$MoAl_3$	4.59	1260	100	Oxidation complete	2
	..	1000	2	0.6	77
NiAl(Al-rich)	5.74(est.)	1150	2	1.0	77
NiAl(Ni-rich)	..	1300	>30	2.5–4.0	79
	5.86	1000	100	1.0	80
NiAl	..	1000	>100	0.004	79
	..	900	>30	~2	79

Low-Temperature Disintegration of $NbAl_3$. The compound, $NbAl_3$, is susceptible to low-temperature disintegration in the temperature range of 550 to 850°C. Rausch[14] measured the oxidation kinetics of arc-melted buttons over the temperature range of 540°C to 1300°C. A maximum in the oxidation rate was observed at 750°C. The powdering during disintegration was preceded by an incubation period, which varied with the temperature. Between 950°C and 1100°C the oxidation rates were at a minimum, and no powdering was observed. At 1300°C oxidation was more rapid, but no disintegration occurred. Wukusick[11] observed spalling of $NbAl_3$ coatings on niobium in the same temperature region as Rausch. The oxidation rate for $NbAl_3$ reached a maximum at 750°C, but Nb_2Al and Nb_3Al exhibited increasing rates of oxidation with increasing temperature.

Wukusick[11] reported that the oxide scale formed from oxidation of $NbAl_3$ after 165 hours at 760°C was a mixture of $Nb_2O_5 \cdot Al_2O_3$ and Al_2O_3. Rausch[14] obtained only α Al_2O_3 diffraction patterns above 950°C.

Preoxidation of $NbAl_3$ at high temperatures showed no effect in the spallation characteristics.[14] Wukusick[11] showed by metallographic examination of coated Nb specimens that the Nb–Al compounds and surface oxides cracked on cooling to room temperature because of differences in expansion coefficients. When the specimens were reheated to 1370°C, the cracks closed because of plasticity of the scale and the mobility of aluminum, but they remained open when treated to only 750°C, exposing fresh Nb–Al compounds. The formation of $Nb_2O_5{\cdot}Al_2O_3$ compounds at lower temperature may have caused spalling of the coating.

Wukusick[89] found that the oxidation rate was faster for $NbAl_3$ compounds, which were slightly depleted in aluminum. Figure 4 shows the effect of Al content on the oxidation behavior. Excess aluminum in $NbAl_3$ was found to extend the life at 700°C. Apparently, excess Al was slowly oxidized until a point was reached where no excess Al existed and spalling occurred. $NbAl_3$ compounds containing an excess of Al formed scales that were gray, indicative of Al_2O_3, whereas compounds that were deficient in Al formed black scale characteristic of niobium oxide. At temperatures of about 650°C, excess Al had no effect on spalling. Because the eutectic temperature between Al and $NbAl_3$ was between 650°C and 700°C, the presence of a liquid phase may have prevented spalling at 700°C by absorption of stresses.

Lever[60] studied a group of fifty alloy additions to ascertain if there was an effect on spalling at low temperatures. The alloy additions were made so that either niobium or aluminum was replaced in $NbAl_3$. The additions, which inhibited spalling between 700°C and 760°C for at least 1000 hours, were Ag, Bi, Co, Cu, Si, Sn, and Ti. Silver inhibited

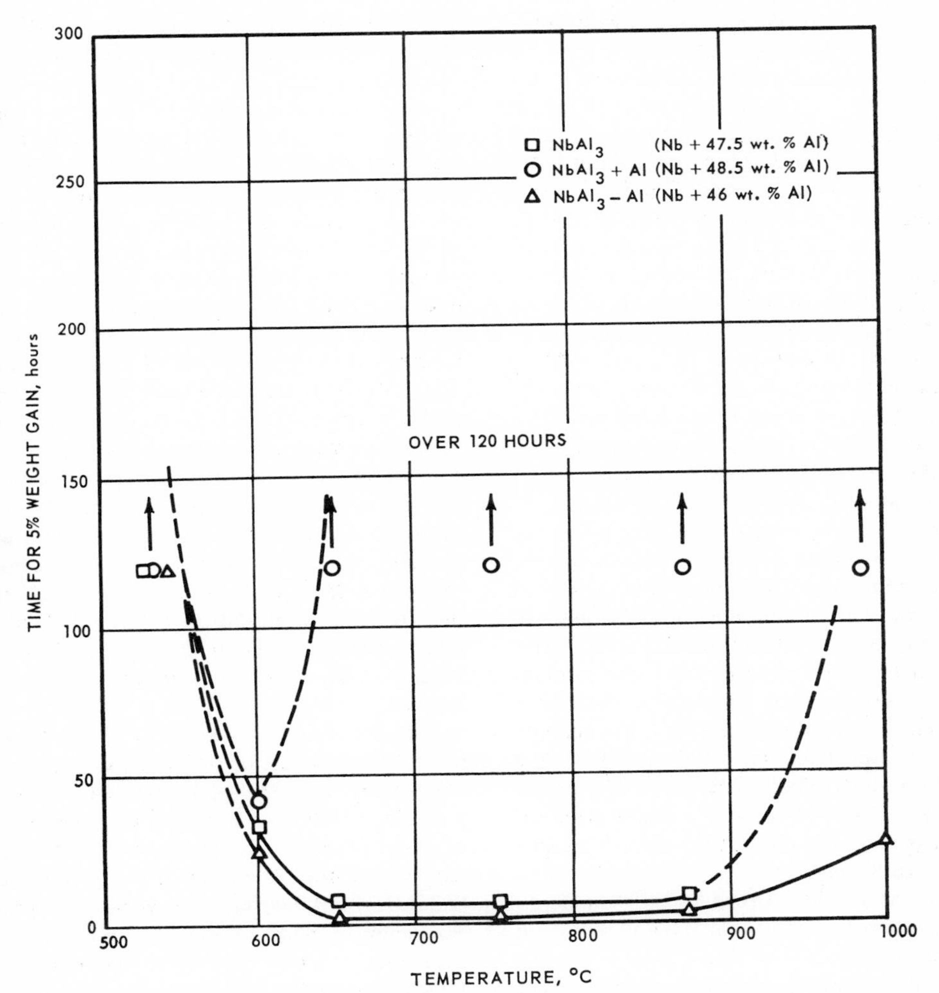

Fig. 4. Oxidation behavior of $NbAl_3$ in air at intermediate temperatures. (Fragmentation occurs at ~1 per cent weight gain).

spalling when 5 atomic per cent was substituted for Nb, but molten Al bubbles were observed to form on the surface. When 10 atomic per cent Ag was substituted for the Al, spalling readily occurred. Other additions, which resulted in the formation of free aluminum, were Bi, Cu, Si, and Sn. The elements, which inhibited spalling without the appearance of free aluminum, were Ti, Co, Si, and Sn, when substituted for aluminum. Most of these additions were not suitable for high-temperature oxidation resistance because of low-melting temperatures or high volatility. The large additions of titanium that were required to prevent spalling changed the basic crystal structure from the $NbAl_3$ type to (Nb, Ti)Al compounds. The high-temperature oxidation of the TiAl and TiAl modified with other metal additions is discussed below.

TiAl Compounds. The compound, TiAl, shows good oxidation resistance to about 1000°C.[92] McAndrew and McPherson (MM)[10] observed oxidation to a depth of about 0.012 inch after 100 hours at 1000°C and to a depth of about 0.018 inch after 50 hours at 1200°C. For a compound of TiAl, which was slightly deficient in Al (35 w/o), the oxidation occurred to a depth of 0.035 inch after 96 hours at 1200°C. X-ray diffraction examination of the scale indicated that both TiO_2 and α-Al_2O_3 phases were present. The scale generally appeared as two layers under polarized light. No evidence of low-temperature disintegration was observed with the TiAl compounds.

MM studied qualitatively the effect of various additives on the oxidation resistance of TiAl. Additions of niobium, tantalum, silver, and silicon improved the oxidation resistance. Chromium additions resulted in fairly protective scales; however, there was a marked tendency to spall when the samples were cooled quickly. Zirconium additions reduced the oxidation resistance, but manganese, tin, molybdenum, nickel, and iron had less deleterious effects. The depth of penetration by oxygen decreased steadily with increasing amounts of niobium, which ranged from 0 to 8 weight per cent after 96 hours at 1200°C. After 140 hours at 1000°C, the depth of oxidation was less than 0.001 inch for all samples with additions of niobium from 1 to 7 per cent. Continued heating of a 5 per cent Nb alloy at 1000°C to a total of 1020 hours caused no greater penetration. Additions of silver up to 7 weight per cent gave good protection at 1000°C, but moderate to severe pitting was observed during oxidation. Tantalum additions showed oxidation levels comparable to the niobium additions.

The TiAl compound has a high solubility for certain metals, particularly niobium. Lever[60] and Wukusick[89] showed improved oxidation resistance of TiAl with additions of niobium near the saturation limit (33–40 weight per cent). The oxygen penetration was about 0.001 inch after 110 hours at 1200°C and after 20 hours at 1425°C. This result was based on arc-melted buttons. Specimens prepared by induction melting often gave lower oxidation resistance, possibly due to higher impurity content.[64] The TiAl compound of composition Ti–32Al–35Nb formed thin adherent scales at 1425°C, but, when the alloy was oxidized between 980°C and 1315°C, a linear oxidation rate was observed and the scale spalled on cooling, leaving a thin adherent subscale.[60] Electron-beam probe analysis of the scale formed after oxidizing for 110 hours at 1200°C and 1315°C showed that the scale was a two-phase mixture of TiO_2 and Al_2O_3.[63] An underlayer of Nb_3Al was formed between the matrix and the scale due to the depletion of titanium and aluminum. The matrix remained unchanged. When it was oxidized for 20 hours at 1425°C, the scale was double-layered. The outer layer was a mixture of TiO_2 and Al_2O_3, but the dense inner layer next to the matrix was only Al_2O_3. The Nb_3Al compound was not present, but the matrix had become a two-phase structure probably because of aluminum depletion. The lowered oxidation resistance of the Ti–32Al–35Nb alloy between 980°C and 1315°C was attributed to the presence of the Nb_3Al layer, which had relatively poor oxidation resistance, and to the absence of the dense Al_2O_3 inner layer, which resisted spalling.

The addition of small amounts of chromium improved the oxidation properties of Ti–32Al–35Nb alloy, but resulted in increased brittleness. The addition of 6 weight per cent chromium in place of the aluminum reduced the tendency of the scale to spall between 980°C and 1315°C. Additions of 6 weight per cent chromium in place of the titanium virtually eliminated spalling even after 500 hours at 1200°C and 1315°C. However, there was some tendency to spall at 1150°C. Upon oxidizing at temperatures above 1300°C, some loss in weight was noted due to volatilization of chromium. Electron-beam probe examination of the scale formed on the matrix with 6 per cent chromium additions showed that the scale consisted of Al_2O_3 only, after oxidation between 1200°C and 1425°C. No titanium, chromium, or niobium was found in the scale. Between 1200°C and 1315°C a nonuniform layer of Nb_3Al was formed, which contained large amounts of chromium. The addition of chromium apparently increased the oxidation resistance of Nb_3Al and promoted the formation of Al_2O_3 in a sublayer at all temperatures. TiO_2 was detected in the scale by X-ray diffraction in the early stages of oxidation, but easily flaked off during further oxidation, and thus it was not seen by the microprobe.

The tendency of Nb–Ti–Al alloy with 6 weight per cent chromium to spall at 1150°C was eliminated entirely, if it was preoxidized at 1370°C for four hours before exposure at the lower temperature.[63] The combination of the TiAl compound modified to give the composition Nb–28Ti–32Al–6Cr and preoxidation at 1370°C resulted in an aluminide compound, which did not disintegrate at low temperatures or spall on cooling after oxidation at temperatures up to 1400°C.

Nickel Aluminide. The compound, NiAl, forms an adherent film on oxidation in air at temperatures up to 1200°C.[82] Mozhukhin, Pivovarov, and Umanskiy[77] have made oxidation measurements on aluminum-rich and nickel-rich NiAl. Comparison of the weight-gain data indicated that the cast aluminum-rich NiAl compound was oxidized to a lesser degree at 1150°C than electrolytic nickel at 1000°C. The cast nickel-rich NiAl oxidized at a rate, which was comparable to the pure nickel.

The nickel-rich compound oxidized linearly with time and the aluminum-rich compound oxidized with about a 0.45 power dependence on the time. Examination of the oxide films by X-ray diffraction showed that both the nickel-rich and aluminum-rich compounds contained $NiO \cdot Al_2O_3$ spinel. The nickel-rich compound contained in addition α-Al_2O_3 and Ni_3Al phases. The authors concluded that disruption of the coating on the nickel-rich compound resulted from the presence of the Ni_3Al phase, which was not adherent to the spinel phase.

Fitzer and Gerasimoff[79] studied the oxidation of Ni–Al compounds prepared by sintering of a powder obtained by reaction in mercury. They found that the highest resistance to scaling was obtained by alloys containing between 80–84 weight per cent nickel, which is in the immiscibility region between the compounds NiAl and Ni_3Al. The oxidation rate of sintered NiAl was somewhat higher than the rate for the cast alloys reported by Mozhukhin et al. Imai and Kumazawa[78] have reported weight-gain data for oxidation of NiAl as only 4 micrograms/cm^2 after several hundred hours at 1000°C, which is considerably below the values obtained by Fitzer and Gerasimoff or Mozhukhin et al. These data may differ appreciably from a lack of knowledge of the true surface area. It is apparent that the composition of NiAl has a strong influence on the oxidation rate and more careful control of the stoichiometry is needed.

5.4 A(III)-B(V) Compounds

The compounds of type A(III)-B(V), in which A is Al, Ga, or In and B is P, As, or Sb, are useful semiconductors and all have the same crystallographic structure. The largely covalent character concentrates the energy in directed bonds, constraining the crystal to a high degree of chemical and structural perfection. Thus there is a low solubility of either element in the compound. Aside from the technological interest, these factors make the A(III)-B(V) compounds useful in a scientific study of the oxidation reactions at the surface.

Crystal Polarity. These compounds form hybridized sp^3 bond orbitals, similar to their group IV neighbors, Si and Ge. Their characteristic zinc blende structure exhibits crystal polarity in the $\langle 111 \rangle$ direction. Thus A atoms terminate at the $\{111\}$ plane (A surface) and B atoms terminate at the $\{\bar{1}\bar{1}\bar{1}\}$ plane (B surface). On the A surface the A atom, which is triply bonded inward to the bulk, has no electrons available for the fourth sp^3 orbital. On the B surface the B atoms, also triply bonded inward, have the four sp^3 orbitals filled, but one is left dangling (unshared). Consequently, the A and B surfaces would be expected to react differently when they are exposed to oxidizing agents.* Because unshared electron pairs are located on B surfaces, oxidation would presumably occur there most readily.

---

* See also Chapter 16 for a discussion of this topic.

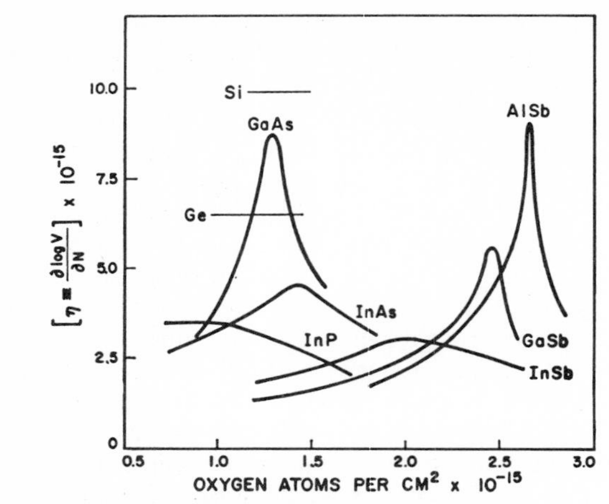

Fig. 5. Transition from "chemisorption" to oxide formation at 26°C.

Oxidation and corrosion studies have tended to support this argument. Gatos and Lavine[65] have measured the relative rates of dissolution of InSb and GaAs in the ⟨111⟩ directions with various oxidizing etchants. At temperatures (~0°C) at which the rate of attack was chemically controlled, the B surfaces had rates of dissolution higher than those of the A surfaces by almost an order of magnitude. The A surfaces formed dislocation etch pits, whereas the B surfaces did not; however, by using corrosion inhibitors, etch pits on both A and B surfaces were developed. Similar etching effects were observed with GaAs.

Faster oxidation of the B surfaces of GaAs was observed at high temperature by Miller, Harper, and Perry.[66] By using the DTA method, the temperature differences between thermocouples attached to A and B surfaces were recorded as a relative measure of the oxidation rate. Oxidation of the A surface appeared about 10°C sooner than the B surface; however, when the temperature reached 800°C, the B surface reacted faster. It was concluded that the relative rates of oxidation below 800°C were controlled by diffusion through an oxide film. Above 800°C the film became nonprotective, and the oxidation was reaction-controlled at the intermetallic surface and was faster at the B surface due to the available unshared electrons of the B atoms. In a study of the oxidation kinetics of InSb at temperatures between 212°C and 494°C, Rosenberg[18] noted a slight but distinct anisotropy in the rate between various crystal faces. For oxidation levels up to 3×10^{17} oxygen atoms/cm² at a temperature of 367°C, the order of decreasing oxidation rate was $\{111\} > \{211\} > \{\bar{1}\bar{1}\bar{1}\} > \{110\} > \{100\}$. The order of reaction between the {111} (A surface) and $\{\bar{1}\bar{1}\bar{1}\}$ (B surface) was in the same direction as observed with GaAs[66] when it was oxidized in the temperature region where a protective film was formed. The rate anisotropy reported by Rosenberg[18] was of a smaller magnitude than that observed in the low-temperature oxidation of metals and of Ge.

Low-Temperature Oxidation of A(III)-B(V) Compounds. The room-temperature oxidation kinetics of vacuum-crushed compounds in the A(III)-B(V) series were measured by Rosenberg.[18] The extent of oxidation was not linear with respect to the logarithm of time for the compounds studied but exhibited regions of negative and positive curvature. Rapid adsorption of a monolayer of oxygen was observed at 26°C on InAs, GaAs, and InP, but two or more layers of oxygen were absorbed rapidly on the antimonides. In all cases the rate apparently diminished as a metastable surface configuration was approached. This rate decreased for given B(V) atoms in the order Al > In > Ga.

Analyzing the rate data in terms of the η-function, Rosenberg showed that the transitions from chemisorption to oxide formation produced maxima in the η-values at particular oxidation levels. As shown in Fig. 5, maxima in η were observed at 2.5×10^{15} and 1.35×10^{15} oxygen atoms/cm² for antimonides and arsenides, respectively. InP exhibited no distinct maximum in η. The magnitude of η at the maximum was a measure of the stability of the configuration, which decreased in the order AlSb ≈ GaAs > GaSb > InAs > InSb. Separate measurements at 71°C showed that, except for InSb, the oxidation level for maximum η did not change with temperature, suggesting that the metastable configuration was a unique system. The metastable configurations corresponded to between three and four oxygen atoms per surface atom for the antimonide and to between one and two atoms per surface atom for the arsenides and InP. Rosenberg noted that the distinct grouping of the antimonides from the arsenide and phosphides had a parallel in the coordination chemistry of the group V elements in that antimony is octahedrally coordinated with oxygen, whereas arsenic and phosphorus are tetrahedrally coordinated. The group III elements generally exhibited octahedral coordination and had little effect on the metastable structure. At larger levels of oxidation, the η-value decreased, indicating a possible return to a diffusion-controlled mechanism and a breakdown of the metastable configuration to the stable oxide.

At liquid nitrogen temperatures, Rosenberg and Menna[24] found that the oxygen uptake of vacuum-crushed InSb consisted of a reversible and an irreversible reaction. The reversible reaction followed a square-root dependence on the pressure, indicating that the adsorption was dissociative. Because dissociative reactions are expected to involve oxygen ions rather than atoms, some difference might be expected between the rate of sorption of oxygen for crystals with different chemical potential of the electron. There was no variation in sorption rates observed by Rosenberg and Menna in the oxidation of intrinsic, Cd-doped and Te-doped single crystals of InSb, hence the reversible reaction for InSb at 78°K indicates some physical adsorption mechanism in spite of the observed pressure dependence. In the reversible reaction, the total amount of gas after 1300 minutes at 78°K was less than a monolayer; however, as the amount of irreversibly adsorbed gas increased, the equilibrium constant for the reversible reaction also increased.

Indium Antimonide. The high-temperature oxidation of one A(III)-B(V) compound, InSb, was studied in detail by Rosenberg and Lavine.[19] The rate was measured as a function of thickness in the range of 5×10^{15} to 5×10^{19} atoms/cm^2 over a temperature range of 212°C to 494°C and over a pressure range of 2.0 microns to 3000 microns. The observed rates plotted in terms of the ϵ-function over the range of thicknesses, at various temperatures as shown in Fig. 6, indicated the following:

1. The positive value of ϵ indicated that the film was increasingly protective with increasing thickness.
2. The composition and structure were not uniquely determined by thickness, because the curves were not superimposable at all temperatures.
3. Between 308°C and 367°C singularities in ϵ occurred at thicknesses of about 5×10^{16} atoms/cm^2. Beyond the maximum, ϵ decreased to between 1 and 2 and then rose slightly. At 430°C and 494°C the curves showed additional singularities, but ϵ eventually approached a constant value of 2.0 (cubic oxidation rate law).

The activation energy of the oxidation rate reached a maximum of 45 kcal/gm atom at about 4×10^{16} atoms/cm^2 and declined to a value of about 25 kcal/gm atom at 1.8×10^{18} atoms/cm^2. If the oxidation rate was measured

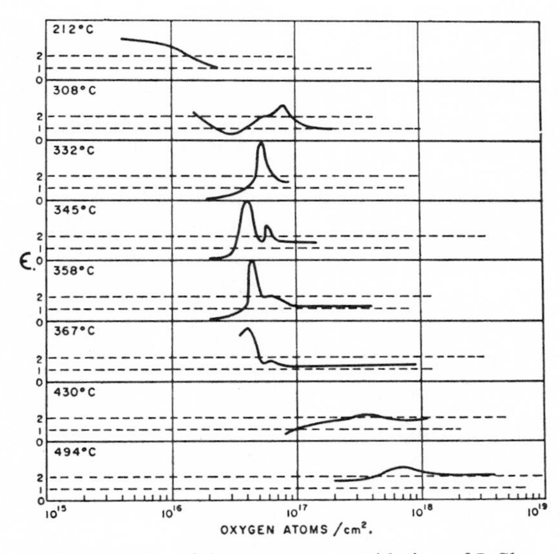

Fig. 6. The high-temperature oxidation of InSb.

on a specimen in which the film was grown to a given thickness at a fixed temperature, a slightly higher activation energy was observed. The oxidation rate varied between a 0.10 and 0.17 power dependence on the oxygen pressure with no systematic trend in thickness or temperature. When InSb was doped with 2×10^{18} atoms/cm^3 of cadmium, the oxidation rates of doped crystals at 400°C were higher than the undoped crystal. Comparison of the oxidation rate of InSb with liquid In metal and Sb metal showed that the parabolic rate constants at 367°C of InSb and In metal differed by a factor of 1.4, but the rate constant for Sb was 10^2 times higher than that for InSb.

The observed singularity of ϵ at 5×10^{16} atoms/cm^2, which is equivalent to fifty monolayer equivalents of oxygen, and the simultaneously abrupt change in activation energy were attributed to a change in composition of the film. Independent dissolution and electron diffraction experiments indicated that the film consisted of In_2O_3 with some dissolved Sb at thicknesses $>10^{17}$ atoms/cm^2. The region of the oxide film ($<10^{17}$ atoms/cm^2) near the surface of the substrate showed, in addition to In_2O_3, free Sb. Rosenberg concluded that antimony oxidized initially, evaporated as $(Sb_2O_3)_2$, leaving behind indium which oxidized to In_2O_3. As In_2O_3 accumulated, the access of oxygen was progressively restricted, as indicated by a rise in ϵ. After the In_2O_3 film was completed, reaction continued only by diffusion through the film. Interstitial migration of trivalent indium ion was the preferred diffusion-controlling process. As the In_2O_3 film grew, some antimony entered the film, but a large fraction remained behind and accumulated at the InSb-oxide interface.

Using the technique of interrupted oxidation, Rosenberg[22] determined the defect concentration, C_0, and diffusion coefficient, D, separately from the kinetics with the following temperature dependence:

$$C_0 = 9.4 \times 10^{22} \exp(-5400/RT) \text{ indium defects/cm}^3$$

$$D = 0.0078 \exp(-31{,}200/RT) \text{ cm}^2/\text{sec}$$

The oxidation rate was found to be influenced by the time and temperature of interruption. When the oxidation reaction was interrupted by removal of oxygen gas, the preferential flow of indium was stopped, but the slower diffusing antimony continued. On reaching the surface, antimony reduced the In_2O_3 to indium metal, and volatilized as $(Sb_2O_3)_2$. When oxygen was readmitted, the accumulated indium was rapidly reoxidized. The amount of accumulated indium increased with the temperature and time of interruption.

6. CORROSION IN MEDIA OTHER THAN OXYGEN

Corrosion data are meager for intermetallic compounds in media other than oxygen. Except for recent studies on the etching behavior of intermetallic semiconductor compounds, there has been little technological interest. Other corrosion media, which may become important in future studies, are aqueous corrosion, liquid metal corrosion, and hydrogen atmosphere corrosion. The latter is not commonly associated with corrosion phenomena, but significant reactions of hydrogen with intermetallic compounds have been observed.

The presence of moisture was known to influence disintegration of many intermetallic compounds many years ago. Tamman and Ruhenbeck[25] noted that some samples of intermetallic compounds disintegrated on storage in laboratory atmosphere over a period of years. Lohberg[26] made further studies on the disintegration phenomena of low-melting compounds of aluminum and magnesium in water and in steam. The rate of attack increased during the course of disintegration and was complete after 20–40 hours. The compounds of Mg_2Cu, $MgCu_2$, MgZn, $MgZn_2$, Al_3Mg_2, Al_2Cu, and AlCu were free of disintegration. Golubev[90] has reviewed the role of intermetallic compounds in selective aqueous corrosion of solid-solution alloys of aluminum and copper. The application of stress changes the rate of attack on the compound dispersed in a solid-solution alloy, which leads to intercrystalline cracking. The author proposes that changes in the bonding type from metallic to atomic as a result of stress influences the electrochemical behavior of the compound and suggests this approach as a means to understanding the corrosion behavior of intermetallic compounds.

Impurities, notably carbon, will promote disintegration on storage in ambient atmosphere, and carbon contamination such as from graphite containers may mislead one about the inherent stability of pure intermetallic compounds. The A(III)-B(V) intermetallic compounds hydrolyze in acid solution forming the hydrides PH_3, AsH_3, and SbH_3. The extent of hydrolysis diminishes with increasing atomic number of the hydride. No hydrolysis is observed in neutral or alkaline solution.

Since 1959 a number of articles have been published on the aqueous corrosion characteristics of the A(III)-B(V) compounds. This subject was treated comprehensively in a chapter of a recent book.[29] Most of this work was performed at the Lincoln Laboratories of MIT by Gatos and co-workers.[28–32,65] The kinetics of dissolution of the semiconductor compounds were chemically controlled, and the rate of attack was a function of the type of crystal surface. The corrosion model based on the crystal polarity of the zinc blende structure in the ⟨111⟩ direction accounts for the observed kinetic behavior of both oxidation and aqueous corrosion. The role of surface damage by cold work and the effects of surface active agents on the etching behavior of the crystal surface have been examined. Because of the atomic configurations in the crystal and the relative reactivity of the group III and group V atoms, dislocation etch pits were different between the $\{111\}$ and $\{\bar{1}\bar{1}\bar{1}\}$ surfaces. Chromic ion solutions develop etch pits associated with In edge dislocations at low-angle boundaries on the Sb surface $\{\bar{1}\bar{1}\bar{1}\}$. Organic sulfides develop pits associated with Sb dislocations on both In $\{111\}$ and Sb $\{\bar{1}\bar{1}\bar{1}\}$ surfaces.

Larrabee[33] showed that metal atoms in the aqueous media readily contaminated semiconductor surfaces. The metals were physically adsorbed or chemically deposited on the crystal surface, depending on the electrochemical potential of the metal-metal ion with respect to the potential of the intermetallic semiconductor. The deposition process was irreversible and resulted in a potentially serious surface contamination which electrically shorted the *p-n* junction of a semiconductor device.

The corrosion of intermetallic compounds by liquid metals is almost unexplored. Bernstein[83] tested a number of bonding agents for alloying or eutectic formation as an aid to making semiconductor devices using the III-V semiconductor compounds. The elements Au, Pb, Sr, Al, and In were found favorable for bonding to the intermetallic semiconductors. In addition, alloys of AuGe and AuSi formed good alloy bonds with all the compounds, except InSb which appeared more sensitive to cracking. Other factors such as doping, solubility, and eutectic formation were considered in the choice of the proper bonding agent.

Another form of corrosion is the attack by hydrogen on the intermetallic compound. Some compounds are known to form hydrides with a unique structure. Many others containing elements which are hydride formers readily dissociate to the hydride of one constituent and the elemental form of the other. Beck[84] has surveyed the hydriding characteristics of a large number of intermetallic compounds. The compounds which are listed in Table 8 have been found to absorb significant amounts of hydrogen. Those marked with an asterisk absorbed more hydrogen than would be possible by the gas-metal reaction of the pure components. Beck[84] stated that the two basic requirements that seemed to determine the formation hydrides of intermetallic compounds were:

1. Tetrahedral interstices should be available in the structure and should be capable of accommodating a hard sphere much greater than 0.46 Å. Smaller interstices lead to lower thermal stability.
2. The hydrogen should be bound to metals as resonating covalent bonds.

The degree of occupancy of lattice sites by hydrogen was found to be dependent on the number of available tetrahedral sites in the proper size range and on the spacing between the tetrahedral sites separated by a plane of close-packed metal atoms. Incorporation of lower valent metals eased the ability to fill smaller-sized interstitial holes.

7. SUMMARY

The corrosion studies of intermetallic compounds have been only a little more than exploratory, and only few systems have been studied in detail. Much more work is needed before some understanding of the present problems is obtained. The present corrosion studies of intermetallic compounds are motivated by two needs:

1. The need for high-melting, hard materials which form protective films to operate at high temperatures in corrosive environments.

TABLE 8

Ternary Hydride Compounds with Hydrogen-to-Metal Ratios Greater than 0.30 [84]

Compound	Maximum Observed H/M	Compound	Maximum Observed H/M
Ti_3Au	1.17	Ti_2Ni	0.97
Ti_3Pt	0.50	Hf_2Co*	1.54
HfCo*	1.49	Ti_2Mn +(10 *a/o* O_2)	0.47
Zr_2Al	0.90	Zr_2Cr (+10 *a/o* O_2)	1.00
Ti_2Ga	0.85	Hf_2Pt	0.44
ThAl	1.25	Ti_2Pd	0.49
ThCo	1.70	Ti_2Pt	0.35
ZrNi*	1.44	Ti_2Cr (+10 *a/o* O_2)	0.39
$ZrMn_2$*	0.69	Zr_3Al_2	0.52
$ThMn_2$	1.19	Nb_3Sn	0.22
$ZrCr_2$*	1.16	V_3Sn	0.06
			0.25
ZrV_2*	1.38	PrSb	0.21
$(Zr_{0.75}V_{0.25})Mo_2$		Y–TiCu	0.21
HfV_2	1.06	$TiCr_2$	0.22
Th_2Al	1.27	$CaAl_2$	0.15
Ti_3Al	0.94	$ZrMo_2$	0.11
			0.27
$Th_6Mn_{23}^*$	0.89	$PrGa_2$	0.11
Th_7Ni_3	(2.6?)	Ti_2Fe (+10 *a/o* O_2)	0.21
Ti_2Cu	1.18	Zr_2Al_3	0.21
Hf_2Mn*	(1.50?)		

* Compounds designated with an asterisk absorbed more hydrogen than would be possible by gas-metal reaction with the pure components.

2. The need for control of surface contamination of the semiconductor compounds to provide better electronic materials.

The studies of the semiconductor intermetallic compounds have used nearly perfect crystalline compounds with controlled defect concentrations. The availability of single crystals of extreme purity and the convenient temperature range for sensitive kinetic measurements make the task more straightforward. In contrast, oxidation studies of high-melting, refractory intermetallic compounds have used polycrystalline materials of unknown purity. As a result the corrosion mechanism for the refractory compounds is less well understood. Future basic studies on the refractory intermetallic compounds would benefit by a choice of materials that can be made with a greater degree of crystalline perfection. Also, more attention must be paid to auxiliary analytical tools to complement weight-gain data. The recent development of the electron-beam microprobe should be a significant aid for interpretation of reactions of intermetallic compounds.

In the present technology, use of intermetallic compounds in semiconductor devices does not require intrinsic protection from corrosive media during operation for the materials are easily encapsulated. Contamination appears most extensively in the preparative stages of device fabrication which presumably can be controlled. Thus it does not appear that the corrosion resistance of a semiconductor is a prime factor relative to the potential electronic properties in selecting a particular compound for a semiconductor application. Thus corrosion and oxidation studies on these compounds will continue to be of academic interest.

In refractory applications, there is a need for intermetallic compounds which are strong and oxidation-resistant at temperatures above 1000°C to replace the iron- and nickel-base solid solution and precipitation-hardened alloys that are presently available. The aluminide, beryllide, and silicide groups of compounds are the most promising candidates. However, without proper design to minimize impact failures their acceptance for applications other than heating elements will be difficult to achieve. The limiting temperatures for oxidation resistance of the common refractory compounds are compared in Fig. 7 from data presented in the preceding tables. Since it is difficult to reduce all the pertinent technology of oxidation resistance to one graph, some information such as spalling after cycling is ignored, if it is indeed even known. The graph is arranged according to the transition element which is combined with aluminum, beryllium, or silicon. In many cases there is more than one component in the binary system, and the temperature limitation applies to the most resistant in the system that is known. The temperature limitation is based on an approximate oxidation rate of 20 mg/cm² for 100 hours. The intermetallic

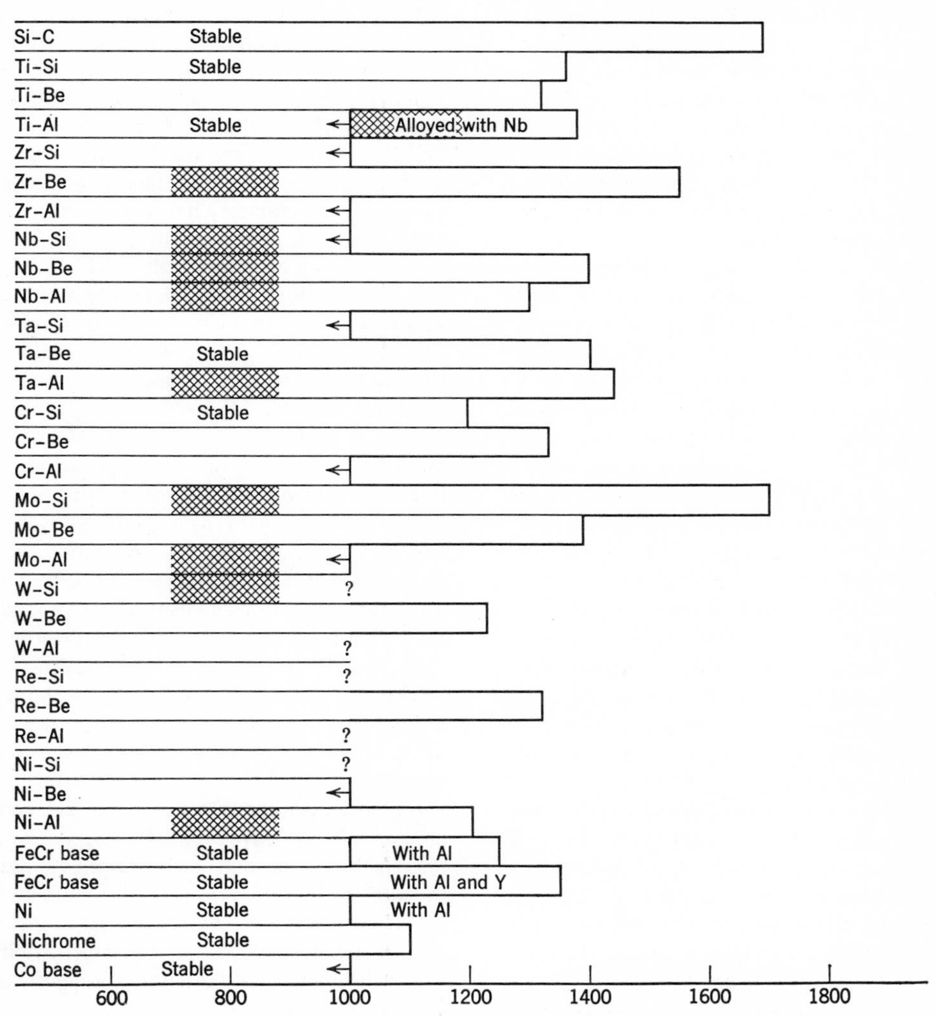

Fig. 7. Comparison of oxidation rates.

compounds with no oxidation resistance above 1000°C are indicated by an arrow. For comparison several oxidation-resistant iron, nickel, and cobalt base alloys are listed. If significant improvement in the oxidation resistance has been observed by minor alloy additions, the bar length has been extended. Regions of known instability, such as by "pest" reaction, are indicated by cross-hatching.

A more popular approach for achieving oxidation resistance in structural materials for use above 1000°C is by applying intermetallic compounds as coating on some of the refractory metals[91] such as niobium, tantalum, molybdenum, and tungsten. The ductility of the base metal provides the impact resistance, and the coating provides the protection from oxidation. Extensive effort is being expended in this direction with some success if the lifetimes are short and the use requires only one cycle. Multiple thermal and mechanical cycling often leads to cracks in the intermetallic coatings and ultimate breakdown of the coating. If the coating can be made self-healing, the presence of cracks during service would not be as critical.

A type of self-healing, oxidation-resistant coating has been demonstrated in recent studies[85–87] at the Naval Research Laboratory by formation of Nb–Zn intermetallic compounds on niobium metal. The Nb–Zn compounds (primarily $NbZn_3$) serve as a reservoir of zinc, which can reform a coating of ZnO if a crack appears. The reaction is described as a dissociation of $NbZn_3$ to $NbZn_2$ and zinc vapor. The zinc vapor will redeposit as the oxide and reheal the crack. The original oxide film is believed to be a porous niobium oxide with the approximate composition NbO_2, in which the pores have filled by ZnO. The lifetime of the coating depends on the amount of zinc available for oxidation and on the rate of its consumption from cracking or from solid-state diffusion through the oxide layer. In a coating of this type the activity of zinc in the intermetallic compound should be high enough to maintain growth and repair of the oxide barrier layer. Not many of the intermetallic compounds have sufficient activity to provide a high-vapor concentration for self-healing. In the case of zinc, the vapor pressures are apparently sufficient to achieve self-healing.

The use of intermetallic compounds such as $NbZn_3$ or $NbAl_3$ on niobium for oxidation protection of the base metal will probably be the major source of oxidation data for intermetallic compounds in the near future.

Solutions to the low-temperature disintegration or "pest" phenomena are of prime importance, if refractory intermetallic compounds are to be useful in the future. The "pest" phenomena so far has been observed in many intermetallic compounds of different crystal type. The reaction occurs over a narrow temperature interval which is specific for each compound and requires the presence of a reactive gas such as nitrogen or oxygen, although the extent of the attack is very small when disintegration starts. The temperature range (500–900°C) at which it occurs makes it difficult to design around the phenomena because most high-temperature materials operating above 1000°C will have significant dwell times at 500–900°C. A study is necessary to determine why certain intermetallic compounds are not susceptible to the "pest" phenomena. The "pest" reaction is reported to be absent in $TaBe_{12}$ but this compound has been shown to exhibit grain-boundary hardening just as with other materials in the beryllide family. It may be that certain structures, although exhibiting local hardening through segregation, can absorb the stresses created and not fragment. Excess aluminum metal, which is molten in the pest region, was found empirically to eliminate disintegration of $NbAl_3$. Unfortunately, molten aluminum is not suitable for high-temperature refractory applications but this observation lends encouragement to solution of the problem by way of alloying or adding dispersed phases. A more direct approach towards solution would be to determine the general mechanism of the pest reaction. The basic kinetic studies by Westbrook and Wood[69] and Westbrook and Seybolt[70] on the kinetics of grain-boundary hardening give indications that grain-boundary segregation of diffused oxygen is responsible. However, the demonstration of a direct link between hardening and disintegration has yet to be accomplished. It is surprising, if segregation is responsible, why is it so damaging to intermetallic compounds and yet not to pure metals and solid solutions in which segregation is not an uncommon occurrence. The marked effect of stoichiometry in intermetallics suggests that a careful analysis of the chemical activities would be fruitful.

REFERENCES

1. Schwartzkopf P., and R. Kieffer, *Refractory Hard Metal*. The Macmillan Co., New York, 1953.
2. Paine R. M., A. J. Stonehouse, and W. W. Beaver, WADC TR 59-29, Parts I and II, PB Rpt. 161, 683 (1960), U.S. Dept. Comm. OTS.
3. Kubaschewski O., and B. E. Hopkins, *Oxidation of Metals and Alloys*. Academic Press, New York, 1962.
4. Hauffe Karl, *Oxydation von Metallen und Metallegierungen*. Springer-Verlag, Berlin, 1956.
5. Roberts M. W., *Quart. Rev.* **16** (1962) 71–79.
6. Popovkin B. A., L. M. Kovba, V. P. Zlomanov, and A V. Novoselova, *Dokl. Akad. Nauk SSSR* **129** (1959) 809–12; *CA* **55**, 25566.
7. Zlomanov V. P., O. I. Tananaeva, A. V. Novoselova, *Zh. Neorgan. Khim.* **6** (1961) 2753; CA **56,** 11195.
8. Korneeva I. V., and A. V. Novoselova, *Zh. Neorgan. Khim.* **5** (1960) 2265; CA **56,** 12518.
9. Jones R. H., *Proc. Phys. Soc.*, 1957 (B) **70** (11), 1025; *Met. Abstr.* **25,** 865.
10. McAndrew J. B., and D. J. McPherson, WADC TR 53-182, Part II, January, 1955. NSA 10-5673.
11. Wukusick C. S., XDC 61-4-54. General Electric Company, Cincinnati, Ohio, April 1961.
12. Jorgensen P. J., M. E. Wadsworth, and I. B. Cutler, *J. Am. Ceram. Soc.* **44** (1961) 258.
13. Fitzer E., *Plansee Proc.* (1955). Pergamon Press, London, 1956.
14. Rausch J. J., ARF 2981-4, Armour Research Foundation, August 31, 1961. NSA 15-31171.
15. Stonehouse A. J., Private Communication.
16. Lewis J. R., *J. Metals* **13** (1961) 357.
17. Lewis J. R., *J. Metals* **13** (1961) 829.
18. Rosenberg A. J., *J. Phys. Chem Solids* **14** (1960) 175.
19. Rosenberg A. J., and M. C. Lavine, *J. Phys. Chem.* **64** (1960) 1135.
20. Wagner C., *Z. Phys. Chem.* **B21** (1933) 25.
21. Rosenberg A. J., *J. Electrochem. Soc.* **107** (1960) 795.
22. Rosenberg A. J., *J. Phys. Chem.* **64** (1960) 1143.
23. Evans U. R., *The Corrosion and Oxidation of Metals*. St. Martins Press, New York, 1960.
24. Rosenberg A. J., and A. A. Menna, Proceedings of the 4th International Symposium on the Reactivity in Solids, J. H. DeBoer, Ed. Elsevier Publish. Co., New York, 1961, p. 164.
25. Tamman G., and A. Ruhenbeck, *Z. Anorg. Allgem. Chem.* **223** (1935) 285.
26. Lohberg K., *Z. Metallk.* **41** (1950) 56.
27. Numata T., *J. Phys. Soc. (Japan)* **17** (1962) 878.
28. Kafalos J. A., H. C. Gatos, and M. J. Button, *J. Am. Chem. Soc.* **79** (1957) 4260.
29. *The Surface Chemistry of Metals and Semiconductors*, H. C. Gatos, Ed. Wiley, New York, 1949.
30. Gatos H. C., and M. C. Lavine, *J. Electrochem. Soc.* **107** (1960) 427, 433.
31. Gatos H. C., M. C. Lavine, and E. P. Warekois, *J. Electrochem. Soc.* **108** (1961) 645.
32. Lavine M. C., H. C. Gatos, and M. C. Finn, *J. Electrochem. Soc.* **108** (1961) 974.
33. Larrabee G. B., *J. Electrochem. Soc.* **108** (1961) 1130.
34. Kieffer R., and F. Benesovsky, Iron and Steel Inst. Spec. Rpt. No. 58, 292 (1956).
35. Kieffer R., and E. Cerwenka, *Z. Metallk.* **43** (1952) 101.
36. Kieffer R., F. Benesovsky, H. Nowotny, and H. Schachner, *Z. Metallk.* **44** (1953) 242.
37. Kieffer R., F. Benesovsky, and E. Gallistl, *Z. Metallk.* **43** (1952) 284.
38. Kieffer R., F. Benesovsky, and R. Machenschalk, *Z. Metallk.* **45** (1954) 493.
39. Kieffer R., F. Benesovsky, and H. Schroth, *Z. Metallk.* **44** (1953) 437.
40. Arbiter W., WADC TR 53-190, Parts 1 and 2, November 1953. PB 111413 and PB 121018 OTS.
41. Kieffer R., F. Benesovsky, and H. Schmid, *Z. Metallk.* **47** (1956) 247.
42. Fitzer E., and J. Schwab, *J. Metallk.* **9** (1955) 1062.
43. Maxwell W. A., NACA RM E52A04, March 14, 1952.
44. Berkowitz WADD TR 60-377, September 1960.
45. Lea A. C., *Trans. Brit. Ceram. Soc.* **40** (1941) 93.
46. Lea A. C., *J. Soc. Glass Technol.* **33** (1949) 27.
47. Elmer T. H., and W. J. Koshuba, NEPA Rpt. No. 1768, Oak Ridge, Tenn., March 1951. NSA 5-3420.
48. Suzuki H., *Yogyo Kyokai Shi* **65** (1957) 88; *Ceram. Abt.*, January 1958, p. 17a.
49. Ervin G., *J. Am. Ceram. Soc.* **41** (1958) 347.

50. Jorgensen P. J., M. E. Wadsworth, and I. B. Cutler, *J. Am. Ceram. Soc.* **42** (1959) 613.
51. Jorgensen P. J., M. E. Wadsworth, and I. B. Cutler, *J. Am. Cer. Soc.* **43** (1960) 209.
52. Engel H. J., and K. Hauffe, *Metall.* **6** (1952) 285. Also Bruce Chalmers, *Progress in Metal Physics*, Vol. IV. Interscience Publishers, New York, 1953, p. 97.
53. Cook M. A., and A. G. Oblad, *Ind. Eng. Chem.* **45** (1953) 1456.
54. Law J. T., *J. Phys. Chem.* **61** (1957) 1200.
55. Higuchi I., T. Ree, and H. Eyring, *J. Am. Chem. Soc.* **77** (1955) 4969.
56. Aitken E. A., and J. P. Smith, GEMP-105. General Electric Co., Cincinnati, Ohio. NSA 17-532 OTS.
57. Aitken E. A., and J. P. Smith, *J. Nucl. Mater.* **6** (1962) 119.
58. Perkins F. C., DRI-2042, Denver Research Inst., June 1, 1962. NSA 17-530 OTS.
59. Klopp W. D., DMIC Report 167, Battelle Memorial Inst., Columbus, Ohio, March 12, 1962. NSA 16-18024 OTS.
60. Lever R. C., APEX-913, General Electric Co., June 20, 1962.
61. Wukusick C. S., GEMP-3A, General Electric Co., Cincinnati, Ohio, July 1, 1961–August 31, 1961, p. 7. NSA 16-2175 OTS.
62. Wukusick C. S., GEMP-5A, General Electric Co., Cincinnati, Ohio, November 15, 1961, p. 13. NSA 16-2177 OTS.
63. Wukusick C. S., GEMP-7A. General Electric Co., Cincinnati, Ohio, January 15, 1962, p. 18. NSA 16-9195 OTS.
64. Wukusick C. S., Private communication.
65. Gatos H. C., and M. C. Lavine, *J. Phys. Chem. Solids* **14** (1960) 169.
66. Miller D. P., J. G. Harper, and T. R. Perry, *J. Electrochem. Soc.* **108** (1961) 1123.
67. Campbell I. E., *High Temperature Technology*. Wiley, New York, 1956.
68. Staritsky E., *Anal. Chem.* **28** (1956) 553.
69. Westbrook J. H., and D. L. Wood, *J. Nucl. Mater.* **12** (1964) 208.
70. Seybolt A. U., and J. H. Westbrook, *Acta Met.* **12** (1964) 449.
71. Ervin G., and M. Nakata, NAA-SR-6493, December 15, 1960.
72. Gulbransen E., and K. F. Andrew, *J. Electrochem. Soc.* **97** (1950) 383.
73. Austerman S. B., NAA-SR-5893, October 1, 1960. NSA 15-18494 OTS.
74. Austerman S. B., NAA-SR-6427, September 75, 1961. NSA 15-29707 OTS.
75. Holt J. B., Private communication.
76. Oishi Y. D., and W. D. Kingery, *J. Chem. Phys.* **33** (1960) 480.
77. Mozhukhin E. I., L. Kh. Pivovarov, and Ya. S. Umanskiy, *J. Appl. Chem.* (*USSR*) **30** (1957) 1593.
78. Imai Y., and M. Kumazawa, *Sci. Rept. Res. Inst. Tohuku Univ.* **11** No. 4, August 1959.
79. Fitzer E., and P. Gerasimoff, *Z. Metallk.* **50** (1959) 187.
80. Muller K., *Wiss. Z. Tech. Univ. Dresden* **9** (1959/60) 1217.
81. Samsonov G. V., and V. S. Sinel'nilova, *Tsvetn. Metal.* **35** No. 11 (1962) 92.
82. Eisenkolb F., and H. E. Rollig, *Neue Huette* **3** (December 1958) 721.
83. Bernstein L., *J. Electrochem. Soc.* **109** (1962) 270.
84. Beck R. L., DRI 2059, Summary Report, October 15, 1962. NSA 17-30969 OTS.
85. NRL Report No. 5550, Progress Rept.-1, July 1, 1960. NSA 15-14722.
86. NRL Report No. 5581, Progress Rept.-2, October, 1960. NSA 15-14723.
87. DMIC Memo 88, OTS PB 161238, March 3, 1961.
88. Krikorian O. H., UCRL 6132, September 6, 1960, NSA 15-1832 OTS.
89. Wukusick C. S., GEMP-218. General Eleetric Co., Cincinnati, Ohio, July 31, 1963.
90. Golubev A. I., *Mezhkriztal Korroziya i Korroziya Metal v Napryazhen Sostoyanii Vsesoyuz Sovet Nauk-Tekh Obshchestv.* **1960** 15–26; CA **55** (1961) 15306d.
91. Lorenz R. H., and A. B. Michael, *J. Electrochem. Soc.* **108** (1961) 885.
92. Samsonov G. V., and V. S. Sinel'nilova, *Tsvetn. Metal.* **35,** No. 11 (1962) 92.

chapter 26

Thermoelectric Properties

IRVING CADOFF
New York University
New York

1. INTRODUCTION

The importance of intermetallic compounds for thermoelectric conversion is a result of their particularly favorable transport properties—Seebeck coefficient (thermoelectric power), electrical conductivity, and thermal conductivity—as compared with metals and elemental semiconductors. Although the thermoelectric effects were discovered in the mid-1800's and the potential use of these effects for energy conversion was proposed in the early 1900's,[1] the significant advances in efficient operation of such devices have occurred only in the last twenty years, paralleling the development of the field of semiconducting intermetallic compounds. In fact, the desire to obtain more efficient thermoelectric materials was one of the main incentives for the expanding research efforts on this class of materials.

As already mentioned, the usefulness of intermetallic compounds is based on their favorable thermoelectric parameters. To develop this point more fully, a brief, nonrigorous* review of the theory of thermoelectric conversion is presented here, from which the relation between the efficiency and the materials parameters is derived. Also included is a discussion of the relation between the thermoelectric parameters and the structure (electronic and physical) of the compounds, from which criteria for compound selection are obtained. The discussion is concluded with a review of those compounds that have been used successfully as well as those that are currently being studied for possible use.

* For a more complete description the reader is referred to other texts listed at the end of the chapter.

2. THERMOELECTRIC CONVERSION

The applications of thermoelectric conversion include both power generation and refrigeration. The idealized thermoelectric circuit shown in Fig. 1 can be used to produce cooling or electrical power. The circuit consists of two dissimilar elements contacted at two junctions. The junction at T_i is shown as a physical one, and the junction at T_o is schematically indicated as electrical. For the general discussion of thermoelectric parameters, a physical contact at T_o will be assumed. For

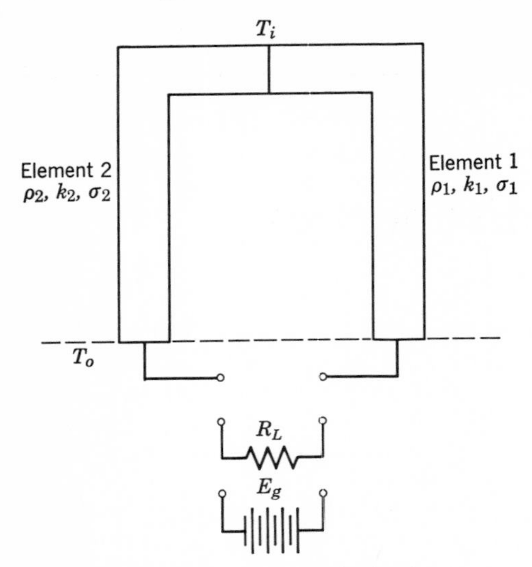

Fig. 1. Thermoelectric circuit.

power generation or refrigeration, the load R or the source E_g will be inserted at the electrical contacts. The parameters of the circuit are:

T_i = source temperature
T_o = sink temperature
k_1, k_2 = specific thermal conductivity
ρ_1, ρ_2 = specific electrical resistivity
S_{12} = relative Seebeck coefficient
π_{12} = relative Peltier coefficient
D_1, D_2 = form factors
= area/length of the thermoelements

The Seebeck effect is measured by the emf, E_{12}, developed in the circuit when $T_i > T_o$ such that

$$E_{12} = S_{12}\,\Delta T$$

and therefore

$$S_{12} = \frac{E_{12}}{\Delta T}$$

or

$$S_{12} = \lim_{\Delta T \to 0} \frac{E_{12}}{\Delta T}. \tag{1}$$

E_{12} and S_{12} are considered positive if the conventional current flows from 1 to 2 at the hot junction.

The Peltier effect is measured by the heat $\dot{Q}$ evolved or absorbed at a junction as electric current, I, passes through it.

$$\dot{Q}_{12} = \pi_{12} I \tag{2}$$

π_{12} is positive if heat is absorbed when current flows from 1 to 2. Although both effects require the contact of two dissimilar elements, the relative coefficient can be expressed in terms of absolute coefficients

$$S_{12} = S_1 - S_2$$

and

$$\pi_{12} = \pi_1 - \pi_2$$

where the subscripts 1 and 2 refer to the elements 1 and 2, respectively.

From thermodynamic arguments (the thermoelectric effects are thermodynamically reversible) William Thomson (Lord Kelvin) derived relations between these effects and predicted the existence of a third effect—the Thomson effect. This effect relates to the reversible absorption or evolution of heat, $\dot{Q}_1$, in a homogeneous conductor carrying a current, I, through a temperature gradient, dT/dx, such that

$$\dot{Q}_1 = \tau_1 I \frac{dT}{dx}$$

The Thomson coefficient τ_1 is defined for one material only and is positive if heat is absorbed when I and dT/dx are in the same direction. The Kelvin analysis also resulted in the following relations among the parameters

$$\pi_{12} = TS_{12} \tag{3}$$

$$\tau_1 - \tau_2 = T\frac{dS_{12}}{dT} \tag{4}$$

and in the definition of an absolute Seebeck coefficient

$$S_1 = \int_0^T \frac{\tau_1}{T}\,dT \tag{5}$$

from which the absolute Peltier coefficient

$$\pi_1 = TS_1 \tag{6}$$

is calculated. Although the Thomson effect is important in relating the Peltier and Seebeck effects, its effect on the operation of a simple thermoelectric circuit can be neglected.

Efficient operation of the thermoelectric circuit depends on optimization of the materials parameters and the circuit parameters. This can be illustrated for the case of refrigeration. If the dc source E_B is placed in the circuit (Fig. 1) in series with the thermoelements, the current I in the circuit can be used to pump heat from T_i to T_o. If $T_i < T_o$, the circuit will operate as a refrigerator. At the cold junction, T_i, the Peltier heat removal (from Eqs. 2 and 6)

$$\dot{Q}_p = \pi_{12} I = S_{12} T_i I \tag{7}$$

must work against the heat supplied by the source $\dot{Q}_a$, thermal conduction heat

$$\dot{Q}_k = (k_1 D_1 + k_2 D_2)\, \Delta T \tag{8}$$

flowing back from T_o, and one half of the Joule heat

$$\dot{Q}_j = I^2 \left(\frac{\rho_1}{D_1} + \frac{\rho_2}{D_2}\right) \tag{9}$$

developed in the thermoelements. A heat balance at T_i shows that

$$\dot{Q}_a = \dot{Q}_p - \tfrac{1}{2}\dot{Q}_j - \dot{Q}_k \tag{10}$$

From a casual examination of this heat balance it can be seen that the most efficient heat removal requires high Peltier absorption, low thermal conduction, and low Joule heating. The effects of thermal conduction and electrical resistance are minimized when

$$\frac{D_1}{D_2} = \left(\frac{\rho_1 k_2}{\rho_2 k_1}\right)^{1/2} \tag{11}$$

A mathematical optimatization of Eq. 10 results in an efficiency term which contains an expression defined as the thermoelectric figure of merit for the circuit.

$$Z_{12} = \frac{S_{12}^2}{[(k_1\rho_1)^{1/2} + (k_2\rho_2)^{1/2}]^2} \tag{12}$$

Examination of this "efficiency" term leads to the same conclusion: for efficient operation, the elements in the circuit should possess high Seebeck coefficient, low thermal conductivity, and low electrical resistivity. The analysis of power generation leads to precisely the same figure of merit.*

For a given material the figure of merit may be written as

$$Z = \frac{S^2}{\rho k} \tag{13}$$

which clearly points up the materials-parameter requirements for thermoelements.

3. THEORY

It should be noted at this point that ρ and k cannot be divorced from the thermoelectric parameters S_1, π, and τ when describing the thermoelectric properties of a material, and in a broad sense they can be considered thermoelectric properties. It was shown earlier that S_1, π, and τ are specifically related properties of a material. S_1, ρ, and k_1 are not directly related but neither are they independent; therefore it is possible to develop relations among these parameters which will point up the conditions for optimization. The factor common to all three is the charge-transport effect, basically related to the concentration of free carriers in the material. It has generally been found that metallic materials are not highly efficient thermoelements, and consequently most of the investigations have been directed toward semiconductors.

In a semiconductor containing only negative carriers, n, the electrical conductivity, σ, may be written as

$$\sigma = \frac{1}{\rho} = ne\mu \tag{14}$$

where μ is the mobility of the carriers. If this semiconductor is nondegenerate, the Seebeck coefficient is given by

$$S = -\frac{k}{e}\left[\delta - \ln\left(\frac{n}{M}\right)\right] \tag{15}$$

where M is an effective density of states parameter and includes an effective mass term m^*, δ is a scattering parameter which is two for lattice scattering and four for impurity scattering, and k is the Boltzmann constant. Figure 2 shows schematically the variation of S, σ, and $S^2\sigma$ with carrier concentration, n. The quantity $S^2\sigma$ is the numerator of the figure of merit

$$Z = \frac{S^2\sigma}{k} \tag{16}$$

which results when σ is substituted for $1/\rho$ in Eq. 13. The position of the broad maximum in the $S^2\sigma$ versus n curve varies somewhat with the crystal lattice, but generally occurs in the range 10^{18} to 10^{21} carriers/cm^3.

The thermal conductivity k is the sum of the lattice conduction k_L and the electronic transport k_e, the latter being related to the electrical conductivity by

$$k_e = \delta \left(\frac{k}{e}\right)^2 T\sigma \tag{17}$$

In Eq. 17, δ is the scattering parameter and T the absolute temperature. The schematic

* Note, however, that the thermocouple, used for temperature measurement, is a type of power generator having somewhat different efficiency considerations. The thermocouple is most efficient if it exhibits a high Seebeck coefficient; optimum power transfer is not required.

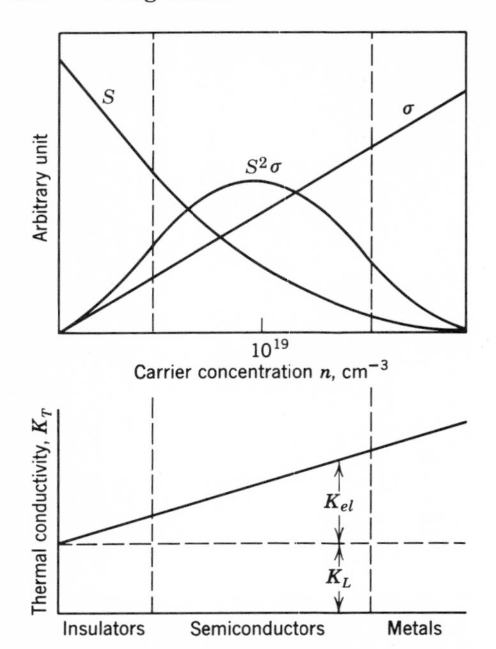

Fig. 2. Schematic dependence of S, σ, and k on carrier concentration.

variation of k with n is also shown in Fig. 2. For metals, $k_e \gg k_L$, and therefore $k \approx k_e$ so that

$$Z_{\text{metals}} = \frac{S^2}{\delta(k/e)^2 T}$$

For metals, $S \sim 10^{-6}$ volts/°K, which yields a figure of merit $Z \sim 0.5 \times 10^{-4}$ °K^{-1} at room temperature. For semiconductors, Goldsmid,[2] Ioffe,[3] and Rittner[4] have shown that the figure of merit is related to a quantity

$$\frac{\mu m^{*3/2}}{k_L} \tag{18}$$

In Eq. 18, the Seebeck coefficient and conductivity are not explicitly present, because they both vary (in opposite manner) with the carrier concentration. Use of this equation requires also that the carrier concentration be optimized, as shown in Fig. 2. This optimum concentration is approximately

$$n_{\text{opt}} = 2.4 \times 10^{10} \left[m^* \frac{T}{300} \right]^{3/2} \text{cm}^{-3} \tag{19}$$

The Seebeck coefficient corresponding to this n_{opt} is about 200×10^{-6} volts/°K with corresponding $Z \sim 10^{-3}$ °K^{-1}.

A high thermoelectric efficiency thus depends on high mobility, high effective mass, and low thermal conductivity. The effective mass term, generally a "density-of-states" mass, is not predictable from the simple physical characteristics of the material. It is a complex parameter which has a wide range of values and temperature dependencies. The effective mass also appears in the description of mobility for a nondegenerate semiconductor

$$\mu = 4/3 \frac{el}{(2\pi m^* kT)^{1/2}} \tag{20}$$

where l is the mean free path of the carrier. Control of the mobility is achieved by control of l rather than m^*. The mean free path is a function of the scattering processes that impede the flow of charge carriers, the principal ones being lattice vibrations, impurity centers, and interactions among carriers of opposite sign. As shown by Eqs. 14 and 17, high mobility will mean higher electrical conductivity, but also higher electron thermal conductivity. For the most part the two are inseparable. The electronic thermal conduction is small compared with S^2, but the efficiency is reduced by a relatively high k_L in nonmetals. The lattice thermal conduction is essentially a transport property similar to electron conduction and is also governed by scattering mechanisms. Here again a mean free path is involved. For optimization of the thermoelectric efficiency, the phonon mean free path should be minimized. k_L is small if the atoms are heavy, the binding is soft, and the structure exhibits anharmonicity. This latter point can be achieved by having unequal masses in the lattice or by using an asymmetrical lattice structure.

Additional theoretical treatment[5] has shown that the figure of merit is related to the bandwidth of a semiconductor, E_G, and that for optimum operation the gap should be in the order of $4kT_h$, where T_h is the hot-junction temperature.

4. CRITERIA FOR MATERIALS SELECTION

As seen from the previous discussion, the principal controllable factors in the efficiency of thermoelectric materials are a low lattice thermal conductivity, high mobility, and an optimum carrier concentration. The problems of selecting favorable materials and of optimizing their properties are inseparable. In looking for useful materials, the first consideration is thermal conductivity. For high Z, k should be less than 0.01 w/cm-°K. The requirements are that the material consist of atoms which are heavy, and that there be at least two elements with their atoms being appreciably different in size and weight. The crystal structure should be complex with a high thermal expansion and low Debye temperature so that large anharmonicity will result.

Goldsmid has suggested ionic binding as a means of reducing k_L, but has also pointed out that ionic crystals generally exhibit reduced mobility. Ioffe also studied the control of lattice thermal conduction and showed that by alloying the materials with elements having similar valence structure, but different atomic mass, the lattice conduction could be reduced appreciably with only a relatively small decrease in carrier mobility.

The electrical characteristics are the second consideration. These relate to the carrier concentration and mobility. The material selected should be a semiconductor with a band gap in the order of $4kT$. A lower limit is set because the number of intrinsic carriers must be small in the operating range. Large intrinsic concentrations (unless the carrier mobilities are very different) will result in a lowering of the Seebeck coefficient and an increase in thermal conduction due to electron-hole pair transport (ambipolar conduction). An upper limit is placed on E_G because lower mobility and higher extrinsic ionization energies are generally associated with "tighter" binding. The optimum carrier concentration of about 10^{19}/cm^3 requires that the material selected have a large solubility for carrier dopants which must be introduced without appreciably reducing the mobility. As will be seen in the next section, extrinsic carriers may be introduced by stoichiometric deviations in compounds as well as by adding doping elements.

The requirements listed above point to intermetallic compounds as the most appropriate class of materials for thermoelements. Table 1 contains a partial listing of some electronic characteristics of semiconducting compounds; the elements Ge and Si and the ionic compound AgI, are included for comparison. A number of interesting features are found in this tabulation which supplement the theoretical evaluation of the materials.

1. For the elements and homologous compounds, E_G and mp decrease with increasing at. wt. but μ increases as at. wt. increases.
2. For an isoelectronic series (InSb, CdTe, AgI) the gap increases with increasing ionic component in binding with, as before, decreasing mobility with increasing E_G.
3. The mp increases with partial ionic bond.
4. The mobilities in III-V structures are slightly higher than their isoelectronic elements.

The lattice conduction properties are similarly listed in Table 2, the group IV elements

TABLE 1

Electronic Properties of Some Semiconducting Compounds

	Class	Mean at. wt.	Crystal Structure[a]	E_G	mp	μ_n
Diamond	E	12	D	~6	...	...
Si	E	28	D	1.1	1400°C	1,200
Ge	E	72.6	D	0.67	1000	3,900
AlP	III–V	29	Z	3	...	...
GaAs	III–V	72.8	Z	1.1	1240	4,000
InSb	III–V	118.3	Z	0.18	535	77,000
CdTe	II–VI	120	Z	1.5	1045	300
AgI	I–VII	117.4	W, Z	2.8	...	30
PbS	IV–VI	120	N	0.30	1100	560
PbSe	IV–VI	143	N	0.22	1065	1,020
PbTe	IV–VI	167.4	N	0.27	904	1,620
Bi_2Te_3	V–VI	160	R	0.15	585	800

[a] D = diamond cubic, Z = zinc blende, W = wurzite, N = NaCl, R = rhombohedral.

TABLE 2

Lattice Conduction of Some Semiconducting Compounds

	Mean at. wt.	k_L w/cm-°K
Diamond	12	1.8
Ge	72.6	0.6
GaSb	95.7	0.25
InSb	118.3	0.16
KCl	37.3	0.06
RbBr	82.7	0.04
RbCl	60.5	0.02
Bi_2Te_3	160	0.016

and ionic compounds are included for illustrative purposes. Here it is clearly seen that k_L decreases with at. wt. for homologous compounds, and that, as an ionic component is introduced, k_L also decreases. However, it may be noted that for RbBr and RbCl, which are not very different in mean at. wt., the lower k is found for the lighter RbCl. This is attributed to the greater mass ratio of Rb/Cl as compared with Rb/Br. Bi_2Te_3 is included for comparison.

From these data we can see why Bi_2Te_3 and PbTe have received the widest attention as thermoelements. The principal limitations on the use of Bi_2Te_3 are derived from its relatively low E_G. The current research areas are predominantly in high-temperature materials, because, as a Carnot engine, the thermoelectric generator will be more efficient as the T_h is increased. As will be seen, the use of higher E_G materials, which will be less efficient than Bi_2Te_3 at room temperature, is necessary. For optimum efficiency over a wide temperature range, cascaded thermoelements of different composition operating in different ranges are used. An alternate possibility is the use of composite thermoelements consisting of high E_G materials at the hot end and low E_G materials at the cold end. This arrangement introduces many complex problems.

5. THERMOELECTRIC PROPERTIES OF INTERMETALLIC COMPOUNDS

The principal thermoelements currently in use are PbTe and Bi_2Te_3. Although other compounds, notably ZnSb, CdSb, and GeTe, also possess useful properties, PbTe and Bi_2Te_3 will be discussed in detail. Comparative physical constants are shown in Table 3. The $Te^{(1)}$–$Te^{(1)}$ layers are weakly bonded with the result that the Bi_2Te_3 crystals cleave readily along these planes. Furthermore, the layered Bi_2Te_3 structure is anisotropic, so that the thermoelectric properties are anisotropic. Dennis[6] has found that the relationship between the Seebeck coefficient parallel to the cleavage plane $S_{\parallel}$ and the coefficient perpendicular to the cleavage plane $S_{\perp}$

$$S_{\parallel} - S_{\perp} = \pm\left(\frac{k}{e}\right)\frac{d(\sigma_{\parallel}/\sigma_{\perp})}{d(\ln kT)}$$

gives a difference, for iodine doped n-type Bi_2Te_3, of about 22 μV/°K in the range of 100 to 300°K where the $S_{\parallel}$ and $S_{\perp}$ are in the range of -100 to -200 μV/°K.

In order to obtain the optimum carrier concentration, doping of the intrinsic material is required. In intermetallic compounds deviations from stoichiometry can be used to control extrinsic carrier concentration. For example, in PbTe excess Pb results in n-type and excess Te in p-type conduction. In Bi_2Te_3, excess Bi results in p-type and excess Te in n-type conduction.

These stoichiometric deviations are common to all intermetallic compounds. Figure 3 illustrates the phase diagram of Bi_2Te_3 in the vicinity of the compound. The features to note are the relatively wide range of stability of the compound about the stoichiometric composition which permits the control of excess carriers over a wide range, and the noncoincidence of the maximum melting point and the stoichiometric point. This is not generally expected but has been found to occur in other systems as well. In the case of Bi_2Te_3 this means that on freezing liquid of stoichiometric proportions the solid to form first will be Bi-rich and p-type because liquid at point A is in equilibrium with solid at point B. As solidification proceeds, the

TABLE 3

Comparison of PbTe and Bi_2Te_3

	Structure	E_G	mp
Bi_2Te_3	Rhombohedral ($Te^{(1)}$–Bi–$Te^{(2)}$–Bi–$Te^{(1)}$ stacking sequence)	0.15 eV	585°C
PbTe	NaCl	0.35 eV	924°C

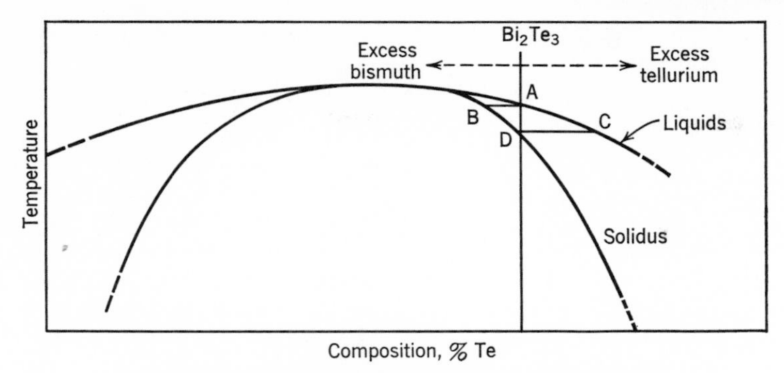

Fig. 3. Partial phase diagram in the vicinity of Bi_2Te_3.

composition of the solid will move along the solidus line from *B* to *D* at which time it is stoichiometric. This corresponds to solidification from a liquid of composition *C*. Further solidification results in formation of Te-rich *n*-type material. This effect poses problems in obtaining uniform material. By normal chemical standards this nonstoichiometry is small, perhaps less than 0.01 per cent, but this represents a large deviation for semiconductor behavior. A similar effect in PbTe is shown in Fig. 4. The resistivity and carrier concentration along the length of an ingot crystallized from a stoichiometric melt are included together with a section of the phase diagram. As shown in this figure, the carrier concentration varies from less than $10^{16}/cm^3$ at the stoichiometric point to $10^{18}/cm^3$ at the seed end. To obtain *n*-type PbTe, it is necessary to start with a melt containing more than 0.42 *a/o* excess Pb.

The type and amount of extrinsic carriers which can be obtained in this manner is limited by the thermodynamics of the system. Additional carriers may be obtained by doping with other elements. In Bi_2Te_3, for example, Pb is a *p*-type dope, whereas Cu, Ag, I, and Br are *n*-type dopants. Figure 5 shows the effect of excess Bi on the thermoelectric properties of Bi_2Te_3. The resistivity abscissa is related to $1/n$, and the relations described previously are apparent: *S* increases as *n* decreases (ρ increases), *k* decreases as *n* decreases, and

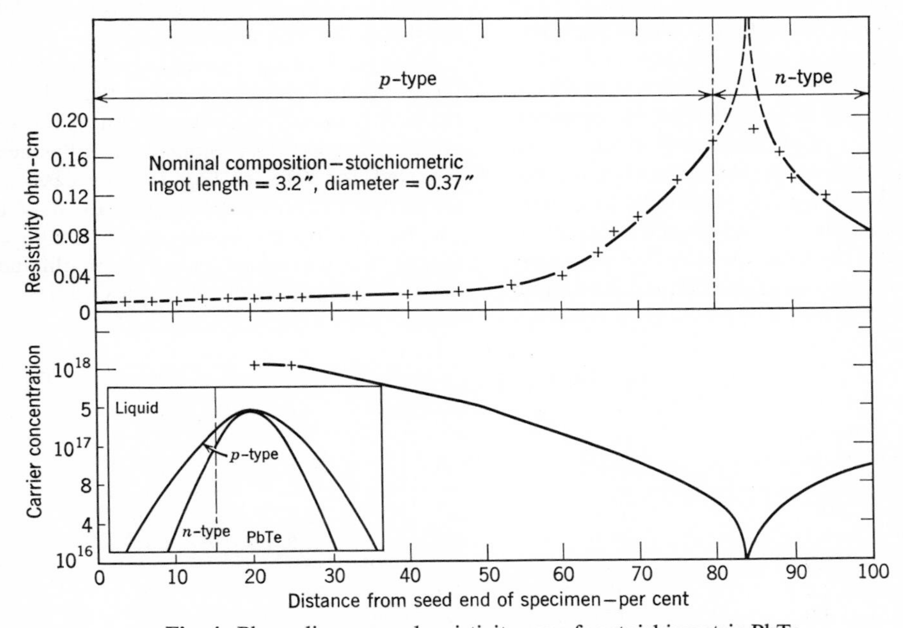

Fig. 4. Phase diagram and resistivity scan for stoichiometric PbTe.

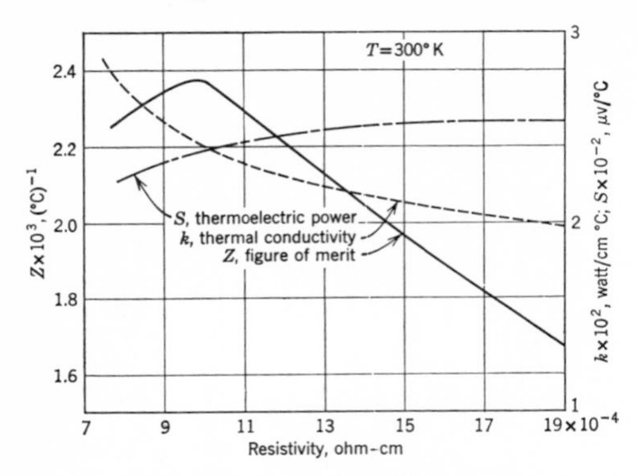

Fig. 5. Properties of Bi_2Te_3.

Z passes through a maximum of 2.4×10^{-3} $°K^{-1}$ corresponding to a resistivity of 1×10^{-3} Ω-cm, with $S_{opt} \cong 230$ μV/°C.

Figure 6 illustrates the effect of doping and temperature on the figure of merit of PbTe. It is shown that the Z falls sharply with temperature due to intrinsic onset, but that the intrinsic effect can be pushed to higher temperatures by doping. In this case Na is added to PbTe to yield *p*-type material. Here, for 1.07 per cent Na the Z is optimized at 300°C. Similar effects are observed in Bi_2Te_3, but the temperature optimization occurs at room temperature. It is thus seen that the figure of merit is optimized not only with respect to carrier concentration but also to temperature. The difference in temperature of optimization between Bi_2Te_3 and PbTe can be attributed to their different values of E_G. The curves shown in Figs. 5 and 6 are typical of all systems; the values at which the optimum values occur will of course vary from system to system. What is important is that the parameters are extremely structure- and temperature-sensitive, and specifying a particular compound is far from sufficient for specifying its thermoelectric behavior. Equations 14 and

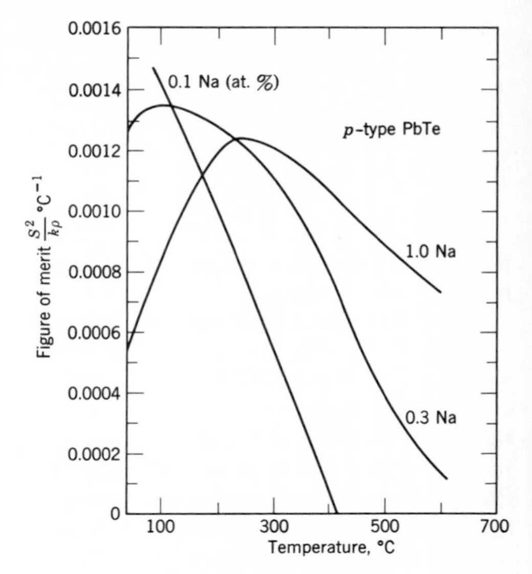

Fig. 6. Properties of PbTe.

15 show the relations between S and n and σ and n, and therefore the $S^2\sigma$ of the Z factor can be optimized and is more or less fixed. As already mentioned, most of the development effort has been directed towards minimizing k, particularly k_L. As suggested by Ioffe, compound alloys should achieve this. Of the alloy systems, those involving $(Bi, Sb)_2(Se, Te)_3$ have been investigated most thoroughly.

Table 4 contains some representative data for the $Bi_2(Se, Te)_3$ alloys.[7]

It is seen that the thermal conductivity for the alloys is significantly lower than that of either compound. However, no real gain in Z appears in these data. Similar work has been reported for $(Bi, Sb)_2Te_3$. In this case the lattice thermal conductivity for an alloy of $\frac{1}{2}Bi_2Te_3 + \frac{1}{2}Sb_2Te_3$ is about $\frac{1}{3}$ that of unalloyed Bi_2Te_3.[8] The maximum figure of merit obtained in $(Bi, Sb)_2Te_3$ of about 3×10^{-3}/°C represents

TABLE 4

Properties of $Bi_2(Se,Te)_3$ Alloys

Composition x in $Bi_2Te_{(3-x)}Se_x$	S μV/°C	σ (Ω cm)$^{-1}$	k mw./cm °C	Z °C^{-1} × 10^3
0	212 *p*	730	19.8	1.66
0.5	290 *p*	200	11.9	1.46
1.0	0	85	15.6	0
2.0	135 *n*	830	14.5	1.05
3.0	70 *n*	1950	27.0	0.35

TABLE 5

Properties of InSb and InAs

Composition	S μV/°C	ρ Ω cm	k w/cm-°C	E_G eV	$T_{meas.}$ °C
InSb	−130	4.5×10^{-4}	0.075	0.27	500
InAs	−123	4.5×10^{-4}	0.075	0.47	750

the highest reproducible value for figure of merit reported to date.

In the PbTe system, alloying with homologous SnTe or PbSe has been found to have similar effects. In pure PbTe the lattice thermal conductivity is about 0.02 w/cm-°C, and the mean free path of the phonons is only about three interatomic distances. Alloying with PbSe reduces the conductivity to a minimum of about 0.009 w/cm-°C for a 50% PbSe–50% PbTe alloy. Also, it has been found that the variation of the thermal conductivity with temperature for a solid solution is much less than that for a pure compound or element.

An additional effect of solution-alloying of the compounds is observed in the mobility. Airapetyants et al.[9] have found anomalous effects in the Sb_2Te_3–Bi_2Te_3, Bi_2S_3–Bi_2Te_3, PbSe–PbTe, and PbTe–SnTe systems. The replacement of the electropositive elements, such as Bi with Sb and Pb with Sn, causes the electron mobility to be reduced to a much greater extent than the hole mobility. Conversely, replacing the electropositive element Te with either S or Se results in a larger decrease in hole mobility. The conclusion reached is that for p-type thermoelements a solution in which the electronegative elements are replaced should be used and for n-type elements replacement of the electropositive element should be used.

With PbTe and Bi_2Te_3 and their alloys as prototypes, the investigations of additional compounds have developed in two directions: into ternary systems for higher-lattice scattering and into higher E_G compounds for higher temperature of operation. The III-V compounds, as seen from Table 1, have E_G values above those for Bi_2Te_3 and PbTe. Although main interest in them has been for "transistor"-type devices, they have also been looked into for thermoelectric applications. Some representative data appear in Table 5.[10,11]

The figure of merit for these compounds is a function of temperature, going through a maximum (the curves are similar in form to those for PbTe) at $Z = 0.5 \times 10^{-3}$ °K^{-1} at 400°C for InSb and $Z = 0.6 \times 10^{-3}$ °K^{-1} at 700°C for InAs. This again points up the relation between Z_{opt} and E_G. For alloys of the form $InAs_{(1-x)}P_x$ the figure of merit varies with x, reaching a maximum of $Z = 0.73 \times 10^{-3}$ °K^{-1} for a alloy fraction $x = 0.1$. The temperature of the maximum remains at 700°C. The effectiveness of phosphorus is attributed to two factors: the energy gap of InAs is increased with added phosphorous and additional scattering centers are introduced. A similar study of the (In, Ga)As system[12] indicated a reduction in k_L from 0.29 w/cm-°C in InAs to 0.05 w/cm-°C for a 50% GaAs–50% InAs alloy. S values in the range of 100 to 400 μV/°K depending on carrier concentration were also reported. The band gap varied on a concave upward curve from 0.35 eV to 1.35 eV. In the GaAs–GaP system the band gap varies from 1.35 eV to 2.24 eV. It is thus possible to obtain any band gap between 0.35 eV and 2.24 eV by proper alloying in the GaAs, GaP, and InAs systems. Silicides and rare-earth compounds are also promising high-temperature materials. Cerium sulfide is reported to have a value of $ZT = 0.3$ at 1500°K.[13] The compound exists over a composition range from Ce_2S_3, with insulating characteristics to Ce_3S_4 with semimetal characteristics. Some preliminary data for silicides appear in Table 6.[14]

In the silicides the resistivity was generally found to increase with temperature, and the

TABLE 6

Properties of Some Silicides

Compound	S, μV/°C	ρ, Ω cm
$CrSi_2$	+200	1.20×10^{-3}
MnSi	+73	0.2×10^{-3}
$MnSi_2$	+170	2.5×10^{-3}
CoSi	−90	0.35×10^{-3}

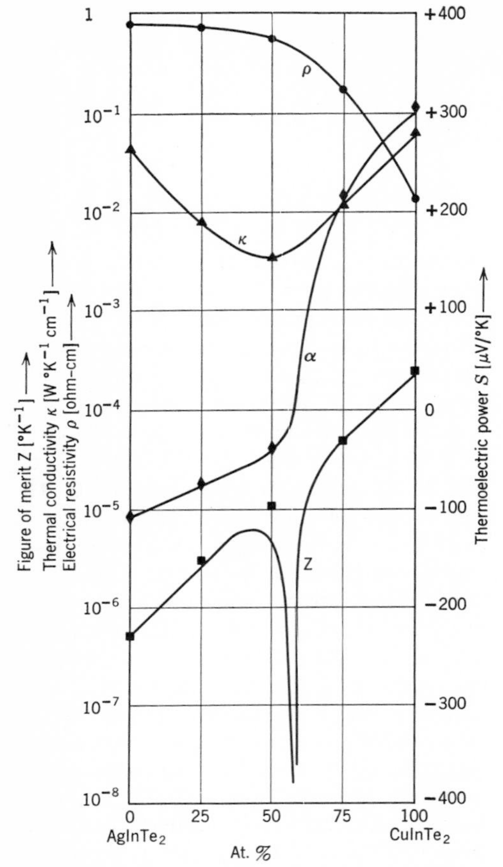

Fig. 7. Thermoelectric properties of $CuInTe_2$–$AgInTe_2$ alloys.

Seebeck coefficient was found to go through a maximum in the range between room temperature and 1000°C. As can be seen, the data are incomplete, particularly with respect to thermal conductivity, the measurement of which presents many difficulties in the temperature range of interest.

The ternary compounds are in general based on the diamond structure and are of the chalcopyrite structure. A typical series of such compounds is: $CuGaSe_2$, $ZnGeSe_2$ (isoelectronic with Ge), $AgInTe_2$, $CdSnTe_2$ (isoelectronic with Sn), and cross-substitutions such as $AgGaSe_2$, $CuInTe_2$, $CuGaTe_2$, $CuGaS_2$. Many such combinations are possible, and almost all are semiconductors. A promising one appears to be $CuGaTe_2$ with room temperature values of $S = 128\ \mu V/°C$, $\rho = 4.1 \times 10^{-3}\ \Omega$ cm, and $k = 0.022$ w/cm-°C. A set of data for a ternary alloy series of $CuInTe_2$ with $AgInTe_2$ is shown in Fig. 7.[15] A characteristic feature is the minimum in the thermal conductivity at $Ag_{1/2}Cu_{1/2}InTe_2$. In the ternary systems behavior similar to that in the binaries is observed with different values. The problems of stoichiometry and doping are more complex the more components in the system, and much additional work is needed in this class before a systematic analysis of their behavior can be made.

As mentioned previously it is now apparent that specifying a compound does not specify its thermoelectric behavior. Much additional investigation is necessary to optimize its properties which are very structure-sensitive. This structure-sensitivity is common to all semiconductors and to a certain extent affects the usefulness of the materials. For example, thermal cycling of the compounds has been found to result in permanent changes in the properties. The properties are also sensitive to mechanical deformation and high-energy irradiations. In all cases the change in properties may be related to generation and interaction of defects in the material. A curve showing the effect of annealing on the Seebeck coefficient of a $Ag_{0.5}Cu_{0.5}InTe_2$ alloy is shown in Fig. 8. There is a marked irreversible change both in sign and magnitude of the coefficient. Balicki et al.[16] reported that nuclear bombardment of Bi_2Te_3 and PbTe resulted in significant changes in S and ρ, ρ being more affected in PbTe than in Bi_2Te_3. p-type Bi_2Te_3 was converted to n type. Flux exposures ranged from 10^{19} to 10^{20} fast neutrons/cm^2. Annealing experiments indicate an almost complete return of properties in the range of 150 to 200°C.

6. SUMMARY

The thermoelectric properties of intermetallic compounds can vary over wide ranges of values depending on the structure and chemistry of the compound. For efficient energy conversion the requirements are narrowed down to produce an optimum value of the figure of merit

$$Z = \frac{S^2\sigma}{k}$$

which, in terms of the basic parameters of the materials, reduces to a term involving

$$\frac{\mu m^*}{k_L}$$

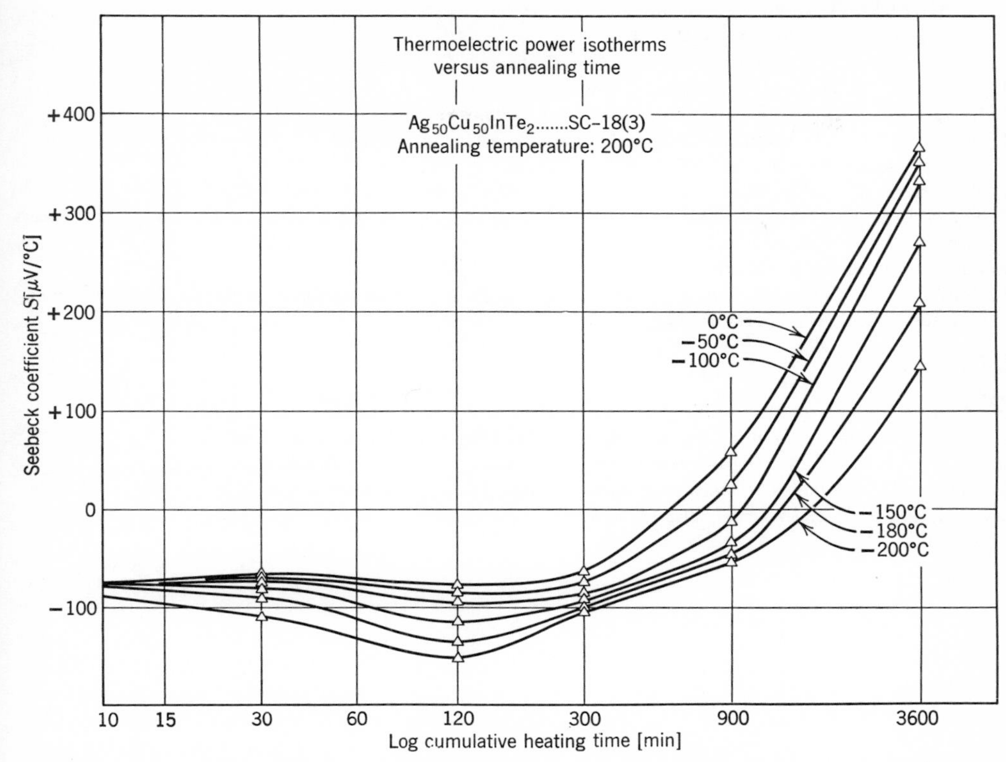

Fig. 8. Effect of cumulative annealing on the Seebeck coefficient of a $Cu_{0.5}Ag_{0.5}InTe_2$ alloy.

where the lattice conduction term k_L becomes a significant variable and the principal control factor. The need for low k_L narrows the useful field of compounds to those possessing a high mean atomic weight. Another important factor is the optimum carrier concentration which relates the Seebeck coefficient and conductivity. These factors and the temperature-stability requirements also depend on the band gap of the compound which should be chosen in the range of $E_G \sim 4kT$, showing that a higher temperature of operation requires higher band gap material. The higher band gap material is generally less efficient than a lower band gap material at lower temperatures, so that for optimum efficiency over a wide temperature range a cascading arrangement of various compounds would be needed.

The investigation of thermoelectric compounds is relatively young, the "classical" material being Bi_2Te_3 which is used for low temperatures and is primarily useful for cooling. The higher band gap PbTe is potentially useful for moderate temperature (~300°C) operation. GeTe appears to be useful in the 500°C range. Many other compounds in various classes are being studied. Those which seem to have most promise for high-temperature use are compounds containing rare-earth elements.

REFERENCES

1. Altenkirch E., *Physik. Z.* **12** (1911) 920.
2. Goldsmid H. J., *J. Electron. Control.* **1** (1958) 218.
3. Ioffe A. F., and A. V. Ioffe, *Dokl. Akad. Nauk SSSR* **97** (1954) 821.
4. Rittner E. S., *J. Appl. Phys.* **30** (1959) 702.
5. Miller R. C., R. R. Heikes, and A. E. Fein, *J. Appl. Phys.* **33** (1962) 1928.
6. Dennis J. H., *Advan. Eng. Conv.* **1** (1961) 99.

7. LaChance M. H., and E. E. Gardner, *Advan. Eng. Conv.* **1** (1961) 133.
8. Wright D. A., *Nature* **181** (1958) 834.
9. Airapetyants S. V., and B. A. Efimova, *Zh. Tekhn. Fiz.* **28** (1958) 1768.
10. Bowers R., R. W. Ure, J. E. Bauerle, and A. J. Cornish, *J. Appl. Phys.* **30** (1959) 930.
11. Bowers R., J. E. Bauerle, and A. J. Cornish, *J. Appl. Phys.* **30** (1959) 1050.
12. Abrahams M. S., R. Braunstein, and F. D. Rosi, *Properties of Elemental and Compound Semiconductors.* Interscience, New York, 1960, p. 275.
13. Kurnick S. W., M. F. Merriam, and R. L. Fitzpatrick, *Advan. Eng. Conv.* **1** (1961) 157.
14. Mayer S. E., and A. I. Mlavsky, *Properties of Elemental and Compound Semiconductors.* Interscience, New York, 1960, p. 261.
15. Zalar S. M., and I. B. Cadoff, *Trans. AIME* **224** (1962) 436.
16. Balicki M., J. C. Corelli, and R. T. Frost, *Metallurgy of Elementary and Compound Semiconductors.* Interscience, New York, 1961, p. 403.

General Texts

Egli P. H. *Thermoelectricity*. Wiley, New York 1960.
Goldsmid H. J. *Applications of Thermoelectricity*. Methuen and Co. London 1960.
Ioffe A. F. *Semiconductor Thermoelements and Thermoelectric Cooling*. Infosearch, 1957.
Cadoff I. and E. Miller, *Thermoelectric Materials and Devices*. Reinhold. New York, 1960.
Heikes R. and R. Ure, *Thermoelectricity; Science and Engineering*. Interscience, New York 1961.

chapter 27

Magnetic Properties

J. S. KOUVEL
General Electric Research and Development Center
Schenectady, New York

1. INTRODUCTION

The past decade or so has seen a prodigious growth of interest in the magnetic properties of intermetallic compounds. This is not to dismiss many earlier events of considerable importance in the history of intermetallic compounds as magnetic materials. On the contrary, starting at the turn of this century with F. Heusler's[1] discovery that certain alloys of magnanese, containing none of the traditional ferromagnetic metals (iron, cobalt, or nickel), were nevertheless ferromagnetic, and subsequently with the recognition that many of them were chemically ordered compounds of fairly definite stoichiometry, the list of known ferromagnetic intermetallics has been increasing steadily. Although some, such as MnBi, were found to have extremely high coercivities (under suitable fabrication conditions), which made them promising as permanent magnet materials, most have been regarded until recently with mild scientific curiosity.

Certainly it is true that no single group of magnetic intermetallic compounds has been able to challenge the success of the mixed iron oxides known as the ferrites in capturing the attention of so many scientists and technologists. The combination of high magnetic permeability and low electrical conductivity has made the ferrites essentially ideal for various types of experiments and devices involving high-frequency magnetic fields. An additional and scientifically more significant advantage of the ferrites and other ionic compounds of transition-group metals has been the relative simplicity of their basic magnetic characteristics. For example, the magnetic moment of each transition-group cation in such a compound is invariably close to that of an integral number of electrons; moreover, the coupling between neighboring cation moments is fairly rigorously understood as a short-range process (called superexchange) involving the excited electronic states of intervening anions. A remarkable feature of these superexchange interactions in ionic materials is that they almost always give rise to an antiparallel rather than parallel (i.e., ferromagnetic) alignment of moments. In cases such as MnO, where there is an equal number of oppositely directed ionic (Mn^{2+}) moments of equal magnitude resulting in a complete cancellation of the spontaneous magnetization, the system is described as antiferromagnetic, whereas in a ferrite such as magnetite (Fe_3O_4) where the antiparallel moments are unequal in magnitude and/or number, the cancellation of the spontaneous magnetization is only partial and the system is called ferrimagnetic.

Shortly after the possibility of an antiferromagnetically ordered state was first postulated by Néel,[2] many materials were found to have a more or less pronounced maximum in their magnetic susceptibility versus temperature characteristics, an anomaly that Néel had predicted to occur at the antiferromagnetic disordering temperature. Most of these materials were ionic, but some, such as CrSb, MnSe, and MnTe, could be more appropriately considered intermetallic. When Néel later extended his hypothesis of antiparallel atomic moments to the ferrimagnetic case,[3] he was concerned with its applicability not only to the ionic ferrites but also to the intermetallic Mn_2Sb, whose ferrimagnetic state had been proposed earlier by Guillaud.[4] Hence the role played by intermetallic compounds in this exciting chapter in the history of magnetism was far from inconspicuous.

However, the high level of scientific interest which magnetic intermetallics are currently enjoying would probably not have been reached if it were not for some important new developments. Foremost among the experimental developments was the use of neutron diffraction in determining the arrangement of atomic moments in magnetic systems. Following its application to MnO,[5] which provided the most definitive confirmation of Néel's hypothesis of the antiferromagnetic state, this technique was soon brought to bear on a wide assortment of magnetic materials, and many intermetallic compounds have since been identified as antiferromagnetic, ferrimagnetic, or simply ferromagnetic. In some unusual cases, the magnetic state is such that the atomic moments are not even aligned collinearly (i.e., parallel or antiparallel) but at some angle other than 0 or 180° with respect to one another; an exotic example is $MnAu_2$, whose Mn-moment directions describe a helix with a pitch many times greater than the linear dimensions of the chemical cell. The variety of magnetic configurations in intermetallic compounds revealed to date by neutron diffraction undoubtedly reflects the diversity of their ordered atomic arrangements, and it will continue to expand as more atomically ordered structures containing transition-group metals are discovered. This wide range of ordered magnetic states in intermetallic compounds in turn gives rise to a corresponding range of properties, which today offers many exciting scientific and technological opportunities, some of them unique to this class of materials. Within the broad spectrum of magnetic properties exhibited by these materials, none is probably more intriguing than the abrupt transition between contrastingly different magnetic states, which the ordered alloys FeRh and MnAs (among others) undergo at a particular critical temperature. This and many other examples of unconventional magnetic behavior in intermetallic compounds will be described later. Although the magnetic properties of most of the compounds to be discussed are in general less spectacular, they do nevertheless raise many questions of considerable importance to our understanding (as well as to the application) of basic magnetic processes.

2. THEORETICAL CONSIDERATIONS

2.1 Basic Concepts

Certain general features of the ordered magnetic states found in intermetallic compounds have had profound theoretical implications pertinent to all magnetic materials. Their significance can best be appreciated within the context of various interpretations of the fundamental magnetic parameters, which for any system can be specified as (*a*) the magnitude of its atomic moments and (*b*) the strength and sign* of the coupling between these moments. All modern theories of magnetism agree that the origin of both (*a*) and (*b*) lies in a quantum-mechanical part of the electron Coulomb energy known as the exchange energy, but they differ in the approximations that must be made in their application

* According to the usual convention, a positive (ferromagnetic) or negative (antiferromagnetic) coupling is one that tends to align magnetic moments parallel or antiparallel, respectively, to each other.

to particular substances. At one extreme, in the atomistic theories, the viewpoint is taken that (*a*) and (*b*) can be considered quite separately. This type of approach is especially valid for ionic crystals, in which all the electrons, including the valence electrons, are tightly bound to the ion cores. The energy levels of the unfilled *d*- or *f*-electron shell of a transition-group ion in such a crystal, although modified in their spacing by interactions with the electrostatic fields produced by neighboring ions, are nevertheless fairly discrete and are therefore occupied, if at all, by an integral number of electrons (per ion). Moreover, according to quantum-mechanical rules, the magnetic moments of these electrons are restricted to certain particular orientations with respect to one another, and consequently the net magnetic moment per ion is equal to a specific number of individual electron moments (or Bohr magnetons). In most cases, in which the electrostatic interactions referred to above are rather weak, the net ionic moment in a crystal is nearly the same as that of the free ion. It is also consistent with this approximation that the effective coupling between the moments of adjacent transition-group ions in a crystal (such as the short-range superexchange mechanism mentioned earlier) is a minor perturbation, which does not appreciably affect the electronic states (or, therefore, the net moments) of the ions.

At the other extreme is the so-called collective electron theory. In this approximation, strong interactions among all the outermost electrons associated with different transition-group atoms give rise to an energy spectrum, in which the *d*-electron levels as well as the *s*-electron levels have been broadened into bands. Furthermore, because the *s* (conduction) and *d*-bands overlap in energy and are both occupied up to the same maximum energy level (known as the Fermi level), the number of electrons in each band will not in general be integral. If there is in addition a sufficiently strong exchange coupling that favors a parallel alignment of the magnetic moments of the *d*-electrons, the energies of those with moments parallel or antiparallel to a reference direction will be lowered or raised, respectively, with regard to the Fermi level. As a result, there will be a net spontaneous magnetization corresponding in general to a nonintegral number of Bohr magnetons per atom, which agrees qualitatively with the situation for the ferromagnetic metals: iron, cobalt and nickel. However, because this theory handles all the *d*-electrons of a material collectively, it makes virtually no distinction between the constituent elements of disordered transition-group alloys and has therefore enjoyed only a limited success in explaining their basic magnetic properties.

However, it is with respect to the magnetic properties of *ordered* alloy systems or intermetallic compounds that the limitations of these extreme theoretical viewpoints are particularly evident. As later discussion of many specific intermetallic compounds will make clear, the magnetic moment of a given transition-group atom such as iron varies appreciably in magnitude not only from one compound to another, but also among crystallographically different sites in the same compound. Any such variability in atomic moment may be taken as evidence of a corresponding variability in the interaction between the *d*-electrons as well as between conduction electrons of neighboring atoms. It is only in special cases—for example, when the intermetallics contain a rare-earth metal, such as gadolinium, whose atomic moment generally has about the same value independent of its environment—that the simple atomistic theory described earlier can be said to apply. Then the *d*- (or *f*-) electrons are fairly tightly coupled to the atomic nuclei and thus produce specific atomic moments through their mutual interactions, whereas most of the coupling between the moments of different atoms is provided by the conduction electrons. However, the occurrence of nonintegral moments in the iron-group compounds should not necessarily be regarded from the opposite viewpoint of the collective electron model. In fact, the variability of the atomic moments that generally accompanies their nonintegrality (as mentioned above) would preclude the validity of this model in the simple form described earlier.

From this qualitative discussion we may conclude that the appropriate magnetic model for all intermetallic compounds is probably intermediate between the simple collective electron and atomistic models. As a deviation from the latter, the correct theory would not strictly localize the *d*- (or *f*-) electrons about the nuclei of the transition-group atoms, but would free them to participate to some extent in exchange interactions involving different atoms. Consequently, the magnitudes of the atomic moments would deviate from integral numbers of Bohr magnetons, and the coupling between adjacent atoms would be largely determined by direct interactions between *d*-electrons. The coupling between more distant atoms, however, would still be the result of indirect interactions between their quasi-localized *d*-electrons via the sea of conduction electrons. Quantitatively, this deviation from the simple atomistic theory can certainly be expected to differ from one intermetallic compound to another, depending sensitively on the crystal structure and composition. It is within this context that the wide variety of intermetallic compounds, if carefully examined for systematic correlations in properties, can offer unusual opportunities for a fundamental understanding of magnetism in solids.

2.2 Categories of Intrinsic Magnetic Behavior

Although, as determined in the preceding section, intermetallic compounds, unlike ionic compounds, cannot be adequately explained by a simple atomistic magnetic model, it is still possible to classify and discuss their intrinsic magnetic properties within the framework of the *molecular field theory*, which is based implicitly on an atomistic model, if allowance is made for nonintegral atomic moments and other deviations from this model. In this theory, the net exchange coupling between a given atom and all its neighbors is approximated by the interaction of its magnetic moment with an effective "molecular" field. The consequences of the molecular field theory are particularly simple when all the interactions in the system are positive (i.e., ferromagnetic), in which case the spontaneous magnetization at 0°K corresponds to a parallel alignment of all the atomic moments and at higher temperatures decreases as the moment directions undergo thermal fluctuation. The manner in which the magnetization is expected to decrease from its value M_0 at 0°K and finally to vanish at the Curie temperature (T_c) is shown in Fig. 1*a*. Above T_c, where the system is magnetically disordered (i.e., paramagnetic), the magnetic susceptibility (χ) can be expected to decrease with rising tempera-

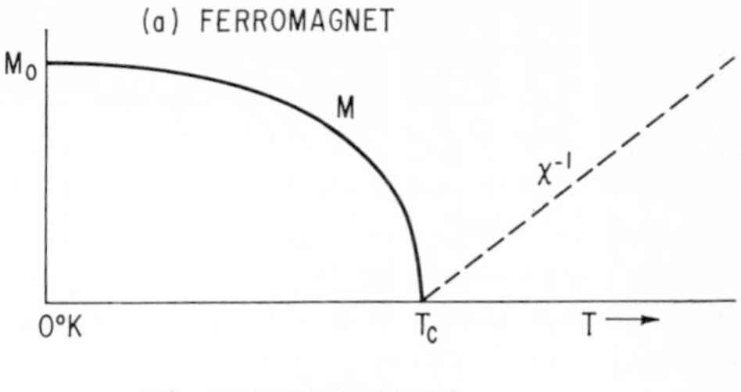

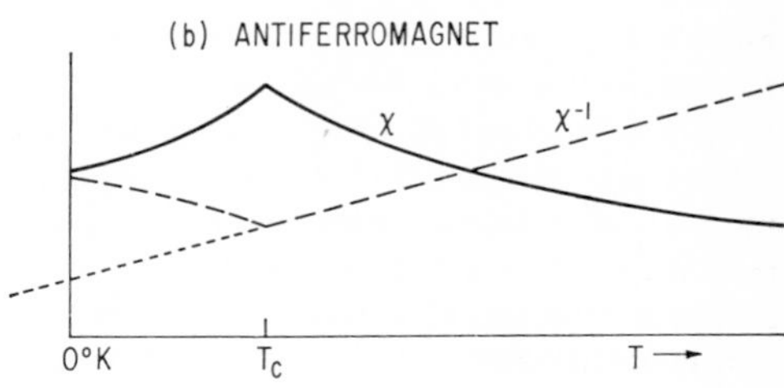

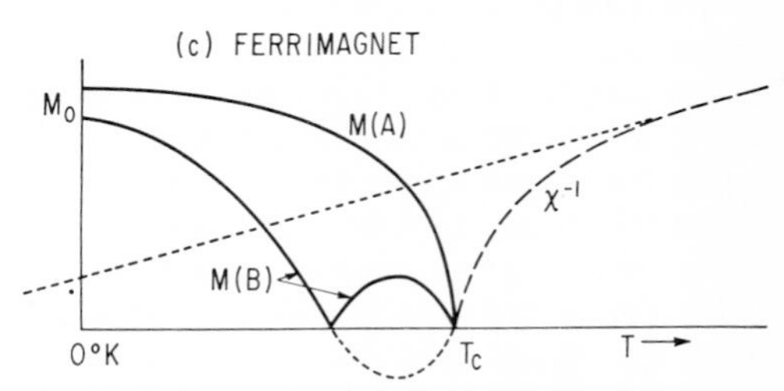

Fig. 1. Typical variation of magnetization (M) or susceptibility (χ) and of χ^{-1} with temperature from 0°K to Curie point (T_c) of ferromagnet, antiferromagnet, and ferrimagnet.

ture according to the following relationship, known as the Curie-Weiss law:

$$\chi = \frac{C}{T - \theta} \tag{1}$$

Thus, as shown in the figure, $1/\chi$ versus T describes a straight line intersecting the temperature axis at $T = \theta$. In the simplest case, when all the atomic moments in the system are equal, the theory predicts that θ and T_c will be identical. However, in many ferromagnetic intermetallic compounds, whose moment-bearing atoms are of different elements and/or are not crystallographically equivalent, θ will not be exactly equal to T_c. In fact, even Eq. 1 will not be strictly valid except at high temperatures, as in the ferrimagnetic case to be discussed later. Deviations from Eq. 1 can also be caused by short-range magnetic ordering.

In the simplest type of antiferromagnetic system, all the magnetic atoms are chemically and crystallographically equivalent and, as a consequence of negative interactions between them, divide in equal numbers into two groups (or sublattices) having oppositely oriented moments. There is therefore no spontaneous magnetization of the system as a whole. Instead, as shown in Fig. 1*b*, there is a susceptibility which, according to the molecular field theory (for a randomly oriented polycrystal), will increase as the temperature rises from 0°K, reaching a peak value at the magnetic-disordering temperature T_c. The susceptibility decrease above T_c will again follow Eq. 1, except that θ will be negative, although still approximately equal to T_c in magnitude. Any inequality between θ and $-T_c$ can be accounted for by the theory as evidence of interatomic interactions within (in addition to those between) the magnetic sublattices.

The simplest type of system that can be called ferrimagnetic is also characterized by two magnetic sublattices with magnetizations antiparallel to one another. Here, however, the sublattices are not equivalent, and therefore the magnitudes of their magnetizations are not the same. Hence the system will have a net spontaneous magnetization, whose variation with temperature will often resemble that of a ferromagnet as illustrated by curve *A* in Fig. 1*c*. Many other unusual forms are theoretically possible for the temperature dependence of the net magnetization of a ferrimagnet and indeed have been encountered experimentally in various ferrimagnetic compounds. The most exotic example is perhaps that demonstrated by curve *B* in Fig. 1*c*, which can be readily understood in terms of the magnetization of one sublattice being larger at low temperatures but decreasing more rapidly than that of the other sublattice. Because the two sublattice magnetizations are opposite in direction, the net magnetization (when measured in a small field) actually reverses its polarity, as shown by dotted curve, and the temperature at which it goes through zero is known as the compensation point. Regardless of any differences in their magnetization-versus-temperature behavior, all ferrimagnetic substances are expected to show a similar temperature dependence of susceptibility above their Curie temperatures. As indicated in Fig. 1*c*, the theoretical $1/\chi$ versus T curve for a ferrimagnet follows a complicated hyperbolic relationship well above T_c before it approaches a straight line which obeys Eq. 1, usually with a negative value for θ approximately equal to T_c. Again, as in the antiferromagnetic case, θ can be expected to deviate from $-T_c$, if there are appreciable magnetic interactions between atoms of the same sublattice.

A more general statement concerning θ in Eq. 1 is that it represents an algebraic sum of all the positive (ferromagnetic) and negative (antiferromagnetic) interactions in the system, in contrast to T_c, which is increased or decreased depending on whether a given interaction helps or hinders the atomic-moment alignment describing the ordered magnetic state. Hence, especially in an intermetallic compound of any complexity, a deviation of θ from $\pm T_c$ is to be expected and, moreover, can be used in determining the magnitudes and signs of the different interactions. Another important distinction is that between the average ferromagnetic (or ferrimagnetic) moment $\bar{\mu}_F$ and the so-called average "effective" paramagnetic moment $\bar{\mu}_P$ of the same material. The former is simply the magnetization M divided by the number of moment-bearing atoms N, both normalized to a unit volume (or mass) of the material. Because there may be several distinct types of magnetic atoms, each having a moment of different magnitude and/or direction and thus constituting a separate magnetic sublattice, we may write

$$\bar{\mu}_F = \frac{M}{N} = \frac{\sum_i n_i \mu_{F_i}}{N} \tag{2}$$

in which the summation over the various sublattices (each having n_i atoms, so that $\sum_i n_i = N$) should be regarded as a *vector* sum of their moments $n_i \mu_{F_i}$. At temperatures well below T_c the individual atomic moments within each sublattice may be expressed as

$$\mu_{F_i} = g_i S_i \mu_B \tag{3}$$

where g_i is the spectroscopic splitting factor, S_i the electron spin quantum number, and μ_B the Bohr magneton. As already discussed, the moment in Bohr magnetons for a transition-group atom in an intermetallic compound can deviate appreciably from an integral value; hence, although $g_i \simeq 2$ from resonance experiments, S_i in Eq. 3 generally will not have a half-integral value and thus will not be obeying simple quantum rules. Rare-earth atoms, however, even in metallic systems retain a half-integral quantum number, which in this case is J_i (replacing S_i in Eq. 3) and contains contributions from the orbital motion as well as the spin of the electrons; moreover, for the rare earths, g_i has a wide range of values. The average paramagnetic moment per atom $\bar{\mu}_P$, on the other hand, is defined by the following expression for the Curie constant C in Eq. 1:

$$C = \frac{N\bar{\mu}_P{}^2}{3k} \tag{4}$$

where k is the Boltzmann constant. Moreover, according to this definition, if there are two or more magnetic sublattices each consisting of n_i atoms with a paramagnetic moment μ_{P_i}, we write as a *scalar* sum

$$\bar{\mu}_P{}^2 = \frac{\sum_i n_i \mu_{P_i}^2}{N} \tag{5}$$

where

$$\mu_{P_i} = g_i\{S_i(S_i + 1)\}^{1/2}\mu_B \tag{6}$$

Comparing these expressions with Eqs. 2 and 3, we observe not only that the ferromagnetic and paramagnetic moments of a given atom, μ_{F_i} and μ_{P_i}, are themselves defined differently, but also that in a multisublattice system they are combined differently with those of other atoms in specifying the average moments $\bar{\mu}_F$ and $\bar{\mu}_P$. These distinctions must be considered in any valid comparison of the magnetic behavior of a material above and below its Curie temperature.

2.3 Anisotropy and Permanent Magnet Properties

In the absence of an external field, the atomic moments in a crystal of any magnetic substance generally prefer to lie along certain crystallographic axes; this preference in orientation is known as the magnetocrystalline (or crystal) anisotropy. For a ferromagnet or ferrimagnet, this anisotropy is defined in terms of the energy required of an external field to rotate the spontaneous magnetization away from one of its so-called easy axes. Theoretically, magnetocrystalline anisotropy has its origins in an effective coupling of the atomic moments, via the orbital motion of the electrons, with the electrostatic field of the crystal. This crystal field can be particularly intense when the local crystalline environment of the magnetic atoms has low symmetry and it is therefore understandable that a large crystal anisotropy is a property of many noncubic intermetallic compounds. However, some materials with cubic crystal symmetry are also strongly anisotropic magnetically. Although they are in the minority, they include several ferromagnetic intermetallic compounds that are technologically important as permanent magnets.

In order that a strong magnetocrystalline anisotropy in a ferromagnet may be useful as a permanent magnet property, it is necessary in response to a reversing magnetic field that the magnetization be constrained to rotate against the anisotropy forces. If this constraint is absent, the magnetization will reverse at lower reverse fields by the process of domain-wall motion. The result as shown schematically in Fig. 2*a*, will be a thin hysteresis loop with a very small intrinsic coercive field H_{ci}, which is typical for any homogeneous bulk specimen. However, if the same material is subdivided into small spherical particles of diameter comparable to the thickness of a domain wall (approximately 1000 Å in most ferromagnetic materials), it becomes energetically unfavorable for any domain walls to exist in the particles. Each particle is then a single domain, whose magnetization can reverse only by a rotation process requiring a reverse external field approximately equal to the anisotropy field $2K/M_s$, where K is the dominant anisotropy coefficient and M_s is the saturation magnetization. Because the anisotropy fields of known materials range up to several thousand oersteds, the hysteresis loop of a compact of single-domain particles can be extremely wide, as shown in Fig. 2*b*. It is not this hysteresis loop of M versus H, however, but rather the loop in which the magnetic flux density B is plotted against H (shown as a dashed curve in the figure) that determines the value of a material as a practical permanent magnet. Because $B = H + 4\pi M$, the coercive field H_c defined at $B = 0$ is in general smaller than H_{ci} defined at $M = 0$ and is limited, in fact, to a maximum value of $4\pi M_R = B_R$, the remanent flux density, regardless of how large H_{ci} may be. Thus the magnetization as well as the crystal anisotropy must be fairly large in order for a material to qualify for use in a fine-particle magnet, which eliminates

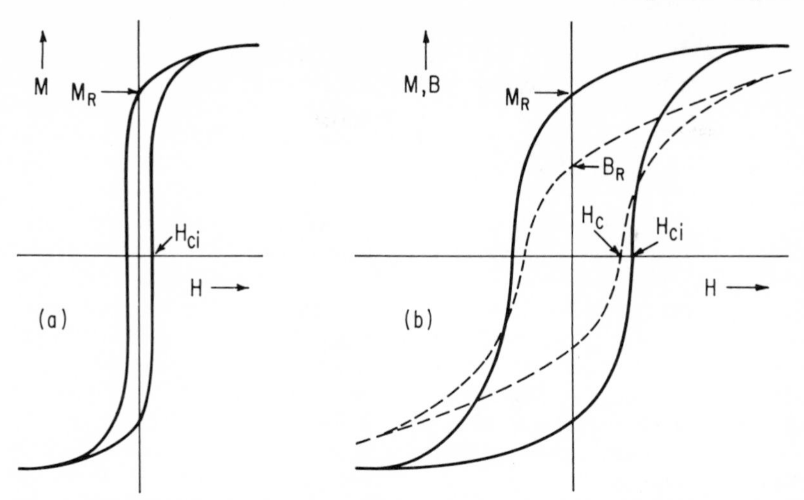

Fig. 2. Typical hysteresis loops of magnetization (M) or magnetic induction (B) versus field (H) for a soft magnetic material (a) and a hard magnetic material (b).

many intermetallic compounds (and other types of materials) from consideration. Many intermetallic systems, however, have atomically ordered ferromagnetic phases that do meet these qualifications, and in some cases an additional advantage can be derived from the fact that the atomic ordering process is inhomogeneous and can be easily controlled. If a suitable heat treatment of such a system can produce very small regions of the ordered phase dispersed in a matrix of a weakly magnetic disordered phase, the result is magnetically analogous to a compact of single-domain particles with desirable permanent-magnet properties.

3. CLASSIFICATION OF COMPOUNDS

The compilation of structures and properties that follows is restricted to compounds that are both "intermetallic" and "magnetic"; the definitions of the two terms here are necessarily arbitrary. As intermetallics we include materials known more appropriately as ordered alloys, but only when there is a very high degree of long-range atomic order; any discussion of a disordered phase will be limited to a cursory comparison of its magnetic properties with those of the fully ordered phase of the same chemical composition. A more difficult problem is the differentiation between intermetallic compounds and other types of compounds. Rather than try to resolve the many subtleties of this problem, we have simply chosen to exclude from consideration all the hydrides, borides, carbides, nitrides, oxides, phosphides, and sulfides (as well as the halides and other ionic salts) and to retain all other compounds, including borderline cases such as the selenides and tellurides. From the magnetic point of view, we confine our attention to those compounds, for which there is experimental evidence at some temperature for a ferromagnetically, ferrimagnetically, or antiferromagnetically ordered state.

All the intermetallic compounds included in this compilation are classified into different general categories according to their constituent elements. First and the most completely treated are the binary compounds of iron-group metals, which are discussed in the alphabetical order of their chemical formula (in which the iron-group element takes preference). The only iron-group compounds excluded from this first category are those containing either uranium-group or rare-earth elements. The latter are incorporated in the discussion of the next two categories: the binary compounds of uranium-group metals (those of magnetic interest being discussed alphabetically) and the binary compounds of rare-earth metals. Finally, there is a short treatment of the ternary compounds of iron-group metals, particularly the Heusler-type compounds.

If it is known, the crystal structure of each compound is described according to type. Hansen[6] has been relied on for most crystal-structure work prior to 1957; for more recent work (which in some cases supersedes the old) reference is made to the original publications. Whenever a compound is known to exist only at nonstoichiometric compositions or when its stoichiometry is in serious doubt, the stoichiometric chemical formula under which it is listed appears in quotation marks. Discussion of pseudobinary systems, that is, solid solutions of two (usually isomorphous) binary compounds, is limited mostly to those whose magnetic properties are unexpectedly different and cannot be interpolated from the properties of the binary compounds.

For convenience, references to each specific compound appear with the discussion of that compound. The general references for Sections 1 to 3 are found at the end of the chapter.

4. BINARY COMPOUNDS OF IRON-GROUP METALS

4.1 Cobalt Compounds

CoFe: Cubic, CsCl(B2)-type crystal structure. *Ferromagnetic*, with T_c well above 1000°K, the critical temperature for atomic ordering. For atomically well-ordered CoFe, saturation magnetization[7] corresponds to average atomic moment of $2.35\mu_B$, and saturation magnetostriction[8] is 9.2×10^{-5}, representing increases of 4 per cent and 40 per cent, respectively, from values for the quenched (partially ordered, bcc alloy. However, more recent measurements give $\lambda_{100} \simeq 14 \times 10^{-5}$ and $\lambda_{111} \simeq 3 \times 10^{-5}$ for the magnetostriction coefficients of ordered CoFe and very similar values for the disordered alloy.[9] Atomic ordering reduces magnetocrystalline anisotropy considerably,[9] probably accounting for large magnetic permeability of the ordered CoFe alloy known commercially as *Permendur*.[10]

CoPt: Tetragonal, CuAu (Ll_0)-type crystal structure. *Ferromagnetic*; properties of fully ordered alloy are not well established, except that saturation induction (B_s) at room temperature, after long anneal below critical ordering temperature, is about 4000 gauss, compared to 7200 gauss for the quenched fcc alloy.[11]

Partially ordered alloy is one of the strongest permanent magnet materials known, especially with respect to coercivity (H_c), which ranges between 3500 and 4700 oersteds (Oe), depending on heat treatment.[12] Good permanent-magnet properties probably derive from coexistence of microscopic ordered and disordered regions,[13] which have different magnetizations, and from a presumably large magnetocrystalline anisotropy. This hypothesis is supported by recent ferromagnetic domain study of partially ordered alloy, showing small regions with domain structure typical of a material of very strong anisotropy and low magnetization.[14]

$CoPt_3$: Cubic, Cu_3Au (Ll_2)-type crystal structure. *Ferromagnetic*, with $T_c \simeq 290$°K, compared to T_c of about 480°K for disordered fcc alloy of the same composition.[15]

4.2 Chromium Compounds

Cr_2As: Tetragonal, Cu_2Sb(C38)-type crystal structure. *Antiferromagnetic* with $T_c \simeq$ 393°K; susceptibility and specific heat were observed to go through maxima.[16] From $1/\chi$ versus T curve well above T_c: $\theta = -2067$°K. Earlier report[17] of weak ferromagnetism in Cr_2As is probably due to presence of some Cr_3As_2.

Cr_3As_2: Tetragonal crystal structure; further identification from X-ray diffraction data was not made.[18] *Ferrimagnetic*, according to magnetic measurements[18] which, extrapolated to 0°K, give $\sigma_s = 21.0$ emu/gm, corresponding to $\mu_F = 0.38\mu_B$ per Cr atom. Magnetic disordering at T_c of 213°K is extremely rapid,

[7] Goldman, J. E., and R. Smoluchowski, *Phys. Rev.* **75** (1949) 310.

[8] Goldman, J. E., and R. Smoluchowski, *Phys. Rev.* **75** (1949) 140.

[9] Hall, R. C., *J. Appl. Phys. Suppl.* **31** (1960) 157.

[10] Bozorth, R. M., *Ferromagnetism*. D. Van Nostrand Co., New York, 1951, pp. 196–199.

[11] Bozorth, R. M., *Ferromagnetism*. D. Van Nostrand Co., New York, 1951, p. 413.

[12] Martin, D. L., and A. H. Geisler, *J. Appl. Phys.* **24** (1953) 498.

[13] Newkirk, J. B., R. Smoluchowski, A. H. Geisler, and D. L. Martin, *J. Appl. Phys.* **22** (1951) 290.

[14] Craik, D. J., and F. Nunez, *Proc. Phys. Soc. (London)* **78** (1961) 225.

[15] Simpson, A. W., and R. H. Tredgold, *Proc. Phys. Soc. (London)* **B67** (1954) 38.

[16] Yuzuri, M., *J. Phys. Soc. Japan* **15** (1960) 2007.

[17] Nowotny, H., and O. Arstad, *Z. Physik. Chem.* **B38** (1938) 461.

[18] Yuzuri, M., *J. Phys. Soc. Japan* **15** (1960) 2007.

suggestive of first-order transition. Above T_c, $1/\chi$ versus T curve is nonlinear in the manner of a ferrimagnet, and its Curie constant at high temperatures gives $\mu_P = 2.0\mu_B$ per Cr atom. From earlier measurements,[19] μ_F and μ_P = 0.68 and $2.3\mu_B$ per Cr atom, respectively.

CrAs: Orthorhombic, MnP (B31)-type crystal structure. *Antiferromagnetic*; susceptibility peak at about 823°K.[20] Earlier report[21] of weak ferromagnetism in CrAs is probably due to presence of some Cr_3As_2.

$CrAu_4$: Tetragonal, Ni_4Mo ($D1_a$)-type crystal structure.[22] *Antiferromagnetic*, judging from susceptibility maximum (χ_{max}). Temperature of χ_{max} varies over wide range, centered around 300°K, depending on thermal history[22]; this is presumably caused by changes in degree of atomic ordering.

$CrGe_2$: Crystal structure unknown. *Ferromagnetism* (or *ferrimagnetism*) with $T_c \simeq$ 100°K observed in Cr–Ge alloys of 50 to 95 at.% Ge and attributed to compound of composition $CrGe_2$, which metallographically appears to be essentially single-phase and whose saturation moment is largest for this composition range.[23] Paramagnetic susceptibility data above T_c give $\theta \simeq 140$°K. Ferromagnetism in Ge-rich alloys was previously associated with $CrGe_3$ compound,[24] whose existence now seems questionable.

CrPt: Tetragonal, CuAu ($L1_0$)-type crystal structure *Antiferromagnetic* structure, determined from neutron diffraction at room temperature,[25] is identical to that of MnNi (see Fig. 18). Chromium atomic moments (of magnitude $2.24 \pm 0.15\mu_B$) are antiparallel to moments of chromium nearest-neighbors and lie perpendicular to tetragonal c-axis of crystal; atomic moment of platinum indeterminate, but probably very small.

$CrPt_3$: Cubic, Cu_3Au ($L1_2$)-type crystal

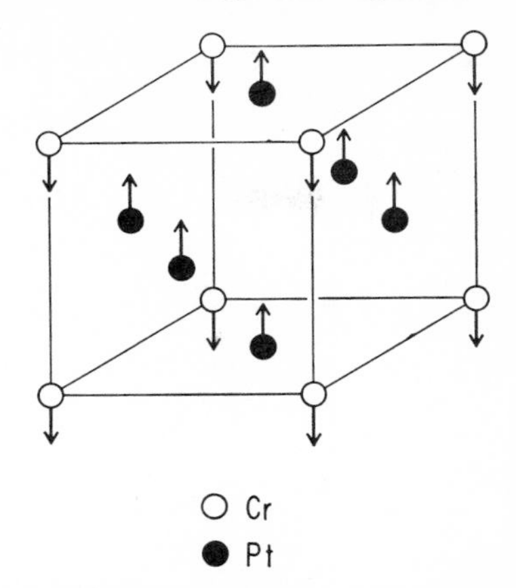

Fig. 3. Ferrimagnetic structure of $CrPt_3$. Absolute moment directions unknown.

structure. *Ferrimagnetic* structure, obtained from room-temperature neutron diffraction measurements,[26] shown in Fig. 3. Atomic moments of chromium and platinum antiparallel to each other ($\mu_{Cr} = 2.33 \pm 0.10\mu_B$, $\mu_{Pt} = 0.27 \pm 0.05\mu_B$), but axis of easy magnetization unknown. Ferromagnetism previously reported for Ct–Pt alloys of 22 to 48.5 at.% Cr[27] probably due to $CrPt_3$ phase or some ordered phase intermediate between $CrPt_3$ and CrPt; from this work,[27] T_c of $CrPt_3 \simeq 450$°K. Recent report[28] on ordered alloy of 30 at.% Cr gives $T_c = 687$°K and σ_s (at 20°K) = 19.1 emu/gm, corresponding to $\mu_F = 0.52\mu_B$ per atom, both of which are larger than values for stoichiometric $CrPt_3$.

CrSb: Hexagonal, NiAs($B8_1$)-type crystal structure. *Antiferromagnetic* structure, determined by neutron diffraction at room temperature,[29] consists of ferromagnetic sheets, parallel to basal plane, of chromium moments (of magnitude $2.7 \pm 0.2\mu_B$) directed along c-axis and alternating in sign, as shown in Fig. 4; the same magnetic structure was deduced more recently from similar data.[30] Broad susceptibility peak observed at about

---

[19] Haraldsen, H., and E. Nygaard, *Z. Electrochem.* **45** (1939) 686.
[20] Yuzuri, M., *J. Phys. Soc. Japan* **15** (1960) 2007.
[21] Nowotny, H., and O. Arstad, *Z. Physik. Chem.* **B38** (1938) 461.
[22] Wachtel, E., and U. Vetter, *Naturwissenschaften* **18** (1961) 156.
[23] Margolin, S. D., and I. G. Fakidov, *Phys. Metals Metallog.* (*USSR*) (*English Transl.*) **9** (1960) 22.
[24] Fakidov, I. G., and N. P. Grazhdankina, *Phys. Metals Metallog.* (*USSR*) (*English Transl.*) **6** (1958) 62.
[25] Pickart, S. J. and R. Nathans, *J. Appl. Phys.* **34** (1963) 1203.
[26] Pickart, S. J., and R. Nathans, *J. Appl. Phys* **34.** (1963) 1203.
[27] Kussmann, A., and E. Friederich, *Physik. Z.* **36** (1935) 185.
[28] Meyer, A. J. P., and R. Asfeld, *Compt. Rend.* **254** (1962) 4266.
[29] Snow, A. I., *Phys. Rev.* **85** (1952) 365.
[30] Takei, W. J., D. E. Cox, and G. Shirane, *Phys. Rev.* **129** (1963) 200.8

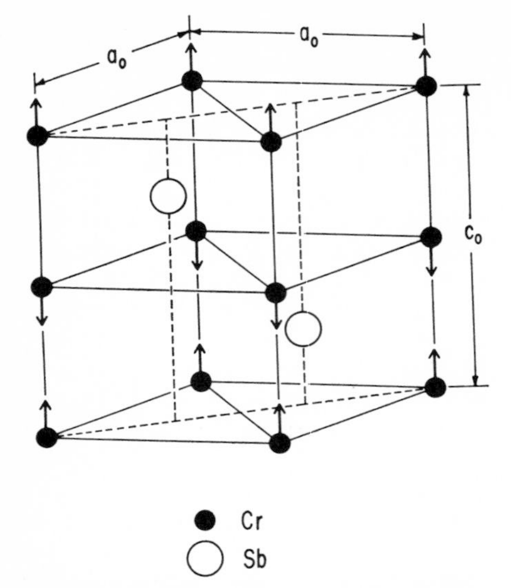

Fig. 4. Antiferromagnetic structure of CrSb.

700°K;[30–33] this temperature was identified as Curie point also from lattice-parameter measurements,[34,35] which indicate rapid decrease of c_0 and a slower increase of a_0 as T drops below T_c. Above T_c, $1/\chi$ versus T curve becomes linear and gives $\theta = -625$°K and $2S = 3.89$ for each Cr atom (assuming $g = 2$).[30]

Torque data on CrSb crystal below T_c are consistent with magnetic structure described above and, when combined with powder-susceptibility data, give 10^5 erg/gm as order-of-magnitude value for uniaxial anisotropy coefficient.[36] Susceptibility measurements on single crystal are also consistent with this magnetic structure and reveal large thermoremanent magnetization for crystal cooled to room temperature from above T_c in a magnetic field;[37] this thermoremanence effect was previously reported for polycrystalline sample.[38]

[31] Foëx, G., and M. Graff, *Compt. Rend.* **209** (1939) 160.
[32] Hirone, T., S. Maeda, I. Tsubokawa, and N. Tsuya, *J. Phys. Soc. Japan* **11** (1956) 1083.
[33] Lotgering, F. K., and E. W. Gorter, *J. Phys. Chem. Solids* **3** (1957) 238.
[34] Snow, A. I., *Rev. Mod. Phys.* **25** (1953) 127.
[35] Willis, B. T. M., *Acta Cryst.* **6** (1953) 425.
[36] Tsubokawa, I., *J. Phys. Soc. Japan* **16** (1961) 277.
[37] Wendling, R., and J. Wucher, *Compt. Rend.* **250** (1960) 2691.
[38] Perakis, N., J. Wucher, G. Parravano, and R. Wendling, *Compt. Rend.* **246** (1958) 3037.

CrSe: Hexagonal, NiAs($B8_1$)-type crystal structure. *Antiferromagnetic* structure deduced from neutron diffraction data at 77 and 4.2°K[39] has chromium moments forming umbrella-like configuration in each basal plane and alternating in direction along c-axis; magnitude of component of each moment in basal plane is $2.90\mu_B$. Susceptibility versus temperature data on polycrystalline specimens[40–43] show anomaly (but not well-defined peak) suggesting that T_c is somewhere in the range 200 to 300°K. Linear $1/\chi$ versus T curve above T_c gives $\theta = -185$°K and $\mu_P = 4.50\mu_B$ per Cr atom.[42] Susceptibilities along a- and c-directions of CrSe single crystal[43] decrease as temperature increases from 4°K and merge at Curie point anomaly at 279°K; at all temperatures below T_c, $\chi_c > \chi_a$. Specific heat measurements locate T_c at 320°K.[43]

Cr_3Se_4: Monoclinic, distorted NiAs($B8_1$)-type crystal structure, with ordering of vacancies on Cr sites.[44] *Antiferromagnetic*; magnetic unit cell was identified from neutron diffraction, but details of magnetic structure were not resolved.[45] It is reported that $T_c =$ 80°K and that paramagnetic $\theta = -3$°K.[45]

CrTe: Hexagonal, NiAs($B8_1$)-type crystal structure. *Ferromagnetic*, with saturation magnetization (extrapolated to 0°K) of about 75 emu/gm,[46,47] corresponding to $\mu_F \simeq 2.4\mu_B$ per Cr atom; the Curie point (T_c) $\simeq 340$°K,[46–50] and from susceptibility data above T_c, the effective paramagnetic moment, $\mu_P \simeq 4.0\mu_B$.[47–49] If g-factor is assumed to be 2, above

[39] Corliss, L. M., N. Elliott, J. M. Hastings, and R. L. Sass, *Phys. Rev.* **122** (1961) 1402.
[40] Haraldsen, H., and A. Neuber, *Z. Anorg. Chem.* **234** (1937) 353.
[41] Tsubokawa, I., *J. Phys. Soc. Japan* **11** (1956) 662.
[42] Lotgering, F. K., and E. W. Gorter, *J. Phys. Chem. Solids* **3** (1957) 238.
[43] Tsubokawa, I., *J. Phys. Soc. Japan* **15** (1960) 2243.
[44] Chevreton, M., and E. F. Bertaut, *Compt. Rend.* **253** (1961) 145.
[45] Bertaut, E. F., A. Delapalme, F. Forrat, G. Roult, F. de Bergevin, and R. Pauthenet, *J. Appl. Phys. Suppl.* **33** (1962) 1123.
[46] Guillaud, C. and S. Barbezat, *Compt. Rend.* **222** (1946) 386.
[47] Lotgering, F. K., and E. W. Gorter, *J. Phys. Chem. Solids* **3** (1957) 238.
[48] Haraldsen, H., and A. Neuber, *Z. Anorg. Chem.* **234** (1937) 353.
[49] Tsubokawa, I., *J. Phys. Soc. Japan* **11** (1956) 662.
[50] Fakidov, I. G., N. P. Grazhdankina, and A. K. Kikoin, *Dokl. Akad. Nauk (SSSR)* **68** (1949) 491.

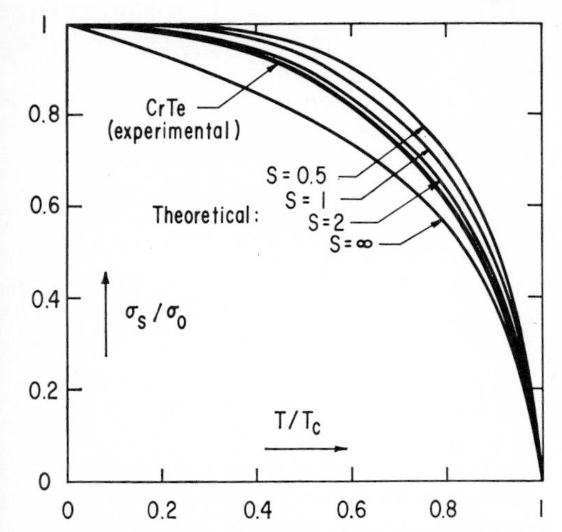

Fig. 5. Normalized saturation magnetization versus temperature curves for CrTe compared to theoretical curves for different values of spin quantum number, S.

values for μ_F and μ_P give 1.2 and 1.6, respectively, when S is the spin quantum number. These values are smaller than $S \simeq 2$ obtained from the best fit of magnetization versus temperature data with simple molecular field theory curves,[51] as shown in Fig. 5. Torque measurements on CrTe single crystal show that c-axis is the easy axis of magnetization and that the anisotropy coefficient (K_1) is about 5.5×10^6 erg/cm³ at 4°K and decreases monotonically to zero at the Curie point.[52]

4.3 Iron Compounds

Fe_3Al: Cubic, BiF_3($D0_3$)-type crystal structure. *Ferromagnetic* structure was obtained from neutron diffraction data[53] at room temperature. As shown in Fig. 6, A and B sites are occupied by Fe atoms with average moments of $1.46 \pm 0.1\mu_B$ and $2.14 \pm 0.1\mu_B$, respectively; C sites are occupied by Al atoms. This difference between $\mu_{Fe(A)}$ and $\mu_{Fe(B)}$ was attributed alternatively to (a) difference in $3d$-electron configuration and hence in intrinsic magnetic moment,[53] (b) larger degree of magnetic disorder of A-site magnetization[54] or (c) some antiferromagnetic alignment of A-site moments.[55] Explanation (a) was favored by later experiments showing that the A and B sites are different in the orbital symmetry of their unpaired spin densities[56] and in their internal (Mössbauer) fields.[57]

Curie temperature $T_c \simeq 750$°K[58,59] and saturation magnetization σ_s (in emu/gm) $\simeq$ 160 (at 81°K) and 140 (at 300°K),[60] all of which are lower than values for disordered bcc alloy. Magnetocrystalline anisotropy coefficient K_1 (in erg/cm³) $\simeq -19 \times 10^4$ at 80°K and -8×10^4 at 300°K, whereas disordered alloy has small positive K_1.[61] Magnetostriction coefficients at room temperature for ordered Fe_3Al[62] are $\lambda_{100} \simeq +7 \times 10^{-5}$ and $\lambda_{111} \simeq +2 \times 10^{-5}$.

$FeBe_2$: Hexagonal, $MgZn_2$(C14)-type crystal structure. *Ferromagnetic*, with Curie point of about 795°K.[63] Large (but quantitatively

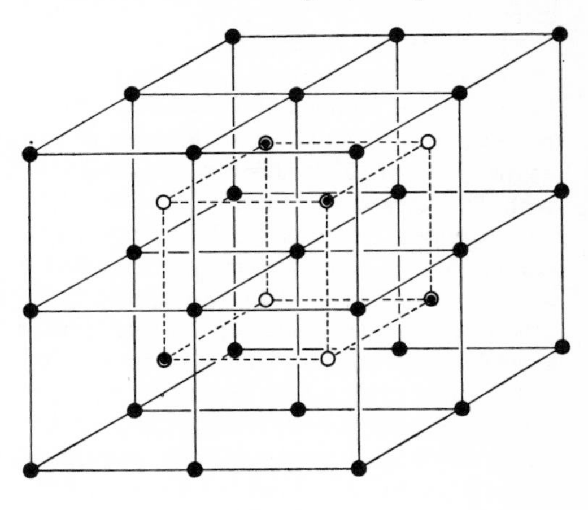

Fig. 6. Crystal structure of Fe_3Al.

[51] Guillaud, C., *Compt. Rend.* **222** (1946) 1224.
[52] Hirone, T., and S. Chiba, *J. Phys. Soc. Japan* **15** (1960) 1991.
[53] Nathans, R., M. T. Pigott, and C. G. Shull, *J. Phys. Chem. Solids* **6** (1958) 38.
[54] Sato, H., and A. Arrott, *J. Appl. Phys.* **29** (1958) 515.
[55] Goodenough, J. B., *Phys. Rev.* **120** (1960) 61.
[56] Pickart, S. J., and R. Nathans, *Phys. Rev.* **123** (1961) 1163.
[57] Ono, K., Y. Ishikawa, and A. Ito, *J. Phys. Soc. Japan* **17** (1962) 1747.
[58] Sykes, C., and H. Evans, *J. Iron Steel Inst.* **131** (1935) 225.
[59] Rassman, G., and H. Wich, *Arch. Eisenhüttenw.* **33** (1963) 115.
[60] Taylor, A., and R. M. Jones, *J. Phys. Chem. Solids* **6** (1958) 16.
[61] Puzei, I. M., *Phys. Metals Metallog.* (*USSR*) (*English Transl.*) **9**, 2 (1960) 103.
[62] Hall, R. C., *J. Appl. Phys.* **30** (1959) 816.
[63] Misch, L., *Z. Physik. Chem.* **B29** (1935) 42.

undetermined) magnetocrystalline anisotropy, easy direction of magnetization lying in basal plane.[63]

$FeBe_5$: Cubic, $MgCu_2$(C15)-type crystal structure. *Ferromagnetic* at liquid air temperature but not at room temperature.[64]

FeCo: See CoFe.

Fe_3Ge: Hexagonal, $Ni_3Sn(D0_{19})$-type crystal structure reported for compositions near $Fe_{3.25}Ge$[65] but later ascribed to stoichiometric compound Fe_3Ge.[66] The latter is *ferromagnetic*, with $T_c = 655°K$, $\mu_F = 2.2\mu_B$ per Fe atom (derived from saturation magnetization extrapolated to 0°K), and $\mu_P = 3.6\mu_B$ per Fe atom (derived from paramagnetic susceptibility above T_c).[66]

Annealing at about 900°K causes transformation to a new ferromagnetic phase reported as fcc, for which μ_F and μ_P are the same as those given above but $T_c = 755°K$.[66]

"Fe_2Ge": Hexagonal, $Ni_2In(B8_2)$-type crystal structure; compound generally has nonstoichiometric composition $Fe_{2-x}Ge$, where x represents Fe atom vacancies only in B sites, as shown in Fig. 22 for Mn_2Sn. *Ferromagnetic*, with T_c decreasing from about 480 to 370°K as x increases from 0.22 to 0.77.[67] For $Fe_{1.67}Ge$ compound, σ_s versus T data give $T_c = 485°K$ and $\sigma_0 = 89$ emu/gm (by extrapolation to 0°K), which corresponds to 1.59 μ_B per Fe atom; from linear part of $1/\chi$ versus T curve well above T_c: $\theta = +540°K$ and $\mu_P = 3.14\ \mu_B$ per Fe atom.[68] These magnetic data were shown to be consistent with moments of 2 and 1 μ_B for Fe atoms on A and B sites, respectively, all moments being aligned ferromagnetically.[68] Similar conclusion was drawn from variation of saturation magnetization (σ_s) with composition (x), in contrast to ferrimagnetic moment alignment deduced for $Mn_{2-x}Sn$[69] (see Mn_2Sn).

FeGe: Hexagonal, CoSn(B35)-type crystal structure.[70] Susceptibility maxima were observed at 410°K and 340°K; the former was tentatively attributed to *antiferromagnetic* ordering of the CoSn compound and the latter to ferromagnetic ordering of an extraneous amount of "Fe_2Ge" phase in the specimen.[70] At higher temperatures, susceptibility follows Curie-Weiss relation with $\theta = +220°K$ and $\mu_P = 3.1\ \mu_B$ per Fe atom.

$FeGe_2$: Tetragonal, $CuAl_2$(C16)-type crystal structure. Weakly *ferrimagnetic*, according to magnetization measurements giving $T_c = 190°K$ and $\sigma_0 = 3.15$ emu/gm (by extrapolation to 0°K), which corresponds to 0.11 μ_B per Fe atom; linear $1/\chi$ versus T curve yields $\theta = +90°K$ and $\mu_P = 2.43\ \mu_B$ per Fe atom.[71] Contrast between low μ_F and high μ_P suggests that magnetic structure of $FeGe_2$ is basically *antiferromagnetic*, possibly similar to that shown in Fig. 9 for $FeSn_2$.

$FeNi_3$: Cubic, $Cu_3Au(L1_2)$-type crystal structure. Neutron diffraction and magnetization data at room temperature for $FeNi_3$ of fairly high degree of atomic order give $\mu_{Fe} = 2.97(\pm 0.15)\ \mu_B$ and $\mu_{Ni} = 0.62\ (\pm 0.05)\ \mu_B$ for the *ferromagnetically* aligned atomic moments; an alternative ferrimagnetic solution (i.e., μ_{Fe} antiparallel to μ_{Ni}) deduce from the same data seems less likely.[72] Although Curie point of ordered phase lies well above order-disorder temperature of about 780°K, magnetic measurements made under rapid heating place it at 863°K, compared to 833°K for the disordered fcc phase.[73] From similar measurements: T_c (ordered) = 954[74] or 983°K,[75] and T_c (disordered) = 871[74] or 858°K;[75] however, these T_c values for ordered $FeNi_3$ are probably too high, because they were obtained by assuming a fixed shape for M_s/M_0 versus T/T_c curve, whereas this curve may be more concave towards the T/T_c-axis for the ordered phase.[76] Saturation magnetization was found to increase by 3 to 6 per cent with atomic ordering,[74,75,77,78] M_s for ordered

64 Misch, L., *Z. Physik. Chem.* **B29** (1935) 42.
65 Shtol'ts, A. K., and P. V. Gel'd, *Fiz. Metal. Metalloved* **12** (1961) 462.
66 Kanematsu, K., K. Yasukochi, and T. Ohoyama, *J. Phys. Soc. Japan* **18** (1963) 920.
67 Lecocq, P., and A. Michel, *Compt. Rend.* **253** (1961) 2235.
68 Yasukochi, K., K. Kanematsu, and T. Ohoyama, *J. Phys. Soc. Japan* **16** (1961) 429.
69 Yasukochi, K., K. Kanematsu, and T. Ohoyama, *J. Phys. Soc. Japan* **17,** Suppl. B-1 (1962) 165.
70 Ohoyama, T., K. Kanematsu, and K. Yasukochi, *J. Phys. Soc. Japan* **18** (1963) 589.
71 Yasukochi, K., K. Kanematsu, and T. Ohoyama, *J. Phys. Soc. Japan* **16** (1961) 429.
72 Shull, C. G., and M. K. Wilkinson, *Phys. Rev.* **97** (1955) 304.
73 Josso, E., *J. Phys. Radium* **12** (1951) 399.
74 Wakelin, R. J., and E. L. Yates, *Proc. Phys. Soc. (London)* **B66** (1953) 221.
75 Taoka, T., and T. Ohtsuka, *J. Phys. Soc. Japan* **9** (1954) 712.
76 Went, J. J., *Physica* **17** (1951) 596.
77 McKeehan, L. W., and E. M. Grabbe, *Phys. Rev.* **55** (1939) 505.
78 Grabbe, E. M., *Phys. Rev.* **57** (1940) 728.

$FeNi_3$ at room temperature being 1007 emu/cm^3.[78]

From magnetization[78] and torque[79] measurements at room temperature, anisotropy coefficient K_1 of ordered $FeNi_3$ crystal is found to be about -3×10^4 erg/cm^3, in contrast to $K_1 \simeq 0$ for the same crystal disordered. This difference in K_1 may account for the larger coercivity and smaller permeability of the ordered phase.[74] Saturation magnetostriction data for $FeNi_3$ crystal at room temperature show an increase of λ_{111} and a decrease of λ_{100} with atomic ordering,[79] net effect for polycrystal being an increase of its positive longitudinal magnetostriction.[73,75]

FePd: Tetragonal, CuAu($L1_0$)-type crystal structure. *Ferromagnetic.* Neutron diffraction and magnetization data, combined and extrapolated to 0°K, give 2.9 $(\pm 0.1)\mu_B$ and 0.30 $(\pm 0.05)\mu_B$ for the atomic moments of iron and palladium, respectively.[80] From earlier magnetic measurements[81] σ_s (at 290°K = 105 emu/gm, corresponding to an average atomic moment of 1.52 μ_B, and T_c = 749°K, compared to 703°K for the disordered fcc alloy.

Partial ordering anneal at 773°K increases the coercive field of FePd alloy from 2 to 260 Oe.[82]

$FePd_3$: Cubic, $Cu_3Au(L1_2)$-type crystal structure. *Ferromagnetic.* According to neutron diffraction and magnetization measurements at 300°K, the atomic moments of iron and palladium are 2.73 $(\pm 0.13)\mu_B$ and 0.51 (± 0.04) μ_B, respectively.[83] Similar data extrapolated to 0°K give $\mu_{Fe} = 3.0(\pm 0.1)\mu_B$ and $\mu_{Pd} = 0.45(\pm 0.05)\mu_B$.[84] From earlier magnetic measurements[85] σ_s (at 290°K) = 45 emu/gm, equivalent to $0.94\mu_B$ per atom, and T_c = 529°K, compared to 497°K for the disordered fcc alloy.

Fe_3Pt: Cubic, $Cu_3Au(L1_2)$-type crystal structure. *Ferromagnetic.* For ordered stoichiometric compound, $T_c \simeq 430$°K,[86,87] compared to 290°K for disordered fcc alloy.[87] Saturation magnetization of ordered Fe_3Pt, extrapolated to 0°K, is about 138 emu/gm, corresponding to 2.2 μ_B per atom.[86]

Large saturation magnetostrictions (10^{-4} in order of magnitude) reported for ordered alloys of about 25 at.% Pt.[88–90] From variation of magnetostriction versus temperature curves with ordering treatment, it was deduced that the pressure dependence of Curie point, unlike the Curie point itself, is lowest for the most highly ordered state.[90]

FePt: Tetragonal, CuAu($L1_0$)-type crystal structure. *Ferromagnetic*; Curie point of ordered stoichiometric compound is about 750°K, compared to 530°K for disordered fcc alloy of same composition.[91,92] Saturation magnetization of ordered FePt is 31.7 emu/gm at 290°K and 33.8 emu/gm at 0°K (by extrapolation);[92] the latter corresponds to 0.75 μ_B per atom and is considerably smaller than σ_0 value for disordered alloy.[91]

Intrinsic coercivity of 1800 Oe and remanent induction (B_R) of 3000 G (gauss) measured in ordered FePt; optimum permanent magnet properties found in ordered 40 at.% Pt alloy.[93] Similar properties observed earlier[91,94] and attributed to large strains at the twin boundaries of ordered tetragonal phase.[94]

$FePt_3$: Cubic, $Cu_3Au(L1_2)$-type crystal structure. *Antiferromagnetic*, judging from susceptibility maximum and negative value of paramagnetic θ of ordered alloys of 72 to 76 at.% Pt; for stoichiometric $FePt_3$, χ_{max} occurs at about 90°K and $\theta \simeq -120$°K.[95] Cold-working of these alloys produces strong ferromagnetism.[95]

---

[79] Bozorth, R. M., and J. G. Walker, *Phys. Rev.* **89** (1953) 624.

[80] Cable, J. W., E. O. Wollan, W. C. Koehler, and M. K. Wilkinson, *J. Appl. Phys. Suppl.* **33** (1962) 1340.

[81] Fallot, M., *Ann. Phys.* **10** (1938) 291.

[82] Jellinghaus, W., *Z. Tech. Physik* **17** (1936) 33.

[83] Pickart, S. J., and R. Nathans, *J. Appl. Phys. Suppl.* **33** (1962) 1336.

[84] Cable, J. W., E. O. Wollan, W. C. Koehler, and M. K. Wilkinson, *J. Appl. Phys. Suppl.* **33** (1962) 1340.

[85] Fallot, M., *Ann. Phys.* **10** (1938) 291.

[86] Fallot, M., *Ann. Phys.* **10** (1938) 291.

[87] Kussmann, A., and G. v. Rittberg, *Z. Metallk.* **41** (1950) 470.

[88] Akulov, N. S., Z. I. Alizade, and K. P. Belov, *Dokl. Akad. Nauk.* (*SSSR*) **65** (1949) 815.

[89] Kussmann, A., and G. v. Rittberg, *Ann. Phys.* **7** (1950) 173.

[90] Belov, K. P., and Z. D. Sirota, *Sov. Phys. JETP* (*English Transl.*) **36(9)** (1959) 752.

[91] Graf, L., and A. Kussmann, *Z. Physik.* **36** (1935) 544.

[92] Fallot, M., *Ann. Phys.* **10** (1938) 291.

[93] Kussmann, A., and G. v. Rittberg, *Ann. Phys.* **7** (1950) 173.

[94] Lipson, H., D. Schoenberg, and G. V. Stupart, *J. Inst. Metals* **67** (1941) 333.

[95] Crangle, J., *J. Phys. Radium* **20** (1959) 435.

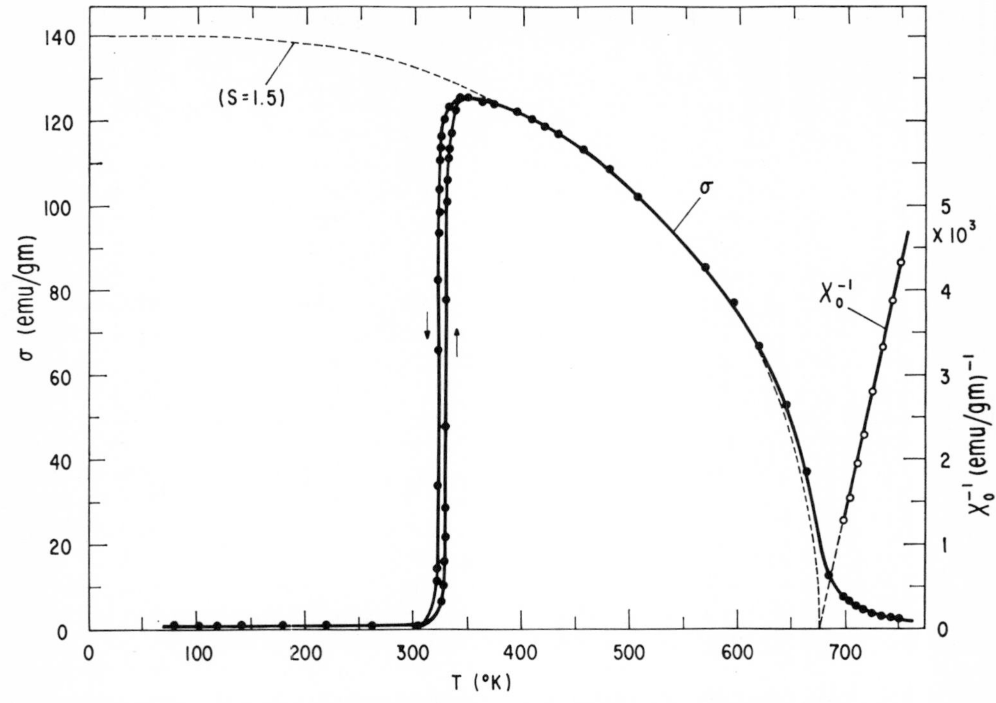

Fig. 7. Magnetization (in 10 kOe field) and inverse susceptibility versus temperature for FeRh.

FeRh: Cubic, CsCl(B2)-type crystal structure; first-order transformation at about 350°K involves abrupt but isotropic change in volume (approximately 1 per cent) with no detectable change in crystal symmetry.[96] This transformation corresponds to transition between *antiferromagnetic* and *ferromagnetic* states. According to neutron diffraction measurements on 53 at.% Rh alloy,[97] antiferromagnetic structure below transition temperature consists of nearest-neighbor iron atoms having antiparallel moments, and rhodium moments, if not zero, arranged in similar configuration; quantitatively, $\mu_{Fe}^2 + \mu_{Rh}^2 = 10.88\ \mu_B^2$, so that $\mu_{Fe} = 3.3\ \mu_B$ if $\mu_{Rh} = 0$. For ferromagnetic structure just above the transition temperature of same alloy,[97] $\mu_{Fe} = 3.04\,(\pm 0.14)\ \mu_B$ and $\mu_{Rh} = 0.62\,(\pm 0.14)\ \mu_B$, resulting in saturation magnetization (σ_s) of about 125 emu/gm. This σ_s value agrees with magnetization data on 52 at.% Rh alloy,[98,99] the most recent of which are plotted versus temperature in Fig. 7. The curve yields σ_0 (extrapolated to 0°K) of 140 emu/gm and ferromagnetic Curie point (T_c) of 675°K; from linear $1/\chi$ versus T curve above T_c, $\mu_P = 3.3\ \mu_B$ per average atom. Recent neutron diffraction and Mössbauer experiments[100] on 48 at.% Rh alloy in ferromagnetic state find that $\mu_{Fe} = 3.2\ \mu_B$ and $\mu_{Rh} = 0.9\ \mu_B$ and that the hyperfine field for excess iron atoms on the Rh sites is larger than that for iron atoms on regular Fe sites, even though the reverse is probably true for the magnitudes of their moments. This suggests that the magnetic polarization of conduction electrons may be opposite to that of *d*-electrons and sensitive to nearest-neighbor environment of iron atoms.[100]

Critical temperature (T_{crit}) for antiferromagnetic-ferromagnetic transition observed to decrease with increasing magnetic field;

[96] de Bergevin, F., and L. Muldawer, *Compt. Rend.* **252** (1961) 1347.

[97] Bertaut, E. F., A. Delapalme, F. Forrat, G. Roult, F. de Bergevin, and R. Pauthenet, *J. Appl. Phys. Suppl.* **33** (1962) 1123.

[98] Kouvel, J. S., and C. C. Hartelius, *J. Appl. Phys. Suppl.* **33** (1962) 1343.

[99] Kouvel, J. S., C. C. Hartelius, and P. E. Lawrence, *Bull. Am. Phys. Soc.* **8,** 1 (1963) 54.

[100] Shirane, G., C. W. Chen, P. A. Flinn, and R. Nathans, *J. Appl. Phys.* **34** (1963) 1044.

$dT_{crit}/dH = -0.8°K/kOe$ (kilo-oersted)[99] or $-1.08°K/kOe$.[101] T_{crit} also found to increase with increasing pressure.[98]

"FeSe": Variants of hexagonal, NiAs($B8_1$)-type crystal structure for compounds of composition $Fe_{1-x}Se$, where x represents Fe atom vacancies located only in every alternate basal-plane layer; ordering of vacancies in these layers defines stoichiometric structure of compositions Fe_7Se_8[102,103] and Fe_3Se_4.[102] The latter is monoclinic (with $a/b = 1.745$, $\beta = 92°$) and is shown in Fig. 8. *Ferrimagnetic* over composition range $\frac{1}{8} \leq x \leq \frac{1}{4}$, judging from hyperbolic form of $1/\chi$ versus T curves above T_c and from low saturation magnetization (at 90°K) of about 35 emu/cm^3, corresponding to only about 0.2 μ_B per Fe atom.[104] Specific heat data[105] show T_c increasing from 447 to 465°K as x increases from $\frac{1}{8}$ to $\frac{1}{4}$. Ferrimagnetic configuration proposed for these compounds consists of ferromagnetic sheets parallel to basal plane with moments in adjacent sheets antiparallel to each other,[104] as indicated in Fig. 8.

Single crystal torque data[106] are interpreted as showing gradual shift of easy direction of magnetization of Fe_7Se_8 from basal plane towards c-axis, as the temperature is decreased below 200°K; easy direction of magnetization of Fe_3Se_4 is in basal plane at all temperatures. It is also deduced from these torque data that saturation magnetization of Fe_7Se_8 and Fe_3Se_4 (at 90°K) are 68 and 80 emu/cm^3, respectively; discrepancy with value of 35 emu/cm^3 obtained for polycrystalline specimens is presumably caused by very large magnetocrystalline anisotropy.

Fe_3Si: Cubic, BiF_3(DO_3)-type crystal structure. *Ferromagnetic*, with Curie point of about 805°K.[107] Saturation magnetizations: σ_s = 126 emu/gm at 290°K and σ_0 = 134 emu/gm at 0°K (by extrapolation).[107] This σ_0 value is equivalent to about 1.6 μ_B per Fe atom, which probably represents an average of different moment values for the two different Fe atom sites (see Fe_3Al).

[101] Flippen, R. B., and F. J. Darnell, *J. Appl. Phys.* **34** (1963) 1094.
[102] Okazaki, A., and K. Hirakawa, *J. Phys. Soc. Japan* **11** (1956) 930.
[103] Okazaki, A., *J. Phys. Soc. Japan* **16** (1961) 1162.
[104] Hirone, T., S. Maeda, and N. Tsuya, *J. Phys. Soc. Japan* **9** (1954) 496.
[105] Hirone, T., and S. Chiba, *J. Phys. Soc. Japan* **12** (1956) 666.
[106] Hirakawa, K., *J. Phys. Soc. Japan* **12** (1957) 929.
[107] Fallot, M., *Ann. Phys.* **6** (1936) 305.

Fe_3Sn: Hexagonal, Ni_3Sn(DO_{19})-type crystal structure. *Ferromagnetic*, with $T_c = 743°K$ and $\mu_F = 1.90\ \mu_B$ per Fe atom.[108]

Fe_3Sn_2: Complex monoclinic crystal structure. *Ferromagnetic*, with $T_c = 612°K$ and $\mu_F = 2.15\ \mu_B$ per Fe atom.[109]

"FeSn": Hexagonal, NiAs($B8_1$)-type crystal structure; however, because this phase exists only at iron-rich compositions, the excess iron atoms are probably in B sites of related Ni_2In structure, as shown in Fig. 22 for Mn_2Sn. *Ferromagnetic*. $Fe_{1.3}Sn$ compound has $T_c = 676°K$ and σ_s (extrapolated to 0°K) = 69 emu/gm, corresponding to $\mu_F = 1.8\ \mu_B$ per Fe atom.[110] Same μ_F, but $T_c = 583°K$ reported for $Fe_{1.6}Sn$ compound.[111]

[108] Jannin, C., P. Lecocq, and A. Michel, *Compt. Rend.* **257** (1963) 1906.
[109] Jannin, C., P. Lecocq, and A. Michel, *Compt. Rend.* **257** (1963) 1906.
[110] Asanuma, M., *J. Phys. Soc. Japan* **15** (1960) 1343.
[111] Jannin, C., P. Lecocq, and A. Michel, *Compt. Rend.* **257** (1963) 1906.

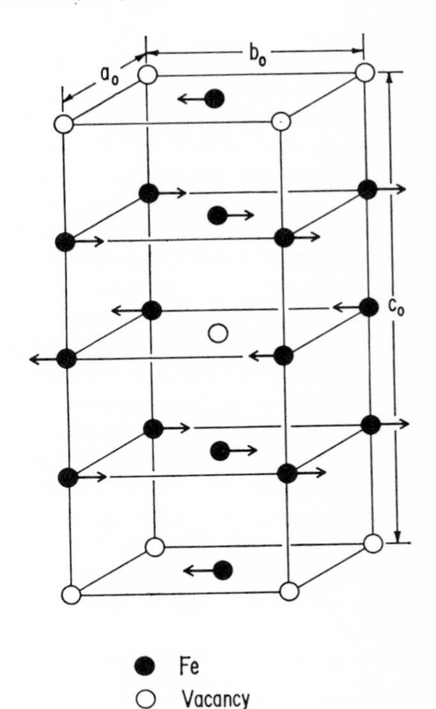

Fig. 8. Ferrimagnetic structure of Fe_3Se_4 resulting from ordering of vacancies in Fe sites. Se atoms omitted for clarity. Absolute moment directions in basal plane are unknown.

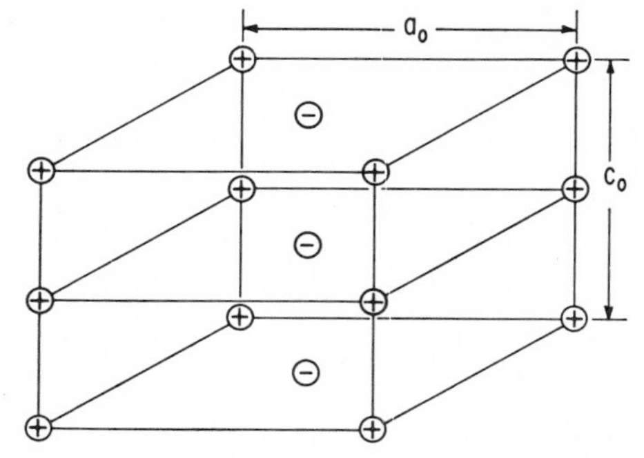

Fig. 9. Antiferromagnetic structure of $FeSn_2$. Sn atoms omitted for clarity. Only relative directions of Fe moments have been determined.

$FeSn_2$: Tetragonal, $CuAl_2$(C16)-type crystal structure. *Antiferromagnetic* configuration, derived from neutron diffraction data at 300°K, is shown in Fig. 9; only the relative directions of the Fe moments (but neither their magnitudes nor their absolute directions relative to crystal axes) were specified.[112] Curie point was placed at 384°K, in agreement with temperature of susceptibility maximum.[113] Linear $1/\chi$ versus T curve above T_c gives $\theta = -230$°K and $\mu_P = 3.36\ \mu_B$ per Fe atom.[113]

"FeTe": Tetragonal crystal structure intermediate between PbO(B10) and Cu_2Sb (C38) types; phase probably limited to compositions close to $FeTe_{0.9}$. *Antiferromagnetic* with Curie point of 63°K identified by sharp maxima in susceptibility[114] and specific heat.[115] Linear $1/\chi$ versus T curve above T_c gives $\theta = -130$°K and $\mu_P = 2.44\ \mu_B$ per Fe atom (or $S = 0.82$ if assumed that $g = 2$).[114] Recent magnetic measurements on $FeTe_{0.8}$ and $FeTe_{0.95}$ show *weak ferrimagnetism* (with $T_c \simeq 480$°K), which was tentatively attributed to crystallographic ordering of excess Fe atoms.[116]

$FeTe_2$: Orthorhombic, marcasite(C18)-type crystal structure. Susceptibility peak[117] at about 83°K indicative of *antiferromagnetic* Curie point, although no specific heat anomaly is observed at this temperature.[118] However, linear $1/\chi$ versus T curve below about 550°K, giving $\theta = -67$°K (and $\mu_P = 3.35\ \mu_B$ per Fe atom), also suggests antiferromagnetism at low temperatures.[119]

Fe_2Ti: Hexagonal, $MgZn_2$(C14)-type crystal structure. *Ferromagnetic*; saturation magnetization measurements at room temperature and 4.2°K give 0.35 μ_B and 0.92 μ_B, respectively, for the average moment per Fe_2Ti molecule.[120] According to neutron diffraction data at room temperature,[121] the two Fe atoms in one kind of site and the six Fe atoms in the other (considering one unit cell) have moments of 0.10 $(\pm 0.03)\mu_B$ and 0.20 $(\pm 0.02)\mu_B$, respectively, with no detectable moment for the Ti atoms. However, subsequent experiments show a temperature-independent susceptibility between 4.2 and 300°K.[122] This apparent contradiction with earlier work is possibily resolved by recent measurements indicating that Fe_2Ti is *antiferromagnetic* in pure state and *ferromagnetic* when contaminated with oxygen.[123]

FeV: Cubic, CsCl(B2)-type crystal structure for metastable phase formed at about 900°K. *Ferromagnetic*, with saturation magnetization (extrapolated to 0°K) ranging from about 20 to 78 emu/gm, as iron concentration is increased from about 38 to 60 at.%; corresponding range in average moment per iron atom is from about 0.50 μ_B to 1.25 μ_P.[124] Neutron diffraction measurements at 77°K on compound of 46.7 at.% Fe (having T_c near 280°K[125]) gave average iron moment of about 0.7 μ_B and essentially no moment for the

[112] Iyengar, P. K., B. A. Dasannacharya, P. R. Vijayaraghavan, and A. P. Roy, *J. Phys. Soc. Japan* **17** (1962) 247.
[113] Kanematsu, K., K. Yasukochi, and T. Ohoyama, *J. Phys. Soc. Japan* **15** (1960) 2358.
[114] Tsubokawa, I., and S. Chiba, *J. Phys. Soc. Japan* **14** (1959) 1120.
[115] Westrum, E. F., Jr., Chien Chou, and F. Gronvold, *J. Chem. Phys.* **30** (1959) 761.
[116] Naya, R., M. Murakami, and E. Hirahara, *J. Phys. Soc. Japan* **15** (1960) 360.
[117] Finlayson, D. M., J. P. Llewellyn, and T. Smith, *Proc. Phys. Soc. (London)* **74** (1959) 75.
[118] Westrum, E. F., Jr., Chien Chou, and F. Gronvold, *J. Chem. Phys.* **30** (1959) 761.
[119] Chiba, S., *J. Phys. Soc. Japan* **10** (1955) 837.
[120] McGuire, T. R., as reported in Reference 121.
[121] Kocher, C. W., and P. J. Brown, *J. Appl. Phys. Suppl.* (1962) 1091.
[122] Piegger, E., and R. S. Craig, *J. Chem. Phys.* **39** (1963) 137.
[123] Wallace, W. E., private communication describing unpublished work of R. S. Craig, G. Stey, and W. Löser.
[124] Nevitt, M. V., and A. T. Aldred, *J. Appl. Phys.* **34** (1963) 463.
[125] Nevitt, M. V., as reported in Reference 126.

vanadium atoms.[126] Saturation magnetizations of disordered bcc phase (quenched from about 1500°K) and stable σ-phase are, respectively, somewhat higher and much lower than those of CsCl-type compounds of similar composition.[124]

Fe_2Zr: Cubic, $MgCu_2$(C15)-type crystal structure. *Ferromagnetic.* From saturation magnetization data, the average moment per Fe_2Zr molecule at room temperature and 4.2°K is 2.56 μ_B and 3.12 μ_B, respectively.[127] Neutron diffraction experiment at room temperature assigns all this moment to the Fe atoms and indicates that the unpaired electrons in this compound have a more compact distribution about the Fe atom nuclei than in α-iron.[128] More recent measurements give slightly higher moments and a Curie point of 628°K.[129]

4.4 Manganese Compounds

"MnAl": Tetragonal, CuAu($L1_0$)-type crystal structure; this phase is metastable, produced only at compositions near 55 at.% Mn by controlled cooling through transformation temperature of about 1000°K.[130–133] *Ferromagnetic*, although neutron diffraction data indicate that excess Mn atoms in "Al" sites have their moments antiparallel to those of Mn atoms in "Mn" sites, so that nonstoichiometric compounds are actually *ferrimagnetic*; tetragonal *c*-axis is easy axis of magnetization.[134] Magnetic measurements place Curie point at about 650°K.[130–132] For saturation magnetization at room temperature, extrapolation to infinite field gives $\sigma_s = 96$ emu/gm, equivalent to saturation induction ($4\pi M_s$) of 6200 G;[131] however, swaging-annealing treatment results in $4\pi M$ of 7100 G at 12 kOe.[135] Extrapolation of magnetization data to infinite field and 0°K gives $\sigma_0 = 100$ emu/gm, corresponding to average moment per Mn atom of 1.40 μ_B, compared to 2.31 μ_B deduced from paramagnetic susceptibility data.[130]

From approach of M versus H curve to saturation, it was estimated that magnetocrystalline anisotropy coefficient is about 10^7 erg/cm³ at room temperature.[131] Intrinsic coercivity (H_{ci}), which is under 1000 Oe for bulk material, increases to over 4000 Oe when material is pulverized, but σ_s is reduced by about 25 per cent; subsequent anneal lowers H_{ci} and raises σ_s.[131] Hence coercive properties are not typical of single-domain particles, but are mostly determined by amount of plastic deformation which affects degree of atomic order.[134] Permanent magnet properties of swaged bar measured parallel to its axis (which has become easy axis of magnetization): $B_R = 4280$ G, $H_c = 2750$ Oe ($H_{ci} = 4600$ Oe), $(BH)_{max} = 3.5 \times 10^6$ G–Oe.[131] Swaging-annealing treatment was reported[135] to give $(BH)_{max}$ of 4.0×10^6 G–Oe.

Mn_2As: Tetragonal, Cu_2Sb(C38)-type crystal structure. *Antiferromagnetic* configuration, according to room-temperature neutron diffraction data,[136] is identical in moment directions to that deduced for low-temperature state for Cr-substituted Mn_2Sb (see Fig. 19*b*), but the magnitudes of Mn moments associated with A and B sites are 3.7 μ_B and 3.5 μ_B, respectively. For $Mn_{2.3}As$ compound with this crystal structure, susceptibility and specific heat maxima were observed at 573°K and identified with antiferromagnetic Curie point; weak ferromagnetism (with T_c = 273°K) detected in compounds with slightly less manganese (e.g., $Mn_{2.1}As$) were attributed to traces of Mn_3As_2 compound with crystal structure different from those of Mn_2As or MnAs.[137]

MnAs: Hexagonal, NiAs($B8_1$)-type crystal structure; recent X-ray diffraction studies,[138,139] however, indicate from about 315 to 400°K an orthorhombic distortion resulting in MnP(B31)

[126] Chandross, R. J., and D. P. Shoemaker, *J. Phys. Soc. Japan* **17** Suppl. B-III (1962) 16.
[127] McGuire, T. R., as reported in Reference 128.
[128] Kocher, C. W., and P. J. Brown, *J. Appl. Phys. Suppl.* **33** (1962) 1091.
[129] Piegger, E., and R. S. Craig, *J. Chem. Phys.* **39** (1963) 137.
[130] Kōno, H., *J. Phys. Soc. Japan* **13** (1958) 1444.
[131] Koch, A. J. J., P. Hokkeling, M. G. v.d. Steeg, and K. J. de Vos, *J. Appl. Phys. Suppl.* **31** (1960) 75.
[132] Köster, W., and E. Wachtel, *Z. Metallk.* **51** (1960) 271.
[133] Kōno, H., *J. Phys. Soc. Japan* **17** (1962) 1092.
[134] Braun, P. B., and J. A. Goedkoop, as reported in Reference 131.
[135] Bohlmann, M. A., *J. Appl. Phys. Suppl.* (1962) 1315.
[136] Austin, A. E., E. Adelson, and W. H. Cloud, *J. Appl. Phys. Suppl.* **33** (1962) 1356.
[137] Yuzuri, M., and M. Yamada, *J. Phys. Soc. Japan* **15** (1960) 1845.
[138] Kornelson, R. O., *Can. J. Phys.* **39** (1961) 1728.
[139] Wilson, R. H., and J. S. Kasper, *Acta Cryst.* **17** (1964) 95.

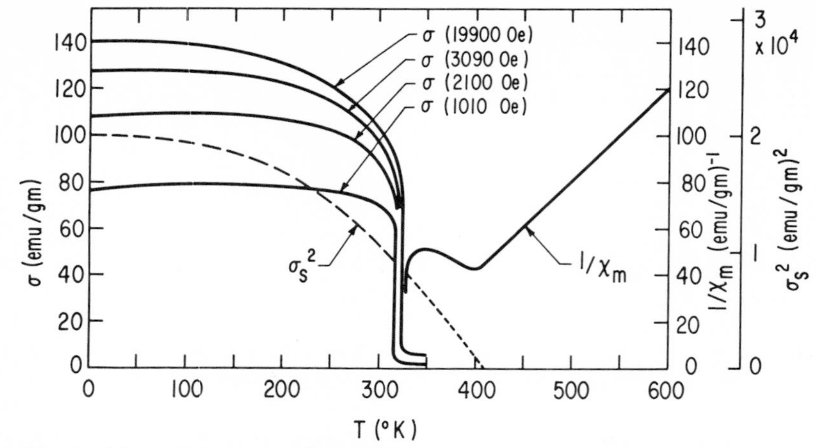

Fig. 10. Magnetization at different fields, saturation magnetization squared, and inverse susceptibility of MnAs as functions of temperature.

type structure. Early discovery[140] that MnAs is *ferromagnetic* was followed by reports that its ferromagnetism disappears suddenly at about 315°K[141,142] and that this magnetic transition is accompanied by latent heat[142] (1.79 cal/gm) and abrupt changes in electrical resistivity[143] and volume[144] (fourhold increase and 2 per cent contraction, respectively, with increasing temperature). This abrupt volume change arises entirely from dimensional changes in the basal plane.[139,144,145] Magnetization versus temperature curves for different fields,[146] reproduced in Fig. 10, illustrate abrupt magnetic transition at 315°K and show that σ_s extrapolated to 0°K gives $\sigma_0 =$ 138 emu/gm, equivalent to 3.4 μ_B per Mn atom. As is also shown in the figure, σ_s^2 versus T curve extrapolates to $\sigma_s = 0$ at about 400°K, at which temperature the $1/\chi$ versus T curve[147] has anomalous minimum (becoming linear at higher temperatures with $\theta \simeq 285$°K), which suggested that MnAs is antiferromagnetic between 315 and 400°K.[146] This hypothesis seemed further supported by anomalies in specific heat,[148] thermal expansion,[145] and magnetocaloric effect[149] observed near 400°K. However, neutron diffraction data showed no new reflections in the 315 to 400°K range that could be ascribed to antiferromagnetic ordering,[150] and, because the anomalies in various properties at 400°K can be attributed to crystallographic changes mentioned earlier, the abrupt transition at 40°K is now interpreted as a first-order transformation from ferromagnetism directly to *paramagnetism*, resulting from strong dependence of exchange interaction energy on lattice spacing.[151,152]

Thermodynamic model based on the latter interpretation[151,152] gives consistent explanation of the temperature hysteresis of the abrupt magnetic transition[153] (about 10°K) and the variation of its critical temperature with field[154,155] and pressure[156] (+0.33°K per

140 Heusler, F., *Z. Angew. Chem.* **17** (1904) 260.
141 Hilpert, S., and T. Dieckmann, *Berlin. Deut. Chem. Ges.* **44** (1911) 2378, 2831.
142 Bates, L. F., *Phil. Mag.* **8** (1929) 714.
143 Bates, L. F., *Phil. Mag.* **17** (1934) 783.
144 Guillaud, C., and J. Wyart, *Compt. Rend.* **219** (1944) 393.
145 Willis, B. T. M., and H. P. Rooksby, *Proc. Phys. Soc. (London)* **B67** (1954) 290.
146 Guillaud, C., *J. Phys. Radium* **12** (1951) 223.
147 Serres, A., *J. Phys. Radium* **8** (1947) 146.
148 Meyer, A. J. P., and P. Taglang, *J. Phys. Radium* **12** (1951) 63S.
149 Meyer, A. J. P., and P. Taglang, *Compt. Rend.* **246** (1958) 1820.
150 Bacon, G. E., and R. Street, *Nature* **175** (1955) 518.
151 Rodbell, D. S., and C. P. Bean, *J. Appl. Phys. Suppl.* **33** (1962) 1037.
152 Bean, C. P., and D. S. Rodbell, *Phys. Rev.* **126** (1962) 104.
153 Bates, L. F., *Proc. Phys. Soc. (London)* **42** (1930) 441.
154 Meyer, A. J. P., and P. Taglang, *J. Phys. Radium* **14** (1953) 82.
155 Rodbell, D. S., and P. E. Lawrence, *J. Appl. Phys. Suppl.* **31** (1960) 275.
156 Rodbell, D. S., *Progress in Very High Pressure Research*, F. Bundy, W. R. Hibbard, and H. M. Strong, eds. Wiley, New York, 1961, pp. 283–286.

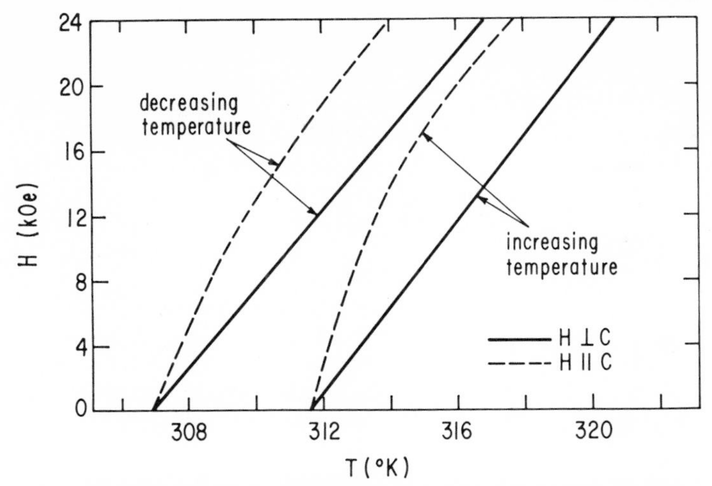

Fig. 11. Critical temperature for first-order magnetic transition in MnAs crystal as function of field applied perpendicular or parallel to *c*-axis. Thermal hysteresis indicated by difference between curves for increasing and decreasing temperature.

kilo-oersted and -12°K per kilobar, respectively); it also explains recent detailed results for temperature-field-pressure characteristics of the abrupt transition in a MnAs single crystal.[157] Further study[158] of single crystal gave curves, reproduced in Fig. 11, for transition temperature as a function of the field applied along easy and hard directions of magnetization (i.e., perpendicular and parallel to *c*-axis, respectively)[150], measured for increasing and decreasing temperature. Note that when the field along the hard direction is large enough for saturation, the curves become parallel to linear curves for field along the easy direction, but are displaced by about -3°K. These results combined those of torque measurements on the same crystal gave uniaxial magnetocrystalline anisotropy coefficients K_1, K_2, and $K_3 = -5.75$, $+1.5$, and -1.15 ($\times$ 10^6 erg/cm^3), respectively.[158]

MnAu: Cubic, CsCl(B2)-type crystal structure, which becomes tetragonally distorted below some temperature T_1 (about 500°K for MnAu, lower for nonstoichiometric compositions), lattice-parameter ratio c/a becoming <1; in compounds with ≤ 50 at.% Au c/a ratio switches to >1 at a still lower temperature T_2 (about 380°K for MnAu composition).[159] *Antiferromagnetic* structure, according to recent neutron diffraction study,[160] consists of ferromagnetic sheets of Mn moments (of magnitude between 4.0 and 4.2 μ_B) aligned antiparallel to those in adjacent sheets; moment directions lie within ferromagnetic sheets, which are always perpendicular to shorter crystal axis, either *a* or *c* (see above). Earlier neutron-diffraction study found Curie point of about 500°K (i.e., $\simeq T_1$) for both slow-cooled and quenched specimens,[161] consistent with temperature of electrical resistivity anomaly[162] and susceptibility maximum[163,164] (above which $1/\chi$ versus T curve is linear with $\theta = -352$°K[164]). However, after 500°K anneal[161] or long room-temperature aging[165] of the MnAu specimen, susceptibility peak is observed at about 370°K (i.e., $\simeq T_2$) as well as at 500°K. Susceptibility maximum found only at 383°K for annealed powder sample, for which X-ray diffraction data show tetragonal structure ($c/a < 1$) and

[157] DeBlois, R. W., and D. S. Rodbell, *Phys. Rev.* **130** (1963) 1347.
[158] DeBlois, R. W., and D. S. Rodbell, *J. Appl. Phys.* **34** (1963) 1101.

[159] Smith, J. H., and P. Gaunt, *Acta Met.* **9** (1961) 819.
[160] Bacon, G. E., *Proc. Phys. Soc. (London)* **79** (1962) 938.
[161] Bacon, G. E., and R. Street, *Proc. Phys. Soc. (London)* **72** (1958) 470.
[162] Giansoldati, A., and J. O. Linde, *J. Phys. Radium* **16** (1955) 341.
[163] Giansoldati, A., *J. Phys. Radium* **16** (1955) 342.
[164] Meyer, A. J. P., *J. Phys. Radium* **20** (1959) 430.
[165] Giansoldati, A., J. O. Linde, and G. Borelius, *J. Phys. Chem. Solids* **9** (1959) 183.

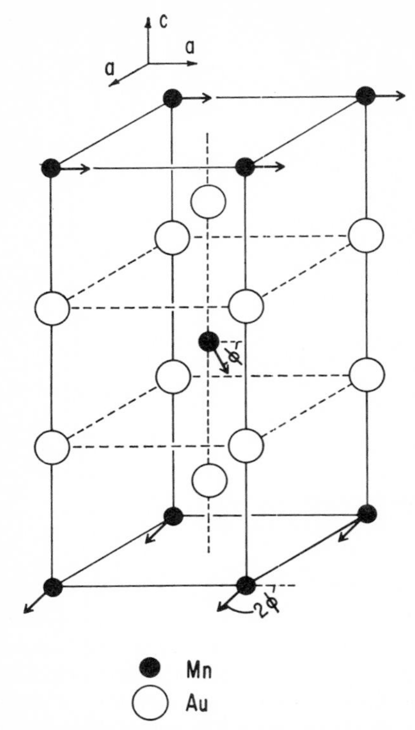

Fig. 12. Spiral magnetic structure of $MnAu_2$. All moments lie in basal plane.

cubic structure below and above this temperature, respectively.[166]

$MnAu_2$: Complex tetragonal crystal structure,[167,168] in which sheets of Mn atoms, parallel to basal plane, are separated by two similar sheets of Au atoms, as shown in Fig. 12. In fields up to 8 kilo-oersted (kOe), the susceptibility is field-independent and has maximum value at about 370°K, suggestive of *antiferromagnetism*; at higher fields and below this temperature, the magnetization rises very rapidly and then approaches *ferromagnetic* saturation.[169] Magnetocaloric effect data[170] also indicate field-induced antiferromagnetic-to-ferromagnetic transition below Curie point of 370°K. Magnetization versus field curves[169] reproduced in Fig. 13 show that the transition at lower temperatures requires higher fields, consistent with other magnetic measurements.[171] Extrapolation to infinite field and 0°K gives $\sigma_0 = 43.35$ emu/gm, or $\mu_F = 3.49\ \mu_B$ per Mn atom;[172] from linear $1/\chi$ versus T curve at high temperatures $\theta = +451$°K and $\mu_P = 5.05\ \mu_B$ per Mn atom.[173] Somewhat smaller value of μ_F (3.38 μ_B per Mn atom) was obtained with pulsed fields up to 300 kOe.[174]

High-field transition (called *metamagnetic*) and other magnetic properties of $MnAu_2$ are explained in terms of a model with two magnetic sublattices coupled antiferromagnetically and with crystal anisotropy comparable in strength to coupling between sublattices.[175] However, subsequent neutron diffraction experiments[176,177] indicate that antiferromagnetic state of $MnAu_2$ is as shown in Fig. 12; Mn moments lie within ferromagnetic sheets, and their directions are rotated by angle ϕ (51° at room temperature) with respect to those in nearest-neighbor sheets. Because rotation of moment direction is progressive along tetragonal *c*-axis, this is called a *spiral* or *helical* structure and is interpreted as a compromise state resulting from conflicting demands of interactions between Mn moments in nearest- and next-nearest-neighbor sheets. Neutron-diffraction study[176,177] also shows that in response to external field the moments turn more readily within rather than out of ferromagnetic basal-plane sheets.

Threshold field for metamagnetic transition is found to decrease with increasing hydrostatic pressure,[178,179] suggesting $MnAu_2$ contracts when transforming from antiferromagnet to ferromagnet.

166 Morris, D. P., and R. P. Preston, *Proc. Phys. Soc.* (*London*) **B69** (1956) 849.
167 Michel, P., *Compt. Rend.* **246** (1958) 2632.
168 Herpin, A., P. Mériel, and A. J. P. Meyer, *Compt. Rend.* **246** (1958) 3170.
169 Meyer, A. J. P., and P. Taglang, *Compt. Rend.* **239** (1954) 961.
170 Meyer, A. J. P., and P. Taglang, *Compt. Rend.* **239** (1954) 1611.
171 Kussman, A., and E. Raub, *Naturwissenschaften* **42** (1955) 411.
172 Meyer, A. J. P., and P. Taglang, *J. Phys. Radium* **17** (1956) 457.
173 Serres, A., as reported in Reference 172.
174 Zavadskii, E. A., and I. G. Fakidov, *Phys. Metal. Metallog.* (*USSR*) (*English Transl.*) **12** (1961) 47.
175 Néel, L., *Compt. Rend.* **242** (1956) 1549.
176 Herpin, A., P. Mériel, and J. Villain, *Compt. Rend.* **249** (1959) 1334; *J. Phys. Radium* **21** (1960) 67.
177 Herpin, A., and P. Mériel, *Compt. Rend.* **250** (1960) 1450; *J. Phys. Radium* **22** (1961) 337.
178 v. Klitzing, K. H., and J. Gielessen, *Z. Physik* **150** (1958) 409.
179 Rodbell, D. S., *Progress in Very High Pressure Research*, F. Bundy, W. R. Hibbard, and H. M. Strong, eds. Wiley, New York, 1961, pp. 283–286.

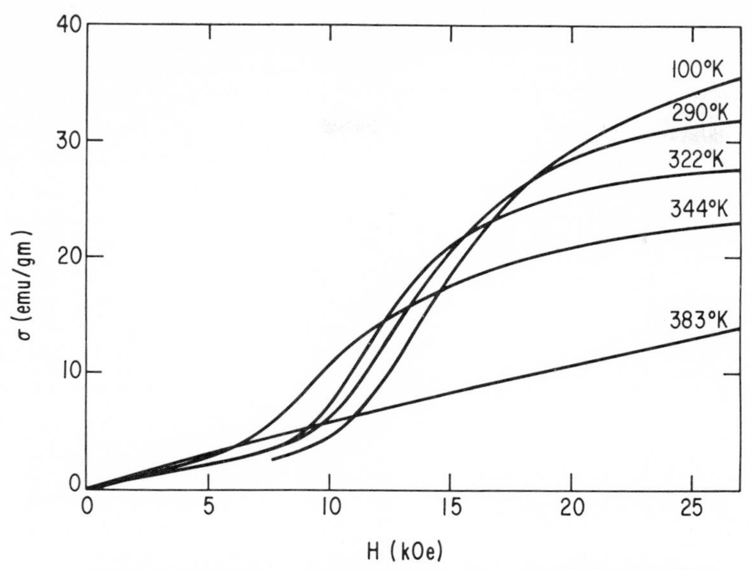

Fig. 13. Magnetization versus field curves for $MnAu_2$ at different temperatures.

$MnAu_3$: Complex crystal structure, reported as tetragonal,[180] but recent electron[181] and X-ray[182] diffraction data suggest that it is orthorhombic. Susceptibility maximum[182–185] and electrical resistivity anomaly[186] at about 150°K indicate *antiferromagnetic* state below this temperature; from linear $1/\chi$ versus T curve above 150°K: $\theta \simeq +200$°K and $\mu_P \simeq$ 4.2 μ_B per Mn atom.[182–184,187] Susceptibility below T_c (150°K) is observed to increase with increasing field (maximum field used $\simeq 21$ kOe); this behavior is interpreted as the beginning of a transition, in which axis of antiparallel moments shifts from being parallel to being perpendicular to applied field, whose threshold value was estimated to be 71.6 kOe at 20°K.[184] This so-called *spin-flop* transition was recently observed in its entirety at 83°K[182] and at 4.2°K.[185] The M versus H data at 4.2°K, reproduced in Fig. 14, show transition occurring over field range centered at 47 kOe; the susceptibility at high fields is about twice that at very low fields. This susceptibility ratio and a similar value obtained for ratio of low-field susceptibility at T_c to that at 4.2°K was shown to be consistent with orthorhombic crystal symmetry.[185]

Magnetic properties are very sensitive to the degree of atomic ordering.[182–185,187] Disordered fcc $MnAu_3$ alloy was reported as ferromagnetic with T_c of 120°K[183,184] but its properties were more recently attributed to complex ferro-antiferromagnetic state,[185] similar to that proposed for disordered $MnNi_3$.

$MnAu_4$: Tetragonal, Ni_4Mo-type crystal structure, according to recent electron diffraction data.[188] *Ferromagnetic*[189] with Curie point of about 360°K;[190] extrapolation of σ_s versus T curves to 0°K gives $\sigma_0 = 27.5$ emu/gm, equivalent to 4.15 μ_B per Mn atom.[190] From linear $1/\chi$ versus T curve above T_c: $\theta = +373$°K and $\mu_P = 4.85\ \mu_B$ per Mn atom.[190]

[180] Raub, E., U. Zwicker, and H. Baur, *Z. Metallk.* **44** (1953) 312.
[181] Watanabe, D., *J. Phys. Soc. Japan* **15** (1960) 1030.
[182] Sato, K., T. Hirone, H. Watanabe, S. Maeda, and K. Adachi, *J. Phys. Soc. Japan* **17** Suppl. B-I (1962) 160.
[183] Meyer, A. J. P., *Compt. Rend.* **244** (1957) 2028.
[184] Meyer, A. J. P., and M.-J. Besnus, *Compt. Rend.* **253** (1961) 2651.
[185] Jacobs, I. S., J. S. Kouvel, and P. E. Lawrence, *J. Phys. Soc. Japan* **17** Suppl. B-I (1962) 157.
[186] Giansoldati, A., and J. O. Linde, *J. Phys. Radium* **16** (1955) 341.
[187] Giansoldati, A., J. O. Linde, and G. Borelius, *J. Phys. Chem. Solids* **11** (1959) 46.
[188] Watanabe, D. *Acta Cryst.* **10** (1957) 483; *J. Phys. Soc. Japan* **15** (1960) 1251.
[189] Kussman, A., and E. Raub, *Z. Metallk.* **47** (1956) 9.
[190] Meyer, A. J. P., *Compt. Rend.* **244** (1957) 2028; *J. Phys. Radium* **20** (1959) 430.

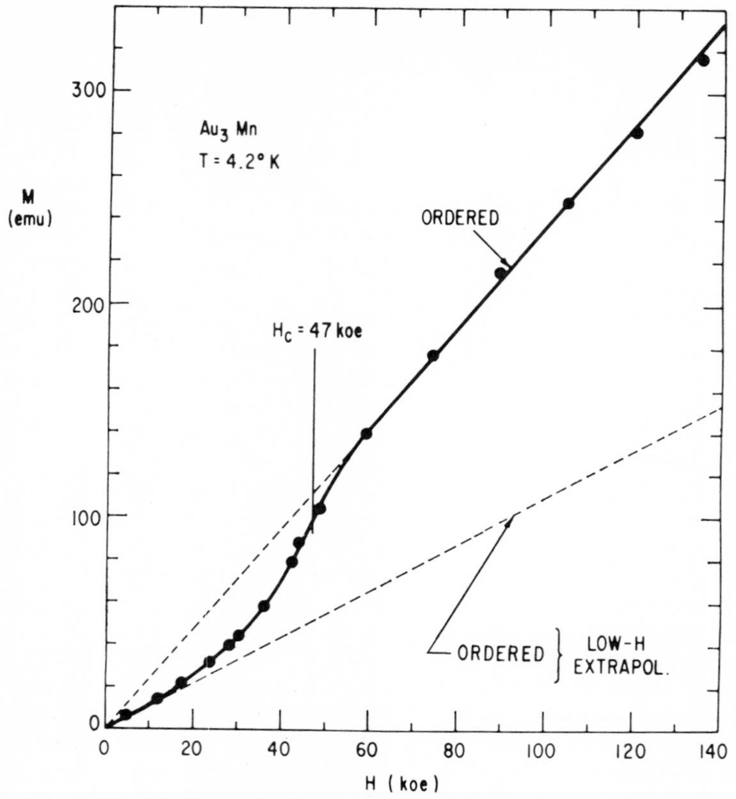

Fig. 14. Magnetization versus field curve for $MnAu_3$ at 4.2°K.

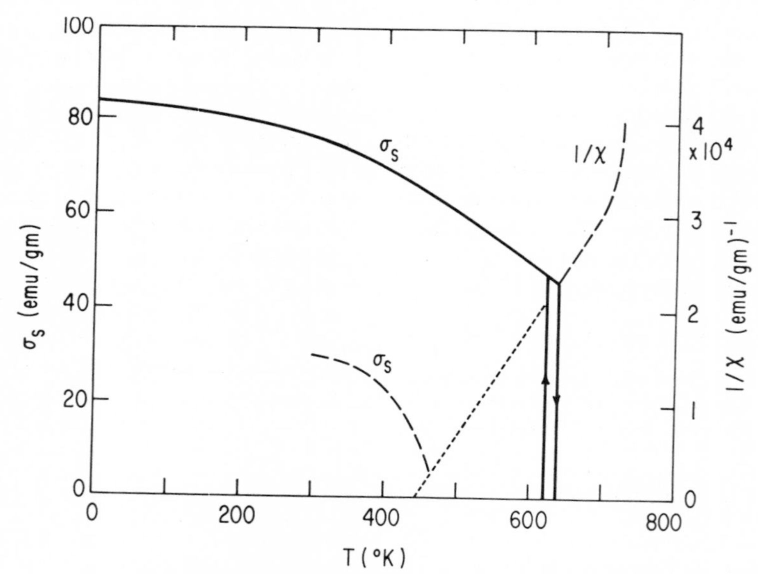

Fig. 15. Saturation magnetization versus temperature for MnBi cooled slowly (solid curve) or quenched from above 630°K (dashed curve). Inverse susceptibility measured above 630°K (and extrapolated to lower temperatures) also shown.

Remanent magnetization is only about 20 per cent of saturation value (σ_s) over temperature range 90 to 270°K; coercivity (H_{ci}) is approximately 250 Oe at 293°K.[190]

Disordered fcc $MnAu_4$ alloy reported to be ferromagnetic with $T_c \simeq 120$°K, similar to disordered $MnAu_3$[190]

MnBi: Hexagonal, NiAs($B8_1$)-type crystal structure, probably transforming above 630°K to related Ni_2In($B8_2$)-type structure, as discussed below. After early reports that MnBi is *ferromagnetic*[191] with Curie point near 630°K,[192,193] it was found that the ferromagnetic moment vanishes abruptly at this temperature and then upon slow cooling reappears abruptly at a temperature 15 to 20°K lower.[194,195] In Fig. 15 derived from Reference 195, solid σ_s versus T curves indicate sharpness and temperature hysteresis of this transition and, when extrapolated to 0°K, give $\sigma_0 = 84$ emu/gm ($\mu_F = 3.95\ \mu_B$ per Mn atom). This μ_F value was also obtained from neutron diffraction data;[196] lower values ranging down to 2.74 μ_B per Mn atom[197] were reported earlier.[193,194,197] Magnetic transition at 630°K is accompanied by discontinuous lattice-parameter changes, a_0 increasing by 1.5 per cent and c_0 decreasing by 3 per cent with increasing temperature, but essentially with no net volume change;[195,198] only c-parameter change was observed in earlier X-ray diffraction experiment.[194] Antiferromagnetic state was proposed for temperatures above 630°K transition, the magnetic disordering temperature being identified with specific heat anomaly at about 720°K.[194] However, magnetic measurements[195] above 630°K and after quenching to 300°K give, respectively, the $1/\chi$ versus T curve and the dashed σ_s versus T curve shown in Fig. 15, which together describe a ferromagnet with a lower T_c and σ_0 than the equilibrium phase below 630°K. For this high-temperature phase (which can be retained down to 300°K only by quenching), neutron diffraction data[196] at 653°K suggest a modified NiAs structure, in which about 10 per cent of the Mn atoms have gone into interstitial sites of the kind occupied in Ni_2In structure; at about 720°K this phase decomposes to α-Mn and molten Bi.

The energy required to saturate a MnBi crystal along hard direction of magnetization was measured as function of temperature;[199] the results are represented by solid curve in Fig. 16. This energy can be equated to $K_1 + K_2$, the sum of first two magnetocrystalline anisotropy coefficients. Above 85°K when the anisotropy energy is positive, the easy axis of magnetization is the c-axis of crystal; at 293°K it was resolved into $K_1 = +8.9 \times 10^6$ and $K_2 = +2.7 \times 10^6$, both in erg/cm^3. Below 85°K, when this anisotropy energy becomes negative, the easy directions of magnetization can be expected to have rotated into the basal plane, but neutron diffraction measurements[196] indicate that this rotation is incomplete, suggesting perhaps that $K_1 + K_2$ is still positive (although small) at very low temperatures, as shown by dashed curve in Fig. 16, and is opposed by small negative shape anisotropy of nonspherical particles in crystal specimen.[200] At room temperature, the magnetostriction measured parallel to 22 kOe field applied in basal plane was -2.5×10^{-4}, and was extrapolated to -8.0×10^{-4} at saturating fields; magnetostrictions measured perpendicular to applied field were somewhat smaller.[201] Recent measurements between 77 and 300°K show that crystal anisotropy and linear magnetostriction of MnBi should be represented by three-coefficient expressions.[202]

Coercive field for MnBi powder was found to be extremely sensitive to average particle size, rising from 600 Oe for 100 μ diameter particles to 12,000 Oe for 3μ diameter particles.[199] This behavior was attributed to single-domain particles, whose intrinsic coercivity would have an upper limit equal to

[191] Heusler, F., *Z. Angew. Chem.* **17** (1904) 260.
[192] Hilpert, S., and T. Dieckmann, *Berlin. Deut. Chem. Ges.* **44** (1911) 2831.
[193] Thielmann, K., *Ann. Phys.* **37** (1940) 41.
[194] Guillaud, C., *J. Phys. Radium* **12** (1951) 143, 223.
[195] Heikes, R. R., *Phys. Rev.* **99** (1955) 446.
[196] Roberts, B. W., *Phys. Rev.* **104** (1956) 607.
[197] Galperin, F., *Dokl. Akad. Nauk.* (*SSSR*) **75** (1950) 647.
[198] Willis, B. T. M., and H. P. Rooksby, *Proc. Phys. Soc.* (*London*) **B67** (1954) 290.
[199] Guillaud, C., Thesis, University of Strasbourg, 1943. Discussed in R. M. Bozorth, *Ferromagnetism*. D. Van Nostrand Co., New York, 1951.
[200] Jacobs, I. S., and C. P. Bean, as reported in Reference 196.
[201] Williams, H. J., R. C. Sherwood, and O. L. Boothby, *J. Appl. Phys.* **28** (1957) 445.
[202] Albert, P. A., and W. J. Carr, Jr., *J. Appl. Phys. Suppl.* **32** (1961) 201.

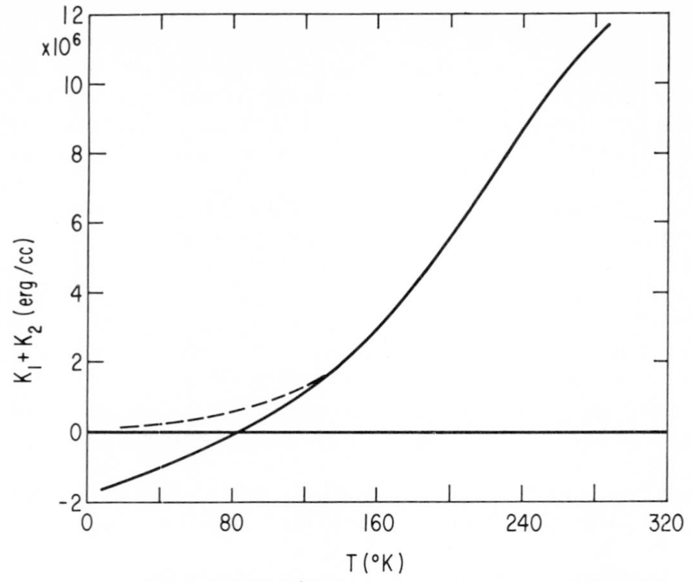

Fig. 16. Magnetic anisotropy energy of MnBi measured as function of temperature (solid curve) and proposed variation of magnetocrystalline anisotropy energy at low temperatures (dashed curve).

$2K_1/M_s$, or about 30,000 Oe.[203] *Permanent-magnet* properties achieved[204] with MnBi powder are $H_c = 3400$ Oe, $B_R = 4300$ G, $(BH)_{\max} = 4.3 \times 10^6$ G–Oe; the fabrication process involves hot-pressing of powder in very high field,[205] which presumably aligns the easy axes of different particles.

Mn_3Ga: Hexagonal, $Ni_3Sn(D0_{19})$-type crystal structure.[206] Weakly *ferrimagnetic*, as indicated by hyperbolic shape of $1/\chi$ versus T curve above T_c (= 743°K) and also by small moment per Mn atom (0.02 μ_B) deduced from saturation magnetization (about 2 emu/gm at 0°K) compared to Mn moment of 3.1 μ_B derived from paramagnetic susceptibility.[206]

"Mn_3Ge": Hexagonal, $Ni_3Sn(D0_{19})$-type crystal structure. This phase is limited to nonstoichiometric compositions near $Mn_{3.3}Ge$ and is retained at low temperatures only by quenching from above about 900°K.[207] Susceptibility of $Mn_{3.4}Ge$ compound rises slowly with increasing temperature up to 350°K, when it begins to drop very rapidly; linear $1/\chi$ versus T curve above 350°K suggests that this is an *antiferromagnetic* Curie point and, moreover, it gives $\theta = +210$°K and $\mu_P = 2.89\ \mu_B$ per Mn atom.[207] Susceptibility maximum previously reported[208] at 139°K for $Mn_{3.25}Ge$ was later attributed to high-temperature Mn_5Ge_2 phase.[207]

Long annealing at 700°K results in transformation to new tetragonal structure (possibly Ti_3Cu-type).[209] Its saturation magnetization extrapolated to 0°K gives $\sigma_0 = 11$ emu/gm ($\mu_F = 0.15\ \mu_B$ per Mn atom) and rises with increasing temperature, reaching maximum of 17 emu/gm at about 970°K before decreasing (presumably because of transformation to hexagonal phase); this unusual temperature dependence of magnetization and the low value for μ_F are considered evidence of *ferrimagnetic* state.[207,209]

Mn_5Ge_2: Hexagonal crystal structure, probably related to $Ni_2In(B8_2)$ type, retained at low temperatures only by quenching from

[203] Guillaud, C., *Compt. Rend.* **229** (1949) 992.
[204] Adams, E., W. M. Hubbard, and A. M. Syeles, *J. Appl. Phys.* **23** (1952) 1207.
[205] Adams, E., *Rev. Mod. Phys.* **25** (1953) 306.
[206] Tsuboya, I., and M. Sugihara, *J. Phys. Soc. Japan* **18** (1963) 143.
[207] Ohoyama, T., *J. Phys. Soc. Japan* **16** (1961) 1995.
[208] Yasukochi, K., T. Ohoyama, and K. Kanematsu, *J. Phys. Soc. Japan* **14** (1959) 1820.
[209] Ohoyama, T., K. Yasukochi, and K. Kanematsu, *J. Phys. Soc. Japan* **16** (1961) 352.

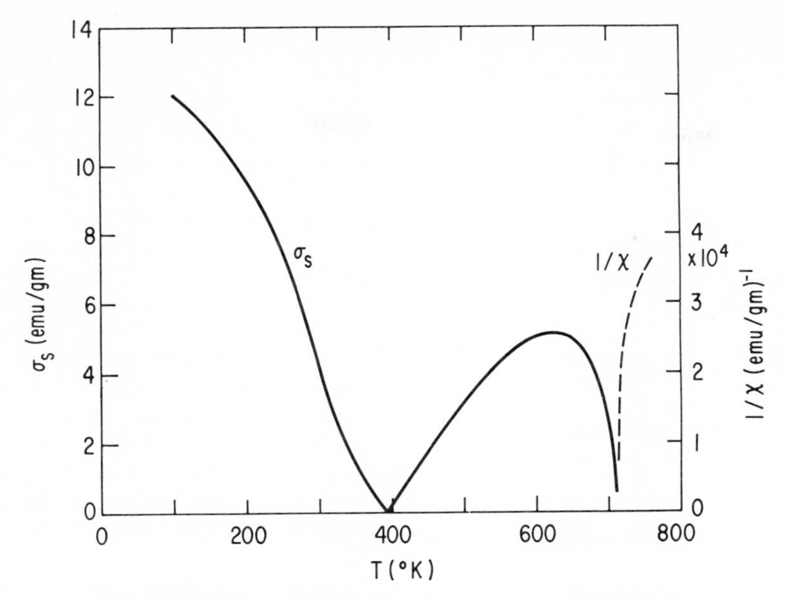

Fig. 17. Saturation magnetization and inverse susceptibility versus temperature for Mn_5Ge_2.

above about 1100°K.[210] Susceptibility maximum at about 150°K is suggestive of *antiferromagnetic* Curie point, although complex χ versus T curve obtained at higher temperatures is not typical for paramagnetic state.[210]

Long annealing at about 770°K results in transformation to new phase of undetermined structure.[210] Its saturation magnetization versus temperature curve,[211] reproduced in Fig. 17, shows σ_s dropping to essentially zero at 395°K, then rising, and finally falling off at Curie point of 710°K. Low-field measurements[211] show reversal of magnetization direction as temperature is varied through 395°K, indicating that this temperature is a *compensation point* for a so-called *N*-type *ferrimagnet*; the curvature of $1/\chi$ versus T curve above T_c, shown in Fig. 17, was also interpreted as ferrimagnetic behavior.[211]

Mn_5Ge_3: Hexagonal, $Mn_5Si_3(D8_8)$-type crystal structure. *Ferromagnetic* (or possibly *ferrimagnetic*); σ_s versus T data[212] give $T_c = 293$°K and by extrapolation to 0°K, $\sigma_0 = 105$ emu/gm ($\mu_F = 1.85\ \mu_B$ per Mn atom), whereas from earlier measurements,[213] $T_c = 320$°K and $\mu_F = 2.5\ \mu_B$ per Mn atom. Above T_c the $1/\chi$ versus T curve becomes linear, giving $\theta = +335$°K and $\mu_P = 2.56\mu_B$ per Mn atom.[212] The Curie point was found to increase with increasing hydrostatic pressure at the rate of 0.43°K per kilobar.[214]

The magnetocrystalline anisotropy is quite large (1.2×10^6 erg/cm^3) even at room temperature; this strong anisotropy is manifested in simple magnetic domain patterns observed.[215] Recent torque data give anisotropies of 0.3 and 4.2 ($\times 10^6$ erg/cm^3) at 300° and 77°K, respectively, the *c*-axis being the easy axis.[216]

Mn_3Ge_2: Crystal structure unknown. Weakly *ferrimagnetic* with T_c near 300°K, but with decreasing temperature, magnetization in fixed field decreases rapidly after reaching maximum value at about 173°K and then levels off below 113°K.[217,218] This critical

[210] Ohoyama, T., *J. Phys. Soc. Japan* **16** (1961) 1995.
[211] Yasukochi, K., K. Kanematsu, and T. T. Ohoyama, *J. Phys. Soc. Japan* **15** (1960) 932.
[212] Fontaine, R., and R. Pauthenet, *Compt. Rend.* **254** (1962) 650.
[213] Castelliz, L., *Z. Metallk.* **46** (1955) 198.
[214] Bloch, D., and R. Pauthenet, *Compt. Rend.* **254** (1962) 1222.
[215] Szczeniowski, S. E., and A. Wrzeciono, *J. Phys. Soc. Japan* **17** Suppl. B-I (1962) 647.
[216] Tawara, Y., and K. Sato, *J. Phys. Soc. Japan* **18** (1963) 773.
[217] Margolin, S. D., and I. G. Fakidov, *Phys. Metal. Metallog* (*USSR*)(*English Transl.*) **7,** 1 (1959) 153.
[218] Fakidov, I. G., and Y. N. Tsiovkin, *Phys. Metal. Metallog.* (*USSR*)(*English Transl.*) **7,** 5 (1959) 47.

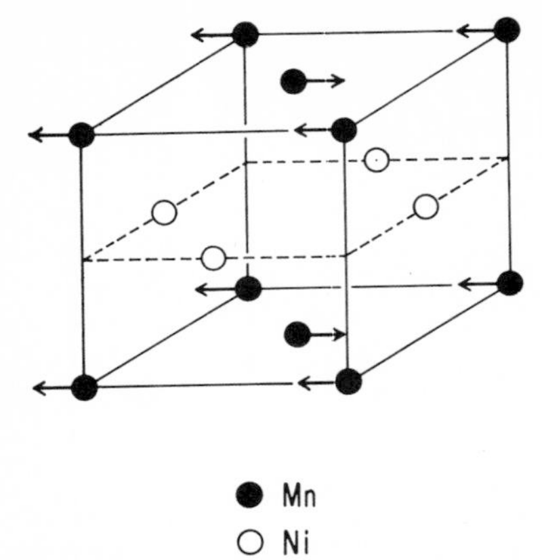

Fig. 18. Antiferromagnetic structure of MnNi. Absolute moment directions in basal plane undetermined.

behavior at $T_{crit} \simeq 150°K$, originally observed in Mn–Ge alloy of 70 at.% Ge[219] and later attributed to Mn_3Ge_2 compound,[217] is considered indicative of first-order transition, whose latent heat and temperature hysteresis were also detected.[217] Linear $1/\chi$ versus T curve above T_c gives $\theta = 300°K$ and $\mu_P = 2.5\ \mu_B$ per Mn atom.[218] Subsequent magnetic measurements[220] confirm these results and find that the magnetization between T_{crit} and T_c has small saturable component (with maximum value 1.56 emu/gm at 180°K) and a component essentially proportional to field, and that below T_{crit} only the latter component prevails. It is suggested that the 150°K transition involves change from an antiferromagnetic state to a weakly ferrimagnetic (or imperfectly antiferromagnetic) state with increasing temperature.[220] The field dependence of T_{crit} was found to be −0.48°K per kilo-oersted.[221]

MnHg: Cubic, CsCl(B2)-type crystal structure, which according to recent X-ray diffraction data[222] becomes tetragonally distorted below 198°K. Susceptibility maximum and electrical resistivity anomaly were also observed[223] at 198°K, suggesting that crystal distortion is associated with long-range *antiferromagnetic* order below this temperature.

Mn_3In: Cubic, γ-brass($D8_2$)-type crystal structure. Recent X-ray diffraction data[224] confirm earlier report[225] that Mn_3In is the only intermetallic compound in Mn–In system. Magnetic measurements[224] indicate Mn_3In is *ferrimagnetic* below T_c of 583°K; its saturation magnetization at 77°K is 3 emu/gm, or $\mu_F = 0.05\ \mu_B$ per Mn atom. Ferrimagnetism, rather than ferromagnetism, is strongly suggested by hyperbolic form of $1/\chi$ versus T curve above T_c and by the fact that the linear high-temperature asymptote of this curve gives a negative θ (−400°K) and $\mu_P = 3.7\ \mu_B$ per Mn atom, or $gS = 2.8$ (assuming $g = 2$), which is considerably larger than μ_F/μ_B. These results,[224] structural and magnetic, disagree with reports that ferromagnetism in this system is attributable to compounds, Mn_4In[226] or Mn_2In[227]

MnNi: Tetragonal, CuAu($L1_0$)-type crystal structure. *Antiferromagnetic* configuration, determined from neutron diffraction data at 77°K[228] is shown in Fig. 18. Nearest-neighbor Mn atomic moments (4.0 μ_B in magnitude) are antiparallel to each other and lie perpendicular to tetragonal c-axis of crystal; atomic moment of Ni indeterminate but probably very small. Essentially no change was observed in the "magnetic" reflection intensities of the neutron diffraction spectrum when the temperature was increased to 600°K, indicating Curie point is well above this temperature. Very high Curie point was also suggested by temperature-independence of susceptibility from 1.8° to 300°K.[228]

$MnNi_3$: Cubic, $Cu_3Au(L1_2)$-type crystal structure. Neutron diffraction and magnetization data for highly ordered $MnNi_3$ at room temperature give $\mu_{Mn} = 3.18(\pm 0.25)\ \mu_B$ and $\mu_{Ni} = 0.30\ (\pm 0.05)\ \mu_B$ for the *ferromagnetically* aligned atomic moments; an alternate but less likely ferrimagnetic solution (with μ_{Mn} antiparallel to μ_{Ni}) is also allowed by this data.[229] The Curie point of the stoichiometric

[219] Margolin, S. D., and I. G. Fakidov, *Phys. Metal. Metallog.* (*USSR*)(*English Transl.*) **5**, 2 (1957) 153.
[220] Fontaine, R., and R. Pauthenet, *Compt. Rend.* **254** (1962) 650.
[221] Flippen, R. B., and F. J. Darnell, *J. Appl. Phys.* **34** (1963) 1094.
[222] Nakogawa, Y., and T. Hori, *J. Phys. Soc. Japan* **17** (1962) 1313.
[223] Nakogawa, Y., and T. Hori, *J. Phys. Soc. Japan* **16** (1961) 1470.
[224] Aoyagi, K., and M. Sugihara, *J. Phys. Soc. Japan* **17** (1962) 1072.
[225] Zwicker, U., *Z. Metallk.* **41** (1950) 399.
[226] Valentiner, S., *Z. Metallk.* **44** (1953) 259.
[227] Goeddel, W. V., and D. M. Yost, *Phys. Rev.* **82** (1951) 555.
[228] Kasper, J. S., and J. S. Kouvel, *J. Phys. Chem. Solids* **11** (1959) 231.
[229] Shull, C. G., and M. K. Wilkinson, *Phys. Rev.* **97** (1955) 304.

$MnNi_3$ compound is about 750°K,[230–235] whereas for the disordered fcc alloy of this composition: T_c = 132°K.[236] For the ordered alloy at room temperature, saturation magnetization (M_s) values ranging from 560 emu/cm³ (or $\sigma_s \simeq 68$ emu/gm) to 850 emu/cm³ (or $\sigma_s \simeq 103$ emu/gm) have been reported,[230–233,235,237,238] probably reflecting different degrees of atomic order; the highest value obtained for M_0 by extrapolation to 0°K is about 870 emu/cm³ [237] ($\sigma_0 \simeq 105$ emu/gm, $\mu_F \simeq 1.1\ \mu_B$ per average atom). Disordered $MnNi_3$ has much lower magnetization with unusual temperature and field dependence, which has been attributed to complex ferro-antiferromagnetic state.[236,239,240]

Recent magnetostriction measurements on ordered $MnNi_3$ crystal[241] give $\lambda_{100} = -3.7 \times 10^{-6}$ and $\lambda_{111} = -0.5 \times 10^{-6}$ at 290°K and much larger negative values for both at lower temperatures; these results also suggest that λ_{100} and λ_{111} are positive above 370°K, consistent with earlier magnetostriction data on polycrystalline specimen.[235]

Coercive fields of almost 100 Oe[242] and large magnetic after effect (or viscosity)[243] observed in partially ordered $MnNi_3$, which according to electron microscopy consists of many small prolate-ellipsoidal single-domain regions,[243] presumably highly ordered and surrounded by disordered material.

[230] Kaya, S., and A. Kussman, *Z. Physik.* **72** (1931) 293.
[231] Kaya, S., and M. Nakayama, *Proc. Math.-Phys. Soc. Japan* **22** (1940) 126.
[232] Thompson, N., *Proc. Phys. Soc.* (*London*) **52** (1940) 217.
[233] Guillaud, C., *Compt. Rend.* **219** (1944) 614.
[234] Köster, W., and W. Rauscher, *Z. Metallk.* **39** (1948) 178.
[235] Taoka, T., and T. Ohtsuka, *J. Phys. Soc. Japan* **9** (1954) 723.
[236] Kouvel, J. S., C. D. Graham, Jr., and J. J. Becker, *J. Appl. Phys.* **29** (1958) 518.
[237] Piercy, G. R., and E. R. Morgan, *Can. J. Phys.* **31** (1953) 529.
[238] Hahn, R., and E. Kneller, *Z. Metallk.* **49** (1958) 426.
[239] Kouvel, J. S., C. D. Graham, Jr., and I. S. Jacobs, *J. Phys. Radium* **20** (1959) 198.
[240] Kouvel, J. S., and C. D. Graham, Jr., *J. Phys. Chem. Solids* **11** (1959) 220.
[241] Yamamoto, M., and T. Nakamichi, *J. Phys. Soc. Japan* **17** (1962) 588.
[242] Volkenstein, N., and A. Komar, *Zh. Eksp. i Teoret. Fiz.* (*SSSR*) **11** (1941) 723.
[243] Taoka, T., *J. Phys. Soc. Japan* **11** (1956) 537.

"Mn_2Pd_3": Complex tetragonal crystal structure, referred to as β_2-phase.[244] Compound of composition $Mn_{0.38}Pd_{0.62}$ found to have susceptibility maximum at 493°K, which was identified as *antiferromagnetic* Curie point.[245] However, field dependence of susceptibility and thermoremanence effects observed below this temperature[246] suggest that the ordered magnetic state is probably *weakly ferrimagnetic*.

$MnPt_3$: Cubic, $Cu_3Au(L1_2)$-type crystal structure. *Ferromagnetic*; neutron diffraction and magnetization data at 77°K give μ_{Mn} = 3.60 (±0.09) μ_B and μ_{Pt} = 0.17 (±0.04) μ_B for the parallel-aligned atomic moments.[247] Curie point is about 370°K for stoichiometric $MnPt_3$ and increases rapidly with increasing per cent of Mn; saturation magnetization (M_s) for $MnPt_3$ composition is about 640 and 320 emu/cm³ at 80°K and 293°K, respectively.[248]

MnRh: Cubic, CsCl(B2)-type crystal structure, recently found to transform below 170°K to a tetragonal, $CuAu(L1_0)$-type structure.[249] Magnetic susceptibility and electrical resistivity measurements indicate that cubic-MnRh is paramagnetic down to the transformation temperature (linear $1/\chi$ versus T curve giving $\theta \simeq -260$°K and $\mu_P \simeq 3.3\ \mu_B$ per atom) and that tetragonal-MnRh is probably *antiferromagnetic* with a Curie point well above the transformation temperature.[249] Recent neutron diffraction experiment[250] at 77°K confirms antiferromagnetism of tetragonal-MnRh; magnetic structure is identical to that shown in Fig. 18 for MnNi with $\mu_{Mn} \simeq 2.3\ \mu_B$ and μ_{Rh} indeterminate but probably very small.

Mn_2Sb: Tetragonal, $Cu_2Sb(C38)$-type crystal structure. Neutron diffraction measurements[251] at 295°K establish *ferrimagnetic* configuration shown in Fig. 19*a*, in which

[244] Raub, E., and W. Mahler, *Z. Metallk.* **45** (1954) 430.
[245] Wendling, R., *Compt. Rend.* **252** (1961) 3207.
[246] Wendling, R., *Compt. Rend.* **253** (1961) 408.
[247] Pickart, S. J., and R. Nathans, *J. Appl. Phys. Suppl.* **33** (1963) (1962) 1336.
[248] Auwärter, M., and A. Kussmann, *Ann. Physik.* **7** (1950) 169.
[249] Kouvel, J. S., C. C. Hartelius, and L. M. Osika, *J. Appl. Phys.* **34** (1963) 1095.
[250] Kasper, J. S., private communication.
[251] Wilkinson, M. K., N. S. Gingrich, and C. G. Shull, *J. Phys. Chem. Solids* **2** (1957) 289.

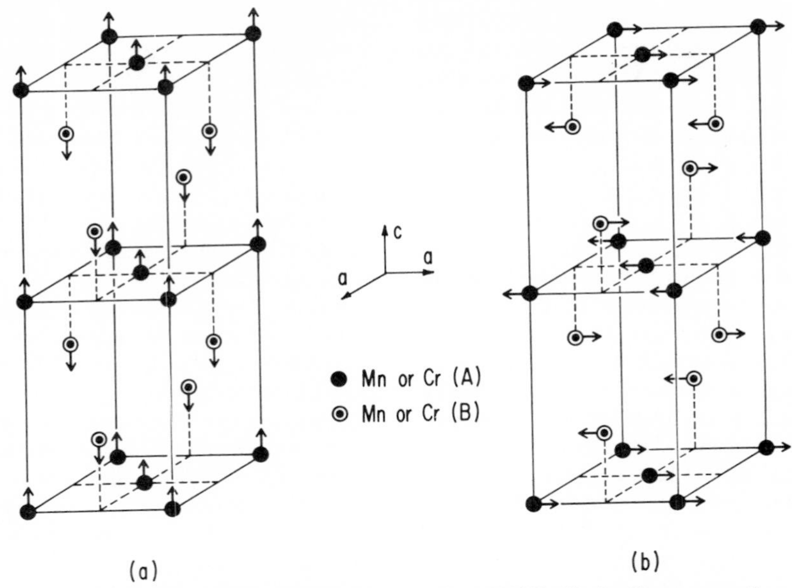

Fig. 19. (*a*) Ferrimagnetic structure of Mn_2Sb at 295°K. (*b*) Antiferromagnetic structure of $Mn_{1.9}Cr_{0.1}Sb_{0.95}In_{0.05}$ below first-order transition temperature; absolute moment directions in basal plane unknown. Sb and In atoms omitted for clarity.

moments of Mn atoms in *A* and *B* sites are unequal in magnitude and antiparallel to each other along tetragonal *c*-axis; similar measurements down to 77°K indicate that below about 240°K the magnetic axis has shifted to basal plane (ferrimagnetic configuration remaining otherwise the same) and that the *B*-site moment is larger and increases more rapidly with decreasing temperature than the *A*-site moment, extrapolation to 0°K giving $\mu_{Mn(A)} = -2.13\ (\pm 0.20)\mu_B$ and $\mu_{Mn(B)} = +3.87\ (\pm 0.20)\ \mu_B$. Recent diffraction experiments with polarized neutrons[252] give comparable values for the moments and indicate that the unpaired spin distribution is aspherical (and, for the *B* site, not centered exactly about the atomic nucleus). These results support earlier suggestion[253] that below Curie point of 550°K Mn_2Sb is ferrimagnetic (rather than ferromagnetic); this suggestion was based on hyperbolic form of $1/\chi$ versus *T* curve above T_c[254] and on unusual temperature dependence of saturation moment (σ_s).[255] Extrapolation of σ_s versus *T* curve to 0°K gave $\sigma_0 = 45.2$ emu/gm, or 0.936 μ_B per average Mn atom,[255] which is somewhat higher than value deduced from neutron diffraction data.

The magnetocrystalline anisotropy (*K*) is predominantly uniaxial and positive (about 2×10^5 emu/cm^3) at room temperature; with decreasing temperature, it decreases to zero at about 240°K, when it reverses in sign and then increases in negative value, reaching about -15×10^5 emu/cm^3 when extrapolated to 0°K.[255] This sign reversal in *K*, which is consistent with shift in magnetic axis already mentioned, was also demonstrated by measurements[256] showing that the initial susceptibility rises to sharp maximum at 240°K and is inversely proportional to $|K|$ above and below this temperature, as predicted theoretically. Coercive field measurements[257] on fine powders of Mn_2Sb give H_c of 16 Oe at 239°K and 280 Oe at 292°K; over this temperature range, H_c was found to be proportional to K/M_s, indicative of single-domain behavior.

Isomorphous mixed compounds of composition $Mn_{2-x}Cr_x\ Sb_{0.95}In_{0.05}$ were recently

---

[252] Alperin, H. A., P. J. Brown, and R. Nathans, *J. Appl. Phys.* **34** (1963) 1201.

[253] Guillaud, C., *Compt. Rend.* **235** (1952) 468.

[254] Serres, A., *J. Phys. Radium* **8** (1947) 146.

[255] Guillaud, C., Thesis, University of Strasbourg, 1943.

[256] Guillaud, C., R. Bertrand, and R. Vautier, *Compt. Rend.* **228** (1949) 1403.

[257] Guillaud, C., *Compt. Rend.* **229** (1949) 818.

found[258] to exhibit abrupt transition at temperature T_s varying from about 125°K for $x = 0.025$ to about 400°K for $x = 0.20$ (whereas T_c decreases only slightly with increasing x), below which magnetization has dropped to very low value. X-ray diffraction study[258] showed large contraction of c_0 and small expansion of a_0 resulting in volume decrease of about 0.18 per cent (for $x = 0.12$ compound) as temperature is lowered through T_s. *Antiferromagnetic* state predicted[258] below T_s was found by neutron diffraction study[259] of $x = 0.1$ compound and is described by configuration shown in Fig. 19*b*, in which magnetic axis lies in basal plane. Moments on adjacent basal-plane sheets of B sites are antiparallel, rather than parallel as in ferrimagnetic structure above T_s (Fig. 19*a*), consistent with proposed "exchange inversion" model[260] in which a net exchange coupling varies rapidly and reverses in sign with changes in atomic spacing. First-order antiferromagnetic-ferrimagnetic transition was also observed in ternary compounds $Mn_{2-x}Cr_xSb$[261] and subsequently in $Mn_{2-x}A_xSb$ ($A \equiv$ V, Co, Cu) and $Mn_2Sb_{1-x}Z_x$ ($Z \equiv$ Ge, As).[262] Moreover, neutron diffraction experiments indicate that the transition in $Mn_{2-x}Cr_xSb$ for $x \leq 0.03$ involves intermediate state with spiral magnetic structure.[263] Critical temperature (T_s) of $Mn_{2-x}Cr_xSb$ compounds was shown to decrease with increasing field; for $x = 0.1$ compound with T_s near room temperature: $dT_s/dH \simeq 0.4$°K/kOe, from which a latent heat of about 0.4 cal/gm was deduced.[264] Pressure dependence of T_s for this compound was measured to be about +3°K/kb, consistent with above latent heat value.[261]

MnSb: Hexagonal, NiAs($B8_1$)-type crystal structure for stoichiometric compound; this phase also exists at Mn-rich compositions in which excess Mn atoms are presumably located in B sites of related Ni_2In structure (see Fig. 22 for Mn_2Sn). Magnetic measurements[265,266] show that stoichiometric MnSb is *ferromagnetic* with T_c of 587°K and a saturation magnetization extrapolated to 0°K of about 110 emu/gm or $\mu_F = 3.5\ \mu_B$ per Mn atom. Linear $1/\chi$ versus T curve[266] above T_c gives $\mu_P = 4.48\ \mu_B$ per Mn atom or $gS = 3.6$ (assuming $g = 2$), in good agreement with μ_F value; earlier measurement, however, gave $\mu_P = 4.10\ \mu_B$ per Mn atom.[267] X-ray diffraction study of thermal expansion shows that a_0 versus T curve has pronounced inflection near T_c, whereas c_0 rises uniformly with increasing temperature.[268]

Recent neutron diffraction and magnetic measurements[269] on annealed MnSb specimen give agreement with above values for μ_F and T_c and indicate that magnetic moments lie in basal plane below 520°K but along *c*-axis at higher temperatures (up to T_c); similar experiments on quenched specimen give μ_{Mn} of 3.37 μ_B and T_c of 565°K and indicate that the shift in moment direction begins below room temperature. The low values obtained for μ_{Mn} and T_c of quenched specimen and the still lower μ_{Mn} value of 3.3 μ_B deduced from previous neutron diffraction data[270] was attributed[269] to material being composed of Mn-rich MnSb phase plus free Sb. Earlier work[265] had already established that excess Mn reduces saturation moment and Curie point of this phase.

MnSe: Cubic, NaCl(B1)-type crystal structure. *Antiferromagnetic* Curie point can be identified with temperature of susceptibility maximum[271–273] which is roughly 130°K and is a sensitive function of thermal history.[273] Specific heat anomaly observed[274] at 247°K may be of crystallographic rather than

[258] Swoboda, T. J., W. H. Cloud, T. A. Bither, M. S. Sadler, and H. S. Jarrett, *Phys. Rev. Letters* **4** (1960) 509.

[259] Cloud, W. H., H. S. Jarrett, A. E. Austin, and E. Adelson, *Phys. Rev.* **120** (1960) 1969.

[260] Kittel, C., *Phys. Rev.* **120** (1960) 335.

[261] Cloud, W. H., T. A. Bither, and T. J. Swoboda, *J. Appl. Phys. Suppl.* **32** (1961) 55.

[262] Bither, T. A., P. H. L. Walter, W. H. Cloud, T. J. Swoboda, and P. E. Bierstedt, *J. Appl. Phys. Suppl.* **33** (1962) 1346.

[263] Bierstedt, P. E., F. J. Darnell, W. H. Cloud, R. B. Flippen, and H. S. Jarrett, *Phys. Rev. Letters* **8** (1962) 15.

[264] Flippen, R. B., and F. J. Darnell, *J. Appl. Phys.* **34** (1963) 1094.

[265] Guillaud, C., *Ann. physique* **4** (1949) 671.

[266] Hirone, T., S. Maeda, I. Tsubokawa, and N. Tsuya, *J. Phys. Soc. Japan* **11** (1956) 1083.

[267] Serres, A., *J. Phys. Radium* **8** (1947) 146.

[268] Willis, B. T. M., and H. P. Rooksby, *Proc. Phys. Soc. (London)* **B67** (1954) 290.

[269] Takei, W. J., D. E. Cox, and G. Shirane, *Phys. Rev.* **129** (1963) 2008.

[270] Pickart, S. J., and R. Nathans, *J. Appl. Phys. Suppl.* **30** (1959) 280.

[271] Squire, C. F., *Phys. Rev.* **56** (1939) 922.

[272] Bizette, H., and B. Tsai, *Compt. Rend.* **212** (1941) 75.

[273] Lindsay, R., *Phys. Rev.* **84** (1951) 569.

[274] Kelley, K. K., *J. Am. Chem. Soc.* **61** (1939) 203.

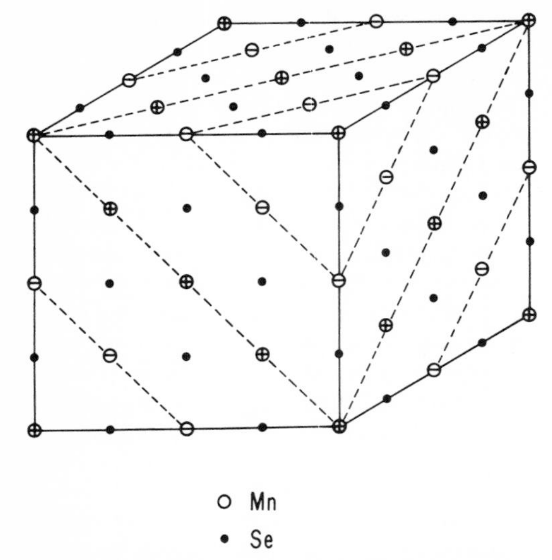

Fig. 20. Antiferromagnetic structure of MnSe; ferromagnetic sheets indicated by dotted lines. Orientation of magnetic axis undetermined.

magnetic origin because various phases (zinc blende and wurtzite,[275] NiAs[276]) have been reported as coexisting with the NaCl phase. Susceptibility data above room temperature gave linear $1/\chi$ versus T curve with $\mu_P = 5.6\ \mu_B$ per Mn atom[277,278] and $\theta = -361°$K[277] or $-297°$K.[278] According to qualitative report[279] on neutron diffraction study below Curie point, MnSe has the same antiferromagnetic structure as its isomorphs, MnO and MnS. This structure, shown in Fig. 20 with orientation of magnetic axis unspecified, suggests that antiferromagnetic coupling between next-nearest-neighbor Mn atoms is dominant; this coupling was estimated to be about 50 per cent stronger than that between each nearest-neighbor Mn atom pair, on basis of T_c and θ values.[278]

Magnetic[280] and neutron diffraction[281] measurements on isomorphous mixed compound $Mn_{0.9}Li_{0.1}Se$ show that it is *ferromagnetic* from about 110°K down to 70°K where it abruptly becomes antiferromagnetic (with configuration different from that of Fig. 20) or possibly weakly ferrimagnetic. The field-dependence of the critical temperature of this abrupt transition was found to be -0.19°K/kOe, from which a latent heat of 0.24 cal/gm was deduced.[282]

$MnSe_2$: Cubic, pyrite(C2)-type crystal structure. Neutron diffraction measurements[283,284] at 4.2°K give *antiferromagnetic* structure whose unit cell is illustrated in Fig. 21, where only the Mn atoms forming a fcc lattice are shown; the magnitude of each Mn moment is $5\ \mu_B$. In this configuration, each Mn atom has eight nearest neighbors with antiparallel moments and four with parallel moments, similar to the simpler antiferromagnetic configuration for $MnTe_2$ (see Fig. 25); unlike the latter, however, each Mn atom has some next-nearest neighbors with antiparallel moments. Susceptibility data[284] indicate Curie point in the vicinity of 77°K; linear $1/\chi$ versus T curve at higher temperatures gives $\theta = -483$°K and $\mu_P = 5.93\ \mu_B$ per Mn atom.

"Mn_3Sn": Hexagonal, $Ni_3Sn(D0_{19})$-type crystal structure; this phase exists only at nonstoichiometric compositions near $Mn_{3.5}Sn$. Unusually pronounced (though broad) susceptibility maximum found in $Mn_{3.67}Sn$ at about 365°K was interpreted tentatively as evidence for *antiferromagnetic* ordering; from linear $1/\chi$ versus T curve at higher temperatures: $\theta = -275$°K and $\mu_F = 3.58\ \mu_B$ per Mn atom.[285]

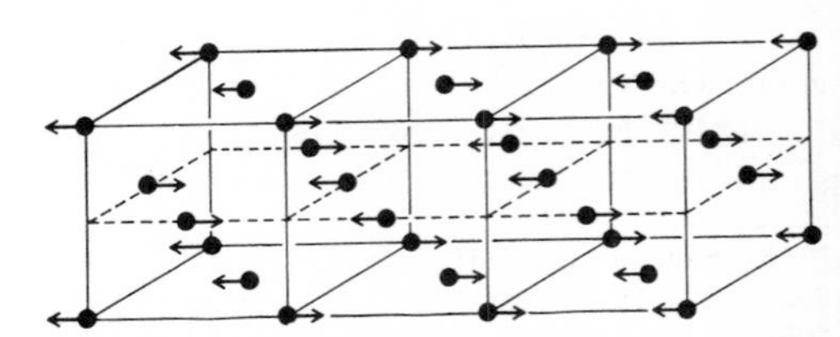

Fig. 21. Antiferromagnetic structure of $MnSe_2$; Se atoms omitted for clarity.

[275] Baroni, A., *Z. Krist.* **99** (1938) 336.
[276] Taylor, A., as reported in Ref. 280.
[277] Serres, A., *J. Phys. Radium* **8** (1947) 146.
[278] Banewicz, J. J., R. F. Heidelberg, and A. H. Luxem, *J. Phys. Chem.* **65** (1961) 615.
[279] Shull, C. G., W. A. Strauser, and E. O. Wollan, *Phys. Rev.* **83** (1951) 333.
[280] Heikes, R. R., T. R. McGuire, and R. J. Happel, Jr., *Phys. Rev.* **121** (1961) 703.
[281] Pickart, S. J., R. Nathans, and G. Shirane, *Phys. Rev.* **121** (1961) 707.
[282] Flippen, R. B., and F. J. Darnell, *J. Appl. Phys*, **34** (1963) 1094.
[283] Corliss, L. M., N. Elliott, and J. M. Hastings, *J. Appl. Phys.* **29** (1958) 391.
[284] Hastings, J. M., N. Elliott, and L. M. Corliss, *Phys. Rev.* **115** (1959) 13.
[285] Yasukōchi, K., K. Kanematsu, and T. Ohoyama, *J. Phys. Soc. Japan* **16** (1961) 1123.

Mn_2Sn: Hexagonal, $Ni_2In(B8_2)$-type crystal structure. This phase extends to nonstoichiometric compositions $Mn_{2-x}Sn$ where x represents Mn atom vacancies in B sites, as shown in Fig. 22; it was also reported[286,287] that some of the "Sn sites" may be occupied by Mn atoms. X-ray diffraction data on $Mn_{1.77}Sn$ compound annealed below 850°K show additional superstructure lines which suggest tripling of a_0 of unit cell.[287] Discovery[288] and other early reports[289–292] of ferromagnetism in various Mn–Sn alloys of unknown structure can probably be attributed to presence of Mn_2Sn phase. Recent measurements[293,294] show that saturation magnetization (extrapolated to 0°K) and Curie temperature increase from about 48 emu/gm and 256°K for $Mn_{1.8}Sn$ to about 68 emu/gm and 269°K for $Mn_{1.5}Sn$; correspondingly, μ_F (per Mn atom) increases from 1.76 μ_B to 2.38 μ_B. This rise in μ_F with increasing x in $Mn_{2-x}Sn$, corroborated by other recent work,[295] was interpreted as indicative of *ferrimagnetic* configuration shown in Fig. 22, in which the moments of Mn atoms in B sites are antiparallel to (and smaller in magnitude than) the Mn moments associated with A sites.[294,295] Ferrimagnetism of these compounds was also suggested by hyperbolic shape of $1/\chi$ versus T curve above T_c.[293,294] Heat treatment of $Mn_{1.77}Sn$ compound at temperatures below 850°K was found to raise both T_c and μ_F.[287]

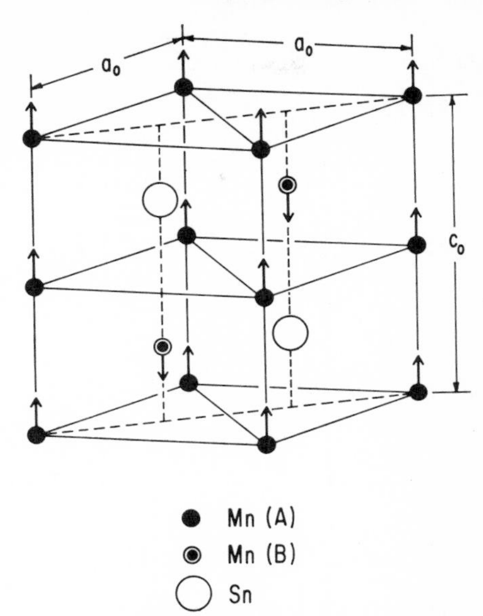

Fig. 22. Ferrimagnetic structure of Mn_2Sn. Orientation of magnetic axis unknown.

$MnSn_2$: Tetragonal, $CuAl_2(C16)$-type crystal structure. Susceptibility maximum was observed at 324°K and identified with *antiferromagnetic* Curie point;[296] however, measurements down to about 77°K showed an unusual rise in susceptibility with decreasing temperature[296,297] and another susceptibility peak at 86°K.[297] At still lower temperatures, the susceptibility drops extremely abruptly at 73°K and then remains constant down to 4.2°K.[297] Recent susceptibility data[298] indicating all these effects are shown in Fig. 23; the linear portion of $1/\chi$ versus T curve above 325°K gives $\theta = +160$°K and $\mu_P = 3.71\ \mu_B$ per Mn atom (or $gS = 2.90$, assuming $g = 2$), which are in fair agreement with previous values[296]: $\theta = +190$°K and $\mu_P = 3.41\ \mu_B$ per Mn atom.

In a recent neutron diffraction study,[299] it was deduced that from 73°K up to about 330°K $MnSn_2$ has the antiferromagnetic structure shown in Fig. 24 with a Mn moment of 2.36 μ_B at 77°K (which differs from the antiferromagnetic structure of its isomorph $FeSn_2$, shown in Fig. 9). An antiferromagnetic configuration was also deduced for temperatures below 73°K,[299] which is a simple modification of that found above 73°K in that the magnitude of the Mn moment is no longer constant but varies sinusoidally along the

[286] Nowotny, H., and K. Schubert, *Z. Metallk.* **37** (1946) 17.
[287] Yasukōchi, K., K. Kanematsu, and T. Ohoyama, *J. Phys. Soc. Japan* **16** (1961) 1123.
[288] Heusler, F., *Z. Angew. Chem.* **17** (1904) 260.
[289] Williams, R. S., *Z, Anorg. Allgem. Chem.* **55** (1907) 1.
[290] Honda, K., *Ann. Physik* **32** (1910) 1027.
[291] Potter, H. H., *Phil. Mag.* **12** (1931) 255.
[292] Guillaud, C., Thesis, University of Strasbourg, 1943.
[293] Asanuma, M., *J. Phys. Soc. Japan* **16** (1961) 1265
[294] Asanuma, M., *J. Phys. Soc. Japan* **17** (1962) 300
[295] Yasukōchi, K., K. Kanematsu, and T. Ohoyama, *J. Phys. Soc. Japan* **17** Suppl. B-I (1962) 165.
[296] Yasukōchi, K., K. Kanematsu, and T. Ohoyama, *J. Phys. Soc. Japan* **16** (1961) 1123.
[297] Kouvel, J. S., and C. C. Hartelius, *Phys. Rev.* **123** (1961) 124.
[298] Kouvel, J. S., and C. C. Hartelius, unpublished data.
[299] Corliss, L. M., and J. M. Hastings, *J. Appl. Phys.* **34** (1963) 1192.

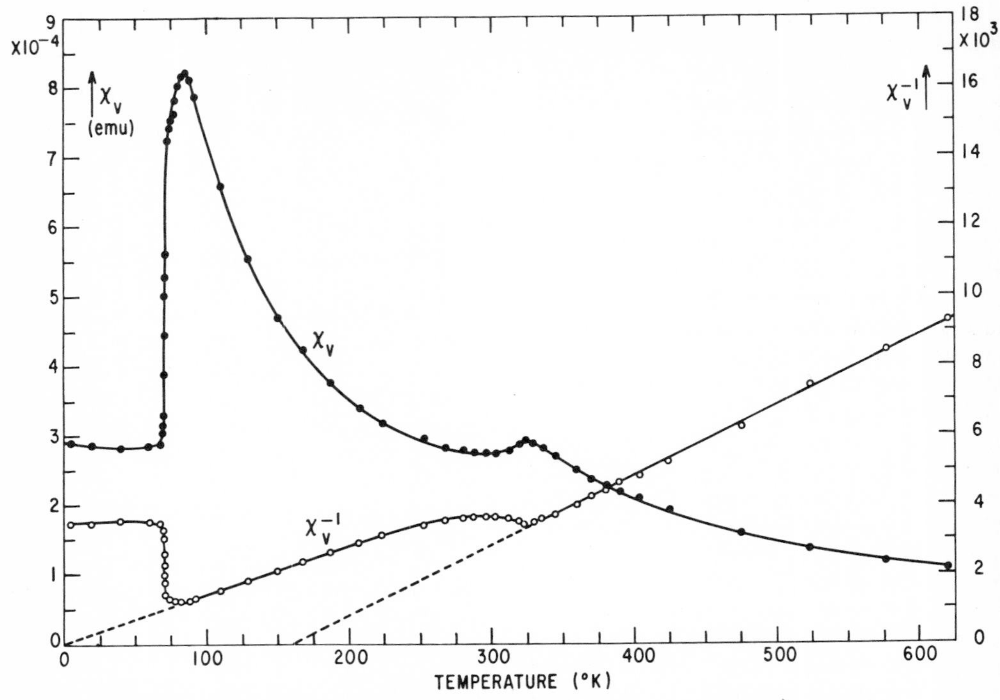

Fig. 23. Magnetic susceptibility, and its reciprocal, versus temperature for $MnSn_2$.

direction perpendicular to the ferromagnetically aligned sheets; the wavelength of the modulation is about 24 [subsequently corrected to four[300]] times the distance between adjacent ferromagnetic sheets, and the amplitude of the modulated moment is 3.11 μ_B at 4.2°K. Thus the abrupt transition at 73°K, which is marked by discontinuous changes in susceptibility, electrical resistivity,[297] and volume,[298] corresponds to a first-order transformation between different antiferromagnetic states. In that case, the rapid rise in susceptibility as the temperature is decreased to 86°K, previously interpreted as paramagnetic behavior,[297] remains an unexplained anomaly.

MnTe: Hexagonal, NiAs($B8_1$) type crystal structure. Susceptibility maximum[301] and anomalies in the temperature dependences of the electrical resistivity[302] and of the specific heat[303] were observed near 307°K, indicating that MnTe is *antiferromagnetic* below this temperature. Subsequent susceptibility measurements[304,305] place χ_{max} at about 325°K, but the more recent study[305] also finds an abrupt change in slope of χ versus T curve at 310°K. Various $1/\chi$ versus T curves obtained at higher temperatures give $\theta = -690$°K,[304] -650°K,[305]

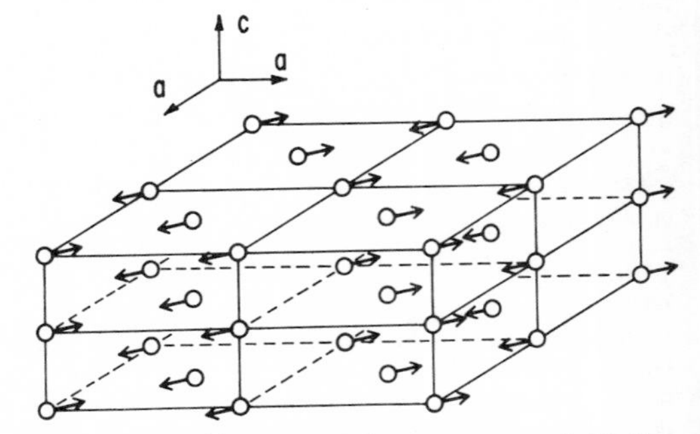

Fig. 24. Antiferromagnetic structure of $MnSn_2$ above first-order transition temperature. Sn atoms omitted for clarity.

[300] Corliss, L. M., and J. M. Hastings, private communication.
[301] Squire, C. F., *Phys. Rev.* **56** (1939) 922.
[302] Squire, C. F., *Phys. Rev.* **56** (1939) 960.
[303] Kelley, K. K., *J. Am. Chem. Soc.* **61** (1939) 203.
[304] Serres, A., *J. Phys. Radium* **8** (1947) 146.
[305] Uchida, E., H. Kondoh, and N. Fukuoka, *J. Phys. Soc. Japan* **11** (1956) 27.

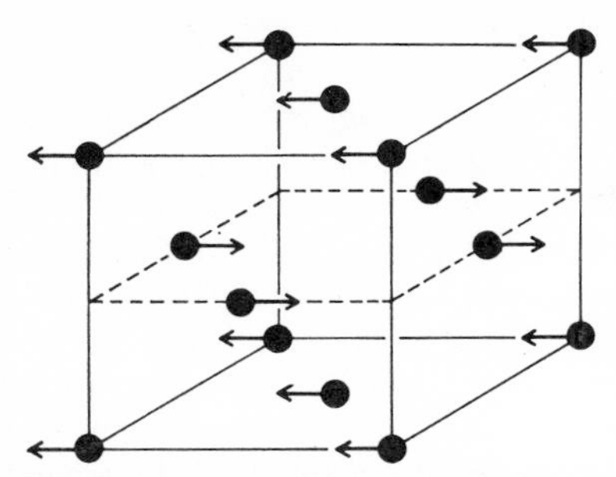

Fig. 25. Antiferromagnetic structure of $MnTe_2$; Te atoms omitted for clarity. Absolute moment directions in plane of ferromagnetic sheets are undetermined.

−584°K,[306] or −585°K,[307] and μ_P (per Mn atom) = 6.1 μ_B,[304] 5.1 μ_B,[305] or 4.81 μ_B.[307] Susceptibility and torque measurements on MnTe single crystal below T_c indicate antiparallel arrangement of atomic moments lying in basal plane.[307]

X-ray diffraction data[308] show that slope of c_0 versus T curve has abrupt change at 310°K, whereas a_0 varies smoothly with temperature; similar thermal expansion anomaly was observed earlier but at 329°K.[309] From values for the abrupt changes in dc_0/dT and in the specific heat at T_c, it was estimated[308] that T_c would increase with increasing pressure at the rate of 2.6°K per kilobar, in good agreement with dT_c/dp value of +2.0 (±0.4)°K/kilobar obtained directly from pressure measurements.[310]

$MnTe_2$: Cubic, pyrite(C2)-type crystal structure. *Antiferromagnetic* configuration determined from neutron diffraction data[311,312] at 4.2°K is illustrated in Fig. 25, where only the Mn atoms, forming a fcc lattice, are shown; each Mn moment is 5 μ_B in magnitude and lies in plane of ferromagnetic sheet. In this configuration, each Mn atom has eight nearest neighbors with antiparallel moments and four with parallel moments, whereas all six next-nearest neighbors have parallel moments, unlike $MnSe_2$ (see Fig. 21). Susceptibility data[312,313] indicate Curie point is below 77°K; from linear $1/\chi$ versus T curves at higher temperatures: $\theta = -528$°K[312] or −430°K,[313] $\mu_P = 6.22\ \mu_B$[312] or 5.68 μ_B[313] per Mn atom.

"MnZn": Cubic, CsCl(B2)-type crystal structure. This phase exists only at compositions near $Mn_{0.6}Zn_{0.4}$; although stable only at high temperatures, it can be retained at room temperature by quenching.[314] *Ferromagnetic*, with Curie point higher than 500°K. At room temperature, the magnetization in 10kOe field is about 40 emu/gm or 0.7 μ_B per Mn atom.[314] This low average moment per Mn atom would suggest *ferrimagnetic* ordering.

$MnZn_3$: Metastable phase with ordered hexagonal crystal structure, probably of $Ni_3Sn(D0_{19})$-type, produced by annealing at 370°K after quenching from 820°K.[315] This hexagonal compound is *ferromagnetic* with T_c higher than 400°K,[315] whereas the disordered (quenched) hcp alloy becomes ferromagnetic below about 250°K.[316] Earlier report[317] of ferromagnetism at room temperature in Mn–Zn alloys of about 30 at.% Mn probably can be attributed to presence of ordered hexagonal phase. Measurements at 98°K and 295°K on single crystal of hexagonal $MnZn_3$ compound show that it is more easily magnetized parallel rather than perpendicular to the c-axis, but the magnetization in both directions is still increasing with increasing field at 12 kOe.[316]

Further annealing at 370°K causes transformation to the stable phase having ordered cubic structure of Cu_3Au ($L1_2$) type, which was reported to be ferromagnetic below 140°K.[315] However, when complete transformation was achieved by annealing at 520°K,[316] the cubic $MnZn_3$ compound was found to exhibit a (nearly) field-independent

[306] Banewicz, J. J., R. F. Heidelberg, and A. H. Luxem, *J. Phys. Chem.* **65** (1961) 615.
[307] Komatsubara, T., M. Murakami, and E. Hirahara, *J. Phys. Soc. Japan* **18** (1963) 356.
[308] Grazhdankina, N. P., and D. I. Gurfel', *Sov. Phys. JETP English Transl.* **8** (1959) 631.
[309] Greenwald, S., *Acta Cryst.* **6** (1953) 396.
[310] Grazhdankina, N. P., *Sov. Phys. JETP English Transl.* **6** (1958) 1178.
[311] Corliss, L. M., N. Elliott, and J. M. Hastings, *J. Appl. Phys.* **29** (1958) 391.
[312] Hastings, J. M., N. Elliott, and L. M. Corliss, *Phys. Rev.* **115** (1959) 13.
[313] Uchida, E., H. Kondoh, and N. Fukuoka, *J. Phys. Soc. Japan* **11** (1956) 27.
[314] Nakagawa, Y., S. Sakai, and T. Hori, *J. Phys. Soc. Japan* **17** Suppl. B-I (1962) 168.
[315] Tezuka, S., S. Sakai, and Y. Nakagawa, *J. Phys. Soc. Japan* **15** (1960) 931.
[316] Nakagawa, Y., S. Sakai, and T. Hori, *J. Phys. Soc. Japan* **17** Suppl. B-I (1962) 168.
[317] Nowotny, H., and H. Bittner, *Monatsh. Chem.* **81** (1950) 887.

susceptibility with a maximum value at 150°K, suggestive of *antiferromagnetic* ordering; linear $1/\chi$ versus T curve above 200°K gave $\theta = +110$°K and $\mu_P = 3.7\ \mu_B$ per Mn atom. X-ray diffraction data showed there is also at 150°K the onset of a tetragonal lattice distortion.[316]

4.5 Other Iron-Group Compounds

Cu_2Sb: Prototype of tetragonal (C38) crystal structure. *Antiferromagnetic*; $T_c \simeq 373$°K determined from susceptibility maximum and specific heat anomaly.[318] $1/\chi$ versus T curve well above T_c gives $\theta = -1694$°K.

Ni_3Fe, see $FeNi_3$.

Ni_2Mg: Prototype of hexagonal (C36)-type crystal structure. *Ferromagnetic* with Curie point near 510°K.[319]

Ni_3Mn, see $MnNi_3$.

NiMn, see MnNi.

Ni_3Pt: Cubic, $Cu_3Au(L1_2)$-type crystal structure. *Ferromagnetic*. Annealed (ordered) stoichiometric Ni_3Pt alloy found to have Curie point of 288°K and σ_s of 22.4 emu/gm at 77°K, extrapolation to 0°K giving $\sigma_0 =$ 32.1 emu/gm or $\mu_F = 0.38\ \mu_B$ per atom; for quenched (disordered) fcc alloy, both T_c and μ_F are somewhat higher.[320]

VSe: Hexagonal, NiAs($B8_1$)-type crystal structure. Susceptibility maximum and specific heat anomaly observed at 163°K were cited as evidence for *antiferromagnetic* ordering; linear $1/\chi$ versus T curve above T_c of 163°K gave $\theta = -2570$°K and $gS = 3.2$ for vanadium (when assumed that $g = 2$).[321]

$ZrZn_2$: Cubic, $MgCu_2$(C15)-type crystal structure. *Ferromagnetic* with Curie point of 35°K and saturation moment (μ_F) of about 0.13 μ_B per molecule.[322]

5. BINARY COMPOUNDS OF URANIUM-GROUP METALS

$PuGe_2$: Tetragonal, α-$ThSi_2$-type crystal structure. *Ferromagnetic*, with Curie point of 34.5°K; magnetization measured at 4°K in 420 Oe field corresponded to 0.144 μ_B per molecule.[323]

UFe_2: Cubic, $MgCu_2$ (C15)-type crystal structure. *Ferromagnetic*. Magnetization measurements down to 104°K place Curie point at 195°K and by extrapolation to 0°K give $\sigma_0 =$ 18.0 emu/gm, corresponding to 1.13 μ_B per UFe_2 molecule.[324] However, from more recent magnetization data[325] taken down to 4.2°K: $T_c = 172$°K and $\sigma_0 = 16.24$ emu/gm (1.02 μ_B per molecule). Neutron diffraction measurements indicate that the magnetization of UFe_2 originates essentially from the ferromagnetic alignment of small Fe moments (i.e., about 0.5 μ_B per Fe atom), with some possibility of much smaller moments on U atoms.[326] Results of Mössbauer experiments also appear consistent with this small atomic moment for Fe in UFe_2.[327]

UGe_2: Orthorhombic crystal structure, probably related to tetragonal α-$ThSi_2$-type structure.[328] *Ferromagnetic*, with Curie temperature of 52.0°K; measurements at 4°K in 840 Oe field gave magnetization equivalent to 0.803 μ_B per molecule.[328]

UMn_2: Cubic, $MgCu_2$(C15)-type crystal structure. Susceptibility maximum was observed at 260°K and considered as evidence for *antiferromagnetic* ordering.[329] However, below 230°K the magnetization-versus-field curves become slightly nonlinear and hysteretic; although this effect diminished at very low temperatures, the susceptibility rose very steeply as the temperature was decreased to 4.2°K. Moreover, when the compound was cooled in a magnetic field, its hysteresis loops at 77°K and 4.2°K were found to be asymmetrical with respect to the origin. All these effects represent departures from ideal antiferromagnetism. From linear $1/\chi$ versus T curve above 260°K: $\theta = -1977$°K and $\mu_P =$ 6.6 μ_B per UMn_2 molecule.

[318] Yuzuri, M., *J. Phys. Soc. Japan* **15** (1960) 2007.
[319] Voss, G., *Z. Anorg. Allgem. Chem.* **57** (1908) 34.
[320] Marian, V., *Ann. physique* **7** (1937) 459.
[321] Tsubokawa, I., *J. Phys. Soc. Japan* **14** (1959) 196.
[322] Matthias, B. T., and R. M. Bozorth, *Phys. Rev.* **109** (1959) 604.
[323] Olsen, C. E., *J. Appl. Phys. Suppl.* **31** (1960) 340.
[324] Komura, S., N. Kunitomi, Y. Hamaguchi, and M. Sakamoto, *J. Phys. Soc. Japan* **16** (1961) 1486.
[325] Lin, S. T., and R. E. Ogilvie, *J. Appl. Phys.* **34** (1963) 1372.
[326] Hamaguchi, S. Komura, N. Kunitomi, and M. Sakamoto, *J. Phys. Soc. Japan* **17** (Suppl. B-III) (1961) 46.
[327] Komura, S., N. Kunitomi, Poh-Kun Tseng, N. Shikazono, and H. Takekochi, *J. Phys. Soc. Japan* **16** (1961) 1479.
[328] Olsen, C. E., *J. Appl. Phys. Suppl.* **31** (1960) 340.
[329] Lin, S. T., and A. R. Kaufmann, *Phys. Rev.* **108** (1957) 1171.

6. BINARY COMPOUNDS OF RARE-EARTH METALS

A majority of the rare earth compounds whose magnetic properties have been studied have one of three types of crystal structures: the cubic $MgCu_2$(C15)-type the hexagonal $MgZn_2$(C14)-type, or the hexagonal $CaCu_5$-type. Moreover, in most of these compounds of composition AB_2 or AB_5, where A is a rare earth element, the other constituent, B, has usually been a transition element of the iron, palladium or platinum groups. The results of several magnetic investigations[330–336] of the AB_2 compounds are summarized in Table 1, where experimental values reported for the Curie points and saturation moments of the ferromagnetic (or ferrimagnetic) compounds are listed. A similar summary of a few magnetic studies[337–339] of the AB_5 compounds appears in Table 2.

Many of the saturation moment values in Tables 1 and 2 become more meaningful when reference is made to the magnetic moment that can be expected from the unfilled $4f$ shell of each rare-earth atom. The angular momenta associated with the spin and the orbital motion of the $4f$ electrons are proportional respectively to the quantum numbers S and L. For the rare earths whose $4f$ shell is less than half-filled (i.e., with <7 electrons), the spin and orbital momenta are antiparallel to each other and the quantum number J representing the total angular momentum equals $L - S$. Conversely, for those with more than seven $4f$ electrons, the spin and orbital momenta are parallel to each other and $J = L + S$. The magnetic moment per atom (in μ_B) is obtained by multiplying J by the appropriate g-factor which is itself a function of S, L, and J.[340] Values for all these quantities representing the lowest energy states of the different rare-earth atoms are listed in Table 3; strictly speaking, these values are appropriate for the trivalent ions, which the rare-earth atoms in an intermetallic compound usually resemble. When the column of gJ values in this table is compared with the moments (in μ_B) shown in Table 1 for the rare-earth-aluminum compounds, the general agreement is quite good. Fairly good agreement is also found with the moment values given in Table 1 for the rare-earth compounds with Os and Ir, suggesting that even though the latter are transition group atoms they contribute very little if anything to the saturation moments.

The situation is clearly very different for the rare-earth-cobalt compounds whose saturation moments are listed in Table 2. In this case the moment values have been reasonably well explained[339] by assuming that the cobalt moments in all these compounds are antiparallel to the rare-earth spin moments and are therefore parallel or antiparallel to the total moments of the rare earths depending on whether $J = L - S$ or $L + S$, respectively. The calculated values listed in Table 3 for the saturation moments (per formula unit) of these ACo_5 compounds were obtained by taking the moment per Co atom to be 1.71 μ_B and combining it in the manner just described with the gJ values for the different rare earths. These calculated values agree rather closely with the experimental values given in Table 2. A similar explanation involving an antiferromagnetic alignment of spin moments has been given for the AFe_2 compounds[333] represented in Table 1 and for AFe_4 compounds of yet undetermined structure.[339] However, the saturation moment values listed in Tables 1 and 2 for the ANi_2 and ANi_5 compounds do not seem to follow an analogous ferrimagnetic scheme, unless the moment associated with the nickel is taken to be very small.[339]

Another indication of ferrimagnetism in the ACo_5 and AFe_4 compounds mentioned above, where A is a rare earth with seven or more $4f$

[330] Compton, V. B., and B. T. Matthias, *Acta Cryst.* **12** (1959) 651.
[331] Bozorth, R. M., B. T. Matthias, H. Suhl, E. Corenzwit, and D. D. Davis, *Phys. Rev.* **115** (1959) 1595.
[332] Hubbard, W. M., E. Adams, and J. V. Gilfrich, *J. Appl. Phys. Suppl.* **31** (1960) 368.
[333] Williams, H. J., and E. A. Nesbit, as reported by G. K. Wertheim, and J. H. Wernick, *Phys. Rev.* **125** (1962) 1937.
[334] Williams, H. J., J. H. Wernick, E. A. Nesbitt, and R. C. Sherwood, *J. Phys. Soc. Japan* **17** Suppl. B-I (1962) 91.
[335] Nesbitt, E. A., H. J. Williams, J. H. Wernick, and R. C. Sherwood, *J. Appl. Phys.* **34** (1963) 1347.
[336] Skrabek, E. A., and W. E. Wallace, *J. Appl. Phys.* **34** (1963) 1356.
[337] Nassau, K., L. V. Cherry, and W. E. Wallace, *J. Phys. Chem. Solids* **16** (1960) 131.
[338] Cherry, L. V., and W. E. Wallace, *J. Appl. Phys. Suppl.* **32** (1961) 340.
[339] Nesbitt, E. A., H. J. Williams, J. H. Wernick, and R. C. Sherwood, *J. Appl. Phys.* **33** (1962) 1674.
[340] Kittel, C., *Introduction to Solid State Physics* 2nd ed., Wiley, New York, 1956, equation (9.17).

TABLE 1

Ferromagnetic Curie Points and Saturation Moments (Per Formula Unit, at 0°K) of AB_2 Compounds of Rare-Earth Elements (A), Having Cubic (c) C15-Type or Hexagonal (h) C14-Type Structure

A \ *B*	Mn[a]	Fe[b]	Ni[c]	Ru[d]	Rh[e]	Os[d]	Ir[d]	Pt[e]	Al[f]
Ce		c 878°K 6.97 μ_B	c NF	c S	c NF	c NF	c NF		c 8°K 2.5 μ_B
Pr			c 8°K 0.73 μ_B	c 40°K 0.7 μ_B	c 8.6°K	? 27°K 1.0 μ_B	c 16°K 2.1 μ_B	c 7.9°K	c 34°K 2.6 μ_B
Nd			c 20°K 1.65 μ_B	c 28°K	c 8.1°K	h 23°K 1.4 μ_B	c 11.8°K 1.6 μ_B	c 6.7°K	c 64°K 2.1 μ_B
Sm		c 674°K	c 77°K 6.30 μ_B[g]			h 34°K 0.2 μ_B	c 37°K 0.3 μ_B		c 123°K 0.1 μ_B
Gd	c ~300°K 2.85 μ_B	c 813°K	c 90°K 6.74 μ_B	h 83°K 6.3 μ_B	c $>$77°K	h 66°K 6.7 μ_B	c 88°K 6.9 μ_B	c $>$77°K	c 176°K 6.9 μ_B
Tb	c F		c 46°K 5.85 μ_B			h 34°K 7.3 μ_B	c 45°K 7.0 μ_B		c 121°K 7.6 μ_B
Dy	c F	c 663°K 5.44 μ_B	c 32°K 7.14 μ_B			h 15°K 6.7 μ_B	c 23°K 7.7 μ_B		c 53°K 8.7 μ_B
Ho	c F	c 608°K 6.02 μ_B	c 23°K 7.19 μ_B			h 9°K 6.0 μ_B	c 12°K 7.5 μ_B		c 27°K 7.9 μ_B
Er		c 473°K 5.02 μ_B	c 14°K 5.70 μ_B	h 8°K 6.1 μ_B		h 3°K 5.4 μ_B	c 3°K 6.1 μ_B		c 21°K 7.1 μ_B
Tm		c 613°K 2.92 μ_B	c 14°K 2.59 μ_B				c 1°K 3.0 μ_B		c 5°K 4.7 μ_B

[a] Values for $GdMn_2$ given in Reference 332; Gd, Tb, Dy, and Ho compounds reported to be ferrimagnetic in Reference 335.

[b] Reference 333.

[c] Reference 336.

[d] Reference 331; somewhat different values given in Reference 330. NF ≡ not ferromagnetic; S ≡ superconducting.

[e] Reference 330. NF ≡ not ferromagnetic.

[f] Reference 334.

[g] W. E. Wallace (private communication) reports that this moment value was obtained for impure $SmNi_2$ specimen and that the correct value is 0.23 μ_B.

TABLE 2

Ferromagnetic Curie Points and Saturation Moments (Per Formula Unit, at Temperatures Indicated) of AB_5 Compounds of Rare-Earth Elements (A), Having Hexagonal $CaCu_5$ Structure

A \ B	Mn[a]	Co[b]	Ni[c]
Y	490°K 2.21 μ_B (78°K) 1.38 μ_B (298°K)	995°K 8.2 μ_B (1.4°K)	~0 μ_B (1.4°K)
Ce		687°K 7.4 μ_B (1.4°K)	
Pr		9.9 μ_B (1.4°K)	
Nd		10.5 μ_B (1.4°K)	2.2 μ_B (1.4°K)
Sm	440°K 1.72 μ_B (78°K) 1.40 μ_B (298°K)	1015°K 8.6 μ_B (1.4°K)	0.7 μ_B (1.4°K)
Gd	465°K 6.23 μ_B (78°K) 2.89 μ_B (298°K)	1030°K 1.3 μ_B (1.4°K)	6.1 μ_B (1.4°K)
Tb	445°K 6.18 μ_B (78°K) 2.66 μ_B (298°K)	0.7 μ_B (1.4°K)	7.0 μ_B (1.4°K)
Dy	430°K 5.34 μ_B (78°K) 2.49 μ_B (298°K)	1125°K 1.5 μ_B (1.4°K)	7.7 μ_B (1.4°K)
Ho	425°K 5.12 μ_B (78°K) 1.99 μ_B (298°K)	1025°K 1.9 μ_B (1.4°K)	7.2 μ_B (1.4°K)
Er	415°K 3.74 μ_B (78°K) 1.63 μ_B (298°K)	1.3 μ_B (1.4°K)	7.7 μ_B (1.4°K)
Tm		1.5 μ_B (1.4°K)	

[a] Reference 338.
[b] Curie points from Ref. 337; moments from Ref. 339.
[c] Reference 339.

electrons (see Table 3), is the fact that their magnetization versus-temperature curves each show a compensation point.[339,341] In the case of $HoCo_5$ a recent neutron diffraction study[342] demonstrated that the cobalt moments are indeed antiparallel to those of holmium, the latter rising rapidly with decreasing temperature and equaling the contribution of the former at about 70°K, which is also the compensation point determined by magnetization measurement.[341] This neutron diffraction study also showed that in the case of YCo_5, where yttrium has no unfilled electron shell

[341] Nesbitt, E. A., H. J. Williams, J. H. Wernick, and R. C. Sherwood, *J. Appl. Phys. Suppl.* **32** (1961) 342.
[342] Ballestracci, R., E. F. Bertaut, J. Coing-Boyat, A. Delapalme, W. James, R. Lemaire, R. Pauthenet, and G. Roult, *J. Appl. Phys.* **34** (1963) 1333.

TABLE 3

A	No. of f electrons	S	L	J	g	gJ	ACo_5 moment $\lvert gJ \pm 5 \times 1.71 \rvert$
Ce	1	$\frac{1}{2}$	3	$\frac{5}{2}$	$\frac{6}{7}$	2.14	10.69
Pr	2	1	5	4	$\frac{4}{5}$	3.20	11.75
Nd	3	$\frac{3}{2}$	6	$\frac{9}{2}$	$\frac{8}{11}$	3.27	11.82
Pm	4	2	6	4	$\frac{3}{5}$	2.40	10.95
Sm	5	$\frac{5}{2}$	5	$\frac{5}{2}$	$\frac{2}{7}$	0.71	9.26
Eu	6	3	3	0	5	0	8.55
Gd	7	$\frac{7}{2}$	0	$\frac{7}{2}$	2	7.00	1.55
Tb	8	3	3	6	$\frac{3}{2}$	9.00	0.45
Dy	9	$\frac{5}{2}$	5	$\frac{15}{2}$	$\frac{4}{3}$	10.00	1.45
Ho	10	2	6	8	$\frac{5}{4}$	10.00	1.45
Er	11	$\frac{3}{2}$	6	$\frac{15}{2}$	$\frac{6}{5}$	9.00	0.45
Tm	12	1	5	6	$\frac{7}{6}$	7.00	1.55
Yb	13	$\frac{1}{2}$	3	$\frac{7}{2}$	$\frac{8}{7}$	4.00	4.55

and therefore no moment, the cobalt moments are ferromagnetically aligned and are 1.74 μ_B per atom (or 8.7 μ_B per formula unit, in fair agreement with the value shown in Table 2). From the variation of the magnetization of the mixed compounds $Gd(Co_xFe_{1-x})_5$ with composition (x), it was concluded that although all the Co moments are antiparallel to the Gd moments in $GdCo_5$ (consistent with the above discussion), this is not true for the Fe moments in $GdFe_5$ some of which are parallel to the Gd moments.[343]

A recent development of considerable interest is the coexistence of ferromagnetism and superconductivity that was found in the mixed compounds $(Ce_{1-x}Gd_x)Ru_2$ having the cubic C15-type structure.[344–346] This coexistence was demonstrated by magnetic hysteresis loop measurements at 1.3°K for the compound with $x \simeq 0.08$. For $x = 0.04$ hysteresis loops at this temperature show superconductivity accompanied only by strong paramagnetism; the compounds with $x \geq 0.1$ are simply ferromagnetic. A ferromagnetic-superconducting state at 1.3°K was also discovered in the isomorphous mixed compound $La_{0.91}Gd_{0.09}Os_2$.[346]

Additional magnetic information on these compounds (much of it previously unpublished) as well as structural and magnetic data on other rare-earth compounds appears in a comprehensive book devoted to these materials.[347]

7. TERNARY COMPOUNDS OF IRON-GROUP METALS

7.1 Heusler-Type Compounds

Since the early discovery by F. Heusler[348] that certain Cu–Mn–Sn and Cu–Mn–Al alloys are *ferromagnetic*, an extensive literature has gradually accumulated on their structural and magnetic properties. This literature has been thoroughly reviewed by Bozorth,[349] and only its salient features will be summarized here. The maximum saturation moments in these two ternary systems were found to occur at the compositions Cu_2MnSn and Cu_2MnAl, and the alloys of this stoichiometry were then shown to have the ordered cubic structure depicted in Fig. 26, which is the prototype of the $L2_1$ type. When fully ordered, both compounds have a saturation magnetization at room temperature of about 7500 emu/cm³, and in both cases extrapolation to 0°K gives a moment per Mn atom of about 4 μ_B. The Curie point of Cu_2MnAl has been estimated to be near 700°K, but this value is very uncertain because atomic disordering begins to take

[343] Hubbard, W. M., and E. Adams, *J. Phys. Soc. Japan* **17** Suppl. B-I (1962) 143.

[344] Matthias, B. T., H. Suhl, and E. Corenzwit, *Phys. Rev. Letters* **1** (1958) 449.

[345] Bozorth, R. M., and D. D. Davis, *J. Appl. Phys. Suppl.* **31** (1960) 321.

[346] Bozorth, R. M., D. D. Davis, and A. J. Williams, *Phys. Rev.* **119** (1960) 1570.

[347] Gschneidner, K. A., *Rare Earth Alloys*. D. Van Nostrand Co., Princeton, New Jersey, 1961.

[348] Heusler, F., *Verhandl. Deut. Physik. Ges.* **5** (1903) 219. *Z. Angew. Chem.* **17** (1904) 260.

[349] Bozorth, R. M., *Ferromagnetism*. D. Van Nostrand Co., New York, 1951, pp. 328–334.

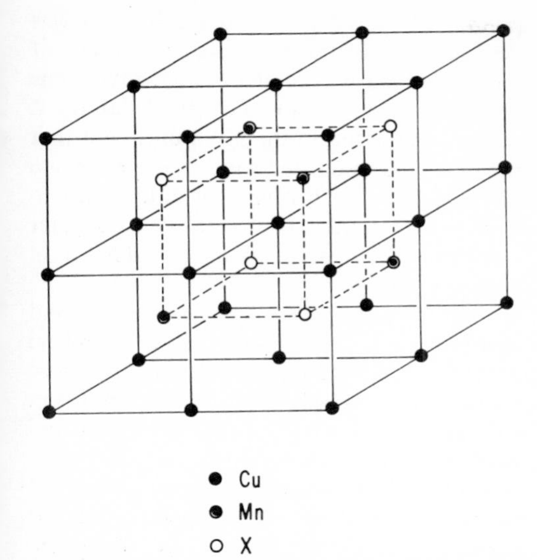

Fig. 26. Crystal structure of Heusler-type alloys ($Cu_2Mn X$).

place at lower temperatures. Recent work[350,351] has revealed a similar but more severe situation in Cu_2MnSn, making its Curie temperature virtually indeterminable. The magnetocrystalline anisotropy (K_1) of Cu_2MnAl was recently found to decrease monotonically from -3700 erg/cm^3 at 77°K to -1500 erg/cm^3 at room temperature.[352]

Ferromagnetism has also been discovered[353] in Cu–Mn–In alloys and later found by X-ray diffraction studies[354,355] to be attributable to a Heusler-type phase of stoichiometry Cu_2MnIn. Magnetic measurements[354] on this compound gave $T_c = 506°K$ and $\sigma_0 = 74.4$ emu/gm, the latter corresponding to 4.06 μ_B per Mn atom (similar to Cu_2MnSn and Cu_2MnAl).

F. Heusler[348] also reported ferromagnetism in Cu–Mn–Sb alloys. However, it was recently discovered[356] that although Cu_2MnSb has a Heusler-type structure it is not ferromagnetic. Instead, as it is cooled below room temperature, its susceptibility rises to a maximum value at 38°K and then decreases, indicative of *antiferromagnetic* order. It was also pointed out[356] that the lattice parameter of Cu_2MnSb is intermediate between those of Cu_2MnAl and Cu_2MnIn, both of which are ferromagnetic; thus the coupling between Mn moments in these compounds is not a simple function of atomic spacing.

The alloy of composition Ag_5MnAl has been found to be *ferromagnetic* with a Curie point near 630°K and a saturation magnetization at room temperature of about 70 emu/cm^3,[357] despite its low magnetization, the very high intrinsic coercivity (about 600 Oe) of this alloy makes it a fairly useful permanent magnet material, and it is known commercially as *Silmanal.* It was thought to derive its high coercivity from a Widmanstätten precipitate presumed to have a ferromagnetic Heusler-type ordered structure.[358] However, subsequent experiments have shown that the ferromagnetic component of Silmanal is a metastable precipitate whose structure was identified alternatively as hexagonal[359] or fcc.[360]

Alloys of composition Au_2MnAl and Au_2MnIn were observed to be *ferromagnetic* with Curie temperatures of about 230°K and 140°K, respectively;[361] recent magnetization measurements on Au_2MnAl gave $\sigma_0 = 20.6$ emu/gm, corresponding to only 2.24 μ_B per Mn atom.[362] Room-temperature X-ray measurements on these two materials were unable to establish that the Mn and Al (or In) atoms had the ordered positions of the cubic Heusler-type structure (Fig. 26), thus allowing the possibility of at least a partially disordered atomic distribution over these lattice sites; below about 235°K the structure of Au_2MnIn is tetragonally distorted.[361]

Ni_2MnIn, Ni_2MnGa, Co_2MnGa, and Pd_2MnSb have been tentatively identified as having the Heusler ($L2_1$) type structure, and *PdMnSb* the related fluorite ($C1_1$) type structure; all these compounds appeared to be

350 Taglang, P., and G. Asch, *Compt. Rend.* **238** (1954) 2500.
351 Taglang, P., and M. Fournier, *J. Phys. Radium* **22** (1961) 295.
352 Aoyagi, K., and M. Sugihara, *J. Phys. Soc. Japan* **16** (1961) 1027.
353 Valentiner, S., *Naturwissenschaften* **34** (1947) 123.
354 Coles, B. R., W. Hume-Rothery, and H. P. Myers, *Proc. Roy. Soc. (London)* **A196** (1949) 125.
355 Hames, F. A., and D. S. Eppelsheimer, *Trans. AIME* **185** (1949) 495.
356 Oxley, D. P., C. T. Slack, R. S. Tebble, and K. C. Williams, *Nature* **197** (1962) 465.
357 Potter, H. H., *Phil. Mag.* **12** (1931) 255.
358 Geisler, A. H., *Trans. ASM* **43** (1951) 90.
359 Hall, E. O., *Phil. Mag.* **4** (1959) 730.
360 Köster, W., and N. K. Anantha Swamy, *Z. Metallk.* **53** (1962) 299.
361 Morris, D. P., R. R. Preston, and I. Williams, *Proc. Phys. Soc. (London)* **73** (1959) 520.
362 Oxley, D. P., R. S. Tebble, and K. C. Williams, *J. Appl. Phys.* **34,** (1963) 1362.

strongly *ferromagnetic*, except Pd_2MnSb which appeared to be only weakly so.[363]

7.2 Other Ternary Compounds

A new ferromagnetic phase has been discovered in alloys of composition

$$Mn_{1-x}Al_{1-y}Z_{x+y}$$

where Z is Fe,[364] Cu[365], Ni[366] or Co[367] and x and y are typically 0.1 to 0.4 and 0 to 0.1, respectively. This phase has the CsCl (B2) type structure and the compounds are therefore pseudobinary. Above their Curie temperatures, the compounds with $y = 0$ have linear $1/\chi$ versus T curves, whereas the curves for those with $y > 0$ are distinctly concave towards the T-axis, suggesting that the former are *ferromagnetic* and the latter *ferrimagnetic*. Moreover, from the variations of saturation magnetization ($\sim 30 - 80$ emu/gm) and Curie point (350—450°K) with composition, it was deduced that of the two equivalent types of lattice sites one is occupied by all the Al atoms and some Mn atoms and the other by all the Z atoms and the rest of the Mn atoms, and that the moments associated with each type of site are parallel to each other but antiparallel to those on the other type of site. This was later confirmed (for Z $\equiv$ Cu) by neutron diffraction experiments.[368]

Another *ferromagnetic* ternary compound reported recently is one of composition *MnAlGe*.[369] It has a tetragonal structure whose c-axis is the easy axis of magnetization. From magnetization curves measured parallel and perpendicular to the c-axis of a single crystal, it was found that the saturation magnetization is 294 emu/cm^3 and the magnetocrystalline anisotropy coefficient K_1 is 5.3×10^6 erg/cm^3 at room temperature. Coercive fields of 1500–2200 Oe were observed in MnAlGe powders embedded in molten wax in a field; the maximum attainable coercivity is estimated to be about 36,000 Oe.

---

[363] Hames, F. A., *J. Appl. Phys. Suppl.* **31** (1960) 370.
[364] Tsuboya, I., and M. Sugihara, *J. Phys. Soc. Japan* **15** (1960) 1534.
[365] Tsuboya, I., and M. Sugihara, *J. Phys. Soc. Japan* **16** (1961) 571; Tsuboya, I., *op. cit.* **16** (1961) 1875.
[366] Tsuboya, I., and M. Sugihara, *J. Phys. Soc. Japan* **16** (1961) 1257; **17,** Suppl. B-I (1962) 172.
[367] Tsuboya, I., and M. Sugihara, *J. Phys. Soc. Japan* **174** (1962) 10.
[368] Katsuraki, H., H. Takada, and K. Suzuki, *J. Phys. Soc. Japan* **18** (1963) 93.
[369] Wernick, J. H., S. E. Haszco, and W. J. Romanow, *J. Appl. Phys.* **32** (1961) 2495.

REFERENCES

1. Heusler F., *Verhandl. Deut. Physik. Ges.* **5** (1903) 219; *Z. Angew. Chem.* **17** (1904) 260.
2. Néel L., *Ann. Physique* **17** (1932) 5.
3. Néel L., *Ann. Physique* **3** (1948) 137.
4. Guillaud C., Thesis, University of Strasbourg, 1943.
5. Shull C. G., and J. S. Smart, *Phys. Rev.* **76** (1949) 1256.
6. Hansen M., *Constitution of Binary Alloys*, McGraw-Hill Book Co., New York, 1958, 2nd ed.

chapter 28

Electronic Properties and Applications

C. Sheldon Roberts

C. SHELDON ROBERTS

Consultant
Los Altos, California

1. INTRODUCTION

The intermetallic compounds of practical value in electronic applications all are essentially electronic semiconductors rather than metallic conductors. Semiconductors make useful electronic devices primarily because of these properties:

(*a*) As temperature or impurity content is changed, electrical conductivity is controlled primarily by the changes in the density of charge carriers rather than by changes in the mobility as in metallic conductors.

(*b*) Conductivity occurs both by electrons and by holes, the conductivity type being controlled by whichever are the *majority carriers*.

(*c*) The occurrence of both *n*- and *p*-type conductivity allows the creation of *p-n junctions* where important nonlinear effects occur.

The preferred raw material for electronic device development is single crystal of extremely high purity. There are some applications in thermoelectric power conversion and electroluminescence in which polycrystalline, lower-purity materials are suitable. In most cases it is preferable to produce relatively high-purity material and then to "dope" it controllably to lower-purity levels with known impurity atoms.

There are certain combinations of electrical properties which are needed for device applications that can not be found in the popular elemental semiconductors germanium and silicon. The compounds allow a greater freedom of choice in these combinations. They offer higher energy gaps, higher carrier mobilities, and special energy transitions which are useful in electroluminescent applications.

Devices are desired to operate at temperatures of several hundred degrees Centigrade rather than up to 100 and 200°C as with germanium and silicon, respectively. Higher energy gaps between the conduction and the valence bands in the crystals are required. Added impurities control the behavior of semiconductor devices by supplying electrons or holes to the crystal. At any given temperature, electrons and holes are also being produced by thermal ionization in the crystal. At a sufficiently high temperature in any semiconductor material, these intrinsic carriers become equal in concentration to the impurity-controlled or *extrinsic* carriers. Then the effect of the impurities is lost, and most semiconductor devices are no longer useful, although no permanent damage results. This temperature limitation is especially important in electronic devices that depend on *p-n* junctions, such as the junction transistor, because of the importance of impurity carrier control in its structure. Large energy gap semiconductors are needed in order to extend the useful extrinsic range of these devices.

In addition, it is often desired to operate electronic devices at very high frequencies or, in the case of computer applications, at very high transient speeds. A controlling electrical property is the *mobility*—the velocity of a current carrier in unit electrical field within the solid. In general, a high mobility is desired. Electron and hole mobilities are often greatly different in compound semiconductors. If a material has both a large energy gap and at least one very high mobility, transistors made from it would operate at very high temperatures and very high frequencies. Unfortunately, ideal combinations are rather hard to find, because, as the energy gap increases from one compound to another, the mobilities tend to decrease.

Most of the important diode and transistor *p-n* junction effects involve minority carrier action with the exception of the tunneling junction effect in tunnel diodes. A vital concern in the design of electronic devices depending on minority carrier effects is the *lifetime* of these carriers. This time constant measures the exponential rate of return to equilibrium when the majority and/or minority carrier concentrations have been increased by phonons (thermal excitation), photons (photo-excitation) or by applied electric fields (injection). Equilibrium in both bulk semiconductors and *p-n* junctions can be considered a dynamic balance between the processes of carrier generation (formation of free electrons and holes) and carrier recombination.

2. THE IMPORTANT COMPOUNDS

The literature on these materials is extremely voluminous. Key review articles and authoritative books dealing with almost all of the most important compounds have been published.[1–8]

The application of compound semiconductors in solid state electronics has become far-reaching, and the number of intermetallic compounds available to serve as electronic semiconductors is enormous. In this chapter only a summary view of a rapidly growing field can be given.

The only intermetallic compounds of adequate potential electronic applications to be discussed in this chapter are listed in Table 1, roughly in the order of their commercial importance. The prominence of the III-V compounds is clear. Those intermetallic semiconductor compounds of great interest in thermoelectric power generation and cooling are excluded because of coverage in Chapter 26.

The following list summarizes the desirable characteristics for intermetallic compound semiconductors:

(*a*) high mobility of at least one carrier;

(*b*) adequately large energy gap for intended application temperature;

(*c*) controllable values of carrier lifetime;

(*d*) chemical stability;

(*e*) low cost;

(*f*) easy adaptability to technologies of impurity control by purification, diffusion, epitaxial growth, and the control of *p-n* junction lateral geometry, that is, masking; and

(*g*) special energy transitions for optical use.

The justification for the order of listing in Table 1 is as follows: The indium and gallium compounds have high electron mobilities and offer a wide spectrum of energy gap values. They all have good chemical stability. Minority carrier lifetimes generally are less than 1 microsecond. They can not be intrinsically cheap, no matter how far expensive crystal growth processes are improved, because gallium and indium are relatively rare metals. They are the only compounds on the list which have shown reasonable adaptability to impurity control methods to date. All six have found a prototype application in at least one electronic device, and the first three are now regularly sold in commercial device form.

TABLE 1

Properties of Selected Intermetallic Compound Semiconductors and Comparison with Silicon and Germanium[a]

Compound	Structure	Melting Point °C	Lattice Parameter, Å	Density g/cm³	Energy Gap at 300°K, eV	Dielectric Constant	Mobilities at 300°K cm²/volt-sec.	
							Electron	Hole
InSb	Zinc blende	525	6.48	5.78	0.17	16.8	78,000	750
InAs	Zinc blende	943	6.06	5.66	0.36	10.2	33,000	460
GaAs	Zinc blende	1237	5.65	5.31	1.38	10.2	8,500	400
GaP	Zinc blende	1500	5.45	4.13	2.24	8.4	110	75
GaSb	Zinc blende	712	6.10	5.62	0.67	13.7	4,000	1400
InP	Zinc blende	1062	5.87	4.79	1.27	10.9	4,600	150
α-SiC	Wurtzite	2800	$c = 15.12$, $a = 3.08$	3.21	3.1	10.2	>100	>20
β-SiC	Zinc blende	2800	4.36	3.21	2.2			
Cs_3Sb	Cubic				1.6			
AlSb	Zinc blende	1054	6.14	4.22	1.60	10.1	200	420
AlAs	Zinc blende	>1600	5.63	3.81	2.2		1,600	200
AlP	Zinc blende	>1500	5.43	2.85	3.0	11.6		
Ge	Diamond	957	5.66	5.32	0.67	16.3	3,800	1820
Si	Diamond	1412	5.43	2.33	1.11	11.7	1,300	500

[a] Values principally from References 6 and 9.

The hexagonal α-SiC offers the more immediate promise of application of the two allotropes because of the larger amount of work which has been applied to its difficult high temperature growth. The cubic form is stable and thus growable only at lower temperatures, below about 2000°C. Mobility is moderately low, but the energy gap is high enough to offer promise of device service above 500°C. Little is known about carrier lifetime, but chemical stability and potentially low cost are outstanding. Adaptability to impurity control has been poor so far, partly because of the small size of available crystals. Crude prototype single-crystal devices have been made, although thermistors and other nonlinear resistors have been made from impure, polycrystalline SiC for many years.

Cesium antimonide enjoys a special photocell application because of its exceptional photoelectric quantum efficiency. The aluminum compounds offer attractive advantages in high energy gap and reasonable mobility combinations, and in potentially low cost. They have been held back from any but crude device applications because of their hygroscopic behavior and difficulties in preparation.

The most rapid technological progress has been made with selected members of the III-V intermetallic compounds. This group has been presented rather understandably in a rhombic diagram by Welker. His original 4 × 4 diagram included the nitrides and the boron compounds. Figure 1 shows the combinations of the three group III elements—aluminum, gallium, and indium—with the three group V

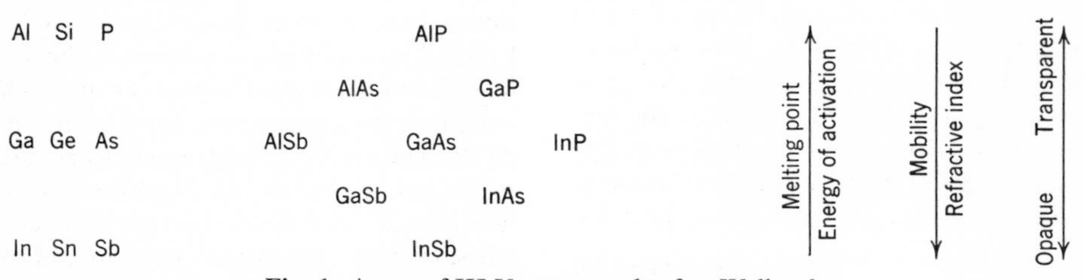

Fig. 1. Array of III-V compounds after Welker.[2]

TABLE 2

Experimentally Determined Electrical Effects of Impurities in Some Compound Semiconductors

Compound	Acceptors	Donors
InSb	Mg, Zn, Cd, Cu[b], Ag[b], Au[b], Si, Ge, Mn	S, Se, Te, Sn, Na
InAs	Mg, Zn, Cd	S, Se, Te, Si, Ge, Sn, Cu
GaAs	Mg, Zn, Cd, Hg, Na, Cu, O[a], Fe[a], Co[a], Ni[a]	S, Se, Te, Si, Ge, Sn, Pb, Li
GaP	Zn	S
GaSb	Mg, Zn, Cd, Si, Ge, Sn, Cu	Se, Te, Li
InP	Zn, Cd	S, Se, Te
α-SiC	B, Al	N

[a] Deep level.
[b] Double, shallow, and deep levels.

elements—phosphorus, arsenic, and antimony. The vertical center column of the rhombus contains those compounds that are isoelectronic with the elemental semiconductors silicon, germanium, and gray tin. The arrows show the direction of change of some electrical and optical properties with position up and down the diagram. Indium antimonide, for example, has an energy gap of 0.17 eV, a melting point of 525°C, an electron mobility of 76,000 cm^2/volt-sec, and is opaque to visible light. Aluminum phosphide has an energy gap of about 3 eV, a melting point above 1500°C, a mobility probably less than 100 cm^2/volt-sec, and is transparent to visible light.

All of these compounds show some ionicity in an essentially covalently bonded crystal of the zinc blende structure. The polarity of this structure leads to differences in X-ray reflection[10] and in other properties between {111} and $\{\bar{1}\bar{1}\bar{1}\}$ faces. Differences in the nature of etch pitting,[11,13,14] faster crystal growth from the melt,[12] slower etch rates, and more noble electrode potentials[14] are observed on the faces where single dangling bonds extend from an outer layer of group III atoms.*

Knowledge of the effects of impurities on the electrical behavior of intermetallic compounds is basic to intelligent design of useful electronic devices with them. The impurity effects in silicon carbide are similar to those in the group IV elemental semiconductors, that is, group III elements are shallow acceptors. Group V elements are shallow donors, and other impurity atoms tend to occupy more or less deep levels. The effects in III-V compounds are considerably more complicated. Group II elements generally act as acceptors and group VI as donors. However, there are exceptions where some of these elements in some of the compounds show the opposite behavior or in fact have two levels of ionization. In general, group IV elements are important carrier-controlling impurities, and they may act either as donors or acceptors in III-V compounds. Table 2 summarizes the currently known impurity effects for the first seven compounds from Table 1.

3. POWER RECTIFIERS

The advantage sought for compound semiconductors has generally been higher temperature operation and/or higher allowable power dissipation than with the well-established silicon rectifiers. The two high energy gap compounds, gallium phosphide and silicon carbide, have been investigated most thoroughly for rectifier application. Davis and co-workers at Westinghouse Electric Corporation have directly compared the performance of power diodes made from the two materials with special emphasis on the phosphide.[15] One of the principal results was that with *p-n* junctions, made by both fusion ("alloying") and by solid-state diffusion techniques, the advantage in lower forward voltage drop at 400°C, and thus less power loss, went to the phosphide. The early work left much to be desired in the level of the reverse breakdown voltage, however. It was found that 500°C operation left silicon carbide as the only material maintaining the rectification ratio of 100, which has

* For a more extended discussion, see Chapter 16, Section 2.

been defined as useful. At lower temperatures, the large forward voltage drop of the silicon carbide units is a real disadvantage, which might be minimized by the use of external heaters or by self-heating according to Davis.

The formation of ohmic contacts between the external metallic conductors and the active regions of the semiconductor is a perennial and important problem in the formation of diodes and transistors. The technical progress of a compound semiconductor is linked with the ease of accomplishing such contacts. The most common metallurgical process for contact formation is that of fusion with a metal or alloy to form a liquid which will freeze to form a junction-free or nonrectifying interface with the semiconductor. Although such contact formation is more difficult in gallium phosphide than in the lower-melting gallium arsenide, it has so far been easier to accomplish ohmic contacts to it than to the high-melting silicon carbide. The control of the purity and perfection of the single-crystal semiconductor itself has been more rapidly accomplished in the case of gallium phosphide. Preparational methods have been advanced in the technology of silicon carbide, but crystals of the dimensions of centimeters and predictable control of impurity content have not yet been achieved.

4. COMPUTER SWITCHING, VARACTOR, AND TUNNEL DIODES, MICROWAVE SOURCES

High speed and a low "on" resistance are the primary requirements for a computer switch. The turn-on and turn-off speeds are complex functions of carrier mobility and minority carrier lifetime. High mobility and low lifetime are desirable. Gallium arsenide satisfied these requirements well. Zinc-diffused junction "mesa" diodes[16] have nanosecond switching times and rectification ratios (at 2V) of 10^{10}. Even shorter switching times have been mentioned for point-contact diodes.[17]

Gallium arsenide has been found to be superior to silicon in varactor (variable capacitance) diode applications.[16] The cutoff frequency is an important figure of merit of these devices. This frequency is directly proportional to the majority carrier mobility in the bulk of the diode and inversely proportional to the dielectric constant. The higher mobility and lower dielectric constant of gallium arsenide have contributed to successful operations in a 2.8 gigacycle variable reactance amplifier.[16]

Microwave mixer diodes are also frequently made from gallium arsenide. Point-contact rectification has traditionally been used in this application. Gallium arsenide point-contact units have functioned as efficient, low-noise first detectors at frequencies greater than 10 gigacycles. Conversion loss is significantly lower than for silicon and germanium.[17,18] Rectification ratios are much poorer for these point-contact diodes than for diffused diodes because of higher reverse leakage and higher forward series resistance.

Tunnel diodes have been of great interest for computer switching, frequency doubling, and high frequency oscillator-amplifier applications. The negative resistance portion of the current-voltage characteristic curve (Fig. 2) is used in the latter. Although the first tunnel diodes were made from germanium, GaAs became a favorite material,[19] and diodes made from it soon came into small-run production. Gallium arsenide tunnel diodes have been actively marketed by several companies in the United States. Experimental investigation of other III-V compounds as tunnel-diode material has also been popular. An InSb tunnel diode made by alloying Cd on *n*-type antimonide has oscillated at 2.3 gc, while cooled to 77°K.[20] Gallium antimonide offers promise for room-temperature tunnel-diode operation if a large voltage swing is not necessary.

The most important electrical parameters of the tunnel diode are the peak current, I_p, the valley current, I_V, their ratio I_p/I_V, the

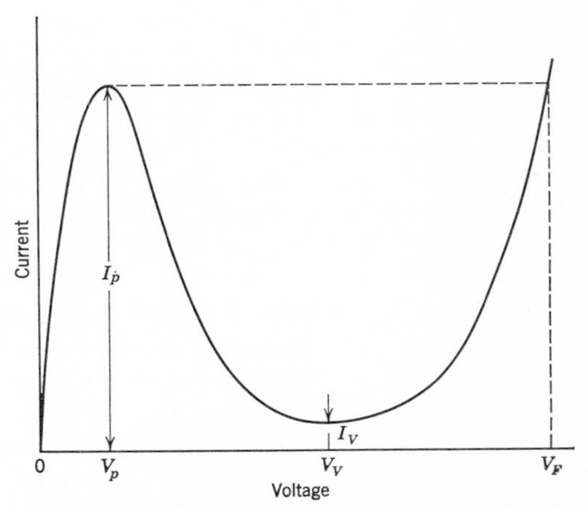

Fig. 2. Tunnel diode current-voltage characteristic.

diode capacitance, and the voltage swing, $(V_F - V_p)$. Gallium arsenide gives high-peak currents and high I_p/I_V ratios apparently because of favorable tunneling probabilities. The capacitance per unit area is lower because of the lower dielectric constant of GaAs as compared to germanium. This is desirable for high-frequency applications, in which device capacitance becomes a limiting factor. The tunnel-diode voltage swing is roughly equal to the band gap of the semiconductor. On this count, gallium arsenide gives a much larger swing than germanium. This is desirable in digital switching applications in which a large output pulse, both in voltage and in power, is desirable.

The tunnel diode performance is clearly outstanding with gallium arsenide. However, it is uncertain what the growth of its applications will be. The severe problems of constructing practical signal-handling circuitry from two-terminal devices have held back the tunnel diode. Unfortunately both for tunnel diodes and for gallium arsenide as a tunnel diode material, production and marketing of the devices was well-advanced in 1960–1961 before a peculiar peak-current degradation phenomenon was discovered. Confidence of electronic system designers in both the device type and the material was damaged until a systematic study was conducted of the conditions under which it occurs.[21] There is no degradation for operation from the origin to the valley voltage, V_V. Operation beyond V_V leads to negligible degradation when $I_F/C < 0.8$ mA/pF where I_F is the forward current past the valley and C is the junction capacitance measured at the valley. Complicated mechanisms involving the motion of impurity atoms have been proposed.

Some of the fabrication and reliability problems with tunnel diodes have resulted from the mechanically weak structure of conventional alloyed and etched junction devices. A new construction has been proposed which will furnish great mechanical improvement without appreciable increase in junction capacitance.[22]

An amazing application of gallium arsenide single crystals is the generation of microwave oscillations from a simple bulk device.[23] A thin crystal element is excited with a pulse through ohmic contacts and breaks into oscillations in the gigacycle frequency range. Theoretical understanding of the effect is still incomplete, but the practical possibility of generating hundreds of milliwatts of power exists. Realization of this promise depends on commercial controllability of this oscillation.

5. TRANSISTORS

Whereas minority carrier lifetime must be controllably low in switching diodes and is of little consequence in the other types, it must be adequately long in high-frequency or high-speed switching transistors to insure a usable current gain. If minority carrier lifetime is too short, those carriers injected by the emitter into the transistor base will never arrive at the collector.

A review article by Jenny[24] came to the conclusion that, considering energy gap and carrier mobility, the greatest potential among the compound semiconductors for combined high temperature and high frequency lay with gallium arsenide. Both bipolar and unipolar transistors were considered.

The advantage of GaAs over germanium was predicted primarily in operating temperature range for bipolar transistors. Some significant operating frequency range benefit was also calculated for bipolar units. Minority carrier lifetime differences were not discussed, although Jenny clearly stated that only low values, relative to silicon and germanium, had been attained. There have not been significant increases of minority carrier lifetime over early values of 10^{-8} sec as gallium arsenide purity (measured by both resistivity and mobility increases) has improved. A similar stagnant situation exists with indium antimonide at about the 10^{-7} sec level. Carrier lifetime might be expected to be intrinsically limited to values less than 10^{-6} sec because of the direct energy transition of carriers across the energy gap in these compounds.

The reported characteristics of gallium arsenide and indium phosphide transistors have been very meager and not extremely encouraging for the future of transistors from these compounds. A current gain of 13 and power gain of 36 dB was obtained in InP using a "surface diffusion" method involving a zinc-doped base and a metal-semiconductor contact emitter.[24] Electron lifetime in the base was estimated to be about 10^{-7} sec. Power gain (about 4 dB) was only obtained in gallium arsenide transistors after GaP was alloyed to form a wide gap emitter. The low electron lifetime of 10^{-8} sec was offset by greatly increased injection efficiency.

Efficient transport across the base of a junction transistor is effected by high minority carrier mobility as well as long minority carrier lifetime. Both parameters are superior in InSb relative to the other compound semiconductors. Thus it is not surprising that by operating at 77°K, at which indium antimonide can behave extrinsically, Henneke was able to observe very high current gains, very fast switching, and excellent high-frequency performance.[25] Of course, the cooling requirements limit the incentive for extended developments in this direction. The results, nevertheless, offer hope for eventually equivalent results in the high-energy gap compounds.

A transistor-like device has been constructed in gallium arsenide by utilizing photons as the carrier between the input and output junctions.[26] This operation is feasible because of the high efficiency of the electroluminescence at GaAs *p-n* junctions, as will be discussed more thoroughly later. The current gain of these units is very low, but the potential speed of the device is great.

6. GALVANOMAGNETIC DEVICES

Indium antimonide and indium arsenide are by far the two most important compounds for these applications. The devices may be considered in two categories—those that depend on magnetoresistance and those that depend on the Hall effect. A relatively large magnetoresistance, that is, a large change in the resistance of a two-terminal device with varying magnetic field, can be produced in a high-mobility semiconductor when the Hall voltage has been almost completely shorted out. Contactless potentiometers, transducers, dc-to-ac converters, rectifiers, and amplifiers may be built with magnetoresistive devices.[24] Hall effect devices, on the other hand, are four-terminal units in which a control current is passed through two terminals or is used to vary the magnetic field. A Hall voltage output, proportional to the control current and to the magnetic field, is obtained at the other two terminals.[28–31]

Hilsum and Rose-Innes, in comparing practical use of magnetoresistance and the Hall effect, find that the power efficiency is greater for the Hall effect at fields below about 2000 Oe and greater for magnetoresistance at higher fields.[6] As a result, they consider the Hall generator, as it is often called, to be more important in fluxmeter applications and for the many other devices in which the magnetic field is signal-produced. Reference to the comprehensive listing of such devices by Grubbs[28] shows three in addition to the fluxmeter which have been produced in prototype or are in steady production either in the United States or in Europe. They are the analog multiplier, the wattmeter, and the amplifier.

Four proposed Hall effect devices depend on a constant magnetic field and on the nonreciprocity of the Hall effect. They are the gyrator, isolator, negative-resistance amplifier, and circulator.[28]

Indium antimonide gives the highest output of any known material in Hall effect devices, because their efficiency is proportional to the square of the carrier mobility. Indium arsenide has a somewhat lower efficiency than the antimonide but is still far superior to other materials. The most important differences between the two materials are in the variations of the Hall constant and resistivity with temperature at low doping levels at which the Hall mobility is still high. These variations are

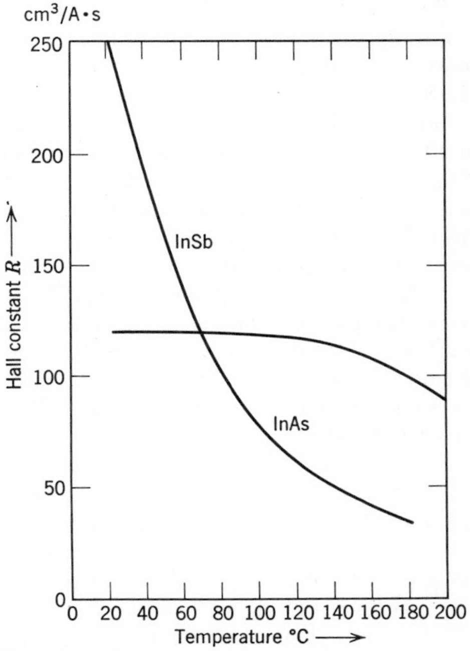

Fig. 3. Hall constant of InSb and InAs as a function of temperature.[29]

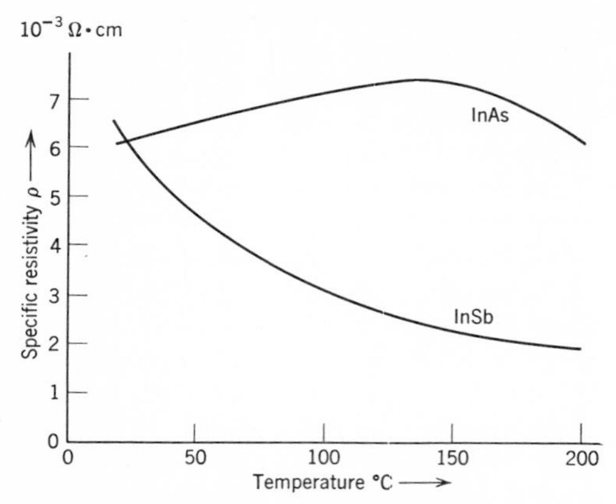

Fig. 4. Resistivity of InSb and InAs as a function of temperature.[29]

illustrated in Figs. 3 and 4. The utility of many of the Hall generator applications, especially the fluxmeter, depends on a lack of drift or sensitivity as well as high linearity. It can be seen that the temperature coefficients are much lower for the arsenide. This is the principal reason why most commercially available Hall effect devices are made with indium arsenide.

7. PHOTOSENSITIVE AND PHOTOVOLTAIC CELLS

Cesium antimonide, Cs_3Sb, is one of the most popular substances for the emitter of photoelectric cathodes.[32–34] The sensitivity shows a threshold at 6800 Å and a maximum at 4000 Å. A high quantum efficiency of 20 to 30 per cent is obtained at 4500 Å. That is to say, only three to five photons in the blue visible range are necessary to cause the emission of one electron. A typical sequence in the formation of this compound emitter is to evaporate antimony onto the intended cathode substrate. This antimony layer is then exposed to cesium vapor at about 150°C until the compound is formed. If monitoring is provided by transmission through a transparent substrate, it is found that maximum photoelectric sensitivity is attained when the layer has a deep red color. The low work function, which contributes to the efficiency of this emitter, is extremely low, as would be expected from the large percentage of the alkali metal cesium.

The outstanding application for photovoltaic cells is in the conversion of solar energy to electrical energy. Those photovoltaic cells of the semiconductor *p-n* junction type used as solar energy converters are known as a solar cells. The power conversion efficiency is the most important characteristic of these units. One of several controlling factors in the efficiency of this conversion is the match between the energy gap of the solid and the energy of the light spectrum. The solar spectrum is known with considerable accuracy now both in outer space and at the earth's surface. The greatest demand for efficient solar cells is for application outside of the earth's atmosphere, at so-called air mass zero. There the solar spectrum contains a large amount of short-wavelength energy which is absorbed in the earth's atmosphere. Thus, although the blue sensitivity of a solar cell is relatively unimportant on the earth's surface, it is of great concern for outer-space applications.

Calculations have been made of the optimum energy gap for the semiconductor to be used in the efficient conversion of solar energy. There is reasonable agreement that from a single band-gap cell the optimum lies at about $E_g = 1.6$ eV, well above that available from any of the elemental semiconductors. Silicon ($E_g = 1.1$ eV) is still the material used for the great majority of solar cells. The components of greatest promise for solar-cell use are gallium arsenide, indium phosphide, and aluminum antimonide.[35,36] Cells are commercially available in gallium arsenide. The conversion efficiency is not quite so high as is regularly attainable in silicon, and the cost per unit cell area is still excessive because of the use of expensive bulk crystal. As was described above, there is much activity in the field of epitaxial film deposition of GaAs. Solar-cell manufacture in this way may eventually allow a compound cell to be competitive with silicon cells.

Some exploratory work on the use of gallium phosphide for solar-cell applications has also been done.[37] The cell open-circuit voltage is greater than a volt because of the large energy gap. The collection efficiency has been so low that the power obtainable is still very low. The temperature coefficient of power output of these early gallium phosphide cells is positive rather than negative, as in the case of silicon and gallium arsenide. The explanation of this effect has not been satisfactory. It is not clear that the temperature coefficient anomaly is intrinsic to the phosphide, but rather it may arise from some peculiar impurity-level effect.

The importance of the temperature coefficient results from the perennial problem of silicon solar-cell heating, both via absorption of infrared energy in the contacts and supports and via joule heating because of finite cell resistance.

8. ELECTROLUMINESCENCE

The relatively new field of optoelectronics concentrates on the transfer of photons to gain advantages of speed, simplification, and isolation over purely electronic systems in computing and control. The electroluminescence which is produced by radiative recombination of injected carriers at semiconductor *p-n* junctions is an important photon source in optoelectronics. A receiver or photovoltaic cell of high conversion efficiency is used as the other side of the couple. The two primary goals in the design of the electroluminescent element are quantum efficiency and optimum light spectrum. An efficiency as near to unity as possible is desired, and the spectral requirements are fixed by the response of the receiver, which is most frequently a silicon photovoltaic cell. This requirement resolves into one for a light source of high quantum efficiency at wavelengths less than about 1 μ, with energy deep into the visible range desirable. For the first time, this appears possible because of the advancing technology of large energy gap III-V intermetallic compounds, particularly GaAs, GaP, and their solid solutions.

The original observations were of radiation from injection and recombination at point or area metal contacts in the compounds GaAs, GaSb, and InP.[38] It was concluded in this early work that the spectral maxima were at the band-gap values, and thus that direct recombination of electron-hole pairs was involved. After several years' lapse of interest, during which diffused *p-n* junctions in GaAs became a reality, a new series of results from GaAs junction diodes at 77°K was published.[39–41] It became clear that the mechanism was not one of simple direct recombination, but that shallow acceptor levels were probably involved. Most exciting of all these results from forward-biased GaAs junctions was the discovery of a high quantum efficiency, apparently close to unity.[40–42] The proof of high frequency modulation capability of this electroluminescence accelerated interest in it. Modulation at 200 Mc was achieved with a limit by the measurement equipment. The *RC* time constant limit of the gallium arsenide diode itself was said to be of the order of 10 gigacycles.[43]

The radiation from GaAs injection electroluminescence has a broad maximum in the infrared around 8500 Å. If the oxygen content of the crystal is high, a peak corresponding to 0.65 eV is observed.[44] It is desirable to shift the electroluminescence used in optoelectronics to shorter wavelengths, both for convenience in avoiding the use of image converters and for more efficient match to receiver sensitivity. The first step in progress in this direction was a series of detailed studies of the injection luminescence in gallium phosphide.[45–49] Although the interpretation of the complex spectrum from GaP is still rather controversial, the following may be a fair summary:

(*a*) The highest efficiency process results from minority carrier injection with two peaks at 5650 Å and 7000 Å.

(*b*) The green peak is associated with zinc doping and the red with zinc plus oxygen. The bounding concentrations have been fairly well defined.[48]

(*c*) The mechanism of the green electroluminescence is probably from the conduction band to the shallow zinc acceptor level. The mechanism of the red electroluminescence is more complicated and may involve zinc as a shallow acceptor and oxygen as a deep donor.[47–49]

(*d*) Lower-efficiency electroluminescence peaks at 5800 Å and 6500 Å by mechanisms which are even more uncertain.[49]

These processes in GaP were all soon recognized to be less efficient than those in GaAs. This difference is explained on the basis of differences in the band structure of the two compounds. The conduction band minimum and the valence band maximum both occur at $k = 0$ in GaAs. Direct transitions, whether they be from band-to-band or from band-to-impurity level can occur without phonon energy loss. The conduction band minimum is not at $k = 0$ for GaP or for $x < \frac{1}{2}$ in the solid solution series $GaAs_xP_{1-x}$. As a result, intensive study of alloys with $x > \frac{1}{2}$ has begun in order to find the best composition for both visible and high-efficiency electroluminescence.

The variation of emission peak wavelength versus x has been published.[50]

Thorough studies of the variation of transmission of light with conductivity type and temperature in gallium arsenide have led to a better understanding of the internal losses of efficiency in electroluminescence diodes.[51] Two excellent reviews of this explosively growing subject have been written by Gershenzon.[52,53]

Accurate measurements of diode electroluminescence has experimentally proved that internal quantum efficiencies of GaAs *p-n* junctions are 50–90 per cent.[54,55] The design of efficient light sources from the material has been resolved principally into design for efficient extraction of the useful light. Commercial electroluminescent lamps are being manufactured and sold by at least two American firms.

9. INJECTION LASERS

One of the most exciting discoveries of recent time in solid-state science was that the radiation from GaAs diodes described in the prèvious section became coherent, narrowband, and directional when a threshold current was exceeded. Several groups of workers in the United States simultaneously and independently (followed closely by workers abroad) recognized the effect as that of stimulated emission.[56–59] This is the phenomenon basic to laser action. The injection lasers offer the possibility of electrical pumping, thus avoiding bulky and complicated optical pumping arrangements. They are small in size and easily adaptable to modulation. They are apparently inferior to the other solid-state lasers and to the gas lasers on the counts of coherence, directionality, and power. These lasers operate efficiently at 77°K with threshold currents of 1000–2000 amperes/cm^2 depending on the optical mode. Continuous operation of small diodes is possible under these conditions. Good directionality has been obtained by the use of cleaved rectangular parallelepipeds.[60] The active region is on the *p*-type side of the junction and is about 10 microns thick.[61] The spatial distribution of radiation has been studied.[62] The volume of the light-emitting region does not change when the current is raised within the region of stimulated emission. Shaping of the diodes has led to some improvements in efficiency.[63] The widespread interest in GaAs lasers has stimulated detailed measurements of refractive index[64] and optical absorption.[65]

The effect has been observed at room temperature with threshold currents approximately ten times those needed at 77°K.[63] At 4.2°K, the threshold current is a factor of fifteen less than that at 77°K.[58] Detailed studies of the coherent output and threshold current in the temperature range 4 to 125°K have been published.[67]

Laser work with $GaAs_xP_{1-x}$ crystals has shown that a *p-n* junction laser can probably be built with output wavelength from below 6200 Å (2.0 eV) to near 8400 Å (1.48 eV).[69]

REFERENCES

General Review

1. Pincherle L., and J. M. Radcliffe, *Advan. Phys.* **5** (1956) 271.
2. Welker H., and H. Weiss, *Solid State Phys.* **3** (1956) 1.
3. Cunnell F. A., and E. W. Saker, "Properties of III-V Compound Semiconductors," *Progress in Semiconductors*, A. F. Gibson, ed., Vol. 2, Wiley, New York, 1957, p. 37.
4. Cunnell F. A., J. T. Edmond, and W. R. Harding, *Solid-State Electron.* **1** (1960) 97.
5. O'Connor J. R., and J. Smiltens, eds., *Silicon Carbide*. Pergamon Press, Oxford, 1960.
6. Hilsum C., and A. C. Rose-Innes, *Semiconducting III-V Compounds*. Pergamon Press, Oxford, 1961.
7. Hulme K. F., and J. B. Mullin, *Solid-State Electron.* **5** (1962) 211.
8. Willardson R. K., and H. L. Goering, *Compound Semiconductors*, Vol. 1. Reinhold Publishing Corporation, New York, 1962.
9. *Selected Constants Relative to Semiconductors*, Vol. 12 of *Tables of Constants and Numerical Data*. Pergamon Press, Oxford, 1961.

Special Properties

10. Warekois E. P., and P. H. Metzger, *J. Appl. Phys.* **30** (1959) 960.
11. White J. G., and W. C. Roth, *J. Appl. Phys.* **30** (1959) 946.

12. Ellis S. G., *J. Appl. Phys.* **30** (1959) 947.
13. Faust J. W., Jr., and A. Sagar, *J. Appl. Phys.* **31** (1960) 331.
14. Gatos H. C., and M. C. Lavine, *J. Electrochem. Soc.* **107** (1960) 427.

Diodes and Transistors

15. Davis R. E., "Gallium Phosphide for Power Rectifiers," *Properties of Elemental and Compound Semiconductors*. Interscience, New York, 1960, p. 295.
16. Lowen J., and R. H. Rediker, *J. Electrochem. Soc.* **107** (1960) 26.
17. Sharpless W., *Bell System Tech. J.* **38** (1959) 259.
18. Jenny D. A., *Proc. IRE* **46** (1958) 717.
19. Holonyak N., and I. A. Lesk, *Proc. IRE* **48** (1960) 1405.
20. Batdorf R. L. et al., *J. Appl. Phys.* **31** (1960) 613.
21. Pikor A., G. Elie, and R. Glicksman, *J. Electrochem. Soc.* **110** (1963) 178.
22. Coupland M. J., C. Hilsum, and R. J. Sherwell, *Solid-State Electron.* **5** (1962) 405.
23. Gunn J. B., *Solid-State Commun.* **1** (1963) 88.
24. Jenny D. A., *Proc. IRE* **47** (1958) 959.
25. Henneke H. L., *Solid-State Electron.* **3** (1961) 159.
26. Rutz R. F., *Proc. IEEE* **51** (1963) 470.

Galvanomagnetic Devices and Photocells

27. Willardson R. K., and A. C. Beer, *Elec. Mf.* **57** (1956) 79.
28. Grubbs W. J., *Bell System Tech. J.* **38** (1959) 853.
29. Kuhrt F., *Siemens-Z.* **28** (1954) 370.
30. Saker E. W., F. A. Cunnell, and J. T. Edmond, *British J. Appl. Phys.* **6** (1955) 217.
31. Schillmann E., *Techn. Rundschau* (October 9, 1957).
32. Gorlich P., *Z. Physik* **101** (1936) 335.
33. Sommer A., *Proc. Phys. Soc.* **55** (1943) 145.
34. Spicer W. E., *Phys. Rev.* **112** (1958) 114.
35. Jenny D. A., J. J. Loferski, and P. Rappaport, *Phys. Rev.* **101** (1956) 1208.
36. Rappaport P., *RCA Review* **20** (1959) 373.
37. Grimmeiss H. G., W. Kischio, and H. Koelmans, *Solid-State Electron.* **5** (1962) 155.

Electroluminescence

38. Braunstein R., *Phys. Rev.* **99** (1955) 1892.
39. Nasledov D. N. et al., *Fiz. Tverd. Tela* **4** (1962) 1062; *Soviet Phys.-Solid State* **4** (1962) 782.
40. Keyes R. J., and T. M. Quist, *Proc. IRE* **50** (1962) 1822.
41. Black J., H. Lockwood, and S. Mayburg, *J. Appl. Phys.* **34** (1963) 178.
42. Galginaitis S. V., *J. Appl. Phys.* **35** (1964) 295.
43. Pankove J. I., and J. F. Berkeyheiser, *Proc. IRE* **50** (1962) 1976.
44. Turner W. S., G. D. Pettit, and N. G. Ainslie, *J. Appl. Phys.* **34** (1963) 3274.
45. Grimmeiss H. G., and A. Rabenau, *J. Appl. Phys.* **32** (1961) 2123.
46. Gershenzon M., and R. M. Mikulyak, *J. Appl. Phys.* **32** (1961) 1338.
47. Gershenzon M., and R. M. Mikulyak, *Solid-State Electron.* **5** (1962) 313.
48. Starkiewicz J., and J. W. Allen, *J. Phys. Chem. Solids* **23** (1962) 881.
49. Ullman F. G., *J. Electrochem. Soc.* **109** (1962) 805.
50. Allen J. W., and M. E. Moncaster, *Phys. Letters* **4** (1963) 27.
51. Hill Dale E., *Phys. Rev.* **133** (1964) A866.
52. Gershenzon M., *Radiative Recombination in the III-V Compounds*. Bell Telephone Laboratories, Murray Hill, New Jersey,
53. Gershenzon M., "Electroluminescence from *p-n* Junctions in Semiconductors," Ch. 11 in *The Luminescence of Inorganic Solids*, Paul Goldberg, ed.
54. Hill D. E., *J. Appl. Phys.* **36** (1965) 3405.
55. Carr W. N., *IEEE Trans. Electron Devices*, *ED*-12 (1965) 531.

Injection Lasers

56. Nathan I., W. P. Dumke, G. Burns, F. H. Dill, Jr., and G. Lasher, *Appl. Phys. Letters* **1** (1962) 62.
57. Hall R. N., G. E. Fenner, J. D. Kingsley, T. J. Soltys, and R. O. Carlson, *Phys. Rev. Letters* **9** (1962) 366.

58. Quist R. M. et al., *Appl. Phys. Letters* **1** (1962) 91.
59. Bolger B. et al., *Phys. Letters* **3** (1963) 252.
60. Burns G., R. A. Laff, S. E. Blum, F. H. Dill, Jr., and M. I. Nathan, *IBM Journal* **7** (1963) 62.
61. Michel A. E., E. J. Walker, and M. I. Nathan, *IBM Journal* **7** (1963) 70.
62. Fenner G. E., and J. D. Kingsley, *J. Appl. Phys.* **34** (1963) 3204.
63. Franklin A. R., and R. Newman, *J. Appl. Phys.* **35** (1964) 1153.
64. Marple D. T. F., *J. Appl. Phys.* **35** (1964) 1241.
65. Turner W. J., and W. E. Reese, *J. Appl. Phys.* **35** (1964) 350.
66. Burns G., and J. I. Nathan, *IBM Journal* **7** (1963) 72.
67. Engeler W. E., and M. Garfinkel, *J. Appl. Phys.* **34** (1963) 2746.
68. Mayburg S., *J. Appl. Phys.* **34** (1963) 3417.
69. Holonyak N., Jr., and S. F. Bevacqua, *Appl. Phys. Letters* **1** (1962) 82.

chapter 29

Superconductive Properties

B. W. ROBERTS
General Electric Research and Development Center
Schenectady, New York

1. INTRODUCTION

The first superconductive alloy that may have contained an intermetallic compound was an amalgam formed from tinfoil and liquid mercury and found to have a critical temperature of 4.2°K by Onnes[1] in 1913. The science of crystal-structure determination was in its infancy, so that the exact composition and the crystal structures present at the time are unknown. Little additional research on alloys and compounds was made until the period of 1928 to 1932. At this time many intermetallic compounds and alloys, as well as carbides, nitrides, and sulfides were studied at three locations: by McLennan and co-workers at Toronto, by W. Meissner and co-workers near Munich, and by De Haas and co-workers at Leiden. Also, during this period compounds containing nonmetals were found to be superconductive, and superconducting alloys were formed of two elements otherwise not superconductive to the lowest temperatures tested.

An extensive study of compounds and materials was not undertaken until the 1950's. In this decade the number of superconductive compounds increased possibly fivefold, a microscopic theory of superconductivity evolved, intense study and development effort were applied to the superconductive cryotron for computer use, and the first practical utilization of the superconductive phenomenon was instigated by the discovery of high-field superconductivity. Many exploratory studies of superconductive materials were made by Matthias and co-workers that led to empirical rules for estimating the probable occurrence of superconductivity in a compound or alloy and to the discovery that certain crystal-structure types had unusual superconductive properties.

A schematic description of some of the properties of superconductors is shown in Fig. 1. The central physical phenomena are the complete loss of electrical resistance* below a critical temperature, T_c, and, while

* Experiment has shown that the resistivity is no more than 10^{-17} of the normal state resistivity of copper at room temperature.

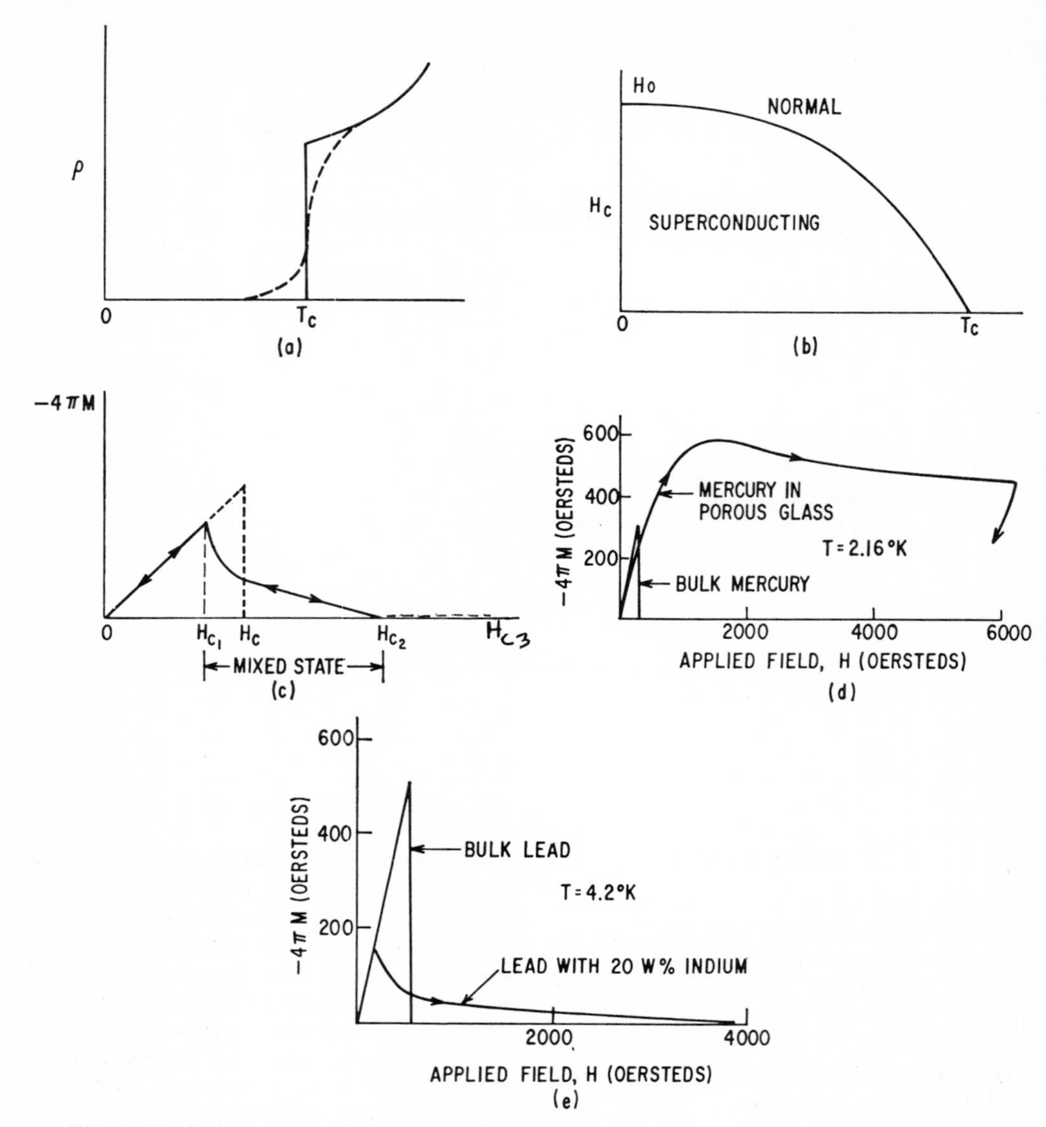

Fig. 1. Physical properties of superconductors. (*a*) Resistivity versus temperature for pure and perfect lattice (solid line). Dashed line indicates impure and/or imperfect material. (*b*) Magnetic field-temperature dependence for type I or "soft" superconductors. (*c*) Magnetization curve for a soft superconductor (dashed line) and magnetization for an Abrikosov-Goodman model superconductor (solid line) after Livingston.[4,152] (*d*) Bulk Hg magnetization behavior compared to filamentary model behavior after Bean.[3] Note irreversibility. (*e*) Example[4] of Abrikosov-Goodman model behavior in Pb and Pb with 20 wt. % In.

in this superconductive state, the exclusion of magnetic flux up to a value H_c. At the field H_c the superconducting state is quenched, and the metal returns to the normal state. Figures 1*a* and *b* illustrate this behavior of super conductors variously described as type I, "soft," or ideal. Another type of behavior known as "mixed-state" or type II superconductivity allows the superconductive state to exist in large magnetic fields. In Fig. 1*c* the magnetization of a superconductor as a function of field is shown for a type II superconductor according to theoretical models of Abrikosov and Goodman. As is shown, at the field H_{c1}, which is lower than H_c, the magnetization curve falls below the perfect diamagnetism of type I superconductivity. The material then enters what is termed the "mixed state," a state pictured as interwoven superconductive and normal regions. At a field H_{c2}, greater than H_c, the sample reverts completely to the normal state. High-field superconductors, instead of reverting to the normal state at a field H_c (usually less than a few thousand oersteds), may carry supercurrents in fields up to several hundred thousand oersteds. The announcement by Kunzler[2] of a Nb_3Sn wire complex that carries large supercurrents in intense magnetic fields has stimulated great activity in both the study of the basic phenomena and the search for and discovery of new superconductive compounds.

A superconducting surface sheath, recently predicted[151] to occur in magnetic fields well above H_{c2}, has been clearly established by new experimental work and reinterpretation of existing data. The superconductive sheath occurs up to 1.69 H_{c2} or H_{c3} (see Fig. 1*c*) when the direction of the magnetic field lies in the plane of the superconductor surface. The sheath may persist below H_{c2} to H_{c1} concurrently with another type surface phenomenon[152] induced by the resistance to passage of magnetic field bundles (fluxons) through the surface.

Another group of high-field superconductive materials are those characterized by a filamentary structure. Recent work on this type by Bean[3] is illustrated in Fig. 1*d*. The sample consisted of a meshwork of mercury filaments formed by forcing the liquid into the pores of unfired Vycor glass. The filaments had diameters of 30 Å to 50 Å. The magnetization for this case breaks away from the linear change observed for bulk mercury below H_c, continues to a maximum, and smoothly decreases to zero in a sufficiently high field found to be many times H_c. In Fig. 1*d* the field has been reversed before the high field has been attained, and irreversibility is evidenced. This hysteresis is produced by supercurrents trapped in the multiply connected meshwork.

Figure 1*e* shows the experimental magnetization measurements of Livingston[4] for Pb containing 20 w/o In. The correspondence with the Abrikosov-Goodman model is striking. There is observed an order of magnitude increase in the maximum field H_{c2} below which supercurrents may still flow.

As these high-field superconductivity models and experimental results indicate, much greater diversity has entered the description and study of superconductive materials beyond the simple parameters T_c and H_0. The magnetic and electrical properties may be controlled by composition and crystal structure, impurities, dislocations, strains, or sample size, shape, and configuration.

The critical temperature has become very important in utilizing high-field superconductive properties because H_c and H_{c2} increase with T_c. Thus materials with large T_c are sought when conductors of supercurrents are needed to produce large magnetic fields. This need has led to extensive search for high T_c superconductive compounds because almost all materials with T_c above 10°K exist as compounds or modifications of compounds.

2. THEORY RELATING TO SUPERCONDUCTIVE COMPOUND FORMATION

An important step in establishing the criteria for the superconductive state in elements, alloys, and compounds was developed by Matthias[5,16] from empirical considerations. In Fig. 2 is sketched Matthias' function $T(n)$ describing the qualitative variance of T_c with the average number of valence electrons per atom in an element, alloy, or compound. The average number of valence electrons per atom for alloys and compounds is the simple average of the valence (counting all electrons outside a filled shell) for each element in the pure state weighted as to composition. For example, $Nb_3^{(val.5)}Sn^{(val.4)}$ has 4.75 valence electrons/atom. For the nontransition metal elements, $T(n)$ was suggested to be a smooth function of the valence electrons/atom, whereas for transition metals a peaked function was given with maxima at three, five, and seven valence electrons/atom. The limiting valences for the occurrence of superconductivity have been extended[16] from two and eight to appreciably greater than one or less than ten, and the maximum at three valence electrons/atom has been eliminated.[38] Matthias also suggested from empirical considerations that T_c would vary as the atomic volume raised to the fourth or fifth power and vary inversely as the atomic mass. The volume and mass correlations were not generally applicable to materials with different crystal structures. It is noted that T_c varies with isotopic mass in the same element as the inverse half-power (or less for some of the transition elements). DeSorbo[146] has demonstrated a relationship between solute and solvent atom size and the effective electron-per-atom ratio upon T_c. The superconductive materials he studied were solid solutions with up to 10 per cent solute atom present.

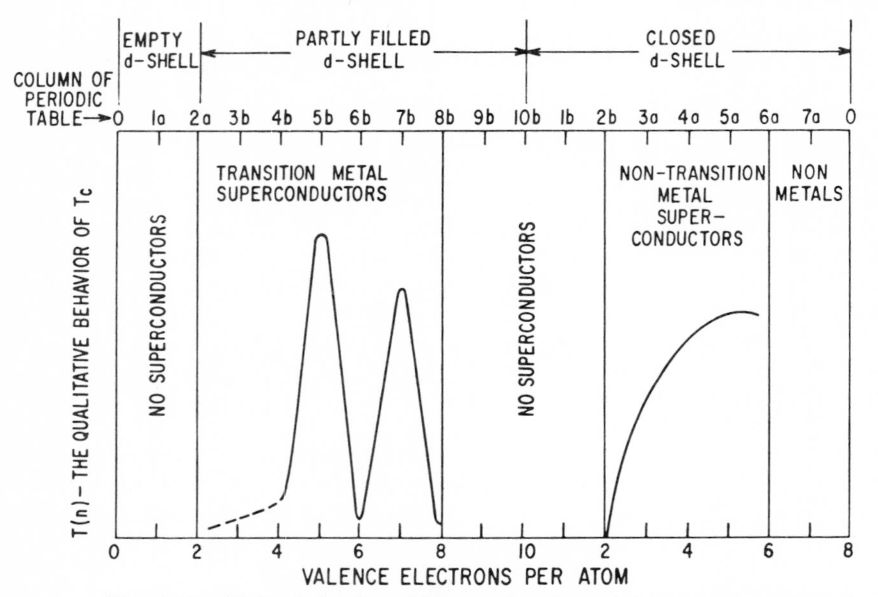

Fig. 2. Qualitative behavior of T_c as a function of the average number of valence electrons per atom as proposed by Matthias.[5,38]

Pines[6] undertook a theoretical study to seek the basis of the empirical regularities demonstrated by Matthias and co-workers. He used the Bardeen, Cooper, and Schrieffer[7] (BCS) microscopic theory of superconductivity as a base. Regularities were noted such as (1) superconductivity is observed only for metals and (2) no ferromagnetic or antiferromagnetic substances superconduct. Pines considered the calculation of the interaction energy V between electrons from first principles. V occurs in the equation[7]

$$kT_c = 1.14\langle\hbar\omega\rangle_{\text{av}} \exp\frac{-1}{N(0)V}$$

Here k is Boltzmann constant, $\langle\hbar\omega\rangle_{\text{av}}$ is the average energy of the phonons that scatter electrons at the Fermi surface, $N(0)$ is the number of electron states per unit energy range at the Fermi surface, and V is the interaction energy between electrons very close to the Fermi surface. V is composed of an attractive interaction induced by phonons and the repulsive screened Coulomb interaction between the electrons. Pines' analysis yielded qualitative agreement with Matthias' empirical regularities. A smooth increase in T_c with the number of valence electrons per atom was predicted for the nontransition elements. The three peaks in T_c as a function of the number of valence electrons per atom were suggested to result from a cyclic variation of the density of states, $N(0)$.

In an important contribution, Morin and Maita[8] evaluated the BCS[7] relationship by measuring the specific heat as a function of temperature for a series of superconductive materials. From an analysis of the specific heat curve, they obtained T_c, θ_D (the Debye temperature), and $N(0)$. Taking $k\theta_D$ to be roughly proportional to $1.14\langle\hbar\omega\rangle_{\text{av}}$, they made an approximate determination of the interaction energy V. For superconductive materials investigated, they found V to be approximately constant. This implies that the large variations observed in T_c must be due to changes in $N(0)$ or in θ_D. Because θ_D is known not to vary rapidly, we must conclude that

$N(0)$ is the variable causing the maxima in the critical temperature function $T(n)$ of Matthais. This variation of the density of states with the number of valence electrons per atom is nicely illustrated in Fig. 3. Also, if V is roughly constant, a plot of $\log(T_c/\theta_D)$ versus $1/N(0)$ should show straight-line dependence. Figure 4*a* is a plot of the data including materials with electron occupation in the 4*d*- and 5*d*-bands for a series of alloys including Nb_3Sn. The approximate constant slope of the datum lines suggests the constancy of V.

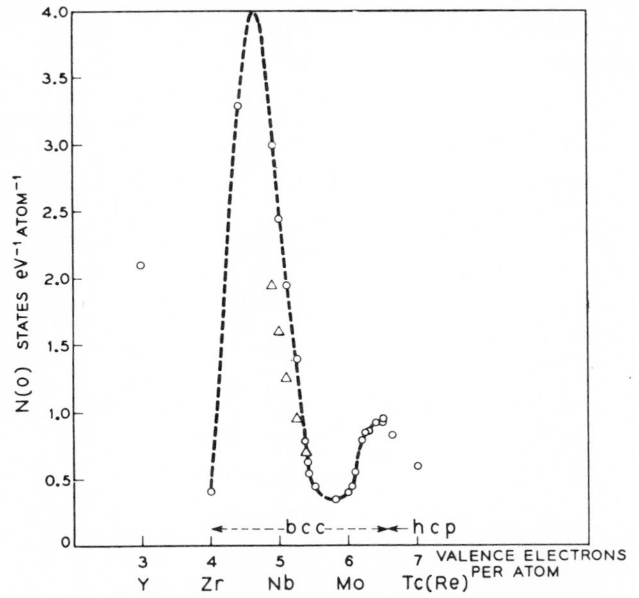

Fig. 3. Plot of $N(0)$, the density of states in number of states of one spin per eV per atom as a function of the number of valence electrons. (From Morin and Maita.[8])

Morin and Maita speculate that, because the upper line in Fig. 4 is not occupied to high $N(0)$ values, discovery of new compound structures or modifications of certain materials may permit attainment of higher T_c than now observed in the five valence electrons/atom peak of $T(n)$ for there the maximum allowable density of states has been nearly attained. The *d*-band should be narrowed possibly by compound formation thus increasing $N(0)$ while maintaining the Fermi level in the center of the peak near seven valence electrons/atom.

The interaction energy has also been considered by Bucher, Heiniger, Muheim, and Müller[9,142] in work on transition-metal alloy systems. They find that evaluation of the specific heat and other parameters indicate some regularities of V to appear if the transition temperatures are considered to be dependent on the electronic specific heat as well as on the position of the alloy components in the periodic system. For transition metals with four to seven valence electrons/atom they have observed, for a given value of the electronic specific heat, lower values of T_c/θ_D if one or both alloy components are 3*d*-elements (Ti, V, Cr, Mn) and higher values if 4*d* or 5*d* elements are involved.

Bucher et al.[9] have also reported study of the binary alloy system, Ti–Zr. Both elements are in the fourth column so that valency changes and consequent band-structure adjustment do not influence $N(0)$ and V. In Fig. 4*b*, T_c is observed to be qualitatively proportional to γ and thus to $N(0)$. However, the calculated V from these data rises steadily from Ti to Zr, which suggests that further study of $N(0)$ and V in alloys and compounds is required before secure generalizations are possible.

Correlations of T_c with other electronic properties have been attempted. The paramagnetic susceptibility of a series of alloys has been measured by Bucher, Heiniger, and Müller[9,10] as well as by Bender et al.[11] They

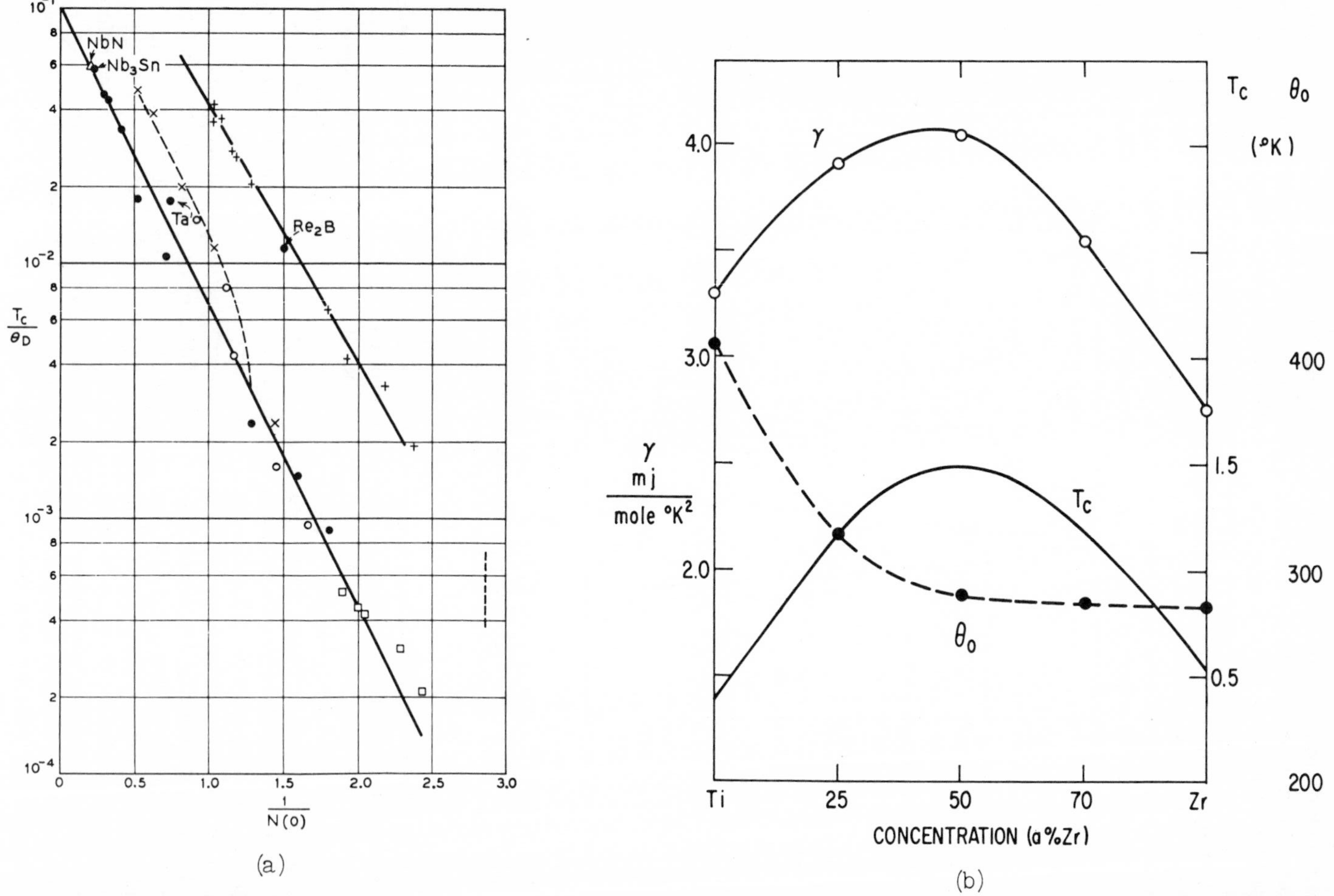

Fig. 4. (*a*) Experimental results for a variety of alloys and compounds plotted according to Eq. 1 from Morin and Maita.[8] Lower curve is for data for second peak in *d*-band; upper curve is for data for third peak in *d*-band. (*b*) Coefficient of the electronic specific heat, γ, Debye temperature at $T = 0$, θ_0, and T_c of hexagonal-close-packed Ti–Zr alloys.[9]

found no relationship between the susceptibility and superconductive properties, and suggested that this may be the result of orbital paramagnetic terms even larger than the spin paramagnetism. Further study has led to a scheme for separating the spin and orbital contributions.[12]

Clogston and Jaccarino[12] and others have discussed Knight shift measurements made by nuclear magnetic resonance (NMR) on a series of V_3X compounds with the A15 or Cr_3O-type structure. They find that the NMR properties and T_c are related to the temperature dependence of the magnetic susceptibility in the normal state. The temperature dependence is believed associated with the vanadium "d" electrons. Blumberg et al.[13] have shown that the quadrupole interaction at the vanadium site in this compound series increases with T_c.

Kikoin and Lazaref[14] noted very early the positive Hall coefficient for superconductive elements. Chapnik[15] has recently compared the Hall coefficient for all of the elements where known and finds that eighteen out of twenty-one known superconductive metals have positive coefficients.* This indicates that the carriers are moving on portions of the Fermi surface such that they appear to be holes. From his analysis he deduced that a criterion for the occurrence of superconductivity is the presence of a sufficient concentration of holes (greater than 10^{19} cm^{-3}). He also proposed that the separation between atoms should lie between 2.6 to 2.9 and 4 angstrom units.

From the discussion above of the physical variables associated with the superconductive state, the most informative measurement would appear to be the specific heat as a function of temperature. This measurement yields information on the density of states and the other necessary parameters to utilize the BCS theory and to obtain the interaction energy associated with the superconductive state. Unfortunately, measurements of the specific heats of superconductive alloys are limited to a relatively small fraction of known superconductive materials. Additional useful measurements are the Hall coefficient and NMR observations that are presently even less available. Thus we generally are limited to data concerning the crystal structure, the valence, and the critical temperature. In a few compounds the critical fields have been measured at least qualitatively and have been reported.[17] In the following sections concerning specific compound series, we shall only consider correlations with the structural parameters and the valence electron-to-atom ratio.

A summation of superconductive materials with known crystal-structure data has been made by Matthias, Geballe, and Compton.[16] Data on all superconductive materials including many negative results have been given by Roberts.[17]

Almost all references to superconductive materials with a component element of D, H, B, C, N, O, and S have been excluded from this report for the purpose of brevity.

3. SUPERCONDUCTIVE BEHAVIOR OF SPECIFIC STRUCTURE TYPES

3.1 Cr_3O(β-W)-Type Phases

The early discovery[34] of the high T_c in Nb_3Sn has led to an extensive search for Cr_3O-type intermetallic compounds. Most of these compounds are superconductive, and a substantial fraction have critical temperatures above 10°K. Because of their importance, the known materials are given in Table 1, beginning with the binary compounds, then the ternary and quaternary alloys, and finally the data on those binaries that exist over an extended composition range plus the mixed binary pairs.

The Cr_3O-type structure,† denoted A15 in the *Strukturbericht* series (see Pearson[45]), is described by space group $O_h{}^3$, $Pm3n$, as having "A" atoms in ordered configurations at positions 000, and $\frac{1}{2}, \frac{1}{2}, \frac{1}{2}$, and "B" atoms in pairs on the faces at positions $\frac{1}{4}, 0, \frac{1}{2}; \frac{1}{2}, \frac{1}{4}, 0; 0, \frac{1}{2}, \frac{1}{4}; \frac{3}{4}, 0, \frac{1}{2}; \frac{1}{2}, \frac{3}{4}, 0; 0, \frac{1}{2}, \frac{3}{4}$. A novel feature of the atomic arrangement is the lines of atoms or "chains," which occur in three orthogonal sets parallel to the unit cell edges, as illustrated in Fig. 5. Several workers have suggested that the occurrence of superconductivity in this intermetallic compound series is associated with the continuity of the lines of atoms. For

* Three elements, Ir, W, and Mo, have been added recently to the list of superconductive metals.

† As pointed out by Nevitt (Chapter 13), there is justification for calling this structure class Cr_3Si rather than Cr_3O.

TABLE 1

Superconductive A15(Cr_3O-Type) Compounds—A. Binary Compounds; B. Ternary and Quaternary Compositions; C. Extended Compositions

Compound	T_c(°K)	a(Å)	Electrons/Atom	Reference
A. Binary Compounds				
$AlMo_3$	0.58	4.950	5.25	16
$AlNb_3$	17.5, 18.0	5.187	4.5	18, 19
$AlNb_3$ (sintered)	17.1	5.186	4.5	20
$AuNb_3$	11.5, 11.0	5.21	6.5(11)[a]	
			4.0(1)	21, 153
AuV_3	0.74	4.883	6.5(11)	
			4 (1)	16
$AuZr_3$	0.92	5.482	5.75(11)	
			3.25(1)	16
$BiNb_3$	2.25	5.320	5.0	148
Cr_3Ir	0.45	4.668	6.75	22
Cr_3Ru (annealed)	2.15	4.683	6.5	22
	3.3			
$GaMo_3$	0.76	4.943	5.25	16, 21
$GaNb_3$	14.5	5.171	4.5	23
GaV_3	16.8	4.816	4.5	24
$GeMo_3$	1.43	4.933	5.5	21
$GeNb_3$	6.9	5.166	4.75	16
GeV_3	6.01	4.769	4.75	25
$InNb_3$	9.2	5.303	4.5	26
$IrMo_3$	8.8, 9.6	4.972	6.75	16, 27
$IrNb_3$	1.7	5.131	6.0	21, 5
$IrTi_3$	5.40	5.009	5.25	28
Mo_3Os	7.20	4.973	6.5	29
Mo_3Si	1.30	4.893	5.5	30
Nb_3Os	1.05	5.121	5.75	16
Nb_3Pt	9.20	5.153	6.25	29, 21
Nb_3Rh	2.5	5.115	6.0	21, 5
Nb_3Sn	18.05	5.289	4.75	29
$Nb_{0.8}Sn_{0.2}$	18.5	5.290	4.8	20
$PbZr_3$	0.76		4.0	16
$PtTi_3$	0.58	5.032	5.5	16
PtV_3	2.83	4.814	6.25	28
$SbTi_3$	5.80	5.217	4.25	28
SbV_3	0.80	4.941	5.0	16
SiV_3	17.0	4.722–8	4.75	32, 25
$Si_{0.263}V_{0.737}$	15.8	4.726	4.74	33
$Si_{0.206}V_{0.794}$	14.5	4.729	4.79	33
$SnTa_3$	6.0	5.276–8	4.75	34
SnV_3	7.0, 3.8	4.94–6	4.75	35, 21
B. Ternary and Quaternary Compositions				
$Al_{0.5}Nb_3Sn_{0.5}$	16.3		4.625	16
$Ga_{0.5}Ge_{0.5}Nb_3$	7.3	5.175	4.625	20
$Ga_{0.5}Si_{0.5}V_3$	8.6–11.9		4.625	16
$Ge_{0.5}Nb_3Sn_{0.5}$	12.6	5.236	4.75	20
$Ir(Ir_{0.33}V_{2.67})$	1.39	4.794	6.33	28
$Mo_{0.15}Si_{0.25}V_{0.60}$	4.54	4.758	4.9	33
$Mo_{0.12}Si_{0.25}V_{0.63}$	5.1		4.87	33
$Mo_{0.05}Si_{0.25}V_{0.70}$	5.59	4.736	4.8	33
$Mo_{0.02}Si_{0.25}V_{0.73}$	10.4		4.77	33
$Mo_{0.009}Si_{0.248}V_{0.743}$	14.0		4.76	33
$Mo_{0.005}Si_{0.25}V_{0.745}$	16.0		4.76	33
$Nb_3Si_{0.5}Sn_{0.5}$	8.3		4.75	36, 37, 138

[a] The alternative valence of 11 for Au rather than the normal value of 1 is considered here and elsewhere as noted.

TABLE 1 *(Continued)*

Compound	T_c(°K)	a(Å)	Electrons/Atom	Reference
$Nb_3Si_{0.6}Sn_{0.4}$	6.5		4.75	36, 138
$Nb_{2.75}SnTa_{0.25}$	17.8		4.75	35
$Nb_{2.5}SnTa_{0.5}$	17.6		4.75	35
$NbSnTa_2$	10.8	5.287	4.75	35
Nb_2SnTa	16.4	5.280	4.75	35
$Nb_2SnTa_{0.5}V_{0.5}$	12.2		4.75	35
$NbSnTaV$	6.2	5.175	4.75	35
$Nb_3Sn_2V_3$	7.4		4.75	36
$Nb_{2.5}SnV_{0.5}$	14.2		4.75	35
Nb_2SnV	9.8	5.171	4.75	35
$NbSnV_2$	5.5	5.115	4.75	35
$Si(V_{0.9}Ru_{0.1})_3$	2.9	4.707	4.97	25, 30
$Si(V_{0.9}Ti_{0.1})_3$	10.9	4.736	4.70	25, 30
$Si(V_{0.9}Cr_{0.1})_3$	11.3	4.697	4.82	25, 30
$Si(V_{0.9}Mo_{0.1})_3$	11.7	4.732	4.82	25, 30
$Si(V_{0.9}Nb_{0.1})_3$	12.8	4.756	4.75	25, 30
$Si(V_{0.9}Zr_{0.1})_3$	13.2	4.724	4.70	25, 30
$Si_{0.9}Ge_{0.1}V_3$	14.0	4.731	4.7	25, 30
$Si_{0.9}Al_{0.1}V_3$	14.05	4.727	4.6	25
$Si_{0.9}B_{0.1}V_3$	15.8	4.720	4.72	25
$Si_{0.9}V_3C_{0.1}$	16.4	4.723	4.75	25
SiV_3(0.25% Fe, Mn)	17.0	4.722	4.8	25
SiV_3(0.4% Fe, Mn)	16.3	4.723	4.8	25
SiV_3(1.3% Fe, Mn)	14.4	4.720	4.8	25
$SiV_{\sim 3}(Al_{0.009})$	16.12	4.727		33
$SiV_{\sim 3}(Al_{0.053})$	<14	4.733		33
$SiV_{\sim 3}(B_{0.088})$	16.2	4.722		33
$SiV_{\sim 3}(Be_{0.06})$	15.6	4.726		33
$SiV_{\sim 3}(Be_{0.002})$	16.6	4.726		33
$SiV_{\sim 3}(C_{0.016})$	16.5	4.724		33
$SiV_{\sim 3}(Ce_{0.034})$	15.32	4.729		33
$SiV_{\sim 3}(Cr_{0.05})$	<14	4.709		33
$SiV_{\sim 3}(La_{0.002})$	16.48	4.727		33
$SiV_{\sim 3}(Mn_{0.007})$	16.25	4.721		33
$SiV_{\sim 3}(O_{0.016})$	15.94	4.726		33
$SiV_{\sim 3}(Re_{0.048})$	<14	4.727		33
(Note: Starting V at 99.4% and Si at 99.8% atomic for series above.)				
$SnTa_2V$	3.7	5.174	4.75	35
$SnTaV_2$	2.8	5.041	4.75	35
	C. Extended Compositions			
$Al_{0.22-0.31}Nb_y$ (sintered)	16–18.0		4.56–4.83	39
$Al_{0.17-0.35}Nb_y$	15.2–17.6		4.66–4.30	39
$Al_{1-x}Nb_3Sb_x$	<4.2–15.7			40
$Al_{0-0.8}Nb_3Sn_{1-0.2}$	14.6–18.1	5.262–5.292		20
$As_xNb_3Sn_{1-x}$	17.91–18.0			41
$Cr_{0.8-0.9}Ir_y$	0.5–0.78			22
$Cr_{0.7-0.85}Ru_y$	1.13–2.15			22
$Ga_{0-0.6}Nb_3Sn_{1-0.4}$	14.0–18.1	5.230–5.288		20
$Ga_xNb_3Sn_{1-x}$	18.0–18.35			41, 16
$Ga_{0.34-0.2}V_y$	5.3–14.5			24
$Ge_{0.14-0.22}Nb_y$	4.9–5.5			20
$Ge_xNb_3Sn_{1-x}$	17.6–18.0			41
$In_{0-0.3}Nb_3Sn_{1-0.7}$	18.0–18.19			41, 42
$Nb_3Sb_{0-0.3}Sn_{1-0.7}$	14.7–18.0			42, 41
$Nb_3Sb_{0-0.7}Sn_{1-0.3}$	6.8–18.0	5.292–5.270		20
$Nb_{0.72-0.84}Sn_y$	18.2–5.6	5.290–5.282		20
$Nb_{3(1-x)}SnTa_{3x}$	14.1–18.0			43, 44
$Nb_{3x}SnTa_{3(1-x)}$	6.0–18.0			44
$Nb_3Sn_{1-x}Tl_x$	18.0–18.16			41

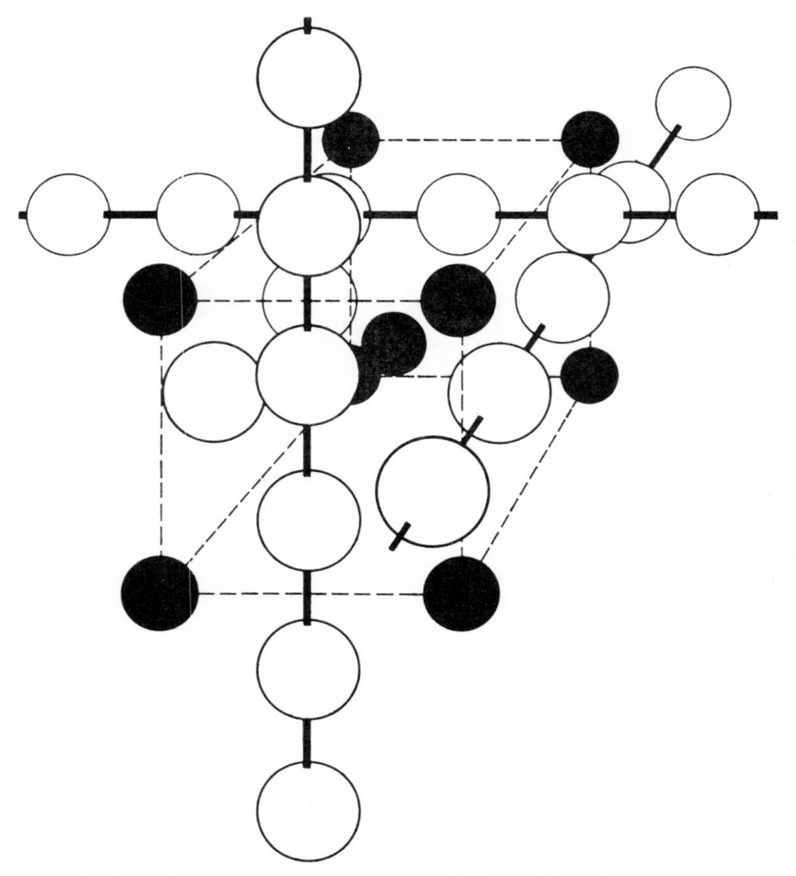

Fig. 5. Sketch of unit cell for A15(Cr_3O)-type structures indicating extended chains or lines of "B" atoms in AB_3.

example, Weger[134] has considered this observation in relation to the band structure. The Nb atoms in Nb_3"B" compounds are approximately 10 per cent closer together than in metallic Nb, and in general the "B" atoms are also 8 to 14 per cent smaller. The coordination schemes and the determination of atomic radii for this structure series has been developed by Geller[46] and discussed in Chapter 13 by Nevitt. Figure 6 shows the mean atomic volumes of those compounds for which both structure data and T_c are known. The majority of the atomic volumes fall in a broad band with highest T_c occurring for large mean atomic volume. The grouping of SiV_3 and modifications illustrates the small atomic volumes for silicon when alloyed in Cr_3O-type intermetallic compounds.[46a] GaV_3 is noted to have a small volume and high T_c. The plotted datum points take no account of the state of ordering in the compounds, because usually this has not been determined. In Fig. 7 is shown T_c versus the average number of valence electrons per unit volume. This correlation shows a threefold peaking of T_c with most of the high-temperature materials centering around 0.26 valence electrons/Å^3. The silicides occur at a higher value, again illustrating the small atomic volume assumed by the silicon atoms. A few examples of molybdenum and vanadium compounds are found with the larger values.

In Fig. 8 is plotted the average number of valence electrons/atom as a function of T_c. This is the most striking correlation for any compound series. It illustrates very clearly the importance of the density of states in controlling the critical temperature and shows Matthias' empirical correlation distinctly. The mean points of the maxima are to the low side

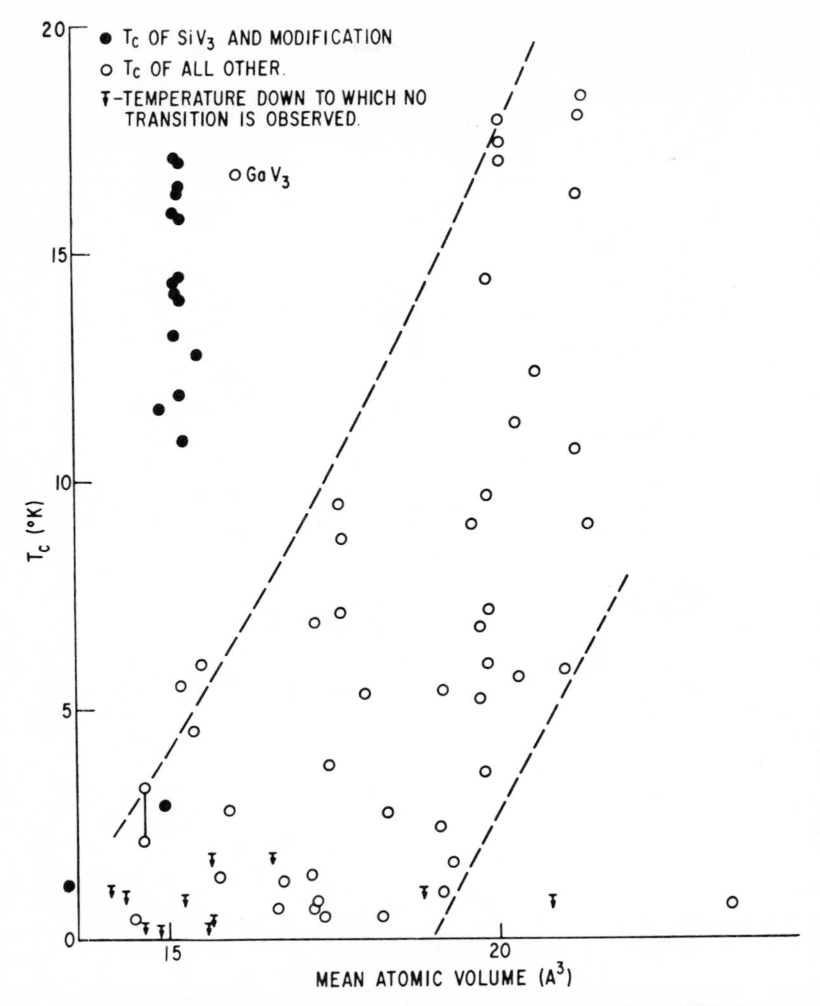

Fig. 6. Critical temperature versus mean atomic volume for Al5(Cr_3O)-type compounds.

of the five and seven peak rule, as has been found in other compounds series as well as in studies of metal solid solutions carried out extensively by Blaugher and Hulm.[47] It is to be noted that only five examples of Cr_3O-type compounds have been found with the average number of valence electrons/atom at four and below, and four of these will fit the plot as well or better if the suggestion by Matthias[87] is followed in which the valence of Au is increased to eleven. The four datum points for optional valence of the Au containing Cr_3O-compounds are marked with tails in Fig. 8 and subsequent figures. Most Cr_3O-type compounds found to date are composed of a transition-nontransition pair or of two transition elements in some cases.

A few studies have been made of the effects of ordering on the superconductive properties. Saur[44] and co-workers have shown that sintered niobium-tin alloys with Cr_3O-type structure give increasing critical temperature as a function of increasing niobium content. If we prepare the compound by vapor deposition, as done by Hanak et al.[48] low critical temperatures are observed. This rapid decrease in critical temperature has been correlated to the high degree of disorder introduced in the vapor deposited Nb–Sn. Figure 9 shows the occupancy of the sites as a function of composition

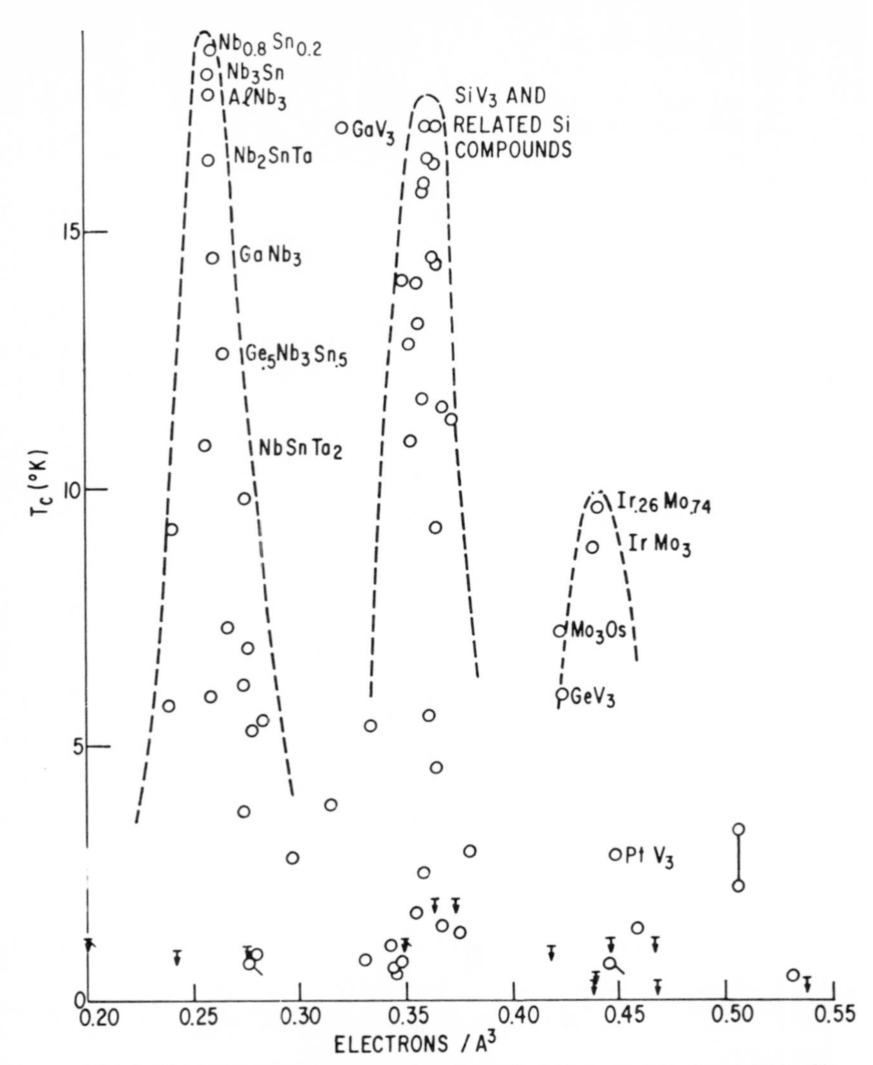

Fig. 7. Critical temperature as a function of electron density for A15(Cr_3O)-type compounds. (*Note*: dashed lines represent apparent envelopes of experimental points only.)

from 75 to 100 a/o niobium for two models. The first model[48] is for an ordered lattice in which the per cent of tin sites occupied by niobium atoms increases from 0 to 100 per cent as the stoichiometric compound composition of Nb_3Sn is exceeded. Niobium sites are not occupied by tin atoms.

The second model considers the occupancy of all sites by a random array of niobium and tin atoms. The per cent of tin sites occupied by niobium atoms varies from 75 to 100 per cent and the niobium sites occupied by tin atoms varies from 25 to 0 per cent as pure niobium is approached. This sketch assumes that the Cr_3O structure exists continuously above 75 per cent niobium. Alloys with "A" component above about 90 per cent atomic content have not been found to have the Cr_3O structure. In Fig. 10 is shown the critical temperature as a function of the per cent of tin sites occupied by niobium as inferred from the known composition and with the assumptions that above 80 per cent the occupancy of the sites in the ordered lattice can be described, as shown in Fig. 9, and that the materials were single-phase. A roughly linear plot of T_c versus

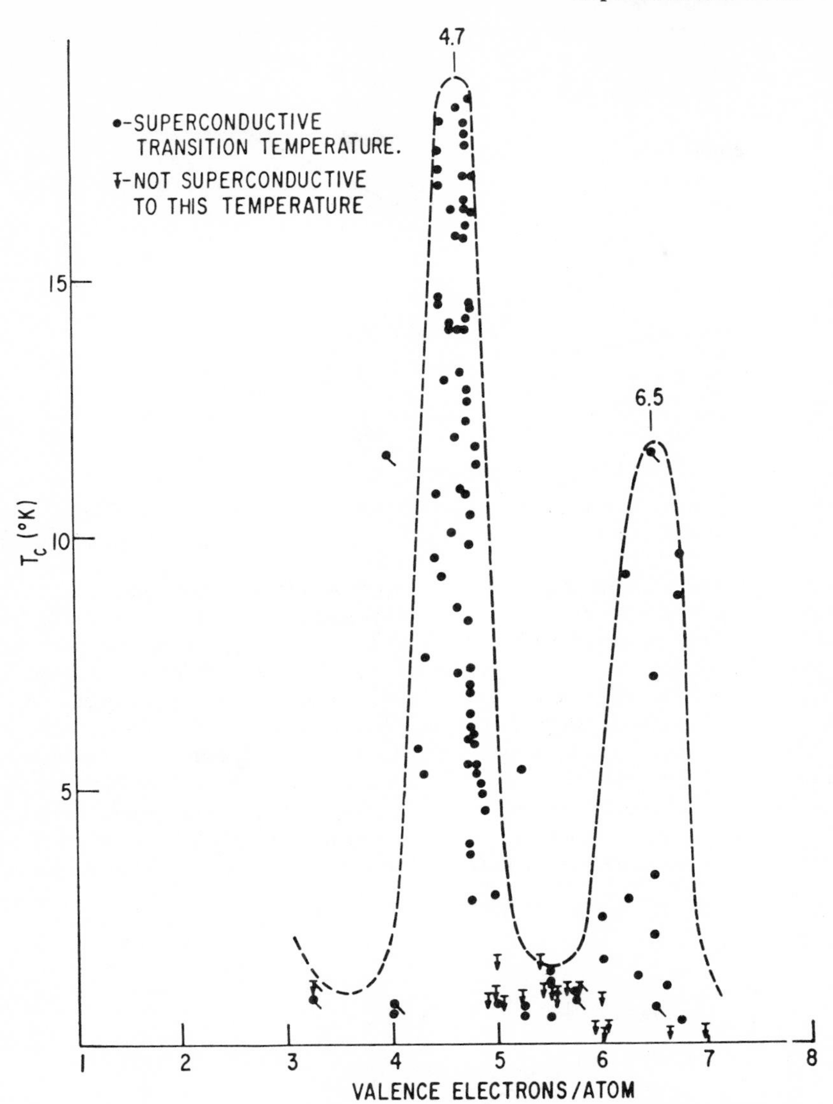

Fig. 8. Dependence of critical temperature on average number of valence electrons per atom for A15(Cr_3O)-type compounds.

the per cent of Sn sites occupied by Nb is now obtained with both the sintered and vapor-deposited samples. Annealing of vapor-deposited films was found to increase T_c, as expected.

Reed and co-workers[20] have also considered the ordering possibilities in a series of Nb–Sn alloys. They found one alloy at a composition denoted by $Nb_{0.8}Sn_{0.2}$, which exhibited a low critical temperature (5.6°K) and which also showed X-ray evidence of a high degree of disorder. This sample had been fired at 1800°C for three hours. A similar sample refired at 1200°C for three hours yielded a critical temperature near the expected value of 18°K. A sample of the same composition annealed for sixteen hours at 1200°C gave a critical temperature of 18.5°K, which is the highest reported critical temperature of any superconductive material.

Disorder may thus be introduced in at least two ways in Nb–Sn alloys: by sintering above certain optimum temperatures and by vapor deposition. The presence of gross disorder

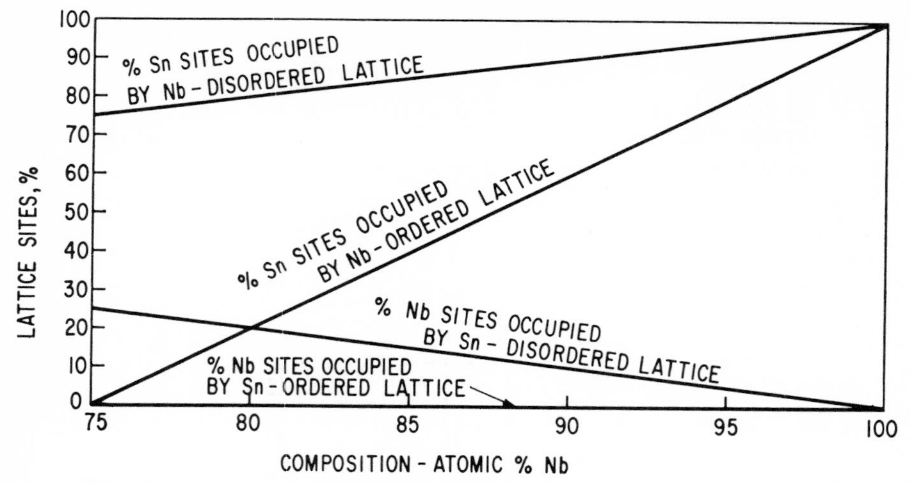

Fig. 9. Lattice site occupancy vs. composition for ordered and disordered Nb–Sn. (By Hanak et al.[48])

decreases T_c in Cr_3O-type alloys, but a small amount of excess Nb on the Sn or "A" sites with the "B" sites fully occupied with Nb appears to raise T_c slightly.

The critical temperature of several A15 compounds has been reduced slightly by neutron irradiation while current carrying capabilities were greatly improved.[154]

The superconductive property data for a variety of ternaries and quaternaries exhibiting the Cr_3O structure (Table 1*B*) will require extensive further analysis to obtain a complete understanding of atom ordering in both the "A" and "B" type sites and the additional possibility of vacancy formation when the "A" sites are not completely occupied.

If a Cr_3O-type ternary alloy series is prepared with compositions lying between two superconductive Cr_3O-type binaries, the critical temperature has been found to be less than that of the binary with the highest T_c, except for very slight increases in T_c found in a few

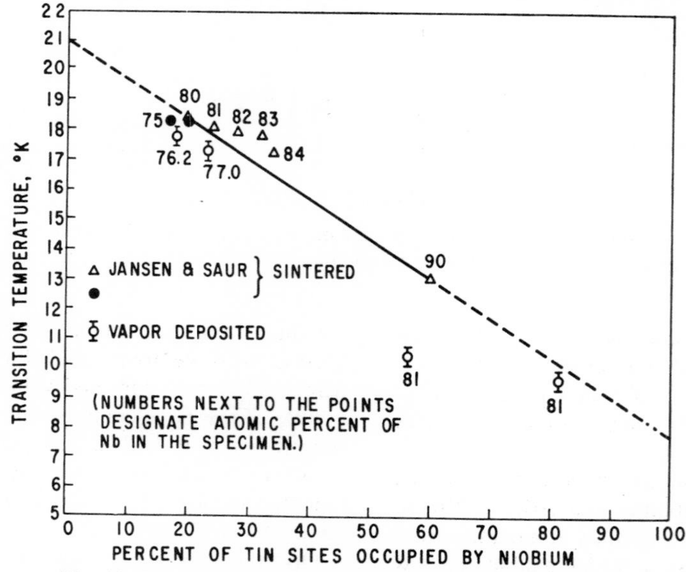

Fig. 10. Transition temperature as a function of per cent Sn sites occupied by Nb atoms for A15(Cr_3O) Nb–Sn. (After Hanak et al.[48])

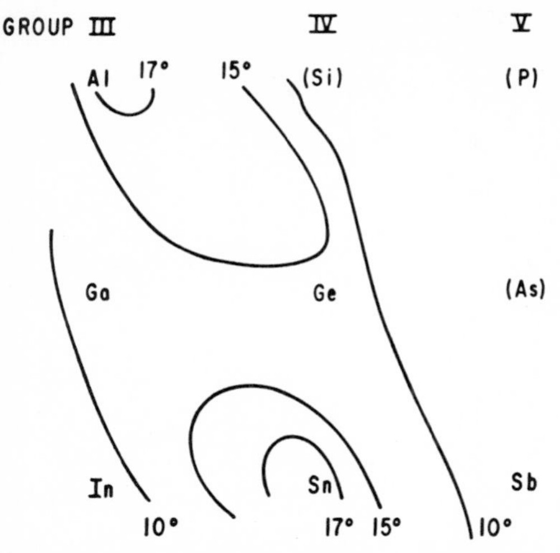

Fig. 11. Iso-T_c lines for Nb_3"B" alloys with A15(Cr_3O)-type structure. (After Reed et al.[20])

cases at compositions near the binary compound. Evidence for this generalization is found in Fig. 11 in which iso-T_c lines for some Nb_3"B" alloys exhibiting the Cr_3O-type structure are plotted.[20] Two high regions in T_c are found around Nb_3Sn and Nb_3Al. In Fig. 12 the ternary representation of $(Nb, Ta, V)_3Sn$ alloy critical temperatures is shown after Cody et al.[35] Each alloy was predominantly Cr_3O-type. This alloy series illustrates the lowering of T_c of the high T_c binary by addition of a third alloying element.

A statistical criterion for high T_c in Cr_3O-type compounds and alloys appears (from the data in Figs. 6 through 12 and related discussion) to be a binary alloy composed of a transition and nontransition element with 4.7 electrons/atom and a mean atomic volume >21 Å^3, which combine to yield a mean electron density of >0.25 electron/Å^3. The alloy would be ordered to the extent that the "A" sites should be fully occupied with one atomic species.

New examples of Cr_3O-type compounds such as Nb_3Bi have been reported to have been prepared under high pressures and temperatures.[26,136,148]

The A15 compound $AuNb_3$ ($T_c = 11.0$°K) forms a solid solution on the body-centered cubic, A2 lattice when rapidly quenched.[153] T_c of the latter is down an order of magnitude to 1.2°K.

3.2 The Sigma-Phase and Related Structure Types*

The σ-Phase ($D8_b$). Sigma phases frequently exist over broad composition ranges.[49] For example, in the Ta–Os and W–Os phase diagrams the σ-phase is found in the composition range 25 to 50 Å/o Os. For the thirty atoms in the unit cell, five nonequivalent crystallographic positions are required. Preferential ordering of sites has been observed in all σ-phases studied in detail. General compositions must lead to one or more of the five positions having mixed occupancy.

Greenfield and Beck[50] have looked for electronic parameters to correlate with the σ-phase structures and have found that the average number of valence electrons/atom was

* These structure types are extensively discussed by Wernick in Chapter 12.

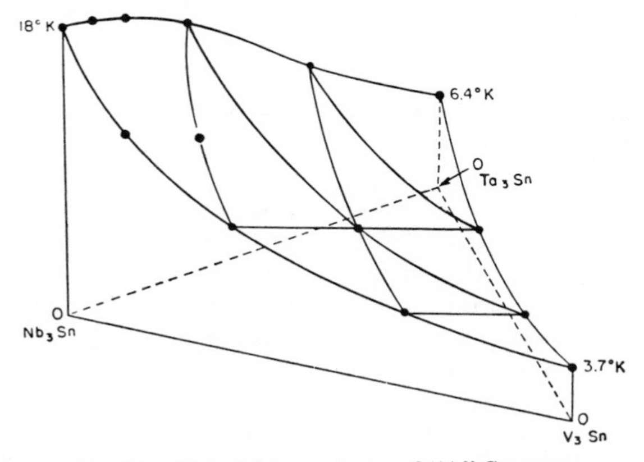

Fig. 12. Critical temperatures of "A"$_3$Sn compounds with A15(Cr_3O)-type structure. (After Cody et al.[35])

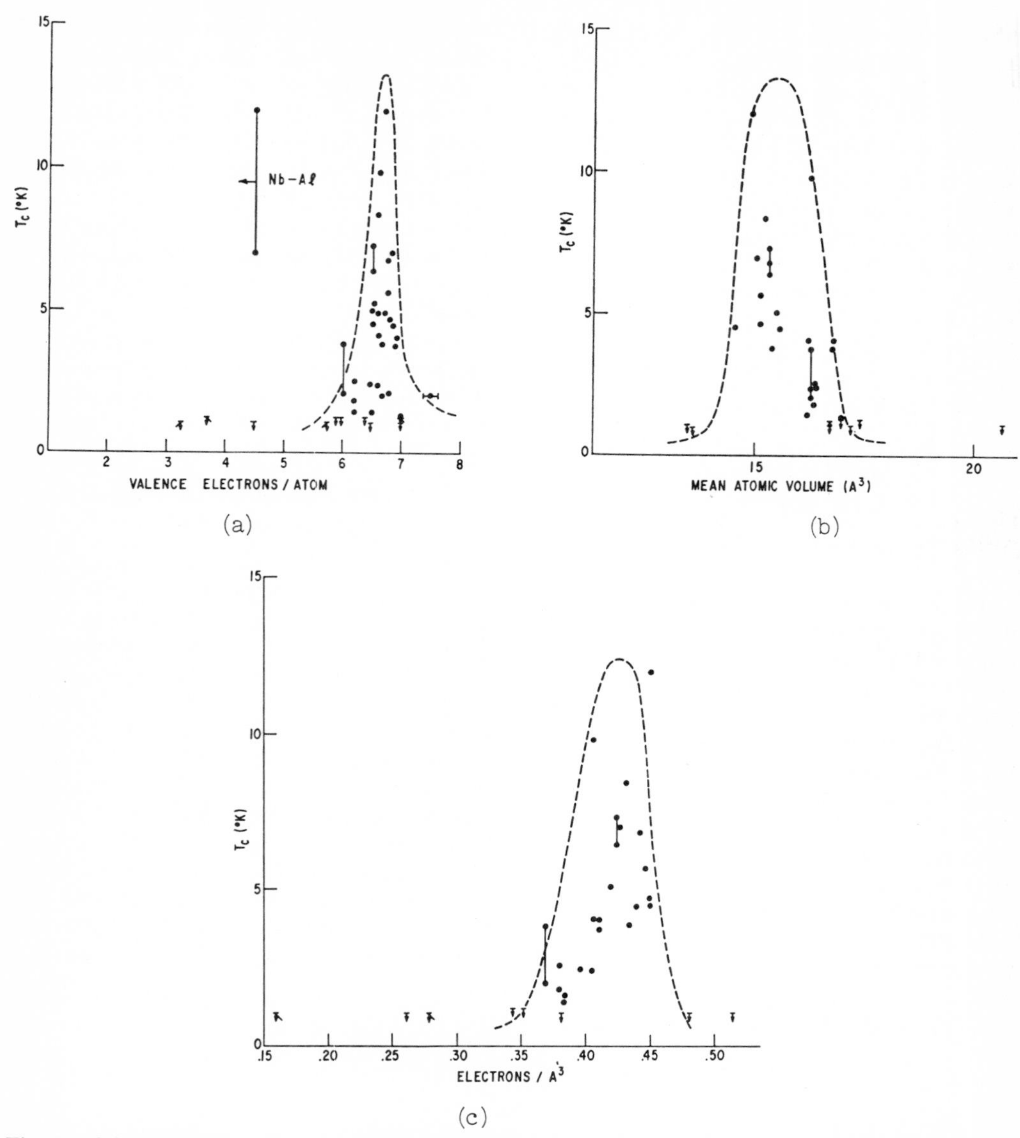

Fig. 13. Critical temperature data for $D8_b$ (σ-phase) compounds: (*a*) versus number of valence electrons per atom, (*b*) versus mean atomic volume, and (*c*) versus density of electrons.

6.93 (9.1 per cent standard deviation) for the mean compositions of σ-phases available. The valence was calculated just as Matthias determined the average valence. When the probable *d*-shell occupancy was considered, they found 3.41 (4.1 per cent standard deviation) and 3.61 (2.8 per cent standard deviation) valence electrons/atom by different schemes. Berman and Shoemaker[51] have suggested that the mean valence electron/atom ratio is between these two values. By taking into account interatomic distances with Pauling's valence scheme, they arrive at 5.76 electrons/atom. They have also discussed the probable valence of atoms associated with each of the five sites, and generalizations of occupancy have been made as well by Kasper and Waterstrat.[52] Unfortunately, relatively few σ-phases have been studied with respect to ordering, and apparently only one of the superconductive binaries, Mo–Re, has been investigated.[137]

The data for σ-phases are given in Figs. 13*a*,

b, and *c* and in Table 2. The T_c versus average valence electron/atom plot shows a pronounced maximum at 6.7. It is to be noted that all T_c of the σ-phases fall in this maximum, except for the Nb–Al alloy that lies between four and five and is reported with a broad superconductive transition. The plot of T_c versus mean atomic volume shows a broad dependence centering upon 15 Å³ to 16 Å³ per atom. The combined parameters of volume and valence show a range of ~0.37 to 0.45 electron/Å³ with the higher T_c associated with the larger values. Bucher, Heiniger, and Müller[55] have given similar correlations with T_c and, in addition, have found T_c times $\bar{M}$ (where $\bar{M}$ is the mean atomic weight) to yield a distinct maximum at 6.7 when plotted against the number of valence electrons/atom. The peak at 6.7 electrons/atom corresponds to the mean valence attributed to all σ-phase data as utilized by Greenfield and Beck.[50]

The α-Mn-Type (A12). This complex cubic structure, which is also generally known as the χ-phase, has four subgroups of atoms with coordinations of 16, 13, and 12. Studies of elemental Mn in the antiferromagnetic state[63] suggest that different atomic states are associated with the various sites. With binary alloys

TABLE 2

Superconductive D8$_b$(σ-Phase) Compounds and Crystal Structure Data

Compound	T_c(°K)	a(Å)	c(Å)	Electrons/Atom	Reference
$Al_{>0.25}Nb_{<0.75}$	7–12			<4.5	18
$Cr_{0.42-0.33}Re_{0.58-0.67}$	~1–2.50	9.26–9.32	4.805–4.845		53, 141
Cr_2Ru	2.02			6.67	22
$Cr_{0.6}Ru_{0.4}$	2.10			6.8	22
$Cr_{0.5}Ru_{0.5}$	1.30			7.0	22
$IrMo_3$	6.8	9.631	4.956	6.75	54
$Ir_{0.37}Nb_{0.63}$	2.40	9.86	5.06	6.48	55
$Ir_{0.4}Nb_{0.6}$	9.8	9.834	5.052	6.6	28
$Ir_{0.28}W_{0.72}$	4.46	9.67	5.00	6.84	28, 56
$Mo_{0.62}Os_{0.38}$	5.65	9.60	4.93	6.76	55
$Mo_{0.5}Re_{0.5}$	6.4, 7.3	9.61	4.98	6.5	55, 57
$Mo_{0.42}Re_{0.58}$	8.4	9.59	4.97	6.58	55
$Mo_{0.6}Ru_{0.4}$	7.0	9.55	4.95	6.8	55, 59
$Mo_{0.3}Tc_{0.7}$	12.0	9.5091	4.9448	6.7	57, 58
$Nb_{0.6}Os_{0.4}$	1.78, 1.85	9.844	5.056	6.2	28, 55
$Nb_?Pd_?$	2.0				59
$Nb_{0.625}Pt_{0.375}$	3.73	9.91	5.12	6.88	28
$Nb_{0.62}Pt_{0.38}$	4.01	9.91	5.13	6.90	55
$Nb_{0.4}Re_{0.6}$[a]	2.5	9.77	5.14	6.2	55
NbRe[b]	2.0–3.8	9.79	5.10	6.0	55
$Nb_{0.6}Rh_{0.4}$	4.04	9.80	5.07	6.6	55
$Os_{0.34}W_{0.66}$	3.81	9.63	4.98	6.68	55
$Os_{0.23-0.33}W_y$	2.5–3.6				60
$Pt_{0.3}Ta_{0.7}$	<1.2–1.5	9.93	5.16	6.5	55
$Re_{0.6}Ta_{0.4}$[c]	1.4	9.77	5.09	6.2	55
$Re_{0.76}V_{0.24}$	4.52	9.45	4.88	6.52	62, 56
$Re_{0.52}W_{0.48}$	5.2			6.52	60
$Re_{0.7}W_{0.3}$	4.9			6.7	60
$Re_{0.5-0.7}W_y$	4.8–5.2				61
$Re_{0.6}W_{0.4}$	4.9			6.6	60
$Re_{0.5}W_{0.5}$	5.03	9.63	5.01	6.5	55
$Rh_{0.4}Ta_{0.6}$	2.35	9.80	5.09	6.6	55
$Ru_{0.4}W_{0.6}$	4.67	9.57	4.96	6.8	55

[a] Plus A12 with a = 9.773Å.
[b] Plus cubic with a = 3.184Å.
[c] Plus A12 with a = 9.783Å.

the valence of different species and the problems of filling the various subgroups in an ordered manner is little understood and has not been studied for compounds found to be superconductive, which are given in Table 3. Almost all of the compounds are transition-metal pairs. Figure 14 shows the three valence and volume correlations with T_c. The peak of the valence electron/atom ratio is near 6.6, although most of the examples fall on the low side. A sharp maximum in T_c is found when plotted against mean atomic volume (Fig. 14*b*). In fact, all superconductive α-Mn type compounds are found within ±4 per cent of the mean atomic value. The number of electrons/unit volume also peaks sharply in T_c at 0.43 electron/atom.

These simple correlations with T_c support the early contention by Beck and co-workers[65] that the α-Mn and other similar materials were "electron compounds." The factors causing the sharply defined mean atomic volume are not known.

Other Phases. No examples of superconductive μ-phases ($D8_5$), *P*-phases, or *R*-phases have been reported.

Only one example of a β-Mn-type (A13) compound has been found to be superconductive: Al_2CMo_3[150] (T_c = 10.0°K, a = 6.86 Å, 4.67 valence electrons/atom).

A single example[155] of a Hume-Rothery phase is now known to be superconductive: $Ag_{0.44}Hg_{0.56}$($D8_{1,2,\text{or }3}$, complex cubic, 52 atoms/unit cell, T_c = 0.64°K, 1.56 electrons/atom).

3.3 The Laves Phases*

The three Laves phases—$MgZn_2$-type, hexagonal, C14; $MgCu_2$-type, cubic, C15; and $MgNi_2$-type, hexagonal, C36—are well described by Wernick[49] and appear to be examples of structure types dependent on packing considerations as well as on electronic properties. The "B" atoms of AB_2 are the smaller atoms and are located at the corners of tetrahedra and joined at the points in C15, at the points and tetrahedra bases in C14, and a mixture of the two schemes in C36. The larger "A" atoms fit into spaces enclosed by the tetrahedra. The three structures are related

* These structure types are extensively discussed by Wernick in Chapter 12.

TABLE 3

Superconductive A12(α-Mn)-Type Compounds and Crystal Structure Data

Compound	T_c(°K)	a(Å)	Electrons/Atom	Reference
$Al_{0.17}Re_{0.83}$	3.35	9.60	6.32	62
$Hf_{0.14}Re_{0.86}$	5.86		6.57	60
$MoRe_3$	9.89, 9.26		6.75	59, 16
$Nb_{0.5}Os_{0.5}$	2.86	9.760	6.5	55
$NbOs_2$	2.52	9.655	7.0	28
$Nb_{0.6}Pd_{0.4}$	2.04–2.47	9.77	7.0	55
$Nb_{0.18}Re_{0.82}$	9.7, 8.89	9.641	6.64	28, 59
$Nb_{0.38}Re_{0.62}$	2.45	9.770	6.24	64, 56
$Nb_{0.32}Re_{0.68}$	4.5	9.730	6.36	64, 56
$Nb_{0.26}Re_{0.74}$	7.2	9.688	6.48	64, 56
$Nb_{0.2}Re_{0.8}$	9.1	9.648	6.6	64, 56
$Nb_{0.14}Re_{0.86}$	8.5	9.610	6.72	64, 56
$Nb_{0.4}Re_{0.6}$	2.36	9.781	6.2	55
$Nb_{0.38}Re_{0.62}$	2.45		6.24	56
$NbTc_3$	10.5	9.625	6.5	57
$Re_{0.64}Ta_{0.36}$	1.46	9.765	6.28	28
Re_3Ta	6.78		6.5	59
$Re_{0.65}Ta_{0.35}$	1.58	9.762	6.3	55
$Re_{0.83}Ti_{0.17}$	6.6	9.587	6.48	28
OsTa	1.95	9.773	6.5	28
$Re_{0.83}Ti_{0.17}$	5.1	9.595	6.49	55
Re_3W	9.0		6.75	59
Re_6Zr	7.40	9.698	6.57	28, 57
Tc_6Zr	9.7	9.636	6.57	57

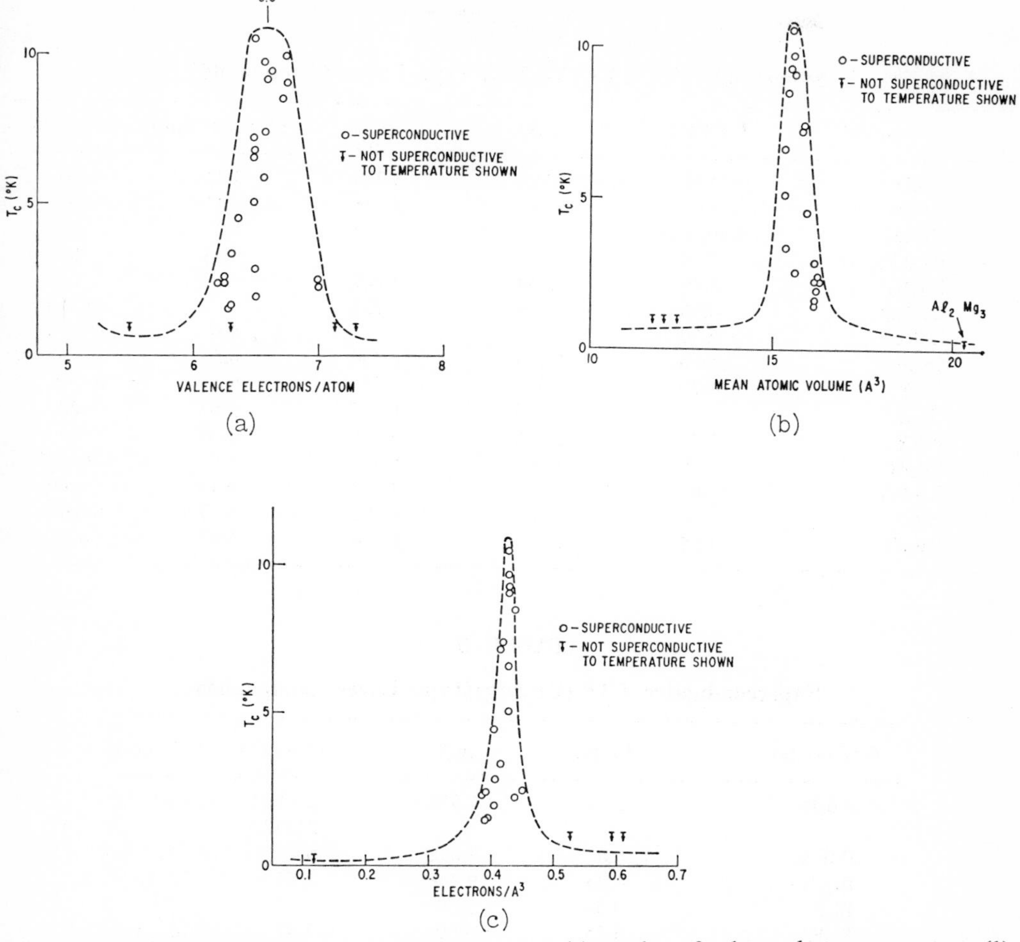

Fig. 14. Data for Al2(α-Mn)-type compounds. T_c versus (*a*) number of valence electrons per atom, (*b*) mean atomic volume, and (*c*) electron density.

by stackings of double layers in different sequences. The "A" atom must be larger than the "B" atom, so that an effective ratio near the ideal of $r_A/r_B = 1.225$ is achieved.

Many of the Laves phases have been studied with compounds made from elements with a low number of valence electrons/atom in the range 1 to ~2, where a systematic valence dependence is found. None of these low-valence compounds or ternary alloys has been reported to be superconductive.

In Table 4, the superconductive $MgZn_2$-type (C14) Laves phases are listed with crystallographic data. The transition metals predominate, with average numbers of valence electrons/atom above six, as shown in Fig. 15. The mean atomic volume peaks at 17.5 Å^3 and the number of electrons per unit volume at 0.36 electron/Å^3.

The $MgCu_2$-type (C15) superconductive compound data are given in Table 5. The mean atomic volume in Fig. 16 is found to be peaked at 17 Å to 20 Å^3 with the maximum T_c at 8.8°K for V_2Zr. The mean atomic volume of the atoms in Bi_2K, Bi_2Rb, and Bi_2Cs is nearly twice that of the majority of superconductive C15 compounds. The Bi atoms in Au_2Bi are found in the opposite position set, and thus Au_2Bi is antiisomorphous to the latter three compounds. It is to be noted that V_2Zr is the only $MgCu_2$-type compound with slightly less than five valence electrons/atom. The plot of electrons/atom and electrons per unit volume are both scattered. An extremely low density

TABLE 4

Superconductive C14 ($MgZn_2$)-Type Laves Compounds

Compound	T_c(°K)	a(Å)	c(Å)	Electrons/Atom	Reference
Al_2Zr	<0.35	5.282	8.748	3.33	16
$HfOs_2$	2.69	5.184	8.468	6.67	66
$HfRe_2$	4.80, 5.61	5.239	9.584	6.0	66
$Ir_{1.5}Os_{0.5}Y$	2.40			6.83	67
$LuOs_2$	3.49	5.254	8.661		66
$LuRu_2$	0.86	5.204	8.725		66, 16
OsReY	2.00			6.0	67, 68
Os_2Sc	4.60	5.179	8.484	6.33	66
Os_2Y	4.7	5.307	8.786	6.33	67, 68
Os_2Zr	3.0	5.219	8.538	6.67	28
Re_2Y	1.83	5.396	8.819	5.67	66
Re_2Zr	6.8, 5.9	5.262	8.593	6.0	16, 139
$Ru_2Sc_{1.2}$	1.67	5.119	8.542	6.12	66, 69
Ru_2Y	1.52	5.256	8.792	6.33	66, 69
Ru_2Zr	1.84	5.144	8.504	6.67	28

TABLE 5

Superconductive C15 (Cu_2Mg)-Type Laves Compounds

Compound	T_c(°K)	a(Å)	Electrons/Atom	Reference
Au_2Bi	1.84	7.958	2.33(1) 9.0(11)	70
$BaRh_2$	6.0	7.852	6.67	71
Bi_2Cs	4.75	9.746	3.67	72, 73
Bi_2K	3.58	9.501	3.67	73
Bi_2Rb	4.25	9.609	3.67	74, 75
$CaIr_2$	4–6.15	7.545	6.67	71
$CaRh_2$	6.40	7.525	6.67	71
$CeRu_2$	4.90	7.535		69, 66
IrOsY	2.60		6.67	68, 67
Ir_2Sc	1.03	7.348	7.0	66
Ir_2Sr	5.7	7.700	6.67	71
Ir_2Th	6.50	7.664		28
Ir_2Y	2.18	7.50–7.52	7.0	66, 67
Ir_2Zr	4.10	7.359	7.33	28
$LaOs_2$	6.5	7.737	6.33	66
$LaRu_2$	1.63	7.702	6.67	69, 66
$Ni_{1.5}V_{0.5}Zr$	0.43	7.068	7.17	16
Pt_2Y	1.57	7.590	7.67	66, 67
Rh_2Sr	6.2	7.706	6.67	71
Ru_2Th	3.56	7.651		28
V_2Zr	8.8	7.439	4.67	28
W_2Zr	2.16	7.621	5.33	28

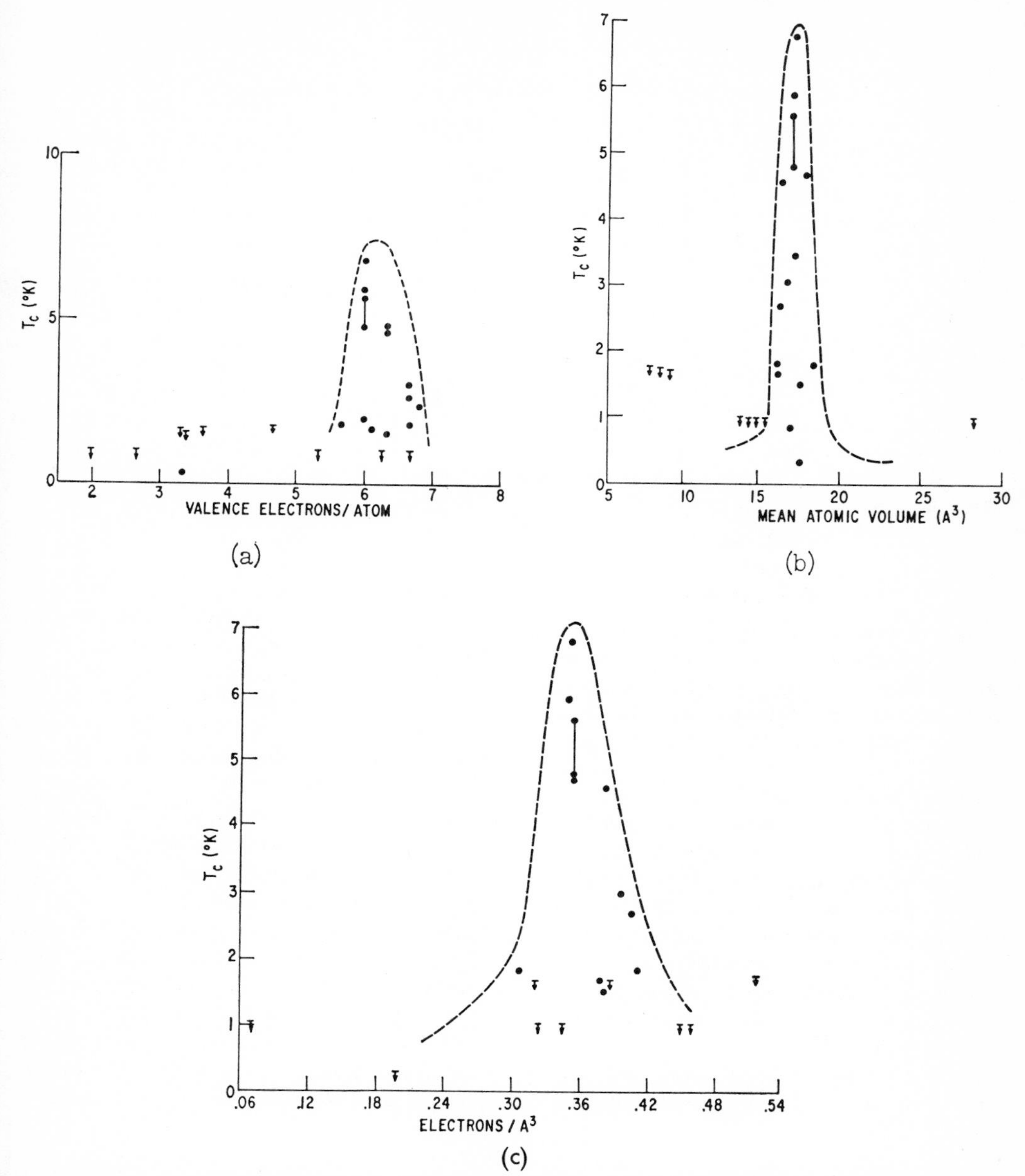

Fig. 15. Data for C14($MgZn_2$)-type Laves phase compounds. T_c versus (*a*) number of valence electrons per atom, (*b*) mean atomic volume, and (*c*) electron density.

of valence electrons (~0.1 electron/Å^3) is observed for the Bi compounds. About one half of the superconductive C15 compounds have 6.67 valence electrons/atom.

$HfMo_2$ is the only example of a $MgNi_2$-type Laves phase (C36) reported tested for superconductivity; no transition was observed[60] down to 1.02°K.

Two of the three Laves phases thus have been shown to exhibit the superconductive state, especially when at least one component is a transition-metal element. The critical temperatures all lie below 10°K. Except for the alloys $Ir_{1.5}Os_{0.5}Y$ (C14) and $Ni_{1.5}V_{0.5}Zr$ (C15), the compositions have been reported as stoichiometric.

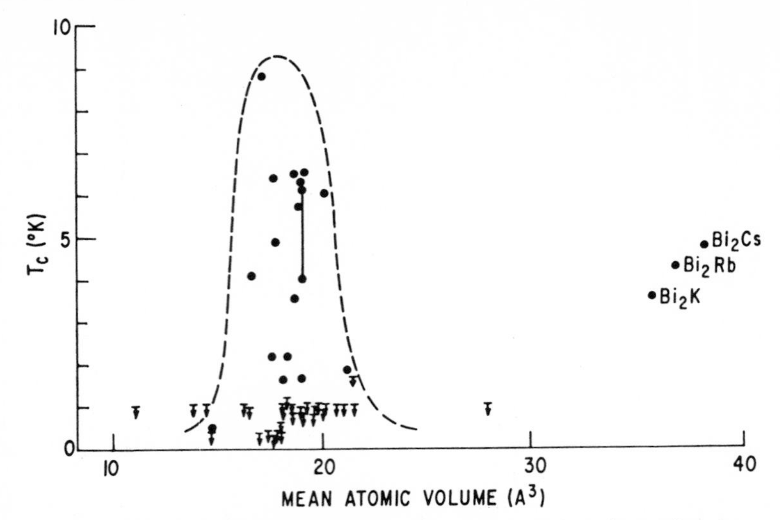

Fig. 16. T_c of C15(Cu_2Mg)-type Laves phase compounds as a function of mean atomic volume.

3.4 Other Compounds

Approximately two dozen crystal structures given *Struckturbericht* types[45] exhibit at least one example of a superconductive compound, and a brief résumé of these is presented in this section. Complete data are given in References 16 and 17.

Bl, NaCl, cubic. Only one system with a B1 superconductive intermetallic compound has been found, although several examples have been tested to about 1°K. InTe[90] (3.45 to 3.20°K, a = 6.16 Å) has been produced by simultaneous application of more than 30 kb pressure and high temperature. The carbides and nitrides of some transition-metal elements form superconductive B1 compounds[17] with T_c as high as 16°K.

B2, CsCl, Cubic. Of twelve reported CsCl-type compounds tested, only three have been found superconductive: $Mg_{0.47}Tl_{0.53}$[76] (2.75°K, a = 3.628 Å), OsTi[16] (0.46°K, a = 3.077 Å), and RuTi[16] (1.07°K, a = 3.067 Å). CoU (1.70°K) has a distorted CsCl-type structure.[77]

B8$_1$, NiAs, Hexagonal. Many NiAs compounds support some form of magnetic behavior at low temperature and, of the remainder, a large fraction is found to be superconductive, although with T_c less than 5°K. The data are given in Table 6. A NiAs superstructure has been reported[83,84] in γ-Bi_2Pd_3 with a broad transition: 3.7 to 4°K. It has also been noted[83] that quenching BiPt may introduce up to 0.8 per cent vacancies. The vacancy concentration may be related to the wide range of T_c observed: 2.4°, 1.2°K.

B18, CuS, Hexagonal. A lone example of this structure type has been found, namely, CuS[85,86] (1.62°K, a = 3.76 Å, c = 16.26 Å).

TABLE 6

Superconductive B8$_1$(NiAs)-Type Compounds

Compound	T_c(°K)	a(Å)	c(Å)	Electrons/Atom	Reference
BiNi	4.25	4.070	5.35	7.5	74
BiPt	1.21 2.4	4.315	5.490	7.5	74
$Bi_{0.1-1}PtSb_{0.9-0}$	1.21–2.05	4.13–4.32	5.47–5.48	7.5	79
BiRh	2.06–2.2	4.094	5.663	7.0	80, 73
PdSb	1.50			7.5	81
PdTe	2.30			8.0	81, 82
PtSb	2.10			7.5	81, 70

TABLE 7

Superconductive B31 (MnP)-Type Compounds

Compound	T_c(°K)	a(Å)	b(Å)	c(Å)	Electrons/Atom	Reference
GeIr	4.70				6.5	81, 82
GePt	0.40	6.088	5.733	3.701	7.0	16
GeRh	0.96	5.70	6.48	3.25	6.5	88, 16
PdSi	0.93	6.133	5.599	3.381	7.0	16
PtSi	0.88	5.932	5.595	3.603	7.0	16

TABLE 8

Superconductive C2 (FeS_2)-Type Compounds

Compound	T_c(°K)	a(Å)	Electrons/Atom	Reference
$AuSb_2$	0.58	6.658	3.67(1) 7.0 (11)	16, 92
$PdSb_2$	1.25	6.459	6.33	16
$Rh_{0.36}Se_{0.64}$	6.0	6.015	7.09	78
$Rh_{0.53}Se_{0.47}$	6.0		7.59	29
$RhTe_2$ (low-temp. mod.)	1.51	6.441	7.0	29, 78

B20, FeSi, Cubic. AuBe[87] (2.64°K) and GaPt[139] (1.74°K, a = 4.91 Å) are examples which have this structure.

B31, MnP, Orthorhombic. Five superconductive examples of this structure type are known and are listed in Table 7. The number of valence electrons/atom is 6.5 or 7.

Cl, CaF_2, Cubic. The reported examples are $CoSi_2$[89] with T_c = 1.22°K and $Ga_{0.7}Pt_{0.3}$[139] with T_c = 2.9°K.

C2, FeS_2, Cubic. Superconductive compounds with this structure are listed in Table 8. Compositions from ~1:1 to 2:1 appear to exist. The electron/atom ratio ranges from 6.33 to 7.59 if the valence for gold is set at eleven, as has been suggested by Matthias.

C2, FeS_2 (Disordered), Cubic. Recent reports of Hulliger and Müller[93,94] on ternary alloys with the cobaltite-type structure give six superconductive examples with T_c under 1.45°K. The valence electron/atom ratio is seven for all. See Table 9.

C6, CdI_2, Trigonal. The compound PdTe has a T_c of 1.53°K when prepared as an annealed single crystal.[95] The structure persists to a composition corresponding to $Pd_{1.75}Te_2$ with T_c increasing to 2.25°K. Unannealed specimens have lower T_c in the range 1.46 to 1.93°K. A substitution of 5 per cent of the Pd with Rh raises T_c to 1.65°K, and 5 and 10 per. cent substitution with Pt raises T_c to 1.71° and 1.65°K, respectively. The valence electron ratio varies in the range of 7.3 to 7.9.

C16, $CuAl_2$, Tetragonal. About one half of the twenty-four known intermetallic compounds of this structure have been tested for

TABLE 9

Superconductive C2 (FeS_2 Disordered)-Type Compounds

Compound	T_c(°K)	a(Å)	Electrons/Atom	Reference
BiPdSe	1.0	6.432	7	93, 94
BiPdTe	1.2	6.656	7	93, 94
BiPtSe	1.45	6.42	7	93, 94
BiPtTe	1.15	6.59	7	93, 94
PdSbSe	1.0	6.323	7	93, 94
PdSbTe	1.2	6.533	7	93, 94

TABLE 10

Superconductive C16 ($CuAl_2$)-Type Compounds

Compound	T_c(°K)	a(Å)	c(Å)	Electrons/Atom	Reference
$AgIn_2$	2.30–2.46	6.883	5.615	2.34	92
$AgTh_2$	2.26	7.56	5.84		28
$AuPb_2$	4.42	7.325	5.655	3.0 (1) 6.33(11)	92
$AuTh_2$	3.08	7.42	5.95		28
$CoZr_2$	6.30	6.367	5.513	5.67	96
$CuTh_2$	3.49	7.28	5.75		28
Pb_2Pd	2.95	6.849	5.833	6.0	92
Pb_2Rh	2.66	6.664	5.865	5.67	92

a transition and eight have been found superconductive up to 6.3°K, as shown in Table 10. The C16 structure may form from elements with highly different radii, the range being R_A/R_B from 1.080 to 1.499.[97] For $AgIn_2$ the number of valence electrons/atom is low, 2.33, and the remaining ratios lie between five and six. Gendron and Jones[92] considered this structure type and noted disagreement with Matthias' generalizations. They suggested that a change in the density of states at the Fermi surface caused by variations in the c/a ratio might be the explanation of the disagreement.

C32, AlB_2, Hexagonal. The compound β-Si_2Th is reported[25] to have T_c equal to 2.41°K.

C_c, Si_2Th, Tetragonal. Three superconductive compounds are known with this structure: Ge_2Y (3.80°K),[98] $LaSi_2$ (~2.5°K),[99] and α-Si_2Th (3.16°K).[30]

$D0_{19}$, Ni_3Sn, Hexagonal. Only Al_3Th (0.75°K, a = 6.500 Å, c = 4.626 Å) is known to be superconductive.[16]

$D0_e$, Ni_3P, Tetragonal. The compound Mo_3P is reported[100] to have a T_c of 5.31°K.

$D1_c$, $PtSn_4$, Orthorhombic. A single compound, $AuSn_4$, has been found superconductive[92] with T_c equal to 2.38°K. The valence electron/atom ratio is 3.4 or 5.4, depending on the valence chosen for Au—one or eleven.

$D2_c$, MnU_6, Tetragonal. Three uranium compounds have been found superconductive:[77] CoU_6 (2.29°K), FeU_6 (3.86°K), and MnU_6 (2.32°K).

$D2_d$, $CaZn_5$, Hexagonal. One example of this structure has been found[135] in Au_5Ba with T_c = 0.35 to 0.7°K, a = 5.69 Å, and c = 4.54 Å. The composition range extended from about $Au_{0.78}Ba_{0.22}$ to $Au_{0.86}Ba_{0.14}$.

$D8_8$, Mn_5Si_3, Hexagonal. The single example of this structure type is Pb_3Zr_5 (4.60°K, a = 8.529 Å, c = 5.864 Å)[16] with a valence electron/atom ratio of four.

$D10_2$, Fe_3Th_7, Hexagonal. The intermetallic compounds with the $D10_2$ structure all have Th for the "B_7" member. The critical temperatures observed are low, and the mean atomic volume is larger than that of most other compounds. The data are given in Table 11.

$E9_3$, Fe_3W_3, Cubic. Two examples of superconductive compounds are known: $CoHf_2$ (0.56°K, a = 12.067 Å)[16] and $CoTi_2$ (3.44°K,

TABLE 11

Superconductive $D10_2$ (Fe_3Th_7)-Type Compounds

Compound	T_c(°K)	a(Å)	c(Å)	Reference
Co_3Th_7	1.83	9.833	6.200	28
Fe_3Th_7	1.86	9.823	6.211	28
Ir_3Th_7	1.52	10.06	6.290	28
Ni_3Th_7	1.98	9.885	6.225	28
Os_3Th_7	1.51	10.02	6.285	28
Rh_3Th_7	2.15	10.031	6.287	16

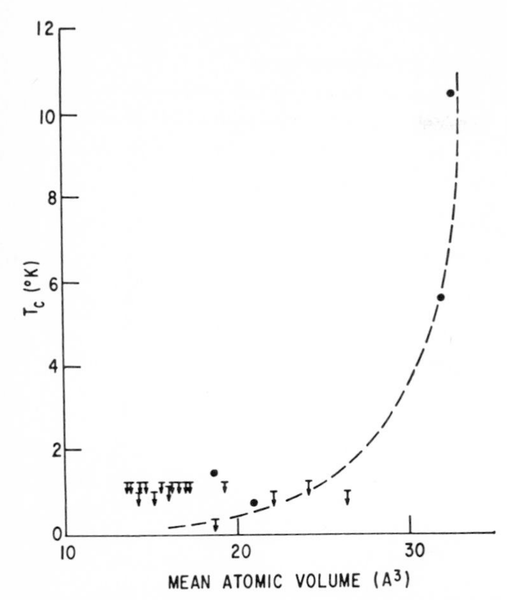

Fig. 17. T_c of $Ll_2(Cu_3Au)$-type compounds versus mean atomic volume.

$a = 11.30$ Å).[28] An interesting series of superconductive oxides[16] also exists with this structural arrangement with T_c as high as 11.8°K.

$L1_0$, CuAu, Tetragonal. The two Bi compounds with Li and Na exhibit this ordered crystal structure. They are α-BiLi (2.47°K, $a = 3.361$ Å, $c = 4.24$ Å)[83] and BiNa (2.25°K, $a = 3.46$ Å, $c = 4.80$ Å).[73] The electron/atom ratio is three for each case.

$L1_2$, Cu_3Au, Cubic. Twenty-three of the more than 100 examples[101] of this intermetallic compound have been tested for superconductive transitions, and only six examples have been found. The notable alloy with high T_c is $InLa_3$[16] at 10.4°K ($a = 5.07$ Å). A second compound of interest is the alloy $Bi_{0.26}Tl_{0.74}$, which appears to exist in both an ordered ($T_c = 4.15$°K) and disordered state ($T_c = 4.4$°K). The states of order have not been demonstrated by diffraction techniques.[102] $AlZr_3$ (0.73°K, $a = 4.37$ Å),[16] Bi_3Sr (5.62°K, $a = 5.042$ Å),[83] and Nb_3Si (1.5°K, $a = 4.211$ Å)[103] are also superconductive.

The valence electron/atom ratio for the observed superconductive $L1_2$ compounds is three to four, but a recent study[62] given in Reference 17 of compounds with ratios near eight yielded no transitions over 1.2°K. Figure 17 suggests that a large atomic volume may be associated with large T_c.

Bismuth Compounds. An extended effort has been made by Alekseevskii, Zhuravlev, Zhdanov, and co-workers[73,83,104] to find and document the Bi-containing superconductive compounds. They have found that superconductive binary compounds are formed with elements not normally found to be superconductive.

In Fig. 18 is plotted an interesting correlation of critical temperature and the shortest interatomic distance.[83] Superconductive compounds are found when the shortest interatomic distance lies between 3.1 Å and 3.8 Å. A broad maximum in T_c is noted. The compounds Bi_2K, Bi_2Rb, and Bi_2Cs show increasing T_c with increasing atomic number,

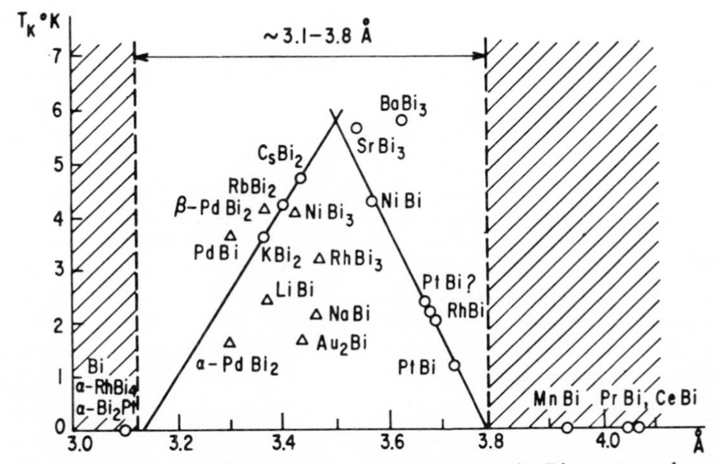

Fig. 18. Plot of shortest interatomic distance in Bi compounds as function of critical temperature. (After Zhuravlev et al.[83])

TABLE 12

Additional Superconductive Compounds with Single Example Structures, Complex Known Structures, and Apparent Compounds with Unsolved Atomic Arrangements

Compound	T_c(°K)	Crystallographic Data				Reference
		a(Å)	b(Å)	c(Å)	β	
Cubic system						
Al_3Mg_2	0.84	28.28				16, 107
$AuZn_3$	1.21	7.893				16
$Nb_{0.26}U_{0.74}$	1.85		(γ-U type)			108
$Rh_{17}S_{15}$	5.8	9.911				78,109
Tetragonal system						
$BaBi_3$	5.69	5.188		5.157		110, 83
$Be_{13}W$	4.1	10.14		4.23		139
Bi_2Pd (β)	4.25	3.362		12.983		111, 112
CdHg	1.77	3.940		2.916		16
Ga_5V_2	3.55		(Mn_2Hg_5 type)			113
InSb	1.6–2.1	5.72		3.18		114
			(A5 disordered?)			
Pb_4Pt	2.80		(Related to C16)			115
$Nb_{0.6}Ru_{0.4}$	1.2	3.140		3.227		116, 117
Nb_3Sn_2	16.6	6.901		9.533		118, 20
Hexagonal system						
$BiIn_2$	5.6		(Disordered)			119
$Bi_2Pt(\beta)$	0.155	6.44		6.25		83
$Bi_4Rh(\gamma)$	2.70					147
Hg_3Li	1.7					120
Hg_2Na	1.62					120
$La_{0.55}Lu_{0.45}$	2.20		(La type)			121, 122
$La_{0.75}Y_{0.25}$	2.50		(La type)			121
SiTa	4.25–4.38					123
Orthorhombic system						
Bi_3Ni	4.06	8.875	4.112	11.477		73
BiPd	3.7	7.203	8.707	10.662		111
$Bi_3Rh(\beta)$	3.2	11.522	9.027	4.24		124
			(like Bi_3Ni)			
Ge_2La	1.49		(Disordered $ThSi_2$ type)			98
Ge_3Rh_5	2.12	5.42	10.32	3.96		88, 81
$NbSn_2$	2.60	9.852	5.645	19.126		125, 140
Monoclinic system						
$Bi_2Pd(\alpha)$	1.70	12.74	4.25	5.665	102°35′	112

whereas BiNi, BiRh, and BiPt show the inverse behavior. The crosshatched areas contain nonsuperconductive compounds to the low interatomic distance side and at least one ferromagnetic compound, BiMn, to the high value side. Along with this atom separation correlation, it is interesting to note that the superconductive Bi structure formed under high pressure would have a shortest interatomic distance of 3.39 Å for bcc packing and 3.5 Å for the fcc atomic arrangement.[83] Bi(III), formed under pressures from 27,000 to 28,400 atm[105] has T_c equal to 7.25°K, which is approximately the T_c for Bi films[106] deposited below about 10°K. No Bi compound has a critical temperature exceeding these values.

A final grouping of intermetallic compounds is presented in Table 12 including unique structures (one example), very complex crystal structures, and those materials that appear to be compounds but have unsolved atomic arrangements at the moment.

Unusual among these compounds is Al_3Mg_3 with a unit cell volume over 20,000 Å and therefore with the order of 1000 atoms/unit cell. The atomic arrangement has not been reported.

Tetragonal InSb is an unusual metastable superconductor formed by the application of high pressures of about 25 kilobars. At room temperature and atmospheric pressure it is a semiconductor. The specimen must be maintained at low temperature after the pressure is released.

A semiconducting compound, $Ge_{0.976}Te$, has been found to be superconductive at very low temperatures—T_c = 0.08 to 0.09°K—by Hulm and co-workers,[145] and similar examples have been found in other systems.[156]

The ferroelectric compound, $SrTiO_3$, when properly doped, has been found to be superconductive by Schooley and co-workers.[157]

No examples of compounds with triclinic symmetry appear to have been reported.

4. HIGH-FIELD MAGNETIC PROPERTIES

The physical properties of high-field superconductive materials is under intense study at many laboratories.[126] Two behavior types have evolved as mentioned in the introduction: first, the type II or Abrikosov-Goodman model materials such as Nb_3Sn and V_3Ga in an inhomogeneous state; second, the filamentary model assemblies characterized most clearly by mercury and lead forced into the pores of unfinished Vycor glass. A review of high-field superconductivity has been given by Bean and Schmitt,[127] in which the physics of the model possibilities has been pursued.

The resistantless current carrying capabilities in the presence of large magnetic fields has brought attention to the intermetallic compounds because of their high critical temperature and proportionately higher H_{c2}, the field up to which a supercurrent is still flowing. Table 13 lists the four Cr_3O-type compounds in which high-field magnetic properties have been studied along with representative data. The T_c values are in the range 16 to 18°K. The highest magnetic field in which a current may flow is predicted to be about 350 kOe for GaV_3. This value has been obtained by a parabolic extrapolation of low-field data.

Nb_3Sn, the most studied compound, has been prepared by several techniques, such as vapor deposition, formation on the interior and exterior of wires and tubes, and direct sintering and melting. The highest field reported for H_{c2} of Nb_3Sn is about 190 kOe. After the original discovery of high current-carrying capability in a straight wire centered with Nb_3Sn and other Nb–Sn compounds, many attempts have been made to design a superconductive coil that would produce as large a fraction of H_{c2} as possible. A magnetic field of 101 kOe has recently been attained by Martin and co-workers[133] with a coil approximately two inches in length and two inches in diameter. The Nb_3Sn was formed by heat treating Sn-coated Nb wire at a selected temperature. The impact on technology of the production of very large magnetic fields with small volume superconductive coils is unpredictable at the moment, but appears to have

TABLE 13

High-Field Superconductive Properties of Intermetallic Compounds

Compound	T_c(°K)	H_{c1} (kOe)	H_{c2} (kOe)	Reference
$AlNb_3$	17.5, 18.0	0.375		19
GaV_3	16.8	0.4	350[a]	24, 19, 128
Nb_3Sn	~18	0.3	183,[a] >190[b]	19, 129, 130, 128, 131
SiV_3	16.9	0.55	156[a]	24, 128, 132

[a] Parabolic extrapolation of H_c versus T.
[b] Pulsed fields.

large potential. A large activity[17] in noncompound superconductive materials, such as Nb–Zr and Mo–Re alloys, also exists because of ease of extrusion and shaping. The alloys are sufficiently ductile for direct wire forming. However, maximum T_c of these alloys is about 12°K, which scales down H_{c2} compared to the compounds and limits high-field current-carrying capabilities. Fields produced with superconductive coils of these alloys have reached 50 to 65 kOe and are in general use in many laboratories.

5. SUMMARY

The known superconductive intermetallic compounds have been presented in a grouping by crystal-structure type. With few exceptions, the critical transition temperatures have been found to follow the pattern predicted by the empirical average valence/atom rules proposed by Matthias. Measurements and interpretation following from the BCS microscopic theory of superconductivity have shown that the density of states at the Fermi surface is an important variable in the occurrence of superconductivity, and also is the physical basis for Matthias' valency rule for maxima in the transition temperature. The determination of the density of states requires measurement of properties, such as the specific heat. Unfortunately, such data on compounds are quite limited. The important results of the available measurements suggest that the eventual complete understanding of the superconductive properties of intermetallic compounds will require more determinations of the density of states and interaction energies, together with extension of the BCS theory.

The readily existing parameters for most compounds are the crystal structure and average valence. With the valence determined directly from the periodic chart and with simple averages for compounds, correlations of T_c with the number of valence electrons/atom, the mean atomic volume, and the valence electron density have been presented. A résumé of the maxima observed in T_c as a function of these correlations for the five crystal-structure types with largest numbers of superconductive examples is given in Table 14.

T_c as a function of mean atomic volume has a maximum near 16 Å^3 in several structures and a range of 14 Å^3 to 22 Å^3, with the notable exception of the three C15 type compounds, Bi_2Cs, Bi_2K, and Bi_2Rb, which have a mean volume of ~37 Å^3. The A15 atomic volumes are evenly distributed over the range 14 Å^3 to 22 Å^3, except for the SiV_3-related compounds, which have a T_c maximum just above 15 Å^3. The highest known T_c of 18.5°K in $Nb_{0.8}Sn_{0.2}$ is associated with the largest mean atomic volume of the A15 compounds. It should be noted that the atomic volumes of most elemental superconductors fall into the 14 Å^3 to 22 Å^3 range and that the large atomic volumes of the superconductive La polymorphs is close to those of the three Bi compounds.

Maxima in T_c are found slightly below five or seven valence electrons/atom for these five structure types. More than 80 per cent of the A15 superconductive materials fall in an envelope centered at 4.7 valence electrons/atom. Only one A15 superconductive compound has an average valence near three, and it contains Au and therefore may be subject to special valence consideration. The superconductive σ-phases ($D8_b$) compounds have

TABLE 14

Maxima in Critical Temperature as Function of Valence and Volume for A12, A15, C14, C15, and $D8_b$ Structures

Struckturbericht Classification	Example	Mean Atomic Volume (Å^3)	Valence Electrons/Atom	Valence Electrons/Å^3	Maximum T_c(°K)
A12	α-Mn	15.5	6.6	0.42_5	9.9
A15 (all)	Cr_3O	15–22	4.7, 6.5	0.26, 0.36	18.5
(high T_c)	or (β-W)	15, 20–22	4.7	0.26	18.5
(V_3Si related)		15	4.7	~0.36	17.0
C14	$MgZn_2$	17	5.8–6.6	0.35–0.41	5.6
C15	$MgCu_2$	17–21, ~37	Some at 6.7	Scattered	8.8
$D8_b$	σ-phases	14–17	6.6	0.37–0.45	12.0

valence electron/atom values near 6.6, except for one compound. The A12 and C14 type superconductive compounds exhibit high T_c when the number of valence electrons/atom is near seven. The presence of these regularities is consistent with the prior designation of these structure types as "electron compounds." For the C14 and C15 Laves phases, in which sphere packing principles as well as valence considerations apply, the maxima in T_c are less distinct.

Additional studies of the electronic density of states and the future development of the band structure of compounds[143,144,149] should lead to a better understanding of the principles of formation of various intermetallic compounds.

The valence electron density shows some regularities. A sharp maximum in T_c is found for A12 type compounds at 0.43 valence electron/Å^3. For large T_c in Cr_3O-type superconductors the density is low. The V_3Si-related compounds have intermediate values, whereas a few Mo compounds have the largest valence electron density. It is noted that the density may vary more than a factor of two for Cr_3O-type compounds. Also, the three maxima in T_c correspond to transition-nontransition binaries with low density, V_3Si and related compounds to intermediate and transition metal binaries to high density of valence electrons.

Atomic positional disorder has been shown to affect superconductive properties. The sintering of Nb–Sn compacts at high temperature and vapor deposition of Nb_3Sn films introduces disorder and may reduce T_c a factor of two or three. On the other hand, the only superconductive compound that is presumed to have the classical Cu_3Au order-disorder transformation, $Bi_{0.26}Tl_{0.74}$, has a T_c 6 per cent larger in the disordered state. Also, the disorder in the Sn-sites introduced by having a few per cent more Nb atoms than required for stoichiometric Nb_3Sn raises T_c about 0.5° to 18.5°K, the highest reported critical temperature. Much additional study, both experimental and theoretical, must be done to understand the detailed effects of order on the superconductive state.

The crystal structure or the repetitive spatial arrangement of atoms appears to play only a secondary role in determining the superconductive properties of a metal. The interrelated controlling factors are the band structure, the valence, and the electron density of states at the Fermi surface.

The interest in superconductive intermetallic compounds will increase with increased technical utilization of the superconductive phenomenon. The highest attainable critical temperature is now more than a scientific curiosity, because the desirable property of the highest magnetic field in which a current may be carried without resistive loss increases with T_c. The brittleness usually associated with intermetallic compounds has been overcome in at least the system Nb_3Sn, which has been used in a superconductive coil to produce a magnetic field over 100 kOe. K. Onnes[31] dreamed of the same feat fifty years ago, utilizing soft elemental superconductors, but found that he could only produce a field of a few hundred oersteds.

ACKNOWLEDGMENTS

Appreciation is extended to W. DeSorbo and R.W. Schmitt for careful reading of the manuscript and for valuable suggestions.

REFERENCES

1. Onnes H. K., *Commun. Phys. Lab. Univ. Leiden*, Rept. No. 133*d* (1913).
2. Kunzler J. E., *Rev. Mod. Phys.* **33** (1961) 499.
3. Bean C. P., *Phys. Rev. Letters* **8** (1962) 250; Bean C. P., and M. V. Doyle, *J. Appl. Phys.* **33** (1962) 3334.
4. Livingston J. D., *Phys. Rev.* **129** (1963) 1943.
5. Matthias B. T., *Progress in Low Temperature Physics*, Vol. II. Interscience Publishers, New York, 1957, p. 138.
6. Pines D., *Phys. Rev.* **109** (1958) 280.
7. Bardeen J., L. N. Cooper, and J. R. Schrieffer, *Phys. Rev.* **108** (1957) 1175.
8. Morin F. J., and J. P. Maita, *Phys. Rev.* **129** (1963) 1115.
9. Bucher E., F. Heiniger, J. Muheim, and J. Müller, *Rev. Mod. Phys.* **36** (1964) 146.

10. Bucher E., F. Heiniger, and J. Müller, *Helv. Phys. Acta* **34** (1961) 843.
11. Bender D., E. Bucher, and J. Müller, *Phys. Kondens. Materie* **1** (1963) 225.
12. Clogston A. M., and V. Jaccarino, *Phys. Rev.* **121** (1961) 1357; A. M. Clogston, A. C. Gossard, V. Jaccarino, and A. Yafet, *Rev. Mod. Phys.* **36** (1964) 170.
13. Blumberg W. E., J. Eisinger, V. Jaccarino, and B. T. Matthias, *Phys. Rev. Letters* **5** (1960) 149.
14. Kikoin I., and B. Lazarev, *J. Exp. Theoret. Phys.* (*USSR*) **3** (1933) 44.
15. Chapnik I. M., *Dokl. Akad. Nauk. SSSR* **141** (1961) 70.
16. Matthias B. T., T. H. Geballe, and V. B. Compton, *Rev. Mod. Phys.* **35** (1963) 1.
17. Roberts B. W., *Progress in Cryogenics*, Vol. IV. Heywood and Co., London, 1964, p. 160.
18. Corenzwit E., *J. Phys. Chem. Solids* **9** (1959) 93.
19. Swartz P. S., *Phys. Rev. Letters* **9** (1962) 448 and personal communication.
20. Reed T. B., H. C. Gatos, W. J. LaFleur, and J. T. Roddy, *Superconductors*. Interscience Publishers, New York, 1962, p. 143. Also *Metallurgy of Advanced Electronic Materials*. Interscience, New York, 1963, pp. 71, 87.
21. Matthias B. T., E. A. Wood, E. Corenzwit, and V. B. Bala, *J. Phys. Chem. Solids* **1** (1956) 188.
22. Matthais B. T., T. H. Geballe, V. B. Compton, E. Corenzwit, and G. W. Hull, Jr., *Phys. Rev.* **128** (1962) 588.
23. Wood E. A., V. B. Compton, B. T. Matthais, and E. Corenzwit, *Acta Cryst.* **11** (1958) 604.
24. Wernick J. H., F. J. Morin, S. L. Hsu, D. Dorsi, J. P. Maita, and J. E. Kunzler, *High Magnetic Fields*, Kolm H., B. Lax, F. Bitter, and R. Mills, eds. MIT Press, Cambridge, Mass., 1962, p. 609. Also see *Phys. Rev. Letters* **5** (1960) 149; *J. Appl. Phys.* **32** (1961) 325.
25. Hardy G. F., and J. K. Hulm, *Phys. Rev.* **93** (1954) 1004.
26. Banus M. D., T. B. Reed, and H. C. Gatos, *J. Phys. Chem. Solids* **23** (1962) 971.
27. Matthias B. T., and E. Corenzwit, *Phys. Rev.* **94** (1954) 1069.
28. Matthias B. T., V. B. Compton, and E. Corenzwit, *J. Phys. Chem. Solids* **19** (1961) 130.
29. Matthias B. T., *Phys. Rev.* **97** (1955) 74.
30. Hardy G. F., and J. K. Hulm, *Phys. Rev.* **89** (1953) 884.
31. Onnes H. K., *Commun. Kamerlingh Onnes Lab. Univ. Leiden* **13,** Suppl. No. 34*b* (1913).
32. Samsonov G. V., *Silicides, Their Utilization and Technology*, Acad. Sci. Ukrainian SSR, Kiev, 1959, p. 21.
33. Alekseevskii N. E., E. M. Savitsky, V. V. Baron, and Yu. V. Efimov, *Doklady Akad. Nauk SSSR* **145** (1962) 82.
34. Matthias B. T., T. H. Geballe, S. Geller, and E. Corenzwit, *Phys. Rev.* **95** (1954) 1435.
35. Cody G. D., J. J. Hanak, G. T. McConville, and F. D. Rosi, *Proc. VII Intern. Conf. Low Temp. Phys.* Univ. of Toronto Press, 1961, p. 382.
36. Arrhenius G. O. S., and M. F. Merriam, unpublished results.
37. Galasso F., B. Bayles, and S. Soehle, *Nature* **198** (1963) 984.
38. Matthias B. T., personal communication (1963). Also *Phys. Rev.* **97** (1955) 74.
39. Raetz K., and E. Saur, *Z. Physik.* **169** (1962) 315.
40. Rothwarf F., C. C. Dickson, E. Parthe, and H. Boller, *Bull. Am. Phys. Soc.* **7** (1962) 322.
41. Hagner R., and E. Saur, *Proc. 8th Intern. Congr. Low Temp. Phys.*, London (1962). Butterworths, Washington, 1963, p. 358.
42. Hagner R., and E. Saur, *Naturwissenschaften* **49** (1962) 444.
43. Jansen H. C., *Z. Physik.* **162** (1961) 275.
44. Jansen H. C., and E. J. Saur, *Proc. VII Intern. Conf. Low Temp. Phys.* Univ. of Toronto Press, 1961, p. 379.
45. Pearson W. B., *Handbook of Lattice Spacing and Structures of Metals*. Pergamon Press, New York, 1958.
46. Geller S., *Acta Cryst.* **9** (1956) 885.
46*a*. Nevitt M. V., *Trans. Met. Soc. AIME* **212** (1958) 350.
47. Hulm J. K., and R. D. Blaugher, *Phys. Rev.* **123** (1961) 1569.
48. Hanak J. J., G. D. Cody, J. L. Cooper, and M. Rayl, Ref. 41, p. 353.
49. Wernick J. H., Chapter 12 of this book.
50. Greenfield P., and P. A. Beck, *Trans. AIME* **200** (1954) 253; **206** (1956) 265.
51. Bergman B. G., and D. P. Shoemaker, *Acta Cryst.* **7** (1954) 857.
52. Kasper J. S., and R. M. Waterstrat, *Acta Cryst.* **9** (1956) 289.
53. Bucher E., F. Heiniger, J. Muheim, and J. Müller, *Mod. Phys.* **36** (1964) 146.
54. Matthias B. T., T. H. Geballe, E. Corenzwit, and G. W. Hull, Jr., *Phys. Rev.* **129** (1963) 1025.

55. Bucher E., F. Heiniger, and J. Müller, *Helv. Phys. Acta* **34** (1961) 843.
56. Bucher E., F. Heiniger, and J. Müller, Ref. 41, p. 153.
57. Compton V. B., E. Corenzwit, J. P. Maita, B. T. Matthias, and F. J. Morin, *Phys. Rev.* **123** (1961) 1567.
58. Darby J. B., Jr., and S. T. Zegler, *J. Phys. Chem. Solids* **23** (1962) 1825.
59. Blaugher R. D., and J. K. Hulm, *J. Phys. Chem. Solids* **19** (1961) 134.
60. Blaugher R. D., A. Taylor, and J. K. Hulm, *IBM J. Res. Dev.* **6** (1962) 116.
61. Hulm J. K., and R. D. Blaugher, *Phys. Rev.* **123** (1961) 1569.
62. Müller J., private communication.
63. Kasper J. S., and B. W. Roberts, *Phys. Rev.* **101** (1956) 537.
64. Bucher E., F. Heiniger, and J. Müller, to be published.
65. Das D. K., S. P. Rideout, and P. A. Beck, *J. Metals* **4** (1952) 1071.
66. Compton V. B., and B. T. Matthias, *Acta Cryst.* **12** (1959) 651.
67. Suhl H., B. T. Matthias, and E. Corenzwit, *J. Phys. Chem. Solids* **11** (1959) 346.
68. Matthias B. T., *J. Appl. Phys.* **31** (1960) 23S.
69. Matthias B. T., H. Suhl, and E. Corenzwit, *Phys. Rev. Letters* **1** (1958) 92, 152E.
70. DeHaas W. J., and T. Jurriaanse, *Proc. Acad. Sci. Amsterdam* **35** (1932) 748.
71. Matthias B. T., and E. Corenzwit, *Phys. Rev.* **107** (1957) 1558.
72. Zhuravlev N. N., *J. Expl. Theoret. Phys.* **34** (1958) 827; Soviet Phys. *JETP* **7** (1958) 571.
73. Alekseevskii N. E., G. S. Zhdanov, and N. N. Zhuravlev, *Vestn. Mosk. Univ. Ser. Mat. Mekh. Astron. Fiz. i Khim.* **14,** No. 3 (1959) 113.
74. Zhuravlev N. N., N. E. Alekseevskii, and G. S. Zhdanov, *Vestn. Mosk. Univ. Ser. Mat. Mekh. Astron. Fiz. i Khim.* **14** (1959) 117–127.
75. Zhuravlev N. N., T. A. Mingazin, and G. S. Zhdanov, *J. Exp. Theoret. Phys.* **34** (1958) 820; also *Trans. Soviet Phys. JETP* **34** (1958) 566–571.
76. Guttman L., and J. W. Stout, *Proc. Low Temp Conf.*, NBS (March 1951), p. 65.
77. Chandrasekhar B. S., and J. K. Hulm, *J. Phys. Chem. Solids* **7** (1958) 259–267.
78. Matthias B. T., E. Corenzwit, and C. E. Miller, *Phys. Rev.* **93** (1954) 1415.
79. Zhuravlev N. N., G. S. Zhdanov, and E. M. Smirnova, *Fiz. Metal i Metalloved.* **13** (1962) 68.
80. Glagoleva V. P., and G. S. Zhdanov, *J. Exp. Theoret. Phys.* **25** (1953) 248–254.
81. Matthias B. T., *Phys. Rev.* **92** (1953) 874.
82. Matthias B. T., *Phys. Rev.* **90** (1953) 487.
83. Zhuravlev N. N., G. S. Zhdanov, and R. N. Kuz'min, *Kristallografiya* **5** (1961) 532.
84. Zhuravlev N. N., *Soviet Phys. Cryst.* **3** (1958) 506.
85. Meissner W., *Z. Physik.* **58** (1929) 570–572.
86. Buckel W., and R. Hilsch, *Z. Physik* **128** (1950) 324–346.
87. Matthias B. T., *J. Phys. Chem. Solids* **10** (1959) 342.
88. Geller S., *Acta Cryst.* **8** (1955) 15.
89. Matthias B. T., *Phys. Rev.* **87** (1952) 380.
90. Geller S., A. Jayaraman, and G. W. Hull, Jr., *Appl. Phys. Letters* **4** (1964) 35.
91. Geller S., and B. T. Matthias, *J. Phys. Chem. Solids* **4** (1958) 156.
92. Gendron M. F., and R. E. Jones, *J. Phys. Chem. Solids* **23** (1962) 405; *Bull. Am. Phys. Soc.* **6** (1961) 122.
93. Hulliger F., and J. Müller, *Phys. Letters* **5** (1963) 226.
94. Hulliger F., *Nature* **198** (1963) 382.
95. Guggenheim J., F. Hulliger, and J. Müller, *Helv. Phys. Acta* **34** (1961) 408; **34** (1961) 410.
96. Matthias B. T., and E. Corenzwit, *Phys. Rev.* **100** (1955) 626; *B.T.M.—Intern. Conf. Low Temp. Phys. Proc.*, Paris (1955), p. 570.
97. Laves F., in *Theory of Alloy Phases.* ASM, Cleveland (1956).
98. Matthias B. T., E. Corenzwit, and W. H. Zachariasen, *Phys. Rev.* **112** (1958) 89.
99. Henry W. E., C. Betz, and H. Muir, *Bull. Am. Phys. Soc.* **7** (1962) 474; **7** (1962) 621.
100. Blaugher R. D., J. K. Hulm, and P. N. Yocum, *Phys. Chem. Solids* **26** (1965) 2037.
101. Nevitt M. V., in *Electronic Structure and Alloy Chemistry of the Transition Elements.* Interscience Publishers, New York, 1963, p. 163.
102. Coles B. R., *IBM J. Res. Dev.* **6** (1962) 68.
103. Galasso F., and J. Pyle, *Acta Cryst.* **16** (1963) 228.
104. Alekseevskii N. E., N. B. Brandt, and T. I. Kostina, *Izv. Akad. Nauk SSSR Ser. Fiz.* **16** (1952) 233–263.

105. Brandt N. B., and N. I. Ginzburg, *J. Exp. Theoret. Phys. (USSR)* **39** (1960) 1554–1556, *Trans. Soviet Phys. JETP* **12** (1961) 1082; also *Fiz. Tverd. Tela SSSR* **3** (1961) 3461, *Trans. Soviet Phys.—Solid State* **3** (1962) 2510.
106. Lazarev B. G., E. E. Semenenko, and A. I. Sudovtsov, *Trans. Soviet Phys. JETP* **13** (1961) 75; *J. Exp. Theoret. Phys. (USSR)* **40** (1961) 105.
107. Meissner W., H. Franz, and H. Westerhoff, *Ann. Phys.* **17** (1933) 593–619.
108. Goodman B. B., J. Hillairet, J. J. Veyssie, and L. Weill, *Proc. VII Intern. Conf. Low Temp. Phys.* Univ. of Toronto Press, 1961, p. 350.
109. Geller S., *Acta Cryst.* **15** (1962) 1198.
110. Buckel W., *Naturwissenschaften* **42** (1955) 451.
111. Zhuravlev N. N., and G. S. Zhdanov, *J. Exp. Theoret. Phys. (USSR)* **25** (1953) 485–490.
112. Alekseevskii N. E., and I. I. Lifanov, *J. Exp. Theoret. Phys.* **30** (1956) 405; *Soviet Phys. JETP* **3** (1956) 294.
113. Müller J., and R. Reinmann, to be published.
114. Darnell A. J., W. F. Libby (H. E. Bommel and B. R. Tittmann), *Science* **139** (1963) 1301, 1302. Geller, S., D. B. McWhan, and G. W. Hull, Jr., *Science* **140** (1963) 62.
115. Rosler U., and K. Schubert, *Z. Metallk.* **42** (1951) 395; *Naturwissenschaften* **38** (1951) 331.
116. Bender D., E. Bucher, and J. Müller, *Phys. Kondens. Materie* **1** (1963) 225.
117. Bucher E., J. Muheim, and J. Müller, Ref. 41, p. 151.
118. Reed T. B., and H. C. Gatos, *J. Appl. Phys.* **33** (1962) 2657.
119. Jones R. E., and W. B. Ittner, *Phys. Rev.* **113** (1959) 1520.
120. Merriam M. F., unpublished results.
121. Anderson G. S., S. Legvold, and F. H. Spedding, *Phys. Rev.* **109** (1958) 243.
122. Anderson G. S., S. Legvold, and F. H. Spedding, *Low Temperature Physics and Chemistry*. Univ. of Wisconsin Press, Madison, Wis., 1958, p. 279.
123. Justi E., *Leitfähigheit und Leitungsmechanismus fester Stöffe*. Göttingen Vandenhoeck und Ruprecht, 1948, pp. 187–270.
124. Kuz'min R. N., and G. S. Zhdanov, *Trans. Soviet Phys. Cryst.* **5** (1961) 830.
125. Van Ooijen D. J., J. H. N. Van Vucht, and W. F. Druyvesteyn, *Phys. Letters* **3** (1962) 128.
126. See *Proc. Intern. Conf. Science of Superconductivity*, Hamilton, N.Y. (1963), *Rev. Mod. Phys.* **36** (1964) 1.
127. Bean C. P., and R. W. Schmitt, *Science* **140** (1963) 26.
128. Wernick J. H., *Superconductors*, Interscience Publishers, New York, 1962, p. 35.
129. Kunzler J. E., E. Buehler, F. S. L. Hsu, and J. H. Wernick, *Phys. Rev. Letters* **6** (1961) 89.
130. Alekseevskii N. E., and N. N. Michailov, *Soviet Phys. JETP* **41** (1961) 1809.
131. Hart H. R., Jr., I. S. Jacobs, C. L. Kolbe, and P. E. Lawrence, in *High Magnetic Fields*, Kohn, Lax, Bitter, and Mills, eds. MIT Press, Cambridge, Mass., 1961, p. 584.
132. Hauser J. J., *Phys. Rev. Letters* **9** (1962) 423.
133. Martin D. L., M. G. Benz, C. A. Bruch, and C. H. Rosner, *Cryogenics* **3** No. 2 (1963) 114.
134. Weger M., *Rev. Mod. Phys.* **36** (1964) 175.
135. Arrhenius G., C. J. Raub, D. C. Hamilton, and B. T. Matthias, *Phys. Rev. Letters* **11** (1963) 313.
136. Holleck H., H. Nowotny, and F. Benesovsky, *Monatsh. Chem.* **94** (1963) 473.
137. Wilson C. G., *Acta Cryst.* **16** (1963) 724.
138. Holleck H., H. Nowotny, and F. Benesovsky, *Monatsh. Chem.* **94** (1963) 359, 477.
139. Alekseevskii N. E., and N. N. Michailov, *J. Exp. Theor. Phys.* **43** (1962) 2110; *Soviet Phys. JETP* **16** (1963) 1493.
140. Gomes de Mesquita A. H., C. Langereis, and J. I. Leenhouts, *Philips Res. Repts.* **18** (1963) 377.
141. Bucher E., F. Heiniger, J. Muheim, and J. Müller, *Rev. Mod. Phys.* **36** (1964) 146.
142. Bucher E., G. Busch, F. Heiniger, and J. Müller, to be published.
143. Merriam M. F., *Rev. Mod. Phys.* **36** (1964) 152.
144. Coles B. R., *Rev. Mod. Phys.* **36** (1964) 139.
145. Hulm J. K., R. A. Hein, J. W. Gibson, and R. C. Miller, *Rev. Mod. Phys.* **36** (1964) 242.
146. DeSorbo W., *Phys. Rev.* **130** (1963) 2177.
147. Zhdanov G. S., N. N. Zhuravlev, R. N. Kuz'min, and A. I. Soklakov, *Kristallografiya* **3** (1958) 373; *Soviet Phys. Cryst. (Engl. Transl.)* **3** (1958) 374.
148. Killpatrick D. H., *J. Metals* **16** (1964) 98.
149. Reed T. B., H. C. Gatos, W. J. LaFleur, and J. T. Roddy, *Metallurgy of Advanced Electronic Materials*. Interscience, New York, 1963, pp. 71, 87.
150. Johnson J., L. Toth, K. Kennedy, and E. R. Parker, *Solid State Commun.* **2** (1964) 123.

151. Saint-James D., and P. G. DeGennes, *Phys. Letters* **7** (1964) 306.
152. Bean C. P., and J. D. Livingston, *Phys. Rev. Letters* **12** (1964) 14.
153. Bucher E., F. Laves, J. Müller, and H. Van Philipsborn, *Phys. Letters* **8** (1964) 27.
154. McEvoy J. P., R. F. Decell, and R. L. Novak, *Appl. Phys. Letters* **4** (1964) 43; Swartz P. S., H. R. Hart, Jr., and R. L. Fleischer, **4** (1964) 71.
155. Merriam M. F., *Phys. Letters* **9** (1964) 100.
156. Geller S., and G. W. Hull, Jr., *Phys. Rev. Letters* **13** (1964) 127.
157. Schooley J. F., W. R. Hosler, and M. L. Cohen, *Phys. Rev. Letters* **12** (1964) 474.

Indexes

The labour and the patience, the judgment and the penetration, which are required to make a good index, is only known to those who have gone through this most painful, but least praised part of a publication. But laborious as it is, I think it indispensably necessary, to manifest the treasures of any multifarious collection, facilitate the knowledge to those who seek it, and invite them to make application thereof.

William Oldys (1696–1761)

Author Index

This index lists all pages of text citation, but omits page references to the bibliographies, which are easily accessible by way of the text entries.

Subject Index

Compound Index

This index lists all compounds or compound systems that are specifically cited in text, figures, or tables. Because authors are not always consistent in the way in which they refer to a given compound, binary and ternary compounds are listed multiply under all possible permutations of the compound formula. Quaternary and higher-order compounds, the multiple indexing of which would be cumbersome, are listed only for that number of permutations that is required to present each component once in the first position. The sequence of compound listings follows the order AB, AB_2, A_2B, A_2BC, AC, and so on. It should be noted also that it has not been possible to treat element number 41 (Cb or Nb) in a consistent manner because of the variability in the practice of the individual chapter authors. In referring to compounds containing this element, please search under both symbols.